FORMULAS AND THEOREMS
IN
PURE MATHEMATICS

FORMULAS AND THEOREMS
IN
PURE MATHEMATICS

BY

GEORGE S. CARR

SECOND EDITION

WITH AN INTRODUCTION BY
JACQUES DUTKA

CHELSEA PUBLISHING COMPANY
NEW YORK, N. Y.

SECOND EDITION,

WITH AN INTRODUCTION BY JACQUES DUTKA

THE FIRST EDITION OF THIS WORK WAS PUBLISHED UNDER THE TITLE: A SYNOPSIS OF ELEMENTARY RESULTS IN PURE MATHEMATICS, AT LONDON AND CAMBRIDGE, IN 1886. IN THE PRESENT EDITION THERE IS A SLIGHT CHANGE OF NOTATION AND THE ERRATA NOTED HAVE BEEN CORRECTED. THE FOLD-OUT PLATES ARE TO BE FOUND AS A SEPARATE BOOKLET UNDER THE BACK COVER OF THE BOOK. PUBLISHED, 1970, AT NEW YORK, N. Y., ON ACID-FREE PAPER

LIBRARY OF CONGRESS CATALOG CARD NUMBER 78-113122

INTERNATIONAL STANDARD BOOK NUMBER 0-8284-0239-6

PRINTED IN THE UNITED STATES OF AMERICA

INTRODUCTION

DURING the nineteenth and well into the twentieth century, mathematical education at the British universities was largely oriented toward preparing students for rigorous competitive examinations which were prerequisites for graduation. At Cambridge, the leading mathematical center of the period, and the model for other British universities, mathematical teaching was dominated by the Mathematical Tripos, the most prestigious examination of the time which had a kind of sporting notoriety even outside the academic community. A candidate for the Tripos studied a prescribed range of subjects, generally under the guidance of a mathematical coach, and was intensively trained during most of his undergraduate years to solve difficult problems against a time limit. Becoming a Wrangler, one of the top places in the Tripos, and thus obtaining a degree with honor, virtually assured the candidate of a successful future career in academic life, the Civil Service, or in other professions. Many of the notable British mathematicians of the time were involved in the system, either as candidates, coaches, or examiners. The most famous coach was E. J. Routh, who is perhaps best remembered today for his distinguished treatises on mechanics which are still worth reading. In the thirty-one year period from 1858 to 1888, Routh, who had himself been a Senior Wrangler, (just ahead of J. C. Maxwell, the great physicist), had some 631 pupils, most of whom became Wranglers, including 27 Senior Wranglers.

In such an era, with its heavy emphasis on achieving high places in examinations by mastering a wide range of subjects in a relatively short time, there was a ready market for review books, specialized tracts, course notes, and collections of mathematical problems. Perhaps the best known of these works, published at a time when interest in the Tripos was near its peak, was (as it was originally titled) *A Synopsis of Elementary Results in Pure Mathematics*, by George Shoobridge Carr, sometime Scholar of Gonville and Caius College, Cambridge.

Carr's book has a considerable amount of material which cannot be readily found in one place. It is a comprehensive and well organized summary, in about 6,000 propositions, of the principal results in many of the leading British (and some French) textbooks of the past century. Concise proofs or indications of proofs of many results are given with numerous cross references. A notable feature of the book which enhances its value as a reference work is the elaborate joint index to its contents and to much of the relevant periodical literature of the nineteenth century. This is no routine compendium, but, as the famous British mathematician, G. H. Hardy, has said, "A book written with some real scholarship and with a style and individuality of its own." One of the distinctions of the book is the decisive role it played in the education of the largely self-taught Indian mathematical genius, Srinivasa Ramanujan, by providing him with a general style of mathematical exposition and with the seeds of many of the ideas which afterwards germinated into some of the most beautiful and striking formal results obtained thus far in the twentieth century.

In Part I of the book, consisting of 256 pages, there are sections on algebra, theory of equations, plane and spherical trigonometry, elementary geometry, and geometrical conics. (The first

32 pages contain a summary of physical data and some abridged tables which have been superseded by later developments.) Part II, with 583 pages, treats the differential and integral calculus, the calculus of variations, differential equations, finite differences, plane and solid coordinate geometry. In addition, there are a very detailed index and many plates. Many subjects are treated at length. This is particularly evident in the section on the integral calculus, in which the formal side is very fully developed—perhaps in accordance with Carr's own interests.

Further details regarding the objectives and subject matter of the book are to be found in the author's preface and in the extensive table of contents.

JACQUES DUTKA

RIVERSIDE RESEARCH INSTITUTE
and
COLUMBIA UNIVERSITY
NEW YORK, N. Y.

PREFACE TO PART I.

The work, of which the part now issued is a first instalment, has been compiled from notes made at various periods of the last fourteen years, and chiefly during the engagements of teaching. Many of the abbreviated methods and mnemonic rules are in the form in which I originally wrote them for my pupils.

The general object of the compilation is, as the title indicates, to present within a moderate compass the fundamental theorems, formulæ, and processes in the chief branches of pure and applied mathematics.

The work is intended, in the first place, to follow and supplement the use of the ordinary text-books, and it is arranged with the view of assisting the student in the task of revision of book-work. To this end I have, in many cases, merely indicated the salient points of a demonstration, or merely referred to the theorems by which the proposition is proved. I am convinced that it is more beneficial to the student to recall demonstrations with such aids, than to read and re-read them. Let them be read once, but recalled often. The difference in the effect upon the mind between reading a mathematical demonstration, and originating one wholly or partly, is very great. It may be compared to the difference between the pleasure experienced, and interest aroused, when in the one case a traveller is passively conducted through the roads of a novel and unexplored country, and in the other case he discovers the roads for himself with the assistance of a map.

In the second place, I venture to hope that the work, when completed, may prove useful to advanced students as an *aide-mémoire* and book of reference. The boundary of mathematical science forms, year by year, an ever widening circle, and the advantage of having at hand some condensed statement of results becomes more and more evident.

To the original investigator occupied with abstruse researches in some one of the many branches of mathematics, a work which gathers together synoptically the leading propositions in all, may not therefore prove unacceptable. Abler hands than mine undoubtedly, might have undertaken the task of making such a digest; but abler hands might also, perhaps, be more usefully employed,—and with this reflection I have the less hesitation in commencing the work myself. The design which I have indicated is somewhat comprehensive, and in relation to it the present essay may be regarded as tentative. The degree of success which it may meet with, and the suggestions or criticisms which it may call forth, will doubtless have their effect on the subsequent portions of the work.

With respect to the abridgment of the demonstrations, I may remark, that while some diffuseness of explanation is not only allowable but very desirable in an initiatory treatise, conciseness is one of the chief requirements in a work intended for the purposes of revision and reference only. In order, however, not to sacrifice clearness to conciseness, much more labour has been expended upon this part of the subject-matter of the book than will at first sight be at all evident. The only palpable result being a compression of the text, the result is so far a negative one. The amount of compression attained is illustrated in the last section of the present part, in which more than the number of propositions usually given in treatises on Geometrical Conics are contained, together with the figures and demonstrations, in the space of twenty-four pages.

The foregoing remarks have a general application to the work as a whole. With the view, however, of making the earlier sections more acceptable to beginners, it will be found that, in those sections, important principles have sometimes been more fully elucidated and more illustrated by examples, than the plan of the work would admit of in subsequent divisions.

A feature to which attention may be directed is the uniform system of reference adopted throughout all the sections. With the object of facilitating such reference, the articles have been numbered progressively from the commencement in large Clarendon figures; the breaks which will occasionally be found in these numbers having been purposely made, in order to leave room for the insertion of additional matter, if it should be required in a future edition, without disturbing the original numbers and references. With the same object, demonstrations and examples have been made subordinate to enunciations and formulæ, the former being printed in small, the latter in bold type. By these aids, the interdependence of propositions is more readily shown, and it becomes easy to trace the connexion between theorems in different branches of mathematics, without the loss of time which would be incurred in turning to separate treatises on the subjects. The advantage thus gained will, however, become more apparent as the work proceeds.

The Algebra section was printed some years ago, and does not quite correspond with the succeeding ones in some of the particulars named above. Under the pressure of other occupations, this section moreover was not properly revised before going to press. On that account the table of errata will be found to apply almost exclusively to errors in that section; but I trust that the list is exhaustive. Great pains have been taken to secure the accuracy of the rest of the volume. Any intimation of errors will be gladly received.

I have now to acknowledge some of the sources from which the present part has been compiled. In the Algebra, Theory of Equations, and Trigonometry sections, I am largely indebted to Todhunter's well-known treatises, the accuracy and completeness of which it would be superfluous in me to dwell upon.

In the section entitled Elementary Geometry, I have added to simpler propositions a selection of theorems from Townsend's Modern Geometry and Salmon's Conic Sections.

In Geometrical Conics, the line of demonstration followed agrees, in the main, with that adopted in Drew's treatise on the subject. I am inclined to think that the method of that author cannot be much improved. It is true that some important properties of the ellipse, which are arrived at in

Drew's Conic Sections through certain intermediate propositions, can be deduced at once from the circle by the method of orthogonal projection. But the intermediate propositions cannot on that account be dispensed with, for they are of value in themselves. Moreover, the method of projection applied to the hyperbola is not so successful; because a property which has first to be proved true in the case of the equilateral hyperbola, might as will be proved at once for the general case. I have introduced the method of projection but sparingly, always giving preference to a demonstration which admits of being applied in the same identical form to the ellipse and to the hyperbola. The remarkable analogy subsisting between the two curves is thus kept prominently before the reader.

The account of the C. G. S. system of units given in the preliminary section, has been compiled from a valuable contribution on the subject by Professor Everett, of Belfast, published by the Physical Society of London.* This abstract, and the tables of physical constants, might perhaps have found a more appropriate place in an after part of the work. I have, however, introduced them at the commencement, from a sense of the great importance of the reform in the selection of units of measurement which is embodied in the C. G. S. system, and from a belief that the student cannot be too early familiarized with the same.

The Factor Table which follows is, to its limited extent, a reprint of Burckhardt's "*Tables des diviseurs*," published in

---

* "Illustrations of the Centimetre-Gramme-Second System of Units." London: Taylor and Francis. 1875.

1814–17, which give the least divisors of all numbers from 1 to 3,036,000. In a certain sense, it may be said that this is the only sort of purely mathematical table which is absolutely indispensable, because the information which it gives cannot be supplied by any process of direct calculation. The logarithm of a number, for instance, may be computed by a formula. Not so its prime factors. These can only be arrived at through the tentative process of successive divisions by the prime numbers, an operation of a most deterrent kind when the subject of it is a high integer.

A table similar to and in continuation of Burckhardt's has recently been constructed for the fourth million by J. W. L. Glaisher, F.R.S., who I believe is also now engaged in completing the fifth and sixth millions. The factors for the seventh, eighth, and ninth millions were calculated previously by Dase and Rosenberg, and published in 1862–65, and the tenth million is said to exist in manuscript. The history of the formation of these tables is both instructive and interesting.*

As, however, such tables are necessarily expensive to purchase, and not very accessible in any other way to the majority of persons, it seemed to me that a small portion of them would form a useful accompaniment to the present volume. I have, accordingly, introduced the first eleven pages of Burckhardt's tables, which give the least factors of the first 100,000 integers nearly. Each double page of the table here printed is

---

* See "*Factor Table for the Fourth Million.*" By James Glaisher, F.R.S. London: Taylor and Francis. 1880. Also *Camb. Phil. Soc. Proc.*, Vol. III., Pt. IV., and *Nature*, No. 542, p. 462.

an exact reproduction, in all but the type, of a single quarto page of Burckhardt's great work.

It may be noticed here that Prof. Lebesque constructed a table to about this extent, on the plan of omitting the multiples of seven, and thus reducing the size of the table by about one-sixth.* But a small calculation is required in using the table which counterbalances the advantage so gained.

The values of the Gamma-Function, pages 30 and 31, have been taken from Legendre's table in his "*Exercices de Calcul Intégral*," Tome I. The table belongs to Part II. of this Volume, but it is placed here for the convenience of having all the numerical tables of Volume I. in the same section.

In addition to the authors already named, the following treatises have been consulted—Algebras, by Wood, Bourdon, and Lefebure de Fourcy; Snowball's Trigonometry; Salmon's Higher Algebra; the Geometrical Exercises in Potts's Euclid; and Geometrical Conics by Taylor, Jackson, and Renshaw.

Articles 260, 431, 569, and very nearly all the examples, are original. The latter have been framed with great care, in order that they might illustrate the propositions as completely as possible.

G. S. C.

HADLEY, MIDDLESEX;
May 23, 1880.

---

* "Tables diverses pour la décomposition des nombres en leurs facteurs premiers." Par V. A. Lebesque. Paris. 1864.

TABLE OF CONTENTS.

PART I.

SECTION I.—MATHEMATICAL TABLES.

SECTION II.—ALGEBRA.

SECTION III.—THEORY OF EQUATIONS.

SECTION IV.—PLANE TRIGONOMETRY.

SECTION V.—SPHERICAL TRIGONOMETRY.

SECTION VI.—ELEMENTARY GEOMETRY.

SECTION VII.—GEOMETRICAL CONICS.

INDEX TO PROPOSITIONS OF EUCLID

REFERRED TO IN THIS WORK.

The references to Euclid are made in Roman and Arabic numerals; *e.g.* (VI. 19).

BOOK I.

I. 4.—Triangles are equal and similar if two sides and the included angle of each are equal each to each.

I. 5.—The angles at the base of an isosceles triangle are equal.

I. 6.—The converse of 5.

I. 8.—Triangles are equal and similar if the three sides of each are equal each to each.

I. 16.—The exterior angle of a triangle is greater than the interior and opposite.

I. 20.—Two sides of a triangle are greater than the third.

I. 26.—Triangles are equal and similar if two angles and one corresponding side of each are equal each to each.

I. 27.—Two straight lines are parallel if they make equal alternate angles with a third line.

I. 29.—The converse of 27.

I. 32.—The exterior angle of a triangle is equal to the two interior and opposite; and the three angles of a triangle are equal to two right angles.

Cor. 1.—The interior angles of a polygon of n sides $= (n-2)\,\pi$.

Cor. 2.—The exterior angles $= 2\pi$.

I. 35 to 38.—Parallelograms or triangles upon the same or equal bases and between the same parallels are equal.

I. 43.—The complements of the parallelograms about the diameter of a parallelogram are equal.

I. 47.—The square on the hypotenuse of a right-angled triangle is equal to the squares on the other sides.

I. 48.—The converse of 47.

BOOK II.

II. 4.—If a, b are the two parts of a right line, $(a+b)^2 = a^2+2ab+b^2$.

If a right line be bisected, and also divided, internally or externally, into two unequal segments, then—

II. 5 and 6.—The rectangle of the unequal segments is equal to the difference of the squares on half the line, and on the line between the points of section; or $(a+b)(a-b) = a^2-b^2$.

II. 9 and 10.—The squares on the same unequal segments are together double the squares on the other parts; or

$$(a+b)^2+(a-b)^2 = 2a^2+2b^2.$$

II. 11.—To divide a right line into two parts so that the rectangle of the whole line and one part may be equal to the square on the other part.

II. 12 and 13.—The square on the base of a triangle is equal to the sum of the squares on the two sides *plus* or *minus* (as the vertical angle is *obtuse* or *acute*), twice the rectangle under either of those sides, and the projection of the other upon it; or $a^2 = b^2+c^2-2bc\cos A$ (702).

BOOK III.

III. 3.—If a diameter of a circle bisects a chord, it is perpendicular to it: and conversely.

III. 20.—The angle at the centre of a circle is twice the angle at the circumference on the same arc.

III. 21.—Angles in the same segment of a circle are equal.

III. 22.—The opposite angles of a quadrilateral inscribed in a circle are together equal to two right angles.

III. 31.—The angle in a semicircle is a right angle.

III. 32.—The angle between a tangent and a chord from the point of contact is equal to the angle in the alternate segment.

III. 33 and 34.—To *describe* or to *cut off* a segment of a circle which shall contain a given angle.

III. 35 and 36.—The rectangle of the segments of any chord of a circle drawn through an *internal* or *external* point is equal to the square of the semi-chord perpendicular to the diameter through the internal point, or to the square of the tangent from the external point.

III. 37.—The converse of 36. If the rectangle be equal to the square, the line which meets the circle touches it.

BOOK IV.

IV. 2.—To inscribe a triangle of given form in a circle.
IV. 3.—To describe the same about a circle.
IV. 4.—To inscribe a circle in a triangle.
IV. 5.—To describe a circle about a triangle.
IV. 10.—To construct two-fifths of a right angle.
IV. 11.—To construct a regular pentagon.

BOOK VI.

VI. 1.—Triangles and parallelograms of the same altitude are proportional to their bases.

VI. 2.—A right line parallel to the side of a triangle cuts the other sides proportionally; and conversely.

VI. 3 and A.—The bisector of the *interior* or *exterior* vertical angle of a triangle divides the base into segments proportional to the sides.

VI. 4.—Equiangular triangles have their sides proportional homologously.

VI. 5.—The converse of 4.

VI. 6.—Two triangles are equiangular if they have two angles equal, and the sides about them proportional.

VI. 7.—Two triangles are equiangular if they have two angles equal and the sides about two *other* angles proportional, provided that the third angles are both greater than, both less than, or both equal to a right angle.

VI. 8.—A right-angled triangle is divided by the perpendicular from the right angle upon the hypotenuse into triangles similar to itself.

VI. 14 and 15.—Equal *parallelograms*, or *triangles* which have two angles equal, have the sides about those angles reciprocally proportional; and conversely, if the sides are in this proportion, the figures are equal.

VI. 19.—Similar triangles are in the duplicate ratio of their homologous sides.

VI. 20.—Likewise similar polygons.

VI. 23.—Equiangular parallelograms are in the ratio compounded of the ratios of their sides.

VI. B.—The rectangle of the sides of a triangle is equal to the square of the bisector of the vertical angle *plus* the rectangle of the segments of the base.

VI. C.—The rectangle of the sides of a triangle is equal to the rectangle under the perpendicular from the vertex on the base and the diameter of the circumscribing circle.

VI. D.—Ptolemy's Theorem. The rectangle of the diagonals of a quadrilateral inscribed in a circle is equal to both the rectangles under the opposite sides.

BOOK XI.

XI. 4.—A right line perpendicular to two others at their point of intersection is perpendicular to their plane.

XI. 5.—The converse of 4. If the first line is also perpendicular to a fourth at the same point, that fourth line and the other two are in the same plane.

XI. 6.—Right lines perpendicular to the same plane are parallel.

XI. 8.—If one of two parallel lines is perpendicular to a plane, the other is also.

XI. 20.—Any two of three plane angles containing a solid angle are greater than the third.

XI. 21.—The plane angles of any solid angle are together less than four right angles.

TABLE OF CONTENTS.

PART II.

SECTION VIII.—DIFFERENTIAL CALCULUS.

SECTION IX.—INTEGRAL CALCULUS.

SECTION X.—CALCULUS OF VARIATIONS.

SECTION XI.—DIFFERENTIAL EQUATIONS.

No. of Article.

SECTION XII. — CALCULUS OF FINITE DIFFERENCES.

SECTION XIII.—PLANE COORDINATE GEOMETRY.

ANALYTICAL CONICS IN CARTESIAN COORDINATES.

ANALYTICAL CONICS IN TRILINEAR COORDINATES.

THEORY OF PLANE CURVES.

SECTION XIV.—SOLID COORDINATE GEOMETRY.

PREFACE TO PART II.

APOLOGIES for the non-completion of this volume at an earlier period are due to friends and enquirers. The labour involved in its production, and the pressure of other duties, must form the author's excuse.

In the compilation of Sections VIII. to XIV., the following works have been made use of:—

Treatises on the Differential and Integral Calculus, by Bertrand, Hymer, Todhunter, Williamson, and Gregory's Examples on the same subjects; Salmon's Lessons on Higher Algebra.

Treatises on the Calculus of Variations, by Jellett and Todhunter; Boole's Differential Equations and Supplement; Carmichael's Calculus of Operations; Boole's Calculus of Finite Differences, edited by Moulton.

Salmon's Conic Sections; Ferrers's Trilinear Coordinates; Kempe on Linkages (*Proc. of Roy. Soc.*, Vol. 23); Frost and Wolstenholme's Solid Geometry; Salmon's Geometry of Three Dimensions.

Wolstenholme's Problems.

The Index which concludes the work, and which, it is hoped, will supply a felt want, deals with 890 volumes of 32 serial publications: of these publications, thirteen belong to Great Britain, one to New South Wales, two to America, four to France, five to Germany, three to Italy, two to Russia, and two to Sweden.

As the volumes only date from the year 1800, the

important contributions of Euler to the "Transactions of the St. Petersburg Academy," in the last century, are excluded. It was, however, unnecessary to include them, because a very complete classified index to Euler's papers, as well as to those of David Bernoulli, Fuss, and others in the same Transactions, already exists.

The titles of this Index, and of the works of Euler therein referred to, are here appended, for the convenience of those who may wish to refer to the volumes.

Tableau général des publications de l'Académie Impériale de St. Pétersbourg depuis sa fondation. 1872. [B.M.C.:* *R.R.* 2050,*e.*]

I. Commentarii Academiæ Scientiarum Imperialis Petropolitanæ. 1726–1746; 14 vols. [B. M. C.: 431,*f.*]

II. Novi Commentarii A. S. I. P. 1747–75, 1750–77; 21 vols. [B. M. C.: 431,*f.*15–17, *g.*1–16, *h.*1, 2.]

III. Acta A. S. I. P. 1778–86; 12 vols. [B. M. C: 431, *h.*3–8; or *T.C.* 8, *a.* 11.]

IV. Nova Acta A. S. I. P. 1787–1806; 15 vols. [B. M. C.: 431, *h.*9–15, *i.*1–8; or *T.C.* 8, *a.*23.]

V. Leonhardi Euler Opera minora collecta, vel Commentationes Arithmeticæ collectæ; 2 vols. 1849. [B. M. C.: 8534, *ee.*]

VI. Opera posthuma mathematica et physica; 2 vols. 1862. [B.M.C.: 8534,*f.*]

VII. Opuscula analytica; 1783–5; 2 vols. [B. M. C.: 50,*i.* 15.]

Analysis infinitorum. [B. M. C.: 529,*b.*11.]

G. S. C.

ENDSLEIGH GARDENS,
LONDON, N.W., 1886.

* British Museum Catalogue.

MATHEMATICAL TABLES.

INTRODUCTION.

The Centimetre-Gramme-Second system of units.

Notation.—The decimal measures of length are the *kilometre, hectometre, decametre, metre, decimetre, centimetre, millimetre.* The same prefixes are used with the *litre* and *gramme* for measures of capacity and weight.

Also, 10^7 metres is denominated a *metre-seven;* 10^{-7} metres, a *seventh-metre;* 10^{15} grammes, a *gramme-fifteen;* and so on.

A gramme-six is also called a *megagramme;* and a millionth-gramme, a *microgramme;* and similarly with other measures.

Definitions.—The C. G. S. system of units refers all physical measurements to the *Centimetre* (cm.), the *Gramme* (gm.), and the *Second* (sec.) as the units of length, mass, and time.

The quadrant of a meridian is approximately a *metre-seven.* More exactly, one metre = 3·2808694 feet = 39·370432 inches.

The *Gramme* is the *Unit of mass*, and the *weight of a gramme* is the *Unit of weight*, being approximately the weight of a cubic centimetre of water; more exactly, 1 gm. = 15·432349 grs.

The *Litre* is a cubic decimetre: but *one cubic centimetre* is the C. G. S. *Unit of volume.*

1 litre = ·035317 cubic feet = ·2200967 gallons.

The *Dyne* (dn.) is the *Unit of force*, and is the force which, in one second generates in a gramme of matter a velocity of one centimetre per second.

The *Erg* is the *Unit of work and energy*, and is the work done by a dyne in the distance of one centimetre.

The absolute *Unit of atmospheric pressure* is one megadyne per square centimetre = 74·964 cm., or 29·514 in. of mercurial column at 0° at London, where g = 981·17 dynes.

Elasticity of Volume = k, is the pressure per unit area upon a body divided by the cubic dilatation.

Rigidity $= n$, is the shearing stress divided by the angle of the shear.

Young's Modulus $= M$, is the longitudinal stress divided by the elongation produced, $= 9nk \div (3k+n)$.

Tenacity is the tensile strength of the substance in dynes per square centimetre.

The *Gramme-degree* is the *Unit of heat*, and is the amount of heat required to raise by 1° C. the temperature of 1 gramme of water at or near 0°.

Thermal capacity of a body is the increment of heat divided by the increment of temperature. When the increments are small, this is the thermal capacity *at* the given temperature.

Specific heat is the thermal capacity of unit mass of the body at the given temperature.

The *Electrostatic unit* is the quantity of electricity which repels an equal quantity at the distance of 1 centimetre with the force of 1 dyne.

The *Electromagnetic unit* of quantity $= 3 \times 10^{10}$ *electrostatic units* approximately.

The *Unit of potential* is the potential of unit quantity at unit distance.

The *Ohm* is the common *electromagnetic unit* of resistance, and is approximately $= 10^9$ *C. G. S. units.*

The *Volt* is the *unit of electromotive force*, and is $= 10^8$ *C. G. S. units of potential.*

The *Weber* is the *unit of current*, being the current due to an electromotive force of 1 Volt, with a resistance of 1 Ohm. It is $= \frac{1}{10}$ *C. G. S. unit.*

Resistance of a Wire = *Specific resistance* × *Length* ÷ *Section.*

Physical constants and Formulæ.

In the latitude of London, $g = 32{\cdot}19084$ feet per second.
$= 981{\cdot}17$ centimetres per second.

In latitude λ, at a height h above the sea level,

$$g = (980{\cdot}6056 - 2{\cdot}5028 \cos 2\lambda - {\cdot}000003\,h) \text{ centimetres per second.}$$

Seconds pendulum $= (99{\cdot}3562 - {\cdot}2536 \cos 2\lambda - {\cdot}0000003\,h)$ centimetres.

THE EARTH.—Semi-polar axis, 20,854895 feet* $= 6{\cdot}35411 \times 10^8$ centims.
Mean semi-equatorial diameter, 20,926202 ,, * $= 6\ 37824 \times 10^8$,,
Quadrant of meridian, $39{\cdot}377786 \times 10^7$ inches* $= 1{\cdot}000196 \times 10^7$ metres.
Volume, 1·08279 cubic centimetre-nines.
Mass (with a density $5\frac{1}{2}$) = Six gramme-twenty-sevens nearly.

* These dimensions are taken from Clarke's "Geodesy," 1880.

Velocity in orbit = 2933000 centims. per sec. Obliquity, 23° 27′ 16″.*
Angular velocity of rotation = 1 ÷ 13713.
Precession, 50″·26.* Progression of Apse, 11″·25. Eccentricity, e = ·01679.
Centrifugal force of rotation at the equator, 3·3912 dynes per gramme.
Force of attraction upon moon, ·2701. Force of sun's attraction, ·5839.
Ratio of g to centrifugal force of rotation, $g : r\omega^2 = 289$.
Sun's horizontal parallax, 8″·7 to 9′.* Aberration, 20″·11 to 20″·79.*
Semi-diameter at earth's mean distance, 16′ 1″·82.*
Approximate mean distance, 92,000000 miles, or 1·48 centimetre-thirteens.†
Tropical year, 365·242216 days, or 31,556927 seconds.
Sidereal year, 365·256374 " 31,558150 "
Anomalistic year, 365·259544 days. Sidereal day, 86164 seconds.

THE MOON.—Mass = Earth's mass × ·011364 = 6·98 10^{25} grammes.
Horizontal parallax. From 53′ 56″ to 61′ 24″.*
Sidereal revolution, 27d. 7h. 43m. 11·46s. Lunar month, 29d. 12h. 44m. 2·87s.
Greatest distance from the earth, 251700 miles, or 4·05 centimetre-tens,
Least " " 225600 " 3·63 " "
Inclination of Orbit, 5° 9′. Annual regression of Nodes, 19° 20′.
RULE.—(*The Year* + 1) ÷ 19. *The remainder is the Golden Number.*
(*The Golden Number* − 1) × 11 ÷ 30. *The remainder is the Epact.*

GRAVITATION.— Attraction between masses m, m' at a distance l $\Big\} = \dfrac{mm'}{l^2 \times 1{\cdot}543 \times 10^7}$ dynes.

The mass which at unit distance (1 cm.) attracts an equal mass with unit force (1 dn.) is $= \sqrt{(1{\cdot}543 \times 10^7)}$ gm. = 3928 gm.

WATER.—Density at 0° C., unity; at 4°, 1·000013 (Kupffer).
Volume elasticity at 15°, $2{\cdot}22 \times 10^{10}$.
Compression for 1 megadyne per sq. cm, $4{\cdot}51 \times 10^{-5}$ (Amaury and Descamps).
The heat required to raise the temperature of a mass of water from 0° to t° is proportional to $t + {\cdot}00002t^2 + {\cdot}0000003t^3$ (Regnault).

GASES.—Expansion for 1° C., ·003665 = 1 ÷ 273.

$$\frac{\text{Specific heat at constant pressure}}{\text{Specific heat at constant volume}} = 1{\cdot}408.$$

Density of dry air at 0° with Bar. at 76 cm. = ·0012932 gm. per cb. cm. (Regnault).
At unit pres. (a megadyne) Density = ·0012759.
Density at press. $p = p \times 1{\cdot}2759 \times 10^{-9}$.

Density of saturated steam at t°, with p taken from Table II., is approximately $\Big\} = \dfrac{{\cdot}7936098p}{(1 + {\cdot}00366t)\,10^9}$.

SOUND.—Velocity = $\sqrt{}$(*elasticity of medium* ÷ *density*).
Velocity in dry air at t° = 33240 $\sqrt{(1 + {\cdot}00366t)}$ centimetres per second.
Velocity in water at 0° = 143000 " "

LIGHT.—Velocity in a medium of absolute refrangibility μ
= $3{\cdot}004 \times 10^{10} \div \mu$ (Cornu).

If P be the pressure in dynes per sq. cm., and t the temperature,
$\mu - 1 = 2903 \times 10^{-13} P \div (1 + {\cdot}00366t)$ (Biot & Arago).

* These data are from the "Nautical Almanack" for 1883.
† Transit of Venus, 1874, "Astrom. Soc. Notices," Vols. 37, 38.

Table I.

Various Measures and their Equivalents in C. G. S. units.

Dimensions.

1 inch = 2·5400 cm.
1 foot = 30·4797 „
1 mile = 160933 „
1 nautical do. = 185230 „

1 sq. inch = 6·4516 sq. cm.
1 sq. foot = 929·01 „
1 sq. yard = 8361·13 „
1 sq. mile = $2·59 \times 10^{10}$ „

1 cb. inch = 16·387 cb. cm.
1 cb. foot = 28316 „
1 cb. yard = 764535 „
1 gallon = 4541 „
= 277·274 cb. in. or the volume of 10 lbs. of water at 62° Fah., Bar. 30 in.

Mass.

1 grain = ·06479895 gm.
1 ounce = 28·3495 „
1 pound = 453·5926 „
1 ton = 1,016047 „
1 kilogramme = 2·20462125 lbs.
1 pound Avoir. = 7000 grains
1 pound Troy = 5760 „

Velocity.

1 mile per hour = 44·704 cm. per sec.
1 kilometre „ = 27·77$\dot{7}$ „

Pressure.

1 gm. per sq. cm. = 981 dynes per sq. cm.
1 lb. per sq. foot = 479 „
1 lb. per sq. in. = 68971 „
76 centimetres of mercury at 0° C. = 1,014,000 „

$$\frac{\text{lbs. per sq. in.}}{\text{gms. per sq. cm.}} = 70·307 = \frac{1}{·014223}$$

Force of Gravity.

upon 1 gramme = 981 dynes
„ 1 grain = 63·56777 „
„ 1 oz. = $2·7811 \times 10^4$ „
„ 1 lb. = $4·4497 \times 10^5$ „
„ 1 cwt. = $4·9837 \times 10^7$ „
„ 1 ton = $9·9674 \times 10^8$ „

Work ($g = 981$).

1 gramme-centimetre = 981 ergs.
1 kilogram-metre = 981×10^5 „
1 foot-grain = $1·937 \times 10^3$ „
1 foot-pound = $1·356 \times 10^7$ „
1 foot-ton = $3·04 \times 10^{10}$ „
1 'horse power' p. sec. = $7·46 \times 10^9$ „

Heat.

1 gramme-degree C. = 42×10^6 ergs.
1 pound-degree = 191×10^8 „
1 pound-degree Fah. = 106×10^8 „

Table II.

Pressure of Aqueous Vapour in dynes per square centim.

Temp.	Pressure.	Temp.	Pressure.
−20°	1236	40°	73200
−15°	1866	50°	122600
−10°	2790	60°	198500
− 5°	4150	80°	472900
0°	6133	100°	1014000
5°	8710	120°	1988000
10°	12220	140°	3626000
15°	16930	160°	6210000
20°	23190	180°	10060000
25°	31400	200°	15600000
30°	42050		

Table III.

Values for the principal Lines of the Spectrum in air at 160° C. with Bar. 76 cm.

	Wave-length in centims.	No. of vibrations per second.
A	$7·604 \times 10^{-5}$	$3·950 \times 10^{14}$
B	6·867 „	4·373 „
C	6·56201 „	4·577 „
D (mean)	5·89212 „	5·097 „
E	5·26913 „	5·700 „
F	4·86072 „	6·179 „
G	4·30725 „	6·973 „
H_1	3·96801 „	7·569 „
H_2	3·93300 „	7·636 „

TABLE IV. *in C. G. S. units.*

	Density Water = 1.	Young's Modulus *M.*	Rigidity *n.*	Elasticity of volume *k.*	Tenacity.	Expansion of Volume per degree C.	Linear Expansion between 0 & 100 C.	Specific Heat between 0 & 100 C.	Relative Conductivity.	Rate of Conduction of Sound in cm. per sec.	Elect. Magn. Specific Resistance at 0° C.
Platinum	21	—	—	—	—	·000027	·000875	·0335	381	$2·69 \times 10^5$	9158
Gold	19·26	—	—	—	—	·000045	·001483	—	1000	1·74 ,,	2081
Mercury	13·596	—	—	$0·542 \times 10^{12}$	—	·000180	—	·0330	—	—	96190
Lead	11·35	$·059 \times 10^{12}$	—	—	$2·28 \times 10^8$	·000086	·002861	—	180	1·23 ,,	19850
Silver	10·47	—	—	—	—	·000061	·00196	·0557	973	2·61 ,,	1521
Copper	8·843	1·234 ,,	$4·47 \times 10^{11}$	1·684 ,,	41·4 ,,	·000054	·00175	·0949	898	3·74 ,,	1615
Brass, drawn	8·471	1·075 ,,	3·66 ,,	—	33·8 ,,	·000054	·00193	—	357	3·56 ,,	—
Iron, cast	7·235	1·349 ,,	5·32 ,,	0·964 ,,	—	·000033	·001111	—	—	4·32 ,,	—
Iron, wrought	7·677	1·963 ,,	7·69 ,,	1·456 ,,	58·6 ,,	·000040	·001258	·1098	374	5·06 ,,	9827
Steel	7·849	2·139 ,,	8·19 ,,	1·841 ,,	79·3 ,,	·000037	001260	—	—	5·22 ,,	—
Tin, cast	7·29	—	—	—	3·17 ,,	·000063	·00227	—	304	—	13360
Zinc, cast	7·19	—	—	—	5·17 ,,	·000088	·00294	·0927	363	—	5690
Glass, flint	2·942	0·603 ,,	2·40 ,,	0·415 ,,	—	·000015	·00081	·1770	—	4·53 ,,	—

TABLE V.

	Greatest distance from Sun. Earth's mean distance = 1.	Least distance from Sun.	Sidereal Revolution in Days.	Inclination of Orbit to Ecliptic.			Time of Rotation.			Diameter in Miles.	Mass.	Density.
				°	′	″	h.	m.	s.			
Sun	—	—	—		—		600	0	0	888000	354936	0·25
Mercury	0·46669	0·30750	87·969	7	0	8	24	5	28	3000	0·118	2·01
Venus	0·72826	0·71840	224·701	3	23	31	23	21	21	7700	0·883	0·97
Earth	1·01678	0·98322	365·256		—		23	56	4	7926	1·000	1·00
Mars	1·66578	1·38160	686·980	1	51	5	24	37	22	4500	0·132	0·72
Jupiter	5·45378	4·95182	4332·585	1	18	40	9	55	26	92000	338·034	0·24
Saturn	10·07328	9·00442	10759·220	2	29	28	10	29	17	75000	101·064	0·13
Uranus	20·07612	18·28916	30686·821		46	30		—		36000	14·789	0·15
Neptune	30·29888	29·77506	60126·722	1	46	59		—		35000	24·648	0·27

Table VI.—*Functions of π and e.*

$\pi = 3{\cdot}1415926$	$\pi^{-1} = {\cdot}3183099$	$e = 2{\cdot}71828183$
$\pi^2 = 9{\cdot}8696044$	$\pi^{-2} = {\cdot}1013212$	$e^2 = 7{\cdot}38905611$
$\pi^3 = 31{\cdot}0062761$	$\pi^{-3} = {\cdot}0322515$	$e^{-1} = 0{\cdot}3678794$
$\sqrt{\pi} = 1{\cdot}7724539$	$200^g \div \pi = 63^g{\cdot}6619772$	$e^{-2} = 0{\cdot}1353353$
$\log_{10}\pi = {\cdot}4971499$	$180^\circ \div \pi = 57^\circ{\cdot}2957795$	$\log_{10} e = 0{\cdot}43429448$
$\log_e \pi = 1{\cdot}1447299$	$= 206264''{\cdot}8$	$\log_e 10 = 2{\cdot}30258509$

Table VII.

No.	Square root.	Cube root.
2	1·4142136	1·2599210
3	1·7320508	1·4422496
4	2·0000000	1·5874011
5	2·2360680	1·7099759
6	2·4494897	1·8171206
7	2·6457513	1·9129312
8	2·8284271	2·0000000
9	3·0000000	2·0800837
10	3·1622777	2·1544347
11	3·3166248	2·2239801
12	3·4641016	2·2894286
13	3·6055513	2·3513347
14	3·7416574	2·4101422
15	3·8729833	2·4662121
16	4·0000000	2·5198421
17	4·1231056	2·5712816
18	4·2426407	2·6207414
19	4·3588989	2·6684016
20	4·4721360	2·7144177
21	4·5825757	2·7589243
22	4·6904158	2·8020393
23	4·7958315	2·8438670
24	4·8989795	2·8844991
25	5·0000000	2·9240177
26	5·0990195	2·9624960
27	5·1961524	3·0000000
28	5·2915026	3·0365889
29	5·3851648	3·0723168
30	5·4772256	3·1072325

Table VIII.

N.	$\log_{10} N$.	$\log_e N$.
2	·3010300	·69314718
3	·4771213	1·09861229
5	·6989700	1·60943791
7	·8450980	1·94591015
11	1·0413927	2·39789527
13	1·1139434	2·56494936
17	1·2304489	2·83321334
19	1·2787536	2·94443898
23	1·3617278	3·13549422
29	1·4623980	3·36729583
31	1·4913617	3·43398720
37	1·5682017	3·61091791
41	1·6127839	3·71357207
43	1·6334685	3·76120012
47	1·6720979	3·85014760
53	1·7242759	3·97029191
59	1·7708520	4·07753744
61	1·7853298	4·11087386
67	1·8260748	4·20469262
71	1·8512583	4·26267988
73	1·8633229	4·29045944
79	1·8976271	4·36944785
83	1·9190781	4·41884061
89	1·9493900	4·48863637
97	1·9867717	4·57471098
101	2·0043214	4·61512052
103	2·0128372	4·63472899
107	2·0293838	4·67282883
109	2·0374265	4·69134788

Note.—The authorities for Table IV. are as follows:—Columns 2, 3, and 4 (Mercury excepted), Everett's experiments (Phil. Trans., 1867); g is here taken = 981·4. The densities in these cases are those of the specimens employed. Cols. 5 and 7, Rankine. Col. 6, Watt's Dict. of Chemistry. Col. 8, Dulong and Petit. Col. 10, Wertheim. Col. 11, Matthiessen.

Table V. is abridged from Loomis's Astronomy.

The values in Table III. are Angström's.

BURCKHARDT'S FACTOR TABLES.

For all numbers from 1 to 99000.

Explanation.—The tables give the least divisor of every number from 1 up to 99000: but numbers divisible by 2, 3, or 5 are not printed. All the digits of the number whose divisor is sought, excepting the units and tens, will be found in one of the three rows of larger figures. The two remaining digits will be found in the left-hand column. The least divisor will then be found in the column of the first named digits, and in the row of the units and tens.

If the number be prime, a cipher is printed in the place of its least divisor.

The numbers in the first left-hand column are not consecutive. Those are omitted which have 2, 3, or 5 for a divisor. Since $2^2.3.5^2 = 300$, it follows that this column of numbers will re-appear in the same order after each multiple of 300 is reached.

Mode of using the Tables.—If the number whose prime factors are required is divisible by 2 or 5, the fact is evident upon inspection, and the division must be effected. The quotient then becomes the number whose factors are required. If this number, being within the range of the tables, is yet not given, *it is divisible by* 3. Dividing by 3, we refer to the tables again for the new quotient and its least factor, and so on.

Examples.—Required the prime factors of 310155.

Dividing by 5, the quotient is 62031. This number is within the range of the tables. But it is not found printed. Therefore 3 is a divisor of it. Dividing by 3, the quotient is 20677. The table gives 23 for the least factor of 20677. Dividing by 23, the quotient is 899.

The table gives 29 for the least factor of 899. Dividing by 29, the quotient is 31, a prime number. Therefore $310155 = 3.5.23.29.31$.

Again, required the divisors of 92881. The table gives 293 for the least divisor. Dividing by it, the quotient is 317. Referring to the tables for 317, a cipher is found in the place of the least divisor, and this signifies that 317 is a prime number.

Therefore $92881 = 293 \times 317$, the product of two primes.

It may be remarked that, to have resolved 92881 into these factors without the aid of the tables by the method of Art. 360, would have involved fifty-nine fruitless trial divisions by prime numbers.

	00	03	06	09	12	15	18	21	24	27	30	33	36	39	42	45	48	51	54	57	60	63	66	69	72	75	78	81	84	87
01	0	7	0	17	0	19	0	11	7	37	0	0	13	47	0	7	0	0	11	0	17	0	7	67	19	13	29	0	31	7
07	0	0	0	0	17	11	13	7	29	0	31	0	0	0	7	0	11	0	0	13	0	7	0	0	0	0	37	11	7	0
11	0	0	13	0	7	0	0	0	0	0	0	7	23	0	0	13	17	19	7	0	0	0	11	0	0	7	73	0	13	31
13	0	0	0	11	0	17	7	0	19	0	23	0	0	7	11	0	0	0	0	29	7	59	17	31	0	11	13	7	47	0
17	0	0	0	7	0	37	23	29	0	11	7	31	0	0	0	0	0	7	0	0	11	0	13	0	7	0	0	0	19	23
19	0	11	0	0	23	7	17	13	41	0	0	0	7	0	0	0	61	0	0	7	13	71	0	11	0	73	7	23	0	0
23	0	17	7	13	0	0	0	11	0	7	0	0	0	0	41	0	7	47	11	59	19	0	37	7	31	0	0	0	0	11
29	0	7	17	0	0	11	31	0	7	0	13	0	19	0	0	7	11	23	61	17	0	0	7	13	0	0	0	11	0	7
31	0	0	0	7	0	0	0	0	11	0	7	0	0	0	0	23	0	7	0	11	37	13	19	29	7	17	41	47	0	0
37	0	0	7	0	0	29	11	0	0	7	0	47	0	31	19	13	7	11	0	0	0	0	0	7	0	0	17	79	11	0
41	0	11	0	0	17	23	7	0	0	0	0	13	11	7	0	19	47	53	0	0	7	17	29	11	13	0	0	7	23	0
43	0	7	0	23	11	0	19	0	7	13	17	0	0	0	0	7	29	37	0	0	0	0	7	53	0	19	11	17	0	7
47	0	0	0	0	29	7	0	19	0	41	11	0	7	0	31	0	37	0	13	7	0	11	17	0	0	0	7	0	0	0
49	7	0	11	13	0	0	43	7	31	0	0	17	41	11	7	0	13	19	0	0	23	7	61	0	11	0	47	29	7	13
53	0	0	0	0	7	0	17	0	11	0	43	7	13	59	0	29	23	0	7	11	0	0	0	17	0	7	0	31	79	0
59	0	0	0	7	0	0	11	17	0	31	7	0	0	37	0	47	43	7	53	13	73	0	0	0	7	0	29	41	11	19
61	0	19	0	31	13	7	0	0	23	11	0	0	7	17	0	0	0	13	43	7	11	0	0	0	53	0	7	0	0	0
67	0	0	23	0	7	0	0	11	0	0	0	7	19	0	17	0	31	0	7	73	0	0	59	0	13	7	0	0	0	11
71	0	7	11	0	31	0	0	13	7	17	37	0	0	11	0	7	0	0	0	29	13	23	7	0	11	67	17	0	43	7
73	0	0	0	7	19	11	0	41	0	47	7	0	0	29	0	17	11	7	13	23	0	0	0	19	7	0	0	11	37	31
77	7	13	0	0	0	19	0	7	0	0	17	11	0	41	7	23	0	31	0	53	59	7	11	0	19	0	0	13	7	67
79	0	0	7	11	0	0	0	0	37	7	0	31	13	23	11	19	7	0	0	0	0	0	0	7	29	11	0	0	61	0
83	0	0	0	0	0	0	7	37	13	11	0	17	29	7	0	0	19	71	0	0	7	13	41	0	0	0	0	7	17	0
89	0	0	13	23	0	7	0	11	19	0	0	0	7	0	0	13	0	0	11	7	0	0	0	29	37	0	7	19	13	11
91	7	17	0	0	0	37	31	7	47	0	11	0	0	13	7	0	67	29	17	0	0	7	0	0	23	0	13	0	7	59
97	0	0	17	0	0	0	7	13	11	0	19	43	0	7	0	0	59	0	23	11	7	0	37	0	0	71	53	7	29	19

	01	04	07	10	13	16	19	22	25	28	31	34	37	40	43	46	49	52	55	58	61	64	67	70	73	76	79	82	85	88
01	0	0	0	7	0	0	0	31	41	0	7	19	0	0	11	43	13	7	0	0	0	37	0	0	7	11	0	59	0	13
03	0	13	19	17	0	7	11	0	0	0	29	41	7	0	13	0	0	11	0	7	17	19	0	47	67	0	7	13	11	0
07	0	11	7	19	0	0	0	0	23	7	13	0	11	0	59	17	7	41	0	0	31	43	19	7	0	0	0	29	47	0
09	0	0	0	0	7	0	23	47	13	53	0	7	0	19	31	11	0	0	7	37	41	13	0	43	0	7	11	0	67	23
13	0	7	23	0	13	0	0	0	7	29	11	0	47	0	19	7	17	13	37	0	0	11	7	0	71	23	41	43	0	7
19	7	0	0	0	0	0	19	7	11	0	0	13	0	0	7	31	0	17	0	11	29	7	0	0	13	19	0	0	7	0
21	11	0	7	0	0	0	17	0	0	7	0	11	61	0	29	0	7	23	0	0	0	0	11	7	0	0	89	0	0	0
27	0	7	0	13	0	0	41	17	7	11	53	23	0	0	0	7	13	0	0	0	11	0	7	0	17	29	0	19	0	7
31	0	0	17	0	11	7	0	23	0	19	31	47	7	29	61	11	0	0	0	7	0	59	53	79	0	13	7	0	19	0
33	7	0	0	0	31	23	0	7	17	0	13	0	0	37	7	41	0	0	11	19	0	7	0	13	0	17	0	0	7	11
37	0	19	11	17	7	0	13	0	43	0	0	7	37	11	0	0	0	0	7	13	17	41	0	31	11	7	0	0	0	0
39	0	0	0	0	13	11	7	0	0	17	43	19	0	7	0	0	11	13	29	0	7	47	23	0	41	0	17	7	0	0

49	0	0	7	0	19	17	0	13	0	7	47	0	23	0	0	0	7	29	31	0	11	0	17	7	0	0	0	73	83	0
51	0	11	0	0	7	13	0	0	0	0	23	7	11	0	19	0	0	59	7	0	0	0	43	11	0	7	0	37	17	53
57	0	0	0	7	23	0	19	37	0	0	7	0	13	0	0	0	0	7	0	0	47	11	29	0	7	13	73	23	43	17
61	7	0	0	0	0	11	37	7	13	0	29	0	0	31	7	59	11	0	67	0	61	7	0	23	17	47	19	11	7	0
63	0	0	7	0	29	0	13	31	11	7	0	0	53	17	0	0	7	19	0	11	0	23	0	7	37	79	0	0	0	0
67	0	0	13	11	0	0	7	0	17	47	0	0	0	7	11	13	0	23	19	0	7	29	67	37	53	11	31	7	13	0
69	13	7	0	0	37	0	11	0	7	19	0	0	0	13	17	7	0	11	0	0	31	0	7	0	0	0	13	0	11	7
73	0	11	0	29	0	7	0	0	31	13	19	23	7	0	0	0	0	0	0	7	0	0	13	11	73	0	7	0	0	19
79	0	0	19	13	7	23	0	43	0	0	11	7	0	0	29	0	13	0	7	0	37	11	0	0	47	7	79	17	23	13
81	0	13	11	23	0	41	7	0	29	43	0	59	19	7	13	31	17	0	0	0	7	0	0	73	11	0	23	7	0	83
87	11	0	0	0	19	7	0	0	13	0	0	11	7	61	41	43	0	17	37	7	23	13	11	19	83	0	7	0	31	0
91	0	0	7	0	13	19	11	29	0	7	0	0	17	0	0	0	7	11	0	43	41	0	0	7	19	0	61	0	11	17
93	0	17	13	0	7	0	0	0	0	11	31	7	0	0	23	13	0	67	7	71	11	43	0	41	0	7	0	0	13	0
97	0	7	0	0	11	0	0	0	7	0	23	13	0	17	0	7	19	0	29	0	0	73	7	47	13	43	11	0	0	7
99	0	0	17	7	0	0	0	11	23	13	7	0	29	0	53	37	0	7	11	17	0	67	13	31	7	0	19	43	0	11

	02	05	08	11	14	17	20	23	26	29	32	35	38	41	44	47	50	53	56	59	62	65	68	71	74	77	80	83	86	89
03	7	0	11	0	23	13	0	7	19	0	0	31	0	11	7	0	0	0	13	0	0	7	0	0	11	0	53	19	7	29
09	11	0	0	0	0	0	7	0	0	0	0	11	13	7	0	17	0	0	71	19	7	23	11	0	31	13	0	7	0	59
11	0	7	0	11	17	29	0	0	7	41	13	0	37	0	11	7	0	47	31	23	0	17	7	13	0	11	0	0	79	7
17	7	11	19	0	13	17	0	7	0	0	0	0	11	23	7	53	29	13	41	61	0	7	17	11	0	0	0	0	7	37
21	13	0	0	19	7	0	43	11	0	23	0	7	0	13	0	0	0	17	7	31	0	0	19	0	41	7	13	53	37	11
23	0	0	0	0	0	0	7	23	43	37	11	13	0	7	0	0	0	0	0	0	7	11	0	17	13	0	71	7	0	0
27	0	17	0	7	0	11	0	13	37	0	7	0	43	0	19	29	11	7	17	0	13	61	0	0	7	0	23	11	0	79
29	0	23	0	0	0	7	0	17	11	29	0	0	7	0	43	0	47	73	13	7	0	0	0	0	17	59	7	0	0	0
33	0	13	7	11	0	0	19	0	0	7	53	0	0	0	11	0	7	0	43	17	23	47	0	7	0	11	29	13	89	0
39	0	7	0	17	0	37	0	0	7	0	41	0	11	0	23	7	0	19	0	0	17	13	7	11	43	71	0	31	53	7
41	0	0	29	7	11	0	13	0	19	17	7	0	23	41	0	11	71	7	0	13	79	31	0	37	7	0	11	19	0	0
47	13	0	7	31	0	0	23	0	0	7	17	0	0	11	0	47	7	0	0	19	0	0	41	7	11	61	13	17	0	23
51	0	19	23	0	0	17	7	0	11	13	0	53	0	7	0	0	0	0	0	11	7	0	13	0	0	23	83	7	41	0
53	11	7	0	0	0	0	0	13	7	0	0	11	0	0	61	7	31	53	0	0	13	0	7	23	29	0	0	0	17	7
57	0	0	0	13	31	7	11	0	0	0	0	0	7	0	0	67	13	11	0	7	0	79	0	17	0	0	7	61	11	13
59	7	13	0	19	0	0	29	7	0	11	0	0	17	0	7	0	0	23	0	59	11	7	19	0	0	0	0	13	7	17
63	0	0	0	0	7	41	0	17	0	0	13	7	0	23	0	11	61	31	7	67	0	0	0	13	17	7	11	0	0	0
69	0	0	11	7	13	29	0	23	17	0	7	43	53	11	41	19	37	7	0	47	0	0	0	67	7	17	0	0	0	0
71	0	0	13	0	0	7	19	0	0	0	0	0	7	43	17	13	11	41	53	7	0	0	0	71	31	19	7	11	13	0
77	0	0	0	11	7	0	31	0	0	13	29	7	0	0	11	17	0	19	7	43	0	0	13	0	0	7	41	0	0	47
81	0	7	0	0	0	13	0	0	7	11	17	0	0	37	0	7	0	0	13	0	11	0	7	43	0	31	0	17	0	7
83	0	11	0	7	0	0	0	0	0	19	7	0	11	47	0	0	13	7	0	31	61	29	0	11	7	43	59	83	19	13
87	7	0	0	0	0	0	0	7	0	29	19	17	13	53	7	0	0	0	11	0	0	7	71	0	0	13	0	0	7	11
89	17	19	7	29	0	0	0	0	0	7	11	37	0	59	67	0	7	17	0	53	19	11	83	7	0	0	0	0	0	89
93	0	0	19	0	0	11	7	0	0	41	37	0	17	7	0	0	11	0	0	13	7	19	61	0	59	0	0	7	0	17
99	13	0	29	11	0	7	0	0	0	0	0	59	7	13	11	0	0	0	41	7	0	0	0	23	0	11	7	37	0	0

	90	93	96	99	102	05	08	11	14	17	20	23	26	29	32	35	38	41	44	47	50	53	56	59	62	65	68	71	74	77
01	0	71	0	0	101	0	7	17	13	0	11	0	0	7	43	23	37	59	0	61	7	11	0	0	17	29	53	7	0	31
07	0	41	13	0	59	7	101	29	11	23	0	31	7	0	47	13	0	0	0	7	43	0	0	0	19	17	7	0	13	0
11	0	0	7	11	0	23	19	41	0	7	0	13	0	0	11	59	7	103	0	47	17	61	67	7	13	11	0	71	23	89
13	0	67	0	23	7	0	11	0	101	13	41	7	0	37	73	0	19	11	7	0	0	0	13	0	31	7	17	109	11	0
17	71	7	59	47	17	13	29	0	7	0	61	109	11	0	0	7	41	19	13	0	0	17	7	11	0	83	67	0	0	7
19	29	0	0	7	11	67	31	0	19	0	7	97	0	0	0	11	13	7	0	41	23	0	0	0	7	0	11	17	0	13
23	7	0	0	0	0	17	79	7	0	19	11	0	13	0	7	0	23	29	0	0	83	7	17	0	0	13	0	0	7	37
29	0	19	0	0	53	0	7	31	11	37	23	0	73	7	0	83	0	71	47	11	7	0	0	17	0	0	0	7	29	0
31	11	7	0	0	13	0	0	0	7	0	53	11	17	67	101	7	0	13	0	0	0	0	7	89	0	61	0	37	0	7
37	7	0	23	19	29	41	0	7	0	11	0	13	0	17	7	0	101	67	0	0	11	7	19	0	13	23	113	0	7	0
41	0	0	31	0	7	83	37	13	17	59	0	7	0	0	0	11	0	79	7	0	13	23	0	19	109	7	11	61	107	113
43	0	0	0	61	0	13	7	11	0	0	0	0	47	7	17	29	109	0	11	23	7	67	0	107	37	71	0	7	0	11
47	83	13	11	7	0	53	0	71	0	17	7	0	0	11	13	19	61	7	0	0	41	103	0	37	7	0	17	13	73	0
49	0	0	0	0	37	7	19	0	107	31	0	53	7	23	0	17	11	0	0	7	101	0	0	41	0	13	7	11	0	0
53	11	47	7	37	0	61	0	19	13	7	17	11	0	0	29	0	7	0	97	0	0	13	11	7	0	0	19	17	31	41
59	0	7	13	23	0	0	0	0	7	11	31	17	0	0	0	7	0	0	19	0	11	0	7	0	71	29	23	0	13	7
61	13	11	0	7	31	59	0	0	73	19	7	47	11	13	89	71	83	7	0	29	0	0	0	11	7	0	13	131	19	0
67	0	17	7	0	0	0	0	13	0	7	11	83	53	0	0	0	7	31	17	0	13	11	0	7	0	0	101	0	0	109
71	47	0	19	13	0	11	7	0	0	79	0	89	0	7	23	41	11	37	29	0	7	19	0	0	53	73	0	7	0	13
73	43	7	17	0	0	97	83	0	7	61	0	0	19	0	13	7	0	0	41	11	0	0	7	0	0	0	47	13	101	7
77	29	0	0	11	43	7	73	0	23	0	13	0	7	19	11	0	0	0	31	7	0	0	61	13	41	11	7	89	0	29
79	7	83	0	17	19	71	11	7	13	0	47	0	31	0	7	37	0	11	0	0	17	7	0	19	73	59	0	41	7	23
83	31	11	23	67	7	19	0	53	0	0	43	7	11	0	37	17	0	13	7	0	0	0	0	11	19	7	0	0	0	0
89	61	41	0	7	0	0	0	67	0	0	7	13	0	31	97	107	17	7	0	23	79	11	29	59	7	53	0	0	0	0
91	0	0	11	97	41	7	0	19	0	13	107	0	7	11	0	0	29	23	43	7	0	0	13	0	11	47	7	0	0	0
97	11	0	0	13	7	0	17	0	0	47	0	7	0	41	0	0	13	0	7	0	31	89	11	17	43	7	61	29	0	13

	91	94	97	100	03	06	09	12	15	18	21	24	27	30	33	36	39	42	45	48	51	54	57	60	63	66	69	72	75	78
01	19	7	89	73	0	0	11	23	7	0	0	0	13	0	47	7	0	11	17	19	0	0	7	0	0	13	0	103	11	7
03	0	0	31	7	0	23	0	17	0	11	7	79	0	0	53	61	0	7	0	113	11	73	41	13	7	0	0	0	23	19
07	7	23	17	0	11	0	13	7	37	0	0	19	97	0	7	11	0	0	89	13	0	7	113	0	23	0	11	0	7	0
09	0	97	7	0	13	103	0	11	17	7	0	0	71	0	0	31	7	13	11	59	29	19	23	7	47	17	37	0	0	11
13	13	0	11	17	0	0	7	0	29	0	0	0	0	7	0	0	0	61	23	0	7	0	19	67	11	37	13	7	83	47
19	11	0	0	43	17	7	61	13	0	53	0	11	7	47	19	0	31	59	0	7	13	17	11	83	0	0	7	67	0	103
21	7	0	0	11	0	13	67	7	41	0	17	0	0	29	7	53	0	0	13	0	0	7	79	37	19	11	0	17	7	71
27	0	11	71	37	23	0	7	103	0	0	67	17	11	7	0	0	19	41	73	0	7	0	0	11	29	13	0	7	17	0
31	23	0	37	7	0	0	17	11	13	0	7	31	29	83	0	43	0	7	11	0	0	13	0	17	7	0	0	0	47	11
33	0	0	0	79	0	7	13	47	19	0	11	0	7	0	67	0	0	43	0	7	37	11	0	0	0	0	7	19	89	17
37	0	0	7	0	0	11	0	17	83	7	53	0	47	0	0	13	7	23	0	37	0	43	0	7	17	127	0	11	13	0
39	13	0	0	0	7	0	0	0	11	0	61	7	0	13	0	23	53	29	7	11	0	0	0	43	0	7	13	0	0	0
43	41	7	0	11	0	29	31	0	7	13	0	23	0	0	11	7	73	0	0	0	19	0	7	61	59	11	0	43	53	7

49	7	11	0	13	79	23	0	7	0	17	0	59	11	0	7	0	13	0	0	31	0	7	0	11	0	0	17	47	7	13
51	0	13	7	19	11	0	47	0	0	7	29	0	41	31	13	11	7	0	0	0	109	0	19	7	83	0	11	13	0	0
57	0	7	11	89	0	0	0	0	7	71	0	0	0	11	19	7	17	53	0	83	23	13	7	0	11	0	31	0	97	7
61	0	0	43	0	13	7	97	0	11	29	0	17	7	37	31	19	23	13	0	7	0	0	0	0	0	0	7	41	17	53
63	7	0	13	29	43	0	19	7	31	0	0	11	0	0	7	13	0	17	0	89	59	7	11	0	0	19	0	61	7	0
67	89	0	0	0	7	0	11	19	43	0	23	7	17	73	0	79	0	11	7	0	29	0	0	0	13	7	19	31	11	17
69	53	17	0	0	0	47	7	59	23	11	43	37	113	7	29	0	61	19	17	0	7	31	13	0	0	79	71	7	0	107
73	0	0	29	7	11	13	0	0	71	31	7	0	53	17	43	11	89	7	13	107	0	0	0	0	7	0	11	23	0	61
79	67	0	7	0	97	59	0	0	0	7	19	0	13	11	17	0	7	109	61	0	43	23	31	7	11	13	0	37	0	19
81	0	19	0	17	7	11	79	29	37	109	13	7	0	103	0	0	11	0	7	23	17	113	43	13	0	7	0	11	0	0
87	0	53	0	7	13	0	0	0	0	0	7	0	19	23	11	0	71	7	29	0	0	17	0	0	7	11	0	59	43	31
91	7	0	0	0	0	0	29	7	67	11	73	0	0	13	7	0	17	31	0	0	11	7	0	0	37	0	13	0	7	0
93	29	11	7	0	19	17	0	23	0	7	89	13	11	0	59	0	7	0	0	53	0	0	17	7	13	0	0	0	73	29
97	17	0	97	23	37	19	7	11	0	0	0	0	67	7	0	0	0	17	11	0	7	0	0	0	19	59	23	7	0	11
99	0	7	41	0	0	13	17	0	7	73	11	29	0	0	0	7	0	79	13	47	0	11	7	17	23	0	89	0	0	7

	92	95	98	1 01	04	07	10	13	16	19	22	25	28	31	34	37	40	43	46	49	52	55	58	61	64	67	70	73	76	79
03	0	13	0	0	101	7	0	89	41	0	0	0	7	0	13	71	11	0	17	7	23	37	0	0	47	0	7	11	29	0
09	0	37	17	11	7	0	101	43	13	0	29	7	0	0	11	0	0	41	7	17	67	13	0	89	61	7	73	19	0	0
11	61	0	0	0	29	0	7	0	17	43	0	0	23	7	0	0	0	11	19	13	7	0	97	0	0	17	0	7	11	0
17	13	31	0	67	11	7	23	0	0	17	19	0	7	13	0	11	107	103	47	7	0	59	0	71	0	73	7	0	79	19
21	0	0	7	29	17	71	103	0	0	7	11	19	0	0	0	0	7	0	0	43	31	11	13	7	0	23	0	0	67	0
23	23	89	11	53	7	0	73	13	59	0	17	7	0	11	31	0	37	0	7	0	13	19	0	23	11	7	29	17	0	0
27	0	7	31	13	0	17	0	47	7	0	0	0	101	0	29	7	13	0	0	11	0	0	7	0	0	43	0	0	0	7
29	11	13	0	7	0	0	41	0	29	79	7	11	0	19	13	0	0	7	0	0	97	53	11	127	7	0	0	13	17	0
33	7	0	0	0	0	0	11	7	0	0	13	83	41	23	7	31	0	11	0	109	0	7	71	13	0	29	0	0	7	79
39	0	0	0	0	11	0	7	17	103	0	0	0	37	7	89	11	101	13	0	0	7	41	47	0	17	19	11	7	31	0
41	0	7	13	0	53	23	61	11	7	0	0	0	0	17	0	7	19	0	11	67	0	0	7	0	41	0	0	0	13	7
47	7	0	43	73	31	11	0	7	19	13	37	0	29	0	7	59	11	0	97	0	79	7	13	67	0	0	0	11	7	131
51	11	0	0	0	7	13	43	0	61	17	0	7	71	0	0	0	0	113	7	0	101	0	11	31	0	7	17	0	19	29
53	19	41	59	11	0	0	7	0	43	0	0	0	0	7	11	17	13	31	0	19	7	103	83	29	0	11	0	7	127	13
57	0	19	0	7	0	31	0	41	0	11	7	29	13	59	0	0	0	7	0	0	11	47	101	107	7	13	37	17	0	0
59	47	11	0	0	0	7	0	37	89	0	13	19	7	0	43	0	17	83	107	7	0	0	0	11	109	0	7	0	0	0
63	59	73	7	0	0	47	13	11	107	7	0	17	19	0	0	0	7	53	11	13	0	79	29	7	101	0	113	97	17	11
69	13	7	71	0	19	11	0	0	7	0	0	0	17	13	0	7	11	0	0	0	0	0	7	19	43	41	13	11	0	7
71	73	17	0	7	37	0	0	83	11	0	7	13	61	0	19	47	0	7	17	11	0	23	59	103	7	31	43	29	41	0
77	0	61	7	0	0	13	11	31	0	7	0	0	79	0	0	23	7	11	13	17	0	37	0	7	0	19	0	0	11	0
81	0	11	41	0	47	0	7	19	0	0	0	23	11	7	13	0	0	73	53	71	7	0	0	11	0	97	19	7	0	0
83	0	7	0	17	11	41	0	0	7	23	71	0	13	0	97	7	0	19	0	0	17	0	7	0	53	13	11	0	0	7
87	37	0	0	61	0	7	0	59	13	0	11	41	7	0	0	17	0	0	19	7	0	11	0	0	0	0	7	0	23	0
89	7	43	11	23	17	0	13	7	0	19	0	0	0	11	7	0	73	0	37	13	0	7	0	0	11	103	23	0	7	0
93	0	53	13	0	7	43	0	0	11	67	19	7	0	79	103	13	17	37	7	11	41	31	23	0	0	7	0	0	13	19
99	17	29	19	7	0	0	11	0	0	13	7	43	0	67	0	0	23	7	0	53	0	19	13	97	7	107	0	127	11	41

	1 80	83	86	89	92	95	1 98	2 01	04	07	10	13	16	19	22	25	28	31	34	37	40	43	46	49	52	55	58	61	64	67
01	47	0	11	41	7	0	0	0	23	127	0	7	0	11	149	0	151	13	7	137	0	19	73	37	11	7	0	43	17	0
07	11	0	23	7	0	0	29	0	0	0	7	11	17	19	53	71	0	7	89	151	0	109	11	0	7	23	131	0	0	17
11	7	0	37	0	0	109	11	7	0	139	0	101	0	0	7	0	0	11	41	131	13	7	0	29	17	97	53	0	7	0
13	0	0	7	0	0	13	0	0	137	7	0	0	0	17	97	47	7	29	13	23	11	41	151	7	19	31	83	0	61	0
17	43	13	0	0	11	29	7	0	17	0	0	0	0	7	13	11	0	0	0	37	7	0	103	0	151	17	11	7	0	0
19	37	7	43	0	0	131	0	11	7	0	0	0	13	23	17	7	19	61	11	0	0	83	7	0	0	13	0	0	29	7
23	67	73	11	127	47	7	43	0	13	17	0	0	7	11	71	101	29	19	59	7	0	13	0	0	11	0	7	151	0	0
29	11	0	13	23	7	59	79	0	31	19	17	7	43	0	0	13	37	101	7	61	0	0	11	97	0	7	23	17	13	0
31	13	23	31	11	0	0	7	41	0	0	0	83	97	7	11	0	17	0	0	19	7	29	0	107	23	11	13	7	0	0
37	17	11	0	29	0	7	83	13	107	89	109	19	7	0	37	31	41	17	23	7	13	0	71	11	0	0	7	59	0	0
41	0	0	7	13	71	0	0	11	0	7	53	0	17	37	23	0	7	73	11	0	29	101	41	7	43	0	0	0	137	11
43	0	13	103	19	7	0	0	0	0	0	11	7	23	0	13	0	53	0	7	0	0	11	19	0	0	7	43	13	31	47
47	0	7	29	0	19	11	89	0	7	0	13	0	0	17	0	7	11	79	0	0	139	97	7	13	0	59	0	11	53	7
49	0	59	17	7	0	113	23	0	11	0	7	37	0	47	19	0	73	7	131	11	0	13	157	61	7	29	0	79	0	23
53	7	0	23	11	13	0	0	7	113	0	37	131	59	29	7	19	0	13	47	0	67	7	89	0	0	11	103	0	7	31
59	0	11	47	0	0	0	7	19	41	0	0	13	11	7	0	17	0	0	0	23	7	0	0	11	13	61	19	7	0	0
61	0	7	0	67	11	31	0	0	7	13	0	41	0	0	113	7	0	19	29	0	0	17	7	109	0	0	11	0	47	7
67	7	0	11	13	0	17	0	7	97	19	0	23	47	11	7	0	13	0	31	0	41	7	17	0	11	37	0	137	7	13
71	17	0	0	61	7	0	31	23	11	0	19	7	13	127	0	0	0	17	7	11	0	0	0	0	37	7	41	0	103	19
73	11	19	71	0	0	23	7	0	59	0	13	11	0	7	0	0	89	0	0	0	7	0	11	13	127	107	0	7	23	41
77	0	17	19	7	37	0	11	0	0	79	7	0	53	0	0	107	0	7	17	13	0	19	0	0	7	0	113	0	11	0
79	101	0	0	0	13	7	103	17	0	11	107	0	7	31	0	67	137	13	53	7	11	0	23	0	17	0	7	47	0	61
83	13	31	7	41	11	0	59	0	0	7	29	0	0	13	0	11	7	97	23	17	0	37	0	7	131	0	11	0	71	0
89	0	7	11	17	0	19	0	13	7	0	0	73	23	11	31	7	47	0	83	0	13	29	7	0	11	0	0	0	0	7
91	79	53	0	7	101	11	0	61	31	17	7	0	109	0	0	19	11	7	13	37	0	0	0	67	7	157	17	11	59	73
97	0	0	7	11	23	0	101	19	103	7	17	0	13	0	11	59	7	0	0	53	0	31	0	7	41	11	19	17	0	127

	1 81	84	87	90	93	96	1 99	2 02	05	08	11	14	17	20	23	26	29	32	35	38	41	44	47	50	53	56	59	62	65	68
01	23	0	0	0	0	17	7	0	13	11	0	0	0	7	29	97	0	0	71	0	7	13	17	23	0	0	59	7	0	0
03	43	7	59	31	97	0	13	89	7	71	47	17	11	0	0	7	37	0	19	13	0	23	7	11	0	0	0	0	17	7
07	19	79	13	83	43	7	17	11	0	0	0	0	7	59	0	13	0	23	11	7	0	0	31	17	0	29	7	73	13	11
09	7	41	53	0	0	0	43	7	0	0	11	79	17	13	7	23	31	0	0	29	0	7	0	89	0	0	13	0	7	17
13	59	0	0	0	7	11	0	17	73	13	43	7	0	0	53	0	11	139	7	0	0	0	13	0	17	7	0	11	0	0
19	0	113	0	7	0	23	0	0	17	109	7	0	37	97	11	0	13	7	29	0	89	0	19	127	7	11	0	157	23	13
21	0	13	97	23	139	7	11	73	0	47	0	31	7	19	13	0	0	11	43	7	0	0	59	131	0	0	7	13	11	0
27	0	0	61	53	7	19	0	113	13	59	37	7	0	0	83	11	101	0	7	0	23	13	79	29	19	7	11	0	41	139
31	0	7	0	0	13	67	19	0	7	37	11	29	31	0	137	7	23	13	0	0	59	11	7	0	73	19	0	17	43	7
33	0	0	11	7	0	29	31	0	0	83	7	0	103	11	23	13	17	7	101	0	0	53	0	0	7	0	0	37	13	0
37	7	103	41	0	61	73	0	7	11	67	23	13	0	0	7	0	0	19	0	11	0	7	29	0	13	31	37	0	7	47
39	11	0	7	79	83	41	127	37	19	7	0	11	0	0	89	0	7	17	0	31	101	0	11	7	0	0	0	19	0	0
43	0	0	0	137	23	13	7	31	0	19	0	41	17	7	0	0	0	11	13	113	7	0	109	79	0	0	0	7	11	17

	1 82	85	88	91	94	1 97	2 00	03	06	09	12	15	18	21	24	27	30	33	36	39	42	45	48	51	54	57	60	63	66	69
49	0	19	0	43	11	7	0	0	0	0	0	89	7	17	0	11	53	67	0	7	19	23	0	37	0	13	7	0	139	0
51	7	0	17	0	37	43	71	7	0	29	13	19	0	0	7	0	59	0	11	17	0	7	53	13	101	113	0	0	7	11
57	67	0	0	17	13	11	7	47	61	0	0	43	0	7	79	139	11	13	0	0	7	37	19	0	0	0	101	7	0	107
61	11	0	73	7	19	0	0	0	29	23	7	11	47	13	59	17	0	7	0	107	37	61	11	19	7	67	13	0	0	0
63	41	37	29	11	17	7	0	23	0	31	0	13	7	0	11	131	0	43	0	7	73	17	0	71	13	11	7	0	101	0
67	37	59	7	23	107	71	41	13	131	7	61	0	0	0	0	19	7	53	0	29	11	43	0	7	0	0	23	0	31	67
69	0	11	137	0	7	13	19	0	67	41	0	7	11	29	0	0	103	0	7	0	0	0	17	11	23	7	0	109	163	97
73	17	7	0	0	0	103	0	11	7	0	31	109	0	0	13	7	0	17	11	0	23	0	7	0	0	0	19	13	0	7
79	7	17	89	0	0	11	0	7	13	0	0	47	29	0	7	0	11	0	17	0	0	7	71	31	41	0	83	11	7	0
81	0	0	7	0	0	0	13	17	11	7	59	0	23	71	0	37	7	31	0	11	0	0	0	7	17	61	0	41	19	0
87	13	7	0	0	0	0	11	0	7	0	0	0	0	13	61	7	127	11	103	0	19	47	7	0	53	17	13	97	11	7
91	0	11	19	17	0	7	0	103	59	13	0	0	7	0	0	0	83	0	31	7	17	19	13	11	0	23	7	61	0	0
93	7	0	0	61	11	47	0	7	0	17	0	0	19	0	7	11	0	0	0	0	13	7	0	23	67	0	11	0	7	0
97	31	53	0	13	7	0	0	0	43	0	11	7	71	19	0	0	13	0	7	23	0	11	137	0	109	7	0	0	0	13
99	0	13	11	71	19	0	7	53	0	0	17	0	0	7	13	0	109	23	0	0	7	0	0	19	11	31	0	7	67	37
	1 82	85	88	91	94	1 97	2 00	03	06	09	12	15	18	21	24	27	30	33	36	39	42	45	48	51	54	57	60	63	66	69
03	109	0	0	7	0	17	83	79	11	0	7	0	0	23	43	73	0	7	0	11	0	107	17	13	7	0	0	29	37	0
09	131	83	7	97	13	0	11	23	37	7	127	137	113	0	0	0	7	11	0	0	43	0	0	7	0	47	31	0	11	71
11	0	107	13	29	7	23	0	19	0	11	0	7	17	0	73	13	0	0	7	0	11	127	43	0	0	7	19	83	13	17
17	0	0	31	7	0	0	37	11	53	13	7	0	0	17	29	0	0	7	11	0	61	0	13	0	7	0	0	0	43	11
21	7	0	11	0	0	13	0	7	17	0	0	0	0	11	7	0	0	0	13	19	53	7	0	0	11	17	0	0	7	0
23	0	0	7	13	0	11	0	0	41	7	19	0	139	0	17	31	7	83	0	47	0	137	103	7	0	29	53	11	79	13
27	11	97	67	31	0	0	7	0	0	17	0	11	13	7	41	0	0	0	0	71	7	0	11	0	47	13	17	7	0	0
29	0	7	19	11	0	109	0	29	7	0	13	0	83	0	11	7	0	41	0	0	0	19	7	13	59	11	0	113	31	7
33	0	43	37	19	0	7	13	0	47	11	17	61	7	0	0	127	31	0	0	7	11	0	19	41	29	0	7	17	0	23
39	13	0	0	0	7	0	29	11	0	0	67	7	0	13	19	0	0	0	7	37	0	53	59	23	0	7	13	0	17	11
41	17	0	83	0	0	19	7	0	0	43	11	13	0	7	0	0	0	17	47	89	7	11	0	31	13	0	0	7	0	29
47	71	17	47	41	0	7	0	0	11	0	0	29	7	0	0	23	19	37	13	7	0	0	0	0	0	0	7	0	0	0
51	0	13	7	11	53	0	0	47	107	7	79	23	0	17	11	0	7	19	67	43	0	0	0	7	31	11	109	13	29	0
53	0	0	17	107	7	0	11	0	19	23	53	7	13	0	0	61	0	11	7	17	79	43	29	0	0	7	0	19	11	0
57	0	7	109	0	0	23	31	0	7	19	29	0	11	0	17	7	0	0	41	0	127	13	7	11	0	43	71	0	19	7
59	19	67	0	7	11	0	13	0	73	0	7	0	0	0	37	11	0	7	59	13	17	41	0	139	7	0	11	43	53	0
63	7	19	13	0	0	0	0	7	0	0	11	0	0	37	7	13	0	61	0	31	19	7	23	0	0	0	67	41	7	59
69	0	31	0	29	0	53	7	0	11	13	0	0	19	7	0	0	17	0	0	11	7	79	13	0	0	73	131	7	0	149
71	11	7	113	19	0	17	0	13	7	67	89	11	0	0	23	7	0	0	0	0	13	0	7	0	0	0	29	0	149	7
77	7	13	43	127	0	0	17	7	23	11	0	0	131	67	7	0	47	97	0	0	11	7	0	17	73	149	89	13	7	53
81	101	17	79	0	7	131	43	89	0	0	13	7	0	41	0	11	0	103	7	0	0	47	139	13	83	7	11	23	0	0
83	47	0	23	0	0	73	7	11	13	0	0	113	79	7	0	0	41	67	11	29	7	13	149	0	17	19	0	7	0	11
87	0	0	11	7	13	47	53	19	137	31	7	0	43	11	113	0	0	7	0	17	149	23	41	89	7	107	19	0	0	0
89	0	29	13	31	0	7	0	0	17	139	61	0	7	0	43	13	11	19	0	7	107	67	0	0	71	17	7	11	13	137
93	11	0	7	17	101	0	71	0	0	7	107	11	0	0	83	23	7	149	19	0	17	0	11	7	13	0	97	0	0	0
99	29	7	0	73	17	13	101	0	7	11	19	0	61	79	149	7	0	0	13	103	11	17	7	113	43	0	0	0	0	7

	2 70	73	76	79	82	85	88	91	94	2 97	3 00	03	06	09	12	15	18	21	24	27	30	33	36	39	42	45	48	51	54	57
01	13	23	7	0	0	11	83	0	0	7	19	157	71	13	41	17	7	47	0	53	61	0	0	7	23	0	13	11	0	19
07	113	7	19	11	67	29	0	13	7	61	37	0	127	31	11	7	17	97	23	0	13	19	7	41	79	11	0	0	0	7
11	0	31	0	13	0	7	47	43	0	11	0	17	7	0	23	0	13	163	0	7	11	0	19	0	0	0	7	0	17	13
13	7	11	53	103	89	0	0	7	67	43	0	0	11	19	7	0	29	17	0	0	0	7	0	11	0	0	31	13	7	71
17	0	59	0	0	7	0	0	11	23	0	13	7	17	43	19	0	0	0	7	0	137	0	0	13	0	7	37	0	107	11
19	41	17	71	0	0	19	7	37	13	113	11	0	67	7	0	43	47	0	17	0	7	11	0	107	19	0	0	7	0	23
23	61	89	23	7	13	11	19	0	0	0	7	0	113	17	0	29	11	7	0	43	0	47	0	0	7	19	97	11	0	139
29	151	0	7	11	0	47	127	0	0	7	0	13	109	157	11	41	7	19	0	23	0	0	0	7	13	11	29	0	71	0
31	0	151	0	17	7	103	11	0	19	13	59	7	0	0	0	0	139	11	7	71	17	0	13	0	0	7	61	19	11	0
37	19	0	29	7	11	0	0	0	0	131	7	23	0	0	0	11	13	7	163	19	0	17	0	0	7	0	11	41	0	13
41	7	19	131	0	31	0	151	7	59	0	11	0	13	0	7	0	17	0	0	29	19	7	0	0	97	13	0	0	7	103
43	0	37	7	0	61	17	0	151	0	7	13	19	0	11	157	0	7	0	0	137	173	0	17	7	11	0	0	113	23	31
47	17	23	0	0	47	0	7	0	11	151	0	0	19	7	0	0	0	17	71	11	7	0	0	83	23	179	0	7	0	0
49	11	7	43	19	13	0	17	103	7	71	151	11	0	0	0	7	0	13	37	0	0	0	7	17	29	0	0	0	0	7
53	13	17	0	0	19	7	11	0	0	0	41	127	7	13	0	139	53	11	17	7	0	0	73	19	0	109	7	0	11	0
59	0	109	17	73	7	0	0	13	89	0	0	7	23	83	0	11	0	0	7	17	13	0	97	29	0	7	11	0	59	0
61	0	0	139	0	59	13	7	11	17	0	23	97	0	7	43	37	151	29	11	181	7	73	41	0	0	17	71	7	0	11
67	0	0	73	0	23	7	0	0	79	17	107	0	7	173	0	0	11	19	0	7	43	61	131	0	0	13	7	11	29	47
71	11	101	7	83	17	0	0	31	13	7	0	11	0	0	0	131	7	53	19	0	0	13	11	7	43	181	0	0	79	0
73	0	31	0	11	7	0	13	0	0	19	17	7	37	47	11	0	0	0	7	13	0	23	151	53	0	7	43	17	19	83
77	0	7	13	101	0	17	67	163	7	11	19	37	0	0	0	7	127	23	47	73	11	0	7	61	151	71	0	29	13	7
79	13	11	89	7	0	0	0	0	41	97	7	17	11	13	31	23	71	7	0	0	19	29	0	11	7	151	13	127	17	37
83	7	139	19	0	0	101	17	7	0	13	67	23	61	0	7	0	0	0	11	0	0	7	13	17	0	0	0	151	7	11
89	103	61	0	13	0	11	7	17	37	0	0	0	0	7	67	31	11	0	53	0	7	173	59	41	17	0	139	7	23	13
91	0	7	0	23	19	0	167	0	7	31	0	0	47	17	13	7	0	0	0	11	0	0	7	19	53	0	23	13	0	7
97	7	0	0	0	0	0	11	7	13	83	0	113	0	139	7	19	167	11	0	0	23	7	31	0	0	29	0	61	7	0

	2 71	74	77	80	83	86	89	92	95	2 98	3 01	04	07	10	13	16	19	22	25	28	31	34	37	40	43	46	49	52	55	58
01	41	11	0	0	7	37	0	0	0	17	31	7	11	29	113	0	19	13	7	0	79	127	67	11	0	7	17	0	131	0
03	0	67	13	41	11	0	7	19	163	0	0	0	0	7	23	11	61	0	0	0	7	0	0	37	0	0	11	7	13	0
07	0	0	103	7	0	0	137	0	19	41	7	13	0	101	0	0	0	7	0	53	0	11	37	31	7	0	67	17	0	61
09	0	0	11	37	0	7	0	0	23	13	0	47	7	11	131	73	17	31	19	7	113	0	13	71	11	53	7	137	0	0
13	19	79	7	109	23	13	29	131	11	7	0	17	0	0	173	101	7	0	13	11	0	0	0	7	0	0	0	23	17	59
19	47	7	53	0	0	0	11	61	7	0	0	19	13	0	0	7	59	11	31	37	0	23	7	0	0	13	0	41	11	7
21	37	17	19	7	127	0	0	0	53	11	7	29	31	67	0	103	137	7	17	23	11	19	0	13	7	89	47	0	0	113
27	0	0	7	0	13	0	0	11	0	7	47	0	0	19	0	0	7	13	11	17	157	0	29	7	0	31	53	0	0	11
31	13	0	11	0	41	0	7	0	0	23	29	0	79	7	17	47	37	167	0	0	7	101	89	0	11	0	13	7	0	0
33	43	7	0	17	29	11	0	23	7	0	0	13	73	0	0	7	11	0	0	0	17	67	7	0	13	59	181	11	0	7
37	11	0	0	23	43	7	19	13	0	0	0	11	7	41	0	17	109	0	0	7	13	29	11	101	0	19	7	167	0	0
39	7	23	0	11	17	13	43	7	109	53	0	61	59	0	7	29	19	103	13	0	31	7	0	0	23	11	0	131	7	0

49	17	0	0	7	0	0	0	11	13	19	7	0	97	61	23	0	43	7	11	107	0	13	0	79	7	0	0	101	19	11
51	19	97	0	0	0	7	13	0	29	0	11	37	7	0	107	31	89	0	43	7	0	11	0	17	0	0	7	0	73	0
57	13	0	41	0	7	0	23	17	11	73	53	7	0	13	0	0	0	0	7	11	71	0	0	0	17	7	13	0	31	23
61	157	7	17	11	79	0	0	29	7	13	0	83	19	89	11	7	31	0	0	17	0	0	7	0	0	11	0	37	43	7
63	23	29	0	7	113	0	11	13	17	0	7	41	0	0	79	0	0	7	0	59	13	109	19	23	7	17	0	179	11	0
67	7	11	0	13	19	109	83	7	0	0	97	0	11	47	7	0	13	41	29	23	17	7	0	11	0	0	73	0	7	13
69	101	13	7	0	11	0	59	0	0	7	0	0	29	0	13	11	7	23	0	0	41	0	0	7	0	37	11	13	0	0
73	29	83	0	67	17	53	7	73	0	0	11	31	0	7	137	19	0	59	0	71	7	11	0	13	37	0	41	7	0	29
79	0	0	0	43	13	7	0	19	11	0	103	29	7	0	0	79	113	13	0	7	0	0	17	53	31	0	7	0	47	0
81	7	0	13	0	101	23	73	7	0	0	0	11	0	0	7	13	0	19	31	131	0	7	11	173	0	79	0	0	7	53
87	31	0	37	0	0	0	7	0	0	11	0	43	17	7	0	0	29	83	0	0	7	0	13	89	137	0	59	7	19	17
91	0	37	0	7	11	13	53	17	127	71	7	0	41	0	0	11	0	7	13	31	0	107	0	73	7	113	11	0	0	19
93	71	19	0	13	0	7	79	11	101	167	109	0	7	17	0	41	13	43	11	7	19	0	47	103	163	0	7	29	0	11
97	0	31	7	0	73	0	107	0	17	7	0	0	13	11	0	29	7	0	37	67	89	19	0	7	11	13	79	47	0	0
99	59	107	0	0	7	11	47	83	0	29	13	7	19	137	17	0	11	0	7	167	0	139	73	13	41	7	31	11	97	0

	2 72	75	78	81	84	87	90	93	96	2 99	3 02	05	08	11	14	17	20	23	26	29	32	35	38	41	44	47	50	53	56	59
03	11	7	0	157	0	0	13	0	7	17	0	11	0	19	31	7	0	0	0	13	0	0	7	67	0	0	17	43	0	7
09	7	0	0	0	0	19	0	7	29	11	17	0	0	13	7	37	0	0	0	0	11	7	0	23	19	61	13	17	7	149
11	0	11	7	0	0	0	67	0	0	7	0	13	11	53	101	19	7	79	0	0	0	23	0	7	13	103	157	0	149	0
17	17	7	0	31	157	13	0	19	7	0	11	0	0	29	89	7	101	17	13	0	59	11	7	109	127	149	19	0	0	7
21	163	13	43	61	97	7	0	109	19	0	47	23	7	0	13	0	11	0	0	7	139	0	31	149	0	0	7	11	179	17
23	7	17	0	0	43	0	0	7	11	23	0	131	13	0	7	0	31	0	17	11	0	7	149	0	29	13	0	0	7	0
27	19	0	0	11	7	23	0	0	13	0	167	7	29	17	11	0	0	0	7	19	149	13	0	0	173	7	0	0	23	37
29	73	0	17	23	0	0	7	139	0	173	19	0	0	7	53	0	0	11	67	13	7	0	0	0	0	0	23	7	11	19
33	113	11	13	7	0	59	0	0	0	37	7	19	11	163	17	13	103	7	0	0	167	0	23	11	7	47	53	89	13	0
39	0	0	7	19	0	29	71	0	107	7	11	0	0	0	149	17	7	73	127	0	43	11	13	7	0	0	37	0	157	83
41	0	0	11	107	7	41	113	13	0	79	0	7	0	11	23	0	179	0	7	0	13	17	43	0	11	7	67	59	29	127
47	11	13	0	7	0	17	31	0	23	0	7	11	109	0	13	53	73	7	0	47	0	0	11	0	7	0	101	13	43	103
51	7	0	0	0	23	0	11	7	149	61	13	137	0	0	7	0	0	11	103	83	41	7	0	13	47	19	0	23	7	0
53	0	59	7	47	37	0	17	149	13	7	0	0	0	0	71	113	7	0	0	31	11	13	97	7	131	23	0	0	101	157
57	97	17	89	37	11	149	7	31	47	29	79	0	59	7	83	11	0	13	17	0	7	23	0	0	0	0	11	7	181	41
59	0	7	13	29	149	0	0	11	7	0	0	0	0	0	163	7	0	0	11	23	79	37	7	0	17	0	0	19	13	7
63	137	43	11	0	0	7	0	0	0	19	53	13	7	11	73	23	0	0	89	7	29	0	0	127	11	0	7	0	19	0
69	11	19	29	17	7	13	41	43	0	23	0	7	0	71	0	0	0	0	7	0	17	0	11	47	0	7	0	113	53	0
71	0	79	47	11	71	0	7	23	0	17	0	19	0	7	11	0	13	0	37	0	7	59	0	0	0	11	17	7	0	13
77	0	11	61	19	0	7	0	29	59	31	13	0	7	0	0	43	0	0	41	7	107	0	19	11	23	83	7	17	0	0
81	0	0	7	0	19	17	13	11	67	7	107	53	0	0	0	61	7	0	11	13	23	0	17	7	29	0	0	0	31	11
83	0	0	0	0	7	107	127	0	0	0	11	7	89	0	19	37	0	13	7	0	83	11	31	0	0	7	0	41	17	0
87	13	7	79	71	61	11	17	0	7	157	31	73	67	13	23	7	11	139	0	0	0	0	7	17	0	43	13	11	127	7
89	29	47	167	7	31	0	19	0	11	0	7	13	17	0	0	83	0	7	97	11	0	0	0	179	7	19	0	43	89	17
93	7	41	0	11	0	0	47	7	23	89	0	0	0	0	7	0	67	29	0	0	13	7	0	31	17	11	19	0	7	0
99	0	11	23	163	0	31	7	0	17	131	41	37	11	7	13	0	0	179	19	0	7	0	109	11	0	17	0	7	29	0

	3 60	63	66	69	72	75	78	81	84	87	90	93	96	3 99	4 02	05	08	11	14	17	20	23	26	29	32	35	38	41	44	47
01	7	31	17	0	0	0	103	7	11	13	43	0	199	0	7	101	0	23	19	11	97	7	13	0	0	41	0	0	7	0
07	0	0	0	13	29	0	7	53	193	0	19	23	0	7	31	0	13	11	47	179	7	0	137	107	0	139	71	7	11	13
11	0	11	31	7	127	0	0	23	71	0	7	19	11	107	79	17	37	7	0	53	43	29	0	11	7	13	193	0	89	0
13	0	0	19	0	11	7	0	0	107	0	13	0	7	167	0	11	0	0	0	7	0	17	43	13	79	53	7	31	23	61
17	0	23	7	19	0	0	13	47	41	7	11	0	173	179	131	31	7	0	83	13	0	11	19	7	23	0	43	157	0	97
19	181	0	11	0	7	17	59	0	103	31	0	7	0	11	37	0	0	13	7	0	0	101	17	167	11	7	29	0	43	197
23	13	7	53	0	0	157	109	67	7	0	0	0	0	13	19	7	0	17	23	11	0	0	7	0	0	71	13	0	31	7
29	7	17	0	0	59	0	11	7	83	0	31	67	23	0	7	0	0	11	17	0	13	7	47	0	139	19	41	0	7	0
31	137	47	7	0	31	13	0	17	0	7	23	37	0	73	0	0	7	0	13	29	11	0	89	7	17	101	53	0	157	41
37	0	7	0	43	23	0	157	11	7	0	103	139	13	0	0	7	97	31	11	0	127	0	7	0	0	13	59	19	37	7
41	23	0	11	17	167	7	79	43	13	19	0	0	7	11	0	71	0	0	29	7	17	13	0	23	11	0	7	37	19	0
43	7	0	0	0	0	11	13	7	37	17	0	0	29	59	7	0	11	0	0	13	0	7	0	0	83	0	17	11	7	101
47	11	19	13	0	7	0	0	37	0	0	0	7	41	43	167	13	0	23	7	109	19	17	11	67	59	7	163	131	13	29
49	13	163	67	11	193	0	7	0	0	0	17	19	31	7	11	23	0	0	181	83	7	0	0	29	61	11	13	7	0	73
53	31	0	0	7	0	17	0	0	0	11	7	23	19	0	0	107	0	7	0	43	11	41	13	0	7	97	0	67	0	0
59	107	103	7	13	19	23	17	11	0	7	139	0	0	31	127	0	7	79	11	0	137	0	29	7	181	43	61	0	23	11
61	0	13	61	23	7	0	0	31	0	83	11	7	17	89	13	47	29	0	7	0	0	11	37	0	0	7	23	13	173	17
67	0	41	37	7	83	0	19	0	11	0	7	0	0	17	67	113	0	7	0	11	23	13	0	0	7	19	0	29	53	89
71	7	37	0	11	13	0	0	7	17	137	89	0	0	0	7	29	23	13	113	0	0	7	71	97	0	11	19	0	7	0
73	0	0	7	0	0	0	11	59	79	7	41	0	97	71	17	13	7	11	67	37	0	0	139	7	109	0	73	163	11	0
77	43	11	0	103	0	53	7	0	109	17	23	13	11	7	0	0	41	0	19	0	7	31	0	11	13	0	17	7	79	0
79	109	7	43	0	11	0	0	73	7	13	0	53	0	0	47	7	0	0	0	41	29	0	7	0	113	0	11	0	19	7
83	0	0	0	31	23	7	43	0	29	0	11	0	7	0	0	0	0	0	13	7	0	11	0	53	0	41	7	17	0	19
89	151	0	19	47	7	0	0	0	11	79	0	7	13	0	0	37	31	0	7	11	0	19	0	0	73	7	0	0	17	0
91	11	151	0	71	89	0	7	181	61	0	13	11	19	7	43	0	103	17	0	23	7	0	11	13	0	0	0	7	0	47
97	0	17	0	0	13	7	0	0	137	11	0	0	7	23	59	0	0	13	17	7	11	0	0	19	29	0	7	193	0	0

	3 61	64	67	70	73	76	79	82	85	88	91	94	3 97	4 00	03	06	09	12	15	18	21	24	27	30	33	36	39	42	45	48
01	13	89	7	163	11	19	151	0	0	7	61	31	29	13	191	11	7	0	47	0	0	109	0	7	19	59	11	0	0	71
03	79	59	17	0	7	31	29	11	139	0	0	7	0	109	41	19	0	0	7	17	71	0	0	0	13	7	43	0	191	11
07	0	7	11	23	0	0	0	13	7	151	0	157	59	11	17	7	19	89	0	97	13	0	7	29	11	0	23	0	0	7
09	0	23	0	7	0	11	167	19	97	197	7	0	0	0	173	0	11	7	13	0	17	0	0	41	7	0	19	11	47	0
13	7	13	0	0	0	29	31	7	19	37	0	11	151	0	7	17	163	0	0	0	23	7	11	0	0	0	0	13	7	41
19	19	79	73	0	67	0	7	0	13	11	0	0	0	7	23	151	17	47	0	19	7	13	0	0	0	53	37	7	0	0
21	41	7	0	0	0	17	13	37	7	0	19	79	11	31	61	7	151	0	0	13	73	59	7	11	0	181	167	0	211	7
27	7	73	19	61	163	191	17	7	59	41	11	89	0	13	7	0	0	0	131	151	103	7	0	17	37	0	13	47	7	23
31	0	17	23	19	7	11	83	0	53	13	109	7	67	0	31	41	11	0	7	59	0	151	13	37	0	7	197	11	0	127
33	23	0	109	29	37	0	7	13	11	0	0	47	0	7	53	179	0	0	41	11	7	0	151	23	17	0	0	7	0	107
37	0	83	17	7	0	61	59	0	89	71	7	113	79	0	11	0	13	7	73	17	29	0	0	0	7	11	53	31	0	13
39	71	13	0	0	0	7	11	0	17	0	0	0	7	0	13	0	0	11	0	7	0	31	79	193	19	17	7	13	11	0

	3 62	65	68	71	74	77	80	83	86	89	92	95	3 98	4 01	04	07	10	13	16	19	22	25	28	31	34	37	40	43	46	49
49	37	7	0	0	13	0	137	23	7	53	11	103	0	29	157	7	0	13	0	0	113	11	7	0	67	0	71	0	0	7
51	0	0	11	7	41	23	0	29	19	0	7	0	127	11	0	13	31	7	37	0	61	0	0	0	7	0	0	17	13	0
57	11	0	7	0	0	0	0	67	0	7	0	11	83	41	0	109	7	0	29	19	0	0	11	7	191	149	113	0	17	31
61	0	19	0	0	0	13	7	0	0	0	0	0	0	7	0	73	0	11	13	41	7	0	61	17	131	0	0	7	11	113
63	29	7	97	13	0	0	0	83	7	11	0	19	17	0	181	7	13	0	89	0	11	0	7	0	103	47	0	0	0	7
67	59	0	0	101	11	7	0	17	0	0	53	61	7	103	37	11	71	29	197	7	149	0	0	0	17	13	7	0	41	0
69	7	0	83	19	0	139	43	7	0	47	13	29	0	17	7	67	53	0	11	149	0	7	19	13	31	0	0	0	7	11
73	61	0	11	131	7	101	13	0	17	0	43	7	31	11	47	89	0	149	7	13	181	0	0	19	11	7	0	0	29	23
79	11	0	0	7	0	41	163	101	173	17	7	11	0	13	149	19	43	7	0	0	0	107	11	23	7	31	13	0	0	0
81	97	191	0	11	29	7	19	0	41	59	0	13	7	149	11	17	107	0	43	7	0	23	179	67	13	11	7	0	109	37
87	0	11	0	0	7	13	0	0	47	37	149	7	11	0	0	23	17	19	7	0	0	0	0	11	43	7	0	67	0	0
91	0	7	0	29	139	0	0	11	7	0	0	17	0	47	13	7	179	157	11	163	31	0	7	41	0	0	0	13	17	7
93	17	0	0	7	61	0	0	149	0	19	7	73	13	0	31	0	0	7	0	0	0	11	0	0	7	13	29	0	19	0
97	7	0	31	0	0	11	0	7	13	97	19	127	17	101	7	0	11	61	0	0	0	7	0	71	0	37	0	11	7	17
99	53	17	7	23	149	0	13	0	11	7	0	0	0	0	71	0	7	0	17	11	19	0	127	7	0	89	23	31	103	59
03	41	173	13	11	113	37	7	0	0	0	197	0	53	7	11	13	131	103	0	0	7	19	23	0	0	11	79	7	13	83
09	0	11	0	43	0	7	191	29	0	13	0	0	7	19	17	0	23	101	0	7	0	0	13	11	83	109	7	59	31	0
11	7	29	131	17	11	43	0	7	0	167	113	0	41	0	7	11	0	109	0	0	13	7	31	19	0	0	11	73	7	97
17	0	13	11	0	17	0	7	0	23	0	0	43	29	7	13	19	0	79	0	167	7	17	47	0	11	0	0	7	0	0
21	29	59	0	7	23	67	193	0	11	0	7	0	0	53	83	43	17	7	0	11	0	101	0	13	7	0	0	23	0	29
23	11	0	23	0	0	7	47	19	13	0	61	11	7	0	0	193	0	31	107	7	0	13	11	29	173	23	7	127	0	167
27	17	0	7	137	13	31	11	0	19	7	0	29	0	0	0	139	7	11	0	0	0	23	113	7	0	73	0	19	11	0
29	0	0	13	107	7	29	17	0	0	11	0	7	0	0	0	13	89	37	7	23	11	71	0	17	137	7	0	97	13	179
33	19	7	0	71	11	97	73	0	7	0	0	13	61	67	0	7	37	0	17	19	157	0	7	0	13	101	11	43	0	7
39	7	61	11	0	29	13	0	7	0	23	0	19	0	11	7	0	0	67	13	17	0	7	0	179	11	191	47	101	7	0
41	0	0	7	13	0	11	109	23	17	7	0	0	0	137	37	131	7	0	0	0	53	19	0	7	0	17	0	11	0	13
47	67	7	0	11	0	0	0	31	7	17	13	71	0	19	11	7	0	173	0	0	83	157	7	13	23	11	17	61	0	7
51	0	0	43	97	17	7	13	0	0	11	0	0	7	0	19	0	0	0	0	7	11	17	73	0	0	67	7	0	0	79
53	7	11	137	53	13	19	0	7	0	0	17	37	11	0	7	83	61	13	23	0	29	7	0	11	19	0	0	17	7	0
57	13	139	0	73	7	17	19	11	29	163	37	7	0	13	23	53	0	0	7	0	0	0	17	103	0	7	13	0	0	11
59	101	0	29	0	47	61	7	89	67	0	11	13	23	7	0	0	19	59	0	0	7	11	0	0	13	0	0	7	17	0
63	0	0	191	7	0	11	17	13	23	47	7	0	0	0	43	0	11	7	61	29	13	31	0	17	7	107	139	11	59	0
69	0	13	7	11	89	179	0	17	0	7	107	0	0	0	11	59	7	41	0	0	43	0	163	7	17	11	127	13	19	193
71	19	0	0	0	7	107	11	0	0	0	173	7	13	17	0	0	67	11	7	19	41	0	43	23	29	7	0	0	11	0
77	0	79	0	7	11	37	13	0	0	0	7	19	0	0	17	11	0	7	71	13	67	0	53	0	7	0	11	199	43	41
81	7	157	13	0	37	0	113	7	47	17	11	0	19	23	7	13	0	0	0	0	0	7	137	29	0	0	17	0	7	31
83	13	0	7	19	0	0	0	131	101	7	163	23	0	11	0	17	7	29	73	0	0	97	19	7	11	0	13	0	0	0
87	131	0	0	41	19	29	7	23	11	13	17	31	0	7	0	0	181	0	0	11	7	37	13	19	0	0	0	7	0	0
89	11	7	37	0	0	23	41	13	7	127	101	11	113	0	19	7	17	0	47	199	13	0	7	0	157	0	0	0	23	7
93	0	23	79	13	0	7	11	0	0	0	0	17	7	0	0	19	13	11	173	7	0	191	59	47	23	0	7	103	11	13
99	0	0	0	0	7	0	31	19	0	59	13	7	17	61	0	11	73	0	7	0	0	41	0	13	0	7	11	29	0	17

	4 50	53	56	59	62	65	68	71	74	77	80	83	86	89	92	95	4 98	5 01	04	07	10	13	16	19	22	25	28	31	34	37
01	11	89	31	197	47	7	17	19	107	0	23	11	7	79	0	59	0	0	13	7	0	29	11	17	0	0	7	0	0	83
07	0	0	59	29	7	0	0	17	0	11	61	7	13	0	0	31	0	89	7	0	11	0	0	0	17	7	0	23	0	43
11	19	7	17	31	11	0	0	0	7	0	41	0	0	59	0	7	0	0	0	17	29	13	7	23	109	0	11	173	0	7
13	0	113	0	7	37	193	13	11	17	0	7	0	173	41	29	67	109	7	11	13	139	23	0	0	7	17	0	0	31	11
17	7	0	11	17	113	181	0	7	0	0	0	19	61	11	7	13	31	23	0	41	17	7	71	193	11	0	0	0	7	0
19	13	0	7	47	0	11	0	0	0	7	31	211	0	13	83	23	7	0	127	67	163	19	41	7	79	29	13	11	0	0
23	11	61	43	19	17	0	7	0	47	13	0	11	0	7	0	0	0	0	0	0	7	17	11	137	0	53	101	7	41	31
29	37	0	103	13	0	7	0	0	43	11	0	31	7	113	19	0	13	0	211	7	11	0	17	0	29	0	7	0	23	13
31	7	11	0	23	83	19	0	7	0	59	43	17	11	167	7	0	0	0	29	97	0	7	0	11	19	131	23	13	7	0
37	29	0	47	71	0	173	7	0	13	0	11	0	17	7	53	0	19	181	31	113	7	11	0	167	0	107	0	7	0	17
41	73	0	0	7	13	11	31	17	0	0	7	0	127	109	41	107	11	7	0	0	43	0	113	0	7	0	53	11	0	61
43	31	0	13	0	131	7	139	0	11	0	107	29	7	17	23	13	0	41	73	7	0	0	43	127	89	0	7	19	13	223
47	107	137	7	11	103	89	79	0	17	7	23	13	0	0	11	0	7	0	61	31	0	0	0	7	13	11	43	0	19	71
49	19	101	191	0	7	0	11	0	23	13	0	7	0	31	17	0	79	11	7	19	71	0	13	0	0	7	41	0	11	59
53	0	7	71	0	23	13	0	61	7	17	29	0	11	0	0	7	0	0	13	0	19	89	7	11	0	0	17	23	0	7
59	7	67	0	0	167	0	47	7	0	163	11	37	13	173	7	0	73	0	0	193	0	7	0	223	0	13	0	17	7	0
61	0	0	7	19	0	101	0	0	31	7	13	137	0	11	0	29	7	103	0	23	0	0	19	7	11	0	0	0	193	37
67	11	7	0	43	13	0	0	101	7	37	71	11	41	23	19	7	47	13	109	0	223	31	7	157	0	0	29	79	127	7
71	13	59	109	0	0	7	11	43	37	23	53	0	7	13	29	19	0	11	41	7	0	47	163	0	167	0	7	0	11	17
73	7	17	0	31	0	0	19	7	29	11	0	13	0	0	7	89	53	131	17	0	11	7	0	0	13	19	37	0	7	0
77	0	0	0	23	7	47	0	13	197	0	131	7	0	17	0	11	0	0	7	0	13	83	31	0	61	7	11	41	53	0
79	61	23	17	0	0	13	7	11	79	0	0	101	0	7	0	43	31	19	11	17	7	191	0	59	23	0	0	7	0	11
83	0	13	11	7	31	37	173	29	103	71	7	0	89	11	13	179	83	7	19	43	23	0	0	227	7	0	0	13	79	0
89	11	0	7	0	41	0	0	0	13	7	19	11	181	0	23	17	7	31	29	0	47	13	11	7	0	43	0	0	89	19
91	67	19	0	11	7	0	13	41	0	0	0	7	23	0	11	101	0	53	7	13	19	17	0	0	0	7	227	43	149	0
97	13	11	0	7	67	17	23	109	0	0	7	0	11	13	0	0	41	7	0	79	37	103	17	11	7	149	13	0	61	23

	4 51	54	57	60	63	66	69	72	75	78	81	84	87	90	93	96	4 99	5 02	05	08	11	14	17	20	23	26	29	32	35	38
01	7	83	23	157	0	0	0	7	0	13	103	29	31	19	7	193	139	17	11	37	137	7	13	149	0	23	0	0	7	11
03	23	0	7	179	19	29	17	13	67	7	11	97	113	0	47	0	7	61	0	101	13	11	149	7	193	41	0	83	0	173
07	43	17	0	13	0	11	7	0	0	0	73	0	53	7	0	113	11	0	17	23	7	0	29	131	19	31	191	7	0	13
09	79	7	43	139	0	127	61	17	7	0	0	0	67	0	13	7	29	23	53	11	0	101	7	0	17	0	157	13	73	7
13	197	0	17	11	29	7	43	31	0	137	13	0	7	23	11	0	19	149	0	7	79	0	0	13	0	11	7	127	59	0
19	0	11	131	17	7	0	0	23	19	0	0	7	11	0	149	29	0	13	7	89	17	0	0	11	113	7	0	19	109	0
21	0	53	13	0	11	23	7	0	0	17	0	41	83	7	31	11	0	0	19	0	7	0	0	0	0	101	11	7	13	107
27	0	0	11	0	0	7	167	83	0	13	17	79	7	11	107	0	0	0	0	7	29	0	13	0	11	0	7	17	0	19
31	0	181	7	191	107	13	71	73	11	7	0	19	0	0	0	31	7	0	13	11	0	0	17	7	43	0	41	0	199	0
33	11	0	19	13	7	0	0	149	0	31	127	7	0	0	0	0	13	191	7	0	0	19	11	61	59	7	43	0	17	13
37	0	7	0	19	0	149	11	0	7	0	37	0	13	0	103	7	0	11	97	29	0	0	7	17	199	13	0	139	11	7
39	0	0	53	7	149	0	73	97	137	11	7	59	17	19	0	0	0	7	0	0	11	0	31	13	7	0	167	0	37	17

49	13	47	11	0	0	0	7	37	17	59	89	0	29	7	61	131	199	109	0	0	7	0	0	23	11	17	13	7	0	0
51	163	7	0	0	0	11	29	0	7	109	179	13	0	181	17	7	11	31	0	211	0	23	7	0	13	37	0	11	0	7
57	7	131	0	11	151	13	0	7	19	0	0	47	0	0	7	17	0	29	13	0	0	7	73	0	41	11	0	19	7	0
61	0	13	67	0	7	29	151	167	199	11	17	7	0	71	13	53	47	0	7	181	11	0	191	79	0	7	211	13	19	0
63	19	11	0	73	71	0	7	151	0	23	0	0	11	7	0	0	17	0	59	19	7	53	37	11	0	13	0	7	29	61
67	31	19	0	7	199	23	67	11	13	151	7	17	0	139	0	0	29	7	11	0	19	13	0	0	7	0	0	0	17	11
69	17	41	37	23	89	7	13	0	0	0	11	19	7	0	0	0	107	17	61	7	0	11	0	0	0	31	7	0	0	103
73	199	37	7	0	79	11	107	41	113	7	67	0	17	31	97	13	7	0	103	0	73	0	23	7	83	0	0	11	13	17
79	0	7	0	11	19	0	109	0	7	13	0	0	0	17	11	7	23	137	37	83	61	0	7	19	0	11	31	0	131	7
81	0	0	17	7	0	0	11	13	0	0	7	0	0	0	19	0	151	7	0	17	13	0	53	0	7	139	0	0	11	0
87	73	13	7	17	11	0	19	0	23	7	0	0	0	191	13	11	7	0	0	151	17	0	0	7	0	19	11	13	41	0
91	0	0	29	0	23	0	7	19	0	83	11	0	97	7	0	17	0	0	0	0	7	11	67	13	0	0	19	7	0	0
93	43	7	11	0	17	53	0	0	7	47	0	71	59	11	0	7	0	19	0	0	0	13	7	113	11	23	197	137	0	7
97	0	0	41	31	13	7	0	0	11	211	0	0	7	29	47	0	17	13	19	7	0	23	0	59	151	0	7	223	0	0
99	7	173	13	0	0	17	43	7	0	19	157	11	0	37	7	13	0	179	0	23	0	7	11	53	61	151	0	0	7	0

	4 52	55	58	61	64	67	70	73	76	79	82	85	88	91	94	4 97	5 00	03	06	09	12	15	18	21	24	27	30	33	36	39
03	17	0	163	0	7	0	11	0	181	0	19	7	37	0	127	23	31	11	7	109	0	0	0	0	13	7	0	151	11	19
09	53	17	19	7	11	13	29	0	47	23	7	179	0	0	0	11	43	7	13	0	41	19	103	107	7	0	11	0	0	31
11	29	71	61	13	0	7	53	11	47	0	37	139	7	67	0	0	13	0	11	7	83	0	197	31	17	0	7	89	0	11
17	103	23	0	107	7	11	0	0	17	0	13	7	0	0	0	83	11	67	7	59	0	0	0	13	23	7	0	11	0	0
21	11	7	0	17	61	19	13	79	7	173	0	11	0	0	73	7	0	0	223	13	17	0	7	0	19	0	37	71	29	7
23	41	0	0	7	13	0	59	37	0	17	7	0	0	0	11	19	0	7	23	0	181	67	29	47	7	11	17	0	0	0
27	7	53	0	193	17	0	31	7	97	11	29	0	157	13	7	0	19	59	0	127	11	7	0	0	103	0	13	0	7	0
29	31	11	7	163	29	83	131	19	0	7	17	13	11	73	0	223	7	0	197	0	0	227	0	7	13	67	19	17	0	199
33	0	0	0	0	59	17	7	11	19	0	139	0	47	7	0	41	0	0	11	31	7	29	17	37	0	0	181	7	0	11
39	19	13	23	29	0	7	17	0	0	0	0	0	7	0	13	0	11	71	79	7	0	0	0	17	41	23	7	11	0	0
41	7	0	0	0	0	43	0	7	11	191	19	0	13	157	7	0	163	0	89	11	0	7	47	23	229	13	29	41	7	17
47	0	37	19	0	0	0	7	113	29	0	0	43	0	7	197	0	0	11	0	13	7	19	139	0	179	0	0	7	11	73
51	37	11	13	7	0	0	0	0	17	0	7	47	11	23	0	13	0	7	0	0	53	0	19	11	7	17	0	31	13	0
53	13	0	0	0	11	7	211	0	0	79	73	23	7	13	17	11	0	43	37	7	107	31	0	0	0	71	7	0	0	163
57	167	0	7	101	0	0	0	23	0	7	11	59	0	0	19	0	7	37	179	0	0	11	13	7	0	0	17	229	0	79
59	0	29	11	31	7	19	0	13	0	199	0	7	0	11	0	17	113	0	7	131	13	47	0	43	11	7	97	0	23	0
63	0	7	0	13	97	101	19	0	7	0	17	0	131	211	0	7	13	0	29	11	0	0	7	0	23	19	47	17	103	7
69	7	0	0	137	31	0	11	7	73	0	13	17	0	0	7	157	0	11	23	0	167	7	0	13	71	0	0	83	7	29
71	17	199	7	0	0	0	103	127	13	7	0	0	0	0	61	71	7	17	0	0	11	13	0	7	137	113	73	19	191	31
77	19	7	13	61	0	29	179	11	7	0	23	31	37	0	0	7	0	0	11	19	47	0	7	0	97	89	0	0	13	7
81	0	19	11	0	53	7	23	0	0	0	0	13	7	11	0	67	61	83	59	7	19	0	29	0	11	47	7	0	0	23
83	7	79	17	0	23	11	197	7	41	13	53	19	0	137	7	0	11	0	0	17	0	7	13	0	31	0	109	11	7	37
87	11	0	0	0	7	13	0	0	43	47	109	7	19	101	17	0	0	0	7	67	0	79	11	23	73	7	0	197	37	0
89	0	0	109	11	0	71	7	0	103	37	43	0	0	7	11	0	13	41	173	0	7	23	19	0	0	11	0	7	53	13
93	0	127	0	7	19	73	0	83	37	11	7	0	13	0	43	17	0	7	163	0	11	0	0	19	7	13	0	107	0	0
99	97	0	7	0	0	53	13	11	0	7	0	23	107	0	0	19	7	101	11	13	43	0	0	7	47	37	29	67	0	11

	5 40	43	46	49	52	55	58	61	64	67	70	73	76	79	82	85	88	91	94	5 97	6 00	03	06	09	12	15	18	21	24	27
01	0	13	0	7	0	0	41	0	0	0	7	0	0	0	11	19	127	7	191	227	29	47	0	0	7	11	23	13	0	0
07	53	11	7	0	0	47	0	19	13	7	109	17	11	79	0	41	7	0	0	0	23	13	0	7	97	0	19	173	17	73
11	0	0	97	43	13	0	7	11	19	0	47	223	53	7	0	0	23	13	11	29	7	41	0	17	0	0	113	7	139	11
13	0	7	13	89	0	43	0	0	7	0	11	37	17	29	23	7	103	0	19	211	0	11	7	0	41	137	0	179	13	7
17	19	29	0	0	0	7	0	17	0	43	23	13	7	0	0	163	11	31	0	7	0	0	0	0	13	227	7	11	0	59
19	7	0	193	0	0	59	0	7	11	13	19	31	157	17	7	139	131	0	0	11	47	7	13	0	29	0	0	0	7	19
23	89	0	0	11	7	13	0	0	17	131	127	7	29	0	11	43	59	0	7	0	193	179	0	0	0	7	211	23	0	0
29	97	11	0	7	0	0	0	37	73	17	7	0	11	53	0	107	89	7	67	0	0	23	19	11	7	13	17	0	163	149
31	71	0	0	163	11	7	31	0	0	0	13	0	7	19	0	11	0	29	103	7	173	0	0	13	0	37	7	0	149	0
37	0	67	11	137	7	19	0	73	0	0	0	7	0	11	0	0	17	13	7	31	0	0	0	0	11	7	0	0	29	43
41	13	7	101	0	37	0	19	31	7	23	0	17	0	13	139	7	29	0	0	11	0	83	7	149	47	19	13	0	17	7
43	11	31	53	7	0	67	0	23	0	179	7	11	59	0	0	0	19	7	0	0	97	0	11	0	7	0	0	0	41	0
47	7	0	0	23	101	0	11	7	47	0	0	0	17	0	7	127	83	11	0	0	13	7	0	59	73	0	23	29	7	17
49	0	17	7	0	0	13	0	0	19	7	89	0	0	167	31	0	7	0	13	149	11	29	0	7	23	61	127	19	197	131
53	191	13	31	179	11	73	7	233	0	19	59	83	0	7	13	11	229	149	0	0	7	0	131	0	0	0	11	7	19	0
59	0	19	11	0	0	7	83	89	13	211	0	41	7	11	17	31	71	0	37	7	19	13	0	47	11	0	7	61	0	97
61	7	0	47	17	73	11	13	7	131	31	43	19	23	149	7	157	11	67	97	13	17	7	0	0	0	0	0	11	7	0
67	13	0	0	11	17	181	7	0	0	0	149	0	0	7	11	0	37	0	0	59	7	17	19	41	197	11	13	7	0	23
71	139	0	23	7	19	61	0	0	149	11	7	103	101	29	0	37	17	7	0	0	11	73	13	19	7	23	0	0	179	41
73	23	11	0	0	31	7	59	13	0	0	0	0	7	0	19	0	113	47	0	7	13	0	17	11	71	67	7	79	0	0
77	17	0	7	13	167	149	71	11	0	7	0	181	137	0	101	19	7	17	11	23	0	173	47	7	29	139	43	97	0	11
79	41	13	0	0	7	0	17	0	0	0	11	7	0	37	13	0	97	23	7	0	73	11	0	17	233	7	0	13	43	67
83	0	7	149	0	59	11	29	19	7	0	13	0	37	23	167	7	11	0	17	191	0	0	7	13	0	0	19	11	0	7
89	7	137	17	11	13	0	0	7	0	109	0	0	0	103	7	41	0	13	19	17	0	7	0	71	167	11	199	0	7	37
91	0	109	7	127	0	23	11	83	17	7	37	29	31	0	71	13	7	11	41	0	0	131	137	7	0	17	59	0	11	0
97	47	7	83	43	11	53	0	0	7	13	0	0	0	59	97	7	0	0	0	0	19	0	7	181	0	31	11	37	0	7

	5 41	44	47	50	53	56	59	62	65	68	71	74	77	80	83	86	89	92	95	5 98	6 01	04	07	10	13	16	19	22	25	28
01	0	0	19	0	17	7	0	43	0	79	11	61	7	31	173	0	0	53	13	7	0	11	101	0	59	229	7	0	0	0
03	7	0	11	13	29	0	0	7	0	43	17	137	19	11	7	0	13	73	157	79	0	7	0	53	11	0	103	17	7	13
07	61	41	227	67	7	17	37	0	11	0	0	7	13	19	199	103	0	0	7	11	0	29	17	0	101	7	31	0	0	181
09	11	0	0	0	19	0	7	0	0	0	13	11	0	7	0	29	0	0	0	0	7	193	11	13	37	0	0	7	17	107
13	53	0	0	7	0	19	11	67	31	0	7	0	0	0	0	0	0	7	0	13	47	0	109	17	7	0	101	0	11	23
19	13	0	7	37	11	0	199	17	0	7	0	67	0	13	29	11	7	0	53	41	79	31	0	7	17	43	11	0	101	0
21	0	0	0	0	7	0	0	11	29	0	239	7	197	17	0	31	0	0	7	163	59	23	41	139	13	7	19	43	103	11
27	113	37	0	7	61	11	0	59	0	0	7	0	0	0	17	23	11	7	13	29	0	0	0	0	7	0	0	11	31	0
31	7	13	229	113	0	0	0	7	0	17	0	11	0	0	7	0	31	61	59	19	157	7	11	0	0	0	17	13	7	83
33	0	29	7	11	0	0	0	53	0	7	19	79	13	131	11	17	7	0	37	0	0	223	0	7	0	11	0	0	0	19
37	43	0	127	47	0	23	7	0	13	11	17	19	0	7	0	191	0	37	29	53	7	13	0	67	83	0	241	7	23	31
39	0	7	19	23	0	0	13	0	7	113	0	71	11	127	227	7	17	0	0	13	0	19	7	11	0	53	23	109	0	7

49	173	0	53	0	7	11	0	0	193	13	0	7	17	0	19	223	11	179	7	97	0	0	13	41	31	7	0	11	0	17
51	0	17	0	0	0	19	7	13	11	139	67	73	0	7	23	39	167	193	17	11	7	61	79	0	19	0	41	7	71	0
57	31	13	17	0	197	7	11	101	23	0	61	0	7	0	13	0	19	11	0	7	43	0	0	0	0	0	7	13	11	239
61	41	11	7	0	23	0	107	127	163	7	13	37	11	0	17	0	7	19	0	31	0	103	0	7	43	197	0	23	73	0
63	0	107	23	17	7	0	191	0	13	101	0	7	47	31	0	11	0	0	7	0	17	13	0	227	0	7	11	19	0	37
67	0	7	0	53	13	0	0	0	7	19	11	0	61	0	0	7	0	13	0	131	0	11	7	79	109	0	0	71	19	7
69	19	0	11	7	17	179	97	0	0	29	7	101	41	11	0	13	109	7	71	19	0	17	67	173	7	83	31	73	13	0
73	7	19	0	0	0	0	223	7	11	0	0	13	0	0	7	23	17	0	41	11	19	7	0	157	13	0	29	0	7	0
79	17	157	0	0	79	13	7	167	29	23	0	229	19	7	0	0	0	11	13	0	7	197	0	103	0	37	0	7	11	227
81	0	7	29	13	0	0	17	23	7	11	211	47	0	241	79	7	13	0	0	233	11	31	7	17	0	0	0	61	0	7
87	7	23	0	31	97	233	0	7	71	163	13	0	0	29	7	0	61	101	11	0	139	7	89	13	17	0	0	199	7	11
91	47	29	11	89	7	0	13	181	0	0	0	7	0	11	0	19	0	211	7	13	23	241	31	0	11	7	0	167	0	61
93	0	0	157	37	13	11	7	41	17	0	0	0	0	7	0	0	11	13	23	101	7	0	0	199	29	17	47	7	53	109
97	11	0	37	7	31	0	0	19	0	0	7	11	29	13	23	79	0	7	61	89	17	0	11	107	7	103	13	0	0	0
99	83	0	0	11	0	7	29	0	0	17	47	13	7	0	11	0	41	19	107	7	37	101	163	0	13	11	7	0	59	31

	5 42	45	48	51	54	57	60	63	66	69	72	75	78	81	84	87	90	93	96	5 99	6 02	05	08	11	14	17	20	23	26	29
03	67	0	7	0	17	53	0	13	23	7	0	0	0	97	0	47	7	31	19	37	11	17	41	7	0	0	0	0	0	0
09	151	7	23	0	67	17	0	11	7	0	19	131	0	0	13	7	0	127	11	139	0	0	7	53	0	23	59	13	137	7
11	23	19	59	7	0	0	79	0	0	0	7	17	13	0	0	0	0	7	0	181	19	11	0	23	7	13	0	0	17	53
17	0	0	7	0	151	0	13	199	11	7	29	113	17	89	0	71	7	23	0	11	0	73	61	7	0	0	0	101	0	17
21	59	0	13	11	157	0	7	17	41	0	0	97	67	7	11	13	0	137	0	0	7	0	0	0	17	11	109	7	13	0
23	13	7	73	199	19	103	11	151	7	0	0	23	53	13	37	7	0	11	109	31	0	29	7	19	239	0	13	0	11	7
27	211	11	109	0	43	7	179	23	17	13	89	0	7	37	0	0	67	41	0	7	229	0	13	11	19	17	7	0	0	0
29	7	31	0	29	11	23	43	7	0	0	151	0	0	0	7	11	0	79	0	0	13	7	59	0	47	0	11	157	7	0
33	193	23	0	13	7	0	137	0	0	17	11	7	151	61	71	0	13	0	7	73	29	11	127	113	23	7	17	83	0	13
39	73	0	29	7	0	139	0	53	11	97	7	163	0	47	0	151	43	7	23	11	59	0	83	13	7	107	0	17	0	0
41	11	0	173	67	0	7	0	103	13	0	0	11	7	53	0	0	17	0	19	7	107	13	11	0	0	29	7	31	37	113
47	17	0	13	0	7	107	41	29	37	11	19	7	0	0	211	13	137	17	7	151	11	191	71	47	43	7	0	0	13	19
51	0	7	0	131	11	197	23	37	7	0	0	13	17	0	0	7	0	0	0	0	0	151	7	0	13	0	11	0	31	7
53	227	17	19	7	23	127	0	11	181	13	7	67	0	0	0	41	0	7	11	167	89	19	13	0	7	37	0	23	0	11
57	7	89	11	19	0	13	29	7	53	0	31	0	47	11	7	0	73	0	13	0	0	7	19	23	11	0	0	127	7	157
59	29	0	7	13	31	11	61	0	0	7	0	0	0	19	53	67	7	0	0	17	0	23	0	7	41	151	229	11	0	13
63	11	0	83	0	37	0	7	157	0	0	173	11	13	7	17	0	0	23	0	61	7	71	11	31	0	13	53	7	223	79
69	0	197	0	43	0	7	13	0	61	11	0	23	7	0	59	17	0	0	0	7	11	37	0	0	0	19	7	47	29	0
71	7	11	37	0	13	43	47	7	0	23	0	0	11	0	7	0	19	13	0	0	0	7	29	11	0	223	0	97	7	0
77	0	0	0	23	29	17	7	0	19	227	11	13	31	7	0	53	0	0	83	37	7	11	17	131	13	163	23	7	233	71
81	17	0	0	7	109	11	0	13	0	19	7	71	0	73	0	43	11	7	37	0	13	29	23	193	7	0	0	11	19	0
83	19	0	71	139	113	7	17	0	11	0	0	89	7	83	233	29	0	43	13	7	23	47	107	17	0	31	7	0	0	0
87	0	13	7	11	0	0	0	113	0	7	0	0	107	31	11	0	7	0	17	223	19	43	0	7	0	11	47	13	0	0
89	233	79	131	229	7	47	11	17	83	0	59	7	13	0	23	0	37	11	7	239	0	0	0	43	17	7	29	89	11	0
93	0	7	17	97	211	0	0	0	7	0	23	0	11	0	29	7	0	0	0	17	0	13	7	11	0	61	31	43	71	7
99	7	71	13	17	19	0	0	7	31	0	11	239	0	0	7	13	113	0	0	0	17	7	0	19	89	29	0	23	7	73

	6 30	33	36	39	42	45	48	51	54	57	60	63	66	69	72	75	78	81	84	87	90	93	96	6 99	7 02	05	08	11	14	17
01	251	7	0	0	19	53	11	0	7	0	13	0	0	149	17	7	0	11	73	23	0	37	7	13	0	0	101	97	11	7
07	7	29	0	0	11	251	229	7	0	0	149	61	43	23	7	11	0	13	67	127	151	7	47	53	0	0	11	211	7	0
11	13	0	0	79	7	31	0	0	149	23	11	7	59	13	0	0	19	0	7	0	0	11	151	0	61	7	13	17	0	0
13	61	0	11	0	157	0	7	19	0	0	251	13	29	7	0	181	17	0	37	0	7	0	67	151	11	107	19	7	0	0
17	29	0	0	7	0	149	0	13	11	0	7	17	0	61	0	107	73	7	31	11	13	0	43	139	7	151	23	19	17	29
19	11	23	113	41	149	7	53	0	0	0	107	11	7	0	0	251	0	17	13	7	0	103	11	29	23	97	7	0	0	0
23	19	13	7	97	0	113	11	0	0	7	103	29	17	0	13	0	7	11	53	19	23	181	0	7	0	109	0	13	11	17
29	0	7	0	0	11	173	241	0	7	0	0	19	0	17	23	7	0	193	41	0	0	13	7	0	0	0	11	0	0	7
31	0	0	17	7	0	47	13	11	59	0	7	113	23	0	0	0	29	7	11	13	0	19	179	0	7	251	193	83	61	11
37	13	0	7	17	0	11	23	53	0	7	0	0	37	13	71	0	7	61	0	0	17	0	83	7	0	0	13	11	0	23
41	11	97	23	43	227	233	7	0	31	13	0	11	103	7	19	17	179	0	89	53	7	0	11	0	0	23	0	7	199	0
43	23	7	31	11	17	19	61	13	7	29	211	0	0	0	11	7	0	83	0	0	13	17	7	23	19	11	0	0	0	7
47	67	0	0	13	41	7	19	0	0	11	0	0	7	0	0	0	13	0	0	7	11	31	257	113	199	19	7	0	37	13
49	7	11	0	0	47	17	0	7	0	37	257	43	11	0	7	31	19	23	0	0	29	7	17	11	0	0	0	13	7	157
53	17	0	53	31	7	0	0	11	29	47	13	7	0	23	109	43	0	17	7	197	199	223	0	13	163	7	0	0	0	11
59	0	17	0	7	13	11	79	23	67	19	7	0	191	0	103	0	11	7	17	29	53	43	41	0	7	37	59	11	19	73
61	19	0	13	167	179	7	37	17	11	0	31	0	7	29	0	13	79	0	223	7	0	139	0	43	17	41	7	0	13	0
67	0	0	0	47	7	0	11	0	17	13	0	7	163	167	137	0	0	11	7	0	0	71	13	31	29	7	0	0	11	43
71	59	7	0	17	0	13	0	0	7	89	0	31	11	193	0	7	67	0	13	0	17	0	7	11	0	0	131	0	0	7
73	0	127	41	7	11	31	29	0	233	17	7	0	61	0	0	11	13	7	0	97	0	173	19	167	7	0	11	103	0	13
77	7	0	37	0	17	0	0	7	41	0	11	0	13	0	7	0	103	79	0	0	67	7	0	19	31	13	0	109	7	0
79	0	61	7	137	0	0	0	0	0	7	13	41	131	11	19	0	7	29	31	109	37	0	59	7	11	163	0	17	0	179
83	199	241	43	109	0	17	7	0	11	157	0	0	0	7	61	19	0	41	0	11	7	0	17	47	67	0	73	7	0	23
89	13	0	0	61	53	7	11	19	43	0	0	197	7	13	0	0	29	11	0	7	59	0	227	17	0	0	7	257	11	0
91	7	0	0	89	239	0	0	7	79	11	29	13	17	31	7	257	0	19	0	0	11	7	0	0	13	73	0	0	7	17
97	0	0	0	0	113	13	7	11	0	19	157	67	0	7	173	23	43	47	11	89	7	29	0	0	0	227	31	7	19	11

	6 31	34	37	40	43	46	49	52	55	58	61	64	67	70	73	76	79	82	85	88	91	94	6 97	7 00	03	06	09	12	15	18
01	89	13	11	7	0	0	0	113	17	29	7	23	0	11	13	0	0	7	0	107	43	0	47	0	7	17	0	13	127	19
03	0	19	0	29	0	7	41	0	31	23	0	0	7	0	17	67	11	241	61	7	19	0	43	0	229	13	7	11	0	59
07	11	163	7	0	107	23	47	197	13	7	0	11	41	37	0	0	7	0	0	83	29	13	11	7	167	0	17	31	23	0
09	223	0	0	11	7	0	13	61	109	0	0	7	19	113	11	17	59	0	7	13	0	31	0	0	0	7	23	0	43	0
13	0	7	13	0	73	0	139	0	7	11	17	0	0	19	83	7	113	0	131	0	11	41	7	53	0	241	0	17	13	7
19	7	0	0	0	0	19	0	7	0	13	37	17	137	29	7	0	23	0	11	0	0	7	13	0	19	0	0	229	7	11
21	17	0	7	73	131	0	0	13	0	7	11	127	0	0	23	19	7	17	0	0	13	11	113	7	0	0	0	67	37	0
27	0	7	0	43	0	0	0	19	7	0	89	181	53	97	13	7	0	0	17	11	0	0	7	239	0	0	19	13	0	7
31	0	137	101	11	23	7	29	37	19	0	13	0	7	17	11	0	0	31	0	7	73	0	103	13	53	11	7	19	233	109
33	7	229	17	0	0	0	11	7	13	43	41	31	0	0	7	47	0	11	19	17	257	7	137	59	61	23	89	0	7	29
37	19	11	0	0	7	109	0	89	0	0	0	7	11	43	17	239	41	13	7	19	47	23	0	11	37	7	0	0	0	0
39	103	0	13	17	11	37	7	0	0	0	19	29	0	7	0	11	0	0	0	23	7	0	0	0	31	0	11	7	13	19

49	0	67	7	19	229	13	107	71	11	7	29	0	0	0	0	61	7	139	13	11	0	37	19	7	103	31	0	0	0	0
51	11	107	37	13	7	17	0	23	0	0	83	7	0	19	47	0	13	131	7	31	0	199	11	0	0	7	0	43	0	13
57	137	23	103	7	139	19	17	0	0	11	7	0	241	0	193	29	0	7	179	37	11	0	79	13	7	0	0	0	163	181
61	7	17	0	29	11	0	13	7	53	67	0	41	101	0	7	11	0	0	17	13	23	7	0	0	71	19	11	0	7	0
63	83	0	7	0	13	0	167	11	0	7	109	0	0	199	31	71	7	13	11	0	0	0	0	7	17	0	29	0	0	11
67	13	0	11	0	191	0	7	0	173	0	127	0	179	7	23	157	0	19	0	17	7	0	0	0	11	0	13	7	59	0
69	181	7	43	79	59	11	0	0	7	199	0	13	23	47	0	7	11	233	191	61	263	127	7	41	13	17	0	11	0	7
73	11	0	0	17	0	7	43	13	23	19	0	11	7	0	89	31	101	67	47	7	13	0	11	79	0	29	7	263	19	41
79	0	13	23	139	7	0	181	29	0	11	0	7	43	0	13	0	0	0	7	0	11	17	0	0	0	7	0	13	31	0
81	23	11	0	0	0	71	7	97	0	0	17	19	11	7	43	53	157	0	0	0	7	0	31	11	0	13	0	7	47	0
87	179	0	227	19	31	7	13	0	0	41	11	17	7	73	79	113	0	23	107	7	43	11	19	109	59	0	7	0	17	0
91	29	173	7	0	19	11	17	109	107	7	0	0	0	23	0	13	7	47	113	0	0	0	101	7	43	223	0	11	13	29
93	13	0	0	107	7	0	103	0	11	131	37	7	17	13	19	139	0	31	7	11	0	0	71	29	0	7	13	0	0	17
97	0	7	131	11	71	31	0	17	7	13	53	29	0	229	11	7	97	163	0	0	0	0	7	191	17	11	0	83	0	7
99	0	0	0	7	0	23	11	13	0	0	7	0	67	17	0	0	53	7	181	0	13	0	223	0	7	19	0	37	11	0
	6 32	35	38	41	44	47	50	53	56	59	62	65	68	71	74	77	80	83	86	89	92	95	6 98	7 01	04	07	10	13	16	19
03	7	11	0	13	0	89	0	7	17	59	239	73	11	0	7	79	13	167	31	0	0	7	29	11	23	17	19	113	7	13
09	31	41	0	0	29	0	7	0	0	17	11	0	0	7	0	0	47	83	19	0	7	11	0	13	181	0	17	7	101	0
11	0	7	11	61	41	163	0	241	7	19	73	227	71	11	0	7	23	0	0	137	67	13	7	0	11	31	0	29	19	7
17	7	19	13	97	37	0	79	7	0	29	23	11	109	41	7	13	17	53	59	0	19	7	11	0	67	0	47	0	7	0
21	191	0	19	37	7	61	11	83	211	0	0	7	0	0	0	241	251	11	7	41	0	19	0	0	13	7	29	73	11	23
23	17	139	0	0	23	59	7	0	137	11	47	0	19	7	191	0	0	17	163	157	7	37	13	0	0	197	0	7	67	71
27	23	0	83	7	11	13	0	0	29	0	7	71	17	19	0	11	59	7	13	0	37	251	0	23	7	107	11	0	41	17
29	53	17	29	13	19	7	0	11	0	0	103	0	7	0	0	89	13	0	11	7	107	23	0	19	0	0	7	0	83	11
33	37	0	7	59	0	19	0	79	0	7	107	0	13	11	0	0	7	23	0	29	0	31	0	7	11	13	251	0	0	0
39	11	7	0	31	0	41	13	223	7	233	0	11	89	0	17	7	19	37	0	13	0	0	7	0	0	127	0	0	71	7
41	0	0	0	7	13	101	193	19	41	23	7	0	0	0	11	0	0	7	83	71	17	197	211	0	7	11	19	0	31	0
47	0	11	7	23	17	0	29	101	0	7	31	13	11	83	0	37	7	41	19	0	0	17	0	7	13	263	23	0	0	0
51	19	103	67	0	0	73	7	11	0	0	97	61	0	7	37	0	17	0	11	19	7	157	23	29	0	139	227	7	137	11
53	43	7	0	0	0	13	0	0	7	101	11	0	0	0	0	7	0	29	13	53	23	11	7	31	47	0	41	0	79	7
57	17	13	0	0	43	7	67	0	0	0	59	19	7	0	13	0	11	17	71	7	0	0	0	0	0	173	7	11	131	47
59	7	0	19	83	73	31	17	7	11	71	173	101	13	239	7	0	0	197	0	11	0	7	0	17	0	13	0	0	7	227
63	41	17	0	11	7	0	0	163	13	0	23	7	0	47	11	0	29	137	7	0	0	13	19	0	31	7	179	0	0	0
69	151	11	13	7	23	239	31	131	97	41	7	0	11	0	19	13	43	7	0	17	113	73	109	11	7	0	0	23	13	79
71	13	151	23	0	11	7	0	0	17	37	0	0	7	13	109	11	0	0	43	7	53	29	107	47	19	17	7	149	0	0
77	0	0	11	29	7	211	59	13	0	17	191	7	0	11	0	0	19	101	7	23	13	41	0	0	11	7	17	137	229	167
81	0	7	127	13	17	0	151	0	7	0	79	139	47	0	0	7	13	19	173	11	29	17	7	0	0	37	0	41	43	7
83	11	13	193	7	0	0	37	151	19	0	7	11	0	23	13	0	103	7	0	101	79	149	11	0	7	0	31	13	97	0
87	7	0	29	0	59	17	11	7	0	19	13	0	211	0	7	53	0	11	0	149	193	7	17	13	0	71	67	0	7	0
89	19	0	7	0	0	67	0	23	13	7	151	17	0	0	0	0	7	0	149	19	11	13	47	7	0	29	0	0	17	193
93	167	19	181	23	11	0	7	0	179	0	0	0	151	7	0	11	149	13	73	0	7	0	37	17	157	0	11	7	0	0
99	0	0	11	43	0	7	0	17	0	31	167	13	7	11	0	151	0	0	0	7	23	79	0	0	11	83	7	0	0	0

	7 20	23	26	29	32	35	38	41	44	47	50	53	56	59	62	65	68	71	74	77	80	83	86	89	92	95	7 98	8 01	04	07
01	89	17	79	0	71	31	7	0	47	11	179	257	19	7	181	113	0	0	17	13	7	0	83	0	0	107	0	7	37	0
07	13	0	17	0	19	7	23	11	37	0	107	0	7	13	0	0	89	83	11	7	0	0	0	19	103	43	7	0	0	11
11	107	167	7	0	179	19	31	37	0	7	0	127	0	11	17	0	7	29	199	0	181	0	13	7	11	23	0	0	191	43
13	23	0	0	17	7	11	223	13	0	0	0	7	83	0	0	19	11	59	7	0	13	71	127	23	113	7	0	11	97	0
17	11	7	0	13	211	0	97	137	7	0	0	11	0	89	199	7	13	67	0	23	0	0	7	53	37	131	0	113	29	7
19	0	13	101	7	17	37	0	19	0	0	7	109	0	31	11	0	0	7	0	0	61	17	29	0	7	11	19	13	137	53
23	7	31	0	0	37	0	0	7	19	11	13	0	47	23	7	59	17	233	139	0	11	7	0	13	227	281	0	19	7	89
29	17	151	59	233	13	0	7	11	263	0	0	0	0	7	31	103	0	13	11	19	7	29	61	0	0	67	0	7	0	11
31	0	7	13	0	67	23	17	0	7	0	11	71	53	0	0	7	0	137	0	0	0	11	7	17	0	0	97	227	13	7
37	7	0	19	0	0	151	47	7	11	13	0	0	43	0	7	0	0	0	211	11	73	7	13	193	17	0	29	127	7	0
41	61	0	17	11	7	13	41	151	0	31	0	7	0	0	11	0	43	0	7	17	0	0	19	0	0	7	0	0	257	263
43	0	73	0	13	0	251	7	0	17	41	101	59	67	7	0	0	13	11	43	0	7	157	0	89	109	17	0	7	11	13
47	0	11	0	7	89	0	0	53	109	0	7	0	11	173	19	41	0	7	0	0	17	0	31	11	7	13	0	0	0	0
49	109	71	0	0	11	7	0	0	0	17	13	151	7	53	0	11	31	179	41	7	0	47	0	13	19	0	7	0	0	0
53	0	0	7	0	17	0	13	29	0	7	11	0	0	151	0	37	7	0	73	13	89	11	0	7	41	19	47	0	43	23
59	13	7	113	0	0	17	0	0	7	0	47	179	0	13	0	7	151	19	29	11	0	127	7	23	0	0	13	71	61	7
61	11	269	0	7	61	0	233	0	19	0	7	11	29	37	0	0	101	7	71	0	251	23	11	281	7	0	0	19	17	0
67	19	0	7	131	41	13	0	0	113	7	271	0	17	0	53	23	7	0	13	19	11	0	97	7	31	251	0	0	67	17
71	97	13	0	43	11	0	7	17	0	0	41	23	31	7	13	11	0	0	0	83	7	109	151	157	17	47	11	7	0	37
73	0	7	0	0	47	29	31	11	7	23	37	19	13	17	89	7	0	229	11	0	101	181	7	151	0	13	0	0	0	7
77	0	157	11	0	0	7	0	0	13	37	193	0	7	11	83	73	59	71	0	7	163	13	29	0	11	17	7	0	23	0
79	7	0	0	19	127	11	13	7	71	0	0	43	0	0	7	0	11	113	0	13	0	7	19	0	0	0	23	11	7	0
83	11	0	13	59	7	0	0	31	211	17	0	7	0	0	0	13	0	79	7	0	113	103	11	19	0	7	17	181	13	0
89	0	191	0	7	83	0	37	0	0	11	7	0	0	0	0	19	23	7	0	107	11	43	13	0	7	0	0	17	0	0
91	0	11	157	47	0	7	19	13	163	29	61	0	7	0	23	191	17	0	0	7	13	277	0	11	37	19	7	0	0	173
97	17	13	139	0	7	0	0	0	23	0	11	7	59	0	13	0	131	17	7	0	29	11	0	197	179	7	109	13	101	43

	7 21	24	27	30	33	36	39	42	45	48	51	54	57	60	63	66	69	72	75	78	81	84	87	90	93	96	7 99	8 02	05	08
01	0	7	0	37	23	11	67	0	7	131	13	0	17	0	41	7	11	0	19	0	0	0	7	13	0	0	0	11	79	7
03	0	17	23	7	0	89	263	0	11	19	7	0	0	0	0	0	53	7	17	11	83	13	211	199	7	23	0	139	19	0
07	7	61	0	11	13	0	0	7	0	239	19	0	0	17	7	0	0	13	179	29	37	7	0	41	71	11	0	0	7	19
09	0	19	7	0	0	0	11	0	0	7	0	73	0	29	137	13	7	11	0	17	19	89	31	7	0	0	41	0	11	0
13	37	11	19	0	167	0	7	47	269	79	31	13	11	7	17	23	0	0	0	0	7	19	0	11	13	0	157	7	0	211
19	41	139	0	0	157	7	193	0	43	23	11	53	7	19	167	17	0	37	13	7	191	11	223	31	0	103	7	97	73	0
21	7	0	11	13	17	83	29	7	0	0	43	199	0	11	7	193	13	31	0	59	0	7	0	19	11	0	229	0	7	13
27	11	23	0	103	0	17	7	199	0	0	13	11	41	7	127	19	43	29	0	223	7	0	11	13	23	0	257	7	0	131
31	17	0	257	7	0	29	11	0	0	0	7	0	0	0	37	0	19	7	31	13	23	107	131	0	7	0	67	0	11	0
33	53	113	0	199	13	7	17	19	73	11	0	241	7	139	0	197	107	13	23	7	11	41	43	17	0	0	7	0	29	0
37	13	17	7	0	11	0	107	61	19	7	227	0	53	13	23	11	7	0	17	277	0	0	0	7	0	97	11	19	0	229
39	0	107	0	0	7	211	0	11	131	67	29	7	23	0	97	173	47	0	7	0	0	0	71	0	13	7	0	0	43	11
43	19	7	11	0	71	0	0	13	7	0	163	37	[illegible]	11	0	7	0	0	0	17	13	47	7	0	11	73	0	[illegible]	[illegible]	7

51	23	53	7	11	0	0	0	41	0	7	223	197	13	59	11	0	7	67	0	127	31	19	61	7	73	11	17	0	109	233
57	59	7	31	43	109	73	13	0	7	0	17	61	11	19	29	7	41	23	0	13	0	67	7	11	0	0	37	17	0	7
61	0	0	13	0	0	7	0	11	0	0	0	59	7	23	19	13	0	0	11	7	47	31	17	173	61	37	7	83	13	11
63	7	233	0	0	0	19	37	7	173	43	11	17	239	13	7	31	0	0	0	0	0	7	79	0	19	29	13	0	7	0
67	0	0	0	31	7	11	17	23	0	13	0	7	0	29	0	0	11	0	7	0	0	0	13	17	0	7	0	11	0	193
69	0	0	53	89	0	23	7	13	11	0	0	163	17	7	0	43	19	0	0	11	7	131	227	37	139	0	211	7	23	17
73	0	23	61	7	239	0	0	17	0	0	7	71	0	127	11	0	13	7	0	43	0	97	37	107	7	11	0	0	197	13
79	89	11	7	0	0	0	29	0	17	7	13	0	11	0	0	0	7	0	23	47	0	0	0	7	0	17	0	0	19	31
81	19	0	73	107	7	0	167	59	13	103	0	7	0	0	17	11	23	109	7	19	37	13	0	31	163	7	11	43	61	29
87	37	173	11	7	0	31	241	0	0	0	7	19	0	11	0	13	167	7	0	71	41	0	0	0	7	0	0	0	13	47
91	7	71	83	0	79	59	23	7	11	0	17	13	19	0	7	53	0	0	0	11	0	7	0	139	13	0	41	17	7	23
93	11	0	7	19	23	0	61	0	97	7	0	11	0	47	79	271	7	37	31	0	0	53	11	7	0	0	167	23	83	41
97	23	0	0	67	19	13	7	0	0	0	29	17	0	7	241	0	37	11	13	61	7	0	0	19	0	0	0	7	11	0
99	17	7	43	13	29	0	0	191	7	11	139	103	229	0	19	7	13	17	73	0	11	23	7	83	0	0	0	59	0	7

	7 22	25	28	31	34	37	40	43	46	49	52	55	58	61	64	67	70	73	76	79	82	85	88	91	94	7 97	8 00	03	06	09
03	103	0	47	41	11	7	43	67	61	0	157	0	7	0	0	11	0	23	71	7	0	29	0	0	271	13	7	131	0	17
09	163	31	11	29	7	0	13	19	0	173	0	7	41	11	109	79	53	97	7	13	197	0	0	239	11	7	19	0	149	0
11	0	59	17	113	13	11	7	0	0	23	0	0	47	7	43	41	11	13	0	17	7	0	53	0	0	79	29	7	0	0
17	257	127	0	11	0	7	0	0	29	19	0	13	7	103	11	0	0	0	0	7	17	0	269	61	13	11	7	0	19	0
21	0	47	7	0	0	0	0	13	71	7	19	0	0	163	0	17	7	167	0	67	11	233	23	7	43	29	0	31	0	19
23	0	11	0	83	7	13	79	0	0	0	0	7	11	0	0	73	0	0	7	29	19	17	0	11	0	7	43	47	37	0
27	0	7	19	0	101	0	0	11	7	31	0	0	191	269	13	7	17	53	11	149	137	19	7	67	0	61	79	13	0	7
29	0	29	67	7	97	17	181	239	37	0	7	47	13	0	23	277	0	7	149	0	0	11	17	53	7	13	191	0	0	0
33	7	0	173	0	0	11	101	7	13	0	23	0	0	19	7	0	11	17	29	0	0	7	31	0	0	71	163	11	7	0
39	29	17	13	11	23	19	7	79	101	137	0	0	181	7	11	13	41	0	17	59	7	0	0	0	19	11	0	7	13	29
41	13	7	23	0	271	37	11	17	7	0	67	0	149	13	0	7	0	11	0	41	0	0	7	29	17	23	13	0	11	7
47	7	0	97	193	11	29	0	7	17	149	47	31	73	0	7	11	0	0	0	23	13	7	37	0	53	17	11	0	7	61
51	0	0	263	13	7	0	0	149	19	241	11	7	101	271	89	23	13	0	7	0	17	11	29	0	0	7	0	19	0	13
53	0	13	11	191	0	131	7	0	0	17	0	0	0	7	13	0	29	103	19	137	7	0	0	0	11	173	17	7	59	0
57	19	37	41	7	17	0	103	0	11	23	7	0	31	0	101	0	251	7	79	11	139	17	0	13	7	0	223	107	0	73
59	11	0	0	149	0	7	31	23	13	0	17	11	7	0	157	59	263	0	0	7	0	13	11	0	181	47	7	17	79	19
63	127	149	7	23	13	17	11	0	197	7	73	19	107	0	0	29	7	11	37	53	61	251	17	7	229	31	23	0	11	0
69	0	7	0	19	11	71	17	31	7	61	0	13	0	59	47	7	0	0	101	0	23	0	7	17	13	0	11	0	0	7
71	0	31	0	7	0	0	0	11	89	13	7	0	17	19	0	0	37	7	11	103	29	0	13	41	7	241	0	179	0	11
77	0	0	7	13	0	11	0	0	53	7	0	0	23	17	31	0	7	0	173	0	0	0	0	7	19	0	0	11	0	13
81	11	181	31	0	197	89	7	0	17	97	83	11	13	7	0	0	0	223	0	29	7	179	11	0	0	13	73	7	0	47
83	41	7	0	11	0	0	23	0	7	167	13	0	0	29	11	7	19	0	131	0	0	0	7	13	61	11	53	31	0	7
87	0	29	23	163	43	7	13	73	0	11	79	131	7	47	0	31	157	19	0	7	11	89	0	0	101	23	7	0	0	109
89	7	11	0	0	13	113	43	7	19	31	0	269	11	61	7	17	127	13	0	167	79	7	0	11	29	73	283	19	7	0
93	13	229	0	53	7	109	0	11	113	19	17	7	29	13	0	41	0	193	7	23	59	0	0	0	0	7	13	17	19	11
99	197	19	269	7	67	11	0	13	0	37	7	17	71	23	227	61	11	7	0	0	13	53	257	29	7	199	173	11	17	107

	8 10	13	16	19	22	25	28	31	34	37	40	43	46	49	52	55	58	61	64	67	70	73	76	79	82	85	88	91	94	97
01	0	11	13	0	7	17	31	0	0	0	167	7	11	59	0	13	239	29	7	277	19	67	17	11	193	7	0	0	13	271
07	59	0	79	7	0	0	17	41	0	13	7	0	19	197	139	37	53	7	71	31	167	11	13	17	7	67	0	0	29	109
11	7	17	0	101	229	11	0	7	239	97	0	59	211	19	7	233	11	0	13	0	0	7	79	0	0	61	0	11	7	283
13	0	31	7	13	19	109	0	17	11	7	29	0	191	0	0	0	7	0	0	11	0	0	0	7	17	0	0	0	0	13
17	0	233	17	11	0	19	7	0	0	0	0	0	13	7	11	0	0	0	103	17	7	0	41	0	19	11	0	7	0	73
19	0	7	0	0	0	179	11	43	7	0	13	0	37	0	31	7	0	11	89	0	173	29	7	13	47	17	0	0	11	7
23	0	11	31	17	0	7	13	101	0	29	73	37	7	163	0	0	19	71	0	7	17	0	0	11	0	0	7	0	223	23
29	13	167	0	0	7	0	113	97	19	101	11	7	0	13	0	31	0	43	7	0	29	11	0	23	83	7	13	19	37	53
31	0	0	11	0	0	0	7	59	0	31	17	13	0	7	29	0	0	0	19	43	7	23	0	0	11	223	211	7	0	61
37	11	163	0	0	0	7	0	0	0	0	19	11	7	157	0	23	0	0	13	7	0	0	11	47	0	29	7	0	17	19
41	0	13	7	67	0	59	11	71	181	7	31	19	53	29	13	113	7	11	0	127	0	167	0	7	0	37	73	13	11	43
43	0	0	19	0	7	197	37	29	0	11	229	7	13	173	0	131	0	0	7	0	11	19	0	0	79	7	0	97	0	17
47	0	7	0	19	11	23	0	17	7	83	0	0	47	0	0	7	0	277	137	223	61	13	7	31	17	0	11	239	23	7
49	0	0	0	7	233	0	13	11	0	89	7	0	0	17	163	0	293	7	11	13	0	113	0	37	7	73	23	59	0	11
53	7	0	11	0	83	31	29	7	17	61	0	67	0	11	7	13	0	101	0	0	263	7	23	281	11	17	0	0	7	0
59	11	0	37	41	43	0	7	137	0	13	0	11	0	7	0	67	23	29	31	101	7	0	11	0	0	19	17	7	0	0
61	103	7	127	11	0	0	41	13	7	0	0	29	31	0	11	7	19	0	0	53	13	199	7	0	0	11	0	163	137	7
67	7	11	0	0	0	0	173	7	19	211	0	239	11	0	7	41	17	199	0	0	83	7	29	11	61	31	0	13	7	0
71	0	0	0	0	7	0	79	11	0	19	13	7	227	31	71	0	43	0	7	0	0	41	0	13	103	7	181	23	17	11
73	17	0	23	0	29	71	7	31	13	0	11	139	0	7	269	83	79	17	43	19	7	11	73	0	41	23	0	7	131	107
77	0	19	0	7	13	11	179	0	0	0	7	0	17	0	53	0	11	7	0	107	19	23	43	0	7	101	31	11	0	17
79	89	17	13	73	0	7	67	223	11	199	83	19	7	0	107	13	157	0	17	7	31	59	0	97	43	283	7	257	13	0
83	0	97	7	11	107	269	0	193	31	7	47	13	19	17	11	23	7	0	197	0	0	0	0	7	13	11	0	101	43	0
89	131	7	0	163	19	13	0	41	7	23	0	0	11	37	17	7	0	79	13	59	73	31	7	11	0	0	103	0	109	7
91	83	199	151	7	11	0	0	23	29	0	7	0	0	0	19	11	13	7	0	229	17	281	0	0	7	0	11	79	0	13
97	0	23	7	167	17	151	19	271	0	7	13	37	0	11	0	0	7	0	67	29	251	17	0	7	11	19	0	191	31	0

	8 11	14	17	20	23	26	29	32	35	38	41	44	47	50	53	56	59	62	65	68	71	74	77	80	83	86	89	92	95	98
01	0	0	0	43	0	0	7	19	11	47	37	0	0	7	197	0	17	0	0	11	7	71	0	0	0	41	19	7	0	89
03	11	7	0	0	13	17	0	0	7	181	31	11	71	167	0	7	0	13	23	61	0	0	7	0	227	251	0	0	37	7
07	13	127	0	0	0	7	11	0	113	43	151	0	7	13	23	0	271	11	19	7	0	0	229	0	233	0	7	37	11	31
09	7	0	101	0	53	0	17	7	37	11	241	13	23	0	7	59	0	0	0	47	11	7	139	17	13	0	67	0	7	0
13	29	17	41	0	7	0	0	13	23	0	19	7	0	151	0	11	53	73	7	0	13	61	239	283	47	7	11	0	0	19
19	0	13	11	7	263	0	283	0	47	79	7	29	0	11	13	0	151	7	241	17	0	19	0	0	7	23	0	13	0	0
21	23	0	71	0	191	7	101	0	17	109	0	0	7	0	41	0	11	151	31	7	0	0	0	23	0	13	7	11	0	0
27	31	107	0	11	7	53	13	0	101	17	0	7	193	0	11	0	29	23	7	13	151	0	37	19	0	7	17	0	0	43
31	0	7	13	0	17	19	127	0	7	11	0	0	0	23	0	7	0	53	0	31	11	17	7	47	19	263	113	0	13	7
33	13	11	37	7	281	0	239	0	103	0	7	23	11	13	0	19	0	7	0	71	0	0	59	11	7	61	13	17	0	0
37	7	31	0	0	137	17	197	7	0	13	0	0	0	0	7	29	19	83	11	0	79	7	13	0	0	151	0	0	7	11
39	41	0	7	0	0	23	0	13	139	7	11	17	101	277	61	0	7	0	0	37	13	11	0	7	0	137	19	233	17	0

	12	15	18	21	24	27	30	33	36	39	42	45	48	51	54	57	60	63	66	69	72	75	78	81	84	87	90	93	96	99
49	19	79	0	11	0	7	109	17	29	191	13	0	7	0	11	41	61	0	23	7	0	157	47	13	17	11	7	31	149	0
51	7	47	29	0	0	0	11	7	13	71	19	79	0	17	7	97	23	11	41	0	0	7	0	191	53	0	0	149	7	19
57	0	0	13	31	11	0	7	0	0	0	23	0	131	7	17	11	43	0	101	0	7	19	127	173	149	0	11	7	13	59
61	277	29	0	7	0	131	23	139	0	17	7	13	0	0	0	0	67	7	0	0	43	11	19	107	7	0	17	0	0	23
63	0	0	11	137	23	7	0	53	0	13	0	0	7	11	0	17	31	0	107	7	101	149	13	83	11	0	7	23	0	73
67	23	41	7	0	31	13	163	0	11	7	17	0	29	257	19	0	7	281	13	11	67	47	0	7	97	0	43	17	0	0
69	11	257	0	13	7	19	29	0	193	0	73	7	103	97	0	0	13	0	7	0	61	23	11	0	19	7	0	0	43	13
73	0	7	0	0	0	47	11	0	7	0	41	17	13	241	59	7	149	11	0	109	179	0	7	29	67	13	193	0	11	7
79	7	59	53	211	11	29	13	7	0	37	0	23	17	149	7	11	127	19	0	13	0	7	61	0	0	71	11	73	7	17
81	0	17	7	79	13	89	0	11	19	7	0	0	149	0	0	47	7	13	11	283	0	0	41	7	31	0	101	19	29	11
87	19	7	17	23	0	11	31	37	7	149	29	13	0	0	103	7	11	0	0	17	0	89	7	59	13	131	23	11	101	7
91	11	19	89	103	47	7	37	13	0	0	0	11	7	0	17	0	0	0	131	7	13	0	11	137	157	31	7	29	0	0
93	7	227	263	11	0	13	149	7	179	43	59	19	0	0	7	67	113	0	13	31	17	7	0	0	37	11	0	0	7	241
97	0	13	157	53	7	41	0	31	0	11	269	7	19	43	13	17	23	0	7	113	11	59	0	37	0	7	0	13	0	0
99	0	11	0	19	17	0	7	0	41	53	0	0	11	7	23	43	0	211	0	67	7	17	19	11	109	13	61	7	0	0

	8 12	15	18	21	24	27	30	33	36	39	42	45	48	51	54	57	60	63	66	69	72	75	78	81	84	87	90	93	96	99
03	0	149	179	7	19	191	0	11	13	0	7	0	137	0	41	0	17	7	11	43	29	13	0	19	7	107	0	0	0	11
09	17	0	7	47	23	11	0	227	0	7	107	0	0	0	223	13	7	17	257	233	37	0	277	7	211	43	0	11	13	0
11	13	37	23	157	7	107	17	0	11	0	0	7	0	13	0	0	0	0	7	11	0	0	0	17	0	7	13	31	0	47
17	241	0	0	7	73	181	11	13	0	31	7	223	89	47	229	0	0	7	37	23	13	0	137	0	7	79	0	0	11	0
21	7	11	17	13	0	0	61	7	0	0	0	0	11	0	7	23	13	37	19	17	0	7	53	11	29	0	0	179	7	13
23	0	13	7	41	11	0	0	97	17	7	0	0	271	23	13	11	7	0	29	0	0	0	31	7	0	17	11	13	19	0
27	43	0	47	17	139	0	7	103	241	23	11	181	0	7	0	59	0	173	0	0	7	11	71	13	0	83	127	7	0	19
29	29	7	11	0	31	0	79	23	7	17	0	137	41	11	0	7	0	131	0	0	19	13	7	0	11	0	17	0	47	7
33	0	0	19	23	13	7	43	167	11	0	131	0	7	0	37	0	227	13	41	7	83	17	0	31	191	89	7	157	0	139
39	0	67	0	0	7	17	11	0	0	0	0	7	43	19	0	83	97	11	7	0	23	0	17	53	13	7	269	41	11	0
41	137	73	223	0	19	97	7	0	0	11	61	17	37	7	43	179	139	0	23	227	7	0	13	19	59	0	0	7	17	53
47	113	0	0	13	29	7	0	11	233	127	0	59	7	0	0	19	13	79	11	7	43	0	107	181	241	0	7	47	157	11
51	31	0	7	113	41	83	53	17	23	7	173	0	13	11	0	0	7	0	73	0	0	29	59	7	11	13	0	199	37	293
53	193	0	0	0	7	11	23	19	0	37	13	7	53	17	0	29	11	0	7	89	0	0	0	13	197	7	19	11	0	23
57	11	7	23	29	0	0	13	0	7	59	109	11	0	31	97	7	47	0	193	13	0	0	7	199	53	17	0	19	0	7
59	23	0	109	7	13	0	0	31	269	13	7	0	0	0	11	191	41	7	19	0	71	0	103	23	7	11	29	193	0	0
63	7	0	71	0	0	0	0	7	0	11	0	103	113	13	7	139	89	67	79	19	11	7	41	131	0	37	13	0	7	0
69	181	0	0	127	0	37	7	11	31	0	17	19	0	7	0	199	0	0	11	0	7	67	0	0	0	29	0	7	0	11
71	67	7	19	0	0	13	0	263	7	131	11	23	0	53	127	7	17	0	13	29	197	11	7	37	0	0	0	0	0	7
77	7	29	41	37	67	23	0	7	11	79	71	83	13	19	7	31	0	17	0	11	0	7	0	0	103	13	281	139	7	0
81	0	23	37	11	7	0	251	199	13	137	271	7	17	103	11	0	59	0	7	0	0	13	0	109	23	7	229	0	0	17
83	0	17	0	0	0	19	7	0	67	0	89	41	29	7	73	109	0	11	17	13	7	0	23	163	19	47	0	7	11	0
87	29	11	13	7	0	0	19	61	53	0	7	251	11	17	0	13	31	7	23	37	191	0	0	11	7	19	0	0	13	29
89	13	83	17	0	11	7	0	0	0	47	31	0	7	13	53	11	19	0	0	7	41	0	179	29	107	0	7	71	0	0
93	0	139	7	0	0	0	0	89	127	7	11	29	23	0	17	0	7	19	0	0	0	11	13	7	0	0	41	0	257	31
99	0	7	0	13	0	0	23	0	7	19	0	31	73	0	193	7	13	0	181	11	0	251	7	89	0	0	139	0	19	7

	9 00	03	06	09	12	15	18	21	24	27	30	33	36	39	42	45	48	51	54	57	60	63	66	69	72	75	78	81	84	87
01	0	73	7	0	11	37	0	31	0	7	0	13	0	0	0	11	7	0	0	0	0	23	0	7	13	0	11	0	19	89
07	0	7	11	0	223	13	0	0	7	0	17	0	0	11	0	7	113	0	13	0	19	193	7	0	11	281	47	17	0	7
11	0	13	19	0	197	7	0	0	11	83	281	23	7	0	13	29	0	0	73	7	67	19	17	0	41	0	7	13	0	0
13	7	0	31	229	53	0	0	7	0	23	47	11	13	0	7	0	59	227	0	0	0	7	11	199	0	13	0	41	7	0
17	0	37	0	0	7	23	11	251	13	0	191	7	179	19	71	47	53	11	7	0	0	13	79	17	67	7	29	59	11	0
19	0	181	0	23	19	71	7	0	0	11	167	0	17	7	0	31	0	73	0	13	7	61	53	19	191	113	23	7	0	17
23	0	41	13	7	11	19	0	17	29	0	7	0	251	0	59	11	0	7	37	0	131	0	23	103	7	0	11	0	13	269
29	197	59	7	79	0	0	229	181	17	7	41	0	0	11	0	0	7	251	0	29	109	0	13	7	11	17	0	0	0	0
31	0	103	0	0	7	11	131	13	0	47	31	7	109	29	17	0	11	0	7	0	13	0	71	0	0	7	19	11	257	0
37	179	13	233	7	0	239	0	199	23	0	7	0	0	0	11	17	0	7	19	0	137	0	41	31	7	11	227	13	173	0
41	7	61	0	211	23	0	0	7	97	11	13	31	29	0	7	0	0	89	0	19	11	7	241	13	0	103	0	17	7	293
43	127	11	7	199	0	31	29	0	13	7	19	269	11	37	73	0	7	0	0	67	0	13	0	7	47	23	0	0	0	19
47	53	167	0	0	13	43	7	11	193	163	0	17	37	7	79	0	0	13	11	0	7	23	127	29	31	0	0	7	17	11
49	17	7	13	103	0	83	53	43	7	137	11	277	71	0	307	7	0	17	31	23	139	11	7	67	79	0	0	61	13	7
53	0	0	269	19	0	7	31	0	59	0	0	13	7	47	0	23	11	0	53	7	0	0	19	0	13	0	7	11	0	17
59	0	0	0	11	7	13	97	157	0	23	0	7	73	17	11	0	29	43	7	31	0	167	163	0	0	7	0	103	0	61
61	113	109	17	13	263	19	7	23	0	0	29	89	229	7	0	0	13	11	0	17	7	173	0	47	19	0	0	7	11	13
67	0	23	71	17	11	7	0	37	0	0	13	73	7	0	107	11	19	59	0	7	17	29	0	13	23	43	7	89	0	283
71	0	0	7	0	107	0	13	61	89	7	11	0	47	0	31	17	7	19	0	13	23	11	0	7	211	0	0	127	59	43
73	0	0	11	29	7	0	0	0	19	113	163	7	283	11	0	0	0	13	7	0	191	17	277	0	11	7	97	19	0	0
77	13	7	0	0	97	0	79	0	7	19	0	0	113	13	23	7	17	0	307	11	29	0	7	37	89	0	13	31	19	7
79	11	0	0	7	37	17	139	0	0	0	7	11	23	0	29	271	79	7	0	19	0	31	11	0	7	0	0	0	0	0
83	7	19	29	37	0	0	11	7	23	31	0	0	0	0	7	0	239	11	0	0	13	7	109	293	0	0	0	47	7	173
89	0	13	23	0	11	67	7	0	0	0	0	47	19	7	13	11	0	0	17	0	7	113	31	0	271	23	11	7	149	223
91	23	7	89	19	0	0	43	11	7	0	127	61	13	193	0	7	31	0	11	0	307	41	7	23	17	13	53	149	0	7
97	7	0	0	0	0	11	13	7	17	71	0	59	43	0	7	0	11	23	29	13	0	7	0	0	149	17	223	11	7	31

	9 01	04	07	10	13	16	19	22	25	28	31	34	37	40	43	46	49	52	55	58	61	64	67	70	73	76	79	82	85	88
01	11	0	13	17	7	139	29	137	233	0	157	7	0	23	181	13	43	31	7	0	17	0	11	0	0	7	47	283	13	0
03	13	0	0	11	0	47	7	0	0	17	0	23	0	7	11	0	0	0	43	0	7	149	0	0	0	11	13	7	137	29
07	0	0	61	7	17	101	73	19	0	11	7	0	83	0	0	89	0	7	0	149	11	17	13	0	7	0	19	0	0	0
09	251	11	0	0	0	7	0	13	79	0	17	29	7	0	0	37	107	19	149	7	13	229	97	11	31	0	7	17	23	0
13	97	23	7	13	127	17	107	11	71	7	0	109	31	41	37	0	7	0	11	0	223	67	17	7	23	0	179	0	29	11
19	227	7	83	0	53	11	17	0	7	101	13	0	0	149	257	7	11	0	23	0	277	0	7	13	307	31	0	11	0	7
21	0	19	257	7	29	0	0	0	11	0	7	103	17	167	0	0	23	7	59	11	19	13	311	0	7	41	181	0	83	17
27	0	31	7	227	271	59	11	0	67	7	23	0	19	17	0	13	7	11	0	79	97	211	197	7	0	233	0	0	11	37
31	193	11	0	29	0	0	7	149	17	0	0	13	11	7	0	173	59	0	0	61	7	0	0	11	13	17	0	7	37	23
33	173	7	41	0	11	43	149	0	7	13	0	233	67	0	17	7	0	0	83	47	251	73	7	19	131	89	11	23	0	7
37	23	0	31	59	149	7	89	0	37	17	11	223	7	271	29	101	139	131	13	7	0	11	0	23	19	163	7	193	211	0
39	7	0	11	13	241	0	0	7	29	263	0	41	0	11	7	17	13	0	0	239	127	7	0	0	11	251	37	31	7	13
43	109	149	103	181	7	113	0	0	11	227	17	7	13	157	0	31	19	23	7	11	79	0	89	53	311	7	0	17	0	97

51	17	29	151	83	13	7	0	0	0	11	0	113	7	163	0	0	0	13	19	7	11	0	31	37	67	0	7	0	139	41
57	89	17	47	23	7	151	0	11	0	0	19	7	29	0	157	103	269	0	7	0	0	0	0	71	13	7	23	0	67	11
61	29	7	11	41	103	71	0	13	7	0	59	19	0	11	127	7	0	0	0	257	13	0	7	31	11	61	0	97	0	7
63	0	61	17	7	211	11	41	257	151	0	7	0	0	0	197	181	11	7	13	17	23	19	0	29	7	127	163	11	0	109
67	7	13	139	19	0	31	0	7	0	0	151	11	41	109	7	137	23	0	227	37	0	7	11	113	0	101	0	13	7	0
69	37	0	7	11	0	29	0	0	0	7	0	151	13	19	11	41	7	47	0	0	17	0	0	7	0	11	313	0	241	0
73	0	0	43	61	0	0	7	53	13	11	23	211	79	7	19	17	73	0	31	0	7	13	29	0	0	0	0	7	0	0
79	31	173	13	0	23	7	19	11	43	131	0	0	7	0	0	13	17	0	11	7	0	0	0	193	0	19	7	23	13	11
81	7	0	23	0	0	17	59	7	0	293	11	0	191	13	7	73	19	151	0	0	0	7	17	0	0	23	13	29	7	61
87	0	41	0	79	0	277	7	13	11	29	0	0	0	7	37	0	43	0	61	11	7	0	0	17	0	0	0	7	311	0
91	0	17	163	7	59	0	67	41	53	19	7	0	71	37	11	23	13	7	17	0	43	47	151	79	7	11	29	227	19	13
93	19	13	0	71	0	7	11	17	0	0	41	0	7	23	13	0	0	11	109	7	29	0	43	151	17	211	7	13	11	0
97	0	11	7	0	0	47	0	0	29	7	13	0	11	73	0	281	7	233	0	17	19	0	0	7	0	151	43	0	0	0
99	0	0	29	0	7	107	197	23	13	0	0	7	97	0	0	11	0	157	7	41	0	13	0	89	173	7	11	0	43	0

	9 02	05	08	11	14	17	20	23	26	29	32	35	38	41	44	47	50	53	56	59	62	65	68	71	74	77	80	83	86	89
03	0	7	0	17	13	0	0	241	7	61	11	0	19	139	67	7	0	13	0	29	17	11	7	0	257	41	23	197	151	7
09	7	29	71	31	17	293	0	7	11	53	83	13	0	0	7	0	0	191	67	11	23	7	131	19	13	199	0	37	7	0
11	11	0	7	179	0	0	101	0	37	7	17	11	0	0	19	53	7	0	23	0	0	103	11	7	29	0	0	17	31	0
17	0	7	197	13	113	41	19	0	7	11	31	17	23	0	263	7	13	0	0	0	11	0	7	0	61	19	0	0	17	7
21	83	131	0	0	11	7	17	19	23	0	73	41	7	0	0	11	0	199	0	7	0	263	0	17	37	13	7	0	0	31
23	7	0	0	293	0	37	23	7	0	43	13	0	17	61	7	0	167	19	11	0	0	7	0	13	0	79	83	0	7	11
27	0	0	11	0	7	29	13	17	0	0	53	7	0	11	0	0	0	0	7	13	41	0	0	0	11	7	61	0	0	0
29	23	0	61	0	13	11	7	127	211	19	0	0	101	7	89	43	11	13	0	0	7	83	37	23	0	0	167	7	19	0
33	11	0	0	7	0	0	0	0	17	199	7	11	103	13	0	61	29	7	0	23	0	37	11	137	7	17	13	107	53	19
39	0	37	7	0	61	199	31	13	0	7	0	89	107	23	0	211	7	0	59	197	11	19	179	7	139	43	17	29	0	0
41	31	11	0	0	7	13	0	107	0	0	0	7	11	47	0	17	101	67	7	37	157	29	113	11	0	7	0	43	0	163
47	0	0	0	7	19	23	83	0	0	41	7	139	13	31	0	0	17	7	101	0	109	11	0	19	7	13	0	0	23	0
51	7	23	47	0	109	11	0	7	13	0	0	17	0	0	7	41	11	97	0	229	29	7	0	0	19	239	71	11	7	53
53	17	83	7	0	0	0	13	0	11	7	0	0	127	0	29	19	7	17	41	11	101	0	23	7	0	67	31	59	47	0
57	43	137	13	11	0	0	7	0	0	0	0	0	17	7	11	13	19	167	23	0	7	0	0	0	41	11	0	7	13	17
59	13	7	43	0	0	89	11	19	7	0	179	0	47	13	59	7	23	11	17	0	0	223	7	0	0	29	13	41	11	7
63	0	11	0	0	0	7	43	0	19	13	0	0	7	17	0	193	0	47	271	7	0	61	13	11	0	59	7	19	0	0
69	19	41	89	13	7	163	23	0	0	31	11	7	37	0	17	97	13	0	7	19	0	11	157	0	29	7	281	0	0	13
71	0	13	11	17	23	0	7	71	0	239	19	137	0	7	13	0	0	283	29	0	7	269	73	0	11	0	101	7	79	19
77	11	53	19	73	17	7	0	0	13	109	37	11	7	41	0	0	31	127	241	7	43	13	11	0	107	0	7	0	101	29
81	0	239	7	19	13	0	11	0	0	7	0	0	269	53	107	0	7	11	163	41	0	0	19	7	43	277	0	131	11	0
83	137	0	13	0	7	17	0	0	0	11	0	7	223	19	0	13	0	0	7	53	11	59	17	157	71	7	43	37	13	31
87	17	7	0	67	11	263	71	0	7	0	0	13	0	97	19	7	0	17	103	0	73	0	7	0	13	0	11	0	29	7
89	0	157	97	7	191	19	17	11	59	13	7	31	0	131	61	0	0	7	11	0	0	0	13	17	7	0	47	0	0	11
93	7	17	11	0	0	13	19	7	0	0	29	173	0	11	7	0	0	0	13	59	0	7	0	83	11	19	233	61	7	0
99	11	0	17	0	0	41	7	0	0	113	79	11	13	7	53	47	61	19	83	17	7	29	11	37	0	13	263	7	229	0

Log Γ (*n*).

n	0	1	2	3	4	5	6	7	8	9
1.00		97497	95001	92512	90030	87555	85087	82627	80173	77727
1.01	9.9975287	72855	70430	68011	65600	63196	60799	58408	56025	53648
1.02	51279	48916	46561	44212	41870	39535	37207	34886	32572	30265
1.03	27964	25671	23384	21104	18831	16564	14305	12052	09806	07567
1.04	05334	03108	00889	$\overline{9}8677$	$\overline{9}6471$	$\overline{9}4273$	$\overline{9}2080$	$\overline{8}9895$	$\overline{8}7716$	$\overline{8}5544$
1.05	9.9883379	81220	79068	76922	74783	72651	70525	68406	66294	64188
1.06	62089	59996	57910	55830	53757	51690	49630	47577	45530	43489
1.07	41469	39428	37407	35392	33384	31382	29387	27398	25415	23449
1.08	21469	19506	17549	15599	13655	11717	09785	07860	05941	04029
1.09	02123	00223	$\overline{9}8329$	$\overline{9}6442$	$\overline{9}4561$	$\overline{9}2686$	$\overline{9}0818$	$\overline{8}8956$	$\overline{8}7100$	$\overline{8}5250$
1.10	9.9783407	81570	79738	77914	76095	74283	72476	70676	68882	67095
1.11	65313	63538	61768	60005	58248	56497	54753	53014	51281	49555
1.12	47834	46120	44411	42709	41013	39323	37638	35960	34288	32622
1.13	30962	29308	27659	26017	24381	22751	21126	19508	17896	16289
1.14	14689	13094	11505	09922	08345	06774	05209	03650	02096	00549
1.15	9.9699007	97471	95941	94417	92898	91386	89879	88378	86883	85393
1.16	83910	82432	80960	79493	78033	76578	75129	73686	72248	70816
1.17	69390	67969	66554	65145	63742	62344	60952	59566	58185	56810
1.18	55440	54076	52718	51366	50019	48677	47341	46011	44687	43368
1.19	42054	40746	39444	38147	36856	35570	34290	33016	31747	30483
1.20	29225	27973	26725	25484	24248	23017	21792	20573	19358	18150
1.21	16946	15748	14556	13369	12188	11011	09841	08675	07515	06361
1.22	05212	04068	02930	01796	00669	$\overline{9}9546$	$\overline{9}8430$	$\overline{9}7318$	$\overline{9}6212$	$\overline{9}5111$
1.23	9.9594015	92925	91840	90760	89685	88616	87553	86494	85441	84393
1.24	83350	82313	81280	80253	79232	78215	77204	76198	75197	74201
1.25	73211	72226	71246	70271	69301	68337	67377	66423	65474	64530
1.26	63592	62658	61730	60806	59888	58975	58067	57165	56267	55374
1.27	54487	53604	52727	51855	50988	50126	49268	48416	47570	46728
1.28	45891	45059	44232	43410	42593	41782	40975	40173	39376	38585
1.29	37798	37016	36239	35467	34700	33938	33181	32439	31682	30940
1.30	30203	29470	28743	28021	27303	26590	25883	25180	24482	23789
1.31	23100	22417	21739	21065	20396	19732	19073	18419	17770	17125
1.32	16485	15850	15220	14595	13975	13359	12748	12142	11540	10944
1.33	10353	09766	09184	08606	08034	07466	06903	06344	05791	05242
1.34	04698	04158	03624	03094	02568	02048	01532	01021	00514	00012
1.35	9.9499515	99023	98535	98052	97573	97100	96630	96166	95706	95251
1.36	94800	94355	93913	93477	93044	92617	92194	91776	91362	90953
1.37	90549	90149	89754	89363	88977	88595	88218	87846	87478	87115
1.38	86756	86402	86052	85707	85366	85030	84698	84371	84049	83731
1.39	83417	83108	82803	82503	82208	81916	81630	81348	81070	80797
1.40	80528	80263	80003	79748	79497	79250	79008	78770	78537	78308
1.41	78084	77864	77648	77437	77230	77027	76829	76636	76446	76261
1.42	76081	75905	75733	75565	75402	75243	75089	74939	74793	74652
1.43	74515	74382	74254	74130	74010	73894	73783	73676	73574	73476
1.44	73382	73292	73207	73125	73049	72976	72908	72844	72784	72728
1.45	72677	72630	72587	72549	72514	72484	72459	72437	72419	72406
1.46	72397	72393	72392	72396	72404	72416	72432	72452	72477	72506
1.47	72539	72576	72617	72662	72712	72766	72824	72886	72952	73022
1.48	73097	73175	73258	73345	73436	73531	73630	73734	73841	73953
1.49	74068	74188	74312	74440	74572	74708	74848	74992	75141	75293

Note.—This table is taken from Vol. II. of Legendre's work, and not from Vol. I., as stated in the Preface: the numbers given in Vol. I. being inaccurate in the seventh decimal place. In Vol. II. the values are given to twelve places of decimals. The figure here printed in the seventh place is

n	0	1	2	3	4	5	6	7	8	9
1.50	9.9475449	75610	75774	75943	76116	76292	76473	76658	76847	77040
1.51	77237	77438	77642	77851	78064	78281	78502	78727	78956	79189
1.52	79426	79667	79912	80161	80414	80671	80932	81196	81465	81738
1.53	82015	82295	82580	82868	83161	83457	83758	84062	84370	84682
1.54	84998	85318	85642	85970	86302	86638	86977	87321	87668	88019
1.55	88374	88733	89096	89463	89834	90208	90587	90969	91355	91745
1.56	92139	92537	92938	93344	93753	94166	94583	95004	95429	95857
1.57	96289	96725	97165	97609	98056	98508	98963	99422	99885	$\bar{0}0351$
1.58	9.9500822	01296	01774	02255	02741	03230	03723	04220	04720	05225
1.59	05733	06245	06760	07280	07803	08330	08860	09395	09933	10475
1.60	11020	11569	12122	12679	13240	13804	14372	14943	15519	16098
1.61	16680	17267	17857	18451	19048	19650	20254	20862	21475	22091
1.62	22710	23333	23960	24591	25225	25863	26504	27149	27798	28451
1.63	29107	29767	30430	31097	31767	32442	33120	33801	34486	35175
1.64	35867	36563	37263	37966	38673	39383	40097	40815	41536	42260
1.65	42989	43721	44456	45195	45938	46684	47434	48187	48944	49704
1.66	50468	51236	52007	52782	53560	54342	55127	55916	56708	57504
1.67	58303	59106	59913	60723	61536	62353	63174	63998	64826	65656
1.68	66491	67329	68170	69015	69864	70716	71571	72430	73293	74159
1.69	75028	75901	76777	77657	78540	79427	80317	81211	82108	83008
1.70	83912	84820	85731	86645	87563	88484	89409	90337	91268	92203
1.71	93141	94083	95028	95977	96929	97884	98843	99805	$\bar{0}0771$	$\bar{0}1740$
1.72	9.9602712	03688	04667	05650	06636	07625	08618	09614	10613	11616
1.73	12622	13632	14645	15661	16681	17704	18730	19760	20793	21830
1.74	22869	23912	24959	26009	27062	28118	29178	30241	31308	32377
1.75	33451	34527	35607	36690	37776	38866	39959	41055	42155	43258
1.76	44364	45473	46586	47702	48821	49944	51070	52200	53331	54467
1.77	55606	56749	57894	59043	60195	61350	62509	63671	64836	66004
1.78	67176	68351	69529	70710	71895	73082	74274	75468	76665	77866
1.79	79070	80277	81488	82701	83918	85138	86361	87588	88818	90051
1.80	91287	92526	93768	95014	96263	97515	98770	$\bar{0}0029$	$\bar{0}1291$	$\bar{0}2555$
1.81	9.9703823	05095	06369	07646	08927	10211	11498	12788	14082	15378
1.82	16678	17981	19287	20596	21908	23224	24542	25864	27189	28517
1.83	29848	31182	32520	33860	35204	36551	37900	39254	40610	41969
1.84	43331	44697	46065	47437	48812	50190	51571	52955	54342	55733
1.85	57126	58522	59922	61325	62730	64140	65551	66966	68384	69805
1.86	71230	72657	74087	75521	76957	78397	79839	81285	82734	84186
1.87	85640	87098	88559	90023	91490	92960	94433	95910	97389	98871
1.88	9.9800356	01844	03335	04830	06327	07827	09331	10837	12346	13859
1.89	15374	16893	18414	19939	21466	22996	24530	26066	27606	29148
1.90	30693	32242	33793	35348	36905	38465	40028	41595	43164	44736
1.91	46311	47890	49471	51055	52642	54232	55825	57421	59020	60622
1.92	62226	63834	65445	67058	68675	70294	71917	73542	75170	76802
1.93	78436	80073	81713	83356	85002	86651	88302	89957	91614	93275
1.94	94938	96605	98274	99946	$\bar{0}1621$	$\bar{0}3299$	$\bar{0}4980$	$\bar{0}6663$	$\bar{0}8350$	$\bar{1}0039$
1.95	9.9911732	13427	15125	16826	18530	20237	21947	23659	25375	27093
1.96	28815	30539	32266	33995	35728	37464	39202	40943	42688	44435
1.97	46185	47937	49693	51451	53213	54977	56744	58513	60286	62062
1.98	63840	65621	67405	69192	70982	72774	74570	76368	78169	79972
1.99	81779	83588	85401	87216	89034	90854	92678	94504	96333	98165

the one nearest to the true value whether in excess or defect. This table, and the table of Least Factors, have each been subjected to two complete and independent revisions before finally printing off.

ALGEBRA.

FACTORS.

1 $a^2-b^2 = (a-b)(a+b).$

2 $a^3-b^3 = (a-b)(a^2+ab+b^2).$

3 $a^3+b^3 = (a+b)(a^2-ab+b^2).$

And generally,

4 $a^n-b^n = (a-b)(a^{n-1}+a^{n-2}b+\ldots+b^{n-1})$ always.

5 $a^n-b^n = (a+b)(a^{n-1}-a^{n-2}b+\ldots-b^{n-1})$ if n be even.

6 $a^n+b^n = (a+b)(a^{n-1}-a^{n-2}b+\ldots+b^{n-1})$ if n be odd.

---

7 $(x+a)(x+b) = x^2+(a+b)x+ab.$

8 $(x+a)(x+b)(x+c) = x^3+(a+b+c)x^2 +(bc+ca+ab)x+abc.$

9 $(a+b)^2 = a^2+2ab+b^2.$

10 $(a-b)^2 = a^2-2ab+b^2.$

11 $(a+b)^3 = a^3+3a^2b+3ab^2+b^3 = a^3+b^3+3ab(a+b).$

12 $(a-b)^3 = a^3-3a^2b+3ab^2-b^3 = a^3-b^3-3ab(a-b).$

Generally

$$(a\pm b)^7=a^7\pm 7a^6b+21a^5b^2\pm 35a^4b^3+35a^3b^4\pm 21a^2b^5+7ab^6\pm b^7.$$

Newton's Rule for forming the coefficients—*Multiply any coefficient by the index of the leading quantity, and divide by the number of terms to that place to obtain the coefficient of the term next following.* Thus $21\times 5\div 3$ gives 35, the following coefficient in the example given above. See also (125).

To square a polynomial—*Add to the square of each term twice the product of that term and every term that follows it.*

Thus,

$$(a+b+c+d)^2$$
$$= a^2+2a(b+c+d)+b^2+2b(c+d)+c^2+2cd+d^2.$$

13 $a^4+a^2b^2+b^4 = (a^2+ab+b^2)(a^2-ab+b^2)$.

14 $a^4+b^4 = (a^2+ab\sqrt{2}+b^2)(a^2-ab\sqrt{2}+b^2)$.

15 $\left(x+\frac{1}{x}\right)^2 = x^2+\frac{1}{x^2}+2,\ \left(x+\frac{1}{x}\right)^3 = x^3+\frac{1}{x^3}+3\left(x+\frac{1}{x}\right)$

16 $(a+b+c)^2 = a^2+b^2+c^2+2bc+2ca+2ab$.

17 $(a+b+c)^3 = a^3+b^3+c^3+3(b^2c+bc^2+c^2a+ca^2$
$+a^2b+ab^2)+6abc$.

Observe that in an algebraical equation *the sign of any letter may be changed throughout*, and thus a new formula obtained, it being borne in mind that an *even* power of a negative quantity is positive. For example, by changing the sign of c in (16), we obtain

$$(a+b-c)^2 = a^2+b^2+c^2-2bc-2ca+2ab.$$

18 $a^2+b^2-c^2+2ab = (a+b)^2-c^2 = (a+b+c)(a+b-c)$
by (1).

19 $a^2-b^2-c^2+2bc = a^2-(b-c)^2 = (a+b-c)(a-b+c)$.

20 $a^3+b^3+c^3-3abc = (a+b+c)(a^2+b^2+c^2-bc-ca-ab)$.

21 $bc^2+b^2c+ca^2+c^2a+ab^2+a^2b+a^3+b^3+c^3$
$= (a+b+c)(a^2+b^2+c^2)$.

22 $bc^2+b^2c+ca^2+c^2a+ab^2+a^2b+3abc$
$= (a+b+c)(bc+ca+ab)$.

23 $bc^2+b^2c+ca^2+c^2a+ab^2+a^2b+2abc = (b+c)(c+a)(a+b)$

24 $bc^2+b^2c+ca^2+c^2a+ab^2+a^2b-2abc-a^3-b^3-c^3$
$= (b+c-a)(c+a-b)(a+b-c)$.

25 $bc^2-b^2c+ca^2-c^2a+ab^2-a^2b = (b-c)(c-a)(a-b)$.

26 $2b^2c^2+2c^2a^2+2a^2b^2-a^4-b^4-c^4$
$= (a+b+c)(b+c-a)(c+a-b)(a+b-c)$.

27 $x^3+2x^2y+2xy^2+y^3 = (x+y)(x^2+xy+y^2)$.

Generally for the division of $(x+y)^n-(x^n+y^n)$ by x^2+xy+y^2 see (545).

MULTIPLICATION AND DIVISION,

BY THE METHOD OF DETACHED COEFFICIENTS.

28 Ex. 1: $(a^4-3a^2b^2+2ab^3+b^4)\times(a^3-2ab^2-2b^3)$.

$$\begin{array}{l} 1+0-3+2+1 \\ \underline{1+0-2-2} \\ 1+0-3+2+1 \\ \quad -2-0+6-4-2 \\ \quad\quad -2-0+6-4-2 \\ \hline 1+0-5+0+7+2-6-2 \end{array}$$

Result $a^7-5a^5b^2+7a^3b^4+2a^2b^5-6ab^6-2b^7$.

Ex. 2: $(x^7-5x^5+7x^3+2x^2-6x-2)\div(x^4-3x^2+2x+1)$.

$$\begin{array}{l} 1+0-3+2+1)\ 1+0-5+0+7+2-6-2\ (1+0-2-2 \\ \qquad\qquad\qquad \underline{-1-0+3-2-1} \\ \qquad\qquad\qquad\quad 0-2-2+6+2-6 \\ \qquad\qquad\qquad\quad\ \underline{+2+0-6+4+2} \\ \qquad\qquad\qquad\qquad -2+0+6-4-2 \\ \qquad\qquad\qquad\qquad \underline{+2+0-6+4+2} \end{array}$$

Result x^3-2x-2.

Synthetic Division.

Ex. 3: Employing the last example, the work stands thus,

$$\begin{array}{r|l} & 1+0-5+0+7+2-6-2 \\ -0 & \quad 0+0+0+0 \\ +3 & \qquad +3+0-6-6 \\ -2 & \qquad\quad -2+0+4+4 \\ -1 & \qquad\qquad -1+0+2+2 \\ \hline & 1+0-2-2 \end{array}$$

Result x^3-2x-2. [See also (248).

Note that, in all operations with detached coefficients, the result must be written out in successive powers of the quantity which stood in its successive powers in the original expression.

INDICES.

29 Multiplication: $a^{\frac{1}{3}} \times a^{\frac{1}{2}} = a^{\frac{1}{3}+\frac{1}{2}} = a^{\frac{5}{6}}$, or $\sqrt[6]{a^5}$;

$$a^{\frac{1}{m}} \times a^{\frac{1}{n}} = a^{\frac{1}{m}+\frac{1}{n}} = a^{\frac{m+n}{mn}}, \text{ or } \sqrt[mn]{a^{m+n}}.$$

Division: $a^{\frac{2}{3}} \div a^{\frac{1}{2}} = a^{\frac{2}{3}-\frac{1}{2}} = a^{\frac{1}{6}}$, or $\sqrt[6]{a}$;

$$a^{\frac{1}{n}} \div a^{\frac{1}{m}} = a^{\frac{1}{n}-\frac{1}{m}} = a^{\frac{m-n}{mn}}, \text{ or } \sqrt[mn]{a^{m-n}}.$$

Involution: $(a^{\frac{2}{3}})^{\frac{1}{2}} = a^{\frac{2}{3}\times\frac{1}{2}} = a^{\frac{1}{3}}$, or $\sqrt[3]{a}$.

Evolution: $\sqrt[7]{a^{\frac{2}{3}}} = a^{\frac{2}{3}\times\frac{1}{7}} = a^{\frac{2}{21}}$, or $\sqrt[21]{a^2}$.

$$a^{-n} = \frac{1}{a^n}, \quad a^0 = 1.$$

HIGHEST COMMON FACTOR.

30 Rule.—To find the highest common factor of two expressions—*Divide the one which is of the highest dimension by the other, rejecting first any factor of either expression which is not also a factor of the other. Operate in the same manner upon the remainder and the divisor, and continue the process until there is no remainder. The last divisor will be the highest common factor required.*

31 Example.—To find the H. C. F. of

$3x^5-10x^3+15x+8$ and $x^5-2x^4-6x^3+4x^2+13x+6$.

	1− 2− 6+ 4+13+ 6	3+0−10+ 0+15+ 8	3
	3	−3+6+18−12−39−18	
1	3− 6−18+12+39+18	2) 6+ 8−12−24−10	
	−3− 4+ 6+12+ 5	3+ 4− 6−12− 5	
	2) −10−12+24+44+18	−3− 9− 9− 3	
	− 5− 6+12+22+ 9	− 5−15−15− 5	
	3	+ 5+15+15+ 5	
5	−15−18+36+66+27		
	+15+20−30−60−25		
	2) 2+ 6+ 6+ 2		
	1+ 3+ 3+ 1		

Result H. C. F. $= x^3+3x^2+3x+1$.

32 Otherwise.—To form the H. C. F. of two or more algebraical expressions—*Separate the expressions into their simplest factors. The H. C. F. will be the product of the factors common to all the expressions, taken in the lowest powers that occur.*

LOWEST COMMON MULTIPLE.

33 *The L. C. M. of two quantities is equal to their product divided by the H. C. F.*

34 Otherwise.—To form the L. C. M. of two or more algebraical expressions—*Separate them into their simplest factors. The L. C. M. will be the product of all the factors that occur, taken in the highest powers that occur.*

EXAMPLE.—The H. C. F. of $a^2(b-x)^5c^7d$ and $a^3(b-x)^2c^4e$ is $a^2(b-x)^2c^4$; and the L. C. M. is $a^3(b-x)^5c^7de$.

EVOLUTION.

To extract the square root of

$$a^2-\frac{3a\sqrt{a}}{2}-\frac{3\sqrt{a}}{2}+\frac{41a}{16}+1.$$

Arranging according to powers of a, and reducing to one denominator, the expression becomes $\dfrac{16a^2-24a^{\frac{3}{2}}+41a-24a^{\frac{1}{2}}+16}{16}$.

35 Detaching the coefficients, the work is as follows:—

$$\begin{array}{l|l}
 & 16-24+41-24+16\ (\,4-3+4 \\
 & 16 \\ \hline
8-3 & -24+41 \\
\ \ -3 & \ \ 24-\ \ 9 \\ \hline
8-6+4 & \qquad 32-24+16 \\
 & \quad -32+24-16 \\ \hline
\end{array}$$

Result $\dfrac{4a-3a^{\frac{1}{2}}+4}{4}=a-\frac{3}{4}\sqrt{a}+1.$

To extract the cube root of

37 $$8x^6-36x^5\sqrt{y}+66x^4y-63x^3y\sqrt{y}+33x^2y^2-9xy^2\sqrt{y}+y^3.$$

The terms here contain the successive powers of x and $\sqrt{y}$; therefore, detaching the coefficients, the work will be as follows:—

I.	II.	III.
6 − 3	12	8 − 36 + 66 − 63 + 33 − 9 + 1 (2 − 3 + 1
− 6	− 18 + 9	− 8
6 − 9 + 1	12 − 18 + 9	− 36 + 66 − 63 + 33 − 9 + 1
	+ 9	+ 36 − 54 + 27
	12 − 36 + 27	12 − 36 + 33 − 9 + 1
	6 − 9 + 1	− 12 + 36 − 33 + 9 − 1
	12 − 36 + 33 − 9 + 1	

Result $2x^2-3x\sqrt{y}+y$.

Explanation.—The cube root of 8 is 2, the first term of the result.

Place $3\times2=6$ in the first column I., $3\times2^2=12$ in column II., and $2^3=8$ in III., changing its sign for subtraction.

$-36\div12=-3$, the second term of the result.

Put -3 in I., $(6-3)\times(-3)$ gives $-18+9$ for II.

$(12-18+9)\times3$ (changing sign) gives $36-54+27$ for III. Then add.

Put twice (-3), the term last found, in I., and the square of it in II. Add the two last rows in I., and the three last in II.

$12\div12$ gives 1, the third term of the result.

Put 1 in col. I., $(6-9+1)\times1$ gives $6-9+1$ for col. II.

$(12-36+33-9+1)\times1$ gives the same for III. Change the signs, and add, and the work is finished.

The foregoing process is but a slight variation of Horner's rule for solving an equation of any degree. See (533).

Transformations frequently required.

38 If $\dfrac{a}{b}=\dfrac{c}{d}$, then $\dfrac{a+b}{a-b}=\dfrac{c+d}{c-d}$ .... [68.

39 If $\left.\begin{array}{l}x+y=a\\ \text{and } x-y=b\end{array}\right\}$, then $\left\{\begin{array}{l}x=\frac{1}{2}(a+b)\\ y=\frac{1}{2}(a-b)\end{array}\right.$

---

40 $$(x+y)^2+(x-y)^2=2(x^2+y^2).$$

41 $$(x+y)^2-(x-y)^2=4xy.$$

42 $$(x+y)^2 = (x-y)^2+4xy.$$

43 $$(x-y)^2 = (x+y)^2-4xy.$$

44 EXAMPLES.

$$\frac{2\sqrt{a^2-b^2}+\sqrt{c^2-x^2}}{2\sqrt{a^2-b^2}-\sqrt{c^2-x^2}} = \frac{3\sqrt{a^2-b^2}+\sqrt{c^2-d^2}}{3\sqrt{a^2-b^2}-\sqrt{c^2-d^2}}$$

$$\frac{\sqrt{c^2-x^2}}{2\sqrt{a^2-b^2}} = \frac{\sqrt{c^2-d^2}}{3\sqrt{a^2-b^2}} \quad\text{[38}$$

$$9(c^2-x^2) = 4(c^2-d^2)$$

$$x = \sqrt{5c^2+4d^2}.$$

To simplify a compound fraction, as

$$\frac{\dfrac{1}{a^2-ab+b^2}+\dfrac{1}{a^2+ab+b^2}}{\dfrac{1}{a^2-ab+b^2}-\dfrac{1}{a^2+ab+b^2}}$$

multiply the numerator and denominator by the L. C. M. of all the smaller denominators.

Result $$\frac{(a^2+ab+b^2)+(a^2-ab+b^2)}{(a^2+ab+b^2)-(a^2-ab+b^2)} = \frac{a^2+b^2}{ab}.$$

QUADRATIC EQUATIONS.

45 If $ax^2+bx+c=0$, $x = \dfrac{-b\pm\sqrt{b^2-4ac}}{2a}$.

46 If $ax^2+2bx+c=0$; that is, if the coefficient of x be an even number, $x = \dfrac{-b\pm\sqrt{b^2-ac}}{a}$.

47 ***Method of solution without the formula.***

Ex.: $$2x^2-7x+3=0.$$

Divide by 2, $$x^2-\frac{7}{2}x+\frac{3}{2}=0$$

Complete the square, $x^2-\frac{7}{2}x+\left(\frac{7}{4}\right)^2=\frac{49}{16}-\frac{3}{2}=\frac{25}{16}$

Take square root, $x-\frac{7}{4}=\pm\frac{5}{4}$

$$x=\frac{7\pm5}{4}=3 \text{ or } \frac{1}{2}.$$

48 Rule for "completing the square" of an expression like $x^2-\frac{7}{2}x$, *Add the square of half the coefficient of x.*

49 The solution of the foregoing equation, employing formula (45), is

$$x=\frac{-b\pm\sqrt{b^2-4ac}}{2a}=\frac{7\pm\sqrt{49-24}}{4}=\frac{7\pm5}{4}=3 \text{ or } \frac{1}{2}.$$

THEORY OF QUADRATIC EXPRESSIONS.

If α, β be the roots of the equation $ax^2+bx+c=0$, then

50 $$ax^2+bx+c=a(x-\alpha)(x-\beta).$$

51 Sum of roots $\alpha+\beta=-\frac{b}{a}$.

52 Product of roots $\alpha\beta=\frac{c}{a}$.

Condition for the existence of equal roots—

53 b^2-4ac must vanish.

54 The solution of equations in one unknown quantity may sometimes be simplified by changing the quantity sought.

Ex. (1): $$2x+\frac{3x-1}{3x+1}+\frac{18x+6}{6x^2+5x-1}=14 \quad\text{.........} (1).$$

$$\frac{6x^2+5x-1}{3x+1}+\frac{6(3x+1)}{6x^2+5x-1}=14$$

Put $$y=\frac{6x^2+5x-1}{3x+1} \quad\text{.........} (2)$$

thus $$y + \frac{6}{y} = 14$$
$$y^2 - 14y + 6 = 0$$

y having been determined from this quadratic, x is afterwards found from (2).

55 Ex. 2: $$x^2 + \frac{1}{x^2} + x + \frac{1}{x} = 4.$$
$$\left(x + \frac{1}{x}\right)^2 + \left(x + \frac{1}{x}\right) = 6$$

Put $x + \frac{1}{x} = y$, and solve the quadratic in y.

56 Ex. 3: $$x^2 + x + \frac{3}{2}\sqrt{2x^2+x+2} = \frac{x}{2} + 1.$$
$$2x^2 + x + 3\sqrt{2x^2+x+2} = 2$$
$$2x^2 + x + 2 + 3\sqrt{2x^2+x+2} = 4$$

Put $\sqrt{2x^2+x+2} = y$, and solve the quadratic
$$y^2 + 3y = 4.$$

57 Ex. 4: $$\sqrt[3]{x^n} + \frac{2}{3\sqrt[3]{x^n}} = \frac{16}{3}x^{-n}.$$
$$x^{\frac{4n}{3}} + \frac{2}{3}x^{\frac{2n}{3}} = \frac{16}{3}$$

A quadratic in $y = x^{\frac{2n}{3}}$.

58 *To find Maxima and Minima values by means of a Quadratic Equation.*

Ex.—Given $$y = 3x^2 + 6x + 7,$$
to find what value of x will make y a maximum or minimum.

Solve the quadratic equation
$$3x^2 + 6x + 7 - y = 0$$

Thus $$x = \frac{-3 \pm \sqrt{3y-12}}{3} \qquad [46$$

In order that x may be a real quantity, we must have $3y$ not less than 12; therefore 4 is a minimum value of y, and the value of x which makes y a minimum is -1.

SIMULTANEOUS EQUATIONS.

General solution with two unknown quantities.

Given

59 $$\left.\begin{aligned} a_1x+b_1y &= c_1 \\ a_2x+b_2y &= c_2 \end{aligned}\right\}, \quad x = \frac{c_1b_2-c_2b_1}{a_1b_2-a_2b_1}, \quad y = \frac{c_1a_2-c_2a_1}{b_1a_2-b_2a_1}.$$

General solution with three unknown quantities.

60 Given
$$\left.\begin{aligned} a_1x+b_1y+c_1z &= d_1 \\ a_2x+b_2y+c_2z &= d_2 \\ a_3x+b_3y+c_3z &= d_3 \end{aligned}\right\}$$

$$x = \frac{d_1(b_2c_3-b_3c_2)+d_2(b_3c_1-b_1c_3)+d_3(b_1c_2-b_2c_1)}{a_1(b_2c_3-b_3c_2)+a_2(b_3c_1-b_1c_3)+a_3(b_1c_2-b_2c_1)}$$

and symmetrical forms for y and z.

Methods of solving simultaneous equations between two unknown quantities x and y.

61 I. ***By substitution.***—*Find one unknown in terms of the other from one of the two equations, and substitute this value in the remaining equation. Then solve the resulting equation.*

Ex.:
$$\left.\begin{aligned} x+5y &= 23 \;\ldots\ldots\; (1) \\ 7y &= 28 \;\ldots\ldots\; (2) \end{aligned}\right\}.$$

From (2), $y=4$. Substitute in (1); thus
$$x+20 = 23, \quad x = 3.$$

62 II. ***By the method of Multipliers.***

Ex.:
$$\left.\begin{aligned} 3x+5y &= 36 \;\ldots\ldots\; (1) \\ 2x-3y &= 5 \;\ldots\ldots\; (2) \end{aligned}\right\}.$$

Eliminate x by multiplying eq. (1) by 2, and (2) by 3; thus
$$\begin{aligned} 6x+10y &= 72 \\ 6x-9y &= 15 \\ 19y &= 57, \text{ by subtraction,} \\ y &= 3 \\ \therefore \quad x &= 7, \text{ by substitution in eq. (2).} \end{aligned}$$

63 III. *By changing the quantities sought.*

Ex. 1:
$$\left.\begin{aligned} x-y &= 2 \;\ldots\ldots\; (1) \\ x^2-y^2+x+y &= 30 \;\ldots\ldots\; (2) \end{aligned}\right\}.$$

Let $x+y=u, \quad x-y=v.$

Substitute these values in (1) and (2),
$$\left.\begin{aligned} v &= 2 \\ uv+u &= 30 \end{aligned}\right\}$$
$$\therefore \quad 2u+u = 30$$
$$u = 10$$
$$\therefore \quad x+y = 10$$
$$x-y = 2$$

From which $x = 6$ and $y = 4$.

---

64 Ex. 2:
$$\left.\begin{aligned} 2\frac{x+y}{x-y}+10\frac{x-y}{x+y} &= 9 \;\ldots\ldots\; (1) \\ x^2+7y^2 &= 64 \;\ldots\ldots\; (2) \end{aligned}\right\}.$$

Substitute z for $\dfrac{x+y}{x-y}$ in (1);
$$\therefore \quad 2z+\frac{10}{z} = 9$$
$$2z^2-9z+10 = 0$$

From which $z = \dfrac{5}{2}$ or 2
$$\frac{x+y}{x-y} = 2 \text{ or } \frac{5}{2}$$

From which $x = 3y$ or $\dfrac{7}{3}y$

Substitute in (2); thus $y = 2$ and $x = 6$

or $y = \dfrac{6}{\sqrt{7}}$ and $x = 2\sqrt{7}$.

---

65 Ex. 3:
$$\left.\begin{aligned} 3x+5y &= xy \;\ldots\ldots\; (1) \\ 2x+7y &= 3xy \;\ldots\ldots\; (2) \end{aligned}\right\}.$$

Divide each equation by xy,
$$\left.\begin{aligned} \frac{3}{y}+\frac{5}{x} &= 1 \;\ldots\ldots\; (3) \\ \frac{2}{y}+\frac{7}{x} &= 3 \;\ldots\ldots\; (4) \end{aligned}\right\}$$

Multiply (3) by 2, and (4) by 3, and by subtraction y is eliminated.

66 IV. *By substituting $y=tx$, when the equations are homogeneous in the terms which contain x and y.*

Ex. 1: $$\left.\begin{aligned} 52x^2+7xy &= 5y^2 \text{ ...... (1)} \\ 5x-3y &= 17 \text{ ...... (2)} \end{aligned}\right\}.$$

From (1), $$52x^2+7tx^2 = 5t^2x^2 \text{..... (3)}$$
and, from (2), $$5x-3tx = 17 \text{ ...... (4)}$$
(3) gives $$52+7t = 5t^2$$
a quadratic equation from which t must be found, and its value substituted in (4).

x is thus determined; and then y from $y = tx$.

67 Ex. 2: $$\left.\begin{aligned} 2x^2+xy+3y^2 &= 16 \text{ ...... (1)} \\ 3y-2x &= 4 \text{ ...... (2)} \end{aligned}\right\}.$$

From (1), by putting $y = tx$,
$$x^2(2+t+3t^2) = 16 \text{ ...... (3)}$$
From (2), $$x(3t-2) = 4 \text{ ...... (4)}$$
squaring, $$x^2(9t^2-12t+4) = 16$$
$$\therefore \quad 9t^2-12t+4 = 2+t+3t^3$$
a quadratic equation for t.

t being found from this, equation (4) will determine x; and finally $y=tx$.

RATIO AND PROPORTION.

68 If $a : b :: c : d$; then $ad = bc$, and $\frac{a}{b} = \frac{c}{d}$;
$$\frac{a+b}{b} = \frac{c+d}{d}; \quad \frac{a-b}{b} = \frac{c-d}{d}; \quad \frac{a+b}{a-b} = \frac{c+d}{c-d}.$$

69 If $\frac{a}{b} = \frac{c}{d} = \frac{e}{f} = \text{\&c.}$; then $\frac{a}{b} = \frac{a+c+e+\text{\&c.}}{b+d+f+\text{\&c.}}$.

General theorem.

70 If $\frac{a}{b} = \frac{c}{d} = \frac{e}{f} = \text{\&c.} = k$ say; then
$$k = \left\{\frac{pa^n+qc^n+re^n+\text{\&c.}}{pb^n+qd^n+rf^n+\text{\&c.}}\right\}^{\frac{1}{n}}$$
where p, q, r, &c. are any quantities whatever. Proved as in (71).

71 Rule.—To verify any equation between such proportional quantities—*Substitute for* a, c, e, *&c. their equivalents* kb, kd, kf, *&c. respectively, in the given equation.*

Ex.—If $a : b :: c : d$, to show that

$$\frac{\sqrt{a-b}}{\sqrt{c-d}} = \frac{\sqrt{a}-\sqrt{b}}{\sqrt{c}-\sqrt{d}}.$$

Put kb for a, and kd for c; thus

$$\frac{\sqrt{a-b}}{\sqrt{c-d}} = \frac{\sqrt{kb-b}}{\sqrt{kd-d}} = \frac{\sqrt{b}\sqrt{k-1}}{\sqrt{d}\sqrt{k-1}} = \frac{\sqrt{b}}{\sqrt{d}}$$

also
$$\frac{\sqrt{a}-\sqrt{b}}{\sqrt{c}-\sqrt{d}} = \frac{\sqrt{kb}-\sqrt{b}}{\sqrt{kd}-\sqrt{d}} = \frac{\sqrt{b}\,(\sqrt{k}-1)}{\sqrt{d}\,(\sqrt{k}-1)} = \frac{\sqrt{b}}{\sqrt{d}}$$

Identical results being obtained, the proposed equation must be true.

---

72 If $\boldsymbol{a : b : c : d : e}$ &c., forming a continued proportion.
Then $\boldsymbol{a : c :: a^2 : b^2}$ the duplicate ratio of $\boldsymbol{a : b}$,
$\boldsymbol{a : d :: a^3 : b^3}$ the triplicate ratio of $\boldsymbol{a : b}$, and so on.
Also $\sqrt{a} : \sqrt{b}$ is the subduplicate ratio of $a : b$,
$a^{\frac{3}{2}} : b^{\frac{3}{2}}$ is the sesquiplicate ratio of $a : b$.

---

73 The fraction $\frac{a}{b}$ is made to approach nearer to unity in value, by adding the same quantity to the numerator and denominator. Thus

$$\frac{a+x}{b+x} \text{ is nearer to 1 than } \frac{a}{b} \text{ is.}$$

---

74 Def.—The ratio compounded of the ratios $a : b$ and $c : d$ is the ratio $ac : bd$.

75 If $a : b :: c : d$, and $a' : b' :: c' : d'$; then, by compounding ratios, $aa' : bb' :: cc' : dd'$.

VARIATION.

---

76 If $a \propto c$ and $b \propto c$, then $(a \pm b) \propto c$ and $\sqrt{ab} \propto c$.

77 If $\left.\begin{array}{l} a \propto b \\ \text{and } c \propto d \end{array}\right\}$, then $ac \propto bd$ and $\frac{a}{c} \propto \frac{b}{d}$.

78 If $a \propto b$, we may assume $a = mb$, where m is some constant.

ARITHMETICAL PROGRESSION.

General form of a series in A. P.

79 $a,\ a+d,\ a+2d,\ a+3d,\ \ldots\ldots\ a+(n-1)\,d.$

a = first term,
d = common difference,
l = last of n terms,
s = sum of n terms; then

80 $$l = a+(n-1)\,d.$$

81 $$s = (a+l)\,\frac{n}{2}.$$

82 $$s = \{2a+(n-1)\,d\}\,\frac{n}{2}.$$

Obtained by writing (79) in reversed order, and adding both series together.

GEOMETRICAL PROGRESSION.

General form of a series in G. P.

83 $a,\ ar,\ ar^2,\ ar^3,\ \ldots\ldots\ ar^{n-1}.$

a = first term,
r = common ratio,
l = last of n terms,
s = sum of n terms; then

84 $$l = ar^{n-1}.$$

85 $$s = a\,\frac{r^n-1}{r-1} \quad \text{or} \quad a\,\frac{1-r^n}{1-r}.$$

If r be less than 1, and n be infinite,

86 $$s = \frac{a}{1-r}, \quad \text{since} \quad r^n = 0.$$

(85) is obtained by multiplying (83) by r, and subtracting one series from the other.

HARMONICAL PROGRESSION.

87 a, b, c, d, &c. are in Harm. Prog. when the reciprocals $\frac{1}{a}, \frac{1}{b}, \frac{1}{c}, \frac{1}{d}$, &c. are in Arith. Prog.

88 Or when $a : b :: a-b : b-c$ is the relation subsisting between any three consecutive terms.

89 n^{th} term of the series $= \frac{ab}{(n-1)\,a-(n-2)\,b}$. [87, 80.

90 Approximate sum of n terms of the Harm. Prog. $\frac{1}{a+d}, \frac{a}{a+2d}, \frac{a^2}{a+3d}$, &c., when d is small compared with a,

$$= \frac{(a+d)^n - a^n}{d\,(a+d)^n}.$$

By taking instead the G. P. $\frac{1}{a+d} + \frac{1}{(a+d)^2} + \frac{1}{(a+d)^3} + \dots$

91 Arithmetic mean between a and $b = \frac{a+b}{2}$.

92 Geometric do. $= \sqrt{ab}$.

93 Harmonic do. $= \frac{2ab}{a+b}$.

The three means are in continued proportion.

PERMUTATIONS AND COMBINATIONS.

94 The number of permutations of n things taken *all* at a time $= n\,(n-1)\,(n-2)\dots 3\,.\,2\,.\,1 \equiv \underline{|n}$ or $n^{(n)}$.

Proof by Induction.—Assume the formula to be true for n things. Now take $n+1$ things. After each of these the remaining n things may be arranged in $\underline{|n}$ ways, making in all $n \times \underline{|n}$ (that is, $\underline{|n+1}$) permutations of $n+1$ things; therefore, &c. See also (233) for the mode of proof by Induction.

---

95 The number of permutations of n things taken r at a time is denoted by $P(n, r)$.

$$\boldsymbol{P(n, r) = n(n-1)(n-2) \ldots (n-r+1) \equiv n^{(r)}}.$$

By (94); for $(n-r)$ things are left out of each permutation; therefore $P(n, r) = |\underline{n} \div |\underline{n-r}$.

Observe that $r =$ the number of factors.

---

96 The number of combinations of n things taken r at a time is denoted by $C(n, r)$.

$$\boldsymbol{C(n, r) = \frac{n(n-1)(n-2) \ldots (n-r+1)}{1.2.3 \ldots r} \equiv \frac{n^{(r)}}{|\underline{r}}}$$

$$\boldsymbol{= \frac{|\underline{n}}{|\underline{r}\ |\underline{n-r}} = C(n, n-r).}$$

For every combination of r things admits of $|\underline{r}$ permutations; therefore $C(n, r) = P(n, r) \div |\underline{r}$.

97 $C(n, r)$ is greatest when $r = \frac{1}{2}n$ or $\frac{1}{2}(n \pm 1)$, according as n is even or odd.

---

98 The number of homogeneous products of r dimensions of n things is denoted by $H(n, r)$.

$$\boldsymbol{H(n, r) = \frac{n(n+1)(n+2) \ldots (n+r-1)}{1.2 \ldots r} \equiv \frac{(n+r-1)^{(r)}}{|\underline{r}}}.$$

When r is $> n$, this reduces to

99 $$\boldsymbol{\frac{(r+1)(r+2) \ldots (n+r-1)}{|\underline{n-1}}}.$$

PROOF.—$H(n, r)$ is equal to the number of terms in the product of the expansions by the Bin. Th. of the n expressions $(1-ax)^{-1}$, $(1-bx)^{-1}$, $(1-cx)^{-1}$, &c.

Put $a = b = c =$ &c. $= 1$. The number will be the coefficient of x^r in $(1-x)^{-n}$. (128, 129.)

100 The number of permutations of n things taken all together, when a of them are alike, b of them alike, c alike, &c.

$$= \frac{\lfloor n}{\lfloor a \lfloor b \lfloor c \ldots \&c.}.$$

For if the a things were all different, they would form $\lfloor a$ permutations where there is now but one. So of b, c, &c.

---

101 The number of combinations of n things r at a time, in which any p of them will always be found, is

$$= C(n-p,\ r-p).$$

For if the p things be set on one side, we have to add to them $r-p$ things taken from the remaining $n-p$ things in every possible way.

---

102 THEOREM: $C(n-1,\ r-1)+C(n-1,\ r) = C(n,\ r)$.

PROOF BY INDUCTION; or as follows: Put one out of n letters aside; there are $C(n-1,\ r)$ combinations of the remaining $n-1$ letters r at a time. To complete the total $C(n,\ r)$, we must place with the excluded letter all the combinations of the remaining $n-1$ letters $r-1$ at a time.

---

103 If there be one set of P things, another of Q things, another of R things, and so on; the number of combinations formed by taking one out of each set is $= PQR \ldots$ &c., the product of the numbers in the several sets.

For one of the P things will form Q combinations with the Q things. A second of the P things will form Q more combinations; and so on. In all, PQ combinations of two things. Similarly there will be PQR combinations of three things; and so on. This principle is very important.

---

104 On the same principle, if p, q, r, &c. things be taken out of each set respectively, the number of combinations will be the product of the numbers of the separate combinations;

that is, $$= C(Pp) \,.\, C(Qq) \,.\, C(Rr) \ldots \&c.$$

105 The number of combinations of n things taken m at a time, when p of the n things are alike, q of them alike, r of them alike, &c., will be the sum of all the combinations of each possible form of m dimensions, and this is equal to the coefficient of x^m in the expansion of

$$(1+x+x^2+\ldots+x^p)(1+x+x^2+\ldots+x^q)(1+x+x^2+\ldots+x^r)\ldots.$$

106 The total number of possible combinations under the same circumstances, when the n things are taken in all ways, 1, 2, 3 ... n at a time

$$= (p+1)(q+1)(r+1)\ldots-1.$$

107 The number of permutations when they are taken m at a time, in all possible ways will be equal to the product of $\lfloor m$ and the coefficient of x^m in the expansion of

$$\left\{1+x+\frac{x^2}{\lfloor 2}+\frac{x^3}{\lfloor 3}+\ldots+\frac{x^p}{\lfloor p}\right\}\left\{1+x+\frac{x^2}{\lfloor 2}+\frac{x^3}{\lfloor 3}+\ldots+\frac{x^q}{\lfloor q}\right\}\ldots$$

......&c.

SURDS.

108 To reduce $\sqrt[3]{2808}$. Decompose the number into its prime factors by (360); thus,

$$\sqrt[3]{2808} = \sqrt[3]{2^3 . 3^3 . 13} = 6\sqrt[3]{13}$$

$$\sqrt[3]{a^{15}b^{10}c^8} = a^{\frac{15}{3}}b^{\frac{10}{3}}c^{\frac{8}{3}} = a^5b^3c^2b^{\frac{1}{3}}c^{\frac{2}{3}} = a^5b^3c^2\sqrt[3]{bc^2}.$$

109 To bring $5\sqrt[4]{3}$ to an entire surd.

$$5\sqrt[4]{3} = \sqrt[4]{5^4 . 3} = \sqrt[4]{1875},$$

$$x^{\frac{2}{3}}y^{\frac{1}{5}}z^{\frac{1}{2}} = x^{\frac{20}{30}}y^{\frac{6}{30}}z^{\frac{15}{30}} = \sqrt[30]{x^{20}y^6z^{15}}.$$

110 ***To rationalise fractions having surds in their denominators.***

$$\frac{1}{\sqrt{7}} = \frac{\sqrt{7}}{7}; \quad \frac{1}{\sqrt[3]{7}} = \frac{\sqrt[3]{49}}{\sqrt[3]{7\times 49}} = \frac{\sqrt[3]{49}}{7}.$$

---

111 $$\frac{3}{9-\sqrt{80}} = \frac{3\,(9+\sqrt{80})}{81-80} = 3\,(9+\sqrt{80})$$

since $(9-\sqrt{80})\,(9+\sqrt{80}) = 81-80$, by (1).

---

112 $$\frac{1}{1+2\sqrt{3}-\sqrt{2}} = \frac{1+2\sqrt{3}+\sqrt{2}}{(1+2\sqrt{3})^2-2} = \frac{1+2\sqrt{3}+\sqrt{2}}{11+4\sqrt{3}}$$

$$= \frac{(1+2\sqrt{3}+\sqrt{2})\,(11-4\sqrt{3})}{73}$$

---

113 $$\frac{1}{\sqrt[3]{3}-\sqrt{2}} = \frac{1}{3^{\frac{1}{3}}-2^{\frac{1}{2}}}.$$

Put $3^{\frac{1}{3}}=x$, $2^{\frac{1}{2}}=y$, and take 6 the L. C. M. of the denominators 2 and 3, then

$$\frac{1}{x-y} = \frac{x^5+x^4y+x^3y^2+x^2y^3+xy^4+y^5}{x^6-y^6}, \text{ by (4);}$$

therefore $$\frac{1}{3^{\frac{1}{3}}-2^{\frac{1}{2}}} = \frac{3^{\frac{5}{3}}+3^{\frac{4}{3}}2^{\frac{1}{2}}+3^{\frac{3}{3}}2^{\frac{2}{2}}+3^{\frac{2}{3}}2^{\frac{3}{2}}+3^{\frac{1}{3}}2^{\frac{4}{2}}+2^{\frac{5}{2}}}{3^{\frac{6}{3}}-2^{\frac{6}{2}}}$$

$$= 3\sqrt[3]{9}+3\sqrt[6]{72}+6+2\sqrt[6]{648}+4\sqrt[3]{3}+4\sqrt{2}.$$

---

114 $\dfrac{1}{\sqrt[3]{3}+\sqrt{2}}$. Here the result will be the same as in the last example if the signs of the even terms be changed. [See 5.

---

115 A surd cannot be partly rational; that is, $\sqrt{a}$ cannot be equal to $\sqrt{b}\pm c$. Proved by squaring.

116 The product of two unlike surds is irrational;

$$\sqrt{7}\times\sqrt{3} = \sqrt{21}, \text{ an irrational quantity.}$$

117 The sum or difference of two unlike surds cannot produce a single surd; that is, $\sqrt{a}+\sqrt{b}$ cannot be equal to $\sqrt{c}$. By squaring.

118 If $a+\sqrt{m} = b+\sqrt{n}$; then $a=b$ and $m=n$.

Theorems (115) to (118) are proved indirectly.

---

119 If $$\sqrt{a+\sqrt{b}} = \sqrt{x}+\sqrt{y};$$

then $$\sqrt{a-\sqrt{b}} = \sqrt{x}-\sqrt{y}.$$

By squaring and by (118).

120 To express in two terms $\sqrt{7+2\sqrt{6}}$.

Let $$\sqrt{7+2\sqrt{6}} = \sqrt{x}+\sqrt{y}$$

then $x+y = 7$ .... by squaring and by (118),

and $x-y = \sqrt{7^2-(2\sqrt{6})^2} = \sqrt{49-24} = 5$, by (119);

$$\therefore\ x = 6 \text{ and } y = 1.$$

Result $\sqrt{6}+1$.

General formula for the same—

121 $$\sqrt{a\pm\sqrt{b}} = \sqrt{\tfrac{1}{2}\left(a+\sqrt{a^2-b}\right)} \pm \sqrt{\tfrac{1}{2}\left(a-\sqrt{a^2-b}\right)}.$$

Observe that no simplification is effected unless a^2-b is a perfect square.

---

122 To simplify $\sqrt[3]{a+\sqrt{b}}$.

Assume $$\sqrt[3]{a+\sqrt{b}} = x+\sqrt{y}$$

Let $$c = \sqrt[3]{a^2-b}$$

Then x must be found by trial from the cubic equation

$$4x^3-3cx = a$$

and $$y = x^2-c$$

No simplification is effected unless a^2-b is a perfect cube.

Ex. 1: $$\sqrt[3]{7+5\sqrt{2}} = x+\sqrt{y}$$

$$c = \sqrt[3]{49-50} = -1$$

$$4x^3+3x = 7;\quad \therefore\ x = 1,\ y = 2.$$

Result $1+\sqrt{2}$.

Ex. 2: $\sqrt[3]{9\sqrt{3}-11\sqrt{2}} = \sqrt{x}+\sqrt{y}$, two different surds.

Cubing, $9\sqrt{3}-11\sqrt{2} = x\sqrt{x}+3x\sqrt{y}+3y\sqrt{x}+y\sqrt{y}$

$$\therefore\ \left.\begin{aligned} 9\sqrt{3} &= (x+3y)\sqrt{x} \\ 11\sqrt{2} &= (3x+y)\sqrt{y} \end{aligned}\right\} \qquad (118)$$

$$\therefore\ x = 3 \text{ and } y = 2.$$

123 To simplify $\sqrt{(12+4\sqrt{3}+4\sqrt{5}+2\sqrt{15})}$.

Assume $\sqrt{(12+4\sqrt{3}+4\sqrt{5}+2\sqrt{15})} = \sqrt{x}+\sqrt{y}+\sqrt{z}$.

Square, and equate corresponding surds.

Result $\sqrt{3}+\sqrt{4}+\sqrt{5}$.

---

124 To express $\sqrt[n]{A\pm B}$ in the form of two surds, where A and B are one or both quadratic surds and n is odd. Take q such that $q(A^2-B^2)$ may be a perfect n^{th} power, say p^n, by (361). Take s and t the nearest integers to $\sqrt[n]{q(A+B)^2}$ and $\sqrt[n]{q(A-B)^2}$, then

$$\sqrt[n]{A\pm B} = \frac{1}{2\sqrt[2n]{q}}\{\sqrt{s+t+2p}\pm\sqrt{s+t-2p}\}.$$

Example: To reduce $\sqrt[5]{89\sqrt{3}+109\sqrt{2}}$.

Here $A = 89\sqrt{3}$, $B = 109\sqrt{2}$,

$A^2-B^2 = 1$; $\therefore p = 1$ and $q = 1$.

$$\left.\begin{aligned}\sqrt[5]{q(A+B)^2} &= 9+f\\ \sqrt[5]{q(A-B)^2} &= 1-f\end{aligned}\right\}, \quad \begin{aligned}&f \text{ being a proper fraction;}\\ &\therefore s = 9,\ t = 1.\end{aligned}$$

Result $\frac{1}{2}(\sqrt{9+1+2}\pm\sqrt{9+1-2}) = \sqrt{3}+\sqrt{2}$.

BINOMIAL THEOREM.

---

125 $(a+b)^n =$

$$a^n+na^{n-1}b+\frac{n(n-1)}{\underline{|2}}a^{n-2}b^2+\frac{n(n-1)(n-2)}{\underline{|3}}a^{n-3}b^3+\&c.$$

126 General or $(r+1)^{\text{th}}$ term,

$$\frac{n(n-1)(n-2)\ldots(n-r+1)}{\underline{|r}}a^{n-r}b^r$$

127 or $$\frac{\underline{|n}}{\underline{|n-r}\,\underline{|r}}a^{n-r}b^r$$

if n be a positive integer.

If b be negative, the signs of the even terms will be changed.

If n be negative the expansion reduces to

128 $$(a+b)^{-n} =$$
$$a^{-n}-na^{-n-1}b+\frac{n(n+1)}{\lfloor 2}a^{-n-2}b^2-\frac{n(n+1)(n+2)}{\lfloor 3}a^{-n-3}b^3+\&c.$$

129 General term,

$$(-1)^r\frac{n(n+1)(n+2)\dots(n+r-1)}{\lfloor r}a^{-n-r}b^r. \quad [\text{See } 98.$$

Euler's proof.—Let the expansion of $(1+x)^n$, as in (125), be called $f(n)$. Then it may be proved by Induction that the equation $$f(m)\times f(n)=f(m+n) \quad \dots\dots\dots\dots\dots\dots (1)$$ is true when m and n are integers, and therefore universally true; because the *form* of an algebraical product is not altered by changing the letters involved into fractional or negative quantities. Hence

$$f(m+n+p+\&c.)=f(m)\times f(n)\times f(p), \&c.$$

Put $m=n=p=\&c.$ to k terms, each equal $\frac{h}{k}$, and the theorem is proved for a fractional index.

Again, put $-n$ for m in (1); thus, whatever n may be,

$$f(-n)\times f(n)=f(0)=1,$$

which proves the theorem for a negative index.

---

130 For the greatest term in the expansion of $(a+b)^n$, take

$$r = \text{the integral part of } \frac{(n+1)b}{a+b} \text{ or } \frac{(n-1)b}{a-b},$$

according as n is positive or negative.

But if b be greater than a, and n negative or fractional, the terms increase without limit.

Examples.

Required the 40th term of $\left(1-\frac{2x}{3}\right)^{42}$.

Here $r=39$; $a=1$; $b=-\frac{2x}{3}$; $n=42$.

By (127), the term will be

$$\frac{\lfloor 42}{\lfloor 3\,\lfloor 39}\left(-\frac{2x}{3}\right)^{39}=-\frac{42.41.40}{1.2.3}\left(\frac{2x}{3}\right)^{39} \text{ by (96).}$$

Required the 31st term of $(a-x)^{-4}$.

Here $r=30$; $b=-x$; $n=-4$.

By (129), the term is

$$(-1)^{30}\frac{4.5.6\ldots30.31.32.33}{1.2.3\ldots30}a^{-34}(-x)^{30}=\frac{31.32.33}{1.2.3}\cdot\frac{x^{30}}{a^{34}} \text{ by (98).}$$

131 Required the greatest term in the expansion of $\frac{1}{(1+x)^6}$ when $x=\frac{14}{17}$.

$\frac{1}{(1+x)^6}=(1+x)^{-6}$. Here $n=6$, $a=1$, $b=x$ in the formula

$$\frac{(n-1)b}{a-b}=\frac{5\times\frac{14}{17}}{1-\frac{14}{17}}=23\tfrac{1}{3}$$

therefore $r=23$, by (130), and the greatest term

$$=(-1)^{23}\frac{6.7.8\ldots28}{1.2.3\ldots23}\left(\frac{14}{17}\right)^{23}=-\frac{24.25.26.27\ 28}{1.2.3.4.5}\left(\frac{14}{17}\right)^{23}.$$

132 Find the first negative term in the expansion of $(2a+3b)^{\frac{17}{3}}$.

We must take r the first integer which makes $n-r+1$ negative; therefore $r>n+1=\frac{17}{3}+1=6\frac{2}{3}$; therefore $r=7$. The term will be

$$\frac{\frac{17}{3}\cdot\frac{14}{3}\cdot\frac{11}{3}\cdot\frac{8}{3}\cdot\frac{5}{3}\cdot\frac{2}{3}\left(-\frac{1}{3}\right)}{\underline{|7}}(2a)^{-\frac{4}{3}}(3b)^7 \text{ by (126)}$$

$$=-\frac{17.14.11.8.5.2.1}{\underline{|7}}\frac{b^7}{(2a)^{\frac{4}{3}}}.$$

133 Required the coefficient of x^{34} in the expansion of $\left(\frac{2+3x}{2-3x}\right)^2$.

$$\frac{(2+3x)^2}{(2-3x)^2}=(2+3x)^2(2-3x)^{-2}=\left(\frac{2+3x}{2}\right)^2\left(1-\frac{3}{2}x\right)^{-2}$$

$$=\left(1+3x+\frac{9}{4}x^2\right)\left\{1+2\left(\frac{3x}{2}\right)+\frac{2.3}{1.2}\left(\frac{3x}{2}\right)^2+\ldots\ldots\right.$$

$$\left.\ldots\ldots+33\left(\frac{3x}{2}\right)^{32}+34\left(\frac{3x}{2}\right)^{33}+35\left(\frac{3x}{2}\right)^{34}+\&c.\ldots\ldots\right\}$$

The three terms last written being those which produce x^{34} after multiplying by the factor $(1+3x+\frac{9}{4}x^2)$; for we have

$$\frac{9}{4}x^2\times33\left(\frac{3x}{2}\right)^{32}+3x\times34\left(\frac{3x}{2}\right)^{33}+1\times35\left(\frac{3x}{2}\right)^{34}$$

giving for the coefficient of x^{34} in the result

$$\frac{297}{4}\left(\frac{3}{2}\right)^{32}+102\left(\frac{3}{2}\right)^{33}+35\left(\frac{3}{2}\right)^{34}=306\left(\frac{3}{2}\right)^{32}$$

The coefficient of x^n will in like manner be $9n\left(\frac{3}{2}\right)^{n-2}$.

134 To write the coefficient of x^{3m+1} in the expansion of $\left(x^2 - \frac{1}{x^2}\right)^{2n+1}$.

The general term is

$$\frac{\underline{|\,2n+1}}{\underline{|\,2n+1-r}\;\underline{|\,r}}\, x^{2(2n-r+1)} \cdot \frac{1}{x^{2r}} = \frac{\underline{|\,2n+1}}{\underline{|\,2n+1-r}\;\underline{|\,r}}\, x^{4n-4r+2}$$

Equate $4n-4r+2$ to $3m+1$, thus

$$r = \frac{4n-3m+1}{4}$$

Substitute this value of r in the general term; the required coefficient becomes

$$\frac{\underline{|\,2n+1}}{\underline{\left|\,\frac{4n+3m+3}{4}\right.}\;\underline{\left|\,\frac{4n-3m+1}{4}\right.}}$$

The value of r shows that there is no term in x^{3m+1} unless $\frac{4n-3m+1}{4}$ is an integer.

135 An approximate value of $(1+x)^n$ when x is small is $1+nx$, by (125), neglecting x^2 and higher powers of x.

136 Ex.—An approximation to $\sqrt[3]{999}$ by Bin. Th. (125) is obtained from the first two or three terms of the expansion of

$$(1000-1)^{\frac{1}{3}} = 10 - \tfrac{1}{3} \cdot 1000^{-\frac{2}{3}} = 10 - \tfrac{1}{300} = 9\tfrac{299}{300} \text{ nearly.}$$

MULTINOMIAL THEOREM.

The general term in the expansion of $(a+bx+cx^2+\&c.)^n$ is

137 $$\frac{n(n-1)(n-2)\dots(p+1)}{\underline{|\,q}\;\underline{|\,r}\;\underline{|\,s}\dots}\, a^p b^q c^r \dots x^{q+2r+3s+\dots}$$

where $$p+q+r+s+\&c. = n,$$

and the number of terms p, q, r, &c. corresponds to the number of terms in the given multinomial.

p is integral, fractional, or negative, according as n is one or the other.

If n be an integer, (137) may be written

138 $$\frac{\underline{|\,n}}{\underline{|\,p}\;\underline{|\,q}\;\underline{|\,r}\dots}\, a^p b^q c^r \dots x^{q+2r+3s}.$$

[Deduced from the Bin. Theor.

Ex. 1.—To write the coefficient of a^3bc^5 in the expansion of $(a+b+c+d)^{10}$. Here put $n=10$, $x=1$, $p=3$, $q=1$, $r=5$, $s=0$ in (138).

Result $$\frac{\underline{|10}}{\underline{|3}\,\underline{|5}} = 7\,.\,8\,.\,9\,.\,10$$

Ex. 2.—To obtain the coefficient of x^8 in the expansion of

$$(1-2x+3x^2-4x^3)^4.$$

Here, comparing with (137), we have $a=1$, $b=-2$, $c=3$, $d=-4$,

$$p+q+\;r+\;s = 4,$$
$$q+2r+3s = 8,$$

1	0	1	2
0	2	0	2
0	1	2	1
0	0	4	0

The numbers 1, 0, 1, 2 are particular values of p, q, r, s respectively, which satisfy the two equations given above.

0, 2, 0, 2 are another set of values which also satisfy those equations; and the four rows of numbers constitute all the solutions. In forming these rows always try the highest possible numbers on the right first.

Now substitute each set of values of p, q, r, s in formula (138) successively as under:

$$\frac{\underline{|4}}{\underline{|2}}\,1^1\,(-2)^0\,3^1\,(-4)^2 = 576$$

$$\frac{\underline{|4}}{\underline{|2}\,\underline{|2}}\,1^0\,(-2)^2\,3^0\,(-4)^2 = 384$$

$$\frac{\underline{|4}}{\underline{|2}}\,1^0\,(-2)^1\,3^2\,(-4)^1 = 864$$

$$\frac{\underline{|4}}{\underline{|4}}\,1^0\,(-2)^0\,3^4\,(-4)^0 = 81$$

Result 1905

Ex. 3.—Required the coefficient of x^4 in $(1+2x-4x^2-2x^3)^{-\frac{1}{2}}$. Here $a=1$, $b=2$, $c=-4$, $d=-2$, $n=-\frac{1}{2}$; and the two equations are

$$p+q+\;r+\;s = -\tfrac{1}{2},$$
$$q+2r+3s = \;\;4,$$

$-\frac{5}{2}$	1	0	1
$-\frac{5}{2}$	0	2	0
$-\frac{7}{2}$	2	1	0
$-\frac{9}{2}$	4	0	0

Employing formula (137), the remainder of the work stands as follows:

$$\left(-\tfrac{1}{2}\right)\left(-\tfrac{3}{2}\right)1^{-\frac{5}{2}}2^1(-4)^0(-2)^1 = -\ 3$$

$$\frac{1}{\lfloor 2}\left(-\tfrac{1}{2}\right)\left(-\tfrac{3}{2}\right)1^{-\frac{5}{2}}2^0(-4)^2(-2)^0 = \quad 6$$

$$\frac{1}{\lfloor 2}\left(-\tfrac{1}{2}\right)\left(-\tfrac{3}{2}\right)\left(-\tfrac{5}{2}\right)1^{-\frac{7}{2}}2^2(-4)^1(-2)^0 = \quad 15$$

$$\frac{1}{\lfloor 4}\left(-\tfrac{1}{2}\right)\left(-\tfrac{3}{2}\right)\left(-\tfrac{5}{2}\right)\left(-\tfrac{7}{2}\right)1^{-\frac{9}{2}}2^4(-4)^0(-2)^0 = \quad \frac{35}{8}$$

Result $22\frac{3}{8}$

139 The number of terms in the expansion of the multinomial $(a+b+c+\text{to } n \text{ terms})^r$ is the same as the number of homogeneous products of n things of r dimensions. See (97 and 98).

The greatest coefficient in the expansion of $(a+b+c+$ to m terms$)^n$, n being an integer, is

140 $$\frac{\lfloor n}{(\lfloor q)^m (q+1)^k}$$ notation of (95) where $qm+k=n$.

Obtained by making the denominator in (138) as small as possible.

LOGARITHMS.

142 $\log_a N = x$ signifies that $a^x = N$, or

DEF.—*The logarithm of a number is the power to which the base must be raised to produce that number.*

143 $$\log_a a = 1, \quad \log 1 = 0.$$

144 $$\log MN = \log M + \log N.$$

$$\log \frac{M}{N} = \log M - \log N.$$

$$\log (M)^n = n \log M.$$

$$\log \sqrt[n]{M} = \frac{1}{n} \log M.$$ [142

145 $$\log_b a = \frac{\log_c a}{\log_c b}.$$

That is—*The logarithm of a number to any base is equal to the logarithm of the number divided by the logarithm of the base*, the two last named logarithms being taken to any the same base at pleasure.

146 $$\log_b a = \frac{1}{\log_a b}.$$ Put $c=a$ in (145).

147 $$\log_{10} N = \frac{\log_e N}{\log_e 10} \text{ by } (145).$$

148 $$\frac{1}{\log_e 10} = \cdot 43429448 \ldots$$

is called the modulus of the common system of logarithms; that is, the factor which will convert logarithms of numbers calculated to the base e into the corresponding logarithms to the base 10. See (154).

EXPONENTIAL THEOREM.

149 $$a^x = 1 + cx + \frac{c^2x^2}{\underline{|2}} + \frac{c^3x^3}{\underline{|3}} + \&\text{c.},$$

where $$c = (a-1) - \tfrac{1}{2}(a-1)^2 + \tfrac{1}{3}(a-1)^3 - \&\text{c.}$$

PROOF: $a^x = \{1+(a-1)\}^x$. Expand this by Binomial Theorem, and collect the coefficients of x; thus c is obtained. Assume c_2, c_3, &c. as the coefficients of the succeeding powers of x, and with this assumption write out the expansions of a^x, a^y, and a^{x+y}. Form the product of the first two series, which product must be equivalent to the third. Therefore equate the coefficient of x in this product with that in the expansion of a^{x+y}. In the identity so obtained equate the coefficients of the successive powers of y to determine c_2, c_3, &c.

Let e be that value of a which makes $c=1$, then

150 $$e^x = 1 + x + \frac{x^2}{\underline{|2}} + \frac{x^3}{\underline{|3}} + \&c.$$

Hence, by putting $x=1$,

151 $$e = 1 + 1 + \frac{1}{\underline{|2}} + \frac{1}{\underline{|3}} + \&c.$$

$= 2{\cdot}718281828$ ...... an incommensurable quantity. See (295).

152 By making $x=1$ in (149), and $x=c$ in (150), we obtain

$a = e^c$; that is, $c = \log_e a$. Therefore by (149)

154 $$\log_e a = (a-1) - \tfrac{1}{2}(a-1)^2 + \tfrac{1}{3}(a-1)^3 - \&c.$$

155 $$\log(1+x) = \quad x - \frac{x^2}{2} + \frac{x^3}{3} - \frac{x^4}{4} + \&c.$$

156 $$\log(1-x) = -x - \frac{x^2}{2} - \frac{x^3}{3} - \frac{x^4}{4} - \&c.$$ [154

157 $$\therefore \log\frac{1+x}{1-x} = 2\left\{x + \frac{x^3}{3} + \frac{x^5}{5} + \&c.\right\}$$

Put $\dfrac{m-1}{m+1}$ for x in (157); thus,

158 $$\log m = 2\left\{\frac{m-1}{m+1} + \frac{1}{3}\left(\frac{m-1}{m+1}\right)^3 + \frac{1}{5}\left(\frac{m-1}{m+1}\right)^5 + \&c.\right\}$$

Put $\dfrac{1}{2n+1}$ for x in (157); thus,

159 $\log(n+1) - \log n$

$$= 2\left\{\frac{1}{2n+1} + \frac{1}{3(2n+1)^3} + \frac{1}{5(2n+1)^5} + \&c.\right\}$$

CONTINUED FRACTIONS AND CONVERGENTS.

160 To find convergents to $3{\cdot}14159 = \dfrac{314159}{100000}$. Proceed as in the rule for H. C. F.

7	100000	314159	3
	99113	300000	
1	887	14159	15
	854	887	
1	33	5289	
	29	4435	
4	4	854	25
	4	66	
		194	
		165	
		29	7
		28	
		1	

The continued fraction is

$$3+\cfrac{1}{7+\cfrac{1}{15+\&c.}};$$

or, as it is more conveniently written,

$$3+\frac{1}{7+}\;\frac{1}{15+}\;\&c.$$

The convergents are formed as follows:—

$$\begin{array}{cccccccc} 3 & 7 & 15 & 1 & 25 & 1 & 7 & 4 \\ \dfrac{3}{1}, & \dfrac{22}{7}, & \dfrac{333}{106}, & \dfrac{355}{113}, & \dfrac{9208}{2931}, & \dfrac{9563}{3044}, & \dfrac{76149}{24239}, & \dfrac{314159}{100000}. \end{array}$$

161 Rule.—Write the quotients in a row, and the first two convergents at sight (in the example 3 and $3+\frac{1}{7}$). Multiply the numerator of any convergent by the next quotient, and add the previous numerator. The result is the numerator of the next convergent. Proceed in the same way to determine the denominator. The last convergent should be the original fraction in its lowest terms.

162 *Formula for forming the convergents.*

If $\dfrac{p_{n-2}}{q_{n-2}}, \dfrac{p_{n-1}}{q_{n-1}}, \dfrac{p_n}{q_n}$ are any consecutive convergents, and a_{n-2}, a_{n-1}, a_n the corresponding quotients; then

$$p_n = a_n p_{n-1} + p_{n-2}, \quad q_n = a_n q_{n-1} + q_{n-2}.$$

The nth convergent is therefore

$$\frac{p_n}{q_n} = \frac{a_n p_{n-1} + p_{n-2}}{a_n q_{n-1} + q_{n-2}} \equiv F_n.$$

The true value of the continued fraction will be expressed by

163 $$F = \frac{a'_n p_{n-1} + p_{n-2}}{a'_n q_{n-1} + q_{n-2}}$$

in which a'_n is the complete quotient or value of the continued fraction commencing with a_n.

164 $p_n q_{n-1} - p_{n-1} q_n = \pm 1$ alternately, by (162).

The convergents are alternately greater and less than the original fraction, and are always in their lowest terms.

165 The difference between F_n and the true value of the continued fraction is

$$< \frac{1}{q_n q_{n+1}} \quad \text{and} \quad > \frac{1}{q_n (q_n + q_{n+1})}$$

and this difference therefore diminishes as n increases. Proved by taking the difference, $\frac{p_n}{q_n} - \frac{a' p_{n+1} + p_n}{a' q_{n+1} + q_n}$ (163).

Also F is nearer the true value than any other fraction with a less denominator.

166 $F_n F_{n+1}$ is greater or less than F^2 according as F_n is greater or less than F_{n+1}.

General Theory of Continued Fractions.

167

First class of continued fraction.	Second class of continued fraction.
$F = \frac{b_1}{a_1 +} \frac{b_2}{a_2 +} \frac{b_3}{a_3 + \&c.}$	$V = \frac{b_1}{a_1 -} \frac{b_2}{a_2 -} \frac{b_3}{a_3 - \&c.}$

a_1, b_1, &c. are taken as positive quantities.

$\frac{b_1}{a_1}$, $\frac{b_2}{a_2}$, &c. are termed *components* of the continued fraction. If the components be infinite in number, the continued fraction is said to be infinite.

Let the successive convergents be denoted by

$$\frac{p_1}{q_1} = \frac{b_1}{a_1}; \quad \frac{p_2}{q_2} = \frac{b_1}{a_1 +} \frac{b_2}{a_2}; \quad \frac{p_3}{q_3} = \frac{b_1}{a_1 +} \frac{b_2}{a_2 +} \frac{b_3}{a_3}; \text{ and so on.}$$

168 The law of formation of the convergents is,

For F,	For V,
$\begin{cases} p_n = a_n p_{n-1} + b_n p_{n-2} \\ q_n = a_n q_{n-1} + b_n q_{n-2} \end{cases}$	$\begin{cases} p_n = a_n p_{n-1} - b_n p_{n-2} \\ q_n = a_n q_{n-1} - b_n q_{n-2} \end{cases}$

Proved by Induction.

The relation between the successive differences of the convergents is, by (168),

169
$$\frac{p_{n+1}}{q_{n+1}} - \frac{p_n}{q_n} = \mp \frac{b_{n+1} q_{n-1}}{q_{n+1}} \left(\frac{p_n}{q_n} - \frac{p_{n-1}}{q_{n-1}} \right)$$

Take the $-$ sign for F, and the $+$ for V.

170
$$p_n q_{n-1} - p_{n-1} q_n = (-1)^{n-1} b_1 b_2 b_3 \ldots b_n \text{ (by 168).}$$

171 The odd convergents for F, $\frac{p_1}{q_1}$, $\frac{p_3}{q_3}$, &c., continually decrease, and the even convergents, $\frac{p_2}{q_2}$, $\frac{p_4}{q_4}$, &c., continually increase. (167.)

Every odd convergent is greater, and every even convergent is less, than all following convergents. (169.)

172 Def.—If the difference between consecutive convergents diminishes without limit, the infinite continued fraction is said to be *definite*. If the same difference tends to a fixed value greater than zero, the infinite continued fraction is *indefinite;* the odd convergents tending to one value, and the even convergents to another.

173 F is definite if the ratio of every quotient to the next component is greater than a fixed quantity.

Apply (169) successively.

174 F is incommensurable when the components are all proper fractions and infinite in number.

Indirectly, and by (168).

175 If a be never less than $b+1$, the convergents of V are all positive proper fractions, increasing in magnitude, p_n and q_n also increasing with n. By (167) and (168).

176 If, in this case, V be infinite, it is also definite, being $=1$, if a always $=b+1$ while b is less than 1, (175); and being less than 1, if a is ever greater than $b+1$. By (180).

177 V is incommensurable when it is less than 1, and the components are all proper fractions and infinite in number.

180 If in the continued fraction V (167), we have $a_n = b_n + 1$ always; then, by (168),

$$p_n = b_1 + b_1 b_2 + b_1 b_2 b_3 + \dots \text{ to } n \text{ terms, and } q_n = p_n + 1.$$

---

181 If, in the continued fraction F, a_n and b_n are constant and equal, say, to a and b respectively; then p_n and q_n are respectively equal to the coefficients of x^{n-1} in the expansions

of $$\frac{b}{1-ax-bx^2} \quad \text{and} \quad \frac{a+bx}{1-ax-bx^2}.$$

For p_n and q_n are the nth terms of two recurring series.

See (168) and (251).

---

182 *To convert a Series into a Continued Fraction.*

The series $$\frac{1}{u} + \frac{x}{u_1} + \frac{x^2}{u_2} + \dots + \frac{x^n}{u_n}$$

is equal to a continued fraction V (167), with $n+1$ components; the first, second, and $n+1^{\text{th}}$ components being

$$\frac{1}{u}, \quad \frac{u^2 x}{u_1 + ux}, \quad \dots \quad \frac{u^2_{n-1} x}{u_n + u_{n-1} x}$$

Proved by Induction.

183 The series

$$\frac{1}{v}+\frac{x}{vv_1}+\frac{x^2}{vv_1v_2}+\ldots+\frac{x^n}{vv_1v_2\ldots v_n}$$

is equal to a continued fraction V (167), with $n+1$ components, the first, second, and $n+1^{\text{th}}$ components being

$$\frac{1}{v},\quad \frac{vx}{v_1+x},\quad \ldots\ldots\quad \frac{v_{n-1}x}{v_n+x}.$$ [Proved by Induction.

184 The sign of x may be changed in either of the statements in (182) or (183).

185 Also, if any of these series are convergent and infinite, the continued fractions become infinite.

---

186 ***To find the value of a continued fraction with recurring quotients.***

Let the continued fraction be

$$x=\frac{b_1}{a_1+}\ldots+\frac{b_n}{a_n+y}\quad\text{where}\quad y=\frac{b_{n+1}}{a_{n+1}+}\ldots+\frac{b_{n+m}}{a_{n+m}+y}$$

so that there are m recurring quotients. Form the nth convergent for x, and the mth for y. Then, by substituting the complete quotients a_n+y for a_n, and $a_{n+m}+y$ for a_{n+m} in (168), two equations are obtained of the forms

$$x=\frac{Ay+B}{Cy+D}\quad\text{and}\quad y=\frac{Ey+F}{Gy+H}$$

from which, by eliminating y, a quadratic equation for determining x is obtained.

---

187 If

$$\frac{b_1}{a_1+}\ldots+\frac{b_n}{a_n+}$$

be a continued fraction, and

$$\frac{p_1}{q_1},\quad\ldots\ldots\quad\frac{p_n}{q_n}$$

the corresponding first n convergents; then $\frac{q_{n-1}}{q_n}$, developed by (168), produces the continued fraction

$$\frac{1}{a_n+}\,\frac{b_n}{a_{n-1}+}\,\frac{b_{n-1}}{a_{n-2}+}\dots+\frac{b_3}{a_2+}\,\frac{b_2}{a_1}$$

the quotients being the same but in reversed order.

INDETERMINATE EQUATIONS.

188 Given $$ax+by=c$$
free from fractions, and α, β integral values of x and y which satisfy the equation, the complete integral solution is given by

$$x=\alpha-bt$$
$$y=\beta+at$$

where t is any integer.

EXAMPLE.—Given $5x+3y=112$.

Then $x=20$, $y=4$ are values;

$$\therefore \quad \left.\begin{aligned} x &= 20-3t \\ y &= 4+5t \end{aligned}\right\}$$

The values of x and y may be exhibited as under:

$t=$	-2	-1	0	1	2	3	4	5	6	7
$x=$	26	23	20	17	14	11	8	5	2	-1
$y=$	-6	-1	4	9	14	19	24	29	34	39

For solutions in positive integers t must lie between $\frac{20}{3}=6\frac{2}{3}$ and $-\frac{4}{5}$; that is, t must be 0, 1, 2, 3, 4, 5, or 6, giving 7 positive integral solutions.

189 If the equation be

$$ax-by=c$$

the solutions are given by

$$x=\alpha+bt$$
$$y=\beta+at$$

EXAMPLE: $4x - 3y = 19.$

Here $x = 10$, $y = 7$ satisfy the equation;

$$\therefore \quad \left.\begin{array}{l} x = 10+3t \\ y = \ \ 7+4t \end{array}\right\} \text{ furnish all the solutions.}$$

The simultaneous values of t, x, and y will be as follows:—

$t = -5$	-4	-3	-2	-1	0	1	2	3
$x = -5$	-2	1	4	7	10	13	16	19
$y = -13$	-9	-5	-1	3	7	11	15	19

The number of positive integral solutions is infinite, and the least positive integral values of x and y are given by the limiting value of t, viz.,

$$t > -\frac{10}{3} \text{ and } t > -\frac{7}{4};$$

that is, t must be -1, 0, 1, 2, 3, or greater.

---

190 If two values, α and β, cannot readily be found by inspection, as, for example, in the equation

$$17x+13y = 14900,$$

divide by the least coefficient, and equate the remaining fractions to t, an integer; thus

$$y + x + \frac{4x}{13} = 1146 + \frac{2}{13} \text{..........................(1);}$$

$$\therefore \quad 4x-2 = 13t.$$

Repeat the process; thus

$$x - \frac{2}{4} = 3t + \frac{t}{4},$$

$$\therefore \quad t+2 = 4u.$$

Put $u = 1,$

$$\therefore \quad t = 2,$$

$$x = \frac{13t+2}{4} = 7 = \alpha;$$

and $y+x+t = 1146$, by (1),

$$\therefore \quad y = 1146-7-2 = 1137 = \beta.$$

The general solution will be

$$x = 7-13t,$$
$$y = 1137+17t.$$

Or, changing the sign of t for convenience,

$$x = 7+13t,$$
$$y = 1137-17t.$$

Here the number of solutions in positive integers is equal to the number of integers lying between $-\frac{7}{13}$ and $\frac{1137}{17}$;

or $-\frac{7}{13}$ and $66\frac{15}{17}$; that is, 67.

191 Otherwise.—Two values of x and y may be found in the following manner:—

Find the nearest converging fraction to $\frac{17}{13}$ by (160).

This is $\frac{4}{3}$. By (164) we have

$$17 \times 3 - 13 \times 4 = -1.$$

Multiply by 14900, and change the signs;

$$\therefore \quad 17(-44700) + 13(59600) = 14900;$$

which shews that we may take $\begin{cases} \alpha = -44700 \\ \beta = 59600 \end{cases}$

and the general solution may be written

$$x = -44700 + 13t,$$
$$y = 59600 - 17t.$$

This method has the disadvantage of producing high values of α and β.

192 The values of x and y, in positive integers, which satisfy the equation $ax \pm by = c$, form two Arithmetic Progressions, of which b and a are respectively the common differences. See examples (188) and (189).

193 Abbreviation of the method in (169).

Example: $11x - 18y = 63.$

Put $x = 9z$, and divide by 9; then proceed as before.

194 ***To obtain integral solutions of $ax + by + cz = d$.***

Write the equation thus

$$ax + by = d - cz.$$

Put successive integers for z, and solve for x, y in each case.

TO REDUCE A QUADRATIC SURD TO A CONTINUED FRACTION.

195 EXAMPLE:

$$\sqrt{29} = 5 + \sqrt{29} - 5 = 5 + \frac{4}{\sqrt{29}+5},$$

$$\frac{\sqrt{29}+5}{4} = 2 + \frac{\sqrt{29}-3}{4} = 2 + \frac{5}{\sqrt{29}+3},$$

$$\frac{\sqrt{29}+3}{5} = 1 + \frac{\sqrt{29}-2}{5} = 1 + \frac{5}{\sqrt{29}+2},$$

$$\frac{\sqrt{29}+2}{5} = 1 + \frac{\sqrt{29}-3}{5} = 1 + \frac{4}{\sqrt{29}+3},$$

$$\frac{\sqrt{29}+3}{4} = 2 + \frac{\sqrt{29}-5}{4} = 2 + \frac{1}{\sqrt{29}+5},$$

$$\sqrt{29}+5 = 10 + \sqrt{29} - 5 = 10 + \frac{4}{\sqrt{29}+5}.$$

The quotients 5, 2, 1, 1, 2, 10 are the greatest integers contained in the quantities in the first column. The quotients now recur, and the surd $\sqrt{29}$ is equivalent to the continued fraction

$$5 + \frac{1}{2+}\,\frac{1}{1+}\,\frac{1}{1+}\,\frac{1}{2+}\,\frac{1}{10+}\,\frac{1}{2+}\,\frac{1}{1+}\,\frac{1}{1+}\,\frac{1}{2+}\,\&c.$$

The convergents to $\sqrt{29}$, formed as in (160), will be

$$\frac{5}{1},\ \frac{11}{2},\ \frac{16}{3},\ \frac{27}{5},\ \frac{70}{13},\ \frac{727}{135},\ \frac{1524}{283},\ \frac{2251}{418},\ \frac{3775}{701},\ \frac{9801}{1820}.$$

196 Note that the last quotient 10 is the greatest and twice the first, that the *second* is the first of the recurring ones, and that the recurring quotients, excluding the last, consist of pairs of equal terms, quotients equi-distant from the first and last being equal. These properties are universal. (See 204—210).

To form high convergents rapidly.

197 Suppose m the number of recurring quotients, or any multiple of that number, and let the mth convergent to $\sqrt{Q}$ be represented by F_m; then the $2m^{\text{th}}$ convergent is given by the formula

$$F_{2m} = \frac{1}{2}\left\{F_m + \frac{Q}{F_m}\right\} \text{ by (203) and (210).}$$

198 For example, in approximating to $\sqrt{29}$ above, there are five recurring quotients. Take $m = 2 \times 5 = 10$; therefore, by

$$F_{20} = \frac{1}{2}\left\{F_{10} + \frac{29}{F_{10}}\right\}$$

$$F_{10} = \frac{9801}{1820}, \text{ the } 10^{\text{th}} \text{ convergent.}$$

Therefore $$F_{20} = \frac{1}{2}\left\{\frac{9801}{1820} + 29 \times \frac{1820}{9801}\right\} = \frac{192119201}{35675640},$$

the 20th convergent to $\sqrt{29}$; and the labour of calculating the intervening convergents is saved.

GENERAL THEORY.

199 The process of (174) may be exhibited as follows :—

$$\frac{\sqrt{Q}+c_1}{r_1} = a_1 + \frac{r_2}{\sqrt{Q}+c_2}$$

$$\frac{\sqrt{Q}+c_2}{r_2} = a_2 + \frac{r_3}{\sqrt{Q}+c_3}$$

$$\ldots \quad \ldots \quad \ldots \quad \ldots$$

$$\ldots \quad \ldots \quad \ldots \quad \ldots$$

$$\frac{\sqrt{Q}+c_n}{r_n} = a_n + \frac{r_{n+1}}{\sqrt{Q}+c_{n+1}}$$

200 Then $$\sqrt{Q} = a_1 + \frac{1}{a_2+}\,\frac{1}{a_3+}\,\frac{1}{a_4+}\&c.$$

The quotients a_1, a_2, a_3, &c. are the integral parts of the fractions on the left.

201 The equations connecting the remaining quantities are

$$c_1 = 0 \qquad\qquad r_1 = 1$$

$$c_2 = a_1 r_1 - c_1 \qquad\qquad r_2 = \frac{Q - c_2^2}{r_1}$$

$$c_3 = a_2 r_2 - c_2 \qquad\qquad r_3 = \frac{Q - c_3^2}{r_2}$$

$$\cdots \quad \cdots \quad \cdots \qquad\qquad \cdots \quad \cdots \quad \cdots$$

$$c_n = a_{n-1} r_{n-1} - c_{n-1} \qquad\qquad r_n = \frac{Q - c_n^2}{r_{n-1}}$$

The nth convergent to $\sqrt{Q}$ will be

202 $$\frac{p_n}{q_n} = \frac{a_n p_{n-1} + p_{n-2}}{a_n q_{n-1} + q_{n-2}}$$ [By Induction.

The true value of $\sqrt{Q}$ is what this becomes when we substitute for a_n the complete quotient $\frac{\sqrt{Q}+c_n}{r_n}$, of which a_n is only the integral part. This gives

203 $$\sqrt{Q} = \frac{(\sqrt{Q}+c_n)\, p_{n-1} + r_n p_{n-2}}{(\sqrt{Q}+c_n)\, q_{n-1} + r_n q_{n-2}}$$

By the relations (199) to (203) the following theorems are demonstrated :—

204 All the quantities a, r, and c are positive integers.

205 The greatest c is c_2, and $c_2 = a_1$.

206 No a or r can be greater than $2a_1$.

207 If $r_n = 1$, then $c_n = a_1$.

208 For all values of n greater than 1, $a - c_n$ is $< r_n$.

209 The number of quotients cannot be greater than $2a_1^2$. The last quotient is $2a_1$, and after that the terms repeat. The first complete quotient that is repeated is $\frac{\sqrt{Q}+c_2}{r_2}$, and a_2, r_2, c_2 commence each cycle of repeated terms.

210 Let a_m, r_m, c_m be the last terms of the first cycle; then a_{m-1}, r_{m-1}, c_{m-1} are respectively equal to a_2, r_2, c_2; a_{m-2}, r_{m-2}, c_{m-2} are equal to a_3, r_3, c_3, and so on. By (187).

EQUATIONS.

Special Cases in the Solution of Simultaneous Equations.

211 First, with two unknown quantities.

$$\left.\begin{aligned} a_1x+b_1y &= c_1 \\ a_2x+b_2y &= c_2 \end{aligned}\right\}. \quad x = \frac{c_1b_2-c_2b_1}{a_1b_2-a_2b_1}, \quad y = \frac{c_1a_2-c_2a_1}{b_1a_2-b_2a_1}.$$

If the denominators vanish, we have

$$\frac{a_1}{a_2} = \frac{b_1}{b_2}, \text{ and } x = \infty, \; y = \infty;$$

unless at the same time the numerators vanish, for then

$$\frac{a_1}{a_2} = \frac{b_1}{b_2} = \frac{c_1}{c_2}; \quad x = \frac{0}{0}; \quad y = \frac{0}{0};$$

and the equations are *not independent*, one being produced by multiplying the other by some constant.

212 Next, with three unknown quantities. See (60) for the equations.

If d_1, d_2, d_3 all vanish, divide each equation by z, and we have three equations for finding the two ratios $\frac{x}{z}$ and $\frac{y}{z}$, two only of which equations are necessary, any one being deducible from the other two if the three be consistent.

213 ***To solve simultaneous equations by Indeterminate Multipliers.***

Ex.—Take the equations

$$\begin{aligned} x + 2y + 3z + 4w &= 27, \\ 3x + 5y + 7z + w &= 48, \\ 5x + 8y + 10z - 2w &= 65, \\ 7x + 6y + 5z + 4w &= 53. \end{aligned}$$

Multiply the first by A, the second by B, the third by C, leaving one equation unmultiplied; and then add the results.

Thus
$$(A+3B+5C+7)\,x+(2A+5B+8C+6)\,y$$
$$+(3A+7B+10C+5)\,z+(4A+B-2C+4)\,w$$
$$=27A+48B+65C+53.$$

To determine either of the unknowns, for instance x, equate the coefficients of the other three separately to zero, and from the three equations find A, B, C. Then

$$x=\frac{27A+48B+65C+53}{A+3B+5C+7}$$

MISCELLANEOUS EQUATIONS AND SOLUTIONS.

214 $$x^6 \pm 1 = 0.$$

Divide by x^3, and throw into factors, by (2) or (3). See also (480).

215 $$x^3 - 7x - 6 = 0.$$

$x=-1$ is a root, by inspection; therefore $x+1$ is a factor. Divide by $x+1$, and solve the resulting quadratic.

216 $$x^3+16x = 455.$$

$$x^4+16x^2 = 455x = 65\times 7x$$

$$x^4+65x^2+\left(\frac{65}{2}\right)^2 = 49x^2+65\times 7x+\left(\frac{65}{2}\right)^2$$

$$x^2+\frac{65}{2} = 7x+\frac{65}{2}$$

$$x^2 = 7x; \qquad \therefore \quad x = 7.$$

Rule.—Divide the absolute term (here 455) into two factors, if possible, such that one of them, minus the square of the other, equals the coefficient of x. See (483) for general solution of a cubic equation.

---

217 $x^4-y^4=14560, \quad x-y=8.$

Put $x=z+v$ and $y=z-v$.

Eliminate v, and obtain a cubic in z, which solve as in (216).

---

218 $x^5-y^5=3093, \quad x-y=3.$

Divide the first equation by the second, and subtract from the result the fourth power of $x-y$. Eliminate (x^2+y^2), and obtain a quadratic in xy.

---

219 ***On forming Symmetrical Expressions.***

Take, for example, the equation

$$(y-c)(z-b)=a^2.$$

To form the remaining equations symmetrical with this, write the corresponding letters in *vertical* columns, observing the circular order in which a is followed by b, b by c, and c by a. So with x, y, and z. Thus the equations become

$$(y-c)(z-b)=a^2$$
$$(z-a)(x-c)=b^2$$
$$(x-b)(y-a)=c^2$$

To solve these equations substitute

$$x=b+c+x', \quad y=c+a+y', \quad z=a+b+z';$$

and multiplying out, and eliminating y and z, we obtain

$$x=\frac{bc(b+c)-a(b^2+c^2)}{bc-ca-ab}$$

and therefore, by symmetry, the values of y and z, by the rule just given.

---

220 $$y^2+z^2+yz=a^2 \quad \ldots\ldots (1),$$
$$z^2+x^2+zx=b^2 \quad \ldots\ldots (2),$$
$$x^2+y^2+xy=c^2 \quad \ldots\ldots (3).$$

$$\therefore \quad 3(yz+zx+xy)^2=2b^2c^2+2c^2a^2+2a^2b^2-a^4-b^4-c^4 \ldots\ldots (4).$$

Now add (1), (2), and (3), and we obtain

$$2(x+y+z)^2-3\,(yz+zx+xy) = a^2+b^2+c^2 \ldots\ldots\ldots (5).$$

From (4) and (5) $(x+y+z)$ is obtained, and then (1), (2), and (3) are readily solved.

221

$$x^2-yz = a^2 \ldots\ldots\ldots\ldots\ldots\ldots\ldots\ldots (1),$$

$$y^2-zx = b^2 \ldots\ldots\ldots\ldots\ldots\ldots\ldots\ldots (2),$$

$$z^2-xy = c^2 \ldots\ldots\ldots\ldots\ldots\ldots\ldots\ldots (3).$$

Multiply (2) by (3), and subtract the square of (1).

Result $\quad x\,(3xyz-x^3-y^3-z^3) = b^2c^2-a^4$

$$\therefore \qquad \frac{x}{b^2c^2-a^4} = \frac{y}{c^2a^2-b^4} = \frac{z}{a^2b^2-c^4} = \lambda \ldots\ldots\ldots\ldots (4).$$

Obtain λ^2 by proportion as a fraction with numerator

$$= x^2-yz = a^2$$

222

$$x = cy+bz \ldots\ldots\ldots\ldots\ldots\ldots\ldots\ldots (1),$$

$$y = az+cx \ldots\ldots\ldots\ldots\ldots\ldots\ldots\ldots (2),$$

$$z = bx+ay \ldots\ldots\ldots\ldots\ldots\ldots\ldots\ldots (3).$$

Eliminate a between (2) and (3), and substitute the value of x from equation (1).

Result $\quad \dfrac{y^2}{1-b^2} = \dfrac{z^2}{1-c^2} = \dfrac{x^2}{1-a^2}$

IMAGINARY EXPRESSIONS.

---

223 The following are conventions :—

That $\sqrt{(-a^2)}$ is equivalent to $a\sqrt{(-1)}$; that $a\sqrt{(-1)}$ vanishes when a vanishes; that the symbol $a\sqrt{(-1)}$ is subject to the ordinary rules of Algebra. $\sqrt{(-1)}$ is denoted by i.

224 If $\alpha + i\beta = \gamma + i\delta$; then $\alpha = \gamma$ and $\beta = \delta$.

225 $\alpha + i\beta$ and $\alpha - i\beta$ are conjugate expressions; their product $= \alpha^2 + \beta^2$.

226 The sum and product of two conjugate expressions are both real, but their difference is imaginary.

227 The modulus is $+\sqrt{\alpha^2+\beta^2}$.

228 If the modulus vanishes, α and β must vanish.

229 If two imaginary expressions are equal, their moduli are equal, by (224).

230 The modulus of the product of two imaginary expressions is equal to the product of their moduli.

231 Also the modulus of the quotient is equal to the quotient of their moduli.

METHOD OF INDETERMINATE COEFFICIENTS.

232 If $A + Bx + Cx^2 + \ldots = A' + B'x + C'x^2 + \ldots$ be an equation which holds for all values of x, the coefficients A, B, &c., not involving x; then $A = A'$, $B = B'$, $C = C'$, &c.; that is, the coefficients of like powers of x must be equal. Proved by putting $x = 0$, and dividing by x alternately. See (234) for an example.

233 METHOD OF PROOF BY INDUCTION.

Ex.—To prove that

$$1 + 2^2 + 3^2 + \ldots + n^2 = \frac{n(n+1)(2n+1)}{6}.$$

Assume $$1 + 2^2 + 3^2 + \ldots + n^2 = \frac{n(n+1)(2n+1)}{6};$$

$$\therefore \quad 1+2^2+3^2+\ldots+n^2+(n+1)^2 = \frac{n(n+1)(2n+1)}{6}+(n+1)^2$$

$$= \frac{n(n+1)(2n+1)+6(n+1)^2}{6} = \frac{(n+1)\{n(2n+1)+6(n+1)\}}{6}$$

$$= \frac{(n+1)(n+2)(2n+3)}{6} = \frac{n'(n'+1)(2n'+1)}{6}$$

where n' is written for $n+1$;

$$\therefore \quad 1+2^2+3^2+\ldots+n'^2 = \frac{n'(n'+1)(2n'+1)}{6}.$$

It is thus proved that *if the formula be true for n it is also true for* $n+1$.

But the formula *is* true when $n = 2$ or 3, as may be shewn by actual trial; therefore it is true when $n = 4$; therefore also when $n = 5$, and so on; therefore universally true.

234 Ex.—The same theorem proved by the method of Indeterminate coefficients.

Assume

$$1+2^2+3^2+\ldots+n^2 = A+Bn+Cn^2+Dn^3+\&c.;$$

$$\therefore 1+2^2+3^2+\ldots+n^2+(n+1)^2 = A+B(n+1)+C(n+1)^2+D(n+1)^3+\&c.;$$

therefore, by subtraction,

$$n^2+2n+1 = B+C(2n+1)+D(3n^2+3n+1),$$

writing no terms in this equation which contain higher powers of n than the highest which occurs on the left-hand side, for the coefficients of such terms may be shewn to be separately equal to zero.

Now equate the coefficients of like powers of n; thus

$$3D = 1, \qquad \therefore \quad D = \frac{1}{3};$$

$$2C+3D = 2, \qquad \therefore \quad C = \frac{1}{2}, \quad \text{and} \quad A = 0;$$

$$B+C+D = 1, \qquad \therefore \quad B = \frac{1}{6};$$

therefore the sum of the series is equal to

$$\frac{n}{6}+\frac{n^2}{2}+\frac{n^3}{3} = \frac{n(n+1)(2n+1)}{6}.$$

PARTIAL FRACTIONS.

In the resolution of a fraction into partial fractions four cases present themselves, which are illustrated in the following examples.

235 First.—When there are no repeated factors in the denominator of the given fraction.

Ex.—To resolve $\frac{3x-2}{(x-1)(x-2)(x-3)}$ into partial fractions.

Assume $$\frac{3x-2}{(x-1)(x-2)(x-3)}=\frac{A}{x-1}+\frac{B}{x-2}+\frac{C}{x-3};$$

$$\therefore\quad 3x-2=A(x-2)(x-3)+B(x-3)(x-1)+C(x-1)(x-2).$$

Since A, B, and C do not contain x, and this equation is true for all values of x, put $x=1$; then

$$3-2=A(1-2)(1-3),\text{ from which } A=\frac{1}{2}.$$

Similarly, if x be put $=2$, we have

$$6-2=B(2-3)(2-1);\qquad\therefore\quad B=-4;$$

and, putting $x=3$,

$$9-2=C(3-1)(3-2);\qquad\therefore\quad C=\frac{7}{2}.$$

Hence $$\frac{3x-2}{(x-1)(x-2)(x-3)}=\frac{1}{2(x-1)}-\frac{4}{x-2}+\frac{7}{2(x-3)}.$$

236 Secondly.—When there is a repeated factor.

Ex.—Resolve into partial fractions $\frac{7x^3-10x^2+6x}{(x-1)^3(x+2)}$.

Assume $$\frac{7x^3-10x^2+6x}{(x-1)^3(x+2)}=\frac{A}{(x-1)^3}+\frac{B}{(x-1)^2}+\frac{C}{(x-1)}+\frac{D}{x+2}.$$

These forms are necessary and sufficient. Multiplying up, we have

$$7x^3-10x^2+6x=A(x+2)+B(x-1)(x+2)+C(x-1)^2(x+2)+D(x-1)^3\quad\text{............ (1).}$$

Make $x=1$; $\therefore$ $7-10+6=A(1+2)$; $\therefore$ $A=1$.

Substitute this value of A in (1); thus

$$7x^3-10x^2+5x-2=B(x-1)(x+2)+C(x-1)^2(x+2)+D(x-1)^3.$$

Divide by $x-1$; thus

$$7x^2-3x+2=B(x+2)+C(x-1)(x+2)+D(x-1)^2\quad\text{......... (2).}$$

Make $x=1$ again, $7-3+2=B(1+2)$; $\therefore$ $B=2$.

Substitute this value of B in (2), and we have

$$7x^2-5x-2=C(x-1)(x+2)+D(x-1)^2.$$

Divide by $x-1$, $\quad 7x+2=C(x+2)+D(x-1)$ .... (3).

Put $x=1$ a third time, $7+2=C(1+2)$; $\therefore$ $C=3$.

Lastly, make $x = -2$ in (3),

$$-14+2 = D(-2-1); \quad \therefore \quad D = 4.$$

Result $$\frac{1}{(x-1)^3} + \frac{2}{(x-1)^2} + \frac{3}{x-1} + \frac{4}{x+2}.$$

---

237 Thirdly.—When there is a quadratic factor of imaginary roots not repeated.

Ex.—Resolve $\dfrac{1}{(x^2+1)(x^2+x+1)}$ into partial fractions.

Here we must assume

$$\frac{1}{(x^2+1)(x^2+x+1)} = \frac{Ax+B}{x^2+1} + \frac{Cx+D}{x^2+x+1};$$

x^2+1 and x^2+x+1 have no real factors, and are therefore retained as denominators. The requisite form of the numerators is seen by adding together two simple fractions, such as $\dfrac{a}{x+b} + \dfrac{c}{x+d}$.

Multiplying up, we have the equation

$$1 = (Ax+B)(x^2+x+1) + (Cx+D)(x^2+1) \quad \ldots\ldots\ldots\ldots\ldots (1).$$

Let $$x^2+1 = 0; \quad \therefore \quad x^2 = -1.$$

Substitute this value of x^2 in (1) repeatedly; thus

$$1 = (Ax+B)x = Ax^2+Bx = -A+Bx;$$

or $$Bx-A-1 = 0.$$

Equate coefficients to zero; $\therefore \quad B = 0,$

$$A = -1.$$

Again let $$x^2+x+1 = 0;$$

$$\therefore \quad x^2 = -x-1.$$

Substitute this value of x^2 repeatedly in (1); thus

$$1 = (Cx+D)(-x) = -Cx^2-Dx = Cx+C-Dx;$$

or $$(C-D)x+C-1 = 0.$$

Equate coefficients to zero; thus $C = 1,$

$$D = 1.$$

Hence $$\frac{1}{(x^2+1)(x^2+x+1)} = \frac{x+1}{x^2+x+1} - \frac{x}{x^2+1}.$$

---

238 Fourthly.—When there is a repeated quadratic factor of imaginary roots.

Ex.—Resolve $\dfrac{40x-103}{(x+1)^2(x^2-4x+8)^3}$ into partial fractions.

Assume

$$\frac{40x-103}{(x+1)^2(x^2-4x+8)^3} = \frac{Ax+B}{(x^2-4x+8)^3} + \frac{Cx+D}{(x^2-4x+8)^2} + \frac{Ex+F}{x^2-4x+8} + \frac{G}{(x+1)^2} + \frac{H}{x+1};$$

$\therefore$

$$40x-103 = \{(Ax+B)+(Cx+D)(x^2-4x+8)+(Ex+F)(x^2-4x+8)^2\}(x+1)^2 + \{G+H(x+1)\}(x^2-4x+8)^3 \quad\ldots\ldots (1).$$

In the first place, to determine A and B, equate x^2-4x+8 to zero; thus $x^2 = 4x-8$.

Substitute this value of x^2 repeatedly in (1), as in the previous example, until the first power of x alone remains. The resulting equation is

$$40x-103 = (17A+6B)\,x-48A-7B.$$

Equating coefficients, we obtain two equations

$$\left.\begin{aligned}17A+6B &= 40\\ 48A+7B &= 103\end{aligned}\right\}, \quad \text{from which} \quad \begin{aligned}A &= 2\\ B &= 1.\end{aligned}$$

Next, to determine C and D, substitute these values of A and B in (1); the equation will then be divisible by x^2-4x+8. Divide, and the resulting equation is

$$0 = 2x+13+\{Cx+D+(Ex+F)(x^2-4x+8)\}(x+1)^2 + \{G+H(x+1)\}(x^2-4x+8)^2 \quad\ldots\ldots (2).$$

Equate x^2-4x+8 again to zero, and proceed exactly as before, when finding A and B.

Next, to determine E and F, substitute the values of C and D, last found in equation (2); divide, and proceed as before.

Lastly, G and H are determined by equating $x+1$ to zero successively, as in Example 2.

CONVERGENCY AND DIVERGENCY OF SERIES.

239 Let $a_1+a_2+a_3+\&c.$ be a series, and a_n, a_{n+1} any two consecutive terms. The following tests of convergency may be applied. The series will converge, if, after any fixed term—

(i.) The terms decrease and are alternately positive and negative.

(ii.) Or if $\frac{a_n}{a_{n+1}}$ is always *greater* than some quantity greater than unity.

(iii.) Or if $\frac{a_n}{a_{n+1}}$ is never less than the corresponding ratio in a known converging series.

(iv.) Or if $\left(\frac{na_n}{a_{n+1}} - n\right)$ is always *greater* than some quantity greater than unity. By 244, & iii.

(v.) Or if $\left(\frac{na_n}{a_{n+1}} - n - 1\right) \log n$ is always *greater* than some quantity greater than unity.

---

240 The conditions of divergency are obviously the converse of rules (i.) to (v.)

---

241 The series $a_1 + a_2 x + a_3 x^2 + \&c.$ converges, if $\frac{a_{n+1}}{a_n}$ is always less than some quantity p, and x less than $\frac{1}{p}$.
By 239 (ii.)

---

242 To make the sum of the last series less than an assigned quantity p, make x less than $\frac{p}{p+k}$, k being the greatest coefficient.

General Theorem.

243 If $\phi(x)$ be positive for all positive integral values of x, and continually diminish as x increases, and if m be any positive integer, then the two series

$$\phi(1) + \phi(2) + \phi(3) + \phi(4) + \ldots\ldots$$

$$\phi(1) + m\phi(m) + m^2\phi(m^2) + m^3\phi(m^3) + \ldots\ldots$$

are either both convergent or both divergent.

---

244 Application of this theorem. To ascertain whether the series
$$\frac{1}{1^p} + \frac{1}{2^p} + \frac{1}{3^p} + \frac{1}{4^p} + \ldots\ldots$$
is divergent or convergent when p is greater than unity.

Taking $m = 2$, the second series in (243) becomes

$$1 + \frac{2}{2^p} + \frac{4}{4^p} + \frac{8}{8^p} + \&c. \ldots\ldots$$

a geometrical progression which converges; therefore the given series converges.

---

245 The series of which $\dfrac{1}{n(\log n)^p}$ is the general term is convergent if p be greater than unity, and divergent if p be not greater than unity. By (243), (244).

---

246 The series of which the general term is

$$\frac{1}{n\lambda(n)\,\lambda^2(n)\ldots\ldots\lambda^r(n)\,\{\lambda^{r+1}(n)\}^p}$$

where $\lambda(n)$ signifies $\log n$, $\lambda^2(n)$ signifies $\log\{\log(n)\}$, and so on, is convergent if p be greater than unity, and divergent if p be not greater than unity. By Induction, and by (243).

---

247 The series $a_1 + a_2 + \&c.$ is convergent if

$$na_n \log(n) \log^2(n) \ldots\ldots \log^r(n)\,\{\log^{r+1}(n)\}^p$$

is always finite for a value of p greater than unity; $\log^2(n)$ here signifying $\log(\log n)$, and so on.

[See Todhunter's Algebra, or Boole's Finite Differences.

EXPANSION OF A FRACTION.

---

248 A fractional expression such as $\dfrac{4x - 10x^2}{1 - 6x + 11x^2 - 6x^3}$ may be expanded in ascending powers of x in three different ways.

First, by dividing the numerator by the denominator in the ordinary way, or by Synthetic Division, as shewn in (28).

Secondly, by the method of Indeterminate Coefficients (232).

Thirdly, by Partial Fractions and the Binomial Theorem.

To expand by the method of Indeterminate Coefficients, proceed as follows:—

$$\text{Assume}\quad \frac{4x-10x^2}{1-6x+11x^2-6x^3} = A+Bx+Cx^2+Dx^3+Ex^4+\&c.\ldots$$

$$\begin{aligned}\therefore\quad 4x-10x^2 = A+ Bx+ & Cx^2+ Dx^3+ Ex^4+ Fx^5+\ldots\\ -6Ax- & 6Bx^2- 6Cx^3- 6Dx^4- 6Ex^5-\ldots\\ & +11Ax^2+11Bx^3+11Cx^4+11Dx^5+\ldots\\ & \quad - 6Ax^3- 6Bx^4- 6Cx^5-\ldots\end{aligned}$$

Equate coefficients of like powers of x, thus

$$\begin{aligned} A &= 0, & & \\ B-6A &= 4, & \therefore\ B &= 4;\\ C-6B+11A &= -10, & \therefore\ C &= 14;\\ D-6C+11B-6A &= 0, & \therefore\ D &= 40;\\ E-6D+11C-6B &= 0, & \therefore\ E &= 110;\\ F-6E+11D-6C &= 0, & \therefore\ F &= 304;\\ \ldots\quad \ldots\quad \ldots\quad \ldots & \ldots & \ldots & \ldots\end{aligned}$$

The formation of the same coefficients by synthetic division is now exhibited, in order that the connexion between the two processes may be clearly seen.

The division of $4x-10x^2$ by $1-6x+11x^2-6x^3$ is as follows:—

$$\begin{array}{r|l} & 0+4-10 \\ +6 & \qquad 24+84+240+660 \\ -11 & \qquad\quad -44-154-440-1210 \\ +6 & \qquad\qquad +24+84+240+660 \\ \hline & 0+4+14+40+110+304+\ \ldots\ldots\ldots \\ & A\ \ B\ \ C\ \ D\ \ E\ \ F \end{array}$$

If we stop at the term $110x^4$, then the undivided remainder will be $304x^5-970x^6+660x^7$, and the complete result will be

$$4x+14x^2+40x^3+110x^4+\frac{304x^5-970x^6+660x^7}{1-6x+11x^2-6x^3}.$$

249 Here the concluding fraction may be regarded as the sum to infinity after four terms of the series, just as the original expression is considered to be the sum to infinity of the whole series.

---

250 If the general term be required, the method of expansion by partial fractions must be adopted. See (257), where the general term of the foregoing series is obtained.

RECURRING SERIES.

$a_0+a_1x+a_2x^2+a_3x^3+$&c. is a recurring series if the coefficients are connected by the relation

251 $$a_n = p_1a_{n-1}+p_2a_{n-2}+ \dots +p_ma_{n-m}.$$

The Scale of Relation is

252 $$1-p_1x-p_2x^2- \dots -p_mx^m.$$

The sum of n terms of the series is equal to

253 [The first m terms

$-p_1x$ (first $m-1$ terms + the last term)

$-p_2x^2$ (first $m-2$ terms + the last 2 terms)

$-p_3x^3$ (first $m-3$ terms + the last 3 terms)

... ...

$-p_{m-1}x^{m-1}$ (first term + the last $m-1$ terms)

$-p_mx^m$ (the last m terms)] $\div\,[1-p_1x-p_2x^2-\dots-p_mx^m]$.

254 If the series converges, and the sum to infinity is required, omit all "the last terms" from the formula.

255 Example.—Required the Scale of Relation, the general term, and the apparent sum to infinity of the series

$$4x+14x^2+40x^3+110x^4+304x^5+854x^6+\dots$$

Observe that six arbitrary terms given are sufficient to determine a Scale of Relation of the form $1-px-qx^2-rx^3$, involving three constants p, q, r, for, by (251), we can write three equations to determine these constants; namely,

$$\left.\begin{aligned} 110 &= 40p+14q+4r \\ 304 &= 110p+40q+14r \\ 854 &= 304p+110q+40r \end{aligned}\right\} \quad \text{The solution gives } p=6,\quad q=-11,\quad r=6.$$

Hence the Scale of Relation is $1-6x+11x^2-6x^3$.

The sum of the series without limit will be found from (254), by putting $p_1=6$, $p_2=-11$, $p_3=6$, $m=3$.

$$\begin{aligned} &\text{The first three terms} &&= 4x+14x^2+40x^3 \\ &-6x\times\text{the first two terms} &&= \quad -24x^2-84x^3 \\ &+11x\times\text{the first term} &&= \qquad\qquad +44x^3 \\ \hline & && \quad 4x-10x^2 \end{aligned}$$

$$\therefore \qquad S = \frac{4x-10x^2}{1-6x+11x^2-6x^3};$$

the meaning of which is, that if this fraction be expanded in ascending powers of x, the first six terms will be those given in the question.

256 To obtain more terms of the series, we may use the Scale of Relation; thus the 7th term will be

$$(6\times 854-11\times 304+6\times 110)\,x^7 = 2440x^7.$$

257 To find the general term, S must be decomposed into partial fractions; thus, by the method of (235),

$$\frac{4x-10x^2}{1-6x+11x^2-6x^3} = \frac{1}{1-3x}+\frac{2}{1-2x}-\frac{3}{1-x}.$$

By the Binomial Theorem (128),

$$\frac{1}{1-3x} = 1+3x+3^2x^2+\ldots\ldots+3^nx^n,$$

$$\frac{2}{1-2x} = 2+2^2x+2^3x^2+\ldots\ldots+2^{n+1}x^n,$$

$$-\frac{3}{1-x} = -3-3x-3x^2-\ldots\ldots-3x^n.$$

Hence the general term involving x^n is

$$(3^n+2^{n+1}-3)\,x^n.$$

And by this formula we can write the "last terms" required in (253), and so obtain the sum of any finite number of terms of the given series. Also, by the same formula we can calculate the successive terms at the beginning of the series. In the present case this mode will be ore expeditious than that of employing the Scale of Relation.

258 If, in decomposing S into partial fractions for the sake of obtaining the general term, a quadratic factor with imaginary roots should occur as a denominator, the same method must be pursued, for the imaginary quantities will disappear in the final result. In this case, however, it is more convenient to employ a general formula. Suppose the fraction which gives rise to the imaginary roots to be

$$\frac{L+Mx}{a+bx+x^2} = \frac{L+Mx}{(p-x)(q-x)},$$

p and q being the imaginary roots of $a+bx+x^2=0$.

Suppose $\qquad p = \alpha+i\beta,$

$\qquad\qquad q = \alpha-i\beta,$ where $i=\sqrt{(-1)}$.

If, now, the above fraction be resolved into two partial fractions in the ordinary way, and if these fractions be expanded separately by the Binomial Theorem, and that part of the general term furnished by these two expansions written out, still retaining p and q, and if the imaginary values of p and q be then substituted, it will be found that the factor will disappear, and that the result may be enunciated as follows.

259 The coefficient of x^{n-1} in the expansion of

$$\frac{L+Mx}{(a^2+\beta^2)^n-2ax+x^2}$$

will be

$$\frac{L}{\beta(a^2+\beta^2)}\{na^{n-1}\beta-C(n,3)\,a^{n-3}\beta^3+C(n,5)\,a^{n-5}\beta^5-\dots\}$$

$$+\frac{M}{\beta(a^2+\beta^2)^{n-1}}\{(n-1)\,a^{n-2}\beta-C(n-1,3)\,a^{n-4}\beta^3$$
$$+C(n-1,5)\,a^{n-6}\beta^5-\dots\}.$$

260 With the aid of the known expansion of $\sin n\theta$ in Trigonometry, this formula for the nth term may be reduced to

$$\sqrt{\frac{(L+Ma)^2+M^2\beta^2}{\beta^2(a^2+\beta^2)^n}}\,.\,\sin(n\theta-\phi),$$

in which $\quad \theta=\tan^{-1}\dfrac{\beta}{a}, \qquad \phi=\tan^{-1}\dfrac{M\beta}{L+Ma}.$

If n be not greater than 100, $\sin(n\theta-\phi)$ may be obtained from the tables correct to about six places of decimals, and accordingly the nth term of the expansion may be found with corresponding accuracy. As an example, the 100th term in the expansion of $\dfrac{1+x}{5-2x+x^2}$ is readily found by this method to be $\dfrac{41824}{10^{41}}\,x^{99}$.

To determine whether a given Series is recurring or not.

261 If certain first terms only of the series be given, a scale of relation may be found which shall produce a recurring series whose first terms are those given. The method is exemplified in (255). The number of unknown coefficients p, q, r, &c. to be assumed for the scale of relation must be equal to half the number of the given terms of the series, if that number be even. If the number of given terms be odd, it may be made even by prefixing zero for the first term of the series.

262 Since this method may, however, produce zero values for one or more of the last coefficients in the scale of relation, it may be advisable in practice to determine a scale from the first two terms of the series, and if that scale does not produce the following terms, we may try a scale determined from the first four terms, and so on until the true scale is arrived at.

If an indefinite number of terms of the series be given, we may find whether it is recurring or not by a rule of Lagrange's.

263 Let the series be

$$S = A + Bx + Cx^2 + Dx^3 + \&c.$$

Divide unity by S as far as two terms of the quotient, which will be of the form $p + qx$, and write the remainder in the form $S'x^2$, S' being another indefinite series of the same form as S.

Next, divide S by S' as far as two terms of the quotient, and write the remainder in the form $S''x^2$.

Again, divide S' by S'', and proceed as before, and repeat this process until there is no remainder after one of the divisions. The series will then be proved to be a recurring series, and the order of the series, that is, the degree of the scale of relation, will be the same as the number of divisions which have been effected in the process.

EXAMPLE.—To determine whether the series 1, 3, 6, 10, 15, 21, 28, 36, 45, ... is recurring or not.

Introducing x, we may write

$$S = 1 + 3x + 6x^2 + 10x^3 + 15x^4 + 21x^5 + 28x^6 + 36x^7 + 45x^8 \ldots$$

Then we shall have $\dfrac{1}{S} = 1 - 3x + \ldots$ with a remainder

$$3x^2 + 8x^3 + 15x^4 + 24x^5 + 35x^6 + \&c.$$

Therefore $\quad S' = 3 + 8x + 15x^2 + 24x^3 + 35x^4 + \&c.,$

$$\frac{S}{S'} = \frac{1}{3} + \frac{x}{9},$$

with a remainder $\frac{1}{9}(x^2+3x^3+6x^4+10x^5+\&c....)$.

Therefore we may take $S''=1+3x+6x^2+10x^3+\&c.$

Lastly $\frac{S'}{S''}=3-x$ without any remainder.

Consequently the series is a recurring series of the third order. It is, in fact, the expansion of $\frac{1}{1-3x+3x^2-x^3}$.

SUMMATION OF SERIES BY THE METHOD OF DIFFERENCES.

264 Rule.—*Form successive series of differences until a series of equal differences is obtained.* Let a, b, c, d, &c. be the first terms of the several series; then the nth term of the given series is

265 $$a+(n-1)\,b+\frac{(n-1)(n-2)}{1\,.\,2}\,c+\frac{(n-1)(n-2)(n-3)}{1\,.\,2\,.\,3}\,d+$$

The sum of n terms

266 $$=na+\frac{n\,(n-1)}{1\,.\,2}\,b+\frac{n\,(n-1)\,(n-2)}{1\,.\,2\,.\,3}\,c+\&c.$$

[Proved by Induction.

Example: a ... 1+ 5+15+35+70+126+...
b ... 4+10+20+35+56+...
c ... 6+10+15+21+...
d ... 4+5+6+...
e ... 1+1+...

The 100th term of the first series

$$=1+99\,.\,4+\frac{99\,.\,98}{1\,.\,2}\,6+\frac{99\,.\,98\,.\,97}{1\,.\,2\,.\,3}\,4+\frac{99\,.\,98\,.\,97\,.\,96}{1\,.\,2\,.\,3\,.\,4}.$$

The sum of 100 terms

$$=100+\frac{100\,.\,99}{1\,.\,2}\,4+\frac{100\,.\,99\,.\,98}{1\,.\,2\,.\,3}\,6+\frac{100\,.\,99\,.\,98\,.\,97}{1\,.\,2\,.\,3\,.\,4}\,4+\frac{100\,.\,99\,.\,98\,.\,97\,.\,96}{1\,.\,2\,.\,3\,.\,4\,.\,5}.$$

267 To interpolate a term between two terms of a series by the method of differences.

Ex.—Given log 71, log 72, log 73, log 74, it is required to find log 72·54. Form the series of differences from the given logarithms, as in (266),

	log 71	log 72	log 73	log 74
a ...	1·8512583	1·8573325	1·8633229	1·8692317
b ...	·0060742	·0059904	·0059088	
c ...	−·0000838	−·0000816		
d ...	−·0000022	considered to vanish.		

Log 72·54 must be regarded as an interpolated term, the number of its place being 2·54.

Therefore put 2·54 for n in formula (265).

Result log 72·54 = 1·8605777.

DIRECT FACTORIAL SERIES.

268 Ex.: $5.7.9+7.9.11+9.11.13+11.13.15+\ldots$

d = common difference of factors,

m = number of factors in each term,

n = number of terms,

a = first factor of first term $-d$.

$$n^{\text{th}} \text{ term} = (a+nd)\,(a+\overline{n+1}\,d)\ \ldots\ldots\ (a+\overline{n+m-1}\,d).$$

269 To find the sum of n terms. Rule.—*From the last term with the next highest factor take the first term with the next lowest factor, and divide by* $(m+1)\,d$. Proof, by Induction.

Thus the sum of 4 terms of the above series will be, putting $d=2$, $m=3$,

$n=4$, $a=3$, $$S=\frac{11\,.\,13\,.\,15\,.\,17-3\,.\,5\,.\,7\,.\,9}{(3+1)\,2}$$

Proved either by Induction, or by the method of Indeterminate Coefficients.

INVERSE FACTORIAL SERIES.

270 Ex.: $\frac{1}{5.7.9}+\frac{1}{7.9.11}+\frac{1}{9.11.13}+\frac{1}{11.13.15}+\dots$

Defining d, m, n, a as before, the

$$n^{\text{th}}\text{ term} = \frac{1}{(a+nd)\,(a+\overline{n+1}\,d)\,\dots\,a+\overline{n+m-1}\,d}$$

271 To find the sum of n terms. RULE.—*From the first term wanting its last factor take the last term wanting its first factor, and divide by* $(m-1)\,d$. Proof, by Induction.

Thus the sum of 4 terms of the above series will be, putting $d=2$, $m=3$, $n=4$, $a=3$,

$$\frac{\frac{1}{5.7}-\frac{1}{13.15}}{(3-1)\,2}$$

Proved either by Induction, or by decomposing the terms, as in the following example.

---

272 Ex.: To sum the same series by decomposing the terms into partial fractions. Take the general term in the simple form

$$\frac{1}{(r-2)\,r\,(r+2)}$$

Resolve this into the three fractions

$$\frac{1}{8\,(r-2)}-\frac{1}{4r}+\frac{1}{8\,(r+2)} \text{ by (235).}$$

Substitute, 7, 9, 11, &c. successively for r, and the given series has for its equivalent the three series

$$\frac{1}{8}\left\{\ \frac{1}{5}+\frac{1}{7}+\frac{1}{9}+\frac{1}{11}+\frac{1}{13}\ \dots\dots\dots\dots\dots\ +\frac{1}{2n+3}\right\}$$
$$+\frac{1}{8}\left\{-\frac{2}{7}-\frac{2}{9}-\frac{2}{11}-\frac{2}{13}-\dots\dots\dots-\frac{2}{2n+3}-\frac{2}{2n+5}\right\}$$
$$+\frac{1}{8}\left\{\ \frac{1}{9}+\frac{1}{11}+\frac{1}{13}+\dots\dots+\frac{1}{2n+3}+\frac{1}{2n+5}+\frac{1}{2n+7}\right\}$$

and the sum of n terms is seen, by inspection, to be

$$\frac{1}{8}\left\{\frac{1}{5}-\frac{1}{7}-\frac{1}{2n+5}+\frac{1}{2n+7}\right\}=\frac{1}{4}\left\{\frac{1}{5.7}-\frac{1}{(2n+5)\,(2n+7)}\right\}$$

a result obtained at once by the rule in (271), taking $\frac{1}{5.7.9}$ for the first term, and $\frac{1}{(2n+3)\,(2n+5)\,(2n+7)}$ for the nth or last term.

273 Analogous series may be reduced to the types in (268) and (270), or else the terms may be decomposed in the manner shewn in (272).

Ex.: $$\frac{1}{1.2.3}+\frac{4}{2.3.4}+\frac{7}{3.4.5}+\frac{10}{4.5.6}+\ldots\ldots$$

has for its general term

$$\frac{3n-2}{n(n+1)(n+2)}=-\frac{1}{n}+\frac{5}{n+1}-\frac{4}{n+2} \text{ by (235)},$$

and we may proceed as in (272) to find the sum of n terms.

The method of (272) includes the method known as 'Summation by Subtraction,' but it has the advantage of being more general and easier of application to complex series.

COMPOSITE FACTORIAL SERIES.

274 If the two series

$$(1-x)^{-5}=1+5x+\frac{5.6}{1.2}x^2+\frac{5.6.7}{1.2.3}x^3+\frac{5.6.7.8}{1.2.3.4}x^4+\ldots$$

$$(1-x)^{-3}=1+3x+\frac{3.4}{1.2}x^2+\frac{3.4.5}{1.2.3}x^3+\frac{3.4.5.6}{1.2.3.4}x^4+\ldots$$

be multiplied together, and the coefficient of x^4 in the product be equated to the coefficient of x^4 in the expansion of $(1-x)^{-8}$, we obtain as the result the sum of the composite series

$$5.6.7.8\times 1.2+4.5.6.7\times 2.3+3.4.5.6\times 3.4$$
$$+2.3.4.5\times 4.5+1.2.3.4\times 5.6=\frac{\underline{|4}\ 2\underline{|11}}{\underline{|7}\ \underline{|4}}.$$

275 Generally, if the given series be

$$P_1Q_1+P_2Q_2+\ldots+P_{n-1}Q_{n-1} \quad\ldots\ldots\ldots\ldots\ldots (1),$$

where $\qquad Q_r=r(r+1)(r+2)\ldots(r+q-1)$,

and $\qquad P_r=(n-r)(n-r+1)\ldots(n-r+p-1)$;

the sum of $n-1$ terms will be

$$\frac{\underline{|p}\ \underline{|q}}{\underline{|p+q+1}}\cdot\frac{\underline{|n+p+q-1}}{\underline{|n-2}}$$

MISCELLANEOUS SERIES.

276 *Sum of the powers of the terms of an Arithmetical Progression.*

$$1+2+3+\ldots+n = \frac{n(n+1)}{2} = S_1$$

$$1+2^2+3^2+\ldots+n^2 = \frac{n(n+1)(2n+1)}{6} = S_2$$

$$1+2^3+3^3+\ldots+n^3 = \left\{\frac{n(n+1)}{2}\right\}^2 = S_3$$

$$1+2^4+3^4+\ldots+n^4 = \frac{n(n+1)(2n+1)(3n^2+3n-1)}{30} = S_4$$

By the method of Indeterminate Coefficients (234).

A general formula for the sum of the rth powers of $1.2.3\ldots n$, obtained in the same way, is

$$S_r = \frac{1}{r+1}n^{r+1}+\tfrac{1}{2}n^r+A_1n^{r-1}+\ldots A_{r-1}n,$$

where A_1, A_2, &c. are determined by putting $p=1, 2, 3$, &c. successively in the equation

$$\frac{1}{2\,\underline{|p+1}}$$

$$= \frac{1}{\underline{|p+2}}+\frac{A_1}{r\,\underline{|p}}+\frac{A_2}{r(r-1)\,\underline{|p-1}}+\ldots\frac{A_p}{r(r-1)\ldots(r-p+1)}$$

277 $a^m+(a+d)^m+(a+2d)^m+\ldots+(a+nd)^m$

$$= (n+1)a^m+S_1ma^{m-1}d+C(m,2)a^{m-2}d^2+C(m,3)a^{m-3}d^3+\&c.$$

By Bin. Theor. and (276).

278 *Summation of a series partly Arithmetical and partly Geometrical.*

Example.—To find the sum of the series $1+3x+5x^2+$ to n terms.

Let $s = 1+3x+5x^2+7x^3+\ldots+(2n-1)\,x^{n-1}$

$\therefore\ sx = x+3x^2+5x^3+\ldots+(2n-3)\,x^{n-1}+(2n-1)\,x^n$

$\therefore$ by subtraction,

$$s\,(1-x) = 1+2x+2x^2+2x^3+\ldots+2x^{n-1}-(2n-1)\,x^n$$

$$= 1+2x\,\frac{1-x^{n-1}}{1-x}-(2n-1)\,x^n$$

$$\therefore\quad s = \frac{1-(2n-1)\,x^n}{1-x}+\frac{2x\,(1-x^{n-1})}{(1-x)^2}$$

---

279 A general formula for the sum of n terms of

$$a+(a+d)\,r+(a+2d)\,r^2+(a+3d)\,r^3+\&c.$$

is
$$S = \frac{a-(a+\overline{n-1}\,d)\,r^n}{1-r}+\frac{dr\,(1-r^{n-1})}{(1-r)^2}$$

Obtained as in (278).

RULE.—*Multiply by the ratio and subtract the resulting series.*

---

280 $$\frac{1}{1-x} = 1+x+x^2+x^3+\ldots+x^{n-1}+\frac{x^n}{1-x}$$

281 $$\frac{1}{(1-x)^2} = 1+2x+3x^2+4x^3+\ldots$$
$$\ldots+nx^{n+1}+\frac{(n+1)\,x^n-nx^{n+1}}{(1-x)^2}$$

---

282 $$(n-1)\,x+(n-2)\,x^2+(n-3)\,x^3+\ldots+2x^{n-2}+x^{n-1}$$
$$= \frac{(n-1)\,x-nx^2+x^{n+1}}{(1-x)^2}$$ By (253).

---

283 $$1+n+\frac{n\,(n-1)}{\lfloor 2}+\frac{n\,(n-1)\,(n-2)}{\lfloor 3}+\&c. = 2^n,$$

$$1-n+\frac{n\,(n-1)}{\lfloor 2}-\frac{n\,(n-1)\,(n-2)}{\lfloor 3}+\&c. = 0.$$

By making $a=b$, $a=-b$ in (125).

284 The series

$$1-\frac{n-3}{2}+\frac{(n-4)(n-5)}{\lfloor 3}-\frac{(n-5)(n-6)(n-7)}{\lfloor 4}+\dots$$

$$\dots+(-1)^{r-1}\frac{(n-r-1)(n-r-2)\dots(n-2r+1)}{\lfloor r}$$

consists of $\frac{n}{2}$ or $\frac{n-1}{2}$ terms, and the sum is given by

$$S=\quad \frac{3}{n} \text{ if } n \text{ be of the form } 6m+3,$$

$$S=\quad 0 \text{ if } n \text{ be of the form } 6m\pm 1,$$

$$S=-\frac{1}{n} \text{ if } n \text{ be of the form } 6m,$$

$$S=\quad \frac{2}{n} \text{ if } n \text{ be of the form } 6m\pm 2.$$

By (545), putting $p=x+y$, $q=xy$, and applying (546).

---

285 The series $n^r-n(n-1)^r+\frac{n(n-1)}{\lfloor 2}(n-2)^r$

$$-\frac{n(n-1)(n-2)}{\lfloor 3}(n-3)^r+\&c....$$

takes the values $0,\quad \lfloor n,\quad \frac{1}{2}n\lfloor n+1,$

according as r is $<n,\ =n,$ or $=n+1.$

Proved by expanding $(e^x-1)^n$, in two ways: first, by the Exponential Theorem and Multinomial; secondly, by the Bin. Th., and each term of the expansion by the Exponential. Equate the coefficients of x^r in the two results.

Other results by putting $r=n+2$, $n+3$, &c.

The series (285), when divided by $\lfloor r$, is, in fact, equal to the coefficient of x^r in the expansion of

$$\left\{x+\frac{x^2}{\lfloor 2}+\frac{x^3}{\lfloor 3}+\dots\right\}^n$$

286 By exactly the same process we may deduce from the function $\{e^x - e^{-x}\}^n$ the result that the series

$$n^r - n(n-2)^r + \frac{n(n-1)}{\lfloor 2}(n-4)^r - \&c.$$

takes the values 0 or $2^n \lfloor n$, according as r is $< n$ or $= n$; this series, divided by $\lfloor r$, being equal to the coefficient of x^r in the expansion of

$$2^n \left\{ x + \frac{x^3}{\lfloor 3} + \frac{x^5}{\lfloor 5} + \ldots \right\}^n$$

POLYGONAL NUMBERS.

287 The nth term of the rth order of polygonal numbers is equal to the sum of n terms of an Arith. Prog. whose first term is unity and common difference $r-2$; that is

$$= \frac{n}{2}\{2+(n-1)(r-2)\} = n + \tfrac{1}{2}n(n-1)(r-2).$$

288 The sum of n terms

$$= \frac{n(n+1)}{2} + \frac{n(n-1)(n+2)(r-2)}{6}$$

By resolving into two series.

Order.	n^{th} term.							
1	1	1	1	1	1	1	1	1
2	n	1	2	3	4	5	6	7
3	$\frac{1}{2}n(n+1)$	1	3	6	10	15	21	28
4	n^2	1	4	9	16	25	36	49
5	$\frac{1}{2}n(3n-1)$	1	5	12	22	35	51	70
6	$(2n-1)n$	1	6	15	28	45	66	91
...	... ...	...	...	...	...	...	...	...
r	$n + \frac{n(n-1)}{2}(r-2)$	1,	r,	$3+3(r-2)$,	$4+6(r-2)$,	$5+10(r-2)$,	$6+15(r-2)$,	&c.

In practice—to form, for instance, the 6th order of polygonal numbers—write the first three terms by the formula, and form the rest by the method of differences.

Ex.: 1 6 15 28 45 66 91 120 ...
5 9 13 17 21 25 29 ...
$[r-2=4]$ 4 4 4 4 4 4 ...

FIGURATE NUMBERS.

289 The nth term of any order is the sum of n terms of the preceding order.

The nth term of the rth order is

$$\frac{n(n+1)\dots(n+r-2)}{\lfloor r-1}=H(n,\ r-1) \qquad \text{By (98).}$$

290 The sum of n terms is

$$\frac{n(n+1)\dots(n+r-1)}{\lfloor r}=H(n,\ r)$$

Order.	Figurate Numbers.	n^{th} term.
1	1, 1, 1, 1, 1, 1	1
2	1, 2, 3, 4, 5, 6	n
3	1, 3, 6, 10, 15, 21	$\frac{n(n+1)}{1.2}$
4	1, 4, 10, 20, 35, 56	$\frac{n(n+1)(n+2)}{1.2.3}$
5	1, 5, 15, 35, 70, 126	$\frac{n(n+1)(n+2)(n+3)}{1.2.3.4}$
6	1, 6, 21, 56, 126, 252	$\frac{n(n+1)(n+2)(n+3)(n+4)}{1.2.3.4.5}$

HYPERGEOMETRICAL SERIES.

291 $$1+\frac{\alpha\,.\,\beta}{1\,.\,\gamma}x+\frac{\alpha(\alpha+1)\,\beta(\beta+1)}{1\,.\,2\,.\,\gamma(\gamma+1)}x^2+\frac{\alpha(\alpha+1)(\alpha+2)\,\beta(\beta+1)(\beta+2)}{1\,.\,2\,.\,3\,.\,\gamma(\gamma+1)(\gamma+2)}x^3+\&c.$$

is convergent if x is < 1,
and divergent if x is > 1; (239 ii.)

and if $x = 1$, the series is
convergent if $\gamma-\alpha-\beta$ is positive,
divergent if $\gamma-\alpha-\beta$ is negative, (239 iv.)
and divergent if $\gamma-\alpha-\beta$ is zero. (239 v.)

Let the hypergeometrical series (291) be denoted by $F(\alpha, \beta, \gamma)$; then, the series being convergent, it is shewn by induction that

292 $$\frac{F(\alpha,\beta+1,\gamma+1)}{F(\alpha,\beta,\gamma)}=\cfrac{1}{1-\cfrac{k_1}{1-\cfrac{k_2}{1-\&c.\dots}}}$$ concluding with $$1-\frac{k_{2r-1}}{1-k_{2r}z_{2r}}$$

where k_1, k_2, k_3, &c., with z_{2r}, are given by the formulæ

$$k_{2r-1}=\frac{(\alpha+r-1)(\gamma+r-1-\beta)\,x}{(\gamma+2r-2)(\gamma+2r-1)}$$

$$k_{2r}=\frac{(\beta+r)(\gamma+r-\alpha)\,x}{(\gamma+2r-1)(\gamma+2r)}$$

$$z_{2r}=\frac{F(\alpha+r,\beta+r+1,\gamma+2r+1)}{F(\alpha+r,\beta+r,\gamma+2r)}$$

The continued fraction may be concluded at any point with $k_{2r}z_{2r}$. When r is infinite, $z_{2r}=1$ and the continued fraction is infinite.

293 Let

$$f(\gamma) \equiv 1+\frac{x^2}{1.\gamma}+\frac{x^4}{1.2.\gamma(\gamma+1)}+\frac{x^6}{1.2.3.\gamma(\gamma+1)(\gamma+2)}+\&c.$$

the result of substituting $\frac{x^2}{\alpha\beta}$ for x in (291), and making $\beta=\alpha=\infty$. Then, by last, or independently by induction,

$$\frac{f(\gamma+1)}{f(\gamma)}=\frac{1}{1+}\,\frac{p_1}{1+}\,\frac{p_2}{1+}\ldots+\frac{p_m}{1+}\&c.$$

$$\text{with } p_m=\frac{x^2}{(\gamma+m-1)(\gamma+m)}$$

---

294 In this result put $\gamma=\frac{1}{2}$ and $\frac{y}{2}$ for x, and we obtain by Exp. Th. (150),

$$\frac{e^y-e^{-y}}{e^y+e^{-y}}=\frac{y}{1+}\,\frac{y^2}{3+}\,\frac{y^2}{5+}\&c. \quad \text{the } r^{\text{th}} \text{ component being } \frac{y^2}{2r-1}$$

Or the continued fraction may be formed by ordinary division of one series by the other.

---

295 $e^{\frac{m}{n}}$ is incommensurable, m and n being integers. From the last and (174), by putting $x=\frac{m}{n}$.

INTEREST.

If r be the Interest on £1 for 1 year,
n the number of years,
P the Principal,
A the amount in n years. Then

---

296 At Simple Interest $A=P(1+nr)$.

297 At Compound Interest $A=P(1+r)^n$. By (84)

298 But if the payments of Interest be made q times a year ... ... $A = P\left(1+\frac{r}{q}\right)^{nq}$

If A be an amount due in n years time, and P the present worth of A. Then

299 At Simple Interest $P = \frac{A}{1+nr}$ By (296)

300 At Compound Interest $P = \frac{A}{(1+r)^n}$ By (297)

301 Discount $= A-P$.

ANNUITIES.

302 The amount of an Annuity of £1 in n years, at Simple Interest... $= n + \frac{n(n-1)}{2}r$ By (82)

303 Present value of same $= \frac{n+\frac{1}{2}n(n-1)r}{1+nr}$ By (299)

304 Amount at Compound Interest ... ... $= \frac{(1+r)^n-1}{(1+r)-1}$ By (85)

Present worth of same $= \frac{1-(1+r)^{-n}}{(1+r)-1}$ By (300)

305 Amount when the payments of Interest are made q times per annum ... ... $= \frac{\left(1+\frac{r}{q}\right)^{nq}-1}{\left(1+\frac{r}{q}\right)^{q}-1}$ By (298)

Present value of same $= \frac{1-\left(1+\frac{r}{q}\right)^{-nq}}{\left(1+\frac{r}{q}\right)^{q}-1}$

306 Amount when the payments of the Annuity are made m times per annum ... ... $= \dfrac{(1+r)^n-1}{m\{(1+r)^{\frac{1}{m}}-1\}}$

Present value of same $= \dfrac{1-(1+r)^{-n}}{m\{(1+r)^{\frac{1}{m}}-1\}}$

307 Amount when the Interest is paid q times and the Annuity m times per annum ... $= \dfrac{\left(1+\frac{r}{q}\right)^{nq}-1}{m\left\{\left(1+\frac{r}{q}\right)^{\frac{q}{m}}-1\right\}}$

Present value of same $= \dfrac{1-\left(1+\frac{r}{q}\right)^{-nq}}{m\left\{\left(1+\frac{r}{q}\right)^{\frac{q}{m}}-1\right\}}$

PROBABILITIES.

309 If all the ways in which an event can happen be m in number, all being equally likely to occur, and if in n of these m ways the event would happen under certain restrictive conditions; then the probability of the restricted event happening is equal to $n \div m$.

Thus, if the letters of the alphabet be chosen at random, any letter being equally likely to be taken, the probability of a vowel being selected is equal to $\frac{5}{26}$. The number of unrestricted cases here is 26, and the number of restricted ones 5.

310 If, however, all the m events are not equally probable, they may be divided into groups of equally probable cases. The probability of the restricted event happening in each group separately must be calculated, and the sum of these probabilities will be the total probability of the restricted event happening at all.

EXAMPLE.—There are three bags A, B, and C.

A	contains	2	white and	3	black balls
B	,,	3	,,	4	,,
C	,,	4	,,	5	,,

A bag is taken at random and a ball drawn from it. Required the probability of the ball being white.

Here the probability of the bag A being chosen $= \frac{1}{3}$, and the subsequent probability of a white ball being drawn $= \frac{2}{5}$.

Therefore the probability of a white ball being drawn from A

$$= \frac{1}{3} \times \frac{2}{5} = \frac{2}{15}$$

Similarly the probability of a white ball being drawn from B

$$= \frac{1}{3} \times \frac{3}{7} = \frac{1}{7}$$

And the probability of a white ball being drawn from C

$$= \frac{1}{3} \times \frac{4}{9} = \frac{4}{27}$$

Therefore the total probability of a white ball being drawn

$$= \frac{2}{15} + \frac{1}{7} + \frac{4}{27} = \frac{401}{945}$$

If a be the number of ways in which an event can happen, and b the number of ways in which it can fail; then the

311 Probability of the event happening $= \dfrac{a}{a+b}$

312 Probability of the event failing $= \dfrac{b}{a+b}$

Thus Certainty $= 1$.

If p, p' be the respective probabilities of two *independent* events; then

313 Probability of both happening $= pp'$.

314 ,, of not *both* happening $= 1-pp'$.

315 ,, of one happening and one failing

$$= p+p'-2pp'.$$

316 ,, of both failing $= (1-p)(1-p')$.

If the probability of an event happening in one trial be p, and the probability of its failing, q; then

317 Probability of the event happening r times in n trials

$$= C(n, r)\, p^r q^{n-r}.$$

318 Probability of the event failing r times in n trials

$$= C(n, r)\, p^{n-r} q^r.$$ [By induction.

---

319 Probability of the event happening *at least* r times in n trials = the sum of the *first* $n-r+1$ terms in the expansion of $(p+q)^n$.

320 Probability of the event failing *at least* r times in n trials = the sum of the *last* $n-r+1$ terms in the same expansion.

---

321 The number of trials in which the probability of the same event happening amounts to p'

$$= \frac{\log(1-p')}{\log(1-p)}$$

From the equation $(1-p)^x = 1-p'$.

---

322 Definition.—When a sum of money is to be received if a certain event happens, that sum multiplied into the probability of the event is termed the expectation.

Example.—If three coins be taken at random from a bag containing one sovereign, four half-crowns, and five shillings, the expectation will be the sum of the expectations founded upon each way of drawing three coins. But this is also equal to the average value of three coins out of the ten; that is, $\frac{3}{10}$ths of 35 shillings, or 10*s.* 6*d.*

---

323 The Probability that, after r chance selections of the numbers 0, 1, 2, 3 ... n, the sum of the numbers drawn will be s, is equal to the coefficient of x^s in the expansion of

$$(x^0+x^1+x^2+\ldots+x^n)^r \div (n+1)^r.$$

324 The probability of the existence of a certain cause of an observed event out of several known causes, one of which *must* have produced the event, is proportional to the *a priori* probability of the cause existing multiplied by the probability of the event happening from it if it does exist.

Thus, if the *a priori* probabilities of the causes be P_1, P_2, ... &c., and the corresponding probabilities of the event happening from those causes Q_1, Q_2 ... &c.; then, the probability of the rth cause having produced the event is

$$\frac{P_r Q_r}{\Sigma\,(PQ)}$$

---

325 If P'_1, P'_2 ... &c. be the *a priori* probabilities of a second event happening from the same causes respectively; then, *after* the first event has happened, the probability of the second happening is
$$\frac{\Sigma\,(PQP')}{\Sigma\,(PQ)}$$

For this is the sum of such probabilities as $\dfrac{P_r Q_r P'_r}{\Sigma\,(PQ)}$, which is the probability of the rth cause existing multiplied by the probability of the second event happening from it.

Ex. 1.—Suppose there are

4 vases containing each 5 white and 6 black balls,
2 vases containing each 3 white and 5 black balls,
and 1 vase containing 2 white and 1 black ball.

A white ball has been drawn, and the probability that it came from the group of 2 vases is required.

Here
$$P_1 = \frac{4}{7},\quad P_2 = \frac{2}{7},\quad P_3 = \frac{1}{7};$$
$$Q_1 = \frac{5}{11},\quad Q_2 = \frac{3}{8},\quad Q_3 = \frac{2}{3}.$$

Therefore, by (324), the probability required is

$$P = \frac{\dfrac{2.3}{7.8}}{\dfrac{4.5}{7.11} + \dfrac{2.3}{7.8} + \dfrac{1.2}{7.3}} = \frac{99}{427}$$

Ex. 2.—After the white ball has been drawn and replaced, a ball is drawn again; required the probability of the ball being black.

Here $$P_1' = \frac{6}{11}, \quad P_2' = \frac{5}{8}, \quad P_3' = \frac{1}{3}.$$

The probability, by (325), will be

$$\frac{\frac{4.5.6}{7.11.11} + \frac{2.3.5}{7.8.8} + \frac{1.2.1}{7.3.3}}{\frac{4.5}{7.11} + \frac{2.3}{7.8} + \frac{1.2}{7.3}} = \frac{58639}{112728}$$

If the probability of the second ball being white is required, $P_1 P_2 P_3$ must be employed instead of $Q_1 Q_2 Q_3$.

326 The probability of one event *at least* happening out of a number of events whose respective probabilities are a, b, c, &c., is

$$\boldsymbol{P_1 - P_2 + P_3 - P_4 + \&c.}$$

where P_1 is the probability of 1 event happening,
P_2 " " 2 "

and so on. For, by (316), the probability is

$$1-(1-a)(1-b)(1-c) \ldots = \Sigma a - \Sigma ab + \Sigma abc -, \ldots$$

327 The probability of the occurrence of r assigned events and no more out of n events is

$$\boldsymbol{Q_r - Q_{r+1} + Q_{r+2} - Q_{r+3} + \&c.},$$

where Q_r is the probability of the r assigned events; Q_{r+1} the probability of $r+1$ events including the r assigned events.

For if a, b, c ... be the probabilities of the r events, and a', b', c' ... the probabilities of the excluded events, the required probability will be

$$abc \ldots (1-a')(1-b')(1-c') \ldots$$
$$= abc \ldots (1 - \Sigma a' + \Sigma a'b' - \Sigma a'b'c' + \ldots)$$

328 The probability of *any* r events happening and no more is

$$\boldsymbol{\Sigma Q_r - \Sigma Q_{r+1} + \Sigma Q_{r+2} - \&c.}$$

NOTE.—If $a = b = c =$ &c.; then $\Sigma Q_r = C(n, r)\, Q_r$, &c.

INEQUALITIES.

330 $\dfrac{a_1+a_2+\ldots+a_n}{b_1+b_2+\ldots+b_n}$ lies between the greatest and least of the fractions $\dfrac{a_1}{b_1}, \dfrac{a_2}{b_2}, \ldots \dfrac{a_n}{b_n}$, the denominators being all of the same sign.

PROOF.—Let k be the greatest of the fractions, and $\dfrac{a_r}{b_r}$ any other; then $a_r < kb_r$. Substitute in this way for each a. Similarly if k be the least fraction.

---

331 $$\frac{a+b}{2} > \sqrt{ab}.$$

---

332 $$\frac{a_1+a_2+\ldots+a_n}{n} > \sqrt[n]{a_1 a_2 \ldots a_n};$$

or, Arithmetic mean > Geometric mean.

PROOF.—Substitute both for the greatest and least factors their Arithmetic mean. The product is thus *increased* in value. Repeat the process indefinitely. The limiting value of the G. M. is the A. M. of the quantities.

---

333 $$\frac{a^m+b^m}{2} > \left(\frac{a+b}{2}\right)^m$$

excepting when m is a positive proper fraction.

PROOF: $a^m+b^m = \left(\dfrac{a+b}{2}\right)^m \{(1+x)^m+(1-x)^m\}$

where $x = \dfrac{a-b}{a+b}$. Employ Bin. Th.

---

334 $$\frac{a_1^m+a_2^m+\ldots+a_n^m}{n} > \left(\frac{a_1+a_2+\ldots+a_n}{n}\right)^m$$

excepting when m is a positive proper fraction.

Otherwise.—*The Arithmetic mean of the mth powers is greater than the mth power of the Arithmetic mean, excepting when m is a positive proper fraction.*

Proof.—Similar to (332). Substitute for the greatest and least on the left side, employing (333).

---

336 If x and m are positive, and x and mx less than unity; then

$$(1+x)^{-m} > 1-mx. \qquad (125, 240).$$

---

337 If x, m, and n are positive, and n greater than m; then, by taking x small enough, we can make

$$1+nx > (1+x)^m.$$

For x may be diminished until $1+nx$ is $> (1-mx)^{-1}$, and this is $> (1+x)^m$, by *last*.

---

338 If x be positive, $\log(1+x) < x$ (150).

If x be positive and > 1, $\log(1+x) > x - \frac{x^2}{2}$ (155, 240).

If x be positive and < 1, $\log \frac{1}{1-x} > x$ (156).

---

339 When n becomes infinite in the two expressions

$$\frac{1.3.5\ldots(2n-1)}{2.4.6\ldots 2n} \text{ and } \frac{3.5.7\ldots(2n+1)}{2.4.6\ldots 2n}$$

the first vanishes, the second becomes infinite, and their product lies between $\frac{1}{2}$ and 1.

Shewn by adding 1 to each factor (see 73), and multiplying the result by the original fraction.

---

340 If m be $> n$, and $n > a$,

$$\left(\frac{m+a}{m-a}\right)^m \text{ is } < \left(\frac{n+a}{n-a}\right)^n$$

341 If a, b be positive quantities,

$$a^a b^b \text{ is } > \left(\frac{a+b}{2}\right)^{a+b}$$

Similarly $$a^a b^b c^c > \left(\frac{a+b+c}{3}\right)^{a+b+c}$$

These and similar theorems may be proved by taking logarithms of each side, and employing the Expon. Th. (158), &c.

SCALES OF NOTATION.

342 If N be a whole number of $n+1$ digits, and r the radix of the scale, $N = p_n r^n + p_{n-1} r^{n-1} + p_{n-2} r^{n-2} + \ldots + p_1 r + p_0$, where p_n, p_{n-1}, ... p_0 are the digits.

343 Similarly a radix-fraction will be expressed by

$$\frac{p_1}{r} + \frac{p_2}{r^2} + \frac{p_3}{r^3} + \&c.$$

where p_1, p_2, &c. are the digits.

EXAMPLES: 3426 in the scale of 7 $= 3.7^3 + 4.7^2 + 2.7 + 6$

·1045 in the same scale $= \frac{1}{7} + \frac{0}{7^2} + \frac{4}{7^3} + \frac{5}{7^4}$

344 Ex.—To transform 34268 from the scale of 5 to the scale of 11.

RULE.—*Divide successively by the new radix.*

```
11 | 34268
    11 | 1343 — t
        11 | 40 — 3
              1 — 9
```

Result $193t$, in which t stands for 10.

345 Ex.—To transform $\cdot t0e1$ from the scale of 12 to that of 7, e standing for 11, and t for 10.

Rule.—*Multiply successively by the new radix.*

$$\begin{array}{r} \cdot t0e1 \\ 7 \\ \hline 5\cdot t657 \\ 7 \\ \hline 6\cdot 1931 \\ 7 \\ \hline 1\cdot 0497 \\ 7 \\ \hline 0\cdot 2971 \end{array} \qquad \text{Result} \quad \cdot 5610\ldots$$

346 Ex.—In what scale does $2t7$ represent the number 475 in the scale of ten?

Solve the equation $\quad 2r^2+10r+7 = 475.$ [178.

Result $\quad r = 13.$

347 The sum of the digits of any number divided by $r-1$ leaves the same remainder as the number itself divided by $r-1$; r being the radix of the scale. (401).

348 The difference between the sums of the digits in the even and odd places divided by $r+1$ leaves the same remainder as the number itself when divided by $r+1$.

THEORY OF NUMBERS.

349 If a is prime to b, $\frac{a}{b}$ is in its lowest terms.

Proof.—Let $\frac{a}{b} = \frac{a_1}{b_1}$, a fraction in lower terms.

Divide a by a_1, remainder a_2 quotient q_1,

b by b_1, remainder b_2 quotient q_1;

and so on, as in finding the H. C. F. of a and a_1, and of b and b_1 (see 30). Let a_n and b_n be the highest common factors thus determined.

Then, because $\frac{a}{b} = \frac{a_1}{b_1}$ $\therefore$ $\frac{a}{b} = \frac{a - q_1 a_1}{b - q_1 b_1} = \frac{a_2}{b_2}$ (70);

and so on; thus $\frac{a}{b} = \frac{a_1}{b_1} = \frac{a_2}{b_2} = \text{\&c.} \ldots\ldots = \frac{a_n}{b_n}$

Therefore a and b are equimultiples of a_n and b_n; that is, a is not prime to b if any fraction exists in lower terms.

---

350 If a is prime to b, and $\frac{a'}{b'} = \frac{a}{b}$; then a' and b' are equimultiples of a and b.

PROOF.—Let $\frac{a'}{b'}$ reduced to its lowest terms be $\frac{p}{q}$. Then $\frac{p}{q} = \frac{a}{b}$, and since p is now prime to q, and a prime to b; it follows, by 349, that $\frac{p}{q}$ is neither greater nor less than $\frac{a}{b}$; that is, it is equal to it. Therefore, &c.

---

351 If ab is divisible by c, and a is not; then b must be.

PROOF.—Let $\frac{ab}{c} = q$ $\therefore$ $\frac{a}{c} = \frac{q}{b}$

But a is prime to c; therefore, by last, b is a multiple of c.

352 If a and b be each of them prime to c, ab is prime to c. By last.

---

353 If $abcd \ldots$ is divisible by a prime, one at least of the factors a, b, c, &c. must also be divisible by it.

Or, if p be prime to all but one of the factors, that factor is divisible by p. (351).

354 Therefore, if a^n is divisible by p, p cannot be prime to a; and if p be a prime it must divide a.

---

355 If a is prime to b, any power of a is prime to any power of b.

Also, if a, b, c, &c. are prime to each other, the product of any of their powers is prime to any other product of their powers.

356 No expression with integral coefficients such as $A+Bx+Cx^2+\dots$ can represent primes only.

Proof.—For it is divisible by x if $A=0$; and if not, it is divisible by A, when $x=A$.

---

357 The number of primes is infinite.

Proof.—Suppose p the greatest prime. Then the product of all primes up to p plus unity is either a prime, or divisible by a prime greater than p.

---

358 If a be prime to b, and the quantities $a, 2a, 3a, \dots (b-1)a$ be divided by b, the remainders will be different.

Proof.—Assume $ma-nb = m'a-n'b$, m and n being less than b, $\therefore \quad \frac{a}{b} = \frac{n-n'}{m-m'}$ Then by (350).

---

359 A number can be resolved into prime factors in one way only. By (353).

360 To resolve 5040 into its prime factors.

Rule.—*Divide by the prime numbers successively.*

$$\begin{array}{r|l} 2\times 5 & 5040 \\ 2 & 504 \\ 2 & 252 \\ 2 & 126 \\ 7 & 63 \\ 3 & 9 \\ & 3 \end{array}$$

Thus $5040 = 2^4 . 3^2 . 5 . 7$.

---

361 Required the least multiplier of 4704 which will make the product a perfect fourth power.

By (196), $4704 = 2^5 . 3 . 7^2$.

Then $2^5 . 3^1 . 7^2 \times 2^3 . 3^3 . 7^2 = 2^8 . 3^4 . 7^4 = 84^4$.

The indices 8, 4, 4 being the least multiples of 4 which are not less than 5, 1, 2 respectively.

Thus $2^3 . 3^3 . 7^2 = 10584$ is the multiplier required.

362 All numbers are of one of the forms $2n$ or $2n+1$

,,	,,	$2n$ or $2n-1$
,,	,,	$3n$ or $3n\pm1$
,,	,,	$4n$ or $4n\pm1$ or $4n+2$
,,	,,	$4n$ or $4n\pm1$ or $4n-2$
,,	,,	$5n$ or $5n\pm1$ or $5n\pm2$

and so on.

363 All square numbers are of the form $5n$ or $5n\pm1$.

Proved by squaring the forms $5n$, $5n\pm1$, $5n\pm2$, which comprehend all numbers whatever.

364 All cube numbers are of the form $7n$ or $7n\pm1$.
And similarly for other powers.

---

365 The highest power of a prime p, which is contained in the product $\underline{|m}$, is the sum of the integral parts of

$$\frac{m}{p},\ \frac{m}{p^2},\ \frac{m}{p^3},\ \&c.$$

For there are $\frac{m}{p}$ factors in $\underline{|m}$ which p will divide; $\frac{m}{p^2}$ which it will divide a second time; and so on. The successive divisions are equivalent to dividing by

$$p^{\frac{m}{p}}\,.\,p^{\frac{m}{p^2}}\ \ldots\ \&c. = p^{\frac{m}{p}+\frac{m}{p^2}+}$$

Example.—The highest power of 3 which will divide $\underline{|29}$. Here the factors 3, 6, 9, 12, 15, 18, 21, 24, 27 can be divided by 3. Their number is $\frac{29}{3}=9$ (the integral part).

The factors 9, 18, 29 can be divided a second time. Their number is $\frac{29}{3^2}=3$ (the integral part).

One factor, 27, is divisible a third time. $\frac{29}{3^3}=1$ (integral part).

$9+3+1=13$; that is, 3^{13} is the highest power of 3 which will divide $\underline{|29}$.

---

366 The product of any r consecutive integers is divisible by $\underline{|r}$.

Proof: $\dfrac{n(n-1)\ldots(n-r+1)}{\underline{|r}}$ is necessarily an integer, by (96).

367 If n be a prime, every coefficient in the expansion of $(a+b)^n$, except the first and last, is divisible by n. By last.

368 If n be a prime the coefficient of every term in the expansion of $(a+b+c\ldots)^n$, except a^n, b^n, &c., is divisible by n.

By last. Put β for $(b+c+\ldots)$.

---

369 *Fermat's Theorem.*—If p be a prime, and N prime to p; then **$N^{p-1}-1$ is divisible by p.**

PROOF: $N^p = (1+1+\ldots)^p = N+Mp$. By last.

370 If p be any number, and if $1, a, b, c, \ldots (p-1)$ be all the numbers less than, and prime to p; and if n be their number, and x any one of them; then x^n-1 is divisible by p.

PROOF.—If $x, ax, bx \ldots (p-1)x$ be divided by p, the remainders will be all different and prime to p (as in 358); therefore the remainders will be $1, a, b, c \ldots (p-1)$; therefore the product $x^n abc \ldots (p-1) = abc \ldots (p-1)+Mp$.

---

371 *Wilson's Theorem.*—If p be a prime, and only then, $1+\underline{|p-1}$ is divisible by p.

Put $p-1$ for r and n in (285), and apply Fermat's Theorem to each term.

372 If p be a prime $=2n+1$, then $(\underline{|n})^2+(-1)^n$ is divisible by p.

PROOF.—By multiplying together equi-distant factors of $\underline{|p-1}$ in Wilson's Theorem, and putting $2n+1$ for p.

---

373 Let $N=a^p b^q c^r \ldots$ in prime factors, the number of integers including 1, which are less than n and prime to it, is

$$N\left(1-\frac{1}{a}\right)\left(1-\frac{1}{b}\right)\left(1-\frac{1}{c}\right)\ldots$$

PROOF.—The number of integers prime to N contained in a^p is $a^p-\dfrac{a^p}{a}$. Similarly in b^q, c^r, &c. Take the product of these.

Also the number of integers less than and prime to $(N \times M \times \&c.)$ is the product of the corresponding numbers for N, M, &c. separately.

374 The number of divisors of N, including 1 and N itself, is $=(p+1)(q+1)(r+1)\,\ldots.$ For it is equal to the number of terms in the product

$$(1+a+\ldots+a^p)(1+b+\ldots+b^q)(1+c+\ldots+c^r)\ldots\ \&c.$$

375 The number of ways of resolving N into two factors is half the number of its divisors (374). If the number be a square the two equal factors must, in this case, be reckoned as two divisors.

376 If the factors of each pair are to be prime to each other, put p, q, r, &c. each equal to one.

377 The sum of the divisors of N is

$$\frac{a^{p+1}-1}{a-1} \cdot \frac{b^{q+1}-1}{b-1} \cdot \frac{c^{r+1}-1}{c-1} \ldots$$

PROOF.—By the product in (374), and by (85).

378 If p be a prime, then the $p-1^{\text{th}}$ power of any number is of the form mp or $mp+1$. By Fermat's Theorem (369).

Ex.—The 12th power of any number is of the form $13m$ or $13m+1$.

379 To find all the divisors of a number; for instance, of 504.

I.	II.						
		1					
504	2	2					
252	2	4					
126	2	8					
63	3	3	6	12	24		
21	3	9	18	36	72		
7	7	7	14	28	56	21	42
		84	168	63	126	252	504

EXPLANATION.—Resolve 504 into its prime factors, placing them in column II.

The divisors of 504 are now formed from the numbers in column II., and placed to the right of that column in the following manner:—

Place the divisor 1 to the right of column II., and follow this rule—*Multiply in order all the divisors which are written down by the next number in column* II., *which has not already been used as a multiplier: place the first new divisor so obtained and all the following products in order to the right of column* II.

---

380 S_r the sum of the rth powers of the first n natural numbers is divisible by $2n+1$.

PROOF: $$x(x^2-1^2)(x^2-2^2)\ldots(x^2-n^2)$$

constitutes $2n+1$ factors divisible by $2n+1$, by (366). Multiply out, rejecting x, which is to be less than $2n+1$. Thus, using (372),

$$x^{2n}-S_1x^{2n-2}+S_2x^{2n-4}-\ldots S_{n-1}x^2+(-1)^n(\lfloor n)^2 = M(2n+1).$$

Put 1, 2, 3 ... $(n-1)$ in succession for x, and the solution of the $(n-1)$ equations is of the form $\quad S_r = M(2n+1)$.

THEORY OF EQUATIONS.

FACTORS OF AN EQUATION.

General form of a rational integral equation of the nth degree.

400 $p_0x^n+p_1x^{n-1}+p_2x^{n-2}+\dots+p_{n-1}x+p_n=0,\ p_r$ ***real.***

The left side will be designated $f(x)$ in the following summary.

---

401 If $f(x)$ be divided by $x-a$, the remainder will be $f(a)$. By assuming $f(x)=P(x-a)+R$.

402 If a be a root of the equation $f(x)=0$, then $f(a)=0$.

---

403 To compute $f(a)$ numerically; *divide $f(x)$ by $x-a$, and the remainder will be $f(a)$.* [401

404 Example.—To find the value of $4x^6-3x^5+12x^4-x^2+10$ when $x=2$.

$$\begin{array}{c|l} & 4-3+12\ +0\ -1\ \ +0\ +10 \\ 2 & \quad\ 8+10+44+88+174+348 \\ \hline & 4+5+22+44+87+174+358 \end{array}$$

Thus $f(2)=358$.

If $a, b, c \dots k$ be the roots of the equation $f(x)=0$; then, by (401) and (402),

405 $$f(x)=p_0(x-a)(x-b)(x-c)\dots(x-k).$$

By multiplying out the last equation, and equating coefficients with equation (400), considering $p_0=1$, the following results are obtained:—

406 $-p_1 =$ the sum of all the roots of $f(x)$.

$p_2 = \left\{\right.$ the sum of the products of the roots taken two at a time.

$-p_3 = \left\{\right.$ the sum of the products of the roots taken three at a time.

... ...

... ...

$(-1)^r p_r = \left\{\right.$ the sum of the products of the roots taken r at a time.

... ...

$(-1)^n p_n =$ product of all the roots.

407 The number of roots of $f(x)$ is equal to the degree of the equation.

---

408 Imaginary roots must occur in pairs of the form

$$\alpha+\beta\sqrt{-1}, \quad \alpha-\beta\sqrt{-1}.$$

The quadratic factor corresponding to these roots will then have real coefficients; for it will be

$$x^2-2\alpha x+\alpha^2+\beta^2. \qquad [405, 226$$

---

409 If $f(x)$ be of an odd degree, it has at least one real root of the opposite sign to p_n.

Thus $x^7-1=0$ has *at least* one positive root.

410 If $f(x)$ be of an even degree, and p_n negative, there is *at least* one positive and one negative root.

Thus x^4-1 has $+1$ and -1 for roots.

---

411 If several terms at the beginning of the equation are of one sign, and all the rest of another, there is one, and only one, positive root.

Thus $x^5+2x^4+3x^3+x^2-5x-4=0$ has only one positive root.

---

412 If all the terms are positive there is no positive root.

413 If all the terms of an even order are of one sign, and all the rest are of another sign, there is no negative root.

414 Thus $x^4-x^3+x^2-x+1=0$ has no negative root.

415 If all the indices are even, and all the terms of the same sign, there is no real root; and if all the indices are odd, and all the terms of the same sign, there is no real root but zero.

Thus $x^4+x^2+1=0$ has no real root, and $x^5+x^3+x=0$ has no real root but zero. In this last equation there is no absolute term, because such a term would involve the zero power of x, which is even, and by hypothesis is wanting.

DESCARTES' RULE OF SIGNS.

416 In the following theorems every two adjacent terms in $f(x)$, which have the same signs, count as one "continuation of sign"; and every two adjacent terms, with different signs, count as one change of sign.

417 $f(x)$, multiplied by $(x-a)$, has an *odd* number of changes of sign thereby introduced, and *one at least*.

418 $f(x)$ cannot have more positive roots than changes of sign, or more negative roots than continuations of sign.

419 When all the roots of $f(x)$ are real, the number of positive roots is equal to the number of changes of sign in $f(x)$; and the number of negative roots is equal to the number of changes of sign in $f(-x)$.

420 Thus, it being known that the roots of the equation

$$x^4-10x^3+35x^2-50x+24=0$$

are all real; the number of positive roots will be equal to the number of changes of sign, which is four. Also $f(-x)=x^4+10x^3+35x^2+50x+24=0$, and since there is no change of sign, there is consequently, by the rule, no negative root.

421 If the degree of $f(x)$ exceeds the number of changes of sign in $f(x)$ and $f(-x)$ together, by μ, there are at least μ imaginary roots.

422 If, between two terms in $f(x)$ of the same sign, there be an odd number of consecutive terms wanting, then there must be at least one more than that number of imaginary roots; and if the missing terms lie between terms of different sign, there is at least one less than the same number of imaginary roots.

Thus, in the cubic equation $x^3+4x-7=0$, there must be two imaginary roots.

And in the equation $x^6-1=0$ there are, for certain, four imaginary roots.

423 If an even number of consecutive terms be wanting in $f(x)$, there is at least the same number of imaginary roots.

Thus the equation $x^5+1=0$ has four terms absent; and therefore four imaginary roots at least.

THE DERIVED FUNCTIONS OF $f(x)$.

Rule for forming the derived functions.

424 *Multiply each term by the index of x, and reduce the index by one; that is, differentiate the function with respect to x.*

Example.—Take

$$\begin{aligned}
f(x) &= x^5+x^4+x^3-x^2-x-1 \\
f^1(x) &= 5x^4+4x^3+3x^2-2x-1 \\
f^2(x) &= 20x^3+12x^2+6x-2 \\
f^3(x) &= 60x^2+24x+6 \\
f^4(x) &= 120x+24 \\
f^5(x) &= 120
\end{aligned}$$

$f^1(x)$, $f^2(x)$, &c. are called the first, second, &c. derived functions of $f(x)$.

425 To form the equation whose roots differ from those of $f(x)$ by a quantity a.

Put $x=y+a$ in $f(x)$, and expand each term by the Binomial Theorem, arranging the results in vertical columns in the following manner:—

$$\begin{aligned}
f(a+y) = \; & (a+y)^5+(a+y)^4+(a+y)^3-(a+y)^2-(a+y)-1 \\
= \; & a^5 + a^4 + a^3 - a^2 - a - 1 \\
& +(5a^4 + 4a^3 + 3a^2 - 2a - 1)\,y \\
& +(10a^3 + 6a^2 + 3a - 1)\,y^2 \\
& +(10a^2 + 4a + 1)\,y^3 \\
& +(5a + 1)\,y^4 \\
& +y^5
\end{aligned}$$

Comparing this result with that seen in (424), it is seen that

426 $$f(a+y) = f(a) + f^1(a)\,y + \frac{f^2(a)}{\underline{|2}}\,y^2 + \frac{f^3(a)}{\underline{|3}}\,y^3 + \frac{f^4(a)}{\underline{|4}}\,y^4 + \frac{f^5(a)}{\underline{|5}}\,y^5$$

so that the coefficient generally of y^r in the transformed equation is $\dfrac{f^r(a)}{\underline{|r}}$.

---

427 To form the equation most expeditiously when a has a numerical value, *divide $f(x)$ continuously by $x-a$, and the successive remainders will furnish the coefficients.*

EXAMPLE.—To expand $f(y+2)$ when, as in (425),

$$f(x) = x^5 + x^4 + x^3 - x^2 - x - 1.$$

Divide repeatedly by $x-2$, as follows :—

	1	+ 1	+ 1	− 1	− 1	− 1	
2		+ 2	+ 6	+ 14	+ 26	+ 50	
	1	+ 3	+ 7	+ 13	+ 25	+ 49	$= f(2)$
2		+ 2	+ 10	+ 34	+ 94		
	1	+ 5	+ 17	+ 47	+ 119		$= f^1(2)$
2		+ 2	+ 14	+ 62			
	1	+ 7	+ 31	+ 109			$= \dfrac{f^2(2)}{\underline{|2}}$
2		+ 2	+ 18				
	1	+ 9	+ 49				$= \dfrac{f^3(2)}{\underline{|3}}$
2		+ 2					
	1	+ 11					$= \dfrac{f^4(2)}{\underline{|4}}$
	1						$= \dfrac{f^5(2)}{\underline{|5}}$

That these remainders are the required coefficients is seen by inspecting the form of the equation (426); for if that equation be divided by $x-a = y$ repeatedly, these remainders are obviously produced when $a = 2$.

Thus the equation, whose roots are each less by 2 than the roots of the proposed equation, is $y^5 + 11y^4 + 49y^3 + 109y^2 + 119y + 49 = 0$.

---

428 To make any assigned term vanish in the transformed equation, a must be so determined that the coefficient of that term shall vanish.

EXAMPLE.—In order that there may be no term involving y^4 in equation (426), we must have $f^4(a) = 0$.

Find $f^4(a)$ as in (424);

thus $$120a + 24 = 0; \quad \therefore \quad a = -\tfrac{1}{5}.$$

The equation in (424) must now be divided repeatedly by $x+\frac{1}{5}$ after the manner of (427), and the resulting equation will be minus its second term.

429 Note, that to remove the second term of the equation $f(x)=0$, the requisite value of a is $=-\frac{p_1}{np_0}$; that is, *the coefficient of the second term, with the sign changed, divided by the coefficient of the first term, and by the number expressing the degree of the equation.*

430 To transform $f(x)$ into an equation in y so that $y=\phi(x)$, a given function of x, *put* $x=\phi^{-1}(y)$, *the inverse function of* y.

EXAMPLE.—To obtain an equation whose roots are respectively three times the roots of the equation $x^3-6x+1=0$. Here $y=3x$; therefore $x=\frac{y}{3}$, and the equation becomes $\frac{y^3}{27}-\frac{6y}{3}+1=0$, or $y^3-54y+27=0$.

431 To transform $f(x)=0$ into an equation in which the coefficient of the first term shall be unity, and the other coefficients the least possible integers.

EXAMPLE.—Take the equation

$$288x^3+240x^2-176x-21=0.$$

Divide by the coefficient of the first term, and reduce the fractions; the equation becomes

$$x^3+\frac{5}{6}x^2-\frac{11}{18}x-\frac{7}{96}=0.$$

Substitute $\frac{y}{k}$ for x, and multiply by k^3; we get

$$y^3+\frac{5k}{6}y^2-\frac{11k^2}{18}y-\frac{7k^3}{96}=0.$$

Next resolve the denominators into their prime factors,

$$y^3+\frac{5k}{2.3}y^2-\frac{11k^2}{2.3^2}y-\frac{7k^3}{2^5.3}=0.$$

The smallest value must now be assigned to k, which will suffice to make each coefficient an integer. This is easily seen by inspection to be $2^2.3=12$, and the resulting equation is $y^3+10y^2-88y-126=0$, the roots of which are connected with the roots of the original equation by the relation

$$y=12x.$$

EQUAL ROOTS OF AN EQUATION.

By expanding $f(x+z)$ in powers of z by (405), and also by (426), and equating the coefficients of z in the two expansions, it is proved that

432 $$f'(x) = \frac{f(x)}{(x-a)} + \frac{f(x)}{(x-b)} + \frac{f(x)}{(x-c)} + \&c.,$$

from which result it appears that, if the roots a, b, c, &c. are all unequal, $f(x)$ and $f'(x)$ can have no common measure involving x. If, however, there are r roots each equal to a, s roots equal to b, t roots equal to c, &c., so that

$$f(x) = p_0(x-a)^r(x-b)^s(x-c)^t\ldots$$

then

433 $$f'(x) = \frac{rf(x)}{x-a} + \frac{sf(x)}{x-b} + \frac{tf(x)}{(x-b)} + \&c.;$$

and the greatest common measure of $f(x)$ and $f'(x)$ will be

444 $$(x-a)^{r-1}(x-b)^{s-1}(x-c)^{t-1}\ldots$$

When $x=a$, $f(x)$, $f'(x)$, ... $f^{r-1}(x)$ all vanish. Similarly when $x=b$, &c.

Practical method of finding the equal roots.

445 Let $f(x) = X_1 X_2^2 X_3^3 X_4^4 X_5^5 \ldots X_m^m$, where

$X_1 \equiv$ product of all the factors like $(x-a)$,
$X_2^2 \equiv$ " " $(x-a)^2$,
$X_3^3 \equiv$ " " $(x-a)^3$.

Find the greatest common measure of $f(x)$ and $f'(x) = F_1(x)$ say,
" " $F_1(x)$ and $F_1'(x) = F_2(x)$,
" " $F_2(x)$ and $F_2'(x) = F_3(x)$,
... ...
... ...

Lastly, the greatest common measure of $F_{m-1}(x)$ and $F'_{m-1}(x) = F_m(x) = 1$.

Next perform the divisions

$$f(x) \div F_1(x) = \phi_1(x) \text{ say},$$
$$F_1(x) \div F_2(x) = \phi_2(x),$$
... ...
... ...
$$F_{m-1}(x) \div 1 = \phi_m(x).$$

And, finally,

$$\phi_1(x) \div \phi_2(x) = X_1,$$
$$\phi_2(x) \div \phi_3(x) = X_2,$$
... ...
... ...
$$\phi_{m-1}(x) \div \phi_m(x) = X_{m-1},$$
$$F_{m-1}(x) = \phi_m(x) = X_m.$$

[T. 82.

The solution of the equations $X_1 = 0$, $X_2 = 0$, &c. will furnish all the roots of $f(x)$; those which occur twice being found from X_2; those which occur three times each, from X_3; and so on.

446 If $f(x)$ has all its coefficients commensurable, X_1, X_2, X_3, &c. have likewise their coefficients commensurable.

Hence, if only one root be repeated r times, that root must be commensurable.

447 In all the following theorems, unless otherwise stated, $f(x)$ is understood to have unity for the coefficient of its first term.

LIMITS OF THE ROOTS.

448 If the greatest negative coefficients in $f(x)$ and $f(-x)$ be p and q respectively, then $p+1$ and $-(q+1)$ are limits of the roots.

449 If x^{n-r} and x^{n-s} are the highest negative terms in $f(x)$ and $f(-x)$ respectively, $(1+\sqrt[r]{p})$ and $-(1+\sqrt[s]{q})$ are limits of the roots.

450 If k be a superior limit to the positive roots of $f\left(\frac{1}{x}\right)$, then $\frac{1}{k}$ will be an inferior limit to the positive roots of $f(x)$.

451 If each negative coefficient be divided by the sum of all the preceding positive coefficients, the greatest of the fractions so formed + unity will be a superior limit to the positive roots.

452 *Newton's method.*—Put $x = h+y$ in $f(x)$; then, by (426),

$$f(h+y) = f(h) + yf'(h) + \frac{y^2}{\underline{|2}} f^2(h) + \ldots + \frac{y^n}{\underline{|n}} f^n(h) = 0.$$

Take h so that $f(h)$, $f'(h)$, $f^2(h) \ldots f^n(h)$ are all positive; then h is a superior limit to the positive roots.

453 According as $f(a)$ and $f(b)$ have the same or different signs, the number of roots intermediate between a and b is even or odd.

454 *Rolle's Theorem.*—One real root of the equation $f'(x)$ lies between every two adjacent real roots of $f(x)$.

455 Cor. 1.—$f(x)$ cannot have more than one root greater than the greatest root in $f'(x)$; or more than one less than the least root in $f'(x)$.

456 Cor. 2.—If $f(x)$ has m real roots, $f^r(x)$ has at least $m-r$ real roots.

457 Cor. 3.—If $f^r(x)$ has μ imaginary roots, $f(x)$ has also μ at least.

458 Cor. 4.—If $\alpha, \beta, \gamma \ldots \kappa$ be the roots of $f'(x)$; then the number of changes of sign in the series of terms

$$f(\infty),\ f(\alpha),\ f(\beta),\ f(\gamma) \ldots f(-\infty)$$

is equal to the number of roots of $f(x)$.

NEWTON'S METHOD OF DIVISORS.

459 To discover the integral roots of an equation.

Example.—To ascertain if 5 be a root of

$$x^4-16x^3+86x^2-176x+105 = 0.$$

If 5 be a root it will divide 105. Add the quotient to the next coefficient. Result, -155.

If 5 be a root it will divide -155. Add the quotient to the next coefficient; and so on.

If the number tried be a root, the divisions will be effectible to the end, and the last quotient will be -1, or $-p_0$, if p_0 be not unity.

$$\begin{array}{r} 5\,)\,\underline{105} \\ 21 \\ \underline{-176} \\ 5\,)\,\underline{-155} \\ -31 \\ \underline{86} \\ 5\,)\,\underline{55} \\ 11 \\ \underline{-16} \\ 5\,)\,\underline{-5} \\ -1 \end{array}$$

460 In employing this method, limits of the roots may first be found, and divisors chosen between those limits.

461 Also, to lessen the number of trial divisors, take any integer m; then any divisor a of the last term can be rejected if $a-m$ does not divide $f(m)$.

In practice take $m = +1$ and -1.

To find whether any of the roots determined as above are repeated, divide $f(x)$ by the factors corresponding to them, and then apply the method of divisors to the resulting equation.

EXAMPLE.—Take the equation

$$x^6+2x^5-17x^4-26x^3+88x^2+72x-144 = 0.$$

Putting $x=1$, we find $f(1) = -24$. The divisors of 144 are

1, 2, 3, 4, 6, 8, 9, 12, 16, 24, &c.

The values of $a-m$ (since $m=1$) are therefore

0, 1, 2, 3, 5, 7, 8, 11, 15, 23, &c.

Of these last numbers only 1, 2, 3, and 8 will divide 24. Hence 2, 3, 4, and 9 are the only divisors of 144 which it is of use to try. The only integral roots of the equation will be found to be ± 2 and ± 3.

---

462 If $f(x)$ and $F(X)$ have common roots, they are contained in the greatest common measure of $f(x)$ and $F(X)$.

463 If $f(x)$ has for its roots a, $\phi(a)$, b, $\phi(b)$ amongst others; then the equations $f(x)=0$ and $f\{\phi(x)\}=0$ have the common roots a and b.

---

464 But, if all the roots occur in pairs in this way, these equations coincide.

For example, suppose that each pair of roots, a and b, satisfies the equation $a+b=2r$. We may then assume $a-b=2z$. Therefore $f(z+r)=0$. This equation involves only even powers of z, and may be solved for z^2.

---

465 Otherwise: Let $ab=z$; then $f(x)$ is divisible by $(x-a)(x-b) = x^2-2rx+z$. Perform the division until a remainder is obtained of the form $Px+Q$, where P and Q only involve z.

The equations $P=0$, $Q=0$ determine z, by (462); and a and b are found from $a+b=2r$, $ab=z$.

RECIPROCAL EQUATIONS.

---

466 A reciprocal equation has its roots in pairs of the form a, $\frac{1}{a}$; also the relation between the coefficients is

$$p_r = p_{n-r}, \text{ or else } p_r = -p_{n-r}.$$

---

467 A reciprocal equation *of an even degree, with its last term positive*, may be made to depend upon the solution of an equation of half the same degree.

468 EXAMPLE: $4x^6 - 24x^5 + 57x^4 - 73x^3 + 57x^2 - 24x + 4 = 0$ is a reciprocal equation of an even degree, with its last term positive.

Any reciprocal equation which is not of this form may be reduced to it *by dividing by* $x+1$ *if the last term be positive; and, if the last term be negative, by dividing by* $x-1$ *or* x^2-1, *so as to bring the equation to an even degree.* Then proceed in the following manner:—

469 First bring together equidistant terms, and divide the equation by x^3; thus

$$4\left(x^3 + \frac{1}{x^3}\right) - 24\left(x^2 + \frac{1}{x^2}\right) + 57\left(x + \frac{1}{x}\right) - 73 = 0.$$

By putting $x + \frac{1}{x} = y$, and by making repeated use of the relation $x^2 + \frac{1}{x^2} = \left(x + \frac{1}{x}\right)^2 - 2$, the equation is reduced to a cubic in y, the degree being one-half that of the original equation.

Put p for $x + \frac{1}{x}$, and p_m for $x^m + \frac{1}{x^m}$.

470 The relation between the successive factors of the form p_m may be expressed by the equation

$$p_m = pp_{m-1} - p_{m-2}.$$

471 The equation for p_m, in terms of p, is

$$p_m = p^m - mp^{m-2} + \frac{m(m-3)}{1.2} p^{m-4} - \dots$$
$$+ (-1)^r \frac{m(m-r-1)\dots(m-2r+1)}{\lfloor r} p^{m-2r} + \dots$$

By (545), putting $q = 1$.

BINOMIAL EQUATIONS.

472 If a be a root of $x^n - 1 = 0$, then a^m is likewise a root where m is any positive or negative integer.

473 If a be a root of $x^n + 1 = 0$, then a^{2m+1} is likewise a root.

474 If m and n be prime to each other, x^m-1 and x^n-1 have no common root but unity.

Take $pm-qn=1$ for an indirect proof.

475 If n be a prime number, and if a be a root of $x^n-1=0$, the other roots are a, a^2, $a^3 \dots a^n$.

These are all roots, by (472). Prove, by (474), that no two can be equal.

476 If n be not a prime number, other roots besides these may exist. The successive powers, however, of some root will furnish all the rest.

477 If $x^n-1=0$ has the index $n=mpq$; m, p, q being prime factors; then the roots are the terms of the product

$$(1+\alpha+\alpha^2+\dots+\alpha^{m-1})(1+\beta+\beta^2+\dots+\beta^{p-1})\times(1+\gamma+\gamma^2+\dots+\gamma^{q-1}),$$

where α is a root of x^m-1,
β ,, x^p-1,
γ ,, x^q-1,

but neither α, β, nor $\gamma=1$. Proof as in (475).

478 If $n=m^3$, and
α be a root of $x^m-1=0$,
β ,, $x^m-\alpha=0$,
γ ,, $x^m-\beta=0$;

then the roots of $x^n-1=0$ will be the terms of the product

$$(1+\alpha+\alpha^2+\dots+\alpha^{m-1})(1+\beta+\beta^2+\dots+\beta^{m-1})\times(1+\gamma+\gamma^2+\dots+\gamma^{m-1}).$$

479 $x^n\pm 1=0$ may be treated as a reciprocal equation, and depressed in degree after the manner of (468).

480 The complete solution of the equation

$$x^n-1=0$$

is obtained by De Moivre's Theorem. (757)

The n different roots are given by the formula

$$x=\cos\frac{2r\pi}{n}\pm\sqrt{-1}\,\sin\frac{2r\pi}{n}$$

in which r must have the successive values 0, 1, 2, 3, &c., concluding with $\frac{n}{2}$, if n be even; and with $\frac{n-1}{2}$, if n be odd.

481 Similarly the n roots of the equation

$$x^n + 1 = 0$$

are given by the formula

$$x = \cos\frac{(2r+1)\pi}{n} \pm \sqrt{-1}\sin\frac{(2r+1)\pi}{n}$$

r taking the successive values 0, 1, 2, 3, &c., up to $\frac{n-2}{2}$, if n be even; and up to $\frac{n-1}{2}$, if n be odd.

482 The number of different values of the product

$$A^{\frac{1}{m}} B^{\frac{1}{m}}$$

is equal to the least common multiple of m and n, when m and n are integers.

CUBIC EQUATIONS.

483 To solve the general cubic equation

$$x^3 + px^2 + qx + r = 0.$$

Remove the term px^2 by the method of (429). Let the transformed equation be $\quad x^3 + qx + r = 0.$

484 *Cardan's method.*—The complete theoretical solution of this equation by Cardan's method is as follows:—

Put $\quad x = y + z \quad$ (i.)

$$y^3 + z^3 + (3yz + q)(y + z) + r = 0.$$

Put $\quad 3yz + q = 0; \quad \therefore \quad y = -\frac{q}{3z}.$

Substitute this value of y, and solve the resulting quadratic in y^3. The roots are equal to y^3 and z^3 respectively; and we have, by (i.),

485 $$x = \left\{-\frac{r}{2} + \sqrt{\frac{r^2}{4} + \frac{q^3}{27}}\right\}^{\frac{1}{3}} + \left\{-\frac{r}{2} - \sqrt{\frac{r^2}{4} + \frac{q^3}{27}}\right\}^{\frac{1}{3}}.$$

The cubic must have one real root at least, by (409).

Let m be one of the three values of $\left\{-\frac{r}{2}+\sqrt{\frac{r^2}{4}+\frac{q^3}{27}}\right\}^{\frac{1}{3}}$, and n one of the three values of $\left\{-\frac{r}{2}-\sqrt{\frac{r^2}{4}+\frac{q^3}{27}}\right\}^{\frac{1}{3}}$.

486 Let 1, α, α^2 be the three cube roots of unity, so that

$$\alpha=-\frac{1}{2}+\frac{1}{2}\sqrt{-3}, \text{ and } \alpha^2=-\frac{1}{2}-\frac{1}{2}\sqrt{-3}.$$ [472

487 Then, since $\sqrt[3]{m^3}=m\sqrt[3]{1}$, the roots of the cubic will be

$$m+n, \quad \alpha m+\alpha^2 n, \quad \alpha^2 m+\alpha n.$$

Now, if in the expansion of

$$\left\{-\frac{r}{2}\pm\sqrt{\frac{r^2}{4}+\frac{q^3}{27}}\right\}^{\frac{1}{3}}$$

by the Binomial Theorem, we put

μ = the sum of the odd terms, and

ν = the sum of the even terms;

then we shall have $m=\mu+\nu$, and $n=\mu-\nu$;

or else $m=\mu+\nu\sqrt{-1}$, and $n=\mu-\nu\sqrt{-1}$;

according as $\sqrt{\frac{r^2}{4}+\frac{q^3}{27}}$ is real or imaginary.

By substituting these expressions for m and n in (487), it appears that—

488 (i.) If $\frac{r^2}{4}+\frac{q^3}{27}$ be positive, the roots of the cubic will be

$$2\mu, \quad -\mu+\nu\sqrt{-3}, \quad -\mu-\nu\sqrt{-3}.$$

(ii.) If $\frac{r^2}{4}+\frac{q^3}{27}$ be negative, the roots will be

$$2\mu, \quad -\mu+\nu\sqrt{3}, \quad -\mu-\nu\sqrt{3}.$$

(iii.) If $\frac{r^2}{4}+\frac{q^3}{27}=0$, the roots are

$$2m, \quad -m, \quad -m;$$

since m is now equal to μ.

---

489 *The Trigonometrical method.*—The equation

$$x^3+qx+r=0$$

may be solved in the following manner, by Trigonometry, when $\frac{r^2}{4}+\frac{q^3}{27}$ is negative.

Assume $x=n\cos\alpha$. Divide the equation by n^3; thus

$$\cos^3\alpha+\frac{q}{n^2}\cos\alpha+\frac{r}{n^3}=0.$$

But $$\cos^3\alpha-\frac{3}{4}\cos\alpha-\frac{\cos 3\alpha}{4}=0.$$ By (657)

Equate coefficients in the two equations; the result is

$$n = \left(-\frac{4q}{3}\right)^{\frac{1}{2}}, \qquad \cos 3a = -4r\left(-\frac{3}{4q}\right)^{\frac{3}{2}},$$

a must now be found with the aid of the Trigonometrical tables.

490 The roots of the cubic will be

$$n \cos a, \quad n \cos\left(\tfrac{2}{3}\pi + a\right), \quad n \cos\left(\tfrac{2}{3}\pi - a\right).$$

---

491 Observe that, according as $\frac{r^2}{4} + \frac{q^3}{27}$ is positive or negative, Cardan's method or the Trigonometrical will be practicable. In the former case, there will be *one real and two imaginary roots;* in the latter case, *three real roots.*

BIQUADRATIC EQUATIONS.

---

492 ***Descartes' Solution.***—To solve the equation

$$x^4 + qx^2 + rx + s = 0 \quad \ldots\ldots\ldots\ldots\ldots\ldots \text{(i.)}$$

the term in x^3 having been removed by the method of (429).

Assume $\quad (x^2+ex+f)(x^2-ex+g) = 0 \ldots\ldots\ldots\ldots\ldots \text{(ii.)}$

Multiply out, and equate coefficients with (i.); and the following equations for determining f, g, and e are obtained

$$g+f = q+e^2, \quad g-f = \frac{r}{e}, \quad gf = s \quad \ldots\ldots\ldots \text{(iii.)}$$

493 $$e^6+2qe^4+(q^2-4s)\,e^2-r^2 = 0 \ldots\ldots\ldots\ldots\ldots \text{(iv.)}$$

---

494 *The cubic in e^2 is reducible by Cardan's method, when the biquadratic has two real and two imaginary roots.* For proof, take $a \pm i\beta$ and $-a \pm \gamma$ as the roots of (i.), since their sum must be zero. Form the sum of each pair for the values of e [see (ii.)], and apply the rules in (488) to the cubic in e^2.

If the biquadratic has all its roots real, or all imaginary, the cubic will have all its roots real. Take $a \pm i\beta$ and $-a \pm i\gamma$ for four imaginary roots of (i.), and form the values of e as before.

---

495 *If a^2, β^2, γ^2 be the roots of the cubic in e^2, the roots of the biquadratic will be* $-\frac{1}{2}(a+\beta+\gamma)$, $\frac{1}{2}(a+\beta-\gamma)$, $\frac{1}{2}(\beta+\gamma-a)$, $\frac{1}{2}(\gamma+a-\beta)$.

For proof, take w, x, y, z for the roots of the biquadratic; then, by (ii.), the sum of each pair must give a value of e. Hence, we have only to solve the symmetrical equations

$$y+z=\alpha, \qquad w+x=-\alpha,$$
$$z+x=\beta, \qquad w+y=-\beta,$$
$$x+y=\gamma, \qquad w+z=-\gamma.$$

---

496 ***Ferrari's solution.***—To the left member of the equation

$$x^4+px^3+qx^2+rx+s=0,$$

add the quantity $ax^2+bx+\frac{b^2}{4a}$, and assume the result

$$=\left(x^2+\frac{p}{2}x+m\right)^2.$$

497 Expanding and equating coefficients, the following cubic equation for determining m is obtained

$$8m^3-4qm^2+(2pr-8s)\,m+4qs-p^2s-r=0.$$

Then x is given by the two quadratics

$$x^2+\frac{p}{2}x+m=\pm\frac{2ax+b}{2\sqrt{a}}$$

498 *The cubic in m is reducible by Cardan's method when the biquadratic has two real and two imaginary roots.* Assume α, β, γ, δ for the roots of the biquadratic; then $\alpha\beta$ and $\gamma\delta$ are the respective products of roots of the two quadratics above. From this find m in terms of $\alpha\beta\gamma\delta$.

---

499 ***Euler's solution.***—Remove the term in x^3; then we have

$$x^4+qx^2+rx+s=0.$$

500 Assume $x=y+z+u$, and it may be shewn that y^2, z^2, and u^2 are the roots of the equation

$$t^3+\frac{q}{2}t^2+\frac{q^2-4s}{16}t-\frac{r^2}{64}=0.$$

501 The six values of y, z, and u, thence obtained, are restricted by the relation $yzu=-\frac{r}{8}$.

Thus $x=y+z+u$ will take four different values.

COMMENSURABLE ROOTS.

502 To find the commensurable roots of an equation.

First transform it by putting $x = \frac{y}{k}$ into an equation of the form $\quad x^n + p_1 x^{n-1} + p_2 x^{n-2} + \dots + p_n = 0,$
having $p_0 = 1$, and the remaining coefficients integers. (431)

503 This equation cannot have a rational fractional root, and the integral roots may be found by Newton's method of Divisors (459).

These roots, divided each by k, will furnish the commensurable roots of the original equation.

504 Example.—To find the commensurable roots of the equation

$$81x^5 - 207x^4 - 9x^3 + 89x^2 + 2x - 8 = 0.$$

Dividing by 81, and proceeding as in (431), we find the requisite substitution to be

$$x = \frac{y}{9}.$$

The transformed equation is

$$y^5 - 23y^4 - 9y^3 + 801y^2 + 162y - 5832 = 0.$$

The roots all lie between 24 and -34, by (451).

The method of divisors gives the integral roots

$$6, \ -4, \text{ and } 3.$$

Therefore, dividing each by 9, we find the commensurable roots of the original equation to be $\quad \frac{2}{3}, \ -\frac{4}{9}, \text{ and } \frac{1}{3}.$

505 To obtain the remaining roots; diminish the transformed equation by the roots 6, -4, and 3, in the following manner (see 427):—

$$\begin{array}{r|l}
 & 1-23-\ \ 9+801+162-5832 \\
6 & \quad\ \ 6-102-666+810+5832 \\
\hline
 & 1-17-111+135+972 \\
-4 & \ \ -\ 4+\ 84+108-972 \\
\hline
 & 1-21-\ 27+243 \\
3 & \quad\ \ 3-\ 54-243 \\
\hline
 & 1-18-\ 81
\end{array}$$

The depressed equation is therefore

$$y^2 - 18y - 81 = 0.$$

The roots of which are $9(1+\sqrt{2})$ and $9(1-\sqrt{2})$; and, consequently, the incommensurable roots of the proposed equation are $1+\sqrt{2}$ and $1-\sqrt{2}$.

INCOMMENSURABLE ROOTS.

506 *Sturm's Theorem.*—If $f(x)$, freed from equal roots, be divided by $f'(x)$, and the last divisor by the last remainder, *changing the sign* of each remainder before dividing by it, until a remainder independent of x is obtained, or else a remainder which cannot change its sign; then $f(x)$, $f'(x)$, and the successive remainders constitute Sturm's functions, and are denoted by $f(x), f_1(x), f_2(x)$, &c. ...... $f_m(x)$.

The operation may be exhibited as follows:—

$$\begin{aligned} f(x) &= q_1 f_1(x) - f_2(x), \\ f_1(x) &= q_2 f_2(x) - f_3(x), \\ f_2(x) &= q_3 f_3(x) - f_4(x), \\ &\dots\quad \dots\quad \dots \\ f_{m-2}(x) &= q_{m-1} f_{m-1}(x) - f_m(x). \end{aligned}$$

507 Note.—Any constant factor of a remainder may be rejected, and the quotient may be set down for the corresponding function.

---

508 An inspection of the foregoing equations shews—

(1) That $f_m(x)$ cannot be zero; for, if it were, $f(x)$ and $f_1(x)$ would have a common factor, and therefore $f(x)$ would have equal roots, by (432).

(2) Two consecutive functions, after the first, cannot vanish together; for this would make $f_m(x)$ zero.

(3) When any function, after the first, vanishes, the two adjacent ones have contrary signs.

509 *If, as x increases, $f(x)$ passes through the value zero, Sturm's functions lose one change of sign.*

For, before $f(x)$ takes the value zero, $f(x)$ and $f_1(x)$ have contrary signs, and afterwards they have the same sign; as may be shewn by making h small, and changing its sign in the expansion of $f(x+h)$, by (426).

510 *If any other of Sturm's functions vanishes, there is neither loss nor gain in the number of changes of sign.*

This will appear on inspecting the equations.

511 Result.—*The number of roots of $f(x)$ between a and b is equal to the difference in the number of changes of sign in Sturm's functions, when $x=a$ and when $x=b$.*

512 Cor.—The total number of roots of $f(x)$ will be found by taking $a = +\infty$ and $b = -\infty$; the sign of each function will then be the same as that of its first term.

When the number of functions exceeds the degree of $f(x)$ by unity, the two following theorems hold :—

513 *If the first terms in all the functions, after the first, are positive; all the roots of $f(x)$ are real.*

514 *If the first terms are not all positive; then, for every change of sign, there will be a pair of imaginary roots.*

For the proof put $x = +\infty$ and $-\infty$, and examine the number of changes of sign in each case. (416).

515 If $\phi(x)$ has no factor in common with $f(x)$, and if $\phi(x)$ and $f'(x)$ take the same sign when $f(x) = 0$; then the rest of Sturm's functions may be found from $f(x)$ and $\phi(x)$, instead of $f'(x)$. For the reasoning in (509) and (510) will apply to the new functions.

516 If Sturm's functions be formed without first removing equal roots from $f(x)$, the theorem will still give the number of distinct roots, without repetitions, between assigned limits.

For if $f(x)$ and $f_1(x)$ be divided by their highest common factor (see 444), and if the quotients be used instead of $f(x)$ and $f_1(x)$ to form Sturm's functions; then, by (515), the theorem will apply to the new set of functions, which will differ only from those formed from $f(x)$ and $f_1(x)$ by the absence of the same factor in every term of the series.

517 Example.—To find the position of the roots of the equation

$$x^4 - 4x^3 + x^2 + 6x + 2 = 0.$$

Sturm's functions, formed according to the rule given above, are here calculated.

$$\begin{aligned} f(x) &= x^4 - 4x^3 + x^2 + 6x + 2 \\ f_1(x) &= 2x^3 - 6x^2 + x + 3 \\ f_2(x) &= 5x^2 - 10x - 7 \\ f_3(x) &= x - 1 \\ f_4(x) &= 12 \end{aligned}$$

The first terms of the functions are all positive; therefore there is no imaginary root.

The changes of sign in the functions, as x passes through integral values, are exhibited in the adjoining table. There are two changes of sign lost while x passes from -1 to 0, and two more lost while x passes from 2 to 3. There

$x =$	-2	-1	0	1	2	3	4
$f(x) =$	+	+	+	+	+	+	+
$f_1(x) =$	−	−	+	+	−	+	+
$f_2(x) =$	+	+	−	−	−	+	+
$f_3(x) =$	−	−	−	+	+	+	+
$f_4(x) =$	+	+	+	+	+	+	+
No. of changes of sign	4	4	2	2	2	0	0

are therefore two roots lying between 0 and -1; and two roots also between 2 and 3.

These roots are all incommensurable, by (503).

---

518 ***Fourier's Theorem.***—Fourier's functions are the following quantities $f(x),\ f'(x),\ f''(x) \ldots\ldots f^n(x)$.

519 Properties of Fourier's functions.—As x increases, Fourier's functions lose *one* change of sign for each root of the equation $f(x) = 0$, through which x passes, and r changes of sign for r repeated roots.

520 If any of the other functions vanish, an *even* number of changes of sign is lost.

---

521 RESULTS.—*The number of real roots of $f(x)$ between α and β cannot be more than the difference between the number of changes of sign in Fourier's functions when $x = \alpha$, and the number of changes when $x = \beta$.*

522 When that difference is *odd*, the number of intermediate roots is *odd*, and therefore *one at least*.

523 When the same difference is *even*, the number of intermediate roots is either *even or zero*.

---

524 Descartes' rule of signs follows from the above for the signs of Fourier's functions, when $x = 0$ are the signs of the terms in $f(x)$; and when $x = \infty$, Fourier's functions are all positive.

---

525 ***Lagrange's method of approximating to the incommensurable roots of an equation.***

Let a be the greatest integer less than an incommensurable root of $f(x)$. Diminish the roots of $f(x)$ by a. Take the reciprocal of the resulting equation. Let b be the greatest integer less than a positive root of this equation. Diminish the roots of this equation by b, and proceed as before.

526 Let a, b, c, &c. be the quantities thus determined; then, an approximation to the incommensurable root of $f(x)$ will be the continued fraction $x = a + \frac{1}{b+}\,\frac{1}{c+}$

527 *Newton's method of approximation.*—If c_1 be a quantity a little less than one of the roots of the equation $f(x) = 0$, so that $f(c_1+h) = 0$; then c_1 is a first approximation to the value of the root. Also because

$$f(c_1+h) = f(c_1) + hf'(c_1) + \frac{h^2}{\underline{|2}} f''(c_1) + \&c. \ldots\ldots (426),$$

and h is but small, a second approximation to the root will be

$$c_1 - \frac{f(c_1)}{f'(c_1)} = c_2.$$

In the same way a third approximation may be obtained from c_2, and so on.

528 *Fourier's limitation of Newton's method.*—To ensure that c_1, c_2, c_3, &c. shall successively increase up to the value c_1+h without passing beyond it, it is necessary for all values of x between c_1 and c_1+h.

(i.) *That $f(x)$ and $f'(x)$ should have contrary signs.*
(ii.) *That $f(x)$ and $f''(x)$ should have the same sign.*

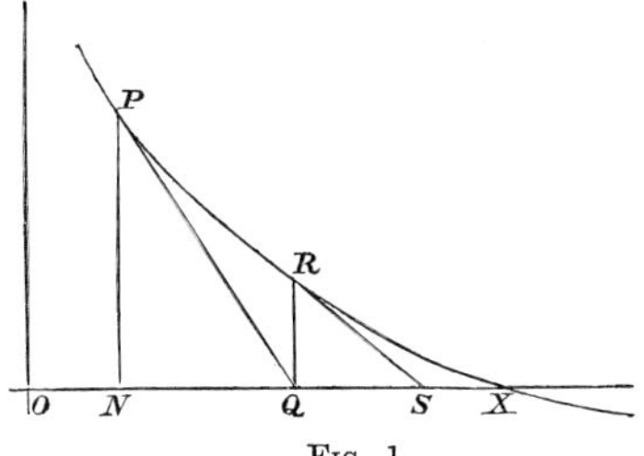

Fig. 1.

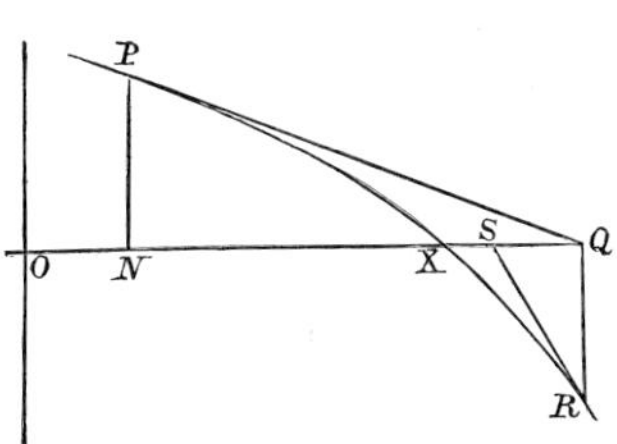

Fig. 2.

A proof may be obtained from the figure. Draw the curve $y = f(x)$. Let OX be a root of the equation, and $ON = c_1$; draw the successive ordinates and tangents NP, PQ, QR, &c. Then $OQ = c_2$, $OS = c_3$, and so on.

Fig. (2) represents $c_2 > OX$, and the subsequent approximations decreasing towards the root.

530 *Newton's Rule for Limits of the Roots.*—Let the coefficients of $f(x)$ be respectively divided by the Binomial coefficients, and let a_0, a_1, $a_2 \ldots a_n$ be the quotients, so that

$$f(x) = a_0x^n + na_1x^{n-1} + \frac{n(n-1)}{1.2} a_2x^{n-2} + \ldots + na_{n-1}x + a_n.$$

Let $A_1, A_2, A_3 \dots A_n$ be formed by the law $A_r = a_r^2 - a_{r-1}a_{r+1}$. Write the first series of quantities over the second, in the following manner:—

$$\begin{array}{llllll} a_0, & a_1, & a_2, & a_3 \dots\dots & a_{n-1}, & a_n, \\ A_0, & A_1, & A_2, & A_3 \dots\dots & A_{n-1}, & A_n. \end{array}$$

Whenever two adjacent terms in the first series have the same sign, and the two corresponding terms below them in the second series also the same sign; let this be called *a double permanence.* When two adjacent terms above have different signs, and the two below the same sign, let this be known as *a variation-permanence.*

531 RULE.—*The number of double permanences in the associated series is a superior limit to the number of negative roots of $f(x)$.*

The number of variation-permanences is a superior limit to the number of positive roots.

The number of imaginary roots cannot be less than the number of variations of sign in the second series.

---

532 ***Sylvester's Theorem.***—Let $f(x+\lambda)$ be expanded by (426) in powers of x, and let the two series be formed as in Newton's Rule (530).

Let $P(\lambda)$ denote the number of double permanences.

Then $P(\lambda) \sim P(\mu)$ is either equal to the number of roots of $f(x)$, or surpasses that number by an even integer.

NOTE.—The first series may be multiplied by $\lfloor n$, and will then stand thus,

$$f^n(\lambda), \quad f^{n-1}(\lambda), \quad \lfloor 2\, f^{n-2}(\lambda), \quad \lfloor 3\, f^{n-3}(\lambda) \dots \lfloor n\, f(\lambda).$$

The second series may be reduced to

$$G_n(\lambda), \quad G_{n-1}(\lambda), \quad G_{n-2}(\lambda) \dots G(\lambda),$$

where $$G_r(\lambda) \equiv \{f^r(\lambda)\}^2 - \frac{n-r+1}{n-r} f^{r-1}(\lambda)\, f^{r+1}(\lambda).$$

---

533 ***Horner's Method.***—To find the numerical values of the roots of an equation. Take, for example, the equation

$$x^4 - 4x^3 + x^2 + 6x + 2 = 0,$$

and find limits of the roots by Sturm's Method or otherwise.

It has been shewn in (517) that this equation has two incommensurable roots between 2 and 3. The process of calculating the least of these roots is here exhibited.

	−4		+1		+6		+2 (2·414213
	2		−4		−6		0
	−2		−3		0	A_1	20000
	2		0		−6		−19584
	0		−3	B_1	−6000	A_2	4160000
	2		4		1104		−2955839
	2	C_1	100		−4896	A_3	12041610000
	2		176		1872		−11437245184
D_1	40		276	B_2	−3024000	A_4	604364816
	4		192		68161		−566003348
	44		468		−2955839	A_5	38361468
	4		208		68723		−28285470
	48	C_2	67600	B_3	−2887116000	A_6	10075998
	4		561		27804704		−8485368
	52		68161		−2859311296	A_7	1590630
	4		562		27895072		
D_2	560		68723	B_4	−2831416224		
	1		563		139948		282843) 1590630 (562372
	561	C_3	6928600		−283001674		1414215
	1		22576		139970		28284) 176415
	562		6951176	B_5	−282861704		169706
	1		22592		700		2828) 6709
	563		6973768		−28285470		5657
	1		22608		700		282) 1052
D_3	5640	C_4	6996376	B_6	−28284770		848
	4		11		21		28) 204
	5644		69974		−2828456		197
	4		11		21		2) 7
	5648		69985	B_7	−2828435		5
	4		11				2
	5652	C_5	69996				
	4	C_6	7				
D_4	5,656						

Root = 2·414213562372.

METHOD.—Diminish the roots by 2 in the manner of (427).

The resulting coefficients are indicated by A_1, B_1, C_1, D_1.

By Newton's rule (527), $-\dfrac{f(c)}{f'(c)}$; that is, $-\dfrac{A_1}{B_1}$ is an approximation to the remaining part of the root. This gives ·3 for the next figure; ·4 will be found to be the correct one. The highest figure must be taken which will not change the sign of A.

Diminish the roots by ·4. This is accomplished most easily by affixing ciphers to A_1, B_1, C_1, D_1, in the manner shewn, and then employing 4 instead of ·4.

Having obtained A_2, and observing that its sign is +, retrace the steps, trying 5 instead of 4. This gives A_2 with a minus sign, thereby proving the existence of a root between 2·4 and 2·5. The new coefficients are A_2, B_2, C_2, D_2.

$-\dfrac{A_2}{B_2}$ gives 1 for the next figure of the root.

Affix ciphers as before, and diminish the roots by 1, distinguishing the new coefficients as A_3, B_3, C_3, D_3.

Note that at every stage of the work A and B must preserve their signs unchanged. If a change of sign takes place it shews that too large a figure has been tried.

To abridge the calculation proceed thus:—After a certain number of figures of the root have been obtained (in this example four), instead of adding ciphers cut off one digit from B_4, two from C_4, and three from D_4. This amounts to the same thing as adding the ciphers, and then dividing each number by 10000.

Continue the work with the numbers so reduced, and cut off digits in like manner at each stage until the D and C columns have disappeared.

A_7 and B_7 now alone remain, and six additional figures of the root are determined correctly by the division of A_7 by B_7.

To find the other root which lies between 2 and 3, we proceed as follows:—After diminishing the roots by 2, try 6 for the next figure. This gives A_2 negative; 7 does the same, but 8 makes A_2 positive. That is to say, $f(2{\cdot}7)$ is negative, and $f(2{\cdot}8)$ positive. Therefore a root exists between 2·7 and 2·8, and its value may be approximated to, in the manner shewn.

Throughout this last calculation A will preserve the negative sign. Observe also that the trial number for the next figure of the root given at each stage of the process by the formula $-\dfrac{f(c)}{f'(c)}$, will in this case be always too great, as in the former case it was always too small.

SYMMETRICAL FUNCTIONS OF THE ROOTS OF AN EQUATION.

NOTATION.—Let a, b, c ... be the roots of the equation $f(x) = 0$.

Let s_m denote $a^m + b^m + \dots$, the sum of the mth powers of the roots.

Let $s_{m,p}$ denote $a^m b^p + b^m a^p + a^m c^p + \dots$ through all the permutations of the roots, two at a time.

Similarly let $s_{m,p,q}$ denote $a^m b^p c^q + a^m b^p d^q + \dots$, taking all the permutations of the roots three at a time; and so on.

534 *SUMS OF THE POWERS OF THE ROOTS.*

$$s_m + p_1 s_{m-1} + p_2 s_{m-2} + \dots + p_{m-1} s_1 + m p_m = 0,$$

where m is less than n, the degree of $f(x)$.

Obtained by expanding by division each term in the value of $f'(x)$ given at (432), arranging the whole in powers of x, and equating coefficients in the result and in the value of $f'(x)$, found by differentiation as in (424).

535 If m be greater than n, the formula will be

$$s_m + p_1 s_{m-1} + p_2 s_{m-2} + \dots + p_n s_{m-n} = 0.$$

Obtained by multiplying $f(x) = 0$ by x^{m-n}, substituting for x the roots a, b, c, &c. in succession, and adding the results.

By these formulæ s_1, s_2, s_3, &c. may be calculated successively.

536 To find the sum of the negative powers of the roots, put m equal to $n-1$, $n-2$, $n-3$, &c. successively in (535), in order to obtain s_{-1}, s_{-2}, s_{-3}, &c.

---

537 To calculate s_r independently.

RULE: $s_r = -r \times$ *coefficient of* x^{-r} *in the expansion of* $\log \dfrac{f(x)}{x^n}$ *in descending powers of* x.

Proved by taking $f(x) = (x-a)(x-b)(x-c)\dots$, dividing by x^n, and expanding the logarithm of the right side of the equation by (156).

---

538 *SYMMETRICAL FUNCTIONS WHICH ARE NOT POWERS OF THE ROOTS.*

These are expressed in terms of the sums of powers of the roots as under, and thence, by (534), in terms of the roots explicitly,

$$s_{m,\,p} = s_m s_p - s_{m+p},$$

539 $$s_{m,\,p,\,q} = s_m s_p s_q - s_{m+p} s_q - s_{m+q} s_p - s_{p+q} s_m + 2 s_{m+p+q}.$$

The last equation may be proved by multiplying $s_{m,\,p}$ by s_q; and expansions of other symmetrical functions may be obtained in a similar way.

---

540 If $\phi(x)$ be a rational integral function of x, then the symmetrical function of the roots of $f(x)$, denoted by

$\phi(a)+\phi(b)+\phi(c)+$&c., is equal to the coefficient of x^{n-1} in the remainder obtained by dividing $\phi(x) f'(x)$ by $f(x)$.

Proved by multiplying the equation (432) by $\frac{\phi(x)}{f(x)}$, and by theorem (401).

---

541 To find the equation whose roots are the squares of the differences of the roots of a given equation.

Let $F(x)$ be the given equation, and S_r the sum of the rth powers of its roots. Let $f(x)$ and s_r have the same meaning with regard to the required equation.

The coefficients of the required equation can be calculated from those of the given one as follows:—

The coefficients of each equation may be connected with the sums of the powers of its roots by (534); *and the sums of the powers of the roots of the two equations are connected by the formula*

542 $$2s_r = nS_{2r} - 2rS_1S_{2r-1} + \frac{2r(2r-1)}{1.2} S_2S_{2r-2} - \ldots + nS_{2r}.$$

Rule.—$2s_r$ *is equal to the formal expansion of* $(S-S)^{2r}$ *by the Binomial Theorem, with the first and last terms each multiplied by* n, *and the indices all changed to suffixes.* As the equi-distant terms are equal we can divide by 2, and take half the series.

Demonstration.—Let $a, b, c \ldots$ be the roots of $F(x)$.

Let $$\phi(x) = (x-a)^{2r} + (x-b)^{2r} + \ldots \qquad \text{(i.)}$$

Expand each term on the right by the Bin. Theor., and add, substituting S_1, S_2, &c. In the result change x into $a, b, c \ldots$ successively, and add the n equations to obtain the formula, observing that, by (i.),

$$\phi(a) + \phi(b) + \ldots = 2s_r.$$

If n be the degree of $F(x)$, then $\frac{1}{2}n(n-1)$ is the degree of $f(x)$. By (96).

543 The last term of the equation $f(x) = 0$ is equal to

$$n^n F(\alpha) F(\beta) F(\gamma) \ldots$$

where $\alpha, \beta, \gamma, \ldots$ are the roots of $F'(x)$. Proved by shewing that $$F'(a) F'(b) \ldots = n^n F(\alpha) F(\beta) \ldots$$

544 If $f(x)$ has negative or imaginary roots, $F(x)$ must have imaginary roots.

545 The sum of the mth powers of the roots of the quadratic equation $$x^2-px+q=0.$$

$$s_m = p^m - mp^{m-2}q + \frac{m(m-3)}{\lfloor 2} p^{m-4}q^2 - \dots$$

$$\dots + (-1)^r \frac{m(m-r-1)\dots(m-2r+1)}{\lfloor r} p^{m-2r}q^r + \&c.$$

By (537) expanding the logarithm by (156).

---

546 The sum of the mth powers of the roots of $x^n-1=0$ is n if m be a multiple of n, and zero if it be not.

By (537); expanding the logarithm by (156).

---

547 If $$\phi(x) = a_0 + a_1x + a_2x^2 + \&c. \quad \dots\dots\dots\dots\dots \text{(i.)},$$
then the sum of the selected terms

$$a_m x^m + a_{m+n} x^{m+n} + a_{m+2n} x^{m+2n} + \&c.$$

will be $$s = \frac{1}{n}\left\{\alpha^{n-m}\phi(\alpha x) + \beta^{n-m}\phi(\beta x) + \gamma^{n-m}\phi(\gamma x) + \&c.\right\}$$

where α, β, γ, &c. are the nth roots of unity.

For proof, multiply (i.) by α^{n-m}, and change x into αx; so with β, γ, &c., and add the resulting equations.

---

548 To approximate to the root of an equation by means of the sums of the powers of the roots.

By taking m large enough, the fraction $\frac{s_{m+1}}{s_m}$ will approximate to the value of the numerically greatest root, unless there be a modulus of imaginary roots greater than any real root, in which case the fraction has no limiting value.

549 Similarly the fraction $\frac{s_m s_{m+2} - s_{m+1}^2}{s_{m-1}s_{m+1} - s_m^2}$ approximates, as m increases, to the *greatest product* of any pair of roots, real or imaginary; excepting in the case in which the product of the pair of imaginary roots, though less than the product of the two real roots, is greater than the square of the least of them, for then the fraction has no limiting value.

550 Similarly the fraction $\dfrac{s_m s_{m+3}-s_{m+1}s_{m+2}}{s_m s_{m+2}-s_{m+1}^2}$ approximates, as m increases, to the sum of the two numerically greatest roots, or to the sum of the two imaginary roots with the greatest modulus.

EXPANSION OF AN IMPLICIT FUNCTION OF x.

Let $y^{\alpha}(Ax^{a}+)+y^{\beta}(Bx^{b}+)+\ldots+y^{\sigma}(Sx^{s}+)=0 \;\ldots\ldots(1)$

be an equation arranged in descending powers of y, the coefficients being functions of x, the highest powers only of x in each coefficient being written.

551 It is required to obtain y in a series of descending powers of x.

First form the fractions

$$-\frac{a-b}{\alpha-\beta},\quad -\frac{a-c}{\alpha-\gamma},\quad -\frac{a-d}{\alpha-\delta}\;\ldots\ldots\;-\frac{a-s}{\alpha-\sigma}\;\ldots\ldots\ldots(2).$$

Let $-\dfrac{a-k}{\alpha-\kappa}=t$ be the greatest of these algebraically, or if several are equal and greater than the rest, let it be the last of such. Then, with the letters corresponding to these equal and greatest fractions, form the equation

$$Au^{\alpha}+\ldots\ldots+Ku^{\kappa}=0\;\ldots\ldots\ldots\ldots\ldots\ldots(3).$$

Each value of u in this equation corresponds to a value of y, commencing with ux^{t}.

Next select the greatest of the fractions

$$-\frac{k-l}{\kappa-\lambda},\quad -\frac{k-m}{\kappa-\mu}\;\ldots\ldots\;-\frac{k-s}{\kappa-\sigma}\;\ldots\ldots\ldots\ldots\ldots(4).$$

Let $-\dfrac{k-n}{\kappa-\nu}=t'$ be the last of the greatest ones. Form the corresponding equation $Ku^{\kappa}+\ldots+Nu^{\nu}=0\;\ldots\ldots\ldots(5)$.

Then each value of u in this equation gives a corresponding value of y, commencing with $ux^{t'}$.

Proceed in this way until the last fraction of the series (2) is reached.

To obtain the second term in the expansion of y, put

$$y = x^t (u+y_1) \text{ in } (1) \ldots\ldots\ldots\ldots\ldots\ldots\ldots (6),$$

employing the different values of u, and again of t' and u, t'' and u, &c. in succession; and in each case this substitution will produce an equation in y and x similar to the original equation in y.

Repeat the foregoing process with the new equation in y, observing the following additional rule :—

When all the values of t, t', t'', &c. have been obtained, the negative ones only must be employed in forming the equations in u. (7).

552 To obtain y in a series of *ascending* powers of x.

Arrange equation (1) so that α, β, γ, &c. may be in *ascending* order of magnitude, and a, b, c, &c. the *lowest* powers of x in the respective coefficients.

Select t, the greatest of the fractions in (2), and proceed exactly as before, with the one exception of substituting the word *positive* for *negative* in (7).

553 Example.—Take the equation

$$(x^3+x^4)+(3x^2-5x^3)\,y+(-4x+7x^2+x^3)\,y^2-y^5=0.$$

It is required to expand y in ascending powers of x.

The fractions (2) are $-\dfrac{3-2}{0-1}$, $-\dfrac{3-1}{0-2}$, $-\dfrac{3-0}{0-5}$; or 1, 1, and $\frac{3}{5}$.

The first two being equal and greatest, we have $t=1$.

The fractions (4) reduce to $-\dfrac{1-0}{2-5}=\frac{1}{3}=t'$.

Equation (3) is $\quad 1+3u-4u^2=0,$

which gives $\quad u=1$ and $-\frac{1}{4}$, with $t=1$.

Equation (5) is $\quad -4u^2-u^5=0,$

and from this $\quad u=0$ and $-4^{\frac{1}{3}}$, with $t'=\frac{1}{3}$.

We have now to substitute for y, according to (6), either

$$x\,(1+y_1),\quad x\,(-\tfrac{1}{4}+y_1),\quad x^{\frac{1}{3}}y,\ \text{or}\ x^{\frac{1}{3}}\,(-4^{\frac{1}{3}}+y_1).$$

Put $y=x\,(1+y_1)$, the first of these values, in the original equation, and arrange in ascending powers of y, thus

$$-4x^4+(-5x^3+)\,y_1+(-4x^3+)\,y_1^2-10x^3y_1^3-5x^3y_1^4-x^5y_1^5=0.$$

The lowest power only of x in each coefficient is here written.

The fractions (2) now become

$$-\frac{4-3}{0-1},\ -\frac{4-3}{0-2},\ -\frac{4-5}{0-3},\ -\frac{4-5}{0-4},\ -\frac{4-5}{0-5};$$

or $$1,\ \tfrac{1}{2},\ -\tfrac{1}{3},\ -\tfrac{1}{4},\ -\tfrac{1}{5}.$$

From these $t=1$, and equation (3) becomes

$$-4-5u=0;\quad \therefore\ u=-\tfrac{4}{5}.$$

Hence one of the values of y_1 is, as in (6), $y_1 = x\left(-\tfrac{4}{5}+y_2\right)$.

Therefore $$y = x\left\{1+x\left(-\tfrac{4}{5}+y_2\right)\right\} = x-\tfrac{4}{5}x^2+\dots$$

Thus the first two terms of one of the expansions have been obtained.

DETERMINANTS.

554 *Definitions.*—The determinant $\begin{vmatrix} a_1 & a_2 \\ b_1 & b_2 \end{vmatrix}$ is equivalent to $a_1b_2-a_2b_1$, and is called a determinant of the second order. A determinant of the third order is

$$\begin{vmatrix} a_1 & a_2 & a_3 \\ b_1 & b_2 & b_3 \\ c_1 & c_2 & c_3 \end{vmatrix} \equiv a_1(b_2c_3-b_3c_2)+a_2(b_3c_1-b_1c_3)+a_3(b_1c_2-b_2c_1).$$

Another notation is $\Sigma \pm a_1b_2c_3$, or simply $(a_1b_2c_3)$.

The letters are named *constituents*, and the terms are called *elements*. The determinant is composed of all the elements obtained by permutations of the suffixes 1, 2, 3.

The coefficients of the constituents are determinants of the next lower order, and are termed *minors* of the original determinant. Thus, the first determinant above is the minor of c_3 in the second determinant. It is denoted by C_3. So the minor of a_1 is denoted by A_1, and so on.

555 A determinant of the nth order may be written in either of the forms below

$$\begin{vmatrix} a_1 & a_2 & \dots & a_r & \dots & a_n \\ b_1 & b_2 & \dots & b_r & \dots & b_n \\ \dots & \dots & & \dots & & \dots \\ l_1 & l_2 & \dots & l_r & \dots & l_n \end{vmatrix} \quad \text{or} \quad \begin{vmatrix} a_{11} & a_{12} & \dots & a_{1r} & \dots & a_{1n} \\ a_{21} & a_{22} & \dots & a_{2r} & \dots & a_{2n} \\ \dots & \dots & & \dots & & \dots \\ a_{n1} & a_{n2} & \dots & a_{nr} & \dots & a_{nn} \end{vmatrix}$$

In the latter, or double suffix notation, the first suffix indicates the row, and the second the column. The former notation will be adopted in these pages.

A *Composite determinant* is one in which the number of columns exceeds the number of rows, and it is written as in the annexed example. Its value is the sum of all the determinants obtained by taking as many columns as there are rows in every possible way.

$$\left\| \begin{matrix} a_1 & a_2 & a_3 \\ b_1 & b_2 & b_3 \end{matrix} \right\|$$

A *Simple determinant* has single terms for its constituents.

A *Compound determinant* has more than one term in some or all of its constituents. See (570) for an example.

For the definitions of *Symmetrical*, *Reciprocal*, *Partial*, and *Complementary* determinants; see (574), (575), and (576).

General Theory.

556 The number of constituents is n^2.

The number of elements in the complete determinant is $\underline{n|}$.

557 The first or leading element is $a_1 b_2 c_3 \dots l_n$. Any element may be derived from the first by permutation of the suffixes.

The sign of an element is + or − according as it has been obtained from the diagonal element by an even or odd number of permutations of the suffixes.

Hence the following rule for determining the sign of an element.

RULE.—*Take the suffixes in order, and put them back to their places in the first element. Let m be the whole number of places passed over; then* $(-1)^m$ *will give the sign required.*

Ex.—To find the sign of the element $a_4 b_3 c_5 d_1 e_2$ of the determinant $(a_1 b_2 c_3 d_4 e_5)$.

	a_4	b_3	c_5	d_1	e_2
Move the suffix 1, three places... ...	1	4	3	5	2
" " 2, three places... ...	1	2	4	3	5
" " 3, one place	1	2	3	4	5

In all, seven places; therefore $(-1)^7 = -1$ gives the sign required.

558 If two suffixes in any element be transposed, the sign of the element is changed.

Half of the elements are plus, and half are minus.

559 The elements are not altered by changing the rows into columns.

If two rows or columns are transposed, the sign of the

determinant is changed. Because each element changes its sign.

If two rows or columns are identical, the determinant vanishes.

560 If all the constituents but one in a row or column vanish, the determinant becomes the product of that constituent and a determinant of the next lower order.

---

561 A cyclical interchange is effected by $n-1$ successive transpositions of adjacent rows or columns, until the top row has been brought to the bottom, or the left column to the right side. Hence

A cyclical interchange changes the sign of a determinant of an even order only.

The rth row may be brought to the top by $r-1$ cyclical interchanges.

---

562 If each constituent in a row or column be multiplied by the same factor, the determinant becomes multiplied by it.

If each constituent of a row or column is the sum of m terms, the compound determinant becomes the sum of m simple determinants of the same order.

Also, if every constituent of the determinant consists of m terms, the compound determinant is resolvable into the sum of m^2 simple determinants.

---

563 To express the minor of the rth row and k^{th} column as a determinant of the $n-1^{th}$ order.

Put all the constituents in the rth row and k^{th} column equal to 0, and then make $r-1$ cyclical interchanges in the rows and $k-1$ in the columns, and multiply by $(-1)^{(r+k)(n-1)}$.

[$\because \; = (-1)^{(r-1+k-1)(n-1)}$.

---

564 To express a determinant as a determinant of a higher order.

Continue the diagonal with constituents of "ones," and fill up with zeros on one side, and with any quantities whatever (α, β, γ, &c.) on the other.

$$\begin{vmatrix} 1 & 0 & 0 & 0 & 0 \\ \alpha & 1 & 0 & 0 & 0 \\ \beta & \epsilon & a & h & g \\ \gamma & \zeta & h & b & f \\ \delta & \eta & g & f & c \end{vmatrix}$$

565 The sum of the products of each constituent of a column by the corresponding minor in another given column is zero. And the same is true if we read 'row' instead of 'column.' Thus, referring to the determinant in (555),

Taking the pth and q^{th} columns,

$$a_pA_q+b_pB_q+\ldots+l_pL_q=0.$$

Taking the a and c rows,

$$a_1C_1+a_2C_2+\ldots+a_nC_n=0.$$

For in each case we have a determinant with two columns identical.

566 In any row or column the sum of the products of each constituent by its minor is the determinant itself. That is,

Taking the pth column,

$$a_pA_p+b_pB_p+\ldots+l_pL_p=\Delta.$$

Or taking the c row,

$$c_1C_1+c_2C_2+\ldots+c_nC_n=\Delta.$$

567 The last equation may be expressed by $\Sigma c_pC_p=\Delta$.

Also, if (a_pc_q) express the determinant $\begin{vmatrix} a_p & a_q \\ c_p & c_q \end{vmatrix}$; then $\Sigma(a_pc_q)$ will represent the sum of all the determinants of the second order which can be formed by taking any two columns out of the a and c rows. The minor of (a_p, c_q) may be written (A_p, C_q), and signifies the determinant obtained by suppressing the two rows and two columns of a_p and c_q. Thus $\Delta=\Sigma(a_p, c_q)(A_p, C_q)$. And a similar notation when three or more rows and columns are selected.

568 *Analysis of a determinant.*

Rule.—*To resolve into its elements a determinant of the nth order. Express it as the sum of n determinants of the $(n-1)^{th}$ order by* (560), *and repeat the process with each of the new determinants.*

Example:

$$\begin{vmatrix} a_1 & a_2 & a_3 & a_4 \\ b_1 & b_2 & b_3 & b_4 \\ c_1 & c_2 & c_3 & c_4 \\ d_1 & d_2 & d_3 & d_4 \end{vmatrix} = a_1\begin{vmatrix} b_2 & b_3 & b_4 \\ c_2 & c_3 & c_4 \\ d_2 & d_3 & d_4 \end{vmatrix} - a_2\begin{vmatrix} b_3 & b_4 & b_1 \\ c_3 & c_4 & c_1 \\ d_3 & d_4 & d_1 \end{vmatrix} + a_3\begin{vmatrix} b_4 & b_3 & b_2 \\ c_4 & c_3 & c_2 \\ d_4 & d_3 & d_2 \end{vmatrix} - a_4\begin{vmatrix} b_1 & b_2 & b_3 \\ c_1 & c_2 & c_3 \\ d_1 & d_2 & d_3 \end{vmatrix}$$

Again,

$$\begin{vmatrix} b_1 & b_2 & b_3 \\ c_1 & c_2 & c_3 \\ d_1 & d_2 & d_3 \end{vmatrix} = b_1\begin{vmatrix} c_2 & c_3 \\ d_2 & d_3 \end{vmatrix} + b_2\begin{vmatrix} c_3 & c_1 \\ d_3 & d_1 \end{vmatrix} + b_3\begin{vmatrix} c_1 & c_2 \\ d_1 & d_2 \end{vmatrix}$$

and so on. In the first series the determinants have alternately plus and minus signs, by the rule for cyclical interchanges (561), the order being even.

569 *Synthesis of a determinant.*

The process is facilitated by making use of two evident rules. Those constituents which belong to the row and column of a given constituent a, will be designated "a's constituents." Also, two pairs of constituents such as a_p, c_q and a_q, c_p, forming the corners of a rectangle, will be said to be "conjugate" to each other.

RULE I.—*No constituent will be found in the same term with one of its own constituents.*

RULE II.—*The conjugates of any two constituents a and b will be common to a's and b's constituents.*

Ex.—To write the following terms in the form of a determinant:

$$abcd+bfgl+f^2h^2+1edf+cghp+1ahr+elpr$$
$$-fhpr-ablr-ach^2-1fhg-bdf^2-efhl-cedp.$$

The determinant will be of the fourth order; and since every term must contain four constituents, the constituent 1 is supplied to make up the number in some of the terms. Select any term, as $abcd$, for the leading diagonal.

Now apply Rule I.,

a is not found with $e, f, f, g, p, 0$ ...(1). c is not found with $f, f, l, r, 1, 0$ ...(3).
b is not found with $e, h, h, p, 1, 0$ ...(2). d is not found with $g, h, h, l, r, 0$ ...(4).

Each constituent has $2(n-1)$, that is, 6 constituents belonging to it, since $n=4$. Assuming, therefore, that the above letters are the constituents of a, b, c, and d, and that there are no more, we supply a sixth zero constituent in each case.

Now apply Rule II.—The constituents common

to a and b are e, p; to a and c—f, f; to b and c—$1, 0$;
to a and d—$g, 0$; to b and d—$h, h, 0$; to c and d—$l, r, 0$.

The determinant may now be formed. The diagonal being $abcd$; place e, p, the conjugates of a and b, either as in the diagram or transposed.

$$\begin{vmatrix} a & e & f & g \\ p & b & 1 & h \\ f & 0 & c & r \\ 0 & h & l & d \end{vmatrix}$$

Then f and f, the conjugates of a and c, may be written.

1 and 0, the conjugates of b and c, must be placed as indicated, because 1 is one of p's constituents, since it is not found in any term with p, and must therefore be in the second row.

Similarly the places of g and 0, and of l and r, are assigned.

In the case of b and d we have $h, h, 0$ from which to choose the two conjugates, but we see that 0 is not one of them because that would assign two zero constituents to b, whereas b has but one, which is already placed.

By similar reasoning the ambiguity in selecting the conjugates l, r is removed.

The foregoing method is rigid in the case of a complete determinant having different constituents. It becomes uncertain when the zero constituents increase in number, and when several constituents are identical. But even then, in the majority of cases, it will soon afford a clue to the required arrangement.

570 *PRODUCT OF TWO DETERMINANTS OF THE nth ORDER.*

$$\overset{(P)}{\begin{vmatrix} a_1 & a_2 & \dots & a_n \\ b_1 & b_2 & \dots & b_n \\ \dots & & & \dots \\ l_1 & l_2 & \dots & l_n \end{vmatrix}} \; \overset{(Q)}{\begin{vmatrix} \alpha_1 & \alpha_2 & \dots & \alpha_n \\ \beta_1 & \beta_2 & \dots & \beta_n \\ \dots & & & \dots \\ \lambda_1 & \lambda_2 & \dots & \lambda_n \end{vmatrix}} = \overset{(S)}{\begin{vmatrix} A_1 & A_2 & \dots & A_n \\ B_1 & B_2 & \dots & B_n \\ \dots & & & \dots \\ L_1 & L_2 & \dots & L_n \end{vmatrix}}$$

The values of $A_1, B_1 \dots L_1$ in the first column of S are annexed. For the second column write b's in the place of a's. For the third column write c's, and so on.

$$\begin{cases} A_1 = a_1\alpha_1 + a_2\alpha_2 + \dots + a_n\alpha_n \\ B_1 = a_1\beta_1 + a_2\beta_2 + \dots + a_n\beta_n \\ \dots\dots\dots\dots\dots\dots \\ L_1 = a_1\lambda_1 + a_2\lambda_2 + \dots + a_n\lambda_n \end{cases}$$

For proof substitute the values of A_1, B_1, &c. in the determinant S, and then resolve S into the sum of a number of determinants by (562), and note the determinants which vanish through having identical columns.

Rule.—*To form the determinant S, which is the product of two determinants P and Q. First connect by plus signs the constituents in the rows of both the determinants P and Q.*

Now place the first row of P upon each row of Q in turn, and let each two constituents as they touch become products. This is the first column of S.

Perform the same operation upon Q with the second row of P to obtain the second column of S; and again with the third row of P to obtain the third column of S, and so on.

571 If the number of columns, both in P and Q, be n, and the number of rows r, and if n be $> r$, then the determinant S, found in the same way from P and Q, is equal to the sum of the $C(n, r)$ products of pairs of determinants obtained by taking any r columns out of P, and the corresponding r columns out of Q.

But if n be $< r$ the determinant S vanishes.

For in that case, in every one of the component determinants, there will be two columns identical.

---

572 The product of the determinants P and Q may be formed in four ways by changing the rows into columns in either or both P and Q.

---

573 Let the following system of n equations in $x_1 x_2 \ldots x_n$ be transformed by substituting the accompanying values of the variables,

$$\begin{array}{ll} a_1x_1+a_2x_2+\ldots+a_nx_n=0, & x_1=\alpha_1\xi_1+\alpha_2\xi_2+\ldots+\alpha_n\xi_n, \\ b_1x_1+b_2x_2+\ldots+b_nx_n=0, & x_2=\beta_1\xi_1+\beta_2\xi_2+\ldots+\beta_n\xi_n, \\ \ldots\ldots\ldots\ldots & \ldots\ldots\ldots\ldots \\ l_1x_1+l_2x_2+\ldots+l_nx_n=0, & x_n=\lambda_1\xi_1+\lambda_2\xi_2+\ldots+\lambda_n\xi_n. \end{array}$$

The eliminant of the resulting equations in $\xi_1 \xi_2 \ldots \xi_n$ is the determinant S in (570), and is therefore equal to the product of the determinants P and Q. The determinant Q is then termed the modulus of transformation.

---

574 ***A Symmetrical determinant*** is symmetrical about the leading diagonal. If the R's form the rth row, and the K's the kth row; then $R_k = K_r$ throughout a symmetrical determinant.

The square of a determinant is a symmetrical determinant.

---

575 ***A Reciprocal determinant*** has for its constituents the first minors of the original determinant, and is equal to its $n-1^{\text{th}}$ power; that is,

$$\begin{vmatrix} A_1 & \ldots & A_n \\ \ldots & \ldots & \ldots \\ \ldots & \ldots & \ldots \\ L_1 & \ldots & L_n \end{vmatrix} = \begin{vmatrix} a_1 & \ldots & a_n \\ \ldots & \ldots & \ldots \\ \ldots & \ldots & \ldots \\ l_1 & \ldots & l_n \end{vmatrix}^{n-1}$$

PROOF.—Multiply both sides of the equation by the original determinant (555). The constituents on the left side all vanish excepting the diagonal of Δ's.

---

576 ***Partial and Complementary determinants.***

If r rows and the same number of columns be selected from a determinant, and if the rows be brought to the top, and the columns to the left side, without changing their order, then the elements common to the selected rows and columns form a Partial determinant of the order r, and the elements *not* found in any of those rows and columns form the Complementary determinant, its order being $n-r$.

Ex.—Let the selected rows from the determinant $(a_1b_2c_3d_4e_5)$ be the second, third, and fifth; and the selected columns be the third, fourth, and fifth. The original and the transformed determinants will be

$$\begin{vmatrix} a_1 & a_2 & a_3 & a_4 & a_5 \\ b_1 & b_2 & b_3 & b_4 & b_5 \\ c_1 & c_2 & c_3 & c_4 & c_5 \\ d_1 & d_2 & d_3 & d_4 & d_5 \\ e_1 & e_2 & e_3 & e_4 & e_5 \end{vmatrix} \quad \text{and} \quad \begin{vmatrix} b_3 & b_4 & b_5 & b_1 & b_2 \\ c_3 & c_4 & c_5 & c_1 & c_2 \\ e_3 & e_4 & e_5 & e_1 & e_2 \\ a_3 & a_4 & a_5 & a_1 & a_2 \\ d_3 & d_4 & d_5 & d_1 & d_2 \end{vmatrix}$$

The partial determinant of the third order is $(b_3c_4e_5)$, and its complementary of the second order is (a_1d_2).

The complete altered determinant is plus or minus, according as the permutations of the rows and columns are of the same or of different class. In the example they are of the same class, for there have been four transpositions of rows, and six of columns. Thus $(-1)^{10} = +1$ gives the sign of the altered determinant.

577 Theorem.—A partial reciprocal determinant of the rth order is equal to the product of the $r-1^{\text{th}}$ power of the original determinant, and the complementary of its corresponding partial determinant.

Take the last determinant for an example. Here $n=5$, $r=3$; and by the theorem,

$$\begin{vmatrix} B_3 & B_4 & B_5 \\ C_3 & C_4 & C_5 \\ E_3 & E_4 & E_5 \end{vmatrix} = \Delta^2 \begin{vmatrix} a_1 & a_2 \\ d_1 & d_2 \end{vmatrix}$$ where B, C, E are the respective minors.

Proof.—Raise the Partial Reciprocal to the original order *five* without altering its value, by (564); and multiply it by Δ, with the rows and columns changed to correspond as in Ex. (576); thus, by (570), we have

$$\begin{vmatrix} B_3 & B_4 & B_5 & B_1 & B_2 \\ C_3 & C_4 & C_5 & C_1 & C_2 \\ E_3 & E_4 & E_5 & E_1 & E_2 \\ 0 & 0 & 0 & 1 & 0 \\ 0 & 0 & 0 & 0 & 1 \end{vmatrix} \begin{vmatrix} b_3 & b_4 & b_5 & b_1 & b_2 \\ c_3 & c_4 & c_5 & c_1 & c_2 \\ e_3 & e_4 & e_5 & e_1 & e_2 \\ a_3 & a_4 & a_5 & a_1 & a_2 \\ d_3 & d_4 & d_5 & d_1 & d_2 \end{vmatrix} = \begin{vmatrix} \Delta & 0 & 0 & b_1 & b_2 \\ 0 & \Delta & 0 & c_1 & c_2 \\ 0 & 0 & \Delta & e_1 & e_2 \\ 0 & 0 & 0 & a_1 & a_2 \\ 0 & 0 & 0 & d_1 & d_2 \end{vmatrix} = \Delta^3 \begin{vmatrix} a_1 & a_2 \\ d_1 & d_2 \end{vmatrix}$$

578 The product of the differences between every pair of n quantities $a_1, a_2 \dots a_n$,

$$\left.\begin{aligned} (a_1-a_2)(a_1-a_3)(a_1-a_4)\dots(a_1-a_n) \\ \times(a_2-a_3)(a_2-a_4)\dots(a_2-a_n) \\ \times(a_3-a_4)\dots(a_3-a_n) \\ \dots\dots\dots\dots\dots\dots \\ \times(a_{n-1}-a_n) \end{aligned}\right\} = \begin{vmatrix} 1 & 1 & 1 & \dots & 1 \\ a_1 & a_2 & a_3 & \dots & a_n \\ a_1^2 & a_2^2 & a_3^2 & \dots & a_n^2 \\ \dots & \dots & \dots & \dots & \dots \\ a_1^{n-1} & a_2^{n-1} & a_3^{n-1} & \dots & a_n^{n-1} \end{vmatrix}$$

Proof.—The determinant vanishes when any two of the quantities are equal. Therefore it is divisible by each of the factors on the left; therefore by their product. And the quotient is seen to be unity, for both sides of the equation are of the same degree; viz., $\frac{1}{2}n(n-1)$.

579 The product of the *squares* of the differences of the same n quantities

$$= \begin{vmatrix} s_0 & s_1 & \dots & s_{n-1} \\ s_1 & s_2 & \dots & s_n \\ \dots & \dots & \dots & \dots \\ s_{n-1} & s_n & \dots & s_{2n-2} \end{vmatrix}$$

Proof.—Square the determinant in (578), and write s_r for the sum of the rth powers of the roots.

580 With the same meaning for $s_1, s_2 \dots$, the same determinant taken of an order r, less than n, is equal to the sum of the products of the squares of the differences of r of the n quantities taken in every possible way; that is, in $C(n, r)$ ways.

Ex.:

$$\begin{vmatrix} s_0 & s_1 \\ s_1 & s_2 \end{vmatrix} = (a_1-a_2)^2+(a_1-a_3)^2+\&c. \equiv \Sigma\,(a_1-a_2)^2,$$

$$\begin{vmatrix} s_0 & s_1 & s_2 \\ s_1 & s_2 & s_3 \\ s_2 & s_3 & s_4 \end{vmatrix} = \Sigma\,(a_1-a_2)^2\,(a_1-a_3)^2\,(a_2-a_3)^2.$$

The next determinant in order

$$= \Sigma\,(a_1-a_2)^2\,(a_1-a_3)^2\,(a_1-a_4)^2\,(a_2-a_3)^2\,(a_2-a_4)^2\,(a_3-a_4)^2.$$

And so on until the equation (579) is reached.

Proved by substituting the values of $s_1, s_2 \dots$ &c., and resolving the determinant into its partial determinants by (571).

---

581 The quotient of

$$\frac{a_0x^m+a_1x^{m-1}+\dots+a_rx^{m-r}+\dots}{b_0x^n+b_1x^{n-1}+\dots+b_rx^{n-r}+\dots}$$

is given by the formula

$$q_0x^{m-n}+q_1x^{m-n-1}+\dots+q_rx^{m-n-r}+\dots,$$

where

$$q_r = \frac{1}{b_0^{r+1}} \begin{vmatrix} b_0 & 0 & 0 & \dots & a_0 \\ b_1 & b_0 & 0 & \dots & a_1 \\ b_2 & b_1 & b_0 & \dots & a_2 \\ \dots & \dots & \dots & \dots & \dots \\ b_r & b_{r-1} & b_{r-2} & \dots\, b_1 & a_r \end{vmatrix}$$

Proved by Induction.

ELIMINATION.

582 *Solution of n linear equations in n variables.*

The equations and the values of the variables are arranged below:

$$a_1x_1+a_2x_2+\ldots+a_nx_n = \xi_1, \qquad x_1\Delta = A_1\xi_1+B_1\xi_2+\ldots+L_1\xi_n$$
$$b_1x_1+b_2x_2+\ldots+b_nx_n = \xi_2, \qquad x_2\Delta = A_2\xi_1+B_2\xi_2+\ldots+L_2\xi_n$$
$$\ldots\ldots\ldots\ldots\ldots\ldots \qquad \ldots\ldots\ldots\ldots\ldots\ldots$$
$$l_1x_1+l_2x_2+\ldots+l_nx_n = \xi_n, \qquad x_n\Delta = A_n\xi_1+B_n\xi_2+\ldots+L_n\xi_n$$

where Δ is the determinant annexed, and A_1, B_1, &c. are its first minors.

$$\begin{vmatrix} a_1 & \ldots & a_n \\ \ldots & \ldots & \ldots \\ l_1 & \ldots & l_n \end{vmatrix}$$

To find the value of one of the unknowns x_r.

RULE.—*Multiply the equations respectively by the minors of the rth column, and add the results. x_r will be equal to the fraction whose numerator is the determinant Δ, with its rth column replaced by $\xi_1, \xi_2 \ldots \xi_n$, and whose denominator is Δ itself.*

583 If $\xi_1, \xi_2 \ldots \xi_n$ and Δ all vanish, then $x_1, x_2 \ldots x_n$ are in the ratios of the minors of any row of the determinant Δ. For example, in the ratios $C_1 : C_2 : C_3 : \ldots : C_n$.

The eliminant of the given equations is now $\Delta = 0$.

584 *Orthogonal Transformation.*

If the two sets of variables in the n equations (582) be connected by the relation

$$x_1^2+x_2^2+\ldots+x_n^2 = \xi_1^2+\xi_2^2+\ldots+\xi_n^2 \quad \ldots\ldots (1),$$

then the changing from one set of variables to the other, by substituting the values of the ξ's in terms of the x's in any function of the former, or *vice versâ*, is called orthogonal transformation.

When equation (1) is satisfied, two results follow.

I. *The determinant* $\Delta = \pm 1$.

II. *Each of the constituents of* Δ *is equal to the corresponding minor, or else to minus that minor according as* Δ *is positive or negative.*

PROOF.—Substitute the values of $\xi_1, \xi_2 \dots \xi_n$ in terms of $x_1, x_2 \dots x_n$ in equation (1), and equate coefficients of the squares and products of the new variables. We get the n^2 equations

$$\left.\begin{array}{r} a_1^2 + b_1^2 \;\; + = 1 \\ a_1a_2 + b_1b_2 + = 0 \\ a_1a_3 + b_1b_3 + = 0 \\ \dots\dots\dots\dots \\ a_1a_n + b_1b_n + = 0 \end{array}\right\} \quad \left.\begin{array}{r} a_2a_1 + b_2b_1 + = 0 \\ a_2^2 + b_2^2 \;\; + = 1 \\ a_2a_3 + b_2b_3 + = 0 \\ \dots\dots\dots\dots \\ a_2a_n + b_2b_n + = 0 \end{array}\right\} \quad \left.\begin{array}{r} a_3a_1 + b_3b_1 + = 0 \\ a_3a_2 + b_3b_2 + = 0 \\ a_3^2 + b_3^2 \;\; + = 1 \\ \dots\dots\dots\dots \\ a_3a_n + b_3b_n + = 0 \end{array}\right\}$$

Also $\Delta = \begin{vmatrix} a_1 b_1 \dots l_1 \\ a_2 b_2 \dots l_2 \\ a_3 b_3 \dots l_3 \\ \dots\dots\dots \\ a_n b_n \dots l_n \end{vmatrix}$ Form the square of the determinant Δ by the rule (570), and these equations show that the product is a determinant in which the only constituents that do not vanish constitute a diagonal of 'ones.' Therefore

$$\Delta^2 = 1 \text{ and } \Delta = \pm 1.$$

Again, solving the first set of equations for a_1 (writing a_1^2 as a_1a_1, &c.), the second set for a_2, the third for a_3, and so on, we have, by (582), the results annexed; which proves the second proposition.

$$\left\{\begin{array}{l} a_1\Delta = A_1 \;\; + A_20 + A_30 + \; = A_1 \\ a_2\Delta = A_10 + A_2 \;\; + A_30 + \; = A_2 \\ a_3\Delta = A_10 + A_20 + A_3 \;\; + \; = A_3 \\ \qquad \&c. \qquad\qquad \&c. \end{array}\right.$$

585 *Theorem.*—The $n-2^{\text{th}}$ power of a determinant of the nth order multiplied by any constituent is equal to the corresponding minor of the reciprocal determinant.

PROOF.—Let ρ be the reciprocal determinant of Δ, and β_r the minor of B_r in ρ. Write the transformed equations (582) for the x's in terms of the ξ's, and solve them for ξ_2. Then equate the coefficient of x_r in the result with its coefficient in the original value of ξ_2.

Thus $\rho\xi_2 = \Delta\,(\beta_1x_1 + \dots + \beta_rx_r + \dots)$, and $\xi_2 = b_1x_1 + \dots + b_rx_r + \dots$;

$$\therefore \;\; \Delta\beta_r = \rho b_r = \Delta^{n-1}b_r \text{ by (575)}; \qquad \therefore \;\; \beta_r = \Delta^{n-2}b_r.$$

586 To eliminate x from the two equations

$$ax^m + bx^{m-1} + cx^{m-2} + \dots = 0 \quad \dots\dots\dots\dots (1),$$

$$a'x^n + b'x^{n-1} + c'x^{n-2} + \dots = 0 \quad \dots\dots\dots\dots (2).$$

If it is desired that the equation should be homogeneous in x and y; put $\frac{x}{y}$ instead of x, and clear of fractions. The following methods will still be applicable.

I. ***Bezout's Method.***—Suppose $m > n$.

RULE.—*Bring the equations to the same degree by multiplying* (2) *by* x^{m-n}. *Then multiply* (1) *by* a', *and* (2) *by* a, *and subtract.*

Again, multiply (1) *by* $a'x+b'$, *and* (2) *by* $(ax+b)$, *and subtract.*

Again, multiply (1) *by* $a'x^2+b'x+c'$, *and* (2) *by* (ax^2+bx+c), *and subtract, and so on until n equations have been obtained. Each will be of the degree* $m-1$.

Write under these the $m-n$ *equations obtained by multiplying* (2) *successively by* x. *The eliminant of the m equations is the result required.*

Ex.—Let the equations be $\begin{cases} a\,x^5+b\,x^4+c\,x^3+d\,x^2+ex+f=0, \\ a'x^3+b'x^2+c'x\;+d' \qquad\qquad\; =0. \end{cases}$

The five equations obtained by the method, and their eliminant, by (583), are, writing capital letters for the functions of a, b, c, d, e, f,

$$\left.\begin{array}{r} A_1x^4+B_1x^3+C_1x^2+D_1x+E_1=0 \\ A_2x^4+B_2x^3+C_2x^2+D_2x+E_2=0 \\ A_3x^4+B_3x^3+C_3x^2+D_3x+E_3=0 \\ a'\,x^4+b'\,x^3+c'\,x^2+d'\,x \qquad =0 \\ a'\,x^3+b'\,x^2+c'\,x+d'\;=0 \end{array}\right\} \quad\text{and}\quad \begin{vmatrix} A_1 & B_1 & C_1 & D_1 & E_1 \\ A_2 & B_2 & C_2 & D_2 & E_2 \\ A_3 & B_3 & C_3 & D_3 & E_3 \\ a' & b' & c' & d' & 0 \\ 0 & a' & b' & c' & d' \end{vmatrix} = 0.$$

Should the equations be of the same degree, the eliminant will be a symmetrical determinant.

---

587 II. ***Sylvester's Dialytic Method.***

RULE.—*Multiply equation* (1) *successively by* x, $n-1$ *times; and equation* (2) $m-1$ *times; and eliminate* x *from the* $m+n$ *resulting equations.*

Ex.—To eliminate x from $\left.\begin{array}{r} ax^3+bx^2+cx+d=0 \\ px+qx+r=0 \end{array}\right\}$.

The $m+n$ equations and their eliminant are

$$\left.\begin{array}{r} px^2+qx+r=0 \\ px^3+qx^2+rx \quad\;\; =0 \\ px^4+qx^3+rx^2 \qquad\;\; =0 \\ ax^3+bx^2+cx+d=0 \\ ax^4+bx^3+cx^2+dx \quad =0 \end{array}\right\} \quad\text{and}\quad \begin{vmatrix} 0 & 0 & p & q & r \\ 0 & p & q & r & 0 \\ p & q & r & 0 & 0 \\ 0 & a & b & c & d \\ a & b & c & d & 0 \end{vmatrix} = 0.$$

588 III. ***Method of elimination by Symmetrical Functions.***

Divide the two equations in (586) respectively by the coefficients of their first terms, thus reducing them to the forms

$$f(x) \equiv x^m + p_1 x^{m-1} + \dots + p_m = 0,$$
$$\phi(x) \equiv x^n + q_1 x^{n-1} + \dots + q_n = 0.$$

RULE.—*Let $a, b, c \dots$ represent the roots of $f(x)$. Form the equation $\phi(a)\,\phi(b)\,\phi(c) \dots = 0$. This will contain symmetrical functions only of the roots $a, b, c \dots$.*

Express these functions in terms of $p_1, p_2 \dots$ by (538), *&c., and the equation becomes the eliminant.*

Reason of the rule.—The eliminant is the condition for a common root of the two equations. That root must make one of the factors $\phi(a), \phi(b) \dots$ vanish, and therefore it makes their product vanish.

589 The eliminant expressed in terms of the roots $a, b, c \dots$ of $f(x)$, and the roots $\alpha, \beta, \gamma \dots$ of $\phi(x)$, will be

$$(a-\alpha)(a-\beta)(a-\gamma) \dots (b-\alpha)(b-\beta)(b-\gamma) \dots \&c.,$$

being the product of all possible differences between a root of one equation and a root of another.

590 The eliminant is a homogeneous function of the coefficients of either equation, being of the nth degree in the coefficients of $f(x)$, and of the mth degree in the coefficients of $\phi(x)$.

591 The sum of the suffixes of p and q in each term of the eliminant $= mn$. Also, if p, q contain z; if p_2, q_2 contain z^2; if p_3, q_3 contain z^3, and so on, the eliminant will contain z^{mn}.

Proved by the fact that p_r is a homogeneous function of r dimensions of the roots $a, b, c \dots$, by (406).

592 If the two equations involve x and y, the elimination may be conducted with respect to x; and y will be contained in the coefficients $p_1, p_2 \dots, q_1, q_2 \dots$.

593 ***Elimination by the Method of Highest Common Factor.***

Let two algebraical equations in x and y be represented by $A = 0$ and $B = 0$.

It is required to eliminate x.

Arrange A and B according to descending powers of x, and, having rejected any factor which is a function of y only, proceed to find the Highest Common Factor of A and B.

The process may be exhibited as follows:

$$\left.\begin{aligned} c_1A &= q_1B + r_1R_1 \\ c_2B &= q_2R_1 + r_2R_2 \\ c_3R_1 &= q_3R_2 + r_3R_3 \\ c_4R_2 &= q_4R_3 + r_4 \end{aligned}\right\}$$

c_1, c_2, c_3, c_4 are the multipliers required at each stage in order to avoid fractional quotients; and these must be constants or functions of y only.

q_1, q_2, q_3, q_4 are the successive quotients.

$r_1R_1, r_2R_2, r_3R_3, r_4$ are the successive remainders; r_1, r_2, r_3, r_4 being functions of y only.

The process terminates as soon as a remainder is obtained which is a function of y only; r_4 is here supposed to be such a remainder.

Now, the simplest factors having been taken for c_1, c_2, c_3, c_4, we see that

$$\left.\begin{array}{llll} 1 \text{ is the H. C. F. of} & c_1 & \text{and} & r_1 \\ d_2 \quad ,, \quad ,, & c_1 & \text{and} & r_2 \\ d_3 \quad ,, \quad ,, & \dfrac{c_1c_2}{d_2} & \text{and} & r_3 \\ d_4 \quad ,, \quad ,, & \dfrac{c_1c_2c_3}{d_2d_3} & \text{and} & r_4 \end{array}\right\}$$

The values of x and y, which satisfy simultaneously the equations $A=0$ and $B=0$, are those obtained by the four pairs of simultaneous equations following:

$$\left.\begin{array}{l} r_1 = 0 \text{ and } B = 0 \,\ldots\ldots\ldots (1) \\ \dfrac{r_2}{d_2} = 0 \text{ and } R_1 = 0 \,\ldots\ldots\ldots (2) \\ \dfrac{r_3}{d_3} = 0 \text{ and } R_2 = 0 \,\ldots\ldots\ldots (3) \\ \dfrac{r_4}{d_4} = 0 \text{ and } R_3 = 0 \,\ldots\ldots\ldots (4) \end{array}\right\}$$

The final equation in y, which gives all admissible values, is

$$\frac{r_1r_2r_3r_4}{d_2d_3d_4} = 0.$$

If it should happen that the remainder r_4 is zero, the simultaneous equations (1), (2), (3), and (4) reduce to

$$r_1=0 \text{ and } \frac{B}{R_3}=0; \quad \frac{r_2}{d_2}=0 \text{ and } \frac{R_1}{R_3}=0; \quad \frac{r_3}{d_3}=0 \text{ and } \frac{R_2}{R_3}=0.$$

594 To find infinite values of x or y which satisfy the given equations.

Put $x = \frac{1}{z}$. Clear of fractions, and make $z = 0$.

If the two resulting equations in y have any common roots, such roots, together with $x = \infty$, satisfy simultaneously the equations proposed.

Similarly we may put $y = \frac{1}{z}$.

PLANE TRIGONOMETRY.

ANGULAR MEASUREMENT.

600 The unit of Circular measure is the angle at the centre of a circle which subtends an arc equal to the radius. Hence

601 Circular measure of an angle $= \dfrac{\text{arc}}{\text{radius}}$.

602 Circular measure of two right angles $= 3{\cdot}14159\ldots \equiv \pi$.

603 The unit of Centesimal measure is a Grade, and is the one-hundredth part of a right angle.

604 The unit of Sexagesimal measure is a Degree, and is the one-ninetieth part of a right angle.

To change degrees into grades, or circular measure, or *vice versâ*, employ one of the three equations included in

605 $$\frac{D}{90} = \frac{G}{100} = \frac{2C}{\pi},$$

where D, G, and C are respectively the numbers of degrees, grades, and units of circular measure in the angle considered.

TRIGONOMETRICAL RATIOS.

606 Let OA be fixed, and let the revolving line OP describe a circle round O. Draw PN always perpendicular to AA'. Then, in all positions of OP,

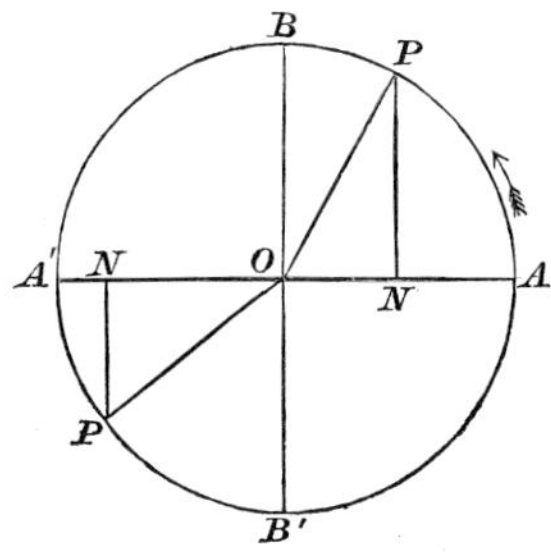

$\dfrac{PN}{OP}$ = the sine of the angle AOP,

$\dfrac{ON}{OP}$ = the cosine of the angle AOP,

$\dfrac{PN}{ON}$ = the tangent of the angle AOP.

607 If P be *above* the line AA', $\sin AOP$ is positive.
If P be *below* the line AA', $\sin AOP$ is negative.

608 If P lies to the *right* of BB', $\cos AOP$ is positive.
If P lies to the *left* of BB', $\cos AOP$ is negative.

---

609 Note, that by the angle AOP is meant the angle through which OP has revolved from OA, its initial position; and this angle of revolution may have any magnitude. If the revolution takes place in the opposite direction, the angle described is reckoned negative.

---

610 The secant of an angle is the reciprocal of its cosine, or

$$\cos A \sec A = 1.$$

611 The cosecant of an angle is the reciprocal of its sine, or

$$\sin A \operatorname{cosec} A = 1.$$

612 The cotangent of an angle is the reciprocal of its tangent, or

$$\tan A \cot A = 1.$$

Relations between the trigonometrical functions of the same angle.

613 $$\sin^2 A + \cos^2 A = 1.$$ [I. 47

614 $$\sec^2 A = 1 + \tan^2 A.$$

615 $$\operatorname{cosec}^2 A = 1 + \cot^2 A.$$

616 $$\tan A = \frac{\sin A}{\cos A}.$$ [606

If $\tan A = \dfrac{a}{b}$.

617 $$\sin A = \frac{a}{\sqrt{a^2+b^2}},$$

$$\cos A = \frac{b}{\sqrt{a^2+b^2}}.$$ [606

618 $$\sin A = \frac{\tan A}{\sqrt{1+\tan^2 A}}, \quad \cos A = \frac{1}{\sqrt{1+\tan^2 A}}$$ [617

619 The Complement of A is $= 90^\circ - A$.

620 The Supplement of A is $= 180^\circ - A$.

621 $\sin(90^\circ - A) = \cos A$,
$\tan(90^\circ - A) = \cot A$,
$\sec(90^\circ - A) = \operatorname{cosec} A$.

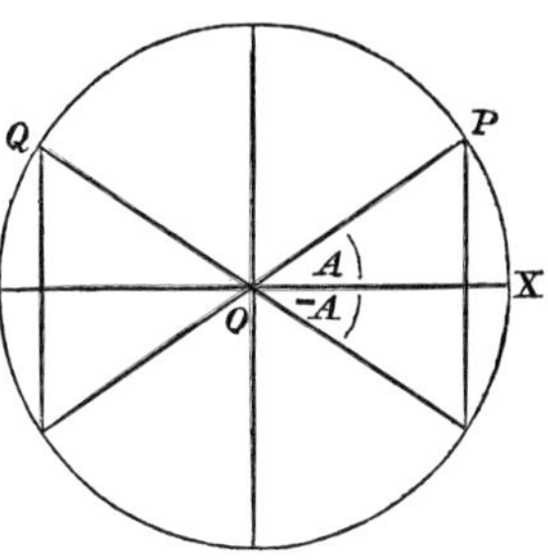

622 $\sin(180^\circ - A) = \sin A$,
$\cos(180^\circ - A) = -\cos A$,
$\tan(180^\circ - A) = -\tan A$.

In the figure

$\angle QOX = 180^\circ - A$. [607, 608

---

623 $\sin(-A) = -\sin A$.

624 $\cos(-A) = \cos A$.

By Fig., and (607), (608).

The secant, cosecant, and cotangent of $180^\circ - A$, and of $-A$, will follow the same rule as their reciprocals, the cosine, sine, and tangent. [610—612

---

625 To reduce any ratio of an angle greater than 90° to the ratio of an angle less than 90°.

RULE.—Determine the sign of the ratio by the rules (607), and then substitute for the given angle the *acute* angle formed by its two bounding lines, produced if necessary.

EX.—To find all the ratios of 660°. Measuring 300° (= 660° − 360°) round the circle from A to P, we find the acute angle AOP to be 60°, and P lies *below* AA', and to the *right* of BB'.

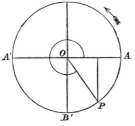

Therefore

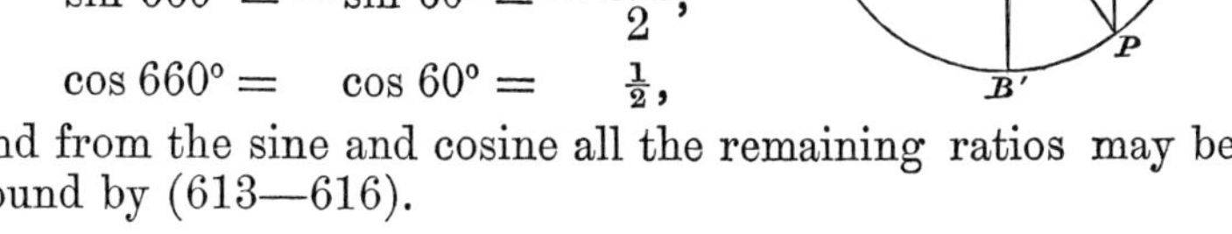

$$\sin 660^\circ = -\sin 60^\circ = -\frac{\sqrt{3}}{2},$$

$$\cos 660^\circ = \cos 60^\circ = \tfrac{1}{2},$$

and from the sine and cosine all the remaining ratios may be found by (613—616).

INVERSE NOTATION.—The angle whose sine is x is denoted by $\sin^{-1} x$.

626 All the angles which have a given sine, cosine, or tangent, are given by the formulæ

$$\sin^{-1} x = n\pi + (-1)^n \theta \quad\text{.............. (1)},$$
$$\cos^{-1} x = 2n\pi \pm \theta \quad\text{........................ (2)},$$
$$\tan^{-1} x = n\pi + \theta \quad\text{........................ (3)}.$$

In these formulæ θ is any angle which has x for its sine, cosine, or tangent respectively, and n is any integer.

$\operatorname{Cosec}^{-1} x$, $\sec^{-1} x$, $\cot^{-1} x$ have similar general values, by (610—612).

These formulæ are verified by taking A, in Fig. 622, for θ, and making n an odd or even integer successively.

FORMULÆ INVOLVING TWO ANGLES, AND MULTIPLE ANGLES.

627 $\sin(A+B) = \sin A \cos B + \cos A \sin B,$

628 $\sin(A-B) = \sin A \cos B - \cos A \sin B,$

629 $\cos(A+B) = \cos A \cos B - \sin A \sin B,$

630 $\cos(A-B) = \cos A \cos B + \sin A \sin B.$

Proofs of (627) to (630).—By (700) and (701), we have

$$\sin C = \sin A \cos B + \cos A \sin B,$$

and $\sin C = \sin(A+B)$, by (622).

To obtain $\sin(A-B)$ change the sign of B in (627), and employ (623), (624), $\cos(A+B) = \sin\{(90°-A)-B\}$, by (621).

Expand by (628), and use (621), (623), (624). For $\cos(A-B)$ change the sign of B in (629).

631 $$\tan(A+B) = \frac{\tan A + \tan B}{1-\tan A \tan B}.$$

632 $$\tan(A-B) = \frac{\tan A - \tan B}{1+\tan A \tan B}.$$

633 $$\cot(A+B) = \frac{\cot A \cot B - 1}{\cot A + \cot B}.$$

634 $$\cot(A-B) = \frac{\cot A \cot B + 1}{\cot B - \cot A}.$$

Obtained from (627—630).

635 $$\sin 2A = 2 \sin A \cos A.$$ [627. Put $B=A$

636 $$\cos 2A = \cos^2 A - \sin^2 A,$$

637 $$= 2\cos^2 A - 1,$$

638 $$= 1 - 2\sin^2 A.$$ [629, 613

639 $$2\cos^2 A = 1 + \cos 2A.$$ [637

640 $$2\sin^2 A = 1 - \cos 2A.$$ [638

641 $$\sin \frac{A}{2} = \sqrt{\frac{1-\cos A}{2}}.$$ [640

642 $$\cos \frac{A}{2} = \sqrt{\frac{1+\cos A}{2}}.$$ [639

---

643 $$\tan \frac{A}{2} = \sqrt{\frac{1-\cos A}{1+\cos A}} = \frac{1-\cos A}{\sin A} = \frac{\sin A}{1+\cos A}.$$ [641, 642, 613

---

646 $$\cos A = \frac{1-\tan^2 \frac{A}{2}}{1+\tan^2 \frac{A}{2}}. \qquad \sin A = \frac{2\tan \frac{A}{2}}{1+\tan^2 \frac{A}{2}}.$$ [643, 613

---

648 $$\cos A = \frac{1}{1+\tan A \tan \frac{A}{2}}.$$

649 $$\sin\left(45^\circ + \frac{A}{2}\right) = \cos\left(45^\circ - \frac{A}{2}\right) = \sqrt{\frac{1+\sin A}{2}}.$$ [641

650 $$\cos\left(45^\circ + \frac{A}{2}\right) = \sin\left(45^\circ - \frac{A}{2}\right) = \sqrt{\frac{1-\sin A}{2}}.$$ [642

651 $$\tan\left(45^\circ + \frac{A}{2}\right) = \sqrt{\frac{1+\sin A}{1-\sin A}} = \frac{1+\sin A}{\cos A} = \frac{\cos A}{1-\sin A}.$$

---

652 $$\tan 2A = \frac{2\tan A}{1-\tan^2 A}.$$ [631. Put $B=A$

653 $$\cot 2A = \frac{\cot^2 A - 1}{2\cot A}.$$

654 $$\tan(45^\circ + A) = \frac{1 + \tan A}{1 - \tan A}.$$

655 $$\tan(45^\circ - A) = \frac{1 - \tan A}{1 + \tan A}.$$ [631, 632

656 $$\sin 3A = 3\sin A - 4\sin^3 A.$$

657 $$\cos 3A = 4\cos^3 A - 3\cos A.$$

658 $$\tan 3A = \frac{3\tan A - \tan^3 A}{1 - 3\tan^2 A}.$$

By putting $B=2A$ in (627), (629), and (631).

659 $$\sin(A+B)\sin(A-B) = \sin^2 A - \sin^2 B = \cos^2 B - \cos^2 A.$$

660 $$\cos(A+B)\cos(A-B) = \cos^2 A - \sin^2 B = \cos^2 B - \sin^2 A.$$

From (627), &c.

661 $$\sin\frac{A}{2} + \cos\frac{A}{2} = \pm\sqrt{1 + \sin A}.$$

Proved by squaring.

662 $$\sin\frac{A}{2} - \cos\frac{A}{2} = \pm\sqrt{1 - \sin A}.$$

663 $$\sin\frac{A}{2} = \frac{1}{2}\left\{\sqrt{1+\sin A} - \sqrt{1-\sin A}\right\}.$$

664 $$\cos\frac{A}{2} = \frac{1}{2}\left\{\sqrt{1+\sin A} + \sqrt{1-\sin A}\right\},$$

when $\frac{A}{2}$ lies between -45° and $+45^\circ$.

665 In the accompanying diagram the signs exhibited in each quadrant are the signs to be prefixed to the two surds in the value of $\sin\frac{A}{2}$ according to the quadrant in which $\frac{A}{2}$ lies.

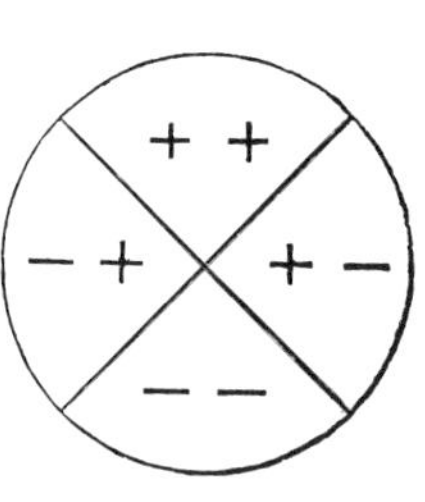

For $\cos\frac{A}{2}$ change the second sign.

Proved by examining the changes of sign in (661) and (662) by (607).

666 $\sin(A+B)+\sin(A-B) = 2\sin A\cos B.$

667 $\sin(A+B)-\sin(A-B) = 2\cos A\sin B.$

668 $\cos(A+B)+\cos(A-B) = 2\cos A\cos B.$

669 $\cos(A-B)-\cos(A+B) = 2\sin A\sin B.$

[627—630

670 $$\sin A+\sin B = 2\sin\frac{A+B}{2}\cos\frac{A-B}{2}.$$

671 $$\sin A-\sin B = 2\cos\frac{A+B}{2}\sin\frac{A-B}{2}.$$

672 $$\cos A+\cos B = 2\cos\frac{A+B}{2}\cos\frac{A-B}{2}.$$

673 $$\cos B-\cos A = 2\sin\frac{A+B}{2}\sin\frac{A-B}{2}.$$

Obtained by changing A into $\frac{A+B}{2}$, and B into $\frac{A-B}{2}$, in (666—669).

It is advantageous to commit the foregoing formulæ to memory, in words, thus—

sin sum + sin difference = 2 sin cos,
sin sum − sin difference = 2 cos sin,
cos sum + cos difference = 2 cos cos,
cos difference − cos sum = 2 sin sin.

sin first + sin second = 2 sin half sum cos half difference,
sin first − sin second = 2 cos half sum sin half difference,
cos first + cos second = 2 cos half sum cos half difference,
cos second − cos first = 2 sin half sum sin half difference.

674 $\sin(A+B+C)$

$$= \sin A\cos B\cos C+\sin B\cos C\cos A + \sin C\cos A\cos B-\sin A\sin B\sin C.$$

675 $\cos(A+B+C)$

$$= \cos A\cos B\cos C-\cos A\sin B\sin C - \cos B\sin C\sin A-\cos C\sin A\sin B.$$

676 $\tan(A+B+C)$

$$= \frac{\tan A+\tan B+\tan C-\tan A\tan B\tan C}{1-\tan B\tan C-\tan C\tan A-\tan A\tan B}.$$

Put $B+C$ for B in (627), (629), and (631).

If $A+B+C=180°$,

677 $\sin A+\sin B+\sin C = 4\cos\frac{A}{2}\cos\frac{B}{2}\cos\frac{C}{2}$.

$\sin A+\sin B-\sin C = 4\sin\frac{A}{2}\sin\frac{B}{2}\cos\frac{C}{2}$.

678 $\cos A+\cos B+\cos C = 4\sin\frac{A}{2}\sin\frac{B}{2}\sin\frac{C}{2}+1$.

$\cos A+\cos B-\cos C = 4\cos\frac{A}{2}\cos\frac{B}{2}\sin\frac{C}{2}-1$.

679 $\tan A+\tan B+\tan C = \tan A\tan B\tan C$.

680 $\cot\frac{A}{2}+\cot\frac{B}{2}+\cot\frac{C}{2} = \cot\frac{A}{2}\cot\frac{B}{2}\cot\frac{C}{2}$.

681 $\sin 2A+\sin 2B+\sin 2C = 4\sin A\sin B\sin C$.

682 $\cos 2A+\cos 2B+\cos 2C = -4\cos A\cos B\cos C-1$.

General formulæ, including the foregoing, obtained by applying (666—673).

If $A+B+C=\pi$, and n be any integer,

683 $4\sin\frac{nA}{2}\sin\frac{nB}{2}\sin\frac{nC}{2}$

$$=\sin\left(\frac{n\pi}{2}-nA\right)+\sin\left(\frac{n\pi}{2}-nB\right)+\sin\left(\frac{n\pi}{2}-nC\right)-\sin\frac{n\pi}{2}.$$

684 $4\cos\frac{nA}{2}\cos\frac{nB}{2}\cos\frac{nC}{2}$

$$=\cos\left(\frac{n\pi}{2}-nA\right)+\cos\left(\frac{n\pi}{2}-nB\right)+\cos\left(\frac{n\pi}{2}-nC\right)+\cos\frac{n\pi}{2}.$$

If $A+B+C=0$,

685 $4\sin\frac{nA}{2}\sin\frac{nB}{2}\sin\frac{nC}{2} = -\sin nA-\sin nB-\sin nC$.

686 $4\cos\frac{nA}{2}\cos\frac{nB}{2}\cos\frac{nC}{2} = \cos nA+\cos nB+\cos nC+1$.

RULE.—*If, in formulæ* (683) *to* (686), *two factors on the left be changed by writing sin for cos, or cos for sin; then, on the right side, change the signs of those terms which do not contain the angles of the altered factors.*

Thus, from (683), we obtain

687 $$4\sin\frac{nA}{2}\cos\frac{nB}{2}\cos\frac{nC}{2}$$
$$= -\sin\left(\frac{n\pi}{2}-nA\right)+\sin\left(\frac{n\pi}{2}-nB\right)+\sin\left(\frac{n\pi}{2}-nC\right)+\sin\frac{n\pi}{2}.$$

A Formula for the construction of Tables of sines, cosines, &c.—

688 $$\sin(n+1)\,\alpha-\sin n\alpha=\sin n\alpha-\sin(n-1)\,\alpha-k\sin n\alpha,$$
where $\alpha=10''$, and $k=2(1-\cos\alpha)=\cdot 0000000023504$.

689 Formulæ for verifying the tables—
$$\sin A+\sin(72^\circ+A)-\sin(72^\circ-A)=\sin(36^\circ+A)-\sin(36^\circ-A),$$
$$\cos A+\cos(72^\circ+A)+\cos(72^\circ-A)=\cos(36^\circ+A)+\cos(36^\circ-A),$$
$$\sin(60^\circ+A)-\sin(60^\circ-A)=\sin A.$$

RATIOS OF CERTAIN ANGLES.

690 $$\sin 45^\circ=\cos 45^\circ=\frac{1}{\sqrt{2}},\quad \tan 45^\circ=1.$$

691 $$\sin 60^\circ=\frac{\sqrt{3}}{2},\quad \cos 60^\circ=\frac{1}{2},\quad \tan 60^\circ=\sqrt{3}.$$

692 $$\sin 15^\circ=\frac{\sqrt{3}-1}{2\sqrt{2}},\quad \cos 15^\circ=\frac{\sqrt{3}+1}{2\sqrt{2}},$$
$$\left.\begin{array}{l}\tan 15^\circ=2-\sqrt{3}\\ \cot 15^\circ=2+\sqrt{3}\end{array}\right\}.$$

693 $$\sin 18^\circ=\frac{\sqrt{5}-1}{4},\quad \cos 18^\circ=\frac{\sqrt{5+\sqrt{5}}}{2\sqrt{2}},$$
$$\tan 18^\circ=\sqrt{\frac{5-2\sqrt{5}}{5}}.$$

694 $$\sin 54^\circ=\frac{\sqrt{5}+1}{4},\quad \cos 54^\circ=\frac{\sqrt{5-\sqrt{5}}}{2\sqrt{2}},$$
$$\tan 54^\circ=\sqrt{\frac{5+2\sqrt{5}}{5}}.$$

695 By taking the complements of these angles, the same table gives the ratios of 30°, 75°, 72°, and 36°.

696 NOTE.—$\sin 15°$ is obtained from $\sin(45°-30°)$, expanded by (628).

697 $\sin 18°$ from the equation $\sin 2x = \cos 3x$, where $x = 18°$.

698 $\sin 54°$ from $\sin 3x = 3\sin x - 4\sin^3 x$, where $x = 18°$.

699 And the ratios of various angles may be obtained by taking the sum, difference, or some multiple of the angles in the table, and making use of known formulæ. Thus

$$12° = 30° - 18°, \quad 7\tfrac{1}{2}° = \frac{15°}{2}, \quad \text{\&c. \&c.}$$

PROPERTIES OF THE TRIANGLE.

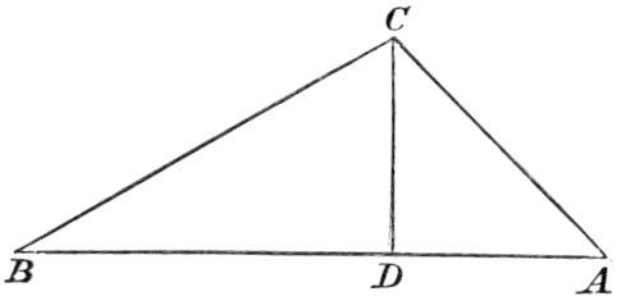

700 $c = a\cos B + b\cos A.$

701 $\dfrac{a}{\sin A} = \dfrac{b}{\sin B} = \dfrac{c}{\sin C}.$

702 $a^2 = b^2 + c^2 - 2bc\cos A.$

[By Euc. II. 5 & 6.

703 $$\cos A = \frac{b^2+c^2-a^2}{2bc}.$$

If $s \equiv \dfrac{a+b+c}{2}$, and Δ denote the area ABC,

704 $$\sin\frac{A}{2} = \sqrt{\frac{(s-b)(s-c)}{bc}}, \quad \cos\frac{A}{2} = \sqrt{\frac{s(s-a)}{bc}}.$$

[641, 642, 703, 9, 10, 1

705 $$\tan\frac{A}{2} = \sqrt{\frac{(s-b)(s-c)}{s(s-a)}}.$$

706 $$\sin A = \frac{2}{bc}\sqrt{s(s-a)(s-b)(s-c)}.$$ [635, 704

707 $$\Delta = \frac{bc}{2}\sin A = \sqrt{s(s-a)(s-b)(s-c)}.$$ [707, 706

708 $$= \tfrac{1}{4}\sqrt{2b^2c^2+2c^2a^2+2a^2b^2-a^4-b^4-c^4}.$$

The Triangle and Circle.

Let

r = radius of inscribed circle.

r_a = radius of escribed circle touching the side a.

R = radius of circumscribing circle.

709 $$r = \frac{\Delta}{s}.$$

[From Fig., $\Delta = \frac{ra}{2} + \frac{rb}{2} + \frac{rc}{2}$.

710 $$r = \frac{a \sin \frac{B}{2} \sin \frac{C}{2}}{\cos \frac{A}{2}}.$$

[By $a = r \cot \frac{B}{2} + r \cot \frac{C}{2}$.

711 $$r_a = \frac{\Delta}{s-a}.$$ [By $\Delta = \frac{r_a b}{2} + \frac{r_a c}{2} - \frac{r_a a}{2}$.

712 $$r_a = \frac{a \cos \frac{B}{2} \cos \frac{C}{2}}{\cos \frac{A}{2}}.$$ [From $a = r_a \tan \frac{B}{2} + r_a \tan \frac{C}{2}$.

---

713 $$R = \frac{a}{2 \sin A} = \frac{abc}{4\Delta}.$$

[By (III. 20) and (706).

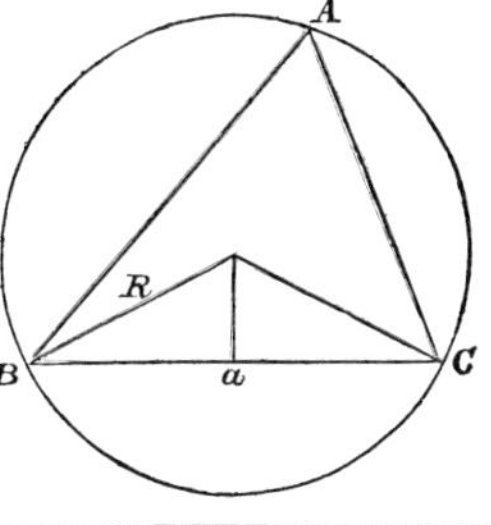

715 $$= \tfrac{1}{4}\sqrt{\left\{(b+c)^2 \sec^2 \frac{A}{2} + (b-c)^2 \operatorname{cosec}^2 \frac{A}{2}\right\}}$$

[702

Distance between the centres of inscribed and circumscribed circles

716 $$= \sqrt{R^2 - 2Rr}.$$ [936

Radius of circle touching b, c and the inscribed circle

717 $$r' = r \tan^2 \tfrac{1}{4}(B+C).$$ [By $\sin \frac{A}{2} = \frac{r-r'}{r+r'}$.

SOLUTION OF TRIANGLES.

Right-angled triangles are solved by the formulæ

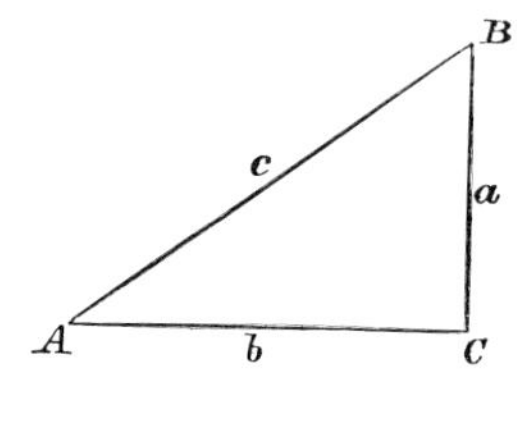

718 $$c^2 = a^2+b^2;$$

719 $$\begin{cases} a = c \sin A, \\ b = c \cos A, \\ a = b \tan A, \end{cases}$$
&c.

Scalene Triangles.

720 Case I.—The equation

$$\frac{a}{\sin A} = \frac{b}{\sin B} \qquad [701$$

will determine any one of the four quantities A, B, a, b when the remaining three are known.

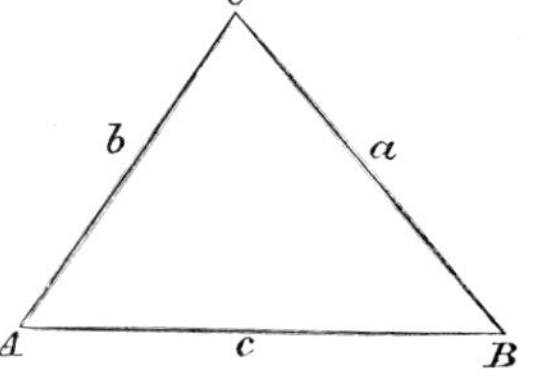

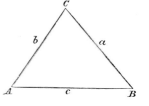

721 *The Ambiguous Case.*

When, in Case I., two sides and an acute angle opposite to one of them are given, we have, from the figure,

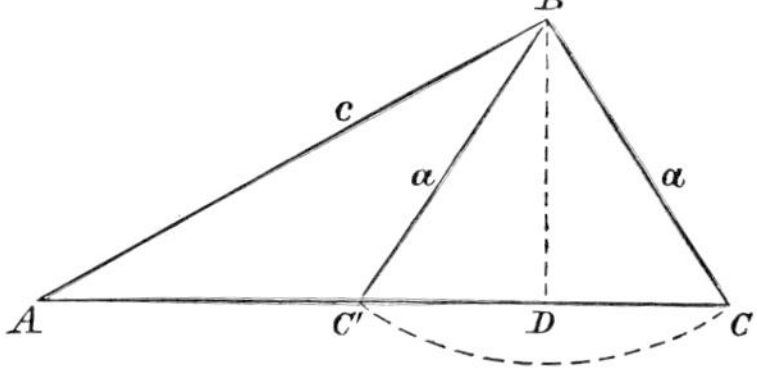

$$\sin C = \frac{c \sin A}{a}.$$

Then C and $180°-C$ are the values of C and C', by (622).

Also $$b = c \cos A \pm \sqrt{a^2-c^2 \sin^2 A},$$

because $$= AD \pm DC.$$

722 When an angle B is to be determined from the equation

$$\sin B = \frac{b}{a} \sin A,$$

and $\frac{b}{a}$ is a small fraction; the circular measure of B may be approximated to by putting $\sin(B+C)$ for $\sin A$, and using theorem (796).

723 CASE II.—When two sides b, c and the included angle A are known, the third side a is given by the formula

$$a^2 = b^2+c^2-2bc\cos A, \qquad [702$$

when logarithms are not used.

Otherwise, employ the following formula with logarithms,

724 $$\tan\frac{B-C}{2} = \frac{b-c}{b+c}\cot\frac{A}{2}.$$

Obtained from $\frac{b-c}{b+c} = \frac{\sin B - \sin C}{\sin B + \sin C}$ (701), and then applying (670) and (671).

$\frac{B-C}{2}$ having been found from the above equation, and $\frac{B+C}{2}$ being equal to $90^\circ - \frac{A}{2}$, we have

725 $$B = \frac{B+C}{2} + \frac{B-C}{2}, \qquad C = \frac{B+C}{2} - \frac{B-C}{2}.$$

B and C having been determined, a can be found by Case I.

---

726 If the logarithms of b and c are known, the trouble of taking out $\log(b-c)$ and $\log(b+c)$ may be avoided by employing the subsidiary angle $\theta = \tan^{-1}\frac{b}{c}$, and the formula

727 $$\tan\tfrac{1}{2}(B-C) = \tan\left(\theta - \frac{\pi}{4}\right)\cot\frac{A}{2}. \qquad [655$$

Or else the subsidiary angle $\theta = \cos^{-1}\frac{c}{b}$, and the formula

728 $$\tan\tfrac{1}{2}(B-C) = \tan^2\frac{\theta}{2}\cot\frac{A}{2}. \qquad [643$$

If a be required without calculating the angles B and C, we may use the formula

729 $$a = \frac{(b+c)\sin\frac{A}{2}}{\cos\frac{1}{2}(B-C)}.$$

From the figure in 960, by drawing a perpendicular from B to EC produced.

---

730 If a be required in terms of b, c, and A alone, and in a form adapted to logarithmic computation, employ the subsidiary angle

$$\theta = \sin^{-1}\left(\frac{4bc}{(b+c)^2}\cos^2\frac{A}{2}\right),$$

and the formula $$a = (b+c)\cos\theta. \qquad [702, 637$$

CASE III.—When the three sides are known, the angles may be found without employing logarithms, from the formula

731 $$\cos A = \frac{b^2+c^2-a^2}{2bc}.$$ [703

732 If logarithms are to be used, take the formulæ for $\sin\frac{A}{2}$, $\cos\frac{A}{2}$, or $\tan\frac{A}{2}$; (704) and (705).

QUADRILATERAL INSCRIBED IN A CIRCLE.

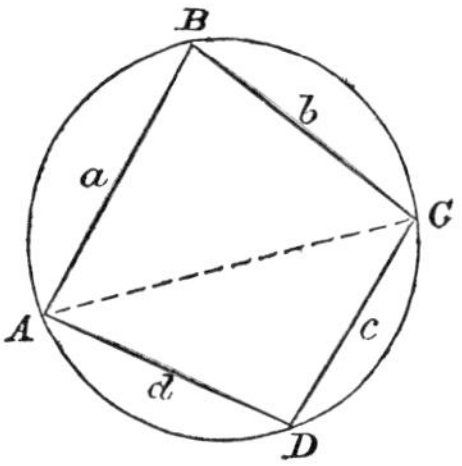

733 $$\cos B = \frac{a^2+b^2-c^2-d^2}{2(ab+cd)}.$$

From $AC^2 = a^2+b^2-2ab\cos B = c^2+d^2+2cd\cos B$, by (702), and $B+D=180°$.

734 $$\sin B = \frac{2Q}{ab+cd}.$$ [613, 733

735 $$Q = \sqrt{(s-a)(s-b)(s-c)(s-d)}$$ $= \text{area of } ABCD$,

and $$s = \tfrac{1}{2}(a+b+c+d).$$

Area $= \frac{1}{2}ab\sin B + \frac{1}{2}cd\sin B$; substitute $\sin B$ from last.

736 $$AC^2 = \frac{(ac+bd)(ad+bc)}{(ab+cd)}.$$ [702, 733

Radius of circumscribed circle

737 $$= \frac{1}{4Q}\sqrt{(ab+cd)(ac+bd)(ad+bc)}.$$ [713, 734, 736

If AD bisect the side of the triangle ABC in D,

738 $$\tan BDA = \frac{4\Delta}{b^2-c^2}.$$

739 $$\cot BAD = 2\cot A + \cot B.$$

740 $$AD^2 = \tfrac{1}{4}(b^2+c^2+2bc\cos A) = \tfrac{1}{2}(b^2+c^2-\tfrac{1}{2}a^2).$$

If AD bisect the angle A of a triangle ABC,

742 $$\tan BDA = \cot\frac{B-C}{2} = \frac{b+c}{b-c}\tan\frac{A}{2}.$$

743 $$AD = \frac{2bc}{b+c}\cos\frac{A}{2}.$$

If AD be perpendicular to BC,

744 $$AD = \frac{bc \sin A}{a} = \frac{b^2 \sin C + c^2 \sin B}{b+c}.$$

745 $$BD \sim CD = \frac{b^2 - c^2}{a} = a\,\frac{\tan B - \tan C}{\tan B + \tan C}.$$

REGULAR POLYGON AND CIRCLE.

Radius of circumscribing circle $= R$.
Radius of inscribed circle $= r$.
Side of polygon $= a$.
Number of sides $= n$.

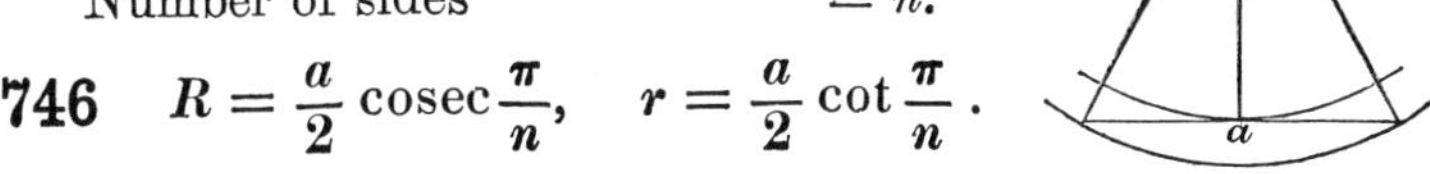

746 $$R = \frac{a}{2}\operatorname{cosec}\frac{\pi}{n}, \quad r = \frac{a}{2}\cot\frac{\pi}{n}.$$

Area of Polygon

748 $$= \tfrac{1}{4} n a^2 \cot\frac{\pi}{n} = \tfrac{1}{2} n R^2 \sin\frac{2\pi}{n} = n r^2 \tan\frac{\pi}{n}.$$

USE OF SUBSIDIARY ANGLES.

749 To adapt $a \pm b$ to logarithmic computation.

Take $$\theta = \tan^{-1}\sqrt{\frac{b}{a}}; \text{ then } a + b = a \sec^2\theta.$$

750 For $a - b$ take $\theta = \tan^{-1}\left(\frac{b}{a}\right)$; thus

$$a - b = \frac{a\sqrt{2}\cos(\theta + 45^\circ)}{\cos\theta}.$$

751 To adapt $a \cos C \pm b \sin C$ to logarithmic computation.

Take $\theta = \tan^{-1}\frac{a}{b}$; then

$$a \cos C \pm b \sin C = \sqrt{(a^2 + b^2)}\sin(\theta \pm C).$$ [By 617

For similar instances of the use of a subsidiary angle, see (726) to (730).

752 To solve a quadratic equation by employing a subsidiary angle.

If $x^2 - 2px + q = 0$ be the equation,

$$x = p\left(1 \pm \sqrt{1 - \frac{q}{p^2}}\right).$$ [By 45

CASE I.—If q be $<p^2$, put $\frac{q}{p^2} = \sin^2\theta$; then

$$x = 2p\cos^2\frac{\theta}{2}, \text{ and } 2p\sin^2\frac{\theta}{2}.$$ [639, 640

CASE II.—If q be $>p^2$, put $\frac{q}{p^2} = \sec^2\theta$; then

$$x = p\,(1 \pm i\tan\theta), \text{ imaginary roots.}$$ [614

CASE III.—If q be negative, put $\frac{q}{p^2} = \tan^2\theta$; then

$$x = \sqrt{q}\cot\frac{\theta}{2} \quad\text{and}\quad -\sqrt{q}\tan\frac{\theta}{2}.$$ [644, 645

LIMITS OF RATIOS.

753 $\frac{\sin\theta}{\theta} = \frac{\tan\theta}{\theta} = 1,$

when θ vanishes.

For ultimately $\frac{PN}{AP} = \frac{AT}{AP} = 1$ [601, 606

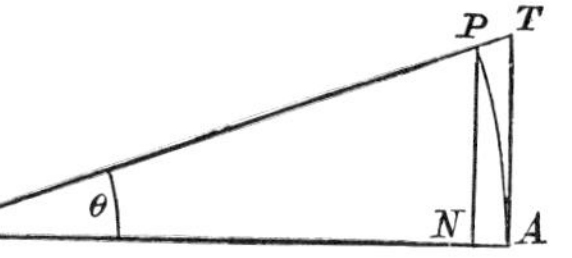

754 $n\sin\frac{\theta}{n} = \theta$ when n is infinite. By putting $\frac{\theta}{n}$ for θ in last.

755 $\left(\cos\frac{\theta}{n}\right)^n = 1$ when n is infinite.

Put $\left(1 - \sin^2\frac{\theta}{n}\right)^{\frac{n}{2}}$, and expand the logarithm by (156).

DE MOIVRE'S THEOREM.

756 $$(\cos\alpha + i\sin\alpha)(\cos\beta + i\sin\beta)\ \dots\ \&c.$$
$$= \cos(\alpha+\beta+\gamma+\dots) + i\sin(\alpha+\beta+\gamma+\dots),$$

where $i = \sqrt{-1}$. Proved by Induction.

757 $$(\cos\theta + i\sin\theta)^n = \cos n\theta + i\sin n\theta.$$

By Induction, or by putting α, β, &c. each $= \theta$ in (756).

Expansion of $\cos n\theta$, &c. in powers of $\sin\theta$ and $\cos\theta$.

758 $$\cos n\theta = \cos^n\theta - C(n,2)\cos^{n-2}\theta\sin^2\theta + C(n,4)\cos^{n-4}\theta\sin^4\theta - \&c.$$

759 $$\sin n\theta = n\cos^{n-1}\theta\sin\theta - C(n,3)\cos^{n-3}\theta\sin^3\theta + \&c.$$

By expanding (757) by Bin. Th., and equating real and imaginary parts.

760 $$\tan n\theta = \frac{n\tan\theta - C(n,3)\tan^3\theta + \&c.}{1 - C(n,2)\tan^2\theta + C(n,4)\tan^4\theta - \&c.}.$$

In series (758, 759), stop at, and exclude, all terms with indices greater than n. Note, *n is here an integer.*

Let s_r = sum of the $C(n, r)$ products of $\tan\alpha$, $\tan\beta$, $\tan\gamma$, &c. to n terms.

761 $$\sin(\alpha+\beta+\gamma+\&c.) = \cos\alpha\cos\beta\ldots(s_1 - s_3 + s_5 - \&c.)$$

762 $$\cos(\alpha+\beta+\gamma+\&c.) = \cos\alpha\cos\beta\ldots(1 - s_2 + s_4 - \&c.)$$

By equating real and imaginary parts in (756).

763 $$\tan(\alpha+\beta+\gamma+\&c.) = \frac{s_1 - s_3 + s_5 - s_7 + \&c.}{1 - s_2 + s_4 - s_6 + \&c.}.$$

Expansions of the sine and cosine in powers of the angle.

764 $$\sin\theta = \theta - \frac{\theta^3}{\lfloor 3} + \frac{\theta^5}{\lfloor 5} - \&c. \qquad \cos\theta = 1 - \frac{\theta^2}{\lfloor 2} + \frac{\theta^4}{\lfloor 4} - \&c.$$

By putting $\frac{\theta}{n}$ for θ in (757) and $n = \infty$, employing (754) and (755).

766 $$e^{i\theta} = \cos\theta + i\sin\theta. \qquad e^{-i\theta} = \cos\theta - i\sin\theta.$$ By (150).

768 $$e^{i\theta} + e^{-i\theta} = 2\cos\theta. \qquad e^{i\theta} - e^{-i\theta} = 2i\sin\theta.$$

770 $$i\tan\theta = \frac{e^{i\theta} - e^{-i\theta}}{e^{i\theta} + e^{-i\theta}}. \qquad \frac{1 + i\tan\theta}{1 - i\tan\theta} = e^{2i\theta}.$$

Expansion of $\cos^n\theta$ and $\sin^n\theta$ in cosines or sines of multiples of θ.

772 $$2^{n-1}\cos^n\theta = \cos n\theta + n\cos(n-2)\theta + C(n,2)\cos(n-4)\theta + C(n,3)\cos(n-6)\theta + \&c.$$

773 When n is even,

$$2^{n-1}(-1)^{\frac{1}{2}n}\sin^n\theta = \cos n\theta - n\cos(n-2)\theta + C(n,2)\cos(n-4)\theta - C(n,3)\cos(n-6)\theta + \&c.$$

774 And when n is odd,

$$2^{n-1}(-1)^{\frac{n-1}{2}}\sin^n\theta = \sin n\theta - n\sin(n-2)\theta + C(n,2)\sin(n-4)\theta - C(n,3)\sin(n-6)\theta + \&c.$$

Observe that in these series the coefficients are those of the Binomial Theorem, with this exception,—*If n be even, the last term must be divided by* 2.

The series are obtained by expanding $(e^{i\theta} \pm e^{i\theta})^n$ by the Binomial Theorem, collecting the equidistant terms in pairs, and employing (768) and (769).

Expansion of $\cos n\theta$ and $\sin n\theta$ in powers of $\sin\theta$.

775 When n is even,

$$\cos n\theta = 1 - \frac{n^2}{\underline{|2}}\sin^2\theta + \frac{n^2(n^2-2^2)}{\underline{|4}}\sin^4\theta - \frac{n^2(n^2-2^2)(n^2-4^2)}{\underline{|6}}\sin^6\theta + \&c.$$

776 When n is odd,

$$\cos n\theta = \cos\theta\left\{1 - \frac{n^2-1}{\underline{|2}}\sin^2\theta + \frac{(n^2-1)(n^2-3^2)}{\underline{|4}}\sin^4\theta - \frac{(n^2-1)(n^2-3^2)(n^2-5^2)}{\underline{|6}}\sin^6\theta + \&c.\right\}$$

777 When n is even,

$$\sin n\theta = n\cos\theta\left\{\sin\theta - \frac{n^2-2^2}{\underline{|3}}\sin^3\theta + \frac{(n^2-2^2)(n^2-4^2)}{\underline{|5}}\sin^5\theta - \frac{(n^2-2^2)(n^2-4^2)(n^2-6^2)}{\underline{|7}}\sin^7\theta + \&c.\right\}$$

778 When n is odd,

$$\sin n\theta = n\sin\theta - \frac{n(n^2-1)}{\underline{|3}}\sin^3\theta + \frac{n(n^2-1)(n^2-3^2)}{\underline{|5}}\sin^5\theta - \frac{n(n^2-1)(n^2-3^2)(n^2-5^2)}{\underline{|7}}\sin^7\theta + \&c.$$

METHOD OF PROOF.

By (758), we may assume, when n is an even integer,

$$\cos n\theta = 1 + A_2\sin^2\theta + A_4\sin^4\theta + \ldots + A_{2r}\sin^{2r}\theta + \ldots$$

Put $\theta + x$ for θ, and in $\cos n\theta\cos nx - \sin n\theta\sin nx$ substitute for $\cos nx$ and $\sin nx$ their values in powers of nx from (764). Each term on the right is of the type $A_{2r}(\sin\theta\cos x + \cos\theta\sin x)^{2r}$. Make similar substitutions for $\cos x$ and $\sin x$ in powers of x. Collect the two coefficients of x^2 in each term by the multinomial theorem (137) and equate them all to the coefficient of x^2 on the left. In this equation write $\cos^2\theta$ for $1-\sin^2\theta$ everywhere, and then equate the coefficients of $\sin^{2r}\theta$ to obtain the relation between the successive quantities A_{2r} and A_{2r+2} for the series (775).

To obtain the series (777) equate the coefficients of x instead of those of x^2.

When n is an odd integer, begin by assuming, by (759),

$$\sin n\theta = A_1\sin\theta + A_3\sin^3\theta + \&c.$$

779 The expansions of $\cos n\theta$ and $\sin n\theta$ in powers of $\cos\theta$ are obtained by changing θ into $\frac{1}{2}\pi-\theta$ in (775) to (778).

780 *Expansion of $\cos n\theta$ in descending powers of $\cos\theta$.*

$$2\cos n\theta = (2\cos\theta)^n - n(2\cos\theta)^{n-2} + \frac{n(n-3)}{\lfloor 2}(2\cos\theta)^{n-4} - \dots + (-1)^r \frac{n(n-r-1)(n-r-2)\dots(n-2r+1)}{\lfloor r}(2\cos\theta)^{n-2r} +$$

up to the last positive power of $2\cos\theta$.

Obtained by expanding each term of the identity

$$\log(1-zx)+\log\left(1-\frac{z}{x}\right) = \log\left\{1-z\left(x+\frac{1}{x}-z\right)\right\}$$

by (156), equating coefficients of z^n, and substituting from (768).

783 $\sin\alpha + c\sin(\alpha+\beta) + c^2\sin(\alpha+2\beta) + \&c.$ to n terms

$$= \frac{\sin\alpha - c\sin(\alpha-\beta) - c^n\sin(\alpha+n\beta) + c^{n+1}\sin\{\alpha+\overline{n-1}\,\beta\}}{1-2c\cos\beta+c^2}.$$

If c be < 1 and n infinite, this becomes

784 $$= \frac{\sin\alpha - c\sin(\alpha-\beta)}{1-2c\cos\beta+c^2}.$$

785 $\cos\alpha + c\cos(\alpha+\beta) + c\cos(\alpha+2\beta) + \&c.$ to n terms $=$ a similar result, changing *sin* into *cos* in the numerator.

786 Similarly when c is < 1 and n infinite.

787 ***Method of summation.***—*Substitute for the sines or cosines their exponential values* (768). *Sum the two resulting geometrical series, and substitute the sines or cosines again for the exponential values by* (766).

788 $c\sin(\alpha+\beta) + \frac{c^2}{\lfloor 2}\sin(\alpha+2\beta) + \frac{c^3}{\lfloor 3}\sin(\alpha+3\beta) + \&c.$ to infinity $= e^{c\cos\beta}\sin(\alpha+c\sin\beta) - \sin\alpha.$

789 $c\cos(\alpha+\beta) + \frac{c^2}{\lfloor 2}\cos(\alpha+2\beta) + \frac{c^3}{\lfloor 3}\cos(\alpha+3\beta) + \&c.$ to infinity $= e^{c\cos\beta}\cos(\alpha+c\sin\beta) - \cos\alpha.$

Obtained by the rule in (787).

790 If, in the series (783) to (789), β be changed into $\beta+\pi$, the signs of the alternate terms will thereby be changed.

Expansion of θ in powers of $\tan\theta$ (Gregory's series).

791 $$\theta = \tan\theta - \frac{\tan^3\theta}{3} + \frac{\tan^5\theta}{5} - \&c.$$

The series converges if $\tan\theta$ be not > 1.

Obtained by expanding the logarithm of the value of $e^{2i\theta}$ in (771) by (158).

Formulæ for the calculation of the value of π by Gregory's series.

792 $$\frac{\pi}{4} = \tan^{-1}\frac{1}{2} + \tan^{-1}\frac{1}{3} = 4\tan^{-1}\frac{1}{5} - \tan^{-1}\frac{1}{239}$$ [791

794 $$= 4\tan^{-1}\frac{1}{5} - \tan^{-1}\frac{1}{70} + \tan^{-1}\frac{1}{99}.$$

Proved by employing the formula for $\tan(A \pm B)$, (631).

To prove that π is incommensurable.

795 Convert the value of $\tan\theta$ in terms of θ from (764) and (765) into a continued fraction, thus $\tan\theta = \frac{\theta}{1-}\,\frac{\theta^2}{3-}\,\frac{\theta^2}{5-}\,\frac{\theta^2}{7-}\,\&c.$; or this result may be obtained by putting $i\theta$ for y in (294), and by (770). Hence

$$1 - \frac{\theta}{\tan\theta} = \frac{\theta^2}{3-}\,\frac{\theta^2}{5-}\,\frac{\theta^2}{7-}\,\&c.$$

Put $\frac{\pi}{2}$ for θ, and assume that π, and therefore $\frac{\pi^2}{4}$, *is* commensurable. Let $\frac{\pi^2}{4} = \frac{m}{n}$, m and n being integers. Then we shall have $1 = \frac{m}{3n-}\,\frac{mn}{5n-}\,\frac{mn}{7n-}\,\&c.$ The continued fraction is incommensurable, by (177). But unity cannot be equal to an incommensurable quantity. Therefore π is *not* commensurable.

796 If $\sin x = n\sin(x+a)$, $x = n\sin a + \frac{n^2}{2}\sin 2a + \frac{n^3}{3}\sin 3a + \&c.$

797 If $\tan x = n\tan y$, $x = y - m\sin 2y + \frac{m^2}{2}\sin 4y - \frac{m^3}{3}\sin 6y + \&c.$, where $m = \frac{1-n}{1+n}$.

These results are obtained by substituting the exponential values of the sine or tangent (769) *and* (770), *and then eliminating x.*

798 Coefficient of x^n in the expansion of $e^{ax}\cos bx = \frac{(a^2+b^2)^{\frac{n}{2}}}{\lfloor n}\cos n\theta$, where $a = r\cos\theta$ and $b = r\sin\theta$.

For proof, substitute for $\cos bx$ from (768); expand by (150); put $a = r\cos\theta$, $b = r\sin\theta$ in the coefficient of x^n, and employ (757).

799 When e is < 1, $\dfrac{\sqrt{1-e^2}}{1-e\cos\theta} = 1+2b\cos\theta+2b^2\cos 2\theta+2b^3\cos 3\theta+\ldots,$ where $b = \dfrac{e}{1+\sqrt{1-e^2}}$.

For proof, put $e = \dfrac{2b}{1+b^2}$ and $2\cos\theta = x+\dfrac{1}{x}$, expand the fraction in two series of powers of x by the method of (257), and substitute from (768).

---

800 $\sin\alpha+\sin(\alpha+\beta)+\sin(\alpha+2\beta)+\ldots+\sin\{\alpha+(n-1)\beta\}$

$$= \frac{\sin\left(\alpha+\frac{n-1}{2}\beta\right)\sin\frac{n}{2}\beta}{\sin\frac{\beta}{2}}.$$

801 $\cos\alpha+\cos(\alpha+\beta)+\cos(\alpha+2\beta)+\ldots+\cos\{\alpha+(n-1)\beta\}$

$$= \frac{\cos\left(\alpha+\frac{n-1}{2}\beta\right)\sin\frac{n\beta}{2}}{\sin\frac{\beta}{2}}.$$

802 If the terms in these series have the signs + and − alternately, change β into $\beta+\pi$ in the results.

For proof, multiply the series by $2\sin\frac{\beta}{2}$, and apply (669) and (666).

---

803 If $\beta = \frac{2\pi}{n}$ in (800) and (801), each series vanishes.

804 Generally, If $\beta = \frac{2\pi}{n}$, and if r be an integer not a multiple of n, the sum of the rth powers of the sines or cosines in (800) or (801) is zero if r be odd; and if r be even it is $= \frac{n}{2^r}C\left(r, \frac{r}{2}\right)$; by (772) to (774).

General Theorem.—Denoting the sum of the series

805 $c+c_1x+c_2x^2+\ldots+c_nx^n$ by $F(x)$;

then $c\cos\alpha+c_1\cos(\alpha+\beta)+\ldots+c_n\cos(\alpha+n\beta) = \frac{1}{2}\{e^{i\alpha}F(e^{i\beta})+e^{-i\alpha}F(e^{-i\beta})\}$,

and

806 $c\sin\alpha+c_1\sin(\alpha+\beta)+\ldots+c_n\sin(\alpha+n\beta) = \frac{1}{2i}\{e^{i\alpha}F(e^{i\beta})-e^{-i\alpha}F(e^{-i\beta})\}$.

Proved by substituting for the sines and cosines their exponential values (766), &c.

Expansion of the sine and cosine in factors.

807 $$x^{2n} - 2x^n y^n \cos n\theta + y^{2n}$$

$$= \left\{x^2 - 2xy\cos\theta + y^2\right\}\left\{x^2 - 2xy\cos\left(\theta + \frac{2\pi}{n}\right) + y^2\right\}\dots$$

to n factors, adding $\frac{2\pi}{n}$ to the angle successively.

PROOF.—By solving the quadratic on the left, we get $x = y(\cos n\theta + i\sin n\theta)^{\frac{1}{n}}$. The n values of x are found by (757) and (626), and thence the factors. For the factors of $x^n \pm y^n$ see (480).

---

808 $$\sin n\phi = 2^{n-1}\sin\phi\,\sin\left(\phi + \frac{\pi}{n}\right)\sin\left(\phi + \frac{2\pi}{n}\right)\dots$$

as far as n factors of sines.

PROOF.—By putting $x = y = 1$ and $\theta = 2\phi$ in the last.

809 If n be even,

$$\sin n\phi = 2^{n-1}\sin\phi\cos\phi\left(\sin^2\frac{\pi}{n} - \sin^2\phi\right)\left(\sin^2\frac{2\pi}{n} - \sin^2\phi\right)\text{\&c.}$$

810 If n be odd, omit $\cos\phi$ and make up n factors, reckoning two factors for each pair of terms in brackets.

Obtained from (808), by collecting equidistant factors in pairs, and applying (659).

---

811 $$\cos n\phi = 2^{n-1}\sin\left(\phi + \frac{\pi}{2n}\right)\sin\left(\phi + \frac{3\pi}{2n}\right)\dots \text{ to } n \text{ factors.}$$

PROOF.—Put $\phi + \frac{\pi}{2n}$ for ϕ in (808).

812 Also, if n be odd,

$$\cos n\phi = 2^{n-1}\cos\phi\left(\sin^2\frac{\pi}{2n} - \sin^2\phi\right)\left(\sin^2\frac{3\pi}{2n} - \sin^2\phi\right)\dots$$

813 If n be even, omit $\cos\phi$.

Proved as in (809).

---

814 $$n = 2^{n-1}\sin\frac{\pi}{n}\sin\frac{2\pi}{n}\sin\frac{3\pi}{n}\dots\sin\frac{(n-1)\pi}{n}.$$

PROOF.—Divide (809) by $\sin\phi$, and make ϕ vanish; then apply (754).

---

815 $$\sin\theta = \theta\left\{1 - \left(\frac{\theta}{\pi}\right)^2\right\}\left\{1 - \left(\frac{\theta}{2\pi}\right)^2\right\}\left\{1 - \left(\frac{\theta}{3\pi}\right)^2\right\}\dots\dots$$

816 $$\cos\theta = \left\{1 - \left(\frac{2\theta}{\pi}\right)^2\right\}\left\{1 - \left(\frac{2\theta}{3\pi}\right)^2\right\}\left\{1 - \left(\frac{2\theta}{5\pi}\right)^2\right\}\dots\dots$$

PROOF.—Put $\phi = \frac{\theta}{n}$ in (809) and (812); divide by (814) and make n infinite.

817 $e^x - 2\cos\theta + e^{-x}$

$$= 4\sin^2\frac{\theta}{2}\left\{1+\frac{x^2}{\theta^2}\right\}\left\{1+\frac{x^2}{(2\pi\pm\theta)^2}\right\}\left\{1+\frac{x^2}{(4\pi\pm\theta)^2}\right\}\cdots$$

Proved by substituting $x = 1+\frac{z}{2n}$, $y = 1-\frac{z}{2n}$, and $\frac{\theta}{n}$ for θ in (807), making n infinite and reducing one series of factors to $4\sin^2\frac{\theta}{2}$ by putting $z = 0$.

De Moivre's Property of the Circle. — Take P any point, and $POB = \theta$ any angle,

$$BOC = COD = \&c. = \frac{2\pi}{n};$$

$$OP = x; \quad OB = r.$$

819 $x^{2n} - 2x^n r^n \cos n\theta + r^{2n}$
$= PB^2\, PC^2\, PD^2 \ldots$ to n factors.

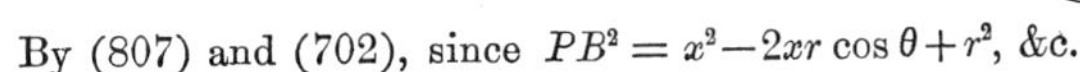

By (807) and (702), since $PB^2 = x^2 - 2xr\cos\theta + r^2$, &c.

820 If $x = r$, $2r^n \sin\frac{n\theta}{2} = PB\,.\,PC\,.\,PD \ldots$ &c.

---

821 *Cotes's properties.*—If $\theta = \frac{2\pi}{n}$,

$$x^n \sim r^n = PB\,.\,PC\,.\,PD \ldots \text{ \&c.}$$

822 $$x^n + r^n = Pa\,.\,Pb\,.\,Pc \ldots \text{ \&c.}$$

ADDITIONAL FORMULÆ.

823 $\cot A + \tan A = 2\,\text{cosec}\,2A = \sec A\,\text{cosec}\,A.$

824 $\text{cosec}\,2A + \cot 2A = \cot A. \quad \sec A = 1 + \tan A \tan\frac{A}{2}.$

826 $$\cos A = \cos^4\frac{A}{2} - \sin^4\frac{A}{2}.$$

827 $$\tan A + \sec A = \tan\left(45^\circ + \frac{A}{2}\right).$$

828 $$\frac{\tan A + \tan B}{\cot A + \cot B} = \tan A \tan B.$$

829 $$\sec^2 A\,\text{cosec}^2 A = \sec^2 A + \text{cosec}^2 A.$$

830 If $A+B+C=\frac{\pi}{2}$,

$$\tan B\tan C+\tan C\tan A+\tan A\tan B=1.$$

831 If $A+B+C=\pi$,

$$\cot B\cot C+\cot C\cot A+\cot A\cot B=1.$$

---

832 $\sin^{-1}\frac{3}{5}+\sin^{-1}\frac{4}{5}=\frac{\pi}{2}.\quad \tan^{-1}\frac{1}{2}+\tan^{-1}\frac{1}{3}=\frac{\pi}{4}.$

In a right-angled triangle ABC, C being the right angle,

833 $$\cos 2B=\frac{a^2-b^2}{a^2+b^2}.\qquad \tan 2B=\frac{2ab}{a^2-b^2}.$$

834 $$\tan\tfrac{1}{2}A=\sqrt{\left(\frac{c-b}{c+b}\right)}.\qquad R+r=\tfrac{1}{2}(a+b).$$

In any triangle,

835 $$\sin\tfrac{1}{2}(A-B)=\frac{a-b}{c}\cos\tfrac{1}{2}C.$$

$$\cos\tfrac{1}{2}(A-B)=\frac{a+b}{c}\sin\tfrac{1}{2}C.$$

836 $$\frac{\sin A-B}{\sin A+B}=\frac{a^2-b^2}{c^2}.\qquad \frac{\tan\frac{1}{2}A+\tan\frac{1}{2}B}{\tan\frac{1}{2}A-\tan\frac{1}{2}B}=\frac{c}{a-b}.$$

837 $$\tfrac{1}{2}(a^2+b^2+c^2)=bc\cos A+ca\cos B+ab\cos C.$$

---

838 Area of triangle $ABC=\frac{1}{2}bc\sin A$

$$=\tfrac{1}{2}a^2\frac{\sin B\sin C}{\sin A}=\tfrac{1}{2}(a^2-b^2)\frac{\sin A\sin B}{\sin(A-B)}.$$

839 $$=\frac{2abc}{a+b+c}\cos\tfrac{1}{2}A\cos\tfrac{1}{2}B\cos\tfrac{1}{2}C.$$

840 $$=\tfrac{1}{4}(a+b+c)^2\tan\tfrac{1}{2}A\tan\tfrac{1}{2}B\tan\tfrac{1}{2}C.$$

With the notation of (709),

841 $$r=\tfrac{1}{2}(a+b+c)\tan\tfrac{1}{2}A\tan\tfrac{1}{2}B\tan\tfrac{1}{2}C.$$

842 $$2Rr=\frac{abc}{a+b+c}.\qquad \Delta=\sqrt{rr_ar_br_c}.$$

843 $$a\cos A+b\cos B+c\cos C=4R\sin A\sin B\sin C.$$

---

844 $R+r = \frac{1}{2}(a\cot A+b\cot B+c\cot C)$ = sum of perpendiculars on the sides from centre of circumscribing circle.

This may also be shown by applying Euc. VI. D. to the circle described on R as diameter and the quadrilateral so formed.

845 $$r_a r_b r_c = abc\cos\tfrac{1}{2}A\cos\tfrac{1}{2}B\cos\tfrac{1}{2}C.$$

846 $$r = \sqrt{(r_b r_c)} + \sqrt{(r_c r_a)} + \sqrt{(r_a r_b)}.$$

847 $$\frac{1}{r} = \frac{1}{r_a} + \frac{1}{r_b} + \frac{1}{r_c}. \qquad \tan\tfrac{1}{2}A = \sqrt{\frac{rr_a}{r_b r_c}}.$$

849 If O be the centre of inscribed circle,

$$OA = \frac{2bc}{a+b+c}\cos\tfrac{1}{2}A.$$

---

850 $$a(b\cos C - c\cos B) = b^2 - c^2.$$

851 $$b\cos B + c\cos C = c\cos(B-C).$$

852 $$a\cos A + b\cos B + c\cos C = 2a\sin B\sin C.$$

853 $$\cos A + \cos B + \cos C = 1 + \frac{2a\sin B\sin C}{a+b+c}.$$

---

854 If $s = \frac{1}{2}(a+b+c)$,

$$1-\cos^2 a-\cos^2 b-\cos^2 c+2\cos a\cos b\cos c$$
$$= 4\sin s\sin(s-a)\sin(s-b)\sin(s-c).$$

855 $$-1+\cos^2 a+\cos^2 b+\cos^2 c+2\cos a\cos b\cos c$$
$$= 4\cos s\cos(s-a)\cos(s-b)\cos(s-c).$$

856 $$4\cos\frac{a}{2}\cos\frac{b}{2}\cos\frac{c}{2}$$
$$= \cos s+\cos(s-a)+\cos(s-b)+\cos(s-c).$$

857 $$4\sin\frac{a}{2}\sin\frac{b}{2}\sin\frac{c}{2}$$
$$= -\sin s+\sin(s-a)+\sin(s-b)+\sin(s-c).$$

---

858 $$\pi^2 = 6\left(1+\frac{1}{2^2}+\frac{1}{3^2}+\dots\right) = 8\left(1+\frac{1}{3^2}+\frac{1}{5^2}+\dots\right).$$

Proved by equating coefficients of $\frac{\sin\theta}{\theta}$ in (764 and 815) or of $\cos\theta$ in (765 and 816).

859 *Examples of the Solution of Triangles.*

Ex. 1: Case II. (724).—Two sides of a triangle b, c, being 900 and 700 feet, and the included angle 47° 25′, to find the remaining angles.

$$\tan\frac{B-C}{2} = \frac{b-c}{b+c}\cot\frac{A}{2} = \frac{1}{8}\cot 23^\circ 42' 30'';$$

therefore $$\log\tan\tfrac{1}{2}(B-C) = \log\cot\frac{A}{2} - \log 8;$$

therefore $$L\tan\tfrac{1}{2}(B-C) = L\cot 23^\circ 42' 30'' - 3\log 2;$$

10 being added to each side of the equation.

$$\begin{array}{lrl} \therefore & L\cot 23^\circ 42' 30'' = & 10{\cdot}3573942\,* \\ & 3\log 2 = & {\cdot}9030900 \\ \therefore & L\tan\tfrac{1}{2}(B-C) = & 9{\cdot}4543042 \end{array} \left\{ \begin{array}{lr} \therefore & \tfrac{1}{2}(B-C) = 15^\circ 53' 19''{\cdot}55\,* \\ \text{and} & \tfrac{1}{2}(B+C) = 66^\circ 17' 30'' \\ \therefore & B = 82^\circ 10' 49''{\cdot}55 \end{array} \right.$$

And, by subtraction, $C = 50^\circ 24' 10''{\cdot}45$

Ex. 2: Case III. (732).—Given the sides a, b, c = 7, 8, 9 respectively, to find the angles.

$$\tan\frac{A}{2} = \sqrt{\frac{(s-b)(s-c)}{s(s-a)}} = \sqrt{\frac{4\,.\,3}{12\,.\,5}} = \sqrt{\frac{2}{10}},$$

therefore $$L\tan\frac{A}{2} = 10 + \tfrac{1}{2}(\log 2 - 1) = 9{\cdot}650515;$$

therefore $$\tfrac{1}{2}A = 24^\circ\ 5'\ 41''{\cdot}43.*$$

$\tfrac{1}{2}B$ is found in a similar manner, and $C = 180^\circ - A - B$.

Ex. 3.—In a right-angled triangle, given the hypotenuse $c = 6953$ and a side $b = 3$, to find the remaining angle.

Here $\cos A = \dfrac{3}{6953}$. But, since A is nearly a right angle, it cannot be determined accurately from $\log\cos A$. Therefore take

$$\sin\frac{A}{2} = \sqrt{\frac{1-\cos A}{2}} = \sqrt{\frac{3475}{6953}},$$

therefore $$L\sin\frac{A}{2} = 10 + \tfrac{1}{2}(\log 3475 - \log 6953) = 9{\cdot}8493913,$$

therefore $$\frac{A}{2} = 44^\circ\ 59'\ 15''{\cdot}52,*$$

therefore $$A = 89^\circ\ 58'\ 31''{\cdot}04 \quad\text{and}\quad B = 0^\circ\ 1'\ 28''{\cdot}96.$$

---

* See Chambers's Mathematical Tables for a concise explanation of the method of obtaining these figures.

SPHERICAL TRIGONOMETRY.

INTRODUCTORY THEOREMS.

870 ***Definitions.***—Planes through the centre of a sphere intersect the surface in *great circles;* other planes intersect it in *small circles.* Unless otherwise stated, all arcs are measured on great circles.

The *poles* of a great circle are the extremities of the diameter perpendicular to its plane.

The sides a, b, c of a spherical triangle are the arcs of great circles BC, CA, AB on a sphere of radius unity; and the angles A, B, C are the angles between the tangents to the sides at the vertices, or the angles between the planes of the great circles. The centre of the sphere will be denoted by O.

The *polar triangle* of a spherical triangle ABC has for its angular points A', B', C', the poles of the sides BC, CA, AB of the *primitive triangle* in the directions of A, B, C respectively (since each great circle has two poles). The sides of $A'B'C'$ are denoted by a', b', c'.

871 The sides and angles of the polar triangle are respectively the supplements of the angles and sides of the primitive triangle; that is,

$$a'+A = b'+B = c'+C = \pi,$$

$$a+A' = b+B' = c+C' = \pi.$$

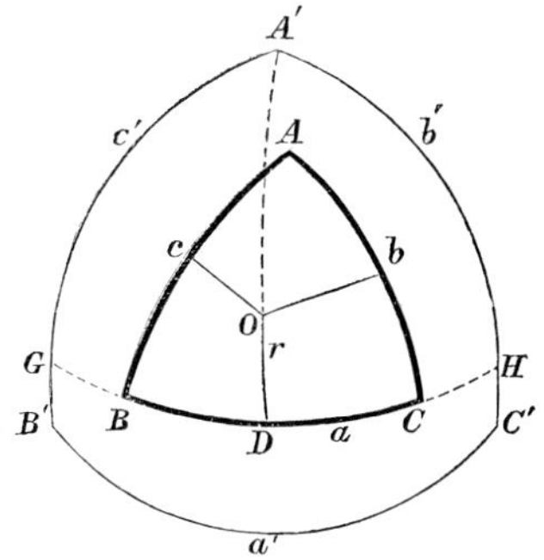

Let BC produced cut the sides $A'B'$, $C'A'$ in G, H. B is the pole of $A'C'$, therefore $BH = \frac{\pi}{2}$. Similarly $CG = \frac{\pi}{2}$, therefore, by addition, $a+GH=\pi$ and $GH=A'$, because A' is the pole of BC.

The polar diagram of a spherical polygon is formed in the same way, and the same relations subsist between the sides and angles of the two figures.

RULE.—*Hence, any equation between the sides and angles of a spherical triangle produces a supplementary equation by changing* a *into* $\pi - \mathrm{A}$ *and* A *into* $\pi - \mathrm{a}$, *&c.*

---

872 The centre of the inscribed circle, radius r, is also the centre of the circumscribed circle, radius R', of the polar triangle, and $r + R' = \frac{1}{2}\pi$.

PROOF.—In the last figure, let O be the centre of the inscribed circle of ABC; then OD, the perpendicular on BC, passes through A', the pole of BC. Also, $OD = r$, therefore $OA' = \frac{1}{2}\pi - r$. Similarly $OB' = OC' = \frac{1}{2}\pi - r$, therefore O is the centre of the circumscribed circle of $A'B'C'$, and $r + R' = \frac{1}{2}\pi$.

---

873 The sine of the arc joining a point on the circumference of a small circle with the pole of a parallel great circle is equal to the ratio of the circumferences, or corresponding arcs of the two circles.

For it is equal to the radius of the small circle divided by the radius of the sphere; that is, by the radius of the great circle.

---

874 Two sides of a triangle are greater than the third. [By XI. 20.

875 The sides of a triangle are together less than the circumference of a great circle. [By XI. 21.

876 The angles of a triangle are together greater than two right angles.

For $\pi - A + \pi - B + \pi - C$ is $< 2\pi$, by (875) and the polar triangle.

---

877 If two sides of a triangle are equal, the opposite angles are equal. [By the geometrical proof in (894).

878 If two angles of a triangle are equal, the opposite sides are equal. [By the polar triangle and (877).

---

879 The greater angle of a triangle has the greater side opposite to it.

PROOF.—If B be $> A$, draw the arc BD meeting AC in D, and make $\angle ABD = A$, therefore $BD = AD$; but $BD + DC > BC$, therefore $AC > BC$.

880 The greater side of a triangle has the greater angle opposite to it. [By the polar triangle and (879).

RIGHT-ANGLED TRIANGLES.

881 *Napier's Rules.*—In the triangle ABC let C be a right angle, then a, $(\frac{1}{2}\pi - B)$, $(\frac{1}{2}\pi - c)$, $(\frac{1}{2}\pi - A)$, and b, are called the five *circular parts*. Taking any part for middle part, Napier's rules are—

I. *sine of middle part = product of tangents of adjacent parts.*

II. *sine of middle part = product of cosines of opposite parts.*

In applying the rules we can take A, B, c instead of their complements, and change sine into cos, or *vice versâ*, for those parts at once. Thus, taking b for the middle part,

$$\sin b = \tan a \cot A = \sin B \sin c.$$

Ten equations in all are given by the rules.

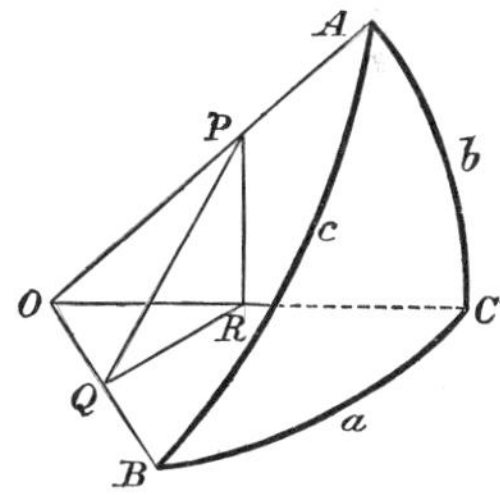

PROOF.—From any point P in OA, draw PR perpendicular to OC, and RQ to OB; therefore PRQ is a right angle, therefore OB is perpendicular to PR and QR, and therefore to PQ. Then prove any formula by proportion from the triangles of the tetrahedron $OPQR$, which are all right-angled. Otherwise, prove by the formulæ for oblique-angled triangles.

OBLIQUE-ANGLED TRIANGLES.

882 $$\cos a = \cos b \cos c + \sin b \sin c \cos A.$$

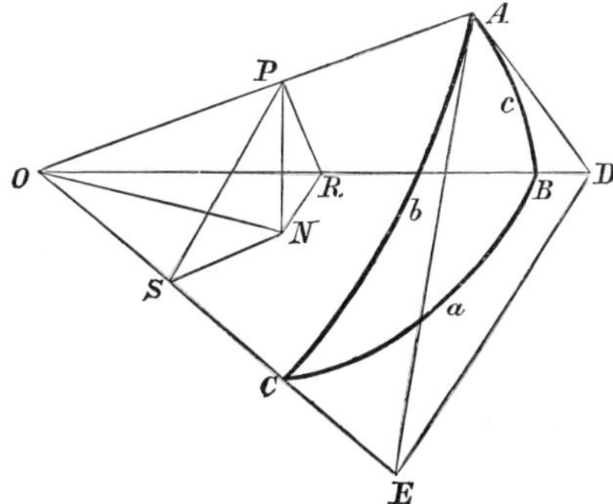

PROOF.—Draw tangents at A to the sides c, b to meet OB, OC in D and E. Express DE^2 by (702) applied to each of the triangles DAE and DOE, and subtract.

If AB and AC are both $> \frac{\pi}{2}$, produce them to meet in A' the pole of A, and employ the triangle $A'BC$.

If AB alone be $> \frac{\pi}{2}$, produce BA to meet BC.

The supplementary formula, by (871), is

883 $$\cos A = -\cos B \cos C + \sin B \sin C \cos a.$$

884 $$\sin\frac{A}{2} = \sqrt{\frac{\sin(s-b)\sin(s-c)}{\sin b \sin c}}.$$

885 $$\cos\frac{A}{2} = \sqrt{\frac{\sin s \sin(s-a)}{\sin b \sin c}}.$$

886 $$\tan\frac{A}{2} = \sqrt{\frac{\sin(s-b)\sin(s-c)}{\sin s \sin(s-a)}} \text{ where } s = \tfrac{1}{2}(a+b+c).$$

PROOF.—$\sin^2\frac{A}{2} = \frac{1}{2}(1-\cos A)$. Substitute for $\cos A$ from (872), and throw the numerator of the whole expression into factors by (673). Similarly for $\cos\frac{A}{2}$.

The supplementary formulæ are obtained in a similar way, or by the rule in (871). They are

887 $$\cos\frac{a}{2} = \sqrt{\frac{\cos(S-B)\cos(S-C)}{\sin B \sin C}}.$$

888 $$\sin\frac{a}{2} = \sqrt{\frac{-\cos S \cos(S-A)}{\sin B \sin C}}.$$

889 $$\tan\frac{a}{2} = \sqrt{\frac{-\cos S \cos(S-A)}{\cos(S-B)\cos(S-C)}}$$
where $S = \frac{1}{2}(A+B+C)$.

890 Let $\sigma = \sqrt{\sin s \sin(s-a)\sin(s-b)\sin(s-c)}$
$= \frac{1}{2}\sqrt{1+2\cos a\cos b\cos c-\cos^2 a-\cos^2 b-\cos^2 c}.$

Then the supplementary form, by (871), is

891 $\Sigma = \sqrt{-\cos S\cos(S-A)\cos(S-B)\cos(S-C)}$
$= \frac{1}{2}\sqrt{1-2\cos A\cos B\cos C-\cos^2 A-\cos^2 B-\cos^2 C}.$

892 $$\sin A = \frac{2\sigma}{\sin b\sin c}. \qquad \sin a = \frac{2\Sigma}{\sin B\sin C}.$$

[By $\sin A = 2\sin\frac{A}{2}\cos\frac{A}{2}$ and (884, 885), &c.

893 The following rules will produce the ten formulæ (884 to 892)—

I. *Write* sin *before each factor in the* s *values of* $\sin\frac{A}{2}$,

$\cos\frac{A}{2}$, $\tan\frac{A}{2}$, $\sin A$, *and* Δ, *in Plane Trigonometry* (704—707), *to obtain the corresponding formulæ in Spherical Trigonometry.*

II. *To obtain the supplementary forms of the five results, transpose large and small letters everywhere, and transpose* sin *and* cos *everywhere but in the denominators, and write minus before* cos S.

894
$$\frac{\sin A}{\sin a} = \frac{\sin B}{\sin b} = \frac{\sin C}{\sin c}.$$

PROOF.—By (882). Otherwise, in the figure of 882, draw PN perpendicular to BOC, and NR, NS to OB, OC. Prove PRO and PSO right angles by I. 47, and therefore $PN = OP \sin c \sin B = OP \sin b \sin C$.

895
$$\cos b \cos C = \cot a \sin b - \cot A \sin C.$$

To remember this formula, take any four consecutive angles and sides (as a, C, b, A), and calling the first and fourth the extremes, and the second and third the middle parts, employ the following rule:—

RULE.—*Product of cosines of middle parts = cot extreme side × sin middle side − cot extreme angle × sin middle angle.*

PROOF.—In the formula for $\cos a$ (882) substitute a similar value for $\cos c$, and for $\sin c$ put $\sin C \frac{\sin a}{\sin A}$.

896 NAPIER'S FORMULÆ.

$$(1)\quad \tan\tfrac{1}{2}(A-B) = \frac{\sin\frac{1}{2}(a-b)}{\sin\frac{1}{2}(a+b)}\cot\frac{C}{2}.$$

$$(2)\quad \tan\tfrac{1}{2}(A+B) = \frac{\cos\frac{1}{2}(a-b)}{\cos\frac{1}{2}(a+b)}\cot\frac{C}{2}.$$

$$(3)\quad \tan\tfrac{1}{2}(a-b) = \frac{\sin\frac{1}{2}(A-B)}{\sin\frac{1}{2}(A+B)}\tan\frac{c}{2}.$$

$$(4)\quad \tan\tfrac{1}{2}(a+b) = \frac{\cos\frac{1}{2}(A-B)}{\cos\frac{1}{2}(A+B)}\tan\frac{c}{2}.$$

RULE.—*In the value of* $\tan\frac{1}{2}(A-B)$ *change* sin *to* cos *to obtain* $\tan\frac{1}{2}(A+B)$. *To obtain* (3) *and* (4) *from* (1) *and* (2), *transpose sides and angles, and change* cot *to* tan.

PROOF.—In the values of $\cos A$ and $\cos B$, by (883), put $m \sin a$ and $m \sin b$ for $\sin A$ and $\sin B$, and add the two equations. Then put $m = \frac{\sin A \pm \sin B}{\sin a \pm \sin b}$, and transform by (670–672).

897 *GAUSS'S FORMULÆ.*

$$(1)\quad \frac{\sin\frac{1}{2}(A+B)}{\cos\frac{1}{2}C} = \frac{\cos\frac{1}{2}(a-b)}{\cos\frac{1}{2}c}.$$

$$(2)\quad \frac{\sin\frac{1}{2}(A-B)}{\cos\frac{1}{2}C} = \frac{\sin\frac{1}{2}(a-b)}{\sin\frac{1}{2}c}.$$

$$(3)\quad \frac{\cos\frac{1}{2}(A+B)}{\sin\frac{1}{2}C} = \frac{\cos\frac{1}{2}(a+b)}{\cos\frac{1}{2}c}.$$

$$(4)\quad \frac{\cos\frac{1}{2}(A-B)}{\sin\frac{1}{2}C} = \frac{\sin\frac{1}{2}(a+b)}{\sin\frac{1}{2}c}.$$

From any one of these formulæ the others may be obtained by the following rule:—

RULE.—*Change the sign of the letter* B *(large or small) on one side of the equation, and write* sin *for* cos *and* cos *for* sin *on the other side.*

PROOF.—Take $\sin\frac{1}{2}(A+B) = \sin\frac{1}{2}A\cos\frac{1}{2}B+\cos\frac{1}{2}A\sin\frac{1}{2}B$, substitute the s values by (884, 885), and reduce.

SPHERICAL TRIANGLE AND CIRCLE.

898 Let r be the radius of the inscribed circle of ABC; r_a the radius of the escribed circle touching the side a, and R, R_a the radii of the circumscribed circles; then

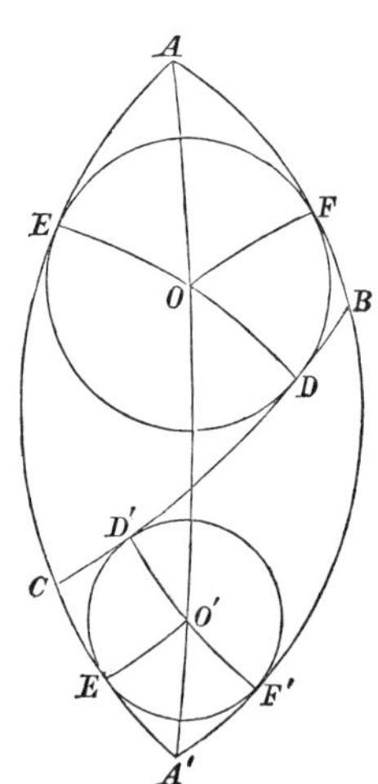

$$(1)\quad \tan r = \tan\tfrac{1}{2}A\sin(s-a) = \frac{\sigma}{\sin s}$$

$$(3)\quad = \frac{2\sin a}{\sin A}\sin\tfrac{1}{2}A\sin\tfrac{1}{2}B\sin\tfrac{1}{2}C$$

$$(4)\quad = \frac{\Sigma}{2\cos\frac{1}{2}A\cos\frac{1}{2}B\cos\frac{1}{2}C}$$

$$= \frac{2\Sigma}{\cos S+\cos(S-A)+\cos(S-B)+\&c.}$$

PROOF.—The first value is found from the right-angled triangle OAF, in which $AF = s-a$. The other values by (884–892).

899 (1) $\tan r_a = \tan\frac{1}{2}A \sin s = \dfrac{\sigma}{\sin(s-a)}$

(3) $$= \frac{2\sin a}{\sin A}\sin\tfrac{1}{2}A\cos\tfrac{1}{2}B\cos\tfrac{1}{2}C$$

(4) $$= \frac{\Sigma}{2\cos\frac{1}{2}A\sin\frac{1}{2}B\sin\frac{1}{2}C}$$

$$= \frac{2\Sigma}{-\cos S-\cos(S-A)+\cos(S-B)+\cos(S-C)}.$$

PROOF.—From the right-angled triangle $O'AF'$, in which $AF'=s$.

NOTE.—The first two values of $\tan r_a$ may be obtained from those of $\tan r$ by interchanging s and $s-a$.

---

900 (1) $\tan R = \dfrac{\tan\frac{1}{2}a}{\cos(S-A)} = \dfrac{-\cos S}{\Sigma}$

(3) $$= \frac{\sin\frac{1}{2}a}{\sin A\cos\frac{1}{2}b\cos\frac{1}{2}c}$$

(4) $$= \frac{2\sin\frac{1}{2}a\sin\frac{1}{2}b\sin\frac{1}{2}c}{\sigma}$$

$$= \frac{-\sin s+\sin(s-a)+\sin(s-b)+\&c.}{2\sigma}$$

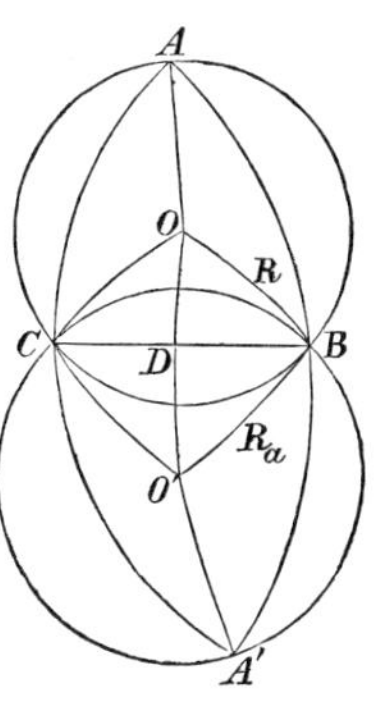

PROOF.—The first value from the right-angled triangle OBD, in which $\angle OBD = S-A$. The other values by the formulæ (887–892).

---

901 (1) $\tan R_a = \dfrac{\tan\frac{1}{2}a}{-\cos S} = \dfrac{\cos(S-A)}{\Sigma}$

(3) $$= \frac{\sin\frac{1}{2}a}{\sin A\sin\frac{1}{2}b\sin\frac{1}{2}c}$$

(4) $$= \frac{2\sin\frac{1}{2}a\cos\frac{1}{2}b\cos\frac{1}{2}c}{\sigma}$$

(5) $$= \frac{\sin s-\sin(s-a)+\sin(s-b)+\sin(s-c)}{2\sigma}.$$

PROOF.—From the right-angled triangle $O'BD$, in which $\angle O'BD = \pi - S$.

SPHERICAL AREAS.

902 area of $ABC = (A+B+C-\pi)\,r^2 = Er^2$,

where $E = A+B+C-\pi$, the *spherical excess*.

PROOF.—By adding the three lunes

$ABDC,\ BCEA,\ CAFB,$

and observing that $ABF = CDE$,

we get $\left(\frac{A}{\pi}+\frac{B}{\pi}+\frac{C}{\pi}\right)2\pi r^2 = 2\pi r^2 + 2ABC.$

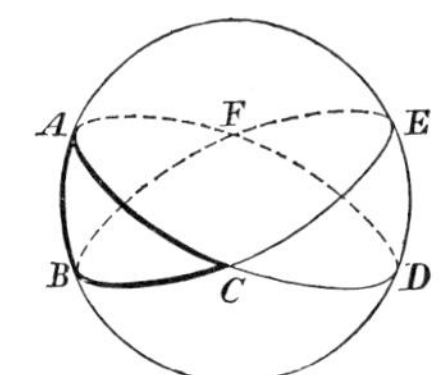

903 *AREA OF SPHERICAL POLYGON.*

n being the number of sides,

$$\begin{aligned}\text{Area} &= \{\text{Interior Angles} - (n-2)\,\pi\}\,r^2\\ &= \{2\pi - \text{Exterior Angles}\}\,r^2\\ &= \{2\pi - \text{sides of Polar Diagram}\}\,r^2.\end{aligned}$$

The last value holds for a curvilinear area in the limit.

PROOF.—By joining the vertices with an interior point, and adding the areas of the spherical triangles so formed.

904 *Cagnoli's Theorem.*

$$\sin\tfrac{1}{2}E = \frac{\sqrt{\{\sin s\,\sin(s-a)\,\sin(s-b)\,\sin(s-c)\}}}{2\cos\frac{1}{2}a\,\cos\frac{1}{2}b\,\cos\frac{1}{2}c}.$$

PROOF.—Expand $\sin[\frac{1}{2}(A+B)-\frac{1}{2}(\pi-C)]$ by (628), and transform by Gauss's equations (897 i., iii.) and (669, 890).

905 *Llhuillier's Theorem.*

$$\tan\tfrac{1}{4}E = \sqrt{\{\tan\tfrac{1}{2}s\,\tan\tfrac{1}{2}(s-a)\,\tan\tfrac{1}{2}(s-b)\,\tan\tfrac{1}{2}(s-c)\}}.$$

PROOF.—Multiply numerator and denominator of the left side by $2\cos\frac{1}{4}(A+B-C+\pi)$ and reduce by (667, 668), then eliminate $\frac{1}{2}(A+B)$ by Gauss's formulæ (897 i., iii.) Transform by (672, 673), and substitute from (886).

POLYHEDRONS.

Let the number of faces, solid angles, and edges, of any polyhedron be F, S, and E; then

906 $$F+S=E+2.$$

PROOF.—Project the polyhedron upon an internal sphere. Let m = number of sides, and s = sum of angles of one of the spherical polygons so formed. Then its area $=\{s-(m-2)\pi\}r^2$, by (903). Sum this for all the polygons, and equate to $4\pi r^2$.

THE FIVE REGULAR SOLIDS.

Let m be the number of sides in each face, n the number of plane angles in each solid angle; therefore

907 $$mF=nS=2E.$$

From these equations and (906), find F, S, and E in terms of m and n, thus,

$$\frac{1}{F}=\frac{m}{2}\left(\frac{1}{m}+\frac{1}{n}-\frac{1}{2}\right),\quad \frac{1}{S}=\frac{n}{2}\left(\frac{1}{m}+\frac{1}{n}-\frac{1}{2}\right),\quad \frac{1}{E}=\frac{1}{m}+\frac{1}{n}-\frac{1}{2}.$$

In order that F, S, and E may be positive, we must have $\frac{1}{m}+\frac{1}{n}>\frac{1}{2}$, a relation which admits of five solutions in whole nnmbers corresponding to the five regular solids. The values of m, n, F, S, and E for the five regular solids are exhibited in the following table:—

	m	n	F	S	E
Tetrahedron	3	3	4	4	6
Hexahedron	4	3	6	8	12
Octahedron	3	4	8	6	12
Dodecahedron	5	3	12	20	30
Icosahedron	3	5	20	12	30

908 The sum of all the plane angles of any polyhedron

$$=2\pi(S-2);$$

Or, *Four right angles for every vertex less eight right angles.*

909 If I be the angle between two adjacent faces of a regular polyhedron,

$$\sin \tfrac{1}{2} I = \cos \frac{\pi}{n} \div \sin \frac{\pi}{m}.$$

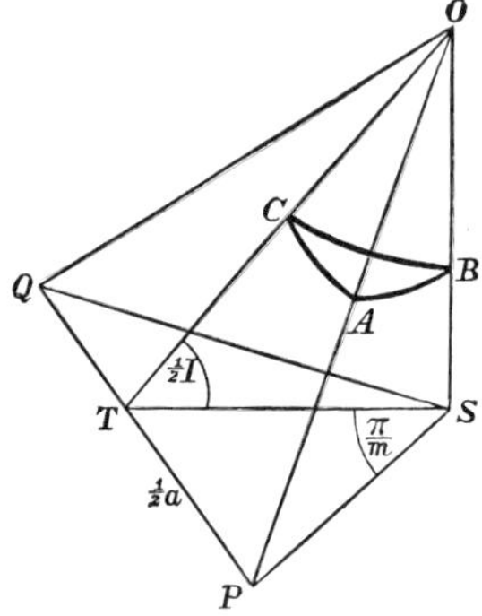

Proof.—Let $PQ = a$ be the edge, and S the centre of a face, T the middle point of PQ, O the centre of the inscribed and circumscribed spheres, ABC the projection of PST upon a concentric sphere. In this spherical triangle,

$$C = \frac{\pi}{2}, \quad A = \frac{\pi}{n}, \quad \text{and} \quad B = \frac{\pi}{m} = PST.$$

Also $\qquad STO = \tfrac{1}{2} I.$

Now, by (881, ii.),

$$\cos A = \sin B \cos BC;$$

that is, $\qquad \cos \frac{\pi}{n} = \sin \frac{\pi}{m} \sin \tfrac{1}{2} I.$

Q. e. d.

If r, R be the radii of the inscribed and circumscribed spheres of a regular polyhedron,

910 $$r = \frac{a}{2} \tan \tfrac{1}{2} I \cot \frac{\pi}{m}, \quad R = \frac{a}{2} \tan \tfrac{1}{2} I \tan \frac{\pi}{n}.$$

Proof. — In the above figure, $OS = r$, $OP = R$, $PT = \frac{a}{2}$; and $OS = PT \cot \frac{\pi}{m} \tan \tfrac{1}{2} I$. Also $OP = PT \operatorname{cosec} AC$, and by (881 i.),

$$\sin AC = \tan BC \cot A = \cot \tfrac{1}{2} I \cot \frac{\pi}{n}, \quad \text{therefore, \&c.}$$

ELEMENTARY GEOMETRY.

MISCELLANEOUS PROPOSITIONS.

920 To find the point in a given line QY, the sum of whose distances from two fixed points S, S' is a minimum.

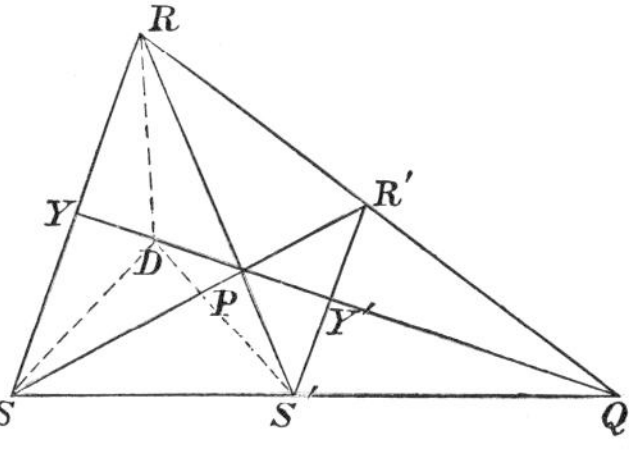

Draw SYR at right angles to QY, making $YR = YS$. Join RS', cutting QY in P. Then P will be the required point.

Proof.—For, if D be any other point on the line, $SD = DR$ and $SP = PR$. But $RD+DS'$ is $> RS'$, therefore, &c. R is called the reflection of the point S, and SPS' is the path of a ray of light reflected at the line QY.

If S, S' and QY are not in the same plane, make SY, YR equal perpendiculars as before, but the last in the plane of S' and QY.

Similarly, the point Q in the given line, the difference of whose distances from the fixed points S and R' is a maximum, is found by a like construction.

The minimum sum of distances from S, S' is given by

$$(SP+S'P)^2 = SS'^2+4SY.S'Y'.$$

And the maximum difference from S and R' is given by

$$(SQ-R'Q)^2 = (SR')^2-4SY.R'Y'.$$

Proved by VI. D., since $SRR'S'$ can be inscribed in a circle.

921 Hence, to find the shortest distance from P to Q *en route* of the lines AB, BC, CD; in other words, the path of the ray reflected at the successive surfaces AB, BC, CD.

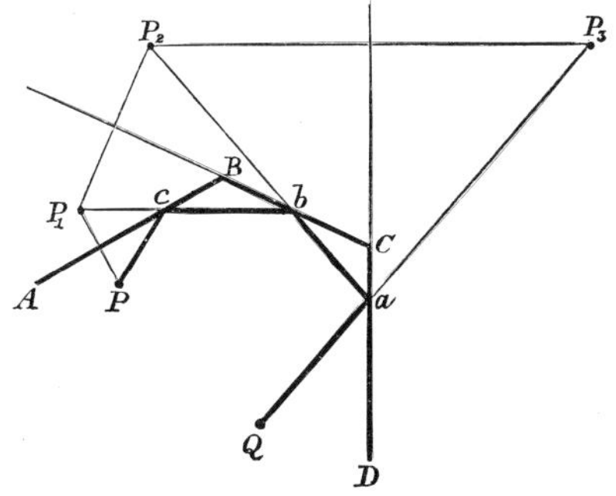

Find P_1, the reflection of P at the first surface; then P_2, the reflection of P_1 at the second surface; next P_3, the reflection of P_2 at the third surface; and so on if there be more surfaces. Lastly, join Q with P_3, the last reflection, cutting CD in a. Join aP_2, cutting BC in b. Join bP_1, cutting AB in c. Join cP. $PcbaQ$ is the path required.

The same construction will give the path when the surfaces are not, as in the case considered, all perpendicular to the same plane.

---

922 If the straight line d from the vertex of a triangle divide the base into segments p, q, and if h be the distance from the point of section to the foot of the perpendicular from the vertex on the base, then

$$b^2+c^2 = p^2+q^2+2d^2+2h\,(p-q). \qquad \text{[II. 12, 13.}$$

The following cases are important:—

(i.) When $p = q$, $\quad b^2+c^2 = 2q^2+2d^2$;

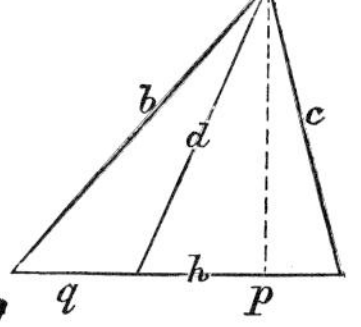

i.e., the sum of the squares of two sides of a triangle is equal to twice the square of half the base, together with twice the square of the bisecting line drawn from the vertex.

(ii.) When $p = 2q$, $\quad 2b^2+c^2 = 6q^2+3d^2$. $\qquad$ (II. 12 or 13)

(iii.) When the triangle is isosceles,

$$b^2 = c^2 = pq+d^2.$$

---

923 If O be the centre of an equilateral triangle ABC and P any point in space. Then

$$PA^2+PB^2+PC^2 = 3\,(PO^2+OA^2).$$

Proof.— $\quad PB^2+PC^2 = 2PD^2+2BD^2. \qquad$ (922, i.)

Also $\quad PA^2+2PD^2 = 6OD^2+3PO^2, \qquad$ (922, ii.)

and $\quad BO = 2OD$;

therefore &c.

Cor.—Hence, if P be any point on the surface of a sphere, centre O, the sum of the squares of its distances from A, B, C is constant. And if r, the radius of the sphere, be equal to OA, the sum of the same squares is equal to $6r^2$.

924 The sum of the squares of the sides of a quadrilateral is equal to the sum of the squares of the diagonals plus four times the square of the line joining the middle points of the diagonals. (922, i.)

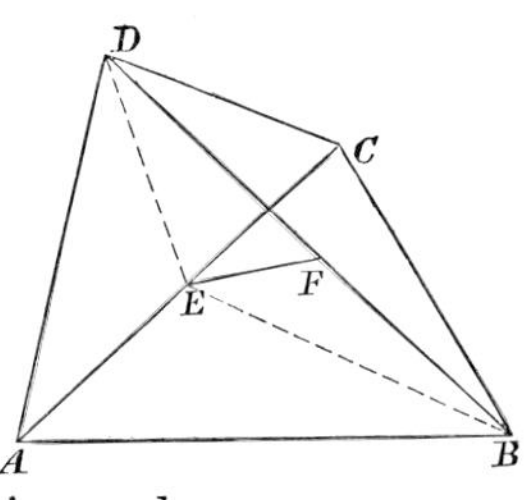

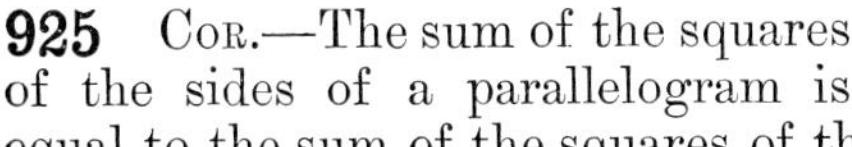

925 Cor.—The sum of the squares of the sides of a parallelogram is equal to the sum of the squares of the diagonals.

926 In a given line AC, to find a point X whose distance from a point P shall have a given ratio to its distance in a given direction from a line AB.

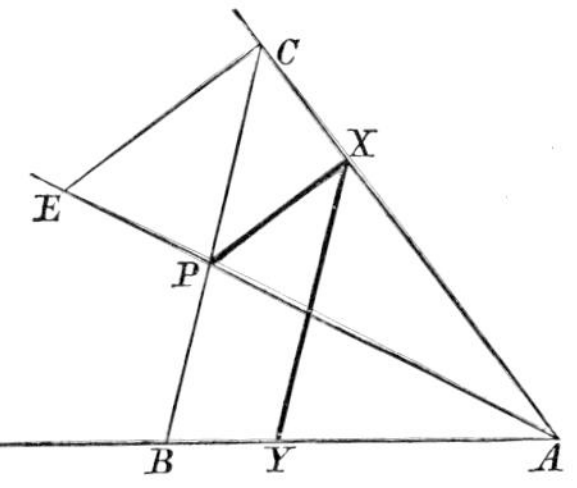

Through P draw BPC parallel to the given direction. Produce AP, and make CE in the given ratio to CB. Draw PX parallel to EC, and XY to CB. There are two solutions when CE cuts AP in two points. [Proof.—By (VI. 2).

927 To find a point X in AC, whose distance XY from AB parallel to BC shall have a given ratio to its distance XZ from BC parallel to AD.

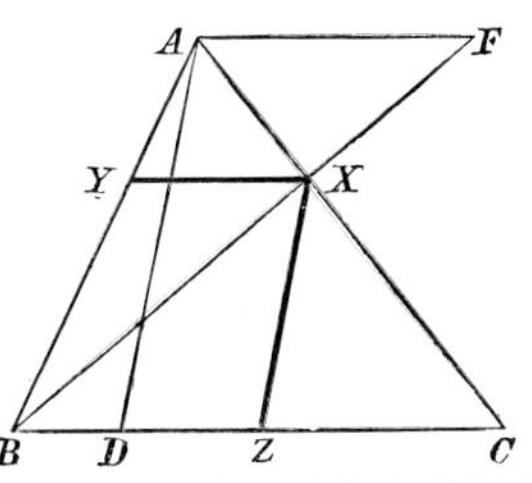

Draw AE parallel to BC, and having to AD the given ratio. Join BE cutting AC in X, the point required. [Proved by (VI. 2).

928 To find a point X on any line, straight or curved, whose distances XY, XZ, in given directions from two given lines AP, AB, shall be in a given ratio.

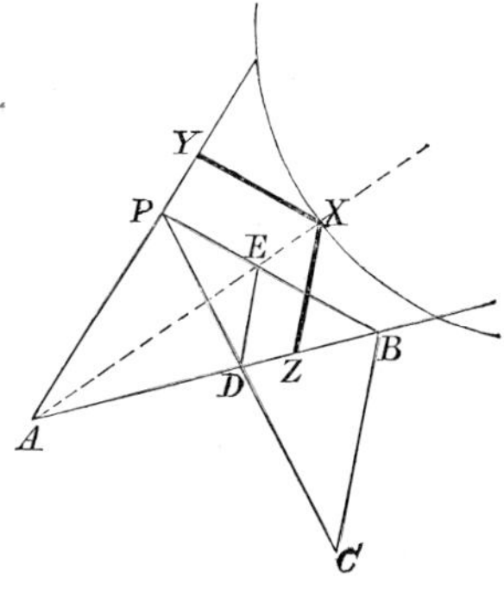

Take P any point in the first line. Draw PB parallel to the direction of XY, and BC parallel to that of XZ, making PB have to BC the given ratio. Join PC, cutting AB in D. Draw DE parallel to CB. Then AE produced cuts the line in X, the point required, and is the locus of such points. [Proof.—By (VI. 2).

929 To draw a line XY through a given point P so that the segments XP, PY, intercepted by a given circle, shall be in a given ratio.

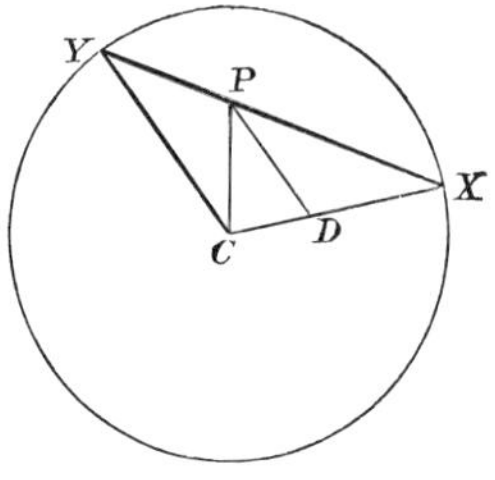

Divide the radius of the circle in that ratio, and, with the parts for sides, construct a triangle PDC upon PC as base. Produce CD to cut the circle in X. Draw XPY and CY.

Then $PD+DC =$ radius,

therefore $PD = DX$.

But $CY = CX$,

therefore PD is parallel to CY (I. 5, 28), therefore &c., by (VI. 2).

930 From a given point P in the side of a triangle, to draw a line PX which shall divide the area of the triangle in a given ratio.

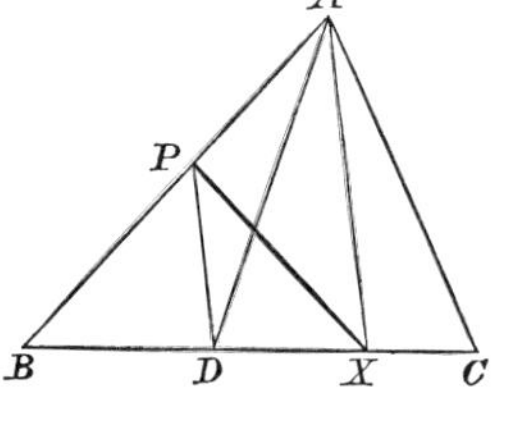

Divide BC in D in the given ratio, and draw AX parallel to PD. PX will be the line required.

$ABD : ADC =$ the given ratio (VI. 1), and $APD = XPD$ (I. 37), therefore &c.

931 To divide the triangle ABC in a given ratio by a line XY drawn parallel to any given line AE.

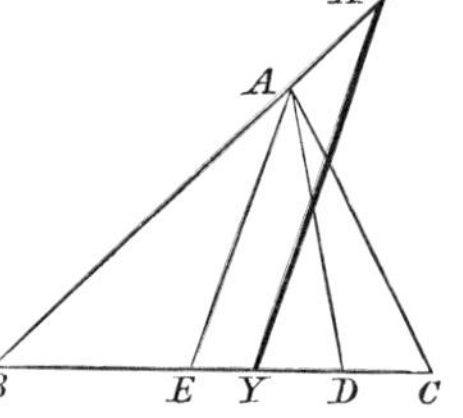

Make BD to DC in the given ratio. Then make BY a mean proportional to BE and BD, and draw YX parallel to EA.

Proof.—AD divides ABC in the given ratio (VI. 1). Now

$ABE : XBY :: BE : BD$, (VI. 19)

or $:: ABE : ABD$;

therefore $XBY = ABD$.

932 If the interior and exterior vertical angles at P of the triangle APB be bisected by straight lines which cut the base in C and D, then the circle circumscribing CPD gives the locus of the vertices of all triangles on the base AB whose sides AP, PB are in a constant ratio.

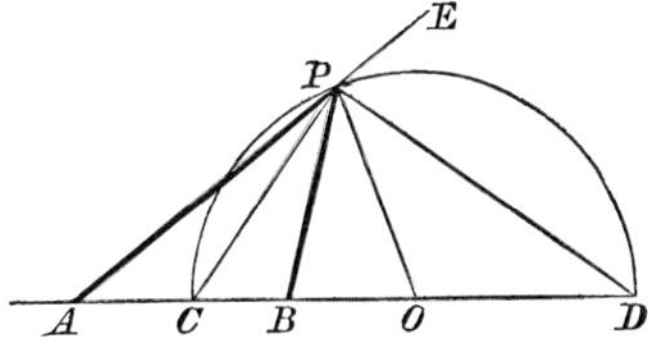

PROOF.—
The $\angle CPD = \frac{1}{2}(APB + BPE)$
$=$ a right angle;
therefore P lies on the circumference of the circle, diameter CD (III. 31). Also
$AP : PB :: AC : CB :: AD : DB$
(VI. 3, and A.), a fixed ratio.

933 AD is divided harmonically in B and C; *i.e.*, $AD : DB :: AC : CB$; or, *the whole line is to one extreme part as the other extreme part is to the middle part.* If we put a, b, c for the lengths AD, BD, CD, the proportion is expressed algebraically by $a : b :: a-c : c-b$, which is equivalent to

$$\frac{1}{a} + \frac{1}{b} = \frac{2}{c}.$$

934 Also $AP : BP = OA : OC = OC : OB$
and $AP^2 : BP^2 = OA : OB,$ (VI. 19)
$AP^2 - AC^2 : CP^2 : BP^2 - BC^2.$ (VI. 3, & B.)

---

935 If Q be the centre of the inscribed circle of the triangle ABC, and if AQ produced meet the circumscribed circle, radius R, in F; and if FOG be a diameter, and AD perpendicular to BC; then

(i.) $\boldsymbol{FC = FQ = FB = 2R \sin \frac{A}{2}}.$

(ii.) $\boldsymbol{\angle FAD = FAO = \frac{1}{2}(B - C)},$
and $\boldsymbol{\angle CAG = \frac{1}{2}(B + C)}.$

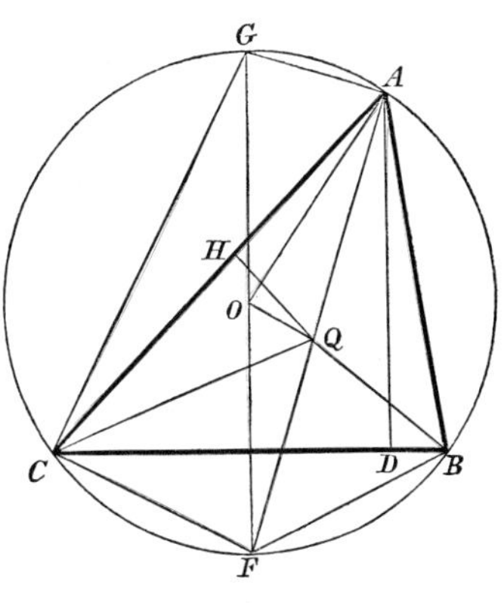

PROOF OF (i.)—
$\angle FQC = QCA + QAC.$
But $QAC = QAB = BCF,$ (III. 21)
$\therefore FQC = FCQ, \quad \therefore FC = FQ.$
Similarly $FB = FQ.$
Also $\angle GCF$ is a right angle, and
$FGC = FAC = \frac{1}{2}A,$ (III. 21)
$\therefore \quad FC = 2R \sin \frac{A}{2}.$

---

936 If R, r be the radii of the circumscribed and inscribed circles of the triangle ABC (see last figure), and O, Q the centres; then $\qquad \boldsymbol{OQ^2 = R^2 - 2Rr}.$

PROOF.—Draw QH perpendicular to AC; then $QH = r$. By the isosceles triangle AOF, $OQ^2 = R^2 - AQ \cdot QF$ (922, iii.), and $QF = FC$ (935, i.), and by similar triangles GFC, AQH, $AQ : QH :: GF : FC$,
therefore $\qquad AQ \cdot FC = GF \cdot QH = 2Rr.$

The problems known as the Tangencies.

937 Given in position any three of the following nine data—viz., three points, three straight lines, and three circles,—it is required to describe a circle passing through the given points and touching the given lines or circles. The following five principal cases occur.

938 I. Given two points A, B, and the straight line CD.

Analysis.—Let ABX be the required circle, touching CD in X. Therefore

$$CX^2 = CA \cdot CB. \qquad \text{(III. 36)}$$

Hence the point X can be found, and the centre of the circle defined by the intersection of the perpendicular to CD through X and the perpendicular bisector of AB. There are two solutions.

Otherwise, by (926), making the ratio one of equality, and DO the given line.

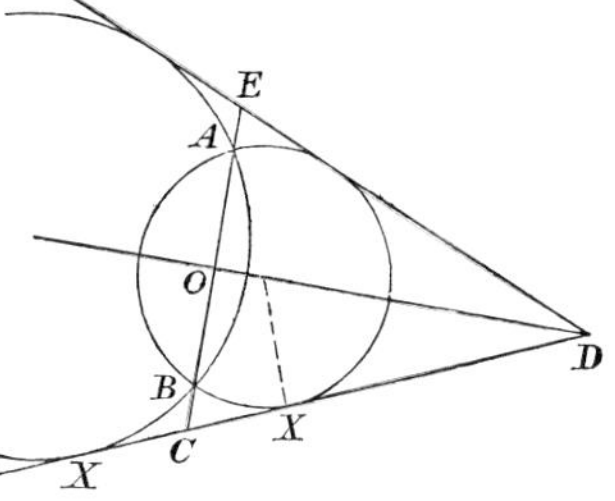

Cor.—The point X thus determined is the point in CD at which the distance AB subtends the greatest angle. In the solution of (941) Q is a similar point in the circumference CD.

(III. 21, & I. 16.)

939 II. Given one point A and two straight lines DC, DE.

In the last figure draw AOC perpendicular to DO, the bisector of the angle D, and make $OB = OA$, and this case is solved by Case I.

940 III. Given the point P, the straight line DE, and the circle ACF.

Analysis.—Let PEF be the required circle touching the given line in E and the circle in F.

Through H, the centre of the given circle, draw $AHCD$ perpendicular to DE. Let K be the centre of the other circle. Join HK, passing through F, the point of contact. Join AF, EF, and AP, cutting the required circle in X. Then

$$\angle DHF = LKF, \qquad \text{(I. 27)}$$

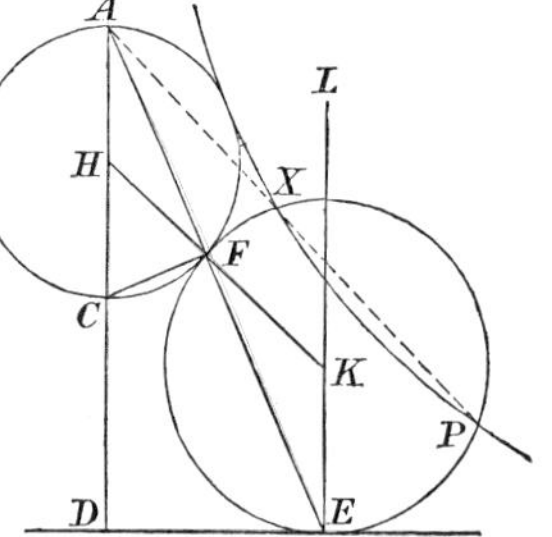

therefore $HFA = KFE$ (the halves of equal angles), therefore AF, FE are in the same straight line. Then, because $AX \cdot AP = AF \cdot AE$, (III. 36) and $AF \cdot AE = AC \cdot AD$ by similar triangles, therefore AX can be found. A circle must then be described through P and X to touch the given line, by Case I. There are two solutions with exterior contact, as appears from Case I. These are indicated in the diagram. There are two more in which the circle AC lies within the described circle. The construction is quite analogous, C taking the place of A.

941 IV. Given two points A, B and the circle CD.

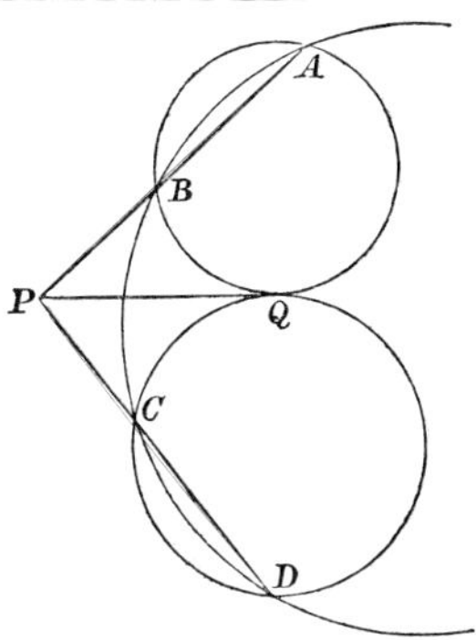

Draw any circle through A, B, cutting the required circle in C, D. Draw AB and DC, and let them meet in P. Draw PQ to touch the given circle. Then, because

$$PC \cdot PD = PA \cdot PB = PQ^2, \quad \text{(III. 36)}$$

and the required circle is to pass through A, B; therefore a circle drawn through A, B, Q must touch PQ, and therefore the circle CD, in Q (III. 37), and it can be described by Case I. There are two solutions corresponding to the two tangents from P to the circle CD.

942 V. Given one point P, and two circles, centres A and B.

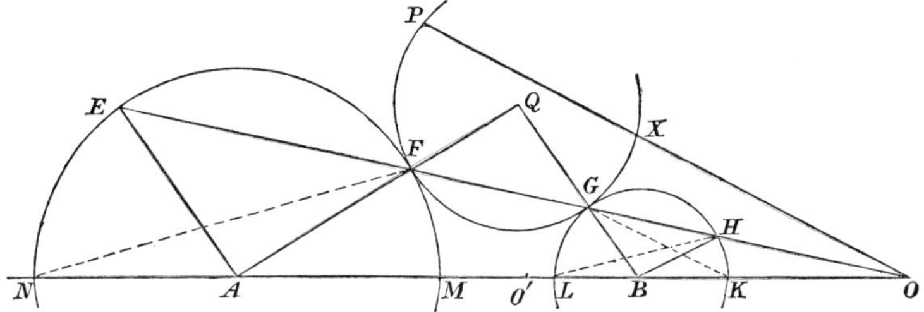

ANALYSIS.—Let PFG be the required circle touching the given ones in F and G. Join the centres QA, QB. Join FG, and produce it to cut the circles in E and H, and the line of centres in O. Then, by the isosceles triangles, the four angles at E, F, G, H are all equal; therefore AE, BG are parallel, and so are AF, BH; therefore $AO : BO :: AF : BH$, and O is a centre of similitude for the two circles. Again, $\angle HBK = 2HLK$, and $FAM = 2FNM$ (III. 20); therefore $FNM = HLK = HGK$ (III. 21); therefore the triangles OFN, OKG are similar; therefore $OF \cdot OG = OK \cdot ON$; therefore, if OP cut the required circle in X, $OX \cdot OP = OK \cdot ON$. Thus the point X can be found, and the problem is reduced to Case IV.

Two circles can be drawn through P and X to touch the given circles. One is the circle PFX. The centre of the other is at the point where EA and HB meet if produced, and this circle touches the given ones in E and H.

943 An analogous construction, employing the internal centre of similitude O', determines the circle which passes through P, and touches one given circle externally and the other internally. See (1047–9).

The centres of similitude are the two points which divide the distance between the centres in the ratio of the radii. See (1037).

944 Cor.—The tangents from O to all circles which touch the given circles, either both externally or both internally, are equal.

For the square of the tangent is always equal to $OK.ON$ or $OL.OM$.

945 The solutions for the cases of three given straight lines or three given points are to be found in Euc. IV., Props. 4, 5.

946 In the remaining cases of the tangencies, straight lines and circles alone are given. By drawing a circle concentric with the required one through the centre of the least given circle, the problem can always be made to depend upon one of the preceding cases; the centre of the least circle becoming one of the *given* points.

947 Definition.—*A centre of similarity of two plane curves is a point such that, any straight line being drawn through it to cut the curves, the segments of the line intercepted between the point and the curves are in a constant ratio.*

948 If AB, AC touch a circle at B and C, then any straight line $AEDF$, cutting the circle, is divided harmonically by the circumference and the chord of contact BC.

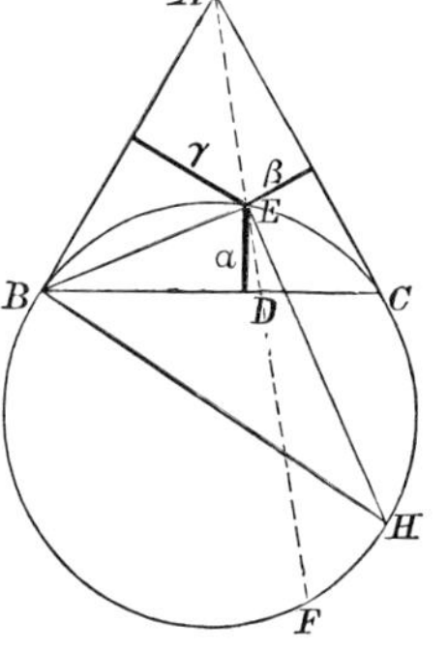

Proof from $AE.AF = AB^2$. (III. 36)

$AB^2 = BD.DC + AD^2$, (923)

and $BD.DC = ED.DF$. (III. 35)

949 If α, β, γ, in the same figure, be the perpendiculars to the sides of ABC from any point E on the circumference of the circle, then $\beta\gamma = \alpha^2$.

Proof.—Draw the diameter $BH = d$; then $EB^2 = \beta d$, because BEH is a right angle. Similarly $EC^2 = \gamma d$. But $EB.EC = \alpha d$ (VI. C.), therefore &c.

950 If FE be drawn parallel to the base BC of a triangle, and if EB, FC intersect in O, then

$$AE : AC :: EO : OB :: FO : OC.$$

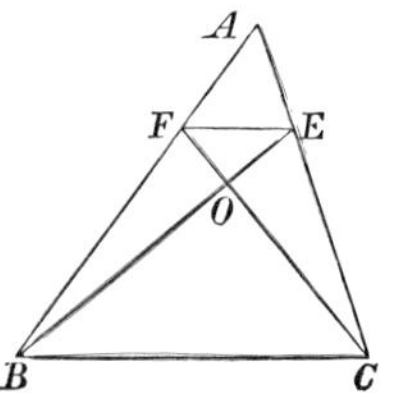

By VI. 2. Since each ratio $= FE : BC$.

Cor.—If $AC = n.AE$, then

$$BE = (n+1)\,OE.$$

951 The three lines drawn from the angles of a triangle to the middle points of the opposite sides, intersect in the same point, and divide each other in the ratio of two to one.

For, by the last theorem, any one of these lines is divided by each of the others in the ratio of two to one, measuring from the same extremity, and must therefore be intersected by them in the same point.

This point will be referred to as the *centroid* of the triangle.

952 The perpendiculars from the angles upon the opposite sides of a triangle intersect in the same point.

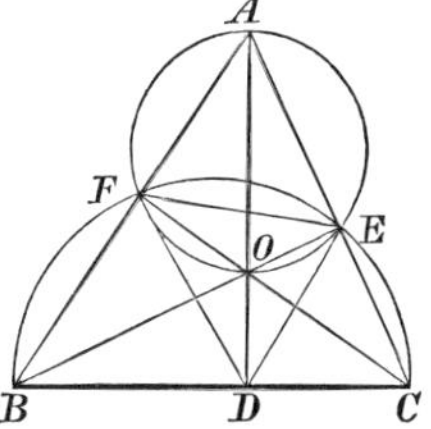

Draw BE, CF perpendicular to the sides, and let them intersect in O. Let AO meet BC in D. Circles will circumscribe $AEOF$ and $BFEC$, by (III. 31);

therefore $\angle FAO = FEO = FCB$; (III. 21)

therefore $\angle BDA = BFC =$ a right angle;

i.e., AO is perpendicular to BC, and therefore the perpendicular from A on BC passes through O.

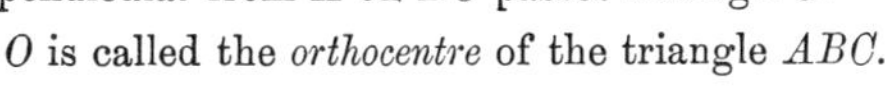

O is called the *orthocentre* of the triangle ABC.

Cor.—The perpendiculars on the sides bisect the angles of the triangle DEF, and the point O is therefore the centre of the inscribed circle of that triangle.

Proof.—From (III. 21), and the circles circumscribing $OEAF$ and $OECD$.

953 If the inscribed circle of a triangle ABC touches the sides a, b, c in the points D, E, F; and if the escribed circle to the side a touches a and b, c produced in D', E', F'; and if

$$s = \tfrac{1}{2}(a+b+c);$$

then

$$\boldsymbol{BF' = BD' = CD = s-c,}$$

and $\boldsymbol{AE' = AF' = s}$;

and similarly with respect to the other segments.

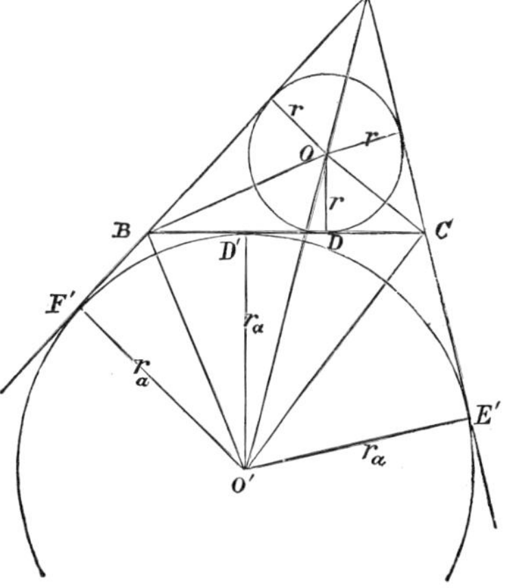

Proof.—The two tangents from any vertex to either circle being equal, it follows that $CD+c =$ half the perimeter of ABC, which is made up of three pairs of equal segments;

therefore $CD = s-c$.

Also

$$AE'+AF' = AC+CD'+AB+BD' = 2s;$$

therefore $AE' = AF' = s$.

The Nine-Point Circle.

954 The *Nine-point* circle is the circle described through D, E, F, the feet of the perpendiculars on the sides of the triangle ABC. It also passes through the middle points of the sides of ABC and the middle points of OA, OB, OC; in all, through nine points.

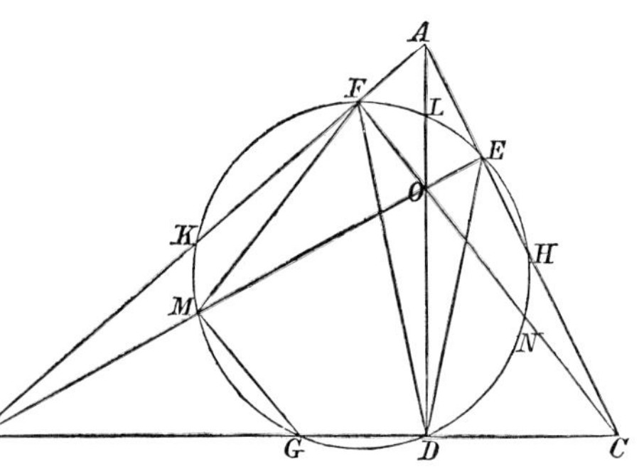

PROOF.—Let the circle cut the sides of ABC again in G, H, K; and OA, OB, OC in L, M, N. $\angle EMF = EDF$ (III. 21) $= 2ODF$ (952, Cor.); therefore, since OB is the diameter of the circle circumscribing $OFBD$ (III. 31), M is the centre of that circle (III. 20), and therefore bisects OB.

Similarly OC and OA are bisected at N and L.

Again, $\angle MGB = MED$ (III. 22) $= OCD$, (III. 21), by the circle circumscribing $OECD$. Therefore MG is parallel to OC, and therefore bisects BC. Similarly H and K bisect CA and AB.

955 The centre of the nine-point circle is the middle point of OQ, the line joining the orthocentre and the centre of the circumscribing circle of the triangle ABC.

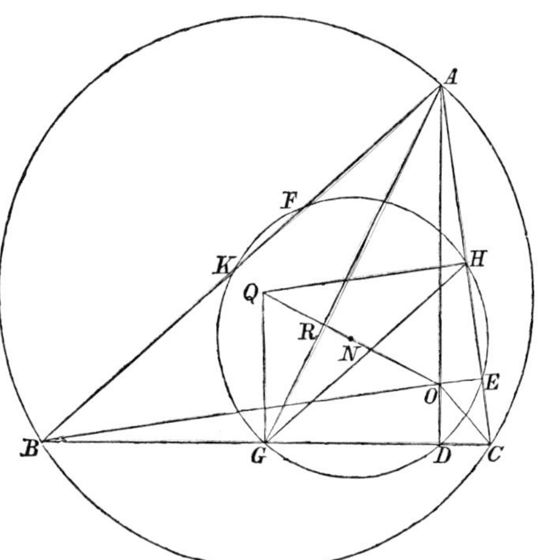

For the centre of the N. P. circle is the intersection of the perpendicular bisectors of the chords DG, EH, FK, and these perpendiculars bisect OQ in the same point N, by (VI. 2).

956 The centroid of the triangle ABC also lies on the line OQ and divides it in R so that $OR = 2RQ$.

PROOF.—The triangles QHG, OAB are similar, and $AB = 2HG$; therefore $AO = 2GQ$; therefore $OR = 2RQ$; and $AR = 2RG$; therefore R is the centroid, and it divides OQ as stated (951).

957 Hence the line joining the centres of the circumscribed and nine-point circles is divided harmonically in the ratio of 2 : 1 by the centroid and the orthocentre of the triangle.

These two points are therefore centres of similitude of the circumscribed and nine-point circles; and any line drawn through either of the points is divided by the circumferences in the ratio of 2 : 1. See (1037.)

958 The lines DE, EF, FD intersect the sides of ABC in the radical axis of the two circles.

For, if EF meets BC in P, then by the circle circumscribing $BCEF$, $PE.PF = PC.PB$; therefore (III. 36) the tangents from P to the circles are equal (985).

959 The nine-point circle touches the inscribed and escribed circles of the triangle.

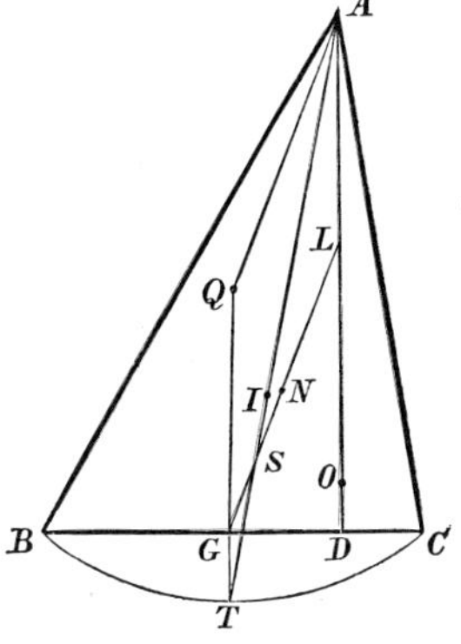

Proof.—Let O be the orthocentre, and I, Q the centres of the inscribed and circumscribed circles. Produce AI to bisect the arc BC in T. Bisect AO in L, and join GL, cutting AT in S.

The N. P. circle passes through G, D, and L (954), and D is a right angle. Therefore GL is a diameter, and is therefore $= R = QA$ (957). Therefore GL and QA are parallel. But $QA = QT$, therefore

$$GS = GT = CT \sin\frac{A}{2} = 2R\sin^2\frac{A}{2}. \quad (935, \text{i.})$$

Also $$ST = 2GS\cos\theta$$

(θ being the angle $GST = GTS$).

N being the centre of the N. P. circle, its radius $= NG = \frac{1}{2}R$; and r being the radius of the inscribed circle, it is required to shew that

$$NI = NG - r.$$

Now $$NI^2 = SN^2 + SI^2 - 2SN.SI\cos\theta. \quad (702)$$

Substitute $$SN = \tfrac{1}{2}R - GS;$$

$$SI = TI - ST = 2R\sin\frac{A}{2} - 2GS\cos\theta;$$

and $GS = 2R\sin^2\frac{1}{2}A$, to prove the proposition.

If J be the centre of the escribed circle touching BC, and r_a its radius, it is shown in a similar way that $NJ = NG + r_a$.

To construct a triangle from certain data.

960 When amongst the data we have the sum or difference of the two sides AB, AC; or the sum of the segments of the base made by AG, the bisector of the exterior vertical angle; or the difference of the segments made by AF, the bisector of the interior vertical angle; the following construction will lead to the solution.

Make $AE = AD = AC$. Draw DH parallel to AF, and suppose EK drawn parallel to AG to meet the base produced in K; and complete the figure. Then BE is the sum, and BD is the difference of the sides.

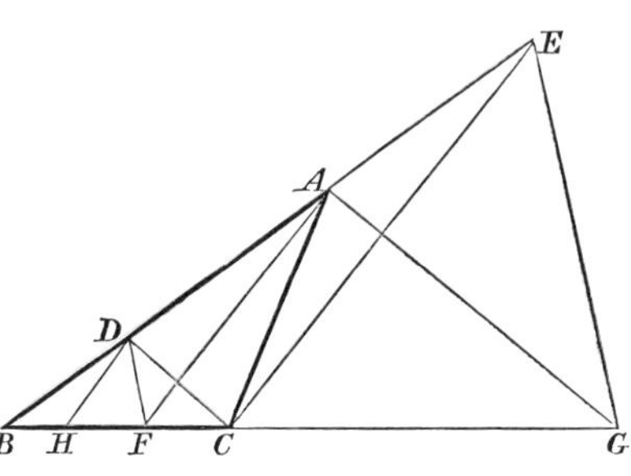

BK is the sum of the exterior segments of the base, and BH is the difference of the interior segments. $\angle BDH = BEC = \frac{1}{2}A$,

$$\angle ADC = EAG = \tfrac{1}{2}(B+C),$$
$$\angle DCB = \tfrac{1}{2}DFB = \tfrac{1}{2}(C-B).$$

961 When the base and the vertical angle are given; the locus of the vertex is the circle ABC in figure (935); and the locus of the centre of the inscribed circle is the circle, centre F and radius FB. When the ratio of the sides is given, see (932).

962 To construct a triangle when its form and the distances of its vertices from a point A' are given.

Analysis.—Let ABC be the required triangle. On $A'B$ make the triangle $A'BC'$ similar to ABC, so that $AB : A'B :: CB : C'B$. The angles ABA', CBC' will also be equal; therefore $AB : BC :: AA' : CC'$, which gives CC', since the ratio $AB : BC$ is known. Hence the point C is found by constructing the triangle $A'CC'$. Thus BC is determined, and thence the triangle ABC from the known angles.

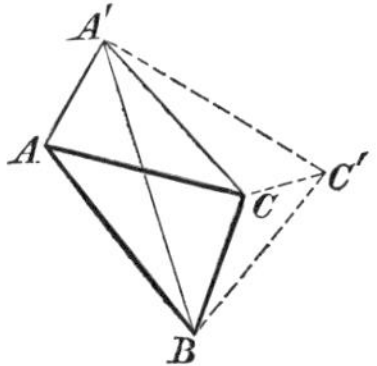

963 To find the locus of a point P, the tangent from which to a given circle, centre A, has a constant ratio to its distance from a given point B.

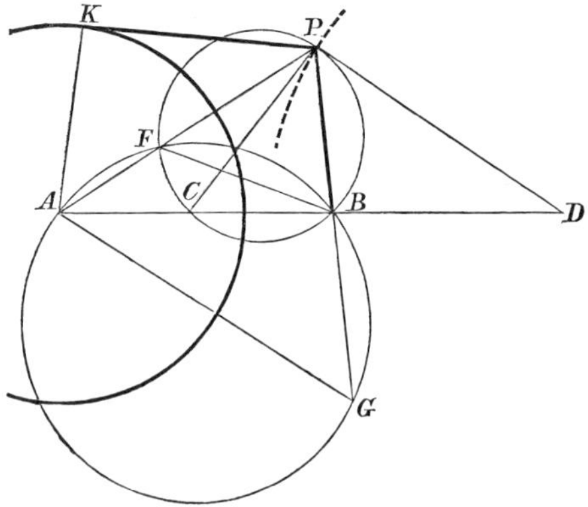

Let AK be the radius of the circle, and $p : q$ the given ratio. On AB take AC, a third proportional to AB and AK, and make

$$AD : DB = p^2 : q^2.$$

With centre D, and a radius

equal to a mean proportional between DB and DC, describe a circle. It will be the required locus.

PROOF.—Suppose P to be a point on the required locus. Join P with A, B, C, and D.

Describe a circle about PBC cutting AP in F, and another about ABF cutting PB in G, and join AG and BF. Then

$$PK^2 = AP^2 - AK^2 = AP^2 - BA \cdot AC \text{ (by constr.)} = AP^2 - PA \cdot AF \text{ (III. 36)}$$
$$= AP \cdot PF \text{ (II. 2)} = GP \cdot PB \text{ (III. 36)}.$$

Therefore, by hypothesis,

$$p^2 : q^2 = GP \cdot PB : PB^2 = GP : PB = AD : DB \text{ (by constr.)};$$

therefore $\angle DPG = PGA$ (VI. 2) $= PFB$ (III. 22) $= PCB$ (III. 21).

Therefore the triangles DPB, DCP are similar; therefore DP is a mean proportional to DB and DC. Hence the construction.

964 COR.—If $p = q$ the locus becomes the perpendicular bisector of BC, as is otherwise shown in (1003).

---

965 To find the locus of a point P, the tangents from which to two given circles shall have a given ratio. (See also 1036.)

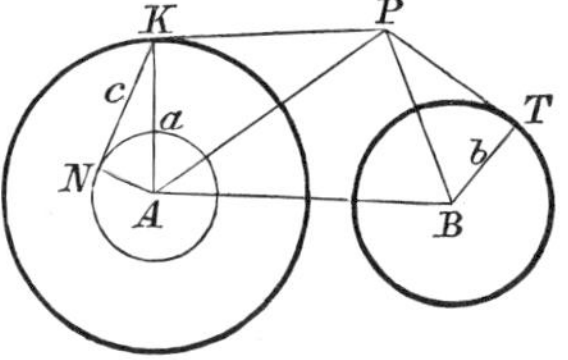

Let A, B be the centres, a, b the radii ($a > b$), and $p : q$ the given ratio. Take c, so that $c : b = p : q$, and describe a circle with centre A and radius $AN = \sqrt{a^2 - c^2}$. Find the locus of P by the last proposition, so that the tangent from P to this circle may have the given ratio to PB. It will be the required locus.

PROOF.—By hypothesis and construction,

$$\frac{p^2}{q^2} = \frac{PK^2}{PT^2} = \frac{c^2}{b^2} = \frac{PK^2 + c^2}{PT^2 + b^2} = \frac{AP^2 - a^2 + c^2}{BP^2} = \frac{AP^2 - AN^2}{BP^2}.$$

COR.—Hence the point can be found on any curve from which the tangents to two circles shall have a given ratio.

---

966 To find the locus of the point from which the tangents to two given circles are equal.

Since, in (965), we have now $p = q$, and therefore $c = b$, the construction simplifies to the following:

Take $AN = \sqrt{(a^2 - b^2)}$, and in AB take $AB : AN : AC$. The perpendicular bisector of BC is the required locus. But, if the circles intersect, then their common chord is at once the line required. See Radical Axis (985).

Collinear and Concurrent systems of points and lines.

967 Definitions.—Points lying in the same straight line are *collinear*. Straight lines passing through the same point are *concurrent*, and the point is called the *focus* of the pencil of lines.

Theorem.—If the sides of the triangle ABC, or the sides produced, be cut by any straight line in the points a, b, c respectively, the line is called a *transversal*, and the segments of the sides are connected by the equation

968 $$(Ab : bC)(Ca : aB)(Bc : cA) = 1.$$

Conversely, if this relation holds, the points a, b, c will be collinear.

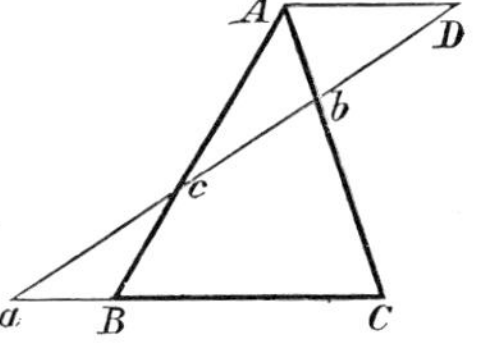

Proof.—Through any vertex A draw AD parallel to the opposite side BC, to meet the transversal in D, then

$Ab : bC = AD : Ca$ and $Bc : cA = aB : AD$

(VI. 4), which proves the theorem.

Note.—In the formula the segments of the sides are estimated positive, independently of direction, the sequence of the letters being preserved the better to assist the memory. A point may be supposed to travel from A over the segments Ab, bC, &c. *continuously*, until it reaches A again.

969 By the aid of (701) the above relation may be put in the form

$$(\sin ABb : \sin bBC)(\sin CAa : \sin aAB)(\sin BCc : \sin cCA) = 1$$

---

970 If O be any focus in the plane of the triangle ABC, and if AO, BO, CO meet the sides in a, b, c; then, as before,

$$(Ab : bC)(Ca : aB)(Bc : cA) = 1.$$

Conversely, if this relation holds, the lines Aa, Bb, Cc will be concurrent.

Proof.—By the transversal Bb to the triangle AaC, we have (968)

$(Ab : bC)(CB : Ba)$
$\times (aO : OA) = 1.$

And, by the transversal Cc to the triangle AaB,

$(Bc : cA)(AO : Oa)$
$\times (aC : CB) = 1.$

Multiply these equations together.

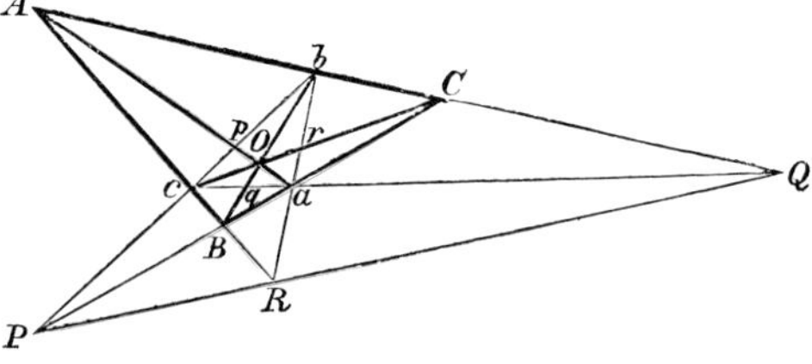

971 If bc, ca, ab, in the last figure, be produced to meet the sides of ABC in P, Q, R, then each of the nine lines in the figure will be divided harmonically, and the points P, Q, R will be collinear.

PROOF.—(i.) Take bP a transversal to ABC; therefore, by (968),

$$(CP : PB)\,(Bc : cA)\,(Ab : bC) = 1;$$

therefore, by (970), $CP : PB = Ca : aB$.

(ii.) Take CP a transversal to Abc, therefore

$$(AB : Bc)\,(cP : Pb)\,(bC : CA) = 1.$$

But, by (970), taking O for focus to Abc,

$$(AB : Bc)\,(cp : pb)\,(bC : CA) = 1;$$

therefore $cP : Pb = cp : pb$.

(iii.) Take PC a transversal to AOc, and b a focus to AOc; therefore, by (968 & 970), $(Aa : aO)\,(OC : Cc)\,(cB : BA) = 1$,

and $(Ap : pO)\,(OC : Cc)\,(cB : BA) = 1$;

therefore $Aa : aO = Ap : pO$.

Thus all the lines are divided harmonically.

(iv.) In the equation of (970) put $Ab : bC = AQ : QC$ the harmonic ratio, and similarly for each ratio, and the result proves that P, Q, R are collinear, by (968).

COR.—If in the same figure qr, rp, pq be joined, the three lines will pass through P, Q, R respectively.

PROOF.—Take O as a focus to the triangle abc, and employ (970) and the harmonic division of bc to show that the transversal rq cuts bc in P.

---

972 If a transversal intersects the sides AB, BC, CD, &c. of any polygon in the points a, b, c, &c. in order, then

$$(Aa : aB)\,(Bb : bC)\,(Cc : cD)\,(Dd : dE)\,\ldots\,\&\text{c.} = 1.$$

PROOF.—Divide the polygon into triangles by lines drawn from one of the angles, and, applying (968) to each triangle, combine the results.

---

973 Let any transversal cut the sides of a triangle and their three intersectors AO, BO, CO (see figure of 970) in the points A', B', C', a', b', c', respectively; then, as before,

$$(A'b' : b'C')\,(C'a' : a'B')\,(B'c' : c'A') = 1.$$

PROOF.—Each side forms a triangle with its intersector and the transversal. Take the four remaining lines in succession for transversals to each triangle, applying (968) symmetrically, and combine the twelve equations.

974 If the lines joining corresponding vertices of two triangles ABC, abc are concurrent, the points of intersection of the pairs of corresponding sides are collinear, and conversely.

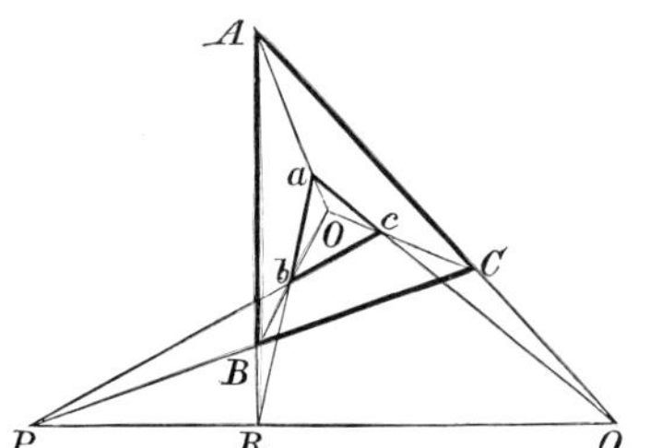

PROOF.—Let the concurrent lines Aa, Bb, Cc meet in O. Take bc, ca, ab transversals respectively to the triangles OBC, OCA, OAB, applying (968), and the product of the three equations shows that P, R, Q lie on a transversal to ABC.

975 Hence it follows that, if the lines joining each pair of corresponding vertices of any two rectilineal figures are concurrent, the pairs of corresponding sides intersect in points which are collinear.

The figures in this case are said to be *in perspective*, or *in homology*, with each other. The point of concurrence and the line of collinearity are called respectively the *centre* and *axis* of perspective or homology. See (1083).

976 *Theorem.*—When three perpendiculars to the sides of a triangle ABC, intersecting them in the points a, b, c respectively, are *concurrent*, the following relation is satisfied; and conversely, if the relation be satisfied, the perpendiculars are concurrent.

$$Ab^2 - bC^2 + Ca^2 - aB^2 + Bc^2 - cA^2 = 0.$$

PROOF.—If the perpendiculars meet in O, then $Ab^2 - bC^2 = AO^2 - OC^2$, &c. (I. 47).

EXAMPLES.—By the application of this theorem, the concurrence of the three perpendiculars is readily established in the following cases:—

(1) When the perpendiculars bisect the sides of the triangle

(2) When they pass through the vertices. (By employing I. 47.)

(3) The three radii of the escribed circles of a triangle at the points of contact between the vertices are concurrent. So also are the radius of the inscribed circle at the point of contact with one side, and the radii of the two escribed circles of the remaining sides at the points of contact beyond the included angle.

In these cases employ the values of the segments given in (953).

(4) The perpendiculars equidistant from the vertices with three concurrent perpendiculars are also concurrent.

(5) When the three perpendiculars from the vertices of one triangle upon the sides of the other are concurrent, then the perpendiculars from the vertices of the second triangle upon the sides of the first are also concurrent.

PROOF.—If A, B, C and A', B', C' are corresponding vertices of the triangles, join AB', AC', BC', BA', CA', CB', and apply the theorem in conjunction with (I. 47).

Triangles of constant species circumscribed to a triangle.

977 Let ABC be any triangle, and O any point; and let circles circumscribe AOB, BOC, COA. The circumferences will be the loci of the vertices of a triangle of constant form whose sides pass through the points A, B, C.

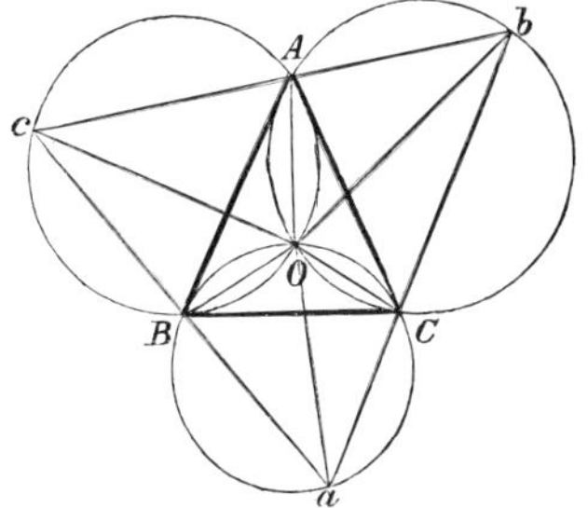

PROOF.—Draw any line bAc from circle to circle, and produce bC, cB to meet in a. The angles AOB, COA are supplements of the angles c and b (III. 22); therefore BOC is the supplement of a (I. 32); therefore a lies on the circle OBC. Also, the angles at O being constant, the angles a, b, c are constant.

978 The triangle abc is a maximum when its sides are perpendicular to OA, OB, OC.

PROOF.—The triangle is greatest when its sides are greatest. But the sides vary as Oa, Ob, Oc, which are greatest when they are diameters of the circles; therefore &c., by (III. 31).

979 To construct a triangle of given species and of given limited magnitude which shall have its sides passing through three given points A, B, C.

Determine O by describing circles on the sides of ABC to contain angles equal to the supplements of the angles of the specified triangle. Construct the figure $abcO$ independently from the known sides of abc, and the now known angles $ObC = OAC$, $OaC = OBC$, &c. Thus the lengths Oa, Ob, Oc are found, and therefore the points a, b, c, on the circles, can be determined.

The demonstrations of the following propositions will now be obvious.

Triangles of constant species inscribed to a triangle.

980 Let abc, in the last figure, be a fixed triangle, and O any point. Take any point A on bc, and let the circles circumscribing OAc, OAb cut the other sides in B, C. Then ABC will be a triangle of constant form, and its angles will have the values $A = Oba + Oca$, &c. (III. 21.)

981 The triangle ABC will evidently be a minimum when OA, OB, OC are drawn perpendicular to the sides of abc.

982 To construct a triangle of given form and of given limited magnitude having its vertices upon three fixed lines bc, ca, ab.

Construct the figure $ABCO$ independently from the known sides of ABC and the angles at O, which are equal to the supplements of the given angles a, b, c. Thus the angles OAC, &c are found, and therefore the angles ObC, &c., equal to them (III. 21), are known. From these last angles the point O can be determined, and the lengths OA, OB, OC being known from the independent figure, the points A, B, C can be found.

Observe that, wherever the point O may be taken, the angles AOB, BOC COA are in all cases either the supplements of, or equal to, the angles c, a, b respectively; while the angles aOb, bOc, cOa are in all cases equal to $C \pm c$, $A \pm a$, $B \pm b$.

983 Note.—In general problems, like the foregoing, which admit of different cases, it is advisable to choose for reference a standard figure which has all its elements of the same affection or sign. In adapting the figure to other cases, all that is necessary is to follow the same construction, letter for letter, observing the convention respecting positive and negative, which applies both to the lengths of lines and to the magnitudes of angles, as explained in (607—609).

Radical Axis.

984 Definition.—The *radical axis* of two circles is that perpendicular to the line of centres which divides the distance between the centres into segments, the difference of whose squares is equal to the difference of the squares of the radii.

Thus, A, B being the centres, a, b the radii, and IP the the radical axis, $$AI^2 - BI^2 = a^2 - b^2.$$

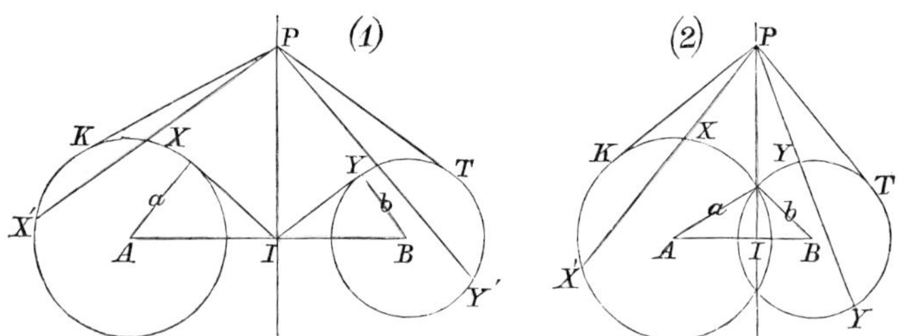

985 It follows that, if the circles intersect, the radical axis is their common chord; and that, if they do not intersect, the radical axis cuts the line of centres in a point the tangents from which to the circles are equal (I. 47).

To draw the axis in this case, see (966).

Otherwise: let the two circles cut the line of centres in C, D and C', D' respectively. Describe any circle through C and D, and another through C' and D', intersecting the former in E and F. Their common chord EF will cut the central axis in the required point I.

PROOF.—$IC.ID = IE.IF = IC'.ID'$ (III. 36); therefore the tangents from I to the circles are equal.

986 *Theorem.*—The difference of the squares of tangents from any point P to two circles is equal to twice the rectangle under the distance between their centres and the distance of the point from their radical axis, or

$$PK^2 - PT^2 = 2AB.PN.$$

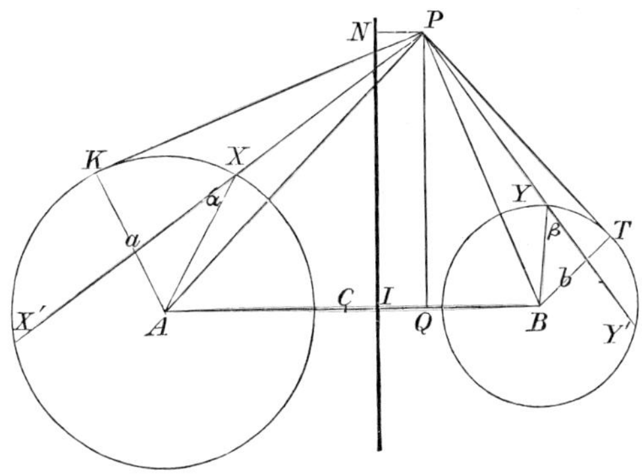

PROOF.

$$PK^2 - PT^2 = (AP^2 - BP^2) - (a^2 - b^2) = (AQ^2 - BQ^2) - (AI^2 - BI^2),$$

by (I. 47) & (984). Bisect AB in C, and substitute for each difference of squares, by (II. 12).

987 COR. 1.—If P be on the circle whose centre is B, then

$$PK^2 = 2AB.PN.$$

988 COR. 2.—If two chords be drawn through P to cut the circles in X, X', Y, Y'; then, by (III. 36),

$$PX.PX' - PY.PY' = 2AB.PN.$$

989 If a variable circle intersect two given circles at constant angles α and β, it will intersect their radical axis at a constant angle; and its radius will bear a constant ratio to the distance of its centre from the radical axis. Or

$$PN : PX = a\cos\alpha - b\cos\beta : AB.$$

PROOF.—In the same figure, if P be the centre of the variable circle, and if $PX = PY$ be its radius; then, by (988),

$$PX(XX' - YY') = 2AB \cdot PN.$$

But $XX' = 2a \cos \alpha$ and $YY' = 2b \cos \beta$;

therefore $PN : PX = a \cos \alpha - b \cos \beta : AB$,

which is a constant ratio if the angles α, β are constant.

990 Also $PX : PN =$ the cosine of the angle at which the circle of radius PX cuts the radical axis. This angle is therefore constant.

991 COR.—A circle which touches two fixed circles has its radius in a constant ratio to the distance of its centre from their radical axis.

This follows from the proposition by making $\alpha = \beta = 0$ or 2π.

If P be on the radical axis; then (see Figs. 1 and 2 of 984)

992 (i.) The tangents from P to the two circles are equal, or $\boldsymbol{PK = PT}$. (986)

993 (ii.) The rectangles under the segments of chords through P are equal, or $\boldsymbol{PX \cdot PX' = PY \cdot PY'}$. (988)

994 (iii.) Therefore the four points X, X', Y, Y' are concyclic (III. 36); and, conversely, if they are concyclic, the chords XX', YY' intersect in the radical axis.

995 DEFINITION.—Points which lie on the circumference of a circle are termed *concyclic*.

996 (iv.) If P be the centre, and if $PX = PY$ be the radius of a circle intersecting the two circles in the figure at angles α and β; then, by (993), $XX' = YY'$, or $a \cos \alpha = b \cos \beta$; that is, *The cosines of the angles of intersection are inversely as the radii of the fixed circles.*

---

997 The radical axes of three circles (Fig. 1046), taken two and two together, intersect at a point called their *radical centre.*

PROOF.—Let A, B, C b the centres, a, b, c the radii, and X, Y, Z the points in which the radical axes cut BC, CA, AB. Write the equation of the definition (984) for each pair of circles. Add the results, and apply (976).

---

998 A circle whose centre is the radical centre of three other circles intersects them in angles whose cosines are inversely as their radii (996).

Hence, if this fourth circle cuts one of the others orthogonally, it cuts them all orthogonally.

999 The circle which intersects at angles α, β, γ three fixed circles, whose centres are A, B, C and radii a, b, c, has its centre at distances from the radical axes of the fixed circles proportional to

$$\frac{b \cos\beta - c \cos\gamma}{BC}, \quad \frac{c \cos\gamma - a \cos\alpha}{CA}, \quad \frac{a \cos\alpha - b \cos\beta}{AB}.$$

And therefore the locus of its centre will be a straight line passing through the radical centre and inclined to the three radical axes at angles whose sines are proportional to these fractions.

Proof.—The result is obtained immediately by writing out equation (989) for each pair of fixed circles.

The Method of Inversion.

1000 Definitions.—Any two points P, P', situated on a diameter of a fixed circle whose centre is O and radius k, so that $OP \cdot OP' = k^2$, are called *inverse points* with respect to the circle, and either point is said to be the *inverse* of the other. The circle and its centre are called the *circle and centre of inversion*, and k the *constant of inversion*.

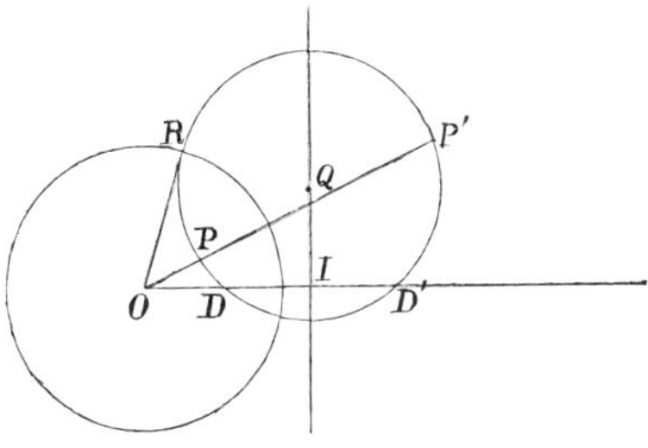

1001 If every point of a plane figure be inverted with respect to a circle, or every point of a figure in space with respect to a sphere, the resulting figure is called the *inverse* or *image* of the original one.

Since $OP : k : OP'$, therefore

1002 $$OP : OP' = OP^2 : k^2 = k^2 : OP'^2.$$

1003 Let D, D', in the same figure, be a pair of inverse points on the diameter OD'. In the perpendicular bisector of DD', take any point Q as the centre of a circle passing through D, D', cutting the circle of inversion in R, and any straight line through O in the points P, P'. Then, by (III. 36), $OP \cdot OP' = OD \cdot OD' = OR^2$ (1000). Hence

1004 (i.) P, P' are inverse points; and, conversely, any two pairs of inverse points lie on a circle.

1005 (ii.) The circle cuts orthogonally the circle of inversion (III. 37); and, conversely, every circle cutting another orthogonally intersects each of its diameters in a pair of inverse points.

1006 (iii.) The line IQ is the locus of a point the tangent from which to a given circle is equal to its distance from a given point D.

1007 Def.—The line IQ is called the *axis of reflexion* for the two inverse points D, D', because there is another circle of inversion, the reflexion of the former, to the right of IQ, having also D, D' for inverse points.

1008 The straight lines drawn from any point P, within or without a circle (Figs. 1 and 2), to the extremities of any chord AB passing through the inverse point Q, make equal angles with the diameter through PQ. Also, the four points O, A, B, P are concyclic, and $QA \cdot QB = QO \cdot QP$.

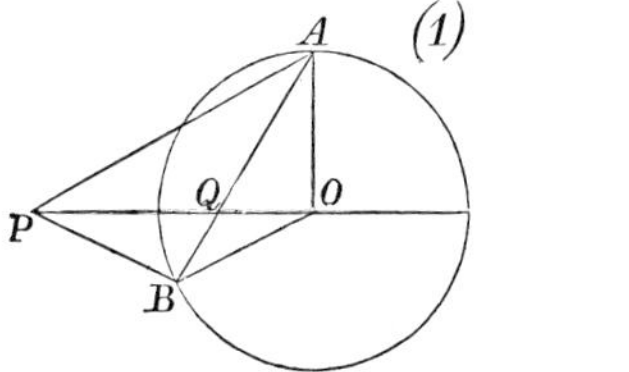

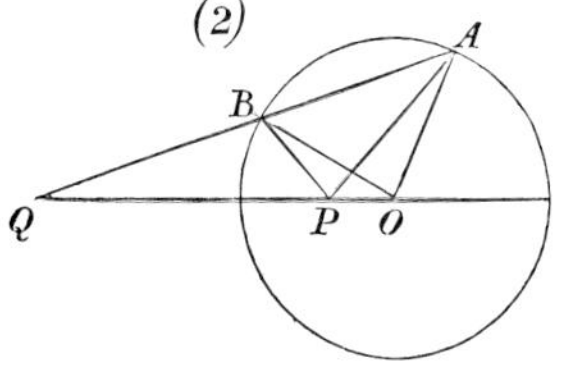

Proof.—In either figure $OP : OA : OQ$ and $OP : OB : OQ$ (1000), therefore, by similar triangles, $\angle OPA = OAB$ and $OPB = OBA$ in figure (1) and the supplement of it in figure (2). But $OAB = OBA$ (I. 5), therefore, &c.

Also, because $\angle OPA = OBA$, the four points O. A, B, P lie on a circle in each case (III. 21), and therefore $QA \cdot QB = QO \cdot QP$ (III. 35, 36).

1009 The inverse of a circle is a circle, and the centre of inversion is the centre of similitude of the two figures. See also (1037).

Proof.—In the figure of (1043), let O be the point where the common tangent RT of the two circles, centres A and B, cuts the central axis, and let any other line through O cut the circles in P, Q P', Q'. Then, in the demonstration of (942), it is shown that OP $OQ' = OQ \cdot OP = k^2$, a constant quantity. Therefore either circle is the inverse of the other, k being the radius of the circle of inversion.

1010 To make the inversions of two given circles equal circles.

Rule.—*Take the centre of inversion so that the squares of the tangents from it to the given circles may be proportional to their radii* (965).

Proof.—(Fig. 1043) $AT : BR = OT : OR = OT^2 : k^2$, since $OT : k : OR$. Therefore $OT^2 : AT = k^2 : BR$, therefore BR remains constant if $OT^2 \propto AT$.

1011 Hence three circles may be inverted into equal circles, for the required centre of inversion is the intersection of two circles that can be drawn by (965).

1012 The inverse of a straight line is a circle passing through the centre of inversion.

Proof.—Draw OQ perpendicular to the line, and take P any other point on it. Let Q', P' be the inverse points. Then $OP \cdot OP' = OQ \cdot OQ'$; therefore, by similar triangles, $\angle OP'Q' = OQP$, a right angle; and OQ' is constant, therefore the locus of P' is the circle whose diameter is OQ'.

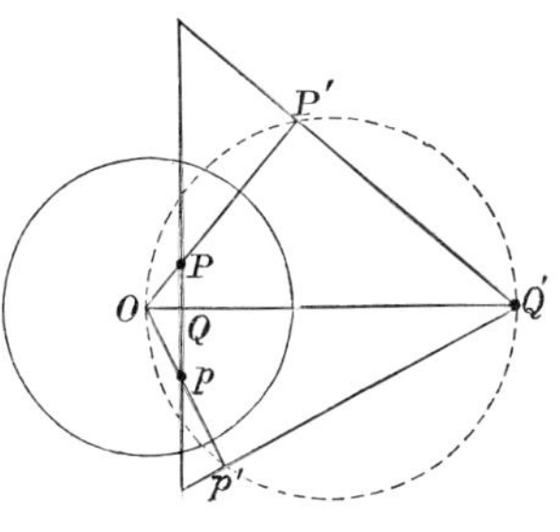

1013 Example.—The inversion of a polygon produces a figure bounded by circular arcs which intersect in angles equal to the corresponding angles of the polygon, the complete circles intersecting in the centre of inversion.

1014 If the extremities of a straight line $P'Q'$ in the last figure are the inversions of the extremities of PQ, then

$$PQ : P'Q' = \sqrt{(OP \cdot OQ)} : \sqrt{(OP' \cdot OQ')}.$$

Proof.—By similar triangles, $PQ : P'Q' = OP : OQ'$ and $PQ : P'Q' = OQ : OP'$. Compound these ratios.

1015 From the above it follows that any homogeneous equation between the lengths of lines joining pairs of points in space, such as $PQ \cdot RS \cdot TU = PR \cdot QT \cdot SU$, the same points appearing on both sides of the equation, will be true for the figure obtained by joining the corresponding pairs of inverse points.

For the ratio of each side of the equation to the corresponding side of the equation for the inverted points will be the same, namely,

$$\sqrt{(OP \cdot OQ \cdot OR \ldots)} : \sqrt{(OP' \cdot OQ' \cdot OR' \ldots)}.$$

Pole and Polar.

1016 Definition.—The *polar* of any point P with respect to a circle is the perpendicular to the diameter OP (Fig. 1012) drawn through the inverse point P'.

1017 It follows that the polar of a point exterior to the circle is the *chord of contact* of the tangents fron the point; that is, the line joining their points of contact.

1018 Also, $P'Q'$ is the polar of P with respect to the circle, centre O, and PQ is the polar of Q'. In other words, *any point* P *lying on the polar of a point* Q', *has its own polar always passing through* Q'.

1019 The line joining any two points P, p is the polar of Q', the point of intersection of their polars.

Proof.—The point Q' lies on both the lines $P'Q'$, $p'Q'$, and therefore has its polar passing through the pole of each line, by the last theorem.

1020 The polars of any two points P, p, and the line joining the points form a *self-reciprocal* triangle with respect to the circle, the three vertices being the poles of the opposite sides. The centre O of the circle is evidently the orthocentre of the triangle (952). The circle and its centre are called the *polar circle* and *polar centre* of the triangle.

If the radii of the polar and circumscribed circles of a triangle ABC be r and R, then

$$r^2 = 4R^2 \cos A \cos B \cos C.$$

Proof.—In Fig. (952), O is the centre of the polar circle, and the circles described round ABC, BOC, COA, AOB are all equal; because the angle BOC is the supplement of A; &c. Therefore $2R \,.\, OD = OB \,.\, OC$ (VI. C) and $r^2 = OA \,.\, OD = OA \,.\, OB \,.\, OC \div 2R$. Also, $OA = 2R \cos A$ by a diameter through B, and (III. 21).

Coaxal Circles.

1021 Definition.—A system of circles having a common line of centres called the *central axis*, and a common radical axis, is termed a *coaxal* system.

1022 If O be the variable centre of one of the circles, and

OK its radius, the whole system is included in the equation

$$OI^2 - OK^2 = \pm \delta^2,$$

where δ is a constant length.

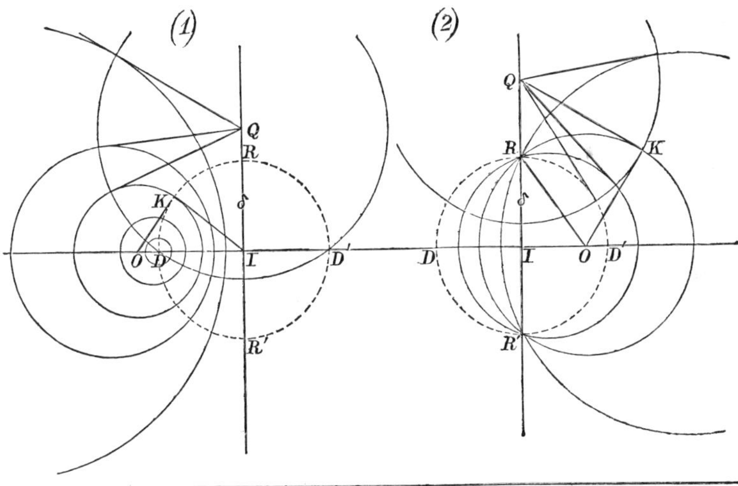

1023 In the first species (Fig. 1),

$$OI^2 - OK^2 = \delta^2,$$

and δ is the length of the tangent from I to any circle of the system (985). Let a circle, centre I and radius δ, cut the central axis in D, D'. When O is at D or D', the circle whose radius is OK vanishes. When O is at an infinite distance, the circle developes into the radical axis itself and into a line at infinity.

The points D, D' are called the *limiting points.*

1024 In the second species (Fig. 2),

$$OR^2 - OI^2 = \delta^2,$$

and δ is half the chord RR' common to all the circles of the system. These circles vary between the circle with centre I and radius δ, and the circle with its centre at infinity as described above. The points R, R' are the *common points* of all circles of this system. The two systems are therefore distinguished as the *limiting points species* and the *common points species* of coaxal circles.

1025 There is a conjugate system of circles having R, R' for limiting points, and D, D' for common points, and the circles of one species intersect all the circles of the conjugate system of the other species orthogonally (1005).

Thus, in figures (1) and (2), Q is the centre of a circle of the opposite species intersecting the other circles orthogonally.

1026 In the first species of coaxal circles, the limiting points D, D' are *inverse* points for every circle of the system, the radical axis being the axis of reflexion for the system.

PROOF.—(Fig. 1) $OI^2 - \delta^2 = OK^2$,

therefore $OD \,.\, OD' = OK^2$, (II. 13)

therefore D, D' are inverse points (1000).

1027 Also, the points in which any circle of the system cuts the central axis are inverse points for the circle whose centre is I and radius δ. [PROOF.—Similar to the last.

1028 PROBLEM.—Given two circles of a coaxal system, to describe a circle of the same system—(i.) to pass through a given point; or (ii.) to touch a given circle; or (iii.) to cut a given circle orthogonally.

1029 I. If the system be of the common points species, then, since the required circle always passes through two known points, the first and second cases fall under the Tangencies. See (941).

1030 To solve the third case, describe a circle through the given common points, and through the inverse of either of them with respect to the given circle, which will then be cut orthogonally, by (1005).

1031 II. If the system be of the limiting points species, the problem is solved in each case by the aid of a circle of the conjugate system. Such a circle always passes through the known limiting points, and may be called a conjugate circle of the limiting points system. Thus,

1032 To solve case (i.)—Draw a conjugate circle through the given point, and the tangent to it at that point will be the radius of the required circle.

1033 To solve case (ii.)—Draw a conjugate circle through the inverse of either limiting point with respect to the given circle, which will thus be cut orthogonally, and the tangent to the cutting circle at either point of intersection will be the radius of the required circle.

1034 To solve case (iii.)—Draw a conjugate circle to touch the given one, and the common tangent of the two will be the radius of the required circle.

1035 Thus, according as we wish to make a circle of the system *touch, or cut orthogonally*, the given circle, we must draw a conjugate circle to *cut orthogonally, or touch it*.

1036 If three circles be coaxal, the squares of the tangents drawn to any two of them from a point on the third are in the ratio of the distances of the centre of the third circle from the centres of the other two.

PROOF.—Let A, B, C be the centres of the circles; PK, PT the tangents from a point P on the circle, centre C, to the other two; PN the perpendicular on the radical axis. By (986),

$$PK^2 = 2AC \,.\, PN \quad \text{and} \quad PT^2 = 2BC \,.\, PN,$$

therefore $$PK^2 : PT^2 = AC : BC.$$

Centres and axes of similitude.

1037 DEFINITIONS.—Let OO' be the centres of similitude (Def. 947) of the two circles in the figure below, and let any line through O cut the circles in P, Q, P', Q'. Then the constant ratio $OP : OP' = OQ : OQ'$ is called the ratio of *similitude* of the two figures; and the constant product $OP \,.\, OQ' = OQ \,.\, OP'$ is called the product of *anti-similitude*. See (942), (1009), and (1043).

The corresponding points P, P' or Q, Q' on the same straight line through O are termed *homologous*, and P, Q' or Q, P' are termed *anti-homologous*.

1038 Let any other line $Opqp'q'$ be drawn through O. Then, if any two points P, p on the one figure be joined, and if P', p', homologous to P, p on the other figure, be also joined, the lines so formed are termed *homologous*. But if the points which are joined on the second figure are anti-homologous to those on the first, the two lines are termed *anti-homologous*. Thus, Pq, $Q'p'$ are anti-homologous lines.

1039 The circle whose centre is O, and whose radius is equal to the square root of the product of anti-similitude, is called the *circle of anti-similitude*.

1040 The four pairs of homologous chords Pp and $P'p'$, Qq and $Q'q'$, Pq and $P'q'$, Qp and $Q'p'$ of the two circles in the figure are parallel. And in all similar and similarly situated figures homologous lines are parallel.

PROOF.—By (VI. 2) and the definition (947).

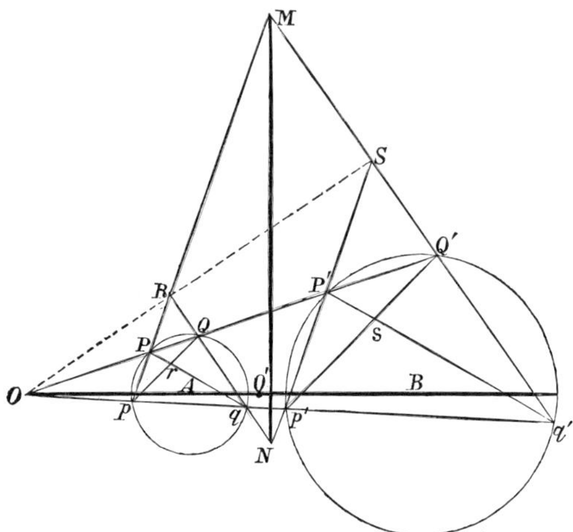

1041 The four pairs of anti-homologous chords, Pp and $Q'q'$, Qq and $P'p'$, Pq and $Q'p'$, Qp and $P'q'$, of the two circles meet on their radical axis.

PROOF.— $$OP \,.\, OQ' = Op \,.\, Oq' = k^2,$$ where k is the constant of inversion; therefore P, p, Q', q' are concyclic; therefore Pp and $Q'q'$ meet on the radical axis. Similarly for any other pair of anti-homologous chords.

1042 COR.—From this and the preceding proposition it follows that the tangents at homologous points are parallel; and that the tangents at anti-homologous points meet on the radical axis. For these tangents are the limiting positions of homologous or anti-homologous chords. (1160)

1043 Let C, D be the inverse points of O with respect to two circles, centres A and B; then the constant product of anti-similitude

$$\boldsymbol{OP' . OQ'} \text{ or } \boldsymbol{OQ . OP' = OA . OD} \text{ or } \boldsymbol{OB . OC}.$$

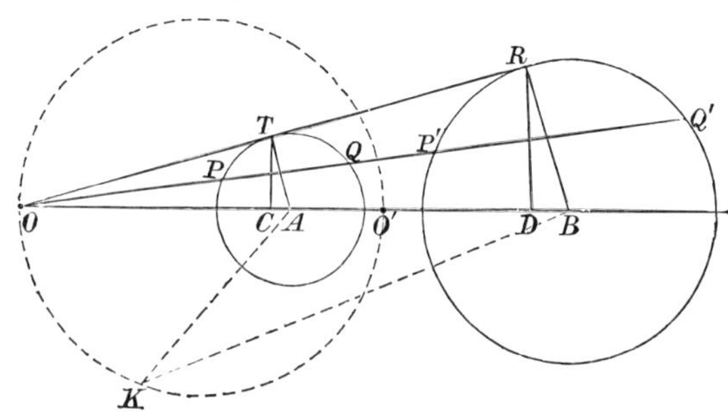

PROOF.—By similar right-angled triangles,

$$OA : OT : OC \quad \text{and} \quad OB : OR : OD;$$

therefore $$OA \,.\, OD = OB \,.\, OC \qquad (1),$$

and also $$OA \,.\, OC = OT^2 = OP \,.\, OQ, \qquad \text{(III. 36)}$$

and $$OB \,.\, OD = OR^2 = OP' \,.\, OQ';$$

therefore $$OA \,.\, OB \,.\, OC \,.\, OD = OP \,.\, OQ \,.\, OP' \,.\, OQ',$$

therefore &c., by (1).

1044 The foregoing definitions and properties (1037 to 1043), which have respect to the external centre of similitude O, hold good for the internal centre of similitude O', with the usual convention of positive and negative for distances measured from O' upon lines passing through it.

1045 Two circles will subtend equal angles at any point on the circumference of the circle whose diameter is OO', where O, O' are the centres of similitude (Fig. 1043). This circle is also coaxal with the given circles, and has been called the *circle of similitude*.

PROOF.—Let A, B be the centres, a, b the radii, and K any point on the circle, diameter OO'. Then, by (932),

$$KA : KB = AO : BO = AO' : BO' = a : b,$$

by the definition (943);

therefore $$a : KA = b : KB;$$

that is, the sines of the halves of the angles in question are equal, which proves the first part. Also, because the tangents from K are in the constant ratio of the radii a, b, this circle is coaxal with the given ones, by (1036, 934).

1046 The six centres of similitude P, p, Q, q, R, r of three circles lie three and three on four straight lines PQR, Pqr, Qpr, Rpq, called *axes of similitude*.

PROOF. — Taking any three of the sets of points named, say P, q, r, they are shewn at once to be collinear by the transversal theorem (968) applied to the triangle ABC.

For the segments of its sides made by the points P, q, r are in the ratios of the radii of the circles.

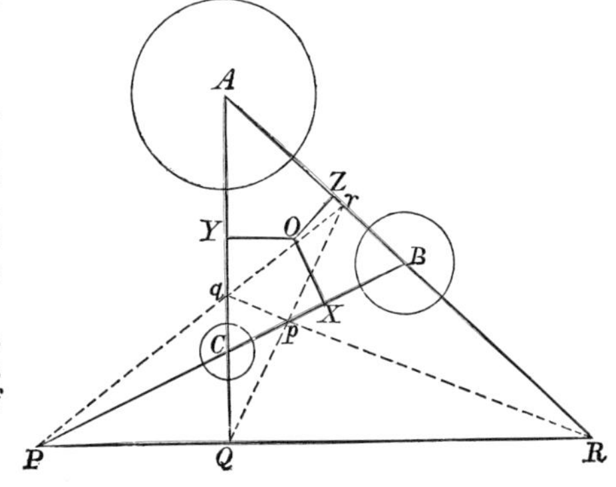

1047 From the investigation in (942), it appears that one circle touches two others in a pair of anti-homologous points, and that the following rule obtains :—

Rule.—*The right line joining the points of contact passes through the external or internal centre of similitude of the two circles according as the contacts are of the same or of different kinds.*

1048 Definition.—Contact of curves is either *internal* or *external* according as the curvatures at the point of contact are in the same or opposite directions.

1049 *Gergonne's method of describing the circles which touch three given circles.*

Take Pqr, one of the four axes of similitude, and find its poles α, β, γ with respect to the given circles, centres A, B, C (1016). From O, the radical centre, draw lines through α, β, γ, cutting the circles in a, a', b, b', c, c'. Then a, b, c and a', b', c' will be the points of contact of two of the required circles.

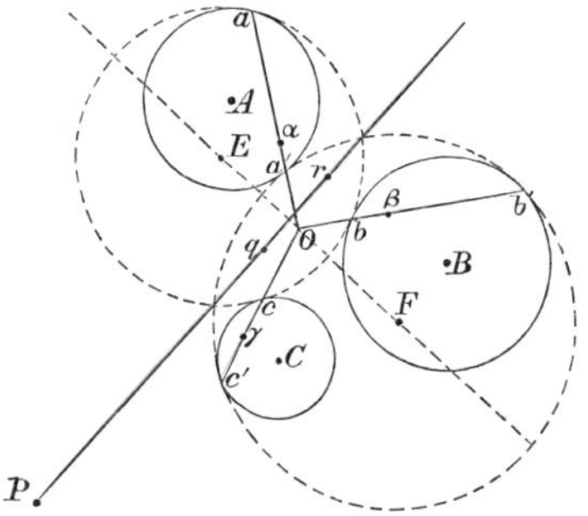

Proof.—*Analysis.*—Let the circles E, F touch the circles A, B, C in a, b, c, a', b', c'. Let bc, $b'c'$ meet in P; ca, $c'a'$ in q; and ab, $a'b'$ in r.

Regarding E and F as touched by A, B, C in turn, Rule (1047) shews that aa', bb', cc' meet in O, the centre of similitude of E and F; and (1041) shews that P, q, and r lie on the radical axis of E and F.

Regarding B and C, or C and A, or A and B, as touched by E and F in turn, Rule (1047) shows that P, q, r are the centres of similitude of B and C, C and A, A and B respectively; and (1041) shews that O is on the radical axis of each pair, and is therefore the radical centre of A, B, and C.

Again, because the tangents to E and F, at the anti-homologous points a, a', meet on Pqr, the radical axis of E and F (1042) ; therefore the point of meeting is the pole of aa' with respect to the circle A (1017). Therefore aa' passes through the pole of the line Pqr (1018). Similarly, bb' and cc' pass through the poles of the same line Pqr with respect to B and C. Hence the construction.

1050 In the given configuration of the circles A, B, C, the demonstration shews that each of the three internal axes of similitude Pqr, Qrp, Rpq (Fig. 1046) is a radical axis and common chord of two of the eight osculating circles which can be drawn. The external axis of similitude PQR is the radical axis of the two remaining circles which touch A, B, and C either all externally or all internally.

1051 The radical centre O of the three given circles is also the common internal centre of similitude of the four pairs of osculating circles. Therefore the central axis of each pair passes through O, and is perpendicular to the radical axis. Thus, in the figure, EF passes through O, and is perpendicular to Pqr.

Anharmonic Ratio.

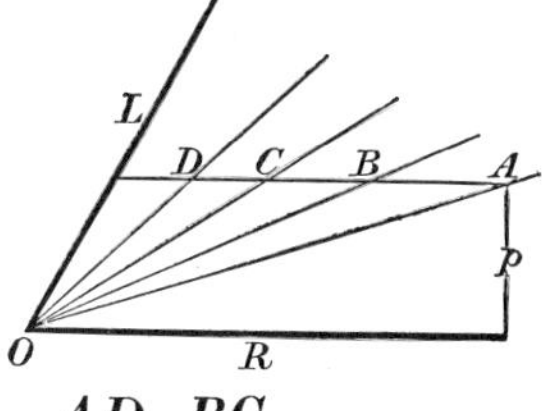

1052 Definition.—Let a pencil of four lines through a point O be cut by a transversal in the points A, B, C, D. The anharmonic ratio of the pencil is any one of the three fractions

$$\frac{AB \cdot CD}{AD \cdot BC} \quad \text{or} \quad \frac{AB \cdot CD}{AC \cdot BD} \quad \text{or} \quad \frac{AD \cdot BC}{AC \cdot BD}.$$

1053 The relation between these three different ratios is obtained from the equation

$$AB \cdot CD + AD \cdot BC = AC \cdot BD.$$

Denoting the terms on the left side by p and q, the three anharmonic ratios may be expressed by

$$p : q, \quad p : p+q, \quad q : p+q.$$

The ratios are therefore mutually dependent. Hence, if the identity merely of the anharmonic ratio in any two systems is to be established, it is immaterial which of the three ratios is selected.

1054 In future, when the ratio of an anharmonic pencil $\{O, ABCD\}$ is mentioned, the form $AB \cdot CD : AD \cdot BC$ will be the one intended, whatever the actual order of the points A, B, C, D may be. For, it should be observed that, by making the line OD revolve about O, the ratio takes in turn each of the forms given above. This ratio is shortly expressed by the notation $\{O, ABCD\}$, or simply $\{ABCD\}$.

1055 If the transversal be drawn parallel to one of the lines, for instance OD, the two factors containing D become infinite, and their ratio becomes unity. They may therefore be omitted. The anharmonic ratio then reduces to $AB : BC$. Thus, when D is at infinity, we may write

$$\{O, ABC\infty\} = AB : BC.$$

1056 The anharmonic ratio

$$\frac{AB \cdot CD}{AD \cdot BC} = \frac{\sin AOB \sin COD}{\sin AOD \sin BOC},$$

and its value is therefore the same for all transversals of the pencil.

Proof.—Draw OR parallel to the transversal, and let p be the perpendicular from A upon OR. Multiply each factor in the fraction by p. Then substitute $p.AB = OA.OB \sin AOB$, &c. (707).

1057 The anharmonic ratio (1056) becomes harmonic when its value is unity. See (933). The harmonic relation there defined may also be stated thus: four points divide a line harmonically when *the product of the extreme segments is equal to the product of the whole line and the middle segment.*

Homographic Systems of Points.

1058 Definition.—If x, a, b, c be the distances of one variable point and three fixed points on a straight line from a point O on the same; and if x', a', b', c' be the distances of similar points on another line through O; then the variable points on the two lines will form two *homographic systems* when they are connected by the anharmonic relation

1059
$$\frac{(x-a)(b-c)}{(x-c)(a-b)} = \frac{(x'-a')(b'-c')}{(x'-c')(a'-b')}.$$

Expanding, and writing A, B, C, D for the constant coefficients, the equation becomes

1060
$$Axx'+Bx+Cx+D = 0.$$

From which

1061
$$x = -\frac{Cx'+D}{Ax'+B}, \quad \text{and} \quad x' = -\frac{Bx+D}{Ax+C}.$$

1062 Theorem.—Any four arbitrary points x_1, x_2, x_3, x_4 on one of the lines will have four corresponding points x'_1, x'_2, x'_3, x'_4 on the other determined by the last equation, and *the two sets of points will have equal anharmonic ratios.*

Proof.—This may be shown by actual substitution of the value of each x in terms of x', by (1061), in the harmonic ratio $\{x_1x_2x_3x_4\}$.

1063 If the distances of four points on a right line from a point O upon it, in order, are α, α', β, β', where α, β; α', β' are the respective roots of the two quadratic equations

$$ax^2+2hx+b = 0, \quad a'x^2+2h'x+b' = 0;$$

the condition that the two pairs of points may be *harmonically conjugate* is

1064
$$ab'+a'b = 2hh'.$$

PROOF.—The harmonic relation, by (1057), is

$$(a-a')(\beta-\beta') = (a-\beta')(a'-\beta).$$

Multiply out, and substitute for the sums and products of the roots of the quadratics above in terms of their coefficients by (51, 52).

1065 If u_1, u_2 be the quadratic expressions in (1063) for two pairs of points, and if u represent a third pair harmonically conjugate with u_1 and u_2, then the pair of points u will also be harmonically conjugate with every pair given by the equation $u_1+\lambda u_2 = 0$, where λ is any constant. For the condition (1064) applied to the last equation will be identically satisfied.

Involution.

1066 DEFINITIONS.—Pairs of *inverse points* PP', QQ', &c., on the same right line, form a system *in involution*, and the relation between them, by (1000), is

$$OP \,.\, OP' = OQ \,.\, OQ' = \&c. = k^2.$$

A O P Q B Q' P'

The radius of the circle of inversion is k, and the centre O is called the *centre of the system*. Inverse points are also termed *conjugate points*.

When two inverse points coincide, the point is called a *focus*.

1067 The equation $OP^2 = k^2$ shows that there are two foci A, B at the distance k from the centre, and on opposite sides of it, real or imaginary according as any two inverse points lie on the same side or on opposite sides of the centre.

1068 If the two homographic systems of points in (1058) be on the *same line*, they will constitute a system *in involution* when $B = C$.

PROOF.—Equation (1060) may now be written

$$Axx' + H(x+x') + B = 0,$$

or
$$\left(x+\frac{H}{A}\right)\left(x'+\frac{H}{A}\right) = \frac{H^2}{A^2} - \frac{B}{A} = k^2,$$

a constant. Therefore $-\frac{H}{A}$ is the distance of the origin O from the centre of inversion. Measuring from this centre, the equation becomes $\xi\xi' = k^2$, representing a system in involution.

1069 *Any four points whatever* of a system in involution on a right line have their anharmonic ratio equal to that of their four conjugates.

PROOF.—Let p, p'; q, q'; r, r'; s, s' be the distances of the pairs of inverse points from the centre.

In the anharmonic ratio of *any* four of the points, for instance $\{pq'rs\}$, substitute $p = k^2 \div p'$, $q' = k^2 \div q$, &c., and the result is the anharmonic ratio $\{p'qr's'\}$.

1070 Any two inverse points P, P' are in harmonic relation with the foci A, B.

A O P B P'

PROOF.—Let p, p' be the distances of P, P' from the centre O; then

$$pp' = k^2, \text{ therefore } \frac{p'}{k} = \frac{k}{p}, \quad \text{therefore } \frac{p'+k}{p'-k} = \frac{k+p}{k-p};$$

that is,

$$\frac{AP'}{BP'} = \frac{AP}{BP}. \quad (933)$$

1071 If a system of points in involution be given, as in (1068), by the equation

$$Axx' + H(x+x') + B = 0 \quad \ldots\ldots\ldots\ldots\ldots\ldots (1);$$

and a pair of conjugate points by the equation

$$ax^2 + 2hx + b = 0 \quad \ldots\ldots\ldots\ldots\ldots\ldots\ldots\ldots (2);$$

the necessary relation between a, h, and b is

1072
$$\boldsymbol{Ab + Ba = 2Hh.}$$

PROOF.—The roots of equation (2) must be simultaneous values of x, x' in (1); therefore substitute in (1)

$$x + x' = -\frac{2h}{a} \quad \text{and} \quad xx' = \frac{b}{a}. \quad (51)$$

1073 COR.—A system in involution may be determined from two given pairs of corresponding points.

Let the equations for these points be

$$ax^2 + 2hx + b = 0 \quad \text{and} \quad a'x^2 + 2hx + b' = 0.$$

Then there are two conditions (1072),

$$Ab + Ba = 2Hh \quad \text{and} \quad Ab' + Ba' = 2Hh',$$

from which A, H, B can be found.

A geometrical solution is given in (985). C, D; C', D' are, in that construction, pairs of inverse points, and I is the centre of a system in involution defined by a series of coaxal circles (1022). Each circle intersects the central axis in a pair of inverse points with respect to the circle whose centre is O and radius δ.

1074 The relations which have been established for a system of *collinear* points may be transferred to a system of concurrent lines by the method of (1056), in which the distance between two points corresponds to the sine of the angle between two lines passing through those points.

The Method of Projection.

1075 Definitions.—The *projection* of any point P in space (Fig. of 1079) is the point p in which a right line OP, drawn from a fixed point O called the *vertex*, intersects a fixed plane called the *plane of projection.*

If all the points of any figure, plane or solid, be thus projected, the figure obtained is called the *projection* of the original figure.

1076 *Projective Properties.*—The projection of a right line is a right line. The projections of a curve, and of the tangent at any point of it, are another curve and the tangent at the corresponding point.

1077 The anharmonic ratio of the segments of a right line is not altered by projection; for the line and its projection are but two transversals of the same anharmonic pencil. (1056)

1078 Also, any relation between the segments of a line similar to that in (1015), in which each letter occurs in every term, is a *projective property.* [Proof as in (1056).

1079 *Theorem.*—Any quadrilateral $PQRS$ may be projected into a parallelogram.

Construction. — Produce PQ, SR to meet in A, and PS, QR to meet in B.

Then, with any point O for vertex, project the quadrilateral upon any plane pab parallel to OAB. The projected figure $pqrs$ will be a parallelogram.

Proof. — The planes OPQ, ORS intersect in OA, and they intersect the plane of projection which is parallel to OA in the lines pq, rs. Therefore pq and rs are parallel to OA, and therefore to each other. Similarly, ps, qr are parallel to OB.

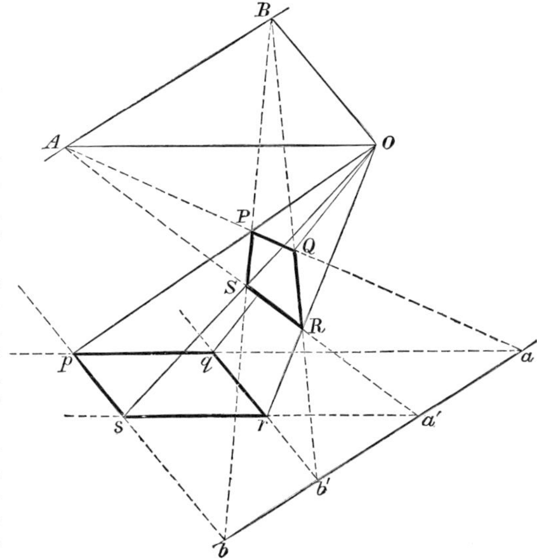

1080 Cor. 1.—The opposite sides of the parallelogram $pqrs$ meet in two points at infinity, which are the projections of the points A, B; and AB itself, which is the third diagonal of the complete quadrilateral $PQRS$, is projected into a line at infinity.

1081 Hence, to project any figure so that a certain line in it may pass to infinity—*Take the plane of projection parallel to the plane which contains the given line and the vertex.*

1082 Cor. 2.—To make the projection of the quadrilateral a rectangle, it is only necessary to make AOB a right angle.

On Perspective Drawing.

1083 Taking the parallelogram $pqrs$, in (1079), for the original figure, the quadrilateral $PQRS$ is its projection on the plane $ABab$. Suppose this plane to be the plane of the paper. Let the planes OAB, pab, while remaining parallel to each other, be turned respectively about the fixed parallel lines AB, ab. In every position of the planes, the lines Op, Oq, Or, Os will intersect the dotted lines in the same points P, Q, R, S. When the planes coincide with that of the paper, $pqrs$ becomes a *ground plan* of the parallelogram, and $PQRS$ is the representation of it in perspective.

AB is then called the *horizontal* line, ab the *picture line*, and the plane of both the *picture plane*.

1084 To find the projection of any point p in the *ground plan.*

Rule.—*Draw* pb *to any point* b *in the picture line, and draw* OB *parallel to* pb, *to meet the horizontal line in* B. *Join* Op, Bb, *and they will intersect in* P, *the point required.*

In practice, pb is drawn perpendicular to ab, and OB therefore perpendicular to AB. The point B is then called the *point of sight*, or *centre of vision*, and O the *station point*.

1085 To find the projection of a point in the *ground plan*, not in the original plane, but at a perpendicular distance c above it.

Rule.—*Take a new picture line parallel to the former, and at a distance above it* $= c \operatorname{cosec} \alpha$, *where* α *is the angle between the original plane and the plane of projection.* For a plane through the given point, parallel to the original plane, will intersect the plane of projection in the new *picture line* so constructed.

Thus, every point of a figure in the *ground plan* is transferred to the drawing.

1086 The whole theory of perspective drawing is virtually included in the foregoing propositions. The original plane is commonly horizontal, and the plane of projection vertical. In this case, $\operatorname{cosec} \alpha = 1$, and the height of the *picture line* for any point is equal to the height of the point itself above the original plane.

The distance BO, when B is the *point of sight*, may be measured along AB, and bp along ab, in the opposite direction; for the line Bb will continue to intersect Op in the point P.

Orthogonal Projection.

1087 Definition.—In orthogonal projection the lines of projection are parallel to each other, and perpendicular to the plane of projection. The vertex in this case may be considered to be at infinity.

1088 The projections of parallel lines are parallel, and the projected segments are in a constant ratio to the original segments.

1089 Areas are in a constant ratio to their projections.

For, lines parallel to the intersection of the original plane and the plane of projection are unaltered in length, and lines at right angles to the former are altered in a constant ratio. This ratio is the ratio of the areas, and is the cosine of the angle between the two planes.

Projections of the Sphere.

1090 In *Stereographic projection*, the vertex is on the surface of the sphere, and the diameter through the vertex is perpendicular to the plane of projection which passes through the other extremity of the diameter. The projection is therefore the *inversion* of the surface of the sphere (1012), and the diameter is the constant k.

1091 In *Globular projection*, the vertex is taken at a distance from the sphere equal to the radius $\div\sqrt{2}$, and the diameter through the vertex is perpendicular to the plane of projection.

1092 In *Gnomonic projection*, which is used in the construction of sun-dials, the vertex is at the centre of the sphere.

1093 *Mercator's projection*, which is employed in navigation, and sometimes in maps of the world, is not a projection at all as defined in (1075). Meridian circles of the sphere are represented on a plane by parallel right lines at intervals equal to the intervals on the equator. The parallels of latitude are represented by right lines perpendicular to the meridians, and at increasing intervals, so as to preserve the actual ratio between the increments of longitude and latitude at every point.

With r for the radius of the sphere, the distance, on the chart, from the equator of a point whose latitude is λ, is $= r \log \tan (45^\circ + \tfrac{1}{2}\lambda)$.

Additional Theorems.

1094 The sum of the squares of the distances of any point P from n equidistant points on a circle whose centre is O and radius r $= n\,(r^2 + OP^2)$.

PROOF.—Sum the values of PB^2, PC^2, &c., given in (819), and apply (803). This theorem is the generalization of (923).

1095 In the same figure, if P be on the circle, the sum of the squares of the perpendiculars from P on the radii OB, OC, &c., is equal to $\frac{1}{2}nr^2$.

PROOF.—Describe a circle upon the radius through P as diameter, and apply the foregoing theorem to this circle.

1096 COR. 1.—The sum of the squares of the intercepts on the radii between the perpendiculars and the centre is also equal to $\frac{1}{2}nr^2$. (I. 47)

1097 COR. 2.—The sum of the squares of the perpendiculars from the equidistant points on the circle to any right line passing through the centre is also equal to $\frac{1}{2}nr^2$.

Because the perpendiculars from two points on a circle to the diameters drawn through the points are equal.

1098 COR. 3.—The sum of the squares of the intercepts on the same right line between the centre of the circle and the perpendiculars is also equal to $\frac{1}{2}nr^2$. (I. 47)

If the radii of the inscribed and circumscribed circles of a regular polygon of n sides be r, R, and the centre O; then,

1099 I. The sum of the perpendiculars from any point P upon the sides is equal to nr.

1100 II. If p be the perpendicular from O upon any right line, the sum of the perpendiculars from the vertices upon the same line is equal to np.

1101 III. The sum of the squares of the perpendiculars from P on the sides is $= n\,(r^2 + \frac{1}{2}OP^2)$.

1102 IV. The sum of the squares of the perpendiculars from the vertices upon the right line is $= n\,(p^2 + \frac{1}{2}R^2)$.

PROOF.—In theorem I., the values of the perpendiculars are given by $r - OP\cos\left(\theta + \frac{2m\pi}{n}\right)$, with successive integers for m. Add together the n values, and apply (803).

Similarly, to prove II.; take for the perpendiculars the values

$$p - R\cos\left(\theta + \frac{2m\pi}{n}\right).$$

To prove III. and IV., take the same expressions for the perpendiculars; square each value; add the results, and apply (803, 804).

For additional propositions in the subjects of this section, see the section entitled *Plane Coordinate Geometry*.

GEOMETRICAL CONICS.

THE SECTIONS OF THE CONE.

1150 Definitions.—A *Conic Section* or *Conic* is the curve AP in which any plane intersects the surface of a right cone.

A *right cone* is the solid generated by the revolution of one straight line about another which it intersects in a fixed point at a constant angle.

Let the axis of the cone, in Fig. (1) or Fig. (2), be in the plane of the paper, and let the cutting plane $PMXN$ be perpendicular to the paper. (*Read either the accented or unaccented letters throughout.*) Let a sphere be inscribed in the cone, touching it in the circle EQF and touching the cutting plane in the point S, and let the cutting plane and the plane of the circle EQF intersect in XM. The following theorem may be regarded as the *defining property* of the curve of section.

1151 ***Theorem.***—The distance of any point P on the conic from the point S, called the *focus*, is in a constant ratio to PM, its distance from the line XM, called the *directrix*, or

$$PS : PM = PS' : PM' = AS : AX = e, \text{ the } eccentricity.$$

[*See next page for the Proof.*]

1152 Cor.—The conic may be generated in a plane from either focus S, S', and either directrix XM, $X'M'$, by the law just proved.

1153 The conic is an *Ellipse*, a *Parabola*, or an *Hyperbola*, according as e is less than, equal to, or greater than unity. That is, according as the cutting plane emerges on both sides of the lower cone, or is parallel to a side of the cone, or intersects both the upper and lower cones.

1154 All sections made by parallel planes are similar; for the inclination of the cutting plane determines the ratio $AE : AX$.

1155 The limiting forms of the curve are respectively—a circle when e vanishes, and two coincident right lines when e becomes infinite.

PROOF OF THEOREM 1151.—Join P, S and P, O, cutting the circular section in Q, and draw PM parallel to NX. Because all tangents from the same point, O or A, to either sphere are equal, therefore $RE = PQ = PS$ and $AE = AS$. Now, by (VI. 2), $RE : NX = AE : AX$ and $NX = PM$; therefore $PS : PM = AS : AX$, a constant ratio denoted by e and called the *eccentricity* of the conic.

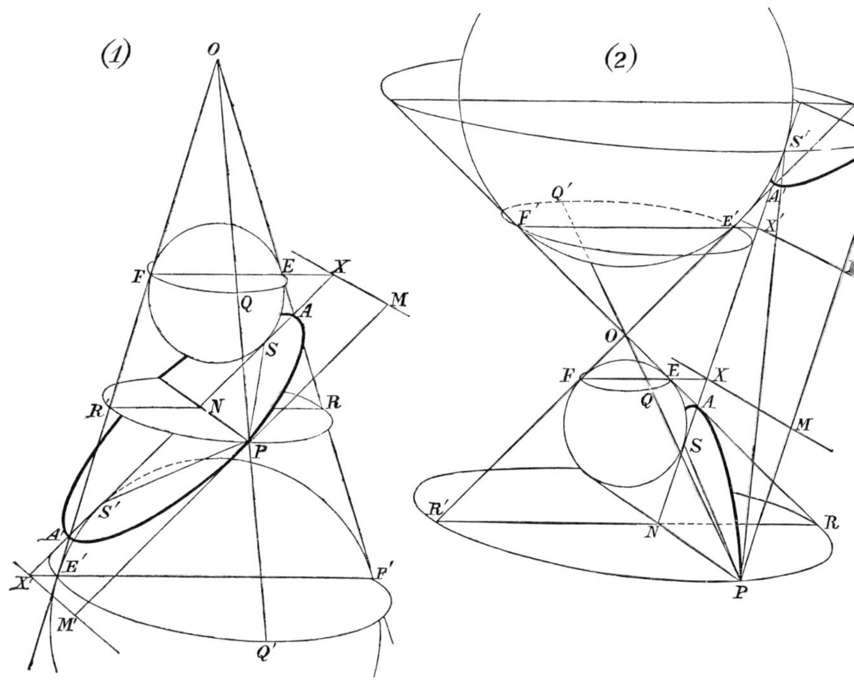

Referring the letters either to the ellipse or the hyperbola in the subjoined figure, let C be the middle point of AA' and N any other point on it. Let DD', RR' be the two circular sections of the cone whose planes pass through C and N; BCB' and PN the intersections with the plane of the conic. In the ellipse, BC is the common ordinate of the ellipse and circle; but, in the hyperbola, BC is to be taken equal to the tangent from C to the circle DD'.

1156 The fundamental equation of the ellipse or hyperbola is

$$PN^2 : AN \cdot NA' = BC^2 : AC^2.$$

PROOF.—$PN^2 = NR \cdot NR'$ and $BC^2 = CD \cdot CD'$ (III. 35, 36). Also, by similar triangles (VI. 2, 6), $NR : CD = AN : AC$ and $NR' : CD' = A'N : A'C$. Multiply the last equations together.

1157 Cor. 1.—PN has equal values at two points equi-distant from AA'. Hence the curve is symmetrical with respect to AA' and BB'.

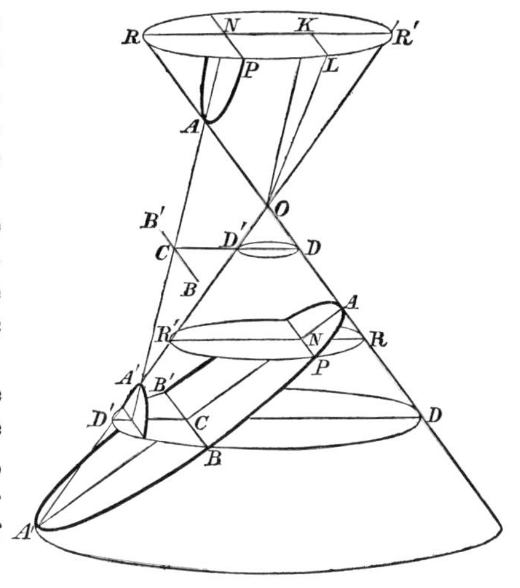

These two lines are called the *major* and *minor axes*, otherwise the *transverse* and *conjugate* axes of the conic.

When the axes are equal, or $BC = AC$, the ellipse becomes a circle, and the hyperbola becomes *rectangular* or *equilateral*.

1158 Any ellipse or hyperbola is the orthogonal projection of a circle or rectangular hyperbola respectively.

Proof.—Along the ordinate NP, measure $NP' = AN.NA'$; therefore by the theorem $PN : P'N = BC : AC$. Therefore a circle or rectangular hyperbola, having AA' for one axis, and having its plane inclined to that of the conic at an angle whose cosine $= BC \div AC$, projects orthogonally into the ellipse or hyperbola in question, by (1089). See Note to (1217).

1159 Hence any *projective property* (1076–78), which is known to belong to the circle or rectangular hyperbola, will also be universally true for the ellipse and hyperbola respectively.

THE ELLIPSE AND HYPERBOLA.

Joint properties of the Ellipse and Hyperbola.

1160 Definitions.—The *tangent* to a curve at a point P (Fig. 1166) is the right line PQ, drawn through an adjacent point Q, in its ultimate position when Q is made to coincide with P.

The *normal* is the perpendicular to the tangent through the point of contact.

In (Fig. 1171), referred to rectangular axes through the centre C (see Coordinate Geometry); the length CN is called the *abscissa*; PN the *ordinate*; PT the *tangent*; PG the *normal*; NT the *subtangent*; and NG the *subnormal*. S, S' are the *foci*; XM, $X'M'$ the *directrices*; PS, PS' the *focal distances*, and a double ordinate through S the *Latus Rectum*.

The *auxiliary circle* (Fig. 1173) is described upon AA' as diameter.

A diameter parallel to the tangent at the extremity of another diameter is termed a *conjugate diameter* with respect to the other.

The *conjugate hyperbola* has BC for its major, and AC for its minor axis (1157).

1161 The following theorems (1162) to (1181) are deduced from the property $PS : PM = e$ obtained in (1151).

The propositions and demonstrations are nearly identical for the ellipse and the hyperbola, any difference in the application being specified.

1162 $\boldsymbol{CS : CA : CX}$, and the common ratio is e.

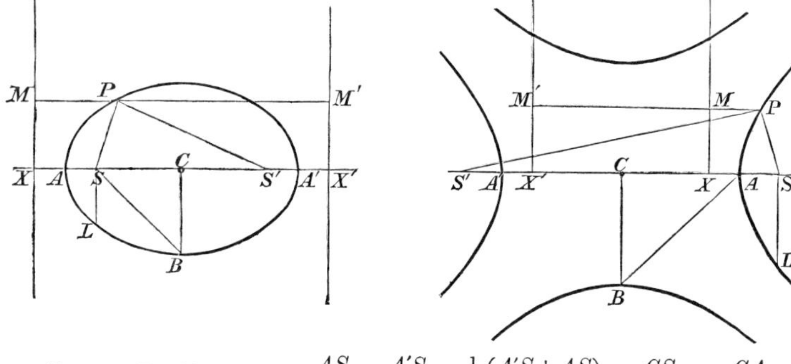

Proof.—By (1151), $e = \dfrac{AS}{AX} = \dfrac{A'S}{A'X} = \dfrac{\frac{1}{2}(A'S \pm AS)}{\frac{1}{2}(A'X \pm AX)} = \dfrac{CS}{CA}$ or $\dfrac{CA}{CX}$.

1163 In the ellipse the sum, and in the hyperbola the difference, of the focal distances of P is equal to the major axis, or

$$PS' \pm PS = AA'.$$

Proof.—With the same figures we have, in the ellipse, by (1151),

$e = \dfrac{PS+PS'}{PM+PM'} = \dfrac{PS+PS'}{XX'}$, and also $e = \dfrac{AS+A'S}{AX+A'X} = \dfrac{AA'}{XX'}$, therefore &c.

For the hyperbola take *difference* instead of *sum*.

1164 $CS^2 = AC^2 - BC^2$ in the ellipse.
[For $BS = AC$, by (1163).

$CS^2 = AC^2 + BC^2$ in the hyperbola.
[By assuming BC. See (1176).

1165 $BC^2 = SL \, . \, AC.$

PROOF.—(Figs. of 1162) $SL : SX = CS : CA$, (1151, 1162)

$\therefore \; SL \, . \, AC = CS \, . \, SX = CS\,(CX \sim CS) = CA^2 \sim CS^2$ (1162) $= BC^2$ (1164).

1166 If a right line through P, Q, two points on the conic, meets the directrix in Z, then SZ bisects the angle QSR.

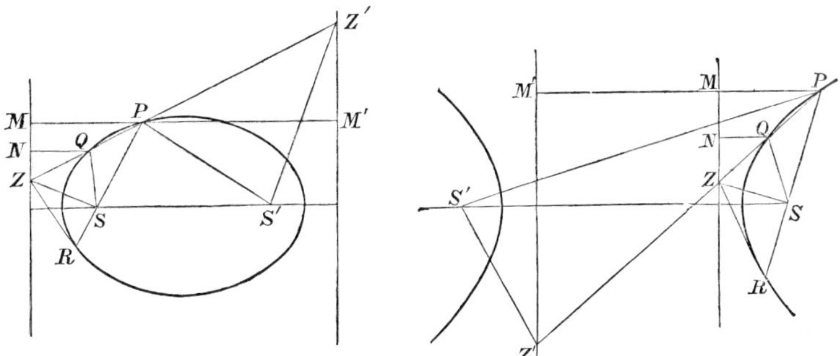

PROOF.—By similar triangles, $ZP : ZQ = MP : NQ = SP : SQ$ (1151), therefore by (VI. A.)

1167 If PZ be a tangent at P, then PSZ and $PS'Z'$ are right angles.

PROOF.—Make Q coincide with P in the last theorem.

1168 The tangent makes equal angles with the focal distances.

PROOF.—In (1166), $PS : PS' = PM : PM'$ (1151) $= PZ : PZ'$; therefore, when PQ becomes the tangent at P, $\angle\, SPZ = S'PZ'$, by (1167) and (VI. 7).

1169 The tangents at the extremities of a focal chord intersect in the directrix.

PROOF.—(Figs. of 1166). Join ZR; then, if ZP is a tangent, ZR is also, for (1167) proves RSZ to be a right angle.

1170 $CN \, . \, CT = AC^2.$

PROOF.—(Figs. 1171.) $\dfrac{TS'}{TS} = \dfrac{PS'}{PS}$ (VI. 3, A.) $= \dfrac{PM'}{PM}$ (1151) $= \dfrac{NX'}{NX}$,

therefore $$\frac{TS'+TS}{TS'-TS} = \frac{NX'+NX}{NX'-NX}, \quad \text{or} \quad \frac{2CT}{2CS} = \frac{2CX}{2CN},$$

therefore $$CN \, . \, CT = CS \, . \, CX = AC^2. \qquad (1162)$$

1171 If PG be the normal,

$$GS : PS = GS' : PS' = e.$$

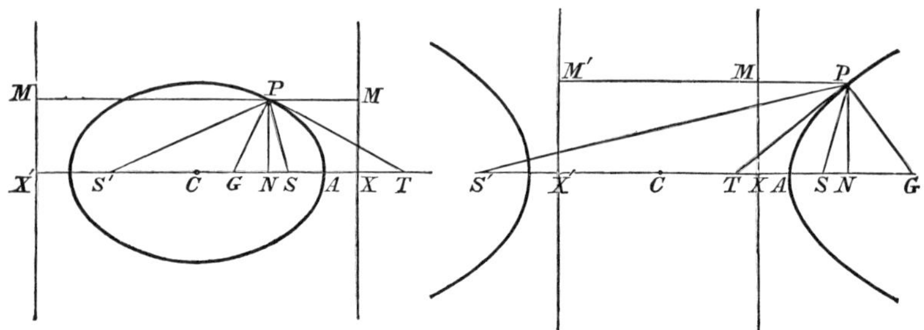

Proof.—By (1168) and (VI. 3, A.),

$$\frac{GS}{PS} = \frac{GS'}{PS'} = \frac{GS'+GS}{PS'+PS} = \frac{2CS}{2CA} = e. \qquad (1162)$$

But, for the hyperbola, change *plus* to *minus*.

1172 The subnormal and the abscissa are as the squares of the axes, or $\quad NG : NC = BC^2 : AC^2.$

Proof.—(Figs. 1171.) Exactly as in (1170), taking the normal instead of the tangent, we obtain $\frac{CG}{CS} = \frac{CN}{CX}$, $\therefore \frac{CN}{CG} = \frac{CX}{CS} = \frac{CA^2}{CS^2}$ (1162),

$$\therefore \frac{CN \sim CG}{CN} = \frac{CA^2 \sim CS^2}{CA^2}, \text{ or } \frac{NG}{NC} = \frac{BC^2}{AC^2} \text{ (1164).}$$

1173 The tangents at P and Q, the corresponding points on the ellipse and auxiliary circle, meet the axis in the same point T. But in the hyperbola, the ordinate TQ of the circle being drawn, the tangent at Q cuts the axis in N.

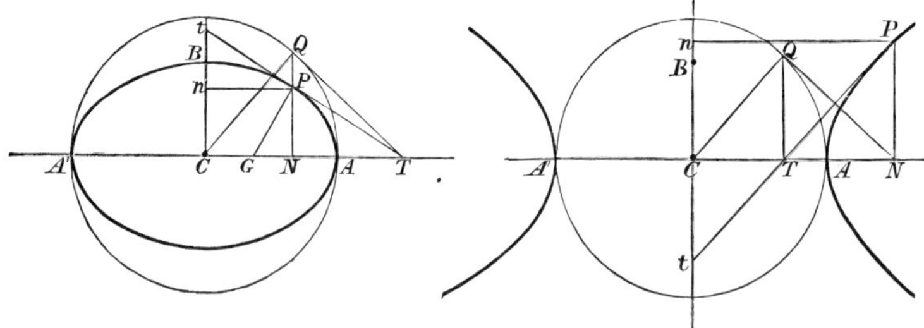

Proof.—For the ellipse: Join TQ. Then $CN \cdot CT = CQ^2$ (1170); therefore CQT is a right angle (VI. 8); therefore QT is a tangent.

For the hyperbola: Interchange N and T.

1174 $$PN : QN = BC : AC.$$

PROOF.—(Figs. 1173). $NG.NT = PN^2$, and $CN.NT = QN^2$. (VI. 8) Therefore $NG : NC = PN^2 : QN^2$; therefore, by (1172).

This proposition is equivalent to (1158), and shows that an ellipse is the orthogonal projection of a circle equal to the auxiliary circle.

1175 COR.—The area of the ellipse is to that of the auxiliary circle as $BC : AC$ (1089).

1176 $$PN^2 : AN.NA' = BC^2 : AC^2.$$

PROOF.—By (1174), since $QN^2 = AN.NA'$ (III. 35, 36). An independent proof of this theorem is given in (1156). The construction for BC in the hyperbola in (1164) is thus verified.

1177 $$Cn.Ct = BC^2.$$

PROOF.—(Figs. 1173.)

$$\frac{Ct}{CT} = \frac{PN}{NT}; \therefore \frac{Cn.Ct}{CN.CT} = \frac{PN^2}{CN.NT} = \frac{PN^2}{QN^2} \text{ (VI. 8)} = \frac{PN^2}{AN.NA'} \text{ (III. 35, 36).}$$

Therefore, by (1170) and (1176), $Cn.Ct : AC^2 = BC^2 : AC^2$.

1178 If SY, $S'Y'$ are the perpendiculars on the tangent, then Y, Y' are points on the auxiliary circle, and

$$SY.S'Y' = BC^2.$$

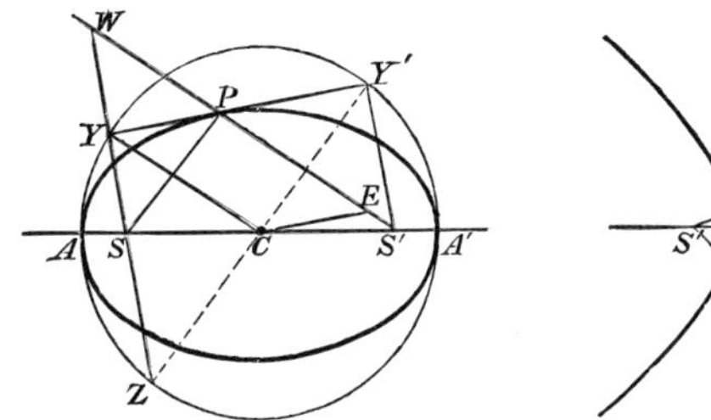

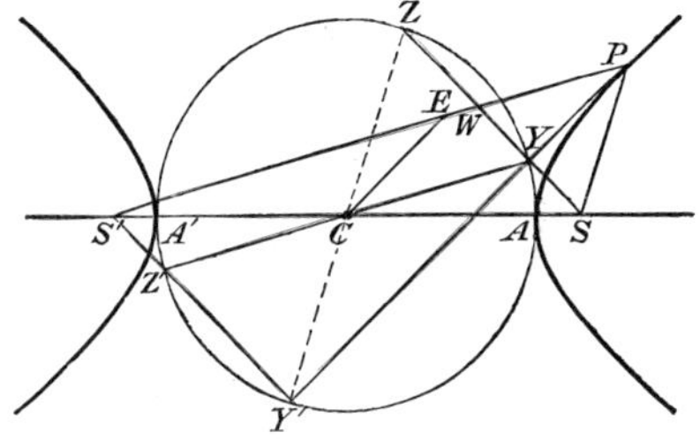

PROOF.—Let PS' meet SY in W. Then $PS = PW$ (1168). Therefore $S'W = AA'$ (1163). Also, $SY = YW$, and $SC = CS'$. Therefore $CY = \frac{1}{2}S'W = AC$. Similarly $CY' = AC$. Therefore Y, Y' are on the circle.

Hence ZY' is a diameter (III. 31), and therefore $SZ = S'Y'$, by similar triangles; therefore $SY.SZ = SA.SA'$ (III. 35, 36) $= CS^2 \sim CA^2$ (II. 5) $= BC^2$ (1164).

1179 COR.—If CE be drawn parallel to the tangent at P, then $$PE = CY = AC.$$

1180 PROBLEM.—To draw tangents from any point O to an ellipse or hyperbola.

CONSTRUCTION.—(Figs. 1181.) Describe two circles, one with centre O and radius OS, and another with centre S' and radius $= AA'$, intersecting in M,

M'. Join MS', $M'S'$. These lines will intersect the curve in P, P', the points of contact. For another method see (1204).

Proof.—By (1163), $PS' \pm PS = AA' = S'M$ by construction. Therefore $PS = PM$, therefore $\angle OPS = OPM$ (I. 8), therefore OP is a tangent by (1168).

Similarly $P'S = P'M'$, and OP' is a tangent.

1181 The tangents OP, OP' subtend equal angles at either focus.

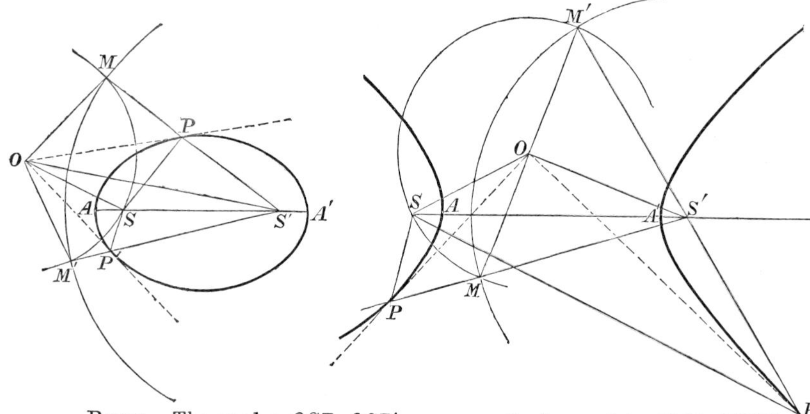

Proof.—The angles OSP, OSP' are respectively equal to OMP, $OM'P'$, by (I. 8), as above; and these last angles are equal, by the triangles $OS'M$, $OS'M'$, and (I. 8). Similarly at the other focus.

Asymptotic Properties of the Hyperbola.

1182 Def. — The *asymptotes* of the hyperbola are the diagonals of the rectangle formed by tangents at the vertices A, A', B, B'.

1183 If the ordinates RN, RM from any point R on an asymptote cut the hyperbola and its conjugate in P, P', D, D',

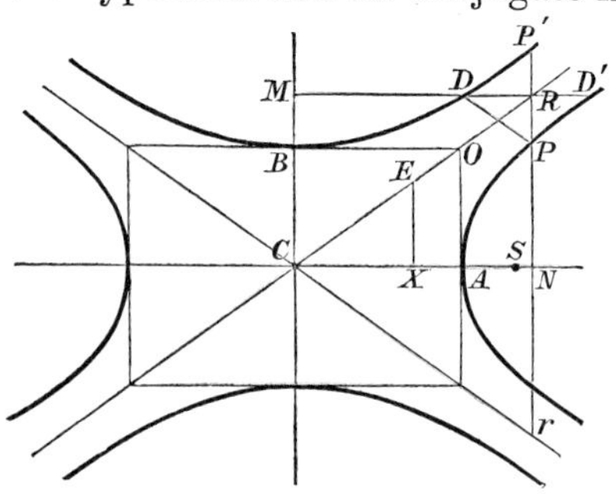

then either of the following pairs of equations will define both the branches of each curve—

$$RN^2 - PN^2 = BC^2 = P'N^2 - RN^2 \quad \text{...........} \quad (1),$$

$$RM^2 - DM^2 = AC^2 = D'M^2 - RM^2 \quad \text{...........} \quad (2).$$

PROOF.—Firstly, to prove (1): By proportion from the similar triangles RNC, OAC, we have $\dfrac{RN^2}{CN^2} = \dfrac{BC^2}{AC^2} = \dfrac{PN^2}{CN^2 - AC^2}$;

by (1176), since $AN.NA' = CN^2 - AC^2$. By (II. 6)

Therefore $\dfrac{RN^2 - PN^2}{AC^2} = \dfrac{BC^2}{AC^2}$, by the theorem (69);

therefore $RN^2 - PN^2 = BC^2$.

Also, by (1176), applied to the conjugate hyperbola, the axes being now reversed, $\dfrac{CN^2}{P'N^2 - BC^2} = \dfrac{AC^2}{BC^2} = \dfrac{CN^2}{RN^2}$, by similar triangles;

therefore $P'N^2 - BC^2 = RN^2$ or $P'N^2 - RN^2 = BC^2$.

Secondly, to prove (2): By proportion from the triangles RMC, OBC, we have $\dfrac{RM^2}{CM^2} = \dfrac{AC^2}{BC^2} = \dfrac{DM^2}{CM^2 - BC^2}$,

by (1176), applied to the conjugate hyperbola, for in this case we should have $BM.MB' = CM^2 - BC^2$.

Therefore $\dfrac{RM^2 - DM^2}{BC^2} = \dfrac{AC^2}{BC^2}$; therefore $RM^2 - DM^2 = AC^2$.

Also, by (1176), since CM, $D'M$ are equal to the coordinates of D',

$$\frac{CM^2}{D'M^2 - AC^2} = \frac{BC^2}{AC^2} = \frac{CM^2}{RM^2}, \text{ by similar triangles;}$$

therefore $D'M^2 - AC^2 = RM^2$ or $D'M^2 - RM^2 = AC^2$.

1184 COR. 1.—If the same ordinates RN, RM meet the other asymptote in r and r', then

$$PR.Pr = BC^2 \text{ and } DR.Dr' = AC^2. \quad \text{(II. 5)}$$

1185 COR. 2.—As R recedes from C, PR and DR continually diminish. Hence the curves continually approach the asymptote.

1186 If XE be the directrix, $CE = AC$.

PROOF.— $CE : CO = CX : CA = CA : CS$ and $CS = CO$. (1164)

1187 PD is parallel to the asymptote.

PROOF.— $\dfrac{RN^2}{RM^2} = \dfrac{BC^2}{AC^2} = \dfrac{RN^2 - PN^2}{RM^2 - DM^2}$ (1183) $= \dfrac{PN^2}{DM^2}$ (69).

Therefore $RN : PN = RM : DM$; therefore, by (VI. 2).

1188 The segments of any right line between the curve and the asymptote are equal, or $QR = qr$.

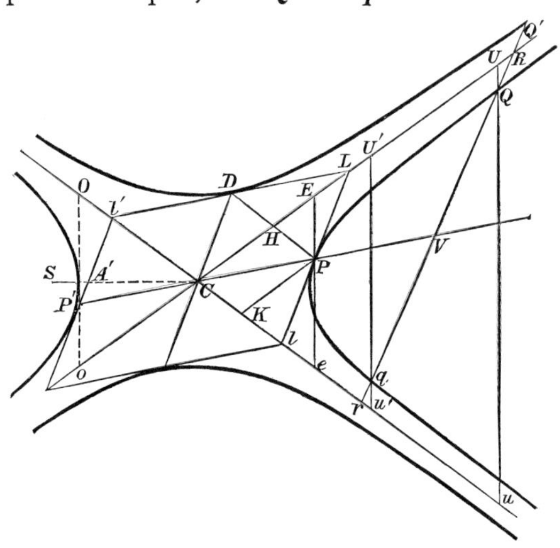

PROOF.— $QR : QU = qR : qU'$
and $Qr : Qu = qr : qu'$ $\Big\}$. Compound the ratios, and employ (1184).

1189 COR. 1.—$PL = Pl$ and $QV = qV$.

1190 COR. 2.—$CH = HL$. Because PD is parallel to lC. (1187)

1191 $QR \,.\, Qr = PL^2 = RV^2 - QV^2 = Q'V^2 - RV^2$.

PROOF.— $QR : QU = PL : PE$
and $Qr : Qu = Pl : Pe$ $\Big\}$ Compound the ratios. Therefore, by (1184), $QR \,.\, Qr = PL \,.\, Pl = PL^2$ (1189).

1192 $4PH \,.\, PK = CS^2$.

PROOF.— $PH : PE = CO : Oo$
and $PK : Pe = Co : Oo$ $\Big\}$ $\therefore PH \,.\, PK : PE \,.\, Pe = CO^2 : Oo^2 = CS^2 : 4BC^2$; therefore, by (1184).

Joint Properties of the Ellipse and Hyperbola resumed.

If PCP' be a diameter, and QV an ordinate parallel to the conjugate diameter CD (Figs. 1195 and 1188).

1193 $QV^2 : PV \,.\, VP' = CD^2 : CP^2$.

This is the fundamental equation of the conic, equation (1176) being the most important form of it.

Otherwise:

In the ellipse, $QV^2 : CP^2 - CV^2 = CD^2 : CP^2.$

In the hyperbola, $QV^2 : CV^2 - CP^2 = CD^2 : CP^2;$

and $Q'V^2 : CV^2 + CP^2 = CD^2 : CP^2.$

PROOF.—(*Ellipse.* Fig. 1195.)—By orthogonal projection from a circle. If C, P, P', D, Q, V are the projections of c, p, p', d, q, v on the circle; $qv^2 = pv \cdot vp'$ and $cd^2 = cp^2$. The proportion is therefore true in the case of the circle. Therefore &c., by (1088).

(*Hyperbola.* Fig. 1188)—

$$\frac{CD^2}{CP^2} = \frac{RV^2}{CV^2} = \frac{PL^2}{CP^2} = \frac{RV^2 \pm PL^2}{CV^2 \pm CP^2} = \frac{QV^2}{CV^2 - CP^2} \text{ or } \frac{Q'V^2}{CV^2 + CP^2}. \quad (1191)$$

1194 The parallelogram formed by tangents at the extremities of conjugate diameters is of constant area, and therefore, PF being perpendicular to CD (Figs. 1195),

$$PF \cdot CD = AC \cdot BC.$$

PROOF.—(*Ellipse.*)—By orthogonal projection from the circle (1089).

(*Hyperbola.* Fig. 1188.)—$CL \cdot Cl = 4PH \cdot PK = CO \cdot Co$ (1192); therefore, by (VI. 15), $\Delta LCl = OCo = AC \cdot BC$.

If PF intersects the axes in G and G',

1195 $$PF \cdot PG = BC^2 \text{ and } PF \cdot PG' = AC^2.$$

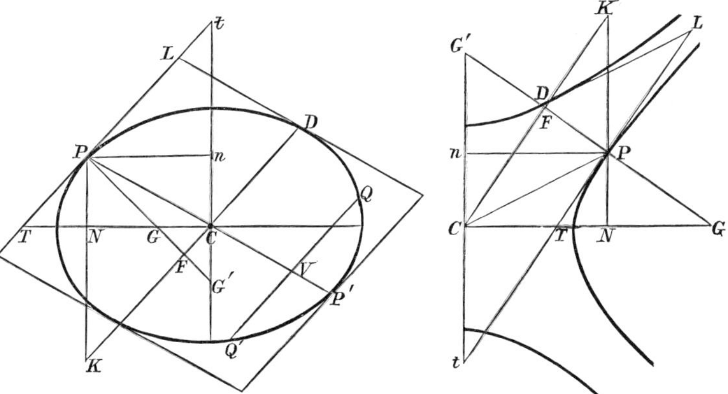

PROOF.— $PF \cdot PG = PK \cdot PN = Cn \cdot Ct = BC^2$ (1177). Similarly for $PF \cdot PG'$.

1197 COR.— $PG \cdot PG' = CD^2 = PT \cdot Pt.$ By (1194)

1198 The diameter bisects all chords parallel to the tangent at its extremity.

PROOF.—(*Ellipse.* Fig. 1195.)—By projection from the circle (1088) $QV = VQ'$. (*Hyperbola.*) By (1189.)

1199 COR. 1.—The tangents at the extremities of any chord meet on the diameter which bisects it.

PROOF.—The secants drawn through the extremities of two parallel chords meet on the diameter which bisects them (VI. 4), and the tangents are the limiting positions of the secants when the parallel chords coincide.

1200 COR. 2.—If the tangents from a point are equal, the diameter through the point must be a principal axis. (I. 8)

1201 COR. 3.—The chords joining any point Q on the curve with the extremities of a diameter PP', are parallel to conjugate diameters, and are called *supplemental chords.*

For the diameter bisecting PQ is parallel to $P'Q$ (VI. 2). Similarly the diameter bisecting $P'Q$ is parallel to PQ.

1202 Diameters are mutually conjugate; If CD be parallel to the tangent at P, CP will be parallel to the tangent at D.

PROOF.—(*Ellipse.* Fig. 1205.)—By projection from the circle (1088).

NOTE.—Observe that, if the ellipse in the figure with its ordinates and tangents be turned about the axis Tt through the angle $\cos^{-1}(BC \div AC)$, it becomes the projection of the auxiliary circle with its corresponding ordinates and tangents.

(*Hyperbola.* Fig. 1188.)—By (1187, 1189) the tangents at P, D meet the asymptotes in the same point L. Therefore they are parallel to CD, CP (VI. 2.)

If QT be the tangent at Q, and QV the ordinate parallel to the tangent at any other point P,

1203 $$CV \,.\, CT = CP^2.$$

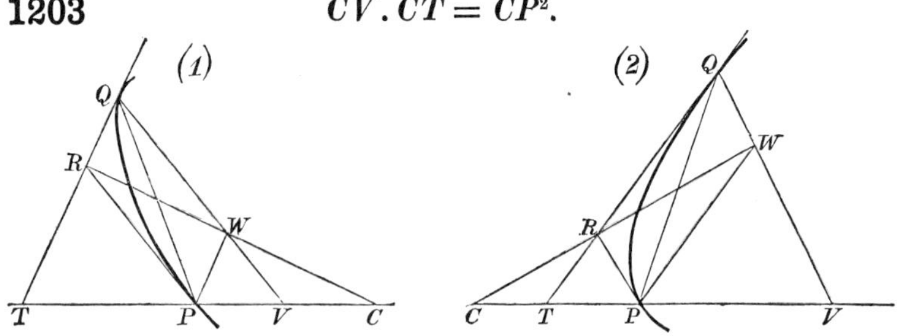

PROOF.—CR bisects PQ (1199). Therefore PW is parallel to QR. Therefore, by (VI. 2), $CV : CP = CW : CR = CP : CT$.

1204 Cor.—Hence, to draw two tangents from a point T, we may find CV from the above equation, and draw QVQ' parallel to the tangent at P to determine the points of contact Q, Q'.

Let PN, DN be the ordinates at the extremities of conjugate diameters, and PT the tangent at P. Let the ordinates at N and R in the ellipse, but at T and C in the hyperbola, meet the auxiliary circle in p and d; then

1205 $$CN = dR, \quad CR = pN.$$

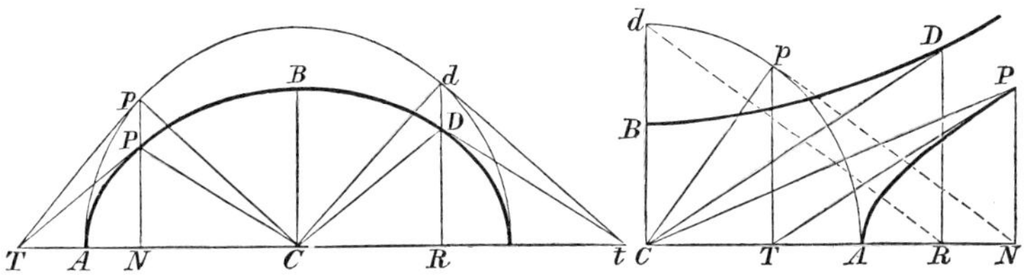

Proof.—(*Ellipse.*) Cp, Cd are parallel to the tangents at d and p (Note to 1202). Therefore pCd is a right angle. Therefore pNC, CRd are equal right-angled triangles with $CN = dR$ and $CR = pN$.

(*Hyperbola.*) $CN \cdot CT = AC^2$ (1170),

and $DR \cdot CT = 2\Delta CDT = 2CDP = AC \cdot BC$ (1194);

$\therefore \quad \frac{CN}{DR} = \frac{AC}{BC} = \frac{pN}{PN}$ (1174), $\quad \therefore \quad \frac{CN}{pN} = \frac{DR}{PN} = \frac{CR}{TN}$ (similar triangles).

But $\frac{CN}{pN} = \frac{pN}{TN}$ (VI. 8); $\therefore CR = pN$. Also $Cp = Cd$; therefore the triangles CpN, dCR are equal and similar; therefore $CN = dR$ and dR is parallel to pN.

1206 Cor.— $$DR : dR = BC : AC.$$

Proof.—(*Ellipse.*) By (1174). (*Hyperbola.*) By the similar right-angled triangles, we have $dR : pN = CR : TN = DR : PN$;

therefore $dR : DR = pN : PN = AC : BC$ (1174).

In the same figures,

1207 (*Ellipse.*) $$CN^2 + CR^2 = AC^2; \quad DR^2 + PN^2 = BC^2.$$

1209 (*Hyperbola.*) $$CN^2 - CR^2 = AC^2; \quad DR^2 - PN^2 = BC^2.$$

Proof.—Firstly, from the right-angled triangle CNp in which $pN = CR$ (1205).

Secondly, In the ellipse, by (1174), $DR^2 + PN^2 : dR^2 + pN^2 = BC^2 : AC^2$, and $dR^2 + pN^2 = AC^2$, by (1205). For the hyperbola, take *difference* of squares.

1211 (*Ellipse.*) $CP^2+CD^2 = AC^2+BC^2$.

1212 (*Hyperbola.*) $CP^2-CD^2 = AC^2-BC^2$.

PROOF.—(Figs. 1205.) By (1205—1210) and (I. 47), applied to the triangles CNP, CRD.

The product of the focal distances is equal to the square of the semi-conjugate diameter, or

1213 $$PS\,.\,PS' = CD^2.$$

PROOF.—(*Ellipse.* Fig. 1171.) $2PS\,.\,PS' = (PS+PS')^2-PS^2-PS'^2$ $= 4AC^2-2CS^2-2CP^2$ (922, i.) $= 2(AC^2+BC^2-CP^2)$ (1164) $= 2CD^2$(1211).

(*Hyperbola.*)—Similarly with $2PS\,.\,PS' = PS^2+PS'^2-(PS'-PS)^2 =$ &c.

1214 The products of the segments of intersecting chords QOq, $Q'Oq'$ are in the ratio of the squares of the diameters parallel to them, or

$$OQ.Oq : OQ'.Oq' = CD^2 : CD'^2.$$

PROOF.—(*Ellipse.*) By projection from the circle (1088); for the proportion is true for the circle, by (III. 35, 36).

(*Hyperbola.* Fig. 1188.) Let O be any point on Qq. Draw IOi parallel to Ee, meeting the asymptotes in I and i; then

$$OR\,.\,Or-OQ\,.\,Oq = QR\,.\,Qr \text{ (II. 5) } = PL^2 \text{ (1191) .... (1).}$$

Now $\dfrac{OR}{OI} = \dfrac{PL}{PE}$, and $\dfrac{Or}{Oi} = \dfrac{Pl}{Pe}$; $\therefore\ \dfrac{OR\,.\,Or}{OI\,.\,Oi} = \dfrac{PL^2}{PE\,.\,Pe} = \dfrac{CD^2}{BC^2}$ (1184).

Therefore $\dfrac{OR\,.\,Or-PL^2}{OI\,.\,Oi-BC^2} = \dfrac{CD^2}{BC^2}$; or, by (1), $\dfrac{OQ\,.\,Oq}{OI\,.\,Oi-BC^2} = \dfrac{CD^2}{BC^2}$.

Similarly for any other chord $Q'Oq'$ drawn through O.

Therefore $OQ\,.\,Oq : OQ'\,.\,Oq' = CD^2 : CD'^2$.

1215 COR.—The tangents from any point to the curve are in the ratio of the diameters parallel to them.

For, when O is without the curve and the chords become tangents, each product of segments becomes the square of a tangent.

1216 If from any point Q on a tangent PT drawn to *any* conic (Fig. 1220), two perpendiculars QR, QL be drawn to the focal distance PS and the directrix XM respectively; then

$$SR : QL = e.$$

PROOF.—Since QR is parallel to ZS (1167), therefore, by (VI. 2),

$$SR : PS = QZ : PZ = QL : PM;$$

therefore $SR : QL = PS : PM = e$.

COR.—By applying the theorem to each of the tangents from Q, a proof of (1181) is obtained.

1217 *The Director Circle.*—The locus of the point of intersection, T, of two tangents always at right angles is a circle called the *Director Circle.*

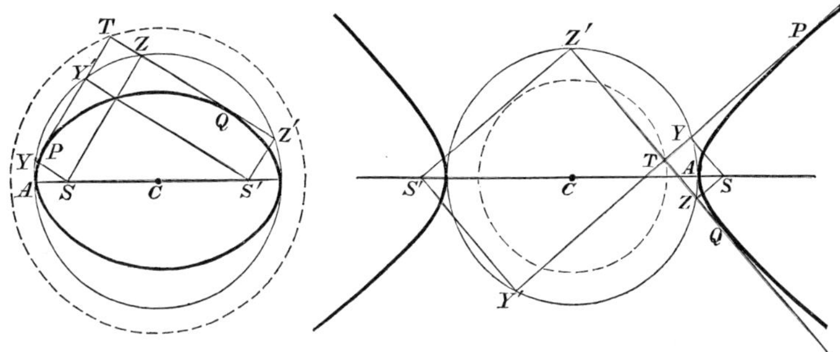

PROOF.—Perpendiculars from S, S' to the tangents meet them in points Y, Z, Y', Z', which lie on the auxiliary circle. Therefore, by (II. 5, 6) and

(III. 35, 36), $TC^2 \sim AC^2 = TZ\,.\,TZ' = SY\,.\,S'Y' = BC^2$. (1178)

Therefore $TC^2 = AC^2 \pm BC^2$, a constant value.

NOTE.—Theorems (1170), (1177), and (1203) may also be deduced at once for the ellipse by orthogonal projection from the circle; and, in all such cases, the analogous property of the hyperbola may be obtained by a similar projection from the rectangular hyperbola if the property has already been demonstrated for the latter curve.

1218 If the points A, S (Fig. 1162) be fixed, while C is moved to an infinite distance, the conic becomes a parabola. Hence, any relation which has been established for parts of the curve which remain finite, when AC thus becomes infinite, *will be a property of the parabola.*

1219 Theorems relating to the ellipse may generally be adapted to the parabola by eliminating the quantities which become infinite, employing the principle that *finite differences may be neglected in considering the ratios of infinite quantities.*

EXAMPLE:—In (1193), when P' is at infinity, VP' becomes $= 2CP$; and in (1213) PS' becomes $= 2CP$. Thus the equations become

$$\frac{QV^2}{PV} = \frac{2CD^2}{CP} \quad \text{and} \quad PS = \frac{CD^2}{2CP}.$$

Therefore $QV^2 = 4PS\,.\,PV$ in the parabola.

THE PARABOLA.

If S be the focus, XM the directrix, and P any point on the curve, the *defining property* is

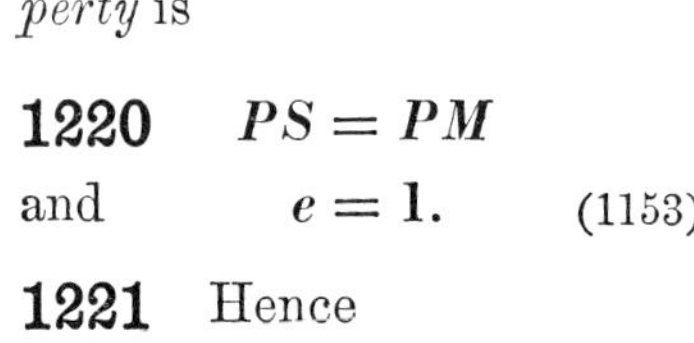

1220 $\quad PS = PM$

and $\quad e = 1. \quad$ (1153)

1221 Hence

$AX = AS.$

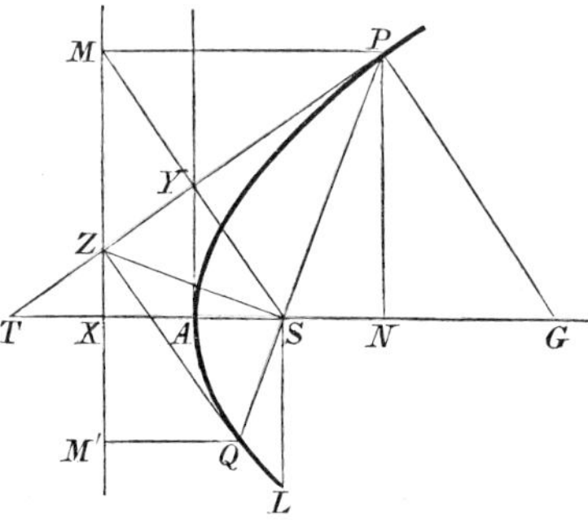

1222 The Latus Rectum $= 4AS$.

PROOF.— $SL = SX$ (1220) $= 2AS$.

1223 If PZ be a tangent at P, meeting the directrix in Z, then PSZ is a right angle.

PROOF.—As in (1167); theorem (1166) applying equally to the parabola.

1224 The tangent at P bisects the angles SPM, SZM.

PROOF.—PZ is common to the triangles PSZ, PMZ; $PS = PM$ and $\angle PSZ = PMZ$ (1223).

1225 COR.— $\quad ST = SP = SG$. $\quad$ (I. 29, 6)

1226 The tangents at the extremities of a focal chord PQ intersect at right angles in the directrix.

PROOF.—(i.) They intersect in the directrix, as in (1169).

(ii.) They bisect the angles SZM, SZM' (1224), and therefore include a right angle.

1227 The curve bisects the sub-tangent. $\quad AN = AT$.

PROOF.— $ST = SP$ (1225) $= PM = XN$, and $AX = AS$.

1228 The sub-normal is half the latus rectum. $\quad NG = 2AS$.

PROOF.—$ST = SP = SG$ and $TX = SN$ (1227). Subtract.

1229 $PN^2 = 4AS . AN.$

PROOF.—$PN^2 = TN . NG$ (VI. 8) $= AN . 2NG$ (1227) $= 4AS . AN$ (1228). Otherwise, by (1176) and (1165); making AC infinite. See (1219).

1230 The tangents at A and P each bisect SM, the latter bisecting it at right angles.

PROOF.—(i.) The tangent at A, by (VI. 2), since $AX = AS$.

(ii.) PT bisects SM at right angles, by (I. 4), since $PS = PM$ and $\angle SPY = MPY$.

1231 COR.— $SA : SY : SP.$ [By similar triangles.

1232 To draw tangents from a point O to the parabola.

CONSTRUCTION.—Describe a circle, centre O and radius OS, cutting the directrix in M, M'. Draw $MQ, M'Q'$ parallel to the axis, meeting the parabola in Q, Q'. Then OQ, OQ' will be tangents.

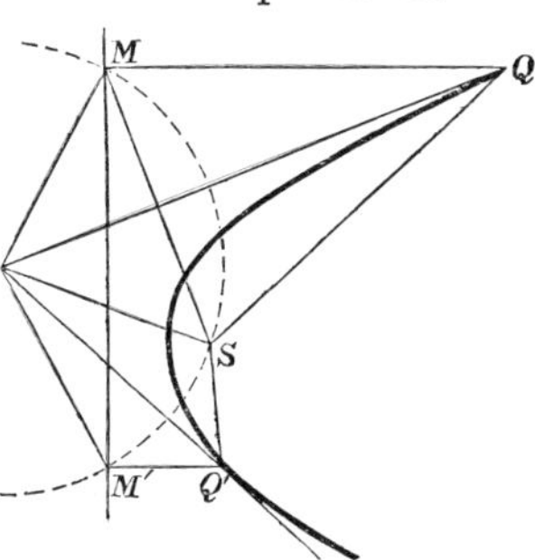

PROOF.—OS, $SQ = OM$, MQ (1220); therefore, by (I. 8), $\angle OQS = OQM$; therefore OQ is a tangent (1224). Similarly OQ' is a tangent.

Otherwise, by (1181). When S' moves to infinity, the circle MM' becomes the directrix.

1233 COR. 1.—The triangles SQO, SOQ' are similar, and

$$SQ : SO : SQ'.$$

PROOF.— $\angle SQO = MQO = SMM' = SOQ'.$ (III. 20)

Similarly $SQ'O = SOQ.$

1234 COR. 2.—The tangents at two points subtend equal angles at the focus; and they contain an angle equal to half the exterior angle between the focal distances of the points.

PROOF.— $\angle OSQ = OSQ'$, by (Cor. 1).

Also $\angle QOQ' = SOQ + SQO = \pi - OSQ = \frac{1}{2}QSQ'.$

1235 DEF.—Any line parallel to the axis of a parabola is called a *diameter*.

1236 The chord of contact QQ' of tangents from any point O is bisected by the diameter through O.

PROOF.—This proposition and the corollaries are included in (1198–1200), by the principle in (1218). An independent proof is as follows.

The construction being as in (1232), we have $ZM = ZM'$; therefore $QV = VQ'$ (VI. 2).

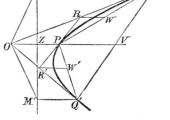

1237 Cor. 1.—The tangent RR' at P is parallel to QQ'; and $OP = PV$.

Proof.—Draw the diameter RW. $QW = WP$; therefore $QR = RO$ (VI. 2). Similarly $Q'R' = R'O$.

1238 Cor. 2.—Hence, the diameter through P bisects *all* chords parallel to the tangent at P.

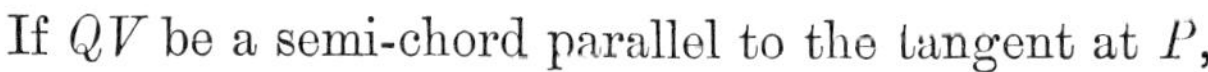

If QV be a semi-chord parallel to the tangent at P,

1239 $$QV^2 = 4PS \,.\, PV.$$

This is the fundamental equation of the parabola, equation (1229) being the most important form of it.

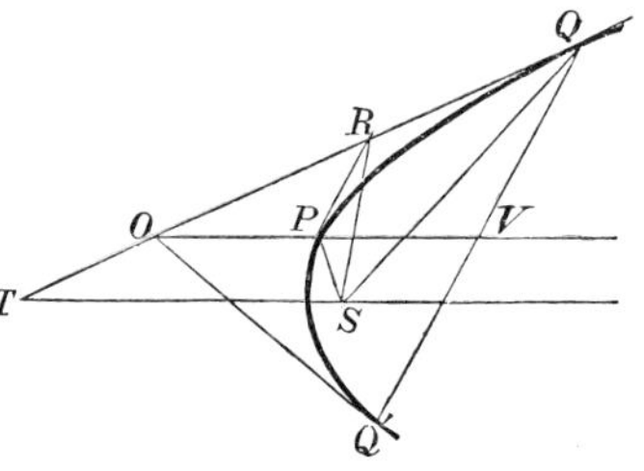

Proof.—Let QO meet the axis in T. By similar triangles (1231),

$\angle SRP = SQR = STQ = POR$;

and $\angle SPR = OPR$ (1224). Therefore $PR^2 = PS.PO = PS.PV$ and $QV = 2PR$.

Otherwise: See (1219), where the equation is deduced from (1193) of the ellipse.

1240 Cor. 1.—If v be any other point, either within or without the curve, on the chord QQ', and pv the corresponding diameter, $$vQ \,.\, vQ' = 4pS \,.\, pv. \qquad \text{(II. 5)}$$

1241 Cor. 2.—The focal chord conjugate to the diameter through P, and called the *parameter* of that diameter, is equal to $4SP$. For PV in this case is equal to PS.

1242 The products of the segments of intersecting chords, QOq, $Q'Oq'$, are in the ratio of the parameters of the diameters which bisect the chords; or

$$OQ.Oq : OQ'.Oq' = PS : P'S.$$

Proof.—By (1240), the ratio is equal to $4PS \,.\, pO : 4P'S \,.\, pO$.

Otherwise: In the ellipse (1214), the ratio is $= \dfrac{CD^2}{CD'^2} = \dfrac{PS \,.\, PS'}{P'S \,.\, P'S'}$ (1213) $= \dfrac{PS}{P'S}$, when S' is at infinity and the curve becomes a parabola (1219).

1243 Cor.—The squares of the tangents to a parabola from any point are as the focal distances of the points of contact.

Proof.—As in (1215). Otherwise, by (1233) and (VI. 19).

1244 The area of the parabola cut off by any chord QQ' is two-thirds of the circumscribed parallelogram, or of the triangle formed by the chord and the tangents at Q, Q'.

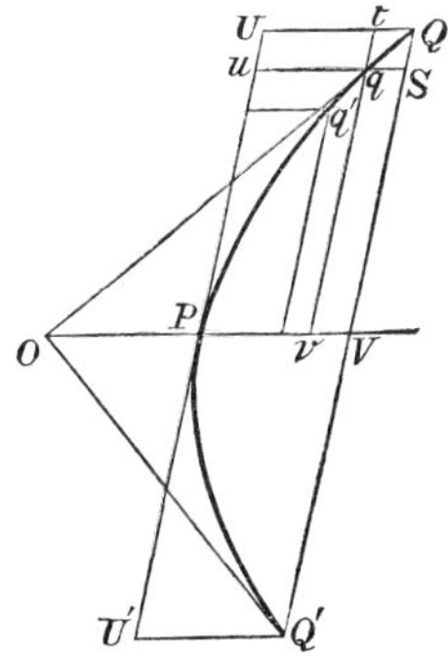

Proof.—Through Q, q, q', &c., adjacent points on the curve, draw right lines parallel to the diameter and tangent at P. Let the secant Qq cut the diameter in O. Then, when q coincides with Q, so that Qq becomes a tangent, we have $OP = PV$ (1237). Therefore the parallelogram $Vq = 2Uq$, by (I. 43), applied to the parallelogram of which OQ is the diagonal. Similarly $vq' = 2uq'$, &c. Therefore the sum of all the evanescent parallelograms on one side of PQ is equal to twice the corresponding sum on the other side; and these sums are respectively equal to the areas PQV, PQU.—(Newton, Sect. I., Lem. II.)

Practical methods of constructing the Conic.

1245 *To draw the Ellipse.*

Fix two pins at S, S' (Fig. 1162). Place over them a loop of thread having a perimeter $SPS' = SS' + AA'$. A pencil point moved so as to keep the thread stretched will describe the ellipse, by (1163).

1246 *Otherwise.*—(Fig. 1173.) Draw PHK parallel to QC, cutting the axes in H, K. $PK = AC$ and $PH = BC$ (1174). Hence, if a ruler PHK moves so that the points H, K slide along the axes, P will describe the ellipse.

1247 *To draw the Hyperbola.*

Make the pin S' (Fig. 1162) serve as a pivot for one end of a bar of any convenient length. To the free end of the bar attach one end of a thread whose length is less than that of the bar by AA'; and fasten the other end of the thread to the pin S. A pencil point moved so as to keep the thread stretched, and touching the bar, will describe the hyperbola, by (1163).

1248 *Otherwise*:—Lay off any scale of equal parts along both asymptotes (Fig. 1188), starting and numbering the divisions from C, in both positive and negative directions.

Join every pair of points L, l, the *product* of whose distances from C is the same, and a series of tangents will be formed (1192) which will define the hyperbola. See also (1289).

1249 *To draw the Parabola.*

Proceed as in (1247), with this difference: let the end of the bar, before attached to S', terminate in a "T-square," and be made to slide along the directrix (Fig. 1220), taking the string and bar of the same length.

1250 *Otherwise*:—Make the same construction as in (1248), and join every pair of points, the algebraic sum of whose distances from the zero point of division is the same.

PROOF.—If the two equal tangents from any point T on the axis (Fig. 1239) be cut by a third tangent in the points R, r; then RQ may be proved equal to rT, by (1233), proving the triangles SRQ, SrT equal in all respects.

1251 COR.—The triangle SRr is always similar to the isosceles triangle SQT.

1252 *To find the axes and centre of a given central conic.*

(i.) Draw a right line through the centres of two parallel chords. This line is a diameter, by (1198); and two diameters so found will intersect in the centre of the conic.

(ii.) Describe a circle having for its diameter any diameter PP' of the conic, and let the circle cut the curve in Q. Then PQ, $P'Q$ are parallel to the axes, by (1201) and (III. 31).

1253 *Given two conjugate diameters*, CP, CD, *in position and magnitude: to construct the conic.*

On CP take $PZ = CD^2 \div CP$; measuring *from* C in the ellipse, and *towards* C in the hyperbola (Fig. 1188). A circle described through the points C, Z, and having its centre O on the tangent at P, will cut the tangent in the points where it is intersected by the axes.

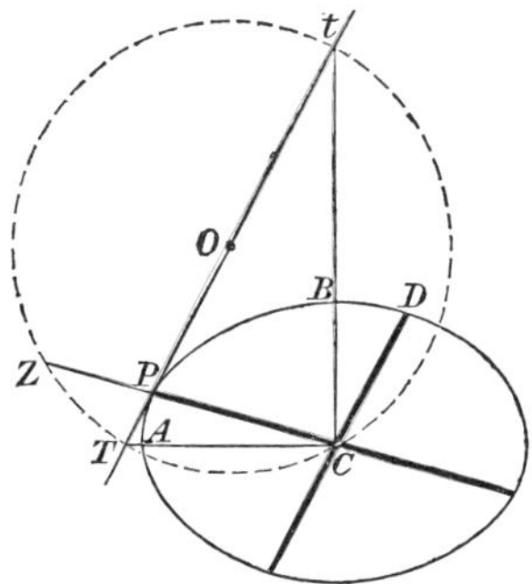

PROOF.—*Analysis*: Let AC, BC cut the tangent at P in T, t. The circle whose diameter is Tt will pass through C (III. 31), and will make

$$CP \,.\, PZ = PT \,.\, Pt \text{ (III. 35, 36)} = CD^2 \text{ (1197)}.$$

Hence the construction.

Circle and Radius of Curvature.

1254 DEFINITIONS.—The circle which has the same tangent with a curve at P (Fig. 1259), and which passes through another point Q on the curve, becomes the *circle of curvature* when Q ultimately coincides with P; and its radius becomes the *radius of curvature*.

1255 Otherwise.—The *circle of curvature* is the circle which passes through *three* coincident points on the curve at P.

1256 Any chord PH of the circle of curvature is called a *chord of curvature* at P.

1257 Through Q draw RQ' parallel to PH, meeting the tangent at P in R, and the circle in Q', and draw QV parallel to PR. RQ is called a *subtense* of the arc PQ.

1258 THEOREM.—*Any chord of curvature PH is equal to the ultimate value of the square of the arc* PQ *divided by the subtense* RQ *parallel to the chord:* and this is also equal to

$$QV^2 \div PV.$$

PROOF.—$RQ' = RP^2 \div RQ$ (III. 36). And when Q moves up to P, RQ' becomes PH; and RP, PQ, and QV become *equal* because *coincident lines.*

1259 In the ellipse or hyperbola, the semi-chords of curvature at P, measured along the diameter PC, the normal PF, and the focal distance PS, are respectively equal to

$$\frac{CD^2}{CP}, \quad \frac{CD^2}{PF}, \quad \frac{CD^2}{AC};$$

the second being the radius of curvature at P.

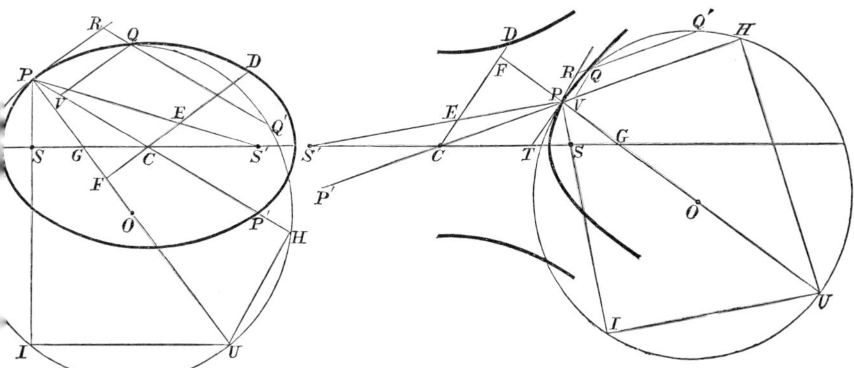

PROOF.—(i.) By (1258), $PH = \frac{QV^2}{PV} = \frac{VP' \cdot CD^2}{CP^2}$ (1193) $= \frac{2CD^2}{CP}$ in the limit when VP' becomes $PP' = 2CP$.

(ii.) By the similar triangles PHU, PFC (III. 31), we have
$$PU \cdot PF = CP \cdot PH = 2CD^2, \text{ by (i.)}$$

(iii.) By the similar triangles PIU, PFE (1168), we have
$$PI \cdot PE = PU \cdot PF = 2CD^2, \text{ by (ii.); and } PE = AC \text{ (1179).}$$

1260 In the parabola, the chord of curvature at P (Fig. 1259) drawn parallel to the axis, and the one drawn through the focus, are each equal to $4SP$, the parameter of the diameter at P (1241).

PROOF.—By (1258). The chord parallel to the axis $= QV^2 \div PV = 4PS$ (1239); and the two chords are equal because they make equal angles with the diameter of the circle of curvature.

1261 COR.—The radius of curvature of the parabola at P (Fig. 1220) is equal to $\mathbf{2SP^2 \div SY}$.

PROOF.—(Fig. 1259.) $\frac{1}{2}PU = \frac{1}{2}PI \sec IPU = 2SP \sec PSY$ (Fig. 1221).

1262 The products of the segments of intersecting chords are as the squares of the tangents parallel to them (1214–15), (1242–43).

1263 The common chords of a circle and conic (Fig. 1264) are equally inclined to the axis; and conversely, if two chords of a conic are equally inclined to the axis, their extremities are concyclic.

PROOF.—The products of the segments of the chords being equal (III. 35, 36), the tangents parallel to them are equal (1262). Therefore, by (1200).

1264 The common chord of any conic and of the circle of curvature at a point P, has the same inclination to the axis as the tangent at P.

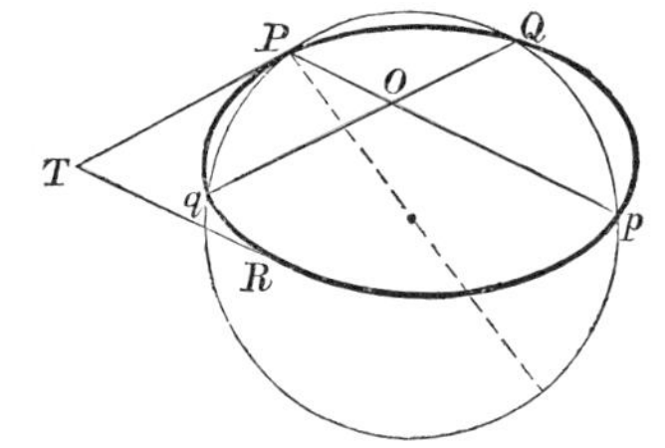

PROOF.—Draw any chord Qq parallel to the tangent at P. The circle circumscribing PQq always passes through the same point p (1263), and does so, therefore, when Qq moves up to P, and the circle becomes the circle of curvature.

1265 PROBLEM.—*To find the centre of curvature at any given point of a conic.*

First Method.—(Fig. 1264.) Draw a chord from the point making the same angle with the axis as the tangent. The perpendicular bisector of the chord will meet the normal in the centre of curvature, by (1264) and (III. 3).

1266 *Second Method.*—Draw the normal PG and a perpendicular to it from G, meeting either of the focal distances in Q. Then a perpendicular to the focal distance drawn from Q will meet the normal in O, the centre of curvature.

PROOF.—(*Ellipse or Hyperbola.*) By (1259), the radius of curvature at P

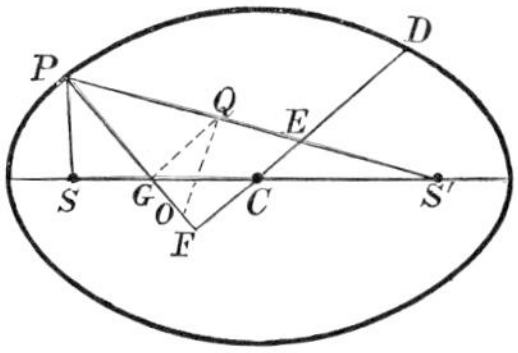

$$= \frac{CD^2}{PF} = \frac{CD^2}{BC^2} PG \text{ (1195)} = \frac{AC^2}{PF^2} PG \text{ (1194)}$$

$$= PG \sec^2 S'PG = PO.$$

For $AC = PE$ (by 1179).

(*Parabola.*) By (1261). The radius of curvature

$$= 2SP^2 \div SY = 2SP \sec SPG = PO.$$

For $2SP = PQ$, because $SP = SG$ (1125).

Miscellaneous Theorems.

1267 In the Parabola (Fig. 1239) let QD be drawn perpendicular to PV, then $QD^2 = 4AS.PV.$ (1231, 1239)

1268 Let RPR' be *any* third tangent meeting the tangents OQ, OQ' in R, R'; the triangles SQO, SRR', SOQ' are similar and similarly divided by SR, SP, SR' (1233–4).

1269 COR.— $OR.OR' = RQ.R'Q'.$

1270 Also, the triangle $PQQ' = 2ORR'.$ (1244)

With the same construction and for *any* conic,

1271 $OQ : OQ' = RQ.RP' : R'Q'.PR.$ (1215, 1243)

1272 Also the angle $RSR' = \frac{1}{2}QSQ'.$ (1181)

1273 Hence, in the Parabola, the points O, R, S, R' are concyclic, by (1234).

1274 In any conic (Figs. 1171), $SP : ST = AN : AT.$

PROOF.— $\frac{SP}{ST} = \frac{S'P}{S'T} = \frac{CA}{CT}$ (69, 1163) $= \frac{CN}{CA}$ (1170) $= \frac{AN}{AT}$ (69).

1275 COR.—If the tangent PT meets the tangent at A in R, then SR bisects the angle PST (VI. 3).

1276 In Figs. (1178), SY', $S'Y$ both bisect the normal PG.

1277 The perpendicular from S to PG meets it in CY.

1278 If CD be the radius conjugate to CP, the perpendicular from D upon CY is equal to BC.

1279 SY and CP intersect in the directrix.

1280 If every ordinate PN of the conic (Figs. 1205) be turned round N, in the plane of the figure, through the same angle PNP', the locus of P' is also a conic, by (1193). The auxiliary circle then becomes an ellipse, of which AC and BC produced are the equi-conjugate diameters.

If the entire figures be thus deformed, the points on the axis AA' remain fixed while PN, DR describe the same angle. Hence CD remains parallel to PT. CP, CD are therefore still conjugate to each other.

Hence, the relations in (1205–6) still subsist when CA, CB are any conjugate radii. Thus universally,

1281 $$PN : CR = DR : CN \quad \text{or} \quad PN \,.\, CN = DR \,.\, CR.$$

1282 If the tangent at P meets any pair of conjugate diameters in T, T', then $PT \,.\, PT'$ is constant and equal to CD^2.

PROOF.—Let CA, CB (Figs. 1205) be the conjugate radii, the figures being deformed through any angle. By similar triangles,

$$\left.\begin{array}{l} PT : PN = CD : DR \\ PT' : CN = CD : CR \end{array}\right\}, \quad \text{therefore } PT \,.\, PT' : PN \,.\, CN = CD^2 : DR \,.\, CR.$$

Therefore $$PT \,.\, PT' = CD^2, \text{ by (1281).}$$

1283 If the tangent at P meets any pair of parallel tangents in T, T', then $PT \,.\, PT' = CD^2$, where CD is conjugate to CP.

PROOF.—Let the parallel tangents touch in the points Q, Q'. Join PQ, PQ', CT, CT'. Then CT, CT' are conjugate diameters (1199, 1201). Therefore $PT \,.\, PT' = CD^2$ (1282).

1284 COR.—$QT \,.\, Q'T' = CD'^2$, where CD' is the radius parallel to QT.

1285 To draw two conjugate diameters of a conic to include a given angle. Proceed as in (1252 ii.), making PP' in this case the chord of the segment of a circle containing the given angle (III. 33).

1286 The focal distance of a point P on any conic is equal to the length QN intercepted on the ordinate through P between the axis and the tangent at the extremity of the latus rectum.

PROOF.—(Fig. 1220). $QN : NX = LS : SX = e$ and $SP : NX = e$.

1287 In the hyperbola (Fig. 1183). $CO : CA = e$. (1162, 1164).

If a right line PKK' be drawn parallel to the asymptote CR, cutting the one directrix XE in K and the other in K'; then

1288 $$SP = PK = e \,.\, CN - AC; \quad S'P = PK' = e \,.\, CN + AC.$$

PROOF.—From $CR = e \,.\, CN$ (1287) and $CE = AC$ (1186).

1289 COR.—Hence the hyperbola may be drawn mechanically by the method of (1249) by merely fixing the cross-piece of the T-square at an angle with the bar equal to BCO.

1290 DEFINITION.—*Confocal* conics are conics which have the same foci.

1291 The tangents drawn to any conic from a point T on a confocal conic make equal angles with the tangent at T.

PROOF.—(Fig. 1217.) Let T be the point on the confocal conic.

$$SY : SZ = S'Z' : S'Y' \quad (1178).$$

Therefore ST and $S'T$ make equal angles with the tangents TP, TQ; and they also make equal angles with the tangent to the confocal at T (1168), therefore &c.

1292 In the construction of (1253), PZ is equal to half the chord of curvature at P drawn through the centre C (1259).

DIFFERENTIAL CALCULUS.

INTRODUCTION.

1400 ***Functions.***—A quantity which depends for its value upon another quantity x is called a *function* of x. Thus, $\sin x$, $\log x$, a^x, a^2+ax+x^2 are all functions of x. The notation $y=f(x)$ expresses generally that y is a function of x. $y=\sin x$ is a particular function.

1401 $f(x)$ is called a *continuous* function between assigned limits, when an indefinitely small change in the value of x always produces an indefinitely small change in the value of $f(x)$.

A *transcendental* function is one which is not purely algebraical, such as the exponential, logarithmic, and circular functions a^x, $\log x$, $\sin x$, $\cos x$, &c.

If $f(x)=f(-x)$, the function is called an *even* function. If $f(x)=-f(-x)$, it is called an *odd* function.

Thus, x^2 and $\cos x$ are even functions, while x^3 and $\sin x$ are odd functions of x; the latter, but not the former, being altered in value by changing the sign of x.

1402 ***Differential Coefficient or Derivative.***—Let y be any function of x denoted by $f(x)$, such that any change in the value of x causes a definite change in the value of y; then x is called the *independent variable*, and y the *dependent variable*. Let an indefinitely small change in x, denoted by dx, produce a corresponding small change dy in y; then the ratio $\frac{dy}{dx}$, in the limit when both dy and dx are vanishing, is called the *differential coefficient*, or *derivative*, of y with respect to x.

1403 THEOREM.—The ratio $dy:dx$ is *definite* for each value of x, and generally *different* for different values.

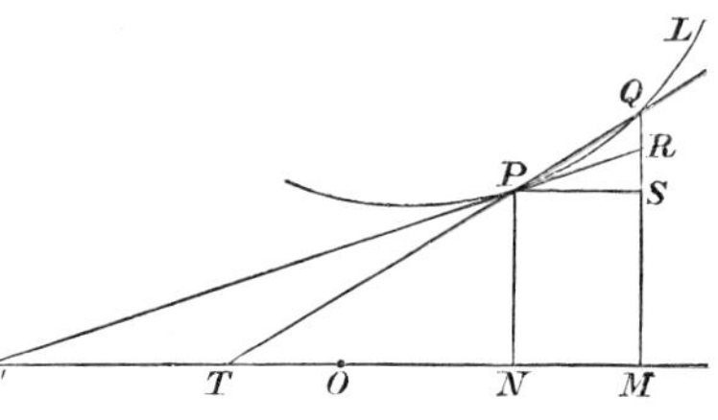

PROOF.—Let an abscissa ON (Defs. 1160) be measured from O equal to x, and a perpendicular ordinate NP equal to y. Then, *whatever may be the form of the function* $y=f(x)$, as x varies, the locus of P will be *some line* PQL. Let $OM=x'$, $MQ=y'$ be values of x and y near to the former values. Let the straight line QP meet the axis in T; and when Q coincides with P, let the final direction of QP cut the axis in T'.

Then $\dfrac{QS}{SP}$ or $\dfrac{y'-y}{x'-x}=\dfrac{PN}{NT}$. And, ultimately, when QS and SP vanish, *they vanish in the ratio of* $PN:NT'$. Therefore $\dfrac{dy}{dx}=\dfrac{PN}{NT'}=\tan PT'N$, a *definite* ratio at each point of the curve, but *different* at different points.

1404 Let NM, the increment of x, be denoted by h; then, when h vanishes, $\dfrac{dy}{dx}=\dfrac{f(x+h)-f(x)}{h}=f'(x)$, a new function of x, called also the *first derived function.* The process of finding its value is called *differentiation.*

1405 *Successive differentiation.*—If $\dfrac{dy}{dx}$ or $f'(x)$ be differentiated with respect to x, the result is the *second differential coefficient* of $f(x)$, or the *second derived function;* and so on to any number of differentiations. These successive functions may be represented in any of the three following systems of notation:—

$$\frac{dy}{dx},\quad \frac{d^2y}{dx^2},\quad \frac{d^3y}{dx^3},\quad \frac{d^4y}{dx^4},\quad \dots\dots\quad \frac{d^ny}{dx^n};$$

$$f'(x),\quad f''(x),\quad f'''(x),\quad f^{\text{iv}}(x),\quad \dots\dots\quad f^n(x);$$

$$y_x,\quad y_{2x},\quad y_{3x},\quad y_{4x},\quad \dots\dots\quad y_{nx}.\text{*}$$

The operations of differentiating a function of x once, twice, or n times, are also indicated by prefixing the symbols

$$\frac{d}{dx},\ \frac{d^2}{dx^2},\ \dots\ \frac{d^n}{dx^n};\ \text{or}\ \frac{d}{dx},\ \left(\frac{d}{dx}\right)^2,\ \dots\ \left(\frac{d}{dx}\right)^n;$$

or, more concisely, $d_x,\ d_{2x},\ \dots\ d_{nx}$.

* See note to (1487).

1406 If, *after* differentiating a function for x, x be made zero in the result, the value may be indicated in any of the following ways: $\frac{dy}{dx_0}$, $f'(0)$, y_{x0}, $\frac{d}{dx_0}$, d_{x0}.

If any other constant a be substituted for x in y_x, the result may be indicated by $y_{x,a}$.

1407 ***Infinitesimals and Differentials.***—The evanescent quantities dx, dy are called *infinitesimals;* and, with respect to x and y, they are called *differentials.* dx^2, d^2y are the *second differentials* of x and y; dx^3, d^3y the third, and so on.

1408 The successive differentials of y are expressed in terms of dx by the equations

$$dy = f'(x)\,dx;\quad d^2y = f''(x)\,dx^2;\quad \text{\&c., and}\quad d^ny = f^n(x)\,dx^n.$$

Since $f'(x)$ is the coefficient of dx in the value of dy, it has therefore been named *the differential coefficient of y or $f(x)$.** For similar reasons $f''(x)$ is called the *second*, and $f^n(x)$ the nth *differential coefficient* of $f(x)$, &c.

1409 Two infinitesimals are of the *same order* when their ratio is neither zero nor infinity.

If dx, dy are infinitesimals of the same order, dx^2, dy^2, and $dx\,dy$ will be infinitesimals of the *second* order with respect to dx, dy; dx^3, dx^2dy, &c. will be of the *third* order, and so on.

dx, dx^2, &c. are sometimes denoted by $\dot{x}$, $\ddot{x}$, &c.

1410 Lemma.—In estimating the ratio of two quantities, any increment of either which is infinitely small in comparison with the quantities may be neglected.

Hence the ratio of two infinitesimals of the same order is not affected by adding to or subtracting from either of them an infinitesimal of a higher order.

Example.— $\frac{dy-dx^2}{dx} = \frac{dy}{dx} - dx = \frac{dy}{dx}$, for dx is zero in comparison with the ratio $\frac{dy}{dx}$. Thus, in Fig. (1403), putting $PS = dx$, $QS = dy$; we have ultimately, by (1258), $QR = k\,dx^2$, where k is a constant. Therefore $\frac{PN}{NT'} = \frac{RS}{PS} = \frac{dy - k\,dx^2}{dx} = \frac{dy}{dx}$ in the limit, by the principle just enunciated; that is, QR *vanishes in comparison with* PS *or* QS *even when those lines themselves are infinitely small.*

* The name is slightly misleading, as it seems to imply that $f'(x)$ is in some sense a coefficient of $f(x)$.

DIFFERENTIATION.

DIFFERENTIATION OF A SUM, PRODUCT, AND QUOTIENT.

Let u, v be functions of x, then

1411 $$\frac{d(u+v)}{dx} = \frac{du}{dx} + \frac{dv}{dx}.$$

1412 $$\frac{d(uv)}{dx} = v\frac{du}{dx} + u\frac{dv}{dx}.$$

1413 $$\frac{d}{dx}\left(\frac{u}{v}\right) = \left(v\frac{du}{dx} - u\frac{dv}{dx}\right) \div v^2.$$

PROOF.—(i.) $d(u+v) = (u+du+v+dv)-(u+v) = du+dv.$

(ii.) $d(uv) = (u+du)(v+dv)-uv = v\,du+u\,dv+du\,dv,$

and, by (1410), $du\,dv$ disappears in the ultimate ratio to dx.

(iii.) $$d\left(\frac{u}{v}\right) = \frac{u+du}{v+dv} - \frac{u}{v} = \frac{v\,du-u\,dv}{(v+dv)\,v},$$

therefore &c., by (1410); $v\,dv$ vanishing in comparison with v^2.

Hence, if u be a constant $= c$,

1414 $$\frac{d(cv)}{dx} = c\frac{dv}{dx} \quad \text{and} \quad \frac{d}{dx}\left(\frac{c}{v}\right) = -\frac{c}{v^2}\frac{dv}{dx}.$$

DIFFERENTIATION OF A FUNCTION OF A FUNCTION.

If y be a function of z, and z a function of x,

1415 $$\frac{dy}{dx} = \frac{dy}{dz} \cdot \frac{dz}{dx}.$$

PROOF.—Since, in all cases, the change dx causes the change dz, and the change dz causes the change dy; therefore the change dx causes the change dy *in the limit.*

Differentiating the above as a product, by (1412), the successive differential coefficients of y can be formed. The first four are here subjoined for the sake of reference. Observe that $(y_z)_x = y_{2z}z_x$.

1416 $y_x = y_z z_x.$

1417 $y_{2x} = y_{2z}z_x^2 + y_z z_{2x}.$

1418 $y_{3x} = y_{3z}z_x^3 + 3y_{2z}z_x z_{2x} + y_z z_{3x}.$

1419 $y_{4x} = y_{4z}z_x^4 + 6y_{3z}z_x^2 z_{2x} + y_{2z}(3z_{2x}^2 + 4z_x z_{3x}) + y_z z_{4x}.$

DIFFERENTIATION OF A COMPOSITE FUNCTION.

If u and v be explicit functions of x, so that $u = \phi(x)$ and $v = \psi(x)$,

1420 $$\frac{dF(u, v)}{dx} = \frac{dF}{du}\frac{du}{dx} + \frac{dF}{dv}\frac{dv}{dx}.$$

Here dF in the first term on the right is the change in $F(u, v)$ produced by du, the change in u; and dF in the second term is the change produced by dv, so that the total change $dF(u, v)$ may be written as in (1408)

$$dF_1 + dF_2 = \frac{dF}{du}du + \frac{dF}{dv}dv.$$

DIFFERENTIATION OF THE SIMPLE FUNCTIONS.

Since $\frac{dy}{dx} = \frac{f(x+h)-f(x)}{h}$ when h vanishes, we have the following rule for finding its value:

1421 Rule.—*Expand* f (x+h) *by some known theorem in ascending powers of* h; *subtract* f (x); *divide by* h; *and in the result put* h *equal to zero.*

The differential coefficients which follow are obtained by the rule and the theorems indicated.

1422 $$y = x^n. \qquad \frac{dy}{dx} = nx^{n-1}.$$

Proof.—Here $\frac{f(x+h)-f(x)}{h} = \frac{(x+h)^n - x^n}{h} = \frac{nx^{n-1}h + C(n, 2)\,x^{n-2}h^2 +}{h}$ (125) $= nx^{n-1} + C(n, 2)\,x^{n-2}h + \ldots = nx^{n-1}$, when h vanishes.

1423 Cor.—

$$\frac{d^ry}{dx^r} = n(n-1)\ldots(n-r+1)\,x^{n-r}. \qquad \frac{d^ny}{dx^n} = \underline{|n}.$$

1424 $$y = \log_a x; \qquad \frac{dy}{dx} = \frac{1}{x \log_e a}.$$

Proof.—By (145), $\frac{\log_a(x+h) - \log_a x}{h} = \frac{1}{x \log_e a}\left\{\log_e\left(1+\frac{h}{x}\right)\right\} \div \frac{h}{x}$.

Expand the logarithm by (155).

1425 Cor.— $$\frac{d^ny}{dx^n} = \frac{(-1)^{n-1}\,\underline{|n-1}}{x^n \log a}.$$ Put $n = -1$ in (1423) and $r = n-1$.

1426 $y = a^x; \qquad \frac{dy}{dx} = a^x \log_e a.$

PROOF.— $\frac{a^{x+h}-a^x}{h} = \frac{a^x(a^h-1)}{h}.$ Expand a^h by (149).

1427 COR.— $\frac{d^ny}{dx^n} = a^x (\log_e a)^n.$

	Function.	*Derivative.*	*Method of Proof by Rule (1421) and Limits (753).*
1428	$\sin x.$	$\cos x.$	Expand by (627, 629), and put $1-\cos h = 2\sin^2\frac{h}{2}.$
1429	$\cos x.$	$-\sin x.$	
1430	$\tan x.$	$\sec^2 x.$	Expand by (631), observing (1410).
1431	$\cot x.$	$-\operatorname{cosec}^2 x.$	By $\cot x = \frac{1}{\tan x}$, and (1415).
1432	$\sec x.$	$\tan x \sec x.$	By $\sec x = \frac{1}{\cos x}$, and (1415).
1433	$\operatorname{cosec} x.$	$-\cot x \operatorname{cosec} x.$	Similarly.
1434	$\left.\begin{array}{l}\sin^{-1} x\\ \cos^{-1} x\end{array}\right\}.$	$\pm\frac{1}{\sqrt{(1-x^2)}}.$	If $\sin^{-1}x = y,\ x = \sin y$, therefore $\frac{dx}{dy} = \cos y = \sqrt{1-x^2};$
1436	$\left.\begin{array}{l}\tan^{-1} x\\ \cot^{-1} x\end{array}\right\}.$	$\pm\frac{1}{1+x^2}.$	therefore $\frac{dy}{dx} = \frac{1}{\sqrt{1-x^2}}.$
1438	$\left.\begin{array}{l}\sec^{-1} x\\ \operatorname{cosec}^{-1} x\end{array}\right\}.$	$\pm\frac{1}{x\sqrt{(x^2-1)}}.$	Similarly for the rest.

EXAMPLES.

1440 $(\sqrt{x})_x = (x^{\frac{1}{2}})_x = \frac{1}{2}x^{-\frac{1}{2}} = \frac{1}{2\sqrt{x}}.$ (1422)

1441 $\left(\frac{1}{x^n}\right)_x = (x^{-n})_x = -nx^{-n-1} = -\frac{n}{x^{n+1}}.$ (1422)

1442 $\{(a+x^2)^3(b+x^3)^2\} = 3(a+x^2)^2\,2x\,(b+x^3)^2 + 2(b+x^3)\,3x^2(a+x^2)^3$
$= 6x(a+x^2)^2(b+x^3)(b+ax+2x^3).$ (1412, '15, '22)

1443 $\left(\frac{e^x-e^{-x}}{e^x+e^{-x}}\right)_x = \frac{(e^x+e^{-x})(e^x+e^{-x})-(e^x-e^{-x})(e^x-e^{-x})}{(e^x+e^{-x})^2} = \frac{4}{(e^x+e^{-x})^2}.$
(1413, 1426)

1444 $d_x(\log\tan x)^2 = 2\log\tan x\,\frac{1}{\tan x}\sec^2 x = \frac{4\log\tan x}{\sin 2x}.$ (1415, '24, '30).

Some differentiations are rendered easier by taking the logarithm of the function. For example,

1445 $y = \sqrt{\dfrac{1-x^2}{(1+x^2)^3}}$; therefore $\log y = \frac{1}{2}\log(1-x^2) - \frac{3}{2}\log(1+x^2)$;

therefore $$\frac{1}{y}\frac{dy}{dx} = \frac{1}{2}\frac{-2x}{(1-x^2)} - \frac{3}{2}\frac{2x}{(1+x^2)};$$

therefore $$\frac{dy}{dx} = y\,\frac{-2x(2-x^2)}{1-x^4} = \frac{-2x(2-x^2)}{(1+x^2)^{\frac{5}{2}}(1-x^2)^{\frac{1}{2}}}.$$

1446 $y = (\sin x)^x$; therefore $\log y = x \log \sin x$;

therefore $$\frac{1}{y}y_x = \log \sin x + \frac{x}{\sin x}\cos x; \qquad (1415,\ '24,\ '28)$$

therefore $$y_x = (\sin x)^x(\log \sin x + x \cot x).$$

Otherwise, by (1420), $y_x = x(\sin x)^{x-1}\cos x + (\sin x)^x \log \sin x$ (1426)

$$= (\sin x)^x (x \cot x + \log \sin x).$$

SUCCESSIVE DIFFERENTIATION.

1460 ***Leibnitz's Theorem.***—If n be any integer,

$$(yz)_{nx} = y_{nx}z + ny_{(n-1)x}z_x + C(n,2)\,y_{(n-2)x}z_{2x} + \ldots$$
$$\ldots + C(n,r)\,y_{(n-r)x}z_{rx} + \ldots + yz_{nx}.$$

Proof.—By Induction (233). Differentiate the two consecutive terms

$$C(n,r)\,y_{(n-r)x}z_{rx} + C(n,r+1)\,y_{(n-r-1)x}z_{(r+1)x},$$

and four terms are obtained, the second and third of which are

$$C(n,r)\,y_{(n-r)x}z_{(r+1)x} + C(n,r+1)\,y_{(n-r)x}z_{(r+1)x}$$
$$= \{C(n,r) + C(n,r+1)\}\,y_{(n-r)x}z_{(r+1)x} = C(n+1,r+1)\,y_{(\overline{n+1}-\overline{r+1})x}z_{(r+1)x},$$

by (102).

This is the general term of the series with n increased by unity. Similarly, by differentiating all the terms the whole series is reproduced with n increased by unity.

DIFFERENTIAL COEFFICIENTS OF THE nth ORDER.

1461 $(\sin ax)_{nx} = a^n \sin(ax + \frac{1}{2}n\pi)$.

1462 $(\cos ax)_{nx} = a^n \cos(ax + \frac{1}{2}n\pi)$.

By Induction and (1428).

1463 $(e^{ax})_{nx} = a^n e^{ax}$. (1426)

1464 $(e^{ax}y)_{nx} = e^{ax}(a + d_x)^n y$,

where, in the expansion by the Binomial Theorem, $d_x^r y$ is to be replaced by y_{rx}. (1460, '63)

1465 $$(e^{ax}\cos bx)_{nx} = r^n e^{ax}\cos(bx+n\phi),$$
where $$a = r\cos\phi \quad\text{and}\quad b = r\sin\phi.$$

Proof.—By Induction. Differentiating once more, we obtain

$$\begin{aligned} &r^n e^{ax}\{a\cos(bx+n\phi) - b\sin(bx+n\phi)\} \\ &\qquad = r^{n+1}e^{ax}\{\cos\phi\cos(bx+n\phi) - \sin\phi\sin(bx+n\phi)\} \\ &\qquad = r^{n+1}e^{ax}\cos(bx+\overline{n+1}\,\phi). \end{aligned}$$

Thus n is increased by one.

1466 $$(x^{n-1}\log x)_{nx} = |\underline{n-1} \div x. \qquad (1460),\ (283)$$

1467 $$\left(\frac{1-x}{1+x}\right)_{nx} = \frac{(-1)^n\, 2}{(1+x)^{n+1}}\,|\underline{n}. \qquad (1423)$$

1468 $$(\tan^{-1}x)_{nx} = (-1)^{n-1}\,|\underline{n-1}\,\sin^n\theta\,\sin n\theta,$$
where $$\theta = \cot^{-1}x.$$

Proof.—By Induction. Differentiating again, we obtain (omitting the coefficient)

$$\begin{aligned} &(n\sin^{n-1}\theta\cos\theta\sin n\theta + n\cos n\theta\sin^n\theta)\,\theta_x \\ &\qquad = n\sin^{n-1}\theta\,(\sin n\theta\cos\theta + \cos n\theta\sin\theta)\,(-\sin^2\theta). \end{aligned}$$

Since, by (1437), $\quad \theta_x = -(1+x^2)^{-1} = -\sin^2\theta.$

Therefore $\quad (\tan^{-1}x)_{(n+1)x} = (-1)^n\,|\underline{n}\,\sin^{n+1}\theta\,\sin(n+1)\,\theta,$

n being increased by one.

1469 $$\left(\frac{1}{1+x^2}\right)_{nx} = (-1)^n\,|\underline{n}\,\sin^{n+1}\theta\,\sin(n+1)\,\theta. \qquad (1436, 1468)$$

1470 $$\left(\frac{x}{1+x^2}\right)_{nx} = (-1)^n\,|\underline{n}\,\sin^{n+1}\theta\,\cos(n+1)\,\theta.$$

Proof.—By (1460), $\quad \left(\dfrac{x}{1+x^2}\right)_{nx} = x\left(\dfrac{1}{1+x^2}\right)_{nx} + n\left(\dfrac{1}{1+x^2}\right)_{(n-1)x}.$

Then by (1469).

1471 *Jacobi's Formula.*

$$d_{(n-1)x}(1-x^2)^{n-\frac{1}{2}} = (-1)^{n-1}\,1.3\ldots(2n-1)\sin(n\cos^{-1}x)\div n.$$

Proof.—Let $y = 1-x^2$; therefore

$$(y^{n+\frac{1}{2}})_{nx} = -(2n+1)\,(xy^{n-\frac{1}{2}})_{(n-1)x}. \quad \text{Also} \quad (y^{n+\frac{1}{2}})_{nx} = (yy^{n-\frac{1}{2}})_{nx}.$$

Expand each of these values by (1460) and eliminate $(y^{n-\frac{1}{2}})_{(n-2)x}$, the derivative of lowest order. Call the result equation (1). Now assume (1471) true for the value n. Differentiate and substitute the result, and also (1471) on the right side of equation (1) to obtain a proof by Induction.

1472 *Theorem.*—If y, z are functions of x, and n a positive integer,

$$zy_{nx} = (yz)_{nx} - n\,(yz_x)_{(n-1)\,x} + C\,(n,\,2)\,(yz_{2x})_{(n-2)\,x} \dots + (-1)^n\,yz_{nx}.$$

PROOF.—By Induction. Differentiate for x, substituting for $z_x y_{nx}$ on the right its value by the formula itself.

PARTIAL DIFFERENTIATION.

1480 If $u = f(x,\,y)$ be a function of two *independent* variables, any differentiation of u with respect to x *requires that y should be considered constant* in that operation, and *vice versâ*. Thus, $\frac{d^2u}{dx^2}$ or u_{2x} signifies that u is to be differentiated successively twice with respect to x, y being considered constant.

1481 The notation $\frac{d^5u}{dx^2\,dy^3}$ or $u_{2x\,3y}$* signifies that u is to be differentiated successively twice for x, y being considered constant, and the result three times successively for y, x being considered constant.

1482 The *order* of the differentiations does not affect the final result, or $u_{xy} = u_{yx}$.

PROOF.—Let $u = f(xy)$; then $u_x = \frac{f(x+h,\,y) - f(x,\,y)}{h}$ in limit. (1484)

$$u_{xy} = \frac{du_x}{dy} = \frac{f(x+h,\,y+k) - f(x,\,y+k) - f(x+h,\,y) + f(x,\,y)}{hk} \text{ in limit.}$$

Now, if u_y had been first formed, and then u_{yx}, the same result would have been obtained. The proof is easily extended. Let $u_x = v$; then $u_{2xy} = v_{xy} = v_{yx} = u_{y2x}$; and so on.

THEORY OF OPERATIONS.

1483 Let the symbols Φ, Ψ, prefixed to a quantity, denote operations upon it of the *same class*, such as multiplication or differentiation. Then the law of the operation is said to be *distributive*, when

$$\Phi\,(x+y) = \Phi\,(x) + \Phi\,(y);$$

* See note to (1487).

that is, the operation may be performed upon an undivided quantity, or it may be distributed by being performed upon parts of the quantity separately with the same result.

1484 The law is said to be *commutative* when

$$\Phi\Psi x = \Psi\Phi x;$$

that is, the order of operation may be changed, Φ operating upon Ψx producing the same result as Ψ operating upon Φx.

1485 $\Phi^m x$ denotes the repetition of the operation Φ m times, and is equivalent to $\Phi\Phi \ldots x$ to m operations. This definition involves the index law,

$$\Phi^m \Phi^n x = \Phi^{m+n} x = \Phi^{n+m} x,$$

which merely asserts, that to perform the operation n times in succession upon x, and afterwards m times in succession upon the result, is equivalent to performing it $m+n$ times in succession upon x.

1486 The three laws of *Distribution*, *Commutation*, and the law of *Indices* apply to the operation of *multiplication*, and also to that of *differentiation* (1411, '12). Therefore any algebraic transformation which proceeds at every step by one or more of these laws *only*, has a valid result when for the operation of *multiplication* that of *differentiation* is substituted.

1487 In making use of this principle, the symbol of differentiation employed is $\frac{d}{dx}$, or simply d_x prefixed to the quantity upon which the operation of differentiating with respect to x is to be performed. The repetition of the operation is indicated by $\frac{d^2}{dx^2}$, $\frac{d^3}{dx^3}$, $\frac{d^5}{dx^2\,dy^3}$, &c., prefixed to the function. An abbreviated notation is d_x, d_{2x}, d_{3x}, $d_{2x\,3y}$, &c. Since $d_x \times d_x = d_x^2$ in the symbolic operation of multiplication, it will be requisite, in transferring the operation to differentiation, to change all such *indices* to *suffixes* when the abbreviated notation is being used.

NOTE.—The notation y_x, y_{2x}, $u_{2x\,3y}$, $d_{2x\,3y}$, &c. is an innovation. It has, however, the recommendations of definiteness, simplicity, and economy of time in writing, and of space in printing. The expression $\frac{d^5u}{dx^2\,dy^3}$ requires at least fourteen distinct types, while its equivalent u_{2x3y} requires but seven. For such reasons I have introduced the shorter notation experimentally in these pages.

All such abbreviated forms of differential coefficients as $y'\,y''\,y'''\ldots$ or $\dot{y}\,\ddot{y}\,\dddot{y}\ldots$, though convenient in practice, are incomplete expressions, because the independent variable is not specified.

The operation $d_{2x\,3y}$, and the derived function $u_{2x\,3y}$, would be more accurately represented by $(d_x^2)_y^3$ and $(u_x^2)_y^3$, the index as usual indicating the repetition of the operation. But the former notation is simpler, and it has the advantage of separating more clearly the index of differentiation from the index of involution.

In the symbols y^2 and y_{2x}, the figure 2 is an index in each case: in the first, it shows the *degree* of involution; in the second, the *order* of differentiation. The index is omitted when the degree or the order is unity, since we write y and y_x.

The suffix takes precedence of the superfix. y_x^2 means the square of y_x. $d_x(y^2)$ would be written $(y^2)_x$ in this notation.

As a concise nomenclature for all fundamental operations is of great assistance in practice, the following is recommended: $\frac{dy}{dx}$ or y_x may be read "*y for x*," as an abbreviation of the phrase, "the differential coefficient of y *for*, or with respect to, x." Similarly, $\frac{d^2y}{dx^2}$, $\frac{d^2u}{dx\,dy}$, $\frac{d^5u}{dx^2\,dy^3}$, or the shortened forms y_{2x}, u_{xy}, $u_{2x\,3y}$, may be read "*y for two x*," "*u for xy*," "*u for two x, three y*," and so forth.

The distinction in meaning between the two forms y_{nx} and y_{xz} is obvious. The first (in which n is numerical and *always an integer*) indicates n successive differentiations for x; the second indicates two successive differentiations for the variables x and z.

The symbols $\frac{dy}{dx_0}$ or y_{x0}, and $\frac{d^2y}{dx_0^2}$ or y_{2x0}, may be read, for shortness, "*y for x zero*," "*y for two x zero*"; $d_{2x\,3y}\,\phi\,(xy)$ can be read "*d* for two x three y of $\phi\,(xy)$."

Although the notation v_x is already employed in a totally different sense in the Calculus of Finite Differences, my own experience is that the double signification of the symbol does not lead to any confusion: and this for the very reason that the two meanings *are* so entirely distinct. Whenever the operation of differentiation is introduced along with the subject of Finite Differences, the notation $\frac{dy}{dx}$ must of course alone be employed.

Thus, in differentiation, we have

1488 THE DISTRIBUTIVE LAW $\quad d_x(u+v) = d_xu + d_xv.$ (1411)

1489 THE COMMUTATIVE LAW $\quad d_x(d_yu) = d_y(d_xu)$

or $\quad d_{xy}u = d_{yx}u.$ (1482)

1490 THE INDEX LAW $\quad d_x^m\,d_x^n\,u = d_x^{m+n}\,u,$

that is, $\quad d_{mx}\,d_{nx}\,u = d_{(m+n)\,x}\,u.$ (1485)

1491 Example.—

$(d_x-d_y)^2 = (d_x-d_y)(d_x-d_y) = d_x d_x - d_x d_y - d_y d_x + d_y d_y = d_{2x} - 2d_{xy} + d_{2y}$.

Here $d_x d_y = d_y d_x$ or $d_{xy} = d_{yx}$, by the *commutative law*. (1489)

$d_x d_x = d_{2x}$ by the *index law*. (1490)

Also $(d_{2x}-2d_{xy}+d_{2y})\,u = d_{2x}u - 2d_{xy}u + d_{2y}u$, by the *distributive law*.

Therefore, finally, $(d_x-d_y)^2 u = d_{2x}u - 2d_{xy}u + d_{2y}u$.

Similarly for more complex transformations.

1492 Thus d_x may be treated as quantitative, and operated upon as such by the laws of Algebra; d_x^n being written d_{nx}, and factors such as $d_x d_y$, in which the independent variables are different, being written d_{xy}, &c.

EXPANSION OF EXPLICIT FUNCTIONS.

TAYLOR'S THEOREM.—EXPANSION OF $f(x+h)$.

1500

$$f(x+h) = f(x) + hf'(x) + \frac{h^2}{1.2}f''(x) + \dots + \frac{h^n}{\lfloor n}f^n(x+\theta h),$$

where θ is some quantity between zero and unity, and n is any integer.

Proof.—(i.) Assume $f(x+h) = A + Bh + Ch^2 + \&c.$

Differentiate both sides of this equation,—first for x, and again for h,—and equate coefficients in the two results.

1501 (ii.) *Cox's Proof.*—Lemma.—If $f(x)$ vanishes when $x = a$, and also when $x = b$, and if $f(x)$ and $f'(x)$ are *continuous* functions between the same limits; then $f'(x)$ vanishes for some value of x between a and b.

For $f'(x)$ must change sign somewhere between the assigned limits (see proof of 1403), and, being *continuous*, it must vanish in passing from plus to minus.

1502 Now, the expression

$$f(a+x) - f(a) - xf'(a) - \dots\dots - \frac{x^n}{\lfloor n}f^n(a)$$

$$-\frac{x^{n+1}}{\lfloor n+1}\frac{\lfloor n+1}{h^{n+1}}\left\{f(a+h) - f(a) - hf'(a) - \dots - \frac{h^n}{\lfloor n}f^n(a)\right\}$$

vanishes when $x = 0$ and when $x = h$. Therefore the differential coefficient with respect to x vanishes for some value of x between 0 and h by the lemma. Let θh be this value. Differentiate, and apply the lemma to the resulting expression, which vanishes when $x = 0$ and when $x = \theta\theta h$. Perform the same process $n+1$ times successively, writing θh for $\theta\theta h$, &c., since θ merely stands for some quantity less than unity. The result shews that

$$f^{n+1}(a+x) - \frac{\lfloor n+1}{h^{n+1}}\left\{f(a+h) - f(a) - hf'(a) - \dots - \frac{h^n}{\lfloor n}f^n(a)\right\}$$

vanishes when $x = \theta h$. Substituting θh and equating to zero, the theorem is proved.

1503 The last term in (1500) is called the remainder after n terms. It may be obtained in either of the subjoined forms, the first being due to Lagrange,

$$\frac{h^n}{\lfloor n} f^n(x+\theta h) \qquad \text{or} \qquad \frac{h^n}{\lfloor n-1}(1-\theta)^{n-1} f^n(x+\theta h).$$

1504 Since the coefficient $\frac{h^n}{\lfloor n}$ diminishes at last without limit as n increases (239, ii.), it follows that *Taylor's series is convergent if* $f^n(x)$ *remains finite for all values of* n.

1505 If in any expansion of $f(x+h)$ in powers of h some index of h be *negative*, then $f(x)$, $f'(x)$, $f''(x)$, &c. all become infinite.

1506 If the least fractional index of h lies between n and $n+1$; then $f^{n+1}(x)$ and all the following differential coefficients become infinite.

PROOF.—To obtain the value of $f^n(x)$, differentiate the expansion n times successively for h, and put $h=0$ in the result.

MACLAURIN'S THEOREM.

Put $x=0$ in (1500), and write x for h; then, with the notation of (1406),

1507 $$f(x) = f(0) + xf'(0) + \frac{x^2}{1.2} f''(0) + \ldots + \frac{x^n}{\lfloor n} f^n(\theta x),$$

where θ, as before, lies between 0 and 1.

Putting $y=f(x)$, this may also be written

1508 $$y = y_0 + x\frac{dy}{dx_0} + \frac{x^2}{1.2}\frac{d^2y}{dx_0^2} + \frac{x^3}{1.2.3}\frac{d^3y}{dx_0^3} + \&c.$$

1509 NOTE.—If any function $f(x)$ becomes infinite with a finite value of x, then $f'(x)$, $f''(x)$, &c. all become infinite. Thus, if $f(x) = \sec^{-1}(1+x)$, $f'(x)$ is infinite when $x=0$ (1438). Therefore $f''(0)$, $f'''(0)$, &c. are all infinite, and $f(x)$ cannot be expanded by this theorem.

Bernoulli's Series.—Put $h=-x$ in (1500); thus,

1510 $$f(0) = f(x) - xf'(x) + \frac{x^2}{1.2} f''(x) - \frac{x^3}{1.2.3} f'''(x) + \&c.$$

1511 If $\phi(y+k)=0$ and $\phi(y)=x$; then

$$k=-xy_x+\frac{x^2}{1.2}y_{2x}-\frac{x^3}{1.2.3}y_{3x}+\&c.$$

PROOF.—Let $y=\phi^{-1}(x)=f(x)$, and let $y+k=f(x+h)$; therefore $x+h=\phi(y+k)=0$.

Therefore $y+k=f(0)=f(x)-xf'(x)+\frac{x^2}{1.2}f''(x)-\&c.$, by (1510); which proves the theorem.

EXPANSION OF $f(x+h,\ y+k)$.

Let $f(xy)=u$. Then, with the notation of (1405),

1512

$$f(x+h,\ y+k)=u+(hu_x+ku_y)+\frac{1}{1.2}(h^2u_{2x}+2hku_{xy}+k^2u_{2y})$$
$$+\frac{1}{1.2.3}(h^3u_{3x}+3h^2ku_{2xy}+3hk^2u_{x2y}+k^3u_{3y})+\&c.$$

1513 The general term is given by $\frac{1}{\underline{|n}}(hd_x+kd_y)^n u$,

where, in the expansion by the Binomial Theorem, each index of d_x and d_y is changed into a suffix; and the coefficients d_x, d_{2x}, &c. are joined to u as symbols of operation (1487); thus u_x^3 is to be changed into u_{3x}.

PROOF.—First expand $f(x+h,y+k)$ as a function of $(x+h)$ by (1500); thus, $f(x+h,y+k)=f(x,y+k)+hf_x(x,y+k)+\frac{1}{1.2}h^2f_{2x}(x,y+k)+\&c.$

Next, expand each term of this series as a function of $(y+k)$. Thus, writing u for $f(xy)$,

$$f(x,y+k)=\quad u\quad+\quad ku_y\quad+\frac{1}{\underline{|2}}k^2u_{2y}\quad+\frac{1}{\underline{|3}}k^3u_{3y}\quad+\frac{1}{\underline{|4}}k^4u_{4y}+\dots$$

$$hf_x(x,y+k)=\quad hu_x+\quad hku_{xy}+\frac{1}{\underline{|2}}hk^2u_{x2y}\quad+\frac{1}{\underline{|3}}hk^3u_{x3y}+\dots\ \dots\ \dots$$

$$\frac{h^2}{\underline{|2}}f_{2x}(x,y+k)=\frac{1}{\underline{|2}}h^2u_{2x}+\frac{1}{\underline{|2}}h^2ku_{2xy}+\frac{1}{\underline{|2}\,\underline{|2}}h^2k^2u_{2x2y}+\dots\ \dots\ \dots\ \dots\ \dots$$

$$\frac{h^3}{\underline{|3}}f_{3x}(x,y+k)=\frac{1}{\underline{|3}}h^3u_{3x}+\frac{1}{\underline{|3}}h^3ku_{3xy}+\dots\ \dots\ \dots\ \dots\ \dots\ \dots\ \dots\ \dots\ \dots$$

$$\frac{h^4}{\underline{|4}}f_{4x}(x,y+k)=\frac{1}{\underline{|4}}h^4u_{4x}+\dots\ \dots\ \dots\ \dots\ \dots\ \dots\ \dots\ \dots\ \dots\ \dots\ \dots$$

The law by which the terms of the same dimension in h and k are formed, is seen on inspection. They lie in successive diagonals; and when cleared of fractions the numerical coefficients are those of the Binomial Theorem.

The theorem may be extended inductively to a function of three or more variables. Thus, if $u = f(x, y, z)$, we have

1514 $$f(x+h, y+k, z+l) = u + (hu_x + ku_y + lu_z)$$
$$+\tfrac{1}{2}(h^2u_{2x} + k^2u_{2y} + l^2u_{2z} + 2klu_{yz} + 2lhu_{zx} + 2hku_{xy}) + \dots,$$
the general term being obtained as before from the expression
$$\frac{1}{\lfloor n}(hd_x + kd_y + ld_z)^n\, u.$$

1515 Cor.—If $u = f(xyz)$ be a function of several independent variables, the term $(hu_x + ku_y + lu_z)$ proves, in conjunction with (1410), that the total change in the value of u, caused by simultaneous small changes in x, y, z, is equal to the sum of the increments of u due to the increments of x, y, z taken separately and *superposed in any order*.

This is known as the principle of *the superposition* of small quantities.

1516 To expand $f(x, y)$ or $f(x, y, z, \dots)$ in powers of x, y, &c., put x, y, z each equal to zero after differentiating in (1512) or (1514), and write $x, y, \dots$ instead of h, k, &c.

1517 Observe that any term in these series may be made the last by writing $x+\theta h$ for x, $y+\theta k$ for y, &c., as in (1500).

SYMBOLIC FORM OF TAYLOR'S THEOREM.

The expansion in (1500) is equivalent to the following

1520 $$f(x+h) = e^{hd_x} f(x).$$

Proof.—By the Exponential Theorem (150), writing the indices of d_x as suffixes (1487),
$$e^{hd_x} f(x) = (1 + hd_x + \tfrac{1}{2}h^2 d_{2x} + \dots) f(x) = f(x) + hf_x(x) + \tfrac{1}{2}h^2 f_{2x}(x) + \dots, \text{ by (1488)}.$$

Cor.— $\Delta f(x) = f(x+h) - f(x) = (e^{hd_x} - 1) f(x)$,

therefore $\Delta^2 f(x) = (e^{hd_x} - 1)^2 f(x)$,

and generally $\Delta^n f(x) = (e^{hd_x} - 1)^n f(x)$,

the index signifying that the operation is performed n times upon $f(x)$.

1521 Similarly $f(x+h, y+k) = e^{hd_x + kd_y} f(x, y)$.

Proof.—
$$e^{hd_x+kd_y} f(x,y) = \{1 + hd_x + kd_y + \tfrac{1}{2}(hd_x + kd_y)^2 + \tfrac{1}{6}(hd_x + kd_y)^3 + \dots\} f(x,y)$$
$$= f(x+h, y+k), \text{ by (150) and (1512).}$$

1522 And, generally, with any number of variables,

$$f(x+h, y+k, z+l \ldots) = e^{hd_x+kd_y+ld_z+\ldots} f(x, y, z \ldots).$$

Cor.—As in (1520),

$$\Delta f(x, y, z \ldots) = (e^{hd_x+kd_y+\ldots}-1) f(x, y, z \ldots).$$

1523 If $u = f(x, y) = \phi(r, \theta)$, where $x = r\cos\theta$, $y = r\sin\theta$; and if $x' = r\cos(\theta+\omega)$, $y' = r\sin(\theta+\omega)$; then $f(x'y')$ is expanded in powers of ω by the formula

$$f(x', y') = e^{\omega(xd_y-yd_x)} f(x, y).$$

Proof.—By (1520), r being constant,

$$\phi(r, \theta+\omega) = e^{\omega d_\theta}\phi(r, \theta) = e^{\omega d_\theta} f(x, y).$$

Now x and y are functions of the single variable θ; therefore

$$u_\theta = u_x x_\theta + u_y y_\theta = u_x(-r\sin\theta) + u_y(r\cos\theta) = xu_y - yu_x.$$

The operation d_θ will be transformed by the same law (1492); therefore

$$d_\theta = xd_y - yd_x; \quad \text{therefore}$$

$$f(x', y') = e^{\omega(xd_y-yd_x)} f(x, y) = 1 + \omega(xu_y - yu_x) + \tfrac{1}{2}\omega^2(x^2u_{2y} - 2xyu_{xy} + y^2u_{2x}) + \&c.$$

1524 Examples.—The Binomial, Exponential, and Logarithmic series for $(1+x)^n$, a^x, and $\log(1+x)$, (125, 149, 155), are obtained immediately by Maclaurin's Theorem (1507); as also the series for $\sin x$ and $\cos x$ (764), and $\tan^{-1}x$ (791). The mode of proceeding, which is the same in all cases, is shewn in the following example; the test of convergency (1504) being applied when practicable.

1525 $$\tan x = x + \frac{1}{3}x^3 + \frac{2}{15}x^5 + \frac{17}{315}x^7 + \frac{62}{2835}x^9 + \&c.$$

Obtained by Maclaurin's theorem, as follows:—Let

$$\left.\begin{aligned} f(x) &= \tan x = y \\ f'(x) &= \sec^2 x = z \end{aligned}\right\} \quad \begin{aligned} &\text{Therefore } y_x = z \text{ and } z_x = 2yz; \\ &y \text{ and } z \text{ being used for shortness.} \end{aligned}$$

$$\begin{aligned}
f''(x) &= 2\sec^2 x \tan x = 2yz, \\
f'''(x) &= 2(zy_x + yz_x) = 2(z^2 + 2y^2z), \\
f^{\text{iv}}(x) &= 2(4yz^2 + 4yz^2 + 4y^3z) = 8(2yz^2 + y^3z), \\
f^{\text{v}}(x) &= 8(2z^3 + 8y^2z^2 + 3y^2z^2 + 2y^4z) = 8(2z^3 + 11y^2z^2 + 2y^4z), \\
f^{\text{vi}}(x) &= 8(12yz^3 + 22yz^3 + \ldots) = 272yz^3 + \ldots, \\
f^{\text{vii}}(x) &= 272z^4 + \ldots \&c.,
\end{aligned}$$

the terms omitted involving positive powers of y, which vanish when x is zero, and which therefore need not be computed if no term of the expansion higher than that containing x^7 is required.

Hence, by making $x=0$, and therefore $y=0$ and $z=1$, we obtain

$$f(0)=0;\; f'(0)=1;\; f''(0)=0;\; f'''(0)=2;\; f^{\text{iv}}(0)=0;\; f^{\text{v}}(0)=16;$$
$$f^{\text{vi}}(0)=0;\; f^{\text{vii}}(0)=272.$$

Thus the terms up to x^7 may be written by substituting these values in (1507).

In a similar manner, may be obtained

1526 $$\sec x = 1+\frac{1}{2}x^2+\frac{5}{24}x^4+\frac{61}{720}x^6+\&c. \ldots\ldots\ldots\ldots$$

Methods of expansion by Indeterminate Coefficients.

1527 RULE I.—*Assume* f(x) = A+Bx+Cx²+&c. *Differentiate both sides of the equation. Then expand* f′(x) *by some known theorem, and equate coefficients in the two results to determine* A, B, C, *&c.*

1528 Ex. $$\sin^{-1}x = x+\frac{1}{2}\frac{x^3}{3}+\frac{1.3}{2.4}\frac{x^5}{5}+\frac{1.3.5}{2.4.6}\frac{x^7}{7}+\&c.$$

Obtained by Rule I. Assume

$$\sin^{-1}x = A+Bx+Cx^2+Dx^3+Ex^4+Fx^5+\ldots\ldots\ldots\ldots\ldots\ldots$$

Therefore, by (1434), $(1-x^2)^{-\frac{1}{2}} = B+2Cx+3Dx^2+4Ex^3+5Fx^4+\ldots\ldots\ldots\ldots$

But, by Bin. Th. (128), $(1-x^2)^{-\frac{1}{2}} = 1+\frac{1}{2}x^2+\frac{1.3}{2.4}x^4+\frac{1.3.5}{2.4.6}x^6+\ldots\ldots\ldots\ldots$

Equate coefficients; therefore $B=1$; $C=0$; $D=\frac{1}{2.3}$; $E=0$; $F=\frac{1.3}{2.4.5}$; &c. By putting $x=0$, we see that $A=f(0)$ always. In this case $A=\sin^{-1}0=0$.

In a similar manner, by Rule I.,

1529 $$e^{\sin x} = 1+x+\frac{x^2}{\lfloor 2}-\frac{3x^4}{\lfloor 4}-\frac{8x^5}{\lfloor 5}-\frac{3x^6}{\lfloor 6}+\&c.\ldots\ldots\ldots\ldots$$

1530 RULE II.—*Assume the series, as before, with unknown coefficients. Differentiate successively until the function reappears. Then equate coefficients in the two equivalent series.*

1531 Ex.—To expand $\sin x$ in powers of x.

Assume $\sin x = A+Bx+Cx^2+Dx^3+Ex^4+Fx^5+\ldots\ldots\ldots\ldots\ldots\ldots\ldots$

Differentiate twice, $\cos x = B+2Cx+3Dx^2+4Ex^3+5Fx^4+\ldots\ldots\ldots$

$$-\sin x = 2C+3.2Dx+4.3Ex^2+5.4Fx^3+\ldots\ldots\ldots\ldots\ldots\ldots$$

Put $x=0$ in the first two equations; therefore $A=0$, $B=1$.

Equate coefficients in the first and third series.

Thus $\quad -2C = A, \quad \therefore C = 0; \quad -3.2D = B, \quad \therefore D = -\frac{1}{2.3};$

$$-4.3E = C, \quad \therefore E = 0; \quad -5.4F = D, \quad \therefore F = \frac{1}{1.2.3.4}, \text{ \&c.}$$

Therefore $\quad \sin x = x - \frac{x^3}{1.2.3} + \frac{x^5}{1.2.3.4.5} - \text{\&c.}$, as in (764).

1532 RULE III.—*Differentiate the equation* y = f (x) *twice with respect to* x, *and combine the results so as to form an equation in* y, y_x, *and* y_{2x}. *Next assume* $y = A + Bx + Cx^2 + \&c.$ *Differentiate twice, and substitute the three values of* y, y_x, y_{2x} *so obtained in the former equation. Lastly, equate coefficients in the result to determine in succession* A, B, C, *&c.*

1533 Ex.—To expand $\sin m\theta$ and $\cos m\theta$ in ascending powers of $\sin\theta$ or $\cos\theta$.

These series are given in (775–779). They may be obtained by Rule III. as follows :—

Put $\quad x = \sin\theta \quad$ and $\quad y = \sin m\theta = \sin(m \sin^{-1} x)$.

Therefore $\quad y_x = \cos(m\sin^{-1}x)\,\frac{m}{\sqrt{1-x^2}} \quad (1434) \text{ .... (i.)}$

$$y_{2x} = -\sin(m\sin^{-1}x)\,\frac{m^2}{1-x^2} + \cos(m\sin^{-1}x)\,\frac{mx}{(1-x^2)^{\frac{3}{2}}}.$$

Therefore, eliminating $\cos(m\sin^{-1}x)$, $\quad (1-x^2)\,y_{2x} - xy_x + m^2y = 0$ ....(ii.)

Let $\quad y = A + A_1x + A_2x^2 + \dots\dots + A_nx^n + \dots$ .... (iii.)

Differentiate twice, and put the values of y, y_x, and y_{2x} in equation (ii.); thus

$$\begin{aligned} 0 = {} & m^2(A + A_1x + A_2x^2 + A_3x^3 + \dots + A_nx^n + \dots\dots\dots\dots) \\ & - x(A_1 + 2A_2x + 3A_3x^2 + \dots + nA_nx^{n-1} + \dots\dots\dots\dots) \\ & + (1-x^2)\{2A_2 + 2.3A_3x + \dots\dots\dots\dots\dots\dots\dots\dots \\ & + (n-1)\,nA_nx^{n-2} + n(n+1)\,A_{n+1}x^{n-1} + (n+1)(n+2)\,A_{n+2}x^n + \dots\}. \end{aligned}$$

Equating the collected coefficients of x^n to zero, we get the relation

$$A_{n+2} = \frac{n^2 - m^2}{(n+1)(n+2)}A_n \text{ .... (iv.)}$$

Now, when $x = 0$, $y = 0$; therefore $A = 0$, by (iii.). And when $x = 0$, $y_x = m$, by (i.); and therefore $A_1 = m$, by differentiating (iii.). The relation (iv.) furnishes the remaining coefficients by making n equal to 0, 1, 2, 3, &c. in succession.

Cos $m\theta$ is obtained in a similar way.

1534 RULE IV.—*Form the equation in* y, y_x, *and* y_{2x}, *as in Rule III. Take the* n^{th} *derivative of this equation by applying Leibnitz's formula* (1460) *to the terms, and an equation in*

$y_{(n+2)x}$, $y_{(n+1)x}$, *and* y_{nx} *is obtained. Put* $x=0$ *in this; and employ the resulting formula to calculate in succession* y_{3x0}, y_{4x0}, *&c. in Maclaurin's expansion* (1507).

1535 Ex. $$e^{a\sin^{-1}x} = 1 + ax + \frac{a^2}{1.2}x^2 + \frac{a(a^2+1)}{1.2.3}x^3 + \frac{a^2(a^2+2^2)}{1.2.3.4}x^4 + \frac{a(a^2+1)(a^2+3^2)}{1.2.3.4.5}x^5 + \&c.$$

Obtained by Rule IV. Writing y for the function, the relation found is

$$(1-x^2)\,y_{2x} - xy_x - a^2y = 0.$$

Differentiating n times, by (1460), we get

$$(1-x^2)\,y_{(n+2)x} - (2n+1)\,xy_{(n+1)x} - (a^2+n^2)\,y_{nx} = 0.$$

Therefore $y_{(n+2)x0} = (a^2+n^2)\,y_{nx0}$, a formula which produces the coefficients in Maclaurin's expansion in succession when y_{x0} and y_{2x0} have been calculated.

1536 ARBOGAST'S METHOD OF EXPANDING $\phi(z)$,

where $$z = a + a_1x + \frac{a_2}{1.2}x^2 + \frac{a_3}{1.2.3}x^3 + \&c. \quad \text{......... (i.)}$$

Let $y = \phi(z)$. When $x = 0$, $y = \phi(a)$; therefore, by Maclaurin's theorem (1508),

$$y = \phi(a) + xy_{x0} + \frac{x^2}{1.2}y_{2x0} + \frac{x^3}{1.2.3}y_{3x0} + \&c. \quad \text{...... (ii.)}$$

Hence, in the values of y_x, y_{2x}, &c., at (1416), x has to be put $= 0$.

Now, when $x=0$, $z=a$; therefore y_z, y_{2z}, &c. become $\phi'(a)$, $\phi''(a)$, &c.; and z_{x0}, z_{2x0}, z_{3x0}, &c. become a_1, a_2, a_3, &c. Hence

$$y_{x0} = \phi'(a)\,a_1,$$
$$y_{2x0} = \phi''(a)\,a_1^2 + \phi'(a)\,a_2,$$
$$y_{3x0} = \phi'''(a)\,a_1^3 + 3\phi''(a)\,a_1a_2 + \phi'(a)\,a_3, \quad \&c.$$

1537 EXAMPLE.—To expand $\log(a+bx+cx^2+dx^3+\&c.)$.

Here $a_1 = b$, $a_2 = 2c$, $a_3 = 6d$, $\phi'(a) = \frac{1}{a}$, $\phi''(a) = -\frac{1}{a^2}$, $\phi'''(a) = \frac{2}{a^3}$.

Therefore $y_{x0} = \frac{b}{a}$, $y_{2x0} = -\frac{b^2}{a^2} + \frac{2c}{a}$, $y_{3x0} = \frac{2b^3}{a^3} - \frac{6bc}{a^2} + \frac{6d}{a}$.

Therefore, substituting in (ii.), we obtain as far as four terms,

$$\log(a+bx+cx^2+\dots) = \log a + \frac{b}{a}x + \left(\frac{c}{a} - \frac{b^2}{2a^2}\right)x^2 + \left(\frac{b^3}{3a^3} - \frac{bc}{a^2} + \frac{d}{a}\right)x^3 +$$

1538 Ex. 2.—To expand $(a+a_1x+a_2x^2+\ldots+a_nx^n)^r$ in powers of x.

Arbogast's method may be employed; otherwise, we may proceed as follows. Assume $(a_0+a_1x+a_2x^2+\ldots a_nx^n)^r = A_0+A_1x+A_2x^2+\ldots\ldots\ldots\ldots\ldots$

Differentiate for x; divide the equation by the result; clear of fractions, and equate coefficients of like powers of x.

BERNOULLI'S NUMBERS.

1539 $$\frac{x}{e^x-1} = 1-\frac{x}{2}+B_2\frac{x^2}{\lfloor 2}-B_4\frac{x^4}{\lfloor 4}+B_6\frac{x^6}{\lfloor 6}-\&c.,$$

where B_2, B_4, &c. are known as Bernoulli's numbers. Their values, as far as B_{18}, are

$$B_2=\frac{1}{6},\quad B_4=\frac{1}{30},\quad B_6=\frac{1}{42},\quad B_8=\frac{1}{30},\quad B_{10}=\frac{5}{66},$$

$$B_{12}=\frac{691}{2730},\quad B_{14}=\frac{7}{6},\quad B_{16}=\frac{3617}{510},\quad B_{18}=\frac{43867}{798}.$$

They are found in succession from the formula

1540 $$nB_{n-1}+C(n,2)\,B_{n-2}+C(n,3)\,B_{n-3}+\ldots$$
$$\ldots+C(n,2)\,B_2-\tfrac{1}{2}n+1=0,$$

the odd numbers B_3, B_5, &c. being all zero.

Proof.—Let $y=\dfrac{x}{e^x-1}$. Then, by (1508),

$$y = y_0+y_{x0}x+y_{2x0}\frac{x^2}{\lfloor 2}+y_{3x0}\frac{x^3}{\lfloor 3}+y_{4x0}\frac{x^4}{\lfloor 4}+\&c.$$

Here $y_{2rx0}=(-1)^{r+1}B_{2r}$. Now $y_0=1$ and $y_{x0}=-\frac{1}{2}$, by (1587). Also $ye^x=y+x$. Therefore, by (1460), differentiating n times,

$$e^x\{y_{nx}+ny_{(n-1)x}+C(n,2)\,y_{(n-2)x}+\ldots+ny_x+y\}=y_{nx}.$$

Therefore $$ny_{(n-1)x0}+C(n,2)\,y_{(n-2)x0}+\ldots+ny_{x0}+y_0=0.$$

Substitute B_{n-1}, B_{n-2}, &c., and we get the formula required.

B_3, B_5, B_7, &c. will all be found to vanish. It may be proved, *à priori*, that this will be the case: for

1541 $$\frac{x}{e^x-1}+\frac{x}{2}=\frac{x}{2}\,\frac{e^x+1}{e^x-1}.$$

Therefore the series (1539) wanting its second term is the expansion of the expression on the right. But that expression is an *even* function of x (1401); changing the sign of x does not alter its value. Therefore the series in question contains *no odd powers of* x after the first.

1542 The connexion between Bernoulli's numbers and the sums of the powers of the natural numbers in (276) is seen by expanding $(1-e^x)^{-1}$ in powers of e^x, and each term afterwards by the Exponential Theorem (150).

1543
$$\frac{x}{e^x+1} = \frac{x}{2} - B_2(2^2-1)\frac{x^2}{\lfloor\underline{2}} + B_4(2^4-1)\frac{x_4}{\lfloor\underline{4}} - B_6(2^6-1)\frac{x^6}{\lfloor\underline{6}} + \&c.$$

1544
$$\frac{e^x-1}{e^x+1} = 2\left\{B_2(2^2-1)\frac{x}{\lfloor\underline{2}} - B_4(2^4-1)\frac{x^3}{\lfloor\underline{4}} + B_6(2^6-1)\frac{x^5}{\lfloor\underline{6}} - \&c.\right\}.$$

PROOF. $\dfrac{x}{e^x+1} = \dfrac{x}{e^x-1} - \dfrac{2x}{e^{2x}-1}$ and $\dfrac{e^x-1}{e^x+1} = 1 - \dfrac{2}{x}\,\dfrac{x}{e^x+1}$, and by (1539).

1545 $$1+\frac{1}{2^{2n}}+\frac{1}{3^{2n}}+\frac{1}{4^{2n}}+\dots\dots = \frac{2^{2n-1}\pi^{2n}}{1.2\dots 2n}B_{2n}.$$

PROOF.—In the expansion of $\dfrac{x}{2}\,\dfrac{e^x+1}{e^x-1}$ (1540) substitute $2i\theta$ for x, and it becomes the expansion of $\theta\cot\theta$ (770). Obtain a second expansion by differentiating the logarithm of equation (815, $\sin\theta$ *in factors*). Expand each term of the result by the Binomial Theorem, and equate coefficients of like powers of θ in the two expansions.

STIRLING'S THEOREM.

1546 $$\phi(x+h)-\phi(x) = h\phi'(x)+A_1h\{\phi'(x+h)-\phi'(x)\} + A_2h^2\{\phi''(x+h)-\phi''(x)\}+\&c.,$$

where $A_{2n} = (-1)^n B_{2n} \div \lfloor\underline{2n}$ and $A_{2n+1} = 0$.

PROOF.—A_1, A_2, A_3, &c. are determined by expanding each function of $x+h$ by (1500), and then equating coefficients of like powers of x. Thus

$$\frac{1}{\lfloor\underline{2}} - A_1 = 0;\quad \frac{1}{\lfloor\underline{3}} - \frac{A_1}{\lfloor\underline{2}} - A_2 = 0;\quad \frac{1}{\lfloor\underline{4}} - \frac{A_1}{\lfloor\underline{3}} - \frac{A_2}{\lfloor\underline{2}} - A_3 = 0;\quad \&c.$$

To obtain the general relation between the coefficients: put $\phi(x) = e^x$, since A_1, A_2, &c. are independent of the form of ϕ. Equation (1546) then produces $$\frac{h}{e^h-1} = 1-A_1h-A_2h^2-A_3h^3-\&c.;$$

and, by (1539), we see that, for values of n greater than zero,
$$A_{2n+1} = 0 \quad\text{and}\quad A_{2n} = (-1)^n B_{2n} \div \lfloor\underline{2n}.$$

BOOLE'S THEOREM.

1547 $$\phi(x+h)-\phi(x) = A_1h\{\phi'(x+h)+\phi'(x)\} + A_2h^2\{\phi''(x+h)+\phi''(x)\}+\&c.$$

PROOF.—A_1, A_2, A_3, &c. are found by the same method as that employed in Stirling's Theorem.

For the general relation between the coefficients, as before, make $\phi(x) = e^x$, and equation (1547) then produces

$$\frac{e^h-1}{e^h+1} = A_1h+A_2h^2+A_3h^3+\&c.;$$

and, by comparing this with (1544), we see that

$$A_{2n} = 0 \quad \text{and} \quad A_{2n-1} = (-1)^{n-1}B_{2n}\frac{2^{2n}-1}{\lfloor 2n}.$$

EXPANSION OF IMPLICIT FUNCTIONS.

1550 Definition.—An equation $f(x, y) = 0$ constitutes y an *implicit* function of x. If y be obtained in terms of x by solving the equation, y becomes an *explicit* function of x.

1551 Lemma.—If y be a function of two independent variables x and z,

$$d_x\{F(y)\,y_z\} = d_z\{F(y)\,y_x\}.$$

Proof.—By performing the differentiations, we obtain

$$F'(y)\,y_xy_z+F(y)\,y_{zx} \quad \text{and} \quad F'(y)\,y_zy_x+F(y)\,y_{xz},$$

which are evidently equal, by (1482).

LAGRANGE'S THEOREM.

1552 Given $y = z+x\phi(y)$, the expansion of $u = f(y)$ in powers of x is

$$f(y) = f(z)+x\phi(z)f'(z)+\ldots+\frac{x^n}{\lfloor n}\frac{d^{n-1}}{dz^{n-1}}\left[\{\phi(z)\}^n f'(z)\right]+$$

Proof.—Expand u as a function of x, by (1507); thus, with the notation of (1406),

$$u = u_0+x\,u_{x0}+\frac{x^2}{\lfloor 2}u_{2x0}+\ \ldots\ +\frac{x^n}{\lfloor n}u_{nx0}+\&c.$$

Here u_0 is evidently $f(z)$.

Differentiating the equation $y = z+x\phi(y)$ for x and z in turn, we have

$$y_x = \phi(y)+x\phi'(y)\,y_x \quad \text{and} \quad y_z = 1+x\phi'(y)\,y_z.$$

Therefore $y_x = \phi(y)\,y_z$; and, since $u_x = f'(y)\,y_x$ and $u_z = f'(y)\,y_z$, therefore also $u_x = \phi(y)\,u_z$ .... (i.)

The following equation may now be proved by induction, equation (i.) being its form when $n = 1$.

Assume that $u_{nx} = d_{(n-1)z}[\{\phi(y)\}^n u_z]$ .... (ii.)

Therefore $u_{(n+1)x} = d_{(n-1)z}d_x[\{\phi(y)\}^n u_z]$ (1482)

$= d_{(n-1)z}d_z[\{\phi(y)\}^n u_x]$ (1551) $= d_{nz}[\{\phi(y)\}^{n+1}u_z]$, by (i.)

Thus, n becomes $n+1$. But equation (ii.) is true when $n=1$; for then it is equation (i.); therefore it is universally true.

Now, since in equations (i.) and (ii.) the differentiations on the right are all effected with respect to z, x may be made zero *before* differentiating instead of *after*. But, when $x=0$, $u_z=f'(z)$ and $\phi(y)=\phi(z)$, therefore equations (i.) and (ii.) give

$$u_{x0}=\phi(z)f'(z); \qquad u_{nx0}=d_{(n-1)z}[\{\phi(z)\}^n f'(z)].$$

1553 Ex. 1.—Given $y^3-ay+b=0$: to expand $\log y$ in powers of $\frac{1}{a}$.

Here $y=\frac{b}{a}+\frac{y^3}{a}$; therefore, in Lagrange's formula,

$$x=\frac{1}{a};\quad z=\frac{b}{a};\quad f(y)=\log y;\quad \phi(y)=y^3;\quad \text{and}\quad y=z+xy^3.$$

Therefore $$u_0=\log z;\quad u_{x0}=z^3\frac{1}{z}=z^2;$$

$$u_{nx0}=d_{(n-1)z}\left(z^{3n}\frac{1}{z}\right)=(3n-1)(3n-2)\ldots(2n+1)z^{2n}.$$

Therefore, substituting the values of x and z, (1552) becomes

$$\log y=\log\frac{b}{a}+\frac{b^2}{a^2}\frac{1}{a}+\ldots+\frac{(3n-1)(3n-2)\ldots(2n+1)}{1.2\ldots n}\frac{b^{2n}}{a^{2n}}\frac{1}{a^n}+\ldots\ldots$$

1554 Ex. 2.—Given the same equation: to expand y^n in powers of $\frac{1}{a}$.

$f(y)$ is now y^n, and, proceeding as in the last example, we find

$$y^n=\frac{b^n}{a^n}\left\{1+n\frac{b^2}{a^2}\frac{1}{a}+\frac{n(n+5)}{1.2}\frac{b^4}{a^4}\frac{1}{a^2}+\frac{n(n+7)(n+8)}{1.2.3}\frac{b^6}{a^6}\frac{1}{a^3}\right.$$
$$\left.+\frac{n(n+9)(n+10)(n+11)}{1.2.3.4}\frac{b^8}{a^8}\frac{1}{a^4}+\&c.\right\}$$

If $n=1$, $$y=\frac{b}{a}\left(1+\frac{b^2}{a^2}\frac{1}{a}+3\frac{b^4}{a^4}\frac{1}{a^2}+12\frac{b^6}{a^6}\frac{1}{a^3}+55\frac{b^8}{a^8}\frac{1}{a^4}+\&c.\ \ldots\ldots\right.$$

CAYLEY'S SERIES FOR $\frac{1}{\phi(z)}$.

1555
$$\frac{1}{\phi(z)}=A-\frac{A^2z}{1.2}[z\phi(z)]_{2z}+\ldots+\frac{A^n(-z)^{n-1}}{1.2\ldots n}\left[z\{\phi(z)\}^{n-1}\right]_{nz}+\ldots,$$

where $A=\frac{1}{\phi(0)}$.

Proof.—Differentiate Lagrange's expansion (1552) for z, noting that $\frac{dy}{dz}=\frac{1}{1-x\phi'(y)}$. Replace x by $\frac{y-z}{\phi(y)}$. Put $f'(y)=\frac{y}{\phi(y)}$; and therefore $f'(z)=\frac{z}{\phi(z)}$, since f is an arbitrary function. Then make $y=0$.

LAPLACE'S THEOREM.

1556 To expand $f(y)$ in powers of x when

$$y = F\{z + x\phi(y)\}.$$

RULE.—*Proceed as in Lagrange's Theorem, merely substituting* F (z) *for* z *in the formula.*

1557 Ex. 3.—To expand e^y in powers of x when $y = \log(z + x\sin y)$.

Here $\qquad f(y) = e^y; \quad F(z) = \log z; \quad \phi(y) = \sin y;$

In the value of u_{nx0} (1552), $\phi(z)$ becomes $\phi\{F(z)\} = \sin\log z$; $f(z)$ becomes $f\{F(z)\} = e^{\log z} = z$; therefore $f'(z) = 1$.

Thus the expansion becomes

$$e^y = z + x\sin\log z + \ldots + \frac{x^n}{\underline{|n}}\, d_{(n-1)z}(\sin\log z)^n.$$

1558 Ex. 4.—Given $\sin y = x\sin(y+a)$: to expand y in powers of x.

Here $\qquad y = \sin^{-1}(x\sin\overline{y+a}), \quad$ with $z = 0$.

$$f(y) = y; \quad F(z) = \sin^{-1} z; \quad \phi(y) = \sin(y+a).$$

$\phi(z)$ in (1552) becomes $\phi\{F(z)\} = \sin(\sin^{-1}z + a)$.

$f(z)$ becomes $f\{F(z)\} = F(z) = \sin^{-1}z$; therefore $F'(z) = (1-z^2)^{-\frac{1}{2}}$.

Thus $\qquad y = x\sin(\sin^{-1}z + a)(1-z^2)^{-\frac{1}{2}}$

$$+\tfrac{1}{2}x^2 d_z\{\sin^2(\sin^{-1}z+a)(1-z^2)^{-\frac{1}{2}}\} + \tfrac{1}{6}x^3 d_{2z}\{\sin^3(\sin^{-1}z+a)(1-z^2)^{-\frac{1}{2}}\} +$$

with z put $=0$ after differentiating. The result is, as in (796),

$$y = x\sin a + \tfrac{1}{2}x^2\sin 2a + \tfrac{1}{3}x^3\sin 3a + \&c.$$

BURMANN'S THEOREM.

1559 To expand one function $f(y)$ in powers of another function $\psi(y)$.

RULE.—*Put* $x = \psi(y)$ *in Lagrange's expansion, and therefore* $\phi(y) = (y-z) \div \psi(y)$; *therefore*

1560 $$f(y) = f(z) + \psi(y)\left\{\frac{y-z}{\psi(y)} f'(y)\right\}_{y=z} + \ldots$$

$$\ldots + \frac{\{\psi(y)\}^n}{\underline{|n}} \frac{d^{n-1}}{dy^{n-1}}\left\{\left(\frac{y-z}{\psi(y)}\right)^n f'(y)\right\}_{y=z} + \&c.$$

Here $y = z$ signifies that *after* differentiating z is to be substituted for y.

1561 COR. 1.—Since $x=\psi(y)$, $y=\psi^{-1}(x)$; therefore (1560) becomes, by writing x for $\psi(y)$,

$$f\{\psi^{-1}(x)\}=f(z)+\ldots+\frac{x^n}{\lfloor n}\frac{d^{n-1}}{dy^{n-1}}\left\{\left(\frac{y-z}{\psi(y)}\right)^n f'(y)\right\}_{,y=z}+\ldots$$

But since the variable y is changed into z after differentiating, it is immaterial what letter is written for y in the second factor of the general term.

1562 COR. 2.—If $f(y)$ be simply y, the equation becomes

$$\psi^{-1}(x)=z+x\left(\frac{y-z}{\psi(y)}\right)_{,y=z}+\ldots+\frac{x^n}{\lfloor n}\frac{d^{n-1}}{dy^{n-1}}\left\{\left(\frac{y-z}{\psi(y)}\right)^n\right\}_{,y=z}+$$

1563 COR. 3.—If $z=0$, so that $y=x\phi(y)$, we obtain the expansion of an inverse function,

$$\psi^{-1}(x)=x\left(\frac{y}{\psi(y)}\right)_{,y=0}+\ldots+\frac{x^n}{\lfloor n}\frac{d^{n-1}}{dy^{n-1}}\left\{\left(\frac{y}{\psi(y)}\right)^n\right\}_{,y=0}+$$

1564 Ex. 5.—The series (1528) for $\sin^{-1}x$ may be obtained by this formula; thus,

Let $\sin^{-1}x=y$, therefore $x=\sin y=\psi(y)$, in (1563); therefore

$$\sin^{-1}x=x\left(\frac{y}{\sin y}\right)_{,\,y=0}+\frac{x^2}{1.2}\left(\frac{y^2}{\sin^2 y}\right)_{y0}+\frac{x^3}{1.2.3}\left(\frac{y^3}{\sin^3 y}\right)_{2y0}+\text{\&c.}$$

1565 Ex. 6.—If $y=\dfrac{x}{1+\sqrt{(1-x^2)}}=\dfrac{1-\sqrt{(1-x^2)}}{x}$, then, by Lagrange's theorem (1552), since $y=\dfrac{x}{2}+\dfrac{y^2}{2}x$, we find

$$y^n=\left(\frac{x}{2}\right)^n+n\left(\frac{x}{2}\right)^{n+2}+\ldots+\frac{n\,\lfloor n+2r-1}{\lfloor r\,\lfloor n+r}\left(\frac{x}{2}\right)^{n+2r}+\ \ldots\ldots\ldots\ldots$$

Put $x=2\sqrt{t}$, thus

1566 $$\left(\frac{1-\sqrt{(1-4t)}}{2}\right)^n=t^n+nt^{n+1}+\ldots+\frac{n\,\lfloor n+2r-1}{\lfloor r\,\lfloor n+r}t^{n+r}+\ \ldots\ldots\ldots\ldots$$

Change the sign of n, thus

1567 $$\left(\frac{1+\sqrt{(1-4t)}}{2}\right)^n=1-nt+\ldots(-1)^r\frac{n\,\lfloor n-r-1}{\lfloor r\,\lfloor n-2r}t^r\pm\ \ldots\ldots\ldots\ldots$$

This last series, continued to $\dfrac{n}{2}+1$ or $\dfrac{n+1}{2}$ terms, according as n is even or odd, is equal to the sum of the two series, as appears by the Binomial theorem.

Also, by Lagrange's Theorem,

1568 $$\log y = \log \frac{x}{2} + \left(\frac{x}{2}\right)^2 + \ldots + \frac{\lfloor 2r-1}{\lfloor r \lfloor r}\left(\frac{x}{2}\right)^{2r} + \ldots\ldots$$

or, by putting $x = 2\sqrt{t}$,

1569 $$\log \frac{1-\sqrt{(1-4t)}}{2t} = t + \ldots + \frac{\lfloor 2r-1}{\lfloor r \lfloor r} t^r + \ldots\ldots$$

1570 Ex. 7.—Given $xy = \log y$; to expand y in powers of x.

The equation can be adapted as follows:

$$y = e^{xy}, \qquad \text{therefore} \quad xy = xe^{xy}.$$

Put $xy = y'$, therefore $y' = xe^{y'}$, from which, by putting $z = 0$ in (1552), y' may be expanded, and therefore y.

Ex. 8.—To expand e^{ay} in powers of ye^{by}.

Here $x = ye^{by}$, $\phi(y) = \frac{y-z}{ye^{by}} = e^{-by}$, if we take $z = 0$. Therefore

1571 $$e^{ay} = 1 + aye^{by} + a(a-2b)\frac{y^2e^{2by}}{1.2} + a(a-3b)^2\frac{y^3e^{3by}}{1.2.3} + \ldots\ldots$$

ABEL'S THEOREM.

1572 If $\phi(x)$ be a function developable in powers of e^x; then

$$\phi(x+a) = \phi(x) + a\phi'(x+b) + \frac{a(a-2b)}{1.2}\phi''(x+2b) + \ldots$$
$$\ldots\ldots + \frac{a(a-rb)^{r-1}}{1.2\ldots r}\phi^r(x+rb) +$$

PROOF.—Let $\phi(y) = A_0 + A_1e^y + A_2e^{2y} + A_3e^{3y} + \ldots\ldots$ (i.)

Put $y = 0, 1, 2, 3$, &c. in (1571), and multiply the results respectively by A_0, A_1e^x, A_2e^{2x}, &c. Then the theorem is proved by equation (i.).

1573 COR.—If $\phi(x) = x^n$, Abel's formula gives

$$(x+a)^n = x^n + na(x+b)^{n-1} + C(n,2)\,a(a-2b)(x+2b)^{n-2} + \ldots$$
$$\ldots\ldots + C(n,r)\,a(a-rb)^{r-1}(x+rb)^{n-r} + \&c.$$

INDETERMINATE FORMS.

1580 Forms $\frac{0}{0}$, $\frac{\infty}{\infty}$. RULE.—*If* $\frac{\phi(\mathrm{x})}{\psi(\mathrm{x})}$ *be a fraction which takes either of these forms when* $\mathrm{x} = \mathrm{a}$; *then* $\frac{\phi(\mathrm{a})}{\psi(\mathrm{a})} = \frac{\phi'(\mathrm{a})}{\psi'(\mathrm{a})}$ or $\frac{\phi^n(\mathrm{a})}{\psi^n(\mathrm{a})}$, *the first determinate fraction obtained by differentiating* *the numerator and denominator simultaneously and substituting* a *for* x *in the result.*

1581 But at any stage of the process the fraction may be reduced to its simplest form before the next differentiation. See example (1589).

PROOF.—(i.) By Taylor's theorem (1500), since $\phi(a) = 0 = \psi(a)$,

$$\frac{\phi(a+h)}{\psi(a+h)} = \frac{\phi(a)+h\phi'(a+\theta h)}{\psi(a)+h\psi'(a+\theta h)} = \frac{\phi'(a+\theta h)}{\psi'(a+\theta h)} = \frac{\phi'(a)}{\psi'(a)},$$

when h vanishes.

(ii.) If $\phi(a) = \psi(a) = \infty$, $\quad \dfrac{\phi(a)}{\psi(a)} = \dfrac{1}{\psi(a)} \div \dfrac{1}{\phi(a)}$,

which is of the first form, and therefore

$$= \frac{\psi'(a)}{\{\psi(a)\}^2} \div \frac{\phi'(a)}{\{\phi(a)\}^2} \;(1414)\; = \frac{\{\phi(a)\}^2}{\{\psi(a)\}^2}\,\frac{\psi'(a)}{\phi'(a)}. \quad \text{Therefore } \frac{\phi(a)}{\psi(a)} = \frac{\phi'(a)}{\psi'(a)}.$$

1582 Vanishing fractions in Algebra are of the indeterminate form just considered, and may be evaluated by the rule, or by rejecting the vanishing factor common to the numerator and denominator.

Ex.—When $x = a$; $\quad \dfrac{x^3-a^3}{x^2-a^2} = \dfrac{0}{0} = \dfrac{(x-a)(x^2+ax+a^2)}{(x-a)(x+a)} = \dfrac{3a^2}{2a} = \dfrac{3}{2}a.$

1583 Form $0 \times \infty$. RULE.—*If* $\phi(x) \times \psi(x)$ *takes this form when* $x = a$, *put* $\phi(a) \times \psi(a) = \phi(a) \div \dfrac{1}{\psi(a)}$, *which is of the form* $\dfrac{0}{0}$.

1584 Forms 0^0, ∞^0, 1^∞. RULE.—*If* $\{\phi(x)\}^{\psi(x)}$ *takes any of these forms when* $x = a$, *find the limit of the logarithm of the expression. For the logarithm* $= \psi(a) \log \phi(a)$, *which, in each case, is of the form* $0 \times \infty$.

1585 Form $\infty - \infty$. RULE.—*If* $\phi(x) - \psi(x)$ *takes this form when* $x = a$, *we have* $e^{\phi(a)-\psi(a)} = \dfrac{e^{-\psi(a)}}{e^{-\phi(a)}} = \dfrac{0}{0}$; *and if the value of this expression be found to be* c, *by* (1580), *the required value will be* log c.

1586 *Otherwise:* $\phi(a) - \psi(a) = \phi(a)\left\{1 - \dfrac{\psi(a)}{\phi(a)}\right\}$, *which is of the form* $\infty \times 0$ (1583).

1587 Ex. 1.—With $x=0$, $y = \frac{x}{e^x-1} = \frac{0}{0} = \frac{1}{e^x}$ (1580) $= 1$.

Also, with $x=0$,

$$y_{x0} = \frac{e^x-1-xe^x}{(e^x-1)^2} = \frac{0}{0} = \frac{e^x-e^x-xe^x}{2(e^x-1)e^x} = \frac{-x}{2(e^x-1)} = \frac{0}{0} = -\frac{1}{2e^x} = -\frac{1}{2}.$$

1588 Ex. 2.—With $x=1$;

$$\cot(\pi x)\log x = \frac{\log x}{\tan(\pi x)} = \frac{0}{0} = \frac{\cos^2(\pi x)}{\pi x} \text{ (1580) } = \frac{1}{\pi}.$$

1589 Ex. 3.—With $x=0$;

$$x^m(\log x)^n = \frac{(\log x)^n}{x^{-m}} = \frac{\infty}{\infty} = \frac{n(\log x)^{n-1}}{-mx^{-m}} = \frac{\lfloor n}{(-m)^n x^{-m}} = 0,$$

by (1581), differentiating n times and reducing the fraction to its simplest form after each differentiation.

1590 Ex. 4.—With $x=0$; $y = (1+ax)^{\frac{1}{x}} = 1^\infty$.

By (1584), $\log y = \frac{\log(1+ax)}{x} = \frac{0}{0} = \frac{a}{1+ax}$ (1580) $= a$; $\therefore\ y = e^a$.

1591 Ex. 5.—With $x=\pi$; $y = (\pi-x)^{\sin x} = 0^0$.

By (1584), $\log y = \sin x \log(\pi-x) = \frac{\log(\pi-x)}{\operatorname{cosec} x} = \frac{-\infty}{\infty} = \frac{\sin^2 x}{(\pi-x)\cos x}$

$$\text{(1580)} = \frac{0}{0} = \frac{2\sin x\cos x}{-\cos x-(\pi-x)\sin x} = 0; \quad \therefore\ y = 1.$$

1592 If $f(x)$ and x become infinite together, then

$$\frac{f(x)}{x} = f'(x) = f(x+1) - f(x).$$

PROOF.— $\frac{f(x)}{x} = \frac{\infty}{\infty} = \frac{f'(x)}{1}$ (1580) $= \frac{f(x+1)-f(x)}{1}$ (1404), since, when $x=\infty$, h may be taken $=1$.

Indeterminate forms involving two variables.

1592 RULE.—*First: If the values* x=a, y=b *make the fraction* $\frac{\phi(x,y)}{\psi(x,y)} = \frac{0}{0}$; *the true value is* $= \frac{\phi_x}{\psi_x}$, *if* ϕ_y *and* ψ_y *both vanish.*

1593 *Secondly: If* $\phi_x : \psi_x = \phi_y : \psi_y = \mathrm{k}$, *the true value of the fraction is* k.

PROOF.—(i.) By (1703) $\frac{\phi(x,y)}{\psi(x,y)} = \frac{\phi_x+\phi_y y_x}{\psi_x+\psi_y y_x}$, and y being an *arbitrary* function of x,—that is, independent of x,—the value of the fraction is indeterminate unless ϕ_y and ψ_y both vanish.

(ii.) If we substitute $\phi_x = k\psi_x$ and $\phi_y = k\psi_y$, the fraction becomes $= k$.

JACOBIANS.

1600 Let u, v, w be n functions of n variables x, y, z $(n=3)$. The following determinant notation is adopted:—

$$\begin{vmatrix} u_x & u_y & u_z \\ v_x & v_y & v_z \\ w_x & w_y & w_z \end{vmatrix} \equiv \frac{d\,(uvw)}{d\,(xyz)}, \qquad \begin{vmatrix} x_u & x_v & x_w \\ y_u & y_v & y_w \\ z_u & z_v & z_w \end{vmatrix} \equiv \frac{d\,(xyz)}{d\,(uvw)}.$$

The first determinant is called the *Jacobian* of u, v, w with respect to x, y, z, and is also denoted by $J\,(uvw)$, or simply by J.

1601 THEOREM.— $\dfrac{d\,(uvw)}{d\,(xyz)} \times \dfrac{d\,(xyz)}{d\,(uvw)} = 1.$

PROOF.—If the product of the two determinants be formed by the rule in (570), first changing the columns into rows in the second determinant (559), the first column of the resulting determinant will be

$$\left.\begin{array}{l} u_x x_u + u_y y_u + u_z z_u = u_u \\ u_x x_v + u_y y_v + u_z z_v = u_v \\ u_x x_w + u_y y_w + u_z z_w = u_w \end{array}\right\}, \begin{array}{l}\text{and the whole}\\ \text{determinant}\\ \text{will be}\end{array} \begin{vmatrix} u_u & v_u & w_u \\ u_v & v_v & w_v \\ u_w & v_w & w_w \end{vmatrix} = \begin{vmatrix} 1 & 0 & 0 \\ 0 & 1 & 0 \\ 0 & 0 & 1 \end{vmatrix} = 1.$$

1602 If u, v, w are n functions of n variables α, β, γ $(n=3)$, and α, β, γ, functions of x, y, z;

$$\frac{d\,(uvw)}{d\,(\alpha\beta\gamma)} \times \frac{d\,(\alpha\beta\gamma)}{d\,(xyz)} = \frac{d\,(uvw)}{d\,(xyz)}.$$

PROOF.—Form the product of the two determinants, changing columns into rows in the second as in (1601). The first column of the resulting determinant will be

$$\left.\begin{array}{l} u_\alpha \alpha_x + u_\beta \beta_x + u_\gamma \gamma_x = u_x \\ u_\alpha \alpha_y + u_\beta \beta_y + u_\gamma \gamma_y = u_y \\ u_\alpha \alpha_z + u_\beta \beta_z + u_\gamma \gamma_z = u_z \end{array}\right\}, \begin{array}{l}\text{and the whole de-}\\ \text{terminant will}\\ \text{be}\end{array} \begin{vmatrix} u_x & v_x & w_x \\ u_y & v_y & w_y \\ u_z & v_z & w_z \end{vmatrix} = \frac{d\,(uvw)}{d\,(xyz)},$$

since rows and columns may be transposed (559).

1603 COR.—If α, β, γ are only given as implicit functions of x, y, z, by the equations $\phi=0$, $\chi=0$, $\psi=0$, involving the six variables; then

$$\frac{d\,(\phi\chi\psi)}{d\,(\alpha\beta\gamma)} \times \frac{d\,(\alpha\beta\gamma)}{d\,(xyz)} = (-1)^3\,\frac{d\,(\phi\chi\psi)}{d\,(xyz)}.$$

PROOF.—By (1737), $\phi_\alpha \alpha_x + \phi_\beta \beta_x + \phi_\gamma \gamma_x = -\phi_x$, where ϕ_x is the partial derivative of ϕ. Thus u_x, v_x, w_x, in the determinant, are now replaced by $-\phi_x$, $-\chi_x$, $-\psi_x$; so with y and z; and by changing the sign of each element, the factor $(-1)^3$ is introduced (562).

1604 If u, v, w, n functions of n variables x, y, z $(n=3)$, be transformed into functions of ξ, η, ζ by the linear substitutions

$$\left.\begin{array}{l} x = a_1\xi + a_2\eta + a_3\zeta \\ y = b_1\xi + b_2\eta + b_3\zeta \\ z = c_1\xi + c_2\eta + c_3\zeta \end{array}\right\};\quad \text{then}\quad \frac{d\,(uvw)}{d\,(\xi\eta\zeta)} = M\,\frac{d\,(uvw)}{d\,(xyz)},\quad \text{or}\quad J' = MJ,$$

where M is the determinant $(a_1 b_2 c_3)$ called the *modulus* of transformation (573).

PROOF.— $J = \begin{vmatrix} u_x & u_y & u_z \\ v_x & v_y & v_z \\ w_x & w_y & w_z \end{vmatrix}$ $M = \begin{vmatrix} a_1 & b_1 & c_1 \\ a_2 & b_2 & c_2 \\ a_3 & b_3 & c_3 \end{vmatrix}$ $J' = \begin{vmatrix} u_\xi & u_\eta & u_\zeta \\ v_\xi & v_\eta & v_\zeta \\ w_\xi & w_\eta & w_\zeta \end{vmatrix}$

Form the product MJ by the rule in (570). The first element of the resulting determinant is $u_x a_1 + u_y b_1 + u_z c_1 = u_x x_\xi + u_y y_\xi + u_z z_\xi = u_\xi$. Similarly for each element. Then transpose rows and columns, and the determinant J' is obtained.

1605 When the modulus is unity, the transformation is said to be *unimodular*.

1606 If, in (1600), $\phi\,(uvw) = 0$, where ϕ is some function; then $J\,(uvw) = 0$; and conversely.

PROOF.—Differentiate ϕ for x, y, and z separately, thus

$$\phi_u u_x + \phi_v v_x + \phi_w w_x = 0;$$

similarly y and z; and the eliminant of the three equations is $J\,(uvw) = 0$.

1607 If $u=0$, $v=0$, $w=0$ be a number of homogeneous equations of dimensions m, n, p in the same number of variables x, y, z; then $J\,(uvw)$ vanishes, and if the dimensions are equal J_x, J_y, J_z also vanish.

PROOF.—By (1624), $\left.\begin{array}{l} xu_x + yu_y + zu_z = mu \\ xv_x + yv_y + zv_z = nv \\ xw_x + yw_y + zw_z = pw \end{array}\right\};\quad \therefore Jx = A_1 mu + B_1 nv + C_1 pw.$

By (582), A_1, B_1, C_1 being the minors of the first column of J. Therefore, if u, v, w vanish, J also vanishes.

Again, differentiating the last equation, $J + xJ_x = A_1 mu_x + B_1 nv_x + C_1 pw_x$. Therefore, if $m = n = p$, $J + xJ_x = m\,(A_1 u_x + B_1 v_x + C_1 w_x) = mJ$. Therefore J_x vanishes when J does.

1608 If $u=0$, $v=0$, $w=0$ are three homogeneous equations of the second degree in x, y, z, their eliminant will be the determinant of the sixth order formed by taking the eliminant of the six equations u, v, w, J_x, J_y, J_z.

PROOF.—J is of the third degree, and therefore J_x, J_y, J_z are of the second degree, and they vanish because u, v, w vanish, by (1607). Hence u, v, w, J_x, J_y, J_z form six equations of the form $(x, y, z)^2 = 0$.

1609 If n variables x, y, z $(n=3)$ are connected with n other variables ξ, η, ζ, by as many equations $u=0, v=0, w=0$; then

$$\frac{dx}{d\xi}\frac{dy}{d\eta}\frac{dz}{d\zeta} = \frac{d(uvw)}{d(\xi\eta\zeta)} \div \frac{d(uvw)}{d(xyz)}.$$

PROOF.—By (562) we have

$$\frac{dx}{d\xi}\frac{dy}{d\eta}\frac{dz}{d\zeta}\begin{vmatrix} u_x & u_y & u_z \\ v_x & v_y & v_z \\ w_x & w_y & w_z \end{vmatrix} = \begin{vmatrix} u_x x_\xi & u_y y_\eta & u_z z_\zeta \\ v_x x_\xi & v_y y_\eta & v_z z_\zeta \\ w_x x_\xi & w_y y_\eta & w_z z_\zeta \end{vmatrix} = \begin{vmatrix} u_\xi & u_\eta & u_\zeta \\ v_\xi & v_\eta & v_\zeta \\ w_\xi & w_\eta & w_\zeta \end{vmatrix}$$

QUANTICS.

1620 DEFINITION.—A ***Quantic*** is a homogeneous function of any number of variables: if of *two*, *three* variables, &c., it is called a *binary*, *ternary* quantic, &c. The following will illustrate the notation in use. The binary quantic

$$ax^3+3bx^2y+3cxy^2+dy^3$$

is denoted by $(a, b, c, d \between x, y)^3$ when the numerical coefficients are those in the expansion of $(x+y)^3$. When the numerical coefficients are all unity, the same quantic is written $(a, b, c, d \between x, y)^3$. When the coefficients are not mentioned, the notation $(x, y)^3$ is employed.

EULER'S THEOREM OF QUANTICS.

If $u=f(x, y)$ be a binary quantic of the nth degree, then

1621 $$xu_x+yu_y = nu.$$

1622 $$x^2u_{2x}+2xyu_{xy}+y^2u_{2y} = n(n-1)\,u.$$

$$\cdots \quad \cdots \quad \cdots \quad \cdots$$

1623 $$(xd_x+yd_y)^r\,u = n(n-1)\,\ldots\,(n-r+1)\,u. \qquad (1492)$$

PROOF.—In (1512) put $h=\alpha x$, $k=\alpha y$; then, because the function is homogeneous, the equation becomes

$$(1+\alpha)^n\,u = u+\alpha\,(xu_x+yu_y)+\tfrac{1}{2}\alpha^2\,(x^2u_{2x}+2xyu_{xy}+y^2u_{2y})+\ \ldots\ldots\ldots\ldots$$

Expand $(1+\alpha)^n$, and equate coefficients of powers of α.

The theorem may be extended to *any* quantic, the quantities on the right remaining unaltered. Thus, in a ternary quantic u of the nth degree,

1624 $$xu_x + yu_y + zu_z = nu; \quad \text{and generally}$$

1625 $$(xd_x + yd_y + zd_z)^r u = n(n-1)\ldots(n-r+1)u.$$

DEFINITIONS.

1626 The ***Eliminant*** of n quantics in n variables is the function of the coefficients obtained by putting all the quantics equal to zero and eliminating the variables (583, 586).

1627 The ***Discriminant*** of a quantic is the eliminant of its first derivatives with respect to each of the variables (Ex. 1631).

1628 An ***Invariant*** is a function of the coefficients of an equation whose value is not altered by linear transformation of the equation, excepting that the function is multiplied by the modulus of transformation (Ex. 1632).

1629 A ***Covariant*** is a quantic derived from another quantic, and such that, when both are subjected to the same linear transformation, the resulting quantics are connected by the same process of derivation (Ex. 1634).

1630 A ***Hessian*** is the Jacobian of the first derivatives of a function.

Thus, the Hessian of a ternary quantic u, whose first derivatives are u_x, u_y, u_z, is

$$\frac{d(u_x u_y u_z)}{d(xyz)} = \begin{vmatrix} u_{2x} & u_{xy} & u_{xz} \\ u_{yx} & u_{2y} & u_{yz} \\ u_{zx} & u_{zy} & u_{2z} \end{vmatrix}.$$

1631 Ex.—Take the *binary cubic* $u = ax^3 + 3bx^2y + 3cxy^2 + dy^3$. Its first derivatives are

$$u_x = 3ax^2 + 6bxy + 3cy^2,$$
$$u_y = 3bx^2 + 6cxy + 3dy^2.$$

Therefore (1627) the *discriminant* of u is the annexed determinant, by (587).

$$\begin{vmatrix} 3a & 6b & 3c & 0 \\ 0 & 3a & 6b & 3c \\ 3b & 6c & 3d & 0 \\ 0 & 3b & 6c & 3d \end{vmatrix}$$

1632 The determinant is also an invariant of u, by (1638); that is, if u be transformed into v by putting $x = \alpha\xi + \beta\eta$ and $y = \alpha'\xi + \beta'\eta$; and, if a corresponding determinant be formed with the coefficients of v, the new determinant will be equal to the original one multiplied by $(\alpha\beta' - \alpha'\beta)^8$.

1633 Again, $u_{2x} = 6ax+6by,\quad u_{2y} = 6cx+6dy,\quad u_{xy} = 6bx+6cy.$
Therefore, by (1630), the Hessian of u is

$$u_{2x}u_{2y}-u_{xy}^2 = (ax+by)(cx+dy)-(bx+cy)^2$$
$$= (ac-b^2)\,x^2+(ad-bc)\,xy+(bd-c^2)\,y^2.$$

1634 And this is also a *covariant;* for, if u be transformed into v, as before, then the result of transforming the Hessian by the same equations will be found to be equal to $v_{2x}v_{2y}-v_{xy}^2$. See (1652).

1635 If a quantic, $u=f(x, y, z\,...)$, involving n variables can be expressed as a function of the second degree in $X_1, X_2\,...\,X_{n-1}$, where the latter are linear functions of the variables, the discriminant vanishes.

PROOF.—Let $u = \phi X_1^2+\psi X_1X_2+\chi X_1X_3+\&c.,$
where $X_1 = a_1x+b_1y+c_1z+\&c.$
The derivatives u_x, u_y, &c. must contain one of the factors $X_1X_2\,...\,X_{n-1}$ in every term, and therefore must have, for common roots, the roots of the simultaneous equations $X_1 = 0,\ X_2 = 0,\ ...\ X_{n-1} = 0$; $n-1$ equations being required to determine the ratios of the n variables. Therefore the discriminant of u, which (1627) is the eliminant of the equations $u_x=0$, $u_y=0$, &c., vanishes, by (588).

1636 COR. 1.—If a binary quantic contains a square factor, the discriminant vanishes; and conversely.

Thus, in Example (1631), if u has a factor of the form $(Ax+By)^2$, the determinant there written vanishes.

1637 COR. 2.—If any quadric is resolvable into two factors, the discriminant vanishes.

An independent proof is as follows:—
Let $u = XY$ be the quadric, where

$$X = (ax+by+cz+...),\qquad Y = (a'x+b'y+c'z+...).$$

The derivatives u_x, u_y, u_z are each of the form $pX+qY$, and therefore have for common roots the roots of the simultaneous equations $X=0$, $Y=0$. Therefore the eliminant of $u_x = 0$, $u_y = 0$, &c. vanishes (1627).

1638 The discriminant of a binary quantic is an invariant.

PROOF.—A square factor remains a square factor after linear transformation. Hence, by (1636), if the discriminant vanishes, the discriminant of the transformed equation vanishes, and must therefore contain the former discriminant as a *factor* (see 1628). Thus the determinant in (1631) is an invariant of the quantic u.

The discriminant of the ternary quadric

1639 $$u = ax^2+by^2+cz^2+2fyz+2gzx+2hxy$$

is the eliminant of the equations

1640 $\left.\begin{array}{l}\frac{1}{2}u_x = ax+hy+gz = 0 \\ \frac{1}{2}u_y = hx+by+fz = 0 \\ \frac{1}{2}u_z = gx+fy+cz = 0\end{array}\right\}$; that is, the determinant $\begin{vmatrix} a & h & g \\ h & b & f \\ g & f & c \end{vmatrix}$

1641 $= abc+2fgh-af^2-bg^2-ch^2 \equiv \Delta.$

1642 The following notation will frequently be employed.

The determinant will be denoted by Δ, and its minors by A, B, C, F, G, H. Their values are readily found by differentiating Δ with respect to a, b, c, f, g, h; thus,

$$A = bc-f^2 = \Delta_a, \quad B = ca-g^2 = \Delta_b, \quad C = ab-h^2 = \Delta_c,$$
$$F = gh-af = \tfrac{1}{2}\Delta_f, \quad G = hf-bg = \tfrac{1}{2}\Delta_g, \quad H = fg-ch = \tfrac{1}{2}\Delta_h.$$

1643 The reciprocal determinant is equal to Δ^2 or

$$\begin{vmatrix} A & H & G \\ H & B & F \\ G & F & C \end{vmatrix} = \begin{vmatrix} a & h & g \\ h & b & f \\ g & f & c \end{vmatrix}^2$$ By (575).

1644 The discriminant of the quaternary quadric

$$u = ax^2+by^2+cz^2+dw^2+2fyz+2gzx+2hxy$$
$$+2pxw+2qyw+2rzw$$

is the eliminant of the equations

1645 $\left.\begin{array}{l}\frac{1}{2}u_x = ax+hy+gz+pw = 0 \\ \frac{1}{2}u_y = hx+by+fz+qw = 0 \\ \frac{1}{2}u_z = gx+fy+cz+rw = 0 \\ \frac{1}{2}u_w = px+qy+rz+dw = 0\end{array}\right\}$; that is, the determinant $\begin{vmatrix} a & h & g & p \\ h & b & f & q \\ g & f & c & r \\ p & q & r & d \end{vmatrix}$

The determinant will be denoted by Δ', and, by decomposing it by (568), we have

1646 $\Delta' = d.\Delta-Ap^2-Bq^2-Cr^2-2Fqr-2Grp-2Hpq.$

1647 $\Delta' = \frac{1}{2}(p\Delta'_p+q\Delta'_q+r\Delta'_r)+d.\Delta.$

1648 THEOREM.—If $\phi(xy)$ be a quantic of an even degree,

$$\phi(d_y, -d_x)\,\phi(x, y)$$

is an invariant of the quantic.

PROOF.—Let the linear substitutions (1628) be

$$x = a\xi + b\eta, \qquad y = a'\xi + b'\eta \quad \text{......................(i.)}$$

Solve for ξ and η. Find ξ_x, ξ_y, η_x, η_y, and substitute in the two equations

$$d_y = d_\xi \xi_y + d_\eta \eta_y; \qquad d_x = d_\xi \xi_x + d_\eta \eta_x.$$

The result is

$$d_y = \{ad_\eta + b(-d_\xi)\} \div M; \quad -d_x = \{a'd_\eta + b'(-d_\xi)\} \div M \text{......(ii.)},$$

where $M = ab' - a'b$, the modulus of transformation. Equations (i.) and (ii.) are parallel, and show that the operations d_y and $-d_x$ can be transformed in the same way as the quantities x and y; that is, if $\phi(x, y)$ becomes $\psi(\xi, \eta)$, then $\phi(d_y, -d_x)$ becomes $\psi(d_\eta, -d_\xi) \div M^n$, where n is the degree of the quantic ϕ. But $\phi(d_y, -d_x)\,\phi(x, y)$ is a function of the coefficients *only* of the quantic ϕ, *since the order of differentiation of each term is the same as the degree of the term;* therefore the function is an invariant, by definition (1628).

1649 EXAMPLE.—Let $\phi(x, y) = ax^4 + bx^3y + exy^3 + fy^4$. The quantic must first be completed; thus, $\phi(xy) = ax^4 + bx^3y + cx^2y^2 + exy^3 + fy^4$, $(c = 0)$; then

$$\phi(d_y, -d_x)\,\phi(x, y) = (ad_{4y} - bd_{3yx} + cd_{2y2x} - ed_{y3x} + fd_{4x})\,\phi(x, y)$$
$$= a.24f - b.6e + c.4c - e.6b + f.24a = 4(12af - 3be + c^2).$$

Therefore $12af - 3be + c^2$ is an invariant of ϕ, and $= (12AF - 3BE + C^2) \div M^4$, where A, B, C, E, F are the coefficients of any equation obtained from ϕ by a linear transformation.

But if the degree of the quantic be odd, these results vanish identically.

1650 Similarly, if $\phi(x, y)$, $\psi(x, y)$ are two quantics of the same degree, the functions

$$\phi(d_y, -d_x)\,\psi(x, y) \quad \text{and} \quad \psi(d_y, -d_x)\,\phi(x, y)$$

are both invariants.

1651 Ex.—If $\phi = ax^2 + 2bxy + cy^2$ and $\psi = a'x^2 + 2b'xy + c'y^2$; then

$$(ad_{2y} - 2bd_{xy} + cd_{2x})(a'x^2 + 2b'xy + c'y^2) = ac' + ca' - 2bb', \text{ an invariant.}$$

1652 A Hessian is a covariant of the original quantic.

PROOF.—Let a ternary quantic u be transformed by the linear substitutions in (1604); so that $u = \phi(x, y, z) = \psi(\xi, \eta, \zeta)$. The Hessians of the two functions are $\dfrac{d(u_x u_y u_z)}{d(xyz)}$ and $\dfrac{d(u_\xi u_\eta u_\zeta)}{d(\xi\eta\zeta)}$ (1630). Now

$$\frac{d(u_\xi u_\eta u_\zeta)}{d(\xi\eta\zeta)} = M\frac{d(u_\xi u_\eta u_\zeta)}{d(xyz)} = M\frac{d(u_x u_y u_z)}{d(\xi\eta\zeta)} = M^2\frac{d(u_x u_y u_z)}{d(xyz)}.$$

The second transformation is seen at once from the form of the determinant by merely transposing rows and columns; the first and third are by theorem (1604). Therefore, by definition (1629), the Hessian of u is a covariant.

1653 ***Cogredients.***—Variables are *cogredient* when they are subjected to the same linear transformation; thus, x, y are cogredient with x', y' when

$$\left.\begin{array}{l} x = a\xi + b\eta \\ y = c\xi + d\eta \end{array}\right\} \quad \text{and} \quad \left.\begin{array}{l} x' = a\xi' + b\eta' \\ y' = c\xi' + d\eta' \end{array}\right\}.$$

1654 ***Emanents.***—If in any quantic $u = \phi(x, y)$, we change x into $x+\rho x'$, and y into $y+\rho y'$, where x', y' are cogredient with x, y; then, by (1512),

$$\phi(x+\rho x', y+\rho y')$$
$$= u+\rho(x'd_x+y'd_y)\,u+\tfrac{1}{2}\rho^2(x'd_x+y'd_y)^2\,u+\&c.,$$

and the coefficients of ρ, ρ^2, ρ^3, ... are called the first, second, third, ... *emanents* of u.

1655 The emanents, of the typical form $(x'd_x+y'd_y)^n u$, are all covariants of the quantic u.

PROOF.—If, in $\phi(x, y)$, we first make the substitutions which lead to the emanent, and afterwards make the cogredient substitutions, we change

x into $x+\rho x'$, and this into $a\xi+b\eta+\rho(a\xi'+b\eta')$.

And if the order of these operations be reversed, we change

x into $a\xi+b\eta$, and this into $a(\xi+\rho\xi')+b(\eta+\rho\eta')$.

The two results are identical, and it follows that, if $\phi(x, y)$ be transformed by the same operations in reversed order, the coefficients of the powers of ρ in the two expansions will be equal, since ρ is indeterminate. Therefore, by the definition (1629), each emanent is a covariant.

1656 For definitions of *contragredients* and *contravariants*, see (1813–4). For other theorems on invariants, see (1794), and the Article on Invariants in Section XII.

IMPLICIT FUNCTIONS.

IMPLICIT FUNCTIONS OF ONE INDEPENDENT VARIABLE.

If y and z be functions of x, the successive application of formula (1420) gives, for the first, second, and third derivatives of the function $\phi(y, z)$ with the notation of (1405),

1700 $$\phi_x(yz) = \phi_y y_x + \phi_z z_x.$$

1701 $$\phi_{2x}(yz) = \phi_{2y} y_x^2 + 2\phi_{yz} y_x z_x + \phi_{2z} z_x^2 + \phi_y y_{2x} + \phi_z z_{2x}.$$

1702 $$\phi_{3x}(yz) = \phi_{3y}y_x^3 + 3\phi_{2yz}y_x^2z_x + 3\phi_{y2z}y_xz_x^2 + \phi_{3z}z_x^3 + 3(\phi_{2y}y_x + \phi_{yz}z_x)\,y_{2x} + 3(\phi_{2z}z_x + \phi_{zy}y_x)\,z_{2x} + \phi_y y_{3x} + \phi_z z_{3x}.$$

By making $z = x$ in the last three formulæ, and consequently $z_x = 1$, $z_{2x} = 0$, or else by differentiating independently, we obtain

1703 $$\phi_x(xy) = \phi_y y_x + \phi_x.$$

1704 $$\phi_{2x}(xy) = \phi_{2y}y_x^2 + 2\phi_{xy}y_x + \phi_{2x} + \phi_y y_{2x}.$$

1705 $$\phi_{3x}(xy) = \phi_{3y}y_x^3 + 3\phi_{2yx}y_x^2 + 3\phi_{y2x}y_x + \phi_{3x}. + 3(\phi_{2y}y_x + \phi_{xy})\,y_{2x} + \phi_y y_{3x}.$$

1706 In these formulæ the notation ϕ_x is used where the differentiation is partial, while $\phi_x(x, y)$ is used to denote the complete derivative of $\phi(x, y)$ with respect to x. Each successive partial derivative of the function $\phi(y, z)$ (1700) is itself treated as a function of y and z, and differentiated as such by formula (1420).

Thus, the differentiation of the product $\phi_y y_x$ in (1700) produces

$$(\phi_y)_x y_x + \phi_y y_{2x} = (\phi_{2y}y_x + \phi_{yz}z_x)\,y_x + \phi_y y_{2x}.$$

The function ϕ_y involves y and z by implication. If it should not in fact contain z, for instance, then the partial derivative ϕ_{yz} vanishes. On the other hand, y_x, y_{2x}, &c. are independent of z; and z_x, z_{2x}, &c. are independent of y.

DERIVED EQUATIONS.

1707 If $\phi(xy) = 0$, its successive derivatives are also zero, and the expansions (1703–5) are then called the *first*, *second*, *and third derived equations* of the primitive equation $\phi(xy) = 0$.

In this case, those equations give, by eliminating y_x,

1708 $$\frac{dy}{dx} = -\frac{\phi_x}{\phi_y}; \qquad \frac{d^2y}{dx^2} = \frac{2\phi_{xy}\phi_x\phi_y - \phi_{2x}\phi_y^2 - \phi_{2y}\phi_x^2}{\phi_y^3}.$$

1710 Similarly, by eliminating y_x and y_{2x}, equation (1705) would give y_{3x} in terms of the partial derivatives of $\phi(xy)$. See the note following (1732).

1711 If $\phi(xy) = 0$ and $\dfrac{dy}{dx} = 0$; $\quad \dfrac{d^2y}{dx^2} = -\dfrac{\phi_{2x}}{\phi_y}$;

1712 and $$\frac{d^3y}{dx^3} = \frac{3\phi_{2x}\phi_{xy} - \phi_{3x}\phi_y}{\phi_y^2}.$$

PROOF.—By (1708), $\phi_x = 0$. Therefore (1704) and (1705) give these values of y_{2x} and y_{3x}.

1713 If ϕ_x and ϕ_y both vanish, y_x in (1708) is indeterminate. In this case it has two values given by the second derived equation (1704), which becomes a quadratic in y_x.

1714 If ϕ_{2x}, ϕ_{xy}, and ϕ_{2y} also vanish, proceed to the third derived equation (1705), which now becomes a cubic in y_x, giving three values, and so on.

1715 Generally, when all the partial derivatives of $\phi(x, y)$ of orders less than n vanish for certain values $x = a$, $y = b$, we have, by (1512), $\phi(a, b)$ being zero,

$$\phi(a+h, b+k) = \frac{1}{\underline{|n}} (hd_x + kd_y)^n \phi(xy)_{,a,b} + \text{ terms of}$$

higher orders which may be neglected in the limit. (x, y are here put $= a, b$ after differentiation.) Now, with the notation of 1406),

$$h\phi_{x,a} + k\phi_{y,b} = 0;$$

therefore

$$\frac{k}{h} = -\frac{\phi_{x,a}}{\phi_{y,b}} = \frac{dy}{dx};$$

the values of which are therefore given by the equation

1716
$$(hd_x + kd_y)^n \phi(x, y)_{,a,b} = 0.$$

1717 If y_x becomes indeterminate through x and y vanishing, observe that $\frac{dy}{dx} = \frac{y}{x}$ in this case, and that the value of the latter fraction may often be more readily determined by algebraic methods.

If x and y in the function $\phi(x, y)$ are connected by the equation $\psi(x, y) = 0$, y is thereby made an implicit function of x, and we have

1718
$$\phi_x(x, y) = \frac{\phi_x\psi_y - \psi_x\phi_y}{\psi_y}.$$

1719
$$\phi_{2x}(x, y) = \{(\phi_{2y}\psi_y - \psi_{2y}\phi_y)\psi_x^2 + (\phi_{2x}\psi_y - \psi_{2x}\phi_y)\psi_y^2 - 2(\phi_{yx}\psi_y - \psi_{yx}\phi_y)\psi_x\psi_y\} \div \psi_y^3.$$

PROOF.—(i.) Differentiate both ϕ and ψ for x, by (1703), and eliminate y_x. (ii.) Differentiate also, by (1704), and eliminate y_x and y_{2x}.

If u, y, z are functions of x, then, as in (1700),

1720
$$\phi_x(uyz) = \phi_u u_x + \phi_y y_x + \phi_z z_x.$$

1721 $$\phi_{2x}(uyz) = \phi_{2u}u_x^2+\phi_{2y}y_x^2+\phi_{2z}z_x^2 +2\phi_{yz}y_xz_x+2\phi_{zu}z_xu_x+2\phi_{uy}u_xy_x +\phi_uu_{2x}+\phi_yy_{2x}+\phi_zz_{2x}.$$

1722 To obtain $\phi_x(xyz)$ and $\phi_{2x}(xyz)$, make $u=x$ in the above equations.

Let $U=\phi(x, y, z, \xi)$ be a function of four variables connected by three equations $u=0$, $v=0$, $w=0$, so that one of the variables, ξ, may be considered independent.

1723 We have, by differentiating for ξ,

$$\left.\begin{array}{ll} \phi_x x_\xi+\phi_y y_\xi+\phi_z z_\xi+\phi_\xi-U_\xi & =0 \\ u_x x_\xi+u_y y_\xi+u_z z_\xi+u_\xi & =0 \\ v_x x_\xi+v_y y_\xi+v_z z_\xi+v_\xi & =0 \\ w_x x_\xi+w_y y_\xi+w_z z_\xi+w_\xi & =0 \end{array}\right\}; \qquad \frac{dU}{d\xi}=\frac{d(\phi uvw)}{d(xyz\xi)}\frac{1}{J}; \quad \text{where } J=\frac{d(uvw)}{d(xyz)}.$$

1724 $$\frac{dx}{d\xi}=-\frac{d(uvw)}{d(\xi yz)}\frac{1}{J}; \quad \frac{dy}{d\xi}=-\frac{d(uvw)}{d(x\xi z)}\frac{1}{J}; \quad \frac{dz}{d\xi}=-\frac{d(uvw)}{d(xy\xi)}\frac{1}{J}.$$

Observe that U_ξ stands for the *complete* and ϕ_ξ for the *partial* derivative of the function U.

Proof.—(i.) U_ξ is found by taking the eliminant of the four equations, separating the determinant into two terms by means of the element $\phi_\xi-U_\xi$, and employing the notation in (1600).

(ii.) x_ξ, y_ξ, and z_ξ are found by solving the last three of the same equations, by (582).

IMPLICIT FUNCTIONS OF TWO INDEPENDENT VARIABLES.

1725 If the equation $\phi(x, y, z)=0$ alone be given, y may be considered an implicit function of x and z. Since x and z are independent, we may make z constant and differentiate for x; thus, for a variation in x *only*, the equations (1703–5) are produced again with $\phi(x, y, z)$ in the place of $\phi(x, y)$.

1726 If x be made constant, z must replace x in those equations as the independent variable.

Again, by differentiating the equation $\phi(xyz)=0$ first for x, making z constant, and the result for z, making x constant, we obtain

1727 $$\phi_{xz}+\phi_{yx}y_z+\phi_{yz}y_x+\phi_{2y}y_xy_z+\phi_yy_{xz}=0.$$

From this and the values of y_x and y_z, by (1708),

1728 $$y_{xz}=\frac{\phi_{yx}\phi_y\phi_z+\phi_{yz}\phi_y\phi_x-\phi_{xz}\phi_y^2-\phi_x\phi_z\phi_{2y}}{\phi_y^3}.$$

1729 If x, y, z in the function $\phi(x, y, z)$ be connected by the relation $\psi(x, y, z)=0$, y may be taken as a function of two independent variables x and z. We may therefore make z constant, and the values of $\phi_x(x, y, z)$ and $\phi_{2x}(x, y, z)$ are identical with those in (1718, '19) if x, y, z be substituted for x, y in each function.

1730 By changing x into z the same formulæ give the values of $\phi_z(x, y, z)$ and $\phi_{2z}(x, y, z)$.

1731 On the same hypothesis, if the value of $\phi_x(x, y, z)$, in forming which z has been made constant, be now differentiated for z while x is made constant, each partial derivative ϕ_x, ψ_y, &c. in (1718) must be differentiated as containing x, y, and z, of which three variables x is now constant and y is a function of z.

The result is

1732 $$\phi_{xz}(x,y,z)=\{(\phi_{xz}\psi_y-\psi_{xz}\phi_y)\psi_y^3-(\phi_{yx}\psi_y-\psi_{yx}\phi_y)\psi_z\psi_y^2$$
$$-(\phi_{yz}\psi_y-\psi_{yz}\phi_y)\psi_x\psi_y^2+(\phi_{2y}\psi_y-\psi_{2y}\phi_y)\psi_x\psi_z\psi_y\}\div\psi_y^3.$$

In a particular instance it is generally easier to apply such rules for differentiating directly to the example proposed, than to deduce the result in a functional form for the purpose of substituting in it the values of the partial derivatives.

1734 EXAMPLE.—Let $\phi(x, y, z)=lx+my+nz$ and $\psi(x, y, z)=x^2+y^2+z^2=1$, x and z being the independent variables; $\phi_x(x, y, z)$ and $\phi_{2x}(x, y, z)$ are required. Differentiating ϕ, considering z constant,

$$\phi_x(x, y, z)=l+m\frac{dy}{dx}=l-m\frac{x}{y}; \quad \text{since } \frac{dy}{dx}=-\frac{\psi_x}{\psi_y}=-\frac{x}{y};$$

$$\phi_{2x}(x, y, z)=-m\frac{y-xy_x}{y^2}=-m\frac{y^2+x^2}{y^3};$$

a result which is otherwise obtained from formula (1719) by substituting the values $\phi_x=l,\quad \phi_y=m,\quad \phi_z=n;\quad \phi_{2x}=\phi_{2y}=\phi_{2z}=0;$
$\psi_x=2x,\quad \psi_y=2y,\quad \psi_z=2z;\quad \psi_{2x}=\psi_{2y}=\psi_{2z}=2.$

Again, to find $\phi_{xz}(x, y, z)$, differentiate for z, considering x constant in the function

$$\frac{d\phi}{dx}=l-m\frac{x}{y}; \quad \text{thus } \frac{d^2\phi}{dx\,dz}=+m\frac{xy_z}{y^2}=-m\frac{xz}{y^3}, \quad \text{since } \frac{dy}{dz}=-\frac{\psi_z}{\psi_y}=-\frac{z}{y}.$$

1735 Let $U = \phi(x, y, z, \xi, \eta)$ be a function of five variables connected by three equations $u=0$, $v=0$, $w=0$; so that *two* of the variables ξ, η may be considered independent. Making η constant, the equations in (1723), and the values obtained for U_ξ, x_ξ, y_ξ, z_ξ, hold good in the present case for the variations due to a variation in ξ, observing that ϕ, u, v, w now stand for functions of η as well as of ξ.

1736 The corresponding values of U_η, x_η, y_η, z_η are obtained by changing ξ into η.

IMPLICIT FUNCTIONS OF n INDEPENDENT VARIABLES.

1737 The same method is applicable to the general case of a function of n variables connected by r equations $u=0$, $v=0$, $w=0$ ... &c.

The equations constitute *any* $n-r$ of the variables we please, *independent*: let these be ξ, η, ζ .... The remaining r variables will be *dependent*: let these be x, y, z ...; and let the function be $U \equiv \phi(x, y, z \ldots \xi, \eta, \zeta \ldots)$.

For a variation in ξ *only*, there will be the derivative of the function U, and r derived equations as under.

1738
$$\begin{aligned} \phi_x x_\xi + \phi_y y_\xi + \phi_z z_\xi + \ldots + \phi_\xi &= U_\xi, \\ u_x x_\xi + u_y y_\xi + u_z z_\xi + \ldots + u_\xi &= 0, \\ v_x x_\xi + v_y y_\xi + v_z z_\xi + \ldots + v_\xi &= 0, \\ \ldots \quad \ldots \quad \ldots \quad \ldots \quad \ldots \quad & \&\text{c.}, \end{aligned}$$

involving the r implicit functions x_ξ, y_ξ, z_ξ, &c. The solution of the r equations, as in (1724), gives

1739
$$\frac{dx}{d\xi} = -\frac{d(uvw\ldots)}{d(\xi yz\ldots)}\frac{1}{J}; \quad \frac{dy}{d\xi} = -\frac{d(uvw\ldots)}{d(x\xi z\ldots)}\frac{1}{J}; \ \&\text{c.},$$

1740 where $J = \dfrac{d(uvw\ldots)}{d(xyz\ldots)}$. Also $\dfrac{dU}{d\xi} = \dfrac{d(\phi uvw)}{d(xyz\xi)}\dfrac{1}{J}$.

The last value being found exactly as in (1723).

1741 With ξ replaced by η we have in like manner the values of x_η, y_η, z_η, U_η; and similarly with each of the independent variables in turn.

1742 If there be n variables and but one equation $\phi(x, y, z \ldots) = 0$, there will be $n-1$ independent and one dependent variable. Let y be dependent. Then for a variation in x only of the remaining variables, the equations (1703–5) apply to the present case, ϕ standing for $\phi(x, y, z \ldots)$. If x be replaced by each of the remaining independents in turn, there will be, in all, $n-1$ sets of derived equations.

CHANGE OF THE INDEPENDENT VARIABLE.

If y be any function of x, and if the independent variable x be changed to t, and if t be afterwards put equal to y, the following formulæ of substitution are obtained, in which $p \equiv y_x$:

1760 $$\frac{dy}{dx} = \frac{y_t}{x_t} = \frac{1}{x_y}.$$

Differentiating these fractions, we get

1762 $$\frac{d^2y}{dx^2} = \frac{y_{2t}x_t - x_{2t}y_t}{x_t^3} = -\frac{x_{2y}}{x_y^3} = pp_y.$$

1765 $$\frac{d^3y}{dx^3} = \frac{x_t\{y_{3t}x_t - x_{3t}y_t\} - 3x_{2t}\{y_{2t}x_t - x_{2t}y_t\}}{x_t^5}.$$

1766 $$= \frac{3x_{2y}^2 - x_y x_{3y}}{x_y^5} = pp_y^2 + p^2p_{2y}.$$

Ex.—If $x = r\cos\theta$ and $y = r\sin\theta$; then

1768 $$\frac{dy}{dx} = \frac{r_\theta \sin\theta + r\cos\theta}{r_\theta\cos\theta - r\sin\theta}; \quad \frac{d^2y}{dx^2} = \frac{r^2 + 2r_\theta^2 - rr_{2\theta}}{(r_\theta\cos\theta - r\sin\theta)^3}.$$

Proof.—Writing θ for t in (1760) and (1762), we have to find x_θ, y_θ, $x_{2\theta}$, $y_{2\theta}$; thus,

$$x_\theta = r_\theta\cos\theta - r\sin\theta; \quad x_{2\theta} = r_{2\theta}\cos\theta - 2r_\theta\sin\theta - r\cos\theta;$$
$$y_\theta = r_\theta\sin\theta + r\cos\theta; \quad y_{2\theta} = r_{2\theta}\sin\theta + 2r_\theta\cos\theta - r\sin\theta.$$

Substituting these values, the above results are obtained.

To change the variable from x to t in $(a+bx)^n y_{nx}$, where $(a+bx) = e^t$, employ the formula

1770 $$(a+bx)^n y_{nx} = b^n(d_t - \overline{n-1})(d_t - \overline{n-2}) \ldots (d_t - 1)\, y_t.$$

in which, multiplication by d_t by the index law signifies the repetition of the operation d_t (1492).

PROOF.— $d_t\{(a+bx)^n y_{nx}\} = \{n(a+bx)^{n-1} b y_{nx} + (a+bx)^n y_{(n+1)x}\} x_t$.

Now $bx_t = e^t = a+bx$. Substitute this, and denote $(a+bx)^n y_{nx}$ by U_n;

therefore $\quad d_t(U_n) = nU_n + \frac{1}{b} U_{n+1}, \quad$ or $U_{n+1} = b(d_t - n) U_n$.

Therefore $U_n = b(d_t - \overline{n-1}) U_{n-1}$ and $U_{n-1} = b(d_t - \overline{n-2}) U_{n-2}$, &c.,

and finally $\quad U_2 = b(d_t - 1) U_1$.

But $\quad U_1 = (a+bx) y_x = bx_t y_x = by_t$.

Therefore $\quad U_n = b^n (d_t - \overline{n-1})(d_t - \overline{n-2}) \dots (d_t - 1) y_t$.

1771 COR.—$(a+x)^n y_{nx}$ and $x^n y_{nx}$ are transformed by the same formula by putting $b=1$.

1772 Let $V \equiv F(x, y)$, where x, y are connected with ξ, η by the equations $u=0$, $v=0$. It is required to change the independent variables x, y to ξ, η in the functions V_x and V_y.

1773 RULE.—To find the value of V_x—*Differentiate* V, u, v, *each with respect to* x, *considering* ξ, η *functions of the independent variables* x, y; *and form the eliminant of the resulting equations;* thus,

1774
$$\left.\begin{aligned} V_\xi \xi_x + V_\eta \eta_x - V_x &= 0 \\ u_\xi \xi_x + u_\eta \eta_x + u_x &= 0 \\ v_\xi \xi_x + v_\eta \eta_x + v_x &= 0 \end{aligned}\right\}; \quad \therefore \quad \begin{vmatrix} V_\xi & V_\eta & -V_x \\ u_\xi & u_\eta & u_x \\ v_\xi & v_\eta & v_x \end{vmatrix} = 0.$$

Similarly, to find V_y.

1775 EX.—Let $x = r\cos\theta$ and $y = r\sin\theta$; then

$$V_x = V_r \cos\theta - V_\theta \frac{\sin\theta}{r}, \qquad V_y = V_r \sin\theta + V_\theta \frac{\cos\theta}{r}.$$

PROOF.— $u \equiv r\cos\theta - x$, $v \equiv r\sin\theta - y$, and the determinant in (1774) takes the form annexed by writing r and θ instead of ξ and η. A similar determinant gives V_y.

$$\begin{vmatrix} V_r & V_\theta & -V_x \\ \cos\theta & -r\sin\theta & -1 \\ \sin\theta & r\cos\theta & 0 \end{vmatrix} = 0.$$

To find V_{2x}, substitute $V_r \cos\theta - V_\theta \frac{\sin\theta}{r}$ in the place of V in the value of V_x; and in differentiating for r and θ, consider V_r and V_θ as functions of both r and θ. Similarly, to find V_{2y} and V_{xy}. Thus,

1777 $$V_{2x} = V_{2r}\cos^2\theta + \left(\frac{1}{r} V_\theta - V_{r\theta}\right) \frac{2\sin\theta\cos\theta}{r} + V_r \frac{\sin^2\theta}{r} + V_{2\theta} \frac{\sin^2\theta}{r^2}.$$

1778 $$V_{2y} = V_{2r}\sin^2\theta - \left(\frac{1}{r} V_\theta - V_{r\theta}\right) \frac{2\sin\theta\cos\theta}{r} + V_r \frac{\cos^2\theta}{r} + V_{2\theta} \frac{\cos^2\theta}{r^2}.$$

By addition these equations give

1779 $$V_{2x} + V_{2y} = V_{2r} + \frac{1}{r} V_r + \frac{1}{r^2} V_{2\theta}.$$

1780 Given $V \equiv f(x, y, z)$ and ξ, η, ζ known functions of x, y, z; V_ξ, V_η, V_ζ are expressed in terms of V_x, V_y, V_z by the formulæ

$$\frac{dV}{d\xi} = \frac{d(V\eta\zeta)}{d(xyz)} \div J, \quad \frac{dV}{d\eta} = \frac{d(\xi V\zeta)}{d(xyz)} \div J, \quad \frac{dV}{d\zeta} = \frac{d(\xi\eta V)}{d(xyz)} \div J.$$

PROOF.—Differentiate V as a function of ξ, η, ζ with respect to independent variables x, y, z. The annexed equations are the result. Solve these by (582) with the notation of (1600).

$$\begin{cases} V_\xi\xi_x + V_\eta\eta_x + V_\zeta\zeta_x = V_x, \\ V_\xi\xi_y + V_\eta\eta_y + V_\zeta\zeta_y = V_y, \\ V_\xi\xi_z + V_\eta\eta_z + V_\zeta\zeta_z = V_z. \end{cases}$$

1781 Given $V \equiv f(x, y, z)$, where x, y, z are involved with ξ, η, ζ in three equations $u = 0$, $v = 0$, $w = 0$, it is required to change the variables to ξ, η, ζ in V_x, V_y, and V_z.

Applying Rule (1773) to the case of three variables, we have

$$\left.\begin{aligned} V_\xi\,\xi_x + V_\eta\eta_x + V_\zeta\zeta_x - V_x &= 0 \\ u_\xi\,\xi_x + u_\eta\eta_x + u_\zeta\zeta_x + u_x &= 0 \\ v_\xi\,\xi_x + v_\eta\eta_x + v_\zeta\zeta_x + v_x &= 0 \\ w_\xi\,\xi_x + w_\eta\eta_x + w_\zeta\zeta_x + w_x &= 0 \end{aligned}\right\}, \quad \therefore \begin{vmatrix} V_\xi & V_\eta & V_\zeta & -V_x \\ u_\xi & u_\eta & u_\zeta & u_x \\ v_\xi & v_\eta & v_\zeta & v_x \\ w_\xi & w_\eta & w_\zeta & w_x \end{vmatrix} = 0.$$

The determinant gives V_x in terms of V_ξ, V_η, V_ζ and the derivatives of u, v, w. V_y and V_z are found in an analogous manner.

1782 Similarly with n equations between $2n$ variables.

1783 Ex.—Given

$$x = r\sin\theta\cos\phi; \quad y = r\sin\theta\sin\phi; \quad z = r\cos\theta.$$

The equations u, v, w become

$$r\sin\theta\cos\phi - x = 0; \quad r\sin\theta\sin\phi - y = 0; \quad r\cos\theta - z = 0.$$

Writing r, θ, ϕ instead of ξ, η, ζ, the determinant becomes

$$\begin{vmatrix} V_r & V_\theta & V_\phi & -V_x \\ \sin\theta\cos\phi & r\cos\theta\cos\phi & -r\sin\theta\sin\phi & -1 \\ \sin\theta\sin\phi & r\cos\theta\sin\phi & r\sin\theta\cos\phi & 0 \\ \cos\theta & -r\sin\theta & 0 & 0 \end{vmatrix} = 0. *$$

From which V_x is obtained. Similarly, V_y and V_z; and, by an exactly similar process, the converse forms for V_r, V_θ, and V_ϕ. The results are

* In writing out a determinant like the above, it will be found expeditious in practice to have the columns written on separate slips of paper in order to be able to transpose them readily. Thus, to find the coefficient of V_θ, bring the second column to the *left* side, and, since this changes the sign of the determinant, transpose *any two* other columns, so that the coefficient of V_θ may be read off in the standard form as the minor of the first element of the determinant.

1784 $$V_x = V_r \sin\theta \cos\phi + V_\theta \frac{\cos\theta \cos\phi}{r} - V_\phi \frac{\sin\phi}{r\sin\theta}.$$

1785 $$V_y = V_r \sin\theta \sin\phi + V_\theta \frac{\cos\theta \sin\phi}{r} + V_\phi \frac{\cos\phi}{r\sin\theta}.$$

1786 $$V_z = V_r \cos\theta - V_\theta \frac{\sin\theta}{r}.$$

1787 $$V_r = V_x \sin\theta \cos\phi + V_y \sin\theta \sin\phi + V_z \cos\theta.$$

1788 $$V_\theta = V_x r\cos\theta \cos\phi + V_y r\cos\theta \sin\phi - V_z r\sin\theta.$$

1789 $$V_\phi = -V_x r\sin\theta \sin\phi + V_y r\sin\theta \cos\phi.$$

1790 To find V_x directly; solve the equations u, v, w, in (1783), for r, θ, and ϕ; *the solution in this case being practicable;* thus,

$$r = \sqrt{(x^2+y^2+z^2)}, \qquad \theta = \tan^{-1}\frac{\sqrt{(x^2+y^2)}}{z}, \qquad \phi = \tan^{-1}\frac{y}{x}.$$

Find r_x, θ_x, ϕ_x from these, and substitute in $V_x = V_r r_x + V_\theta \theta_x + V_\phi \phi_x$. Similarly, V_y and V_z. Also $V_r = V_x x_r + V_y y_r + V_z z_r$. Similarly, V_θ and V_ϕ.

1791 To obtain V_{xy}, substitute the value of V_x in the place of V, in the value of V_y, in (1785), and, in differentiating V_r, V_θ, V_ϕ, consider each of these quantities a function of r, θ, and ϕ.

To change the variables to r, θ, and ϕ, in $V_{2x}+V_{2y}+V_{2z}$, the equations (1783) still subsisting. Result—

1792 $$V_{2x}+V_{2y}+V_{2z}$$
$$= V_{2r} + \frac{2}{r}V_r + \frac{1}{r^2}\left(V_\theta \cot\theta + V_{2\theta} + V_{2\phi}\operatorname{cosec}^2\theta\right).$$

Proof.—Put $r\sin\theta = \rho$, so that $x = \rho\cos\phi$ and $y = \rho\sin\phi$,

therefore, by (1779), $V_{2x}+V_{2y} = V_{2\rho} + \frac{1}{\rho}V_\rho + \frac{1}{\rho^2}V_{2\phi}$ .... (i.).

Also, since $z = r\cos\theta$ and $\rho = r\sin\theta$, we have, by the same formula,

$$V_{2z}+V_{2\rho} = V_{2r} + \frac{1}{r}V_r + \frac{1}{r^2}V_{2\theta} \text{ .... (ii.).}$$

Add together (i.) and (ii.), and eliminate V_ρ by (1776), which gives

$$V_\rho = V_r \sin\theta + V_\theta \frac{\cos\theta}{r}.$$

If V be a function of n variables x, y, z ... connected by the single relation, $x^2+y^2+z^2+\ldots = r^2$ .... (i.).

1793 $$V_{2x}+V_{2y}+V_{2z}+\&c. = V_{2r} + \frac{n-1}{r}V_r.$$

PROOF.— $V_x = V_r r_x = V_r \frac{x}{r}$, since $r_x = \frac{x}{r}$, by differentiating (i.),

therefore $V_{2x} = V_{rx}\frac{x}{r} + V_r\frac{r - xr_x}{r^2} = V_{2r}\frac{x^2}{r^2} + V_r\left(\frac{1}{r} - \frac{x^2}{r^3}\right)$.

Similarly $V_{2y} = V_{2r}\frac{y^2}{r^2} + V_r\left(\frac{1}{r} - \frac{y^2}{r^3}\right)$, &c.

Thus, by addition,

$$V_{2x} + V_{2y} + \&c. = V_{2r}\frac{x^2+y^2+\dots}{r^2} + V_r\left(\frac{n}{r} - \frac{x^2+y^2+\dots}{r^3}\right) = V_{2r} + \frac{n-1}{r}V_r.$$

LINEAR TRANSFORMATION.

1794 If $V \equiv f(x, y, z)$, and if the equations u, v, w in (1781) take the forms

$$\left.\begin{aligned} x &= a_1\xi + b_1\eta + c_1\zeta \\ y &= a_2\xi + b_2\eta + c_2\zeta \\ z &= a_3\xi + b_3\eta + c_3\zeta \end{aligned}\right\}, \quad \text{then} \quad \left\{\begin{aligned} \Delta\xi &= A_1x + A_2y + A_3z, \\ \Delta\eta &= B_1x + B_2y + B_3z, \\ \Delta\zeta &= C_1x + C_2y + C_3z, \end{aligned}\right.$$

by (582), Δ being the determinant $(a_1b_2c_3)$, and A_1 the minor of a_1, &c.

1795 The operations d_x, d_y, d_z will now be transformed by the first set of equations below; and d_ξ, d_η, d_ζ by the second set.

$$\left.\begin{aligned} d_x &= (A_1d_\xi + B_1d_\eta + C_1d_\zeta) \div \Delta \\ d_y &= (A_2d_\xi + B_2d_\eta + C_2d_\zeta) \div \Delta \\ d_z &= (A_3d_\xi + B_3d_\eta + C_3d_\zeta) \div \Delta \end{aligned}\right\}, \quad \left.\begin{aligned} d_\xi &= a_1d_x + a_2d_y + a_3d_z \\ d_\eta &= b_1d_x + b_2d_y + b_3d_z \\ d_\zeta &= c_1d_x + c_2d_y + c_3d_z \end{aligned}\right\}.$$

PROOF.—By $d_x = d_\xi\xi_x + d_\eta\eta_x + d_\zeta\zeta_x$ and $d_\xi = d_xx_\xi + d_yy_\xi + d_zz_\xi$; and the values of ξ_x, x_ξ, &c., from the preceding equations.

1797 From (1795), $V_\xi = a_1V_x + a_2V_y + a_3V_z$. Operating again upon V_ξ, we have

$$V_{2\xi} = (a_1d_x + a_2d_y + a_3d_z)\,V_\xi = a_1(V_\xi)_x + a_2(V_\xi)_y + a_3(V_\xi)_z,$$

and by substituting the value of V_ξ, and similarly with $V_{2\eta}$, $V_{2\zeta}$, we obtain the formulæ,

1798

$$\left.\begin{aligned} V_{2\xi} &= a_1^2V_{2x} + a_2^2V_{2y} + a_3^2V_{2z} + 2a_2a_3V_{yz} + 2a_3a_1V_{zx} + 2a_1a_2V_{xy} \\ V_{2\eta} &= b_1^2V_{2x} + b_2^2V_{2y} + b_3^2V_{2z} + 2b_2b_3V_{yz} + 2b_3b_1V_{zx} + 2b_1b_2V_{xy} \\ V_{2\zeta} &= c_1^2V_{2x} + c_2^2V_{2y} + c_3^2V_{2z} + 2c_2c_3V_{yz} + 2c_3c_1V_{zx} + 2c_1c_2V_{xy} \end{aligned}\right\}.$$

ORTHOGONAL TRANSFORMATION.

1799 If the transformation is orthogonal (584), we have

$$x^2+y^2+z^2 = \xi^2+\eta^2+\zeta^2;$$

and since, by (582, 584), $\Delta=1$, $A_1=a_1$, &c.; equations (1794) now become

1800 $$\left.\begin{aligned} x &= a_1\xi+b_1\eta+c_1\zeta \\ y &= a_2\xi+b_2\eta+c_2\zeta \\ z &= a_3\xi+b_3\eta+c_3\zeta \end{aligned}\right\},\qquad \left.\begin{aligned} \xi &= a_1x+a_2y+a_3z \\ \eta &= b_1x+b_2y+b_3z \\ \zeta &= c_1x+c_2y+c_3z \end{aligned}\right\}.$$

And equations (1795) become

1802 $$\left.\begin{aligned} d_x &= a_1d_\xi+b_1d_\eta+c_1d_\zeta \\ d_y &= a_2d_\xi+b_2d_\eta+c_2d_\zeta \\ d_z &= a_3d_\xi+b_3d_\eta+c_3d_\zeta \end{aligned}\right\},\qquad \left.\begin{aligned} d_\xi &= a_1d_x+a_2d_y+a_3d_z \\ d_\eta &= b_1d_x+b_2d_y+b_3d_z \\ d_\zeta &= c_1d_x+c_2d_y+c_3d_z \end{aligned}\right\}.$$

The double relations between x, y, z and ξ, η, ζ, in the six equations of (1800–1), and the similar relations in (1802–3) between $d_xd_yd_z$ and $d_\xi d_\eta d_\zeta$, are indicated by a single diagram in each case; thus,

1804

	ξ	η	ζ
x	a_1	b_1	c_1
y	a_2	b_2	c_2
z	a_3	b_3	c_3

	d_ξ	d_η	d_ζ
d_x	a_1	b_1	c_1
d_y	a_2	b_2	c_2
d_z	a_3	b_3	c_3

1806 Hence, when the transformation is orthogonal, the quantities x, y, z are *cogredient* with d_x, d_y, d_z, by the definition (1653).

1807 Extending the definition in (1629), it follows that any function $u\equiv\phi(x, y, z)$, when orthogonally transformed, has, for a covariant, the function $\phi(d_x, d_y, d_z)\,u$. That is, if by the transformation,

$$u = \phi(x, y, z) = \psi(\xi, \eta, \zeta),$$

then also $$\phi(d_x, d_y, d_z)\,u = \psi(d_\xi, d_\eta, d_\zeta)\,u.$$

1808 But if u be a quantic, then, as shown in (1648), $\phi(d_x, d_y, d_z)\,u$ is always *a function of the coefficients only* of u, and the *covariant* is, in this case, an *invariant*.

1809 Ex.—Let u or $\phi(x, y, z) = ax^2+by^2+cz^2+2fyz+2gzx+2hxy$,

$\therefore\ \phi(d_x, d_y, d_z)\,u = au_{2x}+bu_{2y}+cu_{2z}+2fu_{yz}+2gu_{zx}+2hu_{xy}$
$= 2\{a^2+b^2+c^2+2f^2+2g^2+2h^2\}$, and this is an invariant of u.

1810 When $V \equiv f(x, y, z)$ is orthogonally transformed,

$$V_{2x}+V_{2y}+V_{2z} = V_{2\xi}+V_{2\eta}+V_{2\zeta}.$$

PROOF.—By adding together equations (1798), and by the relations

$$a_1^2+b_1^2+c_1^2 = 1, \text{ \&c., and } a_1a_2+b_1b_2+c_1c_2 = 0, \text{ \&c.,}$$

established in (584).

1811 If two functions u, v be subjected to the same orthogonal transformation, so that

$$u \equiv \phi(x, y, z) = \Phi(\xi, \eta, \zeta) \quad \text{and} \quad v \equiv \psi(x, y, z) = \Psi(\xi, \eta, \zeta);$$

then $$\phi(d_x, d_y, d_z)\, v = \Phi(d_\xi, d_\eta, d_\zeta)\, v.$$

1812 Ex.—Let $u = ax^2+by^2+cz^2+2fyz+2gzx+2hxy \equiv \phi$

$$= a'\xi^2+b'\eta^2+c'\zeta^2+2f'\eta\zeta+2g'\zeta\xi+2h'\xi\eta \equiv \Phi,$$

and let $$v = x^2+y^2+z^2 = \xi^2+\eta^2+\zeta^2 \equiv \psi \text{ and } \Psi.$$

Then $$\phi(d_x, d_y, d_z)\, v = av_{2x}+bv_{2y}+cv_{2z}+2fv_{yz}+2gv_{zx}+2hv_{xy},$$

and $$\Phi(d_\xi, d_\eta, d_\zeta)\, v = a'v_{2\xi}+b'v_{2\eta}+c'v_{2\zeta}+2f'v_{\eta}+2g'v+2h'v_{\xi\eta}.$$

But $v_{2x} = 2$, and $v_{yz} = 0$, &c. Hence the theorem gives $a+b+c = a'+b'+c'$; in other words, $a+b+c$ is an *invariant*.

1813 *Contragredient.*—When the transformation is not orthogonal, (1795) shows that d_x is not transformed by the *same*, but by a *reciprocal* substitution, in which a_1, b_1, c_1 are replaced by the corresponding minors A_1, B_1, C_1. In this case d_x, d_y, d_z are said to be *contragredient* to x, y, z.

1814 *Contravariant.*—If, in (1629), the quantics are subjected to *a reciprocal* transformation instead of *the same*, we obtain the definition of a *contravariant*.

1815 When z is a function of two independent variables x and y, the following notation is often used:

$$\frac{dz}{dx} \equiv p, \qquad \frac{dz}{dy} \equiv q, \qquad \frac{dp}{dx} = \frac{d^2z}{dx^2} \equiv r,$$

$$\frac{dp}{dy} = \frac{dq}{dx} = \frac{dz}{dx\,dy} \equiv s, \qquad \frac{dq}{dy} = \frac{d^2z}{dy^2} \equiv t.$$

Let $\phi(x, y, z) = 0$. It is required to change the independent variables from x, y to z, y. The formulæ of transformation are

1816 $\dfrac{dx}{dz}=\dfrac{1}{p};\qquad \dfrac{dx}{dy}=-\dfrac{q}{p};\qquad \dfrac{d^2x}{dz^2}=-\dfrac{r}{p^3};$

1819 $\dfrac{d^2x}{dy^2}=\dfrac{2spq-tp^2-rq^2}{p^3};\qquad \dfrac{d^2x}{dy\,dz}=\dfrac{qr-ps}{p^3}.$

PROOF.—Formulæ (1761, 1763) give x_z and x_{2z}, because, since y remains constant, ϕ may be considered a function of only two variables, x and z.

Formulæ (1708–9) give x_y and x_{2y} in terms of partial derivatives of ϕ, since z is now constant, and ϕ may be taken as a function of the two variables x and y.

But $\phi(x, y, z)=0$ is equivalent to $\psi(x, y)-z=0$; and the partial derivatives of ϕ with respect to x and y are the same as those of ψ; and therefore the same as those of z when x and y are the independent variables. Hence z may be written for ϕ in the formulæ.

Lastly, $\dfrac{d^2x}{dy\,dz}=\left(-\dfrac{q}{p}\right)_z=\left(-\dfrac{q}{p}\right)_x\dfrac{dx}{dz}=\dfrac{qp_x-pq_x}{p^2}\,\dfrac{1}{p}=\dfrac{qr-ps}{p^3}.$

The independent variable is here changed from z to x, without reference to the equation $\phi=0$; and this is allowable, because y is constant for the time being in either case.

MAXIMA AND MINIMA.

Maxima and Minima values of a function of one independent variable.

1830 DEFINITION.—A function $\phi(x)$ has a *maximum* value when some value $x=a$ makes $\phi(x)$ *greater* than it is made by any value of x indefinitely near to a. Similarly for a *minimum* value, reading *less* for *greater*.

1831 ILLUSTRATION.—If the ordinate y in the figure be always drawn $=f(x)$, it has maximum values at A, C, E, and minimum at B and D (1403).

NOTE.—For the algebraic determination of maxima and minima values by a quadratic equation, see (58).

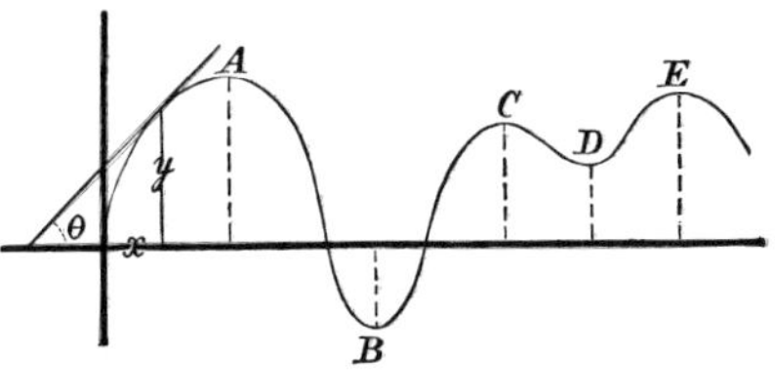

1832 RULE I.—*A function* $\phi(\mathrm{x})$ *is a maximum or minimum when* $\phi'(\mathrm{x})$ *vanishes, and changes its sign as* x *increases from plus to minus or from minus to plus respectively.*

1833 RULE II.—*Otherwise* $\phi(x)$ *is a maximum or minimum when an odd number of consecutive derivatives of* $\phi(x)$ *vanish, and the next is minus or plus respectively.*

PROOF.—(i.) The tangent to the curve in the last figure becomes parallel to the x axis at the points A, B, C, D, E as x increases; therefore, by (1403), $\tan\theta$, which is equal to $f'(x)$, vanishes at those points, while its sign changes in the manner described.

(ii.) Let $f^n(x)$ be the first derivative of $f(x)$ which does not vanish when $x = a$, n being even; therefore, by (1500), $f(a \pm h) = f(a) + \frac{h^n}{\lfloor n} f^n(a+\theta h)$. The last term retains the sign of $f^n(a)$, when h is small enough, whether h be positive or negative, since n is even. Therefore $f(x)$ diminishes for *any* small variation of x from the value a if $f^n(a)$ be negative, but increases if $f^n(a)$ be positive. Hence the rule.

1834 NOTE.—Before applying the rule for discovering a maximum or minimum, we may evidently—

(i.) *reject any constant factor of the function;*
(ii.) *raise it to any constant power, paying attention to sign;*
(iii.) *take its reciprocal; maximum becoming minimum, and vice versâ;*
(iv.) *take the logarithm of a positive function.*

1835 Ex. 1.—Let $\phi(x) = x^7 - 7x^4 - 35x + 1$,

therefore $\phi'(x) = 7x^6 - 28x^3 - 35 = 7(x^3-5)(x^3+1)$.

Also $\phi''(x) = 7(6x^5 - 12x^2)$. Therefore $x = \sqrt[3]{5}$ makes $\phi'(x)$ vanish, and $\phi''(x)$ positive; and therefore makes $\phi(x)$ a minimum.

1836 Ex. 2.—Let $\phi(x) = (x-3)^{14}(x-2)^{11}$. Here

$\phi'(x) = 14(x-3)^{13}(x-2)^{11} + 11(x-3)^{14}(x-2)^{10} = (x-3)^{13}(x-2)^{10}(25x-61)$,

and we know, by (444) or by (1460), that, when $x = 3$, the first *thirteen* derivatives of $\phi(x)$ vanish; and 13 is an odd number. Therefore $\phi(x)$ is either a maximum or minimum when $x = 3$.

To determine which, examine the change of sign in $\phi'(x)$. Now $(x-3)^{13}$ changes from negative to positive as x increases from a value a little less than 3 to a value a little greater, while the other factors of $\phi'(x)$ remain positive. Therefore, by the rule, $\phi(x)$ is a minimum when $x = 3$.

Again, as x passes through the value 2, $\phi'(x)$ does not change sign, 10 being even. Therefore $x = 2$ gives no maximum or minimum value of $\phi(x)$.

Lastly, as x passes through the value $\frac{61}{25}$ the signs of the three factors in $\phi'(x)$ change from $(-)(+)(-)$ to $(-)(+)(+)$; that is, $\phi'(x)$ changes from $+$ to $-$; and, consequently, $\phi(x)$ is a maximum.

1837 Ex. 3.—Let $\phi(x, y) = x^4 + 2x^2y - y^3 = 0$. To find limiting values of y.

Here y is given only as an implicit function of x. Differentiating, in order to employ formulæ (1708, 1711),

$$\phi_x = 4x^3+4xy, \quad \phi_{2x} = 12x^2+4y, \quad \phi_y = 2x^2-3y^2;$$

$y_x = 0$ makes $\phi_x = 0$. Solving this equation with $\phi(x, y) = 0$, we get $x = \pm 1$, $y = -1$ when y_x vanishes.

And then $y_{2x} = -\frac{\phi_{2x}}{\phi_y} = -\frac{12-4}{2-3} = 8$, positive; therefore, when $x = \pm 1$, y has -1 for a minimum value.

Similarly, by making y the independent variable, it may be shown that, when $y = -\frac{8}{9}$, x has both the maximum and minimum values $\pm\frac{4}{9}\sqrt{6}$.

1838 A limiting value of $\qquad \phi(x, y)$,
subject to the condition $\qquad \psi(x, y) = 0 \ldots\ldots\ldots$ (i.),
is obtained from the equation $\qquad \phi_x\psi_y = \phi_y\psi_x \ \ldots\ldots$ (ii.)

Simultaneous values of x and y, found by solving equations (i.) and (ii.), correspond to a maximum or minimum value of ϕ.

Proof.—By (1718), ϕ being virtually a function of x only; and, by (1832), $\phi_x(xy) = 0$.

1839 Ex.—Let $\phi(x, y) = xy$ and $\psi(x, y) = 2x^3 - xy + y^3 = 0 \ldots\ldots$(i.)
Equation (ii.) becomes $\quad y(3y^2-x) = x(6x^2-y)$.
Solving this with (i.), we find $y^3 = 2x^3$ and $x^2(4x - \sqrt[3]{2})$.

Therefore $x = \frac{1}{4}\sqrt[3]{2}$, $y = \frac{1}{4}\sqrt[3]{4}$ are values corresponding to a *maximum* value of ϕ. That it is a maximum, and not a minimum, is seen by inspecting equation (i.)

1840 Most geometrical problems can be treated in this way, and the alternative of maximum or minimum decided by the nature of the case. Otherwise the sign of $\phi_{2x}(xy)$ may be examined by formula (1719) for the criterion, according to the rule.

Maxima and Minima values of a function of two independent variables.

1841 Rule I.—*A function $\phi(x, y)$ is a maximum or minimum when ϕ_x and ϕ_y both vanish and change their signs from plus to minus or from minus to plus respectively, as x and y increase.*

1842 Rule II.—*Otherwise, ϕ_x and ϕ_y must vanish; $\phi_{2x}\phi_{2y} - \phi_{xy}^2$ must be positive, and ϕ_{2x} or ϕ_{2y} must be negative for a maximum and positive for a minimum value of ϕ.*

Proof.—By (1512), writing A, B, C for ϕ_{2x}, ϕ_{xy}, ϕ_{2y}, we have, for small changes h, k in the values of x and y,

$$\phi(x+h, y+k) - \phi(x, y) = h\phi_x + k\phi_y + \tfrac{1}{2}(Ah^2+2Bhk+Ck^2) + \text{terms}$$

which may be neglected, by (1410).

Hence, as in the proof of (1833), in order that changing the sign of h or k shall not have the effect of changing the sign of the right side of the equation, the first powers must disappear, therefore ϕ_x and ϕ_y must vanish. The next term may be written, by completing the square, in the form $\frac{k^2}{2A}\left\{\left(A\frac{h}{k}+B\right)^2+AC-B^2\right\}$; and, to ensure this quantity retaining its sign for *all values of the ratio* $h:k$, $AC-B^2$ must be positive. ϕ will then be a maximum or minimum according as A in the denominator is negative or positive.

It is clear that A and B might have been transposed in the proof. Hence B must have the same sign as A.

1843 A limiting value of $\phi(x, y, z)$,
subject to the condition $\psi(x, y, z) = 0$......... (i.),
is obtained from the two equations

1844 $\phi_x\psi_y = \phi_y\psi_x$ ...... (ii.), $\phi_z\psi_y = \phi_y\psi_z$ ...... (iii.);

1846 or, as they may be written, $\frac{\phi_x}{\psi_x} = \frac{\phi_y}{\psi_y} = \frac{\phi_z}{\psi_z}$ ...... (iv.)

Simultaneous values of x, y, z, found by solving equations (i., ii., iii.), correspond to a maximum or minimum value of ϕ.

PROOF.—By (1841), ϕ being considered a function of two independent variables x and z, and, by (1729, 1730),

$\phi_x(x, y, z) = 0$ gives (ii.), and $\phi_z(x, y, z) = 0$ gives (iii.)

The criterion of maximum or minimum in (1842) may also be applied without eliminating y by employing the values of ϕ_{2x} and ϕ_{2z} in (1719, '30).

1847 Ex.—Let $\phi(x, y, z) = x^2+y^2+z^2$
and $\psi(x, y, z) = ax^2+by^2+cz^2+2fyz+2gzx+2hxy-1 = 0$..(i.)
Equations (ii.) and (iii.) here become

$$\frac{x}{ax+hy+gz} = \frac{y}{hx+by+fz} = \frac{z}{gx+fy+cz} = \frac{1}{R}, \text{ say, (iv.)}$$

Therefore, by proportion (70) and by (i.), $x^2+y^2+z^2 = R^{-1} = \phi$.

From equations (iv.) we have

1848 $$\left.\begin{aligned} ax+hy+gz &= Rx \\ hx+by+fz &= Ry \\ gx+fy+cz &= Rz \end{aligned}\right\}; \quad \therefore \quad \begin{vmatrix} a-R & h & g \\ h & b-R & f \\ g & f & c-R \end{vmatrix} = 0,$$

1849 or $(R-a)(R-b)(R-c)-2fgh$
$-(R-a)f^2-(R-b)g^2-(R-c)h^2 = 0$, or (see 1641)

$$\boldsymbol{R^3-R^2(a+b+c)+R(bc+ca+ab-f^2-g^2-h^2)-\Delta = 0,}$$

This cubic in R is the eliminant of the three equations in x, y, z. It is called the *discriminating cubic* of the quadric (i.), and its roots are the reciprocals of the maxima and minima values of $x^2+y^2+z^2$.

1850 To show that the roots of the discriminating cubic are all real.

Let R_1, R_2 be the roots of the quadratic equation

1851 $$\begin{vmatrix} R-b & f \\ f & R-c \end{vmatrix} \equiv (R-b)(R-c)-f^2 = 0 \qquad \text{(v.)}$$

$R_1 > b$ and c, and b and $c > R_2$.

Make $R = R_1$ in the cubic, and the result is negative, being minus a square quantity, by (v.). Make $R = R_2$, and the result is positive. Therefore the cubic has real roots between each pair of the consecutive values $+\infty$, R_1, R_2, $-\infty$; that is, three real roots. But since the roots are in order of magnitude, the first must be a maximum value of R, the third a minimum, and the intermediate root neither a maximum nor a minimum.

Maxima and Minima values of a function of three or more variables.

1852 Let $\phi(xyz)$ be a function of three variables. Let $\phi_{2x}, \phi_{2y}, \phi_{2z}, \phi_{yz}, \phi_{zx}, \phi_{xy}$ be denoted by a, b, c, f, g, h; and let A, B, C, F, G, H be the corresponding minors of the determinant Δ, as in (1642).

1853 Rule I.—$\phi(\mathrm{x}, \mathrm{y}, \mathrm{z})$ *is a maximum or minimum when* $\phi_{\mathrm{x}}, \phi_{\mathrm{y}}, \phi_{\mathrm{z}}$ *all vanish and change their signs from plus to minus or from minus to plus respectively, as* x, y, *and* z *increase.* Otherwise—

1854 Rule II.—*The first derivatives of* ϕ *must vanish;* A *and its coefficient in the reciprocal determinant of* Δ *must be positive; and* ϕ *will be a maximum or minimum according as* a *is negative or positive.* Or, in the place of A and a, read B and b or C and c.

Proof.—Pursuing the method of (1842), let ξ, η, ζ be small changes in the values of x, y, z. By (1514),

$$\phi(x+\xi, y+\eta, z+\zeta)-\phi(x, y, z)$$
$$= \xi\phi_x+\eta\phi_y+\zeta\phi_z+\tfrac{1}{2}(a\xi^2+b\eta^2+c\zeta^2+2f\eta\zeta+2g\zeta\xi+2h\xi\eta)+\&c.$$

For constancy of sign on the right, ϕ_x, ϕ_y, ϕ_z must vanish. The quadric may then be re-arranged as under by first completing the square of the terms in ξ, and then collecting the terms in ζ, η, and completing the square. It thus becomes $\frac{1}{2a}\left\{(a\xi+h\eta+g\zeta)^2+\frac{(C\eta-F\zeta)^2}{C}+\frac{BC-F^2}{C}\zeta^2\right\}$.

Hence, for constancy of sign for all values of ξ, η, ζ, it is necessary that C and $BC-F^2$ should be positive. This makes B also positive. By symmetry, it is evident that A, B, C, $BC-F^2$, $CA-G^2$, $AB-H^2$ will all be positive. The sign of a in the first factor then determines, as in (1842), whether ϕ is a maximum or a minimum.

1855 The condition may be put otherwise. Since $BC-F^2 = a\Delta$ by (577), the condition that BC^2-F^2 must be positive is equivalent to the condition that a and Δ must have the *same* sign. Hence we have also the following rule :—

1856 RULE III.—ϕ_x, ϕ_y, ϕ_z *must vanish; the second of the four determinants below must be positive, and the first and third must have the same sign: that sign being negative for a maximum and positive for a minimum value of* $\phi(x, y, z)$.

1857 $$\phi_{2x}, \quad \begin{vmatrix} \phi_{2x} & \phi_{xy} \\ \phi_{yx} & \phi_{2y} \end{vmatrix}, \quad \begin{vmatrix} \phi_{2x} & \phi_{xy} & \phi_{xz} \\ \phi_{yx} & \phi_{2y} & \phi_{yz} \\ \phi_{zx} & \phi_{zy} & \phi_{2z} \end{vmatrix}, \quad \begin{vmatrix} \phi_{2x} & \phi_{xy} & \phi_{xz} & \phi_{xw} \\ \phi_{yx} & \phi_{2y} & \phi_{yz} & \phi_{yw} \\ \phi_{zx} & \phi_{zy} & \phi_{2z} & \phi_{zw} \\ \phi_{wx} & \phi_{wy} & \phi_{wz} & \phi_{2w} \end{vmatrix}.$$

1858 The theorem can be extended in a similar manner to $\phi(x, y, z, w, \ldots)$, a function of any number of variables. Form the successive Hessians of ϕ (1630) for one, two, three, &c. variables in order as shown above; then—

1859 RULE.—*In order that* $\phi(x, y, z, w, \ldots)$ *may be a maximum or minimum*, $\phi_x, \phi_y, \phi_z, \phi_w$, &c. *must vanish; the Hessians of an even order must be positive; and those of an odd order must have the same sign, that sign being negative for a maximum and positive for a minimum value of the function* ϕ.

For a demonstration in full, see Williamson's *Diff. Calc.*, 4th Edit., p. 433.

1860 EX.—Required a limiting value of the function

$$u = ax^2+by^2+cz^2+2fyz+2gzx+2hxy+2px+2qy+2rz+d.$$

The condition in the rule produces equations (1), (2), (3). Equation (4) results from Euler's theorem (1624), thus; introducing a fourth variable w, as in (1645), we have $\quad xu_x+yu_y+zu_z+wu_w = 2u$,

which reduces to (4) by means of (1), (2), (3), and the value of u, putting $w = 1$.

1861 $$\left.\begin{aligned} \tfrac{1}{2}u_x &= ax+hy+gz+p = 0 \ldots\ldots (1) \\ \tfrac{1}{2}u_y &= hx+by+fz+q = 0 \ldots\ldots (2) \\ \tfrac{1}{2}u_z &= gx+fy+cz+r = 0 \ldots\ldots (3) \\ & px+qy+rz+d = u \ldots\ldots (4) \end{aligned}\right\}; \quad \therefore \begin{vmatrix} a & h & g & p \\ h & b & f & q \\ g & f & c & r \\ p & q & r & d-u \end{vmatrix} = 0.$$

The determinant is the eliminant of the four equations, by (583), and is equivalent, by the method of (1724, Proof. i.), to $\Delta' - \Delta u = 0$, or $u = \Delta' \div \Delta$ (Notation of 1646).

To determine whether this value of u is a maximum or minimum, either of the conditions in (1854, '6) may be applied; and since, in this example, $u_{2x} = 2a$, $u_{2y} = 2b$, &c., the letters a, b, c, f, g, h may be considered identical with those in the rule.

1862 To determine a limiting value of $\phi(x, y, z, \ldots)$, a function of m variables connected by n equations $u_1 = 0$, $u_2 = 0$, ... $u_n = 0$.

RULE.—*Assume* n *undetermined multipliers* $\lambda_1, \lambda_2, \ldots \lambda_n$ *with the following* m *equations*:—

$$\phi_x + \lambda_1 (u_1)_x + \lambda_2 (u_2)_x + \ldots + \lambda_n (u_n)_x = 0,$$
$$\phi_y + \lambda_1 (u_1)_y + \lambda_2 (u_2)_y + \ldots + \lambda_n (u_n)_y = 0,$$
$$\ldots \quad \ldots \quad \ldots \quad \ldots \quad \ldots \quad \ldots$$

making in all m+n *equations in* m+n *quantities*, x, y, z, ... *and* $\lambda_1, \lambda_2, \ldots \lambda_n$. *The values of* x, y, z, ..., *found from these equations, correspond to a maximum or minimum value of* ϕ.

PROOF.—Differentiate ϕ and $u_1, u_2, \ldots u_n$ on the hypothesis that $x, y, z, \ldots$ are *arbitrary* functions of an independent variable t. Multiply the resulting equations, excepting the first, by $\lambda_1, \lambda_2, \ldots \lambda_n$ in order, and add them to the value of ϕ_t. The coefficients of $x_t, y_t, z_t, \ldots$ may now be equated to zero, since the functions of t are arbitrary, producing the equations in the rule.

1863 Ex. 1.—To find the limiting values of $r^2 = x^2 + y^2 + z^2$ with the conditions $Ax^2 + By^2 + Cz^2 = 1$ and $lx + my + nz = 0$.

Here $m = 3$, $n = 2$; and, choosing λ and μ for the multipliers, the equations in the rule become

$$\left.\begin{array}{l} 2x + 2A\lambda x + \mu l \;\; = 0 \ldots\ldots (1) \\ 2y + 2B\lambda y + \mu m = 0 \ldots\ldots (2) \\ 2z + 2C\lambda z + \mu n \;\; = 0 \ldots\ldots (3) \end{array}\right\}.$$

Multiply (1), (2), (3) respectively by x, y, z, and add; thus μ disappears, and we obtain

$$x^2 + y^2 + z^2 + (Ax^2 + By^2 + Cz^2)\lambda = 0, \qquad \text{therefore } \lambda = -r^2.$$

Substitute this in (1), (2), (3); solve for x, y, z, and substitute their values in
$$lx + my + nz = 0.$$

1864 The result is $\dfrac{l^2}{Ar^2 - 1} + \dfrac{m^2}{Br^2 - 1} + \dfrac{n^2}{Cr^2 - 1} = 0$, a quadratic in r^2.

The roots are the maximum and minimum values of the square of the radius vector of a central section of the quadric $Ax^2 + By^2 + Cz^2 = 1$ made by the plane $lx + my + nz = 0$.

1865 Ex. 2.—To find the maximum value of $u = (x+1)(y+1)(z+1)$, subject to the condition $N = a^x b^y c^z$.

This is equivalent to finding a maximum value of

$$\log(x+1)+\log(y+1)+\log(z+1),$$

subject to the condition $\log N = x\log a + y\log b + z\log c$.

The equations in the rule become

$$\frac{1}{x+1}+\lambda\log a = 0;\qquad \frac{1}{y+1}+\lambda\log b = 0;\qquad \frac{1}{z+1}+\lambda\log c = 0.$$

By eliminating λ, these are seen to be equivalent to equations (1846).

Multiplying up and adding the equations, we find λ, and thence $x+1$, $y+1$, $z+1$; the values of which, substituted in u, give, for its maximum value,

$$u = \{\log(Nabc)\}^3 \div 3\log a^3\log b^3\log c^3.$$

Compare (374), where a, b, c and x, y, z are integers.

Continuous Maxima and Minima.

1866 If ϕ_x and ϕ_y, in (1842), have a common factor, so that

$$\phi_x = P\psi(x, y),\qquad \phi_y = Q\psi(x, y)\ \ldots\ldots\ldots\text{(i.)},$$

where P and Q may also be functions of x and y; then the equation $\psi(x, y) = 0$ determines a continuous series of values of x and y. For all these values ϕ is constant, but, at the same time, it may be a maximum or a minimum *with respect to any other contiguous values of* ϕ, obtained by taking x and y so that $\psi(xy)$ shall not vanish.

1867 In this case, $\phi_{2x}\phi_{2y}-\phi_{xy}^2$ vanishes with ψ, so that the criterion in Rule II. is not applicable.

Proof.—Differentiating equation (i.), we have

$$\left.\begin{aligned}\phi_{2x} &= P_x\psi+P\psi_x\\ \phi_{2y} &= Q_y\psi+Q\psi_y\end{aligned}\right\},\qquad \left.\begin{aligned}\phi_{xy} &= P_y\psi+P\psi_y\\ \phi_{yx} &= Q_x\psi+Q\psi_x\end{aligned}\right\}.$$

If from these values we form $\phi_{2x}\phi_{2y}-\phi_{xy}\times\phi_{yx}$, ψ will appear as a factor of the expression.

1868 Ex.—Take z as $\phi(xy)$ in the equation

$$z^2 = a^2-b^2+2b\sqrt{(x^2+y^2)}-x^2-y^2\ \ldots\ldots\ldots\ldots\ldots\ldots\ \text{(i.)},$$

$$\therefore\qquad zz_x = x\left(\frac{b}{\sqrt{(x^2+y^2)}}-1\right)\quad\text{and}\quad zz_y = y\left(\frac{b}{\sqrt{(x^2+y^2)}}-1\right).$$

The common factor equated to zero gives $x^2+y^2 = b^2$, and therefore $z = \pm a$...(ii.) Here a is a continuous maximum value of z, and $-a$ a continuous minimum.

Equation (i.) represents, in Coordinate Geometry, the surface of an anchor ring, the generating circle of radius a having its centre at a distance b from the axis of revolution Z. Equations (ii.) give the loci of the highest and lowest points of the surface.

For the application of the Differential Calculus to the Theory of Curves, see the Sections on *Coordinate Geometry*.

INTEGRAL CALCULUS.

INTRODUCTION.

1900 The operations of *differentiation* and *integration* are the converse of each other. Let $f(x)$ be the *derivative* of $\phi(x)$; then $\phi(x)$ is called the *integral* of $f(x)$ with respect to x. These converse relations are expressed in the notations of the Differential and Integral Calculus, by

$$\frac{d\phi(x)}{dx} = f(x) \quad \text{and by} \quad \int f(x)\,dx = \phi(x).$$

1901 Theorem.—Let the curve $y = f(x)$ be drawn as in (1403), and any ordinates Ll, Mm, and let $OL = a$, $OM = b$; then the area $LMml = \phi(b) - \phi(a)$.

Proof.—Let ON be any value of x, and PN the corresponding value of y, and let the area $ONPQ = A$; then A *is some function of* x. Also, if $NN' = dx$, the elemental area $NN'P'P = dA$ $= ydx$ in the limit; therefore $\frac{dA}{dx} = y$. Thus A is that function of x whose derivative *for each value* of x is y or $f(x)$; therefore $A = \phi(x) + C$, where C is any constant. Consequently the area $LMml = \phi(b) - \phi(a)$, whatever C may be.

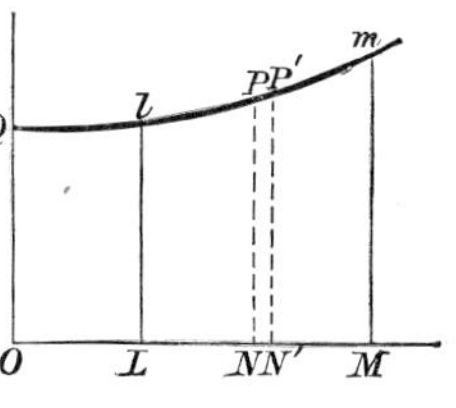

The demonstration assumes that there is only one function $\phi(x)$ corresponding to a given derivative $f(x)$. This may be formally proved.

If possible, let $\psi(x)$ have the same derivative as $\phi(x)$; then, with the same coordinate axes, two curves may be drawn so that the areas defined as above, like $LMml$, shall be $\phi(x)$ and $\psi(x)$ respectively, each area vanishing with x. If these curves do not coincide, then, for a given value of x, they have different ordinates, that is, $\phi'(x)$ and $\psi'(x)$ are different, contrary to the hypothesis. The curves must therefore coincide, that is, $\phi(x)$ and $\psi(x)$ are identical.

1902 Since $\phi(b)-\phi(a)$ is the sum of all the elemental areas like $NN'P'P$ included between Ll and Mm, that is, the sum of the elements ydx or $f(x)\,dx$ taken for all values of x between a and b, this result is written

$$\int_a^b f(x)\,dx = \phi(b)-\phi(a).$$

1903 The expression on the left is termed a *definite integral* because the limits a, b of the integration are assigned.* When the limits are not assigned, the integral is called *indefinite*.

1904 By taking the constant $C=0$ in (1901), we have the area

$$ONPQ = \phi(x) = \int f(x)\,dx.$$

NOTE.—*In practice, the constant should always be added to the result of an integration when no limits are assigned.*

MULTIPLE INTEGRALS.

1905 Let $f(x, y, z)$ be a function of three variables; then the notation

$$\int_{x_1}^{x_2}\int_{y_1}^{y_2}\int_{z_1}^{z_2} f(x, y, z)\,dx\,dy\,dz$$

is used to denote the following operations.

Integrate the function for z between the limits $z=z_1$, $z=z_2$, considering the remaining variables x and y constant. Then, whether the limits z_1, z_2 are constants or functions of x and y, the result will be a function of x and y only. Next, considering x constant, integrate this function for y between the limits y_1 and y_2, which may either be constants or functions of x. The result will now be a function of x only. Lastly, integrate this function for x between the limits x_1 and x_2.

Similarly for a function of any number of variables.

1906 The clearest view of the nature of a multiple integral is afforded by the geometrical interpretation of a triple integral.

---

* The integral may be read "*Sum a to b, $f(x)\,dx$*"; $\int$ signifying "*sum.*"

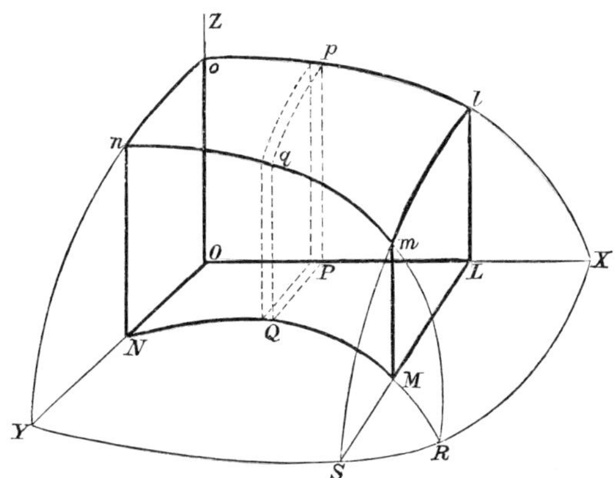

Taking rectangular coordinate axes, let the surface $z=\phi(x,y)$ ($XYom$ in the figure) be drawn, intersected by the cylindrical surface $y=\psi(x)$ ($RMNnm$), and by the plane $x=x_1$ ($LSml$). The volume of the solid $OLMNolmn$ bounded by these surfaces and the coordinate planes will be

$$\int_0^{x_1}\int_0^{\psi(x)}\phi(x,y)\,dx\,dy=\int_0^{x_1}\int_0^{\psi(x)}\int_0^{\phi(x,y)}dx\,dy\,dz.$$

Proof.—Since the volume cut off by any plane parallel to OYZ, and at a distance x from it, varies continuously with x, it must be *some function of x*. Let V be this volume, and let dV be the small change in its value due to a change dx in x. Then, in the limit, $dV=PQqp\times dx$, an element of the solid shown by dotted lines in the figure. Therefore

$$\frac{dV}{dx}=PQqp=\int_0^{\psi(x)}\phi(x,y)\,dy,\ \text{by (1902)},$$

x being constant throughout the integration for y. The result will be a function of x only. Making x then vary from 0 to x, we have, for the whole

volume, $$\int_0^{x_1}\left\{\int_0^{\psi(x)}\phi(x,y)\,dy\right\}dx=\int_0^{x_1}\left\{\int_0^{\psi(x)}\left[\int_0^{\phi(x,y)}dz\right]dy\right\}dx,$$

since $\phi(x,y)=z$. With the notation explained in (1905), the brackets are not required, and the integrals are written as above.

1907 If the solid is bounded by two surfaces $z_1=\phi_1(x,y)$, $z_2=\phi_2(x,y)$, two cylindrical surfaces $y_1=\psi_1(x)$, $y_2=\psi_2(x)$, and two planes $x=x_1$, $x=x_2$, the volume will then be arrived at by taking the difference of two similar integrals at each integration, and will be expressed by the integral in (1905).

If any limit is a constant, the corresponding boundary of the solid becomes a plane.

METHODS OF INTEGRATION.

INTEGRATION BY SUBSTITUTION.

1908 The formula is $\int \phi(x)\,dx = \int \phi(x)\,\frac{dx}{dz}\,dz,$

where z is equal to $f(x)$, some function chosen so as to facilitate the integration.

RULE I.—*Put* x *in terms of* z *in the given function, and multiply the function also by* x_z; *then integrate for* z.

If the limits of the proposed integral are given by $x = a$, $x = b$, *these must be converted into limits of* z *by the equation* $x = f^{-1}(z)$.

The following rule presents another view of the method of substitution, and is useful in practice.

1909 RULE II.—*If* $\phi(x)$ *can be expressed in the form* $F(z)\,z_x$;

then $$\int \phi(x)\,dx = \int F(z)\,z_x dx = \int F(z)\,dz.$$

Ex. 1.—To integrate $\int \frac{dx}{\sqrt{(x^2+a^2)}}$. Substitute $z = x+\sqrt{(x^2+a^2)}$;

therefore $\frac{dz}{dx} = 1+\frac{x}{\sqrt{(x^2+a^2)}} = \frac{x+\sqrt{(x^2+a^2)}}{\sqrt{(x^2+a^2)}} = \frac{z}{\sqrt{(x^2+a^2)}}$,

$$\int \frac{1}{\sqrt{(x^2+a^2)}}\,\frac{dx}{dz}\,dz = \int \frac{dz}{z} = \log z = \log\{x+\sqrt{(x^2+a^2)}\}.$$

Ex. 2. $$\int \frac{5+2x^3}{x+x^4}\,dx = \int \frac{5x^{-6}+2x^{-3}}{x^{-5}+x^{-2}}\,dx = -\log(x^{-5}+x^{-2}).$$

Here $z = x^{-5}+x^{-2}$, $F(z) = -\frac{1}{z}$, $z_x = -(5x^{-6}+2x^{-3})$.

Ex. 3. $$\int \frac{1-cx^2}{1+cx^2}\,\frac{dx}{\sqrt{(1+ax^2+c^2x^4)}} = \int \frac{x^{-2}-c}{x^{-1}+cx}\,\frac{dx}{\sqrt{(c^2x^2+x^{-2}+a)}}$$

$$= -\int \frac{d_x(x^{-1}+cx)\,dx}{(x^{-1}+cx)\sqrt{\{(x^{-1}+cx)^2+a-2c\}}}$$

$$= \frac{1}{\sqrt{(a-2c)}}\log\frac{x\sqrt{(a-2c)}+\sqrt{(1+ax^2+c^2x^4)}}{1+cx^2},$$

or $$-\frac{1}{\sqrt{(2c-a)}}\cos^{-1}\frac{x\sqrt{(2c-a)}}{1+cx^2}.$$

By (1927) or (1926). Here $z = x^{-1}+cx$.

In Examples (2) and (3) the process is analytical, and leads to the discovery of the particular function z, with respect to which the integration is effected. If z be known, Rule I. supplies the direct, though not always the simplest, method of integrating the function.

INTEGRATION BY PARTS.

1910 By differentiating uv with respect to x, we obtain the general formula $\int u_x v\,dx = uv - \int uv_x\,dx.$

The value of the first integral is thus determined if that of the second is known.

RULE.—*Separate the quantity to be integrated into two factors. Integrate one factor, and differentiate the other with respect to* x. If the integral of the resulting quantity is known, or more readily ascertained than that of the original one, the method by "Parts" is applicable.

1911 NOTE.—In subsequent examples, where integration by Parts is directed, the factor which is to be integrated will be indicated. Thus, in example (1951), "By Parts $\int e^{ax}dx$" signifies that e^{ax} is to be integrated and $\sin bx$ differentiated afterwards in applying the foregoing rule. The factor 1 is more frequently integrated than any other, and this step will be denoted by $\int dx$.

INTEGRATION BY DIVISION.

1912 A formula is

$$\int (a+bx^n)^p\,dx = a\int (a+bx^n)^{p-1} + b\int x^n (a+bx^n)^{p-1}\,dx:$$

The expression to be integrated is thus *divided* into two terms, the index p in each being diminished by unity, a step which often facilitates integration.

Similarly, $\int (a+bx^n+cx^m)^p\,dx$

$$= a\,(a+bx^n+cx^m)^{p-1} + bx^n\,(a+bx^n+cx^m)^{p-1} + cx^m\,(a+bx^n+cx^m)^{p-1}.$$

1913 Ex.—To integrate $\int \sqrt{(x^2+a^2)}\,dx.$

By Parts $\int dx$, $\int \sqrt{(x^2+a^2)}\,dx = x\sqrt{(x^2+a^2)} - \int \frac{x^2dx}{\sqrt{(x^2+a^2)}}.$

By Division, $\int \sqrt{(x^2+a^2)}\,dx = \int \frac{a^2dx}{\sqrt{(x^2+a^2)}} + \int \frac{x^2dx}{\sqrt{(x^2+a^2)}}.$

Therefore, by addition,

$$\int \sqrt{(x^2+a^2)}\,dx = \tfrac{1}{2}x\sqrt{(x^2+a^2)} + \tfrac{1}{2}\int \frac{a^2dx}{\sqrt{(x^2+a^2)}}$$

$$= \tfrac{1}{2}x\sqrt{(x^2+a^2)} + \tfrac{1}{2}a^2 \log\{x+\sqrt{(x^2+a^2)}\},$$ by (1909, Ex. 1).

INTEGRATION BY RATIONALIZATION.

1914 In the following example, 6 is the least common denominator of the fractional indices. Hence, by substituting $z=x^{\frac{1}{6}}$, and therefore $x_z=6z^5$, we have

$$\int\frac{x^{\frac{1}{2}}-1}{x^{\frac{1}{3}}-1}\,dx=\int\frac{z^3-1}{z^2-1}\,\frac{dx}{dz}\,dz=6\int\frac{z^8-z^5}{z^2-1}\,dz$$

$$=6\int\left(z^6+z^4-z^3+z^2-z+1-\frac{1}{z+1}\right)dx.$$

Each term of the result is directly integrable by (1922) and (1923). For other examples see (2110).

INTEGRATION BY PARTIAL FRACTIONS.

1915 Rational fractions can always be integrated by first resolving them into partial fractions. The theory of such resolutions will now be given.

1916 If $\phi(x)$ and $F(x)$ are rational algebraic functions of x, $\phi(x)$ being of lowest dimensions, and if $F(x)$ contains the factor $(x-a)$ once, so that

$$F(x)=(x-a)\,\psi(x)\;\ldots\ldots\ldots\ldots\ldots\ldots\;(1);$$

1917 then $\dfrac{\phi(x)}{F(x)}=\dfrac{A}{x-a}+\dfrac{\chi(x)}{\psi(x)}$ and $A=\dfrac{\phi(a)}{F'(a)}\;\ldots\ldots(2).$

PROOF.—Multiply equation (2) by (1), thus

$$\phi(x)=A\psi(x)+(x-a)\,\chi(x).$$

Therefore, putting $x=a$, $\phi(a)=A\psi(a)$. Also, by differentiating (1), and putting $x=a$ afterwards, $F'(a)=\psi(a)$. Therefore $A=\phi(a)\div F'(a)$.

1918 Again, if $F(x)$ contains the factor $(x-a)$, n times, so that

$$F(x)=(x-a)^n\,\psi(x).$$

Assume $$\frac{\phi(x)}{F(x)}=\frac{A_1}{(x-a)^n}+\frac{A_2}{(x-a)^{n-1}}+\ldots+\frac{A_n}{x-a}+\frac{\chi(x)}{\psi(x)}.$$

To determine A_1, A_2 ... A_n. *Multiply by* $(x-a)^n$; *put* $x=a$ *and differentiate, alternately.*

1919 If $F(x)=0$ has a single pair of imaginary roots

$\alpha \pm i\beta$; then, applying (1917), let

$$\frac{\phi(\alpha+i\beta)}{F'(\alpha+i\beta)} = A-iB; \quad \therefore \quad \frac{\phi(\alpha-i\beta)}{F'(\alpha-i\beta)} = A+iB;$$

and the partial fractions corresponding to these roots will be

$$\frac{A-iB}{x-\alpha-i\beta}+\frac{A+iB}{x-\alpha+i\beta}=\frac{2A(x-\alpha)+2B\beta}{(x-\alpha)^2+\beta^2}.$$

For practical methods of resolving a fraction into partial fractions in the different cases which occur, see (235–238).

INTEGRATION BY INFINITE SERIES.

When other methods are not applicable, an integral may sometimes be evaluated by expanding the function in a converging series and integrating the separate terms.

Ex. $$\int \frac{e^{ax}}{x}\,dx = \log x + ax + \frac{a^2x^2}{1.2^2}+\frac{a^3x^3}{1.2.3^2}+\frac{a^4x^4}{1.2.3.4^2}+\&\text{c}. \qquad (150)$$

STANDARD INTEGRALS.

1921 Some elementary integrals are obtained at once from the known derivatives of simple functions. Thus the derivatives (1422–38) furnish corresponding integrals. The following are in constant use:—

1922 $$\int x^m\,dx = \frac{x^{m+1}}{m+1}. \qquad \int \frac{dx}{x} = \log x.$$

1924 $$\int a^x\,dx = \frac{a^x}{\log a}. \qquad \int e^x\,dx = e^x.$$

1926 $$\int \frac{dx}{x\sqrt{(x^2-a^2)}} = \frac{1}{a}\cos^{-1}\frac{a}{x} \text{ or } -\frac{1}{a}\sin^{-1}\frac{a}{x}. \qquad \text{[Subs. } \frac{1}{x}$$

1927 $$\int \frac{dx}{x\sqrt{(a^2\pm x^2)}} = \frac{1}{a}\log\frac{x}{a+\sqrt{(a^2\pm x^2)}}. \qquad \text{[Substitute } \frac{1}{x}$$

1928 $\displaystyle\int \frac{dx}{\sqrt{(x^2 \pm a^2)}} = \log\left\{x + \sqrt{(x^2 \pm a^2)}\right\}.$ (1909, Ex. 1)

1929 $\displaystyle\int \frac{dx}{\sqrt{(a^2 - x^2)}} = \sin^{-1}\frac{x}{a} \quad \text{or} \quad -\cos^{-1}\frac{x}{a}.$ (1434)

1930 $\displaystyle\int \frac{x\,dx}{\sqrt{(a^2 \pm x^2)}} = \pm\sqrt{(a^2 \pm x^2)}.$

1931 By Parts, Division, and adding results (1913), we obtain

$$\int \sqrt{(x^2 \pm a^2)}\,dx = \tfrac{1}{2}x\sqrt{(x^2 \pm a^2)} \pm \tfrac{1}{2}a^2 \log\left\{x + \sqrt{(x^2 \pm a^2)}\right\}.$$

1932 By Parts, Division, and difference of results,

$$\int \frac{x^2\,dx}{\sqrt{(x^2 \pm a^2)}} = \tfrac{1}{2}x\sqrt{(x^2 \pm a^2)} \mp \tfrac{1}{2}a^2 \log\left\{x + \sqrt{(x^2 \pm a^2)}\right\}.$$

1933 $\displaystyle\int \sqrt{(a^2 - x^2)}\,dx = \tfrac{1}{2}a^2 \sin^{-1}\frac{x}{a} + \tfrac{1}{2}x\sqrt{(a^2 - x^2)}.$ [As in (1931)

1934 $\displaystyle\int \frac{x^2\,dx}{\sqrt{(a^2 - x^2)}} = \tfrac{1}{2}a^2 \sin^{-1}\frac{x}{a} - \tfrac{1}{2}x\sqrt{(a^2 - x^2)}.$ [As in (1932)

1935 $\displaystyle\int \frac{dx}{x^2 + a^2} = \frac{1}{a}\tan^{-1}\frac{x}{a} \quad \text{or} \quad -\frac{1}{a}\cot^{-1}\frac{x}{a}$ (1436)

1936 $\displaystyle\int \frac{dx}{x^2 - a^2} = \frac{1}{2c}\log\frac{x - a}{x + a}.$ [By Partial fractions

1937 $\displaystyle\int \frac{dx}{a^2 - x^2} = \frac{1}{2a}\log\frac{a + x}{a - x}.$ [Do.

1938 $\displaystyle\int \sin x\,dx = -\cos x.$ $\displaystyle\int \cos x\,dx = \sin x.$

1940 $\displaystyle\int \tan x\,dx = -\log\cos x.$ $\displaystyle\int \cot x\,dx = \log\sin x.$

1942 $\displaystyle\int \sec x\,dx = \log\tan\left(\frac{\pi}{4} + \frac{x}{2}\right).$ $\displaystyle\int \operatorname{cosec} x\,dx = \log\tan\frac{x}{2}.$

METHOD.—(1940, '2), substitute $\cos x$. (1941, '3), substitute $\sin x$.

1944 $\displaystyle\int \sin^{-1}x\,dx = x\sin^{-1}x + \sqrt{(1 - x^2)}.$

1945 $\displaystyle\int \cos^{-1}x\,dx = x\cos^{-1}x - \sqrt{(1 - x^2)}.$

1946 $\int \tan^{-1} x\,dx = x\tan^{-1}x - \frac{1}{2}\log(1+x^2).$

1947 $\int \cot^{-1} x\,dx = x\tan^{-1}x + \frac{1}{2}\log(1+x^2).$

1948 $\int \sec^{-1} x\,dx = x\sec^{-1}x - \log\{x+\sqrt{(x^2-1)}\}.$

1949 $\int \operatorname{cosec}^{-1} x\,dx = x\operatorname{cosec}^{-1}x + \log\{x+\sqrt{(x^2-1)}\}.$

METHOD.—(1944) to (1949), integrate by Parts, $\int dx$.

1950 $\int \log x\,dx = x\log x - x.$ [By Parts, $\int dx$

1951 $\int \frac{dx}{a+b\cos x} = \frac{2}{\sqrt{(a^2-b^2)}}\tan^{-1}\left\{\tan\frac{x}{2}\sqrt{\left(\frac{a-b}{a+b}\right)}\right\},$

1952 or $\frac{1}{\sqrt{(b^2-a^2)}}\log\frac{\sqrt{(b+a)}+\sqrt{(b-a)}\tan\frac{1}{2}x}{\sqrt{(b+a)}-\sqrt{(b-a)}\tan\frac{1}{2}x},$

according as a is $>$ or $<$ b.

[Subs. $\tan\frac{1}{2}x$, and integrate by (1935 or '37)

VARIOUS INDEFINITE INTEGRALS.

GENERALIZED CIRCULAR FUNCTIONS.

1954 $\int \sin^n x\,dx.$ $\int \cos^n x\,dx.$ $\int \operatorname{cosec}^n x\,dx.$

METHOD.—When n is integral, integrate the expansions in (772–4). Otherwise by successive reduction, see (2060). For $\int \operatorname{cosec}^n dx$, see (2058).

1957 $\int \tan^n x\,dx = \frac{\tan^{n-1}x}{n-1} - \frac{\tan^{n-3}x}{n-3} + \frac{\tan^{n-5}x}{n-5} - \&c.$

PROOF.—By Division; $\tan^n x = \tan^{n-2}x\sec^2 x - \tan^{n-2}x$, the first term of which is integrable; and so on.

1958 $\int \frac{dx}{(a+b\cos x)^n} = \frac{1}{(a^2-b^2)^{n-\frac{1}{2}}}\int (a-b\cos z)^{n-1}\,dz.$

METHOD.—By substituting $\tan\frac{z}{2} = \tan\frac{x}{2}\sqrt{\left(\frac{a-b}{a+b}\right)}$. Similarly with $\sin x$ in the place of $\cos x$, substitute $\frac{1}{2}\pi - x$.

1959 $\int \frac{\cos^p x}{\cos nx}dx.$ $\int \frac{\cos^p x}{\sin nx}dx.$ $\int \frac{\sin^p x}{\cos nx}dx.$ $\int \frac{\sin^p x}{\sin nx}dx.$

METHOD.—By (809 & 812), when p and n are integers, the first two functions can be resolved into partial fractions as under, p being $< n$ in the first and $< n-1$ in the second. The third and fourth integrals reduce to one or other of the former by substituting $\frac{1}{2}\pi - x$.

1963 $$\frac{\cos^p x}{\cos nx} = \frac{1}{n}\,\Sigma_{r=1}^{r=n}(-1)^{r+1}\frac{\sin(2r-1)\theta\cos^p(2r-1)\theta}{\cos x-\cos(2r-1)\theta}, \text{ with } \theta = \frac{\pi}{2n}.$$

1964 $$\frac{\cos^p x}{\sin nx} = \frac{1}{n\sin x}\,\Sigma_{r=1}^{r=n-1}(-1)^{r+1}\frac{\sin^2 r\theta\cos^p r\theta}{\cos x-\cos r\theta}, \text{ with } \theta = \frac{\pi}{n}.$$

The fractions in (1963) are integrated by (1952); those in (1964) by (1990).

Formulæ of Reduction.

1965 $$\int\frac{\cos nx}{\cos^p x}dx = 2\int\frac{\cos(n-1)x}{\cos^{p-1}x}dx - \int\frac{\cos(n-2)x}{\cos^p x}dx$$

1966 $$\int\frac{\cos nx}{\sin^p x}dx = -2\int\frac{\sin(n-1)x}{\sin^{p-1}x}dx + \int\frac{\cos(n-2)x}{\sin^p x}dx$$

1967 $$\int\frac{\sin nx}{\sin^p x}dx = 2\int\frac{\cos(n-1)x}{\sin^{p-1}x}dx + \int\frac{\sin(n-2)x}{\sin^p x}dx$$

1968 $$\int\frac{\sin nx}{\cos^p x}dx = 2\int\frac{\sin(n-1)x}{\cos^{p-1}x}dx - \int\frac{\sin(n-2)x}{\cos^p x}dx$$

PROOF. — In (1965). $2\cos x\cos(n-1)x = \cos nx + \cos(n-2)x$, &c. Similarly in (1966–8).

1969 $$\int\sin^p x\sin nx\,dx$$
$$= -\frac{\sin^p x\cos nx}{p+n} + \frac{p}{p+n}\int\sin^{p-1}x\cos(n-1)x\,dx.$$

1970 $$\int\cos^p x\sin nx\,dx$$
$$= -\frac{\cos^p x\cos nx}{p+n} + \frac{p}{p+n}\int\cos^{p-1}x\sin(n-1)x\,dx.$$

1971 $$\int\sin^p x\cos nx\,dx$$
$$= \frac{\sin^p x\sin nx}{p+n} - \frac{p}{p+n}\int\sin^{p-1}x\sin(n-1)x\,dx.$$

1972 $$\int\cos^p x\cos nx\,dx$$
$$= \frac{\cos^p x\sin nx}{p+n} + \frac{p}{p+n}\int\cos^{p-1}x\cos(n-1)x\,dx.$$

PROOF.—(1969). By Parts, $\int \sin nx\,dx$. In the new integral change $\cos nx \cos x$ into $\cos(n-1)x - \sin nx \sin x$. By successive reduction in this way the integral may be found. Similarly in (1970–2).

Otherwise, expand $\sin^p x$ or $\cos^p x$ in multiple angles by (772–4), and integrate the terms by the following formulæ.

1973—1975

$$\int \sin px \sin nx\,dx = \frac{1}{2}\left(\frac{\sin(p-n)x}{p-n} - \frac{\sin(p+n)x}{p+n}\right).$$

and so with similar forms, by (666–9).

1976 $\int \frac{\cos px\,dx}{\cos nx}$ and $\int \frac{\sin px\,dx}{\cos nx}$ are found from

$$\int \frac{i\cos px - \sin px}{\cos nx}\,dx = 2\int \frac{z^{p+n-1}\,dz}{1+z^{2n}}$$

when p and n are integers, by equating real and imaginary parts after integrating the right side by (2023).

PROOF.—Put $\cos x + i\sin x = z$; therefore $iz\,dx = dz$. Multiplying numerator and denominator of the fraction below by $\cos nx + i\sin nx$, we get

$$\frac{\cos px + i\sin px}{\cos nx} = 2\,\frac{\cos(p+n)x + i\sin(p+n)x}{1+\cos 2nx + i\sin 2nx} = 2\,\frac{z^{p+n}}{1+z^{2n}};$$

therefore $$\int \frac{\cos px + i\sin px}{\cos nx}\,dx = -2i\int \frac{z^{p+n-1}\,dz}{1+z^{2n}}.$$

1978 $\int \frac{\cos px\,dx}{\sin nx}$ and $\int \frac{\sin px\,dx}{\sin nx}$ are found in the same way from $$\int \frac{\cos px + i\sin px}{\sin nx}\,dx = 2\int \frac{z^{p+n-1}\,dz}{1-z^{2n}}.$$

PROOF.—As in (1976), by multiplying numerator and denominator of $\frac{\cos px + i\sin px}{\sin nx}$ by $\cos nx - i\sin nx$.

1980 $$\int \frac{\cos x\,dx}{\sqrt[n]{(\cos nx)}} \quad\text{and}\quad \int \frac{\sin x\,dx}{\sqrt[n]{(\cos nx)}}.$$

Putting $y = \frac{\cos x + i\sin x}{\sqrt[n]{(\cos nx)}}$, we find $\tan nx = i(1-y^n)$, and therefore $y\,dx = \frac{-i\,dy}{2-y^n}$. Hence, multiplying by i, we have

$$\int \frac{i\cos x - \sin x}{\sqrt[n]{(\cos nx)}}\,dx = \int \frac{dy}{2-y^n}.$$

The real part and the coefficient of i in the expansion of the integral on the right by (2021, '2), are the values required.

1982 $\displaystyle\int \frac{dx}{a\cos^2 x + b\sin^2 x} = \frac{1}{\sqrt{(ab)}}\tan^{-1}\left(\tan x\sqrt{\frac{b}{a}}\right).$ [Subs. $\tan x$

1983 $\displaystyle\int \frac{dx}{a+b\tan x} = \frac{1}{a^2+b^2}\{b\log(a\cos x+b\sin x)+ax\}.$ [Subs. $\tan x$

1984 $\displaystyle\int \frac{dx}{a+b\sin^2 x} = \frac{1}{\sqrt{(a^2+ab)}}\tan^{-1}\frac{\sqrt{(a+b)}}{\cot x\sqrt{a}}.$ [Substitute $\cot x$

1985 $\displaystyle\int \frac{\sin x\cos^2 x}{1+a^2\cos^2 x}\,dx = \frac{1}{a^3}\tan^{-1}(a\cos x) - \frac{\cos x}{a^2}.$ [Substitute $a\cos x$

1986 $\displaystyle\int \frac{\cos^3 x\,dx}{1-a^2\cos^2 x} = \frac{1}{a^3\sqrt{(1-a^2)}}\tan^{-1}\frac{a\sin x}{\sqrt{(1-a^2)}} - \frac{\sin x}{a^2}.$ [Subs. $\sin x$

1987 $\displaystyle\int \cos x\sqrt{(1-a^2\sin^2 x)}\,dx = \tfrac{1}{2}\sin x\sqrt{(1-a^2\sin^2 x)} + \frac{1}{2a}\sin^{-1}(a\sin x).$
[Substitute $a\sin x$

1988 $\displaystyle\int \sin x\sqrt{(1-a^2\sin^2 x)}\,dx = -\tfrac{1}{2}\cos x\sqrt{(1-a^2\sin^2 x)}$
$\displaystyle -\frac{1-a^2}{2a}\log\{a\cos x+\sqrt{(1-a^2\sin^2 x)}\}.$ [Subs. $a\cos x$

1989 $\displaystyle\int \sin x\,(1-a^2\sin^2 x)^{\frac{3}{2}}\,dx = -\tfrac{1}{4}\cos x\,(1-a^2\sin^2 x)^{\frac{3}{2}}$
$\displaystyle +\tfrac{3}{4}(1-a^2)\int \sin x\sqrt{(1-a^2\sin^2 x)}\,dx.$ [Subs. $a\cos x$

1990 $\displaystyle\int \frac{dx}{\sin x\,(a+b\cos x)} = \frac{\log\{(a\operatorname{cosec} x+b\cot x)^b\tan^a\tfrac{1}{2}x\}}{a^2-b^2}.$

By $\displaystyle \frac{1}{\sin x\,(a+b\cos x)} = \frac{a-b\cos x}{(a^2-b^2)\sin x} - \frac{b^2\sin x}{(a^2-b^2)(a+b\cos x)}.$

1991 $\displaystyle\int \frac{\tan x\,dx}{\sqrt{(a+b\tan^2 x)}} = \frac{1}{\sqrt{(b-a)}}\cos^{-1}\frac{\cos x\sqrt{(b-a)}}{\sqrt{b}}.$
[Subs. $\cos x\sqrt{(b-a)}$

1992 $\displaystyle\int \frac{\sqrt{(a+b\sin^2 x)}}{\sin x}\,dx = \sqrt{b}\cos^{-1}\frac{\sqrt{b}\cos x}{\sqrt{(a+b)}}$
$\displaystyle -\sqrt{a}\log\{\sqrt{a}\cot x+\sqrt{(a\operatorname{cosec}^2 x+b)}\}.$

METHOD.—By Division (1912), making the numerator rational, and integrating the two fractions by substituting $\cot x$ and $\cos x$ respectively.

1993 $\displaystyle\int \frac{dx}{a+2b\cos x+c\cos 2x} = \frac{c}{m}\left\{\int\frac{dx}{2c\cos x+b-m} - \int\frac{dx}{2c\cos x+b+m}\right\},$
where $m=\sqrt{\{b^2-2c(a-c)\}}$. Then integrate by (1953).

1994 $\displaystyle\int \frac{dx}{a\cos x+b\sin x+c} = \int\frac{d\theta}{\sqrt{(a^2+b^2)}\cos\theta+c}.$ (1953)

METHOD.—Substitute $\theta = x-\alpha$, where $\tan\alpha = \dfrac{b}{a}$.

1995 $$\int \frac{F(\sin x, \cos x)\,dx}{a\cos x + b\sin x + c},$$

F being an integral algebraic function of $\sin x$ and $\cos x$.

METHOD.—Substitute $\theta = x - \alpha$ as in (1994), and the resulting integral takes the form $\int \frac{f(\sin\theta, \cos\theta)\,d\theta}{A\cos\theta + B} = \int \frac{\phi(\cos\theta)\,d\theta}{A\cos\theta + B} + \int \frac{\sin\theta\,\psi(\cos\theta)}{A\cos\theta + B}$, since f contains only integral powers of the sine and cosine, and may therefore be resolved into the two terms as indicated.

To find the first integral on the right, divide by the denominator and integrate each term separately. To find the second integral, substitute the denominator.

1996 $$\int \frac{F(\cos x)\,dx}{(a_1 + b_1\cos x)(a_2 + b_2\cos x)\dots(a_n + b_n\cos x)},$$

where F is an integral function of $\cos x$.

METHOD.—Resolve into partial fractions. Each integral will be of the form $\int \frac{dx}{A + B\cos x}$ (1951).

1997 $$\int \frac{A\cos x + B\sin x + C}{a\cos x + b\sin x + c}\,dx.$$

METHOD.—Let $\phi(x) = a\cos x + b\sin x + c$; $\therefore\ \phi'(x) = -a\sin x + b\cos x$. Assume $A\cos x + B\sin x + C = \lambda\phi(x) + \mu\phi'(x) + \nu$. Substitute the values of $\phi(x)$ and $\phi'(x)$, and equate the coefficients to zero to determine λ, μ, ν. The integral becomes

$$\int \left\{ \lambda + \mu\frac{\phi'(x)}{\phi(x)} + \frac{\nu}{\phi(x)} \right\} dx = \lambda x + \mu\log\phi(x) + \int \frac{\nu}{\phi(x)}\,dx,$$

and the last integral is found by (1994).

EXPONENTIAL AND LOGARITHMIC FUNCTIONS.

1998 $\int e^x F(x)\,dx$ can be found at once when $F(x)$ can be expressed as *the sum of two functions, one of which is the derivative of the other*, for

$$\int e^x \{\phi(x) + \phi'(x)\}\,dx = e^x\phi(x).$$

1999 $\int e^{ax}\cos^n bx\,dx$ and $\int e^{ax}\sin^n bx\,dx$ are respectively $=$

$$\frac{a\cos bx + nb\sin bx}{a^2 + n^2b^2}\,e^{ax}\cos^{n-1}bx + \frac{n(n-1)b^2}{a^2 + n^2b^2}\int e^{ax}\cos^{n-2}bx\,dx.$$

and

$$\frac{a\sin bx - nb\cos bx}{a^2+n^2b^2}\, e^{ax}\sin^{n-1}bx + \frac{n(n-1)b^2}{a^2+n^2b^2}\int e^{ax}\sin^{n-2}bx\,dx.$$

PROOF.—In either case, integrate twice by Parts, $\int e^{ax}\,dx$.

Otherwise, these integrals may be found in terms of multiple angles by expanding $\sin^n x$ and $\cos^n x$ by (772–4), and integrating each term by (1951–2).

2000 $\int e^x \sin^n x \cos^m x\,dx$ is found by expressing $\sin^n x$ and $\cos^n x$ in terms of multiple angles.

Ex.: $\int e^x \sin^5 x \cos^2 x\,dx$. Put $e^{i\theta} = z$ in (768),

$$(2i\sin x)^5 (2\cos x)^2 = (z-z^{-1})^5 (z+z^{-1})^2$$

$$= (z-z^{-1})^3 (z^2-z^{-2})^2 = (z^7-z^{-7}) - 3(z^5-z^{-5}) + (z^3-z^{-3}) + 5(z-z^{-1});$$

$$\therefore \quad 2^6 e^x \sin^5 x \cos^2 x = e^x (\sin 7x - 3\sin 5x + \sin 3x + 5\sin x).$$

Then integrate by (1999).

2001 THEOREM.—Let P, Q be functions of x; and let

$\int P dx = P_1$, $\int P_1 Q_x dx = P_2$, $\int P_2 Q_x dx = P_3$, &c. Then

$$\int PQ^n dx = P_1 Q^n - nP_2 Q^{n-1} + n(n-1) P_3 Q^{n-2} - \dots \pm \lfloor n\, P_{n+1}.$$

PROOF.—Integrate successively by Parts, $\int P\,dx$, &c.

2002 THEOREM.—Let P, Q, as before, be functions of x; and let $P_1 = \left(\frac{P}{Q_x}\right)_x$, $P_2 = \left(\frac{P_1}{Q_x}\right)_x$, $P_3 = \left(\frac{P_2}{Q_x}\right)_x$, &c. Then

$$\int \frac{P}{Q^n}\,dx = -\frac{P}{(n-1)\,Q_x Q^{n-1}} - \frac{P_1}{(n-1)(n-2)\,Q_n Q^{n-2}}$$

$$-\frac{P_3}{(n-1)(n-2)(n-3)\,Q_x Q^{n-3}} \dots\dots + \frac{1}{\lfloor n-1}\int \frac{P_{n-1}}{Q}\,dx.$$

PROOF.—Integrate successively by Parts, $\int -\frac{(n-1)\,Q_x}{Q^n}\,dx$.

EXAMPLES.

2003 $$\int x^{m-1} (\log x)^n\,dx$$

$$= \frac{x^m}{m}\left(l^n - \frac{n}{m} l^{n-1} + \frac{n(n-1)}{m^2} l^{n-2} - \dots + (-1)^n \frac{\lfloor n}{m^n}\right).$$

METHOD.—By (2001). $P_1 = \frac{x^m}{m}$, $P_2 = \frac{x^m}{m^2}$, $P_3 = \frac{x^m}{m^3}$, &c.

2004

$$\int x^{m+nx}\,dx = \int x^m\left\{1+nx\log x+\tfrac{1}{2}(nx\log x)^2+\&c.\right\}dx,$$

[By (149)

and each term of this result can be integrated by (2003).

2005

$$\int\frac{x^{m-1}\,dx}{(\log x)^n}$$

$$= -x^m\left(\frac{l^{-n+1}}{n-1}+\frac{ml^{-n+2}}{(n-1)(n-2)}+\frac{m^2l^{-n+3}}{(n-1)(n-2)(n-3)}+\&c.\right)$$

$$+\frac{m^{n-1}}{\lfloor n-1}\int\frac{x^{m-1}\,dx}{\log x}.$$

METHOD.—By (2002). $Px = x^m$, $P_1 = mx^{m-1}$, $P_2 = m^2x^{m-1}$, &c.

The last method is not applicable when $n=1$. In this case, writing l for $\log x$,

2006 $$\int\frac{x^{m-1}\,dx}{\log x} = \log l + ml + \frac{m^2l^2}{1.2^2}+\frac{m^3l^3}{1.2.3^2}+\frac{m^4l^4}{1.2.3.4^2}+\&c.$$

METHOD: $\dfrac{x^{m-1}}{\log x} = \dfrac{e^{m\log x}}{x\log x}$. Expand the numerator by (150), and integrate.

See also (2161–6) for similar developments of the exponential forms of the same functions.

PARTICULAR ALGEBRAIC FUNCTIONS.

2007

$$\int\frac{dx}{x^n(x-1)^n} = \frac{1}{n-1}\left\{\frac{1}{(1-x)^{n-1}}-\frac{1}{x^{n-1}}\right\}$$

$$+\frac{n}{n-2}\left\{\frac{1}{(1-x)^{n-2}}-\frac{1}{x^{n-2}}\right\}+\ldots+\frac{n(n+1)\ldots2(n-1)}{1.2\ldots(n-1)}\log\frac{x}{1-x}.$$

n being even. (1918)

2008 $$\int\frac{dx}{(1+a^2x^2)\sqrt{(1-x^2)}} = \frac{1}{\sqrt{(1+a^2)}}\cos^{-1}\sqrt{\left(\frac{1-x^2}{1+a^2x^2}\right)}.$$

[Subs. $\dfrac{\sqrt{(1-x^2)}}{x}$

2009 $$\int\frac{dx}{(1-x^2)\sqrt{(1+x^2)}} = \frac{1}{\sqrt{2}}\log\frac{x\sqrt{2}+\sqrt{(1+x^2)}}{\sqrt{(1-x^2)}}.$$

2010 $$\int\frac{dx}{(1+x^2)\sqrt{(x^2-1)}} = \frac{1}{\sqrt{2}}\log\frac{x\sqrt{2}+\sqrt{(x^2-1)}}{\sqrt{(1+x^2)}}.$$

2011 $$\frac{dx}{(c+ex^2)\sqrt{(a+bx^2)}} = \frac{1}{\sqrt{(ace-bc^2)}}\sin^{-1}\sqrt{\left(\frac{x^2(ae-bc)}{aex^2+ac}\right)}, \quad ae > bc$$
$$= \frac{1}{\sqrt{(bc^2-ace)}}\log\frac{c\sqrt{(a+bx^2)}+x\sqrt{(bc^2-ace)}}{\sqrt{(c+ex^2)}}, \quad ae < bc$$

2012 $$\int\frac{dx}{x\sqrt{(a+bx^n)}} = \frac{2}{n\sqrt{a}}\log\frac{\sqrt{(a+bx^n)}-\sqrt{a}}{\sqrt{x^n}}. \qquad \left[\text{Subs. } \frac{1}{\sqrt{x^n}}\right.$$

2013 $$\int\frac{(1+x^2)\,dx}{(1-x^2)\sqrt{(1+x^4)}} = \frac{1}{\sqrt{2}}\log\frac{x\sqrt{2}+\sqrt{(1+x^4)}}{1-x^2}. \qquad \left[\text{Subs. } \frac{x\sqrt{2}}{1-x^2}\right.$$

2014 $$\int\frac{(1-x^2)\,dx}{(1+x^2)\sqrt{(1+x^4)}} = \frac{1}{\sqrt{2}}\sin^{-1}\frac{x\sqrt{2}}{1+x^2}. \qquad \left[\text{Subs. } \frac{x\sqrt{2}}{1+x^2}\right.$$

2015 $$\int\frac{\sqrt{(1+x^4)}}{1-x^4}\,dx = \frac{1}{\sqrt{8}}\left\{\log\frac{\sqrt{(1+x^4)}+x\sqrt{2}}{1-x^2}+\sin^{-1}\frac{x\sqrt{2}}{1+x^2}\right\}.$$

2016 $$\int\frac{x^2dx}{(1-x^4)\sqrt{(1+x^4)}} = \frac{1}{4\sqrt{2}}\left\{\log\frac{\sqrt{(1+x^4)}+x\sqrt{2}}{1-x^2}-\sin^{-1}\frac{x\sqrt{2}}{1+x^2}\right\}.$$

[Substitute $z\sqrt{(1+x^4)} = x\sqrt{2}$ in (2015-6)

2017 $$\int\frac{dx}{(1-x^2)(2x^2-1)^{\frac{1}{4}}} = \tfrac{1}{4}\log\frac{(2x^2-1)^{\frac{1}{4}}-x}{(2x^2-1)^{\frac{1}{4}}+x}-\tfrac{1}{2}\tan^{-1}\frac{x}{(2x^2-1)^{\frac{1}{4}}}.$$

[Substitute $x = z(2x^2-1)^{\frac{1}{4}}$

2018 $$\int\frac{dx}{(1+x^4)\{\sqrt{(1+x^4)}-x^2\}^{\frac{1}{2}}} = \tan^{-1}\frac{x}{\{\sqrt{(1+x^4)}-x^2\}^{\frac{1}{2}}}.$$

[Substitute $\dfrac{x}{\sqrt{\{\sqrt{(1+x^4)}-x^2\}}}$

2019
$$\int\frac{dx}{\sqrt[3]{(1-x^3)}} = \log\sqrt{x+\sqrt[3]{1-x^3}}-\frac{1}{\sqrt{3}}\tan^{-1}\frac{2\sqrt[3]{1-x^3}-x}{x\sqrt{3}}. \qquad \left[\text{Subs. } \frac{\sqrt[3]{(1-x^3)}}{x}\right.$$

2020 $$\int\frac{dx}{(1+x)\sqrt[3]{(1+3x+3x^2)}}$$ reduces to (2019) by substituting $\dfrac{x}{1+x}$.

INTEGRATION OF $\dfrac{x^{l-1}}{x^n \pm 1}$.

If l and n are positive integers, and $l-1 < n$;* then, n being even,

* If $l = n$, the value of the integral is simply $\dfrac{1}{n}\log(x^n-1)$.

2021 $$\int \frac{x^{l-1}\,dx}{x^n-1} = \frac{1}{n}\log(x-1) + \frac{(-1)^l}{n}\log(x+1)$$
$$+\frac{1}{n}\Sigma \cos rl\beta \log(x^2-2x\cos r\beta+1) - \frac{2}{n}\Sigma \sin rl\beta \tan^{-1}\frac{x-\cos r\beta}{\sin r\beta}$$
where $\beta = \frac{\pi}{n}$, and Σ denotes that the sum of all the terms obtained by making $r = 2, 4, 6 \ldots n-2$ successively, is to be taken.

If n be *odd*,

2022 $$\int \frac{x^{l-1}\,dx}{x^n-1} = \frac{1}{n}\log(x-1)$$
$$+\frac{1}{n}\Sigma \cos rl\beta \log(x^2-2x\cos r\beta+1) - \frac{2}{n}\Sigma \sin rl\beta \tan^{-1}\frac{x-\cos r\beta}{\sin r\beta}$$
with $r = 2, 4, 6 \ldots n-1$ successively.

If n be *even*,

2023 $$\int \frac{x^{l-1}\,dx}{x^n+1} = -\frac{1}{n}\Sigma \cos rl\beta \log(x^2-2x\cos r\beta+1)$$
$$+\frac{2}{n}\Sigma \sin rl\beta \tan^{-1}\frac{x-\cos r\beta}{\sin r\beta},$$
with $r = 1, 3, 5 \ldots n-1$ successively.

If n be *odd*,

2024 $$\int \frac{x^{l-1}\,dx}{x^n+1} = \frac{(-1)^{l-1}}{n}\log(x+1)$$
$$-\frac{1}{n}\Sigma \cos rl\beta \log(x^2-2x\cos r\beta+1) + \frac{2}{n}\Sigma \sin rl\beta \tan^{-1}\frac{x-\cos r\beta}{\sin r\beta}$$
with $r = 1, 3, 5 \ldots n-2$ successively.

PROOF.—(2021–4). Resolve $\frac{x^{l-1}}{x^n \pm 1}$ into partial fractions by the method of (1917). We have $\frac{\phi(a)}{F'(a)} = \frac{a^{l-1}}{na^{n-1}} = \frac{a^m}{n}$, since $a^n = \mp 1$. The different values of a are the roots of $x^n \pm 1 = 0$, and these are given by $x = \cos r\beta \pm i \sin r\beta$, with odd or even integral values of r. (See 480, 481; $2r$ and $2r+1$ of those articles being in each case here represented by r.) The first two terms on the right in (2021) arise from the factors $x \pm 1$; the remaining terms from quadratic factors of the type

$$(x-\cos r\beta - i\sin r\beta)(x-\cos r\beta + i \sin r\beta) = (x-\cos r\beta)^2 + \sin^2 r\beta.$$

These last terms are integrated by (1923) and (1935). Similarly for the cases (2022–4).

2025 If, in formulæ (2021–4), $\mp\frac{2}{n}\Sigma\left(\frac{1}{2}\pi-r\beta\right)\sin rl\beta$ be added to the last term for the constant of integration, the integral vanishes with x, and the last term becomes

$$\mp\frac{2}{n}\Sigma\sin rl\beta\tan^{-1}\frac{x\sin r\beta}{1-x\cos r\beta},$$

reading $-$ in (2021–2), and $+$ in (2023–4).

2026 $$\int\frac{x^m\,dx}{a+bx^n}=\frac{1}{a}\left(\frac{a}{b}\right)^{\frac{m+1}{n}}\int\frac{z^m}{1+z^n}\,dz,$$

where $az^n=bx^n$. Then integrate by (2023–4).

2027 $$\int\frac{x^{m-1}+x^{n-m-1}}{x^n+1}\,dx=\frac{4}{n}\Sigma\sin mr\beta\tan^{-1}\frac{x-\cos r\beta}{\sin r\beta},$$

where $\beta=\pi\div n$, and $r=1,3,5,\ldots$ successively up to $n-1$ or $n-2$, according as n is even or odd.

2028

$$\int\frac{x^{m-1}-x^{n-m-1}}{x^n+1}\,dx=-\frac{2}{n}\Sigma\cos mr\beta.\log\left(x^2-2x\cos r\beta+1\right),$$

with the same values of r; but when n is odd, *supply the additional term* $(-1)^m\,2\log(x+1)\div n$.

PROOF.—Follow the method of (2024, *Proof*).

Similar forms are obtainable when the denominator is x^n-1.

2029

$$\int\frac{x^{l-1}\,dx}{x^{2n}-2x^n\cos n\theta+1}=\frac{1}{n\sin n\theta}\Sigma\cos(n-l)\phi.\tan^{-1}\frac{x-\cos\phi}{\sin\phi}$$
$$-\tfrac{1}{2}\Sigma\sin(n-l)\phi.\log\left(x^2-2x\cos\phi+1\right),$$

where $\phi=\theta+\frac{r\pi}{n}$ and $r=0,2,4,\ldots 2(n-1)$ successively. But if the integral is to vanish with x, write $\tan^{-1}\frac{x\sin\phi}{1-x\cos\phi}$, as in (2025).

PROOF.—By the method of (2024).—The factors of the denominator are given in (807), putting y equal to unity.

INTEGRATION OF $\int x^m (a+bx^n)^{\frac{p}{q}}\,dx$.

2035 Rule I.—*When* $\frac{m+1}{n}$ *is a positive integer, integrate by substituting* $z=(a+bx^n)^{\frac{1}{q}}$. *Thus*

$$\int x^m (a+bx^n)^{\frac{p}{q}}\,dx = \frac{q}{nb}\int z^{p+q-1}\left(\frac{z^q-a}{b}\right)^{\frac{m+1}{n}-1} dz.$$

Expand the binomial, and integrate the separate terms by (1922).

2036 But if the positive integer be 1, the integral is known at sight, since m then becomes $=n-1$.

2037 Rule II.—*When* $\frac{m+1}{n}+\frac{p}{q}$ *is a negative integer, substitute* $z=(ax^{-n}+b)^{\frac{1}{q}}$. *Thus*

$$\int x^m (a+bx^n)^{\frac{p}{q}}\,dx = -\frac{q}{na}\int z^{p+q-1}\left(\frac{z^q-b}{a}\right)^{-\frac{m+1}{n}-\frac{p}{q}-1} dz.$$

Expand and integrate as before.

2038 But, if the negative integer be -1, the integral is found immediately by writing it in the form

$$\int x^{m+\frac{np}{q}}(ax^{-n}+b)^{\frac{p}{q}}\,dx = \int x^{-n-1}(ax^{-n}+b)^{\frac{p}{q}}\,dx$$

$$= \frac{q}{na(p+q)}(ax^{-n}+b)^{\frac{p}{q}+1}.$$

EXAMPLES.

2039 To find $\int x^{\frac{1}{2}}(1+x^{\frac{1}{2}})^{\frac{2}{3}}\,dx$. Here $m=\frac{1}{2}$, $n=\frac{1}{2}$, $p=2$, $q=3$, $\frac{m+1}{n}=3$, a positive integer. Therefore, substituting $y=(1+x^{\frac{1}{2}})^{\frac{1}{3}}$, $x=(y^3-1)^2$, $x_y=6y^2(y^3-1)$, and the integral becomes

$$\int (y^3-1)\,y^2\frac{dx}{dy}\,dy = 6\int y^4(y^3-1)^2\,dy,$$

the value of which can be found immediately by expanding and integrating the separate terms.

2040
$$\int x^3(a+bx^4)^{\frac{1}{3}}\,dx = \frac{3}{16b}(a+bx^4)^{\frac{4}{3}}.$$

For $\frac{m+1}{n}=1$ (2036); that is, $m+1=n$, and the factor x^3 is the derivative of $\frac{1}{4}x^4$.

2041 $\int \frac{(1+\sqrt{x})^{\frac{2}{3}}}{x^{\frac{7}{3}}}\,dx$, or $\int x^{-\frac{7}{3}}(1+x^{\frac{1}{2}})^{\frac{2}{3}}\,dx$. Here $m=-\frac{7}{3}$, $n=\frac{1}{2}$, $p=2$, $q=3$, $\frac{m+1}{n}+\frac{p}{q}=-2$, a negative integer. Therefore, substitute $y=(x^{-\frac{1}{2}}+1)^{\frac{1}{3}}$, $x=(y^3-1)^{-2}$, $x_y=-6y^2(y^3-1)^{-3}$. Writing the integral in the form below, and then substituting the values, we have

$$\int x^{-2}(x^{-\frac{1}{2}}+1)^{\frac{2}{3}}\,x_y\,dy = -6\int y^4(y^3-1)\,dy,$$

which can be integrated at once.

2042 $\int \frac{dx}{x(a+bx^n)} = \int x^{-1}(a+bx^n)^{-1}\,dx$. Here $\frac{m+1}{n}+\frac{p}{q}=-1$; therefore, by (2038), the integral

$$= \int x^{-n-1}(ax^{-n}+b)^{-1}\,dx = -\frac{1}{na}\log(ax^{-n}+b).$$

REDUCTION OF $\int x^m(a+bx^n)^p\,dx$.

When neither of the conditions in (2035, 2037) are fulfilled, the integral may be reduced by any of the six following rules, so as to alter the indices m and p, those indices having any algebraic values.

2043 I. *To change* m *and* p *into* m+n *and* p−1.

Integrate by Parts, $\int x^m\,dx$.

2044 II. *To change* m *and* p *into* m−n *and* p+1.

Integrate by Parts, $\int x^{n-1}(a+bx^n)^p\,dx$.

2045 III. *To change* m *into* m+n.

Add 1 *to* p. *Then integrate by Parts,* $\int x^m\,dx$; *and also by Division, and equate the results.*

2046 IV. *To change* m *into* m−n.

Add 1 *to* p, *and subtract* n *from* m. *Then integrate by Parts,* $\int x^m dx$; *and also by Division, and equate the results.*

2047 V. *To change* p *into* p+1.

Add 1 *to* p. *Then integrate by Division, and the new integral by Parts,* $\int x^{n-1} (a+bx^n)^p$.

2048 VI. *To change* p *into* p−1.

Integrate by Division, and the new integral by Parts, $\int x^{n-1} (a+bx^n)^p$.

2049 MNEMONIC TABLE FOR THE SAME RULES.

I.	m+n,	p−1	*By Parts* (m).
II.	m−n,	p+1	*By Parts* (p).
III.	m+n		(p+1), *Parts* (m) *and Division.*
IV.	m−n		(p+1, m−n), *Parts* (m) *and Division.*
V.		p+1	(p+1), *Division, and the new integral by Parts* (p).
VI.		p−1	*Division, and the new integral by Parts* (p).

By applying the rules, *Formulæ of reduction* are obtained. Thus, any of the six values below may be substituted for the integral
$$\int x^m (a+bx^n)^p \, dx.$$

2050—2055

I. $$\frac{x^{m+1}(a+bx^n)^p}{m+1} - \frac{bnp}{m+1}\int x^{m+n}(a+bx^n)^{p-1}\,dx.$$

II. $$\frac{x^{m-n+1}(a+bx^n)^{p+1}}{bn(p+1)} - \frac{m-n+1}{bn(p+1)}\int x^{m-n}(a+bx^n)^{p+1}\,dx.$$

III. $$\frac{x^{m+1}(a+bx^n)^{p+1}}{a(m+1)} - \frac{b(m+n+np+1)}{a(m+1)}\int x^{m+n}(a+bx^n)^p\,dx.$$

IV. $$\frac{x^{m-n+1}(a+bx^n)^{p+1}}{b(m+np+1)} - \frac{a(m-n+1)}{b(m+np+1)}\int x^{m-n}(a+bx^n)^p\,dx.$$

V. $$-\frac{x^{m+1}(a+bx^n)^{p+1}}{an(p+1)} + \frac{m+n+np+1}{an(p+1)}\int x^m(a+bx^n)^{p+1}\,dx.$$

VI. $$\frac{x^{m+1}(a+bx^n)^p}{m+np+1} + \frac{anp}{m+np+1}\int x^m(a+bx^n)^{p-1}\,dx.$$

Examples.

2056 To find $\int \frac{\sqrt{(a^2-x^2)}}{x^3}\,dx$. Apply Rule I. or Formula I.; thus

$$\int x^{-3}(a^2-x^2)^{\frac{1}{2}}\,dx = -\tfrac{1}{2}x^{-2}(a^2-x^2)^{\frac{3}{2}}+\tfrac{1}{2}\int x^{-1}(a^2-x^2)^{-\frac{1}{2}}\,dx \quad (1927).$$

2057 To find $\int \frac{x^4}{(a^2-x^2)^{\frac{3}{2}}}\,dx$. Apply Rule II. or Formula II.; thus

$$\int x^4(a^2-x^2)^{-\frac{3}{2}}\,dx = x^3(a^2-x^2)^{-\frac{1}{2}}-3\int x^2(a^2-x^2)^{-\frac{1}{2}}\,dx \quad (1934).$$

2058 $\int \operatorname{cosec}^m\theta\,d\theta$. Substituting $\sin\theta = x$, the integral becomes

$$\int \sin^{-m}\theta\,\frac{d\theta}{dx}\,dx = \int x^{-m}(1-x^2)^{-\frac{1}{2}}\,dx.$$

Apply Rule III.; thus, increasing p by 1 and integrating, first by Parts $\int x^{-m}\,dx$, and again by Division;

$$\int x^{-m}(1-x^2)^{\frac{1}{2}}\,dx = \frac{x^{1-m}(1-x^2)^{\frac{1}{2}}}{1-m}+\frac{1}{1-m}\int x^{2-m}(1-x^2)^{-\frac{1}{2}}\,dx,$$

$$\int x^{-m}(1-x^2)^{\frac{1}{2}}\,dx = \int x^{-m}(1-x^2)^{-\frac{1}{2}}\,dx-\int x^{2-m}(1-x^2)^{-\frac{1}{2}}\,dx.$$

Equating the results, we obtain

2059
$$\int x^{-m}(1-x^2)^{-\frac{1}{2}}\,dx = \frac{x^{1-m}}{1-m}(1-x^2)^{\frac{1}{2}}+\frac{2-m}{1-m}\int x^{2-m}(1-x^2)^{-\frac{1}{2}}\,dx.$$

By repeating the process, the integral is made to depend finally upon

$$\int x^{-1}(1-x^2)^{-\frac{1}{2}}\,dx \qquad \text{or} \qquad \int (1-x^2)^{-\frac{1}{2}}\,dx,$$

according as m is an odd or even integer (1927, '29).

2060 $\int \sin^m\theta\,d\theta$ is found in a similar manner by Rule IV. The integral to be evaluated is $\int x^m(1-x^2)^{-\frac{1}{2}}\,dx$; and the integral operated upon is $\int x^{m-2}(1-x^2)^{\frac{1}{2}}\,dx$. Otherwise apply Formula IV. See also (1954).

2061 To find $\int \frac{dx}{(x^2+a^2)^r}$. Apply Rule V. $p=-r$, and increasing p by 1, we have, first, by Division,

$$\int (x^2+a^2)^{1-r}\,dx = \int x^2(x^2+a^2)^{-r}\,dx+a^2\int (x^2+a^2)^{-r}\,dx.$$

Integrating the new form by Parts, $\int x(x^2+a^2)^{-r}\,dx$, we next obtain

$$\int x^2(x^2+a^2)^{-r}\,dx = \frac{x(x^2+a^2)^{1-r}}{2(1-r)}-\frac{1}{2(1-r)}\int (x^2+a^2)^{1-r}\,dx.$$

Substituting this value in the previous equation, we have, finally,

2062
$$\int \frac{dx}{(x^2+a^2)^r} = \frac{x}{2(r-1)\,a^2(x^2+a^2)^{r-1}}+\frac{2r-3}{a^2(2r-2)}\int \frac{dx}{(x^2+a^2)^{r-1}}.$$

This equation is given at once by Formula V. Thus r is changed into $r-1$, and by repeating the process of reduction, the original integral is ultimately made to depend upon (1935) for its value if r be an integer.

Another formula for this integral is

2063 $$\int \frac{dx}{(x^2+\beta)^r} = \frac{(-1)^{r-1}}{1.2\ldots(r-1)} \frac{d^{r-1}}{d\beta^{r-1}}\left(\frac{1}{\sqrt{\beta}}\tan^{-1}\frac{x}{\sqrt{\beta}}\right).$$

Proof.—Write β for a^2 in (1935), and differentiate the equation $r-1$ times for β by the principle in (2255).

2064 To find $\int (a^2+x^2)^{\frac{1}{2}n}\,dx$. Apply Rule VI. By Division, we have

$$\int (a^2+x^2)^{\frac{1}{2}n}\,dx = a^2\int (a^2+x^2)^{\frac{1}{2}n-1}\,dx + \int x^2(a^2+x^2)^{\frac{1}{2}n-1}\,dx.$$

The last integral, by Parts, becomes

$$\int x^2(a^2+x^2)^{\frac{1}{2}n-1}dx = \frac{1}{n}x(a^2+x^2)^{\frac{1}{2}n} - \frac{1}{n}\int (a^2+x^2)^{\frac{1}{2}n}.$$

Substituting this value in the previous equation, we obtain

2065 $$\int (a^2+x^2)^{\frac{1}{2}n}\,dx = \frac{x(a^2+x^2)^{\frac{1}{2}n}}{n+1} + \frac{na^2}{n+1}\int (a^2+x^2)^{\frac{1}{2}n-1}\,dx,$$

a result given at once by Formula VI.

If n be an odd integer, we arrive, finally, by successive reduction in this manner, at $\int (a^2+x^2)^{\frac{1}{2}}\,dx$ (1931).

2066 The integral $\int \sin^m\theta \cos^p\theta\, d\theta$ is reducible by the foregoing Rules I. to VI., if, in applying them, n *be always put equal to* 2; *if* p *be changed into* p$\pm$2 *instead of* p$\pm$1; *and if Division be always effected by separating the factor* $\cos^2\theta = 1-\sin^2\theta$.

Proof. $\int \sin^m\theta\cos^p\theta\,d\theta = \int x^m(1-x^2)^{\frac{1}{2}(p-1)}\,dx$, where $x = \sin\theta$. Thus $n = 2$ always, and the index $\frac{1}{2}(p-1)$ is increased by 1 by adding 2 to p.

Thus, Rule I. gives the formula of reduction

2067

$$\int \sin^m\theta\cos^p\theta\,d\theta = \frac{\sin^{m+1}\theta\cos^{p-1}\theta}{m+1} + \frac{p-1}{m+1}\int \sin^{m+2}\theta\cos^{p-2}\theta\,d\theta.$$

But the integral can be found by substitution in the following cases :—

If r be a positive integer,

2068 $\int \cos^{2r+1}x \sin^p x\,dx = \int (1-z^2)^r z^p dz$, where $z = \sin x$.

2069 $\int \sin^{2r+1}x \cos^p x\,dx = -\int (1-z^2)^r z^p dz$, where $z = \cos x$.

If $m+p = -2r$,

2070 $\int \sin^m x \cos^p x\,dx = \int (1+z^2)^{r-1} z^m dz$, where $z = \tan x$.

FUNCTIONS OF $a+bx\pm cx^2$.

The seven following integrals are found either by writing

2071 $$a+bx+cx^2 = \{(2cx+b)^2+4ac-b^2\} \div 4c,$$

and substituting $2cx+b$; or by writing

2072 $$a+bx-cx^2 = \{4ac+b^2-(2cx-b)^2\} \div 4c,$$

and substituting $2cx-b$.

2073 $$\int \frac{dx}{a+bx+cx^2} = \frac{1}{\sqrt{(b^2-4ac)}} \log \frac{2cx+b-\sqrt{(b^2-4ac)}}{2cx+b+\sqrt{(b^2-4ac)}}$$

$$\text{or} \quad \frac{2}{\sqrt{(4ac-b^2)}} \tan^{-1} \frac{2cx+b}{\sqrt{(4ac-b^2)}},$$

according as $b^2 >$ or $< 4ac$ (2071, 1935–6).

2074 $$\int \frac{dx}{a+bx-cx^2} = \frac{1}{b^2+4ac} \log \frac{\sqrt{(b^2+4ac)}+(2cx-b)}{\sqrt{(b^2+4ac)}-(2cx-b)}.$$

(2072, 1937)

2075

$$\int \frac{dx}{\sqrt{(a+bx+cx^2)}} = \frac{1}{\sqrt{c}} \log \{2cx+b+2\sqrt{c}\sqrt{(a+bx+cx^2)}\}.$$

2076 $$\int \frac{dx}{\sqrt{(a+bx-cx^2)}} = \frac{1}{\sqrt{c}} \sin^{-1} \frac{2cx-b}{\sqrt{(4ac+b^2)}}.$$

(2071–2, 1928–9)

2077 $\int\sqrt{(a+bx+cx^2)}\,dx = \frac{1}{4}c^{-\frac{3}{2}}\int\sqrt{(y^2+4ac-b^2)}\,dy.$

2078 $\int\sqrt{(a+bx-cx^2)}\,dx = \frac{1}{4}c^{-\frac{3}{2}}\int\sqrt{(4ac+b^2-y^2)}\,dy,$

where $y=2cx\pm b$. The integrals are given at (1931–3).

2079 $$\int\frac{dx}{(a+bx+cx^2)^p} = 2^{2p-1}c^{p-1}\int\frac{dy}{(y^2+4ac-b^2)^p}.$$

[By (2071), the integral being reduced by (2062–3).

2080 $$\int\frac{(lx+m)\,dx}{(a+bx+cx^2)^p}$$

$$=\frac{l}{2c}\int\frac{(2cx+b)\,dx}{(a+bx+cx^2)^p}+\left(m-\frac{bl}{2c}\right)\int\frac{dx}{(a+bx+cx^2)^p}.$$

The value of the second integral is $(a+bx+cx^2)^{1-p}\div(1-p)$, unless $p=1$, when the value is $\log(a+bx+cx^2)$. For the third, see (2079).

METHOD.—Decompose into two fractions, making the numerator of the first $2cx+b$; that is, the derivative of $a+bx+cx^2$.

2081 $\int\frac{px^2+q}{a+bx^2+cx^4}\,dx$ may be integrated as follows:—

I. If $b^2>4ac$, put α and β for $\frac{-b\pm\sqrt{(b^2-4ac)}}{2c}$, and, by Partial Fractions, the integral is resolved into

$$\frac{1}{c(\alpha-\beta)}\left\{\int\frac{p\alpha+q}{x^2-\alpha}\,dx-\int\frac{p\beta+q}{x^2-\beta}\,dx\right\}. \qquad (1936)$$

II. If $b^2<4ac$, put $\frac{a}{c}=n^2$ and $\frac{2\sqrt{(ac)}-b}{c}=m^2$, and the integral may be decomposed into

2082 $$\frac{1}{2mnc}\left\{\int\frac{(q-pn)\,x+qm}{x^2+mx+n}\,dx-\int\frac{(q-pn)\,x-qm}{x^2-mx+n}\,dx\right\},$$

the value of which is found by (2080).

III. If $b^2=4ac$,

2083 $$\int\frac{dx}{a+bx^2+cx^4}=\frac{x}{2a+bx^2}+\frac{1}{\sqrt{(2ab)}}\tan^{-1}\left(x\sqrt{\frac{b}{2a}}\right). \qquad (2062)$$

2084 $$\int \frac{x\,dx}{a+bx^2+cx^4} = -\frac{2a}{b\,(2a+bx^2)}.$$

2085 $$\int \frac{x^2dx}{a+bx^2+cx^4} = \frac{2a}{b}\left\{\frac{1}{\sqrt{(2ab)}}\tan^{-1}\left(x\sqrt{\frac{b}{2a}}\right) - \frac{x}{2a+bx^2}\right\}$$

( )

REDUCTION OF $\int x^m\,(a+bx^n+cx^{2n})^p\,dx$.

NOTE.—In the following *Formulæ of Reduction*, for the sake of clearness, $x^m\,(a+bx^n+cx^{2n})^p$ is denoted by (m, p), and the integral merely by $\int (m, p)$.

2086 $$(m+1)\int (m, p) = (m+1, p)$$
$$-bnp\int (m+n, p-1) - 2cnp\int (m+2n, p-1) \quad \ldots\ldots\ldots(1).$$

2087 $$bn\,(p+1)\int (m, p) = (m-n+1, p+1)$$
$$-(m-n+1)\int (m-n, p+1) - 2cn\,(p+1)\int (m+n, p) \ \ldots(2).$$

2088 $$2cn\,(p+1)\int (m, p) = (m-2n+1, p+1)$$
$$-(m-2n+1)\int (m-2n, p+1) - bn\,(p+1)\int (m-n, p)\ldots(3).$$

2089 $$(m+np+1)\int (m, p) = (m+1, p)$$
$$+anp\int (m, p-1) - cnp\int (m+2n, p-1)\ldots\ldots\ldots\ldots\ldots\ldots(4).$$

2090 $$(m+2np+1)\int (m, p) = (m+1, p)$$
$$+2anp\int (m, p-1) + bnp\int (m+n, p-1) \quad \ldots\ldots\ldots\ldots\ldots(5).$$

2091 $$b\,(m+np+1)\int (m, p) = (m-n+1, p+1)$$
$$-a\,(m-n+1)\int (m-n, p) - c\,(m+2np+n+1)\int (m+n, p)$$
$$\ldots\ldots(6).$$

2092 $$bn\,(p+1)\int (m, p) = -(m-n+1, p+1)$$
$$+(m+2np+n+1)\int (m-n, p+1) - 2an\,(p+1)\int (m-n, p)$$
$$\ldots\ldots(7).$$

2093 $cn(p+1)\int(m,p) = (m-2n+1,\ p+1)$
$+an(p+1)\int(m-2n,p)-(m+np-n+1)\int(m-2n,\ p+1)$ ......(8).

2094 $an(p+1)\int(m,p) = -(m+1,\ p+1)$
$+(m+np+n+1)\int(m,p+1)+cn(p+1)\int(m+2n,p)$...(9).

2095 $2an(p+1)\int(m,p) = -(m+1,\ p+1)$
$+(m+2np+2n+1)\int(m,\ p+1)-bn(p+1)\int(m+n,\ p)$ ......(10).

2096 $a(m+1)\int(m,p) = (m+1,\ p+1)$
$-b(m+np+n+1)\int(m+n,p)-c(m+2np+2n+1)\int(m+2n,p)$ ......(11).

2097 $c(m+2np+1)\int(m,p) = (m-2n+1,\ p+1)$
$-b(m+np-n+1)\int(m-n,p)-a(m-2n+1)\int(m-2n,p)$ ......(12).

Proof.—By differentiation, we have

2098
$\int(m,p) = m\int(m-1,p)+bnp\int(m+n-1,\ p-1)+2cnp\int(m+2n-1,\ p-1).$
Formulæ (1), (2), and (3) are obtained from this equation by altering the indices m and p, so that each integral on the right, in turn, becomes $\int(m,p)$.

Again, by division,

2099 $\int(m,p) = a\int(m,p-1)+b\int(m+n,p-1)+c\int(m+2n,p-1)$...(A).

And, by changing m into $m-n$, and p into $p+1$,

2100 $\int(m-n,\ p+1) = a\int(m-n,p)+b\int(m,p)+c\int(m+n,p)$ .....(B).

Formulæ (4) to (12) may now be found as follows:—

(4), by eliminating $\int(m+n,\ p-1)$ between (1) and (A);

(5), by eliminating $\int(m+2n,\ p-1)$ between (1) and (A);

(6), by eliminating $\int(m-n,\ p+1)$ between (2) and (B);

(7), by eliminating $\int(m+n,\ p)$ between (2) and (B);

(8), from (4), by changing m into $m-2n$, and p into $p+1$;
(9), from (4), by changing p into $p+1$;
(10), from (5), by changing p into $p+1$;
(11), from (6), by changing m into $m+n$;
(12), from (6), by changing m into $m-n$.

If α and β are real roots of the quadratic equation $a+bx^n+cx^{2n}=0$, then, by Partial Fractions,

2101 $$\int\frac{x^m\,dx}{a+bx^n+cx^{2n}}=\frac{1}{c(\alpha-\beta)}\left\{\int\frac{x^m\,dx}{x^n-\alpha}-\int\frac{x^m\,dx}{x^n-\beta}\right\},$$

and the integrals are obtained by (2021–2).

But, if the roots are imaginary,

2102 $$\int\frac{x^m\,dx}{a+bx^n+cx^{2n}}=\frac{1}{a}\left(\frac{a}{c}\right)^{\frac{m+1}{2n}}\int\frac{z^m\,dz}{1-2z^n\cos n\theta+z^{2n}},\quad(2029)$$

where $\cos n\theta=-\dfrac{b}{2\sqrt{(ac)}}$ and $z=\left(\dfrac{c}{a}\right)^{\frac{1}{2n}}x$.

2103 $\displaystyle\int\frac{x^m\,dx}{(a+bx+cx^2)^p}$ is reduced to (2079–80) by (2097).

2104 $$\int\frac{dx}{(x+h)\sqrt{(a+bx+cx^2)}}=-\int\frac{dy}{\sqrt{(A+By+Cy^2)}},\quad(2075)$$

where $y=(x+h)^{-1}$, $A=c$, $B=b-2ch$, $C=a-bh+ch^2$.

2105 $$\int\frac{dx}{(x+h)\sqrt{x^2-1}}=\frac{1}{\sqrt{1-h^2}}\cos^{-1}\frac{1+hx}{x+h}=\frac{1}{\sqrt{h^2-1}}\cosh^{-1}\frac{1+hx}{x+h}.$$

2106 $$\int\frac{dx}{(x+h)\sqrt{1-x^2}}=\frac{1}{\sqrt{h^2-1}}\sin^{-1}\frac{1+hx}{x+h}.$$ [By (2181).

METHOD.—Substitute $(x+h)^{-1}$, as in (2104). Observe the cases in which $h=1$.

2107 $$\int\frac{dx}{(x+h)^r\sqrt{(a+bx+cx^2)}}=-\int\frac{y^{r-1}\,dy}{\sqrt{(A+By+Cy^2)}},$$

with the same values for A, B, C, and y as in (2104). The integral is reduced by (2097).

2108 $$\int\frac{(lx+m)\,dx}{(x^2+\beta^2)\sqrt{(a+bx+cx^2)}}.$$

METHOD.—Substitute θ by putting $x = \beta \tan(\theta+\gamma)$, and determine the constant γ by equating to zero the coefficient of $\sin 2\theta$ in the denominator. The resulting integral is of the form $\int \frac{L \cos\theta + M \sin\theta}{\sqrt{(P+Q\cos 2\theta)}} d\theta$. Separate this into two terms, and integrate by substituting $\sin\theta$ in the first and $\cos\theta$ in the second.

2109 $$\int \frac{\phi(x)\,dx}{F(x)\sqrt{(a+bx+cx^2)}},$$

where $\phi(x)$ and $F(x)$ are rational algebraic functions of x, the former being of the lowest dimensions.

METHOD.—Resolve $\frac{\phi(x)}{F(x)}$ into partial fractions. The resulting integrals are either of the form (2107), or else they arise from a pair of imaginary roots of $F(x) = 0$, and are of the type $\int \frac{(Ax+B)\,dx}{\{(x-\alpha)^2+\beta^2\}\sqrt{(a+bx+cx^2)}}$. Substitute $x-\alpha$ in this, and the integral (2108) is obtained.

INTEGRATION BY RATIONALIZATION.

In the following articles, F denotes a rational algebraic function. In each case, an integral involving an irrational function of x is, by substitution, made to take the form $\int F(z)\,dz$. This latter integral can always be found by the method of Partial Fractions (1915).

2110 $$\int F\left\{x^n, \left(\frac{a+bx}{f+gx}\right)^{\frac{p}{q}}, \left(\frac{a+bx}{f+gx}\right)^{\frac{r}{s}}, \text{\&c.}\right\} dx.$$

Substitute $\frac{a+bx}{f+gx} = z^l$, where l is the least common denominator of the fractional indices; thus,

$$x = \frac{fz^l - a}{b - gz^l}, \quad \frac{dx}{dz} = \frac{lz^{l-1}(bf-ag)}{(b-gz^l)^2}, \quad \left(\frac{a+bx}{f+gx}\right)^{\frac{p}{q}} = z^{\frac{lp}{q}}, \text{ \&c.},$$

the powers of z being now all integral.

2111 $$\int x^{n-1} F\left\{x^{mn}, \left(\frac{a+bx^n}{f+gx^n}\right)^{\frac{p}{q}}, \left(\frac{a+bx^n}{f+gx^n}\right)^{\frac{r}{s}}, \text{\&c.}\right\} dx.$$

Reduce to the form of (2110) by substituting x^n.

2112 $\int F\{\sqrt{(a+\sqrt{mx+n})}\}\,dx.$ Subs. $\sqrt{(a+\sqrt{mx+n})}$.

2113 $\int F\{x,\ \sqrt{(bx\pm cx^2)}\}\,dx.$ Substitute $x=\dfrac{b}{z^2\mp c}$.

And therefore $\sqrt{(bx\pm cx^2)}=\dfrac{bz}{z^2\mp c},\quad \dfrac{dx}{dz}=-\dfrac{2bz}{(z^2\mp c)^2}.$

2114 $\int F\{x,\ \sqrt{(a+bx+cx^2)}\}\,dx.$

Writing Q for $a+bx+cx^2$, F may always be reduced to the form $\dfrac{A+B\sqrt{Q}}{C+D\sqrt{Q}}$, in which A, B, C, D are constants or rational functions of x. Rationalizing this fraction, it takes the form $L+M\sqrt{Q}$. Thus the integral becomes $\int L\,dx+\int M\sqrt{Q}\,dx$, the first of which two integrals is rational, while the second is equivalent to $\int \dfrac{MQ}{\sqrt{Q}}\,dx$, which is of the form in (2075).

2115 Otherwise.—(i.) When c is positive, the integral may be made rational by substituting

$$x=\frac{a-cz^2}{2cz-b},\quad \frac{dx}{dz}=\frac{2c\,(bz-cz^2-a)}{(2cz-b)^2},\quad \sqrt{(a+bx+cx^2)}=\sqrt{c}\left(\frac{a-cz^2}{2cz-b}+z\right).$$

(ii.) When c is negative, let α, β be the roots of the equation $a+bx-cx^2=0$, which are necessarily real (a, b, and c being now all positive), so that $a+bx-cx^2=c\,(x-\alpha)(\beta-x)$. The integral is now made rational by substituting

$$x=\frac{c\alpha z^2+\beta}{cz^2+1},\quad \frac{dx}{dz}=\frac{2\,(\alpha-\beta)\,cz}{(cz^2+1)^2},\quad \sqrt{(a+bx-cx^2)}=\frac{(\beta-\alpha)\,cz}{cz^2+1}.$$

In each case the result is of the form $\int F(z)\,dz$.

2116 $\int x^mF\{x^n,\ \sqrt{a+bx^n+cx^{2n}}\}\,dx,$

when $\dfrac{m+1}{n}$ is an integer, is reduced to the form (2114) by substituting x^n.

2117

$\int F\{x,\ \sqrt{(a+bx)},\ \sqrt{(f+gx)}\}\,dx.$ Substitute $z^2=\dfrac{a+bx}{f+gx}$, and, therefore, $x = \frac{a-fz^2}{gz^2-b}$, $\sqrt{(a+bx)} = \frac{z\sqrt{(ag-bf)}}{\sqrt{(gz^2-b)}}$,

$$\sqrt{(f+gx)} = \frac{\sqrt{(ag-bf)}}{\sqrt{(gz^2-b)}}, \quad \frac{dx}{dz} = \frac{2(bf-ag)z}{(gz^2-b)^2}.$$

The form $\int F\{z, \sqrt{(gz^2-b)}\}\,dz$ is obtained, which is comprehended in (2114).

2118 $\int x^m F\{x^n, \sqrt{(a+b^2x^{2n})}, bx^n \pm \sqrt{(a+b^2x^{2n})}\}\,dx$,

when $\frac{m+1}{n}$ is an integer, is reduced to the form $\int F(z)\,dz$ by substituting $z = bx^n \pm \sqrt{(a+b^2x^{2n})}$, and therefore

$$x^n = \frac{z^2-a}{2bz}; \qquad \sqrt{(a+b^2x^{2n})} = \frac{z^2+a}{2z};$$

$$x^m \frac{dx}{dz} = \frac{1}{2nb}\left(\frac{z^2-a}{2bz}\right)^{\frac{m+1}{n}-1}\left(\frac{z^2+a}{z^2}\right).$$

2119 $\int x^m (a+bx^n)^{\frac{p}{q}} F(x^n)\,dx$ is rationalized by substituting either $(a+bx^n)^{\frac{1}{q}}$ or $(ax^{-n}+b)^{\frac{1}{q}}$ according as $\frac{m+1}{n}$ or $\frac{m+1}{n}+\frac{p}{q}$ is integral, whether positive or negative.

2120 $\int x^{m-1} F\{x^m, x^n, (a+bx^n)^{\frac{p}{q}}\}\,dx$, when $\frac{m}{n}$ is either a positive or negative integer, is rationalized by substituting $(a+bx^n)^{\frac{1}{q}}$.

INTEGRALS REDUCIBLE TO ELLIPTIC INTEGRALS.

2121 $\int F\{x, \sqrt{(a+bx+cx^2+dx^3+ex^4)}\}\,dx$.

Writing X for the quartic, the rational function F may always be brought to the form $\frac{P+Q\sqrt{X}}{P'+Q'\sqrt{X}}$,

and this again, by rationalizing the denominator, to the form

$M+N\sqrt{X}$, where P, Q, P', Q', M, N are all rational functions of x. $\int M dx$ has already been considered (1915).

$$\int N\sqrt{X}\,dx = \int \frac{NX}{\sqrt{X}}\,dx \text{ or } \int \frac{R\,dx}{\sqrt{X}}, \text{ where } R \text{ is rational.}$$

By substituting $x = \frac{p+qy}{1+y}$, and determining p and q so that the odd powers of y in the denominator may vanish, the last integral is brought to the form

2122 $$\int \frac{R\,dy}{\sqrt{(a+by^2+cy^4)}}.$$

R being a rational function of y, may be expressed as the sum of an odd and an even function; thus the integral is equivalent to the two

2123 $$\int \frac{yF_1(y^2)\,dy}{\sqrt{(a+by^2+cy^4)}} + \int \frac{F_2(y^2)\,dy}{\sqrt{(a+by^2+cy^4)}}.$$

The first integral can be found by substituting $\sqrt{y}$.

The second, by substituting $\frac{a+bx^2}{c+dx^2}$ for y^2, can be made to depend upon three integrals of the forms

2124 $$\int \frac{dx}{\sqrt{1-x^2 . 1-k^2x^2}}, \qquad \int \frac{\sqrt{1-k^2x^2}\,dx}{\sqrt{1-x^2}},$$

$$\int \frac{dx}{(1+nx^2)\sqrt{1-x^2 . 1-k^2x^2}}.$$

By substituting $\phi = \sin^{-1}x$, the above become

2125 $$\int \frac{d\phi}{\sqrt{1-k^2\sin^2\phi}}, \qquad \int \sqrt{1-k^2\sin^2\phi}\,d\phi,$$

$$\int \frac{d\phi}{(1+n\sin^2\phi)\sqrt{1-k^2\sin^2\phi}}.$$

These are the transcendental functions known as *Elliptic Integrals*. They are denoted respectively by

2126 $$F(k, \phi), \qquad E(k, \phi), \qquad \Pi(n, k, \phi).$$

APPROXIMATIONS TO $F(k, \phi)$ AND $E(k, \phi)$ IN SERIES.

When k is less than unity, the values of $F(k, \phi)$ and $E(k, \phi)$, from the origin $\phi = 0$, in converging series, are

2127 $$F(k, \phi) = \phi - \frac{k^2}{2^2} A_2 + \frac{1.3.k^4}{2.4.2^3} A_4 - \frac{1.3.5.k^6}{2.4.6.2^5} A_6 + \dots$$
$$\dots + (-1)^{\frac{1}{2}n} \frac{1.3.5 \dots n-1}{2.4.6 \dots n} \frac{k^n}{2^{n-1}} A_n, \text{ \&c.}$$

2128 $$E(k, \phi) = \phi + \frac{k^2}{2^2} A_2 - \frac{1.k^4}{2.4.2^3} A_4 + \frac{1.3.k^6}{2.4.6.2^5} A_6 - \dots$$
$$+ (-1)^{\frac{1}{2}n+1} \frac{1.3.5 \dots n-3}{2.4.6 \dots n} \frac{k^n}{2^{n-1}} A_n, \text{ \&c.,}$$

where [n being an even integer.

$$A_2 = \frac{\sin 2\phi}{2} - \phi,$$

$$A_4 = \frac{\sin 4\phi}{4} - \frac{4 \sin 2\phi}{2} + 3\phi,$$

$$A_6 = \frac{\sin 6\phi}{6} - \frac{6 \sin 4\phi}{4} + \frac{15 \sin 2\phi}{2} - 10\phi,$$

... ...

$$A_n = \frac{\sin n\phi}{n} - \frac{n \sin (n-2)\phi}{n-2} + \frac{C(n, 2) \sin (n-4)\phi}{n-4}$$
$$- \frac{C(n, 3) \sin (n-6)\phi}{n-6} + \dots + (-1)^{\frac{1}{2}n} \tfrac{1}{2} C(n, \tfrac{1}{2}n)\, \phi.$$

PROOF.—In each case expand by the Binomial Theorem; substitute from (773) for the powers of $\sin\phi$, and integrate the separate terms.

The values of $F(k, \phi)$ and $E(k, \phi)$, between the limits $\phi = 0$, $\phi = \frac{1}{2}\pi$, are therefore

2129
$$F\left(k, \frac{\pi}{2}\right) = \frac{\pi}{2} \left\{ 1 + \left(\frac{1}{2}\right)^2 k^2 + \left(\frac{1.3}{2.4}\right)^2 k^4 + \left(\frac{1.3.5}{2.4.6}\right)^2 k^6 + \text{\&c.} \right\}.$$

2130
$$E\left(k, \frac{\pi}{2}\right) = \frac{\pi}{2} \left\{ 1 - \left(\frac{1}{2}\right)^2 k^2 - \left(\frac{1.3}{2.4}\right)^2 \frac{k^4}{3} - \left(\frac{1.3.5}{2.4.6}\right)^2 \frac{k^6}{5} - \text{\&c.} \right\}.$$

But series which converge more rapidly are

2131
$$F\left(k, \frac{\pi}{2}\right) = \frac{\pi(1+n)}{2} \left\{ 1 + \left(\frac{1}{2}\right)^2 n^2 + \left(\frac{1.3}{2.4}\right)^2 n^4 + \left(\frac{1.3.5}{2.4.6}\right)^2 n^6 + \text{\&c.} \right\}$$

2132

$$E\left(k, \frac{\pi}{2}\right) = \frac{\pi}{2(1+n)}\left\{1+\left(\frac{1}{2}\right)^2 n^2+\left(\frac{1}{2.4}\right)^2 n^4+\left(\frac{1.3}{2.4.6}\right)^2 n^6+\&c.\right\}$$

where $$n = \frac{1-\sqrt{(1-k^2)}}{1+\sqrt{(1-k^2)}}.$$

2133 $\int \frac{F(x)\,dx}{\sqrt{(a+bx+cx^2+bx^3+ax^4)}}$, when $F(x)$ can be expressed in the form $\left(x-\frac{1}{x}\right)f\left(x+\frac{1}{x}\right)$, is integrated by substituting $x+\frac{1}{x}$.

If b is negative, and $F(x)$ of the form $\left(x+\frac{1}{x}\right)f\left(x-\frac{1}{x}\right)$; substitute $x-\frac{1}{x}$.

2134 $$\frac{F(x)\,dx}{\sqrt{(a+bx+cx^2+dx^3+cx^4+bx^5+ax^6)}}.$$

Substitute $$x+\frac{1}{x} = z.$$

Hence $$\frac{dx}{\sqrt{x^3}} = \left(\frac{1}{2\sqrt{z-2}} - \frac{1}{2\sqrt{z+2}}\right)dz,$$

and the integral takes the form

$$\int \frac{P+Q\sqrt{z^2-4}}{\sqrt{\{a(z^3-3z)+b(z^2-2)+cz+d\}}} \cdot \frac{\sqrt{z+2}-\sqrt{z-2}}{2\sqrt{z^2-4}}\,dz,$$

where P, Q are rational functions of z. Writing Z for the cubic in z, we see that the integral depends upon

$$\int \frac{P\,dz}{\sqrt{Z(z\pm 2)}} \quad \text{and} \quad \int \frac{Q(z\pm 2)\,dz}{\sqrt{Z(z\pm 2)}},$$

the radicals in which contain no higher power of z than the fourth. The integrals therefore fall under (2121).

2135 $$\int \frac{F(x)\,dx}{\sqrt{(a+bx^2+cx^4+bx^6+ax^8)}}.$$

Expressing $F(x)$ as the sum of an odd and an even function, as in (2123), the integral is divided into two; and, by substituting x^2, the first of these is reduced to the form in (2121), and the second to the form in (2134) with $a=0$.

2136 $$\int \frac{F(x)\,dx}{\sqrt[3]{(a+bx+cx^2+dx^3)}}.$$

Put $x = y+\alpha$, α being a root of the equation $a+bx+cx^2+dx^3 = 0$; and, in the resulting integral, substitute zy for the denominator. The form finally obtained will be

$$\int (P+Q\sqrt{\alpha+\beta z^3})\,dz,$$

which falls under (2134), P and Q being rational functions of z.

2137 $$\int \frac{F(x)\,dx}{\sqrt[4]{(a+bx^2+cx^4)}}.$$

Expressing $F(x)$ as the sum of an odd and an even function, as in (2123), two integrals are obtained. By putting the denominator equal to z in the first, and equal to xz in the second, each is reducible to an integral of the form

$$\int (P+Q\sqrt{\alpha+\beta z^4})\,dz,$$

which falls under (2121).

2138 $$\int \frac{dx}{\sqrt{(1-x^4)}} = -\frac{1}{\sqrt{2}}\,F\left(\frac{1}{\sqrt{2}},\ \phi\right).$$ [Expand by (2127).

2139 $$\int \frac{dx}{\sqrt{1+x^4}} = \frac{1}{2}\,F\left(\frac{1}{\sqrt{2}},\ \phi\right).$$

Proof.—Substitute $\cos^{-1}x$ in (2138), and $2\tan^{-1}x$ in (2139).

2140
$$\int \frac{dx}{\sqrt{(2ax-x^2)(b-x)}} = \frac{2}{\sqrt{b}}\,F\left(\sqrt{\frac{2a}{b}},\ \phi\right) \text{ or } \sqrt{\frac{2}{a}}\,F\left(\sqrt{\frac{b}{2a}},\ \phi\right),$$
according as b is $>$ or $<$ $2a$.

Proof.—Substitute accordingly, $x = 2a\sin^2\phi$ or $x = b\sin^2\phi$.

2141 $$\int \frac{dx}{\sqrt{(a-x)(x-b)(x+c)}} = -\frac{2}{\sqrt{(a+c)}}\,F\left(\sqrt{\frac{a-b}{a+c}},\ \phi\right).$$

Proof.—Substitute $x = a-(a-b)\sin^2\phi$, x being $<a$ and $>b$.

SUCCESSIVE INTEGRATION.

2148 In conformity with the notation of (1487), let the operation of integrating a function v, once, twice, ... n times for x, be denoted either by

$$\int_x v, \int_{2x} v, \ldots \int_{nx} v, \quad \text{or by} \quad d_{-x},\ d_{-2x},\ \ldots\ d_{-nx},$$

the notation d_{-x} indicating an operation which is the inverse of d_x. Similarly, since y_x, y_{2x}, y_{3x}, &c. denote successive derivatives of y, so y_{-x}, y_{-2x}, y_{-3x}, &c. may be taken to represent the successive integrals of y with respect to x.

2149 Since a constant is added to the result of each integration, every integral of the nth order of a function of a single variable x must be supplemented by the quantity

$$\frac{a_1 x^{n-1}}{\lfloor n-1} + \frac{a_2 x^{n-2}}{\lfloor n-2} + \ldots + a_{n-1}x + a_n = \int_{nx} 0,$$

where a_1, a_2, a_3 ... a_n are arbitrary constants.

Examples.

The six following integrals are obtained from (1922) and (1923).

When p is any positive quantity,

2150
$$\int_{nx} x^p = \frac{x^{p+n}}{(p+1)(p+2)\ldots(p+n)} + \int_{nx} 0.$$

When p is any positive quantity not an integer, or any positive integer greater than n,

2151
$$\int_{nx} \frac{1}{x^p} = \frac{(-1)^n}{(p-1)(p-2)\ldots(p-n)\,x^{p-n}} + \int_{nx} 0.$$

When p is a positive integer not greater than n, the following cases occur—

2152
$$\int_{px} \frac{1}{x^p} = \frac{(-1)^{p-1}}{\lfloor p-1} \log x + \int_{px} 0.$$

2153
$$\int_{(p+1)x} \frac{1}{x^p} = \frac{(-1)^{p-1}}{\lfloor p-1} \cdot (x \log x - x) + \int_{(p+1)x} 0,$$

2154 $$\int_{(p+r)x}\frac{1}{x^p}=\frac{(-1)^{p-1}}{\lfloor p-1}\left\{\int_{(r-1)x}(x\log x)-\frac{x^r}{\lfloor r}\right\}+\int_{(p+r)x}0.$$

For the integral within the brackets, see (2166).

The following formula is analogous to (1461–2)

2155 $$\int_{nx}\left.\begin{matrix}\sin\\\cos\end{matrix}\right\}ax=\frac{1}{a^n}\left.\begin{matrix}\sin\\\cos\end{matrix}\right\}(ax-\tfrac{1}{2}n\pi)+\int_{nx}0,$$

SUCCESSIVE INTEGRATION OF A PRODUCT.

Leibnitz's Theorem (1460) and its analogue in the Integral Calculus are briefly expressed by the two equations

2157 $$D_{nx}(uv)=(d_x+\delta_x)^n uv,\qquad D_{-nx}(uv)=(d_x+\delta_x)^{-n}uv;$$

where D operates upon the product uv, d only upon u, and δ only upon v. Expanding the binomials, we get

2159 $$D_{nx}(uv)=u_{nx}v+nu_{(n-1)x}v_x+\frac{n(n-1)}{1.2}u_{(n-2)x}v_{2x}+\&c.$$

2160 $$D_{-nx}(uv)=u_{-nx}v-nu_{-(n+1)x}v_x+\frac{n(n+1)}{1.2}u_{-(n+2)x}v_{2x}-\&c.$$

PROOF.—The first equation is obtained in (1460). The second follows from the first by the operative law (1483); or it may be proved by Induction, independently, as follows—

Writing it in the equivalent form

$$\int_{nx}(uv)=\int_{nx}uv-n\int_{(n+1)x}uv_x+\frac{n(n+1)}{1.2}\int_{(n+2)x}uv_{2x}-\&c.\ \ldots\ldots\ \text{(i.)},$$

make $n=1$; then

$$\int_x(uv)=\int_x uv-\int_{2x}uv_x+\int_{3x}uv_{2x}-\&c.\ \ldots\ldots\ldots\ldots\ldots\ldots\ \text{(ii.)},$$

a result which may be obtained directly by integrating the left member successively by Parts. Now integrate equation (i.) once more for x, integrating each term on the right as a product by formula (ii.), and equation (i.) will be reproduced with $(n+1)$ in the place of n.

2161 $$\int_{nx}e^{ax}x^m=e^{ax}(a+d_x)^{-n}x^m+\int_{nx}0.$$ Or, by expansion,

2162
$$\int_{nx}e^{ax}x^m=\frac{e^{ax}}{a^n}\left\{x^m-\frac{nm}{a}x^{m-1}+\frac{n(n+1)\,m(m-1)}{1.2\,a^2}x^{m-2}-\&c.\right\}+\int_{nx}0.$$

If m be an integer, the series terminates with $(-1)^m n^{(m)}\div a^m$.

Similarly, by changing the sign of m,

2163
$$\int_{nx} \frac{e^{ax}}{x^m} = \frac{e^{ax}}{a^n} \left\{ \frac{1}{x^m} + \frac{nm}{ax^{m+1}} + \frac{n(n+1)\, m(m+1)}{1.2\, a^2 x^{m+2}} + \&c. \right\} + \int_{nx} 0.$$

PROOF.—Putting $u = e^{ax}$, $v = x^m$ in (2158), the formula becomes

$$\int_{nx} e^{ax} x^m = (d_x + \delta_x)^{-n} e^{ax} x^m = (d_x e^{ax} + e^{ax} \delta_x)^{-n} x^m = e^{ax} (a + \delta_x)^{-n} x^m.$$

Here e^{ax} is written before δ_x within the brackets, because δ does not operate upon e^{ax}. Observe, also, that the index $-n$ affects only the operative symbols d_x and δ_x, but it therefore affects the *results* of those operations. Thus, since $d_x e^{ax}$ produces ae^{ax}, the operation d_x is equivalent to aX, and is retained within the brackets, while the subject e^{ax}, being only now connected as a factor with each term in the expansion of $(a+\delta_x)^{-n}$, may be placed on the left.

2164 $$\int e^{ax} x^m \, dx = \frac{e^{ax}}{a} \left\{ x^m - \frac{m}{a} x^{m-1} + \frac{m(m-1)}{a^2} x^{m-2} - \&c. \right\}$$

2165 $$\int \frac{e^{ax}}{x^m} \, dx = \frac{e^{ax}}{a} \left\{ \frac{1}{x^m} + \frac{m}{ax^{m+1}} + \frac{m(m+1)}{a^2 x^{m+2}} + \&c. \right\}$$

PROOF.—Make $n=1$ in (2162) and (2163).

2166 $$\int_{nx} x^p (\log x)^m = \int_{nx} e^{(p+n)x} x^m.$$ [Subs. $\log x$.

Hence the integral of the logarithmic function may be obtained from that of the equivalent exponential function (2161).

For another method, see (2003–5).

HYPERBOLIC FUNCTIONS.

2180 DEFINITIONS.—The hyperbolic cosine, sine, and tangent are written and defined as follows :—

2181 $$\cosh x = \tfrac{1}{2}(e^x + e^{-x}) = \cos(ix).$$ (768)

2183 $$\sinh x = \tfrac{1}{2}(e^x - e^{-x}) = -i \sin(ix).$$

2185 $$\tanh x = \frac{e^x - e^{-x}}{e^x + e^{-x}} = -i \tan(ix).$$ (770)

By these equations the following relations are readily obtained.

2187 $\cosh 0 = 1; \qquad \sinh 0 = 0; \qquad \cosh \infty = \sinh \infty = \infty.$

2191 $\cosh^2 x - \sinh^2 x = 1.$

2192 $\sinh(x+y) = \sinh x \cosh y + \cosh x \sinh y.$

2193 $\cosh(x+y) = \cosh x \cosh y + \sinh x \sinh y.$

2194 $\tanh(x+y) = \dfrac{\tanh x + \tanh y}{1 + \tanh x \tanh y}.$

2195 $\sinh 2x = 2 \sinh x \cosh x.$

2196 $\cosh 2x = \cosh^2 x + \sinh^2 x.$

2197 $\quad = 2\cosh^2 x - 1 = 1 + 2\sinh^2 x.$

2199 $\sinh 3x = 3 \sinh x + 4 \sinh^3 x.$

2200 $\cosh 3x = 4\cosh^3 x - 3\cosh x.$

2201 $\tanh 2x = \dfrac{2\tanh x}{1+\tanh^2 x}.$

2202 $\tanh 3x = \dfrac{3\tanh x + \tanh^3 x}{1 + 3\tanh^2 x}.$

2203 $\sinh \dfrac{x}{2} = \sqrt{\dfrac{\cosh x - 1}{2}}; \qquad \cosh \dfrac{x}{2} = \sqrt{\dfrac{\cosh x + 1}{2}}.$

2205 $\tanh \dfrac{x}{2} = \sqrt{\dfrac{\cosh x - 1}{\cosh x + 1}} = \dfrac{\cosh x - 1}{\sinh x} = \dfrac{\sinh x}{\cosh x + 1}.$

2208 $\cosh x = \dfrac{1+\tanh^2 \frac{1}{2}x}{1-\tanh^2 \frac{1}{2}x}; \qquad \sinh x = \dfrac{2\tanh \frac{1}{2}x}{1-\tanh^2 \frac{1}{2}x}.$

INVERSE RELATIONS.

2210 Let $u = \cosh x, \quad \therefore x = \cosh^{-1} u = \log\left(u + \sqrt{u^2-1}\right).$

2211 $v = \sinh x, \quad \therefore x = \sinh^{-1} v = \log\left(v + \sqrt{v^2+1}\right).$

2212 $w = \tanh x, \quad \therefore x = \tanh^{-1} w = \frac{1}{2} \log\left(\dfrac{1+w}{1-w}\right).$

GEOMETRICAL INTERPRETATION OF tanh S.

2213 *The tangent of the angle which a radius from the centre of a rectangular hyperbola makes with the principal axis, is equal to the hyperbolic tangent of the included area.*

PROOF.—Let θ be the angle, r the radius, and S the area, in the hyperbola $x^2-y^2=1$ or $r^2=\sec 2\theta$; then

$$S=\int_0^\theta \tfrac{1}{2}\sec 2\theta\, d\theta = \tfrac{1}{2}\log\tan\left(\tfrac{1}{4}\pi+\theta\right). \quad (1942)$$

Therefore $e^{2S}=\dfrac{1+\tan\theta}{1-\tan\theta}$; therefore $\tan\theta=\dfrac{e^S-e^{-S}}{e^S+e^{-S}}=\tanh S$. (2185)

VALUE OF THE LOGARITHM OF AN IMAGINARY QUANTITY.

2214 $$\log(a+ib)=\tfrac{1}{2}\log(a^2+b^2)+i\tan^{-1}\frac{b}{a}.$$

PROOF.— $\log\dfrac{a+ib}{\sqrt{(a^2+b^2)}}=\log\sqrt{\left\{\dfrac{1+i\frac{b}{a}}{1-i\frac{b}{a}}\right\}}=i\tan^{-1}\dfrac{b}{a}$. By (771).

DEFINITE INTEGRALS.

SUMMATION OF SERIES BY DEFINITE INTEGRALS.

2230 $$\int_a^b f(x)\,dx=\left[f(a)+f(a+dx)+\ldots+f(a+n\,dx)\right]dx,$$

where n increases and dx diminishes indefinitely, so that $n\,dx=b-a$ in the limit.

2231 Ex. 1.—To find the sum, when n is infinite, of the series

$\dfrac{1}{n}+\dfrac{1}{n+1}+\dfrac{1}{n+2}\ldots\ldots+\dfrac{1}{n+n}$. Put $n=\dfrac{a}{dx}$; thus,

$$\frac{dx}{a}+\frac{dx}{a+dx}+\frac{dx}{a+2dx}+\ldots\ldots+\frac{dx}{2a}=\int_a^{2a}\frac{dx}{x}=\log 2.$$

2232 Ex. 2.—To find the sum, when n is infinite, of the series

$\dfrac{n}{n^2+1^2}+\dfrac{n}{n^2+2^2}+\dfrac{n}{n^2+3^2}+\ldots\ldots+\dfrac{n}{n^2+n^2}$. Put $n=\dfrac{1}{dx}$, then

$$\frac{dx}{1+(dx)^2}+\frac{dx}{1+(2dx)^2}+\ldots\ldots+\frac{dx}{1+(ndx)^2}=\int_0^1\frac{dx}{1+x^2}=\frac{\pi}{4}. \quad (1935)$$

THEOREMS RESPECTING THE LIMITS OF INTEGRATION.

2233 $$\int_0^a \phi(x)\,dx = \int_0^a \phi(a-x)\,dx.$$ [Substitute $a-x$.

2234 $$\int_0^a \phi(x)\,dx = 2\int_0^{\frac{1}{2}a} \phi(x)\,dx,$$

or *zero*, according as $\phi(x) = \pm\phi(a-x)$ for all values of x between 0 and a.

Ex.— $\int_0^\pi \sin x\,dx = 2\int_0^{\frac{\pi}{2}} \sin x\,dx.$ $\int_0^\pi \cos x\,dx = 0.$

If $\phi(x) = \phi(-x)$, that is, if $\phi(x)$ be an even function (1401) for all values of x between 0 and a.

2236 $$\int_{-a}^0 \phi(x)\,dx = \int_0^a \phi(x)\,dx = \frac{1}{2}\int_{-a}^a \phi(x)\,dx.$$

Ex.— $$\int_{-\frac{\pi}{2}}^0 \cos x\,dx = \int_0^{\frac{\pi}{2}} \cos x\,dx = \frac{1}{2}\int_{-\frac{\pi}{2}}^{\frac{\pi}{2}} \cos x\,dx.$$

If $\phi(x) = -\phi(-x)$, that is, if $\phi(x)$ be an odd function for all values of x between 0 and a.

2238 $$\int_{-a}^0 \phi(x)\,dx = -\int_0^a \phi(x)\,dx \quad \text{and} \quad \int_{-a}^a \phi(x)\,dx = 0.$$

Ex.— $$\int_{-\frac{\pi}{2}}^0 \sin x\,dx = -\int_0^{\frac{\pi}{2}} \sin x\,dx \quad \text{and} \quad \int_{-\frac{\pi}{2}}^{\frac{\pi}{2}} \sin x\,dx = 0.$$

Given $a<c<b$, and that $x=c$ makes $\phi(x)$ infinite, the value of $\int_a^b \phi(x)\,dx$ may be investigated by putting $\mu=0$, after integrating, in the formula

2240 $$\int_a^b \phi(x)\,dx = \int_a^{c-\mu} \phi(x)\,dx + \int_{c+\mu}^b \phi(x)\,dx.$$

If the function $\phi(x)$ changes sign on becoming infinite, this expression, when μ is an indefinitely small quantity, is called the *principal value* of the integral.

Ex.— $\int_{-1}^{1}\frac{dx}{x^3}=\int_{-1}^{-\mu}\frac{dx}{x^3}+\int_{\mu}^{1}\frac{dx}{x^3}=-\frac{1}{2\mu^2}+\frac{1}{2}-\frac{1}{2}+\frac{1}{2\mu^2}=0,$
which is the *principal value*. If, however, μ be made to vanish, the expression takes the indeterminate form $\infty-\infty$.

2241 Given $a<c<b$, the integral $\int_a^b\frac{\psi(x)\,dx}{(x-c)^n}$ will always be finite in value while n is less than unity.

PROOF.—Let μ in (2240) be taken so near to c in value that $\psi(x)$ shall remain finite and of the same sign for all values of x comprised between $c\pm\mu$. Then the part of the integral in which the fraction becomes infinite, and which is omitted in (2240), will be equal to $\int_{c-\mu}^{c+\mu}\frac{dx}{(x-c)^n}$, multiplied by a constant whose value lies between the greatest and least values of $\psi(x)$ which occur between $\psi(c-\mu)$ and $\psi(c+\mu)$. By integration it appears that the last integral is finite in value when n is <1.

2242 $$\int_a^b f(x)\,dx=(b-a)f\{a+\theta(b-a)\},$$

where θ lies between 0 and 1 in value.

The equation expresses the fact that the area in (Fig. 1901), bounded by the curve $y=f(x)$, the ordinates $f(a)$, $f(b)$, and the base $b-a$ is equal to the rectangle under $b-a$ and some ordinate lying in value between the greatest and least which occur in passing from $f(a)$ to $f(b)$.

If $\psi(x)$ does not change sign while x varies from $x=a$ to $x=b$,

2243 $$\int_a^b f(x)\,\psi(x)\,dx=f\{a+\theta(b-a)\}\int_a^b\psi(x)\,dx.$$

2244 If $\phi\left(x,\frac{1}{x}\right)$ is a symmetrical function of x and $\frac{1}{x}$,

$$\int_0^\infty\phi\left(x,\frac{1}{x}\right)\frac{dx}{x}=2\int_0^1\phi\left(x,\frac{1}{x}\right)\frac{dx}{x}.$$

PROOF.—Separate the integral into two parts by the formula $\int_0^\infty=\int_0^1+\int_1^\infty$, and substitute $\frac{1}{x}$ in the last integral.

METHODS OF EVALUATING DEFINITE INTEGRALS.

2245 RULE I.—*Substitute a new variable, and adjust the limits accordingly.*

For examples, see numbers 2291, 2308, 2342, 2345, 2416, 2425, 2457, 2506, 2605, &c.

2246 RULE II.—*Integrate by Parts* (1910), *so as to introduce a known definite integral.*

For examples, see numbers 2283, 2290, 2430, 2453, 2465, 2484–5, 2608–13, 2623, 2625, &c.

2247 RULE III.—*Differentiate or integrate with respect to some quantity other than the variable concerned; if a known integral is thus obtained, evaluate it, and then reverse the operation of differentiation or integration before performed with respect to the secondary variable.*

For examples, see numbers 2346–7, 2364, 2391, 2417, 2421–4, 2426, 2428, 2497–8, 2502–4, 2571, 2575–6, 2591, 2604, 2614, 2617–8, 2632, &c.

2248 RULE IV.—*Substitute imaginary values for constants, and thus transform the expression into one capable of integration.*

For examples, see numbers 2490, 2494, 2577, 2594, 2598, 2603, 2606, 2615, 2641–2.

2249 RULE V.—*Expand the function, if possible, in a finite or converging series, and integrate the separate terms.*

For examples, see numbers 2395–7, 2402–3, 2418–9, 2479, 2506, 2571, 2593, 2598, 2614, 2620, 2625, 2629, 2630–2, 2639.

2250 RULE VI.—*Decompose the integral into a number of partial integrals, and change all these by some substitution into integrals having the same limits. By summing the resulting series, a new integral is obtained which may be a known one.*

For examples, see numbers 2341, 2356–61, 2572, 2638.

2251 RULE VII.—*Separate the function to be integrated into two factors, and replace one of them by its value in the form of a definite integral taken between constant limits with respect to some new variable. The double integral so obtained may frequently be evaluated by changing the order of integration as explained in* (2261).

For examples, see numbers 2507, 2510, 2573, 2619.

2252 RULE VIII.—*Multiply a known definite integral which is discontinuous between certain values of a constant which it* *contains, by some function of that constant, such that the integral of the product with respect to the constant is known. A new definite integral may thus be obtained.*

For examples, see numbers 2518, 2522.

Particular artifices not included in the foregoing rules are employed in 2293, 2305, 2310, 2314–5, 2317, 2367–9, 2404–15, 2422, 2429, 2456, 2495, 2514, 2518, 2585, 2600, 2626, 2635, 2637.

Additional formulæ for integration will be found at 2700, *et seq.*

DIFFERENTIATION UNDER THE SIGN OF INTEGRATION.

Let $u = \int_a^b f(x)\,dx$, where a, b, and $f(x)$ are independent of each other; then

2253 $$\frac{du}{db} = f(b) \qquad \text{and} \qquad \frac{du}{da} = -f(a).$$

PROOF.—Let $u = \phi(b) - \phi(a)$.

Therefore $u_b = \phi'(b) = f(b)$ and $u_a = -\phi'(a) = -f(a)$.

Let $u = \int_a^b f(x, c)\,dx$. Then, when a and b are independent of c,

2255 $$u_c = \int_a^b \{f(x, c)\}_c\,dx \qquad \text{and} \qquad u_{nc} = \int_a^b \{f(x, c)\}_{nc}\,dx.$$

PROOF.— $\frac{du}{dc} = \left\{\int_a^b f(x, c+h)\,dx - \int_a^b f(x, c)\,dx\right\} \div h$

$= \int_a^b \frac{f(x, c+h) - f(x, c)}{h}\,dx$ (since h is constant relatively to x) $= \int_a^b \frac{df(x, c)}{dc}\,dx.$

But if a and b also are functions of c,

2257 $$\frac{du}{dc} = \int_a^b \left\{\frac{df(x, c)}{dc}\right\} dx + f(b, c)\frac{db}{dc} - f(a, c)\frac{da}{dc}.$$

PROOF.—The complete derivative of u with respect to c will now be $u_c + u_b b_c + u_a a_c$. But $u_b = f(b, c)$ and $u_a = -f(a, c)$, by (2253–4).

INTEGRATION BY DIFFERENTIATING UNDER THE SIGN OF INTEGRATION.

2258 Ex. 1.— $\int x^n e^{ax} dx = \int (e^{ax})_{na} dx = d_{na} \int e^{ax} dx$ (2256)

$$= d_{na}(e^{ax}a^{-1}) = e^{ax}(x+d_a)^n a^{-1},$$

by (1464), a and x being transposed.

2259 Ex. 2.— $\int x^n e^{ax} \sin bx\, dx = d_{na} \int e^{ax} \sin bx\, dx.$

The last integral is given in (1999), putting $n=1$.

2260 Ex. 3.—

$$\int x a^{x-1} dx = \int (a^x)_a dx = \left(\int a^x dx\right)_a = \left(\frac{a^x}{\log a}\right)_a = \frac{a^{x-1}}{\log a}\left(x - \frac{1}{\log a}\right).$$

INTEGRATION UNDER THE SIGN OF INTEGRATION.

When the limits are constant,

2261 $$\int_{x_1}^{x_2} \int_{y_1}^{y_2} f(x, y)\, dx\, dy = \int_{y_1}^{y_2} \int_{x_1}^{x_2} f(x, y)\, dy\, dx.$$

That is, the order of integration may be changed.

But an exception to this rule occurs when, at any stage of the integration, an infinite value is produced. The double integrals above may not then have the same value.

APPROXIMATE INTEGRATION.

BERNOULLI'S SERIES.

2262 $$\int_0^a f(x)\, dx = af(a) - \frac{a^2}{1.2} f'(a) + \frac{a^3}{1.2.3} f''(a) - \dots$$

$$+ \frac{(-1)^{n-1} a^n}{1.2 \dots n} f^{n-1}(a) + \frac{(-1)^n}{1.2 \dots n} \int_0^a x^n f^n(x)\, dx.$$

Proof.—Integrate successively by Parts, $\int dx$, $\int x\,dx$, &c. Or change $f'(x)$ into $f(x)$ in (1510).

2263

$$\int_a^b f(x)\, dx = (b-a) f(a) + \frac{(b-a)^2}{1.2} f'(a) + \frac{(b-a)^3}{1.2.3} f''(a) + \&c.$$

PROOF.—Put $f(a)$ for $\phi'(a)$ in the expansion of the right side of equation (1902), by Taylor's theorem (1500); viz.,

$$\int_a^b f(x)\,dx = \phi(b)-\phi(a) = (b-a)\,\phi'(a)+\frac{(b-a)^2}{1.2}\,\phi''(a)+\&c.$$

The following is a nearer approximation :—

Let $(b-a) = nh$, where n is an integer; then

2264 $$\int_a^b f(x)\,dx = h\left\{\tfrac{1}{2}f(b)+\tfrac{1}{2}f(a)+f(a+h)+\ldots+f(b-h)\right\}$$
$$-\frac{h^2B_2}{\lfloor 2}\left\{f'(b)-f'(a)\right\}+\frac{h^4B_4}{\lfloor 4}\left\{f'''(b)-f'''(a)\right\}$$
$$-\frac{h^6B_6}{\lfloor 6}\left\{f^{\text{v}}(b)-f^{\text{v}}(a)\right\}+\&c.$$

PROOF.—Expand $(e^{nhd_x}-1)\div(e^{hd_x}-1)$ by ordinary division, and also by (1539), and operate upon $f(x)$ with each result; thus, after multiplying by h, we obtain, by (1520),

$$h\left\{f(x)+f(x+h)+f(x+2h)+\ldots+f(x+\overline{n-1}\,h)\right\}$$
$$=\left\{f(x+nh)-f(x)\right\}\left(d_{-x}-\frac{h}{2}+\frac{h^2B_2}{1.2}\,d_x-\frac{h^4B_4}{1.2.3.4}\,d_{3x}+\ldots\right),$$

which expression, by changing x into a and $x+nh$ into b, is equivalent to the above, since $d_{-x}\{f(x+nh)-f(x)\} = \int_x^{x+nh} f(x)\,dx$.

2265
$$\int_a^{a+h} f(x)\,dx = \frac{2hf(a-h)}{1.2}+\frac{3^2h^2f'(a-2h)}{1.2.3}+\frac{4^3h^3f''(a-3h)}{1.2.3.4}+\&c.$$

PROOF.—Assume $x = ce^{hx}$. Then x is equal to the coefficient of $\frac{1}{x}$ in the expansion of $-\log\left(1-\frac{ce^{hx}}{x}\right)$. Thus

$$x = c+\frac{2hc^2}{1.2}+\frac{3^2h^2c^3}{1.2.3}+\frac{4^3h^3c^4}{1.2.3.4}+\ldots$$

Substitute d_x for x, and therefore $d_xe^{-hd_x}$ for c in this equation, and operate with it upon $\int\phi(x+h)\,dx$, employing (1520). Finally, write $f(x)$ for $\phi'(x)$, and a for x.

A more general result, obtained in the same way, is

2266 $$\int_a^{a+nh} f(x)\,dx = nhf(a-h)+n(n+2)\frac{h^2}{\lfloor 2}f'(a-2h)$$
$$+(n+3)^2\frac{h^3}{\lfloor 3}f''(a-3h)+\&c.$$

THE INTEGRALS B (l, m) AND $\Gamma(n)$.

EULER'S FIRST INTEGRAL B (l, m).

The three principal forms are—

2280 I. $$B(l,m)=\int_0^1 x^{l-1}(1-x)^{m-1}dx = B(m,l).$$ [By (2233)

2281 II. $$B(l,m)=\int_0^\infty \frac{x^{l-1}}{(1+x)^{l+m}}\,dx.$$ [By substituting $\frac{x}{1-x}$ in I.

2282 III. $$B(l,m)=\int_0^\infty \frac{x^{m-1}}{(1+x)^{l+m}}\,dx.$$ [By substituting $\frac{1-x}{x}$ in I.

When l and m are positive, and l is an integer,

2283 $$B(l, m) = \frac{1^{(l-1)}}{m^{(l)}}.$$

If m be the integer, interchange l and m. If both l and m are integers, the forms are convertible.

Proof.—Integrate (2280) by parts, thus,

$$\int_0^1 x^{l-1}(1-x)^{m-1}dx = \frac{l-1}{m}\int_0^1 x^{l-2}(1-x)^m.$$

Repeat this step successively.

EULER'S SECOND INTEGRAL $\Gamma(n)$.

n being a real and positive quantity,

2284 $$\Gamma(n) = \int_0^\infty e^{-x}x^{n-1}dx = \int_0^1 \left(\log\frac{1}{x}\right)^{n-1} dx.$$

The second form being obtained by substituting e^{-x} in the first.

2286 $$\Gamma(1) = 1, \qquad \Gamma(2) = 1.$$

2288 $$\Gamma(n+1) = n\Gamma(n) = n(n-1)\ldots(n-r)\,\Gamma(n-r).$$

2290 $\Gamma(n+1) = \lfloor n$, when n is an integer.

Proof.—By Parts, $$\int_0^\infty e^{-x}x^{n-1}dx = \left.\frac{x^n}{ne^x}\right]_0^\infty + \frac{1}{n}\int_0^\infty e^{-x}x^n dx.$$

The fraction becomes zero at each limit, as appears by (1580), differentiating the numerator and denominator, each r times, and taking $r > n$.

2291 $$\int_0^\infty e^{-kx}x^{n-1}dx = \frac{\Gamma(n)}{k^n} = \int_0^1 x^{k-1}\left(\log\frac{1}{x}\right)^{n-1}dx.$$

PROOF.—Substitute kx in the first integral, and so reduce it to the form (2284). In the second integral, substitute $-\log x$, reducing it to the former. When n is an integer, (2291) may be obtained by differentiating $n-1$ times for k the equation $\int_0^\infty e^{-kx}dx = \frac{1}{k}$.

When μ is an indefinitely great integer,

2293 $$\Gamma(n) = \frac{1.2.3\ldots\mu}{n(n+1)\ldots(n+\mu)}\mu^n.$$

PROOF.— $\log\frac{1}{x} = \text{lim.}\ \mu(1-x^{\frac{1}{\mu}})$ (1583). Give it this value in (2285), and then substitute $y = x^{\frac{1}{\mu}}$; thus, in the limit,

$\Gamma(n) = \mu^{n-1}\int_0^1 (1-x^{\frac{1}{\mu}})^{n-1}dx = \mu^n\int_0^1 y^{\mu-1}(1-y)^{n-1}dy.$ Then, by (2283), changing μ finally into $\mu+1$ in the fraction.

$\log\Gamma(1+n)$ IN A CONVERGING SERIES.

2294 Let n be <1, μ an indefinitely great integer, and $S_r = 1+\frac{1}{2^r}+\frac{1}{3^r}+\ldots\frac{1}{\mu^r}$, then

$$\log\Gamma(1+n).$$

2295 $$= (\log\mu - S_1)n + \tfrac{1}{2}S_2n^2 - \tfrac{1}{3}S_3n^3 + \tfrac{1}{4}S_4n^4 - \tfrac{1}{5}S_5n^5 + \&c.$$

2296 $$= \tfrac{1}{2}\log\frac{n\pi}{\sin n\pi} + (\log\mu - S_1)\,n - \tfrac{1}{3}S_3n^3 - \tfrac{1}{5}S_5n^5 - \&c.$$

2297 $$= \tfrac{1}{2}\log\frac{n\pi(1-n)}{\sin n\pi(1+n)} + n(1+\log\mu - S_1) + \frac{n^3}{3}(1-S_3) + \frac{n^5}{5}(1-S_5) + \&c.$$

2298 $$= \tfrac{1}{2}\log\frac{n\pi}{\sin n\pi} - \tfrac{1}{2}\log\frac{1+n}{1-n} + \cdot 4227843n$$
$$-\cdot 0673530n^3 - \cdot 0073855n^5 - \cdot 0011927n^7 - \cdot 0002231n^9 - \&c.$$

PROOF.—By (2293), $\Gamma(1+n) = \frac{\mu^n\,1.2.3\ldots\mu}{(n+1)(n+2)\ldots(n+\mu)}$, since $\frac{\mu}{n+\mu+1} = 1$, when $\mu=\infty$. Whence

$$\log\Gamma(1+n) = nl\mu - l(1+n) - l\left(1+\frac{n}{2}\right) - l\left(1+\frac{n}{3}\right) - \ldots - l\left(1+\frac{n}{\mu}\right).$$

Developing the logarithms by (155), the series (2295) is obtained. The next series is deduced from this by substituting

$$S_2 n^2 + \tfrac{1}{2} S_4 n^4 + \tfrac{1}{3} S_6 n^6 + \tfrac{1}{4} S_8 n^8 + \&c. = \log n\pi - \log \sin n\pi,$$

a result obtained from (815) by putting $\theta = n\pi$ and expanding the logarithms by (156).

The series (2297) is deduced from the preceding by adding the expression

$$0 = -\tfrac{1}{2} \log \frac{1+n}{1-n} + n + \frac{n^3}{3} + \frac{n^5}{5} + \&c., \text{ from (157).}$$

2305 $$B(l, m) = \frac{\Gamma(l)\,\Gamma(m)}{\Gamma(l+m)}.$$

PROOF.—Perform the integrations in the double integral

$$\int_0^\infty \int_0^\infty e^{-x(y+1)} x^{l+m-1} y^{l-1}\, dx\, dy,$$

first for x, by formula (2291), and then for y, by (2281), and the result is $B(l, m)\,\Gamma(l+m)$. Again perform the integration, first for y, by (2291), and the result is $\Gamma(l)\,\Gamma(m)$, by (2284).

NOTE.—The double integral may be written by the following rule :—*Write* xy *for* x *in* $\Gamma(l)$, *and multiply by the factors of* $\Gamma(m+1)$. We thus obtain

$$\int_0^\infty \int_0^\infty e^{-xy} (xy)^{l-1} \times e^{-x} x^m\, dx\, dy,$$

which is equivalent to the integral in question.

2306 $$B(l, m)\, B(l+m, n) = B(m, n)\, B(m+n, l) = B(n, l)\, B(n+l, m)$$

2307 $$= \frac{\Gamma(l)\,\Gamma(m)\,\Gamma(n)}{\Gamma(l+m+n)}.$$ [By (2305).

2308 $$\int_0^a x^{l-1}(a-x)^{m-1} dx = a^{l+m-1} B(l, m).$$ [Substitute $\frac{x}{a}$.

If p and q are positive integers, $p<q$, and if $m = \frac{2p+1}{2q}$.

2309 $$\int_0^\infty \frac{x^{2p}}{1+x^{2q}}\, dx = \frac{\pi}{2q \sin m\pi}.$$

2310 $$\int_0^\infty \frac{x^{2p}}{1-x^{2q}}\, dx = \frac{\pi}{2q \tan m\pi}.$$

PROOF.—(i.) In (2023) put $l = 2p+1$, $n = 2q$, and take the value of the integral between the limits $\pm\infty$. The first term becomes $\log 1 = 0$; the second gives the series

$$\frac{\pi}{q}\left\{ \sin \frac{l\pi}{2q} + \sin \frac{3l\pi}{2q} + \ldots + \sin \frac{(2q-1)\,l\pi}{2q} \right\} = \frac{\pi}{q \sin m\pi},$$

by (800). The integral required is one-half of this result, by (2237).

(ii.) (2310) is deduced in a similar manner from (2021).

2311 $\int_0^\infty \frac{x^{m-1}}{1+x}\,dx = \frac{\pi}{\sin m\pi}, \qquad \int_0^\infty \frac{x^{m-1}}{1-x}\,dx = \frac{\pi}{\tan m\pi},$

where m has *any* value between 0 and 1.

PROOF.—By substituting x^{2q} in (2309–10). Also, since $m = \frac{2p+1}{2q}$, by taking the integers p and q large enough, the fraction may, in the limit, be made equal to any quantity whatever lying between 0 and 1 in value.

2313 $\Gamma(m)\Gamma(1-m) = \frac{\pi}{\sin m\pi}, \qquad m$ being <1.

PROOF.—Put $l+m = 1$ in the two values of $B(l, m)$ (2282) and (2305);

thus, $$\Gamma(m)\Gamma(1-m) = \int_0^\infty \frac{x^{m-1}}{1+x}\,dx = \frac{\pi}{\sin m\pi}, \quad \text{by (2311).}$$

2314 COR.— $\Gamma(\tfrac{1}{2}) = \sqrt{\pi}.$

The following is an independent proof:

$$\Gamma(\tfrac{1}{2}) = \int_0^\infty e^{-x}x^{-\frac{1}{2}}\,dx = 2\int_0^\infty e^{-y^2}dy = 2\int_0^\infty e^{-z^2}dz.$$

Now form the product of the last two integrals, and change the variables to r, θ by the equations

$\left.\begin{array}{l} y = r\sin\theta \\ z = r\cos\theta \end{array}\right\}$, from which, by (1609), $dy\,dz = \frac{d(yz)}{d(r\theta)}dr\,d\theta = r\,dr\,d\theta$. Hence

$$\{\Gamma(\tfrac{1}{2})\}^2 = 4\int_0^\infty\int_0^\infty e^{-(y^2+z^2)}dy\,dz = 4\int_0^\infty\int_0^{\frac{1}{2}\pi} e^{-r^2}r\,dr\,d\theta = \pi;$$

the limits for r and θ being obtained from

$$r^2 = y^2+z^2, \quad \tan\theta = \frac{y}{z}.$$

2315 $\Gamma\left(\frac{1}{n}\right)\Gamma\left(\frac{2}{n}\right)\Gamma\left(\frac{3}{n}\right)\ldots\Gamma\left(\frac{n-1}{n}\right) = \sqrt{\frac{(2\pi)^{n-1}}{n}}.$

PROOF.—Multiply the left side by the same factors in reversed order, and apply (2313) thus

$$\Gamma\left(\frac{1}{n}\right)\Gamma\left(1-\frac{1}{n}\right)\Gamma\left(\frac{2}{n}\right)\Gamma\left(1-\frac{2}{n}\right)\ldots\Gamma\left(\frac{n-1}{n}\right)\Gamma\left(1-\frac{n-1}{n}\right)$$

$$= \frac{\pi^{n-1}}{\sin\frac{\pi}{n}\sin\frac{2\pi}{n}\ldots\sin\frac{(n-1)\pi}{n}} = \frac{2^{n-1}\pi^{n-1}}{n}, \quad \text{by (814).}$$

2316 $\frac{n^{nx}}{n\Gamma(nx)}\Gamma(x)\Gamma\left(x+\frac{1}{n}\right)\ldots\Gamma\left(x+\frac{n-1}{n}\right) = \sqrt{\frac{(2\pi)^{n-1}}{n}}.$

PROOF.—Call the expression on the left $\phi(x)$. Change x to $x+r$, where r is any integer, and change each Gamma function by the formula $\Gamma(x+r) = x^{(r)}\Gamma(x)$ (2288). The result after reduction is $\phi(x)$. Hence $\phi(x) = \phi(x+r)$, however great r may be. Therefore $\phi(x)$ is *independent* of x. But, when $x = \frac{1}{n}$, $\phi(x)$ takes the value in question by (2315). Therefore $\phi(x)$ always has that value.

The formula may also be obtained by means of (2293).

NUMERICAL CALCULATION OF $\Gamma(x)$.

2317 All values of $\Gamma(x)$ may be found in terms of values lying between $\Gamma(0)$ and $\Gamma(\frac{1}{2})$.

When x is >1, formula (2289) reduces $\Gamma(x)$ to the value in which x is <1; and when x lies between 1 and $\frac{1}{2}$, formula (2313) reduces the function to the value in which x lies between 0 and $\frac{1}{2}$.

Values of $\Gamma(x)$, when x lies between 0 and 1, can also be made to depend upon values in which x lies between $\frac{1}{6}$ and $\frac{1}{3}$, by the formulæ,

2318 $$\Gamma(x) = 2^{1-2x}\sqrt{\pi}\,\frac{\Gamma(2x)}{\Gamma(\frac{1}{2}+x)}, \quad \Gamma(x) = \frac{\sqrt{\pi}}{2^{1-x}\cos\frac{\pi x}{2}}\,\frac{\Gamma\left(\frac{x}{2}\right)}{\Gamma\left(\frac{1-x}{2}\right)}.$$

PROOF.—To obtain (2318), make $n = 2$ in (2316). To obtain (2319), put $m = \frac{1}{2}(1+x)$ in (2313), and change x into $\frac{1}{2}x$ in (2318), and then eliminate $\Gamma\left(\frac{1+x}{2}\right)$.

Methods of employing the formulæ—

2320 (i.) When x lies between $\frac{2}{3}$ and 1, reduce $\Gamma(x)$ to $\Gamma(1-x)$, by (2313).

2321 (ii.) When x lies between $\frac{1}{3}$ and $\frac{2}{3}$, reduce by (2319), the limits on the right of which will then be $\frac{1}{6}$ and $\frac{1}{3}$.

2322 (iii.) When x lies between 0 and $\frac{1}{6}$, reduce by (2318); $\Gamma(\frac{1}{2}+x)$ will then involve the limits $\frac{1}{2}$ and $\frac{2}{3}$, and will be reducible by case (ii.)

If $2x$ is $<\frac{1}{6}$, reduce $\Gamma(2x)$ by (2318), writing $2x$ for x. If this gives $4x<\frac{1}{6}$, reduce again by the same formula, writing $4x$ for x, and so on.

2323 The figure exhibits the curve whose equation in rectangular coordinates is $y = \Gamma(x)$. Let the unit abscissa be $OA = AB = 1$. Then the ordinates AD, BC are also each $= 1$ by (2286–7).

The minimum value of $\Gamma(x)$ is approximately 0·8556032, corresponding to $x = 1{\cdot}4616321$.

The values of $\log\Gamma(x)$ in the table at page 30 correspond to ordinates taken between $AD = \Gamma(1)$ and $BC = \Gamma(2)$.

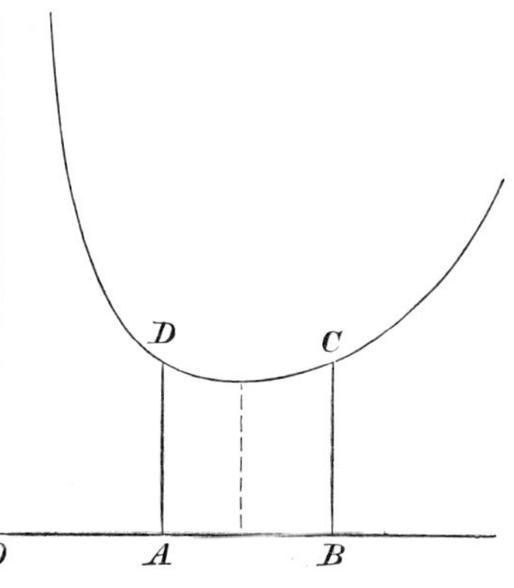

INTEGRATION OF ALGEBRAIC FORMS.

2341
$$\int_0^1 \frac{x^{l-1}+x^{m-1}}{(1+x)^{l+m}}\,dx = B(l,\,m).$$

PROOF.—Add together (2281) and (2282). Separate the resulting integral into $\int_0^1 + \int_1^\infty$, and substitute $\frac{1}{x}$ in the last part.

2342
$$\int_0^1 \frac{x^{l-1}(1-x)^{m-1}}{(x+a)^{l+m}}\,dx = \frac{B(l,m)}{a^m(1+a)^l}.$$
[Substitute $\frac{x+ax}{x+a}$.

2343
$$\int_0^1 x^{l-1}(1-x^n)^{m-1}dx = \frac{1}{n}B\left(\frac{l}{n},\,m\right).$$
[Substitute x^n.

2344
$$\int_0^1 \frac{x^{l-1}(1-x)^{m-1}}{\{ax+b(1-x)\}^{l+m}}\,dx = \frac{1}{a^l b^m}B(l,m).$$
[Subs. $\frac{ax}{(a-b)x+b}$.

The integral is also equivalent to

$$2\int_0^{\frac{\pi}{2}} \frac{\sin^{2l-1}\theta\cos^{2m-1}\theta}{(a\sin^2\theta+b\cos^2\theta)^{l+m}}\,d\theta,$$ and similarly in other cases.

2345
$$\int_0^\infty \frac{x^{n-1}\,dx}{(a+bx)^{n+1}} = \frac{1}{nab^n}.$$
[Substitute $\frac{bx}{a+bx}$.

2346 $$\int_0^\infty \frac{x^{n-1}dx}{(a+bx)^{m+n}} = \frac{1^{(m-1)}}{n^{(m)}} \frac{1}{a^m b^n}.$$

PROOF.—Differentiate (2345) $m-1$ times for a. (2255)

2347 $$\int_0^\infty \frac{x^{m+n-1}}{(1+x)^{n+1}}\,dx = C(m, n)\,\frac{\pi}{\sin m\pi}. \qquad m<1.$$

PROOF.—Substitute $x = ay$ in (2311), and then differentiate n times for a.

2348 $$\int_0^\infty \frac{x^{m-1}dx}{1+x^n} = \frac{\pi}{n}\operatorname{cosec}\frac{m\pi}{n}, \qquad \int_0^\infty \frac{x^{m-1}dx}{1-x^n} = \frac{\pi}{n}\cot\frac{m\pi}{n},$$

where m and n are *any* positive quantities, and m is $< n$.

PROOF.—Change m into $\frac{m}{n}$ in (2311–2), and then substitute $x^{\frac{1}{n}}$.

When n is positive and greater than unity,

2350 $$\int_0^\infty \frac{dx}{1+x^n} = \frac{\pi}{n}\operatorname{cosec}\frac{\pi}{n}. \qquad \int_0^\infty \frac{dx}{1-x^n} = \frac{\pi}{n}\cot\frac{\pi}{n}.$$

PROOF.—Substitute x^m in (2311–2) and change m into $\frac{1}{n}$.

2352 $$\int_0^1 \frac{dx}{\sqrt[n]{1-x^n}} = \frac{\pi}{n}\operatorname{cosec}\frac{\pi}{n}. \qquad \int_0^1 \frac{dx}{\sqrt[n]{1+x^n}} = \frac{\pi}{n}\cot\frac{\pi}{n}.$$

PROOF.—Substitute $\frac{x}{\sqrt[n]{1+x^n}}$ in (2350), and $\frac{x}{\sqrt[n]{1-x^n}}$ in (2351).

When m lies between 0 and 1,

2354 $$\int_0^\infty \frac{x^m dx}{1+x^2} = +\frac{\pi}{2}\sec\frac{m\pi}{2}, \qquad \int_0^\infty \frac{x^m dx}{1-x^2} = +\frac{\pi}{2}\tan\frac{m\pi}{2}.$$

PROOF.—Make $n = 2$ and write $m+1$ for m in (2348–9).

2356 $$\int_0^1 \frac{x^{m-1}+x^{-m}}{1+x}\,dx = \frac{\pi}{\sin m\pi}. \qquad \int_0^1 \frac{x^{m-1}-x^{-m}}{1-x}\,dx = \frac{\pi}{\tan m\pi},$$

where m lies between 0 and 1.

PROOF.—Separate (2311–2) each into two integrals by the formula $\int_0^\infty = \int_0^1 + \int_1^\infty$, and substitute x^{-1} in the last integral.

Otherwise, in (2601) substitute $e^{-\pi x}$, and change a into $\pi a - \frac{1}{2}\pi$.

2358 $\int_0^1 \frac{x^m + x^{-m}}{1+x^2}\,dx = \frac{\pi}{2}\sec\frac{m\pi}{2}. \quad \int_0^1 \frac{x^m - x^{-m}}{1-x^2}\,dx = \frac{\pi}{2}\tan\frac{m\pi}{2}.$

PROOF.—From (2354–5) by the method of (2356).

2360 $$\int_0^1 \frac{x^{m-1} + x^{n-m-1}}{1+x^n}\,dx = \frac{\pi}{n}\operatorname{cosec}\frac{m\pi}{n}.$$

2361 $$\int_0^1 \frac{x^{m-1} - x^{n-m-1}}{1-x^n}\,dx = \frac{\pi}{n}\cot\frac{m\pi}{n}.$$

PROOF.—In the same way, from (2348–9).

2362 $$\int_0^1 \frac{x^{n-m-1} - x^{n+m-1}}{1+x^{2n}}\,dx = \frac{\pi}{2n}\sec\frac{m\pi}{2n}.$$

2363 $$\int_0^1 \frac{x^{n-m-1} - x^{n+m-1}}{1-x^{2n}}\,dx = \frac{\pi}{2n}\tan\frac{m\pi}{2n}.$$

PROOF.—In (2601) substitute $e^{-\frac{\pi x}{2n}}$ and put $a = \frac{m\pi}{2n}$. In (2595) substitute $e^{-\frac{\pi x}{n}}$ and put $a = \frac{m\pi}{n}$.

2364 $$\int_0^\infty \frac{dx}{(x^2+a^2)^n} = \frac{1_2^{(n-1)}}{2_2^{(n-1)}}\,\frac{\pi}{2a^{2n-1}},$$ n being an integer.

PROOF. — By successive reduction by (2062), or by differentiating $\int_0^\infty \frac{dx}{x^2+a^2} = \frac{\pi}{2a}$, $n-1$ times with respect to a^2. (2255)

2365 $$\int_0^1 \frac{x^{m-1}\,dx}{(1+bx)(1-x)^m} = \frac{\pi\operatorname{cosec} m\pi}{(1+b)^m}.$$ [Subs. $\frac{1+bx}{1-x}$. (2311)

2367 $$\int_0^1 \left(\frac{x^{a-1}}{1-x} - \frac{nx^{na-1}}{1-x^n}\right)dx = \log n.$$

PROOF.—The value when $a = 1$ is $\log n$. The difference, when the value is a, is
$$\int_0^1 \frac{x^{a-1}-1}{1-x}\,dx - n\int_0^1 \frac{x^{na-1}-x^{n-1}}{1-x^n}\,dx,$$
which, by substituting x^n in the second integral, is seen to be zero.

$F(x)$ being any integral polynomial,

2368 $\int_{-1}^{+1} \frac{F(x)\,dx}{\sqrt{(1-x^2)}} = A\pi$, where A is equal to the constant term in the product of $F(x)$ and the expansion of $\left(1-\frac{1}{x^2}\right)^{-\frac{1}{2}}$.

PROOF.—By successive reduction by (2053), we know that

$$\int \frac{F(x)\,dx}{\sqrt{(1-x^2)}} = \phi(x)\sqrt{(1-x^2)} + A\int \frac{dx}{\sqrt{(1-x^2)}} \quad\text{..............(1)},$$

where $\phi(x)$ is some integral polynomial and A is a constant. Therefore the integral in question $= A\pi$. To determine A write the last equation thus,

$$\int \frac{F(x)}{ix}\left(1-\frac{1}{x^2}\right)^{-\frac{1}{2}} dx = ix\,\phi(x)\left(1-\frac{1}{x^2}\right)^{-\frac{1}{2}} + A\int \frac{1}{ix}\left(1-\frac{1}{x^2}\right)^{-\frac{1}{2}} dx.$$

Expand each binomial; perform the integrations and equate the coefficients of the two logarithmic terms in the result.

$F(x)$ being an integral polynomial of a degree less than n,

2369 $$\int_a^b \frac{F(x)}{(x-c)^n}\,dx = \frac{1}{\lfloor n-1}\,\frac{d^{n-1}}{dc^{n-1}}\left\{F(c)\log\frac{b-c}{a-c}\right\}.$$

PROOF.— $$\int_a^b \frac{F(x)}{(x-c)^n}\,dx = \frac{1}{\lfloor n-1}\int_a^b \left\{\frac{d^{n-1}}{dc^{n-1}}\left(\frac{F(x)}{x-c}\right)\right\}dx. \qquad (2255)$$

But $\dfrac{F(x)}{x-c} = f(x, c) + \dfrac{F(c)}{x-c}$, where f is of a dimension lower than $n-1$ (421), and therefore $d_{(n-1)\,c}\,f(x, c) = 0$. Hence the integral on the right

$$= \frac{d^{n-1}}{dc^{n-1}}\int_a^b \frac{F(c)}{x-c}\,dx = \frac{d^{n-1}}{dc^{n-1}}\left\{F(c)\log\frac{b-c}{b+c}\right\}.$$

INTEGRATION OF LOGARITHMIC AND EXPONENTIAL FORMS.

2391 $$\int_0^1 x^p \log x\,dx = \frac{-1}{(p+1)^2}. \qquad \int_0^1 \frac{x^p-1\,dx}{\log x} = \log(p+1).$$

PROOF.—These are cases of (2292). Otherwise; to obtain the first integral differentiate, and to obtain the second integrate, the equation $\int_0^1 x^p\,dx = \dfrac{1}{p+1}$ with respect to p (2255 and 2261).

2393 $$\int_0^1 x^p(\log x)^n\,dx = (-1)^n\frac{\Gamma(n+1)}{(p+1)^{n+1}}.$$

PROOF.—See (2292). Otherwise, when n is either a positive or negative integer, the value may be obtained, as in (2391), by performing the differentiation or integration there described, n times successively, and employing formulæ (2166), and (2163) in the case of integration.

2394 $$\int_0^1 \frac{x^p-x^q}{\log x}\,dx = \log\frac{p+1}{q+1}. \qquad \text{[By (2392).}$$

2395 $$\int_0^1 \frac{(\log x)^{2n-1}}{1-x}\,dx = -\frac{2^{2n-2}}{n}B_{2n}\pi^{2n} = -\int_0^\infty \frac{x^{2n-1}}{e^x-1}\,dx.$$

2397 $$\int_0^1 \frac{(\log x)^{2n-1}}{1+x}\,dx = -\frac{2^{2n-1}-1}{2n}B_{2n}\pi^{2n} = -\int_0^\infty \frac{x^{2n-1}}{e^x+1}\,dx.$$

PROOF.—Expand by dividing by $1\mp x$, and integrate by (2393); thus

$$\int_0^1 \frac{(\log x)^{2n-1}}{1-x}\,dx = -\lfloor 2n-1 \left(1+\frac{1}{2^{2n}}+\frac{1}{3^{2n}}+\frac{1}{4^{2n}}+\&c.\right),$$

$$\int_0^1 \frac{(\log x)^{2n-1}}{1+x}\,dx = -\lfloor 2n-1 \left(1-\frac{1}{2^{2n}}+\frac{1}{3^{2n}}-\frac{1}{4^{2n}}+\&c.\right).$$

The first series is summed in (1545). The difference of the two series multiplied by 2^{2n-1} is equal to the first; this gives the value of the second series.

2399

$$\int_0^1 \frac{\log x}{1-x}\,dx = \int_0^1 \frac{\log(1-x)}{x} = -1-\frac{1}{2^2}-\frac{1}{3^2}-\frac{1}{4^2}-\&c. = -\frac{\pi^2}{6}.$$

2400 $$\int_0^1 \frac{\log x}{1+x}\,dx = -1+\frac{1}{2^2}-\frac{1}{3^2}+\frac{1}{4^2}-\&c. = -\frac{\pi^2}{12}.$$

PROOF.—As in (2395–7), making $n = 1$.

The series (2399) may also be summed by equating the coefficients of θ^3 in (764) and (815).

2401 $$\int_0^1 \frac{\log x}{1-x^2}\,dx = -1-\frac{1}{3^2}-\frac{1}{5^2}-\frac{1}{7^2}-\&c. = -\frac{\pi^2}{8}.$$

PROOF.—The integral is half the sum of those in (2399, 2400).

2402 $$\int_0^1 \frac{\log(1+x)}{x}\,dx = \frac{\pi^2}{12}. \qquad \int_0^1 \frac{1}{x}\log\frac{1+x}{1-x}\,dx = \frac{\pi^2}{4}.$$

PROOF.—Expand the logarithms by (155) and (157) and integrate the terms. The series in (2400–1) are reproduced.

2404 Let $$\int_0^a \frac{\log(1-x)}{x}\,dx = \phi(a), \qquad a \text{ being } <1.$$

Substitute $1-x = y$; therefore, writing l for log,

$$\phi(a) = \int_{1-a}^1 \frac{ly\,dy}{1-y} = \int_0^1 \frac{ly\,dy}{1-y} - \int_0^{1-a} \frac{ly\,dy}{1-y}.$$

The second integral by (2399), and the third by Parts, make the right side

$$= -\frac{\pi^2}{6} + la\,l(1-a) - \int_0^{1-a} \frac{l(1-y)}{y}\,dy. \quad \text{Therefore}$$

2405 $\phi(a)+\phi(1-a) = \log a \log(1-a)-\tfrac{1}{6}\pi^2.$

If $a=\frac{1}{2}$, $$\int_0^{\frac{1}{2}} \frac{\log(1-x)}{x}\,dx = \tfrac{1}{2}(\log 2)^2-\frac{\pi^2}{12}.$$

Again, $\phi(x) = \int_0^x \frac{l(1-x)}{x}\,dx$, $\therefore$ (by 2253) $\phi'(x) = \frac{l(1-x)}{x}$ ... (i.)

$$d_x\phi\left(\frac{-x}{1-x}\right) = \frac{-1}{(1-x)^2}\phi'\left(\frac{-x}{x-1}\right) = \frac{1}{x(1-x)}\,l\frac{1}{1-x} \text{ by (i.)},$$

$$\therefore \phi\left(\frac{-x}{1-x}\right) = \int_0^x l\frac{1}{1-x}\,\frac{dx}{x(x-1)} = \int_0^x l\frac{1}{1-x}\,\frac{dx}{1-x}+\int_0^x l\frac{1}{1-x}\,\frac{dx}{x}$$
$$= \tfrac{1}{2}\{l(1-x)\}^2-\phi(x).$$

Put $\frac{x}{1+x}$ for x; then

2406 $$\phi(-x)+\phi\left(\frac{x}{1+x}\right) = \tfrac{1}{2}\{\log(1+x)\}^2.$$

Also, $$\phi(x) = -\int_0^x \left(1+\frac{x}{2}+\frac{x^2}{3}+\frac{x^3}{4}+\&c.\right)$$
$$= -\left(x+\frac{x^2}{2^2}+\frac{x^3}{3^2}+\frac{x^4}{4^2}+\&c.\right).$$

Hence $\phi(x)+\phi(-x) = -\frac{1}{2}\left(x^2+\frac{x^4}{2^2}+\frac{x^6}{3^2}+\&c.\right) = \frac{1}{2}\phi(x^2).$

Eliminate $\phi(-x)$ by (2406); thus

2407 $$\phi\left(\frac{x}{x+1}\right)+\tfrac{1}{2}\phi(x^2)-\phi(x) = \tfrac{1}{2}\{\log(1+x)\}^2.$$

Let $\frac{x}{x+1} = x^2$, and therefore $x = -\frac{1}{2}+\frac{1}{2}\sqrt{5} \equiv \beta$ say,

$\therefore$ by (2407), $\frac{3}{2}\phi(\beta^2)-\phi(\beta) = \frac{1}{2}\{l(1+\beta)\}^2,$

or $\frac{3}{2}\phi(1-\beta)-\phi(\beta) = \frac{1}{2}(l\beta)^2$; $[\because \beta^2 = 1-\beta$ and $1+\beta = \frac{1}{\beta}$;

and by (2405) $\phi(1-\beta)+\phi(\beta) = 2(l\beta)^2-\frac{1}{6}\pi^2,$

$\therefore$ $\phi(\beta) = (l\beta)^2-\frac{1}{10}\pi^2$ and $\phi(1-\beta) = (l\beta)^2-\frac{1}{15}\pi^2$, that is,

2408 $$\int_0^{\frac{\sqrt{5}-1}{2}} \frac{\log(1-x)}{x}\,dx = \left(\log\frac{\sqrt{5}-1}{2}\right)^2-\frac{\pi^2}{10}.$$

2409 $$\int_0^{\frac{3-\sqrt{5}}{2}} \frac{\log(1-x)}{x}\,dx = \left(\log\frac{\sqrt{5}-1}{2}\right)^2-\frac{\pi^2}{15}.$$

2410 Let a be >1, then $\phi(a)$ contains imaginary elements, but its value is determinate. We have

$$\phi(a) = \int_0^1 \frac{l(1-x)}{x}\,dx+\int_1^a \frac{l(1-x)}{x}\,dx = -\frac{\pi^2}{6}+\int_1^a \frac{\pi i+l(x-1)}{x},$$

the integration by 2399, and $l(-1)=\pi i$ by (2214). The last integral

$$=\pi i la+\int_1^a\left\{lx+l\left(1-\frac{1}{x}\right)\right\}\frac{dx}{x}=\pi i la+\frac{1}{2}(la^2)+\int_1^a l\left(1-\frac{1}{x}\right)\frac{dx}{x}.$$

Substitute $\frac{1}{x}=y$ in the last integral, and it becomes

$$\int_{\frac{1}{a}}^1\frac{l(1-y)}{y}dy=\int_0^1\frac{l(1-y)}{y}dy-\int_0^{\frac{1}{a}}\frac{l(1-y)}{y}dy=-\frac{\pi^2}{6}-\phi\left(\frac{1}{a}\right).$$

Hence, when a is >1,

2411 $$\phi(a)+\phi\left(\frac{1}{a}\right)=-\frac{\pi^2}{3}+\pi i\log a+\tfrac{1}{2}(\log a)^2.$$

If $a=2$, this result becomes, by employing (2405),

2412 $$\int_0^2\frac{\log(1-x)}{x}dx=-\frac{\pi^2}{4}+\pi i\log 2.$$

2413 Let $$\psi(x)=\int_0^x\frac{1}{2x}\log\frac{1+x}{1-x}dx.$$

Therefore $$\psi'(x)=\frac{1}{2x}l\left(\frac{1+x}{1-x}\right),\quad d_x\psi\left(\frac{1-x}{1+x}\right)=\frac{lx}{1-x^2},$$

therefore $\psi(x)+\psi\left(\frac{1-x}{1+x}\right)=\int_0^x\left(l\frac{1+x}{1-x}\frac{1}{2x}+\frac{lx}{1-x^2}\right)dx$, therefore

2414 $$\psi(x)+\psi\left(\frac{1-x}{1+x}\right)=\tfrac{1}{2}\log x\log\frac{1+x}{1-x}.$$

The constant vanishes, by (2403) and (2401), putting $x=1$.

Let $x=\frac{1-x}{1+x}$, and therefore $x=\sqrt{2}-1$; then, by (2414),

2415 $$\int_0^{\sqrt{2}-1}\log\frac{1+x}{1-x}\frac{dx}{x}=-\tfrac{1}{2}\{\log(\sqrt{2}-1)\}^2.$$

2416 $$\int_0^1\frac{\log(1+x)}{1+x^2}dx=\frac{\pi}{8}\log 2.$$

Proof.—Substitute $\phi=\tan^{-1}x$; then, by (2233),

$$\int_0^{\frac{\pi}{4}}l(1+\tan\phi)\,d\phi=\int_0^{\frac{\pi}{4}}l\left\{1+\tan\left(\frac{\pi}{4}-\phi\right)\right\}d\phi=\int_0^{\frac{\pi}{4}}l\frac{2}{1+\tan\phi}d\phi.$$

2417 By differentiating or integrating the equations (2341) to (2363) with respect to the index m, the integrals of functions involving $\log x$ are produced; thus, from (2356), by integrating for m between the limits $\frac{1}{2}$ and m, we have

2418 $$\int_0^1 \frac{x^{m-1}-x^{-m}}{(1+x)\log x}\,dx = \log\tan\frac{m\pi}{2}. \qquad (1943)$$

Otherwise, this result may be arrived at by forming the expansion of the fraction in powers of x, and integrating the terms by (2392); the reduction is then effected by 815–6.

In a similar manner, we obtain the more general formula

2419 $$\int_0^1 \frac{(x^{m-1}-x^{n-1})\,dx}{(1+x^p)\log x} = \log\frac{m}{n}\,\frac{n+p}{m+p}\,\frac{m+2p}{n+2p}\,\frac{n+3p}{m+3p}\cdots$$

2420 $$\int_0^1 \frac{x^{m-1}-x^{n-m-1}}{(1+x^n)\log x}\,dx = \log\tan\frac{m\pi}{2n}.$$

PROOF.—Integrate (2360) for m from $2n$ to m.

2421 $$\int_0^1 \frac{x^p-x^r+x^q(r-p)\log x}{(\log x)^2}\,dx$$

$$=(p+1)\log(p+1)-(r+1)\log(r+1)+(r-p)\{1+\log(q+1)\}$$

PROOF.—Integrate (2394) for p between the limits r and p.

2422 $$\int_0^1 \frac{(q-r)\,x^p+(r-p)\,x^q+(p-q)\,x^r}{(\log x)^2}\,dx$$

$$=\log\left\{(p+1)^{(p+1)(q-r)}\,(q+1)^{(q+1)(r-p)}\,(r+1)^{(r+1)(p-q)}\right\}.$$

PROOF.—Write (2421) symmetrically for r, p; p, q and q, r. Multiply the three equations, respectively, by q, r, p, and add, reducing the result by (2394).

2423 $$\int_0^\infty \frac{\log(1+a^2x^2)}{b^2+x^2}\,dx = \frac{\pi}{b}\log(1+ab).$$

PROOF.—Differentiate for a, and resolve into two fractions. Effect the integration for x, and integrate finally with respect to a.

2424

$$\int_0^\infty \log\left(1+\frac{x^2}{a^2}\right)\log\left(1+\frac{b^2}{x^2}\right)\frac{dx}{x^2} = 2\pi\left(\frac{1}{a}+\frac{1}{b}\right)\log\left(1+\frac{b}{a}\right)-\frac{2\pi}{a}.$$

PROOF.—In (2423) put $a=1$, and substitute $\frac{1}{x}=y$; multiply up by b, and integrate for b between limits 0 and $\frac{b}{a}$, and in the result substitute by.

2425 $$\int_0^\infty e^{-kx^2}\,dx = \frac{1}{2}\sqrt{\frac{\pi}{k}}.$$ [Substitute kx^2.

2426 $$\int_0^\infty e^{-kx^2}x^{2n}\,dx = \frac{1.3\ldots(2n-1)}{2^{n+1}}\,k^{-\frac{2n+1}{2}}\sqrt{\pi}.$$

PROOF.—Substitute kx^2. Otherwise, differentiate the preceding equation n times for k.

2427 $$\int_0^\infty \frac{e^{-bx}-e^{-ax}}{x}\,dx = \log a - \log b.$$

2428 $$\int_0^\infty \left(\frac{e^{-bx}-e^{-cx}}{x^2} - \frac{(c-b)\,e^{-ax}}{x}\right)dx = c-b+\log\frac{b^b a^c}{a^b c^c}.$$

PROOF.—Making $n=1$ in (2291), $\int_0^\infty e^{-ax}\,dx = \frac{1}{a}$. Integrate this for a between limits b and a to obtain (2427); and integrate that equation for b between limits b and c to obtain (2428).

2429 $$\int_0^\infty \left(\frac{e^{-bx}-e^{-ax}}{x^2} - \frac{(a-b)\,e^{-ax}}{x}\right)dx = a-b+b\log\frac{b}{a}.$$

PROOF.—Make $c=a$ in (2428).

Otherwise.—Integrating the first term by Parts, the whole reduces to

$$\left[\frac{e^{-ax}-e^{-bx}}{x}\right]_0^\infty + b\int_0^\infty \frac{e^{-ax}-e^{-bx}}{x}\,dx.$$

The indeterminate fraction is evaluated by (1580) and the integral by (2427).

2430 $$\int_0^\infty \left\{\frac{e^{-bx}-e^{-ax}}{x^3} - \frac{(a-b)\,e^{-ax}}{x^2} - \frac{(a-b)^2\,e^{-ax}}{2x}\right\}dx$$
$$= \tfrac{1}{4}\left\{a^2+3b^2-4ab+2b^2(\log a-\log b)\right\}.$$

PROOF.—By two successive integrations by Parts, $\int x^{-3}dx$; &c.,

$$\int_\epsilon^\infty \frac{e^{-x}-e^{-ax}}{x^3}\,dx = \frac{e^{-b\epsilon}-e^{-a\epsilon}}{2\epsilon^2} + \frac{ae^{-a\epsilon}-be^{-b\epsilon}}{2\epsilon} + \frac{1}{2}\int_\epsilon^\infty \frac{b^2e^{-bx}-a^2e^{-ax}}{x}\,dx.$$

Also $$\int_\epsilon^\infty \frac{e^{-ax}}{x^2}\,dx = \frac{e^{-a\epsilon}}{\epsilon} - \int_\epsilon^\infty \frac{ae^{-ax}}{x}\,dx.$$

Substitute these values, and make $\epsilon = 0$. The vanishing fractions are found by (1580), and the one resulting integral is that in (2427).

In a similar manner the value of the subjoined integral may be found.

2431
$$\int_0^\infty \left\{ \frac{e^{-bx}-e^{ax}}{x^4} - \frac{(a-b)\,e^{-ax}}{x^3} - \frac{(a-b)^2\,e^{-ax}}{1.2\,x^2} - \frac{(a-b)^3\,e^{-ax}}{1.2.3\,x^3} \right\} dx.$$

INTEGRATION OF CIRCULAR FORMS.

NOTATION.—Let $a_b^{(n)}$ signify the continued product of n factors in arithmetical progression, the first of which is a, and the common difference of which is b, so that

2451 $$a_b^{(n)} \equiv a\,(a+b)\,(a+2b)\,\ldots\,\{a+(n-1)\,b\}.$$

Similarly, let

2452 $$a_{-b}^{(n)} \equiv a\,(a-b)\,(a-2b)\,\ldots\,\{a-(n-1)\,b\}.$$

These may be read, respectively, "a *to* n *factors, difference* b"; "a *to* n *factors, difference minus* b."

2453 $$\int_0^{\frac{1}{2}\pi} \sin^n x\,dx = \frac{n-1}{n}\int_0^{\frac{1}{2}\pi} \sin^{n-2} x\,dx.$$

PROOF.—By (2048); applying Rule VI., we have, by division,

$$\int \sin^n x\,dx = \int \sin^{n-2} x\,dx - \int \sin^{n-2} x\cos^2 x\,dx,$$

and by Parts, $$\int \sin^{n-2} x\cos^2 x\,dx = \frac{\sin^{n-1} x\cos x}{n-1} + \frac{1}{n-1}\int \sin^n x\,dx.$$

Therefore $$\int_0^{\frac{1}{2}\pi} \sin^{n-2} x\cos^2 x\,dx = \frac{1}{n-1}\int_0^{\frac{1}{2}\pi} \sin^n x\,dx.$$

The substitution of this value in the first equation produces the formula.

If n be an integer, with the notation of (2451),

2454 $$\int_0^{\frac{1}{2}\pi} \sin^{2n+1} x\,dx = \frac{2_2^{(n)}}{3_2^{(n)}} \quad \text{and} \quad \int_0^{\frac{1}{2}\pi} \sin^{2n} x\,dx = \frac{1_2^{(n)}}{2_2^{(n)}}\,\frac{\pi}{2}.$$

PROOF.—By repeated application of formula (2453).

Wallis's Formula.—If m be any positive integer, we have

2456 $$2m\left\{\frac{2_2^{(m-1)}}{1_2^{(m)}}\right\}^2 > \frac{\pi}{2} > (2m-1)\left\{\frac{2_2^{(m-1)}}{1_2^{(m)}}\right\}^2.$$

And since the ratio of these limits to each other constantly approaches unity as m increases, the value of either of them when m is infinite is $\frac{1}{2}\pi$.

Ex.—With $m=4$, $\frac{1}{2}\pi$ lies in magnitude between

$$\frac{2^2.4^2.6^2.8}{1\,.3^2.5^2.7^2} \quad \text{and} \quad \frac{2^2.4^2.6^2.7}{1\,.3^2.5^2.7^2}.$$

PROOF.—Put $2m=n$, then

$$\int_0^{\frac{1}{2}\pi} \sin^{n-1} x\,dx, \quad \int_0^{\frac{1}{2}\pi} \sin^n x\,dx, \quad \text{and} \quad \frac{n-1}{n}\int_0^{\frac{1}{2}\pi} \sin^{n-1} x\,dx$$

are in descending order of magnitude; the first and second because $\sin x$ is <1; the second and third by (2453); then substitute the factorial values by (2454–5).

2457 $$\int_0^{\frac{1}{2}\pi} \tan^{2m-1}\phi\,d\phi = \frac{\pi}{2\sin m\pi}.$$ [Subs. $x=\tan^2\phi$ in (2311).

2458 $$\int_0^{\pi} \sin^n x\,dx = 2\int_0^{\frac{1}{2}\pi} \sin^n x\,dx.$$ [By (2234).

2459 $$\int_0^{\pi} \sin^n x\cos^p x\,dx = 2\int_0^{\frac{1}{2}\pi} \sin^n x\cos^p x\,dx \text{ or zero,}$$

according as p is an even or odd integer. [By (2234).

2461 $$\int_0^{\frac{1}{2}\pi} \sin^n x\cos^p x\,dx = \int_0^{\frac{1}{2}\pi} \sin^p x\cos^n x\,dx.$$ [By (2233).

2462 $$\int_0^{\frac{1}{2}\pi} \sin^n x\cos^p x\,dx = \tfrac{1}{2}B\left(\frac{n+1}{2}, \frac{p+1}{2}\right).$$ [Subs. $\sin^2 x$ (2280).

Let either of the integers n and p, in (2461), be odd, and the other either odd or even; thus, let n be odd and $=2m+1$, then

2463 $$\int_0^{\frac{1}{2}\pi} \sin^{2m+1} x\cos^p x\,dx = \frac{2_2^{(m)}}{(p+1)_2^{(m+1)}}.$$ (2451)

PROOF.—Transposing the indices by (2461), we have, by Parts (2067),

$$\int_0^{\frac{1}{2}\pi} \sin^p x\cos^{2m+1} x\,dx = \frac{2m}{p+1}\int_0^{\frac{1}{2}\pi} \sin^{p+2} x\cos^{2m-1} x\,dx.$$

By repeating the reduction, the integral finally arrived at is

2464 $$\int_0^{\frac{1}{2}\pi} \sin^{p+2m} x\cos x\,dx = \frac{1}{p+2m+1}.$$

If both the indices are even, then

2465 $$\int_0^{\frac{1}{2}\pi} \sin^{2m} x \cos^{2p} x\, dx = \frac{1_2^{(m)}\, 1_2^{(p)}}{2_2^{(m+p)}}\, \frac{\pi}{2}. \qquad (2451)$$

PROOF.—Reduce by Parts as before. The final integral is $\int_0^{\frac{1}{2}\pi} \sin^{2m+2p} x\, dx$, the value of which is given at (2455).

2466 Should either of the indices be a negative integer, the value of the integral is infinite, as the foregoing reduction shows, for the factor zero will then occur somewhere in the denominator.

2467 $$\int_0^{\pi} \sin nx \sin px\, dx = \int_0^{\pi} \cos nx \cos px\, dx = 0,$$

when n and p are unequal integers.

2469 $$\int_0^{\pi} \sin nx \cos px\, dx = \frac{2n}{n^2-p^2} \text{ or } \textit{zero},$$

according as the difference of the integers n and p is *odd* or *even*. [By (1973–5).

2470 $$\int_0^{\pi} \sin^2 nx\, dx = \int_0^{\pi} \cos^2 nx\, dx = \tfrac{1}{2}\pi,$$

when n is an integer.

PROOF.—Express in terms of $\cos 2nx$, and then integrate.

2472 $$\int_0^{\frac{1}{2}\pi} \sin^n x\, dx = \int_0^{\frac{1}{2}\pi} \cos^n x\, dx = \int_0^1 (1-x^2)^{\frac{n-1}{2}}\, dx.$$

The following four integrals (2473–9) all vanish for integral values of n and p excepting in the cases here specified.

2474 $$\int_0^{\pi} \sin^p x \sin nx\, dx = (-1)^{\frac{n-1}{2}}\, C\left(p, \frac{p-n}{2}\right) \frac{\pi}{2^p},$$

when p and n are *both odd*, and n is not greater than p.

2475 But if p be *even*, and n *odd*, the value is

$$(-1)^{\frac{p}{2}} \frac{n}{2^{p-2}} \left\{ \frac{1}{n^2-p^2} - \frac{C(p,1)}{n^2-(p-2)^2} + \frac{C(p,2)}{n^2-(p-4)^2} - \dots \right.$$
$$\left. \dots (-1)^{\frac{p}{2}} \frac{C(p, \frac{1}{2}p)}{2n^2} \right\}.$$

2476 $$\int_0^\pi \sin^p x \cos nx\,dx = (-1)^{\frac{n}{2}}\, C\left(p, \frac{p-n}{2}\right)\frac{\pi}{2^p},$$

when p and n are *both even*, and n is not greater than p.

2477 But if p be *odd* and n *even*, the value is

$$(-1)^{\frac{p-1}{2}}\frac{1}{2^{p-2}}\left\{\frac{p}{p^2-n^2}-\frac{C(p,1)(p-2)}{(p-2)^2-n^2}+\frac{C(p,2)(p-4)}{(p-4)^2-n^2}-\ldots\right.$$
$$\left.\ldots(-1)^{\frac{p-1}{2}}\frac{C\{p, \frac{1}{2}(p-1)\}}{1-n^2}\right\}.$$

2478 $$\int_0^\pi \cos^p x \cos nx\,dx = C\left(p, \frac{p-n}{2}\right)\frac{\pi}{2^p},$$

when p and n are either *both odd* or *both even*, and n is not greater than p.

2479 $\int_0^\pi \cos^p x \sin nx\,dx$, when $p \sim n$ is odd, takes the value

$$\frac{n}{2^{p-2}}\left\{\frac{1}{n^2-p^2}+\frac{C(p,1)}{n^2-(p-2)^2}+\frac{C(p,2)}{n^2-(p-4)^2}+\text{\&c.}\right\},$$

the last term within the brackets being

$\dfrac{C\left(p, \frac{p-1}{2}\right)}{n^2-1}$ when p is *odd* and n *even*, or $\dfrac{C\left(p, \frac{p}{2}\right)}{n^2}$ when p is *even* and n *odd*.

Proof.—(For 2474 to 2479.)—Expand by (772-4), and apply (2467-2470) to the separate terms.

Corollaries.—n being any integer,

2480 $$\int_0^\pi \cos^n x \cos nx\,dx = \frac{\pi}{2^n}, \quad \int_0^{\frac{1}{2}\pi} \cos^n x \cos nx\,dx = \frac{\pi}{2^{n+1}}.$$

2482 $$\int_0^\pi \sin^{2n} x \cos 2nx\,dx = (-1)^n\frac{\pi}{2^{2n}},$$

$$\int_0^\pi \sin^{2n+1} x \sin(2n+1)x\,dx = (-1)^n\frac{\pi}{2^{2n+1}}.$$

2484 $$\int_0^{\frac{1}{2}\pi} \cos^p x \cos nx\,dx = \frac{p(p-1)}{p^2-n^2}\int_0^{\frac{1}{2}\pi} \cos^{p-2} x \cos nx\,dx.$$

2485

$$\int_0^{\frac{1}{2}\pi} \cos^p x \sin nx\,dx = \frac{p(p-1)}{p^2-n^2}\int_0^{\frac{1}{2}} \cos^{p-2} x \sin nx\,dx - \frac{n}{p^2-n^2}.$$

Proof.—(For either formula) By Parts, $\int \cos x\,dx$; and the new integral of highest dimensions in $\cos x$, by Parts, $\int \cos^{p-1} x \sin x\,dx$.

2486 $$\int_0^{\frac{1}{2}\pi} \cos^{n-2} x \cos nx\,dx = 0. \quad \int_0^{\frac{1}{2}\pi} \cos^{n-2} x \sin nx\,dx = \frac{1}{n-1}.$$

Proof.—Make $p=n$ in (2484–5).

When k is a positive integer,

2488 $$\int_0^{\frac{1}{2}\pi} \cos^{n-2k} x \cos nx\,dx = 0,$$

2489 $$\int_0^{\frac{1}{2}\pi} \cos^{n+2k} x \cos nx\,dx = \frac{(n+2k)_{-1}^{(k)}}{1^{(k)}}\,\frac{\pi}{2^{n+2k+1}}.$$

Proof.—The first, by putting $p = n-2,\ n-4,\ \dots\ n-2k$ successively in (2484) and employing (2486). The second, by putting $p = n+2,\ n+4,\ \dots\ n+2k$ successively and employing (2481).

When k is *not* an integer,

2490 $$\int_0^{\frac{1}{2}\pi} \cos^{n-2k} x \cos nx\,dx = 2^{2k-n+1} \sin k\pi\, B(n-2k+1,\ k).$$

Proof.—In (2706) take $f(a) = a^{n-2k}$, and transform by (766). The coefficient of i vanishes by (2239), and the limits are changed by (2237).

2491 $$\int_0^{\frac{1}{2}\pi} \cos^n x \sin nx\,dx = \frac{1}{2^{n+1}}\left(2+\frac{2^2}{2}+\frac{2^3}{3}+\dots+\frac{2^n}{n}\right).$$

Proof.—By successive reduction by (1970), making $m=n$, and the integral definite.

2492 When p and n are integers, one odd and the other even,

$$\int_0^{\frac{1}{2}\pi} \cos^p x \cos nx\,dx = \frac{(-1)^{\frac{1}{2}(p+n\pm1)}\,1^{(p)}}{(n-p)_2^{(p+1)}} \quad \begin{cases} +, \text{ with } p \text{ odd}, \\ -, \text{ with } p \text{ even}. \end{cases}$$

Proof.—Reduce successively by (2484). The final integral, according as p is odd or even, will be

$$\int_0^{\frac{1}{2}\pi} \cos x \cos nx\,dx = \frac{\cos\frac{1}{2}n\pi}{1-n^2} = \frac{(-1)^{\frac{1}{2}n}}{1-n^2} \text{ or } \int_0^{\frac{1}{2}\pi} \cos nx\,dx = \frac{\sin\frac{1}{2}n\pi}{n} = \frac{(-1)^{\frac{1}{2}(n-1)}}{n}.$$

2493 $$\int_0^\pi \cos^p x \cos nx\,dx = \frac{p_{-1}^{(n)}}{1_2^{(n)}}\int_0^\pi \sin^{2n} x \cos^{p-n} x\,dx,$$

where n and p are any integers whatever such that $p-n$ is > 0.

PROOF.—When $p-n$ is odd, each integral vanishes, by (2478) and (2459). When $p-n$ is even, let it $=2k$; then, by (2488),

$$\int_0^{\frac{1}{2}\pi} \cos^{n+2k} x \cos nx\,dx = \frac{(n+2k)_{-1}^{(k)}}{1^{(k)}}\,\frac{\pi}{2^{n+2k+1}} = \frac{(n+2k)_{-1}^{(n)}}{1_2^{(n)}}\,\frac{1_2^{(n)}\,1_2^{(k)}}{2_2^{(n+k)}}\,\frac{\pi}{2}$$

$$= \frac{(n+2k)_{-1}^{(n)}}{1_2^{(n)}}\int_0^{\frac{1}{2}\pi} \sin^{2n} x \cos^{2k} x\,dx, \quad \text{(by 2465).}$$

But $n+2k=p$, and by (2234) the limit may be doubled. Hence the result.

2494 $$\int_0^{\frac{1}{2}\pi} x \cos^{n-2} x \sin nx\,dx = \frac{\pi}{2^n(n-1)}.$$

PROOF.—In (2707), put $k=1$ and $f(x) = x^{n-2}$. Give $e^{in\theta}$ its value from (766). The imaginary term in the result vanishes, and the limits are changed, by (2237). Finally, write x instead of θ.

2495

$$\int_0^\pi f^n(\cos x) \sin^{2n} x\,dx = 1.3\ldots(2n-1)\int_0^\pi f(\cos x)\cos nx\,dx.$$

PROOF.—Let $z = \cos x$. By (1471), we have

$$d_{(n-1)z}(1-z^2)^{n-\frac{1}{2}} = (-1)^{n-1}1.3\ldots(2n-1)\frac{\sin nx}{n} \quad \ldots\ldots\ldots\ldots\ldots \text{(i.)}$$

Also, by integrating n times by Parts,

$$\int_{-1}^1 f^n(z)(1-z^2)^{n-\frac{1}{2}}\,dz = (-1)^n\int_{-1}^1 f(z)\,d_{nz}(1-z^2)^{n-\frac{1}{2}}\,dz$$

$$= -1.3\ldots(2n-1)\int_{-1}^1 f(z)\,d_z\left(\frac{\sin nx}{n}\right)dz, \quad \text{by (i.)}$$

Then substitute $z = \cos x$.

Otherwise.—Let $f(z) = A_0 + A_1 z + A_2 z^2 + \&c. = \Sigma A_p z^p$,

$$\therefore f^n(z) = \Sigma p(p-1)\ldots(p-n+1)A_p z^{p-n},$$

$$\therefore \int_0^\pi f(\cos x)\cos nx\,dx = \Sigma A_p\int_0^\pi \cos^p x \cos nx\,dx$$

$$= \frac{1}{1.3\ldots(2n-1)}\int_0^\pi f^n(\cos x)\sin^{2n} dx, \quad \text{by (2493).}$$

2496 $$\int_0^{\frac{1}{2}\pi} \frac{dx}{a^2\cos^2 x + b^2\sin^2 x} = \frac{\pi}{2ab}. \qquad (1982)$$

2497 $$\int_0^{\frac{1}{2}\pi} \frac{\cos^2 x\,dx}{(a^2\cos^2 x+b^2\sin^2 x)^2} = \frac{\pi}{4a^3b}.$$ [Differentiate (2496) for a.

2498 $$\int_0^{\frac{1}{2}\pi} \frac{\sin^2 x\,dx}{(a^2\cos^2 x+b^2\sin^2 x)^2} = \frac{\pi}{4ab^3}.$$ [Differentiate (2496) for b.

2499 $$\int_0^{\frac{1}{2}\pi} \frac{dx}{(a^2\cos^2 x+b^2\sin^2 x)^2} = \frac{\pi}{4ab}\left(\frac{1}{a^2}+\frac{1}{b^2}\right).$$

[Add together (2497–8)

2500 $$\int_0^{\frac{1}{2}\pi} \frac{dx}{(a^2\cos^2 x+b^2\sin^2 x)^3} = \frac{\pi}{16ab}\left(\frac{3}{a^4}+\frac{2}{a^2b^2}+\frac{3}{b^4}\right).$$

2501 $$\int_0^{\frac{1}{2}\pi} \frac{dx}{(a^2\cos^2 x+b^2\sin^2 x)^4} = \frac{\pi}{32ab}\left(\frac{5}{a^6}+\frac{3}{a\,b^2}+\frac{3}{a^2b}+\frac{5}{b^6}\right).$$

(2500) and (2501) are obtained by repeating upon (2499) the operations by which that integral was itself obtained from (2496).

2502 $$\int_0^1 \frac{\tan^{-1} ax}{x\sqrt{(1-x^2)}}\,dx = \frac{\pi}{2}\log(a+\sqrt{1+a^2}).$$

Proof.—Denote the integral by u.

$$\frac{du}{da} = \int_0^1 \frac{dx}{(1+a^2x^2)\sqrt{1-x^2}} = \frac{\pi}{2}\,\frac{1}{\sqrt{(1+a^2)}}, \qquad \text{[by (2008)}$$

$$\therefore\ u = \frac{\pi}{2}\int_0^a \frac{da}{\sqrt{(1+a^2)}} = \frac{\pi}{2}\log(a+\sqrt{1+a^2}). \qquad (1928)$$

2503 $$\int_0^{\infty} \frac{\tan^{-1} ax}{x(1+x^2)}\,dx = \frac{\pi}{2}\log(1+a).$$

Proof.—Differentiate for a. Integrate for x by partial fractions, and then integrate for a.

2504 $$\int_0^{\infty} \tan^{-1}\frac{x}{a}\tan^{-1}\frac{x}{b}\,\frac{dx}{x^2} = \frac{\pi}{2}\log\left\{\left(1+\frac{a}{b}\right)^{\frac{1}{a}}\left(1+\frac{b}{a}\right)^{\frac{1}{b}}\right\}.$$

Proof.—From (2503) we obtain

$$\int_0^{\infty} \frac{\tan^{-1} x}{x(a^2+x^2)}\,dx = \frac{\pi}{2}\,\frac{\log(1+a)}{a^2}.$$

Integrate for a between limits $\frac{a}{b}$ and ∞, and in the result substitute bx.

2505 $$\int_0^\infty \frac{\tan^{-1}ax - \tan^{-1}bx}{x}\,dx = \frac{\pi}{2}\log\frac{a}{b}.$$

PROOF.—Applying (2700), $\phi(0)$ here vanishes. Also, by Parts, we have

$$\int_{\frac{h}{a}}^{\frac{h}{b}} \frac{\tan^{-1}(bx)}{x}\,dx = \frac{\pi}{2}\int_{\frac{h}{a}}^{\frac{h}{b}} \frac{dx}{x}$$

since bx is infinite and therefore $\tan^{-1}(bx) = \frac{\pi}{2}$ in *every* element of the integral. Hence the required value is

$$\frac{\pi}{2}\int_{\frac{h}{a}}^{\frac{h}{b}} \frac{dx}{x} = \frac{\pi}{2}\log\frac{a}{b}.$$

2506 $$\int_0^\pi \frac{x\sin x}{1+\cos^2 x}\,dx = \frac{\pi^2}{4}.$$

PROOF.—(i.) Substitute $\pi - x$, and the integral is reproduced, and is thus shown to be

$$= \frac{\pi}{2}\int_0^\pi \frac{\sin y}{1+\cos^2 y}\,dy = -\frac{\pi}{2}\left(\tan^{-1}\cos\pi - \tan^{-1}\cos 0\right) = \frac{\pi^2}{4}.$$

(ii.) Otherwise, expand by dividing by the denominator, and integrate each term of the result by Parts. Employing (2478) we obtain the series

$$\pi\left(1 - \frac{1}{3} + \frac{1}{5} - \frac{1}{7} + \&c. \ldots\right) = \frac{\pi^2}{4}.$$ [By (2945).

2507 $$\int_0^\infty \frac{\cos x}{\sqrt{x}}\,dx = \sqrt{\frac{\pi}{2}} = \int_0^\infty \frac{\sin x}{\sqrt{x}}\,dx.$$

PROOF.—By the method of (2251), putting

$$\frac{1}{\sqrt{x}} = \frac{2}{\sqrt{\pi}}\int_0^\infty e^{-xy^2}\,dy,$$

the integral becomes

$$\frac{2}{\sqrt{\pi}}\int_0^\infty\int_0^\infty e^{-xy^2}\cos x\,dx\,dy = \frac{2}{\sqrt{\pi}}\int_0^\infty\int_0^\infty e^{-xy^2}\cos x\,dy\,dx \quad (2261)$$

$$= \frac{2}{\sqrt{\pi}}\int_0^\infty \frac{y^2\,dy}{1+y^4}\ (2610) = \sqrt{\frac{\pi}{2}}. \quad (2348)$$

The second integral is obtained in a similar manner.

2509 $$\int_0^\infty \cos y^2\,dy = \frac{\sqrt{\pi}}{2\sqrt{2}} = \int_0^\infty \sin y^2\,dy.$$

PROOF.—Substitute y^2, and (2507–8) are produced.

When n and p are integers,

2510 $$\int_0^\infty \frac{\sin^n x}{x^p}\,dx = \frac{1}{1.2\,\ldots\,(p-1)}\int_0^\infty\int_0^\infty z^{p-1}e^{-zx}\sin^n x\,dz\,dx.$$

The integration for x in the double integral is given in (2608–9), and the original integral is thus reduced to the integral of a rational fraction.

PROOF.—By the method of (2251), putting

$$\frac{1}{x^p} = \frac{1}{\Gamma(p)}\int_0^\infty e^{-xz}z^{p-1}\,dz. \qquad \text{[By (2291).}$$

2511 $$\int_0^\infty \frac{\sin^2 x}{x^2}\,dx = 2\int_0^\infty \frac{dz}{z^2+4} = \frac{\pi}{2}. \qquad \text{[By (2510).}$$

2512 $$\int_0^\infty \frac{\sin^3 x}{x}\,dx = 6\int_0^\infty \frac{dz}{(z^2+1)(z^2+9)} = \frac{\pi}{4}. \qquad \text{[By (2510) \& (2081).}$$

2513 $$\int_0^\infty \frac{\cos qx - \cos px}{x}\,dx = \log\frac{p}{q}.$$

PROOF.—By (2700). Transforming the numerator by (673), and putting $\frac{1}{2}(p+q) = a$, $\frac{1}{2}(p-q) = b$, this becomes

2514 $$\int_0^\infty \frac{\sin ax \sin bx}{x}\,dx = \frac{1}{2}\log\frac{a+b}{a-b}.$$

2515 $$\int_0^\infty \frac{\cos qx - \cos px}{x^2}\,dx = \frac{\pi}{2}(p-q).$$

PROOF.—Integrate (2572) for r between the limits p and q.

If a and b are positive quantities,

2516 $$\int_0^\infty \frac{\sin ax \cos bx}{x}\,dx = \frac{\pi}{2} \text{ or } 0,$$

according as a is $>$ or $<$ b.

PROOF.—Change by (666), and employ (2572).

2518 $$\int_0^\infty \frac{\sin ax \sin bx}{x^2}\,dx = \frac{\pi a}{2} \quad\text{or}\quad \frac{\pi b}{2},$$

according as a or b is the least of the two numbers.

PROOF.—From (2515), exactly as in (2513).

Otherwise, as an illustration of the method in (2252), as follows. Denoting the integral in (2516) by u, we have, (i.) when b is $> a$,

$$\int_0^b u\,db = \int_0^a \frac{\pi}{2}\,db + \int_a^b 0\,db = \frac{\pi a}{2};$$

that is, $$\frac{\pi a}{2} = \int_0^b \int_0^\infty \frac{\sin ax \cos bx}{x}\,db\,dx = \int_0^\infty \frac{\sin ax \sin bx}{x^2}\,dx. \qquad (2261)$$

(ii.) When b is $< a$, $\int_0^b u\,db = \int_0^b \frac{\pi}{2}\,db = \frac{\pi b}{2}$.

If a is a positive quantity,

2520 $$\int_0^\infty \frac{\sin^2 x \cos ax}{x^2}\,dx = \frac{\pi}{4}(2-a) \quad \text{or} \quad 0,$$

according as a is or is not less than 2.

PROOF.— $\sin^2 x \cos ax = \frac{1}{2}\sin x\,\{\sin(1+a)\,x + \sin(1-a)\,x\}$;
and the result then follows from (2518), the value of the integral being in the two cases $\frac{\pi}{4} - \frac{\pi}{4} = 0$ and $\frac{\pi}{4} - \frac{\pi}{4}(a-1) = \frac{\pi}{4}(2-a)$.

2522 $$\int_0^\infty \frac{\sin^2 x \sin ax}{x^3}\,dx = \frac{\pi}{2} \quad \text{or} \quad \frac{a\pi}{2}\left(1 - \frac{a}{4}\right),$$

according as a is $>$ or $<$ 2.

PROOF.—Denote the integral in (2520) by u; then, when a is > 2, the present integral is equal to

$$\int_0^a u\,da = \int_0^2 \frac{\pi}{4}(2-a)\,da + \int_2^a 0\,da = \frac{\pi}{2}.$$

And, when a is < 2, $\int_0^a u\,da = \int_0^a \frac{\pi}{4}(2-a)\,da = \frac{\pi a}{2} - \frac{\pi a^2}{8}$.

INTEGRATION OF CIRCULAR LOGARITHMIC AND EXPONENTIAL FORMS.

2571 $$\int_0^\infty \frac{e^{-ax} \sin rx}{x}\,dx = \tan^{-1}\frac{r}{a}.$$

PROOF.—Differentiate for r, and integrate by (2584).

Otherwise.—Expand $\sin rx$ by (764), and integrate the terms by (2291). Gregory's series (791) is the result.

2572 $$\int_0^\infty \frac{\sin rx}{x}\,dx = \frac{\pi}{2}.$$

PROOF.—(i.) By making $a = 0$ in (2571).

(ii.) *Otherwise.* By the method of (2250). First, observing that the integral is independent of r, which may be proved by substituting rx, let $r = 1$.

Then $$\int_0^\infty \frac{\sin x}{x}\,dx = \int_0^\pi \frac{\sin x}{x}\,dx + \int_\pi^{2\pi} \frac{\sin x}{x}\,dx + \int_{2\pi}^{3\pi} \frac{\sin x}{x}\,dx + \&c.$$

Now, n being an integer, the general term is either

$$\int_{(2n-1)\pi}^{2n\pi} \frac{\sin x}{x}\,dx = \int_0^\pi \frac{-\sin y\,dy}{(2n-1)\pi + y}, \text{ by substituting } x = (2n-1)\pi + y,$$

or $$\int_{(2n-2)\pi}^{(2n-1)\pi} \frac{\sin x}{x}\,dx = \int_0^\pi \frac{\sin y\,dy}{(2n-1)\pi - y}, \text{ by substituting } x = (2n-1)\pi - y;$$

$$\therefore \int_0^\infty \frac{\sin x}{x}\,dx = \int_0^\pi \sin y \left\{ \frac{1}{\pi - y} - \frac{1}{\pi + y} + \frac{1}{3\pi - y} - \frac{1}{3\pi + y} + \frac{1}{5\pi - y} - \&c. \right\} dy$$

$$= \tfrac{1}{2} \int_0^\pi \sin y \tan \frac{y}{2}\,dy \;\; (2913) \;\; = \int_0^\pi \sin^2 \frac{y}{2}\,dy = \frac{\pi}{2}.$$

2573 $$\int_0^\infty \frac{\cos rx}{1+x^2}\,dx = \frac{\pi}{2} e^{-r} = \int_0^\infty \frac{r \cos x}{r^2 + x^2}\,dx.$$

PROOF.—(i.) By (2251), putting $\dfrac{1}{1+x^2} = 2\displaystyle\int_0^\infty e^{-(1+x^2)y^2} y\,dy$ (2291), the integral takes the form

$$2\int_0^\infty \int_0^\infty \cos rx\, e^{-(1+x^2)y^2} y\,dx\,dy = 2\int_0^\infty \int_0^\infty e^{-y^2} y\, e^{-y^2x^2} \cos rx\,dy\,dx$$

$$= \sqrt{\pi} \int_0^\infty e^{-y^2 - \frac{r^2}{4y^2}}\,dy \;\; (2614) \;\; = \frac{\pi e^{-r}}{2} \;\; (2604).$$

(ii.) *Otherwise.* By the method of (2252), putting $u = \displaystyle\int_0^\infty \frac{\sin ax \cos bx}{x}\,dx$, it follows from (2516) that

$$\int_0^\infty u e^{-a}\,da = \int_0^b 0 e^{-a}\,da + \int_b^\infty \frac{\pi}{2} e^{-a}\,da = \frac{\pi}{2} e^{-b}.$$

Therefore $$\frac{\pi}{2} e^{-b} = \int_0^\infty \int_0^\infty \frac{\sin ax \cos bx}{x} e^{-a}\,da\,dx = \int_0^\infty \frac{\cos bx}{1+x^2}\,dx, \text{ by (2583).}$$

2575 $$\int_0^\infty \frac{x \sin rx}{1+x^2}\,dx = \frac{\pi}{2} e^{-r}.$$

2576 $$\int_0^\infty \frac{\sin rx}{x(1+x^2)}\,dx = \frac{\pi}{2}(1 - e^{-r}).$$

PROOF.—For (2575) differentiate, and for (2576) integrate equation (2573) with respect to r.

2577 $$\int_0^\infty e^{-ax} x^{n-1} \frac{\sin}{\cos}(bx)\,dx = \frac{\Gamma(n)}{(a^2+b^2)^{\frac{1}{2}n}} \frac{\sin}{\cos}\left(n \tan^{-1}\frac{b}{a}\right).$$

PROOF.—By (2291), $\int_0^\infty e^{-kx} x^{n-1}\,dx = \frac{\Gamma(n)}{k^n}$.

Put $k = a+ib$, and $a = r\cos\theta$, $b = r\sin\theta$; thus

$$\int_0^\infty e^{-(a+ib)x} x^{n-1}\,dx = (\cos n\theta - i\sin n\theta)\frac{\Gamma(n)}{r^n},$$

by (757). Substitute on the left side for e^{-ibx} from (767), and equate real and imaginary parts. Otherwise, as in (2259).

2579 $$\int_0^\infty x^{n-1} \frac{\sin}{\cos}(bx)\,dx = \frac{\Gamma(n)}{b^n} \frac{\sin}{\cos}\left(\frac{n\pi}{2}\right).$$

PROOF.—Make $a = 0$ in (2577).

2581 $$\int_0^\infty \frac{\frac{\sin}{\cos}(bx)}{x^m}\,dx = \frac{b^{m-1}\pi}{\Gamma(m)\, 2 \frac{\sin}{\cos}\left(\frac{m\pi}{2}\right)}.$$

PROOF.—Put $n = 1-m$ in (2579), and employ

$$\Gamma n\, \Gamma(1-n) = \frac{\pi}{\sin n\pi} = \frac{\pi}{\sin m\pi}.$$

2583 $$\int_0^\infty e^{-ax}\sin bx\,dx = \frac{b}{a^2+b^2}. \qquad \int_0^\infty e^{-ax}\cos bx\,dx = \frac{a}{a^2+b^2}.$$

PROOF.—Make $n = 1$ in (2577-8).

Otherwise.—Directly from (1999), putting $n = 1$, and $-a$ for a.

2585 $$\int_0^{\frac{1}{2}\pi} \cos^{p-n-1}\theta \sin^{n-1}\theta \frac{\sin}{\cos}(p\theta)\,d\theta = \frac{\Gamma(p-n)\Gamma(n)}{\Gamma(p)} \frac{\sin}{\cos}\left(\frac{n\pi}{2}\right),$$

where n is a positive integer > 1.

PROOF.—In (2577), put $\tan^{-1}\frac{b}{a} = \theta$, thus, writing p for n,

$$\int_0^\infty e^{-ax} x^{p-1} \sin bx\,dx = \frac{\Gamma(p)}{a^p} \cos^p\theta \sin p\theta.$$

Multiply this equation by $b^{n-1}db = a^n \tan^{n-1}\theta \sec\theta\, d\theta$.

and integrate from $b = 0$ to ∞, by (2579). Then the corresponding limits in the integration for θ will be 0 and $\frac{1}{2}\pi$.

2587 $$\int_0^{\frac{1}{2}\pi} \sin^{p-2}\theta \begin{matrix}\sin\\ \cos\end{matrix}(p\theta)\,d\theta = \begin{matrix}-\cos\\ +\sin\end{matrix}\left(\frac{p\pi}{2}\right)\frac{1}{p-1}.$$

PROOF.—Put $n = p-1$ in (2585).

2589 $$\int_0^\infty e^{-x}x^{n-1}\begin{matrix}\sin\\ \cos\end{matrix}\Big\}(x\tan\theta)\,dx = \Gamma(n)\cos^n\theta \begin{matrix}\sin\\ \cos\end{matrix}\Big\}\,n\theta.$$

PROOF.—In (2709), let $\phi(x) = \cos(x\tan\theta)$;

$\therefore$ by (765), $A_0 = 1$, $A_2 = -\dfrac{\tan^2\theta}{1.2}$, $A_4 = \dfrac{\tan^4\theta}{1.2.3.4}$, &c.; A_1, A_3, &c. vanishing. Therefore

$$1 - \frac{a(a+1)}{1.2}\tan^2\theta + \frac{a^{(4)}}{1^{(4)}}\tan^4\theta - \frac{a^{(6)}}{1^{(6)}}\tan^6\theta + \dots = \frac{\int_0^\infty e^{-x}x^{a-1}\cos(x\tan\theta)\,dx}{\int_0^\infty e^{-x}x^{a-1}\,dx}.$$

The series on the left $= \frac{1}{2}(1+i\tan\theta)^{-a} + \frac{1}{2}(1-i\tan\theta)^{-a}$, which by the values (770) and (768) reduces to $\cos a\theta\cos^a\theta$. Then change a into n. Similarly, with *sine* in the place of *cosine*.

2591 $$\int_0^\infty \frac{e^{-ax} - e^{-\beta x}}{x}\sin bx\,dx = \tan^{-1}\frac{\beta}{b} - \tan^{-1}\frac{a}{b}.$$

PROOF.—Integrate (2583) for a between $a = \alpha$ and $a = \beta$.

2592 $$\int_0^\infty e^{-kx}\cos ax\sin^n x\,dx,$$

where n is any positive integer.

See (2717–20) for the values of this integral.

2593 $$\int_0^\infty \frac{e^{ax}+e^{-ax}}{e^{\pi x}-e^{-\pi x}}\sin mx\,dx = \frac{1}{2}\,\frac{e^m - e^{-m}}{e^m + 2\cos a + e^{-m}},$$

a being $< \pi$.

PROOF.—The function expanded by division becomes

$$(e^{ax}+e^{-ax})\sin mx\,(e^{-\pi x} + e^{-3\pi x} + e^{-5\pi x} + \&c.)$$

Multiply in and integrate by (2583). The result is

$$\Sigma\frac{m}{\{(2n-1)\pi - a\}^2 + m^2} + \Sigma\frac{m}{\{(2n-1)\pi + a\}^2 + m^2}.$$

But this series is also produced by differentiating the logarithm of equation (2953). Hence the result.

2594 $$\int_0^\infty \frac{e^{ax}-e^{-ax}}{e^{\pi x}-e^{-\pi x}} \cos mx\, dx = \frac{\sin a}{e^m+2\cos a+e^{-m}}.$$

PROOF.—Change m into $i\theta$ in (2593), thus

$$\int_0^\infty \frac{(e^{ax}+e^{-ax})(e^{\theta x}-e^{-\theta x})}{e^{\pi x}-e^{-\pi x}}\, dx = \frac{\sin\theta}{\cos a+\cos\theta}.$$

Now change a into im and write a instead of θ.

2595 $$\int_0^\infty \frac{e^{ax}-e^{-ax}}{e^{\pi x}-e^{-\pi x}}\, dx = \frac{\tan\frac{1}{2}a}{2}. \quad \int_0^\infty \frac{\sin mx\, dx}{e^{\pi x}-e^{-\pi x}} = \frac{1}{4}\,\frac{e^{\frac{1}{2}m}-e^{-\frac{1}{2}m}}{e^{\frac{1}{2}m}+e^{-\frac{1}{2}m}}.$$

PROOF.—Make $m=0$ in (2594), and $a=0$ in (2593).

2597 $$\int_0^\infty \frac{e^{\pi x}+e^{-\pi x}}{e^{\pi x}-e^{-\pi x}} \sin mx\, dx = \frac{1}{2}\,\frac{e^{\frac{1}{2}m}+e^{-\frac{1}{2}m}}{e^{\frac{1}{2}m}-e^{\ \frac{1}{2}m}}.$$

PROOF.—Make $a=\pi$ in (2593).

2598 $$\int_0^\infty \frac{x^{2n-1}\,dx}{e^{\pi x}-e^{-\pi x}} = \frac{2^{2n}-1}{4n} B_{2n}.$$

PROOF.—Expand $\sin mx$ on the left side of (2596) by (764). The right side is $= -\frac{1}{4}i\tan(\frac{1}{2}im)$ by (770). Expand this by (2917), and equate the coefficients of the same powers of m.

2599 $$\int_0^\infty \frac{e^{ax}-e^{-ax}}{e^{\frac{1}{2}\pi x}+e^{-\frac{1}{2}\pi x}} \sin mx\, dx = \frac{2(e^m-e^{-m})\sin a}{e^{2m}+2\cos 2a+e^{-2m}}.$$

2600 $$\int_0^\infty \frac{e^{ax}+e^{-ax}}{e^{\frac{1}{2}\pi x}+e^{-\frac{1}{2}\pi x}} \cos mx\, dx = \frac{2(e^m+e^{-m})\cos a}{e^{2m}+2\cos 2a+e^{-2m}}.$$

PROOF.—To obtain (2599), put $a+\frac{1}{2}\pi$ and $a-\frac{1}{2}\pi$ successively for a in equation (2593), and take the difference of the results. (2600) is obtained in the same way from (2594).

2601 $$\int_0^\infty \frac{e^{ax}+e^{-ax}}{e^{\frac{1}{2}\pi x}+e^{-\frac{1}{2}\pi x}}\, dx = \sec a.$$

PROOF.—Make $m=0$ in (2600).

2602 $$\int_0^\infty \sin(cx)^2\, dx = \int_0^\infty \cos(cx)^2\, dx = \frac{\sqrt{\pi}}{2c\sqrt{2}}.$$

PROOF.—By (2425) $\int_0^\infty e^{-a^2x}\,dx = \frac{\sqrt{\pi}}{2a}$.

Put $a = \frac{1+i}{\sqrt{2}}c$. Substitute on the left from (766), and equate real and imaginary parts.

2604 $$\int_0^\infty e^{-\left(x^2+\frac{a^2}{x^2}\right)}\,dx = \frac{\sqrt{\pi}}{2}e^{-2a}.$$

PROOF.—Denote the integral by u. Differentiate the equation for a, and substitute $\frac{a}{x}$ in the resulting integral to prove that $\frac{du}{da} = -2u$, and therefore $u = Ce^{-2a}$. When $a=0$, we get

$$\int_0^\infty e^{-x^2}\,dx = C, \quad \therefore \quad C = \tfrac{1}{2}\sqrt{\pi} \ (2425).$$

2605 $$\int_0^\infty e^{-\left(x^2+\frac{a^2}{x^2}\right)k}\,dx = \frac{\sqrt{\pi}}{2\sqrt{k}}e^{-2ak}.$$

PROOF.—Substitute $x\sqrt{k}$, and integrate by (2604).

2606 $$\int_0^\infty e^{-\left(x^2+\frac{a^2}{x^2}\right)\cos\theta}\,\begin{matrix}\cos\\ \sin\end{matrix}\left[\left(x^2+\frac{a^2}{x^2}\right)\sin\theta\right]dx$$
$$= \frac{\sqrt{\pi}}{2}e^{-2a\cos\theta}\,\begin{matrix}\cos\\ \sin\end{matrix}\left(2a\sin\theta+\frac{\theta}{2}\right).$$

PROOF.—In (2605) put $k = \cos\theta + i\sin\theta$; substitute from (766), and equate real and imaginary parts.

2608 $$\int_0^\infty e^{-ax}\sin^{2n+1}x\,dx = \frac{1.2.3\ldots(2n+1)}{(a^2+1)(a^2+3^2)\ldots(a^2+\overline{2n+1}^2)}.$$

2609 $$\int_0^\infty e^{-ax}\sin^{2n}x\,dx = \frac{1.2.3\ldots 2n}{a(a^2+2^2)(a^2+4^2)\ldots(a^2+\overline{2n}^2)}.$$

2610
$$\int_0^\infty e^{-ax}\cos^{2n+1}x\,dx = \frac{a}{a^2+(2n+1)^2}+\frac{a(2n+1)2n}{(a^2+\overline{2n+1}^2)(a^2+\overline{2n-1}^2)}$$
$$+\frac{a(2n+1)2n(2n-1)(2n-2)}{(a^2+\overline{2n+1}^2)(a^2+\overline{2n-1}^2)(a^2+\overline{2n-3}^2)}+\ldots+\frac{a\,|\underline{2n+1}}{(a^2+\overline{2n+1}^2)\ldots(a^2+1)}.$$

2611 $$\int_0^\infty e^{-ax}\cos^{2n}x\,dx = \frac{a}{a^2+(2n)^2}+\frac{a\,2n\,(2n-1)}{(a^2+2n^2)(a^2+\overline{2n-2}^2)}$$

$$+\frac{a\,2n\,(2n-1)\,(2n-2)\,(2n-3)}{(a^2+\overline{2n}^2)\,(a^2+\overline{2n-2}^2)\,(a^2+\overline{2n-4}^2)}+\ldots+\frac{a\,\lfloor 2n}{(a^2+\overline{2n}^2)\ldots(a^2+2^2)}.$$

Proof of (2608-11). — Reduce successively by (1999). The integral part after each reduction disappears between the limits in the cases (2608-9), but not in the cases (2610-1). See also (2721).

2612

$$\int_{-\frac{1}{2}\pi}^{\frac{1}{2}\pi} e^{-ax}\cos^{2n+1}x\,dx = \frac{1.2.3\ldots(2n+1)}{(a^2+1)(a^2+3^2)\ldots(a^2+\overline{2n+1}^2)}\left(e^{\frac{a\pi}{2}}+e^{-\frac{a\pi}{2}}\right)$$

2613

$$\int_{-\frac{1}{2}\pi}^{\frac{1}{2}\pi} e^{-ax}\cos^{2n}x\,dx = \frac{1.2.3\ldots 2n}{a\,(a^2+2^2)(a^2+4^2)\ldots(a^2+\overline{2n}^2)}\left(e^{\frac{a\pi}{2}}-e^{-\frac{a\pi}{2}}\right).$$

Proof.—By successive reduction by (1999).

2614 $$\int_0^\infty e^{-a^2x^2}\cos 2bx\,dx = \frac{\sqrt{\pi}}{2a}e^{-\frac{b^2}{a^2}}.$$

Proof.—Denote the integral by u, then

$$\frac{du}{db} = -\int_0^\infty e^{-a^2x^2}\,2x\sin 2bx\,dx = -\int_0^\infty \frac{2b}{a^2}e^{-a^2x^2}\cos 2bx\,dx = -\frac{2bu}{a^2},$$

the second integration being effected by parts, $\int e^{-a^2x^2}\,2x\,dx$. Therefore $\log u = \log C - \frac{b^2}{a^2}$; and $b=0$ gives $C=\frac{\sqrt{\pi}}{2a}$ (2425).

Otherwise.—Expand the cosine by (765), and integrate the terms of the product by (2426). Thus the general term is

$$(-1)^n\frac{(2b)^{2n}}{\lfloor 2n}\int_0^\infty e^{-a^2x^2}x^{2n}dx = (-1)^n\frac{(2b)^{2n}}{\lfloor 2n}\,\frac{1.3.5\ldots(2n-1)}{2^{n+1}a^{2n+1}}\sqrt{\pi}.$$

$$= (-1)^n\frac{\sqrt{\pi}}{2a}\left(\frac{b}{a}\right)^{2n}\frac{1}{1.2\ldots n},$$ which gives the required result by (150).

2615 $$\int_0^\infty e^{-a^2x^2}\cosh 2bx = \frac{\sqrt{\pi}}{2a}e^{\left(\frac{b}{a}\right)^2}.$$ (2181)

Proof.—Change b into ib in (2614).

2617 $$\int_0^\infty e^{-x^2} x \sin 2bx \, dx = \frac{b\sqrt{\pi}}{2} e^{-b^2}.$$

2618 $$\int_0^\infty e^{-x^2} x^{n+1} \sin\left(2bx + \tfrac{1}{2}n\pi\right) dx = \frac{\sqrt{\pi}}{2^{n+1}} \frac{d^n}{db^n}\left(be^{-b^2}\right).$$

PROOF.—To obtain (2617), put $a = 1$ in (2614), and differentiate for b. To obtain (2618), differentiate, in all, $n+1$ times for b.

2619 $$\int_0^\infty \frac{\cos x - e^{-ax}}{x} \, dx = \log a.$$

PROOF.—By (2251) putting $\frac{1}{x} = \int_0^\infty e^{xy} dy$ (2291), and changing the order of integration, the integral becomes

$$\int_0^\infty \int_0^\infty (\cos x - e^{-ax}) e^{-xy} \, dy \, dx = \int_0^\infty \int_0^\infty (e^{-xy} \cos x - e^{-(a+y)x}) \, dy \, dx$$

$$= \int_0^\infty \left(\frac{y}{1+y^2} - \frac{1}{a+y}\right) dy \text{ (2584, 2291)} = \log a.$$

2620 $$\int_0^\pi \log(1 - 2a\cos x + a^2) \, dx = 0,$$

when a is equal to, or less than, unity; but is equal to $2\pi \log a$, when a is greater than unity.

PROOF.—(i.) $a = 1$. By (2635), since

$$\log 2(1 - \cos x) = \log 4 + 2 \log \sin \tfrac{1}{2}x.$$

(ii.) $a < 1$. By integrating (2922) from 0 to π.

(iii.) $a > 1$. As in (2926), integrating from 0 to π.

2622 $$\int_0^\pi \log(1 - n\cos x) \, dx.$$

When n is less than unity, the values of this integral depend on those of (2620). See (2933).

2623 $$\int_0^\pi \frac{x \sin x \, dx}{1 - 2a\cos x + a^2} = \frac{\pi}{a} \log(1+a), \text{ or } \frac{\pi}{a} \log\left(1 + \frac{1}{a}\right),$$

according as a is less or greater than unity.

PROOF.—Integrate $\int_0^\pi \log(1 - 2a\cos x + a^2) \, dx$ by Parts, $\int dx$, and apply (2620).

2625 $$\int_0^\pi \cos rx \log(1 - 2a\cos x + a^2) \, dx = -\frac{\pi a^r}{r}, \text{ or } -\frac{\pi a^{-r}}{r},$$

according as a is less or greater than unity.

PROOF.—Substitute the value of the logarithm obtained in (2922). The integral of every term of the resulting expansion, excepting the one in which $u=r$, vanishes by (2467).

2627 $$\int_0^\pi \frac{\sin x \sin rx\,dx}{1-2a\cos x+a^2} = \frac{\pi a^{r-1}}{2}, \quad \text{or} \quad \frac{\pi a^{-(r+1)}}{2},$$

according as a is less or greater than unity.

PROOF.—Integrate (2625) by Parts, $\int \cos rx\,dx$.

2629 $$\int_0^\pi \frac{\cos rx\,dx}{1-2a\cos x+a^2} = \frac{\pi a^r}{1-a^2}, \quad a \text{ being } < 1.$$

PROOF.—The fraction $= \cos rx\,(1+2a\cos x+2a^2\cos 2x+2a^3\cos 3x+\ldots)$ $\div(1-a^2)$, by (2919), and the result follows as in (2625).

2630 $$\int_0^\infty \frac{1}{1+x^2}\cdot\frac{dx}{1-2a\cos cx+a^2} = \frac{\pi}{2(1-a^2)}\,\frac{1+ae^{-c}}{1-ae^{-c}}.$$

PROOF.—Expand the second factor by (2919), and integrate the terms by (2573).

2631 $$\int_0^\infty \frac{\log(1-2a\cos cx+a^2)\,dx}{1+x^2} = \pi\log(1-ae^{-c}).$$

PROOF.—Expand the numerator by (2922), and integrate the terms by (2573).

2632 $$\int_0^\infty \frac{x\sin cx\,dx}{(1+x^2)(1-2a\cos cx+a^2)} = \frac{\pi}{2(e^c-a)}.$$

PROOF.—By differentiating (2631) for c.

Otherwise.—Expand by (2921), and integrate the terms by (2574).

2633 $$\int_0^{\frac{1}{2}\pi} \frac{\log(1+c\cos x)}{\cos x}\,dx = \tfrac{1}{2}\left\{\frac{\pi^2}{4}-(\cos^{-1}c)^2\right\}.$$

PROOF.—Put $a=1$ in (1951), and take the integral between the limits 0 and $\frac{1}{2}\pi$, then integrate for b between limits 0 and c; the result is

$$\int_0^{\frac{1}{2}\pi} \frac{\log(1+c\cos x)}{\cos x}\,dx = 2\int_0^c \frac{1}{\sqrt{1-b^2}}\tan^{-1}\sqrt{\frac{1-b}{1+b}}\,db,$$

and the integral on the right is found by substituting $\cos^{-1} b$.

2634 $$\int_0^{\pi} \frac{\log(1+c\cos x)}{\cos x}\,dx = \pi \sin^{-1} c.$$

PROOF.—As in (2633), by taking 0 and π for the limits of x.

2635 $$\int_0^{\frac{1}{2}\pi} \log \sin x\,dx = \frac{\pi}{2}\log\frac{1}{2} = \int_0^1 \frac{\log x}{\sqrt{(1-x^2)}}\,dx.$$

PROOF.—$\int_0^{\frac{1}{2}\pi} \sin x\,dx = \int_0^{\frac{1}{2}\pi} \cos x\,dx$ (2233). Add these integrals and substitute $2x$, applying (2234) to the result.

2637 $$\int_0^{\pi} x \log \sin x\,dx = \frac{\pi^2}{2}\log\tfrac{1}{2}.$$

PROOF.—$\int_0^{\pi} x^2 \log \sin x\,dx = \int_0^{\pi} (\pi-x)^2 \log \sin x\,dx$, by (2233). Equate the difference of these integrals to zero.

2638 $$\int_0^{n\pi} x \log \sin^2 x\,dx = -n^2\pi^2 \log 2, \quad n \text{ being an integer.}$$

PROOF.—Method of (2250),

$$\int_0^{n\pi} xl\sin^2 x\,dx = \int_0^{\pi} xl\sin^2 x\,dx + \int_{\pi}^{2\pi} xl\sin^2 x\,dx + \ldots + \int_{(n-1)\pi}^{n\pi} xl\sin^2 x\,dx$$

$$= \int_0^{\pi} xl\sin^2 x\,dx + \int_0^{\pi} (\pi+y)\,l\sin^2 y\,dy + \ldots + \int_0^{\pi} \{(n-1)\,\pi+y\}\,l\sin^2 y\,dy.$$

Each integral reduces by (2635) and (2637); for example,

$$\int_0^{\pi} (\pi+y)\,l\sin^2 y\,dy = 2\int_0^{\pi} (\pi+y)\,l\sin y\,dy = 2\pi\int_0^{\pi} l\sin y\,dy + 2\int_0^{\pi} y\,l\sin y\,dy$$

$$= -2\pi^2\log 2 - \pi^2\log 2 = -3\pi^2\log 2.$$

The result is $-\{1+3+5+\ldots+(2n-1)\}\,\pi^2\log 2 = -n^2\pi^2\log 2.$

2639 $$\int_0^{\infty} \frac{\sin mx}{e^{2\pi x}-1}\,dx = \frac{1}{2}\left(\frac{1}{e^m-1}+\frac{1}{2}-\frac{1}{m}\right).$$

PROOF.—Develope $\sin mx$ by (764); integrate the terms by (2396), and sum the series by (1539).

2640 $$\int_0^{\infty} \frac{\sin mx}{e^x+1}\,dx = \frac{1}{2m} - \frac{\pi}{e^{m\pi}-e^{-m\pi}}.$$

PROOF.—Develope $\sin mx$ by (764); integrate the terms by (2398). The resulting series is $=\frac{1}{2m}+\frac{\pi}{2i}\operatorname{cosec} im\pi$, by (2918), which is equivalent to the above by (769).

2641 $$\int_0^1 \frac{\cos(m\log x)-\cos(n\log x)}{\log x}\,dx = \tfrac{1}{2}\log\frac{1+m^2}{1+n^2}.$$

2642 $$\int_0^1 \frac{\sin(m\log x)-\sin(n\log x)}{\log x}\,dx = \tan^{-1}m-\tan^{-1}n.$$

PROOF.—Put $p=im$ and $q=in$ in (2394), and equate corresponding parts. See (2214).

2643
$$\int_0^1 \frac{\sin(n\log x)}{\log x}\,dx = \tan^{-1}n. \quad \int_0^1 \frac{\operatorname{vers}(n\log x)}{\log x}\,dx = \tfrac{1}{2}\log\frac{1}{1+n^2}.$$

PROOF.—Put $m=0$ in (2641) and (2642).

MISCELLANEOUS THEOREMS.

FRULLANI'S FORMULA.

2700 $$\int_0^\infty \frac{\phi(ax)-\phi(bx)}{x}\,dx = \phi(0)\log\frac{b}{a}+\int_{\frac{h}{a}}^{\frac{h}{b}}\frac{\phi(bx)}{x}\,dx,$$

h being $=\infty$, and the last term generally $=0$.

PROOF.—In the integral $\int_0^h \frac{\phi(z)-\phi(0)}{z}\,dz$ substitute $z=ax$ and $z=bx$, and equate the results thus,

$$\int_0^{\frac{h}{a}}\frac{\phi(ax)}{x}\,dx-\int_0^{\frac{h}{b}}\frac{\phi(bx)}{x}\,dx=\int_0^{\frac{h}{a}}\frac{\phi(0)}{x}\,dx-\int_0^{\frac{h}{b}}\frac{\phi(0)}{x}\,dx,$$

or $$\int_0^{\frac{h}{a}}\frac{\phi(ax)-\phi(bx)}{x}\,dx-\int_{\frac{h}{a}}^{\frac{h}{b}}\frac{\phi(bx)}{x}\,dx=\int_{\frac{h}{b}}^{\frac{h}{a}}\frac{\phi(0)}{x}\,dx=\phi(0)\log\frac{b}{a}.$$

Then make h infinite. For applications see (2513) and (2505).

2701 $$\int_0^\infty \frac{\phi(ax)-\phi(bx)}{x^2}\,dx = (a-b)\int_0^\infty \frac{\phi'(x)}{x}\,dx$$
$$+\phi'(0)\left(\log\frac{a^a}{b^b}-a+b\right)-(a-b)\frac{\phi(h)}{h}+\int_{\frac{h}{a}}^{\frac{h}{b}}\frac{\phi(bx)}{x^2}\,dx,$$
with $h=\infty$.

PROOF.— $$d_a\left\{\int_0^{\frac{h}{a}}\frac{\phi(ax)}{x^2}\,dx\right\} = \int_0^{\frac{h}{a}}\frac{\phi'(ax)}{x}\,dx-\frac{\phi(h)}{h}$$
$$(2257) = \int_0^h \frac{\phi'(x)}{x}\,dx-\phi'(0)\,la-\frac{\phi(h)}{h},$$
by making $b=1$ in the proof of (2700). Integrate for a between limits a and b, thus $$\int_0^{\frac{h}{a}}\frac{\phi(ax)}{x^2}\,dx-\int_0^{\frac{h}{b}}\frac{\phi(bx)}{x^2}\,dx$$
$$=(a-b)\int_0^h\frac{\phi'(x)}{x}\,dx+\phi'(0)\{a\,la-b\,lb-a+b\}-\frac{(a-b)\,\phi(h)}{h},$$
and the left is $$=\int_0^{\frac{h}{a}}\frac{\phi(ax)-\phi(bx)}{x^2}\,dx-\int_{\frac{h}{a}}^{\frac{h}{b}}\frac{\phi(bx)}{x^2}\,dx.$$

POISSON'S FORMULÆ.

2702 $$\int_0^\pi \frac{f(a+e^{ix})+f(a+e^{-ix})}{1-2c\cos x+c^2}\,dx = \frac{2\pi}{1-c^2}f(a+c),$$
c being <1.

PROOF.—By Taylor's theorem (1500), and by (2919), the fraction is equal to the product of the two expansions
$$2\left\{f(a)+f'(a)\cos x+\frac{1}{1.2}f''(a)\cos 2x+\frac{1}{1.2.3}f'''(a)\cos 3x+\dots\right\}$$
and $$\{1+2c\cos x+2c^2\cos 2x+2c^3\cos 3x+\dots\}$$
divided by $(1-c^2)$. By (2468) the integral of every term of the product vanishes, except when it is of the form $2\int_0^\pi \cos^2 nx$, and this is $=\pi$, by (2471). Hence the result.

2703
$$\int_0^\pi \frac{f(a+e^{ix})+f(a+e^{-ix})}{1-2c\cos x+c^2}(1-c\cos x)\,dx = \pi\{f(a+c)+f(a)\}.$$

2704
$$\int_0^\pi \frac{f(a+e^{ix})-f(a+e^{-ix})}{1-2c\cos x+c^2}\sin x\,dx = \frac{i\pi}{c}\{f(a+c)-f(a)\}.$$

PROOF.—As in (2702), adding unity to each side of (2919), and employing (2921, 2467, 2470).

ABEL'S FORMULA.

Given that $F(x+a)$ can be expanded in powers of e^{-a}, then

2705 $$\int_0^\infty \frac{F(x+iat)+F(x-iat)}{1+t^2}\,dt = \pi F(x+a).$$

PROOF.—Assume $F(x+a) = A+A_1e^{-a}+A_2e^{-2a}+A_3e^{-3a}+\&c.$,

$\therefore\; F(x+iat)+F(x-iat) = 2A+2A_1\cos at+2A_2\cos 2at+\&c.$

Substitute and integrate by (1935) and (2573).

Ex.—Let $F(x) = \dfrac{1}{x}$, then $\displaystyle\int_0^\infty \frac{dt}{(1+t^2)(x^2+a^2t^2)} = \frac{\pi}{2x(x+a)}$.

KUMMER'S FORMULA.

2706 $$\int_{-\frac{1}{2}\pi}^{\frac{1}{2}\pi} f(2x\cos\theta e^{i\theta})\,e^{2ik\theta}\,d\theta = \sin k\pi \int_0^1 (1-z)^{k-1} f(xz)\,dz.$$

PROOF.—If $h = xe^{2i\theta}$, then $x+h = 2x\cos\theta e^{i\theta}$ by (766). Substitute these values in the expansion of $f(x+h)$ by (1500); multiply by $e^{2ik\theta}$ and integrate; thus, after reducing by (769),

$$\int_{-\frac{1}{2}\pi}^{\frac{1}{2}\pi} f(2x\cos\theta e^{i\theta})\,e^{2ik\theta}\,d\theta = \sin k\pi\left\{\frac{f(x)}{k}-\frac{xf'(x)}{k+1}+\frac{x^2f''(x)}{1.2.(k+2)}-\&c.\right\}$$

Again, putting $h = -x\phi$ in (1500), multiplying by $\phi^{k-1}\,d\phi$, and integrating, we have $\int_0^1 \phi^{k-1} f(x-x\phi)\,d\phi =$ the foregoing series within the brackets. Equating the two values and changing ϕ into $1-z$, the formula is obtained.

For an application see (2490).

2707 When k is an integer,

$$\int_{-\frac{1}{2}\pi}^{\frac{1}{2}\pi} f(2x\cos\theta e^{i\theta})\,e^{2ik\theta}\,\theta\,d\theta = \frac{\pi\cos k\pi}{2i}\int_0^1 (1-z)^{k-1} f(xz)\,dz.$$

PROOF.—Divide equation (2706) by $\sin k\pi$, and evaluate the indeterminate fraction by (1580), differentiating with respect to k.

For applications see (2490), (2494).

2708 If X be a function of x so chosen that

$$\int_a^b Xf(x,k)\,dx = C_k\int_a^b Xf(x,0)\,dx \;\ldots\ldots\ldots\ldots \text{(i.)},$$

and if the series

$$A_0f(x,0)+A_1f(x,1)+A_2f(x,2)+\&c.\;\ldots = \phi(x)\ldots\text{(ii.)},$$

where ϕ is a known function, then

$$A_0C_0+A_1C_1+A_2C_2+\&c.\ \ldots = \frac{\int_a^b X\phi(x)\,dx}{\int_a^b Xf(x,\,0)\,dx} \quad \text{......(iii.)}$$

PROOF.—Multiply (ii.) by X, and integrate from a to b, employing (i.)

2709 If the sum of the series

$$A_0+A_1x+A_2x^2+A_3x^3+\&c.\ \ldots = \phi(x)$$

be known, then

$$A_0+A_1\alpha+A_2\alpha(\alpha+1)+A_3\alpha(\alpha+1)(\alpha+2)+\&c.\ldots$$

$$=\frac{\int_0^\infty e^{-x}x^{\alpha-1}\,\phi(x)\,dx}{\int_0^\infty e^{-x}x^{\alpha-1}\,dx}.$$

PROOF.—In (2708) let $X = e^{-x}x^{\alpha-1}$ and $f(x,\,k) = x^k$. Then since, by Parts, we have $\int_0^\infty e^{-x}x^{\alpha+k-1}dx = \alpha(\alpha+1)\ldots(\alpha+k-1)\int_0^\infty e^{-x}x^{\alpha-1}dx$, it follows that $C_k = \alpha(\alpha+1)\ldots(\alpha+k-1)$. Hence, conditions (i.) and (ii.) being fulfilled, result (iii.) is established.

For an application see (2589).

THEOREM.—Let $f(x+iy) = P+iQ$ .... (i.)

2710 Then $$\int_a^b\int_\alpha^\beta \frac{dQ}{dy}\,dx\,dy = \int_\alpha^\beta\int_a^b \frac{dP}{dx}\,dy\,dx \quad \text{........... (ii.)}$$

2711 $$\int_a^b\int_\alpha^\beta \frac{dP}{dy}\,dx\,dy = -\int_\alpha^\beta\int_a^b \frac{dQ}{dx}\,dy\,dx \quad \text{........... (iii.)}$$

PROOF.—Differentiating (i.) independently for x and y,

$$f'(x+iy) = P_x+iQ_x, \quad if'(x+iy) = P_y+iQ_y,$$

$$\therefore\ P_x+iQ_x = Q_y-iP_y, \quad \therefore\ P_x = Q_y \text{ and } Q_x = -P_y.$$

Hence by (2261) the equalities (ii.) and (iii.) are obtained.

Ex.—Let $f(x+iy) = e^{-(x+iy)^2} = e^{-x^2}e^{y^2}(\cos 2xy - i\sin 2xy)$.

Here $P = e^{-x^2}e^{y^2}\cos 2xy$, $Q = -e^{-x^2}e^{y^2}\sin 2xy$, therefore, by (iii.),

$$\int_a^b e^{-x^2}(e^{\beta^2}\cos 2\beta x - e^{\alpha^2}\cos 2\alpha x)\,dx = \int_\alpha^\beta e^{y^2}(e^{-b^2}\sin 2by - e^{-a^2}\sin 2ay)\,dy.$$

Put $a = \alpha = 0$, $b = \infty$; therefore

$$\int_0^\infty e^{-x^2}(e^{\beta^2}\cos 2\beta x - 1)\,dx = 0, \quad \therefore\ e^{\beta^2}\int_0^\infty e^{-x^2}\cos 2\beta x\,dx = \int_0^\infty e^{-x^2}\,dx.$$

CAUCHY'S FORMULA.

2712 Let $\int_0^\infty x^{2n}F(x^2)\,dx = A_{2n}$, n being an integer, then

$$\int_0^\infty x^{2n}F\left\{\left(x-\frac{1}{x}\right)^2\right\}dx$$

$$= A_0+\frac{n(n+1)}{1.2}A_2+\frac{(n-1)^{(4)}}{1^{(4)}}A_4+\frac{(n-2)^{(6)}}{1^{(6)}}A_6+\&c.$$

Proof.—In the integral $\int_{-\infty}^{\infty} z^{2n}F(z^2)\,dz = 2A_{2n}$, substitute $z = x-\frac{1}{x}$, and it becomes $\int_0^\infty \left(x-\frac{1}{x}\right)^{2n}\left(x+\frac{1}{x}\right)F\left\{\left(x-\frac{1}{x}\right)^2\right\}\frac{dx}{x} = 2a_{2n}$ ......(i.)

Let the integral sought be denoted by C_{2n}, then

$$\int_0^\infty \frac{1}{x^{2n+1}}F\left\{\left(x-\frac{1}{x}\right)^2\right\}\frac{dx}{x} = \int_0^\infty x^{2n+1}F\left\{\left(x-\frac{1}{x}\right)^2\right\}\frac{dx}{x} = C_{2n}.$$

This is proved by substituting $\frac{1}{x}$ in the first integral. Therefore by addition

$$\int_0^\infty \left(x^{2n+1}+\frac{1}{x^{2n+1}}\right)F\left\{\left(x-\frac{1}{x}\right)^2\right\}\frac{dx}{x} = 2C_{2n} \quad\text{..............(ii.)}$$

Now, in the expansion of $\cos(2n+1)\theta$ (776), put $2\cos\theta = x+\frac{1}{x}$ and $2i\sin\theta = x-\frac{1}{x}$, where $x = e^{i\theta}$ by (768–9), and multiply the equation by $F\left\{\left(x-\frac{1}{x}\right)^2\right\}\frac{dx}{x}$, and integrate from $x=0$ to $x=\infty$. Then, by (i.) and (ii.), the required result is obtained.

2713 Ex.—Let $F(x) = e^{-ax}$, then

$$A_{2n} = \int_0^\infty x^{2n}e^{-ax^2}\,dx = \frac{1.3\ldots(2n-1)}{2^{n+1}a^{n+\frac{1}{2}}}\sqrt{\pi} \quad\text{and}\quad A_0 = \frac{\sqrt{\pi}}{2\sqrt{a}}.$$

Therefore $$\int_0^\infty x^{2n}e^{-a\left(x^2+\frac{1}{x^2}\right)}\,dx$$

$$= \frac{e^{-2a}\sqrt{\pi}}{2\sqrt{a}}\left\{1+\frac{n(n+1)}{4a}+\frac{(n-1)^{(4)}}{1^{(2)}4^2a^2}+\frac{(n-2)^{(6)}}{1^{(3)}4^3a^3}+\frac{(n-3)^{(8)}}{1^{(4)}4^4a^4}+\&c.\right\}.$$

FINITE VARIATION OF A PARAMETER.

2714 Theorem (2255) may be extended to the case of a finite change in the value of a quantity under the sign of integration.

Let α be independent of a and b, and let Δ be the difference caused by an increase of unity in the value of α, then

$$\int_a^b \Delta\phi(x,\alpha)\,dx = \Delta\int_a^b \phi(x,\alpha)\,dx.$$

2715 Ex. 1. $\int_0^\infty e^{-\alpha x}\,dx = \frac{1}{\alpha}$, $\therefore\ \int_0^\infty \Delta e^{-\alpha x}\,dx = \Delta\frac{1}{\alpha}$, that is

$$\int_0^\infty e^{-\alpha x}(e^{-x}-1)\,dx = -\frac{1}{\alpha(\alpha+1)}.$$

Also, by repeating the operation,

$$\int_0^\infty \Delta^n e^{-\alpha x}\,dx = \Delta^n\frac{1}{\alpha},\quad \text{that is}$$

2716 $$\int_0^\infty e^{-\alpha x}(e^{-x}-1)^n\,dx = \frac{(-1)^n \lfloor n}{\alpha(\alpha+1)\ldots(\alpha+n)}.$$

2717 Ex. 2.—In (2583-4) put k for a and $(2\alpha-m)$ for b, then

$$\int_0^\infty e^{-kx}\,\Delta\sin(2\alpha-m)\,x\,dx = \Delta\frac{2\alpha-m}{k^2+(2\alpha-m)^2}\ \ldots\ldots\ldots\ldots\text{(i.)},$$

$$\int_0^\infty e^{-kx}\,\Delta\cos(2\alpha-m)\,x\,dx = \Delta\frac{k}{k^2+(2\alpha-m)^2}\ \ldots\ldots\ldots\text{(ii.)}.$$

In (ii.) let $m=2p$, an even integer, then

$$\begin{aligned}\Delta^{2p}\cos(2\alpha-2p)\,x &= \cos(2\alpha+2p)\,x-2p\cos(2\alpha+2p-2)\,x+\ldots\\ &\qquad\ldots+\cos(2\alpha-2p)\,x\\ &= \cos 2\alpha x\,[\cos 2px-2p\cos(2p-2)\,x+C(2p,2)\cos(2p-4)\,x-\ldots\\ &\qquad\ldots+\cos 2px]\\ &\quad-\sin 2\alpha x\,[\sin 2px-2p\sin(2p-2)\,x+\ldots\\ &\qquad\ldots-\sin 2px].\end{aligned}$$

The coefficient of $\cos 2\alpha x$, in which equidistant terms are equal, is $=(-1)^p\,2^{2p}\sin^{2p}x$ (773); while the coefficient of $\sin 2\alpha x$ vanishes because the equidistant terms destroy each other. Therefore

$$\Delta^{2p}\cos(2\alpha-2p)\,x = (-1)^p\,2^{2p}\cos 2\alpha x\sin^{2p}x.$$

Hence (ii.) becomes

2718 $$\int_0^\infty e^{-kx}\cos 2\alpha x\sin^{2p}x\,dx = \frac{(-1)^p}{2^{2p}}\,\Delta^{2p}\frac{k}{k^2+(2\alpha-2p)^2}.$$

2719 Again, in (i.) let $m=2p+1$, an odd integer, then

$$\begin{aligned}\Delta^{2p+1}\sin(2\alpha-2p-1)\,x &= \sin(2\alpha+2p+1)\,x-(2p+1)\sin(2\alpha+2p-1)\,x\\ &\quad+C(2p+1,2)\sin(2\alpha+2p-3)\,x-\ldots-\sin(2\alpha-2p-1)\,x\\ &= \sin 2\alpha x\,[\cos(2p+1)\,x-(2p+1)\cos(2p-1)\,x+\ldots-\cos(2p+1)\,x]\\ &\quad+\cos 2\alpha x\,[\sin(2p+1)\,x-(2p+1)\sin(2p-1)\,x+\ldots+\sin(2p+1)\,x].\end{aligned}$$

The coefficient of $\sin 2\alpha x$ vanishes as before, while that of $\cos 2\alpha x$ is

$$=(-1)^p\,2^{2p+1}\sin^{2p+1}x\ \ (774).$$

Therefore equation (i.) becomes

2720

$$\int_0^\infty e^{-kx}\cos 2ax \sin^{2p+1}x\,dx = \frac{(-1)^p}{2^{2p+1}}\,\Delta^{2p+1}\,\frac{2a-2p-1}{k^2+(2a-2p-1)^2}.$$

To compute the right member of equation (2718), we have

$$\Delta^{2p}\frac{k}{k^2+(2a-2p)^2} = k\left[\frac{1}{k^2+(2a+2p)^2}\right.$$

$$\left.-\frac{2p}{k^2+(2a+2p-2)^2}+\frac{C(2p,2)}{k^2+(2a+2p-4)^2}-\ldots+\frac{1}{k^2+(2a-2p)^2}\right].$$

Let $a=0$, then the equidistant terms are equal, and we obtain in this case

2721 $$\Delta^{2p}\frac{k}{k^2+(2a-2p)^2} = \frac{(-1)^p\,1.2\ldots 2p\,.\,2^{2p}}{k\,(k^2+4)(k^2+16)\ldots\{k^2+(2p)^2\}}.$$

Thus formula (2609) is proved.

Similarly, by making $a=0$ in (2720) after expansion, formula (2608) is obtained.

Let p be any integer, and let q and a be arbitrary, but $q<2p$ in (2722), and $<2p+1$ in (2723).

2722

$$\int_0^\infty \frac{\cos 2ax \sin^{2p}x}{x^{q+1}}\,dx = \frac{(-1)^p}{2^{2p}\Gamma(q+1)}\int_0^\infty \Delta^{2p}\,\frac{z^{q+1}}{z^2+(2a-2p)^2}\,dz.$$

2723

$$\int_0^\infty \frac{\cos 2ax \sin^{2p+1}x}{x^{q+1}}\,dx$$

$$= \frac{(-1)^p}{2^{2p+1}\Gamma(q+1)}\int_0^\infty \Delta^{2p+1}\,\frac{(2a-2p-1)\,z^q}{z^2+(2a-2z)^2}\,dz,$$

where Δ has the signification in (2714).

PROOF.—Employing the method of (2510), replace

$$\frac{1}{x^{q+1}} \text{ by } \frac{1}{\Gamma(q+1)}\int_0^\infty e^{-xz}z^q\,dz,$$

q being integral or fractional; therefore

$$\int_0^\infty \frac{\cos 2ax \sin^{2p}x}{x^{q+1}}\,dx = \frac{1}{\Gamma(q+1)}\int_0^\infty\int_0^\infty \cos 2ax \sin^{2p}x\, e^{-zx}z^q\,dz\,dx,$$

by changing the order of integration. Substitute the value in (2718) for the integral containing x, writing the factor z^q under the operator Δ, since it is independent of a.

Similarly, with $2p+1$ in the place of p, we substitute from (2720).

It may be shown that, whenever $a > p$, formula (2722) reduces to

2724

$$\int_0^\infty \frac{\cos 2ax \sin^{2p} x}{x^{q+1}}\,dx = \frac{(-1)^{p+1}\pi}{2^{2p+1}\Gamma(q+1)\sin\frac{q\pi}{2}}\,\Delta^{2p}(2a-2p)^q.$$

For a complete investigation, see Cauchy's "Mémoire de l'Ecole Polytechnique," tome xvii.

2725 Ex.—Let $a = 2,\ p = 1,\ q = \frac{1}{3}$,

$$\int_0^\infty \frac{\cos 4x \sin^2 x}{x^{\frac{4}{3}}}\,dx = \frac{\pi}{8\Gamma\left(\frac{4}{3}\right)\sin\frac{\pi}{6}}\,\Delta^2(2a-2)^{\frac{1}{3}},$$

and $\Delta^2(2a-2)^{\frac{1}{3}} = (2a+2)^{\frac{1}{3}} - 2(2a)^{\frac{1}{3}} + (2a-2)^{\frac{1}{3}} = 6^{\frac{1}{3}} - 2.4^{\frac{1}{3}} + 2^{\frac{1}{3}}$.

FOURIER'S FORMULA.

2726

$$\int_0^h \frac{\sin ax}{\sin x}\,\phi(x)\,dx = \frac{\pi}{2}\,\phi(0),$$

when $a = \infty$ and h is not greater than $\frac{1}{2}\pi$.

Proof.—(i.) Let $\phi(x)$ be a continuous, finite, positive quantity, decreasing in value as x increases from zero to h.

$$\int_0^h \frac{\sin ax}{\sin x}\,\phi(x)\,dx = \int_0^{\frac{\pi}{a}} + \int_{\frac{\pi}{a}}^{\frac{2\pi}{a}} + \int_{\frac{2\pi}{a}}^{\frac{3\pi}{a}} + \dots + \int_{\frac{(r-1)\pi}{a}}^{\frac{r\pi}{a}} + \int_{\frac{r\pi}{a}}^{h} \quad\dots\dots\dots \text{(i.)},$$

$\frac{r\pi}{a}$ being the greatest multiple of $\frac{\pi}{a}$ contained in h. The terms are alternately positive and negative, as appears from the sign of $\sin ax$. The following investigation shows that the terms decrease in value. Take two consecutive terms

$$\int_{\frac{n\pi}{a}}^{\frac{(n+1)\pi}{a}} \frac{\sin ax}{\sin x}\,\phi(x)\,dx, \qquad \int_{\frac{(n+1)\pi}{a}}^{\frac{(n+2)\pi}{a}} \frac{\sin ax}{\sin x}\,\phi(x)\,dx.$$

Substituting $x - \frac{\pi}{a}$ in the second integral, it becomes

$$-\int_{\frac{n\pi}{a}}^{\frac{(n+1)\pi}{a}} \frac{\sin ax}{\sin\left(x+\frac{\pi}{a}\right)}\,\phi\left(x+\frac{\pi}{a}\right)dx,$$

and since ϕ decreases as x increases, an element of this integral is less than the corresponding element of the first integral.

Now, by substituting $ax = y$, we have

$$\int_{\frac{n\pi}{a}}^{\frac{(n+1)\pi}{a}} \frac{\sin ax}{\sin x}\phi(x)\,dx = \int_{n\pi}^{(n+1)\pi} \frac{\sin y}{a\sin\frac{y}{a}}\phi\left(\frac{y}{a}\right)dy = \phi(0)\int_{n\pi}^{(n+1)\pi}\frac{\sin y}{y}\,dy \;\ldots\;\text{(ii.)},$$

when a is infinite, because then $\phi\left(\frac{y}{a}\right) = \phi(0)$ and $a\sin\frac{y}{a} = y$.

Hence the sum of n terms of (i.) may be replaced by $\phi(0)\int_0^{n\pi}\frac{\sin y}{y}\,dy$, which, when n is infinite, takes the value $\phi(0)\,\frac{1}{2}\pi$ by (2572); while the sum of the remaining terms vanishes, because (the signs alternating) that sum is less than the $n+1^{\text{th}}$ term, which itself vanishes when n is infinite.

(ii.) If $\phi(x)$, while always *decreasing*, becomes negative, let C be a constant such that $C+\phi(x)$ remains always positive while x varies from 0 to h. The theorem is true for $C+\phi(x)$, and also for a function constant and equal to C, and it is therefore true for the decreasing function ϕ whatever its sign.

If $\phi(x)$ is a function always *increasing* in value, $-\phi(x)$ is a decreasing function. The theorem applies to the last function, and therefore also to $\phi(x)$.

2727 Cor.—Hence the same integral taken between any two limits lying between zero and $\frac{1}{2}\pi$, vanishes when a is infinite.

2728 $$\int_0^h \frac{\sin ax}{\sin x}\phi(x)\,dx$$

$$= \pi\left\{\tfrac{1}{2}\phi(0)+\phi(\pi)+\phi(2\pi)+\ldots+\phi(n-1)\pi+\phi(n\pi)\right\},$$

when a is an indefinitely great *odd integer*, and $n\pi$ is the greatest multiple of π less than h. But when a is an indefinitely great *even integer*, the second and alternate terms of the series have the minus sign.

Proof. $$\int_0^h \frac{\sin ax}{\sin x}\phi(x)\,dx = \int_0^{n\pi}\frac{\sin ax}{\sin x}\phi(x)\,dx + \int_{n\pi}^{h}\frac{\sin ax}{\sin x}\phi(x)\,dx \ldots\ldots\text{(i.)},$$

decompose the second integral into $2n$ others with the limits 0 to $\frac{1}{2}\pi$, $\frac{1}{2}\pi$ to π, π to $\frac{3}{2}\pi$, ... $(2n-1)\,\frac{1}{2}\pi$ to $n\pi$; and in these integrals put successively $x = y$, $\pi-y$, $\pi+y$, $2\pi-y$, $2\pi+y$, ... $n\pi-y$. The new limits will be 0 to $\frac{1}{2}\pi$, $\frac{1}{2}\pi$ to 0 alternately, with the even terms negative, so that, by changing the signs of the even terms, the limits for each will be 0 to $\frac{1}{2}\pi$. Also, if a is an odd integer, $\frac{\sin ax}{\sin x}$ is changed into $\frac{\sin ay}{\sin y}$ by each substitution, so that (i.) becomes

$$\int_0^{\frac{1}{2}\pi}\frac{\sin ay}{\sin y}\{\phi(y)+\phi(\pi-y)+\phi(\pi+y)+\ldots+\phi(n\pi-y)\}\,dy$$
$$+\int_{n\pi}^{h}\frac{\sin ax}{\sin x}\phi(x)\,dx \;\ldots\ldots\ldots\text{(iii.)}$$

But, when a is even, the substitution of $r\pi \mp y$ for x makes $\frac{\sin ay}{\sin y}$ minus whenever r is odd. The limit of the first part of (iii.) is

$$\frac{\pi}{2}\{\phi(0)+2\phi(\pi)+2\phi(2\pi)+\ldots+2\phi(n-1)\pi+\phi(n\pi)\}, \text{ by (2726).}$$

In the last part of (iii.) put $x=n\pi+y$, and the integral becomes

$$\int_0^{h-n\pi}\frac{\sin ay}{\sin y}\phi(n\pi+y)\,dy=\frac{\pi}{2}\phi(n\pi), \text{ if } h-n\pi \text{ is } \leqq \frac{\pi}{2}, \text{ by (2725).}$$

If $h-n\pi$ lies between $\frac{1}{2}\pi$ and π, decompose the integral into two others; the one with limits 0 to $\frac{1}{2}\pi$ will converge towards $\frac{1}{2}\pi\phi(n\pi)$, while the other with limits $\frac{1}{2}\pi$ to $h-n\pi$ becomes, by putting $y=\pi-z$,

$$\int_{(n+1)\pi-h}^{\frac{1}{2}\pi}\frac{\sin az}{\sin z}\phi[(n+1)\pi-z]\,dz=0,$$

the limit by (2727). Hence the last term of (iii.) is $\frac{1}{2}\pi\phi(n\pi)$. Substituting these values, (2728) is obtained.

2729 Ex.—By (2614), $\int_0^\infty e^{-a^2x^2}\cos 2bx\,dx=\frac{\sqrt{\pi}}{2a}e^{-\frac{b^2}{a^2}}$.

Put $b=0, 1, 2 \ldots n$ successively, and add, after multiplying the first equation by $\frac{1}{2}$, thus

$$\int_0^\infty e^{-a^2x^2}\left(\tfrac{1}{2}+\cos 2x+\cos 4x+\ldots+\cos 2nx\right)dx$$
$$=\frac{\sqrt{\pi}}{2a}\left\{\tfrac{1}{2}+e^{-\frac{1}{a^2}}+e^{-\frac{4}{a^2}}+\ldots+e^{-\frac{n^2}{a^2}}\right\}.$$

The left side $=\frac{1}{2}\int_0^\infty e^{-a^2x^2}\frac{\sin(2n+1)x}{\sin x}dx$, by (801),

and, if $n=\infty$, becomes

$$\frac{\pi}{2}\{\tfrac{1}{2}+e^{-\pi^2a^2}+e^{-4\pi^2a^2}+e^{-9\pi^2a^2}+\ldots\}, \text{ by (2728);}$$

$$\therefore\ \pi\{\tfrac{1}{2}+e^{-\pi^2a^2}+e^{-4\pi^2a^2}+e^{-9\pi^2a^2}+\ldots\}=\frac{\sqrt{\pi}}{a}\{\tfrac{1}{2}+e^{-\frac{1}{a^2}}+e^{-\frac{4}{a^2}}+e^{-\frac{9}{a^2}}+\ldots\}.$$

Put $\pi a=\alpha$ and $\frac{1}{a}=\beta$; therefore

2730 $$\sqrt{\alpha}\{\tfrac{1}{2}+e^{-\alpha^2}+e^{-4\alpha^2}+e^{-9\alpha^2}+\ldots\}$$
$$=\sqrt{\beta}\{\tfrac{1}{2}+e^{-\beta^2}+e^{-4\beta^2}+e^{-9\beta^2}+\ldots\},$$

with the condition $\alpha\beta=\pi$.

2731 $$\int_0^h\frac{\sin ax}{x}\phi(x)\,dx=\frac{\pi}{2}\phi(0),$$

when a is an infinite integer.

Proof.—The integral may be put in the form

$$\int_0^h\frac{\sin ax}{\sin x}\Phi(x)\,dx, \text{ where } \Phi(x)=\frac{\sin x}{x}\phi(x),$$

therefore, by (2726), when h is $\leqq \frac{1}{2}\pi$, and by (2728), if h is $> \frac{1}{2}\pi$, the value is $\frac{1}{2}\pi\Phi(0)$, since in (2728) $\Phi(\pi)$, $\Phi(2\pi)$, &c. all vanish. But $\Phi(0) = \phi(0)$. Hence the theorem is proved.

When a and β are both positive,

2732 $$\int_a^\beta \frac{\sin ax}{x}\phi(x)\,dx = 0 = \int_{-a}^{-\beta} \frac{\sin ax}{x}\phi(x)\,dx.$$

2733 $$\int_{-a}^\beta \frac{\sin ax}{x}\phi(x)\,dx = \pi\phi(0).$$

PROOF.—(i.) $\int_a^\beta = \int_0^\beta - \int_0^a = \frac{\pi}{2}\phi(0) - \frac{\pi}{2}\phi(0)$, by (2729).

(ii.) $\int_{-a}^\beta = \int_{-a}^0 + \int_0^\beta = \frac{\pi}{2}\phi(0) + \frac{\pi}{2}\phi(0)$,

by substituting $-x$ in the second integral.

2734 $$\int_0^a \int_0^h \phi(x)\cos ux\,du\,dx = \frac{\pi}{2}\phi(0), \quad \text{when } a = \infty.$$

PROOF. $\frac{\sin ax}{x} = \int_0^a \cos ux\,du$. Substitute this in (2731).

When a and β are positive, the limit when a is infinite of

2735 $$\int_0^a \int_a^\beta \phi(x)\cos tu\cos ux\,du\,dx,$$

or of $$\int_0^a \int_a^\beta \phi(x)\sin tu\sin ux\,du\,dx,$$

is $\frac{1}{2}\pi\phi(t)$, if t lies between a and β, $\frac{1}{4}\pi\phi(t)$ if $t = a$, and *zero* for any other value of t.

PROOF.—When $a = \infty$ we have, by (668), and integrating with respect to u,

$$\int_a^\beta \int_0^a \phi(x)\cos ux\cos tu\,dx\,du = \tfrac{1}{2}\int_a^\beta \frac{\sin a(x-t)}{x-t}\phi(x)\,dx + \tfrac{1}{2}\int_a^\beta \frac{\sin a(x+t)}{x+t}\phi(x)\,dx$$

$$= \tfrac{1}{2}\int_{a-t}^{\beta-t} \frac{\sin az}{z}\phi(z+t)\,dz + \tfrac{1}{2}\int_{a+t}^{\beta+t} \frac{\sin az}{z}\phi(z-t)\,dz \quad \text{.........(i.)},$$

by substituting $z = x-t$ and $z = x+t$ in the two integrals respectively.

When a is infinite, the limit of each integral is known.

When a and β are positive and t lies between them in value, the limit of (i.) is $\frac{1}{2}\pi\,\phi(t)$, by (2732–3).. (ii.)

When a and β are positive and t does not lie between them, the value is zero, by (2732).. (iii.)

If $a = t$ in (i.), the first integral becomes $= \frac{1}{2}\pi\phi(t)$ by (2731), and the second vanishes as before; so that the value, in this case, is $\frac{1}{4}\pi\,\phi(t)$...(iv.)

The same demonstration applies in the case of (2736), transforming by (669) instead of (668).

Hence, by (ii.), if t be always positive,

2737 $$\int_0^\infty\int_0^\infty \phi(x)\cos tu\cos ux\,du\,dx = \frac{\pi}{2}\phi(t)$$

$$= \int_0^\infty\int_0^\infty \phi(x)\sin tu\sin ux\,du\,dx.$$

2739 Ex.—Let $\phi(x) = e^{-ax}$,

$$\therefore \int_0^\infty\int_0^\infty e^{-ax}\cos tu\cos ux\,du\,dx = \frac{\pi}{2}e^{-at}.$$

Therefore, by (2584), $$\int_0^\infty \frac{a\cos tu}{a^2+u^2}\,du = \frac{\pi}{2}e^{-at},$$

which is equivalent to (2574), with $t = 1$.

The expressions in (2737–8) being even functions of u, we have, supposing t to be always positive,

$$\int_{-\infty}^\infty\int_0^\infty \phi(x)\cos tu\cos ux\,du\,dx = \pi\phi(t) = \int_{-\infty}^\infty\int_0^\infty \phi(x)\sin tu\sin ux\,du\,dx \quad \text{...(i.)}$$

Replacing $\phi(x)$ by $\phi(-x)$, and afterwards substituting $-x$, these equations become

$$\int_{-\infty}^\infty\int_{-\infty}^0 \phi(x)\cos tu\cos ux\,du\,dx = \pi\phi(-t)$$

$$= -\int_{-\infty}^\infty\int_{-\infty}^0 \phi(x)\sin tu\sin ux\,du\,dx \quad \text{.........(ii.)}$$

From (i.) and (ii.), by addition and subtraction, we get

2740 $$\int_{-\infty}^\infty\int_{-\infty}^\infty \phi(x)\cos tu\cos ux\,du\,dx = \pi[\phi(t)+\phi(-t)],$$

2741 $$\int_{-\infty}^\infty\int_{-\infty}^\infty \phi(x)\sin tu\sin ux\,du\,dx = \pi[\phi(t)-\phi(-t)].$$

Whence, by addition,

2742 $$\int_{-\infty}^\infty\int_{-\infty}^\infty \phi(x)\cos u(t-x)\,du\,dx = 2\pi\phi(t),$$

the original formula of Fourier's.

THE FUNCTION $\psi(x)$.

The function $d_x \log \Gamma(x)$ is denominated $\psi(x)$.

2743 $$\psi(x) = \log\mu - \frac{1}{x} - \frac{1}{x+1} - \frac{1}{x+2} - \dots - \frac{1}{x+\mu},$$

when μ is an indefinitely great integer.

PROOF.—By differentiating the logarithm of (2293).

2744 COR. $$\psi(1) = \log\mu - 1 - \frac{1}{2} - \frac{1}{3} - \dots - \frac{1}{\mu+1},$$
when $\mu = \infty$,
$$= -0{\cdot}577215{,}664901{,}532860{,}60 \dots \textit{(Euler)}.$$

All other values of $\psi(x)$, when x is a commensurable quantity, may be made to depend upon the value of $\psi(1)$.

When x is less than 1,

2745 $$\psi(1-x) - \psi(x) = \pi \cot \pi x.$$

PROOF.—Differentiate the logarithm of the equation
$$\Gamma(x)\,\Gamma(1-x) = \pi \div \sin \pi x \quad (2313).$$

2746 $$\psi(x) + \psi\left(x+\frac{1}{n}\right) + \psi\left(x+\frac{2}{n}\right) + \dots + \psi\left(x+\frac{n-1}{n}\right)$$
$$= n\psi(nx) - n\log n.$$

PROOF.—Differentiate the logarithm of equation (2316).

2747 *To compute the value of* $\psi\left(\frac{\mathrm{p}}{\mathrm{q}}\right)$ *when* $\frac{\mathrm{p}}{\mathrm{q}}$ *is a proper fraction.*

Find $\psi\left(\frac{p}{q}\right)$ from the two equations

2748 $$\psi\left(1-\frac{p}{q}\right) - \psi\left(\frac{p}{q}\right) = \pi \cot \frac{p}{q}\pi, \qquad (2745)$$

2749
$$\psi\left(1-\frac{p}{q}\right) + \psi\left(\frac{p}{q}\right) = 2\left\{\psi(1) - \log q + \cos\frac{2p\pi}{q}\log\left(2\operatorname{vers}\frac{2\pi}{q}\right)\right.$$
$$\left. + \cos\frac{4p\pi}{q}\log\left(2\operatorname{vers}\frac{4\pi}{q}\right) + \cos\frac{6p\pi}{q}\log\left(2\operatorname{vers}\frac{6\pi}{q}\right) + \&c.\right\}.$$

The last term within the brackets, when q is odd, is

$$\cos\frac{(q-1)\,p\pi}{q}\log\left(2\operatorname{vers}\frac{(q-1)\,\pi}{q}\right);$$

and when q is even, the last term is $\pm\log 2$ according as p is even or odd.

PROOF.—Equation (2743) may be written

$$\psi(x) = -\frac{1}{x}+l2-\frac{1}{x+1}+l\tfrac{3}{2}-\frac{1}{x+2}+l\tfrac{4}{3}-\dots-\frac{1}{x+\mu}+l\frac{\mu+2}{\mu+1},$$

μ being an indefinitely great integer.

Replace x successively by $\frac{1}{q}$, $\frac{2}{q}$, $\frac{3}{q}$ ... $\frac{q-1}{q}$, 1; where q is any integer; thus

$$\left.\begin{aligned}
\psi\left(\frac{1}{q}\right) &= -\;q\;+l2-\frac{q}{q+1}+l\tfrac{3}{2}-\frac{q}{2q+1}+l\tfrac{4}{3}-\frac{q}{3q+1}+l\tfrac{5}{4}- \\
\psi\left(\frac{2}{q}\right) &= -\frac{q}{2}+l2-\frac{q}{q+2}+l\tfrac{3}{2}-\frac{q}{2q+2}+l\tfrac{4}{3}-\frac{q}{3q+2}+l\tfrac{5}{4}- \\
\dots & \quad \dots \quad \dots \quad \dots \quad \dots \quad \dots \\
\psi\left(\frac{q-1}{q}\right) &= -\frac{q}{q-1}+l2-\frac{q}{2q-1}+l\tfrac{3}{2}-\frac{q}{3q-1}+l\tfrac{4}{3}-\frac{q}{4q-1}+l\tfrac{5}{4}- \\
\psi(1) &= -\;1\;+l2-\;\tfrac{1}{2}\;+l\tfrac{3}{2}-\;\tfrac{1}{3}\;+l\tfrac{4}{3}-\;\tfrac{1}{4}\;+l\tfrac{5}{4}-
\end{aligned}\right\}\dots\dots\text{(i.)}$$

Now, if ϕ be any one of the angles $\frac{2\pi}{q}$, $\frac{4\pi}{q}$, $\frac{6\pi}{q}$ ... $\frac{2(q-1)\pi}{q}$, we shall have

$$1 = \cos q\phi = \cos 2q\phi = \cos 3q\phi = \&\text{c.}\dots\dots\dots\dots\text{(ii.)},$$

$$\cos\phi = \cos(q+1)\phi = \cos(2q+1)\phi = \cos(3q+1)\phi = \&\text{c.}\dots\dots\text{(iii.)},$$

$$\cos\phi+\cos 2\phi+\cos 3\phi+\dots+\cos(q-1)\phi+1 = 0 \text{ by } (803)\dots\dots\text{(iv.)}$$

By means of the relations (ii.) and (iii.), equations (i.) may be written

$$\cos\phi\,\psi\left(\frac{1}{q}\right) = -\;q\;\cos\phi+\cos\phi\,l2-\frac{q}{q+1}\cos(q+1)\phi+\cos\phi\,l\tfrac{3}{2}-,$$

$$\cos 2\phi\,\psi\left(\frac{2}{q}\right) = -\frac{q}{2}\cos 2\phi+\cos 2\phi\,l2-\frac{q}{q+2}\cos(q+2)\phi+\cos 2\phi\,l\tfrac{3}{2}-,$$

$$\cos 3\phi\,\psi\left(\frac{3}{q}\right) = -\frac{q}{3}\cos 3\phi+\cos 3\phi\,l2-\frac{q}{q+3}\cos(q+3)\phi+\cos 3\phi\,l\tfrac{3}{2}-,$$

... ...

$$\cos(q-1)\phi\,\psi\left(\frac{q-1}{q}\right) = -\frac{q}{q-1}\cos(q-1)\phi+\cos(q-1)\phi\,l2$$

$$-\frac{q}{2q-1}\cos(2q-1)\phi+\cos(q-1)\phi\,l\tfrac{3}{2}-,$$

$$\psi(1) = -1 \quad + \quad l2 \quad - \quad \tfrac{1}{2} \quad + \quad l\tfrac{3}{2} \quad -.$$

Upon adding the equations, the coefficient of each logarithm vanishes, by (iv.) The remaining terms on the right form a continuous series, and we have

$$\cos\phi\,\psi\left(\frac{1}{q}\right)+\cos 2\phi\,\psi\left(\frac{2}{q}\right)+\dots+\cos(q-1)\phi\,\psi\left(\frac{q-1}{q}\right)+\psi(1)$$

$$= -q\,\{\cos\phi+\tfrac{1}{2}\cos 2\phi+\tfrac{1}{3}\cos 3\phi+\text{in inf.}\}$$

$$= \tfrac{1}{2}q\log(2-2\cos\phi) \text{ by } (2928)\dots\dots\dots\dots\text{(v.)}$$

Let $\frac{2\pi}{q} = \omega$. Then, by giving to ϕ in equation (v.) its different values ω, 2ω, 3ω ... $(q-1)\,\omega$, we obtain $q-1$ linear equations in the unknown quantities $\psi\left(\frac{1}{q}\right)$, $\psi\left(\frac{2}{q}\right) \dots \psi\left(\frac{q-1}{q}\right)$. To solve these equations for $\psi\left(\frac{p}{q}\right)$ p being an integer less than q, multiply them respectively by

$$\cos p\omega,\ \cos 2p\omega \dots \cos (q-1)\,p\omega,$$

and join to their sum equation (2746), after putting $x = \frac{1}{q}$ and $n = q$.

The coefficient of $\psi\left(\frac{k}{q}\right)$ in the result, k being any integer less than q, is

$$\cos p\omega \cos k\omega + \cos 2p\omega \cos 2k\omega + \dots + \cos (q-1)\,p\omega \cos (q-1)\,k\omega + 1.$$

By expanding each term by (668), we see by (iv.) that this coefficient vanishes excepting for the values $k = q-p$ and $k = p$, in each of which cases it becomes $= \frac{1}{2}q$. Hence, dividing by $\frac{1}{2}q$, we obtain

$$\psi\left(\frac{q-p}{q}\right) + \psi\left(\frac{p}{q}\right) = 2\psi(1) - 2lq + \cos p\omega\, l\,(2-2\cos\omega)$$
$$+ \cos 2p\omega\, l\,(2-2\cos 2\omega) + \dots\dots + \cos (q-1)\,p\omega\, l\,\{2-2\cos(q-1)\,\omega\}.$$

The last term $= \cos p\omega\, l\,(2-2\cos\omega) =$ the third term; the last but one $= \cos 2p\omega\, l\,(2-2\cos 2\omega) =$ the second term, and so on, forming pairs of equal terms. But, if q be even, there is the odd term

$$\cos \tfrac{1}{2}qp\omega \log (2-2\cos \tfrac{1}{2}q\omega) = \pm\, 2\log 2,$$

according as p is even or odd.

EXAMPLES.—By (2748–9) we obtain

2750 $\psi\left(\frac{3}{4}\right) = \psi(1) - 3\log 2 + \frac{\pi}{2}$, $\quad \psi\left(\frac{1}{4}\right) = \psi(1) - 3\log 2 - \frac{\pi}{2}$,

2752 $\psi\left(\frac{2}{3}\right) = \psi(1) - \frac{3}{2}\log 3 + \frac{\pi}{2\sqrt{3}}$, $\quad \psi\left(\frac{1}{3}\right) = \psi(1) - \frac{3}{2}\log 3 - \frac{\pi}{2\sqrt{3}}$,

2754 $\psi\left(\frac{1}{2}\right) = \psi(1) - 2\log(2)$.

DEVELOPMENTS OF $\psi(a+x)$.

When x is any integer,

2755 $$\psi(a+x) = \psi(a) + \frac{1}{a} + \frac{1}{a+1} + \frac{1}{a+2} + \dots + \frac{1}{a+x-1}.$$

PROOF.—By (2289), putting $n = a+x-1$ and $r = x-1$,

$$\Gamma(a+x) = (a+x-1)(a+x-2)\dots(a+2)(a+1)\,a\,\Gamma(a).$$

Differentiate the logarithm of this equation with respect to a.

2756 $$\psi(a+x) = \psi(a) + \frac{x}{a} - \frac{x(x-1)}{2a(a+1)} + \frac{x(x-1)(x-2)}{3a(a+1)(a+2)}$$
$$- \frac{x(x-1)(x-2)(x-3)}{4a(a+1)(a+2)(a+3)} + \&c.$$

If x be a positive integer, the number of terms in this series is finite, and the value of $\psi(a+x)$ can be found from that of $\psi(a)$.

Hence, by this or the preceding formula, in conjunction with (2747), the value of $\psi(N)$, when N is any commensurable quantity, may be found in terms of $\psi(1)$.

PROOF.—Let $\psi(a+x) = A+Bx+Cx(x-1)+Dx(x-1)(x-2)+\&c.$ Change x into $x+1$; then,

$$\Delta\psi(a+x) = \psi(a+x+1)-\psi(a+x) = d_x\{\log\Gamma(a+x+1)-\log\Gamma(a+x)\}$$
$$= d_x\log(a+x)\ (2288) = \frac{1}{a+x},$$

$\Delta x = 1$, $\Delta x(x-1) = 2x$, $\Delta x(x-1)(x-2) = 3x(x-1)$, &c. Therefore

$$\frac{1}{a+x} = B+2Cx+3Dx(x-1)+4Ex(x-1)(x-2)+,$$
$$\Delta\frac{1}{a+x} = 2C+2.3Dx+3.4Ex(x-1)+,$$
$$\Delta^2\frac{1}{a+x} = 2.3D+2.3.4.Ex+,$$
$$\Delta^3\frac{1}{a+x} = 2.3.4E+.$$

Put $x=0$ in each equation to determine the coefficients A, B, C, D, &c.; thus $A = \psi(a)$, $B = \frac{1}{a}$, $2C = \Delta\frac{1}{a} = \frac{1}{a+1}-\frac{1}{a} = -\frac{1}{a(a+1)}$,

$$2.3D = \Delta^2\frac{1}{a} = \Delta\frac{-1}{a(a+1)} = \frac{2}{a(a+1)(a+2)},$$

$$2.3.4E = \Delta^3\frac{1}{a} = \Delta\frac{2}{a(a+1)(a+2)} = -\frac{2.3}{a(a+1)(a+2)(a+3)},$$ and so on.

SUMMATION OF SERIES BY THE FUNCTION $\psi(x)$.

2757 *Formula I.* $$\frac{a}{b}+\frac{a}{b+c}+\frac{a}{b+2c}+\ldots+\frac{a}{b+nc}$$
$$= \frac{a}{b}-\frac{a}{c}\psi\left(\frac{b}{c}+1\right)+\frac{a}{c}\psi\left(\frac{b}{c}+n+1\right).$$

PROOF.—Let S_n denote the n terms of the series to be summed. We have

$$S_{n+1}-S_n = \frac{a}{c}\div\left(\frac{b}{c}+n+1\right) = \frac{a}{c}\left[\psi\left(\frac{b}{c}+n+2\right)-\psi\left(\frac{b}{c}+n+1\right)\right]\ (2288)$$

or $$S_{n+1}-\frac{a}{c}\psi\left(\frac{b}{c}+n+2\right) = S_n-\frac{a}{c}\psi\left(\frac{b}{c}+n+1\right).$$

Hence the difference is independent of n, and therefore

$$S_n-\frac{a}{c}\psi\left(\frac{b}{c}+n+1\right) = S_0-\frac{a}{c}\psi\left(\frac{b}{c}+1\right) = \frac{a}{b}-\frac{a}{c}\psi\left(\frac{b}{c}+1\right).$$

2758 Ex. $1+\frac{1}{3}+\frac{1}{5}+\ldots+\frac{1}{2n+1}=1-\frac{1}{2}\psi\left(\frac{3}{2}\right)+\frac{1}{2}\psi\left(n+\frac{3}{2}\right).$

2759 *Formula II.* $\frac{a}{b}-\frac{a}{b+c}+\frac{a}{b+2c}-\ldots-\frac{a}{b+(2n+1)c}$

$$=\frac{ac}{b(b+c)}+\frac{a}{2c}\left\{\psi\left(\frac{b+3c}{2c}\right)-\psi\left(\frac{b+2c}{2c}\right)+\psi\left(\frac{b+2c}{2c}+n\right)-\psi\left(\frac{b+3c}{2c}+n\right)\right\}.$$

PROOF.—The series is equivalent to

$$\frac{a}{b}+\frac{a}{b+2c}+\frac{a}{b+4c}+\ldots+\frac{a}{b+2nc}-\left\{\frac{a}{b+c}+\frac{a}{b+c+2c}\ldots+\frac{a}{b+c+2nc}\right\},$$

and the result follows by Formula I.

2760 *Formula III.*—

$$\frac{1}{b}-\frac{1}{b+c}+\frac{1}{b+2c}-\ldots\text{ in inf.}$$

$$=\frac{c}{b(b+c)}+\frac{1}{2c}\left\{\psi\left(\frac{b+3c}{2c}\right)-\psi\left(\frac{b+2c}{2c}\right)\right\}.$$

PROOF.—Make $n=\infty$ in Formula II. The last two terms become equal.

2761 Ex. 1.—In (2760) let $b=c=1$, then

$$1-\tfrac{1}{2}+\tfrac{1}{3}-\tfrac{1}{4}+\&c.=\tfrac{1}{2}+\tfrac{1}{2}\left\{\psi(2)-\psi\left(\tfrac{3}{2}\right)\right\}=\log 2.$$

For $\psi(2)=1+\psi(1)$, by (2755); $\psi\left(\frac{3}{2}\right)=2+\psi(1)-2\log 2$, by (2754–5).

2762 Ex. 2.—In (2760) let $b=1$, $c=2$, then

$$1-\tfrac{1}{3}+\tfrac{1}{5}-\tfrac{1}{7}+\&c.=\tfrac{2}{3}+\tfrac{1}{4}\psi\left(\tfrac{7}{4}\right)-\tfrac{1}{4}\psi\left(\tfrac{5}{4}\right)=\frac{\pi}{4}.$$

2763 $$\psi(1+a)=\int_0^1\frac{x^a-1}{x-1}\,dx+\psi(1).$$

PROOF. $\psi(1+a)=\psi(1)+a-\frac{a(a-1)}{2.1.2}+\frac{a(a-1)(a-2)}{3.1.2.3}-\&c.$ [by (2756)

But $\frac{1-(1-x)^a}{x}=a-\frac{a(a-1)}{1.2}x+\frac{a(a-1)(a-2)}{1.2.3}x^2-\&c.,$

therefore $\psi(1+a)=\int_0^1\frac{1-(1-x)^a}{x}\,dx+\psi(1).$

Substitute $1-x$ in the integral.

2764 $$\psi(1+a)-\psi(1+b)=\int_0^1\frac{x^a-x^b}{x-1}\,dx.$$ [By (2763)

Ex.—Put $b=-a$; then

$$\psi(1+a)-\psi(1-a)=\frac{1}{a}+\psi(a)-\psi(1-a)\ (2756)=\frac{1}{a}-\pi\cot\pi a\ (2745).$$

2765 Therefore $$\int_0^1\frac{x^a-x^{-a}}{x-1}\,dx=\frac{1}{a}-\pi\cot\pi a.$$

$\psi(x)$ AS A DEFINITE INTEGRAL INDEPENDENT OF $\psi(1)$.

2766 $$\psi(x)=-\int_0^1\left(\frac{1}{\log u}+\frac{u^{x-1}}{1-u}\right)du.$$

PROOF. $\psi(x)=\log\mu-\frac{1}{x}-\frac{1}{x+1}-\ldots-\frac{1}{x+\mu-1}$ with $\mu=\infty$ (2743).

But $$\frac{1}{x}+\frac{1}{x+1}+\ldots+\frac{1}{x+\mu-1}=\int_0^1\frac{z^{x-1}-z^{x+\mu-1}}{1-z}\,dz,$$
by actual division and integration.

Also $$\log\mu=\int_0^1\left(\frac{1}{1-z}-\frac{\mu z^{\mu-1}}{1-z^\mu}\right)dz\quad(2367).$$

$$\therefore\ \psi(x)=\int_0^1\frac{1-z^{x-1}}{1-z}\,dz+\int_0^1\left\{\frac{z^{x+\mu-1}}{1-z}-\frac{\mu z^{\mu-1}}{1-z^\mu}\right\}dz,\quad \mu=\infty\ \ \ldots\ldots\text{(i.)}$$

Put $z=y^\mu$ in the first integral, therefore

$$\int_0^1\mu\,\frac{1-y^{\mu x-\mu}}{1-y^\mu}\,y^{\mu-1}dy=\mu\int_0^1\frac{y^{\mu-1}-y^{\mu x-1}}{1-y^\mu}\,dy.$$

Replace y by z, and suppress the term common with the second integral of (i.), and we get $\psi(x)=\int_0^1\left\{\frac{z^{x+\mu-1}}{1-z}-\frac{\mu z^{\mu x-1}}{1-z^\mu}\right\}dz.$

Put $z^\mu=u$, and this becomes

$$\psi(x)=\int_0^1\left\{\frac{u^{\frac{x}{\mu}}}{\mu\left(1-u^{\frac{1}{\mu}}\right)}-\frac{u^{x-1}}{1-u}\right\}du.$$

But when $\mu=\infty$ the product $\mu\left(1-u^{\frac{1}{\mu}}\right)$ has $-\log u$ for its limit (1584); and $u^{\frac{x}{\mu}}=1$. Hence the result.

2767 $$\psi(x)=\int_0^\infty\left\{e^{-a}-\frac{1}{(1+a)^x}\right\}\frac{da}{a}.$$

PROOF. $\Gamma(x)=\int_0^\infty e^{-z}z^{x-1}dz$; $\quad d_x\Gamma(x)=\int_0^\infty e^{-z}z^{x-1}\log z\,dz.$

But, by (2427), $$\log z=\int_0^\infty\frac{e^{-a}-e^{-az}}{a}\,da,$$

$$\therefore\ d_x\Gamma(x) = \int_0^\infty\int_0^\infty e^{-z}z^{x-1}\frac{e^{-a}-e^{-az}}{a}\,dz\,da$$

$$= \int_0^\infty\left[e^{-a}\int_0^\infty e^{-z}z^{x-1}\,dz - \int_0^\infty e^{-(1+a)z}z^{x-1}\,dz\right]\frac{da}{a}$$

$$= \Gamma(x)\int_0^\infty\left[e^{-a}-\frac{1}{(1+a)^x}\right]\frac{da}{a}\quad(2291),$$

which establishes the formula since

$$d_x\Gamma(x)\div\Gamma(x) = d_x\log\Gamma(x) = \psi(x).$$

2768 $$\log\Gamma(x) = \int_0^\infty\left[(x-1)\,e^{-\xi}-\frac{e^{-\xi}-e^{-\xi x}}{1-e^{-\xi}}\right]\frac{d\xi}{\xi}$$

2769 $$= \int_0^1\left[\frac{1-z^{x-1}}{1-z}-x+1\right]\frac{dz}{z}.$$

2770 $$\psi(x) = \int_0^\infty\left[\frac{e^{-\xi}}{\xi}-\frac{e^{-\xi x}}{1-e^{-\xi}}\right]d\xi.$$

Proof.—Integrate (2767) for x between the limits 1 and x, observing that $\log\Gamma(1) = 0$; thus

$$\log\Gamma(x) = \int_0^\infty\left\{(x-1)\,e^{-a}-\frac{(1+a)^{-1}-(1+a)^{-x}}{\log(1+a)}\right\}\frac{da}{a}.$$

Subtract from this the equation obtained from it by making $x = 2$, and multiplying the result by $x-1$. We thus obtain

$$\log\Gamma(x) = \int_0^\infty\left[(x-1)(1+a)^{-2}-\frac{(1+a)^{-1}-(1+a)^{-x}}{a}\right]\frac{da}{\log(1+a)}.$$

Substitute $\xi = \log(1+a)$, and (2768) is the result. To obtain (2769), substitute $z = (1+a)^{-1}$. Lastly, (2770) is the result of differentiating (2768) for x.

NUMERICAL CALCULATION OF $\log\Gamma(x)$.

2771 The second member of (2768) can be divided into two parts, one of which appears under a finite form, and the other vanishes with x. If we put

$$P = \left(x-1-\frac{1}{1-e^{-\xi}}\right)\frac{e^{-\xi}}{\xi},\ \text{ and }\ Q = \frac{1}{\xi(1-e^{-\xi})},$$

then $$\log\Gamma(x) = \int_0^\infty(P+Qe^{-\xi x})\,d\xi\ \ldots\ldots\ldots\ldots\ldots\ldots\ldots\ \text{(i.)}$$

If Q be developed in ascending powers of ξ, the terms which contain negative indices are $\frac{1}{\xi^2}+\frac{1}{2\xi} = R$ say.

Put $F(x) = \int_0^\infty (P+Re^{-\xi x})\,d\xi$

$$= \int_0^\infty \left[\left((x-1)-\frac{1}{1-e^{-\xi}}\right)e^{-\xi}+\left(\frac{1}{\xi}+\frac{1}{2}\right)e^{-\xi x}\right]\frac{d\xi}{\xi} \quad \text{.........(ii.)},$$

and $$\varpi(x) = \int_0^\infty (Q-R)\,e^{-\xi x}\,d\xi = \int_0^\infty \left(\frac{1}{1-e^{-\xi}}-\frac{1}{\xi}-\frac{1}{2}\right)e^{-\xi x}\,\frac{d\xi}{\xi} \quad \text{... (iii.)}$$

Then, by (i.), $$\log\Gamma(x) = F(x)+\varpi(x) \quad \text{......................... (iv.)}$$

$F(x)$ can now be calculated in a finite form, and $\varpi(x)$ will have zero for its limit as x increases.

First, to show that $F(\frac{1}{2})$ and $\varpi(\frac{1}{2})$ can be exactly calculated.

$$\varpi(\tfrac{1}{2}) = \int_0^\infty \left(\frac{1}{1-e^{-\xi}}-\frac{1}{\xi}-\frac{1}{2}\right)e^{-\frac{1}{2}\xi}\,\frac{d\xi}{\xi},$$

and, by substituting $\frac{1}{2}\xi$,

$$\varpi(\tfrac{1}{2}) = \int_0^\infty \left(\frac{1}{1-e^{-2\xi}}-\frac{1}{2\xi}-\frac{1}{2}\right)e^{-\xi}\,\frac{d\xi}{\xi} \quad \text{..................(v.)}$$

Again, putting $x=1$ in (iii.), we have

$$\varpi(1) = \int_0^\infty \left(\frac{1}{1-e^{-\xi}}-\frac{1}{\xi}-\frac{1}{2}\right)e^{-\xi}\,\frac{d\xi}{\xi} \quad \text{..................(vi.)};$$

and, by substituting $\frac{1}{2}\xi$,

$$\varpi(1) = \int_0^\infty \left(\frac{1}{1-e^{-2\xi}}-\frac{1}{2\xi}-\frac{1}{2}\right)e^{-2\xi}\,\frac{d\xi}{\xi} \quad \text{..................(vii.)}$$

The difference of (vi.) and (vii.) gives

$$0 = \int_0^\infty \left(\frac{1}{1-e^{-2\xi}}-\frac{2-e^{-\xi}}{2\xi}-\frac{1-e^{-\xi}}{2}\right)e^{-\xi}\,\frac{d\xi}{\xi} \quad \text{...........(viii.)},$$

since $$\frac{e^{-\xi}}{1-e^{-\xi}}-\frac{e^{-2\xi}}{1-e^{-2\xi}} = \frac{e^{-\xi}}{1-e^{-2\xi}}.$$

Subtract (viii.) from (v.), thus

$$\varpi(\tfrac{1}{2}) = \tfrac{1}{2}\int_0^\infty \left(\frac{e^{-\xi}-e^{-2\xi}}{\xi^2}-\frac{e^{-2\xi}}{\xi}\right)d\xi = \frac{1}{2}-\frac{\log 2}{2} \quad (2429).$$

Also, by (iv.), $F(\frac{1}{2})+\varpi(\frac{1}{2}) = l\Gamma(\frac{1}{2}) = \frac{1}{2}l\pi$, $\therefore\ F(\frac{1}{2}) = \frac{1}{2}\log(2\pi)-\frac{1}{2}$...(ix.)

$F(x)$ may now be found by calculating $F(x)-F(\frac{1}{2})$ as follows:—

By (ii.), $$F(x)-F(\tfrac{1}{2}) = \int_0^\infty \left[(x-\tfrac{1}{2})\,e^{-\xi}+\left(\frac{1}{\xi}+\frac{1}{2}\right)(e^{-\xi x}-e^{-\frac{1}{2}\xi})\right]\frac{d\xi}{\xi}$$

$$= \int_0^\infty \frac{e^{-\xi x}-e^{-\frac{1}{2}\xi}+\xi\,(x-\frac{1}{2})\,e^{-\xi}}{\xi^2}\,d\xi+\frac{1}{2}\int_0^\infty \frac{e^{-\xi x}-e^{-\frac{1}{2}\xi}}{\xi}\,d\xi$$

$$= \tfrac{1}{2}-x+(x-\tfrac{1}{2})\log x \quad (2427\text{–}8),$$

$$\therefore\ F(x) = \tfrac{1}{2}\log(2\pi)+(x-\tfrac{1}{2})\log x-x, \quad \text{by (ix.)};$$

$\therefore$ by (iv.) $$\log\Gamma(x) = \tfrac{1}{2}\log(2\pi)+(x-\tfrac{1}{2})\log x-x+\varpi(x) \quad \text{..............(x.)};$$

2772 $$\therefore\ \Gamma(x) = e^{-x}x^{x-\frac{1}{2}}\sqrt{(2\pi)}\,e^{\varpi(x)} \quad \text{........................(xi.)}$$

When x is very large, $e^{\varpi(x)}$ differs but little from unity. For $\varpi(x)$ diminishes without limit as x increases, by the value (iii.)

Replacing $\varpi(x)$ in (x.) by its value (iii.), and observing that

$$\log\Gamma(x+1) = \log x + \log\Gamma(x),$$

we get $$\log\Gamma(x+1) = \tfrac{1}{2}\log(2\pi) + (x+\tfrac{1}{2})\log x - x + \int_0^\infty \left(\frac{1}{1-e^{-\xi}} - \frac{1}{\xi} - \frac{1}{2}\right) e^{-\xi x}\frac{d\xi}{\xi} \quad \text{(xii.)}$$

Now, by (1539),

$$\left(\frac{1}{1-e^{-\xi}} - \frac{1}{\xi} - \frac{1}{2}\right)\frac{1}{\xi} = \frac{B_2}{1.2} - \frac{B_4\xi^2}{1.2.3.4} + \ldots \pm \frac{B_{2n}\xi^{2n-2}}{1\ldots 2n} \mp \frac{\theta B_{2n+2}\xi^{2n}}{1\ldots 2n+2},$$

where θ is <1. Also

$$\int_0^\infty e^{-\xi x}\xi^{2\mu}d\xi = \frac{1\ldots 2\mu}{x^{2\mu+1}}, \qquad \int_0^\infty \theta e^{-\xi x}\xi^{2n}d\xi = \theta_1\frac{1\ldots 2n}{x^{2n+1}}.$$

So that equation (xii.) produces

2773 $$\log\Gamma(x+1) = \frac{\log(2\pi)}{2} + \left(x+\frac{1}{2}\right)\log x - x + \frac{B_2}{1.2x} - \frac{B_4}{3.4x^3} + \ldots \mp \frac{\theta B_{2n+2}}{(2n+1)(2n+2)\,x^{2n+1}}.$$

This series is divergent, the terms increasing indefinitely. The complementary term, which increases with n and is very great when n is very great, is, however, *very small for considerable values of* n. For instance, when $x=10$, the values obtained for $\log\Gamma(11)$, by taking 3, 4, 5, or 6 terms of the series, are respectively,

16·090820096, 16 104415343, 16·104412565, 16·104112563.

CHANGE OF THE VARIABLES IN A DEFINITE MULTIPLE INTEGRAL.

2774 Let x, y, z be connected with ξ, η, ζ by three equations $u=0,\ v=0,\ w=0$.

Then, when the limits of the integral containing the new variables can be assigned independently, we have

$$\int_{x_1}^{x_2}\int_{y_1}^{y_2}\int_{z_1}^{z_2} F(x,y,z)\,dx\,dy\,dz = -\int_{\xi_1}^{\xi_2}\int_{\eta_1}^{\eta_2}\int_{\zeta_1}^{\zeta_2}\Phi(\xi,\eta,\zeta)\frac{\dfrac{d(uvw)}{d(\xi\eta\zeta)}}{\dfrac{d(uvw)}{d(xyz)}}\,d\xi\,d\eta\,d\zeta,$$

where Φ is what F becomes when the values of x, y, z, in terms of ξ, η, ζ, obtained by solving the equations u, v, w, are substituted.

PROOF. $$\iiint F(x, y, z)\, dx\, dy\, dz = \iiint \Phi(\xi, \eta, \zeta) \frac{dx}{d\xi} \frac{dy}{d\eta} \frac{dz}{d\zeta}\, d\xi\, d\eta\, d\zeta.$$

To find x_ξ, consider η and ζ constant, and differentiate the three equations u, v, w for ξ, as in (1723). To find y_η, consider ζ and x constant, and differentiate for η. To find z_ζ, consider x and y constant, and differentiate for ζ. We thus obtain

$$\frac{dx}{d\xi} \frac{dy}{d\eta} \frac{dz}{d\zeta} = -\frac{\dfrac{d(uvw)}{d(\xi yz)} \dfrac{d(uvw)}{d(\eta z\xi)} \dfrac{d(uvw)}{d(\zeta\xi\eta)}}{\dfrac{d(uvw)}{d(xyz)} \dfrac{d(uvw)}{d(yz\xi)} \dfrac{d(uvw)}{d(z\xi\eta)}} = -\frac{\dfrac{d(uvw)}{d(\xi\eta\zeta)}}{\dfrac{d(uvw)}{d(xyz)}},$$

observing that two interchanges of columns in a determinant do not alter its value or sign (559).

Similarly in the case of any number of independent variables.

When, however, the limits in the transformed integral have to be discovered from the given equations, the process is not so simple.

In the first place, we shall show how to change the *order* of integration merely.

2775 Taking a double integral in its most general form, we shall have

$$\int_a^b \int_{\psi(x)}^{\phi(x)} F(x, y)\, dx\, dy = \Sigma \int_a^\beta \int_{\Psi(y)}^{\Phi(y)} F(x, y)\, dy\, dx \quad \text{(i.)}$$

The right member will generally consist of more than one integral, and Σ denotes their sum. The limits of the integration for x may be, one or both, constants, or, one or both, functions of y. Ψ is the inverse of the function ψ, and is obtained by solving the equation $y = \psi(x)$, so that $x = \Psi(y)$. Similarly with regard to ϕ and Φ.

An examination of the solid figure described in (1907), whose volume this integral represents, will make the matter clearer. The integration, the order of which has to be changed, extends over an area which is the projection of the solid upon the plane of xy, and which is bounded by the two straight lines $x = a$, $x = b$, and the two curves $y = \psi(x)$, $y = \phi(x)$.

The summation of the elements $PQqp$ extends from a to b, and includes in the one integral on the left of equation (i.) the whole of the solid in question.

But, on the right, the different integrals represent the summation of elements like $PQqp$, but all parallel to OX, between planes $y = \alpha$, $y = \beta$, &c. drawn through points where the limits of x change their character on account of the boundaries $y = \psi(x)$, $y = \varphi(x)$ not being straight lines parallel to OX.

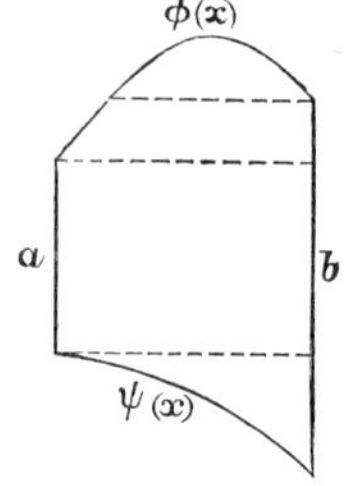

2776 Example.—Let the figure represent the projected area on the xy plane, bounded by the curves $y = \psi(x)$, $y = \phi(x)$, and the straight lines $x = a$, $x = b$. Let $y = \phi(x)$ have a maximum value when $x = c$. The values of y at this point will be $\phi(c)$, and at the points where the straight lines meet the curves the values will be $\phi(a)$, $\phi(b)$, $\psi(a)$, $\psi(b)$.

According to the drawing, the right member of equation (i.) will now stand as follows, U being written for $F(x, y)$,

$$\int_{\psi(b)}^{\psi(a)}\int_{\Psi(y)}^{b} U\,dy\,dx + \int_{\psi(a)}^{\phi(a)}\int_{a}^{b} U\,dy\,dx + \int_{\phi(a)}^{\phi(b)}\int_{\Phi(y)}^{b} U\,dy\,dx + \int_{\phi(b)}^{\phi(c)}\int_{\Phi_1(y)}^{\Phi_2(y)} U\,dy\,dx.$$

The four integrals represent the four areas into which the whole is divided by the dotted lines drawn parallel to the X axis. In the last integral, $\Phi_1(y)$ and $\Phi_2(y)$ are the two values of x corresponding to one of y in that part of the curve $y = \phi(x)$ which is cut twice by any x coordinate.

2777 To change the order of integration in a triple integral, from z, y, x to y, x, z, we shall have an equation of the form

$$\int_{x_1}^{x_2}\int_{\psi_1(x)}^{\psi_2(x)}\int_{\phi_1(x,y)}^{\phi_2(x,y)} F(x, y, z)\,dx\,dy\,dz = \Sigma\int_{z_1}^{z_2}\int_{\Psi_1(z)}^{\Psi_2(z)}\int_{\Phi_1(z,x)}^{\Phi_2(z,x)} F(x, y, z)\,dz\,dx\,dy \quad \text{.........(ii.)}$$

Here the most general form for the integrals whose sum is indicated by Σ is that in which the limits of y are functions of z and x, the limits of x functions of z, and the limits of z constant. Referring to the figure in (1906), the total value of the integral is equivalent to the following. Every element $dx\,dy\,dz$ of the solid described in (1907) is multiplied by $F(xyz)$, a function of the coordinates of the element, and the sum of the products is taken.

This process is indicated by one triple integral on the left of equation (ii.); the limits of the integration for z being two unrestricted curved surfaces $z = \phi_1(x, y)$, $z = \phi_2(x, y)$; the limits for y, two cylindrical surfaces $y = \psi_1(x)$, $y = \psi_2(x)$; and the limits for x, two planes $x = x_1$, $x = x_2$.

But, with the changed order of integration, several integrals may be required. The most general form which any of them can take is that shown on the right of equation (ii.) Solving the equation $z = \phi_1(x, y)$, let $y_1 = \Phi_1(z, x)$, $y_2 = \Phi_2(z, x)$ be two resulting values of y; then the integration for y may be effected between these limits over all parts of the solid where the surface $z = \phi_1(x, y)$ is cut twice by the same y coordinate.

The next integration is with respect to x, and is limited by the cylindrical surface, whose generating lines, parallel to OY, touch the surface $z = \phi_1(x, y)$. At the points of contact, x will have a maximum or minimum value for each value of z; therefore $d_y\phi_1(x, y) = 0$. Eliminating y between this equation and that of the surface, we get $x = \Psi_1(z)$, $x = \Psi_2(z)$ for the limits of x.

Lastly, the result of the previous summations is integrated for z between two parallel planes $z = z_1$, $z = z_2$, drawn so as to include all that portion of the solid over which the limits for x and y, already determined, remain the same.

The remaining integrations will take place between $z = z_2$ and similar successive parallel planes; and, according to the portion of the solid which any two of these planes intercept, the limits of x for that integral will be one or other of the bounding surfaces, curved or plane, the limits of y, one or other of the curved surfaces.

The general problem to change the variables in a multiple integral, and determine the limits from the given equations, may now be solved.

2778 First, in the case of a double integral,

$$\int_{x_1}^{x_2}\int_{\mu_1(x)}^{\mu_2(x)} F(x, y)\, dx\, dy \quad \ldots\ldots\ldots\ldots\ldots\ldots \text{(iii).},$$

to change from x, y to ξ, η, having given the equations $u=0$, $v=0$, involving the four variables.

To change y for η, eliminate ξ between these equations; thus $y=f(x, \eta)$ and $dy = f_\eta(x, \eta)\, d\eta$. Substituting these values, we shall have

$$F(x, y)\, dy = F\{x, f(x, \eta)\} f_\eta(x, \eta)\, d\eta = F_1(x, \eta)\, d\eta.$$

Also, if η_1 corresponds to y_1, the equations $y_1 = \mu_1(x)$ and $y_1 = f(x, \eta_1)$ will give $\eta_1 = \psi_1(x)$. Similarly $\eta_2 = \psi_2(x)$.

Hence the integral (iii.) may now be written

$$\int_{x_1}^{x_2}\int_{\psi_1(x)}^{\psi_2(x)} F_1(x, \eta)\, dx\, d\eta = \Sigma \int_{\eta_1}^{\eta_2}\int_{\Psi_1(\eta)}^{\Psi_2(\eta)} F_1(x, \eta)\, d\eta\, dx \quad \ldots\ldots\ldots \text{(iv.)},$$

the form on the right being obtained *by changing the order of integration*, as explained in (2775).

Next, to change x for ξ, eliminate y between the equations $u=0$, $v=0$; thus, $x = g(\xi, \eta)$ and $dx = g_\xi(\xi, \eta)\, d\xi$. Substituting as before, we shall have

$$F_1(x, \eta)\, dx = F_2(\xi, \eta)\, d\xi.$$

Also, ξ_1 corresponding to x_1, the equations $x_1 = \Psi_1(\eta)$ and $x_1 = g(\xi_1, \eta)$ produce $\xi_1 = m_1(\eta)$, and in the same way $\xi_2 = m_2(\eta)$. Hence, finally,

$$\int_{x_1}^{x_2}\int_{\mu_1(x)}^{\mu_2(x)} F(x, y)\, dx\, dy = \Sigma \int_{\eta_1}^{\eta_2}\int_{m_1(\eta)}^{m_2(\eta)} F_2(\xi, \eta)\, d\eta\, d\xi \quad \ldots\ldots\ldots\ldots \text{(v.)}$$

In the last transformation from x to ξ, the most general form of the integrals which may be included under Σ has been chosen. When any of the limits of x are constants, the process is simplified.

2779 Again, to change the variables from x, y, z to ξ, η, ζ, in the triple integral,

$$\int_{x_1}^{x_2}\int_{\psi_1(x)}^{\psi_2(x)}\int_{\chi_1(x,y)}^{\chi_2(x,y)} F(x, y, z)\, dx\, dy\, dz \ldots\ldots\ldots\ldots\ldots \text{(vi.)},$$

having given the equations $u=0$, $v=0$, $w=0$ between the six variables $x, y, z, \xi, \eta, \zeta$.

First, to change from z to ζ, eliminate ξ and η between the three equations, and let the resulting equation be $z=f(x,y,\zeta)$. From this $dz=f_\zeta(x,y,\zeta)\,d\zeta$; therefore

$$F(x,y,z)\,dz = F\{x,y,f(x,y,\zeta)\}\,f_\zeta(x,y,\zeta)\,d\zeta = F_1(x,y,\zeta)\,d\zeta.$$

Also, if ζ_1 corresponds to the limit z_1, the equations $z_1=\chi_1(x,y)$ and $z_1=f(x,y,\zeta_1)$ give $\zeta_1=\phi_1(x,y)$. Similarly $\zeta_2=\phi_2(x,y)$.

The integral (vi.) may therefore be written

$$\int_{x_1}^{x_2}\int_{\psi_1(x)}^{\psi_2(x)}\int_{\phi_1(x,y)}^{\phi_2(x,y)} F_1(x,y,\zeta)\,dx\,dy\,d\zeta = \Sigma\int_{\zeta_1}^{\zeta_2}\int_{\Psi_1(\zeta)}^{\Psi_2(\zeta)}\int_{\Phi_1(\zeta,x)}^{\Phi_2(\zeta,x)} F_1(x,y,\zeta)\,d\zeta\,dx\,dy \quad \text{.........(vii.)},$$

the last form being the result of changing the order of integration, as explained in (2777). We have now to change from y to η; we therefore eliminate z and ξ from the equations u, v, w, obtaining an equation of the form $y=f(\zeta,x,\eta)$, and proceed exactly as before. The result, as respects the general form of integral in (vii.), will be

$$\int_{\zeta_1}^{\zeta_2}\int_{\Psi_1(\zeta)}^{\Psi_2(\zeta)}\int_{\lambda_1(\zeta,x)}^{\lambda_2(\zeta,x)} F_2(x,\eta,\zeta)\,d\zeta\,dx\,d\eta \quad \text{..................(viii.)}$$

The order of x and η has now to be changed by (2775). Since ζ is a constant with respect to integrations for x and η, $\Psi_1(\zeta)$, $\Psi_2(\zeta)$ will also be constants, while $\lambda_1(\zeta,x)$, $\lambda_2(\zeta,x)$ will be functions of the single variable x.

Suppose $\eta=\lambda_1(\zeta,x)$ gives $x=\Lambda_1(\zeta,\eta)$. Similarly, $x=\Lambda_2(\zeta,\eta)$ may be the other limit.

At the point where $x=\Psi_1(\zeta)$ and $\eta=\lambda_1(\zeta,x)$, we shall obtain by eliminating x, say, $\eta=\mu_1(\zeta)$. Similarly, from $x=\Psi_1(\zeta)$ and $\eta=\lambda_2(\zeta,x)$ suppose, we get $\eta=\mu_2(\zeta)$ for the next limit; then a general form for the transformed integral will be

$$\int_{\zeta_1}^{\zeta_2}\int_{\mu_1(\zeta)}^{\mu_2(\zeta)}\int_{\Lambda_1(\zeta,\eta)}^{\Lambda_2(\zeta,\eta)} F_2(x,\eta,\zeta)\,d\zeta\,d\eta\,dx \quad \text{.....................(ix.)}$$

It now remains to change from the variable x to ξ. Eliminating y and z between the equations u, v, w, we have a result of the form $x=f(\xi,\eta,\zeta)$. Substituting for x and dx as before, we arrive finally at the form

$$\int_{\zeta_1}^{\zeta_2}\int_{\mu_1(\zeta)}^{\mu_2(\zeta)}\int_{\nu_1(\zeta,\eta)}^{\nu_2(\zeta,\eta)} F_3(\xi,\eta,\zeta)\,d\zeta\,d\eta\,d\xi \quad \text{..................... (x.)}$$

It should be noticed that the limits $x=\Lambda_1(\zeta,\eta)$, $x=\Lambda_2(\zeta,\eta)$, in (ix.), are not necessarily different curves. They may, in some of the partial integrals, be different portions of the same curve. This was exemplified in the last integral of (2776).

MULTIPLE INTEGRALS.

The following theorems, (2825) to (2830), which are given for three variables only, hold good for any number.

Let x, y, z be quantities which can take any positive values subject to the condition that their sum is not greater than unity; then

2825 $$\int_0^1\int_0^{1-x}\int_0^{1-x-y} x^{l-1}y^{m-1}z^{n-1}\,dx\,dy\,dz = \frac{\Gamma(l)\,\Gamma(m)\,\Gamma(n)}{\Gamma(l+m+n+1)}.$$

Here $x+y+z=1$ is the limiting equation.

PROOF.—Integrate for z; then for y by (2308); finally for x by (2280), and change to the gamma function by (2305).

2826 $$\iiint \xi^{l-1}\eta^{m-1}\zeta^{n-1}\,d\xi\,d\eta\,d\zeta = \frac{a^l\beta^m\gamma^n}{pqr}\,\frac{\Gamma\left(\frac{l}{p}\right)\Gamma\left(\frac{m}{q}\right)\Gamma\left(\frac{n}{r}\right)}{\Gamma\left(\frac{l}{p}+\frac{m}{q}+\frac{n}{r}+1\right)}.$$

when $\left(\frac{\xi}{a}\right)^p+\left(\frac{\eta}{\beta}\right)^q+\left(\frac{\zeta}{\gamma}\right)^r=1$ is the limiting equation.

PROOF.—Substitute $x=\left(\frac{\xi}{a}\right)^p$, $y=\left(\frac{\eta}{\beta}\right)^q$, $z=\left(\frac{\zeta}{\gamma}\right)^r$, and apply (2825).

2827 When the limiting equation is simply $\xi+\eta+\zeta=h$, the value of the last integral becomes

$$h^{l+m+n}\,\frac{\Gamma(l)\,\Gamma(m)\,\Gamma(n)}{\Gamma(l+m+n+1)}.$$

2828 The value of the same integral, taken between the limits h and $h+dh$ of the sum of the variables, is

$$h^{l+m+n-1}\,\frac{\Gamma(l)\,\Gamma(m)\,\Gamma(n)}{\Gamma(l+m+n)}\,dh.$$

PROOF.—Let u be the value in (2827); then, by Taylor's theorem, the value required is

$$\frac{du}{dh}\,dh = (l+m+n)\,h^{l+m+n-1}\,\frac{\Gamma(l)\,\Gamma(m)\,\Gamma(n)}{\Gamma(l+m+n+1)}\,dh,$$

which reduces to the above, by (2288).

2829 $$\iiint x^{l-1}y^{m-1}z^{n-1}f(x+y+z)\,dx\,dy\,dz$$

$$= \frac{\Gamma(l)\,\Gamma(m)\,\Gamma(n)}{\Gamma(l+m+n)}\int_0^c f(h)\,h^{l+m+n-1}\,dh,$$

if $x+y+z=h$ and h varies from 0 to c. In other words, the variables must take all positive values allowed by the condition that their sum is not greater than c.

PROOF.—For each value of h the integration with respect to x, y, z gives, by (2828), $$f(h)\,h^{l+m+n-1}\frac{\Gamma(l)\,\Gamma(m)\,\Gamma(n)}{\Gamma(l+m+n)}\,dh,$$
the variations of x, y, z not affecting h. This expression has then to be integrated as a function of h from 0 to c.

2830 $$\iiint \xi^{l-1}\eta^{m-1}\zeta^{n-1}f\left\{\left(\frac{\xi}{a}\right)^p+\left(\frac{\eta}{\beta}\right)^q+\left(\frac{\zeta}{\gamma}\right)^r\right\}d\xi\,d\eta\,d\zeta$$
$$=\frac{a^l\beta^m\gamma^n}{pqr}\,\frac{\Gamma\left(\frac{l}{p}\right)\Gamma\left(\frac{m}{q}\right)\Gamma\left(\frac{n}{r}\right)}{\Gamma\left(\frac{l}{p}+\frac{m}{q}+\frac{n}{r}\right)}\int_0^c f(h)\,h^{\frac{l}{p}+\frac{m}{q}+\frac{n}{r}-1}\,dh,$$
with the limiting equation $\left(\frac{\xi}{a}\right)^p+\left(\frac{\eta}{\beta}\right)^q+\left(\frac{\zeta}{\gamma}\right)^r=c.$

PROOF.—From (2829) by substituting $x=\left(\frac{\xi}{a}\right)^p$, &c.

2831 If x, y, z be n variables, taking all positive values subject to the restriction $x^2+y^2+z^2+\ldots\leqq 1$; then
$$\iiint\frac{dx\,dy\,dz\,.\,\&c.}{\sqrt{(1-x^2-y^2-z^2-\&c.)}}=\frac{\pi^{\frac{1}{2}(n+1)}}{2^n\Gamma\left\{\frac{1}{2}(n+1)\right\}}.$$
But, if negative values of the variables are permitted, omit the factor 2^n in the denominator.

PROOF.—In (2830) put $l=m=\&c.=1$; $a=\beta=\&c.=1$; $p=q=\&c.=2$; $f(h)=\frac{1}{\sqrt{(1-h)}}$; $c=1$; and the expression on the right becomes
$$\frac{\{\Gamma(\frac{1}{2})\}^n}{2^n\Gamma(\frac{1}{2}n)}\int_0^1\frac{h^{\frac{1}{2}n-1}}{(1-h)^{\frac{1}{2}}}\,dh.$$
The integral is $=B(\frac{1}{2}n,\frac{1}{2})$ (2280) $=\frac{\Gamma(\frac{1}{2}n)\Gamma(\frac{1}{2})}{\Gamma\{\frac{1}{2}(n+1)\}}$ (2305).
Hence the result.

But if negative values of the variables are allowed under the same restriction, $x^2+y^2+z^2+\ldots\leqq 1$, each element of the integral will occur 2^n times for once under the first hypothesis. Therefore the former result must be multiplied by 2^n.

2832 If n positive variables, x, y, z, &c., are limited by the condition $x^2+y^2+z^2+\&c. \leqq 1$, then

$$\iiint \phi(ax+by+cz+\&c.)\,dx\,dy\,dz\ldots$$
$$= \frac{\pi^{\frac{1}{2}(n-1)}}{2^{n-1}\Gamma\left\{\frac{1}{2}(n+1)\right\}}\int_0^1 \phi(k\xi)(1-\xi^2)^{\frac{1}{2}(n-1)}d\xi,$$

where $k^2 = a^2+b^2+c^2+\&c.$

PROOF.—Change the variables to ξ, η, ζ by the orthogonal transformation (1799), so that

$a^2+b^2+c^2+\&c. = k^2$, and $ax+by+cz+\&c. = k\xi$.

The integral then takes the form

$$\iiint \phi(k\xi)\,d\xi\,d\eta\,d\zeta\ldots \text{ with } \xi^2+\eta^2+\zeta^2+\&c. \not> 1.$$

	x	y	z	...
ξ	$\frac{a}{k}$	$\frac{b}{k}$	$\frac{c}{k}$	...
η	$\frac{a'}{k}$	$\frac{b'}{k}$	$\frac{c'}{k}$	...
ζ	$\frac{a''}{k}$	$\frac{b''}{k}$	$\frac{c''}{k}$	...
...	...	...	...	

Now, integrate for η, ζ, &c., considering ξ constant, by adapting formula (2826). The limiting equation is

$$\left(\frac{\eta}{\sqrt{(1-\xi^2)}}\right)^2+\left(\frac{\zeta}{\sqrt{(1-\xi^2)}}\right)^2+\&c. \text{ to } n-1 \text{ terms,} = 1.$$

Therefore put $l=m=\&c.=1$; $p=q=\&c.=2$; $\alpha=\beta=\&c.=\sqrt{(1-\xi^2)}$.

The result is $\displaystyle\int_0^1 \phi(k\xi)\frac{(1-\xi^2)^{\frac{n-1}{2}}}{2^{n-1}}\frac{\{\Gamma(\frac{1}{2})\}^{n-1}}{\Gamma\{\frac{1}{2}(n-1)+1\}}d\xi,$

which is equivalent to the value above.

2833 With the same limiting equation for n variables and the same value of k,

$$\iiint \frac{\phi(ax+by+cz+\&c.)}{\sqrt{(1-x^2-y^2-z^2-\&c.)}}\,dx\,dy\,dz\ldots$$
$$= \frac{\pi^{\frac{1}{2}n}}{2^{n-1}\Gamma(\frac{1}{2}n)}\int_0^1 \phi(k\xi)(1-\xi^2)^{\frac{1}{2}n-1}d\xi.$$

PROOF.—Making the same orthogonal transformation as in (2832), the integral changes to $\displaystyle\iiint\ldots\frac{\phi(k\xi)\,d\xi\,d\eta\,d\zeta\ldots}{\sqrt{(1-\xi^2-\eta^2-\&c.)}},$

Considering ξ constant, the integration for the remaining variables is effected by (2830). Adapting the integral to that formula, we have

$$\frac{\phi(k\xi)\,d\xi}{\sqrt{(1-\xi^2)}}\iint\ldots\frac{d\eta\,d\zeta\ldots}{\sqrt{\left\{1-\frac{\eta^2}{1-\xi^2}-\frac{\zeta^2}{1-\xi^2}-\&c.\right\}}},$$

with $\displaystyle\left(\frac{\eta}{\sqrt{(1-\xi^2)}}\right)^2+\left(\frac{\zeta}{\sqrt{(1-\xi^2)}}\right)^2+\&c.$, to $n-1$ terms, $\leqq 1$

for the limiting equation.

Here $l = m = \&c. = 1$; $p = q = \&c. = 2$; $\alpha = \beta = \&c. = \sqrt{(1-\xi^2)}$; $f(h) = \dfrac{1}{\sqrt{(1-h)}}$; $c = 1$; and the reductions are similar to those in (2832).

2834 If in (2832–3) negative values of the variables are admitted (since the limiting equation is satisfied by such), each element of the integral with respect to η, ζ, &c. will then occur 2^{n-1} times, and therefore the result in each case must be multiplied by 2^{n-1}, and the limits of the integration for ξ will be -1 and 1 instead of 0 and 1.

EXPANSIONS OF FUNCTIONS IN CONVERGING SERIES.

The expansion of a function by Maclaurin's theorem (1507) can be at once effected if the nth derivative of the function is known, or if merely the value of the same, when the independent variable vanishes, is known. Some n^{th} derivatives of different functions, in addition to those given at (1461–71), are therefore here collected. When the general value would be too complicated, the value for the origin zero alone is given.

DERIVATIVES OF THE nth ORDER.

The following is a general formula for calculating the nth derivative of a function of a function.

If y be a function of z, and z a function of x,

2852 $$\frac{d^n y}{dx^n} = \Sigma \frac{z^r}{\lfloor r} \frac{d^r y}{dz^r} \frac{d^n}{dx^n}\left(\frac{z}{a}-1\right)^r_{a=z},$$

where $r = 1, 2, 3, \dots n$ successively, and a is put $= z$ in each term of the expanded binomial, *after* differentiation.

PROOF.—Assume $y_{nx} = A_1 y_z + \dfrac{A_2}{\lfloor 2} y_{2z} + \dfrac{A_3}{\lfloor 3} y_{3z} + \dots + \dfrac{A_n}{\lfloor n} y_{nz}$.

To determine any coefficient A_r, form r equations from this by making $y = z, z^2, z^3, \dots z^r$ in succession: multiply these r equations respectively by $rz^{-1}, -C(r, 2)\, z^{-2}, C(r, 3)\, z^{-3}, \dots (-1)^{r+1} z^{-r}$, and add the results. All the coefficients excepting A_r disappear. This is shown by differentiating the equation $\quad (1-x)^r = 1 - rx + C(r, 2)\, x^2 - C(r, 3)\, x^3 + \dots \pm x^r$

successively for x, and making x zero after each differentiation. Thus, finally,

$$\frac{A_r}{z^r} = (-1)^{r+1}\left\{\frac{rz_{nx}}{z} - \frac{C(r,2)(z^2)_{nx}}{z^2} + \frac{C(r,3)(z^3)_{nx}}{z^3} - \dots (-1)^{r+1}\frac{(z^r)_{nx}}{z^r}\right\}$$

$$= d_{nx}\left(\frac{z}{a}-1\right),$$

with a put $= z$, after expanding and differentiating the binomial.

2853 EXAMPLES.—The formula may be applied to verify equations (1416–19).

Jacobi's formula (1471) may also be obtained by it.

2854 $$d_{(n+1)x}\sin^{-1}x$$

$$= \frac{1.3\dots(2n-1)}{2^n(1-x)^n(1-x^2)^{\frac{1}{2}}}\left\{1-\frac{n}{2n-1}Z+\frac{1.3\,C(n,2)}{2n-1.2n-3}Z^2\right.$$

$$\left.-\frac{1.3.5\,C(n,3)}{2n-1.2n-3.2n-5}Z^3+\dots\pm Z^n\right\},\quad \text{where } Z=\frac{1-x}{1+x}.$$

PROOF. $(\sin^{-1}x)_{(n+1)x} = \{(1-x)^{-\frac{1}{2}}(1+x)^{-\frac{1}{2}}\}_{nx}$ (1434).
Expand the right member by (1460).

2855 $d_{nx}\tan^{-1}x$. This derivative is obtained in (1468). The following is another method, which also includes the result in (1469).

$$d_x\tan^{-1}x = \frac{1}{1+x^2} = \frac{i}{2}\left\{\frac{1}{x+i}-\frac{1}{x-i}\right\};$$

$$\therefore \text{ by (1425)}\quad d_{nx}\tan^{-1}x = \frac{\lfloor n-1\,(-1)^n}{2i}\left\{\frac{1}{(x+i)^n}-\frac{1}{(x-i)^n}\right\} \quad\dots\dots\dots(1).$$

Put $x=\cot\theta$, therefore $x\pm i = \sqrt{(1+x^2)}\,(\cos\theta\pm i\sin\theta)$, which values substituted in (1) convert the equation, by (757), into

$$(\tan^{-1}x)_{nx} = (-1)^{n-1}\lfloor n-1\,\sin^n\theta\sin n\theta.$$

2856 $$d_{nx}\{e^{x\cos a}\cos(x\sin a)\} = e^{x\cos a}\cos(x\sin a+na).$$

PROOF.—By Induction.

LAGRANGE'S METHOD.

2857 *Lemma.*—The nth derivative of a function $u=f(x)$ will, by Taylor's theorem (1500), be equal to $1.2\dots n$ times the coefficient of h^n in the expansion of $f(x+h)$ in powers of h by any known method.

Let $u = a+bx+cx^2$, and therefore $u_x = b+2cx$; then $d_{rx}(a+bx+cx^2)^n$ is equal to either of the following series, with the notation of (2451–2).

2858 $$1^{(r)}u^{n-r}u_x^r\left\{\frac{n_{-1}^{(r)}}{1^{(r)}}+\frac{n_{-1}^{(r-1)}cu}{1^{(r-2)}u_x^2}+\ldots+\frac{n_{-1}^{(r-p)}c^pu^p}{1^{(p)}1^{(r-2p)}u_x^{2p}}+\ldots\right\};$$

2859 or, putting $2n = m$ and $4ac-b^2 = q^2$,

$$\frac{m_{-1}^{(r)}}{2^r}u^{n-r}\left\{u_x^r+n\frac{r_{-1}^{(2)}}{m_{-1}^{(2)}}q^2u_x^{r-2}+\ldots+C(n,p)\frac{r_{-1}^{(2p)}}{m_{-1}^{(2p)}}q^{2p}u_x^{r-2p}+\ldots\right\}.$$

PROOF.—Changing x into $x+h$ in u^n, it becomes $(u+u_xh+ch^2)^n$. Then, by (2857), $d_{rx}u^n$ will be $= \lfloor r$ times the coefficient of h^r in the expansion of this trinomial. (2858) is the result, and it may be obtained by expanding $\{(u+u_xh)+ch^2\}^n$ as a binomial, and collecting the coefficients of h^r from the subsequent expansions. The value (2859) is found by taking

$$(u+u_xh+ch^2)^n = u^n\left\{\left(1+\frac{u_xh}{2u}\right)^2+\left(\frac{qh}{2u}\right)^2\right\}^n,$$

expanding, collecting coefficients of h^r, and multiplying by $1.2\ldots r$, as before.

2860 Ex.—To find $d_{nx}(a^2+x^2)^n$. Applying formula (2859), we have $u = a^2+x^2$, $u_x = 2x$, $q = 2a$, $r = n$. Therefore

$$d_{nx}(a^2+x^2)^n = (2n)_{-1}^{(n)}\left\{x^n+\frac{n^2(n-1)}{2n(2n-1)}a^2x^{n-2}\right.$$
$$\left.+\frac{n^2(n-1)^2(n-2)(n-3)}{1.2.2n\ldots(2n-3)}a^4x^{n-4}+\&c.\right\}.$$

2861 $$d_{nx}e^{ax^2} = e^{ax^2}\left\{a^n(2x)^n+\ldots+\frac{n_{-1}^{(2r)}}{1^{(r)}}a^{n-r}(2x)^{n-2r}+\&c.\right\},$$

with $r = 1, 2, 3,$ &c. in succession.

PROOF.—By the method of (2857). Putting $e^{a(x+h)^2} = e^{ax^2}e^{2axh}e^{ah^2}$, expand the factors containing h by (150), and from the product of the two series collect the coefficients of h^n.

2862 $$d_{nx}\begin{matrix}\sin\\ \cos\end{matrix}x^2 = (2x)^n\begin{matrix}\sin\\ \cos\end{matrix}\left(x^2+\frac{n\pi}{2}\right)+\ldots$$
$$\ldots+\frac{n_{-1}^{(2r)}}{1^{(r)}}(2x)^{n-2r}\begin{matrix}\sin\\ \cos\end{matrix}\left(x^2+\frac{(n-r)\pi}{2}\right)+\&c.$$

PROOF.— $d_{nx}(\cos x^2 + i\sin x^2) = d_{nx}e^{ix^2}$. Expand the right by (2861), putting $i^{n-r} = e^{i(n-r)\frac{1}{2}\pi}$, since, by (766), $e^{\frac{1}{2}i\pi} = i\sin\frac{\pi}{2} = i$. Also put

$$e^{ix^2+i(n-r)\frac{1}{2}\pi} = \cos\left\{x^2 + \frac{(n-r)\pi}{2}\right\} + i\sin\left\{x^2 + \frac{(n-r)\pi}{2}\right\},$$

and then equate real and imaginary parts.

2864 $$d_{nx}\frac{1}{e^x+1} = (-1)^n\Big[e^{nx} + \{n+1-2^n\}\,e^{(n-1)x}$$
$$+\{C(n+1,2) - 2^n(n+1) + 3^n\}\,e^{(n-2)x}$$
$$+\{C(n+1,3) - 2^nC(n+1,2) + 3^n(n+1) - 4^n\}\,e^{(n-3)x} + \&c.\Big]$$

PROOF.—Let u be the function. By differentiating u it is seen that

$$(e^x+1)^{n+1}u_{nx} = A_ne^{nx} + A_{n-1}e^{(n-1)x} + A_{n-2}e^{(n-2)x} + \ldots + A_1e^x,$$

the A's being constants. To determine their values, expand $u = (e^x+1)^{-1}$, and also $(e^x+1)^{n+1}$, by the Binomial theorem; thus

$$u_{nx} = (-1)^n\{e^{-x} - 2^ne^{-2x} + 3^ne^{-3x} - 4^ne^{-4x} + \&c.\},$$

$$(e^x+1)^{n+1} = e^{(n+1)x} + (n+1)\,e^{nx} + C(n+1,2)\,e^{(n-1)x} + C(n+1,3)\,e^{(n-2)x} + \&c.$$

From the product of the two expansions the coefficients A_n, A_{n-1}, &c. may be selected.

2865 $$d_{nx0}\tan^{-1}x = (-1)^{\frac{n-1}{2}}\,\underline{|n-1}\ \text{ or } zero,$$

according as n is *odd* or even.

PROOF.—By Rule IV. (1534). The first and last differential equations (see Example 1535) are, in this case,

$$(1+x^2)\,y_{2x} + 2xy_x = 0 \ldots\ldots \text{(i.)}; \qquad y_{(n+2)x0} + n(n+1)\,y_{nx0} = 0 \ldots\ldots \text{(ii.)};$$

with $y_{x0} = 1$ and $y_{2x0} = 0$.

Otherwise.—By (1468), putting $x = 0$.

2867 $$d_{nx0}\sin^{-1}x = 1.3^2.5^2\ldots(n-2)^2 \ \text{ or } zero,$$

according as n is *odd* or *even*.

PROOF.—By differentiating (1528).

Otherwise.—As in (2865) where equations (i.) and (ii.) will become in this case $\quad (1-x^2)\,y_{2x} = xy_x \ldots\ldots \text{(i.)} \qquad y_{(n+2)x0} = n^2y_{nx0} \ldots\ldots \text{(ii.)}$

2869 $$d_{nx0}(\sin^{-1}x)^2 = 2.2^2.4^2.6^2\ldots(n-2)^2 \ \text{ or } zero,$$

according as n is *even* or odd.

PROOF.—As in (2865); equations (i.) and (ii.) being identical with those in (2867).

2871 $$d_{nx0}\cos(m\sin^{-1}x)$$
$$=(-1)^{\frac{n}{2}}m^2(m^2-2^2)(m^2-4^2)\dots\left[m^2-(n-2)^2\right],$$
or *zero*; according as n is *even* or *odd* and >0 if even.

2873 $$d_{nx0}\sin(m\sin^{-1}x)$$
$$=(-1)^{\frac{n-1}{2}}m(m^2-1)(m^2-3^2)\dots\left[m^2-(n-2)^2\right],$$
or *zero*; according as n is *odd* or *even* and >1 if odd.

2875 $$d_{nx0}\cos(m\cos^{-1}x)$$
$$=(-1)^{\frac{n-1}{2}}m(m^2-1)(m^2-3^2)\dots\left[m^2-(n-2)^2\right]\sin\frac{m\pi}{2},$$
2876 or
$$=(-1)^{\frac{n}{2}}m^2(m^2-2^2)(m^2-4^2)\dots\left[m^2-(n-2)^2\right]\cos\frac{m\pi}{2},$$
according as n is *odd* and >1, or even and >0.

2877 $$d_{nx0}\sin(m\cos^{-1}x)$$
$$=(-1)^{\frac{n}{2}}m^2(m^2-2^2)(m^2-4^2)\dots\left[m^2-(n-2)^2\right]\sin\tfrac{1}{2}m\pi,$$
2879 or
$$=(-1)^{\frac{n+1}{2}}m(m^2-1)(m^2-3^2)\dots\left[m^2-(n-2)^2\right]\cos\tfrac{1}{2}m\pi,$$
according as n is *even* and >0, or odd and >1.

Observe that, in (2871–3), $\sin^{-1}0=0$, and in (2875–9), $\cos^{-1}0=\frac{\pi}{2}$, are the only values admitted.

PROOF.—For (2871–9). As in (2865); equations (i.) and (ii.) now becoming in each case

$$(1-x^2)\,y_{2x}-xy_x+m^2y=0\dots\dots\text{(i.)}\qquad y_{(n+2)\,x0}=(n^2-m^2)\,y_{nx0}\dots\dots\text{(ii.)}$$

Otherwise.—By the method of (1533).

2880 Let $y=x\cot x$, then

$$n\cos\frac{n-1}{2}\pi=y_0\sin\frac{n}{2}\pi+\dots+y_{rx0}\,C(n,r)\sin\frac{n-r}{2}\pi+\dots$$
$$\dots+y_{(n-1)\,x0}\,n\sin\frac{\pi}{2},$$

with integral values of r, from 0 to $n-1$ inclusive.

2881 Thus, denoting y_{nx0} shortly by y_n, we find, by making $n=1, 2, 3$, &c. successively in the formula,

$$y_2 = -\frac{2}{3},\quad y_4 = -\frac{8}{15},\quad y_6 = -\frac{32}{21},\quad y_8 = -\frac{128}{15}.$$

PROOF.—Take the nth derivative of the equation $x\cos x = y\sin x$ by (1460), reducing the coefficients by (1461-2), and putting x finally $= 0$.

2882 The derivatives of an odd order all vanish. This may be shown independently, as follows :—

Let $y = \phi(x)$, then $\phi(x)$ is an even function of x (1401) ; therefore

$$\phi^{2n+1}(x) = -\phi^{2n+1}(-x);$$

$$\therefore\ \phi^{2n+1}(0) = -\phi^{2n+1}(0);\quad \therefore\ \phi^{2n+1}(0) = 0.$$

2883 $d_{nx0}\{(1+x^2)^{\frac{m}{2}}\sin(m\tan^{-1}x)\} = (-1)^{\frac{n-1}{2}}\, m_{-1}^{(n)}$ or *zero*, according as n is *odd* or *even*.

2885 $d_{nx0}\{(1+x^2)^{\frac{m}{2}}\cos(m\tan^{-1}x)\} = (-1)^{\frac{n-2}{2}}\, m_{-1}^{(n)}$ or *zero*, according as n is *even* or *odd*.

PROOF.—As in (2865). Equations (i.) and (ii.), both for (2883) and (2885), are now $(1+x^2)\,y_{2x} - 2\,(m-1)\,xy_x + m\,(m-1)\,y = 0$..................(i.),

and $y_{(n+2)x0} = -(m-n)\,(m-n-1)\,y_{nx0}$ .... (ii.)

Formula (ii.) gives the factors in succession, starting with $y_0 = 0$, $y_x = m$ in (2883); and with $y_0 = 1$, $y_x = 0$ in (2885).

2887 $d_{nx0}\{(1+x^2)^{-\frac{m}{2}}\cos(m\tan^{-1}x)\} = (-1)^{\frac{n}{2}}\, m^{(n)}$ or *zero*, according as n is *even* or *odd*.

PROOF.—Change the sign of m in (2885).

NOTE.—In formulæ (2883-7) zero is the only admitted value of $\tan^{-1}0$.

2889 $$d_{nx0}\left(\frac{x}{e^x-1}\right) = (-1)^{\frac{n}{2}-1} B_n \text{ or } zero,$$

according as n is *even* or *odd*; by (1539).

2891 When p is a positive integer,

$$d_{nx0}\,(x^p e^{ax}\cos bx) = n_{-1}^{(p)}(a^2+b^2)^{\frac{n-p}{2}}\cos\left\{(n-p)\tan^{-1}\frac{b}{a}\right\}$$

or *zero*, according as n is $>$ or $<$ p.

PROOF.—Put $y = e^{ax}\cos bx$ and $z = x^p$ in (1460), employing (1465).

MISCELLANEOUS EXPANSIONS.

The following series are placed here for the sake of reference, many of them being of use in evaluating definite integrals by Rule V. (2249). Other series and methods of expansion will be found in Articles (125–129), (149–159), (248–295), (756–817), (1460), (1471–1472), (1500–1573). For tests of convergency, see (239–247).

Numerous expansions may be obtained by differentiating or integrating known series or their logarithms. These and other methods are exemplified below.

2911 $$\cot x = \frac{1}{x} - \frac{1}{\pi - x} + \frac{1}{\pi + x} - \frac{1}{2\pi - x} + \frac{1}{2\pi + x} - \frac{1}{3\pi - x} + \frac{1}{3\pi + x} - \&c.$$

PROOF.—By differentiating the logarithm of equation (815).

2912 $$\pi \cot \pi x = \frac{1}{x} + \frac{1}{x-1} + \frac{1}{x+1} + \frac{1}{x-2} + \frac{1}{x+2} + \frac{1}{x-3} + \frac{1}{x+3} + \&c.$$

PROOF.—By changing x into πx in (2911).

2913 $$\tan x = \frac{1}{\frac{1}{2}\pi - x} - \frac{1}{\frac{1}{2}\pi + x} + \frac{1}{\frac{3}{2}\pi - x} - \frac{1}{\frac{3}{2}\pi + x} + \frac{1}{\frac{5}{2}\pi - x} - \frac{1}{\frac{5}{2}\pi + x} + \&c.$$

PROOF.—By changing x into $\frac{1}{2}\pi - x$ in (2911).

2914 $$\operatorname{cosec} x = \frac{1}{x} + \frac{1}{\pi - x} - \frac{1}{\pi + x} - \frac{1}{2\pi - x} + \frac{1}{2\pi + x} + \frac{1}{3\pi - x} - \frac{1}{3\pi + x} - \frac{1}{4\pi - x} + \&c.$$

PROOF.—By adding together equations (2911, 2913), and changing x into $\frac{1}{2}x$.

2915 $$\frac{\pi}{\sin m\pi} = \frac{1}{m} + \frac{1}{1-m} - \frac{1}{1+m} - \frac{1}{2-m} + \frac{1}{2+m} + \frac{1}{3-m} - \frac{1}{3+m} - \&c.$$

PROOF.—By putting $x = m\pi$ in (2914).

2916 $$\cot x = \frac{1}{x} - \frac{2^2 B_2 x}{\lfloor 2} - \frac{2^4 B_4 x^3}{\lfloor 4} - \frac{2^6 B_6 x^5}{\lfloor 6} - \&c.$$

For proof see (1545). The reference in that article (first edition) should be to (1541) not (1540).

2917 $$\tan x = \frac{2^2(2^2-1)}{\lfloor 2} B_2 x + \frac{2^4(2^4-1)}{\lfloor 4} B_4 x^3 + \frac{2^6(2^6-1)}{\lfloor 6} B_6 x^5 + \&c.$$

2918 $$\operatorname{cosec} x = \frac{1}{x} + \frac{2(2-1)}{\lfloor 2} B_2 x + \frac{2(2^3-1)}{\lfloor 4} B_4 x^3 + \frac{2(2^5-1)}{\lfloor 6} B_6 x^5 + \&c.$$

PROOF.—By (2916) and the relations

$$\tan x = \cot x - 2\cot 2x, \qquad \operatorname{cosec} x = \cot \tfrac{1}{2}x - \cot x.$$

2919 $$\frac{1-a^2}{1-2a\cos x + a^2} = 1 + 2a\cos x + 2a^2\cos 2x + 2a^3\cos 3x + \&c.$$

2920 $$\frac{\cos x}{1-2a\cos x + a^2}$$
$$= \frac{a}{1-a^2} + \frac{1+a^2}{1-a^2}(\cos x + a\cos 2x + a^2\cos 3x + a^3\cos 4x + \&c\ldots)$$

2921 $$\frac{\sin x}{1-2a\cos x + a^2} = \sin x + a\sin 2x + a^2\sin 3x + a^3\sin 4x + \&c.$$

PROOF.—By (784–6) making $\alpha = \beta = x$ and $c = a$.

When a is less than unity and either positive or negative,

2922
$$\frac{1}{2}\log(1+2a\cos x+a^2) = a\cos x-\frac{a^2}{2}\cos 2x+\frac{a^3}{3}\cos 3x-\&c.$$

2923 $$\tan^{-1}\frac{a\sin x}{1+a\cos x} = a\sin x-\frac{a^2}{2}\sin 2x+\frac{a^3}{3}\sin 3x-\&c.$$

Proof.—Putting $z = a(\cos x+i\sin x)$, we have

$$\log(1+z) = \log(1+a\cos x+ia\sin x)$$
$$= \tfrac{1}{2}\log(1+2a\cos x+a^2)+i\tan^{-1}\frac{a\sin x}{1+a\cos x} \quad (2214),$$

and also $$\log(1+z) = z-\frac{z^2}{2}+\frac{z^3}{3}-\frac{z^4}{4}+\&c.$$

Substitute the value of z and equate real and imaginary parts.

2924 *Otherwise.*—To obtain (2922),

$$\log(1+2a\cos x+a^2) = \log(1+ae^{ix})+\log(1+ae^{-ix}).$$

Expanding by (154), the series is at once obtained by (768).

2925 *Otherwise.*—Integrate the equation in (786) with respect to a, after changing a and β into x, and c into $-a$.

2926 When a is greater than unity, put

$$\log(1+2a\cos x+a^2) = \log a^2+\log(1+2a^{-1}\cos x+a^{-2}),$$

and the last term can be expanded in a converging series by (2922).

2927
$$\log 2\cos\tfrac{1}{2}x = \cos x-\tfrac{1}{2}\cos 2x+\tfrac{1}{3}\cos 3x-\tfrac{1}{4}\cos 4x+\&c.$$

2928
$$\log 2\sin\tfrac{1}{2}x = -\cos x-\tfrac{1}{2}\cos 2x-\tfrac{1}{3}\cos 3x-\tfrac{1}{4}\cos 4x-\&c.$$

2929 $$\tfrac{1}{2}x = \sin x-\tfrac{1}{2}\sin 2x+\tfrac{1}{3}\sin 3x-\tfrac{1}{4}\sin 4x+\&c.$$

2930 $$\tfrac{1}{2}(\pi-x) = \sin x+\tfrac{1}{2}\sin 2x+\tfrac{1}{3}\sin 3x+\tfrac{1}{4}\sin 4x+\&c.$$

Proof.—(2927–30) Make $a = \pm 1$ in (2922–3).

2931 $$\tfrac{1}{4}\pi = \sin x+\tfrac{1}{3}\sin 3x+\tfrac{1}{5}\sin 5x+\&c.$$

2932 $$\pi = 2\sqrt{2}\,(1+\tfrac{1}{3}-\tfrac{1}{5}-\tfrac{1}{7}+\tfrac{1}{9}+\tfrac{1}{11}-\&c.).$$

Proof.—Add together (2929–30), and put $x = \frac{1}{4}\pi$.

When n is less than unity, and $a = 1\pm\sqrt{(1+n^2)}$,

2933 $$\log(1+n\cos x) = \log(1+2a\cos x+a^2)-\log(1+a^2),$$

and is therefore equal to twice the series in (2922), minus $\log(1+a^2)$. But if a be greater than unity, expand, as in (2926), by

2934 $$\log(1+n\cos x)$$
$$= \log(1+2a^{-1}\cos x + a^{-2}) + \log a^2 - \log(1+a^2).$$

2935
$$(1+2a\cos x)^n = A + A_1\cos x + A_2\cos 2x + A_3\cos 3x + \&c.,$$
where
$$A = 1 + C(n,2)\,2a^2 + \ldots + C(n,2p)\,C(2p,p)\,a^{2p} + \ldots,$$
$$A_1 = 2a\,\{n + C(n,3)\,3a^2 + \ldots + C(n,2p+1)\,C(2p+1,p)\,a^{2p} + \ldots\},$$
and $$A_{r+1} = \frac{A_{r-1}(n-r+1)\,a - rA_r}{(n+r+1)\,a}.$$

If n be a positive integer, the series terminates with the $n+1^{\text{th}}$ term, and the values of A and A_1 are also finite.

Proof.—Differentiate the logarithm of the first equation; multiply up and equate coefficients of $\sin rx$ after transforming by (666); thus A_{r+1} is obtained.

To find A and A_1, expand $(1+2a\cos x)^n$ by the Binomial Theorem, and the powers of $\cos x$ afterwards by (772).

LEGENDRE'S FUNCTION X_n.

2936 $$(1-2ax+a^2)^{-\frac{1}{2}} = 1 + X_1 a + X_2 a^2 + \ldots + X_n a^n +$$
with $$X_n = \frac{1}{2^n \underline{|n}}\,\frac{d^n}{dx^n}(x^2-1)^n.$$

Proof.—Expand by the Binomial Theorem, and in the numerical part of each coefficient of a^n express $1.3.5 \ldots 2n-1$ as $\underline{|2n} \div 2^n \underline{|n}$.

Consecutive functions are connected by the relation

2937 $$d_x X_{n+1} = (2n+1)\,X_n + d_x X_{n-1}.$$

Proof.—Differentiate the factor once under the sign of differentiation in the values of X_{n+1} and X_{n-1} given by the formula for X_n in (2936).

A differential equation for X_n is

2938 $$(1-x^2)\,d_{2x}X_n - 2x\,d_x X_n + n(n+1)\,X_n = 0.$$

2939 When p is any positive integer,

$$1^p+2^p+3^p+\ldots+(n-1)^p$$
$$=\lfloor p\left\{\frac{n^{p+1}}{\lfloor p+1}-\frac{n^p}{2\lfloor p}+\frac{B_2 n^{p-1}}{\lfloor 2\lfloor p-1}-\frac{B_4 n^{p-3}}{\lfloor 4\lfloor p-3}+\frac{B_6 n^{p-5}}{\lfloor 6\lfloor p-5}-\&c.\right.$$

concluding, according as n is even or odd, with

$$(-1)^{\frac{1}{2}p+1}\frac{B_p n}{\lfloor p} \qquad \text{or} \qquad \left.(-1)^{\frac{1}{2}(p+1)}\frac{B_{p-1}n^2}{\lfloor p-1\lfloor 2}\right\}.$$

PROOF. $\frac{e^{nx}-1}{e^x-1}=\frac{e^{nx}-1}{x}\cdot\frac{x}{e^x-1}$. Expand the left side by division, and each term subsequently by (150). Again, expand the first factor of the right side by (150), and the second by (1539), and equate the coefficients of x^p in the two results.

See (276) for the values of the series when p is 1, 2, 3, or 4. But the general formula there is incorrectly printed.

Let the series (2940–4) be denoted by S_{2n}, S'_{2n}, s_{2n}, s'_{2n+1}, as under, n being any positive integer; then

2940 $$S_{2n}\equiv 1+\frac{1}{2^{2n}}+\frac{1}{3^{2n}}+\frac{1}{4^{2n}}+\&c.\ldots=\frac{2^{2n-1}}{\lfloor 2n}\pi^{2n}B_{2n}.$$

2941 $$S'_{2n}\equiv 1-\frac{1}{2^{2n}}+\frac{1}{3^{2n}}-\frac{1}{4^{2n}}+\&c.\ldots=\frac{2^{2n-1}-1}{\lfloor 2n}\pi^{2n}B_{2n}.$$

2942 $$s_{2n}\equiv 1+\frac{1}{3^{2n}}+\frac{1}{5^{2n}}+\frac{1}{7^{2n}}+\&c.\ldots=\frac{2^{2n}-1}{2\lfloor 2n}\pi^{2n}B_{2n}.$$

PROOF.—(i.) S_{2n} is obtained in (1545).

(ii.) $S_{2n}-S'_{2n}=2\left(\frac{1}{2^{2n}}+\frac{1}{4^{2n}}+\&c.\right)=2\frac{S_{2n}}{2^{2n}}$. This give S'_{2n}.

(iii.) $s_{2n}=\frac{1}{2}(S_{2n}+S'_{2n})$.

2943 $$s_{2n}\equiv 1+\frac{1}{3^{2n}}+\frac{1}{5^{2n}}+\frac{1}{7^{2n}}+\&c.=-\frac{\pi d_{(2n-1)x}\cot\pi x}{4^{2n}\lfloor 2n-1},$$
$$x=\tfrac{1}{4}.$$

2944 $$s'_{2n+1}\equiv 1-\frac{1}{3^{2n+1}}+\frac{1}{5^{2n+1}}-\frac{1}{7^{2n+1}}+\&c.=\frac{\pi d_{2nx}\cot\pi x}{4^{2n+1}\lfloor 2n},$$
$$x=\tfrac{1}{4}.$$

PROOF.—By differentiating equation (2912) successively, and putting $x=\frac{1}{4}$ in the result. To compute $d_{nx}\cot\pi x$, see (1525).

2945 The following values have been calculated by formulæ (2940–4).

$$S_2=\frac{\pi^2}{6},\quad S_4=\frac{\pi^4}{90},\quad S_6=\frac{\pi^6}{945},\quad S_8=\frac{\pi^8}{9450};$$

$$S_2'=\frac{\pi^2}{12},\quad S_4'=\frac{7\pi^4}{720},\quad S_6'=\frac{31\pi^6}{30240},\quad S_8'=\frac{127\pi^8}{1209600};$$

$$s_2=\frac{\pi^2}{8},\quad s_4=\frac{\pi^4}{96},\quad s_6=\frac{\pi^6}{960},\quad s_8=\frac{17\pi^8}{161280};$$

$$s_1'=\frac{\pi}{4},\quad s_3'=\frac{\pi^3}{32},\quad s_5=\frac{5\pi^5}{1536},\quad s_7'=\frac{61\pi^7}{184320}.$$

2946 $$\frac{\cos x-\cos a}{1-\cos a}=\left(1-\frac{x^2}{a^2}\right)\left[1-\frac{x^2}{(2\pi-a)^2}\right]\left[1-\frac{x^2}{(2\pi+a)^2}\right]$$
$$\times\left[1-\frac{x^2}{(4\pi-a)^2}\right]\left[1-\frac{x^2}{(4\pi+a)^2}\right]\ldots\ \&c.,$$

2947 $$\frac{\cos x+\cos a}{1+\cos a}=\left[1-\frac{x^2}{(\pi-a)^2}\right]\left[1-\frac{x^2}{(\pi+a)^2}\right]\left[1-\frac{x^2}{(3\pi-a)^2}\right]$$
$$\times\left[1-\frac{x^2}{(3\pi+a)^2}\right]\left[1-\frac{x^2}{(5\pi-a)^2}\right]\ldots\ \&c.$$

PROOF. $\dfrac{\cos x-\cos a}{1-\cos a}=\dfrac{\sin\frac{1}{2}(a+z)\sin\frac{1}{2}(a-x)}{\sin^2\frac{1}{2}a}$. Expand the sines by (815). The two $n+1$th factors of the numerator divided by the corresponding ones of the denominator reduce to

$$\left(1-\frac{2ax+x^2}{4n^2\pi^2-a^2}\right)\left(1+\frac{2ax-x^2}{4n^2\pi^2-a^2}\right)$$
$$=\left(1-\frac{x}{2n\pi-a}\right)\left(1+\frac{x}{2n\pi+a}\right)\left(1+\frac{x}{2n\pi-a}\right)\left(1-\frac{x}{2n\pi+a}\right)$$
$$=\left(1-\frac{x^2}{(2n\pi-a)^2}\right)\left(1-\frac{x^2}{(2n\pi+a)^2}\right).$$

Similarly with (2947) employing (816).

2948 $$\cos x+\tan\frac{a}{2}\sin x=\left(1+\frac{2x}{\pi-a}\right)\left(1-\frac{2x}{\pi+a}\right)\left(1+\frac{2x}{3\pi-a}\right)$$
$$\times\left(1-\frac{2x}{3\pi+a}\right)\left(1+\frac{2x}{5\pi-a}\right)\left(1-\frac{2x}{5\pi+a}\right)\ldots\&\text{c.},$$

2949 $$\cos x-\cot\frac{a}{2}\sin x=\left(1-\frac{2x}{a}\right)\left(1+\frac{2x}{2\pi-a}\right)\left(1-\frac{2x}{2\pi+a}\right)$$
$$\times\left(1+\frac{2x}{4\pi-a}\right)\left(1-\frac{2x}{4\pi+a}\right)\left(1+\frac{2x}{6\pi-a}\right)\ldots\ \&c.$$

PROOF. $\cos x+\tan\dfrac{a}{2}\sin x=\dfrac{\cos(\frac{1}{2}a-x)}{\cos\frac{1}{2}a}$. Expand the cosines by (816), and reduce. Similarly with (2949), employing 815.

2950 $\frac{e^x-e^{-x}}{2} = x\left[1+\left(\frac{x}{\pi}\right)^2\right]\left[1+\left(\frac{x}{2\pi}\right)^2\right]\left[1+\left(\frac{x}{3\pi}\right)^2\right]\ldots\ldots$

2951 $\frac{e^x+e^{-x}}{2} = \left[1+\left(\frac{2x}{\pi}\right)^2\right]\left[1+\left(\frac{2x}{3\pi}\right)^2\right]\left[1+\left(\frac{2x}{5\pi}\right)\right]\ldots\ldots$

PROOF.—Change θ into ix in (815) and (816).

2952 $\frac{e^x-2\cos a+e^{-x}}{2(1-\cos a)}$

$$=\left[1+\left(\frac{x}{a}\right)^2\right]\left[1+\left(\frac{x}{2\pi-a}\right)^2\right]\left[1+\left(\frac{x}{2\pi+a}\right)^2\right]\left[1+\left(\frac{x}{4\pi-a}\right)^2\right]\ldots\ \&c.$$

2953 $\frac{e^x+2\cos a+e^{-x}}{2(1+\cos a)}$

$$=\left[1+\left(\frac{x}{\pi-a}\right)^2\right]\left[1+\left(\frac{x}{\pi+a}\right)^2\right]\left[1+\left(\frac{x}{3\pi-a}\right)^2\right]\left[1+\left(\frac{x}{3\pi+a}\right)^2\right]\ldots\ \&c.$$

PROOF.—Change x into ix in (2946–7).

FORMULÆ FOR THE EXPANSION OF FUNCTIONS IN TRIGONOMETRICAL SERIES.

2955 When x has any value between l and $-l$,

$$\phi(x) = \frac{1}{2l}\int_{-l}^{l}\phi(v)\,dv + \frac{1}{l}\Sigma_{n=1}^{n=\infty}\int_{-l}^{l}\phi(v)\cos\frac{n\pi(v-x)}{l}\,dv\ \ldots\text{(i.)},$$

where n must have all positive integral values in succession from 1 upwards

But, if $x=l$ or $-l$, the left side becomes $l\phi(l)+l\phi(-l)$.

PROOF.—By (2919) we have, when h is <1,

$$\frac{1-h^2}{1-2h\cos\theta+h^2} = 1+2h\cos\theta+2h^2\cos 2\theta+2h^3\cos 3\theta+\&c.\ldots$$

Put $\theta=\frac{\pi(v-x)}{l}$; multiply each side by $\phi(v)$, and integrate for v from $-l$ to l; then make $h=1$. The left side becomes, by substituting $z=v-x$,

$$\int_{-l}^{l}\frac{(1-h^2)\,\phi(v)\,dv}{1-2h\cos\frac{\pi(v-x)}{l}+h^2} = \int_{-l-x}^{l-x}\frac{(1-h^2)\,\phi(x+z)\,dz}{(1-h)^2+4h\sin^2\frac{\pi z}{2l}}.$$

When $h=1$ each element of the integral *vanishes*, excepting for values of v which lie near to x. Therefore the only appreciable value of the integral arises from such elements, and in these z will have values near to zero, both positive and negative, since x has a fixed value between l and $-l$. Let these values of z range from $-\beta$ to α. Then between these small limits we shall have

$$\frac{\sin^2\pi z}{2l} = \frac{\pi^2 z^2}{4l^2}, \text{ and } \phi(x+z)=\phi(x),$$

and the integral takes the form

$$(1-h^2)\,\phi(x)\int_{-\beta}^{\alpha}\frac{dz}{(1-h)^2+\dfrac{h\pi^2z^2}{l^2}}$$

$$=\frac{(1+h)\,l}{\pi\sqrt{h}}\,\phi(x)\left(\tan^{-1}\frac{\alpha\pi\sqrt{h}}{l\,(1-h)}+\tan^{-1}\frac{\beta\pi\sqrt{h}}{l\,(1-h)}\right)=2l\phi(x),$$

when h is made equal to unity, which establishes the formula.

In the case, however, in which $x=l$, $\sin^2\frac{\pi z}{2l}$ vanishes at both limits, that is, when $z=0$ and when $z=-2l$. We have therefore to integrate for z from $-\beta$ to 0, and also from $-2l$ to $-2l+\alpha$, α and β being any small quantities. The first integration gives $l\phi(l)$ as above, putting $\alpha=0$. The second integration, by substituting $y=z+2l$, produces a similar form with limits 0 to α, and with $\phi(x-2l)$ in the place of $\phi(x)$ giving $l\phi(-l)$ when $x=l$. Thus the total value of the integral is $l\phi(l)+l\phi(-l)$. The result is the same when $x=-l$.

That the right side of equation (i.) forms a converging series appears by integrating the general terms by Parts; thus

$$\int_{-l}^{l}\phi(v)\cos\frac{n\pi(v-x)}{l}\,dv=\frac{l}{n\pi}\left\{\phi(v)\sin\frac{n\pi(v-x)}{l}\right\}_{-l}$$

$$-\frac{l}{n\pi}\int_{-l}^{l}\phi'(v)\sin\frac{n\pi(v-x)}{l}\,dv,$$

which vanishes when n is infinite, provided $\phi'(v)$ is not infinite.

Hence the multiplication of such terms by h^n when n is infinite produces no finite result when h is made $=1$, although 1^∞ is a factor of indeterminate value.

2955*a* A function of the form $\phi(x)\cos nx$, with n infinitely great, has been called "*a fluctuating function*," for the reason that between any two finite limits of the variable x, the function changes sign infinitely often, oscillating between the values $\phi(x)$ and $-\phi(x)$. The preceding demonstration shows that the sum of all these values, as x varies continuously between the assigned limits, is zero.

By similar reasoning, the two following equations are obtained.

2956 If x has any value between 0 and l,

$$\phi(x)=\frac{1}{2l}\int_0^l\phi(v)\,dv+\frac{1}{l}\Sigma_1^\infty\int_0^l\phi(v)\cos\frac{n\pi(v-x)}{l}\,dv\ \ldots\ (2).$$

But if $x=0$, write $\tfrac{1}{2}\phi(0)$ on the left; and if $x=l$, write $\tfrac{1}{2}\phi(l)$.

If x has any value between 0 and l,

2957 $$0=\frac{1}{2l}\int_0^l\phi(v)\,dv+\frac{1}{l}\Sigma_1^\infty\int_0^l\phi(v)\cos\frac{n\pi(v+x)}{l}\,dv\ldots(3).$$

But if $x=0$, write $\frac{1}{2}\phi(0)$ on the left; and if $x=l$, write $\frac{1}{2}\phi(l)$.

2958 $$\phi(x)=\frac{1}{l}\int_0^l\phi(v)\,dv+\frac{2}{l}\,\Sigma_1^\infty\cos\frac{n\pi x}{l}\int_0^l\cos\frac{n\pi v}{l}\,\phi(v)\,dv \quad \ldots\ldots(4).$$

This formula is true for any value of x between 0 and l, *both inclusive.*

But if x be $>l$, write $\phi(x\sim 2ml)$ instead of $\phi(x)$ on the left, where $2ml$ is that even multiple of l which is nearest to x in value.

If the sign of x be changed on the right, the left side of the equation remains unaltered in every case.

2959 $$\phi(x)=\frac{2}{l}\,\Sigma_1^\infty\sin\frac{n\pi x}{l}\int_0^l\sin\frac{n\pi v}{l}\,\phi(v)\,dv \quad \ldots\ldots\ldots (5).$$

This formula holds for any value of x between 0 and l *exclusive* of those values.

If x be $>l$, write $\pm\phi(x\sim 2ml)$ instead of $\phi(x)$ on the left, $+$ or $-$ according as x is $>$ or $<$ $2ml$, the even multiple of l which is nearest to x in value.

But if x be 0 or l, or any multiple of l, the left side of this equation vanishes.

If the sign of x be changed on the right, the left side is numerically the same in every case, but of opposite sign.

Proof.—For (2958–9). To obtain (4) take the sum, and to obtain (5) take the difference, of equations (2) and (3). To determine the values of the series when x is $>l$, put $x=2ml\pm x'$, so that x' is $<l$.

Examples.

For all values of x, from 0 to π inclusive,

2960 $$x=\frac{\pi}{2}-\frac{4}{\pi}\left\{\cos x+\frac{\cos 3x}{3^2}+\frac{\cos 5x}{5^2}+\&c.\right\}.$$

Proof.—In formula (4) put $\phi(x)=x$ and $l=\pi$, then

$$\int_0^\pi v\cos nv\,dv=\left[\frac{v\sin nv}{n}+\frac{\cos nv}{n^2}\right]_0^\pi=-\frac{2}{n^2}\text{ or }0,$$

according as n is odd or even.

Similarly, by formula (5), equation (2929) is reproduced.

For all values of x, from $-\frac{1}{2}\pi$ to $\frac{1}{2}\pi$ inclusive.

2961 $$x=\frac{4}{\pi}\left\{\sin x-\frac{\sin 3x}{3^2}+\frac{\sin 5x}{5^2}-\&c.\right\}.$$

Proof.—Change x into $\frac{1}{2}\pi-x$ in (2961).

2962 $\frac{\pi}{2}\frac{e^{ax}-e^{-ax}}{e^{a\pi}-e^{-a\pi}} = \frac{\sin x}{a^2+1} - \frac{2\sin 2x}{a^2+2^2} + \frac{3\sin 3x}{a^2+3^2} - \&c.$

PROOF.—In formula (5) put $\phi(x) = e^{ax}-e^{-ax}$ and $l = \pi$; then

$$\int_0^\pi (e^{av}-e^{-av}) \sin nv\, dv = (-1)^{n+1}\frac{n(e^{a\pi}-e^{-a\pi})}{a^2+n^2}.$$

2963 If $\phi(x)$ be not a continuous function between $x=0$ and $x=l$, let the function be $\phi(x)$ from $x=0$ to $x=a$, and $\psi(x)$ from $x=a$ to $x=l$; then, in formulæ (4) and (5), we shall have $\phi(x)$ or $\psi(x)$ respectively on the left side, according to the situation of x between 0 and a, or between a and l. But, if $x=a$, we must write $\frac{1}{2}\{\phi(a)+\psi(a)\}$ for the left member.

PROOF.—In ascertaining the value of the integral in the demonstration of (2955), we are only concerned with the form of the function *close to* the value of x in question. Hence the result is not affected by the discontinuity unless $x=a$. In this case the integration for z is from $-\beta$ to 0 with $\phi(x)$ for the function, and from 0 to α with $\psi(x)$ for the function, producing $\frac{1}{2}\phi(a)+\frac{1}{2}\psi(a)$.

2964 Hence an expression involving x in an infinite series of sines of consecutive multiples of $\frac{\pi x}{l}$ may be found, such that, when x lies between any of the assigned limits (0 and a, a and b, b and c, ... k and l), the series shall be equal respectively to the corresponding assigned functions

$$f_1(x),\quad f_2(x),\quad f_3(x)\ \dots\ f_n(x),$$

provided that the integrals

$$\int_0^a \sin\left(\frac{n\pi x}{l}\right) f_1(x)\,dx,\qquad \int_a^b \sin\frac{n\pi x}{l} f_2(x)\,dx,\ \dots\ \int_k^l \sin\left(\frac{n\pi x}{l}\right) f_n(x)\,dx$$

can all be determined.

2965 The same is true, reading *cosine* for *sine* throughout, with the additional proviso [as appears from formula (4)] that the integrals

$$\int_0^a f_1(x)\,dx,\qquad \int_a^b f_2(x)\,dx,\ \dots\ \int_k^l f_n(x)\,dx$$

can also be determined.

2966 Ex. 1.—To find in the form of a series of cosines of multiples of x a function of x which shall be equal to the constants α, β, or γ, according as x lies between 0 and a, a and b, or b and π.

Formula (4) produces, putting $l=\pi$,

$$\frac{1}{\pi}\left\{\int_0^a \alpha\,dx+\int_a^b \beta\,dx+\int_b^\pi \gamma\,dx\right\}$$

$$+\frac{2}{\pi}\,\Sigma_{n=1}^{n=\infty}\frac{1}{n}\cos nx\left\{\int_0^a \alpha\cos nx\,dx+\int_a^b \beta\cos nx\,dx+\int_b^\pi \gamma\cos nx\,dx\right\}$$

$$=\frac{1}{\pi}\left\{a(\alpha-\beta)+b(\beta-\gamma)+\pi\gamma\right\}$$

$$+\frac{2}{\pi}\,\Sigma_{n=1}^{n=\infty}\frac{1}{n^2}\cos nx\left\{(\alpha-\beta)\sin na+(\beta-\gamma)\sin nb+\gamma\sin n\pi\right\}.$$

2967 Ex. 2.—To find a function of x having the value c, when x lies between 0 and a, and the value zero when x lies between a and l.

By formula (4), we shall have

$$\int_0^l \cos\frac{n\pi v}{l}\,\phi(v)\,dx = c\int_0^a \cos\frac{n\pi v}{l}\,dv=\frac{cl}{n\pi}\sin\frac{n\pi a}{l},$$

since $\phi(v) = c$ from 0 to a, and zero from a to l.

Therefore $\phi(x)=\dfrac{ca}{l}+\dfrac{2c}{\pi}\left\{\sin\dfrac{\pi a}{l}\cos\dfrac{\pi x}{l}+\dfrac{1}{2}\sin\dfrac{2\pi a}{l}\cos\dfrac{2\pi x}{l}\right.$

$$\left.+\frac{1}{3}\sin\frac{3\pi a}{l}\cos\frac{3\pi x}{l}+\&c.\right\}.$$

When $x=a$, the value is $\frac{1}{2}[\phi(a)+0]=\frac{1}{2}c$, by the rule in (2963). This may be verified by putting $a=-1$ in (2923).

2968 Ex. 3.—To find a function of x which becomes equal to kx when x lies between 0 and $\frac{1}{2}l$, and equal to $k(l-x)$ when x lies between $\frac{1}{2}l$ and l.

By formula (4),

$$\int_0^l \phi(v)\cos\frac{n\pi v}{l}\,dv=\int_0^{\frac{1}{2}l} kv\cos\frac{n\pi v}{l}\,dv+\int_{\frac{1}{2}l}^l k(l-v)\cos\frac{n\pi v}{l}\,dv.$$

This reduces to $\dfrac{kl^2}{\pi^2n^2}\left(2\cos\dfrac{n\pi}{2}-\cos n\pi-1\right)=-\dfrac{4kl^2}{\pi^2n^2}$ or 0,

according as n is, or is not, of the form $4m+2$. Also

$$\int_0^l \phi(v)\,dv=k\int_0^{\frac{1}{2}l} v\,dv+k\int_{\frac{1}{2}l}^l (l-v)\,dv=\frac{kl^2}{4};$$

$$\therefore\quad \phi(x)=\frac{kl}{4}-\frac{8kl}{\pi^2}\left\{\frac{1}{2^2}\cos\frac{2\pi x}{l}+\frac{1}{6^2}\cos\frac{6\pi x}{l}+\frac{1}{10^2}\cos\frac{10\pi x}{l}+\&c.\right\}.$$

APPROXIMATE INTEGRATION.

2991 Let $\int_a^b f(x)\,dx$ be the integral, and let the curve $y=f(x)$ be drawn. By summing the areas of the trapezoids, whose parallel sides are the $n+1$ equidistant ordinates

$y_0, y_1, \ldots y_n$, we find, for a first approximation,

$$\int_a^b f(x)\,dx = \frac{b-a}{2n}\,(y_0+2y_1+2y_2+\ldots+2y_{n-1}+y_n)\ldots\ldots(\text{i.})$$

SIMPSON'S METHOD.

2992 If y_1 be the ordinate intermediate between $y_0=f(a)$ and $y_2=f(b)$, then, approximately,

$$\int_a^b f(x)\,dx = \frac{b-a}{6}\,(y_0+4y_1+y_2)\;\ldots\ldots\ldots\ldots\;(\text{ii.})$$

PROOF.—Take $n=3$ in formula (i.); write y_2 for y_3, and suppose two intermediate ordinates each equal to y_1. The area thus obtained is equal to what it would be if the bounding curve were a parabola having for ordinates y_0, y_1, y_2 parallel to its axis. Otherwise by Cotes's formula (2995).

2993 A closer approximation, in terms of $2n+1$ equidistant ordinates, is given by Simpson's formula,

$$\int_0^1 f(x)\,dx = \frac{1}{6n}\Big[y_0+y_{2n}+4\,(y_1+y_3+\ldots+y_{2n-1})$$
$$+2\,(y_2+y_4+\ldots+y_{2n-2})\Big]\;\ldots\ldots(\text{iii.})$$

PROOF.—We have

$$\int_0^1 f(x)\,dx = \int_0^{\frac{1}{n}} f(x)\,dx + \int_{\frac{1}{n}}^{\frac{2}{n}} f(x)\,dx + \ldots + \int_{\frac{n-1}{n}}^{1} f(x)\,dx.$$

Apply formula (ii.) to each integral and add the results, denoting by y_r the value of y corresponding to $x=\frac{r}{2n}$.

2994 When the limits are a and b, the integral can be changed into another having the limits 0 and 1, by substituting $x=a+(b-a)\,y$.

COTES'S METHOD.

Let n equidistant ordinates, and the corresponding abscissæ, be

$$y_0,\, y_1,\, y_2 \ldots y_{n-1},\, y_n \quad\text{and}\quad 0,\, \frac{1}{n},\, \frac{2}{n} \ldots \frac{n-1}{n},\, 1.$$

2995 A formula for approximation will then be

$$\int_0^1 f(x)\,dx = A_0y_0+A_1y_1+\ldots+A_ry_r+\ldots+A_ny_n \quad (\text{iv.}),$$

where
$$A_r = \frac{(-1)^{n+r}}{\lfloor r \,\lfloor n-r} \int_0^1 \frac{(nx)_{-1}^{(n)}}{nx-r}\,dx. \qquad (2452)$$

Proof.—The method consists in substituting for $f(x)$ the integral function

$$\psi(x) = (-1)^n \left[\frac{(nx)_{-1}^{(n)}}{nx \, |\underline{n}} y_0 \ldots + (-1)^r \frac{(nx)_{-1}^{(n)}}{(nx-r) \, |\underline{r} \, |\underline{n-r}} y_r \ldots \right.$$
$$\left. \ldots + (-1)^n \frac{(nx)_{-1}^{(n)}}{(nx-n) \, |\underline{n}} y_n \right],$$

r taking all integral values from 0 to n inclusive. When $x = r$, we have $\psi(r) = y_r$; so that $\psi(x)$ has $n+1$ values in common with $f(x)$. The approximate value of the integral is therefore $\int_0^1 \psi(x)\, dx$, and may be written as in (iv.)

By substituting $1-x$, it appears that

$$\int_0^1 \frac{(nx)_{-1}^{(n)}}{nx-r} dx = (-1)^n \int_0^1 \frac{(nx)_{-1}^{(n)}}{nx-(n-r)} dx;$$

and therefore $A_r = A_{n-r}$. Consequently it is only necessary to calculate half the number of coefficients in (iv.)

2996 The coefficients corresponding to the values of n from 1 to 10 are as follows. Every number has been carefully verified, and two misprints in Bertrand corrected; namely, 2089 for 2989 in line 8, and 89500 for 89600 in line 11.

$n=1$: $A_0 = A_1 = \frac{1}{2}$.

$n=2$: $A_0 = A_2 = \frac{1}{6}$, $A_1 = \frac{2}{3}$.

$n=3$: $A_0 = A_3 = \frac{1}{8}$, $A_1 = A_2 = \frac{3}{8}$.

$n=4$: $A_0 = A_4 = \frac{7}{90}$, $A \ = A_3 = \frac{16}{45}$, $A_2 = \frac{2}{15}$.

$n=5$: $A_0 = A_5 = \frac{19}{288}$, $A_1 = A_4 = \frac{25}{96}$, $A_2 = A_3 = \frac{25}{144}$.

$n=6$: $A_0 = A_6 = \frac{41}{840}$, $A_1 = A_5 = \frac{9}{35}$, $A_2 = A_4 = \frac{9}{280}$, $A_3 = \frac{34}{105}$.

$n=7$: $A_0 = A_7 = \frac{751}{17280}$, $A_1 = A_6 = \frac{3577}{17280}$,

$A_2 = A_5 = \frac{49}{640}$, $A_3 = A_4 = \frac{2989}{17280}$.

$n=8$: $A_0 = A_8 = \frac{989}{28350}$, $A_1 = A_7 = \frac{2944}{14175}$, $A_2 = A_6 = -\frac{464}{14175}$,

$A_3 = A \ = \frac{5248}{14175}$, $A_4 = -\frac{454}{2835}$.

$n=9$: $A_0 = A_9 = \frac{2857}{89600}$, $A_1 = A_8 = \frac{15741}{89600}$, $A_2 = A_7 = \frac{27}{2240}$,

$A_3 = A_6 = \frac{1209}{5600}$, $A_4 = A_5 = \frac{2889}{44800}$.

$n=10$: $A_0 = A_{10} = \frac{16067}{598752}$, $A_1 = A_9 = \frac{26575}{149688}$, $A_2 = A_8 = -\frac{16175}{199584}$,

$A_3 = A_7 = \frac{5675}{12474}$, $A_4 = A_6 = -\frac{4825}{11088}$, $A_5 = \frac{17807}{24948}$.

GAUSS'S METHOD.

2997 When $f(x)$ is an integral algebraic function of degree $2n$, or lower, Gauss's formula of approximation is

$$\int_0^1 f(x) = A_0 f(x_0) + \ldots + A_r f(x_r) + \ldots + A_n f(x_n) \qquad \text{(v)},$$

where $x_0 \ldots x_r \ldots x_n$ are the $n+1$ roots of the equation

$$\psi(x) \equiv d_{(n+1)x}\{x^{n+1}(x-1)^{n+1}\} = 0 \quad \ldots\ldots\ldots \text{(vi.)},$$

and $$A_r = \int_0^1 \frac{(x-x_0)\ldots(x-x_{r-1})(x-x_{r+1})\ldots(x-x_n)}{(x_r-x_0)\ldots(x_r-x_{r-1})(x_r-x_{r+1})\ldots(x_r-x_n)}\,dx \quad \ldots\text{(vii.)}$$

The formula is evidently applicable to a function of any form which can be expanded in a converging algebraic series not having a fractional index in the first $2n$ terms. The result will be the approximate value of those terms.

Proof.—Let $\psi(x) = (x-x_0)(x-x_1)\ldots(x-x_n)$,

and let $f(x) = Q\psi(x) + R$ .... (viii.),

where $f(x)$ is of the $2n^{\text{th}}$ degree, Q of the $n-1^{\text{th}}$, and R of the nth, since $\psi(x)$ is of the $n+1^{\text{th}}$ degree.

Then the method consists in choosing a function $\psi(x)$ of the $n+1^{\text{th}}$ degree, so that $\int_0^1 Q\psi(x)\,dx$ shall vanish; and a function R of the nth degree, which shall coincide with $f(x)$ when x is any one of the $n+1$ roots of $\psi(x) = 0$.

(i.) To ensure that $\int_0^1 Q\psi(x)\,dx = 0$. We have, by Parts, successively, writing N for $\psi(x)$, and with the notation of (2148),

$$\int_x x^p N = x^p \int_x N - p\int_x \left(x^{p-1}\int_x N\right)$$

$$= x^p \int_x N - px^{p-1}\int_{2x} N + p(p-1)\int_x\left(x^{p-2}\int_{2x} N\right)$$

$$= \qquad \&c. \qquad\qquad \&c.$$

$$= x^p\int_x N - px^{p-1}\int_{2x} N + p(p-1)\,x^{p-2}\int_{3x} N - \ldots \pm \lfloor p \int_{(p+1)x} N \qquad \text{(ix.)}$$

Now $Q\psi(x)$ is made up of terms like $x^p\psi(x)$ with integral values of p from 0 to $n-1$ inclusive. Hence, if the value (vi.) be assumed for $\psi(x)$, we see, by (ix.), that $\int_0^1 Q\psi(x)\,dx$ will vanish at both limits, because the factors x and $x-1$ will appear in every term.

(ii.) Let R be the function on the right of equation (v.) Then, when $x=x_r$, we see, by (vii.), that $A_r=1$, and that the other coefficients all vanish. Hence R becomes $f(x)$ whenever x is a root of $\psi(x)=0$.

The values of the constants corresponding to the first six values of n, according to Bertrand, are as follows. The abscissæ values, only, have been recalculated by the author.

$n=0$:	$x_0=\cdot 5$,	$A_0=1$.	
$n=1$:	$x_0=\cdot 21132487$, $x_1=\cdot 78867513$.	$A_0=A_1=\cdot 5$,	$\log=9\cdot 6989700$;
$n=2$:	$x_0=\cdot 11270167$, $x_1=\cdot 5$; $x_2=\cdot 88729833$,	$A_0=A_2=\frac{5}{18}$, $A_1=\frac{4}{9}$,	$\log=9\cdot 4436975$; $\log=9\cdot 6478175$.
$n=3$:	$x_0=\cdot 06943184$, $x_1=\cdot 33000948$, $x_2=\cdot 66999052$; $x_3=\cdot 93056816$.	$A_0=A_3=\cdot 1739274$, $A_1=A_2=\cdot 3260726$,	$\log=9\cdot 2403681$; $\log=9\cdot 5133143$;
$n=4$:	$x_0=\cdot 04691008$, $x_1=\cdot 23076534$, $x_2=\cdot 5$, $x_3=\cdot 76923466$; $x_4=\cdot 95308992$.	$A_0=A_4=\cdot 1184634$, $A_1=A_3=\cdot 2393143$, $A_2=\cdot 2844444$,	$\log=9\cdot 0735834$; $\log=9\cdot 3789687$; $\log=9\cdot 4539975$;
$n=5$:	$x_0=\cdot 03376524$, $x_1=\cdot 16939531$, $x_2=\cdot 38069041$, $x_3=\cdot 61930959$; $x_4=\cdot 83060469$; $x_5=\cdot 96623476$.	$A_0=A_5=\cdot 0856622$, $A_1=A_4=\cdot 1803808$, $A_2=A_3=\cdot 2339570$,	$\log=8\cdot 9327895$; $\log=9\cdot 2561903$; $\log=9\cdot 3691360$;

As a criterion of the relative degrees of approximation obtained by the foregoing methods, Bertrand gives the following values of

$$\int_0^1 \frac{\log(1+x)}{1+x^2}\,dx = \frac{\pi}{8}\log 2 = \cdot 2721982613.$$

Method of Trapezoids,	$n=10$,	·2712837.
Simpson's method,	$n=10$,	·2722012.
Cotes's ,,	$n=5$,	·2722091.
Gauss's ,,	$n=4$,	·2721980.

For other formulæ of approximation, see also p. 357.

CALCULUS OF VARIATIONS.

FUNCTIONS OF ONE INDEPENDENT VARIABLE.

3028 Let $y = f(x)$, and let V be a known function of x, y, and a certain number of the derivatives y_x, y_{2x}, y_{3x}, &c. The chief object of the Calculus of Variations is to find the *form* of the function $f(x)$ which will make

$$U = \int_{x_0}^{x_1} V\,dx \qquad \text{(i.)}$$

a maximum or minimum. See (3084).

Denote y_x, y_{2x}, y_{3x}, &c. by p, q, r, &c.

For a maximum or minimum value of U, δU must vanish. To find δU, let δy be the change in y caused by a change in the form of the function $y = f(x)$, and let δp, δq, &c. be the consequent changes in p, q, &c.

Now, $$p = y_x.$$

Therefore the new value of p, when a change takes place in the form of the function y, is

$$p + \delta p = (y + \delta y)_x = y_x + (\delta y)_x,$$

therefore $$\delta p = (\delta y)_x; \quad \text{that is,} \quad \delta\left(\frac{dy}{dx}\right) = \frac{d(\delta y)}{dx}.$$

Similarly, $$\delta q = (\delta p)_x,$$
$$\delta r = (\delta q)_x, \text{ \&c.} \qquad \text{(ii.)}$$

Now $\delta U = \int_{x_0}^{x_1} \delta V\,dx$ (1483). Expand by Taylor's theorem, rejecting the squares of δy, δp, δq, &c., and we find

$$\delta U = \int_{x_0}^{x_1} (V_y\,\delta y + V_p\,\delta p + V_q\,\delta q + \ldots)\,dx,$$

or, denoting V_y, V_p, V_q, ... by N, P, Q, ...,

$$\delta U = \int_{x_0}^{x_1} (N\delta y + P\delta p + Q\delta q + \ldots)\,dx, \qquad \text{(iii.)}$$

Integrate each term after the first by Parts, observing that by (ii.) $\int \delta p\, dx = \delta y$, &c., and repeat the process until the final integrals involve $\delta y\, dx$. Thus

$$\int N\,\delta y\, dx \text{ is unaltered,}$$

$$\int P\delta p\, dx = P\delta y - \int P_x \delta y\, dx,$$

$$\int Q\delta p\, dx = Q\delta p - Q_x\delta y + \int Q_{2x}\delta y\, dx,$$

$$\int R\delta r\, dx = R\delta q - R_x\delta p + R_{2x}\delta y - \int R_{3x}\delta y\, dx,$$

$$\ldots \quad \ldots \quad \ldots \quad \ldots \quad \ldots$$

3029 Hence, collecting the coefficients of δy, δp, δq, &c.,

$$\delta U = \int_{x_0}^{x_1} (N - P_x + Q_{2x} - R_{3x} + \ldots)\,\delta y\, dx$$

$$+\delta y_1 (P - Q_x + R_{2x} - \ldots)_1 - \delta y_0 (P - Q_x + R_{2x} - \ldots)_0$$
$$+\delta p_1 (Q - R_x + S_{2x} - \ldots)_1 - \delta p_0 (Q - R_x + S_{2x} - \ldots)_0$$
$$+\delta q_1 (R - S_x + T_{2x} - \ldots)_1 - \delta q_0 (R - S_x + T_{2x} - \ldots)_0 + \&c. \quad \text{(iv.)}$$

The terms affected by the suffixes 1 and 0 must have x made equal to x_1 and x_0 respectively after differentiation.

Observe that P_x, Q_x, &c. are here *complete* derivatives; y, p, q, r, &c., which they involve, being functions of x.

Equation (iv.) is written in the abbreviated form,

3030 $$\delta U = \int_{x_0}^{x_1} K\,\delta y\,\delta x + H_1 - H_0 \quad \ldots\ldots\ldots\ldots \quad \text{(v.)}$$

The condition for the vanishing of δU, that is, for minimum value of U, is

3031 $$K = N - P_x + Q_{2x} - R_{3x} + \&c. = 0 \quad \ldots\ldots\ldots \text{(vi.)},$$

3032 $$\text{and} \quad H_1 - H_0 = 0 \quad \ldots\ldots\ldots\ldots \quad \text{(vii.)}$$

PROOF.—For, if not, we must have

$$\int_{x_0}^{x_1} K\,\delta y\, dx = H_0 - H_1;$$

that is, the integral of an arbitrary function (since y is arbitrary in form) can be expressed in terms of the limits of y and its derivatives; which is impossible. Therefore $H_1 - H_0 = 0$. Also $K = 0$; for, if the integral could vanish without K vanishing, the *form* of the function δy would be restricted, which is inadmissible.

The order of K is twice that of the highest derivative contained in V. Let n be the order of K, then there will be $2n$ constants in the solution of equation (vi.) and the same number of equations for determining them. For there are $2n$ terms in equation (vii.) involving δy_1, δy_0, dp_1, &c. If any of these quantities are arbitrary, their coefficients must vanish in order that equation (vii.) may hold; and if any are not arbitrary, they will be fixed in their values by given equations which, together with the equations furnished by the coefficients which have to be equated to zero, will make up, in all, $2n$ equations.

PARTICULAR CASES.

3033 I.—When V does not involve x explicitly, a first integral of the equation $K=0$ can always be found. Thus, if, for example,

$$V=\phi(y, p, q, r, s),$$

a first integral will be

$$\begin{aligned} V = \; & Pp \\ & + Qp_x - Q_x p \\ & + Rp_{2x} - R_x p_x + R_{2x} p \\ & + Sp_{3x} - S_x p_{2x} + S_{2x} p_x - S_{3x} p + C. \end{aligned}$$

The order of this equation is less by one than that of (vi.)

PROOF.—We have $V_x = Np + Pq + Qr + Rs$.
Substitute the value of N from (vi.), and it will be found that each pair of terms involving P, Q, R, &c. is an exact differential.

3034 II.—When V does not involve y, a first integral can be found at once, for then $N=0$, and therefore $K=0$, and

we have $$P_x - Q_{2x} + R_{3x} - \&c. = 0;$$

and therefore $$P - Q_x + R_{2x} - \&c. = A.$$

3035 III.—When V involves only y and p,

$$V = Pp + A, \qquad \text{by Case I.}$$

3036 IV.—When V involves only p and q,

$$V = Qq + Ap + B. \qquad \text{See also (3046).}$$

PROOF. $K = -P_x + Q_{2x} = 0$, giving, by integration, $P = Q_x + A$.
Also $V_x = Pq + Qr = Aq + Q_x q + Qr$.
Integrating again, we find $V = Qq + Ap + B$,
a reduction from the fourth to the second order of differential equations.

3037 Ex.—To find the brachistochrone, or curve of quickest descent, from a point O taken as origin to a point x_1y_1, measuring the axis of y downwards.*

Velocity at a depth $y = \sqrt{2gy}$.

Therefore time of descent $= \int_{x_0}^{x_1} \frac{\sqrt{1+p^2}}{\sqrt{2gy}}\, dx$.

Here $$V = \sqrt{\frac{1+p^2}{y}} = \frac{p^2}{\sqrt{\{y(1+p^2)\}}} + A, \text{ by Case III.}$$

By reduction, $y(1+p^2) = \frac{1}{A^2} = 2a$, an arbitrary constant.

That is, since $p = \tan\theta$, $y = 2a\cos^2\theta$, the defining property of a cycloid having its vertex downwards and a cusp at the origin

$$H_1 - H_0 \text{ reduces to } \frac{1}{\sqrt{2a}}\{(p\delta y)_1 - (p\delta y)_0\} = 0.$$

If the extreme points are fixed, δy_1 and δy_0 both vanish.

The values x_1, y_1, at the lower point, determine a.

Suppose x_1, but not y_1, is fixed. Then δy_1 is arbitrary; therefore its coefficient in (3) $(P - Q_x + \&c.)_1$ must vanish; that is, $(V_p)_1 = 0$, or $\left\{\frac{p}{\sqrt{y(1+p^2)}}\right\}_1 = 0$, therefore $p_1 = 0$, which means that the tangent at the lower point is horizontal, and the curve is therefore a complete half cycloid.

3038 In the example of the brachistochrone, it is useful to notice that—

(i.) If the extreme points are fixed, δy_0, δy_1 both vanish.

(ii.) If the tangents at the extreme points have fixed directions, δp_0, δp_1 both vanish.

(iii.) If the curvature at each extremity is fixed in value, δp_0, δq_0, δp_1, δq_1 all vanish.

(iv.) If the abscissæ x_0, x_1 only have fixed values, δy_0, δy_1 are then arbitrary, and their coefficients in $H_1 - H_0$ must vanish.

3039 When the limits x_0, x_1 are variable, add to the value of δU in (3029) $$V_1 dx_1 - V_0 dx_0.$$

Proof.—The partial increment of U, due to changes in x_1 and x_0, is

$$\frac{dU}{dx_1}dx_1 + \frac{dU}{dx_0}dx_0 = V_1 dx_1 - V_0 dx_0. \qquad \text{By (2253).}$$

3040 When x_1 and y_1, x_0 and y_0 are connected by given equations, $$y_1 = \psi(x_1), \quad y_0 = \chi(x_0).$$

Rule.—Put

$$\delta y_1 = \{\psi'(x_1) - p_1\}\, dx_1 \quad \text{and} \quad \delta y_0 = \{\chi'(x_0) - p_0\}\, dx_0,$$

* The Calculus of Variations originated with this problem, proposed by John Bernoulli in 1696.

and afterwards equate to zero the coefficients of dx_1 and dx_0, because the values of the latter are arbitrary.

PROOF.—$y_1+\delta y_1$ being a function of x_1,

$$(y_1+\delta y_1)+d_{x_1}(y_1+\delta y_1)\,dx_1 = \psi(x_1+dx_1) = \psi(x_1)+\psi'(x_1)\,dx_1;$$

therefore $\delta y_1+p_1 dx_1 = \psi'(x_1)\,dx_1$, neglecting $\delta p\,dx_1$.

EX.—In the brachistochrone problem (3037), the result thus arrived at signifies that the cycloid is at right angles to each of the given curves at its extremities.

3041 If V involves the limits $x_0, x_1, y_0, y_1, p_0, p_1$, &c., the terms to be added to δU in (3029), on account of the variation of any of these quantities, are

$$dx_1\int_{x_0}^{x_1}\{V_{x_1}+V_{y_1}p_1+V_{p_1}q_1+\dots\}\,dx$$
$$+dx_0\int_{x_0}^{x_1}\{V_{x_0}+V_{y_0}p_0+V_{p_0}q_0+\dots\}\,dx$$
$$+\int_{x_0}^{x_1}\{V_{y_0}\delta y_0+V_{y_1}\delta y_1+V_{p_0}\delta p_0+V_{p_1}\delta p_1+\text{\&c.}\}\,dx.$$

In the last integral, δy_0, δy_1, δp_0, &c. may be placed outside the symbol of integration since, they are not functions of x.

Hence, when V involves the limits $x_0, x_1, y_0, y_1, p_0, p_1$, &c., and those limits are variable, the complete expression for δU is

3042 $$\delta U=\int_{x_0}^{x_1}\{N-P_x+Q_{2x}-R_{3x}+\text{\&c.}\}\,\delta y\,dx$$
$$+\left\{V_1+\int_{x_0}^{x_1}(V_{x_1}+V_{y_1}p_1+V_{p_1}q_1+\dots)\,dx\right\}dx_1$$
$$-\left\{V_0-\int_{x_0}^{x_1}(V_{x_0}+V_{y_0}p_0+V_{p_0}q_0+\dots)\,dx\right\}dx_0$$
$$+\left\{(P-Q_x+R_{2x}-\dots)_1+\int_{x_0}^{x_1}V_{y_1}dx\right\}\delta y_1$$
$$-\left\{(P-Q_x+R_{2x}-\dots)_0-\int_{x_0}^{x_1}V_{y_0}dx\right\}\delta y_0$$
$$+\left\{(Q-R_x+S_{2x}-\dots)_1+\int_{x_0}^{x_1}V_{p_1}dx\right\}\delta p_1$$
$$-\left\{(Q-R_x+S_{2x}-\dots)_0-\int_{x_0}^{x_1}V_{p_0}dx\right\}\delta p_0+\text{\&c.}$$

3043 Also, if $y_1 = \psi(x_1)$ and $y_0 = \chi(x_0)$ be equations restricting the limits, put

$$\delta y_1 = \{\psi'(x_1) - p_1\}\, dx_1 \quad \text{and} \quad \delta y_0 = \{\chi'(x_0) - p_0\}\, dx_0. \qquad (3040)$$

The relation $K = 0$ is unaltered, and, by means of it, the additional integrals which appear in the value of $H_1 - H_0$ become definite functions of x.

3044 Ex.—To find the curve of quickest descent of a particle from some point on the curve $y_0 = \chi(x_0)$ to the curve $y_1 = \psi(x_1)$.

As in (3037), $t = \dfrac{1}{\surd(2g)} \displaystyle\int_{x_0}^{x_1} \sqrt{\frac{1+p^2}{y-y_0}}\, dx$, $V = \sqrt{\dfrac{1+p^2}{y-y_0}}$, and contains y, p, and y_0. Equation (3042) now reduces to

$$\delta U = \int_{x_0}^{x_1} (N - P_x)\, \delta y\, dx + V_1\, dx_1 - \{V_0 - \int_0^{x_1} V_{y_0} p_0\, dx\}\, dx_0$$
$$+ P_1 \delta y_1 - \{P_0 - \int_{x_0}^{x} V_{y_0}\, dx\}\, \delta y_0 \ldots\ldots (1).$$

Now $K = 0$ gives $N - P_x = 0$; therefore $V = Pp + A$ (3035);

therefore $$\sqrt{\frac{1+p^2}{y-y_0}} = \frac{p^2}{\surd(y-y_0)(1+p^2)} + A.$$

Clearing of fractions, and putting $A = \dfrac{1}{\surd(2a)}$, this becomes

$$(y-y_0)(1+p^2) = 2a \ldots\ldots (2).$$

Also $$P = V_p = \frac{p}{\surd(y-y_0)(1+p^2)\}} = \frac{p}{\surd(2a)} \ldots\ldots (3).$$

Hence $$V = \frac{1+p^2}{\surd(2a)};\quad V_{y_0} = -V_y = -N = -P_x \text{ (by } K = 0),$$

therefore $$\int_{x_0}^{x_1} V_{y_0}\, dx = P_0 - P_1 = \frac{p_0 - p_1}{\surd(2a)} \ldots\ldots (4).$$

Substituting the values (2), (3), (4), in (1), the condition $H_1 - H_0$ produces

$$(1+p_1^2)\, dx_1 - (1 + p_0 p_1)\, dx_0 + p_1\, \delta y_1 - p_1 \delta y_0 = 0.$$

Next, put for δy_1 and δy_0 the values in (3040); thus the equation becomes

$$\{1 + p_1 \psi'(x_1)\}\, dx_1 - \{1 + p_1 \chi'(x_0)\}\, dx_0 = 0 \ldots\ldots (5);$$

dx_1, dx_0 being arbitrary, their coefficients must vanish; therefore

$$p_1 \psi'(x_1) = -1 \quad \text{and} \quad p_1 \chi'(x_0) = -1.$$

That is, the tangents of the given curves ψ and χ at the points $x_0 y_0$ and $x_1 y_1$ are both perpendicular to the tangent of the brachistochrone at the point $x_1 y_1$.

Equation (2) shews that the brachistochrone is a cycloid with a cusp at the starting-point, since there $y = y_0$, and therefore $p = \infty$.

OTHER EXCEPTIONAL CASES.

(Continued from 3036.)

3045 V.—Denoting $y, y_x, y_{2x} \dots y_{nx}$ by $y, p_1, p_2 \dots p_n$;
and $V_y, V_{p_1}, V_p \dots V_{p_n}$ by $N, P_1, P_2 \dots P_n$;
let the first m of the quantities y, p_1, p_2, &c. be wanting in the function V; so that

$$V = f(x, p_m, p_{m+1} \dots p_n).$$

Then $K = d_{mx}P_m - d_{(m+1)x}P_{m+1} + \dots (-1)^{n-m} d_{nx}P_n = 0,$

which equation, being integrated m times, becomes

$$P_m - d_xP_{m+1} + d_{2x}P_{m+2} - \dots (-1)^{n-m} d_{(n-m)x}P_n$$
$$= c_0 + c_1x + \dots + c_{m-1}x^{m-1} \quad \dots\dots \text{(i.)},$$

a differential equation of the order $2n-m$.

3046 VI.—Let x also be wanting in V, so that

$$V = f(p_m, p_{m+1} \dots p_n);$$

then $K = 0$ is the same as before, and produces the same differential equation (i.) From that equation take the value of P_m, and substitute it in

$$V_x = P_m p_{m+1} + P_{m+1} p_{m+2} + \dots + P_n p_{n+1}.$$

Each pair of terms, such as $P_{m+2}p_{m+3} - d_{2x}P_{m+2}p_{m+1}$, is an exact differential; and we thus find

$$V = c + P_{m+1}p_{m+1} + (P_{m+2}p_{m+2} - d_xP_{m+2}p_{m+1}) + \dots$$
$$+ (P_np_n - d_xP_np_{n-1} + d_{2x}P_np_{n-2}) - \dots (-1)^{n-m-1} d_{(n-m-1)x}P_np_{m+1}$$
$$+ \int (c_0 + c_1x + \dots + c_{m-1}x^{m-1})\, p_{m+1}\, dx.$$

The resulting equation will be of the order $2n-m-1$, or $m+1$ degrees lower than the original equation.

3047 VII.—If V. be a *linear* function of p_n, that being the highest derivative it contains, P_n will not then contain p_n. Therefore $d_{nx}P_n$ will be, at most, of the order $2n-1$. Indeed, in this case, the equation $K = 0$ cannot be of an order higher than $2n-2$. (*Jellett*, p. 44.)

3048 VIII.—Let p_m be the lowest derivative which V involves; then, if $P_m = f(x)$, and if only the limiting values of x and of derivatives higher than the mth be given, the problem cannot generally be solved. (*Jellett*, p. 49.)

3049 IX.—Let $N = 0$, and let the limiting values of x alone be given; then the equation $K = 0$ becomes

$$P_x - Q_{2x} + R_{3x} - \&c. = 0,$$

or, by integration, $P - Q_x + R_{2x} - \&c. = c$,

and the two conditions furnished by equating to zero the coefficients of δy_1, δy_0, viz.,

$$(P - Q_x + \&c.)_1 = 0, \qquad (P - Q_x + \&c.)_0 = 0,$$

are equivalent to the single equation $c = 0$, and therefore $H_1 - H_0 = 0$ supplies but $2n-1$ equations instead of $2n$, and the problem is indeterminate.

3050 Let $U = \int_{x_0}^{x_1} V\,dx + V'$, where

$$V = F(x, y, p, q \ldots) \quad \text{and} \quad V' = f(x_0, x_1, y_0, y_1, p_0, p_1, \&c.)$$

The condition for a maximum or minimum value of U arising from a variation in y, is, as before, $K = 0$; and the terms to be added to $H_1 - H_0$ are

$$V'_{x_0} dx_0 + V'_{y_0} \delta y_0 + V'_{p_0} \delta p_0 + \ldots + V'_{x_1} dx_1 + V'_{y_1} \delta y_1 + \&c.$$

If the order of V be n, and the number of increments dx_0, δy_0, &c. be greater than $n+1$, the number of independent increments will exceed the number of arbitrary constants in K, and no maximum or minimum can be found.

Generally, U does not in this case admit of a maximum or minimum if either V or V' contains either of the limiting values of a derivative of an order = or > than that of the highest derivative found in V. (*Jellett*, p. 72.)

FUNCTIONS OF TWO DEPENDENT VARIABLES.

3051 Let V be a function of two dependent variables y, z, and their derivatives with respect to x; that is, let

$$\boldsymbol{V = f(x, y, p, q \ldots z, p', q' \ldots)} \quad \ldots\ldots\ldots\ldots\ldots (1),$$

where p, q, ..., as before, are the successive derivatives of y, and p', q', ... those of z.

Then, if the *forms* of the functions y, z vary, the condition for a maximum or minimum value of U or $\int_{x_0}^{x_1} V dx$ is

$$\delta U = \int_{x_0}^{x_1} (K\delta y + K'\delta z)\, dx + H_1 - H_0 + H_1' - H_0' = 0 \;\ldots\; (2).$$

Here K', H' involve $z, p', q', \ldots$, precisely as K, H involve $y, p, q, \ldots$; the values of the latter being given in (3029).

3052 First, if y and z are independent, equation (2) necessitates the following conditions:

$$K = 0, \quad K' = 0, \quad H_1 - H_0 + H_1' - H_0' = 0 \;\ldots\ldots\; (3).$$

The equations $K = 0$, $K' = 0$ give y and z in terms of x, and the constants which appear in the solution must be determined by equating to zero the coefficients of the arbitrary quantities $\delta y_0, \delta y_1, \delta p_0, \delta p_1 \ldots \delta z_0, \delta z_1, \delta p_0', \delta p_1', \ldots,$ which are found in the equation

$$H_1 - H_0 + H_1' - H_0' = 0 \;\ldots\ldots\ldots\ldots\ldots\; (4).$$

3053 The number of equations so obtained is equal to the number of constants to be determined.

PROOF.—Let $V = f(x, y, y_x, y_{2x} \ldots y_{nx}, z, z_x, z_{2x} \ldots z_{mx})$,
K is of order $2n$ in y, and $\therefore$ of form $\phi(x, y, y_x \ldots y_{2nx}, z, z_x \ldots z_{(m+n)x}) \ldots$ (i.),
K' is of order $2m$ in z, and $\therefore$ of form $\phi(x, y, y_x \ldots y_{(m+n)x}, z, z_x \ldots z_{2mx}) \ldots$ (ii.)
Differentiating (i.) $2m$ times, and (ii.) $m+n$ times, $3m+n+2$ equations are obtained, between which, if we eliminate $z, z_x \ldots z_{(3m+n)x}$, we get a resulting equation in y, of order $2(m+n)$, whose solution will therefore contain $2(m+n)$ arbitrary constants. The equations for finding these are also $2(m+n)$ in number, viz., $2n$ in $H_1 - H_0$ and $2m$ in $H_1' - H_0'$.

3054 NOTE.—The number of equations for determining the constants is not generally affected by any auxiliary equations introduced by restricting the limits. For every such equation either removes a term from (4) by annulling some variation (δy, δp, &c.), or it makes two terms into one; in each case diminishing by one the number of equations, and adding one equation, namely itself.

3055 Secondly, let y and z be connected by some equation

$\phi(xyz) = 0$. y and z are then found by solving simultaneously the equations

$$\phi(x, y, z) = 0 \quad \text{and} \quad K : \phi_y = K' : \phi_z.$$

PROOF — $\phi(x, y, z) = 0$, and therefore $\phi(x, y+\delta y, z+\delta z) = 0$, when the forms of y and z vary. Therefore $\phi_y \delta y + \phi_z \delta z = 0$ (1514). Also $K\delta y + K'\delta z = 0$, by (2). Hence the proportion.

3056 Thirdly, let the equation connecting y and z be of the more general form

$$\phi(x, y, p, q \dots z, p', q' \dots) = 0 \dots\dots\dots\dots\dots (5).$$

By differentiation, we obtain

$$\phi_y \delta y + \phi_p \delta p + \phi_q \delta q + \dots \phi_z \delta z + \phi_{p'} \delta p' + \phi_{q'} \delta q' + \dots = 0 \dots (6).$$

If (which rarely happens) this equation can be integrated so as to furnish a value of δz in terms of δy, then $\delta p'$, $\delta q'$, &c. may be obtained, by simple differentiation, in terms of δy, δp. Generally, we proceed as follows :—

$$\delta V = N\delta y + P\delta p + Q\delta q + \dots + N'\delta z + P'\delta p' + Q'\delta q' + \dots \quad \dots (7).$$

Multiply (6) by λ, and add it to (7), thus

$$\delta V = (N + \lambda\phi_y)\,\delta y + (P + \lambda\phi_p)\,\delta p + \dots$$
$$\dots + (N' + \lambda\phi_z)\,\delta z + (P' + \lambda\phi_{p'})\,\delta p' + \dots \quad \dots\dots (8).$$

The expression for δU will therefore be the same as in (2), if we replace N by $N + \lambda\phi_y$, P by $P + \lambda\phi_p$, &c., thus

3057
$$\delta U = \int_{x_0}^{x_1} \Big[\{(N+\lambda\phi_y) - (P+\lambda\phi_p)_x + \dots\}\,\delta y + \{(N'+\lambda\phi_z) - (P'+\lambda\phi_{p'})_x + \dots\}\,\delta z \Big]\,dx$$
$$+ \{P+\lambda\phi_p - (Q+\lambda\phi_q)_x + \dots\}_1\,\delta y_1 - \{P+\lambda\phi_p - (Q+\lambda\phi_q)_x + \dots\}_0\,\delta y_0$$
$$+ \{Q+\lambda\phi_q - (R+\lambda\phi_r)_x + \dots\}_1\,\delta p_1 - \{Q+\lambda\phi_q - (R+\lambda\phi_r)_x + \dots\}_0\,\delta p_0$$
$$\&c. \qquad\qquad \&c.$$
$$+ \text{ similar terms in } P, Q \dots p', q' \dots \&c. \dots (9).$$

3058 To render δU independent of the variation δz, we must then equate to zero the coefficient of δz under the sign of integration; thus

$$N'+\lambda\phi_z-(P'+\lambda\phi_{p'})_x+(Q'+\lambda\phi_{q'})_{2x}-\&c. = 0 \ldots\ldots(10),$$

the equation for determining λ.

3059 Ex. (i.)—Given $V = F(x, y, p, q \ldots z)$, where

$$z = \int v\,dx \quad \text{and} \quad v = F(x, y, p, q \ldots).$$

The equation ϕ is now $z-\int v\,dx = 0$ or $v-z_x = 0$,

$$\phi_y = v_y, \qquad \phi_p = v_p, \qquad \phi_q = v_q, \ \&c.,$$
$$\phi_z = 0, \qquad \phi_{p'} = -1, \qquad \phi_{q'} = 0, \text{ the rest vanishing.}$$

Substituting these values in (9), we obtain

$$\delta U = \int_{x_0}^{x_1} [\{N+\lambda v_y-(P+\lambda v_p)_x+(Q+\lambda v_q)_{2x}-\ldots\}\,\delta y + \{N'+\lambda_x'\}\,\delta z]\,dx$$
$$+\{P+\lambda v_p-(Q+\lambda v_q)_x+\ldots\}_1\,\delta y_1-\{P+\lambda v_p-(Q+\lambda v_q)_x+\ldots\}_0\,\delta y_0$$
$$+\{Q+\lambda v_q-(R+\lambda v_r)_x+\ldots\}_1\,\delta p_1-\{Q+\lambda v_q-(R+\lambda v_r)_x+\ldots\}_0\,\delta p_0+\&c.$$

For the complete variation DU add $V_1 dx_1 - V_0 dx_0$. To reduce the above so as to remove δz, we must put $N'+\lambda_x = 0$, and therefore $\lambda = -\int N'dx$. Let $\lambda = u$ be the solution, u being a function of $x, y, p, q \ldots z$. Substituting this expression for λ, the value of δU becomes independent of δz.

Ex. (ii.)—Similarly, if z in the last example be $= \int_{px} v$ (2148), ϕ becomes $v - z_{px} = 0$; and, to make $N'+\lambda_{px}$ vanish, we must put $\lambda = -\int_{px} N'$.

3061 Ex. (iii.)—Let $U = \int_{x_0} \sqrt{1+y_x^2+z_x^2}\,dx$ .... (1).

Here $N=0$; $N'=0$; $P = \dfrac{p}{\sqrt{1+p^2+p'^2}}$; $P' = \dfrac{p'}{\sqrt{1+p^2+p'^2}}$; $Q=0$; $Q'=0$; and the equations $K=0$, $K'=0$ become

$$P_x = 0, \quad P'_x = 0, \quad \text{or} \quad \frac{p}{\sqrt{1+p^2+p'^2}} = a, \quad \frac{p'}{\sqrt{1+p^2+p'^2}} = b.$$

Solving these equations, we get

$$y_x = m; \quad z_x = n; \quad \text{or} \quad y = mx+A; \quad z = nx+B.$$

3062 *First*, if $x_1, y_1, z_1, x_0, y_0, z_0$ be given, there are four equations to determine m, n, A, and B.

This solves the problem, to find a line of minimum length on a given curved surface between two fixed points on the surface.

3063 *Secondly*, if the limits x_1, x_0 only are given, then the equations

$$(P)_1 = 0, \quad (P)_0 = 0, \quad (P')_1 = 0, \quad (P')_0 = 0,$$

are only equivalent to the two equations $m = 0$, $n = 0$, and A and B remain undetermined.

3064 *Thirdly,* let the limits be connected by the equations

$$\phi(x_1, y_1, z_1) = 0, \quad \psi(x_0, y_0, z_0) = 0.$$

We shall have $(\phi_{x_1}+\phi_{y_1}p_1+\phi_{z_1}p_1')\,dx_1+\phi_{y_1}\delta y_1+\psi_{z_1}\delta z_1 = 0.$

Substitute $\phi_{y_1} = m_1\phi_{x_1}$, $\phi_{z_1} = n_1\phi_{x_1}$, $p_1 = m$, $p_1' = n$; thus

$$(1+mm_1+nn_1)\,dx_1+m_1\delta y_1+n_1\delta z_1 = 0.$$

Eliminate dx_1 by this equation from

$$V_1 dx_1+(P)_1\delta y_1+(P')_1\delta z_1 = 0,$$

and equate to zero the coefficients of δy_1 and δz_1; then

$$m_1 V_1 = (P)_1(1+mm_1+nn_1); \quad n_1 V_1 = (P')_1(1+mm_1+nn_1).$$

Replacing V_1, P_1 by their values, and solving these equations for m and n, we find $m = m_1$, $n = n_1$.

Similarly from the equation $\psi(x_0, y_0, z_0) = 0$ we derive $m = m_0$, $n = n_0$.

Eliminating $x_1, y_1, z_1, x_0, y_0, z_0$ between these equations, and

$$y_1 = mx_1+A; \quad z_1 = nx_1+B; \quad y_0 = mx_0+A; \quad z_0 = nx_0+B;$$

$$\phi(x_1, y_1, z_1) = 0; \quad \psi(x_0, y_0, z_0) = 0;$$

four equations remain for determining m, n, A, and B.

3065 *On determining the constants in the solution of* (3056).

Denoting $p, q, r \ldots$ by $p_1, p_2, p_3 \ldots$, we have

$$V = F(x, y, p_1, p_2 \ldots p_n, z, p_1', p_2' \ldots p_m');$$

and for the limiting equation,

$$\phi(x, y, p_1, p_2, \ldots p_{n'}, z, p_1', p_2', \ldots p_{m'}') = 0.$$

V is of the order n in y and m in z.

ϕ is of the order n' in y and m' in z.

3066 Rule I.—*If* m *be* > m′, *and* n *either* > *or* < n′, *the order of the final differential equation will be the greater of the two quantities* 2(m+n′), 2(m′+n); *and there will be a sufficient number of subordinate equations to determine the arbitrary constants.*

3067 Rule II.—*If* m *be* < m′, *and* n < n′, *the order of the final equation will generally be* 2(m′+n′); *and its solution may contain any number of constants not greater than the least of the two quantities* 2(m′−m), 2(n′−n).

For the investigation, see *Jellett*, pp. 118—127.

3068 If V does not involve x explicitly, a single integral of order $2(m+n)-1$ may be found. The value of V is that given in (3033), with corresponding terms derived from z.

PROOF.— $dV = Ndy + P_1 dp_1 + \dots + P_n dp_n + N'dz + P_1' dp_1' + \dots + P_m' dp_m'$. Substitute for N and N' from the equations $K = 0$, $K' = 0$, as in (3033), and integrate for V.

RELATIVE MAXIMA AND MINIMA.

3069 In this class of problems, a maximum or minimum value of an integral, $U_1 = \int_{x_0}^{x_1} V_1 dx$, is required, subject to the condition that another integral, $U_2 = \int_{x_0}^{x_1} V_2 dx$, involving the same variables, has a constant value.

RULE.—*Find the maximum or minimum value of the function* $U_1 + aU_2$; *that is, take* $V = V_1 + aV_2$, *and afterwards determine the constant* a *by equating* U_2 *to its given value.*

For examples, see (3074), (3082).

GEOMETRICAL APPLICATIONS.

3070 PROPOSITION I.—To find a curve s which will make $\int F(x, y)\, ds$ a maximum or minimum, F being a given function of the coordinates x, y.

The equation (5), in (3056), here becomes

$$p^2 + p'^2 = 1;$$

where $p = x_s$, $p' = y_s$, x and y being the dependent variables, and s the independent variable.

In (3057), we have now, writing u for $F(x, y)$,

$$N = u_x, \quad N' = u_y, \quad \phi_p = 2p, \quad \phi_{p'} = 2p';$$

the rest zero. The equations of condition are therefore

$$u_x - d_s(\lambda x_s) = 0 \quad \text{and} \quad u_y - d_s(\lambda y_s) = 0 \quad \dots\dots (1).$$

Multiplying by x_s, y_s respectively, adding and integrating, the result is

$$\lambda = u,$$

the constant being zero.*

Substituting this value in equations (1), differentiating ux_s and uy_s, and putting $u_s = u_x x_s + u_y y_s$, we get

$$y_s(u_x y_s - u_y x_s) = ux_{2s} \quad \dots\dots (2),$$

$$x_s(u_y x_s - u_x y_s) = uy_{2s} \quad \dots\dots (3).$$

* See Todhunter's "History," p. 405.

Multiplying (2) by y_s, and (3) by x_s, and subtracting, we obtain finally

$$u\,(y_s x_{2s} - x_s y_{2s}) = u_x y_s - u_y x_s, \quad \text{or}$$

3071 $$\frac{u}{\rho} = \frac{du}{dx}\frac{dy}{ds} - \frac{du}{dy}\frac{dx}{ds} \quad \ldots\ldots\ldots\ldots\ldots\ldots \quad (4),$$

ρ being the radius of curvature.

To integrate this equation, the form of u must be known, and, by assigning different forms, various geometrical theorems are obtained.

3072 PROPOSITION II.—To find the curve which will make

$$\int F(x, y)\,ds + \int f(x, y)\,dx \quad \ldots\ldots\ldots\ldots\ldots\ldots \quad (1)$$

a maximum or minimum, the functions F and f being of given form.

Let $$F(x, y) = u \quad \text{and} \quad f(x, y) = v.$$

Equation (1) is equivalent to $\int (u + v x_s)\,ds$.

In (3057) we now have $V = u + vp$; and for ϕ, $p^2 + p'^2 = 1$, as in (3070).

Therefore $$N = u_x + pv_x; \qquad P = V_p = v; \qquad \phi_p = 2p;$$
$$N' = u_y + pv_y; \qquad \phi_{p'} = 2p'; \quad \text{the rest zero.}$$

Therefore, equating to zero the coefficients of δx and δy, the result is the two equations

$$u_x + pv_x - (v + \lambda p)_s = 0,$$
$$u_y + pv_y - (\lambda p')_s = 0;$$

or $$d_s(\lambda x_s) + v_s = u_x + x_s v_x,$$
$$d_s(\lambda y_s) = u_y + x_s v_y.$$

Multiplying by x_s, y_s respectively, adding, and integrating, we obtain, as in (3070), $\lambda = u$, and ultimately,

3073 $$\frac{1}{\rho} = -\frac{1}{u}\left(\frac{du}{dx}\frac{dy}{ds} - \frac{du}{dy}\frac{dx}{ds} + \frac{dv}{dy}\right).$$

3074 Ex.—To find a curve s of given length, such that the volume of the solid of revolution which it generates about a given line may be a maximum.

Here $\int (y^2 x_s - a^2)\,ds$ must be a maximum, by (3069), a^2 being the arbitrary constant. The problem is a case of (3072),

$$u = a^2, \quad u_x = 0, \quad u_y = 0, \quad v = y^2, \quad v_y = 2y.$$

Hence equation (3073) becomes $\dfrac{1}{\rho} = \dfrac{2y}{a^2}$.

Giving ρ its value, $-\dfrac{(1+p^2)^{\frac{3}{2}}}{pp_y}$ $\left(\text{where } p = \dfrac{dy}{dx}\right)$, and integrating, the result

$$\frac{1}{\sqrt{1+p^2}} = \frac{y^2+b^2}{a^2}; \quad \text{from which} \quad x = \int \frac{(y^2+b^2)\,dy}{\sqrt{a^4-(y^2+b^2)}}.$$

FUNCTIONS OF TWO INDEPENDENT VARIABLES.

3075 Let $$V = f(x, y, z, p, q, r, s, t),$$
in which x, y are the independent variables, and p, q, r, s, t stand for $z_x, z_y, z_{2x}, z_{xy}, z_{2y}$ respectively (1815), z being an indeterminate function of x and y.

Let $$U = \int_{x_0}^{x_1}\int_{y_0}^{y_1} V\,dx\,dy,$$
and let the equation connecting x and y at the limits be $\varphi(x, y) = 0$. The complete variation of U, arising solely from an infinitesimal change in the *form* of the function z, is as follows :—

Let V_z, V_p, &c. be denoted by Z, P, Q, R, S, T.

Let
$$\phi = (P - R_x - \tfrac{1}{2}S_y)\,\delta z + \tfrac{1}{2}S\delta q + R\delta p,$$
$$\psi = (Q - T_y - \tfrac{1}{2}S_x)\,\delta z + \tfrac{1}{2}S\delta p + T\delta q,$$
$$\chi = (Z - P_x - Q_y + R_{2x} + S_{xy} + T_{2y})\,\delta z.$$

The variation in question is then

3076 $$\delta U = \int_{x_0}^{x_1}\left(\psi_{y=y_1} - \psi_{y=y_0} + \phi_{y=y_0}\frac{dy_0}{dx} - \phi_{y=y_1}\frac{dy_1}{dx}\right)dx$$
$$+ \left[\int_{y_0}^{y_1}\phi\,dy\right]_{x=x_0}^{x=x_1} + \int_{x_0}^{x_1}\int_{y_0}^{y_1}\chi\,dx\,dy.$$

PROOF.— $$\delta\int_{x_0}^{x_1}\int_{y_0}^{y_1} V\,dx\,dy = \int_{x_0}^{x_1}\int_{y_0}^{y_1}\delta V\,dx\,dy$$
$$= \int_{x_0}^{x_1}\int_{y_0}^{y_1}(Z\delta z + P\delta p + Q\delta q + R\delta r + S\delta s + T\delta t)\,dx\,dy$$
$$= \int_{x_0}^{x_1}\int_{y_0}^{y_1}\left\{\frac{d\phi}{dx} + \frac{d\psi}{dy} + \chi\right\}dx\,dy,$$
as appears by differentiating the values of ϕ and ψ. But
$$\int_{y_0}^{y_1}\frac{d\phi}{dx}\,dy = \frac{d}{dx}\int_{y_0}^{y_1}\phi\,dy + \phi_{y=y_0}\frac{dy_0}{dx} - \phi_{y=y_1}\frac{dy_1}{dx},$$
by (2257), and $$\int_{y_0}^{y_1}\frac{d\psi}{dy}\,dy = \psi_{y=y_1} - \psi_{y=y_0}.$$
Hence the result.

3077 The conditions for a maximum or minimum value of U are, by similar reasoning to that employed in (3032),
$$\phi = 0, \qquad \psi = 0, \qquad \chi = 0.$$

GEOMETRICAL APPLICATIONS.

3078 PROPOSITION I.—To find the surface, S, which will make $\iint F(x, y, z)\, dS$ a maximum or minimum, F being a given function of the coordinates x, y, z. [*Jellett*, p. 276.

Here, putting $F(x, y, z) = u$, $V = u\sqrt{1+p^2+q^2}$;

$$Z = \sqrt{1+p^2+q^2}\frac{du}{dz}; \quad P = \frac{up}{\sqrt{1+p^2+q^2}}; \quad Q = \frac{uq}{\sqrt{1+p^2+q^2}};$$

and V_r, V_s, V_t are all zero.

$$\frac{dP}{dx} = \frac{p}{\sqrt{1+p^2+q^2}}\left(\frac{du}{dx}+p\frac{du}{dz}\right)+u\frac{(1+q^2)\,r-pqs}{(1+p^2+q^2)^{\frac{3}{2}}},$$

$$\frac{dQ}{dy} = \frac{q}{\sqrt{1+p^2+q^2}}\left(\frac{du}{dy}+q\frac{du}{dz}\right)+u\frac{(1+p^2)\,t-pqs}{(1+p^2+q^2)^{\frac{3}{2}}}.$$

The equation $\chi = 0$ or $Z-P_x-Q_y = 0$ gives

$$\frac{(1+q^2)\,r-2pqs+(1+p^2)\,t}{(1+p^2+q^2)^{\frac{3}{2}}}+\frac{1}{u\sqrt{1+p^2+q^2}}\left(p\frac{du}{dx}+q\frac{du}{dy}-\frac{du}{dz}\right) = 0.$$

If R, R' be the principal radii of curvature, and l, m, n the direction cosines of the normal, this equation may be written

3079 $$\frac{1}{R}+\frac{1}{R'}+\frac{1}{u}\left(l\frac{du}{dx}+m\frac{du}{dy}+n\frac{du}{dz}\right) = 0,$$

and according to the nature of the function u different geometrical theorems may be deduced.

3080 PROPOSITION II.—To find the surface S which will make

$$\iint F(x, y, z)\, dS+\iint f(x, y, z)\, dx\,dy$$

a maximum or minimum; F and f being given functions of the coordinates x, y, z.

Let $F(x, y, z) = u$ and $f(x, y, z) = v$. Proceeding throughout as in (3078), we have

$$V = u\sqrt{1+p^2+q^2}+v,$$
$$Z = \sqrt{1+p^2+q^2}\,u_z+v_z,$$

and the remaining equations the same as in that article if we add to the resulting differential equation the term $-\frac{v_z}{u}$ on the left.

This equation may then be put in the form

3081 $$\frac{1}{R}+\frac{1}{R'}=-\frac{1}{u}\left(l\frac{du}{dx}+m\frac{du}{dy}+n\frac{du}{dz}-\frac{dv}{dz}\right),$$

where l, m, n are the direction cosines of the normal to the surface.

3082 Ex.—To find a surface of given area such that the volume contained by it shall be a maximum.

By (3069), the integral $\iint (z-a\sqrt{1+p^2+q^2})\,dx\,dy$

must take a maximum or minimum value. The problem is a case of (3080). We have $u=-a$, $v=z$, $u_x=0$, $u_y=0$, $u_z=0$, $v_z=1$;
and the differential equation of the surface (3081) reduces to

$$(1+q^2)\,r-2pqs+(1+p^2)\,t+\frac{1}{a}(1+p^2+q^2)^{\frac{3}{2}}=0;$$

3083 $$\text{or}\quad \frac{1}{R}+\frac{1}{R'}=\frac{1}{a}.$$

APPENDIX.

ON THE GENERAL OBJECT OF THE CALCULUS OF VARIATIONS.

3084 Definitions.—A function whose form is invariable is called *determinate*, and one whose form is variable, *indeterminate*.

Let du be the increment of a function u due to a change in the magnitude of the independent variable, δu that due to a change in the form of the function, Du the total increment from both causes; then

$$Du = du + \delta u.$$

Thus, in (3042), the terms involving dx_1 and dx_0 constitute du, and the remaining terms δu; the whole variation being Du.

δu is called the *variation* of the function u.

3085 A *primitive* indeterminate function, u, of any number of variables is a function whose variation is of arbitrary but constant form; in other words, $\delta^2 u = 0$.

3086 Let $v = F.u$ be a *derived* function,—that is, a function derived by some process from the function u; F denoting a relation between the forms, but not between the magnitudes, of u and v.

The general object of the Calculus of Variations is to determine the change in a derived function v, caused by a change in the form of its primitive u.

The particular derived functions considered are those whose symbols are d and $\int$, denoting operations of differentiation and integration respectively.

SUCCESSIVE VARIATION.

3087 Let the variation of the variation, or second variation of V due to a change in the form of the involved function, $y = f(x)$, be denoted by $\delta(\delta V)$ or $\delta^2 V$; the third variation by $\delta^3 V$, and so on.

By definition (3085), y being a primitive indeterminate function, and δy its variation, $\delta^2 y = 0$ .... (1).

3088 The second variation of any derivative of y is also zero, *i.e.*, $\delta^2 p$, $\delta^2 q$, &c. all vanish.

PROOF.— $\delta^2(y_{nx}) = \delta(\delta y_{nx}) = \{\delta(\delta y)\}_{nx} = (\delta^2 y)_{nx} = 0$ by (1).

3089 If $V = f(x, y, p, q, r, \text{\&c.} \ldots)$, where y is a primitive indeterminate function of x, then

$$\delta^n V = (\delta y\, d_y + \delta p\, d_p + \delta q\, d_q + \ldots)^n\, V,$$

where, in the formal expansion by the multinomial theorem, δy, δp, &c. follow the law of involution, but the indices of d_y, d_p, &c. indicate repetition of the operation d_y, d_p, &c. upon V.

PROOF.—First, $\delta V = (\delta y\, d_y + \delta p\, d_p + \delta q\, d_q + \ldots)\, V.$

In finding $\delta^2 V$, each product, such as $\delta y\, d_y V$, is differentiated again as a function of y, p, q, &c.; but, since the variations of δy, δp, &c. vanish by (2), it is the same in effect as though δy, δp, &c. were not operated upon at all. They accordingly rank as algebraic quantities merely, and therefore

$$\delta^2 V = (\delta y\, d_y + \delta p\, d_p + \delta q\, d_q + \ldots)^2\, V.$$

Similarly for a third differentiation; and so on.

IMMEDIATE INTEGRABILITY OF THE FUNCTION V.

3090 DEF.—When the function V (3028) is integrable without assigning the value of y in terms of x, and therefore integrable whatever the form of the function y may be, it is said to be *immediately integrable*, or integrable *per se*.

3091 The requisite condition for V to be immediately integrable is that $K=0$ shall be identically true.

PROOF.— $\int_{x_0}^{x_1} V\,dx$ must be expressible in the form

$$\phi\,(x_1 y_1 p_1 q_1 \ldots) - \phi\,(x_0 y_0 p_0 q_0 \ldots),$$

where ϕ is independent of the form of y. Hence, a change in the form of y, which leaves the values at the limits unaltered, will leave

$$\delta \int_{x_0}^{x_1} V\,dx = 0; \quad \text{that is,} \quad \int_{x_0}^{x_1} K\,\delta y = 0.$$

But the last equation necessitates $K=0$, since δy is arbitrary. And $K=0$ must be identically true, otherwise it would determine y as a function of x.

DIFFERENTIAL EQUATIONS.

GENERATION OF DIFFERENTIAL EQUATIONS.

3150 By differentiating ordinary algebraic equations, and eliminating constants or functions, differential equations are produced. Some methods are illustrated in the following examples.

3151 From an equation between two variables and n arbitrary constants, to eliminate the constants.

RULE.—*Differentiate* r *times* ($r<n$), *and from the* $r+1$ *equations any* r *constants may be eliminated, and thus* $C(n, r)$ *different equations of the* r^{th} *order* (3060) *obtained, involving* $\frac{d^r y}{dx^r}, \frac{d^{r-1}y}{dx^{r-1}}$, &c. *Only* $r+1$, *however, of these equations will be independent. By differentiating* n *times and eliminating the constants, a single final differential equation of the* n^{th} *order free from constants may be obtained.*

3152 Ex.—To eliminate the constants a and b from the equation

$$y = ax^2+bx \qquad \text{(i.)}$$

Differentiating, we find

$$\frac{dy}{dx} = 2ax+b \qquad \text{(ii.)}$$

Eliminating a and b in turn, we get

$$x\frac{dy}{dx}+bx = 2y, \quad x\frac{dy}{dx} = ax^2+y \qquad \text{(iii., iv.)}$$

Now, differentiating (iii.) and eliminating b produces the final equation of the second order,

$$x^2\frac{d^2y}{dx^2}-2x\frac{dy}{dx}+2y = 0 \qquad \text{(v.)}$$

The same equation is obtained by differentiating (iv.) and eliminating a.

3153 To eliminate the function ϕ from the equation $z=\phi(v)$, where v is a function of x and y. We have

$$z_x = \phi'(v)\,v_x, \qquad z_y = \phi'(v)\,v_y.$$

Therefore

$$z_x v_y = z_y v_x.$$

3154 To eliminate ϕ from $u = \phi(v)$, where u and v are functions of x, y, z.

Consider x and y the independent variables, and differentiate for each separately, thus

$$u_x + u_z z_x = \phi'(v)(v_x + v_z z_x),$$
$$u_y + u_z z_y = \phi'(v)(v_y + v_z z_y),$$

and, by division, $\phi'(v)$ is eliminated.

3155 To eliminate $\phi_1, \phi_2, \ldots \phi_n$ from the equation

$$F\{x, y, z, \phi_1(\alpha_1), \phi_2(\alpha_2), \ldots \phi_n(\alpha_n)\} = 0,$$

where $\alpha_1, \alpha_2, \ldots \alpha_n$ are known functions of x, y, z.

RULE.—*Differentiate for* x *and* y *as independent variables, forming the derivatives of* F *of each order, up to the* $(2n-1)^{th}$ *in every possible way; that is,* F; F_x, F_y; F_{2x}, F_{xy}, F_{2y}; *&c. There will be* $2n^2$ *unknown functions, consisting of* $\phi_1, \phi_2, \ldots \phi_n$ *and their derivatives, and* $2n^2+n$ *equations for eliminating them.*

3156 To eliminate ξ, $\phi_1(\xi)$, $\phi_2(\xi)$, ... $\phi_n(\xi)$ between the equations

$$F\{x, y, z, \xi, \phi_1(\xi), \phi_2(\xi) \ldots \phi_n(\xi)\} = 0,$$
$$f\{x, y, z, \xi, \phi_1(\xi), \phi_2(\xi) \ldots \phi_n(\xi)\} = 0.$$

RULE.—*Consider* z *and* ξ *functions of the independent variables* x, y, *and form the derivatives of* F *and* f *up to the* $2n-1^{th}$ *order in the manner described in* (3055). *There will be* $4n^2+n$ *functions, and* $4n^2+2n$ *equations for eliminating them.*

3157 To eliminate ϕ from the equation

$$F\{x, y, z, w, \phi(\alpha, \beta)\} = 0,$$

where α, β are known functions of x, y, z, w.

RULE.—*Consider* x, y, z *the independent variables. Differentiate for each, and eliminate* ϕ, ϕ_α, ϕ_β *between the four equations.*

DEFINITIONS AND RULES.

3158 *Ordinary* differential equations involve the derivatives of a *single* independent variable.

3159 *Partial* differential equations involve partial derivatives, and therefore two or more independent variables are concerned.

3160 *The order* of a differential equation is the order of the highest derivative which it contains.

3161 *The degree* of a differential equation is the power to which the highest derivative is raised.

3162 *A Linear* differential equation is one in which the derivatives are all involved in the first degree.

3163 *The complete primitive* of a differential equation is that equation between the primitive variables from which the differential equation may be obtained by the process of differentiation.

3164 *The general solution* is the name given to the complete primitive when it has been obtained by solving the given differential equation.

Thus, reverting to the example in (3051), equation (i.) is the *complete primitive* of (v.) which is obtained from (i.) by differentiation and elimination.

The *differential equation* (v.) being given, the process is reversed. Equations (iii.) and (iv.) are called the *first integrals* of (v.), and equation (i.) the *final integral* or *general solution*.

3165 *A particular solution*, or *particular integral*, of a differential equation is obtained by giving particular values to the arbitrary constants in the general solution.

For the definition of a *singular solution*, see (3068).

3166 To find when two differential equations of the first order have a common primitive.

Rule.—*Differentiate each equation, and eliminate its arbitrary constant. The two results will agree if there is a common primitive, which, in that case, will be found by eliminating* y_x *between the given equations.*

Ex.—Apply the rule to equations (iii.) and (iv.) in (3052).

3167 To find when two solutions of a differential equation, each involving an arbitrary constant, are equivalent.

RULE.—*Eliminate one of the variables. The other will also disappear, and a relation between the arbitrary constants will remain.*

Otherwise, if $V=C$, $v=c$ be the two solutions: V and v being functions of the variables, and C and c constants; then

$$\frac{dV}{dx}\frac{dv}{dy}=\frac{dV}{dy}\frac{dv}{dx}$$

is the required condition.

PROOF.—V must be a function of v. Let $V=\phi(v)$; therefore $V_x=\phi_v v_x$ and $V_y=\phi_v v_y$; then eliminate ϕ_v.

EX.—$\tan^{-1}(x+y)+\tan^{-1}(x-y)=a$ and $x^2+2bx=y^2+1$ are both solutions of $2xy\,y_x=x^2+y^2+1$. Eliminating y, x disappears, and the resulting equation is $b\tan a=1$.

SINGULAR SOLUTIONS.

3168 DEFINITION.—"A *singular solution* of a differential equation is a relation between x and y which satisfies the equation by means of the values which it gives to the differential coefficients y_x, y_{2x}, &c., but is not included in the complete primitive." See examples (3132–3).

3169 To find a singular solution from the complete primitive $\phi(x,y,c)=0$.

RULE I.—*From the complete primitive determine* c *as a function of* x, *by solving the equation* $y_c=0$, *or else by solving* $x_c=0$, *and substitute this value of* c *in the primitive. The result is a singular solution, unless it can also be obtained by giving to* c *a constant value in the primitive.*

3170 *If the singular solution involves* y *only, it results from the equation* $y_c=0$ *only, and if it involves* x *only, it results from* $x_c=0$ *only. If it involves both* x *and* y, *the two equations* $x_c=0$, $y_c=0$ *give the same result.*

3171 *When the primitive equation* $\phi\,(\mathrm{xyc}) = 0$ *is a rational integral function,* $\phi_c = 0$ *may be used instead of* $\mathrm{x}_c = 0$ *or* $\mathrm{y}_c = 0$.

PROOF.—Let $\phi\,(x, y, c) = 0$ be expressed in the form

$$y = f\,(x, c) \quad \text{.......} \quad (1).$$

Then, if c be constant, $\quad y_x = f_x \quad \text{.......} \quad (2);$

and, if c varies, $\quad y_x = f_x + f_c c_x \quad \text{.......} \quad (3).$

When c is constant, the differential equation of which (1) is the primitive is satisfied by the value of y_x in (2). But it will also be satisfied by the same value of y_x when c is variable, provided that either $f_c = 0$ or $f_x = \infty$, and in either case a solution is obtained which is not the result of giving to c a constant value in the complete primitive; that is, it is a singular solution. But $f_c = 0$ is equivalent to $y_c = 0$, and $f_x = \infty$ makes $y_x = \infty$, and therefore $x =$ constant.

GEOMETRICAL MEANING OF A SINGULAR SOLUTION.

3172 Since the process in Rule I. is identical with that employed in finding the envelope of the series of curves obtained by varying the parameter c in the equation $\phi\,(x, y, c) = 0$; the singular solution so obtained is the equation of the envelope itself.

An exception occurs when the envelope coincides with one of the curves of the system.

3173 Ex.—Let the complete primitive be

$$y = cx + \sqrt{1-c^2}, \quad \text{therefore } y_c = x - \frac{c}{\sqrt{1-c^2}}; \quad y_c = 0 \text{ gives } c = \frac{x}{\sqrt{1+x^2}}.$$

Substituting this in the primitive gives $y = \sqrt{1+x^2}$, a singular solution. It is the equation of the envelope of all the lines that are obtained by varying the parameter c in the primitive; for it is the equation of a circle, and the primitive, by varying c, represents all lines which touch the circle. See also (3132–3).

3174 "The determination of c as a function of x by the solution of the equation $y_c = 0$, is equivalent to determining what particular primitive has contact with the envelop at that point of the latter which corresponds to a given value of x.

"The elimination of c between a primitive $y = f\,(x, c)$ and the derived equation $y_c = 0$, does not *necessarily* lead to a singular solution in the sense above explained.

"For it is possible that the derived equation $y_c = 0$ may neither, on the one hand, enable us to determine c as a function of x, so leading to a singular solution; nor, on the other hand, as an absolute constant, so leading to a particular primitive.

"Thus the particular primitive $y = e^{cx}$ being given, the condition $y_c = 0$ gives $e^{cx} = 0$, whence c is $+\infty$ if x be negative, and $-\infty$ if x be positive. It is a dependent constant. The resulting solution $y = 0$ does not then represent an envelope of the curves of particular primitives, nor strictly one of those curves. It represents a curve formed of branches from two of them. It is most fitly characterised as a particular primitive marked by a singularity in the mode of its derivation from the complete primitive."

[*Boole's "Differential Equations," Supplement*, p. 13.

DETERMINATION OF A SINGULAR SOLUTION FROM THE DIFFERENTIAL EQUATION.

3175 RULE II.—*Any relation is a singular solution which, while it satisfies the differential equation, either involves* y *and makes* p_y *infinite, or involves* x *and makes* $\left(\frac{1}{p}\right)_x$ *infinite.*

3176 "One negative feature marks all the cases in which a solution involving y satisfies the condition $p_y = \infty$. It is, that the solution, while expressed by a single equation, is not connected with the complete primitive by a single and absolutely constant value of c.

"The relation which makes p_y infinite satisfies the differential equation only because it satisfies the condition $y_c = 0$, and this implies a connexion between c and x, which is the ground of a real, though it may be unimportant, singularity in the solution itself.

"In the first, or, as it might be termed, the envelope species of singular solutions, c receives an infinite number of different values connected with the value of x by a law. In the second, it receives a finite number of values also connected with the values of x by a law. In the third species, it receives a finite number of values, determinate, but not connected with the values of x."

Hence the general inclusive definition—

3177 "*A singular solution of a differential equation of the first order is a solution the connexion of which with the complete primitive does not consist in giving to* c *a single constant value absolutely independent of the value of* x."

[*Boole's "Differential Equations,"* p. 163, and *Supplement*, p. 19.

RULES FOR DISCRIMINATING A SINGULAR SOLUTION OF THE ENVELOPE SPECIES.

3178 RULE III.—*When* p_y *or* $\left(\frac{1}{p}\right)_x$ *is made infinite by equating to zero a factor having a negative index, the solution "may be considered to belong to the envelope species."*

3179 "In other cases, the solution is deducible from the complete primitive by regarding c as a constant of multiple value,—its particular values being either, 1st, dependent in some way on the value of x, or, 2ndly, independent of x, but still such as to render the property a singular one."

[*Boole's "Differential Equations,"* p. 164.

3180 Rule IV.—*A solution which, while it makes* p_y *infinite and satisfies the differential equation of the first order, does not satisfy all the higher differential equations obtained from it, is a singular solution of the envelope species.*

Ex.: $y_x = my^{\frac{m-1}{m}}$ has the singular solution $y=0$ when m is >1.

Now $$y_{rx} = m(m-1)\,\dots\,(m-r+1)\,y^{\frac{m-r}{m}},$$

and, when r is $>m$, the value $y=0$ makes y_{rx} infinite. The solution is, therefore, by the rule of the envelope species.

3181 Rule V.—"*The proposed solution being represented by* $u=0$, *let the differential equation, transformed by making* u *and* x *the variables, be* $u_x + f(x, u) = 0$. *Determine the integral* $\int_0^u \frac{du}{U}$ *as a function of* x *and* u, *in which* U *is either equal to* $f(x, u)$ *or to* $f(x, u)$ *deprived of any factor which neither vanishes nor becomes infinite when* $u=0$. *If that integral tends to zero with* u, *the solution is singular" and of the envelope species.* [*Boole, Supplement,* p. 30.

3182 Ex.—To determine whether $y=0$ is a singular solution or particular integral of $$y_x = y(\log y)^2.$$

Here $u=y$, and $$\int_0^y \frac{dy}{y(\log y)^2} = -\frac{1}{\log y}.$$

As this tends to zero with y, the solution is singular.

Verification.—The complete primitive is $y = e^{\frac{1}{c-x}}$, and no constant value assigned to c will produce the result $y=0$.

3183 Professor De Morgan has shown that any relation involving both x and y, which satisfies the conditions $p_y=\infty$, $p_x=\infty$, will satisfy the differential equation when it does not make y_{2x}, as derived from it, infinite; that it *may* satisfy it even if it makes y_{2x} infinite; and that, if it does not satisfy the differential equation, the curve it represents is a locus of points of infinite curvature, usually cusps, in the curves of complete primitives. [*Boole, Supplement,* p. 35.

FIRST ORDER LINEAR EQUATIONS.

3184 $$M+N\frac{dy}{dx}=0, \quad \text{or} \quad Mdx+Ndy=0,$$

M and N being either functions of x and y or constants.

SOLUTION BY SEPARATION OF THE VARIABLES.

3185 This method of solution, when practicable, is the simplest, and is frequently involved in other methods.

Ex. $$xy(1+x^2)\,dy=(1+y^2)\,dx,$$

therefore $$\frac{y\,dy}{1+y^2}=\frac{dx}{x(1+x^2)},$$

and each member can be at once integrated.

HOMOGENEOUS EQUATIONS.

3186 Here M and N, in (3084), are homogeneous functions of x and y, and the solution is affected as follows:—

RULE.—*Put* $y=vx$, *and therefore* $dy=vdx+xdv$, *and then separate the variables. For an example, see* (3108).

EXACT DIFFERENTIAL EQUATIONS.

3187 $Mdx+Ndy=0$ is an exact differential when

$$M_y=N_x,$$

and the solution is then obtained by the formula

$$\int Mdx+\int\{N-d_y(\int Mdx)\}\,dy=C.$$

PROOF.—If $V=0$ be the primitive, we must have $V_x=M$, $V_y=N$; therefore $V_{xy}=M_y=N_x$. Also $V=\int Mdx+\phi(y)$, $\phi(y)$ being a constant with respect to x.

Therefore $$N=V_y=d_y\int Mdx+\phi'(y),$$

therefore $$\phi(y)=\int\{N-d_y\int Mdx\}\,dy+C.$$

3188 Ex. $$(x^3-3x^2y)\,dx+(y^3-x^3)\,dy=0.$$

Here $M_y=-3x^2=N_x$. Therefore the solution is

$$C=\frac{x^4}{4}-x^3y+\int\left\{y^3-x^3-d_y\left(\frac{x^4}{4}-x^3y\right)\right\}dy$$

$$=\frac{x^4}{4}-x^3y+\int y^3\,dy=\frac{x^4+y^4}{4}-x^3y.$$

3189 Observe that, if $Mdx+Ndy$ can be separated into two parts, so that one of them is an exact differential, the other part must also be an exact differential in order that the whole may be such.

3190 Also, if a function of x and y can be expressed as the product of two factors, one of which is a function of the integral of the other, the original function is an exact differential.

3191 Ex.— $\frac{1}{y}\cos\frac{x}{y}\,dx-\frac{x}{y^2}\cos\frac{x}{y}\,dy=\cos\frac{x}{y}\cdot\frac{y\,dx-x\,dy}{y^2}=0.$

Here $\frac{x}{y}$ is the integral of the second factor. Hence the solution is

$$\sin\frac{x}{y}=C.$$

INTEGRATING FACTOR FOR $Mdx+Ndy=0$.

When this equation is not an exact differential, a factor which will make it such can be found in the following cases.

3192 I.— *When one only of the functions* $Mx+Ny$ *or* $Mx-Ny$ *vanishes identically, the reciprocal of the other is an integrating factor.*

3193 II.—*If, when* $Mx+Ny=0$ *identically, the equation is at the same time homogeneous, then* $x^{-(n+1)}$ *is also an integrating factor.*

3194 III.—*If neither* $Mx+Ny$ *nor* $Mx-Ny$ *vanishes identically, then, when the equation is homogeneous,* $\frac{1}{Mx+Ny}$ *is an integrating factor; and when the equation can be put in the form* $\phi(xy)\,x\,dy+\chi(xy)\,y\,dx=0$, $\frac{1}{Mx-Ny}$ *is an integrating factor.*

PROOF.—I. and III.—From the identity

$$M\,dx+N\,dy=\tfrac{1}{2}\left\{(Mx+Ny)\,d\log xy+(Mx-Ny)\,d\log\frac{x}{y}\right\},$$

assuming the integrating factor in each case, and deducing the required forms for M and N, employing (3090).

II.—Put $v = \frac{y}{x}$, $M = x^n \phi(v)$, $N = x^n \psi(v)$, and $dy = x\,dv + v\,dx$ in $Mdx + Ndy$ and $Mx + Ny$.

3195 The general form for an integrating factor of $Mdx + Ndy = 0$ is

$$\mu = e^{\int \frac{M_y - N_x}{Nv_x - Mv_y} dv},$$

where v is some chosen function of x and y; and the condition for the existence of an integrating factor under that hypothesis is that

3196 $$\frac{M_y - N_x}{Nv_x - Mv_y} \textit{ must be a function of } v.$$

PROOF.—The condition for an exact differential of $M\mu dx + N\mu dy = 0$ is $(M\mu)_y = (N\mu)_x$ (3087). Assume $\mu = \phi(v)$, and differentiate out; we thus obtain
$$\frac{\phi' v}{\phi v} = \frac{M_y - N_x}{Nv_x - Mv_y}.$$

The following are cases of importance.

3197 I.—If an integrating factor is required which is a function of x only, we put $\mu = \phi(x)$, that is, $v = x$; and the necessary condition becomes

$$\frac{M_y - N_x}{N} \textit{ must be a function of } x \textit{ only.}$$

3198 II.—If the integrating factor is to be a function of xy, the condition becomes, by putting $xy = v$,

$$\frac{M_y - N_x}{Ny - Mx} \textit{ must be a function of } xy \textit{ only.}$$

3199 III.—If the integrating factor is to be a function of $\frac{y}{x}$, the condition is

$$\frac{x^2 (N_x - M_y)}{Mx + Ny} \textit{ must be a function of } \frac{y}{x}.$$

If $Mx + Ny$ vanishes, (3092) must be resorted to.

In this and similar cases, the expression found will be a function of $v = \frac{y}{x}$ if it takes the form $F(v)$ when y is replaced by vx.

3200 IV.—*Theorem.*—The condition that the equation $M dx + N dy = 0$ may have a homogeneous function of x and y of the nth degree for an integrating factor, is

$$\frac{x^2(N_x - M_y) + nNx}{Mx + Ny} = F(u), \quad \text{where } u = \frac{y}{x}.$$

3201 The integrating factor will then be obtained from

$$\mu = x^n e^{\int F(u)\,du}$$

PROOF.—Put $\mu = v = x^n \psi(u)$ in (3097), thus

$$\frac{1}{v} = \frac{M_y - N_x}{Nv_x - Mv_y}.$$

Perform the differentiations, and, by reduction, we get

$$\frac{\psi'(u)}{\psi(u)} = \frac{x^2(N_x - M_y) + nNx}{Mx + Ny}.$$

The right member must be a function of u in order that $\psi(u)$ may be found by integration.

3202 Ex.—To ascertain whether an integrating factor, which is a homogeneous function of x and y, exists for the equation

$$(y^3 + axy^2)\,dy - ay^3\,dx + (x+y)(x\,dy - y\,dx) = 0.$$

Here $\quad M = -(ay^3 + xy + y^2), \qquad N = (y^3 + axy^2 + xy + x^2).$

Substituting in the formula of (3100), we find that, by choosing $n = -3$, the fraction reduces to $\frac{ax^2y^2 - 3xy^3}{y^4}$, and, by putting $y = ux$, it becomes $\frac{a-3u}{u^2}$, a function of u.

$$\int \frac{a-3u}{u^2}\,du = -\frac{a}{u} - 3\log u,$$

$$\therefore \qquad \mu = x^{-3} e^{-\left(\frac{ax}{y} + 3\log\frac{y}{x}\right)} = y^{-3} e^{-\frac{ax}{y}},$$

the integrating factor required. It is homogeneous, and of the degree -3 in x and y, as is seen by expanding the second factor by (150).

3203 If by means of the integrating factor μ the equation $\mu M\,dx + \mu N\,dy = 0$ is found to have $V = C$ for its complete primitive, then the form for all other integrating factors will be $\mu f(V)$, where f is any arbitrary function.

PROOF.—The equation becomes

$$\mu Mf(V)\,dx+\mu Nf(V)\,dy=0.$$

Applying the test of integrability (3087), we have

$$\{\mu Mf(V)\}_y=\{\mu Nf(V)\}_x.$$

Differentiate out, remembering that

$$(\mu M)_y=(\mu N)_x, \qquad V_y=\mu N, \qquad V_x=\mu M,$$

and the equality is established.

3204 GENERAL RULE.—*Ascertain by the determination of an integrating factor that an equation is solvable, and then seek to effect the solution in some more direct way.*

SOME PARTICULAR EQUATIONS.

3205 $$(ax+by+c)\,dx+(a'x+b'y+c')\,dy=0.$$

This equation may be solved in three ways.

I.—Substitute $x=\xi-\alpha, \quad y=\eta-\beta,$

and determine α and β so that the constant terms in the new equation in ξ and η may vanish.

II.—Or substitute $ax+by+c=\xi, \quad a'x+b'y+c'=\eta.$

3206 But if $a:a'=b:b'$, the methods I. and II. fail. The equation may then be written as a function of $ax+by$.

Put $z=ax+by$, and substitute $b\,dy=dz-a\,dx$, and afterwards separate the variables x and z.

3207 III.—A third method consists in assuming

$$(A\eta+C)\,d\xi+(A'\xi+C')\,d\eta=0,$$

and equating coefficients with the original equation after substituting $\qquad \xi=x+m_1y, \qquad \eta=x+m_2y.$

m_1, m_2 are the roots of the quadratic

$$am^2+(b+a')\,m+b'=0.$$

The solution then takes the form

$$\frac{\left\{(am_1-a')(x+m_1y)+cm_1-c'\right\}^{\frac{1}{am_1-a'}}}{\left\{(am_2-a')(x+m_2y)+cm_2-c'\right\}^{\frac{1}{am_2-a'}}}=C.$$

3208 Ex. $(3y-7x+7)\,dx+(7y-3x+3)\,dy=0.$

Put $x=\xi-\alpha$, $y=\eta-\beta$, thus

$$(3\eta-7\xi)\,d\xi+(7\eta-3\xi)\,d\eta=0 \quad\ldots\ldots\ldots\ldots\ldots\ldots\ldots \text{(i.)},$$

with equations for α and β, $7\alpha-3\beta+7=0$; $3\alpha-7\beta+3=0$;

therefore $\alpha=-1, \quad \beta=0\ldots\ldots\ldots\ldots\ldots\ldots\ldots\ldots\ldots\text{(ii.)}$

(i.) being homogeneous, put $\eta=v\xi$, and therefore $d\eta=v\,d\xi+\xi\,dv$ (3086);

$$\therefore \quad (7v^2-7)\,\xi d\xi+(7v-3)\,\xi^2 dv=0, \quad\text{or}\quad \frac{d\xi}{\xi}+\frac{7v-3}{7v^2-7}\,dv=0.$$

The second member is integrated, as in (2080), with $b=0$, and, after reduction, we find $5\log(\eta+\xi)+2\log(\eta-\xi)=C.$

Putting $\xi=x-1$ and $\eta=y$, by (ii.) the complete solution is

$$(y+x-1)^5\,(y-x+1)^2=C.$$

3209 When P and Q are functions of x only, the solution of the equation

$$\frac{dy}{dx}+Py=0 \quad\text{is}\quad y=Ce^{-\int P dx} \quad\ldots\ldots\ldots \text{(i.)}$$

by merely separating the variables.

3210 Secondly, the solution of

$$\frac{dy}{dx}+Py=Q \quad\text{is}\quad y=e^{-\int P dx}\left\{C+\int Qe^{\int P dx}\,dx\right\}.$$

This result is obtained by the method of *variation of parameters.*

Rule.—*Assume equation* (i.) *to be the form of the solution, considering the parameter* C *a function of* x. *Differentiate* (i.) *on this hypothesis, and put the value of* y_x *so obtained in the proposed equation to determine* C.

Thus, differentiating (i.), we get $y_x=C_x e^{-\int P dx}-Py$,

therefore $Q=C_x e^{-\int P dx}$, therefore $C=\int Qe^{\int P dx}\,dx+C'$.

Then substitute this expression for C in equation (i.).

Otherwise, writing the equation in the form $(Py-Q)\,dx+dy=0$, the integrating factor $e^{\int P dx}$ may be found by (3097).

3211 $$y_x+Py=Qy^n$$

is reduced to the last case by dividing by y^n and substituting

$$z=y^{1-n}.$$

3212 $$P_1dx+P_2dy+Q(x\,dy-y\,dx)=0.$$

P_1, P_2 being homogeneous functions of x and y of the pth degree, and Q homogeneous and of the qth degree, is solved by assuming

$$P_1=x^p\phi\left(\frac{y}{x}\right),\qquad P_2=x^p\psi\left(\frac{y}{x}\right),\qquad Q=x^q\chi\left(\frac{y}{x}\right).$$

Put $y=vx$, and change the variables to x and v. The result may be reduced to

$$\frac{dx}{dv}+\frac{x\psi(v)}{\phi(v)+v\psi(v)}=-\frac{\chi(v)\,x^{q-p+2}}{\phi(v)+v\psi(v)},$$

which is identical in form with (3211), and may be solved accordingly.

3213 $$(A_1+B_1x+C_1y)(x\,dy-y\,dx)-(A_2+B_2x+C_2y)\,dy-(A_3+B_3x+C_3y)\,dx=0.$$

To solve this equation, put $x=\xi+\alpha$, $y=\eta+\beta$, and determine α and β so that the coefficients may become homogeneous, and the form of (3212) will be obtained.

RICCATI'S EQUATION.

3214 $$u_x+bu^2=cx^m \quad\ldots\ldots\ldots\ldots\ldots\ldots\ldots\quad (A).$$

Substitute $y=ux$, and this equation is reduced to the form of the following one, with $n=m+2$ and $a=1$. It is solvable whenever $m(2t\pm1)=-4t$, t being 0 or a positive integer.

3215 $$xy_x-ay+by^2=cx^n\quad\ldots\ldots\ldots\ldots\ldots\ldots\ldots\quad (B).$$

I.—This equation is solvable, when $n=2a$, by substituting $y=vx^a$, dividing by x^{2a}, and separating the variables. We thus obtain

$$\frac{dv}{c-bv^2}=x^{a-1}dx.$$

Integrating by (1937) or (1935), according as b and c in equation (B) have the same or different signs, and eliminating v by $y=vx^a$, we obtain the solution

3216 $$y=\sqrt{\frac{c}{b}}\,x^a\,\frac{Ce^{\frac{2x^a\sqrt{(bc)}}{a}}+1}{Ce^{\frac{2x^a\sqrt{(bc)}}{a}}-1}\quad\ldots\ldots\ldots\ldots\ldots\ldots\quad (1),$$

3217 or $y = \sqrt{\left(-\frac{c}{b}\right)} x^a \tan\left\{C - \frac{x^a\sqrt{(-bc)}}{a}\right\}$ .... (2).

3218 II.—Equation (B) may also be solved whenever $\frac{n-2a}{2n} = t$ a positive integer.

Rule.—*Write* z *for* y *in equation* (B), *and* nt+a *for* a *in the second term, and transpose* b *and* c *if* t *be odd.*

Thus, we shall have

$$xz_x - (nt+a)\,z + bz^2 = cx^n \text{ (when } t \text{ is even) ...... (3)},$$
$$xz_x - (nt+a)\,z + cz^2 = bx^n \text{ (when } t \text{ is odd) ...... (4)}.$$

Either of these equations can be solved as in case (I.), when $n = 2(nt+a)$, that is, when $\frac{n-2a}{2n} = t$. z having been determined by such a solution, the complete primitive of (B) will be the continued fraction

$$y = \frac{a}{b} + \frac{x^n}{\frac{n+a}{c}} + \frac{x^n}{\frac{2n+a}{b}} + \ldots + \frac{x^n}{\frac{(t-1)\,n+a}{k}} + \frac{x^n}{z} \ \ldots (5),$$

where k stands for b or c according as t is odd or even.

3219 III.—Equation (B) can also be solved whenever $\frac{n+2a}{2n} = t$ a positive integer. The method and result will be the same as in Case II., *if the sign of* a *be changed throughout and the first fraction omitted from the value of* y. Thus

$$y = \frac{x^n}{\frac{n-a}{c}} + \frac{x^n}{\frac{2n-a}{b}} + \ldots + \frac{x^n}{\frac{(t-1)\,n-a}{k}} + \frac{x^n}{z} \ \ldots\ldots (6).$$

Proof.—Case II.—In equation (B), substitute $y = A + \frac{x^n}{y_1}$, and equate the absolute term to zero. This gives $A = \frac{a}{b}$ or 0.

Taking the first value, the transformed equation becomes

$$x\frac{dy_1}{dx} - (n+a)\,y_1 + cy_1^2 = bx^n.$$

Next, put $y_1 = \frac{n+a}{c} + \frac{x^n}{y_2}$, and so on. In this way the tth transformed equation (3) or (4) is obtained with z written for the tth substituted variable y_t.

Case III.—Taking the second value, $A=0$, the first transformed equation differs from the above only in the sign of a; and consequently the same series of subsequent transformations arises, with $-a$ in the place of a. The successive substitutions produce (5) and (6) in the respective cases for the values of y.

3220 Ex. $$u_x+u^2 = cx^{-\frac{4}{3}}. \qquad (3214)$$

Putting $u=\frac{y}{x}$, $$\frac{du}{dx}=\frac{xy_x-y}{x^2},$$

and the equation is reduced to $xy_x-y+y^2=cx^{\frac{2}{3}}$ of the form (B). Here $a=1$, $b=1$, $n=\frac{2}{3}$, and $\frac{n+2a}{2n}=2$, Case III. By the rule in (3218), changing the sign of a for Case III., equation (3) becomes

$$xz_x-\tfrac{1}{3}z+z^2=cx^{\frac{2}{3}}.$$

Solving as in Case I., we put $z=vx^{\frac{1}{3}}$, &c.; or, employing formula (1) directly,

$$z=\sqrt{c}x^{\frac{1}{3}}\frac{Ce^{6\sqrt{c}x^{\frac{1}{3}}}+1}{Ce^{6\sqrt{c}x^{\frac{1}{3}}}-1}; \text{ and then, by (6), } y=\frac{x^{\frac{2}{3}}}{\frac{1}{3c}+\frac{x^{\frac{2}{3}}}{z}}$$

is the final solution.

FIRST ORDER NON-LINEAR EQUATIONS.

3221 *Type*

$$\left(\frac{dy}{dx}\right)^n+P_1\left(\frac{dy}{dx}\right)^{n-1}+\ldots+P_{n-1}\frac{dy}{dx}+P_n=0 \ldots\ldots(1),$$

where the coefficients $P_1, P_2, \ldots P_n$ may be functions of x and y.

SOLUTION BY FACTORS.

3222 If (1) can be resolved into n equations,

$$y_x-p_1=0, \quad y_x-p_2=0, \quad \ldots \quad y_x-p_n=0 \quad \ldots\ldots (2),$$

and if the complete primitives of these are

$$V_1=c_1, \quad V_2=c_2, \quad \ldots \quad V_n=c_n \ldots\ldots\ldots\ldots(3),$$

then the complete primitive of the original equation will be

$$(V_1-c)(V_2-c)\ldots(V_n-c)=0\ldots\ldots\ldots\ldots(4).$$

PROOF.—Taking $n = 3$, assume the last equation. Differentiate and eliminate c. The result is

$$(V_2 - V_3)^2 (V_3 - V_1)^2 (V_1 - V_2)^2 dV_1 dV_2 dV_3 = 0 \ldots\ldots\ldots\ldots(5).$$

By (2), $dV_1 = \mu_1 (y_x - p_1)\, dx$, &c., where μ_1 is an integrating factor. Substitute these values in (5), rejecting the factors which do not contain differential coefficients, and the result is

$$(y_x - p_1)(y_x - p_2)(y_x - p_3) = 0,$$

which is the differential equation (1).

3223 Ex.—Given $\quad y_x^2 + 3y_x + 2 = 0.$

The component equations are $y_x + 1 = 0$ and $y_x + 2 = 0$, giving for the complete primitive

$$(y + x - c)(y + 2x - c) = 0.$$

SOLUTION WITHOUT RESOLVING INTO FACTORS.

3224 CLASS I.—*Type* $\phi(x, p) = 0$.

When x only is involved with p, and it is easier to solve the equation for x than for p, proceed as follows.

RULE.—*Obtain* $x = f(p)$. *Differentiate and eliminate* dx *by means of* $dy = p\,dx$. *Integrate and eliminate* p *by means of the original equation.*

Similarly, when $y = f(p)$, *eliminate* dy, &c.

3225 Ex.—Given $x = ay_x + by_x^2$, $\quad$ *i.e.*, $x = ap + bp^2 \ldots\ldots\ldots\ldots(1),$

$$dx = a\,dp + 2bp\,dp, \qquad \text{therefore } dy = p\,dx = ap\,dp + 2bp^2 dp,$$

therefore $$y = \frac{ap^2}{2} + \frac{2bp^3}{3} + C.$$

Eliminating p between this equation and (1), the result is the complete primitive

$$(ax + 6by - bc)^2 = (6ay - 4x^2 - ac)(a^2 + 4bx).$$

3226 CLASS II.—*Type*

$$x\phi(p) + y\psi(p) = \chi(p).$$

RULE.—*Differentiate and eliminate* y *if necessary. Integrate and eliminate* p *by means of the original equation.*

If the equation be first divided by $\psi(p)$, the form is simplified into

3227 $$y = x\phi(p) + \chi(p).$$

Differentiate, and an equation is obtained of the form $x_p + Px = Q$, where P and Q are functions of p.

This may be solved by (3210), and p afterwards eliminated.

3228 Otherwise, a differential equation may be formed between y and p, instead of between x and p.

3229 Or, more generally, a differential equation may be formed between x or y and t, any proposed function of p, after which t must be eliminated to obtain the complete primitive.

3230 *Clairaut's equation*, which belongs to this class, is of the form
$$y = px + f(p).$$

RULE.—*Differentiate, and two equations are obtained*—

$$(1) \ldots\ldots p_x = 0, \text{ and } \therefore\ p = c; \qquad (2) \ldots\ldots x + f'(p) = 0.$$

Eliminate p *from the original equation by means of* (1), *and again by means of* (2). *The first elimination gives* $y = cx + f(c)$, *the complete primitive. The second gives a singular solution.*

PROOF.—For, if Rule I. (3169) be applied to the primitive $y = cx + f(c)$, we have $x + f'(c) = 0$; and to eliminate c between these equations is the elimination directed above, c being merely written for p in the two equations.

3231 Ex. 1. $$y = px + x\sqrt{1+p^2}.$$

This is of the form $y = x\phi(p)$, and therefore falls under (3227). Differentiating, we obtain $x\,dp + dx\sqrt{1+p^2} + \dfrac{xp\,dp}{\sqrt{(1+p^2)}} = 0$,

since $dy = p\,dx$; thus
$$\left(\frac{1}{\sqrt{(1+p^2)}} + \frac{p}{1+p^2}\right) dp + \frac{dx}{x} = 0,$$
in which the variables are separated.

Integrating by (1928), and eliminating p, we find for the complete primitive $x^2 + y^2 = Cx$.

3232 Ex. 2. $$y = px + \sqrt{b^2 - a^2p^2}.$$

This is Clairaut's form (3230). Differentiating, we have
$$\frac{dp}{dx}\left\{x - \frac{a^2p}{\sqrt{(b^2-a^2p^2)}}\right\} = 0.$$

The complete primitive is $y = cx + \sqrt{(b^2 - a^2c^2)}$;

and the elimination of p by the other equation gives for the singular solution $a^2y^2 - b^2x^2 = a^2b^2$, an hyperbola and the envelope of the lines obtained by varying c in the complete primitive, which is the equation of a tangent.

3233 Ex. 3.—To find a curve having the tangent intercepted between the coordinate axes of constant length.

The differential equation which expresses this property is

$$\frac{y\sqrt{1+p^2}}{p} - x\sqrt{1+p^2} = a,$$

or $$y = px + \frac{ap}{\sqrt{1+p^2}} \quad\text{.......(1).}$$

Differentiating gives $$\frac{dp}{dx}\left\{x + \frac{a}{(1+p^2)^{\frac{3}{2}}}\right\} = 0 \quad\text{.......(2).}$$

Eliminating p between (1) and (2) gives,

1st, the primitive $$y = cx + \frac{ac}{\sqrt{1+c^2}} \quad\text{.......(3);}$$

2nd, the singular solution $x^{\frac{2}{3}} + y^{\frac{2}{3}} = a^{\frac{2}{3}}$ .... (4).

(3) is the equation of a straight line; (4) is the envelope of the lines obtained by varying the parameter c in equation (3).

3234 Class III.—*Homogeneous in* x *and* y.

Type $$x^n\phi\left(\frac{y}{x}, p\right) = 0.$$

Rule.—*Put* $y = vx$, *and divide by* x^n. *Solve for* p, *and eliminate* p *by differentiating* $y = vx$; *or solve for* v, *and eliminate* v *by putting* $v = \frac{y}{x}$; *and in either case separate the variables.*

3235 Ex. $$y = px + x\sqrt{1+p^2}.$$

Substitute $y = vx$, and therefore $p = v + xv_x$. This gives $v = p + \sqrt{1+p^2}$. Eliminate p between the last two equations, and then separate the variables.

The result is $$\frac{dx}{x} + \frac{2v\,dv}{1+v^2} = 0,$$

from which $$x(v^2+1) = C \quad\text{or}\quad x^2+y^2 = Cx.$$

The same equation is solved in (3131) in another way.

SOLUTION BY DIFFERENTIATION.

3236 To solve an equation of the form

$$F\{\phi(x, y, y_x),\ \psi(x, y, y_x)\} = 0.$$

Rule.—*Equate the functions* ϕ *and* ψ *respectively to arbitrary constants* a *and* b. *Differentiate each equation, and eliminate the constants. If the results agree, there is a common primitive* (3166), *which may be found by eliminating* y_x *between the equations* $\phi = a$, $\phi = b$, *and subsequently eliminating one of the constants by means of the relation* $F(a, b) = 0$.

Ex. $$x - yy_x + f(y^2 - y^2y_x^2) = 0.$$

Here the two equations $x - yy_x = a$, $f(y^2 - y^2y_x^2) = b$,

on applying the test, are found to have a common primitive. Therefore, eliminating y_x, we obtain

$$f\{y^2 - (x-a)^2\} = b.$$

Also, by the given equation, $a + b = 0$.

Hence the solution is $f\{y^2 - (x+b)^2\} = b$.

HIGHER ORDER LINEAR EQUATIONS.

3237 *Type* $\dfrac{d^ny}{dx^n} + P_1\dfrac{d^{n-1}y}{dx^{n-1}} + \ldots + P_{(n-1)}\dfrac{dy}{dx} + P_ny = Q$,

where $P_1 \ldots P_n$ and Q are either functions of x or constants.

LEMMA.—If $y_1, y_2, \ldots y_n$ be n different values of y in terms of x, which satisfy (3237), when $Q = 0$, the solution in that case will be $y = C_1y_1 + C_2y_2 + \ldots + C_ny_n$.

PROOF.—Substitute $y_1, y_2, \ldots y_n$ in turn in the given equation. Multiply the resulting equations by arbitrary constants, $C_1, C_2, \ldots C_n$ respectively; add, and equate coefficients of $P_1, P_2, \ldots P_n$ with those in the original equation.

LINEAR EQUATIONS WITH CONSTANT COEFFICIENTS.

3238 $$y_{nx} + a_1y_{(n-1)x} + \ldots + a_{(n-1)}y_x + a_ny = Q \quad\ldots\ldots\ldots (1).$$

3239 *Case I.—When* $Q = 0$.

The roots of the auxiliary equation

$$m^n + a_1m^{n-1} + \ldots + a_{n-1}m + a_n = 0 \quad\ldots\ldots\ldots\ldots (2)$$

being $m_1, m_2, \ldots m_n$, the complete primitive of the differential equation will be

3240 $$y = C_1e^{m_1x} + C_2e^{m_2x} + \ldots + C_ne^{m_nx} \quad\ldots\ldots\ldots\ldots (3).$$

If the auxiliary equation (2) has a pair of imaginary roots

$(a \pm ib)$, there will be in the value of y the corresponding terms

3241 $$Ae^{ax}\cos bx + Be^{ax}\sin bx \;\ldots\ldots\ldots\ldots\ldots\ldots \;(4).$$

If any real root m' of equation (2) is repeated r times, the corresponding part of the value of y will be

3242 $$(A_0 + A_1x + A_2x^2 + \ldots + A_{r-1}x^{r-1})\,e^{m'x}.$$

And if a pair of imaginary roots occurs r times, substitute for A and B in (3241) similar polynomials of the $r-1^{\text{th}}$ degree in x.

PROOF.—(i.) Substituting $y = Ce^{mx}$ in (1) as a particular solution, and dividing by Ce^{mx}, the auxiliary equation is produced, the roots of which furnish n particular solutions, $y = C_1e^{m_1x}$, $y = C_2e^{m_2x}$, &c., and therefore, by the preceding lemma, the general solution will be equation (2).

(ii.) The imaginary roots $a \pm ib$ give rise to the terms $Ce^{ax+ibx} + C'e^{ax-ibx}$, which, by the Exp. values (766), reduce to

$$(C+C')\,e^{ax}\cos bx + i\,(C-C')\,e^{ax}\sin bx.$$

(iii.) If there are two equal roots $m_2 = m_1$, put at first $m_2 = m_1 + h$. The two terms $C_1e^{m_1x} + C_2e^{(m_1+h)x}$ become $e^{m_1x}(C_1 + C_2e^{hx})$. Expand e^{hx} by (150), and put $C_1 + C_2 = A$, $C_2h = B$ in the limit when $h = 0$, $C_1 = \infty$, $C_2 = -\infty$. By repeating this process, in the case of r equal roots, we arrive at the form

$$(A_0 + A_1x + A_2x^2 + \ldots + A_{r-1}x^{r-1})\,e^{m_1x};$$

and similarly in the case of repeated pairs of imaginary roots.

3243 *Case II.—When* Q *in* (3238) *is a function of* x.

First method.—By variation of parameters.

Putting $Q = 0$, as in Case I., let the complete primitive be

$$y = A\alpha + B\beta + C\gamma + \text{\&c. to } n \text{ terms} \;\ldots\ldots\ldots\; (6),$$

α, β, γ being functions of x of the form e^{mx}. The values of the parameters $A, B, C, \ldots$, when Q has its proper value assigned, are determined by the n equations

3244 $$\begin{array}{llll} A_x\alpha & +B_x\beta & +\text{to } n \text{ terms} & = 0, \\ A_x\alpha_x & +B_x\beta_x & + \quad ,, & = 0, \\ A_x\alpha_{2x} & +B_x\beta_{2x} & + \quad ,, & = 0, \\ \ldots & \ldots & \ldots & \ldots \\ A_x\alpha_{(n-1)x} & +B_x\beta_{(n-1)x} & + \quad ,, & = Q, \end{array}$$

A_x, B_x, &c. being found from these equations, their integrals must be substituted in (6) to form the complete primitive.

PROOF.—Differentiate (6) on the hypothesis that A, B, C, &c. are functions of x; thus

$$y_x = (A\alpha_x + B\beta_x + \ldots) + (A_x\alpha + B_x\beta + \ldots).$$

Now, in addition to equation (1), $n-1$ relations may be assumed between the n arbitrary parameters. Equate then the last term in brackets to zero, and differentiate y, in all, $n-1$ times, equating to zero the second part of each differentiation; thus we obtain

$$\begin{array}{llll} y_x = A\alpha_x + B\beta_x + \&c. & \text{and} & A_x\alpha + B_x\beta + \&c. = 0, \\ y_{2x} = A\alpha_{2x} + B\beta_{2x} + \&c. & \text{and} & A_x\alpha_x + B_x\beta_x + \&c. = 0, \\ \ldots & & \ldots \\ y_{(n-1)x} = A\alpha_{(n-1)x} + B\beta_{(n-1)x} + \&c. & \text{and} & A_x\alpha_{(n-2)x} + B_x\beta_{(n-2)x} + \&c. = 0. \end{array}$$

The n quantities A_x, B_x, &c. are now determined by the $n-1$ equations on the right and equation (1). For, differentiating the value of $y_{(n-1)x}$, we have

$$y_{nx} = \{A\alpha_{nx} + B\beta_{nx} + \&c.\} + \{A_x\alpha_{(n-1)x} + B_x\beta_{(n-1)x} + \&c.\},$$

and if these values of y_x, y_{2x}, ... y_{nx} be substituted in (1), it reduces to

$$A_x\alpha_{(n-1)x} + B_x\beta_{(n-1)x} + \&c. = Q,$$

for the other part vanishes by the hypothetical equation

$$y_{nx} + a_1 y_{(n-1)x} + \ldots + a_n y = 0,$$

since the values of y_x, ... $y_{(n-1)x}$, and the first part of y_{nx}, are the true values in this equation.

3245 *Case II.—Second Method.*—Differentiate and eliminate Q. The resulting equation can be solved as in Case I. Being of a higher order, there will be additional constants which may be eliminated by substituting the result in the given equation.

3246 Ex.—Given $\quad y_{2x} - 7y_x + 12y = x$ .... (1).

1*st Method.*—Putting $x = 0$, the auxiliary equation is $m^2 - 7m + 12 = 0$; therefore $m = 3$ and 4. Hence the complete primitive of $y_{2x} - 7y_x + 12y = 0$ is

$$y = Ae^{2x} + Be^{4x} \ldots\ldots\ldots (2).$$

The corrected values of A and B for the primitive of equation (1) are found from

$$\left.\begin{array}{r} A_x e^{3x} + B_x e^{4x} = 0 \\ 3A_x e^{3x} + 4B_x e^{4x} = x \end{array}\right\}, \quad \therefore\ A_x = -xe^{-3x} \ \text{and}\ A = \frac{3x+1}{9}e^{-3x} + a.$$

$$B_x = xe^{-4x} \ \text{and}\ B = -\frac{4x+1}{16}e^{-4x} + b.$$

Substituting these values of A and B in (2), we find for its complete primitive

$$y = ae^{3x} + be^{4x} + \frac{12x+7}{144}.$$

3247 2*nd Method.* $\quad y_{2x} - 7y_x + 12y = x$ .... (1).

Differentiating to eliminate the term on the right, we get

$$y_{4x}-7y_{3x}+12y_{2x}=0.$$

The aux. equation is $m^4-7m^3+12m^2=0$; therefore $m=4,\ 3,\ 0,\ 0.$

Therefore $$y=Ae^{4x}+Be^{3x}+Cx+D \quad\text{.......................... (2)};$$

$$y_x=4Ae^{4x}+3Be^{3x}+C;\qquad y_{2x}=16Ae^{4x}+9Be^{3x}.$$

Substitute these values in (1); thus $C=\frac{1}{12}$; $D=\frac{7}{144}$;

therefore, substituting in (2), $y=Ae^{4x}+Be^{3x}+\frac{x}{12}+\frac{7}{144}$ as before.

3248 When a particular integral of the linear equation (3238) is known in the form $y=f(x)$, the complete primitive may be obtained by adding to y that value which it would take if Q were zero.

Thus, in Ex. (3247), $y=\frac{x}{12}+\frac{7}{144}$ is a particular integral of (1); and the complementary part $Ae^{4x}+Be^{3x}$ is the value of y when the dexter is zero.

3249 The order of the linear equation (3238) may always be depressed by unity if a particular integral of the same equation, when $Q=0$, be known.

Thus, if $$y_{3x}+P_1y_{2x}+P_2y_x+P_3y=Q \quad\text{.......................... (1)},$$
and if $y=z$ be a particular solution when $Q=0$; let $y=vz$ be the solution of (1). Therefore, substituting in (1),

$$(z_{3x}+P_1z_{2x}+P_2z_x+P_1z)\,v+\&c.=Q,$$

the unwritten terms containing v_x, v_{2x}, and v_{3x}.

The coefficient of v vanishes, by hypothesis; therefore, if we put $v_x=u$, we have an equation of the *second* order for determining u. u being found,

$$v=\int u\,dx+C.$$

3250 The linear equation

$$(a+bx)^n y_{nx}+A\,(a+bx)^{n-1}y_{(n-1)x}+B\,(a+bx)^{n-2}y_{(n-2)x}+\dots$$
$$\dots+Ly=Q,$$

where A, B, ... L are constants, and Q is a function of x, is solved by substituting $a+bx=e^t$, changing the variable by formula (1770), and in the complete primitive putting $t=\log(a+bx)$.

Otherwise, reduce to the form in (3446) by putting $a+bx=\lambda\xi$, and solve as in that article.

HIGHER ORDER NON-LINEAR EQUATIONS.

3251 *Type* $F\left(x,\, y,\, \frac{dy}{dx},\, \frac{d^2y}{dx^2},\, \dots\, \frac{d^ny}{dx^n}\right) = 0.$

SPECIAL FORMS.

3252 $F(x,\, y_{rx},\, y_{(r+1)x} \dots y_{nx}) = 0.$

When the dependent variable y is absent, and y_{rx} is the derivative of lowest order present, the equation may be depressed to the order $n-r$ by putting $y_{rx} = z$. If the equation in z can be solved, the complete primitive will then be

$$y = \int_{rx} z + \int_{rx} 0 \qquad (2149).$$

3253 $F(y,\, y_{rx},\, y_{(r+1)x} \dots y_{nx}) = 0.$

If x be absent instead of y, change the independent variable from x to y, and proceed as before.

Otherwise, change the independent variable to y, and make $p\,(= y_x)$ the dependent variable.

For example, let the equation be of the form

3254 $F(y,\, y_x,\, y_{2x},\, y_{3x}) = 0$ .... (1).

(i.) This may be changed into the form

$F(y,\, x_y,\, x_{2y},\, x_{3y}) = 0$ by (1761, '63, and '66);

and the order may then be depressed to the 2nd by (3252). The solution will thus give x in terms of y.

3255 (ii.) Otherwise, equation (1) may be changed at once into one of the form $F(y,\, p,\, p_y,\, p_{2y}) = 0$, by (1764 and '67), the order being here depressed from the 3rd to the 2nd. If the solution of this equation be $p = \phi\,(y,\, c_1,\, c_2)$, then, since $dy = p\,dx$, we get, for the complete primitive of (1),

$$x = \int \frac{dy}{\phi\,(y,\, c_1,\, c_2)} + c_3.$$

3256 $y_{nx} = F(x).$

Integrate n times successively, thus

$$y = \int_{nx} F(x) + c_1 x^{n-1} + c_2 x^{n-2} + \dots + c_n.$$

3257 $$y_{2x} = F(y).$$

Multiply by $2y_x$ and integrate, thus

$$\left(\frac{dy}{dx}\right)^2 = 2\int F(y)\,dy + c_1, \quad x = \int \frac{dy}{\sqrt{\{2\int F(y)\,dy + c_1\}}} + c_2.$$

3258 $$y_{nx} = F\{y_{(n-1)x}\},$$

an equation between two successive derivatives.

Put $y_{(n-1)x} = z$, then $z_x = F(z)$, from which

$$x = \int \frac{dz}{F(z)} + c \;\ldots\ldots\ldots\ldots\ldots\ldots\ldots\ldots \text{(1)}.$$

If, after integrating, this equation can be solved for z so that $z = \phi(x, c)$, we have $y_{(n-1)x} = \phi(x, c)$, which falls under (3256).

3259 But if z cannot be expressed in terms of x, proceed as follows:—

$$y_{(n-1)x} = z; \quad y_{(n-2)x} = \int z\,dx = \int \frac{dz}{F(z)}\,z;$$

$$y_{(n-3)x} = \int \frac{dz}{F(z)} \int \frac{dz}{F(z)}\,z; \quad \ldots \quad \ldots$$

Finally, $$y = \int \frac{dz}{F(z)} \int \frac{dz}{F(z)} \cdots \int \frac{dz}{F(z)}\,z;$$

the number of integrations and arbitrary constants introduced being $n-1$.

3260 $$y_{nx} = F\{y_{(n-2)x}\}.$$

Put $y_{(n-2)x} = z$; then $z_{2x} = F(z)$, which is (3257), the solution giving x in terms of z and two constants. If z can be found from this in terms of x and the two constants, we get

$$z \quad \text{or} \quad y_{(n-2)x} = \phi(x, c_1, c_2),$$

which may be solved by (3256).

3261 But if z cannot be expressed in terms of x, proceed as in (3259).

DEPRESSION OF ORDER BY UNITY.

3262 When $$F(x, y, y_x, y_{2x}, \ldots) = 0$$

is rendered homogeneous by considering

$$x, y, y_x, y_{2x}, y_{3x}, \text{\&c.}$$

to be of the respective dimensions **1, 1, 0, −1, −2,** &c.; put

$$x = e^\theta, \quad y = ze^\theta, \text{ \&c.}$$

The transformed equation will contain the same power of e in every term, and will reduce to the form

$$F(z, z_\theta, z_{2\theta}, \ldots) = 0,$$

the order of which is depressed by unity by putting $z_\theta = u$.

3263 When the original equation is of the 2nd order, the transformed equation in u and z may be obtained at once by changing x, y, y_x, y_{2x} into $1, z, u+z, u+uu_z$, respectively. The solution is then completed, as in example (3264).

PROOF.—We have $\quad x = e^\theta; \quad y = ze^\theta;$

$$y_x = z_\theta + z; \quad y_{2x} = e^{-\theta}(z_{2\theta}+z_\theta); \quad y_{3x} = e^{-2\theta}(z_{3\theta}-z_\theta); \text{ and so on.}$$

The dimensions of x, y, y_x, &c. with respect to e_θ are 1, 1, 0, −1, −2, &c. Therefore the same power of e^θ will occur in every term of the *homogeneous* equation.

3264 Ex.: $\quad x^4 y_{2x} = (y - xy_x)^3.$

Making the above substitutions for x, y, y_x, and y_{2x}, the equation becomes

$$z_{2\theta} + z_\theta = -z_\theta^3.$$

Put $z_\theta = u$; thus

$$u^3 + u = -u_\theta = -uu_z, \quad \text{therefore } u^2+1 = -u_z, \quad \frac{du}{u^2+1} = -dz,$$

therefore $\tan^{-1} u = a - z$ (1935), therefore $z_\theta = u = \tan(a-z)$,
therefore $d\theta = \cot(a-z)\,dz$, therefore $\theta = -\log b \sin(a-z)$ (1941).

But $\quad \theta = \log x \quad \text{and} \quad z = \frac{y}{x},$

therefore $\quad bx = \operatorname{cosec}\left(a - \frac{y}{x}\right), \quad \text{or} \quad bx = \sec\left(c + \frac{y}{x}\right),$

by altering the arbitrary constant.

3265 When $\quad F(x, y, y_x, y_{2x}, \ldots) = 0$

is made homogeneous by considering x, y, y_x, y_{2x}, &c. to be of the respective dimensions $1, n, n-1, n-2$, &c.; put

$$x = e^\theta, \quad y = ze^{n\theta},$$

and depress the order by putting $z_\theta = u$, as in (3262).

3266 When the original equation is of the 2nd order, the final equation between u and z may be obtained at once by changing

x, y, y_x, y_{2x} into $1, z, u+nz, uu_z+(2n-1)u+n(n-1)z$, respectively.

3267 Ex.: $$y_{2x}y_x = xy \quad\text{.............} \quad (1).$$

With the view of applying (3265), the assumed dimensions of each member of this equation, being equated, give

$$n-2+n-1 = 1+n, \qquad \text{therefore } n = 4.$$

Thus $x = e^\theta$; $y = ze^{4\theta}$; $y_x = e^{3\theta}(z_\theta+4z)$; $y_{2x} = e^{2\theta}(z_{2\theta}+7z_\theta+12z)$.

Substituting in (1), e disappears; and by putting $z_\theta = u$, $z_{2\theta} = uu_z$, the equation is reduced to

$$(u^2+4uz)\,du+(7u^2+40uz+48z^2-z)\,dz = 0,$$

which is linear and of the 1st order. This equation is also obtained at once by the rule in (3266).

3268 When $$F(x, y, y_x, y_{2x}, \ldots) = 0$$

is homogeneous with respect to y, y_x, y_{2x}, &c., put $y = e^{\int u\,dx}$, and remove e as before by division. The equation between u and x will have its order less by unity than the order of F.

3269 Ex.: $$y_{2x}+Py_x+Qy = 0 \quad\text{.............}\quad (1),$$

P and Q being functions of x.

Here $$y = e^{\int u\,dx}; \quad y_x = uy; \quad y_{2x} = (u_x+u^2)\,y.$$

Substituting, the equation becomes $u_x+u^2+Pu+Q = 0$, an equation of the 1st order. If the solution gives $u = \phi(x, c)$, then $\int \phi(x, c)\,dx = \log y$ is the complete primitive of (1).

EXACT DIFFERENTIAL EQUATIONS.

3270 Let $$dU = \phi(x, y, y_x, y_{2x}, \ldots y_{nx})\,dx = 0$$

be an exact differential equation of the nth order. The highest derivative involved will be of the 1st degree.

3271 RULE FOR THE SOLUTION (*Sarrus*).—*Integrate the term involving* y_{nx} *with respect to* $y_{(n-1)x}$ *only, and call the result* U_1. *Find* dU_1, *considering both* x *and* y *as variables.* $dU-dU_1$ *will be an exact differential of the* $n-1^{\text{th}}$ *order.*

Integrate this with respect to $y_{(n-2)x}$ *only, calling the result* U_2, *and so on.*

The first integral of the proposed equation will be

$$U = U_1 + U_2 + \ldots + U_n = C.$$

3272 Ex.: Let $dU = \{y^2 + (2xy-1)\, y_x + xy_{2x} + x^2 y_{3x}\}\, dx = 0.$

Here $U_1 = x^2 y_{2x},\quad dU_1 = (2xy_{2x} + x^2 y_{3x})\, dx;$

$$dU - dU_1 = \{y^2 + (2xy-1)\, y_x - xy_{2x}\}\, dx = 0;$$

$\therefore\ U_2 = -xy_x,\quad dU_2 = -(y_x + xy_{2x})\, dx,\quad dU - dU_1 - dU_2 = (y^2 + 2xyy_x)\, dx;$

therefore $U_3 = xy^2$, and $U = x^2 y_{2x} - xy_x + xy^2 = C$

is the first integral.

3273 Denoting equation (3270) by $dU = V dx$, the series of steps in Rule (3171) involve and amount to the single condition that the equation

$$N - P_x + Q_{2x} - R_{3x} + \&c. = 0,$$

with the notation in (3028), shall be identically true. This then is the condition that V shall be an exact 1st differential.

3274 Similarly, the condition that V shall be an exact 2nd differential is

$$P - 2Q_x + 3R_{2x} - 4S_{3x} + \&c. = 0.$$

3275 The condition that V shall be an exact 3rd differential is

$$Q - 3R_x + \frac{3.4}{1.2} S_{2x} - \frac{3.4.5}{1.2.3} T_{3x} + \&c. = 0,$$

and so on. [Euler, *Comm. Petrop.*, Vol. viii.

MISCELLANEOUS METHODS.

3276 $$y_{2x} + Py_x + Qy_x^3 = 0 \quad\ldots\ldots\ldots\ldots (1),$$

where P and Q are functions of x only.

The solution is $y = \int e^{-\int P dx} \left(2\int Q e^{-2\int P dx} dx\right)^{-\frac{1}{2}} dx.$

Proof. — Put $e^{\int P dx} = z$, and multiply (1) by z; then, since $z_x = Pz$, $zy_x + Qzy_x^3 = 0$. Put $zy_x = u^{-1}$, $\therefore\ uu_x = Qz^{-2}$, $\therefore\ u = \sqrt{(2\int Qz^{-2} dx)}$, &c.

3277 $$y_{2x} + Qy_x^2 + R = 0 \quad\ldots\ldots\ldots\ldots (1),$$

where Q and R are functions of y only.

The solution is $x = \int e^{\int Qd} \left(2\int Re^{2\int Qdy} dy\right)^{-\frac{1}{2}} dy.$

PROOF.—Put $e^{\int Qdy} = z$, and multiply (1) by z.

$\therefore\ (zy_x)_x = Rz, \qquad \therefore\ zy_x (zy_x)_x = Rz^2 y_x, \qquad \therefore\ (zy_x)^2 = 2\int Rz^2 dy$, &c.

3278 $$y_{2x} + Py_x + Qy_x^n = 0,$$
where P, Q involve x only.

Put $y_x = z$, and the form (3211) is arrived at.

3279 $$y_{2x} + Py_x^2 + Qy_x^n = 0.$$
This reduces to the last case by changing the variable from x to y by (1763).

3280 For a few cases in which the equation
$$y_x + Py^2 + Qy + R = 0$$
can be integrated, see De Morgan's "Differential and Integral Calculus," p. 690.

3281 $$y_{2x} = ax + by.$$
Put $ax + by = t$ (1762–3). Result $t_{2x} = bt$. Solve by (3239) or (3257).

3282 $$(1 - x^2)\, y_{2x} - xy_x + q^2 y = 0.$$
Put $\sin^{-1} x = t$, and obtain $y_{2t} + q^2 y = 0$.

Solution, $\quad y = A\cos(q\sin^{-1} x) + B\sin(q\sin^{-1} x)$.

3283 $$(1 + ax^2)\, y_{2x} + axy_x \pm q^2 y = 0.$$
Put $\int \frac{dx}{\sqrt{(1+ax^2)}} = t$, and obtain $y_{2t} \pm q^2 y = 0$ as above.

3284 *Liouville's equation*, $\quad y_{2x} + f(x)\, y_x + F(y)\, y_x^2 = 0.$

Suppress the last term. Obtain a first integral by (3209), and vary the parameter. The complete primitive is
$$\int e^{\int F(y)\,dy} dy = A\int e^{-\int f(x)\,dx} dx + B.$$

3285 *Jacobi's theorem.*—If one of the first integrals of the equation $\quad y_{2x} = f(x, y) \quad$ is $\quad y_x = \phi(x, y, c)$.........(i., ii.), the complete primitive will be

$$\int \phi_c (dy - \phi\, dx) = c.$$

PROOF.—Differentiating (ii.), we obtain $\phi_x + \phi\phi_y = f(x, y)$, and differentiating this for c, $\phi_{xc} + \phi_c\phi_y + \phi\,\phi_{yc} = 0$. But, by (3187), this is the condition for ensuring that $\phi_c dy - \phi_c \phi dx = 0$ shall be an exact differential; therefore ϕ_c is an integrating factor for equation (ii.), $y_x - \phi(x, y, c) = 0$.

Equations involving the arc s, *having given*

3286 $$ds^2 = dx^2 + dy^2 \quad \text{or} \quad s_x = \sqrt{1 + y_x^2}.$$

3287 $$s = ax + by.$$

Here $\sqrt{1 + y_x^2} = a + by_x$. Find y_x from the quadratic equation.

3288 $$x_{2s} = a.$$

Change from s to x by (1763); $\therefore\ -s_x^{-3} s_{2x} = a$, $\therefore\ s_x^{-2} = 2ax + c$,

$$\text{or} \quad 1 + y_x^2 = \frac{1}{2ax + c}, \qquad \therefore\ y = \int \sqrt{\left(\frac{1}{2ax + c} - 1\right)}\, dx + c'.$$

APPROXIMATE SOLUTION OF DIFFERENTIAL EQUATIONS BY TAYLOR'S THEOREM.

3289 The following example will illustrate the method :—

Given $y_{2x} = xy_x + y$, $\qquad \therefore\ y_{3x} = (x^2 + 2)\, y_x + xy$.

Generally, let $y_{nx} = A_n y_x + B_n y$; $\quad y_{(n+1)x} = A_{n+1} y_x + B_{n+1} y$.

But, by differentiation,

$$y_{(n+1)x} = (A_n x + A_n' + B_n)\, y_x + (A_n + B_n')\, y,$$

$$\therefore\ A_{n+1} = A_n x + A_n' + B_n \quad \text{and} \quad B_{n+1} = A_n + B_n'.$$

But $\qquad A_2 = x,\ B_2 = 1,\ \therefore\ A_3 = x^2 + 2,\ B_3 = x;$

$$A_4 = x^3 + 5x, \quad B_4 = x^2 + 3, \ \&c.$$

Now, when $x = a$, let $y = b$ and $y_x = p$; then, by Taylor's theorem (1500),

$$y = a + p(x - a) + (A_2 p + B_2 b)\frac{(x-a)^2}{2!} + (A_3 p + B_3 b)\frac{(x-a)^3}{3!} + \&c.,$$

which converges when $x - a$ is small. [De Morgan, p. 692.

SINGULAR SOLUTIONS OF HIGHER ORDER EQUATIONS.

DERIVATION FROM THE COMPLETE PRIMITIVE.

3301 Let $y_{nx} = \phi(x, y, y_x, y_{2x} \dots y_{(n-1)x})$ .... (1)
be the differential equation, and let its complete primitive be

$$y = f(x, a, b, c \dots s) \quad\quad (2),$$

containing n arbitrary constants.

3302 RULE.—*To find the general singular solution of* (1), *eliminate* abc ... s *between the equations*

$$\mathrm{y} = \mathrm{f}, \quad \mathrm{y_x} = \mathrm{f_x}, \quad \mathrm{y_{2x}} = \mathrm{f_{2x}} \dots, \quad \mathrm{y_{(n-1)x}} = \mathrm{f_{(n-1)x}} \quad\quad (3)$$

and

$$\begin{vmatrix} \mathrm{f_a} & \mathrm{f_{ax}} & \mathrm{f_{a2x}} & \dots & \mathrm{f_{a(n-1)x}} \\ \mathrm{f_b} & \mathrm{f_{bx}} & \mathrm{f_{b2x}} & \dots & \mathrm{f_{b(n-1)x}} \\ \dots & & \dots & & \dots \\ \mathrm{f_s} & \mathrm{f_{sx}} & \mathrm{f_{s2x}} & \dots & \mathrm{f_{s(n-1)x}} \end{vmatrix} = 0 \quad\quad (4).$$

The result is a differential equation of the $\mathrm{n}-1^{\mathrm{th}}$ *order, and the integral of it, containing* n−1 *arbitrary constants, is the singular solution.*

PROOF.—Differentiate (2), considering the parameters $a, b \dots s$ variable, thus $y_x = f_x + f_a a_x + \dots + f_s s_x$. Therefore, as in (3171),

$$y_x = f_x \quad \text{if} \quad f_a\, a_x + f_b\, b_x + \dots f_s\, s_x = 0,$$

$$y_{2x} = f_{2x} \quad \text{if} \quad f_{ax} a_x + f_{bx} b_x + \dots f_{sx} s_x = 0, \text{ as well;}$$

and so on up to $y_{nx} = f_{nx}$. Eliminating $a_x, b_x, \dots s_x$ between the n equations on the right, the determinant equation (4) is produced with the rows and columns interchanged.

3303 Ex.: $y - xy_x + \tfrac{1}{2}x^2 y_{2x} - y_{2x}^2 - (y_x - xy_{2x})^2 = 0$ .... (1).

The complete primitive is $y = \dfrac{ax^2}{2} + bx + a^2 + b^2$ .... (2).

From which $y_x = ax + b$ .... (3),

and the determinant equation is

$$\begin{vmatrix} \tfrac{1}{2}x^2 + 2a, & x \\ x + 2b, & 1 \end{vmatrix} = 0 \quad \text{or} \quad \frac{x^2}{2} + 2bx = 2a \quad\quad (4).$$

Eliminating a and b from (2), (3), and (4), we get the differential equation

$$\frac{4dy+(2x+x^3)\,dx}{\sqrt{(16y+4x^2+x^4)}} = \sqrt{(1+x^2)}\,dx \quad\text{.....................}\quad (5),$$

the integral of which, and the singular solution of (1), is

$$\sqrt{(16y+4x^2+x^4)} = x\sqrt{(1+x^2)}+\log\{x+\sqrt{(1+x^2)}\}+C.$$

[Boole, *Sup.*, p. 49.

3304 Either of the two 'first integrals' (3064) of a second order differential equation leads to the same singular solution of that equation.

3305 The complete primitive of a singular first integral of a differential equation of the second order is itself a singular solution of that equation; but a singular solution of a singular first integral is not generally a solution of the original equation.

Thus the singular first integral (5) of equation (1) in the last example has the singular solution $16y+4x^2+x^4=0$, which is not a solution of equation (1).

DERIVATION OF THE SINGULAR SOLUTION FROM THE DIFFERENTIAL EQUATION.

3306 Rule.—*Assuming the same form* (3173), *a singular solution of the first order of a differential equation of the* nth *order will make* $\dfrac{d(y_{nx})}{d(y_{(n-1)x})}$ *infinite; a singular solution of the second order will make* $\dfrac{d(y_{nx})}{d(y_{(n-1)x})}$, $\dfrac{d(y_{nx})}{d(y_{(n-2)x})}$ *both infinite; and so on.* [Boole, *Sup.*, p. 51.

3307 Ex.—Taking the differential equation (3303) again,

$$y-xy_x+\tfrac{1}{2}x^2y_{2x}-y_{2x}^2-(y_x-xy_{2x})^2=0 \text{.....................} (1),$$

and differentiating for y_x and y_{2x} only,

$$\{\tfrac{1}{2}x^2+2x\,(y_x-xy_{2x})-2y_{2x}\}\,d\,(y_{2x})-\{x+2\,(y_x-xy_{2x})\}\,d\,(y_x)=0.$$

The condition $\dfrac{d\,(y_{2x})}{d\,(y_x)}=\infty$ requires

$$\tfrac{1}{2}x^2+2x\,(y_x-xy_{2x})-2y_{2x}=0.$$

Substituting the value of y_{2x} obtained from this in equation (1), and rejecting the factor (x^2+1), the same singular integral as before is produced (3303, equation 5).

EQUATIONS WITH MORE THAN TWO VARIABLES.

3320 $$Pdx+Qdy+Rdz=0 \quad\ldots\ldots\ldots\ldots\ldots\quad (1).$$

P, Q, R being here functions of x, y, z, the condition that this equation may be an exact differential of a single complete primitive is

3321 $$P(Q_z-R_y)+Q(R_x-P_z)+R(P_y-Q_x)=0.$$

Proof.—Let μ be an integrating factor of $Pdx+Qdy+Rdz=0$. Then $-\mu P\,dx=\mu Q\,dy+\mu R\,dz$, and, by (3187), for an exact differential, we must have $(\mu Q)_z=(\mu R)_y$. Write this symmetrically for P, Q, and R, differentiate out, and add the three equations after multiplying them respectively by P, Q, and R.

To find the single complete primitive of equation (1).

3322 Rule.—*Consider one of the variables* z *constant, and therefore* dz = 0. *Integrate, and add* ϕ (z) *for the constant of integration. Then differentiate for* x, y, *and* z, *and compare with the given equation* (1). *If a primitive exists,* ϕ (z) *will be determined in terms of* z *only by means of preceding equations.*

The complete primitive so obtained is the equation of a system of surfaces, all of the same species, varying in position according to the value assigned to the arbitrary constant.

3323 Ex.: $(x-3y-z)\,dx+(2y-3x)\,dy+(z-x)\,dz=0 \quad\ldots\ldots\ldots\ldots(1).$

Condition (3321) is satisfied; therefore, putting $dz=0$, we have

$$(x-3y-z)\,dx+(2y-3x)\,dy=0.$$

Applying (3187), $M_y=-3=N_x$, and integration gives

$$\tfrac{1}{2}x^2-3xy-zx+y^2+\phi(z)=0.$$

Differentiating now for x, y, and z,

$$(x-3y-z)\,dx+(2y-3x)\,dy+\{\phi'(z)-x\}\,dz=0.$$

Equating coefficients with (1), $\phi'(z)=z$, therefore $\phi(z)=\tfrac{1}{2}z^2+C$. Hence the single complete primitive is

$$x^2+2y^2+z^2-6xy-2zx=C,$$

the equation of a system of surfaces obtained by varying the constant C.

3324 When the equation $Pdx+Qdy+Rdz=0$ is homogeneous, put $x=uz$, $y=vz$. The result, when the coefficient of dz vanishes, is of the form

3325 $$Mdu+Ndv=0,$$

solvable by (3184). Otherwise it is of the form

3326 $$\frac{dz}{z} = M\,du + N\,dv,$$

and the right will be an exact differential if a complete primitive exists.

3327 Ex.: $$yz\,dx + zx\,dy + xy\,dz = 0 \quad\text{..........(1).}$$
Condition (3321) is satisfied. Putting

$$x = uz, \quad y = vz, \quad dx = u\,dz + z\,du, \quad dy = v\,dz + z\,dv,$$

(1) becomes $$\frac{dz}{z} + \frac{du}{3u} + \frac{dv}{3v} = 0,$$

and the solution is $\log(zu^{\frac{1}{3}}v^{\frac{1}{3}}) = C$ or $xyz = C$.

When the equation

$$P\,dx + Q\,dy + R\,dz = 0 \quad\text{..........(1)}$$

has no single primitive:

3328 Rule.—*Assume* $\phi(x, y, z) = 0$ .... (2)

and differentiate; thus

$$\phi_x dx + \phi_y dy + \phi_z dz = 0 \quad\text{.......... (3).}$$

The form of ϕ *being given, eliminate* z *and* dz *from* (1) *by* (2) *and* (3). *The result, being of the form*

$$M\,dx + N\,dy = 0,$$

can be integrated, and the solution taken with (2) *constitutes a solution of equation* (1), *and represents a system of lines (by varying the constant of integration) drawn on the surface* $\phi(x, y, z) = 0$.

3329 Ex.: $(1+2m)\,x\,dx + (1-x)\,y\,dy + z\,dz = 0.$
The condition (3321) not being satisfied, assume $x^2+y^2+z^2 = r^2$ as the function ϕ, therefore $x\,dx + y\,dy + z\,dz = 0$; and by eliminating z and dz, $2m\,dx - y\,dy = 0$, the integration of which gives $y^2 - 4mx = C$, a cylindrical surface intersecting the spherical surface in a system of curves (by varying C), whose projections on the plane of xy are parabolas.

The condition that

3330 $$X\,dx + Y\,dy + Z\,dz + T\,dt = 0,$$

where X, Y, Z, T are functions of x, y, z, t, may be an exact differential, may be shewn, in a manner similar to that of (3321), to be expressed by any three of the equations

3331
$$\begin{aligned} Y(Z_t-T_z)+Z(T_y-Y_t)+T(Y_z-Z_y) &= 0, \\ Z(T_x-X_t)+T(X_z-Z_x)+X(Z_t-T_z) &= 0, \\ T(X_y-Y_x)+X(Y_t-T_y)+Y(T_x-X_t) &= 0, \\ X(Y_z-Z_y)+Y(Z_x-X_z)+Z(X_y-Y_x) &= 0, \end{aligned}$$

the fourth being always deducible from the other equations.

If this condition is fulfilled, the solution of equation (3330) is analogous to (3322).

Integrate as if z and t were constant, and therefore dz and dt zero, adding for the constant of integration $\phi(z, t)$.

Differentiate next for all the variables, and determine ϕ by comparison with the original equation.

3332 If a single primitive does not exist, the solution must be expressed by simultaneous equations in a manner similar to that of (3328).

SIMULTANEOUS EQUATIONS WITH ONE INDEPENDENT VARIABLE.

GENERAL THEORY.

3340 Let the first of n equations between $n+1$ variables be

$$\boldsymbol{P\,dx+P_1\,dy+P_2\,dz+\ldots+P_n\,dw=0} \quad \ldots\ldots\ldots \quad (1),$$

where P, $P_1 \ldots P_n$ may be functions of all the variables.

Let x be the independent variable. The solution depends upon a single differential equation of the nth order between two variables.

Solving the n equations for the ratios $dx : dy : dz :$ &c., let

$$\frac{dx}{Q} = \frac{dy}{Q_1} = \frac{dz}{Q_2} = \ldots = \frac{dw}{Q_n},$$

$$\therefore \quad \frac{dy}{dx} = \frac{Q_1}{Q}, \quad \frac{dz}{dx} = \frac{Q_2}{Q}, \quad \ldots\ldots \quad \frac{dw}{dx} = \frac{Q_n}{Q}.$$

Differentiate the first of these equations $n-1$ times, substituting from the others the values of $z_x \ldots w_x$, and the result is n equations in $y_x, y_{2x} \dots y_{nx}$, and the primitive variables $x, y, z \dots w$.

Eliminate all the variables but x and y, and let the differential equation obtained be

$$F(x, y, y_x \dots y_{nx}) = 0.$$

Find the n first integrals of this, each of the form $F(x, y, y_x \dots y_{(n-1)x}) = C$, and substitute in them the values of $y_x, y_{2x}, \dots y_{nx}$, in terms of $x, y, z \dots w$, found by solving the n equations last mentioned. Thus a system of n primitives is obtained, each of the form $F(x, y, z \dots w) = C$.

3341 The same in the case of three variables.

Here $n = 2$. Let the given equations be

$$P_1 dx + Q_1 dy + R_1 dz = 0,$$
$$P_2 dx + Q_2 dy + R_2 dz = 0.$$

3342 Therefore $\dfrac{dx}{Q_1R_2 - Q_2R_1} = \dfrac{dy}{R_1P_2 - P_2R_1} = \dfrac{dz}{P_1Q_2 - P_2Q_1}$.

From these let $y_x = \phi(x, y, z), \quad z_x = \psi(x, y, z)$.

Therefore $\qquad y_{2x} = \phi_x + \phi_y y_x + \phi_z z_x$.

Substitute the value of z_x, and eliminate z by means of $y_x = \phi(x, y, z)$. An equation of the 2nd order in x, y, y_x, y_{2x} is the result. Let the complete primitive of this be $y = \chi(x, a, b)$. Then we also have $\phi(x, y, z) = d_x\chi(x, a, b)$. These two equations form the complete solution.

FIRST ORDER LINEAR SIMULTANEOUS EQUATIONS WITH CONSTANT COEFFICIENTS.

3343 In equations of this class, the coefficients of the dependent variables are constants, but any function of the independent variable may exist in a separate term.

Such equations may be solved by the method of (3340), but more practically by indeterminate multipliers.

3344 Ex. (1): $\dfrac{dx}{dt} + 7x - y = 0, \qquad \dfrac{dy}{dt} + 2x + 5y = 0.$

Multiply the second equation by m and add. The result may be written

$$\frac{d(x+my)}{dt} + (2m+7)\left\{x + \frac{5m-1}{2m+7}y\right\} = 0 \quad \dots\dots\dots\dots (1).$$

To make the whole expression an exact differential, put $\frac{5m-1}{2m+7}=m$. This gives $m=\frac{-1+i}{2},\quad m'=\frac{-1-i}{2}$ ....(2);

(1) now becomes $\frac{d(x+my)}{dt}+(2m+7)(x+my)=0,$

and the solution is $x+my=ce^{-(2m+7)t}$ and $x+m'y=c'e^{-(2m'+7)t}$. Solving these equations, and substituting the values (2),

$$iy=ce^{-(6+i)t}-c'e^{-(6-i)t}=e^{-6t}\{(c-c')\cos t-i(c+c')\sin t\},$$

$$ix=e^{-6t}\left\{\left(\frac{c-c'}{2}+i\frac{c+c'}{2}\right)\cos t+\left(\frac{c-c'}{2}-i\frac{c+c'}{2}\right)\sin t\right\},$$

or $y=e^{-6t}(C\cos t-C'\sin t),\quad x=\frac{1}{2}e^{-6t}\{(C+C')\cos t+(C-C')\sin t\}$.

3345 Ex. 2: $\quad x_t+5x+y=e^t,\quad y_t+3y-x=e^{2t}$.

Multiply the second equation by m, and add to the first

$$\frac{d(x+my)}{dt}+(5-m)\left\{x+\frac{1+3m}{5-m}y\right\}=e^t+me^{2t}.$$

Put $\frac{1+3m}{5-m}=m$, thus determining two values of m, and put $x+my=z$; thus $z_t+(5-m)z=e^t+me^{2t}$. This is of the form (3210).

NOTE.—The equations of this example, written in the symmetrical form of (3342), would be $\quad \frac{dx}{e^t-5x-y}=\frac{dy}{e^{2t}+x-3y}=dt.$

3346 *General solution by indeterminate multipliers.*

Let
$$\frac{dx}{P_1}=\frac{dy}{P_2}=\frac{dz}{P_3}$$
be given with

$$P_1=a_1x+b_1y+c_1z+d_1,$$
$$P_2=a_2x+b_2y+c_2z+d_2,$$
$$P_3=a_3x+b_3y+c_3z+d_3.$$

Assume a third variable t and indeterminate multipliers l, m, n such that

$$\frac{dt}{t}=\frac{l\,dx+m\,dy+n\,dz}{lP_1+mP_2+nP_3}=\frac{l\,dx+m\,dy+n\,dz}{\lambda(lx+my+nz+r)}\ \ldots\ldots\ldots\ldots\ (1).$$

The last fraction is an exact differential, and, to determine λ, l, m, n, r, we have

$$\begin{aligned}a_1l+a_2m+a_3n&=\lambda l,\\ b_1l+b_2m+b_3n&=\lambda m,\\ c_1l+c_2m+c_3n&=\lambda n,\\ d_1l+d_2m+d_3n&=\lambda r,\end{aligned}\qquad \begin{vmatrix}a_1-\lambda & a_2 & a_3\\ b_1 & b_2-\lambda & b_3\\ c_1 & c_2 & c_3-\lambda\end{vmatrix}=0.$$

The determinant is the eliminant of the first three equations in l, m, n. The roots of this cubic in λ furnish three sets of values of l, m, n, r, which, being substituted in the integral of (1), give rise to three equations involving three arbitrary constants; thus,

$$c_1 t = (l_1 x + m_1 y + n_1 z + r_1)^{\frac{1}{\lambda_1}}, \qquad c_2 t = (l_2 x + m_2 y + n_2 z + r_2)^{\frac{1}{\lambda_2}},$$
$$c_3 t = (l_3 x + m_3 y + n_3 z + r_3)^{\frac{1}{\lambda_3}}.$$

Eliminating t, we find for the solution two equations involving two arbitrary constants.

A similar solution may be obtained when there are more than three variables.

3347 To solve $\dfrac{dx}{P_1 - xP} = \dfrac{dy}{P_2 - yP} = \dfrac{dz}{P_3 - zP} = \&c. \ldots(1),$

where $\quad P = ax + by + c, \quad P_1 = a_1 x + b_1 y + c_1, \quad \&c.$

Assume $\quad p = a\xi + b\eta + c\zeta, \quad p_1 = a_1\xi + b_1\eta + c_1\zeta, \quad \&c.,$

and take
$$\frac{d\xi}{p_1} = \frac{d\eta}{p_2} = \frac{d\zeta}{p} \quad \ldots\ldots (2),$$

the solution of which is known by (3346). Substitute $\xi = x\zeta$, $\eta = y\zeta$, and these equations become

$$\frac{xd\zeta + \zeta dx}{p_1} = \frac{yd\zeta + \zeta dy}{p_2} = \frac{d\zeta}{p},$$

and therefore
$$\frac{\zeta dx}{p_1 - xp} = \frac{\zeta dy}{p_2 - yp} = \frac{d\zeta}{p}.$$

Dividing numerators and denominators by ζ, the first equation in (1) is produced, and therefore its solution is obtained by changing ξ, η in the solution of (2) into $x\zeta$ and $y\zeta$.

Certain simultaneous equations in which the coefficients are not constants may be solved by the method of multipliers. Thus,

3348 Ex. (1): $\quad x_t + P(ax + by) = Q, \quad y_t + P(a'x + b'y) = R,$

P, Q, R being functions of t. Multiply the second equation by m, add, and determine m as in (3344). The solution is obtained from

$$x + my = e^{-(a+ma')\int P dt}\left\{C + \int e^{(a+ma')\int P dt}(Q + mR)\,dt\right\}, \qquad (3210)$$

with two values of m.

3349 Ex. (2): $\quad x_t + \dfrac{2}{t}(x - y) = 1, \quad y_t + \dfrac{1}{t}(x + 5y) = t$

are equations solvable in a similar manner, and the results are

$$x + y = \frac{1}{t^2}\left(C_1 + \frac{t^4}{4} + \frac{t^5}{5}\right), \quad x + 2y = \frac{1}{t^4}\left(C_2 + \frac{t^5}{5} + \frac{t^6}{6}\right).$$

[Boole, p. 307.

REDUCTION OF ORDER IN SIMULTANEOUS EQUATIONS.

3350 Theorem.—n simultaneous equations of any orders between n dependent variables and 1 independent variable are reducible to a system of equations of the *first* order by substituting a new variable for every derivative except the highest.

3351 The number of equations and dependent variables in the transformed system will be equal to the sum of the indices of order of the highest derivatives. This will, therefore, in general be the number of constants introduced in integrating those equations. If, after integrating, all the new variables be eliminated, there will remain n equations in the original variables and the above-named constants. These equations form the complete solution.

In practice, such reduction is unnecessary. The following are methods frequently adopted :—

3352 Rule I.—*Differentiate until by elimination of a variable and its derivatives an equation of a higher order in one dependent variable only is obtained.*

3353 Rule II.—*Employ indeterminate multipliers.*

3354 Ex. (1): $x_{2t} = ax + by, \quad y_{2t} = a'x + b'y.$

By Rule I., differentiating twice for t and eliminating y and y_{2t}, we obtain

$$x_{4t} - (a+b')\,x_{2t} + (ab' - a'b)\,x = 0,$$

which may be solved by (3239).

Otherwise by Rule II., exactly as in (3344), we find

$$a'm^2 + (a-b')\,m - b = 0,$$

and for the exact differential

$$(x+my)_{2t} = (a+ma')\,(x+my),$$

the solution of which, by (3239), is

$$x + my = C_1 e^{\sqrt{(a+ma')}\,t} + C_2 e^{-\sqrt{(a+ma')}\,t}$$

in duplicate with the two values of m.

3355 Ex. (2): $x_{2t} - 2ay_t + bx = 0, \quad y_{2t} + 2ax_t + by = 0.$

Differentiate, and eliminate y, y_t, y_{2t}; thus

$$x_{4t} + 2\,(2a^2+b)\,x_{2t} + b^2 x = 0,$$

and solve by (3239). Otherwise assume

$$x = \xi \cos at + \eta \sin at, \qquad y = \eta \cos at - \xi \sin at,$$

and the given equations reduce to

$$\xi_{2t} = -(a^2+b)\,\xi, \qquad \eta_{2t} = -(a^2+b)\,\eta,$$

which are solved in (3257). [Boole, p. 311.

3356 Ex. (3).—Let $u = 0$, $v = 0$, $w = 0$ be three equations in x, y, z, t, involving derivatives of t up to x_{3t}, y_{6t}, z_{7t}.

To obtain an equation between x and t. Differentiate each equation $6+7 = 13$ times, producing $3+13\times 3 = 42$ equations involving derivatives of t up to x_{16t}, y_{19t}, z_{20t}. Between these 42 equations eliminate $y, y_t, \dots y_{19t}$, $z, z_t, \dots z_{20t}$, in all 41 quantities, and an equation of the 16th order in x and t is the result. [De Morgan.

3357 If a number of equations involve the quantities x, x_{2t}, x_{4t}, &c., y_t, y_{3t}, y_{5t}, &c., all in the first degree, these quantities may be eliminated by assuming

$$x = L\sin pt, \quad y = M\cos pt.$$

3358 If there be n linear homogeneous equations in n variables $x, y, z, \dots$ and their derivatives of the 2nd order only, the equations may be solved by putting

$$x = L\sin pt, \quad y = M\sin pt, \quad z = N\sin pt, \quad \&c.$$

3359 Ex.:
$$x_{2t} = ax+by, \qquad y_{2t} = gx+fy.$$
Putting
$$x = L\sin pt, \qquad y = M\sin pt,$$
$$\left.\begin{array}{r}(a+p^2)L+bM = 0\\ gL+(f+p^2)M = 0\end{array}\right\}, \quad \therefore \begin{vmatrix} a+p^2, & b \\ g, & f+p^2 \end{vmatrix} = 0,$$
p and the ratios $L:M$ are thus found.

Suppose $L = -kb$ and $M = k(p^2+a)$,

then $x = -kb\sin pt$, $y = k(p^2+a)\sin pt$,

and k and t are arbitrary constants.

PARTIAL DIFFERENTIAL EQUATIONS.

3380 An equation is termed *a general primitive* or *a complete primitive* of a partial differential equation, according as the latter is obtained from it by eliminating arbitrary functions or arbitrary constants, as illustrated in (3150–7).

LINEAR FIRST ORDER P. D. EQUATIONS.

3381 To form the P. D. equation from the primitive $u = \phi(v)$, where u and v are functions of x, y, z.

RULE.—*Differentiate for* x *and* y *in turn, and eliminate* $\phi'(v)$. *See* (3054).

Otherwise.—Differentiate the equations $u = a$, $v = b$; thus

$$u_x dx + u_y dy + u_z dz = 0,$$
$$v_x dx + v_y dy + v_z dz = 0.$$

Therefore $\frac{dx}{P} = \frac{dy}{Q} = \frac{dz}{R}$, *where* $P = \frac{d(uv)}{d(yz)}$, *&c.*

Then the P. D. equation will be

$$Pz_x + Qz_y = R.$$

PROOF.—Since z is a function of x and y, $z_x dx + z_y dy = dz$. But $dx = kP$, $dy = kQ$, $dz = kR$, therefore $kPz_x + kQz_y = kR$.

3382 EX.—The general equation of a conical surface drawn through the point (a, b, c) is

$$\frac{y-b}{x-a} = \phi\left(\frac{z-c}{x-a}\right),$$

the form of ϕ being arbitrary.

Considering z as a function of two independent variables x and y, differentiate for x and y in turn, and eliminate ϕ' as in (3154). The result is the partial differential equation

$$(x-a)\, z_x + (y-b)\, z_y + z - c = 0.$$

3383 To obtain the complete primitive; that is, to solve the P. D. equation, $Pz_x + Qz_y = R$, P, Q, R being either functions of x, y, z or constants.

RULE.—*Solve the equations*

$$\frac{dx}{P} = \frac{dy}{Q} = \frac{dz}{R}.$$

Let the two integrals obtained be $u = a$, $v = b$; *then*

$$u = \phi(v)$$

will be the complete primitive.

Propositions (3381) and (3383) extended to any number of variables.

3384 To form the partial differential equation from the primitive

$$\phi(u, v, \dots w) = 0 \quad\dots\dots\dots\dots (1),$$

where $u, v, \dots, w$ are n given functions of n independent variables $x, y, \dots z$ and one dependent t.

RULE.—*Differentiate for all the variables thus,*

$$\phi_u\,du + \phi_v dv + \dots + \phi_w dw = 0 \qquad (2).$$

Therefore, since ϕ *is arbitrary,* du, dv ... dw *must separately vanish, giving rise to the* n *equations*

$$\begin{aligned} du &= u_x dx + u_y dy + \dots + u_t dt = 0, \\ dv &= v_x dx + v_y dy + \dots + v_t dt = 0, \\ &\dots \quad \dots \quad \dots \quad \dots \\ dw &= w_x dx + w_y dy + \dots + w_t dt = 0. \end{aligned}$$

Solving these for the ratios, by (583), *we get*

$$\frac{dx}{P} = \frac{dy}{Q} = \dots \frac{dz}{R} = \frac{dt}{S} \qquad (3),$$

P, Q ... R, S *being functions of the variables or else constants. Now,* t *being a function of all the rest,*

$$t_x dx + t_y dy + \dots + t_z dz = dt \qquad (4),$$

therefore, by (3) *and* (4), *the partial differential equation required is*

3385 $$Pt_x + Qt_y + \dots + Rt_z = S.$$

3386 If $u, v \dots w$ be n functions of n variables, $x, y \dots t$, the condition of interdependence of the functions or existence of some relation expressed by equation (1) is $J(u, v \dots w) = 0$ (see 1606); that is, the eliminant of equations (2) must vanish.

3387 Conversely, to integrate the partial differential equation

$$Pt_x + Qt_y + \dots + Rt_z = S \qquad (1).$$

RULE.—*Solve the system of ordinary equations*

$$\frac{dx}{P} = \frac{dy}{Q} = \&c. = \frac{dz}{R} = \frac{dt}{S} \qquad (2),$$

and let the integrals obtained be u = a, v = b, ... w = k; *then* ϕ(u, v, ... w) = 0 *will be the complete primitive.*

If P, Q ... R, S *are linear functions of the variables, the integrals of equations* (2) *can always be found by the method of* (3346).

NOTE.—Suppose, in equation (1), that any coefficients P, Q vanish; then, by (2), $dx = 0$, $dy = 0$, and therefore the corresponding integrals are $x = a$, $y = b$. The complete primitive thus becomes

$$\phi(x, y, u, v \dots w) = 0.$$

3389 When only one independent variable occurs in the derivatives of the partial differential equation, the equation may be integrated as though the others were constant, adding functions of the remaining variables for the constants of integration.

3390 (Ex. 1): $\frac{dz}{dx} = \frac{y}{\sqrt{y^2-x^2}}$. Integrating for x as though y were constant, the complete primitive is

$$z = y \sin^{-1}\frac{x}{y} + \phi(y).$$

Some equations are reducible to the above class by a transformation. Thus:

3391 Ex. (2): $z_{xy} = x^2+y^2$. Put $z_x = u$,

therefore $u_y = x^2+y^2$, therefore $u = z_x = x^2y+\frac{1}{3}y^3+\phi(x)$,

therefore $$z = \tfrac{1}{3}x^3y+\tfrac{1}{3}xy^3+\int\phi(x)\,dx+\psi(y),$$

or $$z = \tfrac{1}{3}(x^3y+xy^3)+\chi(x)+\psi(y).$$

3392 Ex. (3): $(x-a)\,z_x+(y-b)\,z_y = c-z$.

Solving by (3283), $$\frac{dx}{x-a} = \frac{dy}{y-b} = \frac{dz}{z-c}.$$

The integrals are

$$\left.\begin{array}{l}\log(y-b)-\log(x-a) = \log C\\ \log(z-c)-\log(x-a) = \log C'\end{array}\right\}, \quad \text{or} \quad \frac{y-b}{x-a} = C, \quad \frac{z-c}{x-a} = C',$$

therefore $\frac{y-b}{x-a} = \phi\left(\frac{z-c}{x-a}\right)$ is the complete primitive.

For the converse process in respect of the same equation, see (3382).

3393 Ex. (4).—To find the surface which cuts orthogonally all the spheres whose equations (varying a) are

$$x^2+y^2+z^2-2ax = 0 \quad \dots\dots\dots\dots\dots\dots \quad (1).$$

Let $\phi(x, y, z) = 0$ be the surface. Then

$$(x-a)\,\phi_x+y\phi_y+z\phi_z = 0$$

by the condition of normals at right angles. Substitute the value of a from (1), and divide by ϕ_z; thus,

$$(x^2-y^2-z^2)\,z_x+2xyz_y = 2zx.$$

By (3383), $$\frac{dx}{x^2-y^2-z^2} = \frac{dy}{2xy} = \frac{dz}{2zx},$$

$\frac{dy}{y} = \frac{dz}{x}$ gives $\frac{y}{z} = c$ for one integral.

Substituting $y = cz$, we then have

$$\frac{dx}{x^2-(c^2+1)z^2} = \frac{dz}{2zx},$$

which, being a homogeneous equation in x and z, may be solved by putting $z = vx$ (3186). The resulting integral is $\frac{x^2+y^2+z^2}{z} = C$. Hence the complete primitive is $\frac{x^2+y^2+z^2}{z} = \phi\left(\frac{y}{z}\right)$ and the equation of the surface sought.

3394 Ex. (5).—To find an integrating factor of the equation

$$(x^3y-2y^4)\,dx+(xy^3-2x^4)\,dy = 0 \quad\text{.......}\quad (1).$$

Assuming z for that factor, the condition $(Mz)_y = (Nz)_x$ (3087) produces the P. D. equation

$$(xy^3-2x^4)\,z_x+(2y^4-x^3y)\,z_y = 9\,(x^3-y^3)\,z \quad\text{.......}\quad (2).$$

The system of ordinary equations (3283) is

$$\frac{dx}{xy^3-2x^4} = \frac{dy}{2y^4-x^3y} = \frac{dz}{9\,(x^3-y^3)\,z}.$$

The first of these equations is identical with (1) (and such an agreement always occurs). Its integral is $\frac{x}{y^2}+\frac{y}{x^2} = c.$

Also $$\frac{y\,dx+x\,dy}{xy^4-2x^4y+2xy^4-x^4y} = \frac{dz}{9\,(x^3-y^3)\,z},$$

which reduces to $$\frac{3dx}{x}+\frac{3dy}{y}+\frac{dz}{z} = 0\,;$$

and thus the second integral is $x^3y^3z = c'$.

Hence the complete primitive and integrating factor is

$$z = \frac{1}{x^3y^3}\,\phi\left(\frac{x}{y^2}+\frac{y}{x^2}\right).$$

Any linear P. D. equation may be written as a homogeneous equation with one additional variable; thus, equation (3387) may be written

3395 $$Pu_x+Qu_y+\ldots+Ru_z = Su_t.$$

SIMULTANEOUS LINEAR FIRST ORDER P. D. EQUATIONS.

3396 Prop. I.—*The solution of such equations may be made to depend upon a system of ordinary 1st order differential*

equations having a number of variables exceeding by more than one the number of equations.

Let there be n equations reduced to the homogeneous form (3395) involving one dependent variable P and $n+m$ independent. Select n of the latter, $x, y \dots z$, and let the remaining m be $\xi, \eta \dots \zeta$. From the n equations find $P_x, P_y \dots P_z$ in terms of $P_\xi, P_\eta \dots P_\zeta$, and arrange the results as under:

$$\left.\begin{array}{l} P_x + a_1 P_\xi + b_1 P_\eta \dots + k_1 P_\zeta = 0 \\ P_y + a_2 P_\xi + b_2 P_\eta \dots + k_2 P_\zeta = 0 \\ \dots \qquad \dots \qquad \dots \\ P_z + a_n P_\xi + b_n P_\eta \dots + k_n P_\zeta = 0 \end{array}\right\} \dots\dots\dots\dots (1).$$

Multiply these equations by $\lambda_1, \lambda_2, \dots \lambda_n$ respectively, and add; thus,

$$\lambda_1 P_x + \lambda_2 P_y \dots + \lambda_n P_z + \Sigma(\lambda a) P_\xi + \Sigma(\lambda b) P_\eta \dots + \Sigma(\lambda k) P_\zeta = 0 \quad \dots\dots (2).$$

From this, as in (3387), we have the auxiliary system

$$\frac{dx}{\lambda_1} = \frac{dy}{\lambda_2} \dots = \frac{dz}{\lambda_n} = \frac{d\xi}{\Sigma(\lambda a)} = \frac{d\eta}{\Sigma(\lambda b)} \dots = \frac{d\zeta}{\Sigma(\lambda k)} \dots\dots (3),$$

and, by eliminating $\lambda_1, \lambda_2 \dots \lambda_n$,

$$\left.\begin{array}{l} d\xi - a_1 dx - a_2 dy \dots - a_n dz = 0 \\ d\eta - b_1 dx - b_2 dy \dots - b_n dz = 0 \\ \dots \qquad \dots \qquad \dots \\ d\zeta - k_1 dx - k_2 dy \dots - k_n dz = 0 \end{array}\right\} \dots\dots\dots\dots\dots (4).$$

Then, if $u = a$, $v = b$, &c. be the integrals of (4), they will be values of P satisfying the equivalent system (1), and the integral of that system will be $F(u, v, \dots) = 0$.

3397 Prop. II.—*To integrate a system of linear* 1*st order P. D. equations.*

Let $$\Delta = ad_x + bd_y \dots + kd_z,$$

so that $\Delta P = 0$ represents a homogeneous linear P. D. equation of the 1st order.

Rule.—"*Reduce the equations to the homogeneous form* (1); *express the result symbolically by*

$$\Delta_1 \mathrm{P} = 0, \quad \Delta_2 \mathrm{P} = 0, \quad \dots \Delta_n \mathrm{P} = 0,$$

and examine whether the condition

$$(\Delta_i\Delta_j - \Delta_j\Delta_i)\,P = 0$$

is identically satisfied for every pair of equations of the system. If it be so, the equations of the auxiliary system (Prop. I.) will be reducible to the form of exact differential equations, and their integrals being u = a, v = b, w = c, ..., *the complete value of* P *will be* F(u, v, w, ...), *the form of* F *being arbitrary.*

"*If the condition be not identically satisfied, its application will give rise to one or more new partial differential equations. Combine any one of these with the previous reduced system, and again reduce in the same way.*

"*With the new reduced system proceed as before, and continue this method of reduction and derivation until either a system of P. D. equations arises, between every two of which the above condition is identically satisfied, or, which is the only possible alternative, the system* $P_x = 0$, $P_y = 0$, ... *appears. In the former case, the system of ordinary equations corresponding to the final system of P. D. equations, will admit of reduction to the exact form, and the general value of* P *will emerge from their integrals as above. In the latter case, the given system can only be satisfied by supposing* P *a constant.*"

3398 "Ex.: $P_x + (t+xy+xz)\,P_z + (y+z-3x)\,P_t = 0,$

$$P_y + (xzt+y-xy)\,P_z + (zt-y)\,P_t = 0.$$

Representing these in the form $\Delta_1 P = 0$, $\Delta_2 P = 0$, it will be found that $(\Delta_1\Delta_2 - \Delta_2\Delta_1)\,P = 0$ becomes, after rejecting an algebraic factor, $xP_z + P_t = 0$, and the three equations prepared in the manner explained in the Rule will be found to be

$$P_x + (3x^2+t)\,P_z = 0, \quad P_y + yP_z = 0, \quad P_t + xP_z = 0.$$

No other equations are derivable from these. We conclude that there is but one final integral.

"To obtain it, eliminate P_x, P_y, P_z from the above system combined with

$$P_x dx + P_y dy + P_z dz + P_t dt = 0,$$

and equate to zero the coefficient of P_z in the result. We find

$$dz - (t+3x^2)\,dx - y\,dy - x\,dt = 0,$$

the integral of which is $z - xt - x^3 - \tfrac{1}{2}y^2 = c.$

"An arbitrary function of the first member of this equation is the general value of P." [*Boole, Sup.*, Ch. xxv.

For Jacobi's researches in the same subject, see *Crelle's Journal*, Vol. lx.

NON-LINEAR FIRST ORDER PARTIAL DIFFERENTIAL EQUATIONS.

3399 *Type* $\quad F(x, y, z, z_x, z_y) = 0$ .... (1).

CHARPITS'S SOLUTION.—*Writing* p, q *instead of* z_x *and* z_y, *assume the equations*

$$\frac{dx}{-q_p} = dy = \frac{dz}{q - pq_p} = \frac{dp}{q_x + pq_z} \quad \text{........... (2).}$$

Find a value of p *from these by integration, and the corresponding value of* q *from the given equation, and substitute in the equation*

$$dz = p\,dx + q\,dy \quad \text{..................... (3),}$$

and integrate by (3322) *to obtain the final integral.*

PROOF.—Since $dz = p\,dx + q\,dy$, we have, by the condition of integrability, $p_y = q_x$. Express p_y and q_x on the hypothesis that z is a function of x, y; p a function of x, y, z; q a function of x, y, z, p; considering x constant when finding p_y, and y as constant when finding q_x. Equating the values of p_y and q_x so obtained, the result is the equation

$$Ap_x + Bp_y + Cp_z = D,$$

in which A, B, C, D stand for $-q_p$, 1, $q - pq_p$, $q_x + pq_z$.

Hence, to solve this equation, we have, by (3387), the system of ordinary equations (2).

3400 NOTE.—More than one value of p obtained from equations (2) may give rise to more than one complete primitive.

The first two of equations (2) taken together involve equation (3).

DERIVATION OF THE GENERAL PRIMITIVE AND SINGULAR SOLUTION FROM THE COMPLETE PRIMITIVE.

RULE.—*Let the complete primitive of a P. D. equation of the 1st order be*

$$z = f(x, y, a, b) \quad \text{...................... (1).}$$

3401 *The general primitive is obtained by eliminating* a *between* $\quad z = f\{x, y; a, \phi(a)\}$ and $f_a = 0$ .... (2), *the form of* ϕ *being specified at pleasure.*

3402 *The singular solution is obtained by eliminating* a *and* b *between the complete primitive and the equations*

$$f_a = 0, \quad f_b = 0 \quad \text{...................... (3).}$$

PROOF.—By varying a and b in (1),

$$p = f_x + f_a a_x + f_b b_x, \quad q = f_y + f_a a_y + f_b b_y.$$

Therefore, reasoning as in (3171), we must have

$$f_a a_x + f_b b_x = 0 \quad \text{and} \quad f_a a_y + f_b b_y = 0 \quad \text{.................. (3)},$$

therefore either $f_a = 0$, $f_b = 0$, leading to the singular solution; or, eliminating f_a, f_b, $a_x b_y - a_y b_x = 0$,
and therefore, by (3167), $b = \phi(a)$. Multiply equations (3) by dx, dy respectively, and add, thus $f_a da + f_b db = 0$. Substitute $b = \phi(a)$ in this and in (1), and the equations (2) are the result.

SINGULAR SOLUTION DERIVED FROM THE DIFFERENTIAL EQUATION.

3403 RULE.—*Eliminate* p *and* q *from the differential equation by means of the equations*

$$z_p = 0, \quad z_q = 0.$$

PROOF.—Let the D. E. be $z = f(x, y, p, q)$, and the C. P. $z = F(x, y, a, b)$. Now p and q being implicit functions of a and b, we have, from the first equation, $z_a = z_p p_a + z_q q_a$, $z_b = z_p p_b + z_q q_b$.
Hence the conditions $z_a = 0$, $z_b = 0$ in (3) involve, and are equivalent to, $z_p = 0$, $z_q = 0$.

3404 All possible solutions of a P. D. equation of the 1st order are represented by the complete primitive, the general primitive, and the singular solution. [*Boole*, p. 343.

3405 To connect any given solution with the complete primitive.

Let $z = F(x, y, a, b)$ be the complete primitive, and $z = \phi(x, y)$ some other solution.

Determine the values of a and b which satisfy the three equations $F = \phi$, $F_x = \phi_x$, $F_y = \phi_y$.

If these values are constant, the solution is a particular case of the complete primitive; if they are variable so that one is a function of the other, the solution is a particular case of the general primitive; if they are variable and unconnected, the solution is a singular solution.

3406 COR.—Any two solutions springing from different complete primitives are equivalent.

3407 Ex.: $$z = px+qy+pq \quad\ldots\ldots\ldots\ldots (1).$$

By (3299), $$\frac{dx}{A} = dy = \frac{dz}{C} = \frac{dp}{D} \quad\ldots\ldots\ldots\ldots (2),$$

and we have $$q = \frac{z-px}{p+y};\quad A = -q_p = \frac{xy+z}{(p+y)^2};$$

$$C = q-pq_p = \frac{2pz+yz-p^2x}{(p+y)^2};\quad D = q_x+pq_z = \frac{-p+p}{p+y} = 0.$$

Hence (2) becomes $$\frac{(p+y)^2}{xy+z}dx = dy = \frac{(p+y)^2\,dz}{2pz+yz-p^2x} = \frac{dp}{0};$$

$\therefore\ dp = 0,\ p = a;\ \therefore\ q = \dfrac{z-ax}{a+y}$. Substituting in $dz = p\,dx+q\,dy$,

$$dz = a\,dx + \frac{z-ax}{a+y}dy \quad\ldots\ldots\ldots\ldots (3).$$

By (3322), making z constant, $\dfrac{a\,dx}{z-ax}+\dfrac{dy}{a+y} = 0$,

therefore $$-\log(z-ax)+\log(a+y) = \phi(z) \quad\ldots\ldots\ldots\ldots (4).$$

Differentiate for x, y, z, and equate with (3), thus $\phi'(z) = 0$, therefore $\phi(z) =$ constant (say $-\log b$); therefore, by (4), $z = ax+by+ab$, the C. P. of (1).

3408 To find a singular solution by (3402), we must eliminate a and b between $z_a = 0$, $z_b = 0$; that is, $x+b = 0$ and $y+a = 0$,

therefore $$z = -xy-xy+xy = -xy$$
is the singular solution.

To find the general primitive by (3401), eliminate a between the two equations $z = ax+(y+a)\,\phi(a)$ and $x+(y+a)\,\phi'(a)+\phi(a) = 0$.

NON-LINEAR FIRST ORDER P. D. EQUATIONS WITH MORE THAN TWO INDEPENDENT VARIABLES.

3409 Prop.—To find the complete primitive of the differential equation

$$F(x_1, x_2 \ldots x_n, z, p_1, p_2 \ldots p_n) = 0 \quad\ldots\ldots\ldots (1),$$

where $$p_1 = \frac{dz}{dx_1},\quad p_2 = \frac{dz}{dx_2},\quad \&c.$$

3410 Rule.—*Form the linear P. D. equation in* Φ *denoted by*

$$\Sigma_r\left\{\left(\frac{dF}{dx_r}+p_r\frac{dF}{dz}\right)\frac{d\Phi}{dp_r}-\frac{dF}{dp_r}\left(\frac{d\Phi}{dx_r}+p_r\frac{d\Phi}{dz}\right)\right\} = 0,$$

the summation extending from $r=1$ *to* $r=n$. *From the auxiliary system* (3387) $n-1$ *integrals*

$$\Phi_1 = a_1,\quad \Phi_2 = a_2,\ \ldots\ \Phi_{n-1} = a_{n-1}$$

may be obtained. From these equations, together with (1), *find* $p_1, p_2 \dots p_n$ *in terms of* $x_1, x_2 \dots x_n$, *substitute the values in*

$$dz = p_1 dx_1 + p_2 dx_2 \dots + p_n dx_n,$$

and the integral of this last equation will furnish the solution required in the form

$$f(x_1, x_2 \dots x_n, z, a_1, a_2 \dots a_n) = 0.$$

[*Boole, Diff. Eq.*, Ch. xiv., and *Sup.*, Ch. xxvii.

SECOND ORDER P. D. EQUATIONS.

3420 *Type* $F(x, y, z, z_x, z_y, z_{2x}, z_{xy}, z_{2y}) = 0$.

The derivatives $z_x, z_y, z_{2x}, z_{xy}, z_{2y}$ are briefly denoted by p, q, r, s, t respectively.

z being a function of the two independent variables x and y, the following values are of frequent use

3421 $dz = pdx + qdy; \quad dp = rdx + sdy; \quad dq = sdx + tdy.$

If u be any function of x, y, and z, the complete derivatives of u are indicated by brackets, thus

3422 $(u_x) = u_x + pu_z, \quad (u_y) = u_y + qu_z.$

A linear 2nd order P. D. equation is of the type

3423 $Rr + Ss + Tt = V$ .... (1),

in which R, S, T, V are functions of x, y, z, p, q.

Proposition.—Any P. D. equation of the 2nd order which has a first integral of the form $u = f(v)$, where u and v involve x, y, z, p, q, is of the form

3424 $Rr + Ss + Tt + U(rt - s^2) = V$..............(2),

where R, S, T, U, V are functions of x, y, z, p, q, and

3425 $U = u_p v_q - u_q v_p$ .... (3).

Proof.—Differentiate $u = f(v)$ for x and y separately, considering x, y, z, p, q all involved in u and v, and eliminate $f'(v)$. The result is equation (2), with the values

3426 $\mu R = u_p(v_y) - (u_y)v_p,\quad \mu T = v_q(u_x) - (v_x)u_q,$

$$\mu S = v_p(u_x) - (v_x)u_p - v_q(u_y) + (v_y)u_q,$$

$$\mu U = u_p v_q - u_q v_p,\quad \mu V = (u_y)(v_x) - (u_x)(v_y),$$

with the notation (3422), μ being an undetermined constant.

3427 Cor.—The condition to be fulfilled in order that equation (1) may have a first integral of the form $u = f(v)$ is

$$u_p v_q - u_q v_p = 0.$$

SOLUTION BY MONGE'S METHOD OF

3428 $$Rr + Ss + Tt = V.$$

Rule.—*Write the two equations*

$$Rdy^2 - S\,dx\,dy + Tdx^2 = 0 \quad\ldots\ldots(1),$$

$$Rdp\,dy - Vdx\,dy + Tdq\,dx = 0 \quad\ldots\ldots(2).$$

Resolve (1) *into its factors, producing the two equations*

$$dy - m_1 dx = 0 \quad \textit{and} \quad dy - m_2 dx = 0.$$

From $dy = m_1 dx$ *and equation* (2) *combined, if necessary, with* $dz = p\,dx + q\,dy$, *find two* 1st *integrals* $u = a$, $v = b$; *then* $u = f(v)$ *will be one* 1st *integral of the given equation. Similarly from* $dy = m_2 dx$ *find another* 1st *integral.*

3429 *The final* 2nd *integral may be found from one of the* 1st *integrals by Lagrange's method* (3383).

3430 Otherwise, determine p and q in terms of x, y, z from the two 1st integrals; substitute in $dz = p\,dx + q\,dy$, and then integrate by (3322) to obtain the final integral.

3431 If equation (1) is a perfect square, there will be only one 1st integral, and Lagrange's method only is applicable.

Proof.—By (3427) we may put $u_q = mu_p$, $v_q = mv_p$; and also $dz = p\,dx + q\,dy$ (3321) in the complete derivatives

$$(du) = u_x dx + u_y dy + u_z dz + u_p dp + u_q dq = 0,\ (dv) = \&c. = 0;$$

$\therefore$ by (3422) $$\left.\begin{aligned}(u_x)\,dx + (u_y)\,dy + u_p(dp + m\,dq) = 0\\ (v_x)\,dx + (v_y)\,dy + v_p(dp + m\,dq) = 0\end{aligned}\right\} \quad\ldots\ldots(3).$$

Solving these equations for the ratios $dx : dy : dp + mdq$, we obtain at once

$$\frac{dx}{R} = \frac{dy + m\,dx}{S} = \frac{m\,dy}{T} = \frac{dp + m\,dq}{V} \quad\ldots\ldots(4),$$

with the values of R, S, T, V in (3426).

Equations (1) and (2) are the result of eliminating m from (4). These two equations with $dz = p\,dx + q\,dy$ suffice to determine a first integral of (3428) when it exists in the form $u = f(v)$.

3432 Ex. (i.): $q(1+q)\,r - (p+q+2pq)\,s + p\,(1+p)\,t = 0.$

Solving the quadratic equation (1), we find

$$p\,dx + q\,dy = 0, \quad \text{or} \quad (1+p)\,dx + (1+q)\,dy = 0 \;\ldots\ldots\ldots\ldots(5).$$

First, $\quad dz = p\,dx + q\,dy = 0, \quad \therefore \; z = A.$

Monge's equation (2) is $\quad q(1+q)\,dp\,dy + p\,(1+p)\,dq\,dx = 0,$

which, by $p\,dx = -q\,dy$, gives $\dfrac{dp}{1+p} = \dfrac{dq}{1+q}$; and, integrating, $\dfrac{1+p}{1+q} = B.$

Hence a first integral is $\quad \dfrac{1+p}{1+q} = \phi(z) \ldots\ldots\ldots\ldots\ldots\ldots(6).$

Next, taking the second equation of (5) with

$$dz = p\,dx + q\,dy, \quad dx + dy + dz = 0, \quad \therefore \; x + y + z = C.$$

Also, by (5), equation (2) now reduces to $q\,dp = p\,dq$, and by integration, $p = qD$; therefore the other first integral is $p = q\psi(x+y+z)$.

For the final integral integrate $p - q\psi = 0$; *i.e.*, $z_x - \psi z_y = 0$, by (3383);

$$\therefore \; dx = -\frac{dy}{\psi(x+y+z)} = \frac{dz}{0}, \quad \therefore \; z = A, \quad \text{and} \quad dx = \frac{dx+dy+dz}{1-\psi(x+y+z)}.$$

$$\therefore \; x = \int \frac{d(x+y+z)}{1-\psi(x+y+z)} = F(x+y+z) + B.$$

Hence the second integral is $x - f(x+y+z) = F(z)$.

3433 Ex. (ii.): $\quad z_{2x} - a^2 z_{2y} = 0.$

(i.) Here, in (3428), $R = 1$, $S = 0$, $T = -a^2$, $V = 0$; therefore (1) and (2) become $\quad dy^2 - a^2dx^2 = 0, \quad dp\,dy - a^2dq\,dx = 0.$

From (1) $dy + a\,dx = 0$, giving $y + ax = c$, and converting (2) into $dp + a\,dq = 0$, which gives $p + aq = c'$; therefore a first integral is

$$p + aq = \phi(y+ax) \;\ldots\ldots\ldots\ldots\ldots\ldots (3).$$

Similarly, from (1), $dy - a\,dx = 0$ gives rise to another first integral

$$p - aq = \psi(y-ax) \;\ldots\ldots\ldots\ldots\ldots\ldots (4).$$

Eliminating p and q by means of (3) and (4) from $dz = p\,dx + q\,dy$, we find

$$dz = (2a)^{-1}\{\phi(y+ax)(dy+a\,dx) - \psi(y-ax)(dy-a\,dx)\},$$

therefore, by integrating, $\quad z = \Phi(y+ax) + \Psi(y-ax).$

For the symbolic solution of the same equation, see (3566).

SOLUTION OF THE P. D. EQUATION.

3434 $\quad Rr + Ss + Tt + U(rt - s^2) = V \;\ldots\ldots\ldots\ldots(1).$

Let m_1, m_2 be the roots of the quadratic equation

3435 $\quad m^2 - Sm + RT + UV = 0 \;\ldots\ldots\ldots\ldots (2).$

Let $u_1 = a$, $v_1 = b$, and $u_2 = a'$, $v_2 = b'$ be respectively the solutions of the two systems of ordinary differential equations.

3436 $$\left.\begin{aligned} Udp &= m_2 dy - T dx \\ Udq &= m_1 dx - R dy \\ dz &= p\,dx + q\,dy \end{aligned}\right\} (3), \qquad \left.\begin{aligned} Udp &= m_1 dy - T dx \\ Udq &= m_2 dx - R dy \\ dz &= p\,dx + q\,dy \end{aligned}\right\} (4).$$

Then the first integrals of (1) will be

$$u_1 = f(v_1), \quad u_2 = f(v_2).$$

To obtain a second integral:

3437 1st.—When m_1, m_2 are unequal, assign any particular forms to f_1 and f_2, then substitute the values of p and q, found from these equations in terms of x and y, in $dz = p\,dx + q\,dy$, which integrate. Otherwise, assign the form of one only of the functions f_1, f_2, involving an arbitrary constant C, solve for p and q, and integrate $dz = p\,dx + q\,dy$, adding an arbitrary function of C for the constant of integration.

3438 2ndly.—When m_1, m_2 are equal, and therefore, by (2),

$$S^2 = 4(RT + UV) \quad \ldots\ldots\ldots\ldots\ldots\ldots \quad (5).$$

Equations (3) and (4) coincide, and, since $m = \frac{1}{2}S$, reduce to

3439
$$Udp = \tfrac{1}{2}S\,dy - T\,dx \quad \ldots\ldots\ldots\ldots\ldots\ldots (6),$$
$$Udq = \tfrac{1}{2}S\,dx - R\,dy \quad \ldots\ldots\ldots\ldots\ldots\ldots (7),$$
$$dz = p\,dx + q\,dy \quad \ldots\ldots\ldots\ldots\ldots\ldots (8).$$

Here $p_y = q_x$, and therefore the last equation is integrable if the values of p and q, obtained by integrating (6) and (7), be substituted in it. Let $u = a$, $v = b$ be the integrals of (6) and (7); and let $z = \phi(x, y, a, b, c)$ $\ldots\ldots\ldots\ldots\ldots\ldots$ (9)

be the integral obtained from (8).

The general integral is found by making the parameters a, b, c vary subject to two conditions $b = f(a)$, $c = F(a)$; that is, by differentiating

$$z = \phi\{x, y, a, f(a), F(a)\}$$

for a, and eliminating a.

3440 The general integral therefore represents the envelope of the surface whose equation is (9).

PROOF.—(Boole, *Sup.*, p. 147.) Assuming a 1st integral of the form $u = f(v)$, eliminate μ and v from equations (3426) by multiplying (i.) by $(u_x)u_q$, (ii.) by $(u_y)u_p$, (iv.) by $(u_x)(u_y)$, (v.) by u_pu_q, and adding. Again, eliminate μ and v by multiplying (i.) by $(u_x)^2$, (ii.) by $(u_y)^2$, (iii.) by $(u_x)(u_y)$, (v.) by $(u_x)u_p + (u_y)u_q$, and adding. The two resulting equations are

$$\left.\begin{array}{l} R(u_x)u_q + T(u_y)u_p - U(u_x)(u_y) + Vu_pu_q = 0 \\ R(u_x)^2 + S(u_x)(u_y) + T(u_y)^2 + V\{(u_x)u_p + (u_y)u_q\} = 0 \end{array}\right\} \quad \ldots\ldots(10).$$

Multiply the 2nd of these by m, divide by V, and add to the 1st equation; the result is expressible in two factors either as (11) or (12),

$$\{R(u_x) + m_1(u_y) + Vu_p\}\{m_1(u_x) + T(u_y) + Vu_q\} = 0 \quad \ldots\ldots\ldots(11),$$

$$\{R(u_x) + m_2(u_y) + Vu_p\}\{m_2(u_x) + T(u_y) + Vu_q\} = 0 \quad \ldots\ldots\ldots(12),$$

m_1, m_2 being the roots of the quadratic (2). By equating to zero one factor of (11) and one of (12), we have four systems of two linear 1st order P. D. equations. Taking each system in turn with the equations

$$(u_x) + ru_p + su_q = 0,$$
$$(u_y) + su_p + tu_q = 0,$$

and eliminating (u_x), (u_y), u_p, u_q, we have the determinant annexed for the case in which the 1st factor of (11) and the 2nd of (12) are equated to zero. In this case, and also when the 2nd factor of (11) and the 1st of (12) are chosen, transposing m_1, m_2 in the determinant, the eliminant is equivalent to

$$\begin{vmatrix} R & m_1 & V & 0 \\ m_2 & T & 0 & V \\ 1 & 0 & r & s \\ 0 & 1 & s & t \end{vmatrix} = 0,$$

$$V\{Rr + Ss + Tt + U(rt - s^2) - V\} = 0,$$

having regard to the values of m_1m_2 and $m_1 + m_2$ from (2).

When the 1st factor of both (11) and (12) is taken, the 2nd order P. D. equation produced by the elimination is

$$Vt - R(rt - s^2) = 0,$$

and when the 2nd factor of each is taken, the elimination produces

$$Vr - T(rt - s^2) = 0.$$

Hence the hypothesis of a 1st integral of (1), of the form $u = f(v)$, involves the satisfying one or other of the systems of two simultaneous equations, (13) or (14), below:

$$\left.\begin{array}{l} R(u_x) + m_1(u_y) + Vu_p = 0 \\ m_2(u_x) + T(u_y) + Vu_q = 0 \end{array}\right\} \ldots(13). \qquad \left.\begin{array}{l} R(u_x) + m_2(u_y) + Vu_p = 0 \\ m_1(u_x) + T(u_y) + Vu_q = 0 \end{array}\right\} \ldots(14).$$

Now multiply the 2nd equation of (13) by λ and add it to the 1st. In the result, collect the coefficients of u_x, u_y, u_p, u_q, u_z. The Lagrangean system of auxiliary equations (3387) will then be found to be

$$\frac{dx}{R + \lambda m_2} = \frac{dy}{m_1 + \lambda T} = \frac{dp}{V} = \frac{dq}{\lambda V} = \frac{dz}{Rp + m_1q + \lambda(Tq + m_2p)} = \frac{du}{0}.$$

Eliminating λ, equations (3) are produced. Treating equations (14) in the same way, equations (4) are produced.

POISSON'S EQUATION.

3441 $$P = (rt-s^2)^n Q,$$

where P is a function of p, q, r, s, t, and homogeneous in r, s, t; Q is a function of x, y, z, and derivatives of z, which does not become infinite when $rt-s^2$ vanishes, and n is positive.

RULE.—*Assume* q $= \phi$ (p) *and express* s *and* t *in terms of* $\mathrm{q_p}$ *and* r; *thus,* rt$-$s² *vanishes, and the left side becomes a function of* p, q, *and* $\mathrm{q_p}$.

Solve for a 1*st integral in terms of* p *and* q, *and integrate again for the final solution.*

PROOF.— $s = q_x = q_p p_x = q_p r;\quad t = q_y = q_p p_y = q_p^2 r;$ therefore $rt-s^2 = 0$. Also P is of the form $(r, s, t)^m = r^m(1, q_p, q_p^2)^m$. Hence the equation takes the form $(1, q_p, q_p^2)^m = 0$.

LAPLACE'S REDUCTION OF THE EQUATION.

3442 $$Rr+Ss+Tt+Pp+Qq+Zz = U \;\ldots\ldots\ldots\ldots (1),$$

where R, S, T, P, Q, Z, U are functions of x and y only.

Let two integrals of Monge's equation (3428)

$$Rdy^2 - Sdx\,dy + Tdx^2 = 0$$

be $$\phi(x, y) = a,\quad \psi(x, y) = b.$$

Assume $$\xi = \phi(x, y),\quad \eta = \psi(x, y).$$

3443 To change the variables in equation (1) to ξ and η, we have

$$\begin{aligned}
p &= z_x = z_\xi \xi_x + z_\eta \eta_x;\qquad q = z_y = z_\xi \xi_y + z_\eta \eta_y;\\
r &= z_{2x} = z_{2\xi}\xi_x^2 + 2z_{\xi\eta}\xi_x\eta_x + z_{2\eta}\eta_x^2 + z_\xi \xi_{2x} + z_\eta \eta_{2x};\\
t &= z_{2y} = z_{2\xi}\xi_y^2 + 2z_{\xi\eta}\xi_y\eta_y + z_{2\eta}\eta_y^2 + z_\xi \xi_{2y} + z_\eta \eta_{2y};\\
s &= z_{xy} = z_{2\xi}\xi_x\xi_y + z_{2\eta}\eta_x\eta_y + z_{\xi\eta}(\xi_x\eta_y + \xi_y\eta_x) + z_\xi \xi_{xy} + z_\eta \eta_{xy}.
\end{aligned}\qquad (1701)$$

The transformed equation is of the form

$$z_{\xi\eta} + Lz_\xi + Mz_\eta + Nz = V \;\ldots\ldots\ldots\ldots\ldots\ldots (2),$$

where L, M, N, V are functions of ξ and η. This equation may be written in the form

$$(d_\xi + M)(d_\eta + L)\,z + (N - LM - L_\xi)\,z = V \;\ldots\ldots\ldots (3).$$

If $$N - LM - L_\xi = 0 \;\ldots\ldots\ldots\ldots\ldots\ldots\ldots (4),$$

we shall have

$$(d_\xi + M)\,z' = V \quad\text{with}\quad (d_\eta + L)\,z = z',$$

and the solution by a double application of (3210) is obtained from

$$z = e^{-\int L d\eta}\{\phi(\xi)+\int z' e^{\int L d\eta} d\eta\},$$

$$z' = e^{-\int M d\xi}\{\psi(\eta)+\int V e^{\int M d\xi} d\xi\}.$$

By symmetry, equation (1) is also solvable, if

$$N-LM-M_{\eta}=0 \quad \text{.................... (5).}$$

But if neither of these conditions is found to hold, find z in terms of z' from (3). It will be of the form

$$z = Az'_{\xi}+Bz'+C,$$

where A, B, C contain ξ and η. Substitute this for z in $(d_{\eta}+L)z = z'$, and the result is of the form

$$z'_{\xi\eta}+L'z'_{\xi}+M'z'_{\eta}+N'z' = V'.$$

The same conditions of integrability, if fulfilled for this equation, will lead to a solution of (1), and, if not fulfilled, the transformation may be repeated until one of the equations, similar to (4) or (5), is satisfied.

3444 Cor.—The solution of the equation

$$z_{\xi\eta}+az_{\xi}+bz_{\eta}+abz = V$$

is $$z = e^{-a\eta}\phi(\xi)+e^{-b\xi}\psi(\eta)+e^{-a\eta-b\xi}\iint e^{a\eta+b\xi}V d\eta\, d\xi.$$

3445 For the solution of equation (2), when L, M, V contain also z, see Prof. Tanner, *Proc. Lond. Math. Soc.*, Vol. viii., p. 159.

LAW OF RECIPROCITY. [*Boole*, ch. xv.

3446 Let a differential equation of the 1st order be

$$\phi(x, y, z, p, q) = 0 \quad \text{.................... (1).}$$

Let the result of interchanging x and p, y and q, and of changing z into $px+qy-z$, be

$$\phi(p, q, px+qy-z, x, y) = 0 \quad \text{.............. (2);}$$

then, if $z = \psi(x, y)$ be the solution of either (1) or (2), the solution of the other will be obtained by eliminating ξ and η between the equations

$$x = d_\xi \psi(\xi, \eta), \quad y = d_\eta \psi(\xi, \eta), \quad z = \xi x + \eta y - \psi(\xi, \eta).$$

3447 Ex.—Let $z = pq$ ...... (1), $px + qy - z = xy$ ...... (2), be the two reciprocal equations.

The integral of (2) is $z = xy + xf\left(\frac{y}{x}\right)$, $\therefore\ \psi(\xi\eta) = \xi\eta + xf\left(\frac{\eta}{\xi}\right)$.

ξ, η have now to be eliminated between

$$x = \eta - \frac{\eta}{\xi} f'\left(\frac{\eta}{\xi}\right) + f\left(\frac{\eta}{\xi}\right), \quad y = \xi + f'\left(\frac{\eta}{\xi}\right), \quad z = \xi\eta \text{ ...... (3)}.$$

Each form assigned to f gives a particular integral of (1). If $f\left(\frac{\eta}{\xi}\right) = a\frac{\eta}{\xi} + b$, the equations (3) become $x = \eta + b$, $y = \xi + a$, $z = \xi\eta$, and the elimination produces $z = (x-b)(y-a)$.

3448 In an equation of the 2nd order, the reciprocal equation is formed by making the changes in (3446), and, in addition, changing

$$r \text{ into } \frac{t}{rt-s^2}, \quad s \text{ into } \frac{-s}{rt-s^2}, \quad t \text{ into } \frac{r}{rt-s^2};$$

then, if the 2nd integral of either equation be $z = \psi(x, y)$, that of the other will be found by the same rule.

3449 The above transformation makes any equation of the form

$$\phi(p, q)\, r + \psi(p, q)\, s + \chi(p, q)\, t = 0$$

dependent for solution upon one of the form

$$\chi(x, y)\, r - \psi(x, y)\, s + \phi(x, y)\, t = 0.$$

3450 And, in the same way, an equation of the form

$$Rr + Ss + Tt = V(rt - s^2)$$

is dependent for solution upon one of the form

$$Rr + Ss + Tt = V.$$

See *De Morgan*, Camb. Phil. Trans., Vol. VIII.

SYMBOLIC METHODS.

FUNDAMENTAL FORMULÆ.

Q denoting a function of θ,

3470 $$(d_\theta - m)^{-1} Q = e^{m\theta} \int e^{-m\theta} Q\, d\theta.$$

PROOF.—The right is the value of y in the solution of $d_\theta y - my = Q$ by (3210). But this equation is expressed symbolically by $(d_\theta - m)\,y = Q$ (see 1492), therefore $y = (d_\theta - m)^{-1} Q$.

Let $x = e^\theta$, therefore $d_\theta = xd_x$ and $xd\theta = dx$. Hence (3470) may be written

3471 $$(xd_x - m)^{-1} Q = x^m \int x^{-m-1} Q dx.$$

3472 COR.— $$(d_\theta - m)^{-1}\, 0 = Ce^{m\theta},$$

3473 or $$(xd_x - m)^{-1}\, 0 = Cx^m.$$

Let $F(m)$ denote a rational integral function of m; then, since $d_\theta e^{m\theta} = me^{m\theta}$, $d_{2\theta} e^{m\theta} = m^2 e^{m\theta}$, &c., the operation d_θ is always replaced by the operation $m\times$. Hence, in all cases,

3474 $$F(d_\theta)\, e^{m\theta} = e^{m\theta} F(m).$$

3475 $$F(d_\theta)\, e^{m\theta} Q = e^{m\theta} F(d_\theta + m)\, Q.$$

Formula (2161) is a particular case of this theorem.

3476 $$e^{m\theta} F(d_\theta)\, Q = F(d_\theta - m)\, e^{m\theta} Q.$$

Also, by (3474–6),

3477 $$F(m) = e^{-m\theta} F(d_\theta)\, e^{m\theta}.$$

3478 $$F(d_\theta + m)\, Q = e^{-m\theta} F(d_\theta)\, e^{m\theta} Q.$$

3479 $$F(d_\theta)\, Q = e^{-m\theta} F(d_\theta - m)\, e^{m\theta} Q.$$

To the last six formulæ correspond

3480 $$F(xd_x)\, x^m = x^m F(m).$$

3481 $$F(xd_x)\, x^m Q = x^m F(xd_x + m)\, Q.$$

3482 $$x^m F(xd_x)\, Q = F(xd_x - m)\, x^m Q.$$

3483 $$F(m) = x^{-m} F(xd_x)\, x^m.$$

3484 $$F(xd_x + m)\, Q = x^{-m} F(xd_x)\, x^m Q.$$

3485 $$F(xd_x)\, Q = x^{-m} F(xd_x - m)\, x^m Q.$$

If $U = a + bx + cx^2 + \&c.$, then, by (3480),

3486 $$F(xd_x)\, U = F(0)\, a + F(1)\, bx + F(2)\, cx^2 + \&c.$$

3487 $F^{-1}(xd_x)\,U = F^{-1}(0)\,a + F^{-1}(1)\,bx + F^{-1}(2)\,cx^2 + \&c.$

3488 $F(xd_x, yd_y, zd_z, \dots)\,x^m y^n z^p \dots = F(m, n, p \dots)\,x^m y^n z^p \dots$

3489 $x^n u_{nx} = d_\theta(d_\theta - 1)(d_\theta - 2)\dots(d_\theta - n + 1)\,u,$

or, more succinctly, writing D for d_θ,

$$D(D-1)\dots(D-n+1)\,u \quad \text{or} \quad D_{-1}^{(n)}u \quad (2452).$$

3490 Otherwise $x^n u_{nx} = xd_x(xd_x - 1)\dots(xd_x - n + 1)\,u.$

Proof.—As in (1770). Otherwise, by Induction, differentiating again, and remembering that $x_\theta = x$.

Note.—In the symbolic solution of differential equations, we may either employ the operator xd_x directly, or the operator d_θ after substituting e^θ for x. Formulæ (3480–5) or (3474–9) will be required accordingly.

3491 $$\{\phi(D)\,e^{r\theta}\}^n Q$$

$$= e^{nr\theta}\phi(D+nr)\,\phi(D+\overline{n-1}.r)\dots\phi(D+r)\,Q$$

$$= \phi(D)\,\phi(D-r)\,\phi(D-2r)\dots\phi\{D-(n-1)\,r\}\,e^{nr\theta}Q.$$

Proof.—By repeated application of (3475) or (3476).

For ready reference, formulæ (1520, '21) are reprinted here.

3492 $$f(x+h) = e^{hd_x}f(x).$$

3493 $$f(x+h,\ y+k) = e^{hd_x+kd_y}f(x, y).$$

Let $$a_0 + a_1x + a_2x^2 \dots + a_nx^n = f(x),$$
then, denoting d_θ by D,

3494 $f(D)\,uv = uf(D)\,v + u_\theta f'(D)\,v + \dfrac{u_{2\theta}}{1.2}f''(D)\,v + \&c.,$

where $f'(D)$ means that D is to be written for x after differentiating $f(x)$.

Proof.—Expand uv, $D.uv$, $D^2.uv \dots D^n.uv$ by Leibnitz's theorem (1460); multiply the equations respectively by $a_0, a_1, a_2 \dots a_n$, and add the results.

3495 $uf(D)\,v = f(D).uv - f'(D)\,u_\theta v + \dfrac{1}{1.2}f''D.u_{2\theta}v - \&c.$

PROOF.—Expand $uv_\theta, uv_{2\theta}, uv_{3\theta} \dots uv_{n\theta}$ by theorem (1472), and proceed as in the last.

3496 $$F(d_{2x}) \frac{\sin}{\cos} mx = F(-m^2) \frac{\sin}{\cos} mx.$$

A more general theorem is

3497 $$F(\pi^2)(u_{im}+u_{-im}) = F(-m^2)(u_{im}+u_{-im}),$$

where u and π have the meanings assigned below (3499), and $i = \sqrt{-1}$.

Theorem.—If ϕ and ψ denote any algebraic functions of x and y, it may be shown, by (3474) and (3475), that

3498 $$\psi(d_x+y)\,\phi(x) = \phi(d_y+x)\,\psi(y).$$

3499 Let u, or, more definitely, $u_n = (x, y, z, \dots)^n$, represent a homogeneous function of the nth degree in severable variables, and let

3500 $$\pi = xd_x+yd_y+zd_z+\text{\&c.}$$

Then, by (3480),

3501 $$\pi u = nu, \quad \pi^2 u = n^2 u, \quad \pi^3 u = n^3 u, \quad \text{\&c.}$$

3502 Hence $$F(\pi)\,u = F(n)\,u.$$

REDUCTION OF $F(\pi_1)$ TO $f(\pi)$.

3503 Let u be any implicit function of the variables, and let $\pi = \pi_1+\pi_2$, where π_1 operates only upon x as contained in u, and π_2 only upon x as contained in πu, &c. after repetitions of the operation π. Then

3504 $$\pi_1 u = \pi u, \qquad \pi_1^2 u = (\pi-1)\,\pi u,$$

$$\dots \quad \dots \quad \dots \quad \dots \quad \dots$$

3506 $$\pi_1^r u = (\pi-r+1)\dots(\pi-2)(\pi-1)\,\pi u.$$

PROOF.— $$\pi_1 u = (\pi-\pi_2)\,u = \pi u,$$

since π_2 has here no subject to operate upon.

$$\pi_1^2 u = (\pi-\pi_2)\,\pi u = (\pi-1)\,\pi u,$$

for, πu being of the 1st degree, π_2 and 1 are equivalent as operators. In the next step, π_2 and 2 are equivalent, and so on.

COR.—When u is a homogeneous function, we have, by

(3501), $\pi^r u = n^r u$, therefore π and n are equivalent operators upon u. Hence (3506) may be written

3507 $\pi_1^r u = (n-r+1)\ldots(n-2)(n-1)\,nu = n_{-1}^{(r)} u,$

which is Euler's theorem of homogeneous functions (1625), since in that theorem the operator is confined to u.

3508 As an illustration, let $\pi u = (xd_x + yd_y)\,u = \pi_1 u,$

then $\pi_1^2 u = (x^2 d_{2x} + 2xy\,d_{xy} + y^2 d_{2y})\,u, \qquad \pi^2 u = (\pi_1 + \pi_2)\,\pi u = \pi_1^2 u + \pi_2 \pi u.$

Here $\qquad \pi_2 \pi u = (xd_x + yd_y)(xd_x + yd_y)\,u,$

the operation being confined to x and y in the second factor (3503), and therefore producing $(xd_x + yd_y)\,u$ merely.

Hence $\pi^2 u = (x^2 d_{2x} + 2xy\,d_{xy} + y^2 d_{2y} + xd_x + yd_y)\,u,$ which proves (3505).

If $U = u_0 + u_1 + u_2 + \ldots$, a series of homogeneous functions of dimensions 0, 1, 2, ..., then, by (3502),

3509 $F(\pi)\,U = F(0)\,u_0 + F(1)\,u_1 + F(2)\,u_2 + \ldots,$

3510 $F^{-1}(\pi)\,U = F^{-1}(0)\,u_0 + F^{-1}(1)\,u_1 + F^{-1}(2)\,u_2 + \ldots$

3511 Ex. 1: $\quad a^\pi U = u_0 + au_1 + a^2 u_2 + \ldots,$

3512 $\quad a^{-\pi} U = u_0 + a^{-1} u_1 + a^{-2} u_2 + \ldots$

Ex. 2: If u have the meaning in (3499),

3513

$$F(\pi)\,e^u = F(0)\,1 + F(n)\,u + F(2n)\,\frac{u^2}{1.2} + F(3n)\,\frac{u^3}{1.2.3} + \&c.,$$

and similarly for the inverse operation $F^{-1}(\pi)$.

PROOF.—By (3502) applied to the expansion of the subject by (150).

3514 $$\frac{C(n,m)\,u_n}{m!} = \Sigma\,\frac{x^p d_{px}\, y^q d_{qy}\, z^r d_{rz}\ldots}{p!\;q!\;r!\;\ldots}\,u_n,$$

where $\quad p+q+r+\ldots = m, \quad$ and $\quad p! = 1.2\ldots p.$

PROOF.—Equate coefficients of a^m in the expansion of

$$(1+a)^\pi\,U = (1+a)^{xd_x}\,(1+a)^{yd_y}\,(1+a)^{zd_z}\ldots U,$$

reducing by (3490).

3515 The general symbolic solution of the equation $F(d_\theta)\,u = Q$ is

$$u = F^{-1}(d_\theta)\,Q + F^{-1}(d_\theta)\,0, \quad \text{by (1488–90).}$$

3516 The solution of the equation (3238), viz.,

$$y_{nx}+a_1 y_{(n-1)x}+\dots+a_{(n-1)}y_x+a_n y = Q \quad\dots\dots\dots (1),$$

where Q is a function of x, is most readily obtained by the symbolic method. Thus $m_1, m_2, \dots m_n$ being the roots of the auxiliary equation in (3239), and $A, B, C \dots N$ the numerators of the partial fractions into which $(m^n+a_1 m^{n-1}+\dots+a_n)^{-1}$ can be resolved, the complete primitive will be

3517

$$y = \{A(d_x-m_1)^{-1}+B(d_x-m_2)^{-1}\dots+N(d_x-m_n)^{-1}\}(Q+0),$$

where $$(d_x-m)^{-1}Q = e^{mx}\int e^{-mx}Q\,dx, \quad (3470)$$

and the whole operation upon zero produces, by (3472), for the complementary term,

3518 $$C_1 e^{m_1 x}+C_2 e^{m_2 x}\dots+C_n e^{m_n x}.$$

Proof.—Equation (1) may be written

$$(d_{nx}+a_1 d_{(n-1)x}+a_2 d_{(n-2)x}\dots+a_n)\,y = Q,$$

or $$(d_x-m_1)(d_x-m_2)\dots(d_x-m_n)\,y = Q,$$

$\therefore$ by (3515), $y = \{(d_x-m_1)(d_x-m_2)\dots(d_x-m_n)\}^{-1}(Q+0)$,

which, by partial fractions, is converted into the formula above.

If r of the roots $m_1, m_2, \dots$ are each $= m$, those roots give rise in (3517) to a single term of the form

3519 $$(A+Bd_x+Cd_{2x}\dots+Rd_{rx})\,e^{mx}\int_{rx} e^{-mx}Q.$$

Proof.—By (1918), the r roots equal to m will produce

$$\{A'(d_x-m)^{-r}+B'(d_x-m)^{-r+1}\dots+R'(d_x-m)^{-1}\}\,Q,$$

or $$(A+Bd_x+Cd_{2x}\dots+Rd_{rx})(d_x-m)^{-r}Q.$$

3520 But, by (3470), $(d_x-m)^{-2}Q = (d_x-m)^{-1}e^{mx}\int e^{-mx}Q\,dx$

$$= e^{mx}\int\left\{e^{-mx}e^{mx}\int e^{-mx}Q\,dx\right\}dx = e^{mx}\int_{2x} e^{-mx}Q\,dx, \quad \text{and so on.}$$

3521 Ex. (1): $$y_{3x}-y_{2x}-5y_x-3 = Q.$$

Here $$m^3-m^2-5m-3 = (m-3)(m+1)^2,$$

and $$\frac{1}{(m-3)(m+1)^2} = \frac{1}{16(m-3)}-\frac{1}{16(m+1)}-\frac{1}{4(m+1)^2},$$

therefore $$y = \tfrac{1}{16}(d_x-3)^{-1}Q-\tfrac{1}{16}(d_x+1)^{-1}Q-\tfrac{1}{4}(d_x+1)^{-2}Q$$

$$= \tfrac{1}{16}e^{3x}\int e^{-3x}Q\,dx-\tfrac{1}{16}e^{x}\int e^{-x}Q\,dx-\tfrac{1}{4}e^{-x}\iint e^{x}Q\,dx^2. \quad (3520)$$

3522 Ex. (2): $$u_{2x}+a^2u = Q,$$

therefore $$u = (d_{2x}+a^2)^{-1}Q.$$

Here $$(m^2+a^2)^{-1} = (2ia)^{-1}\{(m-ia)^{-1}-(m+ia)^{-1}\},$$

therefore
$$\begin{aligned} u &= (2ia)^{-1}\{(d_x-ia)^{-1}Q-(d_x+ia)^{-1}Q\} \\ &= (2ia)^{-1}\{e^{iax}\int e^{-iax}Qdx-e^{-iax}\int e^{iax}Qdx\} \qquad (3470) \\ &= a^{-1}\sin ax\int\cos ax\,Q\,dx-a^{-1}\cos ax\int\sin ax\,Q\,dx, \end{aligned}$$

by the exponential values (766–7).

3523 Cor. 1.—The solution of $u_{2x}+a^2u=0$ is
$$u = A\cos ax+B\sin ax.$$

3524 Cor. 2.—The solution of $u_{2x}-a^2u=0$ is
$$u = Ae^{ax}+Be^{-ax}.$$

Change a into ia in the fifth line of (3522), and put $Q=0$.

3525 When Q is a function whose derivatives of the nth and higher orders vanish, proceed as in the following example.

Ex. (3): $$u_{2x}+a^2u = (1+x)^2,$$

therefore
$$\begin{aligned} u &= (d_{2x}+a^2)^{-1}(1+x)^2+(d_{2x}+a^2)^{-1}0 \\ &= (a^{-2}-a^{-4}d_{2x}+a^{-6}d_{4x}-\&c.)(1+2x+x^2)+(d_{2x}+a^2)^{-1}0 \\ &= a^{-2}(1+x)^2-2a^{-4}+A\cos ax+B\sin ax, \end{aligned}$$

the last two terms by (3523).

Exceptional Case of the Inverse Process.

3526 Ex. (4): $$u_{2x}+a^2u = \cos nx,$$

$$\begin{aligned} \therefore\ u &= (d_{2x}+a^2)^{-1}(\cos nx+0) = \tfrac{1}{2}(d_{2x}+a^2)^{-1}(e^{inx}+e^{-inx}+0) \\ &= \tfrac{1}{2}(e^{inx}+e^{-inx})(-n^2+a^2)^{-1}+A\cos ax+B\sin ax \quad \text{by (3474) and (3523)} \\ &= \cos nx\,(a^2-n^2)^{-1}+\&c. \end{aligned}$$

Now, if $n=a$, the first term becomes infinite. In such cases proceed as follows:—

Put $A = A'-(a^2-n^2)^{-1}$, and find the value of $\dfrac{\cos nx-\cos ax}{a^2-n^2}$, when $n=a$. By (1580) it is $=\dfrac{x\sin ax}{2a}$. Thus the solution is

$$u = \frac{x\sin ax}{2a}+A'\cos ax+B\sin ax.$$

The same result is obtained by making $Q=\cos ax$ in the solution of (3522).

For another example, see (3559).

3527 Ex. (5): $y_{2x}-9y_x+20y = x^2e^{3x}$,

therefore $y = \{(d_x-4)(d_x-5)\}^{-1}x^2e^{3x}+\{(d_x-4)(d_x-5)\}^{-1}0$

$= e^{3x}\{(d_x-1)(d_x-2)\}^{-1}x^2+Ae^{4x}+Be^{5x}$. (3475, 3517, 3472)

Now $(m^2-3m+2)^{-1} = \frac{1}{2}\left(1-\frac{3m-m^2}{2}\right)^{-1}$

$$= \tfrac{1}{2}\left\{1+\frac{3m-m^2}{2}+\left(\frac{3m-m^2}{2}\right)^2+\&c.\right\}$$

Hence the solution becomes

$$y = e^{3x}\{\tfrac{1}{2}+\tfrac{3}{4}d_x+\tfrac{7}{8}d_{2x}+\&c.\}\,x^2+Ae^{4x}+Be^{5x}$$

$$= e^{3x}\left\{\frac{x^2}{2}+\frac{3x}{2}+\frac{7}{4}\right\}+Ae^{4x}+Be^{5x}.$$

3528 Ex. (6): $(d_x-a)^n u = e^{ax}$,

therefore $u = (d_x-a)^{-n}e^{ax} = e^{ax}(d_x)^{-n}1$ (3476) $= e^{ax}\left(\frac{x^n}{n\,!}+\int_{nx}0\right)$. (2149)

3529 Ex. (7): $(d_x+a)^2 y = \sin mx$,

therefore $y = (d_x+a)^{-2}\sin mx+(d_x+a)^{-2}0$

$= (d_x-a)^2(d_{2x}-a^2)^{-2}\sin mx+e^{-ax}(d_x)^{-2}0$ [by (3478) with $Q=0$]

$= (-m^2-a^2)^{-2}(d_x-a)^2\sin mx+e^{-ax}(Ax+B)$ (by 3496)

$= (m^2+a^2)^{-2}(-m^2\sin mx-2am\cos mx+a^2\sin mx)+e^{-ax}(Ax+B)$.

REDUCTION OF AN INTEGRAL OF THE nth ORDER.

3530 $$\int_{nx}Q = \frac{1}{n-1\,!}\left\{x^{n-1}\int Q\,dx-(n-1)\,x^{n-2}\int Qx\,dx\right.$$
$$\left.+C(n,2)\,x^{n-3}\int Qx^2dx\,\ldots\pm\int Qx^{n-1}dx,\right.$$

where $n-1\,! = 1.2\,\ldots\,n$.

Proof.—By (3489) $d_{nx}Q = e^{-n\theta}(d_\theta-n+1)(d_\theta-n+2)\,\ldots\,d_\theta Q$ .... (1),

therefore $d_{-nx}Q = \{(d_\theta-n+1)(d_\theta-n+2)\,\ldots\,d_\theta\}^{-1}e^{n\theta}Q$

$= \frac{1}{n-1\,!}\{(d_\theta-n+1)^{-1}-(n-1)(d_\theta-n+2)^{-1}$
$+C(n,2)(d_\theta-n+3)^{-1}-\&c.\}\,e^{n\theta}Q$ (3517)

$= \frac{1}{n-1\,!}\{e^{(n-1)\theta}(d_\theta)^{-1}e^\theta-(n-1)\,e^{(n-2)\theta}(d_\theta)^{-1}e^{2\theta}+\&c.\}\,Q.$

Then replace e^θ by x.

The equation

3531 $$ax^my_{mx}+bx^ny_{nx}+\&c. = A+Bx+Cx^2+\&c. = Q$$

may, by (3489), be transformed into

$$\{a\,(xd_x)_{-1}^{(m)}+b\,(xd_x)_{-1}^{(n)}+\&c.\}\,y=Q \quad \text{or} \quad F(xd_x)\,y=Q.$$

The solution is then obtained from

3532 $$y=F^{-1}(xd_x)\,Q+F^{-1}(xd_x)\,0.$$

The value of the 1st part is given in (3487).

3533 If α, β, γ, &c. are the roots of $F(m)=0$, the second part gives rise to the arbitrary terms

$$C_1x^{\alpha}+C_2x^{\beta}+\&c.$$

3534 If a root α is repeated r times, the corresponding terms are

$$x^{\alpha}\{C_1(\log x)^{r-1}+C_2(\log x)^{r-2}+\ldots+C_r\}.$$

PROOF.—The partial fractions into which $F^{-1}(xd_x)\,0$ can be resolved, as in (3517), are of the type $C\,(xd_x-m)^{-1}0$, m being a root of $F(x)=0$. But $(xd_x-m)^{-1}\,0=Cx^m$ (3473), C being an arbitrary constant.

For a root m repeated r times, the typical fraction is $C\,(xd_x-m)^{-p}$, p being less than r. Now

$$(xd_x-m)^p\,Cx^m\,(\log x)^{p-1}=(d_\theta-m)^p\,Ce^{m\theta}\,\theta^{p-1}=e^{m\theta}\,(d_\theta)^p\,C\theta^{p-1}\ (3475)=0,$$

therefore $$(xd_x-m)^{-p}\,0=Cx^m\,(\log x)^{p-1}.$$

The equation

3535 $$ay_{m\theta}+by_{n\theta}+\&c.=f(e^\theta,\ \sin\theta,\ \cos\theta)$$

is reducible to the form of (3531) by $x=e^\theta$; or, substituting from (768), it may be written

$$F(d_\theta)\,y=\Sigma\,(A_m e^{m\theta}),$$

and the solution will take the form

3536 $$y=\Sigma A_m e^{m\theta}F^{-1}(m)+F^{-1}(d_\theta)\,0,$$

for the last term of which the forms in (3533–4) are to be substituted with x changed to e^θ.

3537 Ex. (1): $$x^3y_{3x}=ax^m+bx^n$$

$$xd_x\,(xd_x-1)(xd_x-2)\,y=ax^m+bx^n,$$

$$\therefore\ y=\{xd_x\,(xd_x-1)(xd_x-2)\}^{-1}\,(ax^m+bx^n)+\{xd_x\,(xd_x-1)(xd_x-2)\}^{-1}0$$

$$=\frac{ax^m}{m\,(m-1)\,(m-2)}+\frac{bx^n}{n\,(n-1)(n-2)}+A+Bx+Cx^2,$$

by (3180) and (3533). A result evident by direct integration.

3538 Ex. (2): $x^2y_{2x}+3xy_x+y=(1-x)^{-2}$. By (3490)

$\{xd_x(xd_x-1)+3xd_x+1\}y=(xd_x+1)^2y=1+2x+3x^2+\&c.,$

$\therefore\ y=(xd_x+1)^{-2}(1+2x+3x^2+)=(0+1)^{-2}+2(1+1)^{-2}x+3(2+1)^{-2}x^2+$

(3480) $=1+\frac{x}{2}+\frac{x^3}{3}+\&c.+\frac{A\log x}{x}+\frac{B}{x}=-\frac{1}{x}\log(1-x)+\&c.$

3539 Ex. (3): $y_{2x}+(4x-1)y_x+(4x^2-2x+2)y=0$.

Let $\pi=d_x+2x$. Then the equation may be written

$$\pi(\pi-1)y=0,\quad \therefore\ y=\{\pi(\pi-1)\}^{-1}0=(\pi-1)^{-1}0-\pi^{-1}0.$$

Let $(\pi-1)^{-1}0=u$, $\therefore\ (\pi-1)u=0$, or $u_x+(2x-1)u=0$, $\therefore\ u=Ae^{x^2-x}$.

3540 The solution of a P. D. equation of the type

$$a\pi_1^m z+b\pi_1^n z+\&c.=u_1+u_2+\&c.,$$

where u_1, u_2, &c. are homogeneous functions of the 1st, 2nd degrees, &c. in x, y, and $\pi_1=xd_x+yd_y$ (3503), is analogous to (3531), and is obtained from that solution by substituting u_1, u_2, &c. for Bx, Cx^2, &c.; and, for such terms as Cx^a, an arbitrary homogeneous function of x and y of the same degree.

3541 Solution of $F(\pi)u=Q$,

where $F(\pi)=\pi^n+A_1\pi^{n-1}+A_2\pi^{n-2}+A_n$,

and $Q=u_0+u_1+u_2+\&c.$,

a series of homogeneous functions of x, y, z, ... of the respective dimensions 0, 1, 2, &c.

Here $u=F^{-1}(\pi)\,Q+F^{-1}(\pi)\,0.$

3542 The value of the 1st term is given in (3510). For the general value of the last term (see Proof of 3533), let $F(m)=0$ have r roots $=m$; then

3543 $C(\pi-m)^{-p}0=C\{u(\log x)^{p-1}+v(\log x)^{p-2}...+w\}$,

where $u, v, \ldots w$ are arbitrary functions of the variables all of the degree m.

3544 Cor.— $(\pi-m)^{-1}0=(x,y,\ldots)^m$,

that is, a single homogeneous function of the variables of the degree m (1620).

3545 Ex.: $x^2z_{2x}+2xyz_{xy}+y^2z_{2y}-a(xz_x+yz_y)+az = u_m+u_n$,

u_m, u_n being homogeneous functions of the mth and n^{th} degrees. The equation may be written $\qquad (\pi_1^2-a\pi_1+a)\,z = u_m+u_n$;

or, by (3505), $\qquad (\pi-a)(\pi-1)\,z = u_m+u_n$,

therefore $\quad z = \{(\pi-a)(\pi-1)\}^{-1}(u_m+u_n)+\{(\pi-a)(\pi-1)\}^{-1}0$

$$= \frac{u_m}{(m-a)(m-1)}+\frac{u_n}{(n-a)(n-1)}+U_a+U_1.$$

The first two terms by formula (3502); the last two terms are arbitrary functions of the degrees a and 1 respectively, and result from formula (3543) by taking $p=1$ and $m=a$ and 1.

3546 To reduce a P. D. equation, when possible, to the symbolic form

$$(\Pi^n+A_1\Pi^{n-1}+A_2\Pi^{n-2}\ldots+A_n)\,u = Q \quad\ldots\ldots\ldots (1),$$

where $\qquad \Pi = Md_x+Nd_y+\&c.$,

and Q, M, N, &c. are any functions of the independent variables.

Consider the case of two independent variables,

$$(Md_x+Nd_y)^2\,u = M^2u_{2x}+2MNu_{xy}+N^2u_{2y}$$
$$+(MM_x+NM_y)\,u_x+(MN_x+NN_y)\,u_y \ldots (2).$$

Here the form of Π is obtainable from the right by considering the terms involving the highest derivatives only, for these terms are algebraically equivalent to $(Md_x+Nd_y)^2$.

The reduction being effected, and the equation being brought to the form of (1); then, if the auxiliary equation

3547 $$m^n+A_1m^{n-1}+A_2m^{n-2}\ldots+A_n = 0 \quad\ldots\ldots\ldots\ldots (3)$$

have its roots a, b, ... all unequal, the solution of (1) will be of the form

3548 $$u = (\Pi-a)^{-1}Q+(\Pi-b)^{-1}Q+\&c. \quad\ldots\ldots\ldots\ldots(4).$$

The terms on the right involve the solution of a series of linear first order P. D. equations, the first of which is

3549 $$Mu_x+Nu_y+\ \ldots\ -au = Q,$$

and the rest involve b, c, &c.

If equal or imaginary roots occur in the auxiliary equation, we may proceed as in the following example.

3550 Ex.:

$$(1+x^2)^2 z_{2x} - 4xy(1+x^2) z_{xy} + 4x^2y^2 z_{2y} + 2x(1+x^2) z_x + 2y(x^2-1) z_y + a^2 z = 0.$$

Here $\Pi = (1+x^2) d_x - 2xy\, d_y$, and the equation becomes $(\Pi^2 + a^2) z = 0$.

Let the variables x, y be now changed to ξ, η, so that $\Pi = d_\xi$. Therefore, since $\Pi(\xi) = 1$, $\quad \Pi(\xi) = (1+x^2)\xi_x - 2xy\,\xi_y = 1$.

Therefore, by (3383), $$\frac{dx}{1+x^2} = \frac{dy}{-2xy} = d\xi,$$

from which, by separating the variables and integrating, we obtain

$$x^2y + y = A \quad\text{..........(1)},$$

and, by (1436), $$\xi = \tan^{-1} x + B \quad\text{..........(2)}.$$

Also, since $\Pi(\eta) = \eta_\xi = 0$, $\quad (1+x^2)\eta_x - 2xy\,\eta_y = 0$.

Therefore $$\frac{dx}{1+x^2} = \frac{dy}{-2xy} = \frac{d\eta}{0},$$

the solution of which is equation (1). Thus

$$\xi = \tan^{-1} x \quad \text{and} \quad \eta = x^2y + y.$$

The transformed equation is now $\quad (d_{2\xi} + a^2) z = 0$,

and the solution, by (3523), is

$$z = \phi(\eta) \cos a\xi + \psi(\eta) \sin a\xi,$$

arbitrary functions of the variable, which is not explicitly involved, being substituted for the constants (3389). Therefore finally,

$$z = \phi(x^2y + y) \cos(a \tan^{-1} x) + \psi(x^2y + y) \sin(a \tan^{-1} x).$$

MISCELLANEOUS EXAMPLES.

3551 $$u_{2x} + u_{2y} + u_{2z} = 0.$$

Put $d_{2y} + d_{2z} = a^2$. Thus $u_{2x} + a^2 u = 0$, the solution of which, by (3523), is $$u = \phi(y, z) \cos ax + \psi(y, z) \sin ax,$$

arbitrary functions of y and z being put for the constants A and B. Expand the sine and cosine by (764–5); replace a^2 by its operative equivalent, and, in the expansion of $\sin ax$, put $a\psi(y, z) = \chi(y, z)$; thus

$$u = \phi(y, z) - \frac{x^2}{2!}(d_{2y} + d_{2z})\,\phi(y, z) + \frac{x^4}{4!}(d_{2y} + d_{2z})\,\phi(y, z) - \&c.$$

$$+ x\chi(y, z) - \frac{x^3}{3!}(d_{2y} + d_{2z})\,\chi(y, z) + \frac{x^5}{5!}(d_{2y} + d_{2z})\,\chi(y, z) - \&c.$$

[See (3626) for another solution.

3552 $$u_x + u_y + u_z = xyz.$$

Here $$u = (d_x + d_y + d_z)^{-1}(xyz + 0)$$

$$= \{d_{-x} - d_{-2x}(d_y + d_z) + d_{-3x}(d_y + d_z)^2 - \dots\}(xyz + 0).$$

Operating upon xyz, we get

$$u = \tfrac{1}{2}x^2yz - \tfrac{1}{6}x^3(z + y) + \tfrac{1}{12}x^4,$$

the rest vanishing. For symmetry, take $\frac{1}{3}$rd of the sum of three such expressions; thus

$$u = \tfrac{1}{3}\left\{\tfrac{1}{12}(x^4+y^4+z^4)-\tfrac{1}{6}(x^3y+xy^3+y^3z+yz^3+z^3x+zx^3)+\tfrac{1}{2}xyz(x+y+z)\right\}.$$

Operating upon zero, we have, in the first place, $d_{-x}0 = \phi(yz)$ instead of a constant, therefore $d_{-2x}0 = x\phi(yz)$, &c.

The result is

$$\{1-x(d_y+d_z)+\tfrac{1}{2}x^2(d_y+d_z)^2-\dots\}\,\phi(yz) = e^{-x(d_y+d_z)}\phi(yz) = \phi(y-x,\,z-x)$$

(3493) the complementary term.

3553 Otherwise, putting $d_y+d_z = \mathfrak{D}$, we have, by (3478),

$$(d_x+\mathfrak{D})^{-1}xyz = e^{-x\mathfrak{D}}d_{-x}e^{x\mathfrak{D}}xyz = e^{-x\mathfrak{D}}d_{-x}\{x(y+x)(z+x)\}, \quad (3493)$$

$$= e^{-x\mathfrak{D}}\{\tfrac{1}{2}x^2yz+\tfrac{1}{3}x^3(y+z)+\tfrac{1}{4}x^4\}$$

$$= \tfrac{1}{2}x^2(y-x)(z-x)+\tfrac{1}{3}x^3(y+z-2x)+\tfrac{1}{4}x^4,$$

which agrees with the former solution.

3554 $$au_x+bu_y+cu_z = xyz.$$

Substitute $x = a\xi$, $y = b\eta$, $z = c\zeta$, and the equation becomes

$$u_\xi+u_\eta+u_\zeta = abc\xi\eta\zeta,$$

which is solved in (3552).

The same methods furnish the solution of

3555 $$au_x+bu_y+cu_z = x^my^nz^p.$$

3556 $$xz_x+yz_y = 2xy\sqrt{a^2-z^2}.$$

Put $$z = a\sin v,$$

$$\therefore\ \pi z = a\cos v\,.\,\pi v, \qquad \therefore\ \pi v = 2xy, \qquad \therefore\ z = a\sin(xy+c).$$

3557 $$axu_x+byu_y+czu_z-nu = 0.$$

Put $$x = \xi^a, \quad y = \eta^b, \quad z = \zeta^c;$$

$$\therefore\ \xi u_\xi+\eta u_\eta+\zeta u_\zeta-nu = 0, \qquad \therefore \text{ by (3544) } u = (x^{\frac{1}{a}}, y^{\frac{1}{b}}, z^{\frac{1}{c}})^n.$$

3558 The solution of any P. D. equation of the type

$$F(xd_x, yd_y, zd_z, \dots)\,u = \Sigma Ax^my^nz^p \dots$$

is, by (3488) and (3557),

$$u = \Sigma\frac{Ax^my^nz^p}{F(m, n, p, \dots)}+\frac{1}{F(xd_x, yd_y, zd_z, \dots)}0.$$

3559 Ex.: $$xu_x+yu_y-au = Q_m,$$
where $Q_m = (x, y)^m$ (1620).

Here $u = (m-a)^{-1}Q_m+U_a$. When $a = m$, this solution becomes indeterminate. In that case, as in (3526), assume

$$U_a = V_a - \frac{Q_a}{m-a}, \qquad \therefore\ u = \frac{Q_m-Q_a}{m-a}+V_a.$$

Differentiate for a, by (1580), putting Q_a first in the form

$$\tfrac{1}{2}\left\{x^a F\left(\frac{y}{x}\right)+y^a f\left(\frac{x}{y}\right)\right\};$$

thus $$u = \tfrac{1}{2}Q_m(\log x+\log y)+V_m.$$

Similarly, the solution of

3560 $$xu_x+yu_y+zu_z-mu = Q_m$$
is $$u = \tfrac{1}{3}Q_m(\log x+\log y+\log z)+V_m.$$

3561 $$xu_x+yu_y+zu_z = c.$$

The solution, by (3560), is
$$u = \tfrac{1}{3}c(\log x+\log y+\log z)+V_0.$$

3562 $$z_{2x}-2az_{xy}+a^2z_{2y} = 0 \quad \text{or} \quad (d_x-ad_y)^2 z = 0.$$

$$z = (d_x-ad_y)^{-2}0 = (d_x-ad_y)^{-1}\phi(y)\,e^{ax\,d_y} \qquad (3472)$$

by putting ad_y for m and $\phi(y)$ for C. The second operation produces, by (3476), $z = e^{ax\,d_y}\{x\phi(y)+\psi(y)\} = x\phi(y+ax)+\psi(y+ax)$. (3492)

3563 $$x^2z_{2x}-y^2z_{2y}+xz_x-yz_y = 0.$$

This reduces to $(xd_x+yd_y)(xd_x-yd_y)z = 0$.
Here $\pi = xd_x+yd_y$, and $m = 0$ in (3544),
therefore $$z = (x, y)^0+\left(x, \frac{1}{y}\right)^0,$$
the second term being obtained by substituting $y^{-1} = y'$, and so converting the second factor into $(xd_x+y'd_{y'})$. The above may also be written

$$z = F\left(\frac{x}{y}\right)+f(xy),$$

F and f being integral algebraic functions.

3564 $$z_{2x}-a^2z_{2y}+2abz_x+2a^2bz_y = 0.$$

Putting $y = a\eta$, this equation is equivalent to
$$(d_x-d_\eta+2ab)(d_x+d_\eta)z = 0;$$

putting $x = \log x'$ and $\eta = \log \eta'$, this gives, by (3544),

$$z = (e^x, e^{-\eta})^{-2ab} + (e^x, e^\eta)^0 = e^{-2abx}\phi(e^{x+\eta}) + \psi(e^{x-\eta})$$
$$= e^{-2abx}F(y+ax) + f(y-ax),$$

the functions being algebraic and integral.

3565 $$u_{2x} - a^2 u_{2y} = \phi(x, y).$$

$\therefore\ u = (d_{2x} - a^2 d_{2y})^{-1}\phi(x, y)$ (3515)

$= (2ad_y)^{-1}\left\{(d_x - ad_y)^{-1} - (d_x + ad_y)^{-1}\right\}\phi(x, y)$ (3470)

$= (2ad_y)^{-1}\left\{e^{axd_y}\int e^{-axd_y}\phi(x, y)\,dx - e^{-axd_y}\int e^{axd_y}\phi(x, y)\,dx\right\}$ (3470)

$= (2a)^{-1}\int\left\{\Phi_1(x, y+ax) - \Phi_2(x, y-ax)\right\}dy,$

since $e^{axd_y}\phi(x, y) = \phi(x, y+ax)$ (3492).

Here $\Phi_1(x, y) = \int\phi(x, y-ax)\,dx + \psi(y),$

$\Phi_2(x, y) = \int\phi(x, y+ax)\,dx + \chi(y).$

3566 If $\phi(x, y) = 0$, the solution therefore becomes

$$u = \psi_1(y+ax) + \chi_1(y-ax)$$ [*Boole*, ch. 16.

For the solution in this case by Monge's method, see (3433).

3567 $$z_x - az_y = e^{mx}\cos ny.$$

$z = (d_x - ad_y)^{-1}e^{mx}\cos ny = e^{axd_y}\int e^{-axd_y}e^{mx}\cos ny\,dx$ (3470)

$= e^{axd_y}\int e^{mx}\cos n(y-ax)\,dx$ (3492), and this by Parts, or by (1999), is

$= e^{axd_y}e^{mx}\left\{m\cos n(y-ax) - an\sin n(y-ax)\right\}(m^2+a^2n^2)^{-1} + e^{axd_y}\phi(y)$

$\therefore\ z = e^{mx}\left\{m\cos ny - an\sin ny\right\}(m^2+a^2n^2)^{-1} + \phi(y+ax)$, by (3492).

3568 $$z_t - az_{2x} = 0.$$

$z = (d_t - ad_{2x})^{-1}0 = e^{atd_{2x}}\phi(\phi)$, by (3472),

$\phi(x)$ taking the place of the constant C.

Therefore $z = \phi(x) + at\phi_{2x} + \tfrac{1}{2}a^2t^2\phi_{4x} + \&c.$ (3492)

Otherwise, to obtain z in powers of x, we have, putting $b^2 = a^{-1}$,

$$z_{2x} - b^2 zt = 0,$$

$\therefore\ z = \{(d_x + bd_t^{\frac{1}{2}})(d_x + bd_t^{\frac{1}{2}})\}^{-1}0 = e^{bxd_t^{\frac{1}{2}}}\phi(t) + e^{-bxd_t^{\frac{1}{2}}}\psi(t)$ (3518);

then expand by (150).

3569 $$z_{2x} + z_{2y} = \cos nx\cos my.$$

$$z = (d_{2x}+d_{2y})^{-1}\cos nx \cos my.$$

Treating d_{2y} and $\cos my$ as constants, we have, by (3526), putting d_{2y} for a^2,

$$z = \cos nx\,(d_{2y}-n^2)^{-1}\cos my + A\cos ax + B\sin ax, \text{ or by (3496),}$$
$$= \cos nx \cos my\,(-m^2-n^2)^{-1} + \phi(y)\cos(xd_y) + \psi(y)\sin(xd_y),$$

A and B becoming $\phi(y)$ and $\psi(y)$.

3570 $$z_{2\phi}+2z_{\phi\psi}+z_{2\psi}+a^2z = \cos(m\phi+n\psi).$$

Therefore $(\pi+ia)(\pi-ia)\,z = \frac{1}{2}\{e^{i(m\phi+n\psi)}+e^{-i(m\phi+n\psi)}\}$,

where $\pi = d_\phi + d_\psi$. Therefore, by (3510), with $x = e^\phi$, $y = e^\psi$,

$$z = \tfrac{1}{2}\left\{\frac{e^{i(m\phi+n\psi)}}{a^2-(m+n)^2}+\frac{e^{-i(m\phi+n\psi)}}{a^2-(m+n)^n}\right\} + (e^\phi, e^\psi)^{-ia} + (e^\phi, e^\psi)^{ia},$$

or $$z = \frac{\cos(m\phi+n\psi)}{a^2-(m+n)^2} + (e^\phi, e^\psi)^{-ia} + (e^\phi, e^\psi)^{ia}.$$

3571 PROP. I.—To transform a linear differential equation of the form

$$(a+bx+cx^2\ldots)\,u_{nx}+(a'+b'x+c'x^2\ldots)\,u_{(n-1)x}+\&c. = Q \ldots (1)$$

into the symbolical form

$$f_0(D)\,u+f_1(D)\,e^\theta u+f_2(D)\,e^{2\theta}u+\&c. = T \;\ldots\ldots\; (2),$$

where Q is a function of x, T a function of θ, $x = e^\theta$ and $D = d_\theta$.

Multiply the equation by x^n; then the 1st term on the left becomes, by (3489),

$$(a+be^\theta+ce^{2\theta}+\ldots)\,D(D-1)\ldots(D-n+1)\,u.$$

This reduces, by the repeated application of formula (3476) with the notation of (2451), to

3572 $$aD^{(n)}u+b\,(D-1)^{(n)}\,e^\theta u+c\,(D-2)^{(n)}\,e^{2\theta}u+\&c.$$

The other terms admit of similar reductions.

3573 Conversely, to bring back an equation from the symbolic form (2) to the ordinary form (1), employ formula (3475) so as to transfer $e^{m\theta}$ to the left of the operative symbol.

3574 Ex.: $$x^2(x^2u_{2x}+7xu_x+5u) = e^{2\theta}\{D(D-1)+7D+5\}\,u$$
$$= e^{2\theta}(D^2+6D+5)\,u = e^{2\theta}(D+1)(D+5)\,u$$
$$= (D-1)(D+3)\,e^{2\theta}u \quad (3476).$$

For the converse reduction, the steps must be retraced, employing (3475). See also example (3578).

3575 Prop. II.—To solve the equation

$$u+a_1\phi(D)\,e^\theta u+a_2\phi(D)\,\phi(D-1)\,e^{2\theta}u\,\dots$$
$$+a_n\phi(D)\,\phi(D-1)\,\dots\,\phi(D-n+1)\,e^{n\theta}u=U,$$

where U is a function of θ.

By (3491)

$$\phi(D)\,\phi(D-1)\,\dots\,\phi(D-n+1)\,e^{n\theta}u=\{\phi(D)\,e^\theta\}^n u.$$

Putting $\rho^n u$ for this, the equation becomes

$$(1+a_1\rho+a_2\rho^2\dots+a_n\rho^n)\,u=U.$$

3576 Therefore

$$u=\{A_1(1-q_1\rho)^{-1}+A_2(1-q_2\rho)^{-1}\dots+A_n(1-q_n\rho)^{-1}\}\,U,$$

where $q_1, q_2 \dots q_n$ are the roots of the equation

$$q^n+a_1q^{n-1}+a_2q^{n-2}\dots+a_n=0,$$

and $$A_r=\frac{q_r^{n-1}}{(q_r-q_1)(q_r-q_2)\dots(q_r-q_n)}.$$

The solution will then be expressed by

$$u=A_1u_1+A_2u_2\dots+A_nu_n,$$

where u_r is given by the solution of the equation

3577 $$u_r-q_r\phi(D)\,e^\theta u_r=U.$$

3578 Ex.:

$$(x^2+5x^3+6x^4)\,u_{2x}+(4x+25x^2+36x^3)\,u_x+(2+20x+36x^2)\,u=20x^3.$$

Putting $x=e^\theta$, and transforming by (3489),

$$(1+5e^\theta+6e^{2\theta})\,D(D-1)\,u+(4+25e^\theta+36e^{2\theta})\,Du+(2+20e^\theta+36e^{2\theta})\,u=20e^{3\theta}.$$

The first term $=D(D-1)\,u+5(D-1)(D-2)\,e^\theta u+6(D-2)(D-3)\,e^{2\theta}u$ by applying (3476). The other terms similarly; thus, after rearrangement,

$$(D+1)(D+2)\,u+5(D+1)^2e^\theta u+6D(D+1)\,e^{2\theta}u=20e^{3\theta}.$$

Operating upon this with $\{(D+1)(D+2)\}^{-1}$, we get

$$u+5\frac{D+1}{D+2}\,e^\theta u+6\frac{D}{D+2}\,e^{2\theta}u=\frac{20e^{3\theta}}{(3+1)(3+2)}=e^{3\theta},\ \text{by (3474)};$$

or $$(1+5\rho+6\rho^2)\,u=e^{3\theta},\quad \text{if } \rho=(D+1)(D+2)^{-1}e^\theta;$$

therefore $$u=\{3(1+3\rho)^{-1}-2(1+2\rho)^{-1}\}\,e^{3\theta}=3y-2z,$$

if $$y=(1+3\rho)^{-1}e^{3\theta}\quad\text{and}\quad z=(1+2\rho)^{-1}e^{3\theta}.$$

Hence $(1+3\rho)\,y = e^{3\theta}$ or $y+3\,(D+1)(D+2)^{-1}\,e^{\theta}y = e^{3\theta}$;

therefore $(D+2)\,y+3\,(D+1)\,e^{\theta}y = e^{3\theta}\,(3+2)$, by (3474),

or $(D+2)\,y+3e^{\theta}\,(D+2)\,y = 5e^{3\theta}$, by (3475);

that is, $(x+3x^2)\,y_x+2\,(1+3x)\,y = 5x^3$.

Similarly $(x+2x^2)\,z_x+2\,(1+2x)\,z = 5x^3$.

Solve these by (3210), and substitute in $u = 3y-2z$.

3579 PROP. III.—To transform the equation

$$u+\phi(D)\,e^{r\theta}u = U \quad \text{into} \quad v+\phi(D+n)\,e^{r\theta}v = V,$$

put $$u = e^{n\theta}v \quad \text{and} \quad U = e^{n\theta}V.$$

PROOF.—By (3474), because $\phi(D)\,e^{(n+r)\theta}v = e^{n\theta}\phi\,(D+n)\,e^{r\theta}v$.

3580 PROP. IV.—To transform the equation

$$u+\phi(D)\,e^{r\theta}u = U \quad \text{into} \quad v+\psi(D)\,e^{r\theta}v = V,$$

put $$u = P_r\frac{\phi(D)}{\psi(D)}\,v \quad \text{and} \quad U = P_r\frac{\phi(D)}{\psi(D)}\,V,$$

3581 where $$P_r\frac{\phi(D)}{\psi(D)} = \frac{\phi(D)\,\phi(D-r)\,\phi(D-2r)\,\ldots}{\psi(D)\,\psi(D-r)\,\psi(D-2r)}.$$

PROOF.—Put $u = f(D)\,v$ in the 1st equation, and $e^{r\theta}f(D)\,v = f(D-r)\,e^{r\theta}v$ (3476). After operating with $f^{-1}(D)$ it becomes

$$v+\phi(D)\,f(D-r)\,f^{-1}(D)\,e^{r\theta}v = f^{-1}(D)\,U,$$

therefore $$\phi(D)\,f(D-r)\,f^{-1}(D) = \psi(D) \quad \text{by hypothesis;}$$

therefore $$f(D) = \frac{\phi(D)}{\psi(D)}\,f(D-r) = \frac{\phi(D)\,\phi(D-r)}{\psi(D)\,\psi(D-r)}\,f(D-2r),$$

and so *in inf.* Also $U = f(D)\,V$.

3582 To make any elementary factor $\chi(D)$ of $\phi(D)$ become, in the transformed equation, $\chi(D\pm nr)$, where r is an integer; take $\psi(D) = \chi(D\pm nr)\,\chi_1(D)$. See example (3589).

3583 To make any factor of $\phi(D)$ of the form $\dfrac{\chi(D)}{\chi(D\pm nr)}$ disappear in the transformed equation, take $\psi(D) = \chi_1(D)$, where $\chi_1(D)$, in each case, denotes the remaining factors of $\phi(D)$. See example (3591).

3584 In the application of Proposition IV., differentiation or integration will be the last operation according as $P_r\frac{\phi(D)}{\psi(D)}$ (3581) has its factors, after reduction, in the numerator or denominator, and therefore according as $\psi(D)$ is formed by algebraically diminishing or increasing the several factors of $\phi(D)$. However, by first employing Proposition III., the given equation may frequently be so prepared that the final operation with Prop. IV. shall be differentiation only. See example (1).

For further investigation, see *Boole's Diff. Eq.*, Ch. 17, and *Supplement*, p. 187.

3585 To reduce an equation of the homogeneous class (3531) to a binomial equation of the same order of the form

$$y_{nx}+qy = X.$$

The general theory of such solutions is as follows. Let the given equation be

$$u+q\left\{(D+a_1)(D+a_2)\ldots(D+a_n)\right\}^{-1}e^{n\theta}u = U \;\ldots\; (1),$$

$a_1, a_2, \ldots a_n$ being in descending order of magnitude. Putting $u = e^{-a_1\theta}v$, by Prop. III.,

$$v+q\left\{D(D-\overline{a_1-a_2})\ldots(D-\overline{a_1-a_n})\right\}^{-1}e^{n\theta}v = e^{a_1\theta}U \ldots (2).$$

To transform these factors, regarded as $\phi(D)$, by Prop. IV. into $\psi(D) = D(D-1)\ldots(D-n+1)$, we convert D into $D+rn$ (3582), r being an integer.

Hence for the pth factor we must have

$$D+rn-a_1+a_p = D-p+1,$$

3586 and therefore $a_1-a_p = rn+p-1$ .... (3).

If this relation holds for each of the constants $a_1 \ldots a_n$, equation (1) is reducible to the form

3587 $\quad y+q\left\{D(D-1)\ldots(D-n+1)\right\}^{-1}e^{n\theta}y = Y$ ...... (4),

which, by (3489), is equivalent to $y_{nx}+qy = Y_{nx} = X$.

y being found in terms of x from the last equation, and, v being $= P_n\frac{\phi(D)}{\psi(D)}y$ (3580), the solution will result from

3588 $\quad u = e^{-a_1\theta}P_n\frac{(D-1)(D-2)\ldots(D-n+1)}{(D-a_1+a^2)\ldots(D-a_1+a_n)}y;$

while U and Y are connected by the same relation as u and y.

3589 Ex. 1: Given $x^3u_{3x}+18x^2u_{2x}+84xu_x+96u+3x^3u=0$.
Putting $x=e^\theta$ and employing (3489), this becomes

$$\{D(D-1)(D-2)+18D(D-1)+84D+96\}\,u+3e^{3\theta}u=0,$$

or
$$(D+8)(D+4)(D+3)\,u+3e^{3\theta}u=0,$$

therefore
$$u+3\{(D+8)(D+4)(D+3)\}^{-1}e^{3\theta}u=0 \;\ldots\ldots\ldots\ldots\ldots\ldots(1).$$

Employing Prop. III., put $u=e^{-8\theta}v$,

therefore (3476) $v+3\{D(D-4)(D-5)\}^{-1}e^{3\theta}v=0 \;\ldots\ldots\ldots\ldots\ldots\ldots(2).$

To transform this by Prop. IV. into

$$y+3\{D(D-1)(D-2)\}^{-1}e^{3\theta}y=0 \;\ldots\ldots\ldots\ldots\ldots\ldots(3),$$

we have

$$P_3\frac{\phi(D)}{\psi(D)}=\frac{D(D-1)(D-2)(D-3)(D-4)(D-5)\ldots}{D(D-4)(D-5)(D-3)(D-7)(D-8)\ldots}=(D-1)(D-2),$$

$$\therefore\; v=(D-1)(D-2)\,y, \qquad \therefore\; u=e^{-8\theta}(D-1)(D-2)\,y \;\ldots\ldots\ldots(4),$$

and the solution is obtained by differentiation only, performed on the value of y as obtained by the solution of (3), that equation being equivalent to

$$D(D-1)(D-2)\,y+3e^{3\theta}y=0, \quad \text{or, by (3489),} \quad y_{3x}+3y=0.$$

If, however, Prop. IV. were used to pass directly from (1) to (3), we should have

$$P_3\frac{\phi(D)}{\psi(D)}=\frac{D(D-1)(D-2)(D-3)(D-4)(D-5)\ldots}{(D+8)(D+4)(D+3)(D+5)(D+1)\,D\ldots}$$

$$=\frac{1}{(D+8)(D+5)(D+4)(D+3)(D+2)(D+1)},$$

and equation (4) would involve integrations of y as high as $D^{-6}y$.

3590 Note.—By the literal application of Rule IV., the right side of equation (3) ought to be $V=\{(D-1)(D-2)\}^{-1}0$; but no such term is required when the original and transformed equations are of the same order, for in such cases the arbitrary constants introduced by the operation upon zero disappear with the terms containing them in the final differentiation. The result is the same as if the operation upon zero had not been performed.

In the following example, V has to be retained.

3591 Ex. 2: $(x-x^3)\,u_{2x}+(2-12x^2)\,u_x-30xu=0 \;\ldots\ldots\ldots\ldots\ldots\ldots(1).$

Multiply by x, transform by (3489), and remove $e^{2\theta}$ to the right of each function of D by (3476), thus

$$u-\frac{(D+4)(D+3)}{D(D+1)}e^{2\theta}u=0 \;\ldots\ldots\ldots\ldots\ldots\ldots(2).$$

Transform this by Prop. IV. into

$$v-\frac{D+3}{D+1}e^{2\theta}v=V \;\ldots\ldots\ldots\ldots\ldots\ldots(3).$$

We have
$$u=P_2\frac{D+4}{D}v=(D+4)(D+2)\,v,$$

$$V=\{(D+4)(D+2)\}^{-1}0=Ae^{-2\theta}+Be^{-4\theta} \quad (3518).$$

The operation upon zero is required in this example (see 3590), because (3) is of a lower order than (2); but only one term of the result need be retained, because only one additional constant is wanted. Hence (3) becomes

$$(D+1)\,v-(D+3)\,e^{2\theta}v = (D+1)\,Ae^{-2\theta} = -Ae^{-2\theta}.$$

Changing again to x, this equation becomes

$$(x^3-x^5)\,v_x-4x^2v+A = 0.$$

The value of v obtained from this by (3210) will contain two arbitrary constants. The solution of (1) will then be given by

$$u = (D+4)(D+2)\,v.$$

3592 Ex. 3: $u_{2x}-n\,(n+1)\,x^{-2}u-q^2u = 0,$ [*Boole*, p. 424.

n being an integer.

Multiplying by x^2 and employing (3489), this becomes

$$u-q^2\,\{(D+n)(D-n-1)\}^{-1}\,e^{2\theta}u = 0.$$

This is changed by Prop. III. into

$$v-q^2\,\{D\,(D-2n-1)\}^{-1}\,e^{2\theta}v = 0, \quad \text{with } u = e^{-n\theta}v,$$

and this, by Prop. IV., into

$$y-q^2\,\{D\,(D-1)\}^{-1}\,e^{2\theta}y = 0 \quad \text{or} \quad y_{2x}-q^2y = 0 \quad (3489).$$

y being found from this by (3524), we then have

$$u = e^{-n\theta}P_2\frac{D-1}{D-2n-1}\,y = e^{-n\theta}\,(D-1)^{(n)}_{-2}\,y = x^{-n}\,(xd_x-1)^{(n)}_{-2}\,y.$$

But, by (3484), $\quad F\,(xd_x-m) = x^m F\,(xd_x)\,x^{-m},$

$$\therefore \quad u = x^{-n}\,x\,(xd_x)\,x^{-1}.\,x^3\,(xd_x)\,x^{-3}\,\ldots\,x^{2n-1}\,(xd_x)\,x^{-2n+1}\,y,$$

or

$$u = x^{-(n+1)}\,(x^3d_x)^n\,x^{-2n+1}\,y$$
$$= x^{-n-1}\,(x^3d_x)^n\,x^{-2n+1}\,(Ae^{qx}+Be^{-qx}) \quad (3525).$$

This may be evaluated by substituting $z = x^{-2}$. (See *Educ. Times Reprint*, Vol. XVII., p. 77.)

3593 Ex. 4: $u_{2x}-a^2u_{2y}-n\,(n+1)\,x^{-2}\,u = 0.$

The solution is derived from that of Example (2), by putting $q = ad_y$, and arbitrary functions of y *after* the exponentials instead of A and B; thus

$$u = x^{-n-1}\,(x^3d_x)^n\,x^{-2n+1}\,\{e^{axd_y}\,\phi\,(y)+e^{-axd_y}\,\psi\,(y)\}$$
$$= x^{-n-1}\,(x^3d_x)^n\,x^{-2n+1}\,\{\phi\,(y+ax)+\psi\,(y+ax)\}, \quad \text{by (3492).}$$

[*Boole*, p. 425.

3594 $\quad (1+ax^2)\,u_{2x}+axu_x\pm n^2u = 0.$

To solve this equation or its symbolical equivalent obtained by (3489), viz.,

3595 $$u+\frac{a\,(D-2)^2\pm n^2}{D\,(D-1)}\,e^{2\theta}\,u = 0.$$

Substitute $t = \int \frac{dx}{\sqrt{(1+ax^2)}}$ in the solution of $u_{2t}\pm n^2u = 0$, by (3523–4).

3596 Similarly, to solve the equation

$$(x^2+a)\,x^2u_{2x}+(2x^2+a)\,xu_x \pm n^2u = 0,$$

or, the same in its symbolical form,

3597 $$u+\frac{(D-1)(D-2)}{aD^2\pm n^2}\,e^{2\theta}u = 0.$$

Substitute $t = \int \frac{dx}{x\sqrt{(x^2+a)}}$ in the solution of $u_{2t}\pm n^2u = 0$.
(3596) is obtainable from (3593) by changing θ into $-\theta$.

3598 *Pfaff's equation,*

$$(a+bx^n)\,x^2u_{2x}+(c+ex^n)\,xu_x+(f+gx^n)\,u = Q.$$

When $Q = 0$, the symbolical form becomes

$$u+\frac{b\,(D-n)(D-n-1)+e\,(D-n)+g}{aD\,(D-1)+cD+f}\,e^{n\theta}u = 0 \quad\text{.......... (1).}$$

If n be not $= 2$, substitute $2\theta' = n\theta$, and therefore $2d_\theta = nd_{\theta'}$.

3599 Thus $$u+\frac{b\,(D-a_1)(D-a_2)}{a\,(D-\beta_1)(D-\beta_2)}\,e^{2\theta'}u = 0 \quad\text{.................... (2).}$$

where a_1, a_2 are the roots of the equation

$$b\,(\tfrac{1}{2}na-n)(\tfrac{1}{2}na-n-1)+e\,(\tfrac{1}{2}na-n)+g = 0 \quad\text{..............(3),}$$

and β_1, β_2 are the roots of

$$a\,\tfrac{1}{2}n\beta\,(\tfrac{1}{2}n\beta-1)+c\,\tfrac{1}{2}n\beta+f = 0.$$

Four cases occur—

3600 I.—If a_1-a_2 and $\beta_1-\beta_2$ are odd integers, (2) can be reduced by Prop. IV. (3581) to the form $v+\frac{b\,(D-a_1)(D-a_1-1)}{a\,(D-\beta_1)(D-\beta_1-1)}\,e^{2\theta'}v = 0$,

and then resolved into two equations of the first order.

3601 II.—If any one of the four quantities $a_1-\beta_1$, $a_1-\beta_2$, $a_2-\beta_1$, $a_2-\beta_2$ is an even integer, (2) can be reduced by Prop. IV. to an equation of the first order.

3602 III.—If $\beta_1-\beta_2$ and $a_1+a_2-\beta_1-\beta_2$ are both odd integers, then, by Props. III. and IV., (2) is reducible to (3595).

3603 IV.—If a_1-a_2 and $a_1+a_2-\beta_1-\beta_2$ are both odd integers, (2) is reducible in like manner to (3597). [*Boole*, p. 428.

Note.—The integers may be either positive or negative, and when even may be zero.

SOLUTION OF LINEAR DIFFERENTIAL EQUATIONS BY SERIES.

3604 Case I.—Solution of the linear differential equation

$$f_0(D)\,u - f_1(D)\,e^{r\theta}u = 0 \quad \text{or} \quad f_0(xd_x)\,u - f_1(xd_x)\,x^r u = 0,$$

in which $f_0(D)$, $f_1(D)$ are polynomial expressions of the form $a_0 + a_1D + a_2D^2 \ldots + a_nD^n$ and $f_0(D) = (D-a)(D-b)(D-c)\ldots$.

3605 Let $\phi(D) = f_1(D) \div f_0(D)$, and let

3606 $$\Phi(a) = 1 + \phi(a+r)\,x^r + \phi(a+2r)\,\phi(a+r)\,x^{2r} + \phi(a+3r)\,\phi(a+2r)\,\phi(a+r)\,x^{3r} + \&c.$$

Then the solution will be

3607 $$u = Ax^a\Phi(a) + Bx^b\Phi(b) + Cx^c\Phi(c) + \&c.$$

Proof.—Operating with $f_0^{-1}(D)$ and writing ρ for $\phi(D)\,e^{r\theta}$,

$$u - \rho u = f_0^{-1}(D)\,0 = Ae^{a\theta} + Be^{b\theta} + \&c. \quad (3518)$$

Therefore, by (3515), $u = (1-\rho)^{-1}(Ae^{a\theta} + Be^{b\theta} + \ldots)$

$$= (1+\rho+\rho^2+\ldots)\,Ae^{a\theta} + (1+\rho+\rho^2+\ldots)\,Be^{b\theta} + \&c.$$

Now in each term substitute for ρ^n the value in (3491), and remove D by formula (3474).

Case II.—Solution of

3608 $$f_0(D)\,u + f_1(D)\,e^{\theta}u + f_2(D)\,e^{2\theta}u \ldots + f_n(D)\,e^{n\theta}u = 0 \quad \ldots\ldots(1),$$

where $$f_0(D) = (D-a)(D-b)(D-c)\ldots$$

Let $\Psi(a) = 1 + F_1(a+1)\,x + F_2(a+2)\,x^2 + \&c.$,

where the coefficients $F_1(a+1)$ or v_1, $F_2(a+2)$ or v_2, &c. are determined in succession by the formula

3609 $$f_0(m)\,v_m + f_1(m)\,v_{m-1} \ldots + f_n(m)\,v_{m-n} = 0 \text{ and } v_0 = 1 \quad \ldots\ldots(2).$$

The solution will then be expressed by

3610 $$u = Ax^a\Psi(a) + Bx^b\Psi(b) + Cx^c\Psi(c) + \&c.$$

Proof.—From (1)

$$u = \left\{1 + \phi_1(D)\,e^{\theta} \ldots + \phi_n(D)\,e^{n\theta}\right\}^{-1} f_0^{-1}(D)\,.\,0 \quad \ldots\ldots\ldots\ldots\ldots(3),$$

where $$\phi_r(D) = f_r(D) \div f_0(D).$$

Here $f_0^{-1}(D)\,0 = \{(D-a)(D-b)\ldots\}^{-1}\,0 = Ae^{a\theta} + Be^{b\theta} + \ldots$ (3518);

and $\left\{1 + \phi_1(D)\,e^{\theta} \ldots + \phi_n(D)\,e^{n\theta}\right\}^{-1} = 1 + F_1(D)\,e^{\theta} + F_2(D)\,e^{2\theta} + \ldots$

To determine F_1, F_2, &c., operate upon each side with $\{1+\phi(D)e^{\theta}+\&c.\}$, and equate coefficients of powers of e; thus formula (2) is obtained. (3) now becomes

$$u = \{1+F_1(D)e^{\theta}+F_2(D)e^{2\theta}+\dots\}(Ae^{a\theta}+Be^{b\theta}+\dots) \quad\dots\dots\dots (4).$$

Multiply out; apply (3474), and put x for e^{θ}.

3611 Ex.: $\quad x^2u_{2x}-(a+b-1)\,xu_x+abu-qx^2u = 0,$

or, by (3489), $\quad (D-a)(D-b)\,u-qe^{2\theta}u = 0.$

Here $\quad f_0(D) = (D-a)(D-b),\;\; f_1(D) = 0,\;\; f_2(D) = -q.$

Therefore (2) becomes $\quad (m-a)(m-b)\,v_m = qv_{m-2},$

therefore F_1, F_3, &c. vanish, and $F_0(a) = 1$,

$$F_2(a+2) = \frac{qF_0(a)}{(a+2-a)(a+2-b)} = \frac{q}{2(a+2-b)},$$

$$F_4(a+4) = \frac{qF_2(a+2)}{(a+4-a)(a+4-b)} = \frac{q^2}{4.2\,(a+4-b)(a+2-b)}.$$

Therefore $\Psi(a) = 1+\dfrac{qx^2}{2(a+b-2)}+\dfrac{q^2x^4}{4.2\,(a-b+4)(a-b+2)}+\dots$

Similarly we find $F_2(b+2)$, $F_4(b+2)$, &c., and thence $\Psi(b)$; and, substituting in (3610), we have

$$u = Ax^a+\frac{Aqx^{a+2}}{2(a+b-2)}+\frac{Aq^2x^{a+4}}{4.2\,(a-b+4)(a-b+2)}+\dots$$
$$+Bx^b+\frac{Bqx^{b+2}}{2(b-a+2)}+\frac{Bq^2x^{b+4}}{4.2\,(b-a+4)(b-a+2)}+\dots$$

3612 The solution is arrived at more quickly by formula (3607). We have

$$\phi(D) = \frac{q}{(D-a)(D-b)},$$

$$\therefore\; \phi(a+2) = \frac{q}{2(a+2-b)},\qquad \phi(a+4) = \frac{q}{4(a+4-b)},\quad \&c.,$$

producing the same series by the value of $\Phi(a)$. Similarly with $\Phi(b)$.

3613 When $f_0(D)$ has r factors each $= D-a$, the corresponding part of the value of u in equation (4) will produce

3614 $\quad A_0+A_1\log x+A_2(\log x)^2\dots+A_{r-1}(\log x)^{r-1},$

where the coefficients A_0, A_1, ... are each of the form

$$C_0x^a+C_1x^{a+1}+C_2x^{a+2}+\dots$$

3615 But if any one of the quantities $F_r(a+r) = 0$ (3608), then $C_r = 0$ also.

PROOF.—$f_0^{-1}(D)$ now contains a term of the form

$$e^{a\theta}(c_0+c_1\theta \dots + c_r\theta^{r-1}) = e^{a\theta}v, \text{ say.}$$

The corresponding part of u in (4) is

$$\left\{1+F_1(D)\,e^{\theta}+\dots\right\} e^{a\theta} v$$

$$= \left\{e^{a\theta}+e^{(a+1)\theta}F_1(D+a+1)+e^{(a+2)\theta}F_2(D+a+2)+\dots\right\} v \quad \text{by (3475).}$$

Expand each function F by Taylor's theorem in powers of D, operate upon v, and arrange the result according to powers of θ.

In practice, proceed as in the following example.

3616 Ex.: $$xu_{2x}+u_x+q^2xu = 0.$$

Multiplying by x and changing by (3489), this becomes

$$D^2u+q^2e^{2\theta}u = 0. \quad D^2u = 0 \text{ gives } u = A+B\theta.$$

Substitute this value and operate with D, considering A and B as variables, and equate to zero the coefficients of the powers of θ; thus

$$D^2A+q^2e^{2\theta}A+2DB = 0, \quad D^2B+q^2e^{2\theta}B = 0.$$

Then change D into m, and $e^{r\theta}A$ into a_{m-r}, to obtain the relations

$$m^2a_m+q^2a_{m-2}+2mb_m = 0; \quad m^2b_m+q^2b_{m-2} = 0,$$

which determine the constants successively in terms of a_0 and b_0 (which are arbitrary) in the equation

$$u = a_0+a_2x^2+a_4x^4+\dots+\log x\,(b_0+b_2x^2+b_4x^4+\dots),$$

which thus becomes the solution sought. [*Boole*, *Diff. Eq.*, p. 439.

SOLUTION BY DEFINITE INTEGRALS.*

3617 *Laplace's method.*—The solution of the equation

$$x\phi(d_x)\,u+\psi(d_x)\,u = 0 \quad \dots\dots\dots\dots \quad (1)$$

is

$$u = C\int\left\{e^{xt+\int\frac{\psi t}{\phi t}dt}(\phi t)^{-1}\right\}dt \quad \dots\dots\dots \quad (2),$$

the limits being determined by

$$e^{xt+\int\frac{\psi t}{\phi t}dt} = 0 \quad \dots\dots\dots\dots\dots \quad (3).$$

PROOF.—Assume $u = e\int e^{xt}Tdt$, and substitute in (1), putting $\phi(d_x)\,e^{xt} = \phi(t)\,e^{xt}$ (3474), thus

$$\int xe^{xt}\phi(t)\,Tdt+\int e^{xt}\psi(t)\,Tdt = 0.$$

* This method of solution is merely indicated here, and the reader is referred to Boole's *Diff. Eq.*, Ch. xviii., for a complete investigation.

Integrating the first term by parts, this becomes

$$e^{xt}\phi(t)\,T - \int e^{xt}\left[d_t\left\{\phi(t)\,T\right\} - \psi(t)\,T\right]dt = 0 \text{ .... (4),}$$

an equation which is satisfied by equating each term to zero. The second term thus produces a value of T by integration by (3209), and this value substituted in the first term, and in the value of u, gives the results (3) and (2).

3618 Ex. (1): $\quad xu_{2x} + au_x - q^2xu = 0$* .... (4).

Here $\phi(d_x) = d_{2x} - q^2$, $\psi(d_x) = ad_x$, $\phi(t) = t^2 - q^2$, $\psi(t) = at$. Hence (2) and (3) become

$$u = C\int e^{xt}(t^2-q^2)^{\frac{a}{2}-1}dt; \quad e^{xt}(t^2-q^2)^{\frac{a}{2}} = 0;$$

a being positive, the limits are $t = \pm q$, and, putting $t = q\cos\theta$, we find

$$u = C\int_0^\pi e^{qx\cos\theta}\sin^{a-1}\theta\,d\theta \text{ .... (5).}$$

3619 The solution in series by (3608) is as follows. Equations (1) and (2) of that article are in this case

$$D(D+a-1)\,u - q^2e^{2\theta}u = 0 \quad \text{and} \quad m(m+a-1)\,v_m - q^2v_{m-2} = 0.$$

Thus, a in (3608) $= 0$, and $b = 1-a$. Therefore (3610) becomes

$$u = A\left\{1 + \frac{q^2x^2}{2(a+1)} + \frac{q^4x^4}{2.4(a+1)(a+3)} + \&c.\right\}$$

$$+Bx^{1-a}\left\{1 + \frac{q^2x^2}{2(3-a)} + \frac{q^4x^4}{2.4(3-a)(5-a)} + \&c.\right\} \text{ ...... (6).}$$

Both series are convergent by (239 ii.).

The results deduced by Boole are these—

3620 (5) is equivalent to the particular integral represented by the first series of (6).

3621 A second particular integral, by assuming $u = e^{(1-a)\theta}v$, is found to be, when $2-a$ is positive,

$$u = C_2x^{1-a}\int_0^\pi e^{qx\cos\theta}\sin^{1-a}\theta\,d\theta \text{ .... (7).}$$

3622 When a lies between 0 and 2, the complete integral is

$$u = C_1\int_0^\pi e^{qx\cos\theta}\sin^{a-1}\theta\,d\theta + C_2x^{1-a}\int_0^\pi e^{qx\cos\theta}\sin^{1-a}\theta\,d\theta \text{ .... (8).}$$

---

* The method by definite integrals is elucidated by Boole chiefly in the solution of this important equation.

3623 But, if $a=1$, the solution becomes

$$u=\int_0^\pi e^{qx\cos\theta}\left\{A+B\log(x\sin^2\theta)\right\}d\theta \quad\ldots\ldots\ldots\ldots\ldots\ldots (9).$$

3624 If a does not lie between 0 and 2, then, if a be negative, put $a=a'-2n$, and replace the first term of (8) by

$$C_1(xd_x+a'-1)(xd_x+a'-3)\ldots(xd_x+a'-2n+1)\int_0^\pi e^{qx\cos\theta}\sin^{a'-1}d\theta \ldots (10),$$

the transformation being effected by (3580).

3625 And if a be positive and >2, put $u=e^{(1-a)\theta}v=x^{1-a}v$. This converts a into $2-a$, a negative quantity, and the case is reduced to the last one.

3626 Ex. (2).—To solve by this method the P. D. equation

$$u_{2x}+u_{2y}+u_{2z}=0 \ldots\ldots\ldots\ldots\ldots\ldots\ldots\ldots (1)$$

when $r=\sqrt{(x^2+y^2)}$.

Eliminating x and y, (1) becomes

$$ru_{2r}+u_r+ru_{2z}=0 \quad\ldots\ldots\ldots\ldots\ldots\ldots\ldots\ldots\ldots (2).$$

Now the solution of this equation is number (9) of Example (1), if we change x into r, q into id_z, and A and B into arbitrary functions of z. We thus obtain

$$u=\int_0^\pi e^{r\cos\theta\, id_z}\left\{\phi(z)+\psi(z)\log(r\sin^2\theta)\right\}d\theta \quad\ldots\ldots\ldots\ldots\ldots (3),$$

or, by (3492),

$$u=\int_0^\pi \phi\left\{z+ir\cos\theta\right\}d\theta+\int_0^\pi \psi(z+ir\cos\theta)\log(r\sin^2\theta)\,d\theta \ldots\ldots (4).$$

See (3551) for another solution.

3627 If u be the potential of an attracting mass at an external point, and if $u=F(z)$ when $r=0$; then, since $\log r=\infty$, $\psi(z)$ must vanish;

therefore $$F(z)=\int_0^\pi \phi(z)\,d\theta=\pi\phi(z).$$

Hence (4) reduces to $$u=\frac{1}{\pi}\int_0^\pi F\left\{z+ir\cos\theta\right\}d\theta.$$

Parseval's Theorem.

3628 If, for all values of u,

$$A+Bu+Cu^2+\ldots=\phi(u)$$

and $$A'+B'u^{-1}+C'u^{-2}+\ldots=\psi(u) \quad\ldots\ldots\ldots\ldots (1),$$

then

$$AA' + BB' + \dots = \frac{1}{2\pi}\int_0^\pi \left[\phi(e^{i\theta})\,\psi(e^{i\theta}) + \phi(e^{-i\theta})\,\psi(e^{-i\theta})\right] d\theta.$$

PROOF.—Form the product of equations (1), and in it put $u = e^{i\theta}$ and $e^{-i\theta}$ separately, and add the results. Multiply by $d\theta$, integrate from 0 to π, and divide by 2π.

P. D. EQUATIONS WITH MORE THAN TWO INDEPENDENT VARIABLES.

3629 By means of Fourier's theorem (2742), the solution of the equation

$$u_{2t} - h^2(u_{2x} + u_{2y} + u_{2z}) = 0$$

may be deduced by a general method in the form

$$u = (1+d_t)\iiiint\!\!\iint e^{i(A+Bht)}\,\psi(a, b, c)\,da\,db\,dc\,d\lambda\,d\mu\,d\nu,$$

the limits of each integration being $-\infty$ to ∞, and the function ψ being arbitrary and different in the two terms arising from the operator $(1+d_t)$.

Boole, Ch. xviii., and more fully in Cauchy's *Exercice d'Analyse Mathématique*, Tom. I., pp. 53 et 178.

3630 Poisson's solution of the same equation in the form of a double integral is

$$u = (1+dt)\int_0^\pi\int_0^{2\pi} t\sin\xi\,\psi(x+ht\sin\xi\sin\eta,\ y+ht\sin\xi\cos\eta,\ z+ht\cos\xi)\,d\xi\,d\eta$$

with the same latitude in the interpretation of ψ.

[Gregory's *Examples*, p. 504.

DIFFERENTIAL RESOLVENTS OF ALGEBRAIC EQUATIONS.

3631 THEOREM I. (*Boole*).—"If $y_1, y_2 \dots y_n$ are the n roots of the equation

$$y^n - ay^{n-1} + 1 = 0 \quad \dots\dots\dots\dots\dots\dots\dots \quad (1),$$

and if the mth power of any one of these roots be represented by u, and if $a = e^\theta$, then u as a function of θ satisfies the differential equation

$$u - \left(\frac{n-1}{n}D + \frac{m}{n} - 1\right)^{(n-1)} \left(\frac{D}{n} - \frac{m}{n} - 1\right) \left[D^{(n)}\right]^{-1} e^{n\theta} u = 0,$$

and the complete integral of the same will be

$$u = C_1 y_1^m + C_2 y_2^m + \ldots + C_n y_n^m.$$

3632 "Cor. I.—If $m = -1$ and if n be > 2, the differential equation

$$D^{(n-1)} u - \frac{1}{n}\left(\frac{n-1}{n}D - \frac{1}{n} - 1\right)^{(n-1)} e^{n\theta} u = 0$$

has for its general integral

$$u = C_1 y_1^{-1} + C_2 y_2^{-1} \ldots + C_{n-1} y_{n-1}^{-1},$$

$y\,, y\ \ldots\ y_{n-1}$ being any $n-1$ roots of (1).

"If θ be changed into $-\theta$, and therefore D into $-D$, the above results are modified as follows :—

3633 "Cor. II.—The differential equation

$$u - (D-1)^{(n)} \left[\left(\frac{n-1}{n}D - \frac{m}{n}\right)^{(n-1)} \left(\frac{D}{n} + \frac{m}{n}\right)\right]^{-1} e^{n\theta} u = 0$$

has for its complete integral

$$u = C_1 y_1^m + C_2 y_2^m \ldots + C_n y_n^m,$$

$y_1, y_2 \ldots y_n$ being the roots of the equation

$$a y^n - y^{n-1} + a = 0 \quad \ldots\ldots\ldots\ldots\ldots\ldots \quad (2).$$

3634 "Cor. III.—The differential equation

$$u - n\,(D-2)^{(n-1)} \left[\left(\frac{n-1}{n}D + \frac{1}{n}\right)^{(n-1)}\right]^{-1} e^{n\theta} u = 0,$$

supposing $n > 2$ has for its complete integral

$$u = C_1 y_1^{-1} + C_2 y_2^{-1} \ldots + C_{n-1} y_{n-1}^{-1},$$

$y_1, y_2 \cdots y_{n-1}$ being any $n-1$ roots of (2).

" *Theorem II. (Harley).*—

3635 " The differential equation

$$a^r (xd_x)^{(n)} u - \left(\frac{n-r}{n} xd_x + \frac{m}{n} - 1\right)^{(n-r)} \left(\frac{r}{n} xd_x - \frac{m}{n} - 1\right)^{(r)} x^r u = 0$$

is satisfied by the mth power of any root of the equation

$$y^n - xy^{n-r} + a = 0,$$

u being considered as a function of x.

3636 " Cor.—The differential equation

$$n^n \left(\frac{n-r}{r} xd_x - \frac{m}{r}\right)^{(n-r)} (xd_x)^{(r)} u - (n-1)^r \left(\frac{n}{r} xd_x - \frac{m}{r} - 1\right)^{(n)} x^r u = 0$$

is satisfied by the mth power of any root of the equation

$$y^n - ny^{n-r} + (n-1)\, x = 0."$$

[*Boole, Diff. Eq.*, Sup. 191—199.

3637 See also *Boole, Phil. Trans.*, 1864; *Harley, Proc. of the Lit. and Phil. Soc. of Manchester*, Vol. II.; *Rawson, Proc. of the Lond. Math. Soc.*, Vol. 9.

CALCULUS OF FINITE DIFFERENCES.

INTRODUCTION.

3701 In this branch of pure mathematics a function $\phi(x)$ is denoted by u_x, and $\phi(x+h)$ consequently by u_{x+h}. The increment h is commonly unity. If Δx denotes the increment h, and Δu_x the consequent increase in the value of u_x, we have

3702 $$\Delta u_x = u_{x+\Delta x} - u_x.$$

3703 When Δx diminishes without limit, the value of

$$\frac{\Delta u_x}{\Delta_x} \quad \text{or} \quad \frac{u_{x+\Delta x} - u_x}{\Delta x} \quad \text{is} \quad \frac{du_x}{dx}.$$

3704 The repetition of the operation Δ is indicated as follows:

$$\Delta\Delta u_x = \Delta^2 u_x, \quad \Delta\Delta^2 u_x = \Delta^3 u_x, \quad \text{and so on.}$$

3705 Ex.: Let $u_x = x^2$,

$$\begin{array}{rcccccc} x = 1 & 2 & 3 & 4 & 5 & & \ldots \\ x^2 = 1 & 4 & 9 & 16 & 25 & & \ldots \\ \Delta x^2 = 3 & 5 & 7 & 9 & \ldots & & \ldots \\ \Delta^2 x^2 = 2 & 2 & 2 & \ldots & \ldots & & \ldots \end{array}$$

FORMULÆ FOR FIRST AND n^{th} DIFFERENCES.

If $$u_x = ax^n + bx^{n-1} + cx^{n-2} + \&c.,$$

3706 $$\Delta^r u_x = an(n-1)\ldots(n-r+1)\,x^{n-r} + c_1 x^{n-r-1} + \&c.,$$

3707 $$\Delta^n u_x = an(n-1)\ldots 3.2.1.$$

3708 Hence the nth difference of a rational integral function of the nth degree is constant.

3709 So also $\quad \Delta^n x^n = 1.2.3 \ldots n.$

3710 NOTATION.—Factorial terms are denoted as follows :—

$$u_x u_{x-1} \ldots u_{x-m+1} \equiv u_x^{(m)},$$

3711
$$\frac{1}{u_x u_{x+1} \ldots u_{x+m-1}} \equiv u_x^{(-m)}.$$

3712 Thus $\quad x(x-1)\ldots(x-m+1) \equiv x^{(m)},$

3713
$$\frac{1}{x(x+1)\ldots(x+m-1)} \equiv x^{(-m)}.$$

Hence $\underline{|m}$, $m!$, and $m^{(m)}$ are equivalent symbols.

3714 According to (2452), $x^{(m)}$ would here be denoted by $x_{-1}^{(m)}$. The suffix, however, being omitted, it may be understood that the common difference of the factors is always -1.

3715 $\Delta x^{(m)} = m x^{(m-1)}, \quad \Delta^n x^{(m)} = m^{(n)} x^{(m-n)}, \quad \Delta^m x^{(m)} = m^{(m)},$
and, if $m<n$, $\Delta^n x^{(m)} = 0$, since $\Delta c = 0$ if $c =$ constant.

3718 $\Delta x^{(-m)} = -m x^{(-m-1)}, \qquad \Delta^n x^{(-m)} = (-m)^{(n)} x^{(-m-n)}.$

3720
$$\Delta u_x^{(m)} = (u_{x+1} - u_{x-m+1})\, u_x^{(m-1)},$$
$$\Delta u_x^{(-m)} = (u_x - u_{x+m})\, u_x^{(-m-1)}.$$

3722 Ex.:
$$\Delta(ax+b)^{(m)} = am(ax+b)^{(m-1)}, \qquad \Delta(ax+b)^{(-m)} = -am(ax+b)^{(-m-1)}.$$

3724 $\Delta \log u_x = \log\left\{1 + \frac{\Delta u_x}{u_x}\right\}, \quad \Delta \log u_x^{(m-1)} = \log \frac{u_{x+1}}{u_{x-m+1}}.$

3726 $\Delta a^x = (a-1)\, a^x, \qquad \Delta^n a^{mx} = (a^m - 1)^n a^{mx}.$

3728 $\Delta^n \frac{\sin}{\cos}(ax+b) = \left(2\sin\frac{a}{2}\right)^n \frac{\sin}{\cos}\left\{ax+b+\frac{n(a+\pi)}{2}\right\}.$

PROOF.— $\quad \Delta \sin(ax+b) = \sin(ax+a+b) - \sin(ax+b)$
$$= 2\sin\frac{a}{2}\sin\left(ax+b+\frac{a+\pi}{2}\right).$$

That is, Δ is equivalent to adding $\frac{a+\pi}{2}$ to the angle and multiplying the sine by $2\sin\frac{a}{2}$.

3729 Conversely, the same formula holds if the sign of n be changed throughout.

EXPANSION BY FACTORIALS.

3730 If $\Delta^n\phi(0)$ denote the value of $\Delta^n\phi(x)$ when $x=0$, then $\phi(x)=\phi(0)+\Delta\phi(0)\,x+\dfrac{\Delta^2\phi(0)}{2}\,x^{(2)}+\dfrac{\Delta^3\phi(0)}{2.3}\,x^{(3)}+\&c.$

3731 If $\Delta x=h$ instead of unity, the same expansion holds good if for $\Delta^n\phi(0)$ we write $(\Delta^n\phi(x)\div h^n)_{x=0}$; that is, making $x=0$ after reduction.

PROOF.—Assume $\phi(x)=a_0+a_1x+a_2x^{(2)}+a_3x^{(3)}+\&c.$
Compute $\Delta\phi(x)$, $\Delta^2\phi(x)$, &c., and put $x=0$ to determine a_0, a_1, a_2, &c.

GENERATING FUNCTIONS.

3732 If $u_x t^x$ be the general term in the expansion of $\phi(t)$, then $\phi(t)$ is called the *generating function* of u_x or $\phi(t)=Gu_x$.

EX.: $(1-t)^{-2}=G(x+1)$, for $x+1$ is the coefficient of t^x in the expansion.

3733 $$Gu_x=\phi(t),\quad Gu_{x+1}=\frac{\phi(t)}{t},\ \dots\ Gu_{x+n}=\frac{\phi(t)}{t^n}.$$

3734 $$G\Delta u_x=\left(\frac{1}{t}-1\right)\phi(t),\ \dots\ G\Delta^n u_x=\left(\frac{1}{t}-1\right)^n\phi(t).$$

PROOF.— $G\Delta u_x=Gu_{x+1}-Gu_x$, &c.

THE OPERATIONS E, Δ, AND d_x.

3735 E denotes the operation of increasing x by unity,

$$Eu_x=u_{x+1}=u_x+\Delta u_x=(1+\Delta)\,u_x.$$

The symbols E and Δ both follow the laws of *distribution, commutation, and repetition* (1488–90).

3736 $$E=1+\Delta=e^{d_x}\quad\text{or}\quad e^D.*$$

PROOF.— $Eu_x=u_{x+1}=u_x+d_xu_x+\frac{1}{2}d_{2x}u_x+\dfrac{1}{2.3}d_{3x}u_x+\&c.$

$$=(1+d_x+\tfrac{1}{2}d_{2x}+\frac{1}{2.3}d_{3x}+\&c.)\,u_x=e^{d_x}u_x.$$

By (1520), Δx being unity.

* The letter d is reserved as a symbol of differentiation only, and the suffix attached to it indicates the independent variable. See (1487).

3737 Hence $\Delta = e^D - 1$ and $D = \log E$.

3739 Consistently with (3735) E^{-1} denotes the diminishing x by unity; thus $E^{-1}u_x = u_{x-1}$.

For $Eu_{x-1} = u_x$, $\therefore\ u_{x-1} = E^{-1}u_x$.

---

u_{x+n} *in terms of* u_x *and successive differences.*

3740 $u_{x+n} = u_x + n\Delta u_x + C(n, 2)\,\Delta^2 u_x + C(n, 3)\,\Delta^3 u_x + \&c.$

PROOF.—(i.) By induction, or (ii.) by generating functions, or (iii.) by the symbolic law:

(ii.) $$Gu_{x+n} = \left(\frac{1}{t}\right)^n \phi(t) = \left\{1 + \left(\frac{1}{t} - 1\right)\right\}^n \phi(t).$$

Expand by the Binomial theorem, and apply (3734).

(iii.) $$u_{x+n} = E^n u_x = (1+\Delta)^n u_x.$$

Apply the laws in (3735) by expanding the binomial and distributing the operation upon u_x.

Conversely to express $\Delta^n u_x$ in terms of u_x, u_{x+1}, u_{x+2}, &c.

3741 $\Delta^n u_x = u_{x+n} - nu_{x+n-1} + C(n, 2)\,u_{x+n-2} \dots (-1)^n u_x$.

PROOF.— $\Delta^n u_x = (E-1)^n u_x$ (3736).

Expand, and apply (3735) as before. Putting $x = 0$, we also have

3742 $\Delta^n u_0 = u_n - nu_{n-1} + C_{n,2}u_{n-2} \dots (-1)^n u_0$.

3743 $$\Delta^n x^m = (x+n)^m - n(x+n-1)^m + C(n, 2)(x+n-2)^m - \&c.$$

3744 $$\Delta^n 0^m = n^m - n(n-1)^m + C(n, 2)(n-2)^m - C(n, 3)(n-3)^m + \&c.$$

3745 Ex.: By (3717) $\Delta^n 0^n = n!$ Hence a proof of theorem (285) is obtained.

3746 $$\Delta^n u_x v_x = (EE' - 1)^n u_x v_x,$$

where E operates only upon u_x and E' only upon v_x.

PROOF. $\Delta u_x v_x = u_{x+1}v_{x+1} - u_x v_x = Eu_x . E'v_x - u_x v_x = (EE'-1)\,u_x v_x$.

Applications of (3746).

3747 Ex. (1): $\Delta^n u_x v_x = (-1)^n (1 - EE')^n u_x v_x$.

Expand the binomial, and operate upon the subjects u_x, v_x; thus

3748 $\Delta^n u_x v_x = (-1)^n \left\{ u_x v_x - nu_{x+1}v_{x+1} + C(n, 2)\,u_{x+2}v_{x+2} - \&c. \right\}$.

3749 Ex. (2): To expand $a^x \sin x$ by successive differences of $\sin x$.

$$\Delta^n a^x \sin x = \left\{ E(1+\Delta')-1 \right\}^n a^x \sin x = \left\{ \Delta + E\Delta' \right\}^n a^x \sin x$$

$$= \left\{ \Delta^n + n\Delta^{n-1}E\Delta' + C(n,2)\,\Delta^{n-2}E^2\Delta'^2 + \&c. \right\} a^x \sin x$$

$$= \Delta^n a^x.\sin x + n\Delta^{n-1}a^{x+1}\Delta \sin x + C(n,2)\,\Delta^{n-2}a^{x+2}\Delta^2 \sin x + \&c.$$

$$= a^x \left\{ (a-1)^n \sin x + n(a-1)^{n-1}a\Delta \sin x + C(n,2)(a-1)^{n-2}a^2\Delta^2 \sin x + \&c. \right\},$$

by (3727), while $\Delta^r \sin x$ is known from (3728).

3750 Ex. (3): To expand $\Delta^n u_x v_x$ in differences of u_x and v_x alone: put $E = 1+\Delta$, $E' = 1+\Delta'$ in (3746), thus

$$\Delta^n u_x v_x = (\Delta + \Delta' + \Delta\Delta')^n u_x v_x,$$

which must be expanded.

$\Delta^n u_x$ *in differential coefficients of* u_x.

3751 $$\Delta^n u_x = d_x^n u_x + A_1 d_x^{n+1} u_x + A_2 d_x^{n+2} u_x + \&c.$$

PROOF.— $\Delta^n u_x = (e^{d_x} - 1)^n u_x$ (3737).

Expand by (150) and (125) as if d_x were a quantitative symbol. See also (3761).

$\frac{d^n u}{dx^n}$ *in successive differences of* u.

3752 $$\frac{d^n u}{dx^n} = \{\log(1+\Delta)\}^n u.$$

The expansion by (155) and (125) will present a series of ascending differences of u.

PROOF.— $e^{d_x} = 1 + \Delta$, $\therefore\ d_x = \log(1+\Delta)$.

3753 Ex.: If $n = 1$, $\frac{du}{dx} = \Delta u - \frac{\Delta^2 u}{2} + \frac{\Delta^3 u}{3} - \frac{\Delta^4 u}{4} + \&c.$

If C be a constant,

3754 $$\phi(D)\,C = \phi(\Delta)\,C = \phi(0)\,C \quad \text{and} \quad \phi(E)\,C = \phi(1)\,C,$$

Since every term of $\phi(D)$, or of $\phi(\Delta)\,C$, operating upon C, produces 0; and every term of $\phi(E)$ operating upon C produces C.

HERSCHEL'S THEOREM.

3757 $$\phi(e^t) = \phi(E)\,e^{0.t}$$

3758 $$= \phi(1) + \phi(E)\,0.t + \phi(E)\,0^2.\frac{t^2}{1.2} + \&c.$$

PROOF.—Let $\phi(e^t) = A_0 + A_1 e^t \dots + A_n e^{nt}$

$$= A_0 e^{0.t} + A_1 E e^{0.t} \dots + A_n E^n e^{0.t} = (A_0 + A_1 E \dots + A_n E_n)\, e^{0.t}$$

$$= \phi(E)\, e^{0.t} = \phi(E) \left\{ 1 + 0.t + \frac{0^2 t^2}{1.2} + \&c. \right\},$$

and $\phi(E)\,1 = \phi(1)$ by (3756).

A THEOREM CONJUGATE TO MACLAURIN'S (1507).

3759 $\phi(t) = \phi(D)\, e^{0.t}$

3760 $$= \phi(0) + \phi(d_0)\, 0.t + \phi(d_0)\, 0^2 . \frac{t^2}{1.2} + \&c.$$

PROOF.— $\phi(t) = \phi(\log e^t) = \phi(\log E)\, e^{0.t}$ (3757)

$$= \phi(D)\, e^{0.t}\ (3738) = \phi(D) \left\{ 1 + 0.t + \ + \&c. \right\},$$

and $\phi(D)\,1 = \phi(0)$ (3754).

n being a positive integer,

3761

$$\Delta^n u = \frac{d^n u}{dx^n} + \frac{\Delta^n 0^{n+1}}{1.2 \dots (n+1)} \frac{d^{n+1} u}{dx^{n+1}} + \frac{\Delta^n 0^{n+2}}{1.2 \dots (n+2)} \frac{d^{n+2}}{dx^{n+2}} +$$

PROOF.—By (3758), putting $\phi(e^t) \equiv (e^t - 1)^n$,

$$(e^t - 1)^n = (E-1)^n 0.t + (E-1)^n 0^2 . \frac{t^2}{1.2} + \&c. = \Delta^n 0.t + \Delta^n 0^2 \frac{t^2}{1.2} + \&c.$$

Put $t = d_x$, and employ (3736) and (3737).

INTERPOLATION.

Approximate value of u$_x$ *in terms of* n *particular equidistant values.*

3762 If u_x is an integral algebraic function of the degree $n-1$, $\Delta^n u_x$ vanishes, and therefore by making $x = 0$, and writing x for n in (3740),

3763 $$u_x = u_0 + x\Delta u_0 + C_{x,2}\Delta^2 u_0 \dots + C_{x,n-1}\Delta^{n-1} u_0.$$

This is formula (265). The given values are u_0, Δu_0, $\Delta^2 u_0$, &c., corresponding to a, b, c, ...

3764 For an application of the formula to the problem of interpolation, see (267), in which example $x = 1{\cdot}54$ and $u_x = \log 72{\cdot}54$.

3765 When the term to be interpolated is one of a set of equidistant terms, employ (3741). $\Delta^n u_x = 0$, as in (3762); therefore

3766 $$u_n - nu_{n-1} + C_{n,2}u_{n-2} - C_{n,3}u_{n-3} \ldots + (-1)^n u_0 = 0.$$

3767 Ex.: From sin 0, sin 30°, sin 45°, and sin 60°, to deduce the value of sine 15°.

The formula gives $\sin 0 - 4\sin 15° + 6\sin 30° - 4\sin 45° + \sin 60° = 0$,

or $$-4\sin 15° + 3 - 2\sqrt{2} + \tfrac{1}{2}\sqrt{3} = 0,$$

from which $$\sin 15° = \tfrac{1}{8}(6 - 4\sqrt{2} + \sqrt{3}) = \cdot 2594.$$

The true value is ·2588; the error ·0006.

LAGRANGE'S INTERPOLATION FORMULA.

3768 Let $a, b, c, \ldots k$ be n values of x, not equidistant, for which the values of u_x are known; then generally

3769
$$u_x = u_a \frac{(x-b)(x-c)\ldots(x-k)}{(a-b)(a-c)\ldots(a-k)} + u_b \frac{(x-a)(x-c)\ldots(x-k)}{(b-a)(b-c)\ldots(b-k)}$$
$$\ldots\ldots\ldots\ldots + u_k \frac{(x-a)(x-b)(x-c)\ldots}{(k-a)(k-b)(k-c)\ldots}.$$

Proof.—Assume $u_x = A(x-b)(x-c)\ldots(x-k)$
$+B(x-a)(x-c)\ldots(x-k) + C(x-a)(x-b)(x-d)\ldots(x-k) + \&c.$,
and determine A, B, C, &c. by making $x = a, b, c$, &c., in turn.

If the values of $a, b, c, \ldots k$ are $0, 1, 2, \ldots n-1$, (3769) reduces to

3770 $$u_x = u_{n-1}\frac{x(x-1)\ldots(x-n+2)}{1.2.3\ldots(n-1)}$$
$$-u_{n-2}\frac{x(x-1)\ldots(x-n+3)(x-n+1)}{1.1.2.3\ldots(n-2)}$$
$$+u_{n-3}\frac{x(x-1)\ldots(x-n+4)(x-n+2)(x-n+1)}{2.1.1.2.3\ldots(n-3)} - \&c., \text{ or}$$

3771
$$u_x = \frac{x^{(n)}}{(n-1)!}\left\{\frac{u_{n-1}}{x-n+1} - \frac{C_{n-1,1}u_{n-2}}{x-n+2} + \frac{C_{n-1,2}u_{n-3}}{x-n+3} - \&c.\right\}.$$

MECHANICAL QUADRATURE.

The area of a curve whose equation is $y = u_x$ in terms of $n+1$ equidistant ordinates $u, u_1, \dots u_n$, is approximately

3772
$$nu + \frac{n^2}{2}\Delta u + \left(\frac{n^3}{3} - \frac{n^2}{2}\right)\frac{\Delta^2 u}{1.2} + \left(\frac{n^4}{4} - n^3 + n^2\right)\frac{\Delta^3 u}{3!}$$
$$+ \left(\frac{n^5}{5} - \frac{3n^4}{2} + \frac{11n^3}{3} - 3n^2\right)\frac{\Delta^4 u}{4!}$$
$$+ \left(\frac{n^6}{6} - 2n^5 + \frac{35n^4}{4} - \frac{50n^3}{3} + 12n^2\right)\frac{\Delta^5 u}{5!}$$
$$+ \left(\frac{n^7}{7} - \frac{15n^6}{6} + 17n^5 - \frac{225n^4}{4} + \frac{274n^3}{3} - 60n^2\right)\frac{\Delta^6 u}{6!}.$$

PROOF.—The area is $= \int_0^n u_x dx$. Take the value of u_x in terms of $u_0, u_1 \dots u_{n-1}$ from (3763) and integrate.

3773 When $n = 2$, $\int_0^2 u_x dx = \dfrac{u + 4u_1 + u_2}{3}$.

3774 $n = 3$, $\int_0^3 u_x dx = \dfrac{3}{8}(u + 3u_1 + 3u_2 + u_3)$,

3775 $n = 4$, $\int_0^4 u_x dx = \dfrac{14(u+u_4) + 64(u_1+u_3) + 24u_2}{45}$.

3776
$$n = 6, \quad \int_0^6 u_x dx = \frac{3}{10}\{u + u_2 + u_4 + u_6 + 5(u_1 + u_5) + 6u_3\}.$$

In the last formula, which is due to Mr. Weddle, the coefficient of $\Delta^6 u$ is taken as $\frac{3}{10}$ instead of $\frac{41}{140}$, its true value. These results are obtained from (3772) by substituting for each Δ its value from (3742).

COTES'S AND GAUSS'S FORMULÆ.

3777 These give the area of the curve directly in terms of fixed abscissæ.

They are obtained by integrating Lagrange's value of u_x (3769–71), and are fully discussed in articles (2995–7).

LAPLACE'S FORMULA.

3778 $$\int_0^n u_x dx = \frac{u_0}{2} + u_1 + u_2 \ldots + \frac{u_n}{2}$$
$$-\frac{1}{12}(\Delta u_n - \Delta u_0) + \frac{1}{24}(\Delta^2 u_n - \Delta^2 u_0) - \&c.,$$

the coefficients being those in the expansion of
$$t\left\{\log(1+t)\right\}^{-1}.$$

PROOF.— $$\Delta w_x = d_x \left\{\frac{\Delta}{\log(1+\Delta)}\right\} w_x, \quad \text{by (3736)},$$
$$= d_x \left\{1 + \frac{\Delta}{2} - \frac{\Delta^2}{12} + \frac{\Delta^3}{24} - \frac{19}{720}\Delta^4 + \&c.\right\} w_x.$$

Hence, putting $u_x = \Delta w_x$,
$$\int_0^1 u_x dx = \frac{u_0 + u_1}{2} - \frac{\Delta^2 u_0}{12} + \frac{\Delta^3 u_0}{24} - \&c.,$$
$$\int_1^2 u_x dx = \frac{u_1 + u_2}{2} - \frac{\Delta^2 u_1}{12} + \frac{\Delta^3 u_1}{24} - \&c.,$$
and so on; then add together the n equations.

3779 Formula (3778) contains Δu_n, $\Delta^2 u_n$, &c., which cannot be found from $u_0, u_1 \ldots u_n$.

The following formula does not involve differences higher than Δu_{n-1}.

3780 $$\int_0^n u_x dx = \frac{u_0}{2} + u_1 + u_2 \ldots + \frac{u_n}{2}$$
$$-\frac{1}{12}(\Delta u_{n-1} - \Delta u_0) - \frac{1}{24}(\Delta^2 u_{n-1} - \Delta^2 u_0) - \&c.$$

PROOF.—In the proof of (3778), change $\dfrac{\Delta}{\log(1+\Delta)}$ into $E\dfrac{-\Delta E^{-1}}{\log(1-\Delta E^{-1})}$, and put $E^{-1}w_x = w_{x-1}$ (3739) after expansion, and proceed as before.

SUMMATION OF SERIES.

3781 *Definition:* $\Sigma u_x = u_a + u_{a+1} + u_{a+2} \ldots + u_{x-1}$.

3782 *Theorem:* $\Sigma u_x = \Delta^{-1} u_x + C$,

where C is constant for all the assigned values of x.

PROOF.—Let $\phi(x)$ be such that $\Delta\phi(x) = u_x$, then $\phi(x) = \Delta^{-1}u_x$, therefore $u_a = \phi(a+1) - \phi(a)$. Write thus, and add together the values of $u_a, u_{a+1}, \dots u_{x-1}$. Therefore, by (3781), $\Sigma u_x = \phi(x) - \phi(a) = \Delta^{-1}u_x - \phi(a)$, and $\phi(a)$ is constant with respect to x.

Taken between the limits $x=a$, $x=b-1$, we have the notation,

3783 $$\sum_{x=a}^{x=b-1} u_x \quad \text{or} \quad \Sigma_a^{b-1} u_x = \Sigma u_b - \Sigma u_a = \Delta^{-1}u_b - \Delta^{-1}u_a.$$

Functions integrable in finite terms :

3784 *Class I.* $$\Sigma x^{(m)} = \frac{x^{(m+1)}}{m+1} + C.$$

3785 $$\Sigma (ax+b)^{(m)} = \frac{(ax+b)^{(m+1)}}{a(m+1)} + C.$$

3786 *Class II.* $$\Sigma x^{(-m)} = \frac{x^{(-m+1)}}{-m+1} + C.$$ By (3718), and notation (3711).

3787 $$\Sigma (ax+b)^{(-m)} = C - \frac{(ax+b)^{(-m+1)}}{a(m-1)}.$$

Formulæ (3785) and (3786) are equivalent to the rules (269) and (271). They are the direct results of theorem (3782).

3788 $$\Sigma a^x = \frac{a^x}{a-1}.$$ [By (3726).

Class III.—If u_x be a rational integral function,

3789 $$\Sigma_a^{a+x-1} u_x = \{x + C_{x,2}\Delta + C_{x,3}\Delta^2 + \dots\} u_a.$$

PROOF.—By (3735) and (3736),

$$u_a + u_{a+1} \dots + u_{a+x-1} = (1+E+E^2 \dots + E^{x-1})\, v_a = \frac{E^x-1}{E-1} u_a = \frac{(1+\Delta)^x - 1}{\Delta} u_a$$

= the expansion above.

3790 The formula has been given at (266) and an example of its application. The series there summed is $1+5+15+35+70+126+$ to 100 terms. The function u_x which gives rise to these terms is found by (3763) to be

$$u_x = (x^4 + 10x^3 + 35x^2 + 50x + 24) \div 24.$$

3791 If this function be presented as u_x, and $\Sigma_0^{x-1} u_x$ be required, we first find $u_0=1$, $u_1=5$, $u_2=15$, &c.; then the differences $\Delta u_0, \Delta^2 u_0, \ldots \Delta^4 u_4 = 1, 4, 6, 4, 1$, and then, by (3789), the required sum, as in the example referred to.

3792 For another example, let $\Sigma_1^n x^3 = 1+2^3 \ldots +n^3$ be required.

Here $\Delta x^3 = 3x^2+3x+1, \quad \Delta^2 x^3 = 6x+6, \quad \Delta^3 x^3 = 6,$

therefore $\Delta 0^3 = 1, \quad \Delta^2 0^3 = 6, \quad \Delta^3 0^3 = 6 \ldots\ldots\ldots (1),$

x^3 may now be expressed in factorials, and the summation may then be effected by (3784). First, by (3730),

$$x^3 = x+3x(x-1)+x(x-1)(x-2);$$

therefore, by (3784), $\Sigma_1^n x^3 = \Sigma (n+1)^3$ (3783),

$$\Sigma_1^n x^3 = \frac{n(n+1)}{2} + \frac{3(n+1)n(n-1)}{3} + \frac{(n+1)n(n-1)(n-2)}{4} = \frac{n^2(n+1)^2}{4}.$$

3793 Otherwise, by (3789), taking $a=0$, we have

$u_x = x^3, \quad u_0 = 0, \quad \Delta u_0 = 1, \quad \Delta^2 u_0 = 6, \quad \Delta^3 u_0 = 6,$ as above.

Therefore

$$\Sigma_0^{n-1} u_x = \frac{n(n-1)}{2} + \frac{6n(n-1)(n-2)}{1.2.3} + \frac{6n(n-1)(n-2)(n-3)}{1.2.3.4} = \frac{n^2(n-1)^2}{4},$$

therefore, changing n into $n+1$, $\Sigma_0^n u_x = \dfrac{(n+1)^2 n^2}{4}$.

3794 *Class IV.*—When the general term of a series is a rational fraction of the form

$$\frac{A+Bx+Cx^2+}{u_x u_{x+1} \ldots u_{x+m}}, \quad \text{where } u_x = ax+b,$$

and the degree of the numerator is not higher than $x+m-2$; resolve the numerator into

$$A'+B'u_x+C'u_x u_{x+1}+\ldots+D'u_x u_{x+1} u_{x+2} \ldots u_{x+m-2},$$

by (3730). The fraction then separates into a series of fractions with constant numerators which can be summed by (3787).

3795 If the factors $u_x \ldots u_{x+m}$ are not consecutive, introduce the missing ones in the denominator and numerator, and then resolve the fraction as in the foregoing rule.

3796 Ex.: To sum the series $\dfrac{1}{1.4} + \dfrac{1}{2.5} + \dfrac{1}{3.6} +$ to n terms.

The nth term is $\dfrac{1}{n(n+3)} = \dfrac{(n+1)(n+2)}{n(n+1)(n+2)(n+3)}$

$$= \frac{n(n+1)+2n+2}{n(n+1)(n+2)(n+3)} = \frac{1}{(n+2)(n+3)} + \frac{2}{(n+1)(n+2)(n+3)} + \frac{2}{n(n+1)(n+2)(n+3)}.$$

The sum of n terms is, therefore, by the rule (271),

$$\left(\frac{1}{3}-\frac{1}{n+3}\right)+\frac{1}{2}\left(\frac{2}{2\;3}-\frac{2}{(n+2)(n+3)}\right)+\frac{1}{3}\left(\frac{2}{2.3}-\frac{2}{(n+1)(n+2)(n+3)}\right)$$
$$= \frac{11}{18}-\frac{3n^2+12n+11}{3(n+1)(n+2)(n+3)}.$$

If the form in (3787) is used, the total constant part C is determined finally by making $n=0$, which gives $C=\frac{11}{18}$.

3797 *Theorem.* $\quad f(E)\,a^x\phi(x) = a^x f(aE)\,\phi(x)$,

f being an algebraic function.

Proof.—Let $a = e^m$, then the left

$$= f(E)\,e^{mx}\phi(x) = f(e^D)\,e^{mx}\phi(x) \;\; (3736) \;\; = e^{mx}f(e^{D+m})\,\phi(x) \;\; (3475)$$
$$= a^x f(aE)\,\phi(x).$$

Class V.—If $\phi(x)$ be a rational integral function,

3798

$$\Sigma a^x\phi(x) = \frac{a^x}{a-1}\left\{\phi(x)-\frac{a}{a-1}\Delta\phi(x)+\frac{a^2}{(a-1)^2}\Delta^2\phi(x)-\&c.\right\}.$$

The upper limit is understood to be $x-1$, and a constant is to be added, (3781–2).

Proof.— $\Sigma a^x\phi(x) = \Delta^{-1}a^x\phi(x) = (E-1)^{-1}a^x\phi(x) = a^x(aE-1)^{-1}\phi(x)$

$$(3797) \;\; = a^x\left\{a(1+\Delta)-1\right\}^{-1}\phi(x) = \frac{a^x}{a-1}\left(1+\frac{a\Delta}{a-1}\right)^{-1}\phi(x).$$

Then expand the binomial.

$\Sigma a^x\phi(\text{x})$ *in successive derivatives of* $\phi(\text{x})$.

3799 $\quad \Sigma a^x\phi(x) = \dfrac{a^x}{a-1}\left\{1+A_1\phi'(x)+\dfrac{A_2}{1.2}\phi''(x)+\&c.\right\},$

where $\quad A_n = \left(\dfrac{aE-1}{a-1}\right)^{-1}0^n = \left(1+\dfrac{a\Delta}{a-1}\right)^{-1}0^n.$

Proof.—By (3757), $\psi(e^D) = \psi(E)\,e^{0.D}$; therefore (see last proof) $a^x(aE-1)^{-1}\phi(x)$ (putting $E=e^D$) $= a^x(aE-1)^{-1}e^{0.D}\phi(x)$

$$= \frac{a^x}{a-1}\left(1+\frac{a\Delta}{a-1}\right)^{-1}\left\{1+0.D+\frac{0^2}{1.2}D^2+\&c.\right\}\phi(x).$$

3800 Ex.: To sum the series $2.1+4.8+8.27+16.64+$ to n terms.

We require $2^x x^3+\Sigma 2^x x^3 = 2^x x^3+2^x(x^3-2\Delta x^3+2^2\Delta^2x^3-2^3\Delta^3x^3)$

$$= 2^x x^3+2^x\left\{x^3-2(3x^2+3x+1)+4(6x+6)-8.6\right\}$$
$$= 2^x\left\{2x^3-6x^2+18x-26\right\}.$$

3801 If $\Delta^{-n}u_x$ be known for all integral values of n, and if v_x be rational and integral,

$$\Sigma u_xv_x = \Sigma^1 u_x.v_{x-1}-\Sigma^2 u_x.\Delta v_{x-2}+\Sigma^3 u_x.\Delta^2 v_{x-3}-\&c.$$

Proof. $\Sigma u_xv_x = (EE'-1)^{-1}u_xv_x = (\Delta E'+\Delta')^{-1}u_xv_x$ (3746) and (3736). Expand the binomial operator, observing (3738).

3802

$$\Sigma^n u_xv_x = u_x\Sigma^n v_x - n\Delta u_x\Sigma^{n+1}v_{x+1}+C_{n,2}\Delta^2u_x\Sigma^{n+2}v_{x+2}-\&c.$$

Proof. $\Sigma^n u_xv_x = (\Delta'+\Delta E')^{-n}u_xv_x$, as in (3801),

$$= \Delta^{-n}v_xu_x - n(\Delta^{-n-1}E)v_x\Delta u_x + C_{n,2}(\Delta^{-n-2}E^2)v_x\Delta^2u_x - \&c.,$$

producing the above by (3735) and (3782).

Observe that, in (3801) and (3802), two forms are obtainable in each case by expanding the binomial operator from either end of the series.

3803 Ex.: To sum the series $\sin a+2^2\sin 2a+3^2\sin 3a+$ to x terms.

The sum is $= x^2\sin ax+\Sigma x^2\sin ax$. Taking $u_x=\sin ax$ and $v_x = x^2$, we know $\Delta^{-n}\sin ax$, by (3729); therefore (3801) gives

$$\Sigma x^2\sin ax = (2\sin\tfrac{1}{2}a)^{-1}\sin\left\{ax-\tfrac{1}{2}(a+\pi)\right\}(x-1)^2$$
$$-(2\sin\tfrac{1}{2}a)^{-2}\sin\left\{ax-(a+\pi)\right\}(2x-3)+(2\sin\tfrac{1}{2}a)^{-3}\sin\left\{ax-\tfrac{3}{2}(a+\pi)\right\}2.$$

APPROXIMATE SUMMATION.

3820 The most useful formula is the following

$$\Sigma u_x = C+\int u_xdx-\frac{u_x}{2}+\frac{B_2}{2\,!}\frac{du_x}{dx}-\frac{B_4}{4\,!}\frac{d^3u_x}{dx^3}+\&c.$$

$$= C+\int u_xdx-\frac{u_x}{2}+\frac{1}{12}\frac{du_x}{dx}-\frac{1}{720}\frac{d^3u_x}{dx^3}+\frac{1}{30240}\frac{d^5u_x}{dx^5}$$
$$-\frac{1}{1209600}\frac{d^7u_x}{dx^7}+\frac{1}{47900160}\frac{d^9u^x}{dx^9}-\&c.$$

Proof.— $\Sigma u_x = (e^D-1)^{-1}u_x$. Expand by (1539) with D in the place of x.

Ex. 1: The value of Σx^p at (2939) is given at once by the formula.

3821 Ex. 2: To sum the series $1+\frac{1}{2}+\frac{1}{3}\,...+\frac{1}{x}$ approximately,

$$\frac{1}{x}+\Sigma\frac{1}{x}=\frac{1}{x}+C+\log x-\frac{1}{2x}-\frac{1}{12x^2}+\frac{1}{120x^4}-\&c.$$

Put $x=10$ to determine the constant; thus

$$1+\frac{1}{2}+\frac{1}{3}\,...+\frac{1}{10}=C+\log 10+\frac{1}{20}-\frac{1}{1200}+\&c.,$$

from which $C=\cdot 577215$, and the required sum is

$$\cdot 577215+\log x+\frac{1}{2x}-\frac{1}{12x^2}+\frac{1}{120x^4}-\&c.$$

3822 Ex. 3: $$1+\frac{1}{2^3}+\frac{1}{3^3}+\frac{1}{4^3}+\&c.,$$

$$\Sigma\frac{1}{x^3}=C-\frac{1}{2x^2}-\frac{1}{2x^3}-\frac{3B_2}{2x^4}+\frac{5B_4}{2x^6}-\frac{7B_6}{2x^8}+\frac{9B_8}{2x^{10}},$$

$$\therefore\quad \Sigma_1^\infty\frac{1}{x^3}=\frac{1}{2}+\frac{1}{2}+\frac{1}{4}-\frac{1}{12}+\frac{1}{12}-\frac{3}{20}+\frac{5}{12}.$$

The convergent part of this series, consisting of the first five terms, is an approximation to the sum of all the terms.

3823 A much nearer approximation is obtained in this and analogous cases by starting with the summation formula at a more advanced term.

E.g.: $$1+\frac{1}{2^3}+\frac{1}{3^3}+\frac{1}{4^3}+\Sigma_5^\infty\frac{1}{x^3}$$

$$=\frac{2035}{1728}+\frac{1}{2.5^2}+\frac{1}{2.5^3}+\frac{3B_2}{2.5^4}-\frac{5B_4}{2.5^6}+\&c.$$

$$=\frac{2035}{1728}+\frac{1}{50}+\frac{1}{250}+\frac{1}{2500}-\frac{1}{187500}+\&c.$$

The converging part now consists of a far greater number of terms than before, and the convergence at first is much more rapid.

3824 Ex. 4: The series for $\log\Gamma(x+1)$ at (2773) can be obtained by the above formula when x is an integer. For, in that case,

$$\log\Gamma(x+1)=\log 1+\log 2+\log 3\,...+\log x=\log x+\Sigma\log x,$$

and (3820) gives the expansion in question, the constant being determined by making x infinite.

3825 Formula (3820) may also be used to find $\int u_x dx$ by the process of summation, and thus answers the purpose of Laplace's formula (3778).

$\Sigma^n u_x$ *in a series of derivatives of* u_x.

3826 *Lemma.*—

$$\overline{n-1}!\,(e^t-1)^{-n} = (-1)^{n-1}(d_t+n-1)(d_t+n-2)\ldots(d_t+1)\{e^t-1\}^{-1}.$$

PROOF.—Put v_n for $\overline{n-1}!\,(e^t-1)^{-n}$. Then

$$v_{n+1} = -(d_t+n)\,v_n = (d_t+n)(d_t+n-1)\,v_{n-1}.$$

$\Sigma^n u_x$ may now be developed.

3827 Ex.—To develope $\Sigma^3 u_x$, (*Boole*, p. 97)

$$2\,(e^t-1)^{-3} = (d_t+2)(d_t+1)\left\{e^t-1\right\}^{-1}$$

$$= (d_{2t}+3d_t+2)\left\{\frac{1}{t}-\frac{1}{2}+A_1t+A_2t^2+\&c.\right\}$$

with $A_{2r}=0$, and

$$A_{2r+1} = (-1)^r B_{2r+1} \div (2r+2)\,! = \frac{2}{t^3}-\frac{3}{t^2}+\frac{2}{t}+(2A_2+3A_1-1)$$

$$+\Sigma_1^\infty\left\{(r+2)(r+1)\,A_{r+2}+3\,(r+1)\,A_{r+1}+2A_r\right\}t^r.$$

Therefore, changing t into d_x, we get

$$\Sigma^3 u_x = \iiint u_x dx - \frac{3}{2}\iint u_x dx + \int u_x dx - \frac{3}{8}\,u_x + \frac{19}{240}\,\frac{du_x}{dx} - \&c.$$

3828 $\Sigma^n u_x$ *in a series of derivatives of* $u_{x-\frac{n}{2}}$.

Let $x^n \operatorname{cosec}^n x = 1 - C_2x^2 + C_4x^4 - \&c.$, then

$$\Sigma^n u_x = D^{-n}\left\{1+C_2\left(\frac{D}{2}\right)^2+C_4\left(\frac{D}{2}\right)^4+\&c.\right\}u_{x-\frac{n}{2}}.$$

[*Boole*, p. 98.

3829

$$\phi(0)-\phi(1)+\phi(2)-\&c. = \frac{1}{2}\left\{1-\frac{\Delta}{2}+\frac{\Delta^2}{4}-\&c.\right\}\phi(0).$$

By this formula, a series of the given type may often be transformed into one much more convergent.

PROOF.—The left $= \dfrac{1}{1+E}\,\phi(0) = \dfrac{1}{2+\Delta}\,\phi(0) = \dfrac{1}{2}\,\dfrac{1}{1+\frac{1}{2}\Delta}\,\phi(0)$, the expansion of which is the series on the right.

3830 Ex.—To sum $1-\frac{1}{2}+\frac{1}{3}-\frac{1}{4}+\&c.$ Summing the first six terms, it becomes $\frac{37}{60}+\frac{1}{7}-\frac{1}{8}+\&c.$ Taking $\phi(0)=(0+7)^{-1}$,

$$\frac{1}{7}-\frac{1}{8}+\&c. = \frac{1}{2}\left\{\frac{1}{7}+\frac{1}{2.7.8}+\frac{2}{4.7.8.9}+\frac{2.3}{8.7.8.9.10}+\&c.\right\}.$$

The sum after six terms converges rapidly by this formula, and more rapidly than if the formula had been applied to the series from its commencement.

PLANE COORDINATE GEOMETRY.

SYSTEMS OF COORDINATES.

CARTESIAN COORDINATES.

4001 In this system (Fig. 1)* the position of a point P in a plane is determined by its distances from two fixed straight lines OX, OY, called axes of coordinates. These distances are measured parallel to the axes. They are the abscissa PM or ON denoted by x, and the ordinate PN denoted by y. The axes may be rectangular or oblique. The abscissa x is reckoned positive or negative according to the position of P to the right or left of the y axis, and the ordinate y is positive or negative according as P lies above or below the x axis conformably to the rules (607, '8).

4002 These coordinates are called *rectangular* or *oblique* according as the axes of reference are or are not at right angles.

POLAR COORDINATES.

4003 The polar coordinates of P (Fig. 1) are r, the radius vector, and θ, the inclination of r to OX, the initial line, measured as in Plane Trigonometry (609).

4004 To change rectangular into polar coordinates, employ the equations $x = r\cos\theta$, $y = r\sin\theta$.

4005 To change polar into rectangular coordinates, employ

$$r = \sqrt{x^2+y^2}, \qquad \theta = \tan^{-1}\left(\frac{y}{x}\right).$$

* See the end of the volume.

TRILINEAR COORDINATES.

4006 The trilinear coordinates of a point P (Fig. 2) are α, β, γ, its perpendicular distances from three fixed lines which form the *triangle of reference*, ABC, hereafter called the *trigon*. These coordinates are always connected by the relation

4007 $$\mathfrak{a}\alpha+\mathfrak{b}\beta+\mathfrak{c}\gamma = \Sigma,$$

4008 or $$\alpha \sin A+\beta \sin B+\gamma \sin C = \text{constant},$$

where $\mathfrak{a}, \mathfrak{b}, \mathfrak{c}$ are the sides of the trigon, and Σ is twice its area.

4009 If x, y are the Cartesian coordinates of the point $\alpha\beta\gamma$, the equations connecting them with the trilinear coordinates are, by (4094),

$$\alpha = x \cos \alpha+y \sin \alpha-p_1,$$
$$\beta = x \cos \beta+y \sin \beta-p_2,$$
$$\gamma = x \cos \gamma+y \sin \gamma-p_3.$$

4010 Here α has two significations. On the left, it is the length of the perpendicular from the point in question upon the side AB of the trigon. On the right, it is the inclination of that perpendicular to the x axis of Cartesian coordinates. Similarly β and γ.

4011 The angles α, β, γ are connected with the angles A, B, C by the equations

$$\gamma-\beta = \pi-A, \qquad \alpha-\gamma = \pi-B, \qquad \alpha-\beta = \pi+C,$$

only two of which are independent.

4012 p_1, p_2, p_3 are the perpendiculars from the origin upon the sides of the triangle ABC.

AREAL COORDINATES.

If A, B, C (Fig. 2) be the trigon as before, the areal coordinates α', β', γ' of the point P are

4013 $$\alpha' = \frac{\alpha}{p_1} = \frac{PBC}{ABC}, \quad \beta' = \frac{\beta}{p_2} = \frac{PCA}{ABC}, \quad \gamma' = \frac{\gamma}{p_3} = \frac{PAB}{ABC}.$$

The equation connecting the coordinates is now

4014 $$\alpha'+\beta'+\gamma'=1.$$

4015 To convert any homogeneous trilinear equation into the corresponding areal equation.

4016 Substitute $\mathfrak{a}\alpha=\Sigma\alpha',\quad \mathfrak{b}\beta=\Sigma\beta',\quad \mathfrak{c}\gamma=\Sigma\gamma'.$

Also any relation between the coefficients l, m, n in the equation of a right line in trilinears will be adapted to areals by substituting $l\mathfrak{a}$, $m\mathfrak{b}$, $n\mathfrak{c}$ for l, m, n. Similarly for a, b, c, f, g, h, in the general equation of a conic (4656), substitute $a\mathfrak{a}^2$, $b\mathfrak{b}^2$, $c\mathfrak{c}^2$, $f\mathfrak{b}\mathfrak{c}$, $g\mathfrak{c}\mathfrak{a}$, $h\mathfrak{a}\mathfrak{b}$.

In either the trilinear or areal systems, a point is determined if the ratios only of the coordinates are known.

Thus, if $\alpha:\beta:\gamma=P:Q:R$, then, with trilinear coordinates,

4017 $$\alpha=\frac{P\Sigma}{\mathfrak{a}P+\mathfrak{b}Q+\mathfrak{c}R};\quad \text{and, with areal,}\quad \alpha=\frac{P}{P+Q+R}.$$

TANGENTIAL COORDINATES.

4019 In this system the position of a straight line is determined by coordinates, and the position of a point by an equation. If $l\alpha+m\beta+n\gamma=0$ be the trilinear equation of a straight line EDF (Fig. 3); then, making α, β, γ constant, and l, m, n variable, the equation becomes the tangential equation of the point O (α,β,γ); whilst l, m, n are the coordinates of some right line passing through that point.

Let λ, μ, ν (Fig. 3) be the perpendiculars from A, B, C upon EDF, and let p_1, p_2, p_3 be the perpendiculars from A, B, C upon the opposite sides of the trigon; then, by (4624), we have

4020 $$R\lambda=lp_1,\qquad R\mu=mp_2,\qquad R\nu=np_3,$$

where $R=\sqrt{(l^2+m^2+n^2-2mn\cos A-2nl\cos B-2lm\cos C)}$.

Hence the equation of the point O becomes

4021
$$\lambda\frac{\alpha}{p_1}+\mu\frac{\beta}{p_2}+\nu\frac{\gamma}{p_3}=0\quad\text{or}\quad \lambda\frac{\sin\theta_1}{\rho_1}+\mu\frac{\sin\theta_2}{\rho_2}+\nu\frac{\sin\theta_3}{\rho_3}=0,$$

where $\rho_1=OA$, $\theta_1=\angle BOC$, &c., and $2\Delta BOC=\rho_2\rho_3\sin\theta$.

Formula (4021) shows that, when the perpendiculars λ, μ, ν are taken for the coordinates of the line, the coefficients become the areal coordinates of the point referred to the same trigon.

4023 Any homogeneous equation in l, m, n as tangential coordinates is expressed in terms of λ, μ, ν by substituting for $l, m, n, \frac{\lambda}{p_1}, \frac{\mu}{p_2}, \frac{\nu}{p_3}$ respectively. By (4020).

4024 An equation in λ, μ, ν of a degree higher than the first represents a curve such that λ, μ, ν are always the perpendiculars upon the tangent. The *curve* must therefore be the envelope of the line (λ, μ, ν).

TWO-POINT INTERCEPT COORDINATES.

Let $\lambda = AD$, $\mu = BE$ (Fig. 4) be variable distances from two fixed points A, B measured along two fixed parallel lines, then

4025 $$a\lambda + b\mu + c = 0$$

is the equation of a fixed point O through which the line DE always passes. This may easily be proved directly, but we shall show that it is a particular case of the system of three-point tangential coordinates.

Let one of the vertices (C) of the trigon in that system be at infinity (Fig. 3). Then equation (4022) becomes

$$\frac{\lambda \sin\theta_1}{\rho_1} + \frac{\mu \sin\theta_2}{\rho_2} + \sin COE \sin\theta_3 = 0.$$

For $\nu : \rho_3 = \sin COE$ always. Divide by $\sin COE$; then $\lambda \div \sin COE = AD$, &c., and the equation becomes

$$\frac{\sin\theta_1}{\rho_1} AD + \frac{\sin\theta_2}{\rho_2} AE + \sin\theta_3 = 0.$$

The only variables are AD and AE. Calling these λ and μ, the equation may be written $$a\lambda + b\mu + c = 0,$$
the form taken by $a\lambda + b\mu + c'\nu = 0$ when $\nu = \infty$ and c' vanishes.

ONE-POINT INTERCEPT COORDINATES.

4026 Let a, b be the Cartesian coordinates of the point O (Fig. 5); and let the reciprocals of the intercepts on the axes of any line DOE passing through O be $\xi = \frac{1}{AE}$, $\eta = \frac{1}{AD}$. Then, by (4053),

4027 $$a\xi + b\eta = 1$$

is the equation of the point O, the variables being ξ, η.

This is a case of the system of three-point tangential coordinates in which two of the vertices (B, C) of the trigon are at infinity. Equation (4022)

now becomes $$\frac{\lambda \sin\theta_1}{\rho_1} + \sin BOD \sin\theta_2 + \sin COE \sin\theta_3 = 0,$$

or $$\frac{\sin\theta_1}{\rho_1} + \frac{\sin\theta_2}{AD} + \frac{\sin\theta_3}{AE} = 0,$$

which is of the form $a\xi + b\eta = 1$.

TANGENTIAL RECTANGULAR COORDINATES.

4028 This name has been given to the system last described when the two fixed lines are at right angles (Fig. 6).

The coordinates ξ, η, which are defined as the reciprocals of the intercepts of the line they determine, have now also the following values.

4029 Let x, y be the rectangular coordinates of the pole of the line in question with respect to a circle whose centre is the origin and whose radius is k; then

$$\xi = \frac{x}{k^2} \quad \text{and} \quad \eta = \frac{y}{k^2},$$

since $x.OM = y.ON = k^2$; for M, N are the poles of $y = 0$, $x = 0$.

4030 The equation of a point P on NM whose rectangular coordinates are $OR = a$, $OS = b$, is

$$a\xi + b\eta = 1, \text{ by (4053)},$$

this equation being satisfied by the coordinates of all lines passing through that point.

4031 In all these systems an equation of a higher degree in ξ, η represents a curve the coordinates of whose tangents satisfy the equation.

ANALYTICAL CONICS

IN

CARTESIAN COORDINATES.

LENGTHS AND AREAS.

Coordinates of the point dividing in the ratio $n : n'$ the right line which joins the two points xy, $x'y'$.

4032 $$\xi = \frac{nx'+n'x}{n+n'}, \qquad \eta = \frac{ny'+n'y}{n+n'}.$$

PROOF.—(Fig. 7.) $\xi = x + AC = x + \frac{n}{n+n'}(x'-x)$. Similarly for η.

4033 If $n = n'$, $\quad \xi = \frac{x+x'}{2}, \qquad \eta = \frac{y+y'}{2}.$

4034 Length of the line joining the points xy, $x'y'$

$$= \sqrt{(x-x')^2+(y-y')^2}.$$

The same with oblique axes

4035 $$\sqrt{(x-x')^2+(y-y')^2+2(x-x')(y-y')\cos\omega}.$$

PROOF.—By (Fig. 7), Euc. I. 47, and (702).

Area A of a triangle in terms of the coordinates of its angular points x_1y_1, x_2y_2, x_3y_3.

4036 $$A = \tfrac{1}{2}\left\{x_1y_2 - x_2y_1 + x_2y_3 - x_3y_2 + x_3y_1 - x_1y_3\right\}.$$

PROOF.—(Fig. 8.) By considering the three trapezoids formed by y_1, y_2, y_3 and the sides of the triangle, we have

$$A = \tfrac{1}{2}(y_1+y_2)(x_2-x_1) + \tfrac{1}{2}(y_2+y_3)(x_3-x_2) - \tfrac{1}{2}(y_3+y_1)(x_3-x_1).$$

Area of the triangle contained by the x axis and the lines

$$y = m_1x + c_1, \qquad y = m_2x + c_2, \tag{4052}$$

4037 $$A = \frac{(c_1-c_2)^2}{2(m_1-m_2)} = \frac{(B_1C_2-B_2C_1)^2}{2B_1B_2(A_1B_2-A_2B_1)}. \tag{4056}$$

PROOF.—(Fig. 9.) Area $= \frac{1}{2}(c_1-c_2)p$, and p is found from $pm_2 - pm_1 = c_1 - c_2$. The sign of the area is not regarded.

COR.—Area of the triangle contained by the lines

$$y = m_1x + c_1, \qquad y = m_2x + c_2, \qquad y = m_3x + c_3,$$

4038 $$A = \tfrac{1}{2}\left\{\frac{(c_1-c_2)^2}{m_1-m_2} + \frac{(c_2-c_3)^2}{m_2-m_3} + \frac{(c_3-c_1)^2}{m_3-m_1}\right\}.$$

4039 $$= \frac{\{c_1(m_2-m_3)+c_2(m_3-m_1)+c_3(m_1-m_2)\}^2}{2(m_1-m_2)(m_2-m_3)(m_3-m_1)}.$$

4040 $$= \frac{(B_1C_2-B_2C_1)^2}{2B_1B_2(A_1B_2-A_2B_1)} + \&c.$$

4041 $$= \frac{\text{Square of Determinant } (A_1B_2C_3)}{2(A_1B_2-A_2B_3)(A_2B_3-A_3B_2)(A_3B_1-A_1B_3)}.$$

PROOF.—(Fig. 10.) $ABC = AEF + CDE - BED$. Employ (4037).

Area of Polygon of **n** *sides.*

First in terms of the coordinates of the angular points x_1y_1, x_2y_2, ... x_ny_n.

4042 $$2A = (x_1y_2-x_2y_1)+(x_2y_3-x_3y_2)+\ldots+(x_ny_1-x_1y_n)$$
$$= x_1(y_2-y_n)+x_2(y_3-y_1)+\ldots\ldots+x_n(y_1-y_{n-1}).$$

Secondly, when the equations to the sides are given, as in (4037).

4043 $$2A = \frac{(c_1-c_2)^2}{m_1-m_2} + \frac{(c_2-c_3)^2}{m_2-m_3} + \ldots + \frac{(c_n-c_1)^2}{m_n-m_1}.$$

4044 Also three values similar to (4039, '40, '41).

PROOF.—By (4367), adding the component triangles.

4047 Each expression for the area of a triangle or polygon will be adapted to oblique axes by multiplying by $\sin\omega$.

TRANSFORMATION OF COORDINATES.

4048 To transform the origin to the point hk.

Put $x = x'+h, \qquad y = y'+k.$

To transform to rectangular axes inclined at an angle θ to the original axes.

4049 Put

$$x = x'\cos\theta - y'\sin\theta, \quad y = y'\cos\theta + x'\sin\theta. \quad \text{(Fig. 11.)}$$

PROOF.—Consider x' as $\cos\phi$ and y' as $\sin\phi$. Then $x = \cos(\phi+\theta)$ and $y = \sin(\phi+\theta)$ (627, '9).

Generally (Fig. 12), let ω be the angle between the original axes; and let the new axes of x and y make angles α and β respectively with the old axis of x.

4050 Put $x\sin\omega = x'\sin(\omega-\alpha)+y'\sin(\omega-\beta)$

and $y\sin\omega = x'\sin\alpha + y'\sin\beta.$

PROOF.—(Fig. 12.) The coordinates of P referred to the old axes being $OC = x$, $PC = y$, and referred to the new axes, $OM = x'$, $PM = y'$, we have, by projecting OCP and OMP at right angles first to CP and then to OC,

$$CD = MF - ME, \qquad PN = ML + PK,$$

which are equivalent to the above equations.

To change Rectangular coordinates into Polar, hk being the pole O, α the inclination of the initial line to the x axis (Fig. 13), and xy the point P.

4051 Put $x = h + r\cos(\theta+\alpha), \quad y = k + r\sin(\theta+\alpha).$

THE RIGHT LINE.

EQUATIONS OF THE RIGHT LINE.

4052 $y = mx + c$ ....(1),

4053 $\dfrac{x}{a} + \dfrac{y}{b} = 1$ ....(2),

4054 $$x\cos\alpha+y\sin\alpha = p \quad\ldots\ldots\ldots\ldots\ldots\ldots (3).$$

4055 $$Ax+By+C=0 \quad\ldots\ldots\ldots\ldots\ldots\ldots (4).$$

PROOF.—(Fig. 14.) Let AB be the line. Take any point P upon it, coordinates $ON=x$, $PN=y$. Then, in (1), $m=\tan\theta$, where $\theta=BAX$, the inclination to the X axis; therefore $mx=-OC$, and c is the intercept OB. In (2), a, b are the intercepts OA, OB. In (3), $p=OS$, the perpendicular from O upon the line; $\alpha=\angle AOS$.

$$p = OR+LP = x\cos\alpha+y\sin\alpha.$$

(4) is the general equation.

4056 $$m=\tan\theta=-\frac{A}{B}=-\frac{b}{a}=-\cot\alpha.$$

4060 $$\sin\theta=\frac{A}{\sqrt{A^2+B^2}},\qquad \cos\theta=-\frac{B}{\sqrt{A^2+B^2}}.$$

4062 $$p=c\sin\alpha=\frac{c}{\sqrt{1+m^2}}=-\frac{C}{\sqrt{A^2+B^2}}.$$

Oblique Axes.

Equations (4052, '53, '55) hold for oblique axes, but (4054) must be written

4065 $$x\cos\alpha+y\cos\beta=p. \qquad \text{(Fig. 14)}$$

4066 $$\tan\theta=\frac{m\sin\omega}{1+m\cos\omega}=\frac{A\sin\omega}{A\cos\omega-B},$$

ω being the angle between the axes.

PROOF.—From $m=\sin\theta\div\sin(\omega-\theta)$.

4068 $$p=\frac{c\sin\omega}{\sqrt{1+2m\cos\omega+m^2}}=\frac{C\sin\omega}{\sqrt{A^2+B^2-2AB\cos\omega}}.$$

PROOF.—From $p=c\sin(\omega-\theta)$ and (4066).

The equations of two lines being given in the forms (4052) or (4055), the angle, ϕ, between them is given by

4070 $$\tan\phi=\frac{m-m'}{1+mm'} \quad\text{or}\quad \frac{AB'-A'B}{AA'+BB'}.$$

PROOF.—(Fig. 15) $\tan\phi=\tan(\theta-\theta')$. Expand by (632).

To oblique axes:

4072 $$\tan\phi = \frac{(m-m')\sin\omega}{1+(m+m')\cos\omega+mm'}.$$

PROOF.—As in the last, employing (4066).

Equation of a line passing through $x'y'$:

4073 $$y-y' = m(x-x'),$$ (Fig. 8)

4074 or $$y-mx = y'-mx',$$

4075 or $$Ax+By = Ax'+By'.$$

PROOF.—From Figure (13), m being $=\tan\theta$.

Condition of parallelism of two lines:

4076 $$m = m', \quad \text{or} \quad AB' = A'B.$$

Hence the equations differ by a constant.

Condition of perpendicularity:

4078 $$mm' = -1 \quad \text{or} \quad AA'+BB' = 0.$$ (4070)

The same to oblique axes:

4080 $$1+(m+m')\cos\omega+mm' = 0.$$ (4072)

4081 or $$AA'+BB' = (AB'+A'B)\cos\omega,$$

4082 or $$m' = -\frac{1+m\cos\omega}{m+\cos\omega}.$$

A line passing through the points x_1y_1, x_2y_2:

4083 $$\frac{y-y_1}{x-x_1} = \frac{y_1-y_2}{x_1-x_2} = m.$$

PROOF.—(Fig. 16.) By the similar right-angled triangles PCA, ADB.

4084 Or $$y = mx+\frac{x_1y_2-x_2y_1}{x_1-x_2},$$

4085 or $$(x-x_1)(y-y_2) = (x-x_2)(y-y_1).$$

PROOF.—This equation represents a straight line because it is of the first degree; and the coordinates of each of the given points satisfy the equation.

A line passing through $x'y'$ and perpendicular to a given line (m):

4086 $$y-y' = -\frac{1}{m}(x-x').$$ (4073, '78)

4087 or $$Bx-Ay = Bx'-Ay'.$$

The two lines passing through $x'y'$ and making an angle $\beta\ (=\tan^{-1}m_2)$ with a given line (m_1):

4088 $$\frac{y-y'}{x-x'} = \frac{m_1-m_2}{1+m_1m_2} \quad \text{and} \quad \frac{m_1+m_2}{1-m_1m_2}.$$ (4073, '70)

A line passing through hk and dividing the line which joins x_1y_1 and x_2y_2 in the ratio $n_1 : n_2$:

4089 $$\frac{y-k}{x-h} = \frac{n_1(y_2-k)+n_2(y_1-k)}{n_1(x_2-h)+n_2(x_1-h)}.$$ (4073, '32)

Coordinates of the point of intersection of two lines:

4090 $$x = \frac{c_1-c_2}{m_2-m_1} = \frac{B_1C_2-B_2C_1}{A_1B_2-A_2B_1}.$$

4092 $$y = \frac{c_1m_2-c_2m_1}{m_2-m_1} = -\frac{A_1C_2-A_2C_1}{A_1B_2-A_2B_1}.$$ (4116)

Length of the perpendicular from a point $x'y'$ upon a given line

4094 $$= x'\cos\alpha + y'\sin\alpha - p.$$

PROOF.—Let AB (Fig. 14) be the line, and Q the point $x'y'$. Then, by (4054), $x'\cos\alpha+y'\sin\alpha = OT$, the perpendicular from O upon a parallel line through Q, and $p = OS$.

Otherwise, the same perpendicular

4095 $$= \frac{Ax'+By'+C}{\sqrt{A^2+B^2}}.$$ (4060, '61, '94)

The same with oblique axes

4096 $$= \frac{(Ax'+By'+C)\sin\omega}{\sqrt{(A^2+B^2-2AB\cos\omega)}},$$

obtained in a similar way from (4065–69).

Condition of three lines intersecting in one point:

4097 $$c_1m_2-c_2m_1+c_2m_3-c_3m_2+c_3m_1-c_1m_3=0.$$

The area in 1009 must vanish.

4098 *Otherwise.*—If certain values of the constants l, m, n make the expression

$$l(A_1x+B_1y+C_1)+m(A_2x+B_2y+C_2)+n(A_3x+B_3y+C_3)$$

vanish identically, the three lines indicated intersect in one point.

PROOF.—By (4099), for then values of x and y which make (1) and (2) vanish also make (3) vanish.

A line passing through the point of intersection of the lines $Ax+By+C=0$ and $A'x+B'y+C'=0$ is

4099 $$Ax+By+C=k(A'x+B'y+C'),$$

4100 or $$l(Ax+By+C)-m(A'x+B'y+C')=0,$$

k, l, and m being any constants.

4101 RULE.—*If the equation of a right line contains a third variable* k *in the first degree, the line always passes through a fixed point.*

PROOF.—For the values of x and y, which satisfy simultaneously the given equations, also satisfy (4099), whatever k may be. See (4604).

4102 If in the equation of a line $Ax+By+C=0$, the coefficients A, B, C involve x', y', the coordinates of a point which moves along a fixed right line, then the first line passes through some fixed point.

PROOF.—By means of the equation of the fixed line, y' may be eliminated, and x' then remains a third variable in the first degree (4101).

4103 To find the point in which the line $Ax+By+C$ intersects the line joining the points xy, $x'y'$; substitute

$$Ax+By+C \text{ for } n, \text{ and } Ax'+By'+C \text{ for } n' \text{ in (4032).}$$

PROOF.—By (4095), since the segments intercepted are in the ratio of the perpendiculars from xy, $x'y'$ upon the line $Ax+By+C$.

Equations of the line with l, m for direction-ratios, hk a fixed point on the line, and r the distance of the variable point xy from hk.

4104 $$r = \frac{x-h}{l} = \frac{y-k}{m},$$

4105 where $l = \frac{\sin(\omega-\theta)}{\sin\omega}$, $m = \frac{\sin\theta}{\sin\omega}$, (Oblique)

4106 or $l = \cos\theta$, $m = \sin\theta$. (Rectangular)

Polar Equation of a Straight Line.

4107 $$r\cos(\theta-\alpha) = p.$$

(Fig. 17.) Here p is the perpendicular to the line from the pole O, and α is the inclination of p to the initial line OA.

When the line passes through the pole, the equation is

4108 $$\theta = \text{constant}.$$

A line passing through the two points $r_1\theta_1$, $r_2\theta_2$.

4109 $$rr_1\sin(\theta-\theta_1)+r_1r_2\sin(\theta_1-\theta_2)+r_2r\sin(\theta_2-\theta) = 0.$$

Proof.—(Fig. 18.) $\Delta POA+AOB-POB = 0$. Then by (707).

EQUATIONS OF TWO OR MORE RIGHT LINES.

The homogeneous equation of the nth degree,

4110 $$x^n+p_1x^{n-1}y+p_2x^{n-2}y^2+\ldots+p_ny^n = 0,$$

represents n right lines, real or imaginary, passing through the origin.

For it is resolvable into n factors of the form $(x-ay)$, by (405).

For the case of two right lines represented by the general equation of the second degree, see (4469).

Equation of two right lines through the origin:

4111 $$ax^2+2hxy+by^2 = 0.$$

If ϕ be the angle between the lines,

4112 $$\tan\phi = \frac{\sqrt{(h^2-ab)}}{a+b} \quad \text{or} \quad \frac{2\sin\omega\sqrt{(h^2-ab)}}{a+b-2h\cos\omega},$$

according as the axes are rectangular or oblique.

PROOF.—Assume $(y-m_1x)(y-m_2x)=0$, and apply (4088).

Equation of the bisectors of the angle ϕ:

4113 $$hx^2-(a-b)\,xy-hy^2=0.$$

PROOF.—Let $y=\mu x$ be a bisector ($\mu=\tan\psi$); then, since $2\psi=\theta_1+\theta_2$,

$$\frac{2\mu}{1-\mu^2}=\frac{m_1+m_2}{1-m_1m_2}=\frac{2h}{a-b}, \text{ by (4111); and } \mu=\frac{y}{x}.$$

The roots of this equation are always real.

GENERAL METHODS.

APPLICABLE TO ALL EQUATIONS OF PLANE CURVES.

4114 Let $F(x,y)=0$ ......(i.) and $f(x,y)=0$ ......(ii.) be the equations of two curves of any degree.

4115 To find the intercepts on the x and y axes.

Put $y=0$ *in* (i.), *then* x *becomes the intercept on the* x *axis. Similarly, put* $x=0$ *for the intercept on the* y *axis.*

4116 To find the points of intersection of (i.) and (ii.).

Solve as simultaneous equations. Each pair of values of x *and* y *so obtained gives a point of intersection. Imaginary values give an imaginary point.*

4117 To determine equation (i.) so that the line may pass through certain fixed points, x_1y_1, x_2y_2, &c.

Substitute x_1y_1, x_2y_2, *&c. for* xy *successively, so forming as many equations as there are points. From these equations the* constants *in* (i.) *must be determined in terms of* x_1, y_1, x_2, y_2, *&c.*

4118 *The number of arbitrary points cannot exceed the number of constants in the equation.*

4119 Condition that (i.) and (ii.) may touch.

At a point of contact two or more points of intersection must coincide, and therefore the equation for x *or* y, *obtained as in* (4116), *must have two or more equal roots for each point of contact. The contact is said to be of the second order when there are three coincident points; of the third order when there are four, and so on.*

4120 To find the equation of the tangent at a point $x'y'$ on the curve $f(x, y) = 0$.

Form the equation to the secant through two adjacent points x_1y_1, x_2y_2 (4083), *and determine the limiting value of* $\frac{y_1-y_2}{x_1-x_2}$ *when the points coincide by means of the equations* $f(x_1, y_1) = 0$, $f(x_2, y_2) = 0$.

4121 *Otherwise* $\qquad m = \frac{dy}{dx},$ *by* (5101).

4122 For the equation of the normal, change m of the tangent into $-\frac{1}{m}$ (4086).

4123 To express the equation of the tangent, or normal, in terms of m and the constants of the curve.

From the equation of the tangent or normal, the equation to the curve, and the equation furnished by the value of m, *eliminate* x' y', *the coordinates of the point of contact of the tangent.*

THEORY OF POLES AND POLARS.

4124 Let $F(x', y', x, y) = 0$ represent the equation to the tangent of a curve at the point $x'y'$.

Then $F(x, y, x', y') = 0$, the equation obtained by interchanging the constants x', y' with the variables x, y, represents the polar of any fixed point $x'y'$ not on the curve.

Let x_1y_1, x_2y_2 (Fig. 19) be points A, B on the curve, and let the tangents at those points intersect in $x'y'$. Consider the equations

$F(x_1, y_1, x, y) = 0 \dots (1)$, $\quad F(x_2, y_2, x, y) = 0 \dots (2)$, $\quad F(x, y, x', y') = 0 \dots (3)$.

Here (1), (2) are the tangents, and (3) is *some straight line or curve* according to the dimensions of x and y. Also (3) passes through the points of contact x_1y_1, x_2y_2, and may therefore be called the *curve of contact;* or, if a right line, the *chord of contact* of tangents drawn from x', y', *i.e.*, the polar.

4125 Hence the coordinates of the points of contact of tangents from an external point $x'y'$ will be determined by solving (3) and the equation of the curve simultaneously.

4126 Again, let $x'y'$ (Figs. 20 and 21) be any point P not on the curve. Then, from the equations

$$F(x', y', x, y) = 0 \ldots (4), \quad F(x, y, x_3, y_3) = 0 \ldots (5), \quad F(x, y, x_4, y_4) = 0 \ldots (6),$$

we see that (4) is some straight line; that, if x_3y_3 and x_4y_4 are any two points upon it, (5) and (6) are the curves of contact of tangents from those points; and that these curves of contact pass through the point $x'y'$.

4127 If the points x_3y_3, x_4y_4 are taken at A, B, where (4) intersects the curve, (5) and (6) then become curves touching the given curve at A and B, and passing through $x'y'$. We may call these lines the *curve tangents* from $x'y'$.

4128 Lastly, let $x'y'$ in (3) be a point within the given curve (Fig. 22), then the equations

$$F(x, y, x', y') = 0 \ldots (7), \quad F(x_3, y_3, x, y) = 0 \ldots (8), \quad F(x_4, y_4, x, y) = 0 \ldots (9)$$

show that (7) is the locus of a point, the curve tangents from which have their chord of contact always passing through a fixed point. When $x'y'$ is without the curve, as in Fig. (19), the same definition applies to every part of the locus (3) from which tangents can be drawn.

4129 If the given curve be of a degree higher than the second, the line of contact of the tangents from a point is a curve, and the line of contact of the curve tangents from a point is a straight line (Figs. 19 and 20). A similar converse relation is exhibited in Figures (21) and (22).

If the curve be of the second degree, equations (3) and (4) become identical. The line of contact or the polar is always in this case a straight line, and so is the locus (7):

Figures (19) and (20) now become identical, as also (21) and (22).

4130 The polar of the point of intersection of two right lines with regard to a conic passes through their poles.

PROOF.—As in (4124). Let (1) and (2) be the two lines, (x_1y_1), (x_2y_2) their poles, and $x'y'$ their point of intersection.

4131 To find the ratio in which the line joining two given points xy, $x'y'$ is cut by the curve $f(\text{x}, \text{y}) = 0$.

Substitute for x *and* y, *the supposed coordinates of the point of intersection, the values*

$$\frac{nx'+n'x}{n+n'}, \quad \frac{ny'+n'y}{n+n'}, \qquad \text{by (4032)}$$

and determine the ratio n : n′ *from the resulting equation. The real roots of this equation correspond to the real points of intersection.*

4132 To form the equation of all the tangents that can be drawn to the curve from a point $x'y'$.

Express the condition for equal roots of the equation in (4131), *and consider xy a variable point.*

4133 To form the equation of the lines drawn from $x'y'$ to all the points of intersection of two curves.

Substitute $nx'+n'x$, $ny'+n'y$ for x *and* y *in both curves, and eliminate the ratio $n : n'$.*

Proof.—Take any other point xy on the line through $x'y'$ and a point of intersection. The ratio $n : n'$ (4131) is the same for each curve, and therefore may be eliminated.

4134 To find the length, $r = AP$ or AP' (Fig. 23), of the segment intercepted between the point A or $x'y'$ and the curve $f(x, y) = 0$ on a straight line drawn from A at an inclination θ to the X axis. That is, to form the polar equation with $x'y'$ for the pole and the initial line parallel to the x axis.

Substitute for x *and* y, *the assumed coordinates of the point of intersection, the values* x = ON *or* ON′, y = PN *or* P′N′,

that is, $\qquad \text{x} = x'+r\cos\theta, \qquad \text{y} = y'+r\sin\theta,$

and determine r from the resulting equation. That is, put $a = 0$ *in* (4051).

The real values of r are the distances of the points of intersection from $x'y'$.

4135 When an equation has been obtained for determining x the length of a line, important results may frequently be arrived at by applying theorem (406) respecting the sum and product of the roots.

THE CIRCLE.

Equation with the centre for origin.

4136 $$x^2+y^2 = r^2.$$ (Fig. 24.)

Equations of the tangent at the point P or $x'y'$.

4137 $$y-y' = -\frac{x'}{y'}(x-x').$$ (4120)

4138 $$xx'+yy' = r^2.$$

Also, by (4124), the polar of $x'y'$, any point not on the curve.

4139 $$y = mx+r\sqrt{1+m^2}; \qquad m = -\frac{x'}{y'}.$$ (4123)

4140 $$x\cos\alpha+y\sin\alpha = r,$$

α being the inclination to the x axis of the radius to the point $x'y'$.

Equation of the circle with a, b for the coordinates of the centre Q. (Fig. 24.)

4141 $$(x-a)^2+(y-b)^2 = r^2.$$

Tangent at $x'y'$, or Polar,

4142 $$(x-a)(x'-a)+(y-b)(y'-b) = r^2,$$ (4138)

4143 or $$(x-a)\cos\alpha+(y-b)\sin\alpha = r,$$

α being the inclination of the radius to the point $x'y'$.

General equation of the circle :

4144 $$x^2+y^2+2gx+2fy+c = 0.$$

4145 Centre $(-g,\ -f)$. Radius $\sqrt{(g^2+f^2-c)}$.

PROOF.—By equating coefficients with (4141).

Equation of the circle with oblique axes : (Fig. 25.)

4146 $$(x-a)^2+(y-b)^2+2(x-a)(y-b)\cos\omega = r^2,$$ (702)

4147 or $$x^2+2xy\cos\omega+y^2-2(a+b\cos\omega)x \\ -2(b+a\cos\omega)y \\ +a^2+2ab\cos\omega+b^2 = r^2.$$

General Equation.

4148 $$x^2+2xy\cos\omega+y^2+2gx+2fy+c=0.$$

The coordinates of the centre are

4149 $$a=\frac{f\cos\omega-g}{\sin^2\omega},\qquad b=\frac{g\cos\omega-f}{\sin^2\omega}.$$

4150 $$\text{Radius}=\frac{\sqrt{\{g^2-2fg\cos\omega+f^2-c\sin^2\omega\}}}{\sin\omega}.$$

PROOF.—By equating coefficients with (4147).

Polar Equation.

4151 $$r^2+l^2-2rl\cos(\theta-\alpha)=c^2,$$ (Fig. 26)

4152 or $$r^2-2l\cos\alpha\, r\cos\theta-2l\sin\alpha\, r\sin\theta+l^2-c^2=0.$$

PROOF.—By (702), the coordinates of P being r and θ.

General form of the polar equation:—

4153 $$r^2+2gr\cos\theta+2fr\sin\theta+c=0.$$

4154 $$\tan\alpha=\frac{f}{g}.\qquad l=\sqrt{g^2+f^2}.$$

PROOF.—By equating coefficients with (4152).

4156 Equation of the circle passing through the three points x_1y_1, x_2y_2, x_3y_3.

$$(x^2+y^2)\begin{vmatrix}x_1 & y_1 & 1\\ x_2 & y_2 & 1\\ x_3 & y_3 & 1\end{vmatrix}-(x_1^2+y_1^2)\begin{vmatrix}x_2 & y_2 & 1\\ x_3 & y_3 & 1\\ x & y & 1\end{vmatrix}+(x_2^2+y_2^2)\begin{vmatrix}x_3 & y_3 & 1\\ x & y & 1\\ x_1 & y_1 & 1\end{vmatrix}-(x_3^2+y_3^2)\begin{vmatrix}x & y & 1\\ x_1 & y_1 & 1\\ x_2 & y_2 & 1\end{vmatrix}$$

PROOF.—Eliminate g, f, and c from (4144) by (4117).

Equation of the chord joining x_1y_1, x_2y_2, two points on the circle $x^2+y^2=r^2$:

4157 $$x(x_1+x_2)+y(y_1+y_2)=x_1x_2+y_1y_2+r^2,$$ (4083, 4136)

4158 or $$x\cos\tfrac{1}{2}(\theta_1+\theta_2)+y\sin\tfrac{1}{2}(\theta_1+\theta_2)=r\cos\tfrac{1}{2}(\theta_1-\theta_2),$$

where $\quad r\cos\theta_1=x_1,\qquad r\sin\theta_1=y_1,\quad$ &c.

4159 NOTE.—The coordinates x, y of a point on the circle $x^2+y^2=r^2$ may often be expressed advantageously in this way in terms of θ, a single variable.

4160 Let $$S=(x-a)^2+(y-b)^2-r^2=0$$ be any circle (Fig. 27). Then, if xy be a point P outside the circle, S becomes the square of the tangent from P. If xy be a point P' within the circle, S becomes minus the square of the ordinate drawn through P' at right angles to the radius through P'.

CO-AXAL CIRCLES.

(See also 984 and 1021.)

4161 If $$S=x^2+y^2+2gx+2fy+c=0,$$ $$S'=x^2+y^2+2g'x+2f'y+c'=0$$ be two circles, the equation to the radical axis is $$S-S'=0.$$

If $x=0$ be taken for the radical axis, the equation to any circle (radius r) of the system of coaxal circles (1021) is

4162 $$x^2+y^2-2kx\pm\delta^2=0 \quad\text{and}\quad k^2-r^2=\pm\delta^2,$$ + in Figure (1), − in Figure (2). Here $\delta=IR$ a constant, and $k=IO$ a variable.

4164 The polar of $x'y'$ for any circle of the system passes through the intersection of $$xx'+yy'\pm\delta^2=0 \quad\text{and}\quad x+x'=0.$$

PROOF.—Its equation is $xx'+yy'-k(x+x')\pm\delta^2=0$ (4121). Then by (4099).

4165 When $k=\delta$, then $k=ID=ID'$. D and D' are Poncelet's limiting points.

4166 The polar of D with respect to any of the circles passes through D', and *vice versâ*, by (4164).

4167 Tangents from any point on the radical axis to all circles of the system are equal (4160, '61).

4168 The radical axes of three circles, S_1, S_2, S_3, meet in a point called their radical centre.

4169 The reciprocals with respect to the origin D or D' of the system of co-axal circles are all confocal conics (4558).

The equation of the circle, centre Q, cutting the system of circles orthogonally is, putting $IQ = h$,

4170 $$x^2+y^2-2hy-\delta^2 = 0. \qquad (1230, 1236)$$

This circle passes through D and D'.

The common tangents to the two circles

$$(x-a)^2+(y-b)^2 = r^2 \quad \text{and} \quad (x-a')^2+(y-b')^2 = r'^2.$$

(See also 1037.)

The equation for α in (4143) is

4171 $$(a-a')\cos\alpha+(b-b')\sin\alpha+r\mp r' = 0.$$

Proof.—Assume (4143) in a, b, r, α, and also in a', b', r', α' as coinciding lines. Then $\tan\alpha = \tan\alpha'$; therefore $\alpha' = \alpha$ or $\pi+\alpha$. Take the difference of the two equations.

The chords of contact are

4172 $$(a-a')(x-a)+(b-b\,)(y-b\,)+r(r\mp r') = 0,$$

4173 $$(a-a')(x-a')+(b-b')(y-b')+r(r\mp r') = 0,$$

with $-$ for exterior tangents, $+$ for transverse.

Proof.—For these are straight lines, and they pass through the points of contact of each pair of tangents respectively, by (4171).

The centres of similitude O, Q are the intersections of the external and transverse tangents respectively.

4174 Coordinates of O, $$\frac{a'r-ar'}{r-r'}, \quad \frac{b'r-br'}{r-r'}.$$

4175 Coordinates of Q, $$\frac{a'r+ar'}{r+r'}, \quad \frac{b'r+br'}{r+r'}.$$

Proof.—By equating coefficients in (4172) and (4142), the polar of O or Q.

4176 The six centres of similitude of three circles lie on four straight lines called axes of similitude. See the figure of (1046).

PROOF.—The coordinates of the three centres of the forms (4174, '75) will in each case satisfy equation (4083).

4177 The equation of the external axis of similitude is, in determinant notation (554),

$$(1r_2b_3)\,x-(1r_2a_3)\,y+(r_1b_2a_3)=0.$$

PROOF.—By forming the equation of the right line passing through two of the centres of similitude whose coordinates are as in (4174).

4178 The remaining three axes are found by changing in turn the signs of r_1, r_2; r_2, r_3; r_3, r_4.

4179 If one of the circles touches the other two, one axis of similitude passes through the points of contact.

4180 The angle θ, at which the circle $F(x, y)=0$, radius r (Fig. 29), intersects the circle whose centre is hk, and radius R is given by the equation

$$R^2-2Rr\cos\theta=F(h, k).$$

PROOF: $\theta=OQP$, $R^2-2Rr\cos\theta+r^2=PT^2+r^2=F(h, k)+r^2$, (702) and (4160).

4181 COR. 1.—If the circles are given by the equations

$$x^2+y^2+2g'x+2f'y+c'=0, \quad x^2+y^2+2gx+2fy+c=0,$$

the equation for $\cos\theta$ becomes, since $h=-g$, $k=-f$,

$$2Rr'\cos\theta=2gg'+2ff'-c-c'. \qquad (4145)$$

4182 COR. 2.—The condition that the two circles may cut orthogonally is

$$2gg'+2ff'-c-c'=0.$$

4183 COR. 3.—By solving three such equations, we can find the circle cutting three given circles orthogonally (4186).

4184 COR. 4.—The condition that four circles may have a common orthogonal circle is the determinant equation

$$\begin{vmatrix} c & g & f & 1 \\ c_1 & g_1 & f_1 & 1 \\ c_2 & g_2 & f_2 & 1 \\ c_3 & g_3 & f_3 & 1 \end{vmatrix}=0.$$

4185 Cor. 5.—If the circle $x^2+y^2+2Gx+2Fy+C=0$ cuts three other circles at the same angle θ, we have, by (4081), three equations to determine G, F, C. The resulting determinant equation may be written

$$\begin{vmatrix} x^2+y^2 & -x & -y & 1 \\ c_1 & g_1 & f_1 & 1 \\ c_2 & g_2 & f_2 & 1 \\ c_3 & g_3 & f_3 & 1 \end{vmatrix} + 2R\cos\theta \begin{vmatrix} 0 & -x & -y & 1 \\ r_1 & g_1 & f_1 & 1 \\ r_2 & g_2 & f_2 & 1 \\ r_3 & g_3 & f_3 & 1 \end{vmatrix} = 0.$$

4186 The first determinant, put $=0$, is the orthogonal circle (4183), and the second, expanded, is the axis of similitude.

4187 The locus of the centre of a circle cutting three given circles at equal angles is a perpendicular from their radical centre on any of the four axes of similitude.

Proof.—By eliminating R and $\cos\alpha$ between three equations, like (4180).

4188 Each of these four perpendiculars contains the centres of two circles touching the three given circles.

Proof.—Consider $\alpha=0$ or $180°$, in (4180).

To draw the eight circles which touch three given circles, see (946) and (1049).

4189 The equation of the fourth degree of two of the touching circles is

$$\overline{23}\sqrt{S_1}\pm\overline{31}\sqrt{S_2}\pm\overline{12}\sqrt{S_3}=0,$$

where $\overline{23}$ signifies the length of the common tangent of the second and third circles, &c.

Proof.—By first showing that, if four circles are all touched by another circle, the relation

4190 $$\overline{12}.\overline{34}\pm\overline{14}.\overline{23}\pm\overline{31}.\overline{24}=0$$

will subsist, and then supposing the fourth circle to reduce to a point.

THE PARABOLA.

4200 Def.—A conic is the locus of a point which moves in one plane so that its distance from a fixed point S, the *focus*, is in a constant ratio (e) to its distance from a fixed right line XM (the *directrix*).

When $e =$ unity, the curve is a parabola. (See also p. 248, *et seq.*)

Equation of the Parabola with origin of coordinates at the vertex A.

4201 $y^2 = 4ax.$

Here $a = AS,\ x = AN,$
$y = PN.$

Proof.—Geometrically, at (1229). Analytically, from
$PS^2 = y^2+(x-a)^2 = PM^2 = (x+a)^2.$

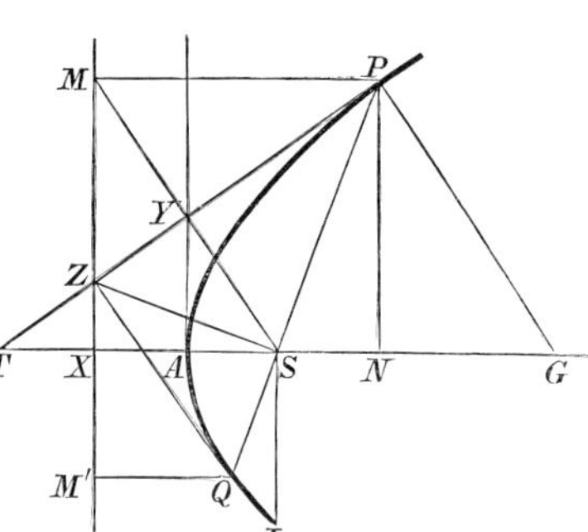

The equations with the origin at S and X respectively are

4202 $y^2 = 4a(x+a),$
$y^2 = 4a(x-a).$ (4048)

Equations of the tangent at $x'y'$:

4204 $$y-y' = \frac{2a}{y'}(x-x').$$ (4120)

4205 $$yy' = 2a(x+x').$$

4206 $$y = mx+\frac{a}{m},\quad m = \frac{2a}{y'}.$$ (4123)

4207 (4204) is also the polar of any point $x'y'$, by (4124). Its intercepts are $-x'$ and $\frac{1}{2}y'$.

Equations of the normal at $x'y'$:

4208 $$y-y' = -\frac{y'}{2a}(x-x').$$ (4122)

4209 $$y'x+2ay-(y'x'+2ay') = 0.$$

4210 $$y = mx-2am-am^3.$$ (4123)

Equation of the parabola with a diameter and tangent for axes of coordinates.

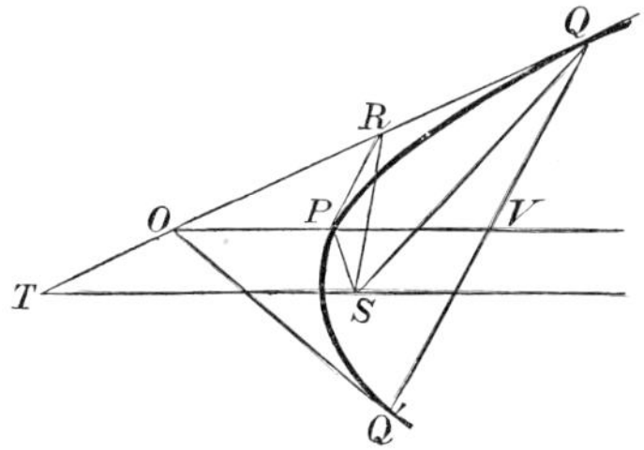

4211 $$y^2 = 4a'x,$$

where

4212 $$a' = a\,\text{cosec}^2\,\theta = SP; \qquad x = PV; \qquad y = QV.$$

PROOF.—Geometrically, at (1239). Otherwise, let $VQ = -VQ'$ be equal roots of opposite signs of the quadratic (4221), V being the point $x'y'$, therefore y^2 or $r^2 = (y'^2 - 4ax')\,\text{cosec}^2\,\theta = 4a\,\text{cosec}^2\,\theta . x$, since $y'^2 = 4a \times$ abscissa of P.

4213 Equations (4204–10) hold good for these axes, with a' written for a in each.

For the polar equation of the parabola, see (4336).

4214 Quadratic for $n_1 : n_2$, the ratio of the segments into which the line joining two given points x_1y_1, x_2y_2 is divided by the parabola $y^2 - 4ax = 0$,

$$n_1^2\,(y_2^2 - 4ax_2) + 2n_1n_2\,\{y_1y_2 - 2a\,(x_1 + x_2)\} + n_2^2\,(y_1^2 - 4ax_1) = 0.$$
(4131)

Equation of a pair of tangents from any point $x'y'$:

4215 $$(y'^2 - 4ax')(y^2 - 4ax) = \{yy' - 2a\,(x + x')\}^2 = 0.$$

The condition for equal roots in (4214).

Quadratics for the coordinates of the points of contact of tangents from $x'y'$:

4216 $$ax^2 - (y'^2 - 2ax')\,x - ax'^2 = 0.$$

4217 $$y^2 - 2yy' + 4ax' = 0.$$

PROOF.—Solve simultaneously the equations of the curve and the polar (4205) and (4125).

Coordinates of the point of intersection of tangents at x_1y_1 and x_2y_2:

4218 $$x = \frac{y_1y_2}{4a}, \qquad y = \frac{y_1 + y_2}{2}.$$

Quadratic for m of the tangent from $x'y'$:

4220 $$m^2x'-my'+a=0.$$ (4206)

4221 General polar equation of the parabola, or quadratic for r, the segment intercepted between a point, $x'y'$, and the curve on a line drawn from that point at an inclination θ to the x axis (4134),

$$r^2\sin^2\theta+2r(y'\sin\theta-2a\cos\theta)+y'^2-4ax'=0.$$

Quadratics for the coordinates of the points of intersection of the line $Ax+By+C$ and the parabola $y^2=4ax$: (4116)

4222 $$A^2x^2-2(2B^2a-AC)x+C^2=0.$$

4223 $$Ay^2+4Bay+4Ca=0.$$

Length of intercepted chord,

4224 $$4\sqrt{\{(B^2a^2-ACa)(A^2+B^2)\}}\div A^2.$$ (4034)

Equation of the secant through x_1y_1, x_2y_2, two points on the parabola:

4225 $$y(y_1+y_2)=y_1y_2+4ax,$$ (4083)

4226 or $$y(m_1+m_2)=2a+2m_1m_2x.$$

4227 The subtangent $NT=2x$. Fig. of (4201)

4228 The subnormal $NG=2a$.

PROOF.—Put $y=0$ in (4205) and (4208).

4229 The tangent $PT^2=4x(a+x)$.

4230 The normal $PG^2=4a(a+x)$.

The perpendicular p from the focus upon the tangent at xy:

4231 $$p=\sqrt{a(x+a)}=\sqrt{aa'}.$$ (4212), (4095)

The part of the normal intercepted by the curve is equal to

4233 $$\frac{4a(1+m^2)^{\frac{3}{2}}}{m^2} = \frac{4a}{\sin^2\theta\cos\theta}.$$ (4221), (4135)

4234 The minimum normal $= 6a\sqrt{3}$ and $m = \sqrt{2}$.

Length of a chord through the focus

4235 $$= \frac{4a}{\sin^2\theta} = 4a'.$$ (4212)

Coordinates of its extremities, with the focus for origin:

4237 $$x = -\frac{2a\cos\theta}{\cos\theta\mp 1}, \qquad y = -\frac{2a\sin\theta}{\cos\theta\mp 1}.$$

Coordinates of its centre:

4239 $$x = \frac{2a\cos^2\theta}{\sin^2\theta}, \qquad y = 2a\cot\theta.$$

THE ELLIPSE AND HYPERBOLA.

(See also p. 233, *et seq.*)

4250 Referring to the definition (4200); when e is less than unity, the conic is an *ellipse;* when greater than unity, an *hyperbola.*

Equation of the ellipse with the origin of coordinates at X and $SX = p$.

4251 $$y^2+(x-p)^2 = e^2x^2.$$

PROOF.—By the definition in (4200).

Abscissæ of vertices: (Supply A' in the following figure.)

4252 $$XA = \frac{p}{1+e}, \qquad XA' = \frac{p}{1-e}.$$ (4115)

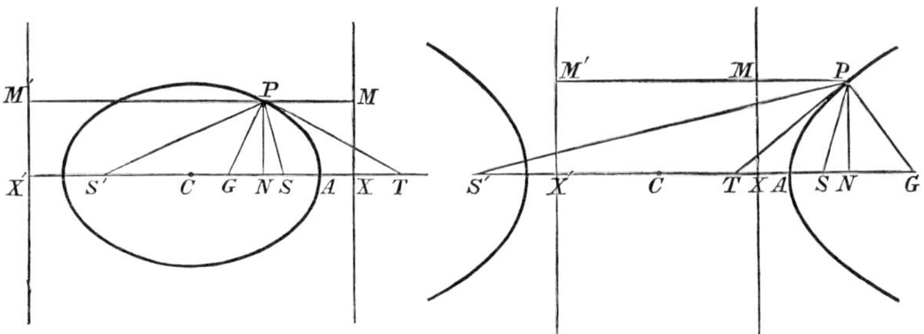

4254 Focal distances of vertices:

$$SA = \frac{ep}{1+e}, \qquad SA' = \frac{ep}{1-e}. \qquad (4251)$$

4256 $$SL \equiv l = ep = a(1-e^2) = \frac{b^2}{a}. \qquad (4251)$$

4260 $$b^2 = a^2(1-e^2); \qquad e^2 = \frac{a^2-b^2}{a^2}.$$

4262 $$CX = \frac{p}{1-e^2} = \frac{a}{e}.$$

4264 $$CA = \frac{ep}{1-e^2} = a.$$

4266 $$CS = \frac{e^2p}{1-e^2} = ae. \qquad (4252)$$

4268 If $b = a\tan\alpha$, then $e = \sec\alpha$ in the hyperbola.

Equation with the origin at A:

4269 $$y^2 = \frac{b^2}{a^2}(2ax-x^2) \qquad \text{Ell.}$$

4270 $$y^2 = \frac{b^2}{a^2}(2ax+x^2) \qquad \text{Hyp.}$$

Proof.—By (4200), $y^2+(x-SA)^2 = e^2(x+AX)^2$, &c.

Equations with the origin at the centre C:

4271 $$y^2 = \frac{b^2}{a^2}(a^2-x^2) = (1-e^2)(a^2-x^2). \qquad (4269)$$

4273 $$\frac{x^2}{a^2}+\frac{y^2}{b^2}=1.$$

PROOF.—By (4200), $y^2+(x+CS)^2=e^2(x+CX)^2$, &c.

4274 $$PN:QN::b:a.$$ (4271)

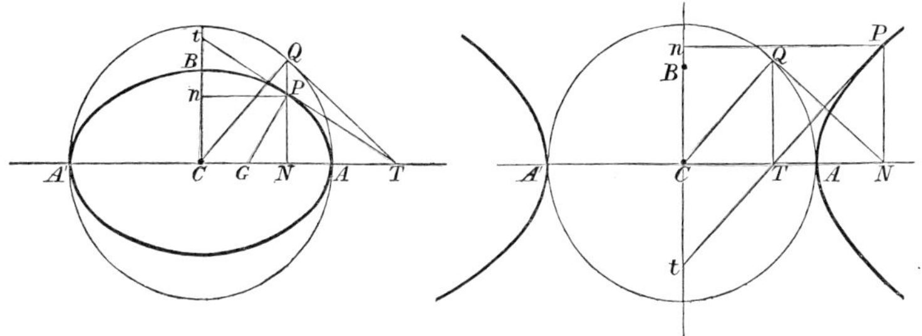

4275 DEF.—QCN is the eccentric angle, ϕ, of the point P.

x and y in terms of the eccentric angle :

4276 $x=a\cos\phi,\qquad y=b\sin\phi.$ (Ell.)

4278 $x=a\sec\phi,\qquad y=b\tan\phi.$ (Hyp.)

Five forms of the equation of the tangent or polar of the point $x'y'$:

4280 $$y-y'=-\frac{b^2x'}{a^2y'}(x-x').$$ (4120)

4281 $$\frac{xx'}{a^2}+\frac{yy'}{b^2}=1.$$

4282 $$y=mx+\sqrt{a^2m^2+b^2}.$$ (4123)

4283 $$\frac{x\cos\phi}{a}+\frac{y\sin\phi}{b}=1.\quad\text{(Ell.)}$$ (4276)

4284 $$\frac{x\sec\phi}{a}-\frac{y\tan\phi}{b}=1.\quad\text{(Hyp.)}$$ (4278)

4285 $$x\cos\gamma+y\sin\gamma=\sqrt{a^2\cos^2\gamma+b^2\sin^2\gamma},$$
γ being the inclination of p. (4054) & (4372)

Five forms of the equation of the normal at $x'y'$:

4286 $$y-y' = \frac{a^2y'}{b^2x'}(x-x').$$ (4122)

4287 $$\frac{a^2x}{x'}-\frac{b^2y}{y'} = a^2-b^2, \quad \text{or} \quad hx-ky = a^2-b^2,$$

where h and k are the intercepts of the tangent.

4289 $$y = mx-\frac{m(a^2-b^2)}{\sqrt{a^2+b^2m^2}}.$$ (4123)

4290 $$ax\sec\phi-by\operatorname{cosec}\phi = a^2-b^2.$$ (4276)

4291 $$x_1x-y_1y = (x_1x'-y_1y'),$$ (4352)

where x_1y_1 is the extremity of the conjugate diameter.

Intercepts of the tangent or polar on the axes:

4292 $$\frac{a^2}{x} \quad \text{and} \quad \frac{b^2}{y}.$$ (4115), (4281)

Intercepts of the normal: (4287)

4294 On the x axis, $\frac{a^2-b^2}{a^2}x$ or e^2x.

4296 On the y axis, $-\frac{a^2-b^2}{b^2}y$ or $-\frac{e^2}{1-e^2}y$.

Focal distances r, r' of a point xy on the curve:

4298 $(a \pm ex)$ in Ell.

4299 $(ex \pm a)$ in Hyp.

Proof.—From $r^2 = (ae \pm x)^2+y^2$, and (4272).

Perpendiculars from the foci upon the tangent:

4300 $$p = b\sqrt{\frac{r}{r'}}, \quad p' = b\sqrt{\frac{r'}{r}}.$$ (4095, 4282)

4302 $\therefore$ (p. 588) $\sin SPT = \dfrac{p}{r} = \dfrac{p'}{r'} = \dfrac{b}{\sqrt{rr'}} = \dfrac{b}{b'}.$ (4365)

4306 $$b^2 = pp'.$$ (4300)

Segments of tangent and normal:

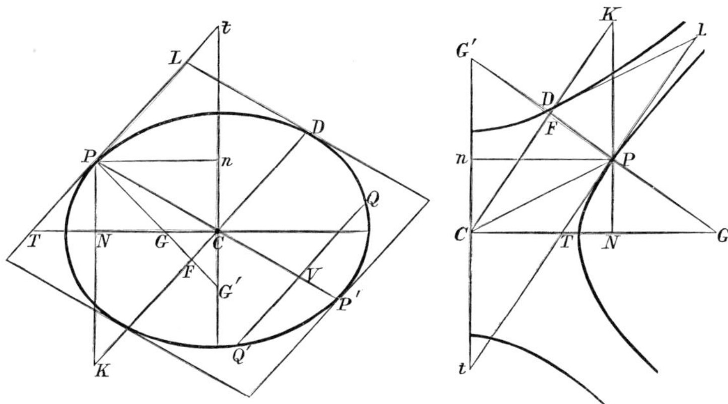

4307 $PT = \dfrac{ay}{bx}\sqrt{rr'},\qquad Pt = \dfrac{bx}{ay}\sqrt{rr'}.$ (4292)

4309 $PG = \dfrac{b}{a}\sqrt{rr'} = \dfrac{bb_1}{a},\qquad PG' = \dfrac{a}{b}\sqrt{rr'} = \dfrac{ab_1}{b}.$ (4294)

Right Line and Ellipse.

Quadratic for the ratio $n_1 : n_2$, in which the line joining two given points x_1y_1, x_2y_2 is cut by the ellipse (4131).

4310
$$\frac{n_1^2}{n_2^2}\left(\frac{x_2^2}{a^2}+\frac{y_2^2}{b^2}-1\right)+2\frac{n_1}{n_2}\left(\frac{x_1x_2}{a^2}+\frac{y_2y_1}{b^2}-1\right)+\left(\frac{x_1^2}{a^2}+\frac{y_1^2}{b^2}-1\right)=0.$$

Equation of the two tangents drawn from $x'y'$:

4311 $$\left(\frac{x'^2}{a^2}+\frac{y'^2}{b^2}-1\right)\left(\frac{x^2}{a^2}+\frac{y^2}{b^2}-1\right)-\left(\frac{xx'}{a^2}+\frac{yy'}{b^2}-1\right)^2=0.$$

Proof.—By the condition for equal roots of (4310).

Quadratic for abscissæ of points of contact of the tangent from $x'y'$:

4312 $$x^2(b^2x'^2+a^2y'^2)-2a^2b^2xx'+a^4(b^2-y'^2)=0. \qquad (4282, 4125)$$

Quadratic for m of the tangent from xy:

4313 $$m^2(x^2-a^2)-2mxy+y^2-b^2=0. \qquad (4282)$$

General polar equation of the ellipse, or quadratic for r, the segment intercepted between the point $x'y'$ and the curve on the right line drawn from that point at an inclination θ to the major axis and x axis of coordinates.

4314 $$(a^2\sin^2\theta+b^2\cos^2\theta)\,r^2 + 2r(a^2y'\sin\theta+b^2x'\cos\theta)+(a^2y'^2+b^2x'^2-a^2b^2)=0.$$

4315 *Length of intercepted chord = difference of roots.*

4316 *Distance to middle point of chord = half sum of roots.*

4317 *Rectangle under segments = products of roots.*

Cor.—If two chords be drawn to a conic at two constant inclinations to the major axis, the ratio of the rectangles under their segments is invariable.

For, if $x'y'$ be their point of intersection, the ratio in question becomes $a^2\sin^2\theta+b^2\cos^2\theta : a^2\sin^2\theta'+b^2\cos^2\theta'$, which is constant if θ and θ' are constant.

Locus of centres of parallel chords:

4318 $$a^2y\sin\theta+b^2x\cos\theta=0. \qquad (4314)$$

Quadratic for abscissæ of points of intersection of the line $Ax+By+C=0$ and the ellipse $b^2x^2+a^2y^2-a^2b^2=0$. (4116)

4319 $$(A^2a^2+B^2b^2)\,x^2+2ACa^2x+C^2a^2-B^2a^2b^2=0.$$

4320 $$x=\frac{-ACa^2\pm Bab\sqrt{A^2a^2+B^2b^2-C^2}}{A^2a^2+B^2b^2}.$$

4321 For the ordinates transpose A, B and a, b.

Length of intercepted chord:

4322 $$\frac{2ab\sqrt{(A^2+B^2)(A^2a^2+B^2b^2-C^2)}}{A^2a^2+B^2b^2}.$$ (4034)

Hence the condition that the line may touch the ellipse is

4323 $$A^2a^2+B^2b^2=C^2.$$

The chord through two points x_1y_1, x_2y_2 is

4324 $$\frac{x(x_1+x_2)}{a^2}+\frac{y(y_1+y_2)}{b^2}=\frac{x_1x_2}{a^2}+\frac{y_1y_2}{b^2}+1;$$

or, denoting the points by their eccentric angles α, β, the chord joining $\alpha\beta$ is

4325 $$\frac{x}{a}\cos\frac{\alpha+\beta}{2}+\frac{y}{b}\sin\frac{\alpha+\beta}{2}=\cos\frac{\alpha-\beta}{2}.$$

The coordinates of the pole of the chord or intersection of tangents at x_1y_1, x_2y_2 (or $\alpha\beta$ as above).

4326 $$x=\frac{x_1y_2+x_2y_1}{y_1+y_2}=\frac{a^2(y_1-y_2)}{x_2y_1-x_1y_2}=a\frac{\cos\frac{1}{2}(\alpha+\beta)}{\cos\frac{1}{2}(\alpha-\beta)}.$$

4329 $$y=\frac{x_1y_2+x_2y_1}{x_1+x_2}=\frac{b^2(x_1-x_2)}{x_1y_2-x_2y_1}=b\frac{\sin\frac{1}{2}(\alpha+\beta)}{\cos\frac{1}{2}(\alpha-\beta)}.$$

The following relations also subsist

4332 $$\frac{a^2b^2}{b^2x^2+a^2y^2}=\frac{a^2\sin\alpha\sin\beta}{a^2-x^2}=\frac{b^2\cos\alpha\cos\beta}{b^2-y^2}.$$

$$=\frac{b(\sin\alpha+\sin\beta)}{2y}=\frac{a(\cos\alpha+\cos\beta)}{2x},$$

"which are of use in finding the locus of (x, y) when α, β are connected by some fixed equation."

(Wolstenholme's Problems, p. 116.)

4334 If $\alpha, \beta, \gamma, \delta$ are the eccentric angles of the feet of the four normals drawn to an ellipse from a point xy, then

$$\alpha+\beta+\gamma+\delta = 3\pi \text{ or } 5\pi.$$

PROOF.—Equation (4290) gives the following biquadratic in $z = \tan\frac{1}{2}\phi$,

$$byz^4+2(ax+a^2-b^2)z^3+2(ax-a^2+b^2)z-by = 0.$$

Let a, b, c, d be the roots. Eliminate d from $ab+ac+\&c. = 0$ and $abcd = -1$ (406). Thus $ab+bc+ca = \frac{1}{bc}+\frac{1}{ca}+\frac{1}{ab}$; from which, since $a = \tan\frac{1}{2}\alpha$, &c., we get

$$\sin(\beta+\gamma)+\sin(\gamma+\alpha)+\sin(\alpha+\beta) = 0;$$

and, since

$$1-(ab+ac+\&c.)+abcd = 0,$$

$$\tan\tfrac{1}{2}(\alpha+\beta+\gamma+\delta) = \infty, \quad \therefore \quad \alpha+\beta+\gamma+\delta = 3\pi \text{ or } 5\pi.$$

4335 The points on the curve where it is met by the normals drawn from a fixed point $x'y'$ are determined by the intersections of the curve and the hyperbola

$$a^2x'y-b^2y'x = c^2xy. \tag{4287}$$

POLAR EQUATIONS OF THE CONIC.

The focus S being the pole (Fig. of 4201), the equation of any conic is

4336 $$r(1+e\cos\theta) = l,$$

θ being measured from A, the nearest vertex.

For the parabola, put $e = 1$.

PROOF.—

$r = SP$; $\theta = ASP$; $l = SL$; $r = e(SX+SN)$ (4200) $= l+er\cos\theta$.

The secant through two points, P, P', on the curve, whose angular coordinates are $\alpha+\beta$ and $\alpha-\beta$ (Fig. 28), is

4337 $$r\{e\cos\theta+\sec\beta\cos(\alpha-\theta)\} = l.$$

PROOF.—Let $ASQ = \alpha$, $PSQ = P'SQ = \beta$.

Analytically. Take (4109) for the equation of PP'. Eliminate r_1 and r_2 by (4336), and substitute 2α for $\theta_1+\theta_2$ and 2β for $\theta_1-\theta_2$.

Geometrically. Let PP' cut the directrix in Z; then QSZ is a right angle, by (1166). Take C any point in PP'; $SC = r$; $ASC = \theta$. Draw CD, CE, CF, CG parallel to SL, SP, SQ, SX, and DH parallel to XL. Then

$$l = SL = SH+HL.$$

$$SH = \frac{SL}{SX}SD = er\cos\theta. \quad \frac{CE}{CG} = \frac{SP}{PM} = \frac{SL}{SX} = \frac{HL}{DX} = \frac{HL}{CG},$$

$$\therefore \; HL = CE = r\sin CSF\sec\beta = r\cos(\alpha-\theta)\sec\beta,$$

$$\therefore \; l = er\cos\beta+r\sec\beta\cos(\alpha-\theta).$$

The equation of the tangent at the point α is, consequently,

4338 $$r\left\{e\cos\theta+\cos(\alpha-\theta)\right\}=l.$$

A Focal Chord.

4339 Length $$=\frac{2l}{1-e^2\cos^2\theta}.$$ (4336)

Coordinates of the extremities, the centre C being the origin:

4340 $$x=\frac{a(e\pm\cos\theta)}{1\pm e\cos\theta},\qquad y=\frac{l\sin\theta}{1\pm e\cos\theta}.$$

4342 The lines joining the extremities of two focal chords meet in the directrix. [By (4337)

Polar equation with vertex for pole:

4343 $$r^2(1-e^2\cos^2\theta)=2l\cos\theta.$$ (4200)

Polar equation with the centre for pole:

4344 $$r^2(a^2\sin^2\theta+b^2\cos^2\theta)=a^2b^2,$$

4345 or $$r\sqrt{(1-e^2\cos^2\theta)}=b.$$

PROOF.—By (4273). Otherwise, by (4314), with $x'=y'=0$.

CONJUGATE DIAMETERS.

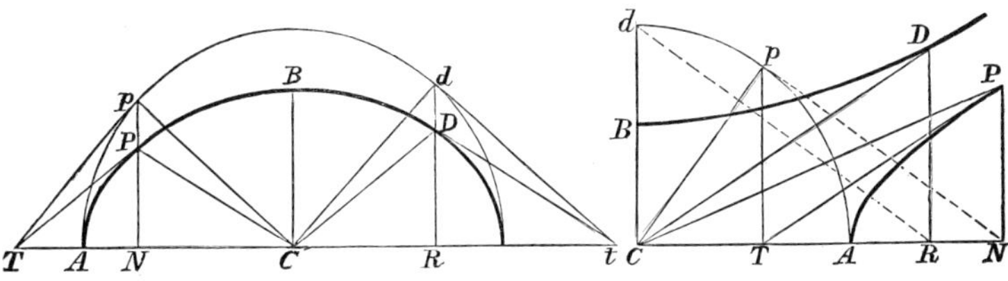

Equation of the ellipse referred to conjugate diameters for coordinate axes:

4346 $$\frac{x^2}{a'^2}+\frac{y^2}{b'^2}=1,$$

where

4347 $$a'^2 = \frac{a^2b^2}{a^2\sin^2\alpha + b^2\cos^2\alpha}, \qquad b'^2 = \frac{a^2b^2}{a^2\sin^2\beta + b^2\cos^2\beta}.$$

Here $a' = CD$, $b' = CP$, α is the angle DCR, and β the angle PCR.

PROOF.—Apply (4050) to the equation (4273), putting $\omega = \frac{\pi}{2}$ and $\tan\alpha\tan\beta = -\frac{b^2}{a^2}$, by (4351).

When $a' = b'$, $\alpha+\beta = \pi$, and equation (4346) becomes

4349 $$x^2+y^2 = a'^2 = \tfrac{1}{2}(a^2+b^2).$$

Let the coordinates of D be x', y', and those of P x, y; the equation of the diameter CP conjugate to CD is

4350 $$\frac{xx'}{a^2} + \frac{yy'}{b^2} = 0.$$

4351 $$\tan\alpha\tan\beta \quad\text{or}\quad mm' = -\frac{b^2}{a^2}. \qquad (4318)$$

xy in terms of $x'y'$, &c.

4352 $$x = -\frac{a}{b}y', \qquad y = \frac{b}{a}x'. \quad \text{Ell.}$$

4354 $$x = \frac{a}{b}y', \qquad y = \frac{b}{a}x'. \quad \text{Hyp.}$$

PROOF.—Solve (4350) with (4273).

4356 $$x = dR, \qquad x' = pN. \qquad (4274, 4352)$$

4358 $$x^2+x'^2 = a^2, \qquad y'^2+y^2 = b^2. \quad \text{Ell.} \qquad (4352)$$

4360 $$x^2-x'^2 = a^2, \qquad y'^2-y^2 = b^2. \quad \text{Hyp.} \qquad (4354)$$

4362 $$a^2+b^2 = a'^2+b'^2. \quad \text{Ell.} \qquad (4358)$$

4363 $$a^2-b^2 = a'^2-b'^2. \quad \text{Hyp.} \qquad (4360)$$

4364 $$a'^2 = b^2+e^2x^2. \qquad (4271, \text{'61})$$

4365 $$b'^2 = a^2-e^2x^2 = rr'. \qquad (4298)$$

The perpendicular from the centre upon the tangent at xy is given by

4366 $$\frac{1}{p^2} = \frac{x^2}{a^4} + \frac{y^2}{b^4}.$$ (4281, 4064)

The area of the parallelogram $PCDL$ (Fig. of 4307) is

4367 $$pa' = ab = a'b' \sin \omega,$$

where $p = PF$, $a' = CD$, $b' = CP$, $\omega = \angle PCD$.

PROOF.—From (4366), and (4352), and $a'^2 = x'^2 + y'^2$.

Other values of p^2:

4369 $$p^2 = \frac{a^2b^2}{a'^2} = \frac{a^2b^2}{a^2+b^2-b'^2}.$$ (4362)

4371 $$p^2 = a^2 \sin^2 \theta + b^2 \cos^2 \theta.$$

PROOF.—From (4344, '67), putting $r = a'$.

4372 $$p^2 = a^2 \cos^2 \gamma + b^2 \sin^2 \gamma,$$ (4371)

γ being the inclination of p.

4373 $$p^2 = a^2 (1 - e^2 \sin^2 \gamma).$$ (4372, 4260)

Equations to the tangents at P and P', the coordinates of D being x', y':

4374 $$xy' - yx' = \pm ab \quad (4073) \quad m = \frac{y'}{x'}.$$

DETERMINATION OF VARIOUS ANGLES.

4375 $$pCd = \frac{\pi}{2}.$$ Fig. p. 595. (4356)

4377 $$\tan PCD = -\frac{a^2b^2}{c^2xy},$$ (4070, 4352-3)

where $c = \sqrt{(a^2 - b^2)} = CS$.

4378 $$\tan(SPT) = \frac{b^2}{cy} = \frac{1+e\cos\theta}{e\sin\theta},$$ (4070, 4256, 4336)

where $\theta = PST$. [See figure on page 588.

If ψ be the inclination of the tangent to the x axis,

4380 $$\tan\psi = -\frac{b^2x}{a^2y} = \frac{e+\cos\theta}{\sin\theta}.$$ (4280)

Proof: $\psi = \theta + SPT$. Then by (631) and (4379).

4382 $$\tan SPS' = \frac{2b^2cy}{b^4-c^2y^2}.$$ (652, 4378)

4383 $$\tan APA' = -\frac{2b^2}{ae^2y}, \quad \tan CPG = \frac{e^2xy}{b^2}.$$

If OP, OP' are tangents to an ellipse,

4385 $$\cos POP' = \frac{CO^2-a^2-b^2}{OS.OS'}.$$

Proof.—By figure and construction of (1180), $POP' = MOS'$. Therefore

$$\cos POP' = \frac{OS^2+OS'^2-S'M^2}{2OS.OS'} = \frac{2CO^2+2CS^2-4a^2}{2OS.OS'} = \&c.$$

If x', y' are the coordinates of O,

4386 $$\tan POP' = \frac{2\sqrt{(b^2x'^2+a^2y'^2-a^2b^2)}}{x'^2+y'^2-a^2-b^2}.$$

Proof.—By (4311), taking terms of the second degree for the two parallel lines through the origin and $\tan\phi$ from (4112).

It is worthy of remark that the substitutions (4276–8) may also be usefully employed when the axes of reference are conjugate diameters: though, in that case, the geometrical signification of ϕ no longer exists.

THE HYPERBOLA
REFERRED TO THE ASYMPTOTES.

4387 $$xy = \tfrac{1}{4}(a^2+b^2).$$

PROOF.—By (4273) and (4050). Here $x = CK$, $y = PK$.

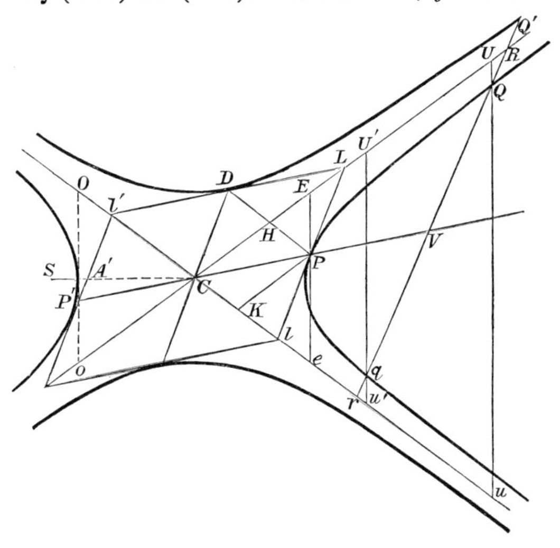

Equations of the tangent at P, (x', y').

4388 $$xy'+x'y = \tfrac{1}{2}(a^2+b^2). \qquad (4120)$$

4389 $$y = mx+\sqrt{m(a^2+b^2)}. \qquad (4123)$$

4390 $$m = -\frac{y'}{x'}.$$

4391 Intercepts on the axes $Cl = 2x'$, $CL = 2y'$.

THE RECTANGULAR HYPERBOLA.

4392 Here $a = b$, $e = \sqrt{2}$; and the equation with the ordinary axes is

4393 $$x^2-y^2 = a^2. \qquad (4273)$$

4394 Tangent $$xx'-yy' = a^2. \qquad (4281)$$

Equation with the asymptotes for axes:

4395 $$2xy = a^2. \quad (4387)$$

4396 Tangent $$xy'+x'y = a^2. \quad (4388)$$

THE GENERAL EQUATION.

The general equation of the second degree is

4400 $$ax^2+2hxy+by^2+2gx+2fy+c = 0,$$

4401 or $ax^2+by^2+cz^2+2fyz+2gzx+2hxy = 0$, with $z = 1$.

The equation will be denoted by u or $\phi(x, y) = 0$.

THE ELLIPSE AND HYPERBOLA.

When the general equation (4400), taken to rectangular axes of coordinates, represents a central conic, the coordinates of the centre, O' (Fig. 30), are

4402 $$x' = \frac{hf-bg}{ab-h^2} = \frac{G}{C}, \quad y' = \frac{gh-af}{ab-h^2} = \frac{F}{C}. \quad (4665)$$

PROOF.—By changing the origin to the point $x'y'$ and equating the new g and f each to zero (4048).

For the case in which $ab = h^2$, see (4430).

4404 The transformed equation is $ax^2+2hxy+by^2+c' = 0$,

4405 where $c' = ax'^2+2hx'y'+by'^2+2gx'+2fy'+c$.

4406 $$= gx'+fy'+c.$$

4407 $$= \frac{abc+2fgh-af^2-bg^2-ch^2}{ab-h^2} = \frac{\Delta}{C}. \quad (4466)$$

The inclination θ of the principal axis of the conic to the x axis is given by

4408 $$\tan 2\theta = \frac{2h}{a-b}.$$

PROOF.—(Fig. 30.) By turning the axes in (4404) through the angle θ (4049) and equating the new h to zero.

The transformed equation now becomes

4409 $$a'x^2+b'y^2+c'=0,$$

4410 in which $a'=\frac{1}{2}\left\{a+b+\sqrt{4h^2+(a-b)^2}\right\},$

4411 $$b'=\frac{1}{2}\left\{a+b-\sqrt{4h^2+(a-b)^2}\right\},$$

a' and b' are found from the two equations

4412 $$a'+b'=a+b, \qquad a'b'=ab-h^2.$$ [See (4418).

The semi-axes and excentricity are

4414 $\sqrt{-\frac{c'}{a'}}$, $\sqrt{-\frac{c'}{b'}}$, and $e=\sqrt{\left(1-\frac{a'}{b'}\right)}$. (4273) (4261)

For the coordinates of the foci, see (5008).

4416 NOTE.—If θ be the acute angle determined by equation (4408), we have to choose between θ and $\theta+\frac{\pi}{2}$ for the inclination in question, since $\tan 2\theta$ is also equal to $\tan(2\theta+\pi)$.

RULE.*—*For the ellipse, the inclination of the major axis to the* x *axis of coordinates will be the acute angle* θ *or* $\theta+\frac{1}{2}\pi$, *according as* h *and* c′ *have the same or different signs. For the hyperbola, read "different or the same."*

PROOF.—Let the transformed equation (4409) be written in terms of the semi-axes p, q; thus $q^2x^2+p^2y^2=p^2q^2$, representing an ellipse. Now turn the axes back again through the angle $-\theta$, and we get

$$(q^2\cos^2\theta+p^2\sin^2\theta)\,x^2-(p^2-q^2)\sin 2\theta\,xy+(q^2\sin^2\theta+p^2\cos^2\theta)\,y^2=p^2q^2.$$

Comparing this with the identical equation (4404), $ax^2+2hxy+by^2=-c'$, we have $\qquad (p^2-q^2)\sin 2\theta=-2h, \qquad p^2q^2=-c'$;

$$\therefore \qquad \sin 2\theta=\frac{2h}{c'}\cdot\frac{p^2q^2}{p^2-q^2}.$$ Hence θ is $<\frac{\pi}{2}$

when h and c' have the same sign, p being $>q$. A similar investigation applies to the hyperbola by changing the sign of q^2.

---

* This rule and the demonstration of it are due to Mr. George Heppel, M.A., of Hammersmith.

INVARIANTS OF THE CONIC.

4417 Transformation of the origin of coordinates alone does not alter the values of a, h, or b, whether the axes be rectangular or oblique. This is seen in (4404).

When the axes are rectangular, turning each through an angle θ does not affect the values of

4418 $$ab-h^2, \quad a+b, \quad g^2+f^2, \quad \text{or} \quad c.$$

When the axes are oblique (inclination ω), transformation in any manner does not affect the values of the expressions

4422 $$\frac{ab-h^2}{\sin^2\omega} \quad \text{and} \quad \frac{a+b-2h\cos\omega}{\sin^2\omega}.$$

These theorems may be proved by actual transformation by the formulæ in (4048–50). For other methods and additional invariants of the conic, see (4951).

4424 If the axes of coordinates are oblique, equation (4400) is transformed to the centre in the same way, and equations (4402–6) still hold good. If the final equation referred to axes coinciding with those of the conic be

4425 $$a'x^2+b'y^2+c'=0,$$

and θ the inclination of the new axis of x to the old one, we shall have c' unaltered,

4426 $$\tan 2\theta = \frac{2h\sin\omega-a\sin 2\omega}{2h\cos\omega-a\cos 2\omega-b};$$

4427 $$a' = \frac{a+b-2h\cos\omega+\sqrt{Q}}{2\sin^2\omega}; \qquad b' = \frac{a+b-2h\cos\omega-\sqrt{Q}}{2\sin^2\omega};$$

where $Q = a^2+b^2+2ab\cos 2\omega+4h(a+b)\cos\omega+4h^2$.

Proof.—(4404) is now transformed by the substitutions in (4050), putting $\beta = \theta+90°$, and equating the new h to zero to determine $\tan 2\theta$. a' and b' are most readily found from the invariants in (4422). Thus, putting the new $h=0$ and the new $\omega = 90°$,

$$a+b' = \frac{a+b-2h\cos\omega}{\sin^2\omega} \quad \text{and} \quad a'b' = \frac{ab-h^2}{\sin^2\omega},$$

equations which determine a' and b'.

The eccentricity of the general conic (4400) is given by the equation

4429 $$\frac{e^4}{1-e^2} = \frac{(a+b-2h\cos\omega)^2}{(ab-h^2)\sin^2\omega} - 4.$$

PROOF.—By (4415), and the invariants in (4422).

THE PARABOLA.

4430 When $ab-h^2=0$, the general equation (4400) represents a parabola.

For x', y' in (4402) then become infinite and the curve has no centre, or the centre may be considered to recede to infinity.

Turn the axes of coordinates at once through an angle θ (4049), and in the transformed equation let the new coefficients be a', $2h'$, b', $2g'$, $2f'$, c'. Equate h' to zero; this gives (4408) again, $\tan 2\theta = \frac{2h}{a-b}$. If θ be the acute angle determined by this equation, we can decide whether θ or $\theta+\frac{1}{2}\pi$ is the angle between the x axis and the axis of the parabola by the following rule.

4431 RULE.—*The inclination of the axis of the parabola to the* x *axis of coordinates will be the acute angle* θ *if* h *has the opposite sign to that of* a *or* b, *and* $\theta+\frac{1}{2}\pi$ *if it has the same sign.*

PROOF.—Since $ab-h^2=0$, a and b have the same sign. Let that sign be positive, changing signs throughout if it is not. Then, for a point at infinity on the curve, x and y will take the same sign when the inclination is the acute angle θ, and opposite signs when it is $\theta+\frac{1}{2}\pi$. But, since $ax^2+by^2=+\infty$, we must have $2hxy=-\infty$, the terms of the first degree vanishing in comparison. Hence the sign of h determines the angle as stated in the rule.

4432 $$\sin\theta = \sqrt{\frac{b}{a+b}}, \qquad \cos\theta = \sqrt{\frac{a}{a+b}}.$$

PROOF.—From the value of $\tan 2\theta$ above, θ being the acute angle obtained, and from $h^2 = ab$.

4434 Also $a'=0$ and $b'=a+b$.

For $a'b'=ab-h^2=0$, and we ensure that a' and not b' vanishes by (4431). Also $a'+b'=a+b$ (4412).

4436 $$g' = g\cos\theta + f\sin\theta = \frac{g\sqrt{a}+f\sqrt{b}}{\sqrt{(a+b)}}.$$

4438 $$f' = f\cos\theta - g\sin\theta = \frac{f\sqrt{a}-g\sqrt{b}}{\sqrt{(a+b)}}.$$

4440 But if h has the same sign as a and b, change θ into $\theta+\frac{1}{2}\pi$. (4431)

PROOF.—By (4418, 4432–3).

The coordinates of the vertex are

4441 $$x' = \frac{f'^2 - b'c}{2b'g'}, \qquad y' = -\frac{f'}{b'}.$$

Obtained by changing the origin to the point $x'y'$ and equating to zero the coefficient of y and the absolute term. The coefficient of x then gives the latus rectum of the parabola; viz.:

4443 $$L = -\frac{2g'}{b'} = -2\frac{g\sqrt{a}+f\sqrt{b}}{\sqrt{(a+b)^3}}.$$ (4437)

METHOD WITHOUT TRANSFORMATION OF THE AXES.

4445 Let the general equation (4400) be solved as a quadratic in y. The result may be exhibited in either of the forms

4446 $$y = \alpha x + \beta \pm \sqrt{\mu\,(x^2 - 2px + q)},$$

4447 $$y = \alpha x + \beta \pm \sqrt{\mu\,\{(x-p)^2 + (q-p^2)\}},$$

4448 $$y = \alpha x + \beta \pm \sqrt{\mu\,(x-\gamma)(x-\delta)},$$

4449 where $$\alpha = -\frac{h}{b}, \qquad \beta = -\frac{f}{b}, \qquad \mu = \frac{h^2 - ab}{b^2}.$$

4452 $$p = \frac{hf - bg}{ab - h^2} \text{ or } \frac{G}{C}, \text{ (1642)} \qquad q = \frac{bc - f^2}{ab - h^2} \text{ or } \frac{A}{C}.$$

4454 $$q - p^2 = \frac{b\,(abc + 2fgh - af^2 - bg^2 - ch^2)}{(ab - h^2)^2} = \frac{b\Delta}{C^2}.$$

4456 $$\gamma \text{ and } \delta = p \pm \sqrt{(p^2 - q)}.$$

4458 Here $y = \alpha x + \beta$ is the equation to the diameter DD

(Fig. 31), γ and δ are the abscissæ of D and D', its extremities, the tangents at those points being parallel to the y axis. The surd $= PN = P'N$ when $x = OM$. The axes may be rectangular or oblique.

When $ab-h^2 = 0$, equation (4446) becomes

4459 $$y = \alpha x + \beta \pm \frac{1}{b}\sqrt{q'-2p'x},$$

4460 where $p' = bg - hf,\quad q' = f^2 - bc.$

4462 In this case, $\dfrac{q'}{2p'}$ is the abscissa of the extremity of the diameter whose equation is $y = \alpha x + \beta$ and the curve has infinite branches.

RULES FOR THE ANALYSIS OF THE GENERAL EQUATION.

First examine the value of ab$-$h², *and, if this is not zero, calculate the numerical value of* c' (4407), *and proceed as in* (4400) et seq. *If* ab$-$h² *is zero, find the values of* p′ *and* q′ (4459). *The following are the cases that arise.*

4464 $ab-h^2$ positive—Locus an ellipse.

Particular Cases.

4465 $\Delta = 0$—Locus the point $x'y'$.

See (4402). For, by (4404), the conjugate axes vanish.

4466 $b\Delta$ positive—No locus.

By (4447–54), since $q-p^2$ is then positive.

4467 $h = 0$ and $a = b$—Locus a circle.

By (4144). In other cases proceed as in (4400–14).

4468 $ab-h^2$ negative—Locus an hyperbola.

Particular Cases.

4469 $\Delta = 0$—Locus two right lines intersecting in the point $x'y'$.

By (4447), since $q-p^2$ then vanishes. In this case solve as in (4447).

4470 $b\Delta$ negative—Locus the conjugate hyperbola.

4471 $a+b=0$—Locus the rectangular hyperbola.

By (4414), since $a' = -b'$.

4472 $a=b=0$—Locus an hyperbola, with its asymptotes parallel to the coordinate axes. The coordinates of the centre are now $-\frac{f}{h}$ and $-\frac{g}{h}$, by (4402). Transfer the origin to the centre, and the equation becomes

4473 $$xy = \frac{2fg-ch}{2h^2}.$$

In other cases proceed as in (4400–14).

4474 $ab-h^2=0$—Locus a parabola.

Particular Cases.

4475 $p'=0$—Locus two parallel right lines. By (4459).

4476 $p'=q'=0$—Locus two coinciding right lines. By (4459).

4477 $p'=0$ and q' negative—No locus.

By (4459). In other cases proceed as in (4430–43).

Ex. 1.: $$2x^2-2xy+y^2+3x-y-1=0.$$

Here the values of a, h, b, g, f, c are respectively $2, -1, 1, \frac{3}{2}, -\frac{1}{2}, -1,$

$$ab-h^2=1; \quad c' = \frac{\Delta}{C} = \frac{abc+2fgh-af^2-bg^2-ch^2}{ab-h^2} = -\frac{9}{4} \quad (4406)$$

The locus is therefore an ellipse, none of the exceptions (4465–7) occurring here. The coordinates of the centre, by (4402), are

$$x' = \frac{hf-bg}{ab-h^2} = -1, \qquad y' = \frac{gh-af}{ab-h^2} = -\frac{1}{2}.$$

Hence the equation transformed to the centre is

$$2x^2-2xy+y^2-\frac{9}{4}=0.$$

Turning the axes of coordinates through an angle θ so that $\tan 2\theta = -2$ (4408), we find the new a and b from

$$a'+b'=3, \qquad a'b'=1; \quad (4412)$$

therefore $a' = \frac{1}{2}(3-\sqrt{5}),\quad b' = \frac{1}{2}(3+\sqrt{5}),$
and the final equation becomes $2(3-\sqrt{5})x^2+2(3+\sqrt{5})y^2 = 9.$

The inclination of the major axis to the original x axis of coordinates is the *acute* angle $\frac{1}{2}\tan^{-1}(-2)$, by the rule in (4416).

Ex. (2): $12x^2+60xy+75y^2-12x-8y-6 = 0.$

The values of a, h, b, g, f, c are respectively 12, 30, 75, -6, -4, -6,

$$ab-h^2 = 0;\qquad p' = bg-hf = -330;\qquad q' = f^2-bc. \qquad (4460)$$

Since p' does not vanish (4475–7), the locus is a parabola. Proceeding, therefore, by (4430–43), we have

$$\tan 2\theta = \frac{2h}{a-b} = -\frac{20}{21};\qquad \sin\theta = \frac{5}{\sqrt{29}},\qquad \cos\theta = \frac{2}{\sqrt{29}}. \qquad (4432)$$

By the rule (4431), we must take $\theta+\frac{1}{2}\pi$ for the angle, instead of θ. Therefore

$$g' = -g\sin\theta+f\cos\theta = \frac{-g\sqrt{b}+f\sqrt{a}}{\sqrt{(a+b)}} = \frac{22}{\sqrt{29}},$$

$$f' = -f\sin\theta-g\cos\theta = \frac{-f\sqrt{b}-g\sqrt{a}}{\sqrt{(a+b)}} = \frac{32}{\sqrt{29}}.$$

and $b' = a+b = 87$ (4435).

Consequently the transformed equation is

$$87y^2+\frac{44}{\sqrt{29}}x+\frac{64}{\sqrt{29}}y-6 = 0.$$

The coordinates of the vertex are computed by (4441), and the final equation with the vertex for origin is $y^2 = \dfrac{44}{87\sqrt{29}}x.$

Ex. (3): $x^2+6xy+9y^2+5x+15y+6 = 0.$

The values of a, h, b, g, f, c are respectively 1, 3, 9, $\frac{5}{2}$, $\frac{15}{2}$, 6,

$$ab-h^2 = 0 \quad\text{and}\quad p' = bg-hf = 0,$$

therefore, by (4475–7), if there is a locus at all, it consists of two parallel or coinciding lines. Solving the equation therefore as a quadratic in y, we obtain it in the form $(x+3y+2)(x+3y+3) = 0,$
the equation of two parallel right lines.

The equation of the tangent or polar of $x'y'$ is

4478 $u_{x'}x+u_{y'}y+u_{z'}z = 0 \quad\text{or}\quad u_x x'+u_y y'+u_z z' = 0;$

(4401, 1405) obtained by (4120) in the form

4479 $(ax'+hy'+g)x+(hx'+by'+f)y+gx'+fy'+c = 0,$

4480 or $(ax+hy+g)x'+(hx+by+f)y'+gx+fy+c = 0,$

4481 or

$$axx'+h(xy'+x'y)+byy'+g(x+x')+f(y+y')+c=0.$$

When the curve passes through the origin, the tangent at the origin is

4482 $$gx+fy=0. \qquad (4479)$$

And the normal at the same point is

4483 $$fx-gy=0.$$

4484 Intercepts of the curve on the axes, $-\frac{2g}{a},\ -\frac{2f}{b}$.

4486 Length of normal intercepted between the origin and the chord

$$=\frac{\sqrt{g^2+f^2}}{a+b}. \qquad (4483\text{–}4)$$

Right Line and Conic with the general Equation.

4487 Quadratic for $n:n'$, the ratio in which the line joining xy, $x'y'$ is cut by the curve.

Let the equation of the curve (4400) be denoted by $\phi(x, y)=0$, and the equation of the tangent (4479) by $\psi(x, y, x', y')=0$; then the quadratic required will be found, by the method of (4131), to be

$$n^2\phi(x', y')+2nn'\psi(x, y, x', y')+n'^2\phi(x, y)=0.$$

The equation of the tangents from $x'y'$ is

4488 $$\phi(x', y')\,\phi(x, y)=\{\psi(x, y, x', y')\}^2.$$

PROOF.—By the condition for equal roots in (4487).

COR.—The equation of two tangents through the origin is

4489 $$Bx^2-2Hxy+Cy^2=0. \qquad (4665)$$

The equation of the asymptotes of u (4400) is

4490 $$au_y^2+2hu_xu_y+bu_x^2=0.$$

The equation of the equi-conjugates of the conic $ax^2+2hxy+by^2=1$ is

4491 $(a+b)(ax^2+2hxy+by^2) = 2(ab-h^2)(x^2+y^2).$

PROOF.—When the conic is $ax^2+by^2=1$, the similar equation is

$$(a+b)(ax^2+by^2) = 2ab(x^2+y^2) \quad \text{or} \quad (ax^2-by^2)=0,$$

given by the intersections of the conic and a circle. Transformation of the axes then produces the above by the invariants in (4418).

4492 When the coordinate axes are oblique, the equation becomes

$$(a-b)(ax^2-by^2)+2x(hx+by)(h-a\cos\omega)+2y(ax+hy)(h-b\cos\omega)=0.$$

General polar equation:

4493 $(a\cos^2\theta+2h\sin\theta\cos\theta+b\sin^2\theta)\,r^2$
$$+2(g\cos\theta+f\sin\theta)\,r+c=0.$$

Polar equation with (x, y) for the pole: (4134)

4494 $(a\cos^2\theta+2h\sin\theta\cos\theta+b\sin^2\theta)\,r^2$
$$+2\{(ax+hy+g)\cos\theta+(by+hx+f)\sin\theta\}\,r+F(xy)=0.$$

Equation of the line through $x'y'$ parallel to the conjugate diameter:

4495 $(x-x')(ax'+hy'+g)+(y-y')(hx'+by'+f)=0.$

PROOF.—By the condition for equal roots of opposite signs (4494).

Equation of the conic with the origin at the extremity of the major axis, L being the latus rectum.

4496 $y^2 = Lx-(1-e^2)\,x^2.$ (4269, '59)

Equation when the point ab is the focus and $Ax+By+C=0$ the directrix:

4497 $\sqrt{\{(x-a)^2+(y-b)^2\}} = e\,\dfrac{Ax+By+C}{\sqrt{\{A^2+B^2\}}}.$ (4200, 4095)

INTERCEPT EQUATION OF A CONIC.

The equation of a conic passing through four points whose intercepts on oblique axes of coordinates are s, s' and t, t', is

4498 $$\frac{x^2}{ss'}+2hxy+\frac{y^2}{tt'}-x\left(\frac{1}{s}+\frac{1}{s'}\right)-y\left(\frac{1}{t}+\frac{1}{t'}\right)+1=0.$$

Equation of a conic touching oblique axes in the points whose intercepts are s and t:

4499 $$\frac{x^2}{s^2}+2hxy+\frac{y^2}{t^2}-\frac{2x}{s}-\frac{2y}{t}+1=0,$$

4500 or $$\left(\frac{x}{s}+\frac{y}{t}-1\right)^2+vxy=0.$$

Comparing with the general equation (4400), we have

4501 $$s=-\frac{c}{g},\quad t=-\frac{c}{f},\quad v=2h-\frac{2}{st}=2\,\frac{hc^2-gf}{c^2}.$$

Perpendicular p from xy, any point on the curve, to the chord of contact:

4505 $$p^2=\frac{vs^2t^2xy\sin^2\omega}{s^2+t^2-2st\cos\omega}.$$ (4096, 4500)

Equation of the tangent at $x'y'$:

4507 $$2\left(\frac{x'}{s}+\frac{y'}{t}-1\right)\left(\frac{x}{s}+\frac{y}{t}-1\right)+v\,(xy'+x'y)=0.$$

4508 The equation of the director-circle is

$$(1+\tfrac{1}{4}stv)(x^2+y^2+2xy\cos\omega)-h\,(x+y\cos\omega)-k\,(y+x\cos\omega)+hk\cos\omega=0.$$

The parabola with the same coordinate axes as in (4499):

4509 $$\left(\frac{x}{s}+\frac{y}{t}-1\right)^2=\frac{4xy}{st}\quad\text{or}\quad\sqrt{\frac{x}{s}}+\sqrt{\frac{y}{t}}=1.$$

PROOF.—From (4500), putting $h=-\frac{1}{st}$ (4474), and therefore $v=-\frac{4}{st}$.

Equation of the tangent at x', y':

4510 $$\frac{x}{\sqrt{(sx')}}+\frac{y}{\sqrt{(ty')}}=1, \qquad (4509)$$

4511 or $$y = mx+\frac{mst}{ms+t}, \qquad m=-\sqrt{\frac{ty'}{sx'}}. \qquad (4123)$$

Equation of the normal at $x'y'$:

4512 $$y = mx+\frac{s^2t-m^3st^2}{(mt+s)^2}, \qquad m=\sqrt{\frac{sx'}{ty'}}. \qquad (4122)$$

4513 Normal through the origin $x\sqrt[3]{s}=y\sqrt[3]{t}$.

The equations of two diameters are, with any axes,

4514 $$\frac{x}{s}-\frac{y}{t}=1 \quad \text{and} \quad \frac{x}{s}-\frac{y}{t}=-1.$$

PROOF.—Diameter through $0t$, $\frac{y-t}{x}=\frac{t}{s}$ by the property $OR=RQ$, in the figure of (4211).

Coordinates of the focus:

4516 $$x=\frac{st^2}{s^2+t^2+2st\cos\omega}, \qquad y=\frac{s^2t}{s^2+t^2+2st\cos\omega}. \qquad (5009)$$

Equation of the directrix:

4518 $$x(s+t\cos\omega)+y(t+s\cos\omega)=st\cos\omega.$$

PROOF.—Expand (4509), and form the equation of the polar of the focus by (4479) and (4516).

When the axes are also rectangular, the latus rectum

4519 $$L=\frac{4s^2t^2}{(s^2+t^2)^{\frac{3}{2}}}. \qquad (4095,\ 4516\text{-}8)$$

4520 Locus of the centre of the conic which touches the axes at the points $s0$, $0t$:

$$tx=sy. \qquad (4500,\ 4402)$$

4521 To make the conic pass through a point $x'y'$; substitute $x'y'$ in (4500), and determine v.

SIMILAR CONICS.

4522 DEFINITION.—If two radii, drawn from two fixed points, maintain a constant ratio and a constant mutual inclination, they will describe *similar curves.*

4523 If the proportional radii be always parallel, the curves are also *similarly situated.*

If there be two conics (1) and (2), with equations of the form (4400), then—

The condition of their being similar and similarly situated is

4524 $$\frac{a}{a'} = \frac{h}{h'} = \frac{b}{b'}.$$

PROOF.—By (4404), changing to polar coordinates, $r : r' =$ constant.

The condition of similarity only is

4525 $$\frac{(a+b)^2}{h^2-ab} = \frac{(a'+b')^2}{h'^2-a'b'}; \qquad (4418\text{–}9)$$

or, with oblique axes,

4526 $$\frac{(a+b-2h\cos\omega)^2}{h^2-ab} = \frac{(a'+b'-2h'\cos\omega)^2}{h'^2-a'b'}. \qquad (4422\text{–}3)$$

CIRCLE OF CURVATURE.

CONTACT OF CONICS.

4527 DEF.—When two points of intersection of two curves coincide on a common tangent, the curves have a contact of the *first order*; when three such points coincide, a contact of the *second order;* and so on. To osculate, is to have a contact higher than the first.

4528 The two conics (Fig. 32) whose equations are

$$ax^2+2hxy+by^2+2gx = 0\ldots\ldots\ldots\ldots\ldots\ldots(1),$$

$$a'x^2+2h'xy+b'y^2+2g'x = 0\ldots\ldots\ldots\ldots\ldots\ldots(2),$$

touch the y axis at the origin, O, by (4482). Eliminate the third terms from (1) and (2), and we obtain $x = 0$, the line through two coincident points, and

4529 $$(ab'-a'b)\,x+2\,(hb'-h'b)\,y-2\,(bg'-b'g) = 0,$$

the equation to LM, the line passing through the two remaining points of intersection of (1) and (2). (4099)

Again, eliminate the last terms from (1) and (2), and we obtain

4530 $$(ag'-a'g)\,x^2+2\,(hg'-h'g)\,xy+(bg'-b'g)\,y^2 = 0,$$

the equation of the two lines OL, OM. [By (4111) and (4099)

4531 If the points L, M coincide, the conics have contact of the first order. The condition for this is that (4530) must have equal roots; therefore

4532 $$(ag'-a'g)(bg'-b'g) = (hg'-h'g)^2.$$

4533 If the conics (1) and (2) are to osculate, M must coincide with O. Therefore, in (4529), $bg' = b'g$.

If in (4532) $bg' = b'g$, the conics have a contact of the third order.

CIRCLE OF CURVATURE.

(See also 1254 *et seq.*)

The radius of curvature at the origin for the conic

$$ax^2+2hxy+by^2+2gx = 0,$$

the axes of coordinates including an angle ω, is

4534 $$\rho = -\frac{g}{b\sin\omega}.$$

Proof.—The circle touching the curve at the origin is

$$x^2+2xy\cos\omega+y^2-2rx\sin\omega = 0,$$

by (4148), and the geometry of the figure, $2r\sin\omega$ being the intercept on the x axis. The condition of osculating (4533) gives the value of ρ.

ρ is positive when the convexity of the curve is towards the y axis.

Radius of curvature for a central conic at the extremity P of a semi-diameter a', the conjugate being b'.

4535 $$\rho = \frac{b'^2}{a' \sin\omega} = \frac{b'^2}{p} = \frac{a^2b^2}{p^3} = \frac{b'^3}{ab}.$$ (4367)

PROOF.—Take the equation and figure of (4346) ($a' = CP$). Transform to parallel axes through P. Then by (4534).

The same in terms of x, y, the coordinates of the point P.

4539 $$\rho = \frac{(b^4x^2+a^4y^2)^{\frac{3}{2}}}{a^2b^2}.$$

PROOF.—By (5138), or from (4538) and the value of b at (4365).

The coordinates of the centre of curvature O for P, the point xy, are

4540 $$\xi = \frac{c^2x^3}{a^4}, \quad \eta = -\frac{c^2y^3}{b^4}, \quad \text{where } c^2 = a^2-b^2.$$

PROOF.—(Fig. 33.) From $\frac{OD}{OP} = \frac{x-\xi}{\rho} = \frac{x}{PG'}$ and $\frac{PD}{OP} = \frac{y-\eta}{\rho} = \frac{y}{PG}$, with the values of ρ, PG, and PG' at (4535) and (4309).

Radius of curvature for the parabola.

Taking the diameter and tangent through the point for axes,

4542 $$\rho = \frac{2a'}{\sin\theta} = \frac{2a}{\sin^3\theta} = \frac{2SP^2}{SY}.$$ (Fig. of 4201)

By (4534), and equation (4211).

Coordinates of the centre of curvature at xy (rectangular axes):

4545 $$\xi = 3x+2a, \quad \eta = -\frac{y^3}{4a^2}.$$

PROOF.—From $y-\eta : \rho = y : PG$ and $\rho = 2a \operatorname{cosec}^3\theta$, $PG = 2a \operatorname{cosec}\theta$ and $y = 2a\cot\theta$.

The evolute of a central conic (Fig. 33):

4547 $$(ax)^{\frac{2}{3}}+(by)^{\frac{2}{3}} = (a^2-b^2)^{\frac{2}{3}},$$

4548 or $$(a^2x^2+b^2y^2-c^4)^3+27a^2b^2c^4x^2y^2 = 0,$$

where $c^2 = a^2-b^2$.

PROOF.—Substitute for x, y in the equation of the conic (4273) their values in terms of ξ, η from (4540). Otherwise as in (4958), or by the method of (5157).

The curve has cusps at L, H, M, and K.

The evolute of the parabola:

4549 $$a\left(\frac{y}{2}\right)^2 = \left(\frac{x-2a}{3}\right)^3.$$

PROOF.—As in (4548), from the equations (4201) and (4545).

CONFOCAL CONICS.

4550 $$\frac{x^2}{a^2}+\frac{y^2}{b^2}=1 \quad \text{and} \quad \frac{x^2}{a'^2}+\frac{y^2}{b'^2}=1$$

are confocal conics, if

$$a^2-a'^2 = b^2-b'^2,$$

or the sign of b'^2 may be changed.

For the confocal of the general conic, see (5007).

4551 Confocal conics intersect, if at all, at right angles.

PROOF.—If u, u' are the two conics in (4550), changing the sign of b'^2 to make the second conic an hyperbola, $u-u'=0$ will be satisfied at their point of intersection; and (by $a^2-a'^2=b^2+b'^2$) this proves the tangents at that point to be at right angles (4078, 4280).

Otherwise geometrically by (1168).

4552 Tangents from a point P on one conic to a confocal conic make equal angles with the tangent at P. [Proof at (1291)

4553 The locus of the pole of the line $Ax+By+C$ with respect to a series of confocal conics in which $a^2-b^2=\lambda$, is the right line perpendicular to the given one,

$$BCx-ACy+AB\lambda = 0.$$

PROOF.—The pole of the line for any of the conics being xy; $Aa^2=-Cx$ and $Bb^2=-Cy$ (4292); also $a^2-b^2=\lambda$. Eliminate a^2 and b^2.

4554 COR.—If the given line touch one of the conics, the locus is the normal at the point of contact.

4555 ***Graves' Theorem.*** — The two tangents drawn to an ellipse from a point on a confocal ellipse together exceed the intercepted arc by a constant quantity.

PROOF.—(Fig. 132.) Let P, P' be consecutive points on the confocal from which the tangents are drawn. Let fall the perpendiculars PN, $P'N'$. From (1291), it follows that $\angle PP'N = P'PN'$, and therefore $P'N = PN'$. The increment in the sum of the tangents in passing from P to P' is

$$RR' - QQ' + P'N - PN' = RR' - QQ'.$$

But this is also the increment in the arc QR, which proves the theorem.

4556 If the tangents are drawn from a confocal hyperbola, as in (Fig. 133), the difference of the tangents PQ, PR is equal to the difference of the arcs QT, RT.

The proof is quite similar to the foregoing.

4557 At the intersection of two confocal conics, the centre of curvature of either is the pole of its tangent with respect to the other.

PROOF.—Take $\frac{x^2}{a^2}+\frac{y^2}{b^2}=1$ (i.) and $\frac{x^2}{a'^2}-\frac{y^2}{b'^2}=1$ (ii.) for confocal conics. At the point of intersection, $x'^2=\frac{a^2a'^2}{c^2}$ and $y'^2=-\frac{b^2b'^2}{c^2}$ (where $c^2=a^2-b^2$); by $a^2-a'^2=b^2+b'^2$. The coordinates of the centre of curvature of $x'y'$ in (i.) are $x''=\frac{c^2x'^3}{a^4}$, $y''=-\frac{c^2y'^3}{b^4}$ (4540-1). The polar of this point with respect to (ii.) will be $\frac{xx''}{a'^2}+\frac{yy''}{b'^2}=1$. Substitute the values of x'', y''; and we see, by the values of x', y', that this is also the tangent of (i.) at P.

4558 A system of coaxal circles (4161), reciprocated with respect to one of the limiting points D or D', becomes a system of confocal conics.

PROOF.—The origin D is one common focus of the reciprocal conics, by (4844). The polar of D with respect to any of the circles is the *same* line, by (4166). D and its polar (both fixed) reciprocate (4858) into the line at infinity and *its* polar, which is the centre of the conic. The centre and one focus being the same for all, the conics are confocal.

ANALYTICAL CONICS

IN

TRILINEAR COORDINATES.

THE RIGHT LINE.

For a description of this system of coordinates, see (4006). The square of the distance between two points $\alpha\beta\gamma$, $\alpha'\beta'\gamma'$ is, with the notation of (4008),

4601
$$\frac{\mathfrak{abc}}{\Sigma^2}\left\{\mathfrak{a}(\beta-\beta')(\gamma-\gamma')+\mathfrak{b}(\gamma-\gamma')(\alpha-\alpha')+\mathfrak{c}(\alpha-\alpha')(\beta-\beta')\right\},$$

4602
$$=\frac{\mathfrak{abc}}{\Sigma^2}\left\{\mathfrak{a}\cos A\,(\alpha-\alpha')^2+\mathfrak{b}\cos B\,(\beta-\beta')^2+\mathfrak{c}\cos C\,(\gamma-\gamma')^2\right\}.$$

PROOF.—Let P, Q be the points. By drawing the coordinates $\beta\gamma$, $\beta'\gamma'$, it is easily seen, by (702), that

$$PQ^2=[(\beta-\beta')^2+(\gamma-\gamma')^2+2(\beta-\beta')(\gamma-\gamma')\cos A]\operatorname{cosec}^2 A \ldots\ldots(1).$$

Now, by (4007), $\mathfrak{a}(\alpha-\alpha')+\mathfrak{b}(\beta-\beta')+\mathfrak{c}(\gamma-\gamma')=0$,

from which $\mathfrak{b}(\beta-\beta')^2=-\mathfrak{a}(\alpha-\alpha')(\beta-\beta')-\mathfrak{c}(\beta-\beta')(\gamma-\gamma')$,

and a similar expression for $\mathfrak{c}(\gamma-\gamma')^2$. Substitute these values of the square terms in (1), reducing by (702).

Coordinates of the point which divides the straight line joining the points $\alpha\beta\gamma$, $\alpha'\beta'\gamma'$ in the ratio $l:m$:

4603 $$\frac{l\alpha+m\alpha'}{l+m},\quad \frac{l\beta+m\beta'}{l+m},\quad \frac{l\gamma+m\gamma'}{l+m}.$$ By (4032).

ABC being the triangle of reference, and $\alpha=0$, $\beta=0$, $\gamma=0$ the equations of its sides, the equation of a line passing through the intersection of the lines $\alpha=0$, $\beta=0$ is

4604 $$l\alpha-m\beta=0 \quad\text{or}\quad \alpha-k\beta=0.$$

PROOF.—For this is the locus of a point whose coordinates α, β are in the constant ratio $m : l$ or k (4099).

When l and m have the same sign, the line divides the external angle C of the triangle ABC; when of opposite sign, the internal angle C.

The general equation of a straight line is

4605 $$l\alpha+m\beta+n\gamma=0,$$

and it may be referred to as the line (l, m, n).

PROOF: $l\alpha+m\beta=0$ is *any* line through the point C, and $(l\alpha+m\beta)+n\gamma=0$ is any line through the intersection of the former line and the line $\gamma=0$ (4604), and therefore any line whatever according to the values of the arbitrary constants l, m, n.

The same straight line in Cartesian coordinates is

4606 $$(l\cos\alpha+m\cos\beta+n\cos\gamma)\,x+(l\sin\alpha+m\sin\beta+n\sin\gamma)\,y-(lp_1+mp_2+np_3)=0.$$

PROOF.—By substituting the values of α, β, γ at (4009).

Or, if the equations of the sides of ABC are given in the form $A_1x+B_1y+C_1=0$, &c., the line becomes

4607 $$(lA_1+mA_2+nA_3)\,x+(lB_1+mB_2+nB_3)\,y+lC_1+mC_2+nC_3=0.$$

PROOF.—By (4095), the denominators like $\sqrt{(A_1^2+B_1^2)}$ being included in the constants l, m, n.

4608 If $u=0$, $v=0$, $w=0$ are the general equations of the lines α, β, γ, then it is obvious that $lu+mv=0$ is, like (4604), a line passing through the intersection of u and v, and $lu+mv+nw=0$ represents any straight line whatever.

To make an equation such as $\alpha=p$ (a constant) homogeneous in α, β, γ; multiply by the equation $\Sigma=\mathfrak{a}\alpha+\mathfrak{b}\beta+\mathfrak{c}\gamma$ (4007), thus

$$(\mathfrak{a}p-\Sigma)\,\alpha+\mathfrak{b}p\beta+\mathfrak{c}p\gamma=0,$$

which is of the same form as (4605).

4610 The point of intersection of the lines

$$l\alpha+m\beta+n\gamma=0 \quad \text{and} \quad l'\alpha+m'\beta+n'\gamma=0$$

is determined by the ratios

$$\frac{\alpha}{mn'-m'n}=\frac{\beta}{nl'-n'l}=\frac{\gamma}{lm'-l'm} \quad \text{and (4017).}$$

The values of α, β, γ are therefore

4611 $$\frac{\Sigma\,(mn'-m'n)}{D},\quad \frac{\Sigma\,(nl'-n'l)}{D},\quad \frac{\Sigma\,(lm'-l'm)}{D},$$

where $D=\mathfrak{a}\,(mn'-m'n)+\mathfrak{b}\,(nl'-n'l)+\mathfrak{c}\,(lm'-l'm)$.

PROOF.—By (4017), or by solving the three equations

$$\mathfrak{a}\alpha+\mathfrak{b}\beta+\mathfrak{c}\gamma=\Sigma,\qquad l\alpha+m\beta+n\gamma=0,\qquad l'\alpha+m'\beta+n'\gamma=0.$$

The equation

4612 $$\mathfrak{a}\alpha+\mathfrak{b}\beta+\mathfrak{c}\gamma=0 \quad \text{or} \quad \alpha\sin A+\beta\sin B+\gamma\sin C=0$$

represents a straight line at infinity.

PROOF.— The coordinates of its intersection with any other line $l\alpha+m\beta+n\gamma=0$ are infinite by (4611).

4613 NOTE: $\mathfrak{a}\alpha+\mathfrak{b}\beta+\mathfrak{c}\gamma=\Sigma$, a quantity not zero. The equation $\mathfrak{a}\alpha+\mathfrak{b}\beta+\mathfrak{c}\gamma=0$ is therefore in itself impossible, and so is a line infinitely distant. The two conceptions are, however, *together consistent*; the one involves the other. And if, in the equation $l\alpha+m\beta+n\gamma=0$, the ratios $l:m:n$ approach the values $\mathfrak{a}:\mathfrak{b}:\mathfrak{c}$, the line it represents recedes to an unlimited distance from the trigon.

4614 The equation corresponding to (4612) in Cartesian coordinates is $0x+0y+C=0$, the intercepts on the axes being both infinite. Cartesian coordinates may therefore be regarded as trilinear with the x and y axes for two sides of the trigon and the other side at an infinite distance.

4615 The condition that three points $\alpha_1\beta_1\gamma_1$, $\alpha_2\beta_2\gamma_2$, $\alpha_3\beta_3\gamma_3$ may lie on the same straight line is the determinant equation,

$$\begin{vmatrix} \alpha_1 & \beta_1 & \gamma_1 \\ \alpha_2 & \beta_2 & \gamma_2 \\ \alpha_3 & \beta_3 & \gamma_3 \end{vmatrix}=0.$$

PROOF.—For it is the eliminant of the three simultaneous equations,

$$l\alpha_1+m\beta_1+n\gamma_1=0,\qquad l\alpha_2+m\beta_2+n\gamma_2,\qquad l\alpha_3+m\beta_3+n\gamma_3=0.\quad (583)$$

4616 Cor.—The above is also the equation of a straight line passing through two of the fixed points if the third point be considered variable.

4617 Similarly, the condition that the three following straight lines may pass through the same point, is the determinant equation on the right,

$$\begin{aligned} l_1\alpha+m_1\beta+n_1\gamma &= 0 \\ l_2\alpha+m_2\beta+n_2\gamma &= 0 \\ l_3\alpha+m_3\beta+n_3\gamma &= 0, \end{aligned} \qquad \begin{vmatrix} l_1 & m_1 & n_1 \\ l_2 & m_2 & n_2 \\ l_3 & m_3 & n_3 \end{vmatrix} = 0.$$

4618 The condition of parallelism of the two straight lines

$$\begin{aligned} l\alpha+m\beta+n\gamma &= 0, \\ l'\alpha+m'\beta+n'\gamma &= 0, \end{aligned} \qquad \begin{vmatrix} l & m & n \\ l' & m' & n' \\ \mathfrak{a} & \mathfrak{b} & \mathfrak{c} \end{vmatrix} = 0,$$

is the determinant equation

Proof.—By taking the line at infinity (4612) for the third line in (4617).

4619 Otherwise the equations of two parallel lines differ by a constant (4076). Thus

$$l\alpha+m\beta+n\gamma+k(\mathfrak{a}\alpha+\mathfrak{b}\beta+\mathfrak{c}\gamma) = 0 \qquad (4007)$$

or

$$(l+k\mathfrak{a})\,\alpha+(m+k\mathfrak{b})\,\beta+(n+k\mathfrak{c})\,\gamma = 0$$

represents any line parallel to $l\alpha+m\beta+n\gamma=0$ by varying the value of k.

The condition of perpendicularity of the two lines in (4618) is

4620 $$ll'+mm'+nn'-(mn'+m'n)\cos A-(nl'+n'l)\cos B - (lm'+l'm)\cos C = 0,$$

4621 or $$l'(l-m\cos C-n\cos B)+m'(m-n\cos A-l\cos C) + n'(n-l\cos B-m\cos A) = 0.$$

Proof.—Transform the two equations into Cartesians, by (4606), and apply the test $AA'+BB'=0$ (4078), remembering that

$$\cos(\beta-\gamma) = -\cos A, \text{ \&c. } (4011).$$

When the second line is AB or $\gamma = 0$, the condition is

4622 $$n = m \cos A + l \cos B.$$

It also appears, by (4676), that (4620) is the condition that the two lines may be conjugate with respect to the conic whose tangential equation is

4623 $$l^2 + m^2 + n^2 - 2mn \cos A - 2nl \cos B - 2lm \cos C = 0.$$

The length of the perpendicular from a point $\alpha'\beta'\gamma'$ to the line $l\alpha + m\beta + n\gamma = 0$:

4624 $$\frac{l\alpha' + m\beta' + n\gamma'}{\sqrt{\{l^2 + m^2 + n^2 - 2mn \cos A - 2nl \cos B - 2lm \cos C\}}}.$$

PROOF.—By (4095) the perpendicular is equal to the form in (4606), with x', y' in the place of x, y, divided by the square root of sum of squares of coefficients of x and y. The numerator $= l\alpha' + m\beta' + n\gamma'$. The denominator reduces by $\cos(\beta - \gamma) = -\cos A$, &c.

4625 Equation of the same perpendicular:

$$\begin{vmatrix} \alpha & \alpha' & l - m \cos C - n \cos B \\ \beta & \beta' & m - n \cos A - l \cos C \\ \gamma & \gamma' & n - l \cos B - m \cos A \end{vmatrix} = 0.$$

PROOF. — This is the eliminant of the three conditional equations $L\alpha + M\beta + N\gamma = 0$, $L\alpha' + M\beta' + N\gamma' = 0$, and equation (4621).

4626 Equation of a line drawn through $\alpha'\beta'\gamma'$ parallel to the line (l, m, n):

$$\begin{vmatrix} \alpha & \alpha' & \mathfrak{c}m - \mathfrak{b}n \\ \beta & \beta' & \mathfrak{a}n - \mathfrak{c}l \\ \gamma & \gamma' & \mathfrak{b}l - \mathfrak{a}m \end{vmatrix} = 0.$$

PROOF. — It is the eliminant of the three conditional equations $l\alpha + m\beta + n\gamma = 0$, $l\alpha' + m\beta' + n\gamma' = 0$, and the equation at (4618).

4627 The tangent of the angle between the lines (l, m, n) and (l', m', n') is

$$\frac{(mn' - m'n) \sin A + (nl' - n'l) \sin B + (lm' - l'm) \sin C}{ll' + mm' + nn' - (mn' + m'n) \cos A - (nl' + n'l) \cos B - (lm' + l'm) \cos C}$$

PROOF.—By (4071) applied to the transformed equations of the lines, (4606), observing (4007).

EQUATIONS OF PARTICULAR LINES AND COORDINATE RATIOS OF PARTICULAR POINTS IN THE TRIGON.

4628 Bisectors of the angles A, B, C:

$$\beta-\gamma = 0, \qquad \gamma-\alpha = 0, \qquad \alpha-\beta = 0.$$

4629 Centre of inscribed circle (or *in-centre*)* $\mathbf{1 : 1 : 1}$.

The coordinates are obtained from their mutual ratios by the formula (4017).

4630 Bisectors of the angles A, $\pi-B$, $\pi-C$:

$$\beta-\gamma = 0, \qquad \gamma+\alpha = 0, \qquad \alpha+\beta = 0.$$

Centre of the escribed circle which touches the side a (or *a ex-circle*) $\mathbf{-1 : 1 : 1}$.

4631 Bisectors of sides drawn through opposite vertices:

$$\beta \sin B = \gamma \sin C, \qquad \gamma \sin C = \alpha \sin A, \qquad \alpha \sin A = \beta \sin B.$$

4632 Point of intersection (or *mass-centre*):

$$\operatorname{cosec} A : \operatorname{cosec} B : \operatorname{cosec} C.$$

PROOF.—Assume $m\beta-n\gamma = 0$, by (4604), as the form of the equation of a line through A, and determine the ratio $m : n$ from the value of $\gamma : \beta$ when $\alpha = 0$.

The coordinates of the point of intersection may be found by (4610), or thus:

$$\alpha : \beta = \sin B : \sin A = \operatorname{cosec} A : \operatorname{cosec} B,$$
$$\beta : \gamma = \sin C : \sin B = \operatorname{cosec} B : \operatorname{cosec} C,$$

therefore $\alpha : \beta : \gamma = \operatorname{cosec} A : \operatorname{cosec} B : \operatorname{cosec} C.$

4633 Perpendiculars to sides drawn through opposite vertices:

$$\beta \cos B = \gamma \cos C, \qquad \gamma \cos C = \alpha \cos A, \qquad \alpha \cos A = \beta \cos B.$$

4634 Orthocentre: $\sec A : \sec B : \sec C$.

---

* This nomenclature is suggested by Professor Hudson, who proposes the following:—"*In-circle, circum-circle, a ex-circle ... mid-circle* for inscribed circle, circumscribed circle, circle escribed to the side a, and nine-point circle; also *in-centre, circum-centre, a ex-centre, ... mid-centre,* for the centres of these circles; and *in-radius, circum-radius, a ex-radius, ... mid-radius,* for their radii; *central line,* for the line on which the circum-centre, mid-centre, ortho-centre, and mass-centre lie; and *central length* for the distance between the circum-centre and the ortho-centre."

If the Cartesian coordinates of A, B, C be x_1y_1, x_2y_2, x_3y_3, the coordinates of the centre of the inscribed circle are

4635 $$x = \frac{ax_1+bx_2+cx_3}{a+b+c}, \qquad y = \frac{ay_1+by_2+cy_3}{a+b+c}.$$

PROOF.—By (4032). Find the coordinates of D where the bisector of the angle A cuts BC in the ratio $b:c$ (VI. 3), and then the coordinates of E where the bisector of B cuts AD in the ratio $b+c:a$.

4636 For the coordinates of the centre of the a ex-circle, change the sign of a in the above values of x and y.

4637 The coordinates of the mass-centre are

$$x = \tfrac{1}{3}(x_1+x_2+x_3), \qquad y = \tfrac{1}{3}(y_1+y_2+y_3).$$

4638 The coordinates of the orthocentre are obtained from the equations of the perpendiculars from x_2y_2, x_3y_3, viz.,

$$(x_1-x_3)\,x+(y_1-y_3)\,y = x_2(x_1-x_3)+y_2(y_1-y_3),$$
$$(x_1-x_2)\,x+(y_1-y_2)\,y = x_3(x_1-x_2)+y_3(y_1-y_2).$$

Perpendicular bisector of the side AB:

4639 $$\alpha\sin A-\beta\sin B+\gamma\sin(A-B) = 0,$$

4640 or $$\alpha\cos A-\beta\cos B-\frac{c}{2}\sin(A-B) = 0,$$

4641 or

$$\left(\alpha+\frac{a\sin B\sin C}{2\sin A}\right)\cos A-\left(\beta+\frac{b\sin C\sin A}{2\sin B}\right)\cos B = 0.$$

4642 Centre of circumscribed circle (or *circum-centre*):

$$\cos A : \cos B : \cos C.$$

PROOF.—A line through the intersection of γ and $\alpha\sin A-\beta\sin B$ (4631) is of the form $\alpha\sin A-\beta\sin B+n\gamma = 0$, and, by (4622),

$$n = -\sin B\cos A+\sin A\cos B = \sin(A-B).$$

Otherwise, by (4633) and (4619),

$$\alpha\cos A-\beta\cos B+k = 0$$

is any line perpendicular to AB; and the constant k is found by giving $\alpha:\beta$ the value which it has at the centre of AB.

4643 Centre of the nine-point circle (or *mid-centre*):

$$\cos(B-C) : \cos(C-A) : \cos(A-B).$$

PROOF.—By (955) the coordinates are the arithmetic means of the corresponding coordinates of the orthocentre and circum-centre. Therefore, by (4634, '42) and (4017),

$$\alpha = k\left\{\frac{\sec A}{\sin A\sec A+\sin B\sec B+\sin C\sec C}+\frac{\cos A}{\sin A\cos A+\sin B\cos B+\sin C\cos C}\right\},$$

which reduces to $\cos(B-C)\times$ constant.

4644 Ex. 1.—In any triangle ABC (Fig. of 955), the mass-centre R, the orthocentre O, and the circum-centre Q lie on the same straight line;* for the coordinates of these points given at (4632, '34, '42), substituted in (4615), give for the value of the determinant

$$\operatorname{cosec} A\,(\sec B\cos C-\cos B\sec C)+\&c.,$$

which vanishes.

Similarly, by the coordinates in (4643), it may be shown that the mid-centre N lies on the same line.

Equation of the central line:

Ex. 2.—To find the line drawn through the orthocentre and mass-centre of ABC. The coordinates of these points are given at (4632, '34). Substituting in the determinant (4616) and reducing, the equation becomes

$$\alpha\sin 2A\sin(B-C)+\beta\sin 2B\sin(C-A)+\gamma\sin 2C\sin(A-B)=0.$$

Ex. 3.—Similarly, from (4629, '42), the line drawn through the centres of the inscribed and circumscribed circles is

$$\alpha(\cos B-\cos C)+\beta(\cos C-\cos A)+\gamma(\cos A-\cos B)=0.$$

Ex. 4.—A parallel to AB drawn through C:

$$\alpha\sin A+\beta\sin B=0.$$

For this is a line through $\alpha\beta$, by (4604), and the equation differs only by a constant from $\gamma=0$, for it may be written

$$(\alpha\sin A+\beta\sin B+\gamma\sin C)-\gamma\sin C=0.$$

Ex. 5.—A perpendicular to BC drawn through C is

$$\alpha\cos C+\beta=0.$$

For a perpendicular is $\beta\cos B-\gamma\cos C=0$ (4633) .... (1),
and a line through C is of the form $l\alpha+m\beta=0$. Hence, by (4619), the constant $k(\alpha\sin A+\beta\sin B+\gamma\sin C)$ must be added to (1) so as to eliminate γ. Thus

$$\beta\sin C\cos B+\alpha\sin A\cos C+\beta\sin B\cos C=0,$$

$$\beta\sin(B+C)+\alpha\sin A\cos C=0 \quad\text{or}\quad \beta+\alpha\cos C=0.$$

---

* The central line. See note to (4629).

ANHARMONIC RATIO.

For the definition, see (1052).

4648 The three ratios of that article are the values of the ratio $k : k'$ in the three following pencils of four lines respectively—

$$\alpha = 0, \quad \alpha - k\beta = 0, \quad \beta = 0, \quad \alpha + k'\beta = 0 \ldots \text{ (i.) (Fig. 34)},$$

$$\alpha = 0, \quad \alpha - k\beta = 0, \quad \alpha - k'\beta = 0, \quad \beta = 0 \ldots \text{ (ii.) (Fig. 35)},$$

$$\alpha = 0, \quad \beta = 0, \quad \alpha + k\beta = 0, \quad \alpha + k'\beta = 0 \ldots \text{(iii.) (Fig. 36)}.$$

4649 The anharmonic ratio (i.) becomes harmonic when $k = k'$. Hence the lines $\alpha + k\beta$, $\alpha - k\beta$ form a harmonic pencil with the lines α, β, the first dividing the external and the second the internal angle between α and β (Fig. 37).

4650 Similarly, the anharmonic ratio of four lines whose equations are

$$\alpha - \mu_1\beta = 0, \quad \alpha - \mu_2\beta = 0, \quad \alpha - \mu_3\beta = 0, \quad \alpha - \mu_4\beta = 0,$$

is the fraction $$\frac{(\mu_1 - \mu_2)(\mu_3 - \mu_4)}{(\mu_1 - \mu_4)(\mu_2 - \mu_3)}.$$

PROOF.—Let OL be the line $\alpha = 0$, and OR, $\beta = 0$.

$\mu_1 - \mu_2 =$ difference of perpendiculars from A and B upon OL, divided by p.

Similarly, $\mu_3 - \mu_4$, &c. These differences are proportional to the segments AB, CD, AD, BC, and p is a common divisor.

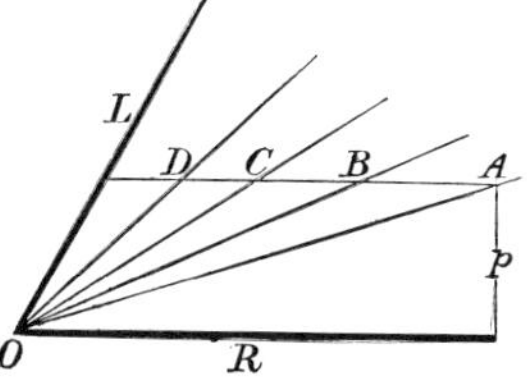

4651 *Homographic* pencils of lines are those which have the same anharmonic ratio. Thus the two pencils

$$\alpha - \mu_1\beta, \quad \alpha - \mu_2\beta, \quad \alpha - \mu_3\beta, \quad \alpha - \mu_4\beta,$$

and $$\alpha' - \mu_1\beta', \quad \alpha' - \mu_2\beta', \quad \alpha' - \mu_3\beta', \quad \alpha' - \mu_4\beta',$$

are homographic pencils.

THE COMPLETE QUADRILATERAL.

4652 Def.—Any four right lines together with the three, called diagonals, which join the points of intersection, make a figure called a complete quadrilateral.

4653 Let O be any point in the plane of the trigon ABC. Draw AOa, BOb, COc, and complete the figure. The equations of the different lines may be written as under, with the aid of proposition (4604), the ratios $l : m : n$ being arbitrary and dependent upon the position of O.

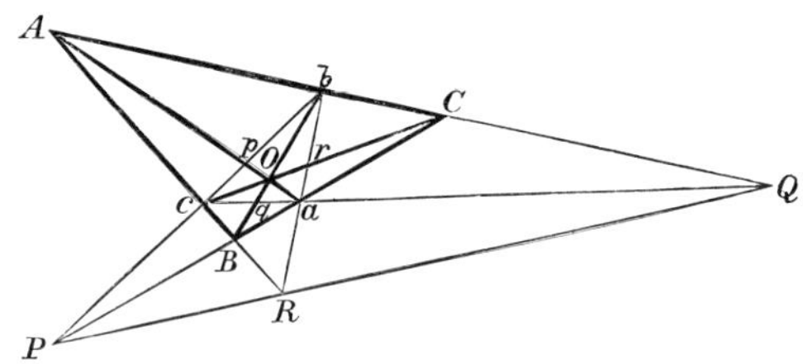

$$Aa,\quad m\beta - n\gamma = 0, \qquad AP,\quad m\beta + n\gamma = 0,$$
$$Bb,\quad n\gamma - l\alpha = 0, \qquad BQ,\quad n\gamma + l\alpha = 0,$$
$$Cc,\quad l\alpha - m\beta = 0, \qquad CR,\quad l\alpha + m\beta = 0;$$

$$bc,\quad m\beta + n\gamma - l\alpha = 0, \qquad OP,\quad m\beta + n\gamma - 2l\alpha = 0,$$
$$ca,\quad n\gamma + l\alpha - m\beta = 0, \qquad OQ,\quad n\gamma + l\alpha - 2m\beta = 0,$$
$$ab,\quad l\alpha + m\beta - n\gamma = 0, \qquad OR,\quad l\alpha + m\beta - 2n\gamma = 0,$$

$$PQR,\quad l\alpha + m\beta + n\gamma = 0.$$

Proof.—Aa, Bb, Cc are concurrent by addition. bc is concurrent with Bb and β, and with Cc and γ, by (4604). AP and OP are each concurrent with bc and α. PQR is concurrent with each pair of lines bc and α, ca and β, ab and γ. Similarly for the rest.

4654 Every pencil of four lines in the above figure (supplying AP, BQ, CR) is a harmonic pencil.

Proof.—By the test in (4649), the alternate pairs of equations being the sum and difference of the other two in every case.

Otherwise by projection. Let $PQRS$ be the quadrilateral, with diagonals RP, QS meeting in C. (Supply the lines AC, BC in the figure.) Taking the plane of projection parallel to OAB, the figure projects into the parallelogram $pqrs$; the points A, B pass to infinity, and therefore the lines AC, BC become lines harmonically divided by the sides of the parallelogram, the centre, and the points at infinity.

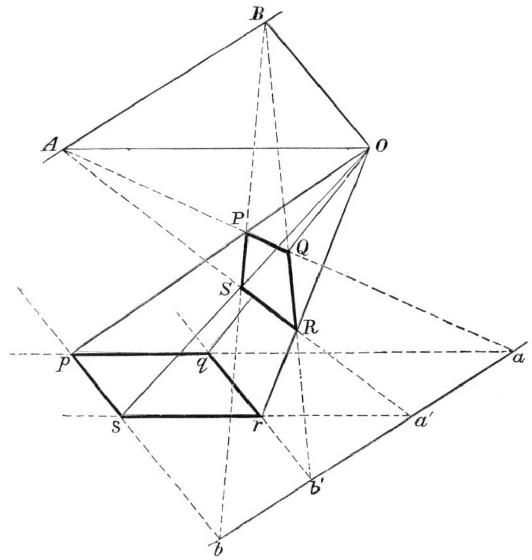

4655 Theorem (974) may be proved by taking α, β, γ for the lines BC, CA, AB, and $l'\alpha+m\beta+n\gamma$, $l\alpha+m'\beta+n\gamma$, $l\alpha+m\beta+n'\gamma$ for bc, ca, ab, the last form being deduced from the preceding by the concurrence of Aa, Bb, and Cc.

THE GENERAL EQUATION OF A CONIC.

The general equation of the second degree is

4656 $$a\alpha^2+b\beta^2+c\gamma^2+2f\beta\gamma+2g\gamma\alpha+2h\alpha\beta=0.$$

This equation will be denoted by $\phi(\alpha,\beta,\gamma)=0$ or $u=0$.

Equation of the tangent or polar:

4657 $$u_{\alpha'}\alpha+u_{\beta'}\beta+u_{\gamma'}\gamma=0 \quad \text{or} \quad u_\alpha\alpha'+u_\beta\beta'+u_\gamma\gamma'=0,$$

the two forms being equivalent and the notation being that of (1405). The first equation written in full is

4659
$$(a\alpha'+h\beta'+g\gamma')\,\alpha+(h\alpha'+b\beta'+f\gamma')\,\beta+(g\alpha'+f\beta'+c\gamma')\,\gamma=0.$$

PROOF.—By the methods in (4120). Otherwise by (4678); let $\alpha\beta\gamma$ be on the curve; then $\phi(\alpha, \beta, \gamma) = 0$. Next let the point where the line cuts the curve move up to $\alpha\beta\gamma$. Then the line becomes a tangent and the ratio $n : n'$ vanishes; the condition for this gives equation (4658).

COR.—The polars of the vertices of the triangle of reference are

4660 $$a\alpha+h\beta+g\gamma = 0, \quad h\alpha+b\beta+f\gamma = 0, \quad g\alpha+f\beta+c\gamma = 0.$$

4661 The condition that u may break up into two linear factors representing two right lines is, by (4469), $\Delta = 0$, where

4662 $$\Delta = abc+2fgh-af^2-bg^2-ch^2. \qquad (4454)$$

4663 The general tangential equation of the conic (4656) expresses the condition that the line $\lambda\alpha+\mu\beta+\nu\gamma$ may touch the curve and is the determinant equation annexed. The same written in full is

$$\begin{vmatrix} a & h & g & \lambda \\ h & b & f & \mu \\ g & f & c & \nu \\ \lambda & \mu & \nu & \end{vmatrix} = 0.$$

4664 $$(bc-f^2)\lambda^2+(ca-g^2)\mu^2+(ab-h^2)\nu^2$$
$$+2(gh-af)\mu\nu+2(hf-bg)\nu\lambda+2(fg-ch)\lambda\mu = 0,$$

4665 or $$A\lambda^2+B\mu^2+C\nu^2+2F\mu\nu+2G\nu\lambda+2H\lambda\mu = 0;$$

writing, as in (1642),

$$A = bc-f^2, \qquad B = ca-g^2, \qquad C = ab-h^2,$$
$$F = gh-af, \qquad G = hf-bg, \qquad H = fg-ch.$$

The tangential equation will be denoted by $\Phi(\lambda, \mu, \nu) = 0$ or $U = 0$, to correspond with (4656).

PROOF.—The determinant is the eliminant of the equation of the line $\lambda\alpha+\mu\beta+\nu\gamma = 0$, and the three equations obtained by equating λ, μ, ν to the coefficients of α, β, γ in (4659).

Otherwise.—Assume $\lambda\alpha+\mu\beta+\nu\gamma = 0$ for the tangent. Substitute the value of the ratio $\beta : \gamma$ obtained from it in the equation of the curve, and express the condition for equal roots (4119).

4666 Conversely, if the line $\lambda\alpha+\mu\beta+\nu\gamma$ has the coefficients λ, μ, ν connected by the equation of the second degree $U = 0$ (4664), then the envelope of the line is the conic in the

$$\begin{vmatrix} A & H & G & \alpha \\ H & B & F & \beta \\ G & F & C & \gamma \\ \alpha & \beta & \gamma & \end{vmatrix} = 0.$$

determinant form annexed corresponding to (4663), or in full

4667 $(BC-F^2)\,\alpha^2+(CA-G^2)\,\beta^2+(AB-H^2)\,\gamma^2$
$+2\,(GH-AF)\,\beta\gamma+2\,(HF-BG)\,\gamma\alpha+2\,(FG-CH)\,\alpha\beta=0.$

4668 or $\Delta\,(a\alpha^2+b\beta^2+c\gamma^2+2f\beta\gamma+2g\gamma\alpha+2h\alpha\beta)=0.$

PROOF.—Eliminate ν from $U=0$ and the given line. The result is of the form $L\lambda^2+2R\lambda\mu+M\mu^2=0$, and therefore the envelope is $LM=R^2$, by (4792). This produces equation (4667). The coefficients are the first minors of the reciprocal determinant of Δ (1643), and therefore, by (585), are equal to $a\Delta$, $b\Delta$, &c.

4669 The condition that U may consist of two linear factors is, as in (4661), $D=0$, where

4670 $D=ABC+2FGH-AF^2-BG^2-CH^2=\Delta^2.$ (1643)

In this case U becomes the equation of two points, since the line $\lambda\alpha+\mu\beta+\nu\gamma$ must pass through one or other of two fixed points. See (4913).

4671 The coordinates of the pole of $\lambda\alpha+\mu\beta+\nu\gamma$ are as

$$A\lambda+H\mu+G\nu : H\lambda+B\mu+F\nu : G\lambda+F\mu+C\nu,$$

4672 or $$U_\lambda : U_\mu : U_\nu .$$

PROOF.—By (4659) we have the equations in the margin, the solution of which gives the ratios of $\alpha:\beta:\gamma$.

$$a\alpha+h\beta+g\gamma=k\lambda$$
$$h\alpha+b\beta+f\gamma=k\mu$$
$$g\alpha+f\beta+c\gamma=k\nu.$$

4673 $$\frac{\alpha}{A\lambda+H\mu+G\nu}=\frac{\beta}{H\lambda+B\mu+F\nu}=\frac{\gamma}{G\lambda+F\mu+C\nu}=\frac{k}{\Delta}.$$

Hence the tangential equation of the pole of $\lambda'\alpha+\mu'\beta+\nu'\gamma$, *i.e.*, the condition that $\lambda\alpha+\mu\beta+\nu\gamma$ may pass through the pole; or, in other words, that the two lines may be mutually conjugate, is

4674 $\lambda U_{\lambda'}+\mu U_{\mu'}+\nu U_{\nu'}=0$ or $\lambda' U_\lambda+\mu' U_\mu+\nu' U_\nu=0,$

the two forms being equivalent, and each

4676 $$=A\lambda\lambda'+B\mu\mu'+C\nu\nu'$$
$$+F\,(\mu\nu'+\mu'\nu)+G\,(\nu\lambda'+\nu'\lambda)+H\,(\lambda\mu'+\lambda'\mu).$$

The coordinates of the centre α_0, β_0, γ_0 are in the ratios

4677 $A\mathfrak{a}+H\mathfrak{b}+G\mathfrak{c} : H\mathfrak{a}+B\mathfrak{b}+F\mathfrak{c} : G\mathfrak{a}+F\mathfrak{b}+C\mathfrak{c},$

where $\mathfrak{a}$, $\mathfrak{b}$, $\mathfrak{c}$ are the sides of the trigon.

PROOF.—By (4671), since the centre is the pole of the line at infinity $a\alpha+b\beta+c\gamma=0$ (4612).

The quadratic for the ratio $n:n'$ of the segments into which the line joining two given points $\alpha\beta\gamma$, $\alpha'\beta'\gamma'$ is divided by the conic is, with the notation of (4656–7),

4678
$$\phi(\alpha',\beta',\gamma')\,n^2+(\phi_\alpha\alpha'+\phi_\beta\beta'+\phi_\gamma\gamma')\,nn'+\phi(\alpha,\beta,\gamma)\,n'^2=0.$$

PROOF.—By the method of (4131).

The equation of the pair of tangents at the points where γ meets the general conic u (4656), is

4679 $$au_\beta^2+2hu_\alpha u_\beta+bu_\alpha^2=0.$$

PROOF.—The point $\alpha'\beta'$, where γ meets the curve, is found from $a\alpha'^2+2h\alpha'\beta'+b\beta'^2=0$ [$\gamma=0$ in (4656)]. The tangent at such a point is $u_\alpha\alpha'+u_\beta\beta'=0$ (4658). Eliminate α', β'.

The equation of a pair of tangents from $\alpha'\beta'\gamma'$ is

4680 $$4\phi(\alpha'\beta'\gamma')\,\phi(\alpha\beta\gamma)=(\phi_\alpha\alpha'+\phi_\beta\beta'+\phi_\gamma\gamma')^2.$$

PROOF.—By the condition for equal roots of (4678).

By actual expansion the equation becomes

4681
$$\begin{aligned}(B\gamma^2+C\beta^2-2F\beta\gamma)\,\alpha'^2+(C\alpha^2+A\gamma^2-2G\gamma\alpha)\,\beta'^2+(A\beta^2+B\alpha^2-2H\alpha\beta)\,\gamma'^2\\ +2\,(-A\beta\gamma+H\gamma\alpha+G\alpha\beta-F\alpha^2)\,\beta'\gamma'\\ +2\,(H\beta\gamma-B\gamma\alpha+F\alpha\beta-G\beta^2)\,\gamma'\alpha'\\ +2\,(G\beta\gamma+F\gamma\alpha-C\alpha\beta-H\gamma^2)\,\alpha'\beta'=0.\end{aligned}$$

In which either α', β', γ' or α, β, γ may be the variables, for the forms are convertible.

Otherwise the equation of the two tangents is

4682 $$\Phi(\beta\gamma'-\beta'\gamma,\ \gamma\alpha'-\gamma'\alpha,\ \alpha\beta'-\alpha'\beta)=0.$$ (4665)

PROOF.—By substituting $\beta\gamma'-\beta'\gamma$, &c. for λ, μ, ν in (4664), the condition that the line joining $\alpha'\beta'\gamma'$ to any point $\alpha\beta\gamma$ on either tangent (see 4616) should touch the conic is fulfilled. The expansion produces the preceding equation (4681).

The equation of the asymptotes is

4683 $$\phi(\alpha,\beta,\gamma)=\phi(\alpha_0,\beta_0,\gamma_0)=k\Sigma \quad \ldots\ldots\ldots\ldots (1),$$

where α_0, β_0, γ_0 are the coordinates of the centre.

Otherwise the equation, in a form homogeneous in α, β, γ, is

4684 $$(\mathfrak{a}\alpha_0+\mathfrak{b}\beta_0+\mathfrak{c}\gamma_0)\,\phi(\alpha,\beta,\gamma) = k(\mathfrak{a}\alpha+\mathfrak{b}\beta+\mathfrak{c}\gamma)^2 \quad\ldots\ldots (2),$$

where $\mathfrak{a}, \mathfrak{b}, \mathfrak{c}$ are the sides of the trigon.

And, finally, if the tangential equation (4664) be denoted by $\Phi(\lambda, \mu, \nu) = 0$, the equation of the asymptotes may be presented in the form

4685 $$\Phi(\mathfrak{a}, \mathfrak{b}, \mathfrak{c})\,\phi(\alpha,\beta,\gamma) = (\mathfrak{a}\alpha+\mathfrak{b}\beta+\mathfrak{c}\gamma)^2\Delta \quad\ldots\ldots\ldots (3).$$

Proof.—(i.) The asymptotes are identical with a pair of tangents from the centre; therefore, put $\alpha_0, \beta_0, \gamma_0$ for α', β', γ' in (4680); thus

$$\phi(\alpha,\beta,\gamma)\,\phi(\alpha_0,\beta_0,\gamma_0) = k^2(\mathfrak{a}\alpha+\mathfrak{b}\beta+\mathfrak{c}\gamma)^2 = k^2\Sigma^2 \quad\ldots\ldots\ldots\ldots (4),$$

since the polar becomes the line at infinity.

Now, multiplying the three equations in (4672) by α, β, γ respectively, and adding, we obtain $\phi(\alpha,\beta,\gamma) = k(\lambda\alpha+\mu\beta+\nu\gamma)$, and therefore

$$\phi(\alpha_0,\beta_0,\gamma_0) = k(\mathfrak{a}\alpha+\mathfrak{b}\beta+\mathfrak{c}\gamma) = k\Sigma \quad\ldots\ldots\ldots\ldots\ldots\ldots\ldots\ldots (5),$$

since the line at infinity (4612) is the pole of the centre.

From (4) and (5), by eliminating k, equation (1) is produced; and by dividing (4) by (5), we get equation (2).

Again, taking the values of α, β, γ from (4673), we have

$$\frac{\lambda\alpha+\mu\beta+\nu\gamma}{k} = \frac{\Phi(\lambda,\mu,\nu)}{\Delta}, \text{ and therefore } \frac{a\alpha_0+b\beta_0+c\gamma_0}{k} = \frac{\Phi(a,b,c)}{\Delta}.$$

By the last equation, (2) is converted into (3). See also (4966).

Cor.—Since the centre $(\alpha_0, \beta_0, \gamma_0)$ is on the asymptotes, we have

4686 $$\phi(\alpha_0,\beta_0,\gamma_0) = \Sigma^2\Delta \div \Phi(\mathfrak{a}, \mathfrak{b}, \mathfrak{c}).$$

4687 The semi-axes of the general conic (4656) are the values of r obtained from the quadratic

$$\begin{vmatrix} \left(a+\dfrac{\mathfrak{a}s\cos A}{r^2}\right), & h, & g, & \mathfrak{a} \\ h, & \left(b+\dfrac{\mathfrak{b}s\cos B}{r^2}\right), & f, & \mathfrak{b} \\ g, & f, & \left(c+\dfrac{\mathfrak{c}s\cos C}{r^2}\right), & \mathfrak{c} \\ \mathfrak{a}, & \mathfrak{b}, & \mathfrak{c}, & \end{vmatrix} = 0,$$

where $\mathfrak{a}, \mathfrak{b}, \mathfrak{c}$ are the sides of the trigon, and

$$s = \mathfrak{abc}\Delta \div \Phi(\mathfrak{abc}).$$

PROOF.—The centre being $\alpha_0\beta_0\gamma_0$, put $\alpha-\alpha_0=x$, $\beta-\beta_0=y$, $\gamma-\gamma_0=z$. $\alpha\beta\gamma$ being a point on the conic, and r the radius to it from the centre, we have, by (4602),

$$r^2=\frac{\mathfrak{abc}}{\Sigma^2}(x^2\mathfrak{a}\cos A+y^2\mathfrak{b}\cos B+z^2\mathfrak{c}\cos C) \quad\dots\dots\dots\dots\dots (1).$$

Also (4656), $\quad \phi(\alpha,\beta,\gamma)=\phi(\alpha_0+x,\ \beta_0+y,\ \gamma_0+z)=0.$

Expand and write l, m, n for $a\alpha_0+h\beta_0+g\gamma_0$, $h\alpha_0+b\beta_0+f\gamma_0$, $g\alpha_0+f\beta_0+c\gamma_0$. The terms in x, y, z become

$$lx+my+nz=l(\alpha-\alpha_0)+\&c.=\Sigma-\Sigma=0\ (4007)\dots\dots\dots\dots(2),$$

and we obtain $\phi(x,y,z)=-\phi(\alpha_0,\beta_0,\gamma_0)=\Sigma^2\Delta\div\Phi(\mathfrak{a},\mathfrak{b},\mathfrak{c})$ (4686) ......(3).

The maximum and minimum values of r^2 and therefore of

$$x^2\mathfrak{a}\cos A+y^2\mathfrak{b}\cos B+z^2\mathfrak{c}\cos C \quad\dots\dots\dots\dots\dots\dots (4)$$

are required, subject to the equations (2) and (3). By the method of undetermined multipliers (1862), the quadratic above is found.—*Ferrers's Tril. Coord.*, Ch. 4, Art. 18.

4688 The area of the conic $=\dfrac{\pi\Sigma\mathfrak{abc}\Delta}{\{\Phi(\mathfrak{a},\mathfrak{b},\mathfrak{c})\}^{\frac{3}{2}}}.$

PROOF.—If the roots of the quadratic (4687) are $\pm r_1^{-1}$, $\pm r_2^{-2}$, the area will be πr_1r_2. The coefficient of r^{-4} reduces by trigonometry to $-\Sigma^2s^2$, and the absolute term is $-\Phi(\mathfrak{a},\mathfrak{b},\mathfrak{c})$. Hence the product of the roots is found.

4689 The conic will be an ellipse, hyperbola, or parabola, according as $\Phi(\mathfrak{a},\mathfrak{b},\mathfrak{c})$ (4664) is positive, negative, or zero.

PROOF.—The squares of the semi-axes have opposite signs in the hyperbola. Therefore the product of the roots of the quadratic (4687) must for an hyperbola be negative, and therefore Φ negative in (4688).

$\Phi(\mathfrak{a},\mathfrak{b},\mathfrak{c})=0$ makes the curve touch the line at infinity (4664), a property which distinguishes the parabola.

The condition that the general conic (4656) may be a rectangular hyperbola is

4690 $\quad a+b+c=2f\cos A+2g\cos B+2h\cos C.$

PROOF.—Let the asymptotes be

$$l\alpha+m\beta+n\gamma=0,\qquad l'\alpha+m'\beta+n'\gamma=0.$$

Forming the product, equating coefficients with (4685), and denoting $\phi(\mathfrak{a},\mathfrak{b},\mathfrak{c})$ by ϕ, we get the proportions

$$\frac{ll'}{a\phi-\mathfrak{a}^2\Delta}=\frac{mm'}{b\phi-\mathfrak{b}^2\Delta}=\frac{nn'}{c\phi-\mathfrak{c}^2\Delta}=\frac{mn'+m'n}{2(f\phi-\mathfrak{bc}\Delta)}$$

$$=\frac{nl'+n'l}{2(g\phi-\mathfrak{ca}\Delta)}=\frac{lm'+l'm}{2(h\phi-\mathfrak{ab}\Delta)}.$$

We may therefore substitute these denominators in (4620) for the condition of perpendicularity of the asymptotes. The result reduces to the equation above, by (837).

For another method, see (5002).

4691 The general conic (4656) will become a circle when the following relation exists between the coefficients:

$$\begin{aligned} & b\sin^2 C + c\sin^2 B - 2f\sin B\sin C \\ = \; & c\sin^2 A + a\sin^2 C - 2g\sin C\sin A \\ = \; & a\sin^2 B + b\sin^2 A - 2h\sin A\sin B. \end{aligned}$$

PROOF.—Equate coefficients of the equation of the conic (4656) with those of the circle in (4751).

4692 The equation of the pair of lines drawn from a point $\alpha'\beta'\gamma'$ to the points of intersection of the conic ϕ and the line $L \equiv \lambda\alpha + \mu\beta + \nu\gamma = 0$ is, writing L' for $\lambda\alpha' + \mu\beta' + \nu\gamma'$, with the notation of (4656–7),

$$L'^2\phi(\alpha, \beta, \gamma) - 2LL'(\phi_\alpha\alpha' + \phi_\beta\beta' + \phi_\gamma\gamma') + L^2\phi(\alpha', \beta', \gamma') = 0.$$

PROOF.—By the method of (4133).

4693 The *Director-Circle* of the conic, that is, the locus of intersection of tangents at right angles, is, in Cartesians,

$$C(x^2+y^2) - 2Gx - 2Fy + A + B = 0.$$

PROOF.—Let the equation of a tangent through xy be

$$m\xi - \eta + (y - mx) = 0.$$

Therefore in the tangential equation (4665) put $\lambda = m$, $\mu = -1$, $\nu = y - mx$, and apply the condition, *Product of roots of quadratic in* $m = -1$ (4078).

The equation of the same circle in trilinears is

4694 $$\begin{aligned} & (B + C + 2F\cos A)\,\alpha^2 + (C + A + 2G\cos B)\,\beta^2 + (A + B + 2H\cos C)\,\gamma^2 \\ & + 2\,(A\cos A - H\cos B - G\cos C - F)\,\beta\gamma \\ & + 2\,(-H\cos A + B\cos B - F\cos C - G)\,\gamma\alpha \\ & + 2\,(-G\cos A - F\cos B + C\cos C - H)\,\alpha\beta = 0; \end{aligned}$$

or, in the form of (4751),

4695
$$(\mathfrak{a}\alpha + \mathfrak{b}\beta + \mathfrak{c}\gamma)\left(\frac{B + C + 2F\cos A}{\mathfrak{a}}\,\alpha + \&c.\right) = \frac{\Phi(\mathfrak{a}, \mathfrak{b}, \mathfrak{c})}{\mathfrak{abc}}\,(\mathfrak{a}\beta\gamma + \mathfrak{b}\gamma\alpha + \mathfrak{c}\alpha\beta).$$

PROOF.—The equation of a pair of tangents (4681) through a point $\alpha\beta\gamma$ in trilinears, when the tangents are at right angles, represents the limiting case of a rectangular hyperbola. Therefore the equation referred to must have the coefficients of α'^2, β'^2, &c. connected by the relation in (4690), which thus becomes the equation of the locus of the point $\alpha\beta\gamma$; *i.e.*, the director-circle.

4696 When the general conic is a parabola, $C=0$ in (4693) and $\Phi(a, b, c) = 0$ in (4695), by (4430) and (4689), and these equations then represent the directrix.

PARTICULAR CONICS.

4697 A conic circumscribing the quadrilateral, the equations of whose sides are $\alpha=0$, $\beta=0$, $\gamma=0$, $\delta=0$, (Fig. 38)

$$\alpha\gamma = k\beta\delta.$$

PROOF.—This is a curve of the second degree, and it passes through the points where α meets β and δ, and also where γ meets β and δ.

4698 The circumscribing circle is $\alpha\gamma = \pm\beta\delta$; $+$ or $-$, as the origin of coordinates lies without or within the quadrilateral.

PROOF.—Transform (4697) into Cartesians (4009); equate coefficients of x and y and put the coefficients of xy equal to zero.

4699 A conic having α and γ for tangents and β for the chord of contact: (Fig. 39)

$$\alpha\gamma = k\beta^2.$$

PROOF.—Make δ coincide with β in (4698).

4700 A conic having two common chords α and β with a given conic S: (Fig. 40)

$$S = k\alpha\beta.$$

4701 A conic having a common chord of contact α with a given conic S: (Fig. 41)

$$S = k\alpha^2.$$

4702 COR.—If RPQ be drawn always parallel to a given line, $PN^2 \propto RP.PQ$, by (4317).

4703 A conic having a common tangent T at a point $x'y'$ and a common chord with the conic S: (Fig. 42)

$$S = T(lx+my+nz).$$

4704 A conic osculating S at the point $x'y'$ where T touches at one extremity of the common chord $l(x-x')+m(y-y')$: (Fig. 43)

$$S = T(lx+my-lx'-my').$$

4705 A conic having common tangents T, T' at common points with the conic S: (Fig. 44)

$$S = kTT'.$$

4706 A conic having four coincident points with the conic S at the point where T touches: (Fig. 45)

$$S = kT^2.$$

4707 The conics $S+L^2 = 0$, $S+M^2 = 0$, $S+N^2 = 0$,

(Fig 46) having respectively L, M, N for common chords of contact with the conic S, will have the six chords of intersection

$$L \pm M = 0, \qquad M \pm N = 0, \qquad N \pm L = 0,$$

passing three and three through the same points.

PROOF.—From $(S+M^2)-(S+N^2) = (M+N)(M-N)$, &c.

By supposing one or more of the conics to become right lines, various theorems may be obtained.

4709 The diagonals of the inscribed and circumscribed quadrilaterals of a conic all pass through the same point and form a harmonic pencil.

PROOF.—(Fig. 47.) By (4707), or by taking $LM = R^2$ and $L'M' = R'^2$ for the equations of the conic by (4784).

4710 If three conics have a chord common to all, the other three chords common to pairs pass through the same point.

PROOF.—(Fig. 48.) Take S, $S+LM$, $S+LN$ for the conics, L being the chord common to all; then M, N, $M-N$ are the other common chords.

4711 The hyperbola $xy = (0x+0y+p)^2$

is of the form (4699), and has for a chord of contact at infinity $0x+0y+p=0$, x, y being the tangents from the centre.

4712 The parabola $y^2 = (0x+0y+p)\,x$

has the tangent at infinity $0x+0y+p=0$.

4713 So the general equation of a parabola may be put in the form of (4699). Thus

$$(\alpha x+\beta y)^2+(2gx+2fy+c)(0x+0y+1)=0.$$

Here $\alpha x+\beta y$ is the chord of contact, that is, a diameter; $2gx+2fy+c$ is the finite tangent at its extremity, and $0x+0y+1$ the tangent at the other extremity, supposed at infinity.

4714 The general conic may be written

$$(ax^2+2hxy+by^2)+(2gx+2fy+c)(0x+0y+1)=0.$$

For this is of the form $\alpha\gamma+k\beta\delta$, δ being at infinity.

4715 The conics S and $S-k\,(0x+0y+1)^2$

have double contact at infinity, and are similar.

4716 The parabolas S and $S-k^2$

have a contact of the third order at infinity.

Proof.—For S and $S-(0x+0y+k)^2$ have the line at infinity for a chord of contact; and, by (4712), this chord of contact is also a tangent to both curves.

4717 All circles are said to pass through the same two imaginary points at infinity (see 4918) and through two real or imaginary finite points.

Proof.—The general equation of the circle (4144) may be written

$$(x+iy)(x-iy)+(2gx+2fy+c)(0x+0y+1)=0;$$

and this is of the form (4697). Here the lines $x\pm iy$ intersect $0x+0y+1$ in two imaginary points which have been called the *circular points at infinity*, and $2gx+2fy+c$ in two finite points P, Q; and these points are all situated on the locus $x^2+y^2+2gx+2fy+c=0$.

4718 Concentric circles touch in four imaginary points at infinity.

PROOF.—The centre being the origin, equation (4136) may be written $(x+iy)(x-iy) = (0x+0y+r)^2$, which, by (4699), shows that the lines $x \pm iy$ have each double contact with the (*supplementary*) curve at infinity, and the variation of r does not affect this result. Compare (4711).

4719 The equation of any conic may be put in the form

$$x^2+y^2 = e^2\gamma^2.$$

Here $x=0$, $y=0$ are two sides of the trigon intersecting at right angles in the focus; $\gamma=0$, the third side, is the directrix, and e is the eccentricity.

The conic becomes a circle when $e=0$ and $\gamma=\infty$, so that $e\gamma=r$, the radius, (4718).

4720 Two imaginary tangents drawn through the focus are, by (4699),

$$(x+iy)(x-iy) = 0.$$

These tangents are identical with the lines drawn through the two circular points at infinity (see 4717). Hence, if two tangents be drawn to the conic from each of the circular points at infinity, they will intersect in two imaginary points, and also in two real points which are the foci of the conic.

All confocal conics, therefore, have four imaginary common tangents, and two opposite vertices of the quadrilateral formed by the tangents are the foci of the conics.

4721 If the axes are oblique, this universal form of the equation of the conic becomes

$$x^2+2xy\cos\omega+y^2 = e^2\gamma^2.$$

The two imaginary tangents through the focus must now be written

$$\{x+y(\cos\omega+i\sin\omega)\}\{x+y(\cos\omega-i\sin\omega)\} = 0.$$

4722 Any two lines including an angle θ form, with the lines drawn from the two circular points at infinity to their point of intersection, a pencil of which the anharmonic ratio is $e^{i(\pi-2\theta)}$.

PROOF.—Take the two lines for sides β, γ of the trigon. The equation of the other pair of lines to the circular points will be obtained by eliminating α between the equations of the line at infinity and the circum-circle,

viz., $$\mathfrak{a}\alpha+\mathfrak{b}\beta+\mathfrak{c}\gamma=0 \quad\text{and}\quad \frac{\mathfrak{a}}{\alpha}+\frac{\mathfrak{b}}{\beta}+\frac{\mathfrak{c}}{\gamma}=0. \qquad (4738)$$

The result is $$\beta^2+2\beta\gamma\cos\theta+\gamma^2=0;$$

or, in factors, $$(\beta+e^{i\theta}\gamma)(\beta+e^{-i\theta}\gamma)=0.$$

The anharmonic ratio of the pencil formed by the four lines β, $\beta+e^{i\theta}\gamma$, γ, $\beta+e^{-i\theta}\gamma$ is, by (4648, i.),

$$-e^{i\theta}:e^{-i\theta}=-e^{i2\theta}=e^{i(\pi-2\theta)}.$$

4723 COR.—If $\theta=\frac{1}{2}\pi$, the lines are at right angles, and the four lines form a harmonic pencil. [*Ferrers' Tril. Coords.*, Ch. VIII.

THE CIRCUMSCRIBING CONIC OF THE TRIGON.

4724 The equation of this conic (Fig. 49) is

$$l\beta\gamma+m\gamma\alpha+n\alpha\beta=0 \quad\text{or}\quad \frac{l}{\alpha}+\frac{m}{\beta}+\frac{n}{\gamma}=0.$$

PROOF.—The equation is of the second degree, and it is satisfied by $\alpha=0$, $\beta=0$ simultaneously. It therefore passes through the point $\alpha\beta$. Similarly through $\beta\gamma$ and $\gamma\alpha$.

The tangents at A, B, and C are

4726 $$\frac{m}{\beta}+\frac{n}{\gamma}=0,\qquad \frac{n}{\gamma}+\frac{l}{\alpha}=0,\qquad \frac{l}{\alpha}+\frac{m}{\beta}=0.$$

PROOF.—By writing (4724) in the form

$$m\gamma\alpha+\beta(l\gamma+n\alpha)=0,$$

$l\gamma+n\alpha=0$ is seen, by (4697), to be the tangent at $\alpha\gamma$; for the intersections of α and γ, with the curve, now coincide, and δ (now $l\gamma+n\alpha$) passes through the two coincident points.

4729 The tangent, or polar, of the point $\alpha'\beta'\gamma'$ is, by (4659),

$$(m\gamma'+n\beta')\,\alpha+(n\alpha'+l\gamma')\,\beta+(l\beta'+m\alpha')\,\gamma=0.$$

4730 The tangents at A, B, C (Fig. 49) meet the opposite sides respectively in P, Q, R on the right line

$$\frac{\alpha}{l}+\frac{\beta}{m}+\frac{\gamma}{n}=0. \qquad \text{By (4604).}$$

4731 The line $\frac{\alpha}{l}-\frac{\beta}{m}$ passes through (D), the intersection of the tangents at A and B.

4732 The diameter through the intersection of the tangents at A and B is

$$n\mathfrak{a}\alpha - n\mathfrak{b}\beta + (l\mathfrak{a} - m\mathfrak{b})\gamma = 0.$$

PROOF.—The coordinates of the point of intersection are $l : m : -n$, by (4726–7), and the coordinates of the centre of AB are $\mathfrak{b} : \mathfrak{a} : 0$. The diameter passes through these points, and its equation is given by (4616).

4733 The coordinates of the centre of the conic are as

$$l(-l\mathfrak{a} + m\mathfrak{b} + n\mathfrak{c}) : m(l\mathfrak{a} - m\mathfrak{b} + n\mathfrak{c}) : n(l\mathfrak{a} + m\mathfrak{b} - n\mathfrak{c}).$$

PROOF.—By (4610), the point being the intersection of two diameters like (4732). Otherwise, by (4677).

4734 The secant through $(\alpha_1\beta_1\gamma_1)$, $(\alpha_2\beta_2\gamma_2)$, any two points on the conic, and the tangent at the first point are respectively,

$$\frac{l\alpha}{\alpha_1\alpha_2} + \frac{m\beta}{\beta_1\beta_2} + \frac{n\gamma}{\gamma_1\gamma_2} = 0 \quad \text{and} \quad \frac{l\alpha}{\alpha_1^2} + \frac{m\beta}{\beta_1^2} + \frac{n\gamma}{\gamma_1^2} = 0.$$

PROOF.—The first is a right line, and it is satisfied by $\alpha = \alpha_1$, &c., and also by $\alpha = \alpha_2$, &c., by (4725). The second equation is what the first becomes when $\alpha_2 = \alpha_1$, &c. For the tangential equation, see (4893).

4735 The conic is a parabola when

$$l^2\mathfrak{a}^2 + m^2\mathfrak{b}^2 + n^2\mathfrak{c}^2 - 2mn\mathfrak{b}\mathfrak{c} - 2nl\mathfrak{c}\mathfrak{a} - 2lm\mathfrak{a}\mathfrak{b} = 0,$$

4736 or $\sqrt{(l\mathfrak{a})} + \sqrt{(m\mathfrak{b})} + \sqrt{(n\mathfrak{c})} = 0.$

PROOF.—Substitute the coordinates of the centre (4733) in $\mathfrak{a}\alpha + \mathfrak{b}\beta + \mathfrak{c}\gamma = 0$, the equation of the line at infinity (4612).

Otherwise, the conic must touch the line at infinity; therefore put $\mathfrak{a}, \mathfrak{b}, \mathfrak{c}$ for λ, μ, ν in (4893).

4737 The conic is a rectangular hyperbola when

$$l\cos A + m\cos B + n\cos C = 0,$$

and in this case it passes through the orthocentre of the triangle.

PROOF.—By (4690), and the coordinates of the orthocentre (4634).

THE CIRCUMSCRIBING CIRCLE OF THE TRIGON.

4738 $$\beta\gamma\sin A + \gamma\alpha\sin B + \alpha\beta\sin C = 0,$$

or $$\frac{\sin A}{\alpha} + \frac{\sin B}{\beta} + \frac{\sin C}{\gamma} = 0.$$

PROOF.—The values of the ratios $l : m : n$, in (4724), may be found geometrically from the equations of the tangents (4726–8).
For the coordinates of the centre, see (4642).

THE INSCRIBED CONIC OF THE TRIGON.

4739 $$l^2\alpha^2+m^2\beta^2+n^2\gamma^2-2mn\beta\gamma-2nl\gamma\alpha-2lm\alpha\beta=0.$$

4740 or $$\sqrt{(l\alpha)}+\sqrt{(m\beta)}+\sqrt{(n\gamma)}=0.$$

PROOF.—(Fig. 50.) The first equation may be written
$$n\gamma\,(n\gamma-2l\alpha-2m\beta)+(l\alpha-m\beta)^2=0.$$
By (4699) this represents a conic of which the lines γ and $n\gamma-2l\alpha-m\beta$ are the tangents at F and f, and $l\alpha-m\beta$ the chord of contact. Similarly, it may be written so as to shew that α and β touch the conic.

4741 The three pairs of tangents at F, f, &c., are

$$\left.\begin{matrix}2m\beta+2n\gamma-l\alpha\\ \text{and } \alpha\end{matrix}\right\},\quad \left.\begin{matrix}2n\gamma+2l\alpha-m\beta\\ \text{and } \beta\end{matrix}\right\},\quad \left.\begin{matrix}2l\alpha+2m\beta-n\gamma\\ \text{and } \gamma\end{matrix}\right\},$$

and they have their three points of intersection P, Q, R on the right line $l\alpha+m\beta+n\gamma$. By (4604).

4742 The coordinates of the centre of the conic are as
$$n\mathfrak{b}+m\mathfrak{c} : l\mathfrak{c}+n\mathfrak{a} : m\mathfrak{a}+l\mathfrak{b}.$$

PROOF.—By putting α and β = zero alternately in (4739), we find, for the coordinates of the points of contact,
$$\text{at } D,\ \beta=\frac{2\Delta n}{n\mathfrak{b}+m\mathfrak{c}};\quad \text{and at } E,\ \alpha=\frac{2\Delta n}{n\mathfrak{a}+l\mathfrak{c}};$$
therefore the equation of the diameter through C bisecting DE is, by (4603),
$$\frac{\alpha}{n\mathfrak{b}+m\mathfrak{c}}=\frac{\beta}{l\mathfrak{c}+n\mathfrak{a}}.$$
Similarly the diameter bisecting DF is $\dfrac{\gamma}{m\mathfrak{a}+l\mathfrak{b}}=\dfrac{\alpha}{n\mathfrak{b}+m\mathfrak{c}}$.
Therefore the point of intersection, or centre, is defined by the ratios given above.

Otherwise, by (4677), and the values in (4665), writing for a, b, c, f, g, h the coefficients in (4739).

4743 The secant through $\alpha_1\beta_1\gamma_1$, $\alpha_2\beta_2\gamma_2$ any two points on the curve.
$$\alpha\sqrt{l}\,(\sqrt{\beta_1\gamma_2}+\sqrt{\beta_2\gamma_1})+\beta\sqrt{m}\,(\sqrt{\gamma_1\alpha_2}+\sqrt{\gamma_2\alpha_1})$$
$$+\gamma\sqrt{n}\,(\sqrt{\alpha_1\beta_2}+\sqrt{\alpha_2\beta_1})=0.$$

PROOF.—Put $\alpha_1\beta_1\gamma_1$ for $\alpha\beta\gamma$, and shew that the expression vanishes by (4740).

4744 The tangent at the point $\alpha_1\beta_1\gamma_1$:

$$\alpha\sqrt{\frac{l}{\alpha_1}}+\beta\sqrt{\frac{m}{\beta_1}}+\gamma\sqrt{\frac{n}{\gamma_1}}=0.$$

PROOF.—Put $\alpha_2=\alpha_1$, &c., in (4743), and divide by $2\sqrt{(\alpha_1\beta_1\gamma_1)}$.

4745 The equation of the polar must be obtained from (4739) by means of (4659).

4746 The conic is a parabola when

$$\frac{l}{a}+\frac{m}{b}+\frac{n}{c}=0.$$

PROOF.—Similar to that of (4736).

THE INSCRIBED CIRCLE OF THE TRIGON.

4747
$$\alpha^2\cos^4\frac{A}{2}+\beta^2\cos^4\frac{B}{2}+\gamma^2\cos^4\frac{C}{2}$$
$$-2\beta\gamma\cos^2\frac{B}{2}\cos^2\frac{C}{2}-2\gamma\alpha\cos^2\frac{C}{2}\cos^2\frac{A}{2}-2\alpha\beta\cos^2\frac{A}{2}\cos^2\frac{B}{2}.$$

4748 or $\cos\frac{A}{2}\sqrt{\alpha}+\cos\frac{B}{2}\sqrt{\beta}+\cos\frac{C}{2}\sqrt{\gamma}=0.$

4749 The a-escribed circle: (4629)

$$\cos\frac{A}{2}\sqrt{-\alpha}+\sin\frac{B}{2}\sqrt{\beta}+\sin\frac{C}{2}\sqrt{\gamma}.$$

PROOF.—At the point of contact where $\gamma=0$, we have, in (4740), geometrically, r being the radius of the circle,

$$l:m=\beta:\alpha=r\cot\frac{A}{2}\sin A:r\cot\frac{B}{2}\sin B=\pm\cos^2\tfrac{1}{2}A:\cos^2\tfrac{1}{2}B;$$

+ for the inscribed; − for the escribed circle and $\pi-B$ instead of B.

4750 The tangent at $\alpha'\beta'\gamma'$, by (4744), is

$$\cos\frac{A}{2}\frac{\alpha}{\sqrt{\alpha'}}+\cos\frac{B}{2}\frac{\beta}{\sqrt{\beta'}}+\cos\frac{C}{2}\frac{\gamma}{\sqrt{\gamma'}}=0.$$

The polar is obtained as in (4745).

GENERAL EQUATION OF THE CIRCLE.

4751
$$(l\alpha+m\beta+n\gamma)(\alpha\sin A+\beta\sin B+\gamma\sin C)$$
$$+k(\beta\gamma\sin A+\gamma\alpha\sin B+\alpha\beta\sin C)=0.$$

PROOF.—The second term is the circumscribing circle (4738), and the first is linear by (4609); therefore the whole represents a circle. By varying k, a system of circles is obtained whose radical axis (4161) is the line $l\alpha+m\beta+n\gamma$, the circumscribing circle being one of the system.

4752 If $l'\alpha+m'\beta+n'\gamma$ be the radical axis of a second system of circles represented by a similar equation, the radical axis of any two circles of the two systems defined by k and k' will be

$$k'(l\alpha+m\beta+n\gamma)-k(l'\alpha+m'\beta+n'\gamma)=0.$$

PROOF.—By eliminating the term

$$\beta\gamma\sin A+\gamma\alpha\sin B+\alpha\beta\sin C.$$

4753 To find the coefficient of x^2+y^2 in the circle when only the trilinear equation is given.

RULE.—*Make* α, β, γ *the coordinates of a point from which the length of the tangent is known, and divide by the square of that length; or, if the point be within the circle, substitute "half the shortest chord through the point" for "the tangent."*

PROOF.—If $S=0$ be the equation of the circle, and m the required coefficient; then, for a point not on the curve, $S\div m=$ square of tangent or semi-chord, by (4160).

THE NINE-POINT CIRCLE.

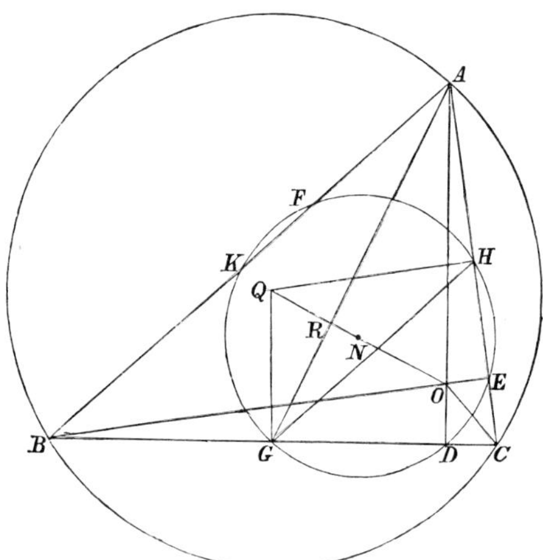

4754 $\alpha^2\sin 2A+\beta^2\sin 2B+\gamma^2\sin 2C$

$$-2(\beta\gamma\sin A+\gamma\alpha\sin B+\alpha\beta\sin C)=0.$$

PROOF.—The equation represents a circle because it may be expressed in the form

$$(\alpha\cos A+\beta\cos B+\gamma\cos C)(\alpha\sin A+\beta\sin B+\gamma\sin C)$$
$$-2\,(\beta\gamma\sin A+\gamma\alpha\sin B+\alpha\beta\sin C)=0.$$

See Proof of (4751). Now, when $\alpha=0$, the equation becomes

$$(\beta\sin B-\gamma\sin C)(\beta\cos B-\gamma\cos C)=0,$$

which shews, by (4631, '3), that the circle bisects BC and passes through D, the foot of the perpendicular from A.

4754*a* The equation of the nine-point circle in Cartesian coordinates, with the side BC and perpendicular on it from A for x and y axes respectively, is

$$x^2+y^2-R\sin(B-C)\,x-R\cos(B-C)\,y=0,$$

where R is the radius of the circum-circle.

THE TRIPLICATE-RATIO CIRCLE.

4754*b* *Let the point S (Fig. 165) be chosen, so that its trilinear coordinates are proportional to the sides of the trigon. Draw lines through S parallel to the sides, then the circle in question passes through the six points of intersection, and the intercepted chords are in the *triplicate-ratio* of the sides.

[The following abbreviations are used, a, b, c, and not $\mathfrak{a}, \mathfrak{b}, \mathfrak{c}$, being in this article written for the sides of the trigon ABC.]

$$K\equiv a^2+b^2+c^2;\quad \lambda\equiv\sqrt{(b^2c^2+c^2a^2+a^2b^2)};\quad \Delta\equiv ABC;$$

$$\mu\equiv\frac{\lambda}{K};\quad \omega=\angle BFD=DE'F',\ \&c.;\quad \theta=DFD'=DE'D',\ \&c.$$

By hypothesis, $$\frac{\alpha}{a}=\frac{\beta}{b}=\frac{\gamma}{c}=\frac{2\Delta}{a^2+b^2+c^2}\ (4007)=\frac{2\Delta}{K}\ \ldots\ldots\ldots\ldots\ldots\ldots(1),$$

$$\frac{BD'}{BF}=\frac{a}{c}=\frac{\alpha}{\gamma}\ (1)=\frac{BF'}{BD},$$

therefore $BF\,.\,BF'=BD\,.\,BD'$, therefore F, F', D, D' are concyclic.

If AS, BS, CS produced meet the opposite sides in l, m, n,

$$\frac{Bn}{An}=\frac{a\sin BCn}{b\sin ACn}=\frac{a\alpha}{b\beta}=\frac{a^2}{b^2},\ \text{by (1)}\ \ldots\ldots\ldots\ldots\ldots\ldots\ (2).$$

---

* The theorems of (1 to 36) are for the most part due to Mr. R. Tucker, M.A. The original articles will be found in *The Quarterly Journal of Pure and Applied Mathematics*, Vol. XIX., No. 76, and Vol. XX., Nos. 77 and 78.

Other and similar investigations have been made by MM. Lemoine and Taylor and Prof. Neuberg, *Mathesis*, 1881, 1882, 1884.

$$SF' = BD = \frac{\gamma}{\sin B} = \frac{2c\Delta}{K\sin B}\ (1) = \frac{ac^2}{K}.\quad \text{Similarly } BF' = \frac{ca^2}{K},\ \&c. \ldots (3).$$

$$DD' = DP\frac{\sin A}{\sin C} = \frac{ca^2}{K}\cdot\frac{a}{c} = \frac{a^3}{K},\ \&c. \ldots\ldots\ldots\ldots\ldots (4).$$

$$BD' = BD + DD' = \frac{a\,(c^2+a^2)}{K},\ \&c. \ldots\ldots\ldots\ldots\ldots\ldots (5).$$

$$FD = \surd(BD^2 + BF^2 - 2BD.BF\cos B) = \frac{c\lambda}{K},\ \text{by (2) and (5)} \ldots\ldots (6).$$

Hence DEF and $D'E'F'$ are triangles similar to ABC, and they are equal to each other because $ESF = E'SF = E'SF'$, &c. (Euc. I. 37.)

$$DF' = \surd(BD^2 + BF'^2 - 2BD.BF'\cos B) = \frac{abc}{K} \ldots\ldots\ldots\ldots (7).$$

Hence $$DF' = FE' = ED'.$$

$$D'F = \frac{b}{a}BD' = \frac{b\,(c^2+a^2)}{K}\ \&c. \ldots\ldots\ldots\ldots\ldots\ldots\ldots (8).$$

$$\cos\omega = \frac{BF^2 + FD^2 - BD^2}{2BF.FD} = \frac{a^2+c^2+b^2}{2\lambda}\ (5\ \&\ 6) = \frac{K}{2\lambda} \ldots\ldots\ldots (9).$$

$$\sin\omega = \sqrt{\left(1 - \frac{K^2}{4\lambda^2}\right)} = \frac{2\Delta}{\lambda}\ (708) \ldots\ldots\ldots\ldots\ldots (10).$$

$$\cos\theta = \cos(A-\omega),\ \&c. = \frac{a^2\cos A + bc}{\lambda} \ldots\ldots\ldots\ldots\ldots (11).$$

$$AFE' + BDF' + CED' = \frac{AF.AE'\sin A}{2} + \&c. = \mu^2\Delta = DEF,\ \text{by (6)} \ldots (12).$$

Or, geometrically, by Euclid I. 37.

Radius of T. R. circle, $\rho = \mu R$, by (6) (R = circum-radius)......(13).

The trilinear equation of the T. R. circle is

$$abc\,(\alpha^2+\beta^2+\gamma^2) = \frac{abc}{K}(a\alpha+b\beta+c\gamma)^2 + a^3\beta\gamma + b^3\gamma\alpha + c^3\alpha\beta \ldots\ldots (14),$$

or $$(b^2+c^2)\,\alpha^2 + (c^2+a^2)\,\beta^2 + (a^2+b^2)\,\gamma^2 = \left\{(a^2+b^2)\,(a^2+c^2) + b^2c^2\right\}\frac{\beta\gamma}{bc}$$
$$+\left\{(b^2+c^2)(b^2+a^2) + c^2a^2\right\}\frac{\gamma\alpha}{ca} + \left\{(c^2+a^2)(c^2+b^2) + a^2b^2\right\}\frac{\alpha\beta}{ab} \ldots (15).$$

Obtained by substituting the trilinear coordinates of D, E, F, through which points the circle passes, in (4751), to determine the ratios $l : m : n$ and k. The coordinates of D are

$$0,\quad \frac{a\,(a^2+b^2)\sin C}{K},\quad \frac{ac^2\sin B}{K}.$$

Similarly those of E and F.

THE SEVEN-POINT CIRCLE.*

4754*c* Let lines be drawn through A, B, C (Fig. 165) parallel to the sides of the triangles DEF, $D'E'F'$, as in the figure, intersecting each other in P, P', L, M, N. Let Q be the circum-centre; then the seven points P, P', L, M, N, Q, S all lie on the circumference of a circle concentric with the T. R. circle. (16)

The proof depends on Euclid III. 21, and the similar triangles DEF, $D'E'F'$.

The radius ρ' of the seven-point circle is

$$\rho' = \frac{\rho}{\lambda}\sqrt{(K^2-3\lambda^2)} = \frac{2PP'\sin 2\omega}{1-3\tan^2\omega} \quad \dots\dots\dots\dots\dots \begin{cases}(17),\\(18),\end{cases}$$

obtained from $\rho'^2 = \rho^2 + SD^2 - 2\rho SD\cos(B-TDD')$.

Expand and substitute $\cos TDD' = \dfrac{DD'}{2\rho} = \dfrac{a^3}{2\rho K}$, by (3) and (5),

$\sin TDD' = \cos\theta$ (11), $\cos B = \dfrac{c^2+a^2-b^2}{2ca}$, $\sin B = \dfrac{2\Delta}{ac}$, $\cos A = \dfrac{b^2+c^2-a^2}{2bc}$.

$3\rho+\rho' = R^2$, by (17) and (13); $\dfrac{\rho'}{\rho} = \sqrt{\dfrac{\cos 3\omega}{\cos\omega}}$, by (17) and (9) ... $\begin{cases}(19),\\(20).\end{cases}$

The trilinear equation of the seven-point circle is

$$abc\,(\alpha^2+\beta^2+\gamma^2) = a^3\beta\gamma + b^3\gamma\alpha + c^3\alpha\beta \quad \dots\dots\dots\dots\dots\dots (21),$$

or

$$a\beta\gamma + b\gamma\alpha + c\alpha\beta = \frac{1}{K}(bc\alpha + ca\beta + ab\gamma)(a\alpha + b\beta + c\gamma) \quad \dots\dots\dots (22).$$

If the coordinates of P are α_1, β_1, γ_1, and those of P' α_1', β_1', γ_1'; then

$$\alpha_1\alpha_1' = \beta_1\beta_1' = \gamma_1\gamma_1' \quad \dots\dots\dots\dots\dots\dots\dots\dots\dots\dots (23).$$

The equation of STQ is, by (4615),

$$\alpha\sin(B-C) + \beta\sin(C-A) + \gamma\sin(A-B) \quad \dots\dots\dots\dots\dots (24).$$

And the equation of PP' is

$$\frac{\alpha}{a}(a^4-b^2c^2) + \frac{\beta}{b}(b^4-c^2a^2) + \frac{\gamma}{c}(c^4-a^2b^2) = 0 \quad \dots\dots\dots\dots (25).$$

The point S has been called the Symmedian point of the triangle. It has also this property. *The line joining the mid-*

---

* This circle was discovered by M. H. Brocard, and has been called "The Brocard Circle," the points P, P' being called the Brocard points.

point of any side to the mid-point of the perpendicular on that side passes through S.

PROOF.—Let X, Y, Z (Fig. 166) be the feet of the perpendiculars; x, y, z the mid-points of the same, and X', Y', Z' the mid-points of the sides. Now the trilinear coordinates of X', S, and x in order are proportional to

$$\begin{vmatrix} 0, & c, & b \\ a, & b, & c \\ 1, & \cos C, & \cos B \end{vmatrix}.$$

This determinant vanishes; therefore the three points are on the same right line, by (4615).

That the three lines $X'x$, $Y'y$, $Z'z$ are concurrent appears at once by (970), since $CX = 2Y'x$, &c.

The Symmedian point may also be defined as the intersection of the three lines drawn from A, B, C to the corresponding vertices of the triangle formed by tangents to the circum-circle at A, B, C.

Let $B\alpha$, $C\beta$, $A\gamma$ be taken $= CX$, AY, BZ respectively. Then $A\alpha$, $B\beta$, $C\gamma$ meet in a point Σ, by (976), and this point by similarity of figure is the Symmedian point of the triangle formed by lines through A, B, C parallel to the sides BC, CA, AB.

If the sides of $X'Y'Z'$ be bisected, similar reasoning shews that σ, the Symmedian point of the triangle $X'Y'Z'$, lies on $S\Sigma$.

It can also be shewn that, if $A'B'C'$ be any triangle having its sides parallel to those of ABC and its vertices on SA, SB, SC, the sides of the two triangles intersect in six points on a circle whose centre lies midway between the circum-centres of the same triangles. When $A'B'C'$ shrinks to the point S, the circle becomes the T. R. circle.

A more general theorem respecting the triangle and circle is the following—

Take ABC any triangle, and let $DD'EE'FF'$ be the points in order, in which any circle cuts the sides.

Let
$$\left.\begin{matrix} BD = pc, & CE = qa, & AF = rb \\ CD' = p'b, & AE' = q'c, & BF' = r'a \end{matrix}\right\} \quad \text{(26)}.$$

From $BD.BD' = BF.BF'$, &c., Euclid III. 35, we can write three equations which are satisfied by the values

$$p = r' = tac, \qquad q = p' = tab, \qquad r = q' = tbc \quad \text{(27)},$$

and from these equations it appears that

$$DF = \sigma c; \quad D'F' = \sigma a, \text{ \&c., where } \sigma = \sqrt{(t^2\lambda^2 - tK + 1)} \quad \text{(28)},$$

so that DEF and $D'E'F'$ are both similar to ABC.

Also $DF' = tabc$, therefore $DF' = FE' = ED'$ .... (29).

From $\sin BFD = \dfrac{tac \sin B}{\sigma}$ we can obtain

$$\cot BFD = \cot\phi = \mp \frac{tK-2}{4t\Delta} \quad \text{.................... (30).}$$

The radius of the circle $= \sigma R$ .... (31),
and the coordinates of its centre are

$$\alpha = R\left\{\cos A + \frac{t\,(Ka^2 - a^4 - b^4 - c^4)}{2bc}\right\}. \quad \text{Similarly } \beta \text{ and } \gamma \text{......(32).}$$

The equation of the circle is

$$a\beta\gamma + b\gamma\alpha + c\alpha\beta = t\,(a\alpha + b\beta + c\gamma)\left\{abc\,(1 - ta^2) + \&c.\right\} \quad \text{...... (33),}$$

or
$$tabc\left\{\alpha^2(1-ta^2) + \beta^2(1-tb^2) + \gamma^2(1-tc^2)\right\}$$
$$= \alpha\beta\gamma\left\{(1-tb^2)(1-tc^2) + t^2b^2c^2\right\} + \&c. \text{......(34).}$$

When $t = 0$, $\sigma = 1$ and the circle is the circum-circle .... (35).

When $tK = 1$, $\sigma = t\lambda = \dfrac{\lambda}{K}$ and the circle is the T. R. circle ...... (36).

CONIC AND SELF-CONJUGATE TRIANGLE.

When the sides of the trigon are the polars of the opposite vertices, the general equation of the conic takes the form

4755 $$l^2\alpha^2 + m^2\beta^2 - n^2\gamma^2 = 0.$$

Proof.—(Fig. 51.) The equation may be written in any one of the three ways,

$$l^2\alpha^2 = (n\gamma + m\beta)(n\gamma - m\beta), \qquad m^2\beta^2 = (n\gamma + l\alpha)(n\gamma - l\alpha),$$
$$n^2\gamma^2 = (l\alpha + im\beta)(l\alpha - im\beta).$$

Hence, by (4699), α or BC is the chord of contact of the tangents $n\gamma \pm m\beta$ (AQ, AS) drawn from A, and β is the chord of contact of the tangents $n\gamma \pm l\alpha$ (BR, BP) drawn from B. Hence α, β are the polars of A, B respectively; and therefore γ or AB is the polar of C (4130). Also γ may be considered to be the chord of contact of the imaginary tangents $l\alpha \pm im\beta$ drawn from C.

4756 If the points of intersection of α and β with the conic be joined, the equations of the sides of the quadrilateral so formed are

$$QR, \quad l\alpha + m\beta + n\gamma = 0, \qquad SP, \quad l\alpha + m\beta - n\gamma = 0,$$
$$PQ, \quad -l\alpha + m\beta + n\gamma = 0, \qquad RS, \quad l\alpha - m\beta + n\gamma = 0.$$

Hence QR, SP and PQ, RS intersect on the line γ in A' and B'.

4757 Each pencil of four lines in the diagram is a harmonic pencil, by the test in (4649).

4758 The triangle $A'B'C$ is also self-conjugate with regard to the conic.

PROOF.—The equations of its sides CB', CA', $A'B'$ are

$$l\alpha - m\beta = 0, \qquad l\alpha + m\beta = 0, \qquad \gamma = 0.$$

Denote these by α', β', γ, and put α, β in (4755) in terms of α', β'. The equation referred to $A'B'C$ thus becomes $\alpha'^2 + \beta'^2 - 2n^2\gamma^2 = 0$, which is of the same form as (4755).

4759 It is clear that the triangles AQS and BPR, formed by a pair of tangents and the chord of contact in each case, are also self-conjugate.

4760 Taking $A'B'C$ for the trigon, and denoting the sides by α, β, γ, the equations of the sides RS, PQ, QR, SP of the quadrilateral become respectively

$$n\gamma \pm l\alpha = 0, \qquad m\beta \pm n\gamma = 0.$$

Ex.—As an example of (4611), we may find the coordinates of P from the equations

$$\left.\begin{aligned} \mathfrak{a}\alpha + \mathfrak{b}\beta + \mathfrak{c}\gamma &= \Sigma \\ 0 + m\beta - n\gamma &= 0 \\ -l\alpha + 0 + n\gamma &= 0 \end{aligned}\right\} \text{ from which } \left\{\begin{aligned} \alpha &= \Sigma lm \div (\mathfrak{a}mn + \mathfrak{b}nl + \mathfrak{c}lm) \\ \beta &= \Sigma mn \div (\mathfrak{a}mn + \mathfrak{b}nl + \mathfrak{c}lm) \\ \gamma &= \Sigma nl \div (\mathfrak{a}mn + \mathfrak{b}nl + \mathfrak{c}lm). \end{aligned}\right.$$

To obtain the coordinates of Q, R, and S, change the signs of m, n, and l respectively.

ON LINES PASSING THROUGH IMAGINARY POINTS.

4761 LEMMA I.—The right line passing through two conjugate imaginary points is real, and is identical with the line passing through the points obtained by substituting unity for $\sqrt{-1}$ in the given coordinates.

PROOF.—Let $(a+ia', b+ib')$ be one of the imaginary points, and therefore $(a-ia', b-ib')$ the conjugate point. The equation of the line passing through them is, by (4083) and reducing, $b'x - a'y + a'b - ab' = 0$, which is real.

But this is also the line obtained by taking for the coordinates of the points $(a+a', b+b')$ and $(a-a', b-b')$.

LEMMA II.—If P, S and Q, R are two pairs of conjugate imaginary points, the lines PS and QR are real, as has just been shown, and, therefore, also their point of intersection is real. The other pairs of lines PQ, RS and PR, QS are imaginary. But the points of intersection of each pair are real, and are identical with the points which are obtained by substituting unity for $\sqrt{-1}$ in the given coordinates, and drawing the six lines accordingly.

PROOF.—Let the coordinates of the four points be as under—

$$P \ldots\ldots a+ia',\quad b+ib',\qquad Q \ldots\ldots \alpha+i\alpha',\quad \beta+i\beta',$$
$$S \ldots\ldots a-ia',\quad b-ib',\qquad R \ldots\ldots \alpha-i\alpha',\quad \beta-i\beta'.$$

The equations of PR and QS, by (4083), are $L+iM$ and $L-iM$, where

$$L=(b-\beta)\;x-(a-\alpha)\;y+a\beta-\alpha b+a'\beta'-\alpha'b',$$
$$M=(b'+\beta')\,x-(a'-\alpha')\,y+a'\beta-\alpha'b-a\beta'+\alpha b'.$$

Now the lines $L\pm iM=0$ intersect in the same real point as the lines $L\pm M=0$, because the values $L=0$, $M=0$ satisfy both equations simultaneously. Hence, to determine this point, we have only to take i as unity in the given coordinates.

LEMMA III.—If P, S are real points, and Q, R a pair of conjugate imaginary points, the lines PS and QR are both real, by Lemma I., and consequently their point of intersection is real. The remaining pairs of lines PQ, RS and PR, QS and their points of intersection are all imaginary. But the line joining these two imaginary points of intersection is real, and is identical with the line obtained by substituting unity for $\sqrt{-1}$ in the given coordinates and drawing the six lines accordingly.

PROOF.—Let the coordinates of the four points be as under—

$$P \ldots\ldots x_1y_1,\qquad Q \ldots\ldots \alpha+i\alpha',\quad \beta+i\beta',$$
$$S \ldots\ldots x_2y_2,\qquad R \ldots\ldots \alpha-i\alpha',\quad \beta-i\beta'.$$

Since the coordinates of R are obtained from those of Q by merely changing the sign of i, the equations of the four imaginary lines will take the forms

$$PQ \ldots\ldots A-iB,\qquad SQ \ldots\ldots C-iD,$$
$$PR \ldots\ldots A+iB,\qquad SR \ldots\ldots C+iD.$$

Now let the coordinates of the point of intersection of PQ and SR be $L+iM$, $L'+iM'$, then will $L-iM$, $L'-iM'$ be the coordinates of the intersection of PR and SQ, for the equations of this pair of lines are got from those of PQ and SR by merely changing the sign of i. The points of intersection are therefore conjugate imaginary points, and the line joining them is real, by Lemma I. Also, since that line is obtained by writing 1 for i in the coordinates of those points, it will also be obtained by writing 1 for i in the original coordinates of Q and R and constructing the figure as before.

4762 To find a common pole and polar of two given conics:

(i.) If the conics intersect in four real points P, Q, R, S, construct the complete quadrilateral (4652). Then $A'B'C$ (Fig. 51) is a self-conjugate triangle for each conic, by (4758), and therefore each vertex and the opposite side form a common pole and polar to the conics.

(ii.) If the conics do not intersect at all in real points, the triangle $A'B'C$ is still real, by Lemma II. (4761), and can be constructed in the manner shown.

(iii.) If two of the points (P, S) are real, and two (Q, R) imaginary, then, by Lemma III., the vertex A' and the side $B'C$ are real, and may be constructed, and they form a common pole and polar of the given conics.

Returning to the triangle of reference ABC,

4763 Let $l\alpha = n\gamma\cos\phi$, $m\beta = n\gamma\sin\phi$; then the chord joining two points ϕ_1, ϕ_2 is

$$l\alpha\cos\tfrac{1}{2}(\phi_1+\phi_2)+m\beta\sin\tfrac{1}{2}(\phi_1+\phi_2) = n\gamma\cos\tfrac{1}{2}(\phi_1-\phi_2),$$

and therefore the tangent at the point ϕ' is

4764
$$l\alpha\cos\phi'+m\beta\sin\phi' = n\gamma.$$

4765 Putting $l^2 = L$, $m^2 = M$, $n^2 = -N$, the conic (4755) becomes

$$L\alpha^2+M\beta^2+N\gamma^2 = 0 \quad\text{.....................(1).}$$

4766 The tangent or polar of $\alpha'\beta'\gamma'$ is

$$L\alpha\alpha'+M\beta\beta'+N\gamma\gamma' = 0 \quad\text{................. (2).}$$

4767 Hence the pole of $\lambda\alpha+\mu\beta+\nu\gamma = 0$

is
$$\left(\frac{\lambda}{L}, \frac{\mu}{M}, \frac{\nu}{N}\right) \quad\text{........................ (3).}$$

4768 The tangential equation is

$$\frac{\lambda^2}{L}+\frac{\mu^2}{M}+\frac{\nu^2}{N} = 0 \quad\text{.....................(4),}$$

and this is the condition that the conic (1) may be touched by the four lines

$$\lambda\alpha\pm\mu\beta\pm\nu\gamma = 0.$$

4769 In like manner,

$$L\alpha'^2+M\beta'^2+N\gamma'^2=0 \quad\text{.....................(5)}$$

is the condition that (1) may pass through the four points

$$(\alpha',\ \pm\beta',\ \pm\gamma').$$

4770 The locus of the pole of the line $\lambda\alpha+\mu\beta+\nu\gamma$ with respect to such conics is

$$\frac{\lambda\alpha'^2}{\alpha}+\frac{\mu\beta'^2}{\beta}+\frac{\nu\gamma'^2}{\gamma}=0.$$

PROOF.—By (3), if (α, β, γ) be the pole, $\alpha=\frac{\lambda}{L}$ &c., $\therefore\ L=\frac{\lambda}{\alpha}$, in (5), the equation of condition.

4771 The locus of the pole of the line $l\alpha+m\beta+n\gamma$, with respect to the conics which touch the four lines $\lambda\alpha\pm\mu\beta\pm\nu\gamma$ is

$$\frac{\lambda^2\alpha}{l}+\frac{\mu^2\beta}{m}+\frac{\nu^2\gamma}{n}=0.$$

PROOF.—By (3), if (α, β, γ) be the pole, $\alpha=\frac{l}{L}$ &c., $\therefore\ L=\frac{l}{\alpha}$, &c., in (4), the equation of condition.

4772 The locus of the centre of the conic is given in each case (4770, '1) by taking the line at infinity

$$\alpha\sin A+\beta\sin B+\gamma\sin C$$

for the fixed line, since its pole is the centre.

4773 Thus the locus of the centre of the conic passing through the four points $(\alpha'\pm\beta'\pm\gamma')$ is

$$\frac{\alpha'^2\sin A}{\alpha}+\frac{\beta'^2\sin B}{\beta}+\frac{\gamma'^2\sin C}{\gamma}=0.$$

4774 The coordinates of the centre of the conic (1) are given by

$$\frac{L\alpha}{\mathfrak{a}}=\frac{M\beta}{\mathfrak{b}}=\frac{N\gamma}{\mathfrak{c}}.$$

PROOF.—Let the conic cut the side a in the points $(0\beta_1\gamma_1)$, $(0\beta_2\gamma_2)$. The right line from A bisecting the chord will pass through the centre of the conic, and its equation will be $\beta:\gamma=\beta_1+\beta_2:\gamma_1+\gamma_2$. Now $\beta_1+\beta_2$ is the sum of the roots of the quadratic in β obtained by eliminating γ and α from the equations $L\alpha^2+M\beta^2+N\gamma^2=0$, $\alpha=0$, and $\mathfrak{a}\alpha+\mathfrak{b}\beta+\mathfrak{c}\gamma=\Sigma$. Similarly for $\gamma_1+\gamma_2$ eliminate α and β. The equation of the diameter through A being found, those through B and C are symmetrical with it.

4775 The condition that the conic (1) may be a parabola is

$$\frac{\mathfrak{a}^2}{L}+\frac{\mathfrak{b}^2}{M}+\frac{\mathfrak{c}^2}{N}=0.$$

PROOF.—This is, by (4), the condition of touching the line at infinity

$$\mathfrak{a}\alpha+\mathfrak{b}\beta+\mathfrak{c}\gamma=0.$$

4776 The condition that (1) may be a rectangular hyperbola is $L+M+N=0$, and in this case the curve passes through the centres of the inscribed and escribed circles of the trigon.

PROOF.—By (4690), (a, b, c are now L, M, N). (1) is now satisfied by $\alpha=\pm\beta=\pm\gamma$, the four centres in question.

4777 Circle referred to a self-conjugate triangle:

$$\alpha^2\sin 2A+\beta^2\sin 2B+\gamma^2\sin 2C=0.$$

PROOF.—The line joining A to the centre is $\frac{M\beta}{\mathfrak{b}}=\frac{N\gamma}{\mathfrak{c}}$ (4774). Therefore $\frac{M}{\mathfrak{b}\cos B}=\frac{N}{\mathfrak{c}\cos C}$, the condition of perpendicularity to a by (4622). Similarly $\frac{N}{\mathfrak{c}\cos C}=\frac{L}{\mathfrak{a}\cos A}$, therefore (1) takes the form above.

IMPORTANT THEOREMS.

CARNOT'S THEOREM.

4778 If A, B, C (Fig. 52) are the angles of a triangle, and if the opposite sides intersect a conic in the pairs of points a, a'; b, b'; c, c'; then

$$Ac.Ac'.Ba.Ba'.Cb.Cb'=Ab.Ab'.Bc.Bc'.Ca.Ca'.$$

PROOF.—Let α, β, γ be the semi-diameters parallel to BC, CA, AB; then, by (4317), $Ab.Ab' : Ac : Ac' = \beta^2 : \gamma^2$. Compound this with two similar ratios.

4779 COR.—If the conic touches the sides in a, b, c, then

$$Ac^2.Ba^2.Cb^2=Ab^2.Bc^2.Ca^2.$$

4780 The reciprocal of Carnot's theorem is: If A, B, C (Fig. 52) are the sides of a triangle, and if pairs of tangents from the opposite angles are a, a'; b, b'; c, c'; then

$$\sin(Ac)\sin(Ac')\sin(Ba)\sin(Ba')\sin(Cb)\sin(Cb')$$
$$=\sin(Ab)\sin(Ab')\sin(Bc)\sin(Bc')\sin(Ca)\sin(Ca'),$$

where (Ac) signifies the angle between the lines A and c.

PROOF.—Reciprocating the former figure with respect to any origin O, let A, B, C (*i.e.*, RQ, QP, PR) be the polars of the vertices A, B, C. Then, by (4130), Q, R will be the poles of AB, AC; and b, b', the polars of the points b, b', will intersect in R and touch the reciprocal conic. Similarly, c, c' will intersect in Q. A, b' are perpendicular to OA, Ob', and therefore $\angle Ab' = \angle AOb'$, and so of the rest.

PASCAL'S THEOREM.

4781 The opposite sides of a hexagon inscribed to a conic meet in three points on the same right line.

PROOF.—(Fig. 53.) Let $\alpha, \beta, \gamma, \gamma', \beta', \alpha'$ be the consecutive sides of the hexagon, and let u be the diagonal joining the points $\alpha\alpha'$ and $\gamma\gamma'$. The equation of the conic is either $\alpha\gamma - k\beta u = 0$ or $\alpha'\gamma' - k'\beta' u = 0$, and, since these expressions vanish for all points on the curve, we must have $\alpha\gamma - k\beta u = \alpha'\gamma' - k'\beta' u$ for *any* values of the coordinates. Therefore $\alpha\gamma - \alpha'\gamma' = u(k\beta - k'\beta')$. Therefore the lines α, α' and also γ, γ' meet on the line $k\beta - k'\beta'$; and β, β' evidently meet on that line.

Otherwise, by projecting a hexagon inscribed in a circle with its opposite sides parallel upon any plane not parallel to that of the circle. The line at infinity, in which the pairs of parallel sides meet, becomes a line in which the corresponding sides of a hexagon inscribed in a conic meet at a finite distance (1075 *et seq.*).

4782 With the same vertices there are sixty different hexagons inscribable in any conic, and therefore sixty different Pascal lines corresponding to any six points on a conic.

PROOF.—Half the number of ways of taking in order five vertices B, C, D, E, F after A is the number of different hexagons that can be drawn, and the demonstration in (4781) applies equally to all.

BRIANCHON'S THEOREM.

4783 The three diagonals of a hexagon circumscribed to a conic pass through the same point (Fig. 54).

PROOF.—Let the three conics $S+L^2$, $S+M^2$, $S+N^2$, in (4707), become three pairs of right lines, then the three lines $L-M$, $M-N$, $N-L$ become the diagonals of a circumscribing hexagon.

Pascal's and Brianchon's theorems may be obtained, the one from the other, by reciprocation (4840).

THE CONIC REFERRED TO TWO TANGENTS AND THE CHORD OF CONTACT.

Let $L=0$, $M=0$, $R=0$ (Fig. 55) be the sides of the trigon; L, M being tangents and R the chord of contact.

4784 The equation of the conic is $\mathbf{LM=R^2}$. (4699)

4785 The lines AP, BP, and CP are respectively

$$\mu L=R, \qquad \mu R=M, \qquad \mu^2 L=M. \qquad \text{[By (4604).}$$

Since the point P on the curve is determined by the value of μ, it is convenient to call it the point μ.

4788 The points μ and $-\mu$ (P and Q) are both on the line $\mu^2L=M$ drawn through C.

4789 The secant through the points μ, μ' (P, P') is

$$\mu\mu' L-(\mu+\mu')R+M=0.$$

PROOF. — Write it $\mu(\mu'L-R)-(\mu'R-M)$, and, by (4604), it passes through the point μ'. Similarly through μ. Otherwise, determine the coordinates of the intersection of $\mu L-R$ and $\mu R-M$, and of $\mu'L-R$ and $\mu'R-M$ by (4610), and the equation of the secant by (4616).

4790 COR.—The tangents at the points μ and $-\mu$ (P, Q) are therefore

$$\mu^2L\mp 2\mu R+M=0.$$

4791 These tangents intersect on R. [Proof by subtraction.

4792 *Theorem.*—If the equation of a right line contains an indeterminate μ in the second degree, it may be written as above, and the line must therefore touch the conic $\mathbf{LM=R^2}$.

4793 The polar of the point (L', M', R') is

$$LM'-2RR'+L'M=0.$$

PROOF.—For $\mu+\mu'$ and $\mu\mu'$, in (4789), put the values of the sum and product of the roots of $\mu^2L'-2\mu R'+M'=0$ (4790).

4794 Similarly the polar of the point of intersection of $aL-R$ and $bR-M$ is

$$abL-2aR+M=0.$$

4795 The line CE joining the vertex C to the intersection of two tangents at μ and μ', or at $-\mu$ and $-\mu'$, is

$$\mu\mu'L-M=0.$$

Otherwise, if two tangents meet on any line $aL-M$, drawn through C, the product of their μ's is equal to a.

PROOF.—Eliminate R from the equations of the two tangents (4790).

4796 The chords PQ', $P'Q$ and the line CE all intersect in the same point on R.

PROOF.—The equations of PQ', $P'Q$ are, by (4789),

$$\mu\mu'L\pm(\mu-\mu')R-M=0,$$

and, by addition and subtraction, we obtain $\mu\mu'L-M=0$ (4795), or $R=0$.

4797 The lines $\mu\mu'L+M$ (CD) and R intersect on the chord PP' which joins the points μ, μ'; or—The extremities of any chord passing through the intersection of $aL+M$ and R have the product of their μ's equal to a.

4798 The chord joining the points $\mu\tan\phi$, $\mu\cot\phi$ touches a conic having the same tangents L, M and chord of contact R.

PROOF.—The equation of the chord is, by (4789),

$$\mu^2L-\mu R(\tan\phi+\cot\phi)+M=0,$$

and this touches the conic $LM\sin^2 2\phi=R^2$ at the point μ, by (4792).

4799 The tangents at the points $\mu\tan\phi$, $\mu\cot\phi$ intersect on the conic $LM=R^2\sin^2 2\phi$.

PROOF.—Write the equations of the two tangents, by (4790), and then eliminate μ.

4800 Ex. 1.—To find the locus of the vertex of a triangle circumscribing a fixed conic and having its other vertices on two fixed right lines.

Take $LM=R^2$ for the conic (Fig. 56), $aL+M$, $bL+M$ for the lines CD, CE. Let one tangent, DE, touch at the point μ; then, by (4795), the others,

PD, PE, will touch at the points $\frac{a}{\mu}$, $\frac{b}{\mu}$, and therefore, by (4790), their equations will be

$$\frac{a^2}{\mu^2}L-\frac{2a}{\mu}R+M,\qquad \frac{b^2}{\mu^2}L-\frac{2b}{\mu}R+M.$$

Eliminate μ, and the locus of P is found to be $(a+b)^2LM = 4abR^2$.
[Salmon, Art. 272.

4801 Ex. 2.—To find the envelope of the base of a triangle inscribed in a conic, and whose sides pass through fixed points P, Q.

(Fig. 57.) Take the line through P, Q for R; $LM-R^2$ for the conic; $aL-M$, $bL-M$ for the lines joining P and Q to the vertex C. Let the sides through P and Q meet in the point μ on the conic; then, by (4797), the other extremities will be at the points $-\frac{a}{\mu}$ and $-\frac{b}{\mu}$, and therefore, by (4789), the equation of the base will be $abL+(a+b)\,\mu R+\mu^2M=0$. By (4792), this line always touches the conic $4ab\,LM = (a+b)^2\,R^2$. [*Ibid.*

4802 Ex. 3.—To inscribe a triangle in a conic so that its sides may pass through three fixed points. (See also 4823.)

We have to make the base $abL+(a+b)\,\mu R+\mu^2M$ (4801) pass through a third fixed point. Let this point be given by $cL = R$, $dR = M$. Eliminating L, M, R, we get $ab+(a+b)\,\mu c+\mu^2cd = 0$, and since, at the point μ, $\mu L = R$, $\mu^2L = M$, that point must be on the line $abL+(a+b)\,cR+cdM$. The intersections of this line with the conic give two solutions by two positions of the vertex. [*Ibid.*

RELATED CONICS.

4803 A conic having double contact with the conics S and S' (Fig. 58) is

$$\mu^2E^2-2\mu\,(S+S')+F^2=0,$$

where E, F are common chords of S and S', so that $S-S'=EF$.

Proof.—The equation may be written in either of the ways

$$(\mu E+F)^2 = 4\mu S \quad\text{or}\quad (\mu E-F)^2 = 4\mu S',$$

showing that $\mu E\pm F$ are the chords of contact AB, CD. There are three such systems, since there are three pairs of common chords.

4804 Cor. 1.—A conic touching four given lines A, B, C, D, the diagonals being E, F (Fig. 59):

$$\mu^2E^2-2\mu\,(AC+BD)+F^2=0.$$

Here $S = AC$ and $S' = BD$, two pairs of right lines.

Otherwise, if L, M, N be the diagonals and $L \pm M \pm N$ the sides, the conic becomes

4805 $$\mu^2 L^2 - \mu (L^2 + M^2 - N^2) + M^2 = 0.$$

For this always touches

$(L^2+M^2-N^2)^2 - 4L^2M^2$ or $(L+M+N)(M+N-L)(N+L-M)(L+M-N)$.

[*Salmon*, Art. 287.]

4806 Cor. 2.—A conic having double contact with two circles C, C' is

$$\mu^2 - 2\mu (C+C') + (C-C')^2 = 0.$$

4807 The chords of contact become

$$\mu + C - C' = 0 \quad \text{and} \quad \mu - C + C' = 0.$$

4808 The equation may also be written

$$\sqrt{C} \pm \sqrt{C'} = \sqrt{\mu},$$

which signifies that the sum or difference of the tangents drawn from any point on the conic to the circles is constant.

ANHARMONIC PENCILS OF CONICS.

4809 The anharmonic ratio of the pencil drawn from any point on a conic through four fixed points upon it is constant.

Proof.—Let the vertices of the quadrilateral in Fig. (38) be denoted by A, B, C, D, and let P be the fifth point. Multiplying the equation of the conic (4697) by the constants AB, CD, BC, DA, we have

$$k\frac{AB.CD}{BC.DA} = \frac{AB\alpha.CD\gamma}{BC\beta.DA\delta} = \frac{PA.PB \sin APB.PC.PD \sin CPD}{PB.PC \sin BPC.PD.PA \sin DPA}$$
$$= \frac{\sin APB.\sin CPD}{\sin BPC.\sin DPA}.$$

Compare (1056).

4810 If the fifth point be taken for origin in the system (4784, Fig. 55), and if the four lines through it be

$$L-\mu_1 R, \quad L-\mu_2 R, \quad L-\mu_3 R, \quad L-\mu_4 R,$$

the anharmonic ratio of the pencil is, by (4650),

$$= \frac{(\mu_1-\mu_2)(\mu_3-\mu_4)}{(\mu_1-\mu_4)(\mu_2-\mu_3)}.$$

4811 Cor. 1.—If four lines through any point, taken for the vertex LM, meet the conic in the points $\mu_1, \mu_2, \mu_3, \mu_4$, the anharmonic ratio of these points, with any fifth point on the conic, is equal to that of the points $-\mu_1, -\mu_2, -\mu_3, -\mu_4$, in which the same lines again meet the conic.

4812 Cor. 2.—The reciprocal theorem is—If from four points upon any right line four tangents be drawn to a conic, the anharmonic ratio of the points of section with any fifth tangent is equal to the corresponding ratio for the other four tangents from the same points.

4813 The anharmonic ratio of the segments of any tangent to a conic made by four fixed tangents is constant.

Proof.—Let $\mu, \mu_1, \mu_2, \mu_3, \mu_4$ (Fig. 60) be the points of contact. The anharmonic ratio of the segments is the same as that of the pencil of four lines from LM to the points of section; that is, of $\mu\mu_1L-M$, $\mu\mu_2L-M$, $\mu\mu_3L-M$, $\mu\mu_4L-M$, a pencil homographic (4651) with that in (4810).

4814 If P, P' are the polars of a point with respect to the conics S, S', then $P+kP'$ will be the polar of the same point with respect to the conic $S+kS'$.

4815 Hence the polar of a given point with regard to a conic passing through four given points (the intersections of S and S') always passes through a fixed point, by (4101).

If Q, Q' are the polars of another point with respect to the same conics, $Q+kQ'$ is the polar with respect to $S+kS'$.

4816 Hence the polars of two points with regard to a system of conics through four points form two homographic pencils (4651).

4817 The locus of intersections of corresponding lines of two homographic pencils having fixed vertices (Fig. 61) is a conic passing through the vertices; and, conversely, if the conic be given, the pencils will be homographic.

Proof.—For eliminating k from $P+kP'=0$, $Q+kQk'$, we get $PQ'=P'Q$.

4818 Cor.—The locus of the pole of the line joining the two points in (4816) is a conic.

Proof.—For the pole is the intersection of $P+kP'$ and $Q+kQ'$.

4819 The right lines joining corresponding points AA', &c. (Fig. 62) of two homographic systems of points lying on two right lines, envelope a conic.

Proof.—This is the reciprocal theorem to (4817); or it follows from (4813).

4820 If two conics have double contact (Fig. 63), the anharmonic ratio of the points of contact A, B, C, D of any four tangents to the inner conic is the same as that of each set of four points (a, b, c, d) or (a', b', c', d') in which the tangents meet the other conic.

Proof.—By (4798). The μ's for the points on the latter conic will be equal to the μ's of the points of contact multiplied by $\tan\phi$ for one set, and by $\cot\phi$ for the other, and therefore the ratio (4810) will be unaltered.

4821 Conversely, if three chords of a conic aa', bb', cc' be fixed, and a fourth dd' moves so that $\{abcd\} = \{a'b'c'd'\}$, then dd' envelopes a conic having double contact with the given one.

For theorems on a right line cut in involution by a conic, see (4824-8).

CONSTRUCTION OF CONICS.

THEOREMS AND PROBLEMS.

4822 If a polygon inscribed to a conic (Fig. 64) has all its sides but one passing through fixed points $A, B, \ldots Y$, the remaining side az will envelope a conic having double contact with the given one.

Proof.—Let $a, b, \ldots z$ be the vertices of the polygon, and a, a', a'', a''' four successive positions of a. Then, by (4811),

$$\{a, a', a'', a'''\} = \{b, b', b'', b'''\} = \&c. = \{z, z', z'', z'''\}.$$

Therefore, by (4821), the side az envelopes a conic, &c.

4823 Poncelet's construction for inscribing in a conic a polygon having its n sides passing through n given points.

Inscribe three polygons, each of n+1 *sides, so that* n *of each may pass through the fixed points, and let the remaining sides be* a'z', a''z'', a'''z''', *denoted in figure* (65) *by* AD, CF, EB. *Let* MLN, *the line joining the intersections of opposite sides of the hexagon* ABCDEF (4781), *meet the conic in* K; *then* K *will be a vertex of the required polygon.*

PROOF.— $\{D.KACE\} = \{A.KDFB\}$, each pencil passing through K, P, N, L; therefore the anharmonic ratio $\{KACE\} = \{KDFB\}$ for any vertex on the conic, by (4809); *i.e.*, $\{Ka'a''a'''\} = \{Kz'z''z'''\}$. But, if az be the remaining side of a fourth polygon inscribed like the others, we have by (4811), as in (4822), $\{aa'a''a'''\} = \{zz'z''z'''\}$. Hence K is the point where a and z coincide.

4824 *Lemma.*—A system of conics passing through four fixed points meets any transversal in a system of points in involution (1066).

PROOF.—Let u, u' be two conics passing through the four points; then $u+ku'$ will be any other. Take the transversal for x axis, and put $y=0$ in each conic, and let their equations thus become $ax^2+2gx+c=0$ and $a'x^2+2g'x+c'=0$. These determine the points where the transversal meets u and u'. It will then meet $u+ku'$ in two points given by $ax^2+2gx+c+k(a'x^2+2g'x+c')=0$, and these points are in involution with the former, by (1065).

Geometrically (Fig. 66),

$$\{a.AdbA'\} = \{c.AdbA'\} \quad (4809),$$

therefore $\{ACBA'\} = \{AB'C'A'\} = \{A'C'B'A\}$, therefore by (1069).

4825 COR. 1.—One of the conics of the system resolves itself into the two diagonals ac, bd. Hence the points B, B', C, C' are in involution with D, D', where the transversal cuts the diagonals.

4826 COR. 2.—A transversal meets a conic and two tangents in four points in involution, so as to meet the chord of contact in one of the foci of the system.

For, in (Fig. 66), if b coincides with c, and a with d, the transversal meets the tangents in C, C', while B, B', D, D', all coincide in F (Fig. 67), one of the foci on the chord of contact.

4827 The reciprocal theorem to (4824) is—Pairs of tangents from any point to a system of conics touching four fixed lines, form a system in involution (4850).

4828 The condition that $\lambda x + \mu y + \nu z$ may be cut in involution by three conics is the vanishing of the determinant

$$\begin{vmatrix} A_1 & H_1 & B_1 \\ A_2 & H_2 & B_2 \\ A_3 & H_3 & B_3 \end{vmatrix} = \begin{vmatrix} a_1 & b_1 & c_1 & 2f_1 & 2g_1 & 2h_1 \\ a_2 & b_2 & c_2 & 2f_2 & 2g_2 & 2h_2 \\ a_3 & b_3 & c_3 & 2f_3 & 2g_3 & 2h_3 \\ \lambda & 0 & 0 & 0 & \nu & \mu \\ 0 & \mu & 0 & \nu & 0 & \lambda \\ 0 & 0 & \nu & \mu & \lambda & 0 \end{vmatrix}.$$

where A_1, H_1, B_1 belong to the first conic and have the values in (4988).

Proof.—The quadratic $A_1x^2 + 2Hxy + B_1y^2 = 0$, obtained in (4987), determines the pair of points of intersection with the first conic. The similar equation for the third conic will have $A_3 = A_1 + \lambda A_2$, &c., if the points are all in involution (1065). The third equation is therefore derived from the other two; therefore the determinant vanishes, by (583).

By expanding and dividing by ν^3, the second determinant above of the sixth order is obtained.

Newton's Method of Generating a Conic.

4829 Two constant angles aPb, aQb (Fig. 68) move about fixed vertices P, Q. If a moves on a fixed right line, b describes a conic which passes through P and Q.

Proof.—Taking four positions of a, we have (see 1054),

$$\{P.bb'b''b'''\} = \{P.aa'a''a'''\} = \{Q.aa'a''a'''\} = \{Q.bb'b''b'''\}.$$

Therefore, by (4817), the locus of b is a conic.

Maclaurin's Method of Generating a Conic.

4830 The vertex V of a triangle (Fig. 69), whose sides pass through fixed points A, B, C, and whose base angles move on fixed lines Oa, Ob, describes a conic passing through A and B.

Proof.—The pencils of lines through A and B in the figure are both homographic with the pencil through C, and are therefore homographic with each other. Therefore the locus of V is a conic, by (4817).

Otherwise, let α, β, γ be the sides of ABC; $l\alpha+m\beta+n\gamma$, $l'\alpha+m'\beta+n'\gamma$ the fixed lines Oa, Ob; and $\alpha=\mu\beta$ the moving base ab.

Then the equations of the sides will be

$$(l\mu+m)\,\beta+n\gamma=0, \quad (l'\mu+m)\,\alpha+n'\mu\gamma=0.$$

Eliminate μ; then $lm'\alpha\beta=(m\beta+n\gamma)(l'\alpha+n'\gamma)$, the conic in question, by (4697).

4831 Given five points, to find geometrically any number of points on the circumscribing conic, and to find the centre.

Let A, B, C, D, E (*Fig.* 70) *be the five points. Draw any line through* A *meeting* CD *in* P. *Draw* PQ *through the intersection of* AB *and* DE *meeting* BC *in* Q; *then* QE *will meet* PA *in* F, *a sixth point on the curve, as is evident from Pascal's theorem* (4781).

To find the centre, choose AP *in the above construction parallel to* CD, *and find two diameters, as in* (1252).

4832 To find the points of contact of a conic with five right lines.

Let ABCDE (*Fig.* 71) *be the pentagon. Join* D *to the intersection of* AC *and* BE. *This line will pass through the point of contact of* AB, *and so on.*

Proof.—By (4783), supposing two sides of the hexagon to become one straight line.

4833 To describe a conic, given four points upon it and a tangent.

Let a, a′, b, b′ (*exterior letters in Fig.* 52) *be the four points. Then, if* AB *is a tangent,* c, c′ *coincide, and Carnot's theorem* (4778) *gives the ratio* $\text{Ac}^2 : \text{Bc}^2$. *Then by* (4831). *Since there are two values of this ratio,* $\pm$ (Ac : Bc), *two conics may be drawn as required.*

4834 To describe a conic, given four tangents and a point.

Let a, a′, b, b′ (*interior letters in Fig.* 52) *be the four tangents. Then, if* Q *be the given point on the curve, the lines* c, c′ *must coincide in direction, and* (4780) *gives the ratio* $\sin^2(\text{Ac}) : \sin^2(\text{Bc})$, *by which the direction of a fifth tangent through* Q *is determined. Then by* (4832). *The two values* $\pm$ (sin Ac : sin Bc) *furnish two solutions.*

Otherwise by (4804), *determining* μ *by the coordinates of the given point.*

4835 To describe a conic, given three points and two tangents.

Let A, A′, A″ *be the points* (*Fig.* 67, *supplying obvious letters*). *Let the two tangents meet* AA′ *in the points* C, C′. *Find* F, F′, *the foci of the system* AA′, CC′ *in involution* (1066) *determining the centre by* (985). *Similarly, find* G, G′, *the foci of a system on the line* AA″. *Then, by* (4826), *the chord of contact of the tangents may be any of the lines* FG, FG′, F′G, F′G′. *There are accordingly four solutions, and the construction of* (4831) *determines the conic.*

4836 To describe a conic, given two points and three tangents.

Let AB, BC, CA (*Fig.* 167) *be the tangents, and* P, P′ *the points. Draw a transversal through* PP′ *meeting the three tangents in* Q, Q′, Q″. *Find* F, *a focus of the system* PP′, QQ′ *in involution* (1066, 985); G *a focus for* PP′, QQ″, *and* H *for* PP′, Q′Q″. *Construct a triangle with its sides passing through* F, G, H, *and with its vertices* L, M, N *on* BC, CA, AB, *by the method of* (4823), *which is equally applicable to a rectilineal figure as to a conic.* L, M, N *will be the points of contact. The reason for the construction is contained in* (4826). *There will, in general, be four solutions.*

If the conic be a parabola, the foregoing constructions can be adapted by considering one tangent at infinity always to be given.

4837 To draw a parabola through four given points a, a', b, b'.

This is problem 4833 with the tangent at infinity.

In figure (52), suppose cc' to coincide and AB to remove to infinity so as to become the tangent at c, the opposite vertex at infinity of a parabola, and therefore to be perpendicular to the axis. Cc then becomes a diameter of the parabola, and Carnot's theorem (4778) shows that

$$\frac{Ca.Ca'}{Cb.Cb'} = \frac{Ac^2}{Ab^2}.\frac{Ba^2}{Bc^2} = \frac{\sin^2 ACc}{\sin^2 BCc},$$

since the points C, a, a', b, b' are all on the axis of the parabola relatively to the infinite distance of AcB. This result, however, is at once obtained from equation (4221), $Ca.Ca' : Cb.Cb'$ being the ratio of the products of the roots of two similar quadratics. Thus a diameter of the parabola can be drawn through C by the known ratio of the sines of ACc and BCc.

Next, describe a circle round three of the given points a, a', b. By the property (1263) and the known direction of the axis, the other point in which the circle cuts the parabola can be found.

Five points being known, we can, by Pascal's theorem, as in (4831), obtain two parallel chords, and then find P, the extremity of their diameter, by the proportion, square of ordinate $\propto$ abscissa (1239).

Lastly, draw the diameter and tangent at P, and then, by equality of angles (1224), draw a line from P which passes through the focus. By obtaining in the same way another pair of parallel chords, a second line through the focus is found, thus determining its position.

4838 To draw a parabola when four tangents are given.

This is effected by the construction of (4832, Fig. 71). Let AB, BC, AE, ED be the four tangents, and CD the tangent at infinity. Then any line drawn to C will be parallel to BC, and any line to D will be parallel to ED.

4839 To draw a parabola, given three points and one tangent.

This is effected by the construction of (4835, Fig. 67). Let bC' be the tangent at ∞; then the centre of involution O must be at C, so that $CC.CC' = 0.\infty = CA.CA' = CF^2$, determining F. F', another point on the chord of contact, being found by joining AA'' or $A'A''$, FF' will be the diameter through a, since the other point of contact b is at infinity.

4840 To draw a parabola, given one point and three tangents.

This is the case of (4834), in which one of the given tangents b' is at infinity. R must therefore be at infinity, and QR, PR and the tangent b, since they all join R to finite points, must be parallel. The ratio found determines another tangent, and the case is reduced to that of (4838).

4841 To draw a parabola, given two points and two tangents.

This is problem (4836). Suppose AC in that construction to be the tangent at infinity. F, G, H will be determined as in (4839) by mean proportionals. The chords LM, NM will become parallel, since M is at infinity; and we have to draw LN and the parallel lines from L and N to pass through F, G, H in their new positions, so that the vertices L, N may lie on BC and AB.

Otherwise by (4509), the intercepts s and t can readily be found from the two equations furnished by the given points.

4842 To describe a conic touching three right lines and touching a given conic twice.

Let AD, CF, EB (*Fig.* 65) *be the three lines as they cut the given conic. Join* AB, AF, BC, BE, *and determine* K *by the Pascal line* MLN. K *will be one point of contact of the two conics, by* (4822) *and the proof in* (4823), *since* AD, CF, EB, *and the tangent at* K *are four positions of the "remaining side" in that proposition. The problem is thus reduced to*

(4834), *since four tangents and* K *the point of contact of one of them are now known.*

4843 To describe a conic touching each of two given conics twice, and passing through a given point or touching a given line.

Proceed by (4803), *determining* μ *by the last condition.*

To describe a conic touching the conics $S+L^2$, $S+M^2$, $S+N^2$ (4707) and touching S twice. [*Salmon*, Art. 387.

THE METHOD OF RECIPROCAL POLARS.

DEF.—The polar reciprocal of a curve is the envelope of the polars of all the points on the curve, or it is the locus of the poles of all tangents to the curve, taken in each case with respect to an arbitrary fixed origin and circle of reciprocation.

4844 Thus, in figure (72), to the points P, Q, R on one curve correspond the tangents qr, rp, and chord of contact pq on the reciprocal curve; and to the points p, q, r correspond the tangents QR, RP, and chord PQ.

The angle between the tangents at P and Q is evidently equal to the angle pOq, since Op, Oq, Or are respectively perpendicular to QR, RP, PQ.

4845 THEOREM.—*The distance of a point from a line is to its distance from the origin as the distance of the pole of the line from the polar of the point is to its distance from the origin.*

PROOF.—(Fig. 73.) Take O for origin and centre of auxiliary circle, PT the polar of c, pt the polar of C, CP perpendicular on polar of c, cp perpendicular on polar of C. Then

$$\left.\begin{aligned} r^2 &= OC.Ot = Oc.OT \\ \text{and}\quad & OC.Om = Oc.OM \end{aligned}\right\}. \quad \text{Therefore, by subtraction, } OC.mt = Oc.MT, \text{ or } OC.cp = Oc.CP;$$

that is, $CP : CO :: cp : cO$. Q. E. D.

COR.—By making CP constant, we see that the reciprocal of a circle is a conic having its focus at the origin and its directrix the polar of the circle's centre.

GENERAL RULES FOR RECIPROCATING.

4846 *A point becomes the polar of the point, and a right line becomes the pole of the line.**

4847 *A line through a fixed point becomes a point on a fixed line.*

4848 *The intersection of two lines becomes the line which joins their poles.*

4849 *Lines passing through a fixed point become the same number of points on a fixed line, the polar of the point.*

4850 *A right line intersecting a curve in* n *points becomes* n *tangents to the reciprocal curve passing through a fixed point.*

4851 *Two lines intersecting on a curve become two points whose joining line touches the reciprocal curve.*

4852 *Two tangents and the chord of contact become two points on the reciprocal curve and the intersection of the tangents at those points.*

4853 *A pole and polar of any curve become respectively a polar and pole of the reciprocal curve; and a point of contact and tangent become respectively a tangent and its point of contact.*

4854 *The locus of a point becomes the envelope of a line.*

4855 *An inscribed figure becomes a circumscribed figure.*

4856 *Four points connected by six lines or a quadrangle become four lines intersecting in six points or a quadrilateral.*

4857 *The angle between two lines is equal to the angle subtended at the origin by the corresponding points.* (4844)

4858 *The origin becomes a line at infinity, the polar of the origin.*

4859 *Two lines through the origin become two points at infinity on the polar of the origin.*

4860 *Two tangents through the origin to a curve become two points at infinity on the reciprocal curve.*

4861 *The points of contact of such tangents become asymptotes of the reciprocal curve.*

4862 *The angle between the same tangents is equal to the angle between the asymptotes.* (4857)

---

* That is, with respect to the circle of reciprocation, and so throughout with the exception of (4853).

4863 *According as the tangents from the origin to a conic are real or imaginary, the reciprocal curve is an hyperbola or ellipse.*

4864 *If the origin be taken on the conic, the reciprocal curve is a parabola.*

For, by (4860, '1), the asymptotes are parallel and at infinity.

4865 *A trilinear equation is converted by reciprocation into a tangential equation.*

Thus $\alpha\gamma = k\beta\delta$ is a conic passing through four of the intersections of the lines $\alpha, \beta, \gamma, \delta$. Reciprocating, we get a tangential equation of the same form $AC = kBD$, and this is a conic touching four of the lines which join the points whose tangential equations are $A = 0$, $B = 0$, $C = 0$, $D = 0$. See (4907).

4866 The equation of the reciprocal of the conic $a^2y^2 + b^2x^2 = a^2b^2$ with the same origin and axes is

$$a^2x^2 + b^2y^2 = k^4,$$

where k is the radius of the auxiliary circle whose centre is the centre of the conic.

Proof.—Let p be the perpendicular on the tangent, θ its inclination; then $k^4r^{-2} = p^2 = a^2\cos^2\theta + b^2\sin^2\theta$ (4732).

4867 The same when the origin of reciprocation is the point $x'y'$,

$$(xx' + yy' + k^2)^2 = a^2x^2 + b^2y^2.$$

Proof: $k^2r^{-1} = p = \sqrt{a^2\cos^2\theta + b^2\sin^2\theta} - (x'\cos\theta + y'\sin\theta)$.

4868 The reciprocal curve of the general conic (4656), the auxiliary circle being $x^2 + y^2 = k^2$ or $x^2 + y^2 + z^2 = 0$ in trilinears, will be symmetrically

$$A\xi^2 + B\eta^2 + C\zeta^2 + 2F\eta\zeta + 2G\zeta\xi + 2H\xi\eta = 0,$$

replacing ζ by $-k^2$.

Proof.—Let $\xi\eta$ be a point on the reciprocal curve, then the polar of $\xi\eta$, namely, $x\xi + y\eta - k^2 = 0$, must touch the conic, by (4853). Therefore, by (4665), we must substitute ξ, η, $-k^2$ for λ, μ, ν in the tangential equation $A\lambda^2 + \&c. = 0$.

4869 From the reciprocal of a curve with respect to the origin of coordinates, to deduce the reciprocal with respect to an origin $x'y'$, substitute in the given reciprocal equation

$$\frac{k^2x}{xx' + yy' + k^2} \text{ for } x \quad \text{and} \quad \frac{k^2y}{xx' + yy' + k^2} \text{ for } y.$$

PROOF.—Let P be the perpendicular from the origin on the tangent and $PR = k^2$. The perpendicular from $x'y'$ is $P - x'\cos\theta - y'\sin\theta$,

$$\therefore\ \frac{k^2}{\rho} = \frac{k^2}{R} - x'\cos\theta - y'\sin\theta, \qquad \therefore\ \frac{k^2}{R} = \frac{xx'+yy'+k^2}{\rho};$$

$$\therefore\ R\cos\theta = \frac{k^2\rho\cos\theta}{xx'+yy'+k^2}.$$

TANGENTIAL COORDINATES.

4870 By employing these coordinates, theorems which are merely the reciprocals of those already deduced in trilinears may be proved independently. See (4019) for a description of this system.

The following proposition serves to transform by reciprocation the whole system of trilinear coordinates of points and equations of right lines and curves, into tangential coordinates of right lines and equations of points and curves.

THEOREM OF TRANSFORMATION.

4871 Given the trilinear equation of a conic (4656), the tangential equation of the reciprocal conic in terms of λ, μ, ν, the perpendiculars from three fixed points A', B', C' upon the tangent (Fig. 74) will be as follows, O being the origin of reciprocation and $OA', OB', OC' \equiv p, q, r$:—

4872 $$\frac{a\lambda^2}{p^2}+\frac{b\mu^2}{q^2}+\frac{c\nu^2}{r^2}+\frac{2f\mu\nu}{qr}+\frac{2g\nu\lambda}{rp}+\frac{2h\lambda\mu}{pq}=0.$$

PROOF.—Let $\alpha=0$, $\beta=0$, $\gamma=0$ be the sides of the original trigon ABC. The poles of these lines will be A', B', C', the vertices of the trigon for the reciprocal curve. Let RS be the polar of a point P on the given conic; α, β, γ the perpendiculars from P upon BC, CA, AB; *i.e.*, the trilinear coordinates of P. Let λ, μ, ν be the perpendiculars from A', B', C' upon RS; *i.e.*, the tangential coordinates of the polar of P referred to A', B', C'. Then, by (4845), $\frac{\alpha}{OP} = \frac{\lambda}{OA'}$, $\frac{\beta}{OP} = \frac{\mu}{OB'}$, $\frac{\gamma}{OP} = \frac{\nu}{OC'}$. Substitute these values of α, β, γ in (4656) and divide by OP^2.

4873 The angular relation between the trigons ABC and $A'B'C'$ is

$$B'OC' = \pi - A, \quad C'OB' = \pi - B, \quad A'OB' = \pi - C.$$

4874 If ABC be self-conjugate with regard to the circle of reciprocation, it will coincide with $A'B'C'$.

4875 Now let O be the circum-centre (4629) of $A'B'C'$ (Fig. 74), then it will be the in-centre of ABC, and, by (4873),

$$2A' = \pi - A, \quad 2B' = \pi - B, \quad 2C' = \pi - C.$$

Also $p = q = r$ in (4872), which becomes $\phi(\lambda, \mu, \nu) = 0$, so that the conic and its reciprocal are represented by the *same* equation. Consequently any relation in trilinear coordinates has its interpretation in tangential coordinates. We have then the following rule:—

4876 RULE.—*To convert any expression in trilinears into tangentials, consider the origin of the former as the in-centre of the trigon, change* α, β, γ *into* λ, μ, ν, *and interpret the result by the rules for reciprocating* (4846–65). *If the angles of the original trigon are involved, change these by* (4875) *into the angles of the reciprocal trigon, of which the origin will now be the circum-centre.*

4877 Referring trilinears and tangentials to the same trigon ABC, the equation of a point, as shown in (4021), becomes

$$\frac{\alpha}{p_1}\lambda + \frac{\beta}{p_2}\mu + \frac{\gamma}{p_3}\nu = 0;$$

4878 or, by multiplying by $\frac{1}{2}\Sigma$,

$$BOC\lambda + COA\mu + AOB\nu = 0. \qquad \text{(Fig. 3)}$$

The equation of a point can generally be obtained directly from the figure by means of this formula.

EQUATIONS IN TANGENTIAL COORDINATES.

For direct demonstrations of the following theorems, the reader may consult *Ferrers' Trilinear Coordinates*, Chap. VII.

4879 The point dividing AB in the ratio $\mathfrak{a} : \mathfrak{b}$, that is, the intersection with the internal or external bisector of C, is

$$\mathfrak{a}\lambda \pm \mathfrak{b}\mu = 0. \qquad \text{Centre of } AB \quad \lambda + \mu = 0.$$

The point O in (4878) is now on the side AB.

4881 Mass-centre, $\lambda + \mu + \nu = 0.$ [For $BOC = COA = AOB$.

4882 In-centre, $\mathfrak{a}\lambda+\mathfrak{b}\mu+\mathfrak{c}\nu = 0.$

4883 a ex-centre, $-\mathfrak{a}\lambda+\mathfrak{b}\mu+\mathfrak{c}\nu = 0.$ [By (4878), for $\frac{\pm BOC}{\mathfrak{a}} = \frac{COA}{\mathfrak{b}} = \frac{AOB}{\mathfrak{c}}$.

4884 Circum-centre $\lambda \sin 2A+\mu \sin 2B+\nu \sin 2C = 0.$

PROOF.—For $BOC = \frac{1}{2}R^2 \sin 2A$, &c. in (4878). *Otherwise.*—By reciprocation (4876), $\alpha \sin A+\beta \sin B+\gamma \sin C = 0$ is the line at infinity referred to the trigon ABC; therefore

$$\lambda \sin A+\mu \sin B+\nu \sin C = 0$$

is the equation of the pole of that line referred to $A'B'C'$; that is,

$$\lambda \sin 2A'+\mu \sin 2B'+\nu \sin 2C', \text{ by (4875).}$$

4885 Foot of perpendicular from C upon AB,

$$\lambda \tan A+\mu \tan B = 0.$$

4886 Orthocentre $\lambda \tan A+\mu \tan B+\nu \tan C = 0.$

4887 Inscribed conic of ABC, [Proof below.

$$L\mu\nu+M\nu\lambda+N\lambda\mu = 0.$$

4888 Point of contact with AB,

$$M\lambda+L\mu = 0.$$

4889 In-circle (4629),

$$(\mathfrak{s}-\mathfrak{a})\,\mu\nu+(\mathfrak{s}-\mathfrak{b})\,\nu\lambda+(\mathfrak{s}-\mathfrak{c})\,\lambda\mu = 0.$$

4890 Point of contact with AB, $(\mathfrak{s}-\mathfrak{b})\,\lambda+(\mathfrak{s}-\mathfrak{a})\,\mu = 0.$

4891 $\mathfrak{a}$ ex-circle, $(\mathfrak{s}-\mathfrak{b})\,\lambda\mu+(\mathfrak{s}-\mathfrak{c})\,\nu\lambda-\mathfrak{s}\mu\nu = 0.$

PROOF.—Since the coordinates of AB of the trigon are 0, 0, ν, the equation of the inscribed conic must be satisfied when any two of the coordinates λ, μ, ν vanish, therefore it must be of the form (4887). Otherwise by reciprocating (4724).

If the circle touches AB in D (Fig. 3), $\lambda : -\mu = AD : BD = \mathfrak{s}-\mathfrak{a} : \mathfrak{s}-\mathfrak{b}$ (Fig. of 709), which proves (4890).

(4888) is the equation of the point of contact, because the line $(0, 0, \nu)$ passes through it and also touches the conic (4887).

(4889) is the in-circle by (4887) and (4890) and what precedes.

4892 Circumscribed conic, [By (4876) applied to (4739, '40).

$$L^2\lambda^2+M^2\mu^2+N^2\nu^2-2MN\mu\nu-2NL\nu\lambda-2LM\lambda\mu = 0, \quad (4740)$$

4893 or $\sqrt{(L\lambda)}+\sqrt{M\mu}+\sqrt{N\nu}=0.$

4894 Tangent at A, $M\mu = N\nu$.

4895 Circum-circle

$$\mathfrak{a}^4\lambda^2+\mathfrak{b}^4\mu^2+\mathfrak{c}^4\nu^2-2\mathfrak{b}^2\mathfrak{c}^2\mu\nu-2\mathfrak{c}^2\mathfrak{a}^2\nu\lambda-2\mathfrak{a}^2\mathfrak{b}^2\lambda\mu = 0\,;$$

4896 or $\mathfrak{a}\sqrt{\lambda}+\mathfrak{b}\sqrt{\mu}+\mathfrak{c}\sqrt{\nu}=0.$

PROOF.—By (4876) applied to (4747, '8), and by $\cos\frac{A}{2} = \sin A'$ (4875)

4897 Relation between the coordinates of any right line:

$$\mathfrak{a}^2(\lambda-\mu)(\lambda-\nu)+\mathfrak{b}^2(\mu-\nu)(\mu-\lambda)+\mathfrak{c}^2(\nu-\lambda)(\nu-\mu) = \Sigma^2.$$

4898 Coordinates of the line at infinity:

$$\lambda = \mu = \nu.$$

PROOF.—The trilinear coordinates of the origin and centre of the reciprocal conic are $\alpha = \beta = \gamma$, (4876). It is also self-evident.

4899 The point $l\lambda+m\mu+n\nu = 0$ will be at infinity when $l+m+n = 0$.

PROOF.—By (4876), for the line $l\alpha+m\beta+n\gamma = 0$ will pass through the origin $\alpha=\beta=\gamma$ when $l+m+n = 0$.

4900 A curve will be touched by the line at infinity when the sum of the coefficients vanishes.

PROOF.—By (4876), for this is the condition that the origin in trilinears, $\alpha = \beta = \gamma$ shall be on the curve.

4901 The equation of the centre of the conic $\phi(\lambda, \mu, \nu)$ is

$$\phi_\lambda+\phi_\mu+\phi_\nu = 0,$$

4902 or $(a+h+g)\,\lambda+(h+b+f)\,\mu+(g+f+c)\,\nu = 0.$

PROOF.—The coordinates of the in-centre of ABC (4876) are $\alpha'=\beta'=\gamma'$, therefore the polar of this point with regard to the conic $\phi(\alpha, \beta, \gamma)$ is $\phi_\alpha+\phi_\beta+\phi_\gamma = 0$ (4658). This point and polar reciprocate into a polar and point, of which the former, being the reciprocal of the in-centre, or origin, is the line at infinity, and therefore the latter is the centre of $\phi(\lambda, \mu, \nu)$, while its equation is as stated.

4903 The equation of the two points in which the line (λ', μ', ν') cuts the conic is

$$4\phi(\lambda', \mu', \nu')\,\phi(\lambda, \mu, \nu) = (\phi_\lambda\lambda' + \phi_\mu\mu' + \phi_\nu\nu')^2. \qquad (4680)$$

4904 The coordinates of the asymptotes are found from the equations

$$\phi(\lambda, \mu, \nu) = 0 \quad \text{and} \quad \phi_\lambda + \phi_\mu + \phi_\nu = 0.$$

PROOF.—These are the conditions that the line (λ, μ, ν) should touch the curve and also pass through the centre (4901).

4905 The equation of the two circular points at infinity is

$$\mathfrak{a}^2(\lambda-\mu)(\lambda-\nu) + \mathfrak{b}^2(\mu-\nu)(\mu-\lambda) + \mathfrak{c}^2(\nu-\lambda)(\nu-\mu) = 0.$$

PROOF.—Put $\lambda' = \mu' = \nu'$ in (4903) to make the line at infinity, and for the conic take the in-circle (4889).

4906 The general equation of a circle is

$$\mathfrak{a}^2(\lambda-\mu)(\lambda-\nu) + \mathfrak{b}^2(\mu-\nu)(\mu-\lambda) + \mathfrak{c}^2(\nu-\lambda)(\nu-\mu) = (l\lambda + m\mu + n\nu)^2 \ldots\ldots (1),$$

where $l\lambda + m\mu + n\nu = 0$ is the equation of the centre.

PROOF.—The general equation of a conic in trilinears may, by (4601), be put in the form

$$\mathfrak{a}(\beta-\beta_0)(\gamma-\gamma_0) + \mathfrak{b}(\gamma-\gamma_0)(\alpha-\alpha_0) + \mathfrak{c}(\alpha-\alpha_0)(\beta-\beta_0) = (l\alpha + m\beta + n\gamma)^2,$$

where $l\alpha + m\beta + n\gamma = 0$ is the directrix, and $\alpha_0\beta_0\gamma_0$ the focus. Now let the focus be the in-centre of the trigon, and therefore $\alpha_0 = \beta_0 = \gamma_0 = \frac{1}{2}\Sigma\mathfrak{s}^{-1}$ (709). By this relation and $\mathfrak{a}\alpha + \mathfrak{b}\beta + \mathfrak{c}\gamma = \Sigma$, the equation is expressed as

$$\mathfrak{a}(\mathfrak{s}-\mathfrak{a})(\alpha-\beta)(\alpha-\gamma) + \&c. = (l'\alpha + m'\beta + n'\gamma)^2,$$

or

$$(\alpha-\beta)(\alpha-\gamma)\cos^2\tfrac{1}{2}A + \&c. = (l'\alpha + m'\beta + n'\gamma)^2.$$

Reciprocating by (4876), this becomes

$$(\lambda-\mu)(\lambda-\nu)\sin^2 A' + \&c. = (l\lambda + m\mu + n\nu)^2,$$

the constant factor introduced on the right being involved in l, m, n; and $\sin A' = \cos\frac{1}{2}A$, by (4875). And we know that this is a circle by (4845 Cor.), and that the directrix of the conic reciprocates into the centre of the circle.

Otherwise.—The left side of (1) represents the two circular points at infinity (4905), and, if for the right we take the equation of a point, the whole represents a conic, as in (4909), of the form $AC = B^2$. In this case, A, C, the points of contact of tangents from B, being the circular points, the conic must be a circle with $B = 0$ for its centre.

Abridged Notation.

4907 Let $A = 0$, $B = 0$, $C = 0$, $D = 0$ (Fig. 75) be the tangential equations of the four points of a quadrangle, where $A \equiv a_1\lambda + b_1\mu + c_1\nu$, $B \equiv a_2\lambda + b_2\mu + c_2\nu$, and so on. Then the equation of the inscribed conic will be $AC = kBD$.

Proof.—The equation is of the second degree in λ, μ, ν; therefore the line (λ, μ, ν) touches a conic. The coordinates of one line that touches this conic are determined by the equations $A = 0$, $B = 0$. That is, the line joining the two points A, B touches the conic, and so of the rest.

4908 If the points B, D coincide (Fig. 76), the equation becomes $AC = kB^2$; and $A = 0$, $C = 0$ are the points of contact of tangents from the point $B = 0$.

4909 Referring the conic to the trigon ABC (Fig. 78), and taking $AC = k^2B^2$ for its equation, let a tangent ef be drawn, and let $Ae : eB = k : m$. The equations of the points e and f will be

$$mA + kB = 0, \qquad mkB + C = 0.$$

Proof.—The first equation corresponds to (4879). For the equation of f, eliminate A from $mA + kB = 0$ and $AC = k^2B^2$.

4910 Let e, h (Fig. 77) be two points on AB whose equations are $mA + kB = 0$, $m'A + kB = 0$. The equation of the point p, in which tangents from e and h intersect, is

$$\mathbf{mm'A + (m+m')\,kB + C = 0.}$$

Proof.—The equation may be put in the form

$$(mA + kB)(m'A + kB) = 0,$$

because $k^2B^2 = AC$ if the line touches the conic. The equation being of the first degree in A, B, C, must represent *some* point. That is, the relation between λ, μ, ν involved in it makes the straight line $\lambda\alpha + \mu\beta + \nu\gamma$ pass through a certain point. But the equation is satisfied when $mA + kB = 0$, a relation which makes the straight line pass through e. Hence a tangent through e passes through a certain fixed point. Similarly, by $m'A + kB = 0$, another tangent passes through h and the same fixed point.

4911 Cor.—Let $m' = m$, then the equation of the point of contact of the tangent joining the points $ma + kB$ and $mkB + C$ (4909) (e and f, Fig. 78) will be

$$\mathbf{m^2A + 2mkB + C = 0.}$$

4912 If in Fig. (78) the trilinear coordinates of the points

A, B, C are x_1, y_1, z_1, x_2, y_2, z_2, x_3, y_3, z_3, the coordinates of the point of contact p of the tangent defined by m will be

$$m^2x_1+2mkx_2+x_3, \quad m^2y_1+2mky_2+y_3, \quad m^2z_1+2mkz_2+z_3,$$

and the tangent at p divides the two fixed tangents in the ratios $k:m$ and $mk:1$, by (4909).

4913 NOTE.—The equation U or $\Phi(\lambda, \mu, \nu) = 0$ (4665) expresses the condition that $\lambda\alpha+\mu\beta+\nu\gamma$ shall touch a certain conic. When U is about to break up into two factors, the minor axis of the conic diminishes (Fig. 79). Every tangent that can now be drawn to the conic passes very nearly through one end or other of the major axis. Ultimately, when the minor axis vanishes, the condition of the line touching the conic becomes the condition of its passing through one or other of two fixed points A, B. In this case, U consists of two factors, which, put equal to zero, are the equations of those points. The conic has become a straight line, and this line is touched at every point by a single tangent.

4914 If U and U' (Fig. 80) be two conics in tangential coordinates, $kU+U'$ is then a conic having for a tangent every tangent common to U and U'; and $kU+AB$ is a conic having in common with U the two pairs of tangents drawn from the points A, B.

The conic U' in this case merges into the line AB, or, more strictly, the two points A, B, as explained in (4913).

4915 If either $kU+U'$ or $kU+AB$ breaks up into two factors, it represents two points which are the opposite vertices of the quadrilateral formed by the four tangents.

ON THE INTERSECTION OF TWO CONICS.

INTRODUCTORY THEOREM.

*Geometrical meaning of $\sqrt{(-1)}$.**

4916 In a system of rectangular or oblique plane coordinates, let the operator $\sqrt{-1}$ prefixed to an ordinate y denote the turning of the ordinate about its foot as a centre through a right angle in a plane perpendicular to the plane of xy. The repetition of this operation will turn the ordinate

* [The fiction of imaginary lines and points is not ineradicable from Geometry. The theory of Quaternions removes all imaginariness from the symbol $\sqrt{-1}$, and, as it appears that a partial application of that theory presents the subject of Projection in a much clearer light, I have here introduced the notion of the multiplication of vectors at right angles to each other.]

through another right angle in the same plane so as to bring it again into the plane of xy. The double operation has converted y into $-y$. But the two operations are indicated algebraically by $\sqrt{-1}\,.\sqrt{-1}\,.y$ or $(\sqrt{-1})^2y=-y$, which justifies the definition.

It may be remarked, in passing, that *any operation* which, being performed twice in succession upon a quantity, changes its sign, offers a consistent interpretation of the multiplier $\sqrt{-1}$.

4917 With this additional operator, borrowed from the Theory of Quaternions, equations of plane curves may be made to represent more extended loci than formerly. Consider the equation $x^2+y^2=a^2$. For values of $x<a$, we have $y=\pm\sqrt{a^2-x^2}$, and a circle is traced out. For values of $x>a$, we may write $y=\pm i\sqrt{x^2-a^2}$, where $i\equiv\sqrt{-1}$. The ordinate $\sqrt{x^2-a^2}$ is turned through a right angle by the vector i, and this part of the locus is consequently an equilateral hyperbola having a common axis with the circle and a common parameter, but having its plane at right angles to that of the circle. Since the foot of each ordinate remains unaltered in position, we may, for convenience, leave the operation indicated by i unperformed and draw the hyperbola in the original plane. In such a case, the circle may be called the *principal*, and the hyperbola the *supplementary*, curve, after Poncelet. When the coordinate axes are rectangular, the supplementary curve is not altered in any other respect than in that of position by the transformation of all its ordinates through a right angle; but, if the coordinate axes are oblique, there is likewise a change of figure precisely the same as that which would be produced by setting each ordinate at right angles to its abscissa in the xy plane.

In the diagrams, the supplementary curve will be shown by a dotted line, and the unperformed operation indicated by i must always be borne in mind. For, on account of it, there can be no geometrical relations between the principal and supplementary curves excepting those which arise from the possession of one common axis of coordinates. This law is in agreement with the algebraic one which applies to the real and imaginary parts of the equation $x^2-(iy)^2=a^2$. When y vanishes, $x=a$ in both curves.

If either the ellipse $b^2x^2+a^2y^2=a^2b^2$ or the hyperbola $b^2x^2-a^2y^2=a^2b^2$ be taken for the principal curve, the other will be the supplementary curve.

It is evident that, by taking different conjugate diameters for coordinate axes, the same conic will have corresponding different supplementary curves. The phrase, "supplementary conic on the diameter DD," for example, will refer to that diameter which forms the common axis of the principal and supplementary conic in question.

4918 Let us now take the circle $x^2+y^2=a^2$ and the right line $x=b$. When b is $>a$, the line intersects the supplementary right hyperbola in two points whose ordinates are $\pm i\sqrt{b^2-a^2}$. By increasing b without limit, we get a pair of, so-called, *imaginary points at infinity*. These lie on the asymptotes of the hyperbola, and the equation of the asymptotes is $(x+iy)(x-iy)=0$.

We can now give a geometrical interpretation to the statements in (4720). The two lines drawn from the focus of the conic $b^2x^2+a^2y^2=a^2b^2$ to the "circular points at infinity" make angles of 45° with the major axis, and they touch the conic in its supplementary hyperbola $b^2x^2-a^2(iy)^2=a^2b^2$. An independent proof of this is as follows.

Draw a tangent from S (Fig. 81) to the supplementary hyperbola, and let x, y be the coordinates of the point of contact P. Then

$$x = \frac{a^2}{CS} \text{ (1170)} = \frac{a^2}{\sqrt{(a^2-b^2)}}; \quad \text{and} \quad y = \frac{b}{a}\sqrt{(x^2-a^2)} = \frac{b^2}{\sqrt{(a^2-b^2)}},$$

by the value of x. Also

$$SN = x - CS = \frac{a^2}{\sqrt{(a^2-b^2)}} - \sqrt{a^2-b^2} = \frac{b^2}{\sqrt{(a^2-b^2)}}.$$

Therefore $y = SN$, therefore SP makes an angle of 45° with CN.

The following results are required in the theory of projection, and are illustrated in figures (82) to (86). Two ellipses are taken in each case for principal curves, and the supplementary hyperbolas are shown by dotted lines. As the planes of the principal and supplementary curves are really at right angles, the intersections of the solid lines with the dotted are only apparent. The intersections of the solid lines are *real* points, while the intersections of the dotted lines represent the *imaginary* points.

4919 Two conics may intersect—

(i.) *in four real points* (Fig. 82);

(ii.) *in two real and two imaginary points* (Fig. 83);

(iii.) *in four imaginary points* (Fig. 84).

[When the two hyperbolas in figures (83) and (84) are similar and similarly situated, two of their points of intersection recede to infinity (Figs. 85 and 86). Hence, and by taking the dotted lines for principal, and the solid for supplementary, curves, we also have the cases]

(iv.) *in two real finite points and two imaginary points at infinity;*

(v.) *in two imaginary finite points and two imaginary points at infinity;*

(vi.) *in two imaginary finite points and two real points at infinity;*

(vii.) *in two real finite points and two real points at infinity.*

4920 Given two conics not intersecting, or intersecting in but two points, to draw the two supplementary curves which have a common chord of intersection conjugate to the diameters upon which they are described, or in other words, to find the imaginary common chord of the conics.

Poncelet has shewn by geometrical reasoning (*Propriétés des Projectives*, p. 31) that such a chord must exist. The following is a method of determining its position—

Let $(abcfgh)(xy1)^2 = 0$ and $(a'b'c'f'g'h')(xy1)^2 = 0$ ....(i.)

be the equations of the conics C, C' (Fig. 89), the coordinate axes being rectangular. Suppose PQ to be the common chord sought. Then the diameters AB, $A'B'$ conjugate to PQ bisect it in D, and the supplementary curves on those diameters intersect in the points P, Q. Now, let the coordinate axes be turned through an angle θ, so that the y axis may become parallel to PQ, and therefore also to the tangents at A, B, A', B'. This is accomplished by substituting for x and y, in equations (i.), the values

$$x\cos\theta - y\sin\theta \quad \text{and} \quad y\cos\theta + x\sin\theta.$$

Let the transformed equations be denoted by $(ABCFGH)(xy1)^2 = 0$ and $(A'B'C'F'G'H')(xy1)^2 = 0$, in which the coefficients are all functions of θ, excepting c, which is unaltered. Solving each of these equations as a quadratic in y, the solutions take the forms

$$y = \alpha x + \beta \pm \sqrt{\mu(x^2 - 2px + q)}, \quad y = \alpha' x + \beta' \pm \sqrt{\mu'(x^2 - 2p'x + q')} \ldots \text{(ii.)},$$

with the values of α, β, μ, p, q given in (4449–53), if for small letters we substitute capitals. Thus, α, β, μ, p, q are obtained in terms of θ and the original coefficients a, h, b, f, g, h.

Now, the coordinates of D being $\xi = ON$, $\eta = DN$, we have $\eta = \alpha\xi + \beta$ and $\eta = \alpha'\xi + \beta'$, therefore $\alpha\xi + \beta = \alpha'\xi + \beta'$ .... (iii.).

The surd in equations (ii.) represents the ordinate of the conic conjugate to the diameter AB or $A'B'$. For values of x in the diagram $> OM$ and $< OR$, the factor $\sqrt{-1}$ appears in this surd, indicating an ordinate of the supplementary curve on AB or $A'B'$. Hence, equating the values of the common ordinate PD, we have

$$\mu(\xi^2 - 2p\xi + q) = \mu'(\xi^2 - 2p'\xi + q') \ldots\ldots\ldots\ldots \text{(iv.)}.$$

Eliminating ξ between equations (iii.) and (iv.), we obtain an equation for determining θ; which angle being found, we can at once draw the diameters AB, $A'B'$.

THE METHOD OF PROJECTION.

4921 PROBLEM.—Given any conic and a right line in its plane and any plane of projection, to find a vertex of projection such that the line may pass to infinity while the conic is projected into a hyperbola or ellipse according as the right line does or does not intersect the given conic; and at the same time to give any assigned proportion and direction to the axes of the projected conic.

Analysis.—Let $HCKD$ be the given conic, and BB the right line, in Fig. (87) not intersecting, and in Fig. (88) intersecting the conic. Draw HK the diameter of the conic conjugate to BB. Suppose O to be the required vertex of projection. Draw any plane $ECGD$ parallel to OBB, intersecting the given conic in CD and the line HK in F, and draw the plane OHK cutting the former plane in E, F, G and the line BB in A; and let the curve $ECGD$ be the conical projection of $HCKD$ on the plane parallel to OBB.

By similar triangles,

$$\frac{EF}{HF}=\frac{OA}{HA} \text{ and } \frac{FG}{FK}=\frac{OA}{AK}, \quad \therefore \frac{EF.FG}{HF.FK}=\frac{OA^2}{HA.AK} \quad\ldots\ldots\ldots(1).$$

Let α, β be the semi-diameters of the given conic parallel to HK and CD; then

$$\frac{CF^2}{HF.FK}=\frac{\beta^2}{\alpha^2}, \quad \therefore \frac{CF^2}{EF.FG}=\frac{\beta^2.HA.AK}{\alpha^2.OA^2} \quad\ldots\ldots\ldots\ldots\ldots(2).$$

Now, since parallel sections of the cone are similar, if the plane of $HCKD$ moves parallel to itself, the ratio on the right remains constant; therefore, by (1193), the section $ECGK$ is an ellipse in Fig. (87) and an hyperbola in Fig. (88). Let a, b be the semi-diameters of this ellipse or hyperbola parallel to EG and CD, that is, to OA and BB; then, by (2),

$$\frac{b^2}{a^2}=\frac{\beta^2}{\alpha^2}\,\frac{HA.AK}{OA^2}, \quad \therefore OA^2=\frac{a^2\beta^2}{b^2\alpha^2}HA.AK \quad\ldots\ldots\ldots\ldots\ldots(3).$$

But $\frac{\beta^2}{\alpha^2}HA.AK=AB^2$, where AB in Fig. (88) is the ordinate at A of the given conic, but in Fig. (87) the ordinate of the conic supplementary to the given one on the diameter conjugate to BB. Therefore

$$AO^2=\frac{a^2}{b^2}AB^2 \quad\ldots\ldots\ldots\ldots\ldots\ldots\ldots\ldots\ldots\ldots\ldots (4).$$

Hence AO, AB are parallel and proportional to a and b. And, since AB is given in magnitude and direction, we have two constants at our disposal, namely, the ratio of the semi-conjugate diameters a and b and the angle between them, or, which is the same thing, the eccentricity and the direction of the axes of the ellipse or hyperbola on the plane of projection.

4922 The construction will be as follows:—

Determine the point A *as the intersection of* BB *with the diameter* HK *conjugate to it. Choose any plane of projection, and in a plane through* BB, *parallel to it, measure* AO *of the length given by equation* (3) *or* (4), *making the angle* BAO *equal to the required angle between* a *and* b. O *will be the vertex of projection, and any plane* LMN *parallel to* OBB *will serve for the plane of projection.*

4923 Cor. I.—If $AO=AB$, the projected curve in Fig. (88) will in every case be a right hyperbola.

4924 Cor. 2.—If BAO is a right angle, the axes of the projected ellipse or hyperbola are parallel and proportional to AO and AB. Hence, in this case, the eccentricity of the hyperbola will be $e = OB : OA$.

4925 Cor. 3.—If $AO = AB$ and $BAO =$ a right angle, the ellipse becomes a circle and the right hyperbola in Cor. 1 has its axes parallel to AO and AB.

4926 To project a conic so that a given point in its plane may become the centre of the projected curve.

Take for the line BB *the polar of the given point, and construct as in* (4922). *For, if* P *be the given point, and* BB *its polar* (*Fig.* 87 *or* 88), p *the projection of* P *will have its polar at infinity, and will therefore be the centre of the projected ellipse or hyperbola, according as* P *is within or without the original conic.*

4927 To project two intersecting conics into two similar and similarly situated hyperbolas of given eccentricity.

Take the common chord of the conics for the line BB (*Fig.* 88), *and project each conic as in* (4922), *employing the same vertex and plane of projection. Then, since the point* A *and the lines* AB *and* AO *are the same for each projection, corresponding conjugate diameters of the hyperbolas are parallel and proportional to* AO *and* AB; *therefore, &c.*

4928 To project two non-intersecting conics into similar and similarly situated ellipses of given eccentricity.

Take the common chord of a certain two of the supplementary curves of the conics (4920), *in other words, the imaginary common chord of the conics, for the line* BB, *and proceed as in* (4927).

4929 To project two conics having a common chord of contact into two concentric, similar and similarly situated hyperbolas.

Take the common chord for the line BB, *and construct as in* (4922). *The common pole of the conics projects into a common centre and the common tangents into common asymptotes.*

4930 To project any two conics into concentric conics.

Find the common pole and polar of the given conics by (4762), *and take the common polar for the line* BB *in the construction of* (4922). *The common pole projects into a common centre.*

4931 Ex. 1. — Given two conics having double contact with each other, any chord of one which touches the other is cut harmonically at the point of contact and where it meets the common chord of contact of the conics.

[*Salmon's Conic Sections*, Art. 354.

Let AB be the common chord of contact, PQ the other chord touching the inner conic at C and meeting AB produced in D. By (4929), project AB, and therefore the point D, to infinity. The conics become similar and similarly situated hyperbolas, and C becomes the middle point of PQ (1189). The theorem is therefore true in this case. Hence, by a converse projection, the more general theorem is inferred.

4932 Ex. 2.—Given four points on a conic, the locus of the pole of any fixed line is a conic passing through the fourth harmonic to the point in which this line meets each side of the given quadrilateral. [*Ibid.*, Art. 354.

Let the fixed line meet a side AB of the quadrilateral in D, and let $ACBD$ be in harmonic ratio. Project the fixed line, and therefore the point D, to infinity. C becomes the middle point of AB (1055), and the pole of the fixed line becomes the centre of the projected conic. Now, it is known that the locus of the centre is a conic passing through the middle points of the sides of the quadrilateral. Hence, projecting back again, the more general theorem is inferred.

4933 Ex. 3. — If a variable ellipse be described touching two given ellipses, while the supplementary hyperbolas of all three have a common chord AB conjugate to the diameters upon which they are described; the locus of the pole of AB with respect to the variable ellipse is an hyperbola whose supplementary ellipse touches the four lines CA, CB, $C'A$, $C'B$, where C, C' are the poles of AB with respect to the fixed ellipses.

(*Salmon*, Art. 355.)

Proof.—Project AB to infinity and the three ellipses into circles. The poles P, C, C' become the centres p, c, c' of the circles. The locus of p is a hyperbola whose foci are c, c'. But the lines Ac, Bc now touch the supplementary ellipse of this hyperbola (4918). Therefore, projecting back again, we get AC, BC touching the supplementary ellipse of the conic which is the locus of P. Similarly, AC', BC' touch the same ellipse.

4934 Any two lines at right angles project into lines which cut harmonically the line joining the two fixed points which are the projections of the circular points at infinity.

Proof.—This follows from (4723).

4935 The converse of the above proposition (4931), which is the theorem in Art. 356 of *Salmon*, is not universally true in any real sense. If the lines drawn through a given point to the two circular points at infinity form a harmonic pencil with two other lines through that point, the latter two are not necessarily at right angles, as the theorem assumes.

The following example from the same article is an illustration of this—

Ex.—Any chord BB (Fig. 88) of a conic $HCKD$ is cut harmonically by any line $PKAH$ through P, the pole of the chord, and the tangent at K.

The ellipse BKB here projects into a right hyperbola; B, B project to infinity. The harmonic pencil formed by PK and the tangent at K, KB and KB projects into a harmonic pencil formed by pk and the tangent at k, kb and kb, where b, b are the circular points at infinity: but pk is not at right angles to the tangent at k of the right hyperbola. The harmonic ratio of the latter pencil can, however, be independently demonstrated, and that of the former can then be inferred. (Note that k is G in figure 88.)

If we may suppose the ellipse to project into an imaginary circle having points at infinity, the imaginary radius of that circle may be supposed to be at right angles to the imaginary tangent. The right hyperbola, however, is the real projection which takes the place of the circle in this and all similar instances; and it is only in the case of principal axes that the radius is at right angles to the tangent.

INVARIANTS AND COVARIANTS.

4936 Let $u \equiv (a\,b\,c\,f\,g\,h \between x\,y\,z)^2$, $u' \equiv (a'b'c'f'g'h' \between x\,y\,z)^2$ be two conics as in (4401) with the notation of (1620).

The three values of k, for which $ku+u'=0$ represents two right lines, are the roots of the cubic equation

4937 $$\Delta k^3+\Theta k^2+\Theta' k+\Delta' = 0,$$

4938 where $\Delta \equiv abc+2fgh-af^2-bg^2-ch^2$,

4939 $\Theta \equiv Aa'+Bb'+Cc'+2Ff'+2Gg'+2Hh'$,

and $A \equiv bc-f^2$, $F \equiv gh-af$, &c. (4665)

For the values of Δ' and Θ' interchange a with a', b with b', &c.

PROOF.—The discriminant of $ku+u'$, which must vanish (4661), is evidently the determinant here written, and it is equivalent to the cubic in question.

$$\begin{vmatrix} ka+a', & kh+h', & kg+g' \\ kh+h', & kb+b', & kf+f' \\ kg+g', & kf+f', & kc+c' \end{vmatrix}$$

4940 Δ, Θ, Θ', and Δ' are invariants of the conic $ku+u'$.

That is, if the axes of coordinates be transformed in any manner, the ratios of the four coefficients in (4937) are unaltered.

PROOF.—The transformation is effected by a linear substitution, as in (1794). Let u, u' thus become v, v'. Then $ku+u'$ becomes $kv+v'$, and k is unaltered. If the equation $ku+u'=0$ represents two right lines, it will continue to do so after transformation; but the condition for this is the vanishing of the cubic in k; and k being constant, the ratios of the coefficients must be unalterable.

4941 The equation of the six lines which join the four points of intersection of the conics u and u' is

$$\Delta u'^3-\Theta u'^2u+\Theta' u'u^2-\Delta' u^3=0.$$

PROOF.—Eliminate k from (4937) by $ku+u'=0$.

4942 The condition that the conics u and u' may touch is

$$(\Theta\Theta'-9\Delta\Delta')^2=4(\Theta^2-3\Delta\Theta')(\Theta'^2-3\Delta'\Theta),$$

4943 or $4\Delta\Theta'^3+4\Delta'\Theta^3+27\Delta^2\Delta'^2-18\Delta\Delta'\Theta\Theta'-\Theta^2\Theta'^2.$

PROOF.—Two of the four points in (3941) must coincide. Hence two out of the three pairs of lines must coincide. The cubic (4937) must therefore have two equal roots. Let α, α, β be the roots; then the condition is the result of eliminating α and β from the equations

$$\Delta(2\alpha+\beta)=-\Theta,\quad \Delta(\alpha^2+2\alpha\beta)=\Theta',\quad \Delta\alpha^2\beta=-\Delta' \ (406).$$

4944 The expression (4943) is the last term of the equation whose roots are the squares of the differences of the roots of the cubic in k, and when it is positive, the cubic in k has two imaginary roots; when it is negative, three real roots; and when it vanishes, two equal roots.

PROOF.—By (543) or (579). The last term of $f(x)$ in (543) is now $=27F(a)F(\beta)$, a, β being the roots of $3\Delta x^2+2\Theta x+\Theta'=0$. When this term is positive, $f(x)$ has a real negative root (409), and therefore $F(x)$ has then two imaginary roots; for, if $(a-b)^2=-c$, $a-b=ic$, and a and b are both imaginary. When the last term of $f(x)$ is negative, all the roots of $f(x)$ are positive, and therefore the roots of $F(x)$ are all real.

INVARIANTS OF PARTICULAR CONICS.

4945 When $u=ax^2+by^2+cz^2$ and $u'=x^2+y^2+z^2$,

$$\Delta=abc,\quad \Theta=bc+ca+ab,\quad \Theta'=a+b+c,\quad \Delta'=1.$$

4946 When $u = (abcfgh \between xyz)^2$ and $u' = x^2+y^2+z^2$,

$$\Theta = A+B+C, \quad \Theta' = a+b+c, \quad \Delta' = 1.$$

4947 When $u = x^2+y^2-r^2$ and $u' = (x-\alpha)^2+(y-\beta)^2-s^2$,

$$\Delta = -r^2, \quad \Delta' = -s^2,$$

$$\Theta = \alpha^2+\beta^2-2r^2-s^2, \quad \Theta' = \alpha^2+\beta^2-r^2-2s^2.$$

4948 The cubic for k reduces to

$$(k+1)\left\{s^2k^2+(r^2+s^2-\alpha^2-\beta^2)\,k+r^2\right\} = 0.$$

4949 When

$$u = b^2x^2+a^2y^2-a^2b^2 \quad \text{and} \quad u' = (x-\alpha)^2+(y-\beta)^2-r^2,$$

$$\Delta = -a^4b^4, \quad \Theta = a^2b^2\left\{\alpha^2+\beta^2-a^2-b^2-r^2\right\},$$

$$\Theta' = a^2\beta^2+b^2\alpha^2-a^2b^2-r^2(a^2+b^2), \quad \Delta' = -r^2.$$

4950 When $u = y^2-4mx$ and $u' = (x-\alpha)^2+(y-\beta)^2-r^2$,

$$\Delta = -4m^2, \quad \Theta = -4m(\alpha+m), \quad \Theta' = \beta^2-4m\alpha-r^2, \quad \Delta' = -r^2.$$

4951 When $u = (abcfgh \between xyz)^2$ and $u' = x^2+2xy\cos\omega+y^2$,

$$\Delta, \quad \Delta' = 0, \quad \Theta = c(a+b)-f^2-g^2+2(fg-ch)\cos\omega,$$

$$\Theta' = c\sin^2\omega.$$

Hence the following are invariants of the general conic, the inclination of the coordinate axes being ω.

4952
$$\frac{abc+2fgh-af^2-bg^2-ch^2}{c\sin^2\omega} = \frac{\Delta}{\Theta'} \quad \ldots\ldots\ldots\ldots (1),$$

4953
$$\frac{c(a+b)-f^2-g^2+2(fg-ch)\cos\omega}{c\sin^2\omega} = \frac{\Theta}{\Theta'} \ldots\ldots\ldots (2),$$

4954
$$\frac{ab-h^2}{\sin^2\omega} \ldots\ldots (3), \quad \text{and} \quad \frac{a+b-2h\cos\omega}{\sin^2\omega} \ldots\ldots (4).$$

For these are what (1) and (2) become when the axes are transformed so as to remove f and g.

If the origin be unaltered, c is invariable, and transformation of the axes will then leave invariable

4956 $$\frac{2fgh-af^2-bg^2}{\sin^2\omega} \quad \text{and} \quad \frac{f^2+g^2-2fg\cos\omega}{\sin^2\omega},$$

as appears by subtracting (3) from (1) and (2) from (4).

4958 Ex. (i.)—To find the evolute of the conic $b^2x^2+a^2y^2 = a^2b^2$. See also (4547).

Proof.—Denote the conic by u, and by u' the hyperbola $c^2xy+b^2y'x-a^2x'y$ (4335), which intersects u in the feet of the normals drawn from $x'y'$. Two of these normals must always coincide if $x'y'$ is to be on the evolute. u and u' must therefore touch. We have

$$\Delta = -a^4b^4, \quad \Theta = 0, \quad \Theta' = -a^2b^2(a^2x^2+b^2y^2-c^4), \quad \Delta' = -2a^2b^2c^2xy.$$

Substitute in (4942), and the equation of the evolute is found to be

$$(a^2x^2+b^2y^2-c^4)^3+27a^2b^2c^4x^2y^2 = 0.$$

4959 Ex. (ii.)—Similarly the evolute of the parabola is obtained from

$$u = y^2-4mx, \quad u' = 2xy+2(2m-x')y-4my',$$

$$\Delta = -4m^2, \quad \Theta = 0, \quad \Theta' = -4(2m-x), \quad \Delta' = 4my,$$

producing the equation $27my^2 = 4(x-2m)^3$. See also (4549).

4960 Ex. (iii.)—The locus of the centre of a circle of radius R, touching the conic $b^2x^2+a^2y^2-a^2b^2$, is called a *parallel* to the conic. Its equation is

$$\begin{aligned}
&R^8c^4-2R^6c^2\left\{c^2(a^2+b^2)+(a^2-2b^2)x^2+(2a^2-b^2)y^2\right\} \\
&+R^4\left\{c^4(a^4+4a^2b^2+b^4)-2c^2(a^4-a^2b^2+3b^4)x^2+2c^2(3a^4-a^2b^2+b^4)y^2\right. \\
&\qquad \left.+(a^4-6a^2b^2+6b^4)x^4+(6a^4-6a^2b^2+b^4)y^4+(6a^4-10a^2b^2+6b^4)x^2y^2\right\} \\
&+R^2\left\{-2a^2b^2c^4(a^2+b^2)+2c^2(3a^4-a^2b^2+b^4)x^2-2c^2(a^4-a^2b^2+3b^4)y^2\right. \\
&\qquad -(6a^4-10a^2b^2+6b^4)(b^2x^4+a^2y^4)+(4a^6-6a^4b^2-6a^2b^4+4b^6)x^2y^2 \\
&\qquad +2(a^2-2b^2)b^2x^6+2(b^2-2a^2)a^2y^6-2(a^4-a^2b^2+3b^4)x^4y^2 \\
&\qquad \left.-2(3a^4-a^2b^2+b^4)x^2y^4\right\} \\
&+(b^2x^2+a^2y^2-a^2b^2)^2\left\{(x-c)^2+y^2\right\}\left\{(x+c)^2+y^2\right\} = 0.
\end{aligned}$$

Proof.—If the curves in (4949) be made to touch, $\alpha\beta$ will be a point on the curve parallel to u at a distance r. Therefore put the values of Δ, Θ, Θ', and Δ' in equation (4942).

[*Salmon*, p. 325.

4961 When u' of (4936) represents two right lines, Δ' vanishes, and

4962 $\Theta' = 0$ is the condition that the two lines should intersect on u;

4963 $\Theta = 0$ is the condition that the two lines should be conjugate with regard to u.

PROOF.—Transform $u' = 0$ into $2xy = 0$, so that the axes x, y are the right lines. This will not affect the invariants (4940). We now have, by (4937),
$$\Delta' = 0, \quad \Theta = 2(fg - ch), \quad \Theta' = -c.$$
$c = 0$ makes u pass through the origin xy; $fg = ch$ makes x and y conjugate. For in (4671), if $\lambda x + \mu y + \nu$ becomes $y = 0$, then $\lambda = \nu = 0$, and the pole is given by $H : B : F$. But $x \equiv a = 0$ at the pole, therefore $H \equiv fg - ch = 0$.

4964 The condition that either of the lines in u' should touch u is, by (4943),
$$\Theta^2 = 4\Delta\Theta' \quad \text{or} \quad AB = 0,$$
with the above values of Θ and Θ'.

4965 The equation of the two tangents to u, when $\lambda x + \mu y + \nu$ is the chord of contact, is, with the notation of (4665),
$$u\Phi(\lambda, \mu, \nu) = (\lambda x + \mu y + \nu z)^2 \Delta.$$

PROOF.—The conic of double contact with u, $ku + (\lambda x + \mu y + \nu)^2$ (4699), must now become two right lines. In (4937) $\Delta' = 0$ and $\Theta' = 0$, therefore $k\Delta + \Theta = 0$. But $\Theta = \Phi(\lambda, \mu, \nu)$. Hence eliminate k.

4966 COR.—Taking the line at infinity $\mathfrak{a}x + \mathfrak{b}y + \mathfrak{c}z$, we obtain the equation of the asymptotes (4685).

The invariant Θ of the conic $ku + u'$ vanishes—

4967 (i.) Whenever an inscribed triangle of u' is self-conjugate to u.

4968 (ii.) Whenever a circumscribed triangle of u is self-conjugate to u'.

4969 Θ' vanishes under similar conditions, transposing u and u' in (i.) and (ii.)

PROOF.—(i.) u becomes $ax^2 + by^2 + cz^2$ (4765), and $f = g = h = 0$. Therefore Θ in (4937) vanishes if $a' = b' = c' = 0$; *i.e.*, if u' is of the form $f'yz + g'zx + h'xy$ (4724).

(ii.) In this case, $f' = g' = h' = 0$ and Θ vanishes if $bc = f^2$, &c., *i.e.*, if the line $x = 0$ touches u, &c.

4970 If u, u' be two conics, and if $\Theta^2 = 4\Delta\Theta'$, any triangle inscribed in u' will circumscribe u, and conversely.

Proof.—Let $u = x^2+y^2+z^2-2yz-2zx-2xy$ and $u' = 2fyz+2gzx+2hxy$, both referred to the same triangle, (4739) and (4724). Then

$$\Delta = -4, \quad \Theta = 4(f+g+h), \quad \Theta' = -(f+g+h)^2, \quad \Delta' = 2fgh;$$

therefore $\Theta^2 = 4\Delta\Theta'$, a relation independent of the axes of reference (4940).

4971 Ex. (i.)—The locus of the centre of a circle of radius r, circumscribing a triangle which circumscribes the conic $b^2x^2+a^2y^2 = a^2b^2$, is

$$(x^2+y^2-a^2-b^2+r^2)^2+4\left\{b^2x^2+a^2y^2-a^2b^2-r^2(a^2+b^2)\right\} = 0,$$

from $\Theta^2 = 4\Delta\Theta'$ and the values in (4949).

4972 Ex. (ii.)—The distance between the centres of the inscribed and circumscribed circles of a triangle is thus found, by employing the values of Θ, Θ', and Δ in (4947), to be $D = \sqrt{(r'^2 \pm 2rr')}$, as in (936).

4973 The tangential equation of the four points of intersection of the two conics $u = 0$, $u' = 0$ is

$$\mathbf{V}^2 = 4UU',$$

with the meanings

4974 $$U \equiv (A\,B\,C\,F\,G\,H \between \lambda\mu\nu)^2; \qquad (4664)$$

$$U' \equiv (A'B'C'F'G'H' \between \lambda\mu\nu)^2.$$

4976 $$\mathbf{V} \equiv (A''B''C''F''G''H'' \between \lambda\mu\nu)^2.$$

4977 $$A \equiv bc-f^2, \text{ \&c.}; \quad A' \equiv b'c'-f'^2, \text{ \&c.},$$

as in (4665), and

4978 $A'' \equiv bc'+b'c-2ff'$, $\qquad F'' \equiv gh'+g'h-af'-a'f$,

4979 $B'' \equiv ca'+c'a-2gg'$, $\qquad G'' \equiv hf'+h'f-bg'-b'g$,

4980 $C'' \equiv ab'+a'b-2hh'$, $\qquad H'' \equiv fg'+f'g-ch'-c'h$.

Proof.—The tangential equation is the condition that $\lambda\alpha+\mu\beta+\nu\gamma$ may pass through one of the four points of intersection of u and u'. The tangential equation of the conic $u+ku'$ is obtained by putting $a+ka'$ for a, &c. in U (4665), and is $U+k\mathbf{V}+k^2U' = 0$. The tangential equation of the envelope of the system is $\mathbf{V}^2 = 4UU'$ (4911). This is the condition that the line (λ, μ, ν) may pass through the consecutive intersections of the conics obtained by varying k. But these conics always intersect in the same four points. The above is therefore the tangential equation of the four points.

4981 The equation of the four common tangents of two conics u, u' is

$$\mathbf{F}^2 = 4\Delta\Delta' uu',$$

where $$\mathbf{F} \equiv (a''b''c''f''g''h'' \between \alpha\beta\gamma)^2,$$

and $$a'' \equiv BC' + B'C - 2FF', \text{ \&c.},$$

$$f'' \equiv GH' + G'H - AF' - A'F, \text{ \&c.},$$

as in (4978–81).

PROOF.—This is the reciprocal of the last theorem. $U + kU'$ is a conic touching the four common tangents of the conics U and U'. The trilinear equation formed from this will, by (4667), be $u\Delta + k\mathbf{F} + k^2u'\Delta' = 0$. The envelope of this system of conics is the equation above, which must therefore represent the four common tangents.

The curve $\mathbf{F}$ passes through the points of contact of u and u' with the locus represented by (4981).

4982 Hence the eight points of contact of the two conics with their common tangents lie on the curve $\mathbf{F}$.

4983 The reciprocal theorem from equation (4973) is—The eight tangents at the intersections of the conics envelope the conic $\mathbf{V}$.

4984 $\mathbf{F} = 0$ is the locus of a point from which the tangents to the two given conics u, u' form a harmonic pencil.

PROOF. — Putting $\gamma = 0$ in (4681), we get a quadratic of the form $a\alpha^2 + 2h\alpha\beta + b\beta^2 = 0$, which determines the two points in which the line γ is cut by tangents from α', β', γ'. Let the similar quadratic for the second conic be $a'\alpha^2 + 2h'\alpha\beta + b'\beta^2 = 0$. Then, by (1064), $ab' + a'b = 2hh'$ is the condition that the four points may be in harmonic relation. This equation will be found to produce $\mathbf{F} = 0$.

4985 The actual values of a, h, b, suppressing the accents on α', β', γ', are

$$C\beta^2 + B\gamma^2 - 2F\beta\gamma, \quad G\beta\gamma + F\gamma\alpha - C\alpha\beta - H\gamma^2,$$
$$A\gamma^2 + C\alpha^2 - 2G\gamma\alpha;$$

and similarly for a', h', b', with A' written for A, &c.

4986 If the *anharmonic* ratio of the pencil of four tangents be given, the locus of the vertex will be $\mathbf{F}^2 = kuu'$. If the given ratio be infinity or zero, the locus becomes the four common tangents in (4981).

4987 $\mathbf{V}=0$ is the envelope of a conic every tangent of which is cut harmonically by the two conics u, u'; *i.e.*, the equation is the condition that $\lambda\alpha+\mu\beta+\nu\gamma$ should be cut harmonically by the two conics.

PROOF.—Eliminate γ between the line (λ, μ, ν), and the conics u and u' separately, and let $A\alpha^2+2H\alpha\beta+B\beta^2=0$ and $A'\alpha^2+2H'\alpha\beta+B'\beta^2=0$ stand for the resulting equations. Then, by (1064), $AB'+A'B=2HH'$ produces the equation $\mathbf{V}=0$, which, by (4666), is the envelope of a conic.

4988 The actual values of A, H, B are respectively

$$a\nu^2+c\lambda^2-2g\nu\lambda, \quad h\nu^2-g\mu\nu-f\nu\lambda+c\lambda\mu, \quad c\mu^2+b\nu^2-2f\mu\nu;$$

and similarly for A', H', B', with a' for a, &c.

4989 $\mathbf{F}^2=4\Delta\Delta'uu'$ is a *covariant* (1629) of the conics u, u'.

For the four common tangents are independent of the axes of reference.

4990 $U=0$ and $\mathbf{V}=0$ (4973) are both *contravariants* (1814) of u and u'.

PROOF.—For $U=0$ is the condition that $\lambda\alpha+\mu\beta+\nu\gamma=0$ shall touch the conic u; and $\mathbf{V}=0$ is the condition that the same line shall be cut harmonically by u and u'; and if all the equations be transformed by a reciprocal substitution (1813, '14), the right line and the conditions remain unaltered.

4991 Any conic covariant with u and u' can be expressed in terms of u, u', and $\mathbf{F}$; and the tangential equation can be expressed in terms of U, U' and $\mathbf{V}$.

4992 Ex. (1).—The polar reciprocal of u with respect to u' is

$$\Theta u'=\mathbf{F}.$$

PROOF.—Referring u, u' to their common self-conjugate triangle,

$$u=ax^2+by^2+cz^2, \quad u'=x^2+y^2+z^2,$$
$$\mathbf{F}=a(b+c)x^2+b(c+a)y^2+c(a+b)z^2.$$

The polar of ξ, η, ζ with respect to u' is $\xi x+\eta y+\zeta z$, and the condition that this may touch u is $bc\xi^2+ca\eta^2+ab\zeta^2=0$ (4664), or, which is the same thing, $(bc+ca+ab)(x^2+y^2+z^2)=\mathbf{F}$ or $\Theta u'=\mathbf{F}$ (4945).

4993 Ex. (2).—The enveloping conic $\mathbf{V}$ in (4987) may also be written

$$\Theta u'+\Theta' u=\mathbf{F}.$$

Proof.—With the same assumptions as in Ex. (1), **V** in (4973) becomes $(b+c)\lambda^2+(c+a)\mu^2+(a+b)\nu^2=0$. The trilinear equation is, therefore, by (4667),

$$(c+a)(a+b)x^2+(a+b)(b+c)y^2+(b+c)(c+a)z^2=0,$$

or $$(bc+ca+ab)(x^2+y^2+z^2)+(a+b+c)(ax^2+by^2+cz^2)=\mathbf{F}.$$

4994 Ex. (3).—The condition that **F** may become two right lines is

$$\Delta\Delta'(\Theta\Theta'-\Delta\Delta')=0.$$

Proof.—Referring to Ex. (1), $A=bc$, $B=ca$, $C=ab$, $F=G=H=0$, $A'=B'=C'=1$; therefore, in (4981), $a''=B+C=a(b+c)$, &c. Hence

the discriminant Δ of $\mathbf{F}=abc(b+c)(c+a)(a+b)$,

or $abc\left\{(a+b+c)(bc+ca+ab)-abc\right\}=$ the above, by (4945).

4995 To reduce the two conics u, u' to the forms

$$x^2+y^2+z^2=0, \qquad \alpha x^2+\beta y^2+\gamma z^2=0.$$

By (4945), α, β, γ will be the roots of the cubic

$$\Delta k^3-\Theta k^2+\Theta' k-\Delta'=0 \quad \text{.................} \quad (1),$$

and x^2, y^2, z^2 will be found in terms of u, u' and **F**, by solving the three equations $x^2+y^2+z^2=u$, $\alpha x^2+\beta y^2+\gamma z^2=u'$ and (by 4994), $\alpha(\beta+\gamma)x^2+\beta(\gamma+\alpha)y^2+\gamma(\alpha+\beta)z^2=\mathbf{F}$ ...... (2).

4996 Ex. (1): Given $x^2+y^2+2y+2x+3=0$; $x^2+2y^2+4y+2x+6=0$; to be reduced as above. To compute the invariants, we take

		a	b	c	f	g	h	
	=	1	1	3	1	1	0	in the first equation.
	and =	1	2	6	2	1	0	in the second.
therefore		A	B	C	F	G	H	
	=	2	2	1	−1	−1	1	in the first equation.
	and =	8	5	2	−2	−2	2	in the second.

Therefore (4938, '9) $\Delta=1$, $\Theta=6$, $\Theta'=11$, $\Delta'=6$. The roots of equation (1) are now 1, 2, 3. Therefore (2) becomes $5X^2+8Y^2+9Z^2=\mathbf{F}$. Computing **F** also by (4981) with the above values of A, B, &c., we get the three equations as under, introducing z for the sake of symmetry,

$$X^2+Y^2+Z^2=x^2+y^2+3z^2+2yz+2zx,$$
$$X^2+2Y^2+3Z^2=x^2+2y^2+6z^2+4yz+2zx,$$
$$5X^2+8Y^2+9Z^2=5x^2+8y^2+22z^2+16yz+10zx,$$

The solution gives $X=x+1$, $Y=y+1$, $Z=1$, and the equations in the forms required are $(x+1)^2+(y+1)^2+1=0$, $(x+1)^2+2(y+1)^2+3=0$.

4997 Ex. (2).—To find the envelope of the base of a triangle inscribed in a conic u' so that two of its sides touch u.

Let $u = x^2+y^2+z^2-2yz-2zx-2xy-2hkxy,$

and $u' = 2fyz+2gzx+2hxy,$

x and y being the sides touched by u. Then $u+ku'$ will be a conic touched by the third side z. By finding the invariants, it appears that $\Theta^2-4\Delta\Theta' = 4\Delta\Delta'k$, whence k is determined, and the envelope becomes

$$(\Theta^2-4\Delta\Theta')\,u'+4\Delta\Delta'u = 0.$$

Compare (4970).

4998 The tangential equation of the two circular points at infinity (4717) is

$$\lambda^2+\mu^2 = 0.$$

PROOF.—This is the condition that $\lambda x+\mu y+\nu$ should pass through either of those points, since $x \pm iy = c$ is the general form of such a line.

4999 $U = 0$ being the tangential equation of a conic, the discriminant of $kU+U'$ is

$$\Delta^2k^3+\Delta\Theta'k^2+\Delta'\Theta k+\Delta'^2.$$

PROOF.—The discriminant of $kU+U'$ is identical in form with (4937), but the capitals and small letters must be interchanged. Let then the discriminant be $\overline{\Delta}k^3+\overline{\Theta}k^2+\overline{\Theta}'k+\overline{\Delta}' = 0$. We have

$$\overline{\Delta} = \Delta^2 \text{ (4670)}, \ \overline{\Theta} = (BC-F^2)\,A'+\&c. = A'a\Delta+\&c. \text{ (4668)} = \Delta\Theta'.$$

Similarly $\overline{\Theta}' = \Delta'\Theta$, $\overline{\Delta}' = \Delta'^2$.

5000 If Θ, Θ' be the invariants of any conic U and the pair of circular points $\lambda^2+\mu^2$ (4998); then $\Theta = 0$ makes the conic a parabola, and $\Theta' = 0$ makes it an equilateral hyperbola.

PROOF.—The discriminant of $kU+\lambda^2+\mu^2$ is $k^2\Delta^2+k\,(a+b)\,\Delta+ab-h^2$. For, as above, $\overline{\Delta} = \Delta^2$; $\overline{\Theta} = A'a\Delta+B'b\Delta = (a+b)\,\Delta$ since $A' = B' = 1$, C' &c. $= 0$; $\overline{\Theta}' = (A'B'-H^2)\,C = C = ab-h^2$; and $\overline{\Delta}' = 0$. The rest follows from the conditions (4471) and (4474).

5001 The tangential equation of the circular points is, in trilinear notation (see the note at 5030),

$$\lambda^2+\mu^2+\nu^2-2\mu\nu\cos A-2\nu\lambda\cos B-2\lambda\mu\cos C.$$

PROOF: $\lambda^2+\mu^2 = 0$, in Cartesians, shows that the perpendicular let fall from any point whatever upon any line passing through one of the points is infinite. Therefore, by (4624).

5002 The conditions in (4689) and (4690), which make the general conic a parabola or equilateral hyperbola, may be obtained by forming Θ and Θ' for the conic and equation (5001) and applying (5000).

5003 If $\Theta'^2 = 4\Theta$, the conic passes through one of the circular points.

5004 When u' in (4984) reduces to $\lambda^2+\mu^2$, that is, to the circular points at infinity, **F** becomes the locus of intersection of tangents to u at right angles, and produces the equations of the director-circle (4693) and (4694).

5005 The tangential equation of a conic confocal with U is

$$kU+\lambda^2+\mu^2 = 0;$$

5006 And if the left side, by varying k, be resolved into two factors, it becomes the equation of the foci of the system.

PROOF.—Since $\lambda^2+\mu^2$ represents the two circular points at infinity (4998), $kU+\lambda^2+\mu^2 = 0$, by (4914), is the tangential equation of a conic touched by the four imaginary tangents of U from those points. But these tangents intersect in two pairs in the foci of U (4720); and, for the same reason, in the foci of $kU+\lambda^2+\mu^2$, which must therefore have the same foci.

If $kU+\lambda^2+\mu^2$ consists of two factors, it represents two points which, by (4913), are the intersections of the pairs of tangents just named, and are therefore the foci.

5007 The general Cartesian equation of a conic confocal with $u = 0$ (4656) is

$$k^2\Delta u+k\left\{C(x^2+y^2)-2Gx-2Fy+A+B\right\}+1 = 0.$$

PROOF.—(5005) must be transformed. Written in full, by (4664), it becomes $(kA+1)\lambda^2+(kB+1)\mu^2+kC\nu^2$. Hence, by (4667), the trilinear equation will be

$$\left\{(kB+1)kC-k^2F^2\right\}\alpha^2+\&c. = k^2(BC-F^2)\alpha^2+kC\alpha^2+\&c.$$
$$=k^2a\Delta\alpha^2+kC\alpha^2+\&c., \qquad (4668)$$

and so on, finally writing x, y, 1 for α, β, γ.

TO FIND THE FOCI OF THE GENERAL CONIC (4656).

(First Method.)

5008 *Substitute in* $kU+\lambda^2+\mu^2$ *either root of its discriminant* $k^2\Delta^2+k(a+b)\Delta+ab-h^2 = 0$ (5000), *and it becomes resolvable into two factors* $(\lambda x_1'+\mu y_1'+\nu)(\lambda x_2+\mu y_2+\nu)$. *The foci are* x_1y_1 *and* x_2y_2, *real for one value of* k *and imaginary for the other.*

PROOF.—By (5006) the two factors represent the two foci, consequently the coordinates of the foci are the coefficients of λ, μ, ν in those factors.

(Second Method.)

5009 *Let* x y *be a focus; then, by* (4720), *the equation of an imaginary tangent through that point is* $(\xi-x)+i(\eta-y)=0$ *or* $\xi+i\eta-(x+iy)=0$. *Therefore substitute, in the tangential equation* (4665), *the coefficients* $\lambda=1$, $\mu=i$, $\nu=-(x+iy)$, *and equate real and imaginary parts to zero. The resulting equations for finding* x *and* y *are, with the notation of* (4665),

5010 $$2(Cx-G)^2=\Delta\left[a-b+\sqrt{\{4h^2+(a-b)^2\}}\right].$$

5011 $$2(Cy-F)^2=\Delta\left[b-a+\sqrt{\{4h^2+(a-b)^2\}}\right].$$

5012 If the conic is a parabola, $C=0$, and the coordinates of the focus are given by

$$(F^2+G^2)\,x=FH+\tfrac{1}{2}(A-B)\,G,$$
$$(F^2+G^2)\,y=GH-\tfrac{1}{2}(A-B)\,F.$$

5013 Ex.—To find the foci of $2x^2+2xy+2y^2+2x=0$. By the first method, we have

$$\left.\begin{array}{lllllll} & a, & b, & c, & f, & g, & h \\ & =2, & 2, & 0, & 0, & 1, & 1 \\ \text{and} & A, & B, & C, & F, & G, & H \\ & =0, & -1, & 3, & 1, & -2, & 0 \end{array}\right\}$$

from which $\Delta=-2$. The quadratic for k is $k^2\Delta^2+4k\Delta+3=(2k-3)(2k-1)=0$, therefore $k=\frac{3}{2}$ or $\frac{1}{2}$.

Taking $\frac{3}{2}$, $kU+\lambda^2+\mu^2=\frac{3}{2}(-\mu^3+3\nu^2+2\mu\nu-4\nu\lambda)+\lambda^2+\mu^2=0$,

or $$2\lambda^2-12\nu\lambda-\mu^2+9\nu^2+6\mu\nu=0.$$

Solving for λ, this is thrown into the factors

$$\left\{2\lambda+\mu\sqrt{2}-3(2+\sqrt{2})\,\nu\right\}\left\{2\lambda-\mu\sqrt{2}-3(2-\sqrt{2})\,\nu\right\}.$$

Therefore the coordinates of the foci, after rationalizing the fractions, are

$$-\frac{2-\sqrt{2}}{3},\quad -\frac{\sqrt{2}-1}{3}\quad\text{and}\quad -\frac{2+\sqrt{2}}{3},\quad \frac{\sqrt{2}+1}{3}.$$

5014 Otherwise, by the second method, equations (5010, '1) become, in this instance, $(3x+2)^2=\pm 2$, $(3y-1)^2=\pm 2$, the solution of which produces the same values of x and y.

5015 *When the axes are oblique, the coordinates* x, y *of a focus are found from the equations*

$$\{C(x+y\cos\omega)-F\cos\omega-G\}^2=\tfrac{1}{2}\Delta\left(\sqrt{I^2-4J}+2a-I\right)$$
$$(Cy-F)^2\sin^2\omega=\tfrac{1}{2}\Delta\left(\sqrt{I^2-4J}-2a+I\right),$$

where I *and* J *are the invariants* (4955) *and* (4954) *respec-*

tively. The equations may be solved for $\mathrm{x}' = \mathrm{x}+\mathrm{y}\cos\omega$ *and* $\mathrm{y}' = \mathrm{y}\sin\omega$, *which are the rectangular coordinates of the focus with the same origin and* x *axis.*

PROOF.—Following the method of (5009), the imaginary tangent through the focus is, by (4721), $\xi - x + (\eta - y)(\cos\omega + i\sin\omega)$. The two equations obtained from the tangential equation are, writing Δa for $BC-F^2$, &c. (4668),

$X^2 - Y^2 = -\Delta\,(a+b-2h\cos\omega - 2a\sin^2\omega)$, $XY = \Delta\,(h\sin\omega - a\sin\omega\cos\omega)$;
where $X = C\,(x+y\cos\omega) - F\cos\omega - G$ and $Y = (Cy-F)\sin\omega$.

5016 *If the equation of the conic to oblique axes be*

$$ax^2+2hxy+by^2+c = 0,$$

the equations for determining the foci reduce to

$$\frac{y\,(x+y\cos\omega)}{a\cos\omega - h} = \frac{x(y+x\cos\omega)}{b\cos\omega - h} = \frac{c}{ab-h^2}.$$

5017 The condition that the line $\lambda x+\mu y+\nu z$ may touch the conic $u+(\lambda' x+\mu' y+\nu' z)^2$ is

$$U+\phi\,(\mu\nu'-\mu'\nu,\ \nu\lambda'-\nu'\lambda,\ \lambda\mu'-\lambda'\mu) = 0. \qquad (4656, 4936, '74)$$

5018 or $$(\Delta+U')\,U = \Pi^2, \qquad (4938)$$

where $$2\Pi \equiv \lambda' U_\lambda + \mu' U_\mu + \nu' U_\nu. \qquad (4674)$$

PROOF.—Put $a+\lambda'^2$ for a, &c. in U of (4664). The second form follows from the first through the identity

$$\Delta\phi\,(\mu\nu'-\mu'\nu,\ \&c.) = UU' - \Pi^2.$$

5019 Otherwise, let $P' \equiv u_{x'}x + u_{y'}y + u_{z'}z$, the polar of x', y', z' (4659), then the condition that P' may touch $u+P''^2$ becomes, in terms of the coordinates of the poles,

5020 $$(1+u'')\,u' = \phi_{x'}x'' + \phi_{y'}y'' + \phi_{z'}z''. \qquad \text{(See 4657).}$$

PROOF.—If we put $u_{x'}, u_{y'}, u_{z'}$, from (4659), for λ, μ, ν in U to obtain the condition of touching, the result is $\Delta u'$; and similar substitutions made in Π give $\Delta\,(\phi_{x'}x''+\&c.)$, therefore (5018) becomes $(1+u'')\,u' = (\phi_{x'}x''+\&c.)$.

5021 The condition that the conics

$$u+(\lambda' x+\mu' y+\nu' z)^2, \quad u+(\lambda'' x+\mu'' y+\nu'' z)^2$$

may touch each other is

$$(\Delta+U')(\Delta+U'') = (\Delta\pm\Pi)^2. \qquad (4938\text{–}74)$$

PROOF.—Make one of the common chords

$$(\lambda' x+\mu' y+\nu' z) \pm (\lambda'' x+\mu'' y+\nu'' z)$$

touch either conic by substituting $\lambda' \pm \lambda''$ for λ, &c. in (5018). The result is $(\Delta+U')(U' \pm 2\Pi+U'') = (U' \pm \Pi)^2$, which reduces to the form above.

5022 The condition, in terms of the coordinates of the poles of the two lines, is found from the last, as in (5019), and is

$$(1+u')(1+u'') = \{1 \pm (\phi_{x'} x''+\phi_{y'} y''+\phi_{z'} z'')\}^2.$$

5023 The Jacobian, J, of three conics u, v, w, is the locus of a point whose polars with respect to the conics all meet in a point. Its equation is

$$\begin{vmatrix} a_1x+h_1y+g_1z, & a_2x+h_2y+g_2z, & a_3x+h_3y+g_3z \\ h_1x+b_1y+f_1z, & h_2x+b_2y+f_2z, & h_3x+b_3y+f_3z \\ g_1x+f_1y+c_1z, & g_2x+f_2y+c_2z, & g_3x+f_3y+c_3z \end{vmatrix} = 0.$$

PROOF.—The equation is the eliminant of the equations of the three polars passing through a point $\xi\eta\zeta$, viz., $u_x\xi+u_y\eta+u_z\zeta=0$, $v_x\xi+v_y\eta+v_z\zeta=0$, $w_x\xi+w_y\eta+w_z\zeta=0$. See (4657) and (1600).

5024 The equation of a conic passing through five points $\alpha_1\beta_1\gamma_1$, $\alpha_2\beta_2\gamma_2$, &c. is the determinant equation annexed; and the equation of a conic touching five right lines $\lambda_1\mu_1\nu_1$, $\lambda_2\mu_2\nu_2$, &c. is the same in form, λ, μ, ν taking the place of α, β, γ.

$$\begin{vmatrix} \alpha^2 & \beta^2 & \gamma^2 & \beta\gamma & \gamma\alpha & \alpha\beta \\ \alpha_1^2 & \beta_1^2 & \gamma_1^2 & \beta_1\gamma_1 & \gamma_1\alpha_1 & \alpha_1\beta_1 \\ \alpha_2^2 & \beta_2^2 & \gamma_2^2 & \beta_2\gamma_2 & \gamma_2\alpha_2 & \alpha_2\beta_2 \\ \alpha_3^2 & \beta_3^2 & \gamma_3^2 & \beta_3\gamma_3 & \gamma_3\alpha_3 & \alpha_3\beta_3 \\ \alpha_4^2 & \beta_4^2 & \gamma_4^2 & \beta_4\gamma_4 & \gamma_4\alpha_4 & \alpha_4\beta_4 \\ \alpha_5^2 & \beta_5^2 & \gamma_5^2 & \beta_5\gamma_5 & \gamma_5\alpha_5 & \alpha_5\beta_5 \end{vmatrix} = 0.$$

PROOF.—The determinant is the eliminant of six equations of the type (4656) in the one case and (4665) in the other. By (583).

5025 If three conics have a common self-conjugate triangle, their Jacobian is three right lines.

PROOF.—The Jacobian of $a_1x^2+b_1y^2+c_1z^2$, $a_2x^2+b_2y^2+c_2z^2$, $a_3x^2+b_3y^2+c_3z^2$ is, by (5023), $xyz=0$.

For the condition that three conics may have a common point, see *Salmon's Conics*, 6th edit., Art. 389*a*, and *Proc. Lond. Math. Soc.*, Vol. IV., p. 404, *J. J. Walker, M.A.*

5026 A system of two conics has four covariant forms u, u', $\mathbf{F}$, J, connected by the equation

$$J^2 = \mathbf{F}^3 - \mathbf{F}^2(\Theta u' - \Theta' u) + \mathbf{F}(\Delta'\Theta u^2 + \Delta\Theta' u'^2)$$
$$+ \mathbf{F}uu'(\Theta\Theta' - 3\Delta\Delta') - \Delta\Delta'^2 u^3 - \Delta'\Delta^2 u'^3$$
$$+ \Delta' u^2 u'(2\Delta\Theta' - \Theta^2) + \Delta u'^2 u(2\Delta'\Theta - \Theta'^2).$$

PROOF.—Form the Jacobian of u, u', and $\mathbf{F}$. This will be the equation of the sides of the common self-conjugate triangle (4992, 5025). Compare the result with that obtained by the method of (4995).

5027 By parity of reasoning, there are four contravariant forms U, U' $\mathbf{V}$, $\mathbf{T}$ where $\mathbf{T}$ is the tangential equivalent of J, and represents the vertices of the self-conjugate triangle. Its square is expressed in terms of U, U', $\mathbf{V}$ and the invariants precisely as J^2 is expressed in (5026).

5028 The locus of the centre of a conic which always touches four given lines is a right line.

PROOF.—Let $U = 0$, $U' = 0$ be the tangential equations of two fixed conics, each touching the four lines; then, by (4914), $U + kU' = 0$ is another conic also touching the four lines. The coordinates of its centre will be $\frac{G+kG'}{C+kC'}$ and $\frac{F+kF'}{C+kC'}$, by (4402). The point is thus seen, by (4032), to lie on the line joining the centres of the two fixed conics and to divide that line in the ratio $kC' : C$.

5029 To find the locus of the focus of a conic touching four given lines.

In the equations (5010, '1) for determining the coordinates of the focus, write $A + kA'$ for A, &c., and eliminate k. The result in general is a cubic curve. If Σ, Σ' be parabolas, $\Sigma + k\Sigma'$ is a parabola having three tangents in common with Σ and Σ'. If $C = C' = 0$ the locus becomes a circle. If the conics be concentric, they touch four sides of a parallelogram, and the locus is a rectangular hyperbola.

Note on Tangential Coordinates.

5030 It must be borne in mind that a tangential equation in trilinear notation (that is, when the variables are the coefficients of α, β, γ in the tangent line $l\alpha + m\beta + n\gamma$) will not agree with the equation of the same locus expressed in the tangential coordinates λ, μ, ν of (4019). Thus, to convert equation (5001), which, for distinctness, will now be written

$$l^2 + m^2 + n^2 - 2mn\cos A - 2nl\cos B - 2lm\cos C = 0$$

into tangential coordinates, we must substitute, by (4023), $\mathfrak{a}\lambda$, $\mathfrak{b}\mu$, $\mathfrak{c}\nu$ for l, m, n. The equation then becomes

$$\mathfrak{a}^2\lambda^2 + \mathfrak{b}^2\mu^2 + \mathfrak{c}^2\nu^2 - 2\mathfrak{bc}\cos A\mu\nu - 2\mathfrak{ca}\cos B\nu\lambda - 2\mathfrak{ab}\cos C\lambda\mu = 0.$$

Put $2\mathfrak{bc}\cos A = \mathfrak{b}^2 + \mathfrak{c}^2 - \mathfrak{a}^2$, &c., and the result is the equation as presented in (4905).

THEORY OF PLANE CURVES.

TANGENT AND NORMAL.

5100 Let P (Fig. 90) be a point on the curve AP; PT, PN, PG, the tangent, ordinate, and normal intercepted by the x axis of coordinates. See definitions in (1160). Let $\angle PTX = \psi$.

5101 $\tan\psi = \frac{dy}{dx}$, by (1403); $\sin\psi = \frac{dy}{ds}$; $\cos\psi = \frac{dx}{ds}$.

5104 *Sub-tangent* $NT = yx_y$, *Sub-normal* $NG = yy_x$.

5106 $PT = y\sqrt{(1+x_y^2)}$, $PT' = x\sqrt{(1+y_x^2)}$.

5108 $PG = y\sqrt{(1+y_x^2)}$, $PG' = x\sqrt{(1+x_y^2)}$.

Let $OP = r$ (Fig. 91), $u = r^{-1}$, $AOP = \theta$, $OPT = \phi$; Arc $AP = s$. Then, by infinitesimals,

5110 $\sin\phi = r\frac{d\theta}{ds}$, $\cos\phi = \frac{dr}{ds}$, $\tan\phi = r\frac{d\theta}{dr}$.

5113 $(dx)^2+(dy)^2 = (ds)^2$, $s_x = \sqrt{(1+y_x^2)}$.

5114 $$\tan\psi = \frac{r_\theta \sin\theta + r\cos\theta}{r_\theta\cos\theta - r\sin\theta}. \quad (1768)$$

5115 *Intercepts of Normal* $OG = r\frac{dr}{dx}$, $OG' = r\frac{dr}{dy}$. (Fig. 90)

PROOF: $OG = \frac{r \sin OPG}{\sin PGN} = \frac{r \cos\phi}{\sin NPT} = r\frac{r_s}{x_s} = r\frac{dr}{dx}$.

5116 $s_\theta = \sqrt{(r^2+r_\theta^2)}, \quad s_r = \sqrt{(1+r^2\theta_r^2)}$.

PROOF.—By $r\theta_s = \sin\phi$ and $\tan\phi = r\theta_r$ (5110).

EQUATIONS OF THE TANGENT AND NORMAL.

The equation of the curve being $y=f(x)$ or $u \equiv \phi(x, y) = 0$, the equation of the tangent at xy is

5118 $$\eta - y = \frac{dy}{dx}(\xi - x), \tag{4120}$$

5119 or $$\xi y_x - \eta = xy_x - y,$$

5120 or $$\xi u_x + \eta u_y = xu_x + yu_y. \tag{1708}$$

5121 If $\phi(x, y) = v_n + v_{n-1} + \ldots + v_0$, where v_n is a homogeneous function of x and y of the nth degree, the constant part forming the right member of equation (5120) takes the value

$$-v_{n-1} - 2v_{n-2} - \ldots - (n-1)\,v_1 - nv_0.$$

By Euler's theorem (1621) and $\phi(x, y) = 0$.

The equation of the normal at xy is

5122 $$\eta - y = -\frac{dx}{dy}(\xi - x), \tag{4122}$$

5123 or $$\xi x_y + \eta = xx_y + y,$$

5124 or $$\xi u_y - \eta u_x = xu_y - yu_x. \tag{1708}$$

POLAR EQUATIONS OF THE TANGENT AND NORMAL.

Let r, θ be the coordinates of P (Fig. 91), and R, Θ those of S, any point on the tangent at P; and let $u = r^{-1}$, $U = R^{-1}$, $\tau = \Theta - \theta$; the polar equation of the tangent at P will be

5125 $$R = \frac{r^2}{d_\theta(r \sin\tau)}, \quad \text{or} \quad U = u\cos\tau + u_\theta \sin\tau.$$

The polar equation of the normal is

5127 $$R = \frac{rr_\theta}{d_\theta(r\cos\tau)}, \quad \text{or} \quad U = u\cos\tau - u^2\theta_u \sin\tau.$$

PROOF.—From $\frac{r}{R} = \frac{OP}{OS} = \frac{\sin OSP}{\sin OPS} = \frac{\sin(\phi-\tau)}{\sin\phi}$, and from $\tan\phi = r\theta_r$ (5112). Similarly for the normal.

Let $OY \equiv p$ be the perpendicular from the pole upon the tangent, then

5129 $$p = r\sin\phi = (u^2+u_\theta^2)^{-\frac{1}{2}}.$$ (5112)

5131 $$\frac{xy_x-y}{\sqrt{(1+y_x^2)}} = \frac{x\phi_x+y\phi_y}{\sqrt{(\phi_x^2+\phi_y^2)}}.$$ (4064 & 5119, '20)

OS, drawn at right angles to r to meet the tangent, is called the *polar sub-tangent.*

5133 Polar sub-tangent $= r^2\theta_r$. (5112)

RADIUS OF CURVATURE AND EVOLUTE.

Let ξ, η be the centre of curvature for a point xy on the curve, and ρ the radius of curvature; then

5134 $$(x-\xi)^2+(y-\eta)^2 = \rho^2 \quad\ldots\ldots\ldots\ldots\ldots\ldots (1).$$

5135 $$(x-\xi)+(y-\eta)\,y_x = 0 \quad\ldots\ldots\ldots\ldots\ldots\ldots (2),$$

$$1+y_x^2+(y-\eta)\,y_{2x} = 0 \quad\ldots\ldots\ldots\ldots\ldots\ldots (3).$$

PROOF.—(2) and (3) are obtained from (1) by differentiating for x, considering ξ, η constants.

The following are different values of ρ:

5137 $$\rho = \frac{(1+y_x^2)^{\frac{3}{2}}}{y_{2x}} = \frac{(\phi_x^2+\phi_y^2)^{\frac{3}{2}}}{2\phi_{xy}\phi_x\phi_y-\phi_{2x}\phi_y^2-\phi_{2y}\phi_x^2}$$

5139 $$= \frac{(x_t^2+y_t^2)^{\frac{3}{2}}}{y_{2t}x_t-x_{2t}y_t} = \frac{1}{y_{2s}x_s-x_{2s}y_s}$$

5141 $$= \frac{1}{\sqrt{(x_{2s}^2+y_{2s}^2)}} = \frac{x_s}{y_{2s}} = -\frac{y_s}{x_{2s}}$$

5144 $$= \frac{(r^2+r_\theta^2)^{\frac{3}{2}}}{r^2+2r_\theta^2+rr_{2\theta}} = \frac{(u+u_\theta^2)^{\frac{3}{2}}}{u^3(u+u_{2\theta})}$$

5146 $$= s_\psi = p+p_{2\psi} = rr_p.$$

PROOFS.—For (5137), eliminate $x-\xi$ and $y-\eta$ between equations (1), (2), and (3).

(5138) is obtained from the preceding value by substituting for y_x and

y_{2x} the values (1708, '9). The equation of the curve is here supposed to be of the form $\phi(x, y) = 0$.

For (5139); change the variable to t. For (5140); make $t = s$.

For (5141–3); let $PQ = QR = ds$ (Fig. 92) be equal consecutive elements of the curve. Draw the normals at P, Q, R, and the tangents at P and Q to meet the normals at Q and R in T and S. Then, if PN be drawn parallel and equal to QS, the point N will ultimately fall on the normal QO. Now the difference of the projections of PT and PN upon OX is equal to the projection of TN. Projection of $PT = ds\,x_s$; that of PN or $QS = ds\,(x_s + x_{2s}\,ds)$ (1500); therefore the difference $= ds\,x_{2s}\,ds = TN\cos\alpha$. But $TN : ds = ds : \rho$, therefore $\rho x_{2s} = \cos\alpha$. Similarly $\rho y_{2s} = \sin\alpha$.

For (5144); change (5137) to r and θ, by (1768, '9).

(5145) is obtained from $\rho = rr_p = -\frac{u_p}{u^3}$ and (5129); or change (5144) from r to u by $r = u^{-1}$.

(5146.) In Fig. (93), $PQ = \rho$, $PP' = ds$, and $PQP' = d\psi$.

(5147.) In Fig. (93), let PQ, $P'Q$ be consecutive normals; PT, $P'T'$ consecutive tangents; OT, OT', ON, ON' perpendiculars from the origin upon the tangents and normals. Then, putting p for $OT = PN$, q for $PT = ON$, and $d\psi$ for $\angle\, TPT' = PQP'$, &c., we have

$$q = \frac{dp}{d\psi},\quad QN = \frac{dq}{d\psi},\quad \text{and}\quad \rho = PQ = p + QN = p + p_{2\psi}.$$

(5148.) $dp = r\cos\phi\, d\psi$ and $\cos\phi = r_s$. Eliminate $\cos\phi$.

5149 Def.—The *evolute* of a curve is the locus of its centre of curvature. Regarding the evolute as the principal curve, the original curve is called its *involute*.

5150 The normal of any curve is a tangent to its evolute.

Proof.—By differentiating equation (5135) on the hypothesis that ξ and η are variables dependent upon x, and combining the result with (3), we obtain $y_x \eta_\xi = -1$.

In (Fig. 94), the normal at P of the curve AP touches the evolute at Q. Otherwise the evolute is the envelope of the normals of the given curve.

If xy and $\xi\eta$ are the points P, Q, we have the relations

5151 $$\frac{\xi - x}{\rho}\, d\xi + \frac{\eta - y}{\rho}\, d\eta = d\rho,\qquad \frac{d\xi}{\xi - x} = \frac{d\eta}{\eta - y} = \frac{d\rho}{\rho}.$$

Proof.—Take $Qn = d\xi$ and $ns = d\eta$, then $Qs = d\rho$. The projection of Qn, ns gives $d\rho$ in (5151) and proportion gives (5152).

5153 The evolute and involute are connected by the formulæ below, in which r', p', s' in the evolute correspond to r, p, s in the involute.

5154 $$\rho \pm s' = \text{constant};\quad p'^2 = r^2 - p^2;\quad r'^2 = r^2 + \rho^2 - 2p\rho.$$

PROOF.—From Fig. (94), $d\rho = \pm ds'$, &c., s being the arc RQ measured from a fixed point R. Hence, if a string is wrapped upon a given curve, the free end describes an involute of the curve. (5155, '6) from Fig. (93).

5157 To obtain the equation of the evolute; eliminate x and y from equations (5135, '6) and the equation of the curve.

5158 To obtain the polar equation of the evolute; eliminate r and p from (5156) and (5157) and the given equation of the curve $r = F(p)$.

5159 Ex.—To find the evolute of the catenary $y = \frac{c}{2}\left(e^{\frac{x}{c}} + e^{-\frac{x}{c}}\right)$. Here $y_x = \frac{1}{2}\left(e^{\frac{x}{c}} - e^{-\frac{x}{c}}\right) = \frac{\sqrt{(y^2-c^2)}}{c}$; $y_{2x} = \frac{1}{2c}\left(e^{\frac{x}{c}} + e^{-\frac{x}{c}}\right) = \frac{y}{c^2}$; so that equations (5135, '6) become

$$(x-\xi) + (y-\eta)\frac{\sqrt{(y^2-c^2)}}{c} = 0 \quad \text{and} \quad 1 + \frac{y^2-c^2}{c^2} + (y-\eta)\frac{y}{c^2} = 0.$$

From these we find $y = \frac{\eta}{2}$, $x = \xi - \frac{\eta}{4c}\sqrt{(\eta^2-4c^2)}$. Substituting in the equation of the curve, we obtain the required equation in ξ and η.

INVERSE PROBLEM AND INTRINSIC EQUATION.

An inverse question occurs when the arc is a given function of the abscissa, say $s = \phi(x)$; the equation of the curve in rectangular coordinates will then be

5160 $$y = \int \sqrt{(s_x^2 - 1)}\, dx.$$ [From (5113).

5161 The *intrinsic equation* of a curve is an equation independent of coordinate axes. Let $y = \phi(x)$ be the ordinary equation, taking for origin a point O on the curve (Fig. 95), and the tangent at O for x axis. Let $s = \text{arc } OP$, and ψ the inclination of the tangent at P; then the intrinsic equation of the curve is

5162 $$s = \int \sec\psi\, x_\psi\, d\psi;$$

where x_ψ is found from $\tan\psi = \phi'(x)$.

To obtain the Cartesian equation from the intrinsic equation:

5163 Let $s = F(\psi)$ be the intrinsic equation. Eliminate ψ between this and the equations

$$x = \int \cos\psi\, ds, \qquad y = \int \sin\psi\, ds.$$

5165 The intrinsic equation of the evolute obtained from the intrinsic equation of the curve, $s = F(\psi)$, is

$$\frac{ds}{d\psi} + s' = l, \quad \text{a constant.} \qquad (5154)$$

5166 The intrinsic equation of the involute obtained from $s' = F(\psi)$, the equation of the curve, is

$$s = \int \{l - F(\psi)\}\, d\psi.$$

For $d\psi$ is the same for both curves (Fig. 94), ψ only differing by $\frac{1}{2}\pi$, and $s = \int \rho\, d\psi$.

ASYMPTOTES.

5167 DEF.—An asymptote of a curve is a straight line or curve which the former continually approaches but never reaches. (*Vide* 1185).

GENERAL RULES FOR RECTILINEAR ASYMPTOTES.

5168 RULE I.—*Ascertain if* y_x *has a limiting value when* $x = \infty$. *If it has, find the intercept on the* x *or* y *axis, that is,* $x - yx_y$ *or* $y - xy_x$ (5104).

There will be an asymptote parallel to the y *axis when* y_x *is infinite, and the* x *intercept finite, or one parallel to the* x *axis when* y_x *is zero and the* y *intercept finite.*

5169 RULE II.—*When the equation of the curve consists of homogeneous functions of* x *and* y, *of the* m^{th}, n^{th}, *&c. degrees, so that it may be written*

$$x^m \phi\left(\frac{y}{x}\right) + x^n \chi\left(\frac{y}{x}\right) + \&c. = 0 \quad \text{.........} \quad (1);$$

put $\mu x+\beta$ *for* y *and expand* $\phi\left(\mu+\frac{\beta}{x}\right)$, &c., *by* (1500). *Divide* (1) *by* x^m, *and make* x *infinite; then* $\phi(\mu)=0$ *determines* μ.

Next, put this value of μ *in* (1), *divide by* x^{m-1}, *and make* x *infinite; thus* $\beta\phi'(\mu)+\chi(\mu)=0$ *determines* β. *Should the last equation be indeterminate, then*

$$\tfrac{1}{2}\beta^2\phi''(\mu)+\beta\chi'(\mu)+\psi(\mu)=0$$

gives two values for β, *and so on.*

When n *is* $<m-1$, $\beta=0$, *and when* n *is* $>m-1$, $\beta=\infty$.

5170 Rule III.—*If* $\phi(x, y)=0$ *be a rational integral equation, to discover asymptotes parallel to the axes, equate to zero the coefficients of the highest powers of* x *and* y, *if those coefficients contain* y *or* x *respectively.*

To find other asymptotes—Substitute $\mu x+\beta$ *for* y *in the original equation, and arrange according to powers of* x. *To find* μ, *equate to zero the coefficient of the highest power of* x. *To find* β, *equate to zero the coefficient of the next power of* x, *or, if that equation be indeterminate, take the next coefficient in order, and so on.*

5171 Rule IV.—*If the polar equation of the curve be* $r=f(\theta)$ *and if* $r=\infty$ *makes the polar subtangent* $r^2\theta_r=c$, *a finite quantity, there is an asymptote whose equation is* $r\cos(\theta-a)=c$; *where* $a\pm\frac{1}{2}\pi=f^{-1}(\infty)=$ *the value of* θ *of the curve when* r *is infinite.*

5172 *Asymptotic curves.*—In these the difference of corresponding ordinates continually diminishes as x increases.

As an example, the curves $y=\phi(x)$ and $y=\phi(x)+\frac{a}{x}$ are asymptotic.

5173 Ex. 1.—To find the asymptotes of the curve

$$(a+3x)(x^2+y^2)=4x^3 \quad\ldots\ldots\ldots\ldots (1).$$

The coefficient of y^2, $a+3x=0$, gives an asymptote parallel to the y axis.

Putting $y=\mu x+\beta$, (1) becomes

$$(a+3x)(x^2+\mu^2x^2+2\mu\beta x+\beta^2)-4x^3=0 \quad\ldots\ldots\ldots (2).$$

The coefficient of x^3, $3(1+\mu^2)-4=0$ gives $\mu=\pm\frac{1}{\sqrt{3}}$. Substituting this value of μ in (2), the coefficient of x^2 becomes $\frac{4a}{\sqrt{3}}\pm\frac{6\beta}{\sqrt{3}}$; and this, equated to zero, gives $\beta=\mp\frac{2a}{3\sqrt{3}}$. Hence the equations of two more asymptotes are $3y\sqrt{3}=\pm(3x-2a)$.

Ex. 2.—To find an asymptote of the curve $r\cos\theta = a\cos 2\theta$. Here

$$r^2\frac{d\theta}{dr} = \frac{a^2\cos 2\theta}{a\cos 2\theta\sin\theta - 2a\sin 2\theta\cos\theta}.$$

When $r=\infty$, $\theta=\frac{1}{2}\pi$, and $r^2\theta_r = -a$. Hence the equation of the asymptote is $r\cos\theta = -a$.

SINGULARITIES OF CURVES.

5174 *Concavity and Convexity.*—A curve is reckoned convex or concave towards the axis of x according as yy_{2x} is positive or negative.

POINTS OF INFLEXION.

5175 *A point of inflexion* (Fig. 96) exists where the tangent has a limiting position, and therefore where y_x takes a maximum or minimum value.

5176 Hence y_{2x} must vanish and change sign, as in (1832).

5177 Or, more generally, an even number of consecutive derivatives of $y=\phi(x)$ must vanish, and the curve will pass from positive to negative, or from negative to positive, with respect to the axis of x, according as the next derivative is negative or positive. [See (1833).

MULTIPLE POINTS.

5178 *A multiple point*, known also as a *node* or *crunode*, exists when y_x has more than one value, as at B (Fig. 98). If $\phi(x, y)=0$ be the curve, ϕ_x and ϕ_y must both vanish, by (1713). Then, by (1704), two values of y_x determining a *double point*, will be given by the quadratic

$$\phi_{2y}y_x^2 + 2\phi_{xy}y_x + \phi_{2x} = 0 \quad\ldots\ldots\ldots\ldots\ldots\ldots (1).$$

5179 If ϕ_{2x}, ϕ_{2y}, ϕ_{xy} also vanish; then, by (1705), three values of y_x, determining a *triple point*, will be obtained from the cubic

$$\phi_{3y}y_x^3 + 3\phi_{2yx}y_x^2 + 3\phi_{y2x}y_x + \phi_{3x} = 0 \quad\ldots\ldots\ldots\ldots (2).$$

5180 Generally, when all the derivatives of ϕ of an order less than n vanish, the equation for determining y_x (put $= z$) may be written

$$(zd_y + d_x)^n \phi(x, y) = 0.$$

PROOF.—Let ab be the multiple point. Then, by (1512),

$$\phi(a+h, b+k) = \frac{1}{n!}(hd_x + kd_y)^n \phi(x, y)$$

+ terms of higher order which vanish when h and k are small.

And $\frac{k}{h} = -\frac{\phi_x}{\phi_y} = \frac{dy}{dx}$ in the limit.

CUSPS.

5181 When two branches of a curve have a common tangent at a point, but do not pass through the point, they form a *cusp*, termed also a *spinode* or *stationary point.*

5182 In the *first species*, or *ceratoid cusp* (Fig. 100), the two values of y_{2x} have opposite signs.

5183 In the *second species*, or *ramphoid cusp* (Fig. 101), they have the same sign.

CONJUGATE POINTS.

5184 A *conjugate point*, or *acnode*, is an isolated point whose coordinates satisfy the equation of the curve. A necessary condition for the existence of a conjugate point is that ϕ_x and ϕ_y must both vanish.

PROOF.—For the tangent at such a point may have any direction, therefore $\frac{\phi_x}{\phi_y}$ is indeterminate (1713).

5185 There are four species of the triple point according as it is formed by the union of

(i.) three crunodes, as in (Fig. 102);
(ii.) two crunodes and a cusp, as in (Fig. 103);
(iii.) a crunode and two cusps, as in (Fig. 104);
(iv.) when only one real tangent exists at the point.

5186 Ex.—The equation $y^2 = (x-a)(x-b)(x-c)$* when $a < b < c$ represents a curve, such as that drawn in (Fig. 97).

* Salmon's *Higher Plane Curves*, Arts. 39, 40.

When $b = c$ the curve takes the form in (Fig. 98). But if, instead, $b = a$, the oval shrinks into a point A (Fig. 99).

If $a = b = c$ the point A becomes a cusp, as in (Fig. 100).

A geometrical method of investigating singular points.

5187 Describe an elementary circle of radius r round the point x, y on the curve $\phi(x, y) = 0$, intersecting the curve in the point $x+h, y+k$. Let $h = r\cos\theta$, $k = r\sin\theta$. Expand $\phi(x+h, y+k) = 0$ by (1512), and put $\phi_x = K\sin\gamma$, $\phi_y = K\cos\gamma$. We thus obtain

$$K\sin(\gamma+\theta)+\frac{r}{2}(\phi_{2x}\cos^2\theta+2\phi_{xy}\sin\theta\cos\theta+\phi_{2y}\sin^2\theta)+\frac{R}{r}=0$$

R being put for the rest of the expansion ....(1),

According as the quadratic in $\tan\theta$,

$$\phi_{2x}+2\phi_{xy}\tan\theta+\phi_{2y}\tan^2\theta = 0,$$

has real, equal, or imaginary roots; *i.e.*, according as $\phi_{xy}^2-\phi_{2x}\phi_{2y}$ is positive, zero, or negative, xy will be a crunode, a cusp, or an acnode. By examining the sign of R, the species of cusp and character of the curvature may be determined.

Figures (105) and (106), according as R and ϕ_{2y} have opposite or like signs, show the nature of a crunode; and figures (107) and (108) show a cusp.

PROOF.—At an ordinary point the circle cuts the curve at the two points given by $\theta = -\gamma$, $\theta = \pi-\gamma$. But, if ϕ_x and ϕ_y both vanish, there is a singular point. Writing A, B, C for $\phi_{2x}, \phi_{xy}, \phi_{2y}$, equation (1) now becomes

$$C\cos^2\theta\left\{\tan^2\theta+\frac{2B}{C}\tan\theta+\frac{A}{C}\right\}+\frac{2R}{r^2}=0 \text{ ....(2).}$$

(i.) If $B^2 > AC$, this may be put in the form

$$C\cos^2\theta(\tan\theta-\tan\alpha)(\tan\theta-\tan\beta)+\frac{2R}{r^2}=0,$$

and the points of intersection with the circle are given by $\theta = \alpha, \beta, \pi+\alpha$, and $\pi+\beta$. (Figs. 105 and 106.)

(ii.) When $B^2 = AC$, we may write equation (1)

$$C\cos^2\theta(\tan\theta-\tan\alpha)^2+\frac{2R}{r^2}=0.$$

If R and C have opposite signs, there is a cusp with α for the inclination of the tangent (Fig. 107). So also, if R and C have the same sign, the inclination and direction being $\pi+\alpha$ (Fig. 108). The cusps exist in this case because R changes its sign when π is added to θ, R being a homogeneous function of the third degree in $\sin\theta$ and $\cos\theta$.

(iii.) If $B^2 < AC$, there are no real points of intersection, and therefore xy is an acnode.

CONTACT OF CURVES.

5188 A contact of the nth order exists between two curves when n successive derivatives, $y_x, \dots y_{nx}$ or $r_\theta, \dots r_{n\theta}$, correspond. The curves cross at the point if n be even. No curve can pass between them which has a contact of a lower order with either.

Ex.—The curve $y = \phi(x)$ has a contact of the nth order, at the point where $x = a$, with the curve $y = \phi(a) + (x-a)\phi'(a) + \dots + \frac{(x-a)^n}{n!}\phi^n(a)$.

5189 Cor.—If the curve $y = f(x)$ has n parameters, they may be determined so that the curve shall have a contact of the $(n-1)^{th}$ order with $y = \phi(x)$.

A contact of the first order between two curves implies a common tangent, and a contact of the second order a common radius of curvature.

Conic of closest contact with a given curve.

5190 *Lemma.*—In a central conic (Fig. of 1195),

$$\tan CPG = \frac{1}{3}\frac{d\rho}{ds}.$$

Proof.—Putting $PCT \equiv \theta$, $CPT \equiv \phi$, $CP \equiv r$, $CD \equiv R$, we have, by (1211), $r^2 + R^2 = a^2 + b^2$, $\therefore\ rr_s = -RR_s$ .... (i.).

Also $Rr\sin\phi = ab$, by (1194), $\therefore\ Rr^2\theta_s = ab$, by (5110) ....(ii.).

Now $\tan CPG = -\cot\phi = -\frac{r_\theta}{r}$ (5112) $= \frac{-r_s}{r\theta_s} = \frac{R^2R_s}{ab}$, by (i.) and (ii.).

But $\rho = \frac{R^3}{ab}$ (4538), $\therefore\ \frac{1}{3}\frac{d\rho}{ds} = \frac{R^2R_s}{ab} = \tan CPG$.

5191 To find the conic having a contact of the fourth order with a given curve at a given point P.

If O be the conic's centre, the radius $r \equiv OP$, and the angle ν between r and the normal are found from the equations

$$\tan\nu = \frac{1}{3}\frac{d\rho}{ds}, \qquad \frac{\cos\nu}{r} = \frac{1}{\rho} - \frac{d\nu}{ds},$$

and these determine the conic.

PROOF.—In Fig. 93, let O be the centre of the conic and P the point of contact. The five disposable constants of the general equation of a conic will be determined by the following five data: two coordinates of O, a common point P, a common tangent at P, and the same radius of curvature PQ.

Since $\nu = POT$, $d\theta = POP'$, $d\psi = TOT'$, and $ds = PP'$, we have, in passing from P to P', $d\nu = P'OT' - POT = d\psi - d\theta$. Now $rd\theta = ds\cos\nu$, therefore $\frac{\cos\nu}{r} = \frac{d\psi - d\nu}{ds\cos\theta} = \frac{1}{\rho} - \frac{d\nu}{ds}$; and $\tan\nu$ has been found in the lemma.

The squares of the semi-axes of the same conic are the roots of the equation

$$(9+b^2-3ac)^3 x^2 - 9a^2(18+2b^2-3ac)(9+b^2-3ac)\,x + 729a^4 = 0,$$

a, b, c being written for ρ, ρ_s, ρ_{2s}. The eccentricity is found from

$$\frac{9(e^2-2)^2}{1-e^2} = \frac{(18+2b^2-3ac)^2}{9+b^2-3ac}.$$

Also the equation of the conic referred to the tangent and normal at the point is

$$Ax^2 + 2Bxy + Cy^2 = 2y,$$

where $A = \frac{1}{\rho}$, $B = -\frac{\rho_s}{3\rho}$, $C = \frac{1}{\rho} + \frac{2\rho_s^2}{9\rho} - \frac{\rho_{2s}}{3}$.

Ed. Times, Math. Reprint, Vol. XXI., p. 87, where the demonstrations by Prof. Wolstenholme will be found.

ENVELOPES.

5192 An envelope of a curve is the locus of the ultimate intersections of the different curves of the same species, got by varying continuously a parameter of the curve; and the envelope touches all the intersecting curves so obtained.

5193 RULE.—*If* F(x, y, a) = 0 *be a curve having the parameter* a, *the envelope is the curve obtained by eliminating* a *between the equations*

$$F(x, y, a) = 0 \quad \text{and} \quad d_a F(x, y, a) = 0.$$

PROOF.—Let a change to $a+h$. The coordinates of the point of intersection of $F(x, y, a) = 0$ and $F(x, y, a+h) = 0$ satisfy the equation

$$\frac{F(x, y, a+h)-F(x, y, a)}{h} = 0, \quad i.e., \quad \frac{dF(x, y, a)}{da} = 0. \qquad (1404)$$

5194 If $F(x, y, a, b, c, \ldots) = 0$ be the equation of a curve having n parameters $a, b, c, \ldots$ connected by $n-1$ equations, then, by varying the parameters, a series of intersecting curves may be obtained. The envelope of these curves will be found by differentiating all the equations with respect to a, b, c, &c., and eliminating $da, db, \ldots$ and $a, b, \ldots$

5195 Ex.—In (2) of (5135), we have the equation of the normal of a curve at a given point xy; ξ, η being the variable coordinates, and x, y the parameters connected by the equation of the curve $F(x, y) = 0$. By differentiating for x and y, (5136) is found, and the elimination as directed in (5157) produces the equation of the evolute which, by (5194), is the envelope of the curve.

INTEGRALS OF CURVES AND AREAS.

FORMULÆ FOR THE LENGTH OF AN ARC s.

5196 $$s = \int ds = \int \sqrt{(1+y_x^2)}\, dx = \int \sqrt{1+x_y^2}\, dy \qquad (5113)$$

5200 $$= \int \sqrt{(x_t^2+y_t^2)}\, dt = \int \sqrt{(r^2+r_\theta^2)}\, d\theta \qquad (5116)$$

5201 $$= \int \sqrt{(r^2\theta_r^2+1)}\, dr = \int \frac{r\,dr}{\sqrt{(r^2-p^2)}}. \qquad (5111)$$

5203 *Legendre's formula*, $s = p_\psi + \int p\,d\psi$.

5204 The whole contour of a closed curve $= \int_0^{2\pi} p\,d\psi$.

PROOF.—In figure (93), let P, P' be an element ds of the curve; $PT, P'T'$ tangents, and OT, OT' the perpendiculars upon them from the origin; $OT \equiv p$, $PT \equiv q$. Then $ds+P'T'-PT = TL$, i.e., $ds+dq = p\,d\psi$; therefore $s+q = \int p\,d\psi$. But $q\,d\psi = -dp$; therefore $s = p_\psi + \int p\,d\psi$. Also, in integrating all round the curve, $P'T'-PT$ taken for every point vanishes in the summation, or $\int dq = 0$. Therefore $\int ds = \int_0^{2\pi} p\,d\psi$.

FORMULÆ FOR PLANE AREAS.

5205 If $y = \phi(x)$ be the equation of a curve, the area bounded by the curve, two ordinates ($x = a$, $x = b$), and the x axis, is, as in (1902),

$$A = \int_a^b \phi(x)\,dx.$$

5206 With polar coordinates the area included between two radii ($\theta = \alpha$, $\theta = \beta$) and the curve is

$$A = \tfrac{1}{2}\int_\alpha^\beta r^2 d\theta = \tfrac{1}{2}\int p\,ds = \tfrac{1}{2}\int \frac{pr\,dr}{\sqrt{(r^2-p^2)}}.$$

PROOF.—From figure (91) and the elemental area OPP'.

5209 The area bounded by two circles of radii a, b, and the two curves $\theta = \phi(r)$, $\theta = \psi(r)$ (Fig. 109).

$$A = \int_a^b \int_{\phi(r)}^{\psi(r)} r\,dr\,d\theta = \int_a^b r\,\{\psi(r) - \phi(r)\}\,dr.$$

Here $r\{\psi(r) - \phi(r)\}\,dr$ is the elemental area between the dotted circumferences.

5211 The area bounded by two radii of curvature, the curve, and its evolute (Fig. 110).

$$A = \tfrac{1}{2}\int \rho^2 d\psi = \tfrac{1}{2}\int \rho\,ds.$$

PROOF.—From figure (93) and the elemental area QPP'.

INVERSE CURVES.

The following results may be added to those given in Arts. (1000–15).

5212 Let r, r' be corresponding radii of a curve and its inverse, so that $rr' = k^2$; s, s' corresponding arcs, and ϕ, ϕ' the angles between the radius and tangents, then

$$\frac{ds}{ds'} = \frac{r}{r'} \quad \text{and} \quad \phi = \phi'.$$

PROOF.—Let PQ be the element of arc ds, $P'Q'$ the element ds', and O the origin.

Then $OP.OP' = OQ.OQ'$, therefore OPQ, $OQ'P'$ are similar triangles; therefore $PQ : P'Q' :: OP : OQ' = r : r'$; also $\angle OPQ = OQ'P'$.

5214 If ρ, ρ' be the radii of curvature,

$$\frac{r}{\rho}+\frac{r'}{\rho'}=2\sin\phi.$$

PROOF.—From $p=r\sin\phi$, $p'=r'\sin\phi$, we have

$$p'=k^2\frac{p}{r^2}, \text{ therefore } \frac{dp'}{dr}=k^2\frac{r^2p_r-2rp}{r^4} \quad\text{.................. (i.).}$$

Also
$$r=\frac{k^2}{r'}, \text{ therefore } \frac{dr}{dr'}=-\frac{k^2}{r'^2} \quad\text{.................. (ii.).}$$

Now $\rho'=r'\frac{dr'}{dp'}$ (5148), therefore $\frac{1}{\rho'}=\frac{1}{r'}\frac{dp'}{dr}\frac{dr}{dr'}=\frac{2p}{rr'}-\frac{r}{\rho r'}$, by (i.) and (ii.).

Therefore
$$\frac{r'}{\rho'}+\frac{r}{\rho}=\frac{2p}{r}=2\sin\phi.$$

5215 To find the equation of the inverse of a curve in rectangular coordinates, substitute

$$\frac{k^2x}{x^2+y^2} \quad\text{and}\quad \frac{k^2y}{x^2+y^2}$$

for x and y in the equation of the given curve.

5216 The inverse of the algebraic curve

$$u_n+u_{n-1}+u_{n-2}+\ldots+u_1+u_0=0,$$

where u_n is a homogeneous function of the nth degree, will be

$$k^{2n}u_n+k^{2(n-1)}u_{n-1}(x^2+y^2)+k^{2(n-2)}u_{n-2}(x^2+y^2)^2+\ldots$$
$$\ldots+u_0(x^2+y^2)^n=0.$$

5217 The inverse of the conic $u_2+u_1+u_0=0$ is

$$k^4u_2+k^2u_1(x^2+y^2)+u_0(x^2+y^2)^2=0.$$

5218 If the origin be on the curve, this equation becomes

$$k^2u_2+u_1(x^2+y^2)=0.$$

5219 The angle ϕ will also be unaltered in any curve, $r=f(\theta)$, if the inversion be effected by putting

$$r=kr'^n \quad\text{and}\quad \theta=n\theta'.$$

PROOF.—

$$\tan\phi'=r'\theta'_{r'} \text{ (5112) }=r'\theta'_r r_{r'}=r'\frac{\theta_r}{n}knr'^{n-1}=kr'^n\theta_r=r\theta_r=\tan\phi.$$

PEDAL CURVES.

5220 The locus of the foot of the perpendicular from the origin upon the tangent is called a *pedal curve*. The pedal of the pedal curve is called the second pedal, and so on. Reversing the order, the envelope of the right lines drawn from each point of a curve at right angles to the radius vector is called the *first negative pedal*, and so on.

5221 The pedal and the reciprocal polar are inverse curves (1000, 4844.)

AREA OF A PEDAL CURVE.

5222 Let C, P, Q be the respective areas of a closed curve, the pedal of the curve, and the pedal of the evolute; then

$$P-Q=C,\quad P+Q=\tfrac{1}{2}\int r^2 d\psi,\quad 2P=C+\tfrac{1}{2}\int r^2 d\psi.$$

PROOF.—With figure (93) and the notation of (5204), we have, by (5206), $P=\frac{1}{2}\int p^2 d\psi$, $Q=\frac{1}{2}\int q^2 d\psi$; therefore $P+Q=\frac{1}{2}\int (p^2+q^2)\,d\psi=\frac{1}{2}\int r^2 d\psi$. Also, taking two consecutive positions of the triangle $OPT\equiv A$, we get $OPT-OP'T'=\delta A=\delta C+\delta Q-\delta P$. Therefore, integrating all round,

$$\int_A^A dA=0=C+Q-P.$$

5225 *Steiner's Theorem.*—If P be the area of the pedal of a closed curve when the pole is the origin, and P' the area of the pedal when the pole is the point xy,

$$P'-P=\tfrac{1}{2}\pi(x^2+y^2)-ax-by,$$

where $a=\int_0^{2\pi} p\cos\theta\, d\theta$ and $b=\int_0^{2\pi} p\sin\theta\, d\theta$;

θ being the inclination of p.

PROOF.—(Fig. 111.) Let LM be a tangent, S the point xy, perpendiculars $OM\equiv p$ and $SR\equiv p'$. Draw SN perpendicular to OM, and let $ON=p_1$;

then $P'=\frac{1}{2}\int p'^2 d\psi=\frac{1}{2}\int (p-p_1)^2\,d\psi=\frac{1}{2}\int p^2 d\psi+\frac{1}{2}\int p_1^2 d\psi-\int pp_1\, d\psi$

$$=P+\frac{\pi}{2}OS^2-\int_0^{2\pi} p\,(x\cos\theta+y\sin\theta)\,d\theta,\text{ by (4094), and } d\theta=d\psi.$$

And $\frac{1}{2}\int_0^{2\pi} p_1^2 d\psi=$ twice the area of the circle whose diameter is OS.

5226 Cor. 1.—If P' be given, the locus of xy is a circle whose equation is (5225), and the centre of this circle is the same for all values of P', the coordinates of the centre being $\frac{a}{\pi}$ and $\frac{b}{\pi}$.

5227 Cor. 2.—Let Q be the fixed centre referred to, and let $QS = c$. Let P'' be the area of the pedal whose origin is Q; then
$$P' - P'' = \tfrac{1}{2}\pi c^2.$$
For a and b must vanish in (5225) when the origin is at the centre Q, and x^2+y^2 then $= c^2$.

5228 Cor. 3.—Hence P'' is the minimum value of P'.*

ROULETTES.

5229 Def.—A *Roulette* is the locus of a point rigidly connected with a curve which rolls upon a fixed right line or curve.

AREA OF A ROULETTE.

5230 When a closed curve rolls upon a right line, the area generated in one revolution by the normal to the roulette at the generating point is twice the area of the pedal of the rolling curve with respect to the generating point.

Proof.—(Fig. 112.) Let P be the point of contact of the rolling curve and fixed straight line, Q the point which generates the roulette. Let R be a consecutive point, and when R comes into contact with the straight line, let $P'Q'$ be the position of RQ. Then PQ is a normal to the roulette at Q, and P is the instantaneous centre of rotation. Draw QN, QS perpendiculars on the tangents at P and R. The elemental area $PQQ'P'$, included between the two normals QP, $Q'P'$, is ultimately equal to $PQR+QRQ'$. But $PQR = dC$, an element of the area of the curve swept over by the radius vector QP or r round the pole Q; and $QRQ' = \frac{1}{2}r^2 d\psi$; therefore, whole area of roulette $= C + \frac{1}{2}\int_0^{2\pi} r^2 d\psi = 2P$, by (5224).

5231 Hence, by (5228), there is one point in any closed curve for which the area of the corresponding roulette is a

* For a discussion of the pedal curves of an ellipse by the Editor of the *Educ. Times* and others, see *Reprint*, Vol. I., p. 23; Vol. XVI., p. 77; Vol. XVII., p. 92; and Vol. XX., p. 106.

minimum. Also the area of the roulette described by any other point, distant c from the origin of the minimum roulette, exceeds the area of the latter by πc^2.

5232 When the line rolled upon is a curve, the whole area generated in one revolution of the rolling curve becomes

$$C+\tfrac{1}{2}\int_0^{2\pi} r^2\left(1+\frac{\rho}{\rho'}\right)d\psi,$$

where ρ, ρ' are the radii of curvature of the rolling and fixed curves, and C is the area of the former.

PROOF.—(Fig. 113.) Instead of the angle $d\psi$, we now have the sum of the angles of contingence at P of the rolling curve and fixed curve, viz.,

$$d\psi+d\psi' = d\psi\left(1+\frac{d\psi'}{d\psi}\right) = d\psi\left(1+\frac{\rho}{\rho'}\right),$$

since $\rho d\psi = ds = \rho' d\psi'$, by (5146).

LENGTH OF THE ARC OF A ROULETTE.

5233 If σ and ζ be corresponding arcs of the roulette and the pedal whose origin is the generating point; then, when the fixed line is straight, $\sigma = \zeta$; and when it is a curve,

5234
$$\int d\sigma = \int\left(1+\frac{\rho}{\rho'}\right)d\zeta.$$

PROOF.—(Fig. 112.) Let R be the point which has just left the straight line, Q the generating point, N, S consecutive points on the pedal curve. Draw the circle circumscribing $RQNS$, of which $RQ = r$ is a diameter, and let the diameter which bisects NS meet the circle in K. Then, when the points P, R, P' coincide, KN and RQ are diameters, and $SKN = SPN = d\psi = QRQ'$; therefore SN or $d\zeta = rd\psi = QQ'$ or $d\sigma$. When the fixed line is a curve, $d\sigma = rd\psi\left(1+\frac{\rho}{\rho'}\right)$, as in (5232).

RADIUS OF CURVATURE OF A ROULETTE.

5235 Let α (Fig. 113) be the angle between the generating line r and the normal at the point of contact; ρ, ρ' the radii of curvature of the fixed and rolling curves, and R the radius of curvature of the roulette; then,

$$R = \frac{r^2}{\dfrac{\rho\rho'}{\rho+\rho'}\cos\alpha - r}.$$

PROOF.— Let consecutive normals of the roulette meet in O; then $OQ = R$, $PQ = r$, $MPT = \alpha$.

$$\frac{R-r}{R} = \frac{PM}{QQ'} = \frac{ds\cos\alpha}{d\sigma}, \quad \text{and} \quad d\sigma = r(d\psi + d\psi') = r\left(\frac{ds}{\rho} + \frac{ds}{\rho'}\right),$$

from which R is obtained. If the curvature of the roulette is convex towards P (Fig. 114), we must write $R+r$ instead of $R-r$ above.

5236 The curvature is convex towards P when R is positive, that is, when the carried point Q falls within the circle whose diameter measured on the normal of the rolling curve $= \dfrac{\rho\rho'}{\rho+\rho'}$. When Q falls without this circle, the curvature is concave; and when Q falls upon the circumference, the point is one of inflexion. The circle has for this reason been called the *circle of inflexions.*

5237 In figure (163) let $PA = \rho$, $PB = \rho'$, $PQ = r$, $OQ = R$, as in (5235). Draw PCD, the circle of inflexions, with its diameter $PC = \dfrac{\rho\rho'}{\rho+\rho'}$, and therefore $PD = \dfrac{\rho\rho'}{\rho+\rho'}\cos\alpha$. From these values and proportion it follows that $BC : BP : BA$ and $QD : QP : QO$. Also, if the circle on diameter $PE = PC$ be drawn, $AE : AP : AB$ and $OF : OP : OQ$.

5238 A simple construction for the centre of curvature of the roulette is the following. (Fig. 164, with letters as in 5237.) At P draw a perpendicular to PQ to meet QB in N. Join NA, which will meet QP produced in O, the required point.

PROOF.—From equation (5235), assuming O to be the centre of curvature, we can deduce the relation $(BA : AP)(PO : OQ)(QN : NB) = 1$, therefore, by (968), A, O, N are collinear points.

THE ENVELOPE OF A CARRIED CURVE.

5239 When a curve is rigidly connected with a rolling curve, it will have an envelope. The path of its point of contact with the envelope is a tangent to both curves, and therefore the normal, common to the carried curve and its envelope, passes through the point of contact P of the rolling and fixed curve.

5240 The centre of curvature of the envelope is obtained as follows.

In Fig. (163), from P draw a normal to the carried curve meeting it in Q, and let S on PQ be the centre of curvature of the envelope for the point Q; and O that of the carried curve. Then PS is found from

$$\frac{1}{\rho}+\frac{1}{\rho'}=\cos\alpha\left(\frac{1}{PS}+\frac{1}{PO}\right).$$

5241 When the envelope is a right line, the centre of curvature lies on the circle of inflexions (5236). When the carried curve is a right line, the same point lies on the circle PEF (Fig. 163), and if the right line always passes through a fixed point, that point lies on the circle PEF.

5242 If p be the perpendicular from a fixed point upon a carried right line whose inclination to a fixed line is ψ; the radius of curvature of the envelope is $\rho=p+p_{2\psi}$, by (5147).

INSTANTANEOUS CENTRE.

5243 When a plane figure moves in any manner in its own plane, the *instantaneous centre of rotation* is the intersection of the perpendiculars at two points to the directions in which the points are moving; and a line from the instantaneous centre to any point of the figure is the normal to the path of that point.

Ex.—Let a triangle ABC slide with its vertices A, B always upon the right lines OA, OB. The perpendiculars at A, B to OA, OB meet in Q, the instantaneous centre, and QC is the normal at C to the locus of C.

Since AB and the angle AOB are of constant magnitude, OQ, the diameter of the circle circumscribing $OAQB$, is of constant magnitude. Hence the locus of the instantaneous centre Q is a circle of centre O and radius OQ.

5244 *Holditch's Theorem.*—If a chord of a given length LM moves completely round a closed curve, the area enclosed between the curve and the locus of a point P on the chord is equal to $\pi cc'$ where $c=LK$, $c'=MK$.

5245 If the ends of LM move on different closed curves whose areas are λ, μ, while the area described by K is κ, then

$$\kappa=\frac{\lambda c'+\mu c}{c+c'}-\pi cc'.$$

Proof.—(5244). Let the innermost oval in figure (134) be the envelope of LM, ϵ its area, and E the point of contact. Let $EL\equiv l$, $EM\equiv m$,

$EK \equiv k$, $l+m = a = c+c'$; θ, the inclination of LM. Then, integrating in every case from 0 to 2π,

$$\left.\begin{array}{l} \tfrac{1}{2}\int l^2 d\theta = \lambda - \epsilon \\ \tfrac{1}{2}\int m^2 d\theta = \mu - \epsilon \end{array}\right\} \quad \therefore \ \tfrac{1}{2}\int (l^2 - m^2)\, d\theta = \frac{a}{2}\int (l-m)\, d\theta = \lambda - m.$$

$$\text{Also } \frac{a}{2}\int (l+m)\, d\theta = \pi a^2,$$

$\therefore\ a\int l d\theta = \pi a^2 + \lambda - \mu$ ...... (i.). Similarly $c\int l d\theta = \pi c^2 + \lambda - \kappa$ ...... (ii.), the last being obtained from $\tfrac{1}{2}\int (l^2 - k^2)\, d\theta = \lambda - \kappa$. κ is then found by eliminating the integral between (i.) and (ii.).

(5245.) If the curves λ, μ coincide, $\lambda = \mu$ and therefore $\lambda - \kappa = \pi c c'$.

TRAJECTORIES.

5246 Def.—A *trajectory* is a curve which cuts according to a given law a system of curves obtained by varying a single parameter.

The differential equation of the trajectory which cuts at a constant angle β the system of curves represented by $\phi(x, y, c) = 0$ is obtained by eliminating c between the equations

$$\phi(x, y, c) = 0 \quad \text{and} \quad \tan\beta = \frac{\phi_x + \phi_y y_x}{\phi_y - \phi_x y_x},$$

the derivatives of ϕ being partial, and y_x referring to the trajectory.*

Proof. — At a point of intersection we have for the given curve $m = -\phi_x \div \phi_y$, and for the trajectory $m' = y_x$. Employ (4070).

If the trajectory is to be orthogonal, $\tan\beta = \infty$, and the second equation becomes

$$\phi_y - \phi_x y_x = 0.$$

Ex.—To find the curve which cuts at a constant angle all right lines passing through the origin.

Let $y = cx$ represent these lines by varying c; then, writing n for $\tan\beta$, the two equations become $y - cx = 0$ and $n(1 + c y_x) = y_x - c$. Eliminating c, $x y_x - y = n(y y_x + x)$. Divide by $x^2 + y^2$ and integrate; thus

$$\tan^{-1}\frac{y}{x} = n \log \sqrt{(x^2+y^2)} + C,$$

which is equivalent to $r = a e^{\frac{\theta}{n}}$, the equation of the logarithmic spiral (5289).

* For a very full investigation of this problem, see Euler, *Novi Com. Petrop.*, Vol. xiv., p. 46, xvii., p. 205; and *Nova Acta Petrop.*, Vol. i., p. 3.

CURVES OF PURSUIT.

5247 DEF.—A *curve of pursuit* is the locus of a point which moves with uniform velocity towards another point while the latter describes a known curve also with uniform velocity.

Let $f(x, y) = 0$ be the known curve, xy the moving point upon it, $\xi\eta$ the pursuing point, and $n:1$ the ratio of their velocities. The differential equation of the path of $\xi\eta$ is obtained by eliminating x and y between the equations

$$f(x, y) = 0 \;\ldots\ldots\;(\text{i.}), \qquad y-\eta = \eta_\xi(x-\xi) \;\ldots\ldots\;(\text{ii.}),$$

$$\sqrt{(x_\xi^2+y_\xi^2)} = n\sqrt{(1+\eta_\xi^2)} \;\ldots\ldots\;(\text{iii.}).$$

PROOF.—(ii.) expresses the fact that xy is always in the tangent of the path of $\xi\eta$.

(iii.) follows from $1:n = \sqrt{(d\xi^2+d\eta^2)} : \sqrt{(dx^2+dy^2)}$; the elements of arc described being proportional to the velocities.

EX.—The simplest case, being the problem usually presented, is that in which the point xy moves in a right line. Let $x = a$ be this line, and let the point $\xi\eta$ start from the origin when the point xy is on the x axis. The equations (i.), (ii.), (iii.) now become, since $x_\xi = 0$,

$$x = a, \quad y = \eta+\eta_\xi(a-\xi), \quad y_\xi = n\sqrt{(1+\eta_\xi^2)}.$$

From the second $y_\xi = \eta_{2\xi}(a-\xi)$, therefore $(a-\xi)\,\eta_{2\xi} = n\sqrt{(1+\eta_\xi^2)}$.

Putting $\eta_\xi = p$, $$\frac{dp}{\sqrt{(1+p^2)}} = \frac{n\,d\xi}{(a-\xi)}.$$

Integrating by (1928), we find

$$\log(p+\sqrt{1+p^2}) = -n\log(a-\xi)+n\log a,$$

so that p and ξ vanish together at the origin;

therefore $\sqrt{1+p^2}+p = \left(\dfrac{a}{a-\xi}\right)^n$, and therefore $\sqrt{1+p^2}-p = \left(\dfrac{a-\xi}{a}\right)^n$;

therefore $$\frac{d\eta}{d\xi} = \frac{1}{2}\left\{\left(\frac{a}{a-\xi}\right)^n - \left(\frac{a-\xi}{a}\right)^n\right\},$$

therefore $$\eta = \frac{1}{2}\left\{\frac{a^n(a-\xi)^{1-n}}{n-1} + \frac{a^{-n}(a-\xi)^{1+n}}{n+1}\right\} + \frac{an}{1-n^2},$$

the equation of the required locus, the constant being taken so that $\xi = \eta = 0$ together. If, however, $n = 1$, the integral is

$$\eta = \frac{\xi^2-2a\xi}{4a} - \frac{a}{2}\log\frac{a-\xi}{a}.$$

CAUSTICS.

5248 DEF.—If right lines radiating from a point be reflected from a given plane curve, the envelope of the reflected rays is called the *caustic by reflexion* of the curve.

Let $\phi(x, y) = 0$, $\psi(x, y) = 0$ be the equations of the tangent and normal of the curve, and let hk be the radiant point; then the equation of the reflected ray will be

$$\phi(h, k)\,\psi(x, y) + \psi(h, k)\,\phi(x, y) = 0,$$

and the envelope obtained by varying the coordinates of the point of incidence, as explained in (5194), will be the caustic of the curve.

Ex.—To find the caustic by reflexion of the circle $x^2+y^2 = r^2$, the radiant point being hk.

Taking for the tangent and normal, as in (4140), $x\cos\alpha+y\sin\alpha = r$, and $x\sin\alpha-y\cos\alpha = 0$, the reflected ray is

$$(h\cos\alpha+k\sin\alpha-r)(x\sin\alpha-y\cos\alpha)$$
$$+(h\sin\alpha-k\cos\alpha)(x\cos\alpha+y\sin\alpha-r) = 0.$$

Reducing this to the form

$$A\cos 2\alpha+B\sin 2\alpha+C\sin\alpha-D\cos\alpha = 0,$$

and differentiating for α,

$$-2A\sin 2\alpha+2B\cos 2\alpha+C\cos\alpha+D\sin\alpha = 0.$$

The result of eliminating α is

$$\{4(h^2+k^2)(x^2+y^2)-r^2(x+h)^2-r^2(y+k)^2\}^3 = 27(kx-hy)^2(x^2+y^2-h^2-k^2)^2,$$

the envelope and caustic required.

5249 *Quetelet's Theorem.*—The caustic of a curve is the evolute of the locus of the image of the radiant point with respect to the tangent of the curve.

Thus, in the Fig. of (1178), if S be the radiant point, W is the image in the tangent at P. The locus of W is, in this case, a circle, and the evolute and caustic reduce to the single point S'.

Since the distance of the image from the radiant point is twice the perpendicular on the tangent, it follows that the locus of the image will always be got by substituting $2r$ for r in the polar equation of the pedal, or $\frac{2k^2}{r}$ for r in the polar equation of the reciprocal of the given curve with respect to the radiant point and a circle of radius k.

TRANSCENDENTAL AND OTHER CURVES.

THE CYCLOID.* (Fig. 115)

5250 DEF.—A *cycloid* is the roulette generated by a circle rolling upon a right line, the carried point being on the circumference. When the carried point is without the circumference, the roulette is called a *prolate cycloid;* and, when it is within, a *curtate cycloid.*

5251 The equations of the cycloid are

$$x = a(\theta + \sin\theta), \qquad y = a(1 - \cos\theta),$$

where θ is the angle rolled through, and a the radius of the generating circle.

PROOF.—(Fig. 115.) Let the circle KPT roll upon the line DE, the point P meeting the line at D and again at E. Arc $KP = KD$; therefore arc $PT = AK = OT$. Also $\theta = PCT$, the angle rolled through from A, the centre of the base ED. Then

$$x = OT + TN = a\theta + a\sin\theta; \quad y = PN = a - a\cos\theta.$$

5253 If s be the arc OP and ρ the radius of curvature at P,

$$s = 2PT = \sqrt{(8ay)}, \qquad \rho = 2PK.$$

PROOF.—(i.) The element $Pp = Bh = 2(OB - Ob)$ ultimately; therefore, by summation, $s = 2OB$. Also $OB = PT = \sqrt{(TK.TR)} = \sqrt{(2ay)}$.

(ii.) Let two consecutive normals at P and p intersect in L. Then PL, pl are parallel to BA, bA; therefore PLp is similar to BAi. But $Pp = 2Bi$; therefore ρ or $PL = 2BA = 2PK$.

5255 COR.—The locus of L, that is the evolute of the cycloid, consists of two half-cycloids as shown in the diagram.

5256 The area of a cycloid is equal to three times the area of the generating circle, and the curve length is four times the diameter of the same circle.

PROOF.—(i.) Area $PpnN = PprR = BbqQ$ ultimately. Therefore, by summation, $DE.AO - \text{cycloid} = \pi a^2$. But $DE.AO = 2\pi a.2a = 4\pi a^2$; therefore cycloid $= 3\pi a^2$.

(ii.) Total curve length $= 8a$, by (5253).

---

* The earliest notice of this curve is to be found in a MSS. by Cardinal de Cusa, 1454 See *Leibnitz, Opera,* Vol. III., p. 95.

5257 The intrinsic equation of the cycloid is

$$s = 4a \sin \psi.$$

PROOF: $s = 2PT = 4a \sin PKT$, and $PKT = PTN = \psi$.*

THE COMPANION TO THE CYCLOID.

5258 This curve is the locus of the point R in Fig. (115). Its equation is

$$y = a\left(1-\cos\frac{x}{a}\right).$$

PROOF.—From $x = a\theta$ and $y = a(1-\cos\theta)$.

5259 The locus of S, the intersection of the tangents at P and B, is the involute of the circle ABO.

PROOF: $BS = BP = \text{arc } OB$.

PROLATE AND CURTATE CYCLOIDS. (5250)

5260 The equations in every case are

$$x = a(\theta+m\sin\theta), \qquad y = a(1-m\cos\theta).$$

The cycloid is prolate when m is > 1 (Fig. 116), and curtate when m is < 1 (Fig. 117), m being the ratio of CP to the radius a.

EPITROCHOIDS AND HYPOTROCHOIDS. (Fig. 118)

5262 These curves are the roulettes formed by a circle rolling upon the convex or concave circumference respectively of a fixed circle, and carrying a generating point either within or without the rolling circle.

The equations of the epitrochoid are

5263 $$x = (a+b)\cos\theta - mb\cos\frac{a+b}{b}\theta,$$

5264 $$y = (a+b)\sin\theta - mb\sin\frac{a+b}{b}\theta,$$

* For other properties, see Pascal, *Histoire de la Roulette;* Carlo Dati, *History of the Cycloid;* Wallis, *Traité de Cycloide;* Groningius, *Historia Cycloidis, Bibliotheca Univ.;* and Lalouère, *Geometria promota in septem de Cycloide libris;* Bernoulli, *Op.*, Vol. IV., p. 98; Euler, *Comm. Pet.*, 1766; and Legendre, *Exercice du Calcul. Int.*, Tom. II, p. 491.

where a, b are the radii of the fixed and rolling circle (Fig. 118), θ is the angle OCX, Q is the generating point initially in contact with the x axis, and m is the ratio $OQ : b$. The dotted line shows the curve described. For the hypotrochoid change the sign of b.

PROOF: $x = CN + MQ$; $CN = (a+b)\cos\theta$;
$MQ = OQ \cos OQM = -OQ\cos(\phi+\theta)$, where $\phi = POR$ and $b\phi = a\theta$.

5265 The length of the arc of an epitrochoid is

$$s = (a+b)\int\left\{1+m^2-2m\cos\frac{a\theta}{b}\right\}^{\frac{1}{2}} d\theta,$$

which is expressed as an elliptic integral $E(k, \phi)$ by substituting $a\theta = 2b\phi$.

For the arc of a hypotrochoid, change the sign of b.

PROOF: $s = \int s_\theta d\theta = \int \sqrt{(x_\theta^2+y_\theta^2)}\, d\theta$ (5113). Find x_θ and y_θ from (5263–4).

EPICYCLOIDS AND HYPOCYCLOIDS. (Fig. 118)

5266 For the equations of these curves make $m = 1$, in (5263, '4). P is then the generating point, and the curve is shown by a solid line in Figure (118).*

5267 If ψ be the inclination of the tangent at a point P on any of these curves,

$$\tan\psi = -\frac{\cos\theta - m\cos\frac{a+b}{b}\theta}{\sin\theta - m\sin\frac{a+b}{b}\theta} = \tan\frac{a+2b}{2b}\theta, \text{ if } m = 1.$$

5268 Hence, in the epicycloid, $\psi = \frac{a+2b}{2b}\theta$,

and the equation of the tangent is

$$x\sin\frac{a+2b}{2b}\theta - y\cos\frac{a+2b}{2b}\theta = (a+2b)\sin\frac{a}{2b}\theta.$$

5269 The equation of the normal will be

$$x\cos\frac{a+2b}{2b}\theta + y\sin\frac{a+2b}{2b}\theta = a\cos\frac{a}{2b}\theta.$$

* Prof. Wolstenholme has investigated these curves considered as the envelopes of a chord whose extremities move on a fixed circle with uniform velocities in the ratio $m : n$ or $m : (-n)$.—*Proc. Lond. Math. Soc.*, Vol. IV., p. 321.

5270 The length of the arc of an epicycloid or hypocycloid included between two successive cusps is

$$\frac{8b}{a}(a \pm b), \text{ and the included area is } \frac{\pi b^2}{a}(3a \pm 2b).$$

PROOF.—Putting $m = 1$ into (5265) and $a\theta = b\phi$, the length becomes

$$\frac{2b}{a}(a \pm b)\int_0^{2\pi} \sin\frac{\phi}{2}\,d\phi = \frac{8b}{a}(a \pm b).$$

Otherwise by (5234); the pedal being the cardioid whose perimeter $= 8a$ (5333).

(ii.) The area, by (5232), is $\pi b^2 + \frac{1}{2}\int_0^{2\pi} 4b^2 \sin^2\frac{\phi}{2}\left(1+\frac{b}{a}\right)d\phi$; since, in Fig. (118), $d\psi$ of (5232) $= dPOR = d\phi$ and $r = PR = 2b\sin\frac{\phi}{2}$.

5271 The evolute of an epicycloid is a similar epicycloid.

PROOF.—The equation of the tangent referred to an x axis drawn through the *summit* of the curve will be (by turning axes through an angle $b\pi \div a$),

$$x\cos\frac{a+2b}{2b}\theta + y\sin\frac{a+2b}{2b}\theta = (a+2b)\cos\frac{a}{2b}\theta.$$

Comparing this with (5270), which is the equation of the tangent of the evolute, we see that the epicycloid and its evolute are similar curves having their parameters in the ratio $a+2b : a$; and that the radius drawn through a cusp of either of the curves passes through a summit of the other.

5272 When $b = -\frac{1}{2}a$, the hypocycloid becomes a straight line, namely, a diameter of the fixed circle.

THE CATENARY. (Fig. 119)

5273 *Characteristic.*—The perpendicular TP from the foot of the ordinate upon the tangent is of a constant length c, and therefore equal to OA, the perpendicular from the origin on the tangent at the vertex. c is the parameter of the curve. The equation is

5274 $$y = \frac{c}{2}\left(e^{\frac{x}{c}} + e^{-\frac{x}{c}}\right).$$

PROOF: $\tan PCT = \frac{dx}{dy} = \frac{c}{\sqrt{(y^2-c^2)}}$, $\therefore x = c\{\log(y+\sqrt{y^2-c^2}) - \log c\}$ (1928), since $x = 0$ when $y = c$. Therefore

$$e^{\frac{x}{c}} = \frac{1}{c}\{y+\sqrt{(y^2-c^2)}\} \quad \text{therefore } e^{-\frac{x}{c}} = \frac{1}{c}\{y-\sqrt{(y^2-c^2)}\}.$$

5275 If $s \equiv \text{arc}\, AC$, $\quad s = \frac{c}{2}\left(e^{\frac{x}{c}} - e^{-\frac{x}{c}}\right) = CP.$

PROOF: $\quad s = \int \sqrt{(1+y_x^2)}\, dx$ (5197)

$$= \int \sqrt{\left(1+\frac{y^2-c^2}{c^2}\right)}\, dx = \int \frac{y}{c} dx = \frac{c}{2}\left(e^{\frac{x}{c}} - e^{-\frac{x}{c}}\right) = \sqrt{(y^2-c^2)} = CP.$$

5276 The area $OACT = cs.$ (5205)

5277 The radius of curvature at $C = \frac{y^2}{c}$, and is therefore equal to the tangent intercepted by the axis of x.

PROOF: $\cos\psi = \frac{c}{y}$, $\therefore\ -\sin\psi\psi_s = -\frac{c}{y^2}y_s$, $\therefore\ \rho = s_\psi = \frac{y^2}{c}$ (5146).

5278 The catenary derives its name from a chain, which, when suspended from its extremities, takes the form of this curve.

For the equation of the evolute of the catenary, see (5159).

THE TRACTRIX. (Fig. 119)

5279 *Characteristic.*—The length of the tangent intercepted by the x axis is constant. This curve is the involute of the catenary, being the locus of P in Figure (119).

The equation of the tractrix is

5280 $x = c \log\left\{c + \sqrt{(c^2-y^2)}\right\} - c\log y - \sqrt{(c^2-y^2)}.$

PROOF.—Let the tangent $PT = c$, then the differential equation of the curve is therefore $yx_y = -\sqrt{c^2-y^2}$. Substitute $z = \sqrt{c^2-y^2}$, and integrate by (1937).

5281 The area included by the four branches $= \pi c^2$.

PROOF.—Area $= 4\int y\,dx = -4\int_0^c \sqrt{c^2-y^2}\,dy = \pi c^2$, by (1933).

THE SYNTRACTRIX.

5282 This curve is the locus of a point Q on the tangent of the tractrix in Fig. (119). Let QT be equal to a given constant length d; then the equation of the syntractrix will be

5283 $x = c \log\left\{d + \sqrt{(d^2-y^2)}\right\} - c\log y - \sqrt{(d^2-y^2)}.$

THE LOGARITHMIC CURVE.* (Fig. 120)

5284 *Characteristic.*—The subtangent is constant.

The equation of the curve is either

5285 $$y = ae^{\frac{x}{n}}, \quad \text{or} \quad x = n \log \frac{y}{a},$$

where $n = NT$, the constant subtangent, and a is the intercept on the y axis.

5287 If n be an even integer, y may take negative values. The most general form of the equation may perhaps be assumed to be

$$y = e^{\frac{x}{n}}\left(\cos \frac{2r\pi}{n} + i \sin \frac{2r\pi}{n}\right).\dagger$$

THE EQUIANGULAR SPIRAL. (Fig. 121)

5288 *Characteristic.*—The angle OPS between the tangent and radius is constant. The equation of the curve is either

5289 $$r = ae^{\frac{\theta}{n}} \quad \text{or} \quad \theta = n \log \frac{r}{a}.$$

5291 $$\tan \phi = n, \qquad s = r \sec \phi,$$

measuring s from the pole.

PROOF.—By (5112) and (5200).

5293 Hence the length of the spiral measured from the pole O to a point P (Fig. 121) is equal to PS, the intercept on the tangent made by the polar subtangent OS.

5294 The locus of S is a similar spiral, and is also an involute of the original curve.

5295 The pedal curve, which is the locus of Y, is also a similar equiangular spiral.

PROOF.—The constancy of the angle ϕ makes the figure $OPYS$ always similar to itself. Therefore P, Y, and S describe similar curves. Hence, if ST is the tangent to the locus of S, $OST = \phi = OPS$; therefore PST is a right angle; therefore the locus of S is an involute of the original spiral. ‡

---

* Originated by James Gregory, *Geometriæ Pars Universalis*, 1668.

† See Euler, *Anal. Infin.*, Vol. II., p. 290; Vincent, *Ann. de Gergonne*, Vol. XV., p. 1; Gregory, *Camb. Math. Journal*, Vol. I., pp. 231, 264; Salmon, *Higher Plane Curves*, p. 274.

‡ For additional properties, see Bernoulli, *Opera*, p. 497.

THE SPIRAL OF ARCHIMEDES.* (Fig. 122)

5296 *Characteristic.*—The distance from the pole is proportional to the angle described. Hence the equation is

5297 $$r = a\theta. \quad \text{Also} \quad \tan\phi = \theta.$$ By (5112).

5299 The intercept, PQ, on any radius between two successive convolutions of the spiral, is constant and $= 2a\pi$.

5300 The area swept over by any radius is one third of the corresponding circular sector of that radius.

5301 This curve is one of the class the general equation of which is

$$r = a\theta^n, \quad \text{with} \quad \tan\phi = \frac{\theta}{n}.$$

THE HYPERBOLIC OR RECIPROCAL SPIRAL. (Fig. 123)

5302 The equation is $r = \dfrac{a}{\theta}$.

5303 An asymptote is the line $y = a$. (5171)

5304 The spiral is also an asymptote to itself.

For when the radius is of the first order of smallness, the distance between two successive convolutions is of the second order. Hence the distance to the pole measured along the curve is infinite.

The area between the radiants r_1, r_2 is $= \frac{1}{2}a(r_1 - r_2)$.

5305 The equation of the *Lituus* is $r = \dfrac{a}{\sqrt{\theta}}$.

THE INVOLUTE OF THE CIRCLE. (Fig. 124)

5306 The equation is

$$\sqrt{(r^2 - a^2)} = a\left(\theta + \cos^{-1}\frac{a}{r}\right).$$

PROOF: $\phi = OPY = \cos^{-1}\dfrac{a}{r}$ and $\sqrt{(r^2-a^2)} = BP = \text{arc}\,AB = a(\theta+\phi)$.

5307 The pedal of the involute is the spiral of Archimedes.

* Invented by Conon, B.C. 250.

PROOF.—Let r', θ' be the coordinates of Y on the pedal curve. Then $r' = BP = \text{arc } AB = a(\theta' + \frac{1}{2}\pi)$. (See 5297).

5308 The reciprocal of the involute is the hyperbolic spiral.

PROOF.—(Fig. 124.) Let P' on OY correspond to P, and let r', θ' be the polar coordinates of P'. Then $r' = OP' = \dfrac{a^2}{OY}$.

But $\quad OY = BP = \text{arc } AB = a(\theta' + \frac{1}{2}\pi), \quad \therefore\ r' = \dfrac{a}{\theta' + \frac{1}{2}\pi}$. See (5302).

THE CISSOID.* (Fig. 125)

5309 *Characteristic.*—A line drawn from the end, O, of a fixed diameter of a circle to the end, Q, of any perpendicular ordinate intersects the parallel ordinate equidistant from the centre in a point, P, whose locus is the cissoid. The equation of the curve is

5310 $$y^2(2a-x) = x^3 \quad \text{and} \quad \frac{dy}{dx} = \frac{(6a-2x)\sqrt{x}}{2\sqrt{(2a-x)^3}}.$$

PROOF.—By similar triangles, $y : x = \sqrt{(2ax-x^2)} : 2a-x$. Two mean proportionals between the radius a and CS are given by the curve, for it appears that $a^2 : CT^2 :: CT : CS$, and therefore $a : CT : \sqrt{CS.CT} : CS$.

5311 The tangent of the circle at B, the other end of the diameter, is an asymptote to both branches of the cissoid.

5312 The area between the curve and its asymptote is equal to three times the area of the circle.

PROOF: In $\int_0^{2a} y\,dx$ substitute $x = 2a\sin^2\theta$.

THE CASSINIAN OR OVAL OF CASSINI. (Fig. 126)

5313 *Characteristic.*—The product $PA.PB$ of the distances of any point on the curve from two fixed points A, B is constant; the equation is consequently

$$\{y^2+(a+x)^2\}\{y^2+(a-x)^2\} = m^4$$

or
$$(x^2+y^2+a^2)^2 - 4a^2x^2 = m^4,$$

where $2a = AB$. The equation in polar coordinates is

$$r^4 - 2a^2r^2\cos 2\theta + a^4 - m^4 = 0.$$

---

* Diocles, A.D. 500.

5314 If a be $> m$, there are two ovals, as shown in the figure. In that case, the last equation shows that if OPP' meets the curve in P and P', we have $OP.OP' = \sqrt{(a^4-m^4)}$; and therefore the curve is its own inverse with respect to a circle of radius $= \sqrt[4]{(a^4-m^4)}$.

5315 O being the centre, the normal PG makes the same angle with PB that OP does with PA.

PROOF.—From $(r+dr)(r'-dr') = m^2$ and $rr' = m^2$; therefore $rdr' = r'dr$ or $r : r' = dr : dr' = \sin\theta : \sin\theta'$, if θ, θ' be the angles between the normal and r, r'. But OP divides APB in a similar way in reverse order.

5316 Let $OP = R$, then the normal PG, and the radius of curvature at P, are respectively equal to

$$\frac{m^2R}{R^2+a^2} \quad \text{and} \quad \frac{2m^2R^3}{3R^4+a^4+m^4}.^*$$

THE LEMNISCATE.† (Fig. 126)

5317 *Characteristic.*—This curve is what a Cassinian becomes when $m = a$. The above equations then reduce to

$$(x^2+y^2)^2 = 2a^2(x^2-y^2) \quad \text{and} \quad r^2 = 2a^2\cos 2\theta.$$

5318 The lemniscate is the pedal of the rectangular hyperbola, the centre being the pole.

5319 The area of each loop $= a^2$. (5206)

THE CONCHOID.‡ (Fig. 127)

5320 *Characteristic.*—If a radiant from a fixed point O intersects a fixed right line, the directrix, in R, and a constant length, $RP = b$, be measured in either direction along the radiant, the locus of P is a *conchoid.* If $OB \equiv a$, be the perpendicular from O upon the directrix, the equation of the curve with B for the origin or O for the pole is

5321 $$x^2y^2 = (a+y)^2(b^2-y^2) \quad \text{or} \quad r = a\sec\theta \pm b.$$

---

* B. Williamson, M.A., *Educ. Times Math.*, Vol. xxv., p. 81.
† Bernoulli, *Opera*, p. 609.
‡ Nicomedes, about A.D. 100.

5323 When $a < b$, there is a loop; when $a = b$, a cusp; and when $a > b$, there are two points of inflexion.

5324 To draw the normal at any point of the curve, erect perpendiculars, at R to the directrix, and at O to OP. They will meet in S the instantaneous centre, and SP will be the normal at P (5242).

5325 To trisect a given angle BON by means of this curve, make $AB = 2ON$, and draw the conchoid, thus determining Q; then $AON = 3AOQ$.

PROOF.—Bisect QT in S; $QT = AB = 2ON$, therefore $SN = SQ = ON$; therefore $NOS = NSO = 2NQO = 2AOQ$.

5326 The total area of the conchoid between two radiants each making an angle θ with OA is

$$a^2 \tan\theta + 2b^2\theta + 3a\sqrt{(b^2-a^2)} \quad \text{or} \quad a^2\tan\theta + 2b^2\theta,$$

according as b is or is not $>a$.

$$\left.\begin{array}{c}\text{The area above the directrix}\\ \text{between the same radiants}\end{array}\right\} = 2ab\log\tan\left(\frac{\pi}{4}+\frac{\theta}{2}\right) + b^2\theta.$$

The area of the loop which exists when b is $>a$ is

$$b^2\cos^{-1}\frac{a}{b} - 2ab\log\frac{a+\sqrt{(b^2-a^2)}}{a-\sqrt{(b^2-a^2)}} + a\sqrt{(b^2-a^2)}.$$

THE LIMAÇON.* (Fig. 128) (Fig. 129)

5327 *Characteristic.*—As in the conchoid, if, instead of the fixed line for directrix, we take a fixed circle upon OB as diameter. This curve is also the inverse of a conic with respect to the focus. The equation, with OB for the initial line and axis of x is

5328 $r = a\cos\theta \pm b$ or $(x^2+y^2-ax)^2 = b^2(x^2+y^2)$,

where $a = OB$, $b = PQ$.

5330 With $b > a$, O is a conjugate point.
With $b \leqq a$, O is a node.

5331 The area $= \pi\left(\frac{1}{2}a^2 + b^2\right)$.

When $a = 2b$, the limaçon has been called the *trisectrix*.

---

* Blaise Pascal, 1643.

THE VERSIERA.* (Fig. 130)

(*Or Witch of Agnesi.*)

5335 *Characteristic.*—If upon a diameter OA of a circle as base a rectangle of variable altitude be drawn whose diagonal cuts the circle in B, the locus of P, the point in which the perpendicular from B meets the side parallel to OA, is the curve in question. Its equation is

5336 $$xy = 2a\sqrt{(2ax-x^2)},$$

where $a = OC$ the radius.

5337 There are points of inflexion where $x = \frac{3}{2}a$.
The total area is four times the area of the circle.

THE QUADRATRIX.† (Fig. 131)

5338 *Characteristic.*—The curve is the locus of the intersection, P, of the radius OD and the ordinate QN, when these move uniformly, so that $x : a :: \theta : \frac{1}{2}\pi$, where $x = ON$, $a = OA$, and $\theta = BOD$. The equation is

$$y = x\tan\left(\frac{a-x}{a}\cdot\frac{\pi}{2}\right).$$

5339 The curve effects the quadrature of the circle, for $OC : OB :: OB : \text{arc } ADB$.

Proof: $OC : OB :: CP : BD$. But $CP = x$ in the limit when it is small, therefore $CP : BD :: a : ADB$.

5340 The area enclosed above the x axis $= 4a^2\pi^{-1}\log 2$.

Proof.—In the integral $\int x\tan\left(\frac{a-x}{a}\frac{\pi}{2}\right)dx$ substitute $\pi(a-x) = 2ay$, and integrate $\int y\tan y\,dy$ by parts, using (1940). The integrated terms produce $\log\cos\frac{1}{2}\pi - \log\cos\frac{1}{2}\pi$ at the limit $\frac{1}{2}\pi$, which vanishes though of the form $\infty-\infty$. The remaining integral is $\int\log\cos y\,dy$, and will be found at (2635).

THE CARTESIAN OVAL. (Fig. 134)

5341 *Characteristic.*—The sum or difference of certain fixed multiples of the distances of a point R on the curve from two

* Donna Maria Agnesi, *Instituzioni Analitiche*, 1748, Art. 238. † Dinostratus, 370 B.C.

fixed points A, B, called the foci, is constant. The equations of the inner and outer ovals are respectively

5342 $$mr_1+lr_2=nc_3, \quad mr_1-lr_2=nc_3,$$

where $r_1=AP$, $r_2=BP$, $c_3=AB$, and $n>m>l$.

5343 To draw the curve, put $\frac{l}{m}=\mu$ and $\frac{nc_3}{m}=a$; therefore $r_1\pm\mu r_2=a$, where a is $>AB$ and $\mu<1$ (1). Describe the circle centre A, and radius $AR=a$. Draw any radiant AQ, and let P, Q be the points in which it cuts the ovals, then, by (1),

5344 $$PR=\mu PB \quad \text{and} \quad QR=\mu QB \quad\text{......................(2).}$$

Hence, by (932), we can draw the circle which will cut AR in the required points P, Q. Thus any number of points on the oval may be found.

5345 By (2) and Euc. VI. 3, it follows that the chord RBr bisects the angle PBQ.

Draw Ap through r, and let PB, QB produced meet Ar in p and q. The triangles PBR, qBr are similar, therefore $qr=\mu qB$; therefore q is on the inner oval. Similarly p is on the outer oval. By Euc. VI. B., $PB.QB=PR.QR+BR^2$; therefore, by (2), $(1-\mu^2)\,PB.QB=BR^2$. Combining this with $PB:Bq=BR:Br$, from similar triangles, we get

5346 $$BQ.Bq=\frac{BR.Br}{1-\mu^2}=\frac{a^2-c_3^2}{1-\mu^2} \quad\text{........................(3).}$$

5347 Draw QC to make $\angle BQC=BAq$; therefore, A, Q, C, q being concyclic, we have, by (3),

$$\boldsymbol{BQ.Bq=AB.BC=\frac{a^2-c_3^2}{1-\mu^2}} \quad\text{..............(4).}$$

Hence C can be found if a, μ, and the points A, B are given. C is the third focus of the ovals, and the equation of either oval may be referred to any two of the three foci.

Putting $BC=c_1$, $AC=c_2$, $AB=c_3$, the equation between l, m, n is obtained from (4) thus: $c_3c_1(1-\mu^2)=a^2-c_3^2$; therefore $c_3(c_3+c_1)=a^2+\mu^2c_1c_3$. But $c_3+c_1=c_2$, $a=\frac{nc_3}{m}$, $\mu=\frac{l}{m}$, and the result is

5348 $$\boldsymbol{l^2c_1+n^2c_3=m^2c_2} \quad\text{or}\quad \boldsymbol{l^2BC+m^2CA+n^2AB=0}\text{...(5),}$$

where $CA=-AC$.

Putting r_1, r_2, r_3 for PA, PB, PC, the equations of the curves are as follows—

	Inner Oval.	Outer Oval.
5349	$mr_1+lr_2 = nc_3$ ... (6),	$mr_1-lr_2 = nc_3$ ... (7),
5351	$nr_1+lr_3 = mc_2$ ... (8),	$nr_1-lr_3 = mc_2$ ... (9),
5353	$mr_3-nr_2 = lc_1$ ...(10),	$nr_2-mr_3 = lc_1$ ...(11).

That (6) and (7) are equations of the curve has been shown. To deduce the other four, we have $\angle APB = AqB = ACQ$ (5347); therefore ACQ, APB are similar triangles. But, by (6), $mAP+lBP = nAB$, therefore $mAC+lCQ = nAQ$ or $nAQ-lCQ = mAC$, which is equation (9). Again, ABQ, APC are similar. But, by (7), $mAQ-lBQ = nAB$; therefore $mAC-lCP = nAP$ or $nAP+lCP = mAC$, which is equation (8).

Equations (10) and (11) are obtained by taking (6) from (8) and (7) from (9), and employing (5).

5355 $$AP.AQ = AB.AC = \text{constant}.$$

Proof.—Since A, Q, C, q are concyclic, $\angle QCA = QqA = ABB$; therefore P, Q, C, B are concyclic; therefore $AP.AQ = AB.AC = \text{constant}$ (12).

5356 $$CP.CP' = CA.CB = \text{constant}.$$

Proof: $\angle PCB = PQB = Bpq = BCq$. Hence, if CP meets the inner oval again in P', CBq, CBP' are similar triangles. Again, because $\angle BPC = BQC = BAq = BAP'$, the points A, B, P', P are concyclic; therefore $CP.CP' = CA.CB = \text{constant}$. Q. E. D.

Hence, by making P, P' coincide, we have the theorem:—

5357 The tangent from the external focus to a series of triconfocal Cartesians is of constant length, and $= \sqrt{(CB.CA)}$.

5358 To draw the tangents to the ovals at P and Q. Describe the circle round $PQCB$, and produce BR to meet the circumference in T; then TP, TQ are the normals at P and Q.

The proof is obtained from the similar triangles TQR, TBQ, which show that $\sin TQA : \sin TQB = l : m$, by (2), and from differentiating equation (7), which produces $\frac{dr_1}{ds} : \frac{dr_2}{ds} = l : m.$*

5359 The *Semi-cubical parabola* $y^2 = ax^3$ is the evolute of a parabola (4549). The length of its arc measured from the origin is
$$s = \frac{8}{27a}\left\{\left(1+\frac{9}{4}ax\right)^{\frac{3}{2}}-1\right\}.$$

---

* For the length of an arc of a Cartesian oval expressed by Elliptic Functions, see a paper by S. Roberts, M.A., in *Proc. Lond. Math. Soc.*, Vol. v., p. 6.

5360 The *Folium of Descartes*, $x^3-3axy+y^3=0$, has two infinite branches, and the asymptote $x+y+a=0$.

For the lengths of arcs and for areas of conics, see (6015), *et seq.*

LINKAGES AND LINKWORK.

5400 A *plane linkage*, in its extended sense, consists of a series of triangles in the same plane connected by hinges, so as to have but one degree of freedom of motion; that is, if any two points of the figure be fixed, and a third point be made to move in some path, every other point of the figure will, in general, also describe a definite path. With two points actually fixed, the linkage is commonly called a *piece-work*, and if straight bars take the place of the triangles, it is called a *link-work*.

THE FIVE-BAR LINKAGE.

5401 Mr. Kempe's fundamental five-bar linkage is shown in Figure (135). A, B, D' are fixed pivots indicated by small circles. C, D, B', C', in the same plane, are moveable pivots indicated by dots. The lengths of the bars AB, BC, CD, DA are denoted by a, b, c, d. The lengths of AB', $B'C'$, $C'D'$, $D'A$ are proportional to the former, and are equal to ka, kb, kc, kd, respectively. Hence $ABCD$, $AB'C'D'$ are similar quadrilaterals, and $\angle AD'C'=ADC$. P being any assigned point on BC and $BP=\lambda$, P' must be taken on $D'C'$ so that $D'P'=\lambda\dfrac{cd}{ab}$. Draw PN, $P'N'$ perpendiculars to AB. Then, throughout the motion of the linkage in one plane, NN' is a constant length.

PROOF: $NN'=BD'-(BN+N'D')$. But $BD'=a-kd$, and

$$BN+N'D'=\lambda\cos B-\lambda\frac{cd}{ab}\cos D=\frac{\lambda}{2ab}(2ab\cos B-2cd\cos D)$$

$$=\frac{\lambda}{2ab}(a^2+b^2-c^2-d^2)\ (702).\ \text{Hence}$$

5402 $$NN' = a - kd - \frac{\lambda}{2ab}(a^2+b^2-c^2-d^2).$$

5403 Case I.—(Fig. 136.) If $\lambda = \dfrac{(a-kd)\,ab}{a^2+b^2-c^2-d^2}$, then $NN' = \dfrac{BD'}{2}$; consequently, if the bars $PO = BB$ and $P'O = P'D'$ be added, the point O will move in the line AB.

If, in this case, $d = ka$ and $b = c$, then $\lambda = b$ and P coincides with C, P' with C', and B' with D, O as before moving in the line AB.

5404 Case II.—(Fig. 137.) If, in Case I., $kd = a$ and $a^2+b^2 = c^2+d^2$, λ is indeterminate; that is, P may then be taken anywhere on BC. D' coincides with B, and $NN' = 0$.

PP' is now always perpendicular to AB. If the bars PO, $P'O$ be added, of lengths such that $PO^2 - P'O^2 = PB^2 - P'B^2$, O will move in the line AB. If, on the other side of PP', bars $PO' = P'B$ and $P'O' = PB$ be attached, then O' will move in a perpendicular to AB through B.

5405 Case III.—(Fig. 138.) If, in Case I., $kd = a$, $b = d$, and $c = -a$, the figure $ABCD$ is termed a contra-parallelogram. $BP = \lambda$ is indeterminate, $BC' = kc = -\dfrac{a^2}{d}$ and $BP' = -\lambda$. Hence BC' and BP' are measured in a reversed direction; PP' is always perpendicular to AB, and if any two equal bars PO, $P'O$ are added, O will move in the line AB.

5406 If three or more similar contra-parallelograms be added to the linkage in this way, as in Figure (139), having the common pivot B and the bars BA, BC, BE, BG in geometrical progression; then, if the bars BA, BG are set to any angle, the other bars will divide that angle into three or more equal parts.

5407 If, in Figure (138), AD be fixed and DC describe an angle ADC, then $B'C'$ describes an equal angle in the opposite direction. Mr. Kempe terms such an arrangement a *reversor*, and the linkage in Figure (139) a *multiplicator*. With the aid of these, and with a *translator* (Fig. 140), for moving a bar AB anywhere parallel to itself, he shows that any plane curve of the nth degree may, theoretically, be constructed by link-work.*

5408 CASE IV.—(Fig. 141.) If, in the original linkage (Fig. 135) $kd = a$, D' coincides with B. Then, if the bars RPO, $RP'O'$ be added by pivots at P, P', and R; and if $OP = PR = BP'$ and $O'P' = P'R = BP$; the points O, O' will move in perpendiculars to AB. For by projecting the equal lines upon AB, we get $NL = BN'$ and $BN = N'L'$, therefore $BL = BL' = NN' =$ a constant, by (5402).

5409 CASE V.—(Fig. 142.) Make $ka = d$ and $\lambda = b$. Then B' coincides with D, P with C, and P' with C'. Replace $D'C'$, $C'D$ by the bars DK, KD' equal and parallel to the former. Also add the bars $CO = DK$ and $OK = CD$. Draw the perpendiculars from O, C and C' to AB. Then by projection, $NL = N'D'$; therefore $BL = BN + NL = BN + N'D' = BD' - NN' =$ constant. Hence the point O will move perpendicularly to AB.

5410 CASE VI.—(Fig. 143.) In the last case take $k = 1$. Therefore $d = a$, D' coincides with B, $BK = BC$, and $CDKO$ is a rhombus. This is Peaucellier's linkage.

5411 CASE VII.—(Fig. 144.) In the fundamental linkage (Fig. 135), transfer the fixed pivots from A, B to P, S, adding the bar SA, so that $PBSA$ shall be a parallelogram. Then, since NN' is constant (5402), the point P' will move perpendicularly to the fixed line PS.

5412 Join AC cutting PS in U, and draw UV parallel to AD. Then $UV : AD = PU : AB = CP : CB =$ constant; therefore PU and UV are constant lengths. Hence it follows that the parallelism of AB to itself may be secured by a fixed pivot at U and a bar UV instead of the pivot S and bar SA.

5413 In Case VII. (Fig. 144), with fixed pivots P and S

---

* *Proc. of the Lond. Math. Soc.*, Vol. VII., p. 213.

and bar SA, make $b=a$, $d=c$, $ka=d$, $\lambda=b$. Then B' coincides with D, N' with N, P with C and L, and P' with C'; and we have Figure 145. DC, DC' are equal, and they are equally inclined to AB or CS; because, in similar quadrilaterals, it is obvious that AB and CD and the homologous sides DC' and AD' include equal angles. Therefore CC' is perpendicular to CS, and C' moves in that perpendicular only.

5414 If two equal linkages like that in (5413), Figure (145), but with the bars AS, CS removed, be joined at D (Fig. 146) and constructed so that $CD\gamma'$, $\gamma DC'$ form two rigid bars, then AB, $\alpha\beta$ will always be in one straight line. Let A, B be made fixed pivots, then, while C describes a circle, the motion of the bar $\alpha\beta$ will be that of a carpenter's plane.

5415 On the other hand, if the linkage of Figure (145), with AS and CS removed as before, be united to a similar inverted linkage (Fig. 147), with DC, DC' common, then, with fixed pivots A, B, D', the motion of the bar $\alpha\beta$ will be that of a lift, directly to and from AB.

5416 The crossing of the links may be obviated by the arrangement in Figure (148). Here the bars $C'\beta$, $C'D$, $C'D'$ are removed, and the bars FD, FE, FG added in parallel ruler fashion.

5417 Case VIII.—(Fig. 149.) In Case VII., substitute the pivot U and the bar UV for S and SA. Make $d=a$, and therefore $k=1$. Then $b'=b$ and $c'=c$, making $BCDC'$ a contra-parallelogram; D' coincides with B, and B' with D. The bars AB, AD are now superfluous. Take $BP=\lambda$; then $BP'=\lambda\frac{c}{b}$; therefore PP' is parallel to CC', therefore to BD, therefore to PV (5412); therefore V, P, P' are always in one right line. P', as in Case VII., moves perpendicularly to PU and AB. This arrangement is Hart's *five-bar linkage*.

5418 When a point P (Fig. 152) moves in a right line PS, it is easy to connect to P a linkage which will make another point move in any other given line we please in the same plane. Let QR be such a line cutting PS in Q. Make Q a fixed pivot, and let OQ, OP, OR be equal bars on a free pivot O. Then, if the angle POR be kept constant by the tie-bar PR, PQR, being one half of POR (Euc. III. 21), will also be constant, and therefore, while P describes one line, R describes the other.

If the bar PO carries a plane along with it, every point in that plane on the circumference of the circle PQR will move in a right line passing through Q.

THE SIX-BAR INVERTOR.*

5419 If in the linkwork (5410, Fig. 143) the bar AD be removed, and D be made to describe any curve, O will describe the inverse curve, just as, when D described a circle, O moved in a right line which is the inverse of a circle.

PROOF.—Let BOD and CK intersect in E. Then $BO.OD = BE^2 - OE^2 = BC^2 - OC^2 =$ a constant called the *modulus* of the cell.

THE EIGHT-BAR DOUBLE INVERTOR.

5420 Two jointed rhombi (Fig. 150) having a common diameter AB form a double *Peaucellier cell* termed positive or negative according as P or Q is made the fulcrum. We have $PQ.PR = PQ.QS = AP^2 - AQ^2$, the constant modulus of the cell.

THE FOUR-BAR DOUBLE INVERTOR.

5421 If, on the bars of a contra-parallelogram $ABCD$ (Fig. 151) four points p, q, r, s be taken in a line parallel to AC or BD, then in every deformation of the linkage, the points p, q, r, s will lie in a right line parallel to AC; and $pq \cdot pr = pq \cdot qs =$ a constant modulus. Thus, if p be a fulcrum and r describes a curve, q will describe the inverse curve. If q be the fulcrum, p will describe the inverse curve.

PROOF.—Let $Ap = mAB$, therefore $pq = mBD$, and $pr = (1-m)\,AC$, therefore $pq \cdot pr = m\,(1-m)\,AC.BD = m\,(1-m)(AD^2 - AB^2) =$ constant.

* Since the curve described is the inverse and not the polar reciprocal of the guiding curve, it seems better to call this linkage an *invertor* rather than a *reciprocator*.

THE QUADRUPLANE, OR VERSOR INVERTOR.

5422 Let the bars of the contra-parallelogram invertor (5421, Fig. 151) carry planes, and let P, Q, R, S be points in the planes similarly situated with respect to the bars which contain p, q, r, s respectively, so that $\angle PAp = QAq$ and $AP : Ap = AQ : Aq$; and similarly at C. Then, if P be the fulcrum and R traces a curve, Q will trace the inverse curve and the angle QPR will be constant.

PROOF.—Let $PA = nAB$ and $PB = n'AB$, therefore, by similar triangles, PAQ, BAD, $PQ = nBD$. Also, by the triangles PBR, ABC, $PR = n'AC$; therefore $PQ.PR = nn'AC.BD = \frac{PA.PB}{AB^2}(AD^2 - AB^2)$, a constant.

Again, the inclination of PQ to BD = that of AP to AB, which is constant. Similarly, by the triangles PBR, ABC, the inclination of PR to AC = that of BR to BC, which is also constant; therefore QPR, the sum of these two inclinations, is a constant angle.

THE PENTOGRAPH, OR PROPORTIONATOR.

5423 Let $ABCD$ (Fig. 153) be a jointed parallelogram, A, B fixed pivots, q a tracer placed at any assigned point in BC produced; then a pencil at p will evidently reproduce any figure traced by q diminished in linear proportions in the ratio of Bq to BC.

THE PLAGIOGRAPH, OR VERSOR PROPORTIONATOR.

5424 In the same figure, make an angle $qBQ = pDP$, $BQ = Bq$, and $DP = Dp$, and let a tracer Q and pencil P be rigidly connected to the arms BC and DC. Then P will produce a similar reduced figure as before, but no longer similarly situated. It will be turned round through an angle QBq. This is Prof. Sylvester's *Plagiograph*.

PROOF.—Let $BC = k.Bq$; therefore $AD = kBQ$, $DP = kAB$, and $\angle ABQ = PDA$; therefore (Euc. VI. 6) $AP = kAQ$. Also PAQ is a constant angle, for $PAQ = BAD - BAQ - PAD = BAD - BAQ - BQA = BAD - (\pi - ABQ) = BAD - \pi + ABC + QBq = QBq$.

THE ISOKLINOSTAT,* OR ANGLE-DIVIDER.

5425 This linkage (Fig. 154) accomplishes the division of an angle into any desired number of equal parts. The dia-

* Invented and so named by Prof. Sylvester.

gram shows the trisection of an angle by it. A number of equal bars are hinged together end to end, and also pivoted on their centres to the same number of equal bars which radiate, fan-like, from a common pivot. The alternate radial bars make equal angles with each other.

The same thing is accomplished in a different way by Kempe's Multiplicator (5406, Fig. 139).

A LINKAGE FOR DRAWING AN ELLIPSE.

5426 In the arrangement of (5413, Fig. 145) the locus of any point P, on DC', excepting D and C', is an ellipse.

PROOF.—Take CS, CC' for x and y axes; P the point xy; $SCD = \theta$, and therefore $CDC' = 2\theta$; $PD = h$. Then we have $x = (c-h)\cos\theta$, $y = (c+h)\sin\theta$, therefore $\frac{x^2}{(c-h)^2}+\frac{y^2}{(c+h)^2} = 1$ is the equation of the locus. Any point on a plane carried by DC' also describes an ellipse round C; but if the point lies on a circle whose centre is D and radius DC, the ellipse becomes a right line passing through C, as appears from (5418).

A LINKAGE FOR DRAWING A LIMAÇON, AND ALSO A BICIRCULAR QUARTIC.*

5427 (Fig. 155.) Let four bars AP', AQ', BC, CD be pivoted at A, B, C, D, and let $AB = BC = BQ' = a$; $AD = DC = DP' = b$. Take a fulcrum F on BC, a tracer at P, and a follower at Q, so that PQ is parallel to BD. Let $FP = \rho$, $FQ = r$; then, if P traces out a circle passing through F, Q will describe a limaçon.

PROOF.—Let $BQ = ma$, therefore $PD = mb$; $r = 2m.BN$, $\rho = (m+1).DN + (1-m)BN$. Also $BN^2 - DN^2 = a^2-b^2$. Eliminate BN and DN, and the equation between r and ρ is

$$r^2+(1-m)\,r\rho - m\rho^2 = m\,(m+1)^2(a^2-b^2) = k^2.$$

If P describes the circle $\rho = c\cos\theta$, Q describes the locus

$$r^2+(1-m)\,cr\cos\theta - mc^2\cos^2\theta = k^2,$$

which is the inverse of a conic, that is, a limaçon (5327).

If C be made the fulcrum, the equation reduces to $r^2-\rho^2 = 4(a^2-b^2)$.

5428 With the same fulcrum F, drawing FH parallel to AC, if a tracer at H describes the circle, then a follower at K on CD will trace out a bicircular quartic.

* W. Woolsey Johnson, *Mess. of Math.*, Vol. v., p. 159.

PROOF.—Draw FL, LK parallel to BA, AD. Let $FH = \rho$, $FK = r$, $CK = \beta$, $CF = a = nFB$, and therefore $CL = n\rho$. Now

$$2(a^2+\beta^2) = r^2+n^2\rho^2+\frac{(a^2-\beta^2)^2}{r^2}.$$

Therefore, if H moves on the circle $\rho = c\cos\theta$, K will describe the curve

$$r^4+n^2c^2r^2\cos^2\theta-2(a^2+\beta^2)r^2+(a^2-\beta^2)^2 = 0,$$

or $$(x^2+y^2)^2+(n^2c^2-2a^2-2\beta^2)x^2-2(a^2+\beta^2)y^2+(a^2-\beta^2)^2 = 0.$$

A LINKAGE FOR SOLVING A CUBIC EQUATION.*

5429 Let the three-bar linkwork (Fig. 156) have the bars AB, DC produced to cross each other. Let $AB = AD = a$, $BC = b$, $CD = c$; and let b and c be adjustable lengths.

Suppose $x^3-qx+r = 0$ a given cubic equation.

Make $c = \frac{1}{2}\sqrt{\left(q+\frac{r}{a}\right)}$, $b = \frac{1}{2}\sqrt{\left(q-\frac{r}{a}\right)}$; then deform the quadrilateral until $EC = CD$; DE will then be equal to a real root of the cubic.

PROOF: $$\cos E = \frac{x^2+c^2-b^2}{2cx} = \frac{(x+a)^2+4c^2-a^2}{4c(x+a)},$$

from which $$x^3-2(c^2+b^2)x+2a(c^2-b^2) = 0.$$

Equate coefficients with the given cubic.†

ON THREE-BAR MOTION IN A PLANE.

5430 If a triangle ABC (Fig. 157) be connected by the bars AO, BO' to the fulcra O, O', the locus of C is called a three-bar curve.

OA, $O'B$ meet in Q, the instantaneous centre of rotation of the triangle, since QA, QB are perpendicular to the movements of A and B respectively. Therefore CQ is the normal to the locus of C.

5431 If a triangle similar to ABC be placed upon OO' (homologous to AB), the circum-circle of the triangle will pass through the node, and the vertices of the triangle are called the foci of the curve.

* M. Saint Loup, *Comptes Rendus*, 1874.

† The foregoing account of linkages is taken chiefly from a paper by A. B. Kempe, F.R.S., in the *Proc. of the Royal Soc.* for 1875, Vol. XXIII. Other results by the same author will be found in the *Proc. of the Lond. Math. Soc.*, Vol. IX., p. 133; and by H. Hart, M.A., *ibid.*, Vol. VI., p. 137, and Vol. VIII., p. 286. See also *The Messenger of Mathematics*, Vol. V.

Figures (158) and (159) exhibit different varieties of the curve according to the relative proportions between the lengths of the bars.*

MECHANICAL CALCULATORS.

The Mechanical Integrator.†

5450 This instrument computes not only the area of any closed plane curve, but the moment and also the moment of inertia of the area about a fixed line. The principle of its action is shown in Figure (160). OP is a bar carrying a tracer at P, and a roller A at some point of its length. The end O is constrained to move in the fixed line ON. *When the tracer* P *moves round a closed curve, the length* OP *multiplied by the entire advance recorded by the roller is equal to the area of the curve.*

PROOF.—Let the motion of the tracer from P to a consecutive point Q be decomposed into PP' and $P'Q$ parallel and perpendicular to ON. Let $OP = a$ and $PON = \theta$. When the pointer moves from P to P', the roll accomplished is $PP' \sin\theta$. The roll due to the motion from P' to Q will be neutralized by the exactly equal and opposite roll in the motion of the pointer from q to p', since the bar will there have again the same inclination. Consequently the product of the entire roll and the length a is equal to the sum of such terms as $aPP' \sin\theta$. But this is the area $OPP'O' = NPP'N'$. The algebraic addition of such rectangles gives the entire area, and the instrument effects this, for the area SN is subtracted, by the motion of the roller, from the area QN which is added.

5451 The instrument itself is shown in Figure (161). A frame moving parallel to OX by means of the guide BB carries two equal horizontal wheels geared to a central wheel which has two circumferences, such that its rate of angular motion is half that of the lower wheel and one third of that of the upper. The latter wheels carry two rollers, M and I, on horizontal axles; and the middle wheel carries an arm OP, a pointer at P, and a roller A. In the initial position, the

* The curve is a tricircular trinodal sextic, and is completely discussed by S. Roberts, F.R.S., and Prof. Cayley, in the *Proc. of the Lond. Math. Soc.*, Vol. VII., pp. 14, 136.

† Invented and manufactured by Mr. J. Amsler-Laffon, of Schaffhausen. The demonstrations (which in clearness and elegance cannot be surpassed) of the action of this instrument, and of the Planimeter which follows, were communicated to the author by Mr. J. Macfarlane Gray, of the Board of Trade.

rollers A and I are parallel, while M is at right angles to A. The frame is thus supported above the paper on the three rollers; and if the arm OP be moved through an angle AOA', the axles of the rollers M and I will describe twice and three times that angle respectively. Putting $OP = a$ as above, and A, M, and I for the linear circumferential advances recorded by the three equal rollers respectively, we have the following results—

I.—*The area traced out by the pointer* P $= \boldsymbol{Aa}$.

II.—*The moment of the area about* OX $= \boldsymbol{M}\dfrac{\boldsymbol{a}^2}{\mathbf{4}}$.

III.—*The moment of inertia about* OX $= (\mathbf{3}\boldsymbol{A}+\boldsymbol{I})\dfrac{\boldsymbol{a}^3}{\mathbf{12}}$.

PROOF.—I. Since O moves in the line OX, while the pointer P moves round a curve, the roller A will, as shown above, make the rolling $\Sigma h \sin\theta$, where $h = PP'$ in Figure (160), and the area of the curve $= a\Sigma h \sin\theta$ or $a \times$ roll.

II. The moment of the area about OX

$$= \Sigma\left(ah \sin\theta \times \frac{a \sin\theta}{2}\right) = \frac{a^2}{4}(\Sigma h - \Sigma h \cos 2\theta).$$

Now Σh vanishes when P returns to the starting point, and $-\Sigma h \cos 2\theta$ is the roll recorded by M. For, when OP makes an angle θ with OX, the axis of M will make an angle $-(90°+2\theta)$ with OX. In this position, while P makes a parallel movement h, the roll produced thereby in M will be $-h \sin(90°+2\theta)$ $= -h \cos 2\theta$. Therefore $\dfrac{a^2}{4} \times$ roll of $M =$ moment of area.

III. Lastly, the moment of inertia of the area about OX

$$= \Sigma\left(ah \sin\theta \times \frac{a^2 \sin^2\theta}{3}\right) = \frac{a^3}{12}\Sigma(3h \sin\theta - h \sin 3\theta).$$

Now, when OP makes an angle θ with OX, the axis of I makes -3θ; therefore $-\Sigma h \sin 3\theta$ is the entire roll of I. Hence the moment of inertia

$$= \frac{a^3}{4} \times \text{roll of } A + \frac{a^3}{12} \times \text{roll of } I.$$

The Planimeter. (Fig. 162)

5452 This instrument * is a simpler form of area computer. O is a fixed pivot; OA, AP are two rods having a free pivot at A; C is the roller, and P the pointer. *The area of a closed curve traced by the pointer is equal to the total roll multiplied by the length* AP.

* Like *The Integrator*, the invention of Mr. Amsler.

PROOF.—Decompose the elementary motion PQ of the pointer into PP', effected with a constant radius OP, and $P'Q$ along the radius OP', and so all round the curve. The roll of C accomplished while P moves from P' to Q will be neutralized by the equal contrary roll when P moves from q to p' on the radius $Op' = OP$. Thus the total roll recorded will be the sum of the rolls due to the movements PP', QQ', &c.

Draw OB perpendicular to AP, and, when P comes to R, let B' be the altered position of B. The area $PQSR = \frac{1}{2}(OP^2 - OR^2)\,\omega$, where $\omega = POQ$. But $OP^2 = OA^2 + PA^2 - 2PA.AC - 2PA.BC$ (Euc. II. 13); therefore, since BC is the only varying length on the right, we have $PQSR = PA(BC \sim BC')\,\omega$. But $BC\omega$ is the roll of C due to the angular motion ω of the rigid frame OAP, and the subtraction of the area OSR from OPQ is effected by the instrument, since when the pointer moves from S to R the direction of the roll must be reversed. Hence the total area $= PA \times$ the total recorded roll.

APPENDIX ON BIANGULAR COORDINATES.*

5453 In the figure of (1178), the biangular coordinates of a point P are defined to be $\theta = PSS'$ and $\phi = PS'S$, or $\alpha = \cot\theta$ and $\beta = \cot\phi$.

5454 The equation of a right line YY' is

$$a\alpha + b\beta = 1,$$

where $a = \cot SYS'$ and $b = \cot SY'S'$.

PROOF.—Supplying the ordinate PN in the figure and denoting the angle $S'SY$ by ψ, the equation is obtained from $CN\cos\psi + PN\sin\psi = p$ the perpendicular on the tangent, $SS'\sin\psi = YY'$ and $SS'\cos\psi = SY - S'Y'$.

5455 $$\cot\psi = a - b.$$

5456 Equation of a line through C: $\alpha - \beta = \text{const.}$

5457 Equation of the line at infinity: $\alpha + \beta = 0$.

5458 Let $SS' = c$, then the distance between two points $\alpha_1\beta_1$, $\alpha_2\beta_2$ is

$$= c^2\left\{\left(\frac{1}{\alpha_1+\beta_1} - \frac{1}{\alpha_2+\beta_2}\right)^2 + \left(\frac{\alpha_1}{\alpha_1+\beta_1} - \frac{\alpha_2}{\alpha_2+\beta_2}\right)^2\right\}.$$

* *Quarterly Journal of Mathematics*, Vols. 9 and 13; W. Walton, M.A.

5459 The equation of a line through the two points is

$$\frac{\alpha-\alpha_1}{\beta-\beta_1}=\frac{\alpha_1-\alpha_2}{\beta_1-\beta_2}.$$

5460 The length of the perpendicular from $\alpha'\beta'$ upon the line $a\alpha+b\beta=1$ is

$$P=\frac{c}{\alpha'+\beta'}\cdot\frac{a\alpha'+b\beta'-1}{\sqrt{\{(a-b)^2+1\}}}.$$

5461 Cor.—The perpendiculars from the poles S, S' are therefore

$$SY=\frac{bc}{\sqrt{\{(a-b)^2+1\}}},\quad S'Y'=\frac{ac}{\sqrt{\{(a-b)^2+1\}}}.$$

5463 When the point $\alpha'\beta'$ is on SS' at a distance h from S,

$$p=\frac{(a-b)\,h+bc}{\sqrt{\{(a-b)^2+1\}}}.$$

With two lines $a\alpha+b\beta=1$, $a'\alpha+b'\beta=1$, the condition

5464 of parallelism is $a-b=a'-b'$,

5465 of perpendicularity $(a-b)(a'-b')+1=0$.

5466 The equation of the line bisecting the angle between the same lines is

$$\frac{a\alpha+b\beta-1}{\sqrt{\{(a-b)^2+1\}}}=\frac{a'\alpha+b'\beta-1}{\sqrt{\{(a'-b')^2+1\}}}.$$

5467 The equation of the tangent at a point $\alpha'\beta'$ on the curve $F(\alpha,\beta)=0$ is

$$(\alpha-\alpha')\,F_{\alpha'}+(\beta-\beta')\,F_{\beta'}=0.$$

5468 And the equation of the normal is

$$\frac{\alpha-\alpha'}{(\alpha'\beta'-1)\,F_{\beta'}+(1+\alpha'^2)\,F_{\alpha'}}=\frac{\beta-\beta'}{(\alpha'\beta'-1)\,F_{\alpha}+(1+\beta'^2)\,F_{\beta'}}.$$

5469 The equation of a circle through S, S' is

$$\alpha\beta-1=m\,(\alpha+\beta),$$

where $m=\cot SPS'$ the angle of the segment.

5470 If C be the centre, the equation becomes

$$\alpha\beta = 1.$$

5471 And, in this case, the equations of the tangent and normal at $\alpha'\beta'$ are respectively

$$\frac{\alpha}{\alpha'}+\frac{\beta}{\beta'} = 2 \quad \text{and} \quad \alpha-\beta = \alpha'-\beta'.$$

5472 The equation of the radical axis of two circles whose centres are S, S', and radii a, b, is

$$(c^2-a^2+b^2)\,\alpha = (c^2+a^2-b^2)\,\beta.$$

PROOF.—By equating the tangents from $\alpha\beta$ to the two circles, their lengths being respectively

$$\frac{c^2(1+\alpha^2)}{(\alpha+\beta)^2}-a^2 \quad \text{and} \quad \frac{c^2(1+\beta^2)}{(\alpha+\beta)^2}-b^2, \text{ by (5458).}$$

5473 To find the equation of the asymptotes of a curve when they exist,—

Eliminate α and β between the equations of the line at

infinity $\qquad \alpha+\beta = 0,$

the curve $\qquad F(\alpha, \beta) = 0,$

and the tangent $\quad (\alpha-\alpha')\,F_{\alpha'}+(\beta-\beta')\,F_{\beta'} = 0.$

Ex.—The hyperbola $\alpha^2+\beta^2 = m^2$ has, for the equation of its asymptotes, $\alpha-\beta = \pm m\sqrt{2}$.

SOLID COORDINATE GEOMETRY.

SYSTEMS OF COORDINATES.

CARTESIAN OR THREE-PLANE COORDINATES.

5501 The position of a point P in this system (Fig. 168) is determined by its distances, $x = PA$, $y = PB$, $z = PC$, from three fixed planes YOZ, ZOX, XOY, the distances being measured parallel to the mutual intersections OX, OY, OZ of the planes, which intersections constitute the axes of coordinates. The point P is referred to as the point xyz, and in the drawing x, y, z are all reckoned positive, ZOX being the plane of the paper and P being situated in front of it, to the right of YOZ and above XOY. If P be taken on the other side of any of these planes, its coordinate distance from that plane is reckoned negative.

FOUR-PLANE COORDINATES.

5502 In this system the position of a point is determined by four coordinates α, β, γ, δ, which are its perpendicular distances from four fixed planes constituting a tetrahedron of reference. The system is in Solid Geometry precisely what trilinear coordinates are in Plane. The relation between the coordinates of a point corresponding to (4007) in trilinears is

5503 $$A\alpha+B\beta+C\gamma+D\delta = 3V,$$

where A, B, C, D are the areas of the faces of the tetrahedron of reference, and V is its volume.

TETRAHEDRAL COORDINATES.

5504 In this system the coordinates of a point are the volumes of the pyramids of which the point is the vertex and the faces of the tetrahedron of reference the bases: viz., $\frac{1}{3}A\alpha$, $\frac{1}{3}B\beta$, $\frac{1}{3}C\gamma$, $\frac{1}{3}D\delta$. The relation between them is

5505 $$\alpha'+\beta'+\gamma'+\delta' = V.$$

POLAR COORDINATES.

5506 Let O be the origin (Fig. 168), XOZ the plane of reference in rectangular coordinates, then the polar coordinates of a point P are r, θ, ϕ, such that $r = OP$, $\theta = \angle POZ$, and $\phi = \angle XOC$ between the planes of XOZ and POZ.

THE RIGHT LINE.

5507 The coordinates of the point dividing in a given ratio the distance between two given points are as in (4032), with a similar value for the third coordinate ζ.

5508 The distance P, Q between the two points xyz, $x'y'z'$ is

$$PQ = \sqrt{\{(x-x')^2+(y-y')^2+(z-z')^2\}}. \quad \text{(Euc. I. 47)}.$$

5509 The same with oblique axes, the angles between the axes being λ, μ, ν.

$$PQ = \sqrt{\{(x-x')^2+(y-y')^2+(z-z')^2+2(y-y')(z-z')\cos\lambda}$$
$$\overline{+2(z-z')(x-x')\cos\mu+2(x-x')(y-y')\cos\nu\}}. \quad \text{(By 702)}.$$

5510 The same in polar coordinates, the given points being $r\theta\phi$, $r'\theta'\phi'$,

$$PQ = \sqrt{\left[r^2+r'^2-2rr'\{\cos\theta\cos\theta'+\sin\theta\sin\theta'\cos(\phi-\phi')\}\right]}.$$

Proof.—Let P, Q be the points, O the origin. Describe a sphere cutting OP, OQ in B, C and the z axis in A; then, by (702), $PQ^2 = OP^2+OQ^2 - 2OP.OQ\cos POQ$ and $\cos POQ$, or $\cos a$ in the spherical triangle ABC, is given by formula (882), since $b=\theta$, $c=\theta'$, and $A=\phi-\phi'$.

DIRECTION RATIOS.

5511 Through any point Q on a right line QP (Fig. 169), draw lines QL, QM, QN parallel to the axes, and through any other point P on the line draw planes parallel to the coordinate planes cutting the lines just drawn in L, M, N; then the direction ratios of the line OP are

5512 $$l = \frac{QL}{QP}, \quad m = \frac{QM}{QP}, \quad n = \frac{QN}{QP}.$$

The angles PQL, PQM, PQN are denoted by α, β, γ; and the angles YOZ, ZOX, XOY between the axes by λ, μ, ν.

5513 When λ, μ, ν are right angles, the axes are called rectangular, and the direction-ratios are called direction-cosines, being in that case severally equal to $\cos\alpha$, $\cos\beta$, $\cos\gamma$.

5514 When L, M, N are the direction-ratios (or numbers proportional to them) of a line which passes through a point abc, the line may be referred to as the line (LMN, abc), or, if direction only is concerned, merely the line LMN.

EQUATIONS BETWEEN THE CONSTANTS OF A LINE.

5515 The relation between the constants of a line with rectangular axes is

$$l^2 + m^2 + n^2 = 1;$$

and with oblique axes, it is

5516 $$l \cos\alpha + m \cos\beta + n \cos\gamma = 1.$$

PROOF.—The first by (Euc. I. 47). The second by projecting the bent line $QLCP$ (Fig. 169) upon PQ, thus $PQ = QL\cos\alpha + LC\cos\beta + CP\cos\gamma$, and $QL = PQ.l$, &c., by (5512).

5517 Also, when the axes are oblique,

$$\cos\alpha = l + m\cos\nu + n\cos\mu,$$
$$\cos\beta = m + n\cos\lambda + l\cos\nu,$$
$$\cos\gamma = n + l\cos\mu + m\cos\lambda.$$

PROOF.—By projecting QP in figure (169) and the bent line $QLCP$ upon each axis in turn, and equating results; thus $PQ\cos\alpha = QL + LC\cos\beta + CP\cos\gamma$, applying (5512).

5518 The relation between l, m, n and λ, μ, ν is

$$l^2 + m^2 + n^2 + 2mn\cos\lambda + 2nl\cos\mu + 2lm\cos\nu = 1.$$

PROOF.—By eliminating $\cos\alpha$, $\cos\beta$, $\cos\gamma$ between (5516) and (5517).

5519 The relation between $\cos\alpha$, $\cos\beta$, $\cos\gamma$ and λ, μ, ν is

$$\begin{aligned}\cos^2\alpha\sin^2\lambda+\cos^2\beta\sin^2\mu+\cos^2\gamma\sin^2\nu\\ +2\cos\beta\cos\gamma(\cos\mu\cos\nu-\cos\lambda)\\ +2\cos\gamma\cos\alpha(\cos\nu\cos\lambda-\cos\mu)\\ +2\cos\alpha\cos\beta(\cos\lambda\cos\mu-\cos\nu)\\ =1-\cos^2\lambda-\cos^2\mu-\cos^2\nu+2\cos\lambda\cos\mu\cos\nu.\end{aligned}$$

Proof.—By eliminating l, m, n between the four equations in (5516) and (5517).

5520 The angle θ between two right lines lmn, $l'm'n'$, the axes being rectangular:

$$\cos\theta = ll'+mm'+nn'.$$

Proof.—In Figure (169), let QP be a segment of the line lmn. The projection of QP upon the line $l'm'n'$ will be $QP\cos\theta$. And this will also be equal to the projection of the bent line $QLCP$, upon $l'm'n'$, for, if planes be drawn through Q, L, C, and P, at right angles to the second line $l'm'n'$, the segment on that line intercepted between the first and last plane will be $= QP\cos\theta$, and the three segments which compose this will be severally equal to $QL.l'$, $LC.m'$, $CP.n'$, the projections of QL, LC, CP. Then, by (5512), $QL = QP.l$, &c.

5521 $\sin^2\theta = (mn'-m'n)^2+(nl'-n'l)^2+(lm'-l'm)^2.$

Proof.—From

$$1-\cos^2\theta = (l^2+m^2+n^2)(l'^2+m'^2+n'^2)-(ll'+mm'+nn')^2 \quad (5515, \text{'20}).$$

5522 With oblique axes,

$$\cos\theta = ll'+mm'+nn'+(mn'+m'n)\cos\lambda + (nl'+n'l)\cos\mu+(lm'+l'm)\cos\nu.$$

Proof.—As in (5520), substituting from (5517) the values of $\cos\alpha$, &c.

EQUATIONS OF THE RIGHT LINE.

5523 $\dfrac{x-a}{L} = \dfrac{y-b}{M} = \dfrac{z-c}{N}$ or $\dfrac{x-a}{l} = \dfrac{y-b}{m} = \dfrac{z-c}{n}$.

Here abc is a datum point on the line, and if r be put for the value of each of the fractions, r is the distance to a variable point xyz. L, M, N are proportional to the direction ratios of the line, which ratios must therefore have the values

5524 $$l = \frac{L}{\sqrt{(L^2+M^2+N^2)}}, \quad m = \frac{M}{\sqrt{(L^2+M^2+N^2)}}, \quad n = \frac{N}{\sqrt{(L^2+M^2+N^2)}}.$$

5525 NOTE.—Instead of a, b, c in the equation we may use $kL+a$, $kM+b$, $kl+c$, where k is an arbitrary constant.

5526 The equations of a line may also be written in the forms $$x = \lambda z + \alpha, \qquad y = \mu z + \beta.$$

5527 These are the equations of the traces on the planes of xz and yz, and are equivalent to

$$\frac{x-\alpha}{\lambda} = \frac{y-\beta}{\mu} = \frac{z-0}{1}.$$

5528 If the line is determined as the intersection of the two planes $Ax+By+Cz = D$ and $A'x+B'y+C'z = D'$, we may write equations (5523) by taking

$$L = BC'-B'C, \quad M = CA'-C'A, \quad N = AB'-A'B,$$

$$a = \frac{DB'-D'B}{N}, \quad b = \frac{DA'-D'A}{N}, \quad c = 0.$$

PROOF.—Eliminate z between the equations of the planes, then the reciprocals of the coefficients of x and y will be L and M.

5529 The projection of the line joining the points xyz and abc upon the line lmn is

$$l(x-a)+m(y-b)+n(z-c).$$

5530 Hence, when the line passes through abc, the square of the perpendicular from xyz upon it is equal to

$$(x-a)^2+(y-b)^2+(z-c)^2-\{l(x-a)+m(y-b)+n(z-c)\}^2.$$

5531 Condition of parallelism of two lines LMN, $L'M'N'$:

$$L : L' = M : M' = N : N',$$

5532 Condition of perpendicularity:

$$LL'+MM'+NN'=0. \qquad (5520)$$

5533 Condition of the intersection of the lines $(LMN,\ abc)$ and $(L'M'N',\ a'b'c')$ (5514):

$$(a-a')(MN'-M'N)+(b-b')(NL'-N'L)$$
$$+(c-c')(LM'-L'M)=0.$$

PROOF.—Eliminate x, y, z between the equations

$$\frac{x-a}{L}=\frac{y-b}{M}=\frac{z-c}{N}=r \quad\text{and}\quad \frac{x-a'}{L'}=\frac{y-b'}{M'}=\frac{z-c'}{N'}=r',$$

by subtracting in pairs, and then eliminate r and r'.

5534 The shortest distance between the same lines is

$$\frac{(a-a')(MN'-M'N)+(b-b')(NL'-N'L)+(c-c')(LM'-L'M)}{\sqrt{\{(MN'-M'N)^2+(NL'-N'L)^2+(LM'-L'M)^2\}}}$$

PROOF.—Assume λ, μ, ν for the dir-cos. of the shortest distance. Then, by projecting the line joining abc, $a'b'c'$ upon the shortest distance, we get $p=(a-a')\lambda+(b-b')\mu+(c-c')\nu$. Also, by (5520), $L\lambda+M\mu+N\nu=0$ and $L'\lambda+M'\mu+N'\nu=0$, giving the ratios $\lambda:\mu:\nu=MN'-M'N:NL'-N'L:LM'-L'M$; and (5524) then gives the values of λ, μ, ν.

5535 The equation of the line of shortest distance between the lines $(lmn,\ abc)$ and $(l'm'n',\ a'b'c')$ is given by the intersection of the two planes

$$l(x-a)+m(y-b)+n(z-\gamma)=\frac{u+u'\cos\theta}{\sin^2\theta} \quad\ldots\ldots\text{(i.)},$$

$$l'(x-a')+m'(y-b')+n'(z-\gamma')=\frac{u'+u\cos\theta}{\sin^2\theta} \quad\ldots\text{(ii.)},$$

where
$$u=l(a'-a)+m(b'-b)+n(c'-c),$$
$$u'=l'(a-a')+m'(b-b')+n'(c-c'),$$

and
$$\cos\theta=ll'+mm'+nn'.$$

PROOF.—(Fig. 170.) Let O be the point xyz on the line of shortest distance AB; P, Q the points abc, $a'b'c'$ on the given lines AP, BQ. Draw BR and PR parallel to AP and AB; RT perpendicular to BQ; and QN, TM perpendicular to BR. Then $\angle RBQ=\theta$, $RN=u$, $QT=u'$, therefore $NM=u'\cos\theta$ and $RM=RN+NM=u+u'\cos\theta$, and in the right-angled triangle RTB, $RM\operatorname{cosec}^2\theta=RB$, the projection of OP upon AP, that is, the left member of equation (i.). Similarly for equation (ii.). It should be observed that (i.) and (ii.) represent planes through AB respectively perpendicular to the given lines AP and BQ.

5536 Otherwise, the line of shortest distance is the intersection of the two planes whose equations are

$$\frac{l'(x-a)+m'(y-b)+n'(z-c)}{l(x-a)+m(y-b)+n(z-c)} = \cos\theta$$

$$= \frac{l(x-a')+m(y-b')+n(z-c')}{l'(x-a')+m'(y-b')+n'(z-c')}.$$

For these equations state that $\cos\theta$ is the ratio of the projections of OP or of OQ upon the given lines, and this fact is apparent from the figure.

5537 Equations of the line passing through the two points abc, $a'b'c'$:

$$\frac{x-a}{a-a'} = \frac{y-b}{b-b'} = \frac{z-c}{c-c'}.$$

5538 A line passing through the point abc and intersecting at right angles the line lmn:

$$\frac{x-a}{L} = \frac{y-b}{M} = \frac{z-c}{N},$$

where $L = lm(b-b')+nl(c-c')-(m^2+n^2)(a-a')$,

and symmetrical values exist for M and N.

PROOF.—The condition of perpendicularity to lmn is

$$Ll+Mm+Nn = 0; \qquad (5520)$$

and the condition of intersecting the line is

$$(a-a')(Mn-mN)+(b-b')(Nl-nL)+(c-c')(Lm-lM) = 0.$$

These equations determine the ratios $L : M : N$.

5539 Equations of the line passing through the point abc, parallel to the plane $Lx+My+Nz = D$, and intersecting the line $(l'm'n', a'b'c')$:

$$\frac{x-a}{l} = \frac{y-b}{m} = \frac{z-c}{n},$$

where l, m, n are found, as in the last, from

$$Ll+Mm+Nn = 0, \quad \text{and}$$

$$(a-a')(mn'-m'n)+(b-b')(nl'-n'l)+(c-c')(lm'-l'm) = 0.$$

5540 Equations of the bisector of the angle between the two lines $l_1m_1n_1$, $l_2m_2n_2$:

$$\frac{x}{l_1+l_2} = \frac{y}{m_1+m_2} = \frac{z}{n_1+n_2}.$$

PROOF.—Taking the intersection of the lines for origin, let $x_1y_1z_1$, $x_2y_2z_2$ be points on the given lines equidistant from the origin; then, if xyz be a point on the bisector midway between the former points, $x = \frac{1}{2}(x_1+x_2)$, &c. (4033); and the direction-cosines of a line through the origin are proportional to the coordinates.

5541 The equations of a right line in four plane coordinates are

$$\frac{\alpha-\alpha'}{L} = \frac{\beta-\beta'}{M} = \frac{\gamma-\gamma'}{N} = \frac{\delta-\delta'}{R} \quad \ldots\ldots\ldots\ldots \quad \text{(i.)},$$

where $\alpha\beta\gamma\delta$ is a variable point, and $\alpha'\beta'\gamma'\delta'$ a fixed point on the line. The relation between L, M, N, R is

5542 $$AL+BM+CN+DR = 0 \quad \ldots\ldots\ldots\ldots \quad \text{(ii.)}.$$

PROOF.—For, since equation (5503) holds for $\alpha\beta\gamma\delta$ and also for $\alpha'\beta'\gamma'\delta'$, we have $A(\alpha-\alpha')+B(\beta-\beta')+C(\gamma-\gamma')+D(\delta-\delta') = 0$.
Substitute from (i.) $\alpha-\alpha' = rL$, $\beta-\beta' = rM$, &c.

5543 In tetrahedral coordinates the same equation (i.) subsists, but the relation between L, M, N, R becomes, by changing $A\alpha$ into α, &c.,

5544 $$L+M+N+R = 0.$$

THE PLANE.

5545 General equation of a plane:

$$Ax+By+Cz+D = 0.$$

5546 Equation in terms of the intercepts on the axes:

$$\frac{x}{a}+\frac{y}{b}+\frac{z}{c} = 1.$$

5547 Equation in terms of p, the perpendicular from the origin upon the plane, and l, m, n, the direction-cosines of p:

$$lx+my+nz = p.$$

PROOF.—If P be any point xyz upon the plane, and O the origin, the projection of OP upon the normal through O is p itself; but this projection is $lx+my+nz$, as in (5520).

5548 The values of l, m, n, p for the general equation (5545) are

$$l = \frac{A}{\sqrt{(A^2+B^2+C^2)}}, \text{ \&c.,} \quad p = \frac{-D}{\sqrt{(A^2+B^2+C^2)}}.$$

PROOF.—Similar to that for (4060-2): by equating coefficients in (5545) and (5547) and employing $l^2+m^2+n^2=1$.

5550 The equation of a plane in four-plane coordinates is

$$l\alpha+m\beta+n\gamma+r\delta = 0,$$

with
$$l : m : n : r = \frac{\alpha_1}{p_1} : \frac{\beta_1}{p_2} : \frac{\gamma_1}{p_3} : \frac{\delta_1}{p_4},$$

where α_1, β_1, γ_1, δ_1 are the perpendiculars upon the plane from A, B, C, D, the vertices of the tetrahedron of reference, and p_1, p_2, p_3, p_4 are the perpendiculars from the same points upon the opposite faces of the tetrahedron.

PROOF.—Put $\gamma = \delta = 0$ for the point where the plane cuts an edge of the tetrahedron, and then determine the ratio $l : m$ by proportion.
See *Frost* and *Wolstenholme*, Art. 81.

5551 The equation of a plane in tetrahedral coordinates is also of the form in (5550), but the ratios are, in that case,

$$l : m : n : r = \alpha_1 : \beta_1 : \gamma_1 : \delta_1.$$

The relation between the three-plane and four-plane coordinates is
$$\alpha = p-lx-my-nz.$$

5552 The equation of a plane in polar coordinates is

$$r\left\{\cos\theta\cos\theta'+\sin\theta\sin\theta'\cos(\phi-\phi')\right\} = p.$$

PROOF.—Here p is the perpendicular from the origin on the plane, and p, θ', ϕ' the polar coordinates of the foot of the perpendicular. Then, if ψ is the angle between p and r, we have $p = r\cos\psi$ and $\cos\psi$ from (882).

5553 The angle θ between two planes

$$lx+my+nz = p \quad \text{and} \quad l'x+m'y+n'z = p'$$

is given by formula (5520), and the conditions of parallelism and perpendicularity by (5531) and (5532), since the mutual inclination of the planes is the same as that of their normals.

5554 The length of the perpendicular from the point $x'y'z'$ upon the plane $Ax+By+Cz+D=0$ is

$$\pm\frac{Ax+By+Cz+D}{\sqrt{(A^2+B^2+C^2)}}=p-lx'-my'-nz'.$$

Proof.—As in (4094).

5556 The same in oblique coordinates

$$=\frac{(Ax+By+Cz+D)}{\rho}=p-x\cos\alpha-y\cos\beta-z\cos\gamma,$$

where ρ is found from (5519) by putting A, B, C for $\rho\cos\alpha$, $\rho\cos\beta$, $\rho\cos\gamma$. This gives

5558 $$\rho^2=\frac{\left\{\begin{array}{l}A^2\sin^2\lambda+B^2\sin^2\mu+C^2\sin^2\nu+2BC(\cos\mu\cos\nu-\cos\lambda)\\ +2CA(\cos\nu\cos\lambda-\cos\mu)+2AB(\cos\lambda\cos\mu-\cos\nu)\end{array}\right\}}{1-\cos^2\lambda-\cos^2\mu-\cos^2\nu+2\cos\lambda\cos\mu\cos\nu}.$$

5559 The distance r of the point $x'y'z'$ from the plane $Ax+By+Cz+D=0$, measured in the direction lmn, the axes being oblique:

$$r=-\frac{Ax'+By'+Cz'+D}{Al+Bm+Cn}.$$

Proof.—By determining r from the simultaneous equations of the line and the plane, viz.,

$$\frac{x-x'}{l}=\frac{y-y'}{m}=\frac{z-z'}{n}=r \quad \text{and} \quad Ax+By+Cz+D=0.$$

Otherwise, by dividing the perpendicular from $x'y'z'$ (5554) by the cosine of its inclination to lmn, viz., $\dfrac{Al+Bm+Cn}{\sqrt{(A^2+B^2+C^2)}}$.

EQUATIONS OF PLANES UNDER GIVEN CONDITIONS.

5560 A plane passing through the point abc and perpendicular to the direction lmn:

$$l(x-a)+m(y-b)+n(z-c)=0.$$

5561 A plane passing through two points abc, $a'b'c'$:

$$\lambda\frac{x-a}{a-a'}+\mu\frac{y-b}{b-b'}+\nu\frac{z-c}{c-c'}=0,$$

5562 with $$\lambda+\mu+\nu=0.$$

PROOF.—By eliminating n between the equations

$$l(x-a)+m(y-b)+n(z-c)=0,\quad l(a-a')+m(b-b')+n(c-c')=0,$$

and altering the arbitrary constant.

5563 A plane passing through the point of intersection of the three planes $u=0$, $v=0$, $w=0$:

$$lu+mv+nw=0.$$

5564 A plane passing through the line of intersection of the two planes $u=0$, $v=0$:

$$lu+mv=0.$$

5565 A plane passing through the two points given by $u=0$, $v=0$, $w=0$ and $u=a$, $v=b$, $w=c$:

$$lu+mv+nw=0 \quad\text{with}\quad la+mb+nc=0.$$

5566 The equation of a plane passing through the three points $x_1y_1z_1$, $x_2y_2z_2$, $x_3y_3z_3$ or A, B, C, is given by the determinant annexed, in which the coefficients of x, y, z represent twice the projections of the area ABC upon the coordinate planes.

$$\begin{vmatrix} x & y & z & 1 \\ x_1 & y_1 & z_1 & 1 \\ x_2 & y_2 & z_2 & 1 \\ x_3 & y_3 & z_3 & 1 \end{vmatrix}=0.$$

PROOF.—The determinant is the eliminant of $Ax+By+Cz=1$, and three similar equations. Expanded it becomes

$$x(y_2z_3-y_3z_2+y_3z_1-y_1z_3+y_1z_2-y_2z_1)+y(\&c.)+z(\&c.)+x_1y_2z_3-\&c.=0.$$

Hence, by (4036), we see that the coefficients are twice the projections of ABC, as stated.

5567 The sum of squares of the coefficients is equal to four times the square of the area ABC.

PROOF.—For, if l, m, n are the dir-cos. of the plane, and $ABC=S$, the coefficients are $=2Sl$, $2Sm$, $2Sn$, by projection.

5568 The determinant (x_1, y_2, z_3), that is, the absolute term in equation (5566), represents six times the volume of the tetrahedron $OABC$, where O is the origin.

PROOF.—Writing the equation of the plane ABC, $Ax+By+Cz+D=0$, we have for the perpendicular from the origin, disregarding sign,

$$p = \frac{D}{\sqrt{(A^2+B^2+C^2)}} = \frac{D}{2S} \text{ (5567)},$$

therefore $D = 2pS = 6 \times$ the tetrahedron $OABC$.

5569 If xyz be a fourth point, P, not in the plane of ABC, the determinant in (5566) represents six times the volume of the tetrahedron $PABC$.

PROOF.—By the last theorem the four component determinants represent six times $(OBCP+OCAP+OABP+OABC)$ for an origin O within the tetrahedron.

5570 A plane passing through the points abc, $a'b'c'$, and parallel to the direction lmn:

$$\left\| \begin{matrix} \frac{x-a}{a-a'} & \frac{y-b}{b-b'} & \frac{z-c}{c-c'} \\ \frac{l}{a-a'} & \frac{m}{b-b'} & \frac{n}{c-c'} \end{matrix} \right\| = 0.$$

PROOF. — Eliminate λ, μ, ν between the equations (5561–2) and $\frac{l\lambda}{a-a'} = \frac{m\mu}{b-b'}, + \frac{n\nu}{c-c'} = 0$, the condition of perpendicularity between lmn and the normal of the plane (5561).

5571 A plane passing through the point abc and parallel to the lines lmn, $l'm'n'$:

$$\begin{vmatrix} x-a & l & l' \\ y-b & m & m' \\ z-c & n & n' \end{vmatrix} = 0.$$

PROOF.—The equation is of the form $\lambda(x-a)+\mu(y-b)+\nu(z-c)=0$, and the conditions of perpendicularity between the normal of the plane and the given lines are $l\lambda+m\mu+n\nu=0$, $l'\lambda+m'\mu+n'\nu=0$. Form the eliminant of the three equations.

5572 A plane equidistant from the two right lines (abc, lmn) and $(a'b'c', l'm'n')$:

$$\begin{vmatrix} x-\frac{1}{2}(a+a') & l & l' \\ y-\frac{1}{2}(b+b') & m & m' \\ z-\frac{1}{2}(c+c') & n & n' \end{vmatrix} = 0.$$

By (5571).

5573 A plane passing through the line (abc, lmn) and perpendicular to the plane $l'x+m'y+n'z = p$:

The equation is that in (5571).

For proof, assume λ, μ, ν for dir-cos. of the normal of the required plane, and write the conditions that the plane may pass through abc and that the normal may be perpendicular to the given line and to the normal of the given plane.

TRANSFORMATION OF COORDINATES.

5574 To change any axes of reference to new axes parallel to the old ones:

Let the coordinates of the new origin referred to the old axes be a, b, c; xyz and $x'y'z'$, the same point referred to the old and new axes respectively; then

$$x = x'+a, \quad y = y'+b, \quad z = z'+c.$$

5575 To change rectangular axes of reference to new rectangular axes with the same origin:

Let OX, OY, OZ be the original axes, and OX', OY', OZ' the new ones,

$l_1\,m_1\,n_1$	the dir-cos. of	OX'	referred to	OX, OY, OZ,
$l_2\,m_2\,n_2$	do.	OY'	do.	do.
$l_3\,m_3\,n_3$	do.	OZ'	do.	do.

xyz, $\xi\eta\zeta$ the same point referred to the old and new axes respectively. Then the equations of transformation are

5576
$$x = l_1\xi+l_2\eta+l_3\zeta \quad \text{(i.)},$$
$$y = m_1\xi+m_2\eta+m_3\zeta \quad \text{(ii.)},$$
$$z = n_1\xi+n_2\eta+n_3\zeta \quad \text{(iii.)}.$$

And the nine constants are connected by the six equations

5577
$$l_1^2+m_1^2+n_1^2 = 1 \ \ldots \text{(iv.)}, \quad l_2l_3+m_2m_3+n_2n_3 = 0 \ \ldots \ \text{(vii.)},$$
$$l_2^2+m_2^2+n_2^2 = 1 \ \ldots \ \text{(v.)}, \quad l_3l_1+m_3m_1+n_3n_1 = 0 \ \ldots \text{(viii.)},$$
$$l_3^2+m_3^2+n_3^2 = 1 \ \ldots \text{(vi.)}, \quad l_1l_2+m_1m_2+n_1n_2 = 0 \ \ldots \ \text{(ix.)},$$

so that three constants are independent.

PROOF.—By (5515) and (5532), since OX', OY', OZ' are mutually at right angles.

5578 The relations (iv. to ix.) may also be expressed thus—

$$\frac{l_1}{m_2n_3-m_3n_2}=\frac{m_1}{n_2l_3-n_3l_2}=\frac{n_1}{l_2m_3-l_3m_2}=\pm 1 \;\ldots\ldots\; \text{(x.)},$$

$$\frac{l_2}{m_3n_1-m_1n_3}=\frac{m_2}{n_3l_1-n_1l_3}=\frac{n_2}{l_3m_1-l_1m_3}=\pm 1 \;\ldots\ldots\; \text{(xi.)},$$

$$\frac{l_3}{m_1n_2-m_2n_1}=\frac{m_3}{n_1l_2-n_2l_1}=\frac{n_3}{l_1m_2-l_2m_1}=\pm 1 \;\ldots\ldots\; \text{(xii.)}.$$

Obtained by eliminating the third term from any two of equations (VII.—IX.). Also, by squaring each fraction in (x.) and adding numerators and denominators, we get

$$\frac{l_1^2+m_1^2+n_1^2}{(l_2^2+m_2^2+n_2^2)(l_3^2+m_3^2+n_3^2)-(l_2l_3+m_2m_3+n_2n_3)}=1, \text{ by (5577).}$$

5579 If the transformation above is *rotational*, that is, if it can be effected by a rotation about a fixed axis, the position of that axis and the angle of rotation θ are found from the equations $\qquad 2\cos\theta = l_1+m_2+n_3-1,$

5580 $$\frac{\cos^2\alpha}{m_2+n_3-l_1-1}=\frac{\cos^2\beta}{n_3+l_1-m_2-1}=\frac{\cos^2\gamma}{l_1+m_2-n_3-1},$$

where α, β, γ are the angles which the axis makes with the original coordinate axes.

PROOF.—(Fig. 171.) Let the original rectangular axes and the axis of rotation cut the surface of a sphere, whose centre is the origin O, in the points x, y, z, and I respectively. Then, if the altered axes cut the sphere in ξ, η, ζ, we shall have $\theta = \angle xI\xi$ in the spherical triangle; $Ix = I\xi = \alpha$; $Iy = I\eta = \beta$; $Iz = I\zeta = \gamma$, and by (882) applied to the isosceles spherical triangles $xI\xi$, &c., $l_1 = \cos x\xi = \cos^2\alpha+\sin^2\alpha\cos\theta$, $m_2 = \cos y\eta = \cos^2\beta+\sin^2\beta\cos\theta$, $n_3 = \cos z\zeta = \cos^2\gamma+\sin^2\gamma\cos\theta$. From these $\cos\theta$, $\cos\alpha$, $\cos\beta$, and $\cos\gamma$ are found.

5581 Transformation of rectangular coordinates to oblique:

Equations (i. to vi.) apply as before, but (vii. to ix.) no longer hold, so that there are now six independent constants.

THE SPHERE.

5582 The equation of a sphere when the point abc is the centre and r is the radius,

$$(x-a)^2+(y-b)^2+(z-c)^2 = r^2.$$

5583 The general equation is

$$x^2+y^2+z^2+Ax+By+Cz+D = 0.$$

The coordinates of the centre are then $-\frac{A}{2},\ -\frac{B}{2},\ -\frac{C}{2}$; and the radius $= \frac{1}{2}\sqrt{(A^2+B^2+C^2-4D)}$.

PROOF.—By equating coefficients with (5582).

5584 If xyz be a point not on the sphere, the value of $(x-a)^2+(y-b)^2+(z-c)^2-r^2$ is the product of the segments of any right line drawn through xyz to cut the sphere.

PROOF.—From Euc. III., 35, 36.

THE RADICAL PLANE.

5585 The radical planes of the two spheres whose equations are $u = 0,\ u' = 0$, is

$$u-u' = 0.$$

5586 The radical planes of three spheres have a common section, and the radical planes of four spheres intersect in the same point.

PROOF.—By adding their equations, and by the principle of (4608) extended to the equations of planes.

POLES OF SIMILITUDE.

5587 DEF.—A *pole of similitude* is a point such that the tangents from it to two spheres are proportional to the radii.

5588 The *external* and *internal poles of similitude* are the vertices of the common enveloping cones.

5589 The locus of the pole of similitude of two spheres is a sphere whose diameter contains the centres and is divided harmonically by them.

CYLINDRICAL AND CONICAL SURFACES.

5590 Def.—A *conical surface* is generated by a right line which passes through a fixed point called the vertex and moves in any manner.

5591 If the point be at infinity, the line moves always parallel to itself and generates a *cylindrical surface.*

5592 Any section of the surface by a plane may be taken for the *guiding curve.*

5593 To find the equation of a cylindrical or conical surface.

Rule.—*Eliminate* xyz *from the equations of the guiding curve and the equations* $\frac{\mathrm{x}-\alpha}{\mathrm{l}} = \frac{\mathrm{y}-\beta}{\mathrm{m}} = \frac{\mathrm{z}}{\mathrm{n}}$ *of any generating line; and in the result put for the variable parameters of the line their values in terms of* x, y, *and* z.

5594 Ex. 1.—To find the equation of the cylindrical surface whose generating lines have the direction lmn, and whose guiding curve is given by $b^2x^2+a^2y^2 = a^2b^2$ and $z=0$.

At the point where the line $\frac{x-\alpha}{l} = \frac{y-\beta}{m} = \frac{z}{n}$ meets the ellipse, $z=0$, $x=\alpha$, $y=\beta$. Therefore $b^2\alpha^2+a^2\beta^2 = a^2b^2$. Substitute in this, for the variable parameters, α, β, $\alpha = x-\frac{lz}{n}$, $\beta = y-\frac{mz}{n}$; and we get, for the cylindrical surface $b^2(nx-lz)^2+a^2(ny-mz)^2 = a^2b^2n^2$.

5595 A conical surface whose vertex is the origin and guiding curve the ellipse $b^2x^2+a^2y^2 = a^2b^2$, $z=c$, is

$$\frac{x^2}{a^2}+\frac{y^2}{b^2}-\frac{z^2}{c^2} = 0.$$

Proof.—Here the generating line is $\frac{x}{l} = \frac{y}{m} = \frac{z}{n}$. At the point of intersection of the line and curve $z=c$, $x=\frac{lc}{n}$, $y=\frac{mc}{n}$; $\therefore \frac{b^2l^2c^2}{n^2}+\frac{a^2m^2c^2}{n^2}=a^2b^2$. Substitute for the variable parameters $l:m:n$ the values $x:y:z$, and the result is obtained.

CIRCULAR SECTIONS.

5596 RULE.—*To find the circular sections of a quadric curve, express the equation in the form* $\mathrm{A}\,(\mathrm{x}^2+\mathrm{y}^2+\mathrm{z}^2+\mathrm{c}^2)+\&\mathrm{c}.=0$. *If the remaining terms can be resolved into two factors, the circular sections are defined by the intersection of a sphere and two planes.*

5597 Generally the two quadrics

$$ax^2+by^2+cz^2+2fyz+2gzx+2hxy=1$$

and $(a+\lambda)\,x^2+(b+\lambda)\,y^2+(c+\lambda)\,z^2+2fyz+2gzx+2hxy=1$

have the same circular sections.

PROOF.—Let r,ρ be coincident radii of the two surfaces having lmn for a common direction. Then $\frac{1}{r^2}=al^2+bm^2+cn^2+2fmn+2gnl+2hlm$ and $\frac{1}{\rho^2}$ = the same $+\lambda$. Therefore, if r has a constant value throughout any section, ρ is also constant throughout that section.

5598 EX.—An oblique circular cone whose vertex is the point $a, 0, b$, and guiding curve the circle $x^2+y^2=c^2$; $z=0$; is

$$(az-bx)^2+b^2y^2=c^2\,(z-b)^2.$$

The equation may be written

$$b^2\,(x^2+y^2+z^2-c^2)=z\,\{2abx+(b^2+c^2-a^2)\,z-2bc^2\},$$

and therefore the cone has two series of parallel circular sections, $z=k$ and $2abx+(b^2+c^2-a^2)\,z-2bc^2=p^2$ (5583). (*Frost* and *Wolstenholme.*)

CONICOIDS.

5599 DEFS.—A *conicoid* is a surface every plane section of which is a conic.

The varieties are the *ellipsoid*, the *one-fold* and *two-fold hyperboloids*, the *elliptic* and *hyperbolic paraboloids*, the *spheroid of revolution*, the *cone*, and the *cylinder*.

In any of the following equations of a conicoid, by making one of the variables constant, the equation of a section parallel to a coordinate plane is obtained, and the equation of the surface is by that means verified. Thus, in the equations of (5600) or (5617), Figs. (172) and (173), if z be put $= ON$, we get the equation of the elliptic section RPQ, the semi-axes of which are $NQ = \frac{a}{c}\sqrt{(c^2-ON^2)}$ and $NR = \frac{b}{c}\sqrt{(c^2-ON^2)}$, a, b, c being the principal semi-axes of the conicoid; that is, OA, OB, OC in the figure.

THE ELLIPSOID.

5600 The equation referred to the principal axes of the figure is

$$\frac{x^2}{a^2}+\frac{y^2}{b^2}+\frac{z^2}{c^2}=1. \qquad \text{(Fig. 172)}$$

5601 There are two planes of circular section whose equations are

$$x^2\left(\frac{1}{b^2}-\frac{1}{a^2}\right)-z^2\left(\frac{1}{c^2}-\frac{1}{b^2}\right)=0,$$

with $a>b>c$.

Proof: $$x^2\left(\frac{1}{a^2}-\frac{1}{r^2}\right)+y^2\left(\frac{1}{b^2}-\frac{1}{r^2}\right)+z^2\left(\frac{1}{c^2}-\frac{1}{r^2}\right)=0$$

is a cone having a common section with the conicoid and a sphere of radius r. If the common section be plane, one of the three terms must vanish in order that the rest may be resolved into two factors.

Since $a>b>c$, the only possible solution for real factors is got by making $r=b$.

5602 Sections by planes parallel to the above are also circles, and any other sections are ellipses.

5603 The umbilici of the ellipsoid (see 5777) are the points whose coordinates are

$$x=\pm a\sqrt{\frac{a^2-b^2}{a^2-c^2}}, \quad y=0, \quad z=\pm c\sqrt{\frac{b^2-c^2}{a^2-c^2}}.$$

Proof.—The points of intersection of the planes (5601) and the ellipsoid (5600) on the xz plane are given by $x'=\pm a\sqrt{\frac{b^2-c^2}{a^2-c^2}}$, $z'=\pm c\sqrt{\frac{a^2-b^2}{a^2-c^2}}$. Since, by (5602) the vanishing circular sections are at the points in the xz plane conjugate to x' and z', we have, by (4352), $x=-\frac{a}{c}z'$, $z=\frac{c}{a}x'$.

5604 If $a = b$, in (5600), the figure becomes a spheroid, and every plane parallel to xy makes a circular section. Hence the spheroid is a surface of revolution. It is called *prolate* or *oblate* according as the ellipse is made to revolve about its major or minor axis.

THE HYPERBOLOID.

5605 The equation of a one-fold hyperboloid referred to its principal axes is

$$\frac{x^2}{a^2}+\frac{y^2}{b^2}-\frac{z^2}{c^2}=1. \qquad \text{(Fig. 173)}$$

5606 The planes of circular section, when $a > b > c$, are all parallel to one or other of the planes whose equations are

$$y^2\left(\frac{1}{b^2}-\frac{1}{a^2}\right)-z^2\left(\frac{1}{c^2}+\frac{1}{a^2}\right)=0.$$

PROOF.—As in (5601), putting $r = a$.

5607 The generating lines of this surface belong to two parallel systems (i.) and (ii.) below, with all values of θ.

5608

$$\left.\begin{aligned}\frac{x}{a}&=\cos\theta+\frac{z}{c}\sin\theta\\ \frac{y}{b}&=\sin\theta-\frac{z}{c}\cos\theta\end{aligned}\right\}\ldots\text{(i.)},\qquad \left.\begin{aligned}\frac{x}{a}&=\cos\theta-\frac{z}{c}\sin\theta\\ \frac{y}{b}&=\sin\theta+\frac{z}{c}\cos\theta\end{aligned}\right\}\ldots\text{(ii.)}.$$

For the coordinates which satisfy either pair of equations, (i.) or (ii.), satisfy also the equation of the surface. The equations may also be put in the forms

5610
$$\frac{x-a\cos\theta}{a\sin\theta}=\frac{y-b\sin\theta}{-b\cos\theta}=\pm\frac{z}{c}.$$

5612 If $z = 0$, $x = a\cos\theta$ and $y = b\sin\theta$. Hence θ is the eccentric angle of the point in which the lines (i.) and (ii.) intersect in the xy plane.

5613 Any two generating lines of opposite systems intersect, but no two of the same system do.

5614 If two generating lines of opposite systems be drawn through the two points in the principal elliptic section whose eccentric angles are $\theta+\alpha$, $\theta-\alpha$, α being constant, the coordinates of the point of intersection will be

$$x = a\cos\theta\sec\alpha,\quad y = b\sin\theta\sec\alpha,\quad z = \pm c\tan\alpha,$$

and the locus of the point, as θ varies, will be the ellipse

5615 $$\frac{x^2}{a^2\sec^2\alpha}+\frac{y^2}{b^2\sec^2\alpha}=1;\quad z=\pm c\tan\alpha.$$

PROOF.—From (i.) and (ii.), putting $\theta\pm\alpha$ for θ.*

5616 The asymptotic cone is the surface given in (5595).

PROOF.—Any plane through the z axis whose equation is $y = mx$ cuts the hyperboloid and this cone in an hyperbola and its asymptotes respectively.

5617 The equation of a two-fold hyperboloid is

$$\frac{x^2}{a^2}-\frac{y^2}{b^2}-\frac{z^2}{c^2}=1. \qquad \text{(Fig. 174)}$$

and the equation of its asymptotic cone is

5618 $$\frac{x^2}{a^2}-\frac{y^2}{b^2}-\frac{z^2}{c^2}=0.$$

PROOF.—Any plane through the x axis, whose equation is $y = mz$, cuts the hyperboloid and this cone in an hyperbola and its asymptotes respectively.

There are two surfaces, one the image of the other with regard to the plane of yz. One only of these is shown in the diagram.

5619 The planes of circular section when b is $>c$ are all parallel to one or other of the planes whose joint equation is

$$x^2\left(\frac{1}{a^2}+\frac{1}{b^2}\right)-z^2\left(\frac{1}{c^2}-\frac{1}{b^2}\right)=0.$$

PROOF.—As in (5601), putting $r^2=-b^2$.

5620 If $b=c$, the figure becomes an hyperboloid of revolution.

THE PARABOLOID.

5621 This surface is generated by a parabola which moves with its vertex always on another parabola; the axes of the two curves being parallel and their planes at right angles.

* The surface of a one-fold hyperboloid, as generated by right lines, may frequently be seen in the foot-stool or work-basket constructed entirely of straight rods of cane or wicker.

The paraboloid is *elliptic* or *hyperbolic* according as the axes of the two parabolas extend in the same or opposite directions.

5622 The equation of the elliptic paraboloid is

$$\frac{y^2}{b}+\frac{z^2}{c}=x, \qquad \text{(Fig. 175)}$$

b and c being the *latera recta* of the two parabolas.

PROOF: $QM^2=b.OM$; $PN^2=c.QN$; $\therefore\ \frac{QM^2}{b}+\frac{PN^2}{c}=OM+QN=x.$

If $b=c$, the figure becomes the *paraboloid of revolution.*

5623 Similarly the equation of the hyperbolic paraboloid is

$$\frac{y^2}{b^2}-\frac{z^2}{c^2}=x. \qquad \text{(Fig. 176)*}$$

5624 The equations of the generating lines of this surface are

$$\frac{y}{\sqrt{b}}\pm\frac{z}{\sqrt{c}}=m \quad \text{and} \quad \frac{y}{\sqrt{b}}\mp\frac{z}{\sqrt{c}}=\frac{x}{m},$$

the upper signs giving one system of generators and the lower signs another system.

5625 The equations of the asymptotic planes are

$$\frac{y}{\sqrt{b}}\pm\frac{z}{\sqrt{c}}=0.$$

CENTRAL QUADRIC SURFACE.

TANGENT AND DIAMETRAL PLANES.

5626 Taking the equation of a central quadric $\frac{x^2}{a^2}+\frac{y^2}{b^2}+\frac{z^2}{c^2}=1$ to include both the ellipsoid and the two hyperboloids

* The curvature of this surface is *anticlastic*, a sort of curvature which may be seen in the saddle of a mountain; for instance, on the smooth sward of some parts of the Malvern Hills, Worcestershire.

according to the signs of b^2 and c^2, the equation of the tangent plane at xyz is

$$\frac{\xi x}{a^2}+\frac{\eta y}{b^2}+\frac{\zeta z}{c^2}=1. \qquad \text{By (5679).}$$

5627 If p be the length of the perpendicular from the origin upon the tangent plane at xyz,

$$\frac{1}{p^2}=\frac{x^2}{a^4}+\frac{y^2}{b^4}+\frac{z^2}{c^4}.$$

PROOF.—From (5549) applied to (5626).

5628 The length of the perpendicular let fall from any point $\xi\eta\zeta$ upon the tangent plane at xyz is

$$p\left(\frac{\xi x}{a^2}+\frac{\eta y}{b^2}+\frac{\zeta z}{c^2}-1\right). \qquad (5554 \;\&\; 5627)$$

5629 Direction cosines of the normal of the tangent plane at xyz,

$$l=\frac{px}{a^2},\quad m=\frac{py}{b^2},\quad n=\frac{pz}{c^2}.$$

PROOF.—By (5548) applied to (5626) and the value in (5627).

5630 If l, m, n are the direction cosines of p,

$$p=lx+my+nz \quad \text{and} \quad p^2=a^2l^2+b^2m^2+c^2n^2.$$

PROOF.—(5630) By projecting the three coordinates x, y, z upon p.
(5631) By substituting the values of x, y, z, obtained from (5629), in (5630).

5632 The equation of the normal at xyz is

$$(\xi-x)\,\frac{a^2}{x}=(\eta-y)\,\frac{b^2}{y}=(\zeta-z)\,\frac{c^2}{z},$$

since the dir-cos. are the same as those of the tangent plane at (5626).

5633 Each term of the above equations

$$=p\sqrt{\{(\xi-x)^2+(\eta-y)^2+(\zeta-z)^2\}}$$

or p multiplied into the length of the normal.

PROOF.—Each term squared $= \dfrac{(\xi-x)^2}{\dfrac{x^2}{a^4}} = \dfrac{(\eta-y)^2}{\dfrac{y^2}{b^4}} = \dfrac{(\zeta-z)^2}{\dfrac{z^2}{c^4}}$.

Add numerators and denominators, and employ (5627).

5634 Equation (5631) is the condition that the plane $lx+my+nz = p$ may touch the conicoid; and if $p = 0$, we have for the condition of the plane $lx+my+nz = 0$ touching the cone $\dfrac{x^2}{a^2}+\dfrac{y^2}{b^2}+\dfrac{z^2}{c^2} = 0$,

5635 $$a^2l^2+b^2m^2+c^2n^2 = 0.$$

5636 The section of the quadric made by a diametral plane conjugate to the diameter through the point xyz has for its

equation $$\frac{\xi x}{a^2}+\frac{\eta y}{b^2}+\frac{\zeta z}{c^2} = 0.$$ By (5688).

5637 Hence the relation between the direction cosines of two conjugate diameters is

$$\frac{ll'}{a^2}+\frac{mm'}{b^2}+\frac{nn'}{c^2} = 0.$$

ECCENTRIC VALUES OF THE COORDINATES.

5638 These are defined to be

$$x = a\lambda, \quad y = b\mu, \quad z = c\nu, \quad \text{with} \quad \lambda^2+\mu^2+\nu^2 = 1.$$

5640 λ, μ, ν are the dir-cos. of a line called the *eccentric line;* and $\xi = r\lambda$, $\eta = r\mu$, $\zeta = r\nu$ are the coordinates of the corresponding point upon an auxiliary sphere of radius r.

5641 The eccentric lines of two conjugate semi-diameters are at right angles. By (5637).

5642 The sum of the squares of three conjugate semi-diameters is constant and $= a^2+b^2+c^2$.

PROOF.—Let a', b', c' be the semi-diameters, and $x_1y_1z_1$, $x_2y_2z_2$, $x_3y_3z_3$ their extremities. Put the eccentric values in the equations $x_1^2+y_1^2+z_1^2 = a'^2$, &c., and add. By (5641), $\lambda_1^2+\lambda_2^2+\lambda_3^2 = 1$, &c.

5643 The sum of the squares of the reciprocals of the same is also constant.

PROOF.—Put $r_1 \cos \alpha_1$, $r_1 \cos \beta_1$, $r_1 \cos \gamma_1$ for x_1, y_1, z_1 in the equation of the quadric. So for x_2, y_2, z_2 and x_3, y_3, z_3. Divide by r_1, r_2, r_3, and add the results.

5644 The sum of squares of reciprocals of perpendiculars on three conjugate tangent planes is constant.

PROOF.—For each perpendicular take (5627), and substitute the eccentric values as in (5642).

5645 The sum of the squares of the areas of three conjugate parallelograms is constant.

PROOF.—By the constant volume of the parallelopiped $p_1A_1 = p_2A_2 = p_3A_3$, (5648) and by (5644).

5646 The sum of the squares of the projections of three conjugate semi-diameters upon a fixed line or plane is constant.

PROOF.—With the same notation as in (5642), let (lmn) be the given line. Substitute the eccentric values (5638) in $(lx_1+my_1+nz_1)^2+(lx_2+my_2+nz_2)^2+(lx_3+my_3+nz_3)^2$. In the case of the plane we shall have

$$a'^2-(lx_1+my_1+nz_1)^2+\&c.$$

5647 COR. — The extremities of three conjugate semi-diameters being $x_1y_1z_1$, $x_2y_2z_2$, $x_3y_3z_3$, it follows that, by projecting upon each axis in turn,

$$x_1^2+x_2^2+x_3^2 = a^2; \quad y_1^2+y_2^2+y_3^2 = b^2; \quad z_1^2+z_2^2+z_3^2 = c^2.$$

5648 The parallelopiped contained by three conjugate semi-diameters is of constant volume $= abc$.

PROOF.—By (5568), the volume $= \begin{vmatrix} x_1 & y_1 & z_1 \\ x_2 & y_2 & z_2 \\ x_3 & y_3 & z_3 \end{vmatrix} = abc \begin{vmatrix} \lambda_1 & \mu_1 & \nu_1 \\ \lambda_2 & \mu_2 & \nu_2 \\ \lambda_3 & \mu_3 & \nu_3 \end{vmatrix}$

by the eccentric values (5638). But the last determinant $=1$ by (584, I.).

5649 COR.—If a', b', c' are the semi-conjugate diameters, ω the angle between a' and b', and p the perpendicular from the origin upon the tangent plane parallel to $a'b'$, the volume of the parallelopiped is $pa'b' \sin \omega = abc$.

5650 Hence the area of a central section in the plane of $a'b'$

$$= \pi a'b' \sin\omega = \pi \frac{abc}{p}.$$

5651 Quadratic for the semi-axis of a central section of the quadric $\frac{x^2}{a^2}+\frac{y^2}{b^2}+\frac{z^2}{c^2}=1$ made by the plane $lx+my+nz=0$:

$$\frac{a^2l^2}{a^2-r^2}+\frac{b^2m^2}{b^2-r^2}+\frac{c^2n^2}{c^2-r^2}=0.$$

PROOF.—The equation is the condition, by (5635), that the plane $lx+my+nz=0$ may touch the cone

$$x^2\left(\frac{1}{a^2}-\frac{1}{r^2}\right)+y^2\left(\frac{1}{b^2}-\frac{1}{r^2}\right)+z^2\left(\frac{1}{c^2}-\frac{1}{r^2}\right)=0,$$

as in the Proof of (5600). For another method, see (1863).

5652 When the equation of the quadric is presented in the form

$$ax^2+by^2+cz^2+2fyz+2gzx+2hxy=1,$$

the quadratic for r^2 takes the form of the determinant equation annexed. Or, by expanding, and writing Δ' for the same determinant, with the fraction $\frac{1}{r^2}$ erased, the equation becomes

$$\begin{vmatrix} a-\frac{1}{r^2} & h & g & l \\ h & b-\frac{1}{r^2} & f & m \\ g & f & c-\frac{1}{r^2} & n \\ l & m & n & \end{vmatrix}=0.$$

$$\Delta'r^4+\left\{(b+c)l^2+(c+a)m^2+(a+b)n^2-2fmn-2gnl-2hlm\right\}r^2$$
$$-l^2-m^2-n^2=0.$$

PROOF.—The equation of the cone of intersection of the sphere and quadric now becomes

$$\left(a-\frac{1}{r^2}\right)x^2+\left(b-\frac{1}{r^2}\right)y^2+\left(c-\frac{1}{r^2}\right)z^2+2fyz+2gzx+2hxy=0,$$

and the condition of touching (5700) produces the determinant equation.

5654 To find the axes of a non-central section of the quadric $\frac{x^2}{a^2}+\frac{y^2}{b^2}+\frac{z^2}{c^2}=1$.

Let PNQ (Fig. 177) be the cutting plane. Take a parallel central section BOC, axes OB, OC, and draw NP, NQ parallel to them. These will be the axes of the section PNQ, and NQ will be found from the equation $\frac{ON^2}{OA^2}+\frac{NQ^2}{OC^2}=1$.

5655 The area of the same section

$$=\frac{\pi abc}{p}\left(1-\frac{p'^2}{p^2}\right),$$

where p' and p are the perpendiculars from O upon the cutting plane and the parallel tangent plane.

PROOF.—The area $=\pi NP.NQ=\pi\frac{NQ^2}{OC^2}.OB.OC$

$$=\pi\left(1-\frac{ON^2}{OA^2}\right)OB.OC=\frac{\pi abc}{p}\left(1-\frac{p'^2}{p^2}\right), \text{ by (5650).}$$

SPHERO-CONICS.

DEF.—A *sphero-conic* is the curve of intersection of the surface of a sphere with any conical surface of the second degree whose vertex is the centre of the sphere.

Properties of cones of the second degree may be investigated by sphero-conics, and are analogous to the properties of conics.

A collection of formulæ will be found at page 562 of Routh's *Rigid Dynamics*, 3rd edition.

CONFOCAL QUADRICS.

5656 DEFINITION.—The two quadrics whose equations are

$$\frac{x^2}{a^2}+\frac{y^2}{b^2}+\frac{z^2}{c^2}=1 \quad \text{and} \quad \frac{x^2}{a^2+\lambda}+\frac{y^2}{b^2+\lambda}+\frac{z^2}{c^2+\lambda}=1,$$

are confocal. We shall assume $a>b>c$.

5657 As λ decreases from being large and positive, the third axis of the confocal ellipsoid diminishes relatively to the others until $\lambda = -c^2$, when the surface merges into the focal ellipse on the xy plane,

$$\frac{x^2}{a^2-c^2}+\frac{y^2}{b^2-c^2}=1.$$

λ still diminishing, a series of one-fold hyperboloids appear until $\lambda = -b^2$, when the surface coincides with the focal hyperbola on the zx plane,

$$\frac{x^2}{a^2-b^2}-\frac{z^2}{b^2-c^2}=1.$$

The surface afterwards developes into a series of two-fold hyperboloids until $\lambda = -a^2$, when it becomes an imaginary focal ellipse on the yz plane.

5658 Through any point xyz three confocal quadrics can be drawn according to the three values of λ furnished by the second equation in (5656). That equation, cleared of fractions, becomes

5659
$$\lambda^3+(a^2+b^2+c^2-x^2-y^2-z^2)\,\lambda^2$$
$$\{b^2c^2+c^2a^2+a^2b^2-(b^2+c^2)\,x^2-(c^2+a^2)\,y^2-(a^2+b^2)\,z^2\}\,\lambda$$
$$+a^2b^2c^2-b^2c^2x^2-c^2a^2y^2-a^2b^2z^2=0.$$

These three confocals are respectively an ellipsoid, a one-fold hyperboloid, and a two-fold hyperboloid. See Figure (178); P is the point xyz; the lines of intersection of the ellipsoid with the two hyperboloids are DPE and FPG, and the two hyperboloids themselves intersect in HPK.

PROOF.—Substitute for λ successively in (5659) a^2, b^2, c^2, $-\infty$; and the left member of the equation will be found to take the signs $+$, $-$, $+$, $-$ accordingly. Consequently there are real roots between a^2 and b^2, b^2 and c^2, c^2 and $-\infty$.

5660 Two confocal quadrics of different species cut each other everywhere at right angles.

PROOF.—Let a, b, c; a', b', c' be the semi-axes of the two quadrics; then, at the line of intersection of the surfaces, we shall have

$$x^2\left(\frac{1}{a^2}-\frac{1}{a'^2}\right)+y^2\left(\frac{1}{b^2}-\frac{1}{b'^2}\right)+z^2\left(\frac{1}{c^2}-\frac{1}{c'^2}\right)=0,$$

which, since $a'^2-a^2 = b'^2-b^2 = c'^2-c^2 = \lambda$, becomes the condition of perpendicularity of the normals by the values in (5629). Thus, in (Fig. 178), the tangents at P to the three lines of intersection of the surfaces are mutually at right angles.

5661 If P be the point of intersection of three quadrics $a_1b_1c_1$, $a_2b_2c_2$, $a_3b_3c_3$ confocal with the quadric abc; the squares of the semi-axes, d_2, d_3, of the diametral section conjugate to P in the first quadric are (considering $a_1 > a_2 > a_3$, and writing the suffixes in circular order)

$$d_3^2 = a_1^2 \sim a_2^2 \qquad d_2^2 = a_1^2 \sim a_3^2,$$

In the second, $$d_1^2 = a_2^2 \sim a_3^2 \qquad d_3^2 = a_2^2 \sim a_1^2,$$

in the third, $$d_2^2 = a_3^2 \sim a_1^2 \qquad d_1^2 = a_3^2 \sim a_2^2.$$

Or, if for a_1^2, a_2^2, a_3^2 we put $a^2+\lambda_1$, $a^2+\lambda_2$, $a^2+\lambda_3$, the above values may be read with λ in the place of a and the same suffixes.

PROOF.—Put $a_1^2-a_2^2 = \mu$; then

$$\frac{x^2}{a_1^2}+\frac{y^2}{b_1^2}+\frac{z^2}{c_1^2} = 1 \quad \text{and} \quad \frac{x^2}{a_1^2-\mu}+\frac{y^2}{b_1^2-\mu}+\frac{z^2}{c_1^2-\mu} = 1$$

are confocal quadrics. Take the difference of the two equations, and we obtain, at a common point $x'y'z'$, $\frac{x'^2}{a^2(a^2-\mu)}+\&c. = 0$. Comparing this with (5651), the quadratic for the axes of the section of the quadric by the plane $lx+my+nz = 0$, we see that, if l, m, n have the values $\frac{x'}{a^2}$, &c., μ is identical with r^2; the plane is the diametral plane of P; and the two values of μ are the squares of its axes. Let d_3^2, d_2^2 be these values; then, since there are but three confocals, the two values of μ must give the remaining confocals, *i.e.*, $a_1^2-d_3^2 = a_2^2$ and $a_1^2-d_2^2 = a_3^2$.

The six axes of the sections are situated as shown in the diagram (Fig. 179). Either axis of any of the three sections is equal to one of the axes in one of the other sections, but the equal axes are not those which coincide. O is supposed to be the centre of the conicoids, and the three lines are drawn from O parallel to the three tangents at P to the lines of intersection.

5662 Coordinates of the point of intersection of three confocal quadrics in terms of the semi-axes:

$$x^2 = \frac{a_1^2 a_2^2 a_3^2}{(a_1^2 - b_1^2)(a_1^2 - c_1^2)}, \qquad y^2 = \frac{b_1^2 b_2^2 b_3^2}{(a_1^2 - b_1^2)(b_1^2 - c_1^2)},$$

$$z^2 = \frac{c_1^2 c_2^2 c_3^2}{(a_1^2 - c_1^2)(b_1^2 - c_1^2)}.$$

The denominators may be in terms of any of the confocals since $a_1^2 - b_1^2 = a_2^2 - b_2^2 = a_3^2 - b_3^2$, &c.

PROOF.—The equation of a confocal may be written $\frac{x^2}{a_1^2} + \frac{y^2}{a_1^2 - h^2} + \frac{z^2}{a_1^2 - k^2} = 1$, producing a cubic in a_1^2, the product of whose roots a_1^2, a_2^2, a_3^2 gives x^2.

5663 The perpendiculars from the origin upon the tangent planes of the three confocal quadrics being p_1, p_2, p_3:

$$p_1^2 = \frac{a_1^2 b_1^2 c_1^2}{(a_1^2 - a_2^2)(a_1^2 - a_3^2)}, \qquad p_2^2 = \frac{a_2^2 b_2^2 c_2^2}{(a_1^2 - a_2^2)(a_2^2 - a_3^2)},$$

$$p_3^2 = \frac{a_3^2 b_3^2 c_3^2}{(a_1^2 - a_3^2)(a_2^2 - a_3^2)}.$$

PROOF.—By (5649), $p_1 d_2 d_3 = a_1 b_1 c_1$; then by the values in (5661).

RECIPROCAL AND ENVELOPING CONES.

5664 DEF.—A right line drawn through a fixed point always perpendicular to the tangent plane of a cone generates the reciprocal cone.

The enveloping cone of a quadric is the locus of all tangents to the surface which pass through a fixed point called the vertex.

5665 The equations of a cone and its reciprocal are respectively

$$Ax^2 + By^2 + Cz^2 = 0 \ldots\ldots \text{(i.)}, \quad \text{and} \quad \frac{x^2}{A} + \frac{y^2}{B} + \frac{z^2}{C} = 0 \ldots\ldots \text{(ii.)}.$$

PROOF.—The equations of the tangent plane of (i.) at any point xyz, and of the perpendicular to it from the origin, are

$$Ax\xi + By\eta + Cz\zeta = 0 \ldots\ldots\ldots \text{(iii.)}, \quad \text{and} \quad \frac{\xi}{Ax} = \frac{\eta}{By} = \frac{\zeta}{Cz} \ldots\ldots\ldots \text{(iv.)}.$$

Eliminate x, y, z between (i.), (iii.), and (iv.).

5667 The reciprocals of confocal cones are concyclic; that is, have the same circular section; and the reciprocals of concyclic cones are confocal.

PROOF.—A series of concyclic cones is given by

$$Ax^2+By^2+Cz^2+\lambda\,(x^2+y^2+z^2)=0$$

by varying λ; and the reciprocal cone is

$$\frac{x^2}{A+\lambda}+\frac{y^2}{B+\lambda}+\frac{z^2}{C+\lambda}=0. \qquad (5665)$$

5668 The reciprocals of the enveloping cones of the series of confocal quadrics $\frac{x^2}{a^2+\lambda}+\frac{y^2}{b^2+\lambda}+\frac{z^2}{c^2+\lambda}=1$, with fgh for the common vertex, P, of the cones, are given by the equation

$$\boldsymbol{a^2x^2+b^2y^2+c^2z^2-(fx+gy+hz)^2+\lambda\,(x^2+y^2+z^2)=0.}$$

PROOF.—Let lmn be the direction of the perpendicular p from the origin upon the tangent plane drawn from P to the quadric. Equate the ordinary value of p^2 at (5631) with that found by projecting OP upon p; thus

$$(a^2+\lambda)\,l^2+(b^2+\lambda)\,m^2+(c^2+\lambda)\,n^2=(fl+gm+hn)^2.$$

Now p generates with vertex O a cone similar and similarly situated to the reciprocal cone with vertex P, and l, m, n are proportional to x, y, z, the coordinates of any point on the former cone. Therefore, by transferring the origin to P, the equation of the reciprocal cone is as stated.

5669 COR.—These reciprocal cones are concyclic; and therefore the enveloping cones are confocal (5667).

5670 The reciprocal cones in (5668) are all coaxal.

PROOF.—Transform the cone given by the terms in (5668) without λ to its principal axes; and its equation becomes $Ax^2+By^2+Cz^2=0$. Now, if the whole equation, including terms in λ, be so transformed, $x^2+y^2+z^2$ will not be altered. Therefore we shall obtain

$$(A+\lambda)\,x^2+(B+\lambda)\,y^2+(C+\lambda)\,z^2=0,$$

a series of coaxal cones.

5671 The axes of the enveloping cone are the three normals to the three confocals passing through its vertex.

PROOF.—The enveloping cone becomes the tangent plane at P for a confocal through P, and one axis in this case is the normal through P. Also this axis is common to all the enveloping cones with the same vertex, by (5670). But there are three confocals through P (5658), and therefore three normals which must be the three axes of the enveloping cone.

5672 The equation of the enveloping cone of the quadric

$\frac{x^2}{a^2+\lambda}+\frac{y^2}{b^2+\lambda}+\frac{z^2}{c^2+\lambda}=1$ is, when transformed to its principal axes,

$$\frac{x^2}{\lambda-\lambda_1}+\frac{y^2}{\lambda-\lambda_2}+\frac{z^2}{\lambda-\lambda_3}=0 \quad \text{or} \quad \frac{x^2}{\lambda'}+\frac{y^2}{\lambda'+d_3^2}+\frac{z^2}{\lambda'+d_2^2}=0,$$

where λ_1, λ_2, λ_3 are the values of λ for the three confocals through P, the vertex, and d_2, d_3 are the semi-axes of the diametral section of P in the first confocal (5661).

Proof.—Transform equation (5668) of the reciprocal of the enveloping cone to its principal axes, as in (5670). Let λ_1, λ_2, λ_3 be the values of λ which make the quadric become in turn the three confocal quadrics through P. Then the reciprocal $(A+\lambda)\,x^2+(B+\lambda)\,y^2+(C+\lambda)\,z^2=0$ must become a right line in each case because the enveloping cone becomes a plane. Therefore one coefficient of x^2, y^2, or z^2 must vanish. Hence $A+\lambda_1=0$, $B+\lambda_2=0$, $C+\lambda_3=0$. Therefore the reciprocal cone becomes

$$(\lambda-\lambda_1)\,x^2+(\lambda-\lambda_2)\,y^2+(\lambda-\lambda_3)\,z^2=0,$$

and therefore the enveloping cone is

$$\frac{x^2}{\lambda-\lambda_1}+\frac{y^2}{\lambda-\lambda_2}+\frac{z^2}{\lambda-\lambda_3}=0.$$

THE GENERAL EQUATION OF A QUADRIC.

5673 This equation will be referred to as $f(x, y, z)=0$ or $U=0$, and, written in full, is

$$ax^2+by^2+cz^2+2fyz+2gzx+2hxy+2px+2qy+2rz+d=0.$$

By introducing a fourth quasi variable $t=1$, the equation may be put in the homogeneous form

5674
$$ax^2+by^2+cz^2+dw^2+2fyz+2gzx+2hxy \\ +2pxt+2qyt+2rzt=0,$$

abbreviated into

$$(a, b, c, d, f, g, h, p, q, r \between x, y, z, t)^2=0,$$

as in (1620).

Transforming to an origin $x'y'z'$ and coordinate axes parallel to the original ones, by substituting $x'+\xi$, $y'+\eta$, $z'+\zeta$ for x, y, and z, the equation becomes, by (1514),

5675 $$a\xi^2+b\eta^2+c\zeta^2+2f\eta\zeta+2g\zeta\xi+2h\xi\eta$$
$$+\xi U_x+\eta U_y+\zeta U_z+U=0,$$
where $U=f(x', y', z')$ (omitting the accents).

5676 The quadratic for r, the intercept between the point $x'y'z'$ and the quadric surface measured on a right line drawn from $x'y'z'$ in the direction lmn, is

$$r^2(al^2+bm^2+cn^2+2fmn+2gnl+2hlm)$$
$$+r(lU_x+mU_y+nU_z)+U=0.$$

Obtained by putting $\xi=rl$, $\eta=rm$, $\zeta=rn$ in (5674).

5677 The tangents from any external point to a quadric are proportional to the diameters parallel to them.

PROOF.—From (5676), as in (1215) and (4317).

5678 The equation of the tangent plane at a point xyz on the quadric is

$$(\xi-x)U_x+(\eta-y)U_y+(\zeta-z)U_z=0$$

5679 or $$\xi U_x+\eta U_y+\zeta U_z+\tau U_t=0,$$

with τ and t made equal to unity after differentiating.

PROOF.—From (5676). Since xyz is a point on the surface, one root of the quadratic vanishes. In order that the line may now *touch* the surface, the other root must also vanish; therefore $lU_x+mU_y+nU_z=0$. Put $rl=\xi-x$, $rm=\eta-y$, $rn=\zeta-z$; $\xi\eta\zeta$ being now a variable point on the line, and therefore on the tangent plane.

5680 Again, $xU_x+yU_y+zU_z+tU_t=2U$, by (1624),
therefore $xU_x+yU_y+zU_z=-tU_t$,
which establishes the second form (5679).

5681 Equation (5679) also represents the polar plane of any point xyz not lying on the quadric surface. Written in full it becomes

$$\begin{aligned}&\xi(ax+hy+gz+p)\\&+\eta(hx+by+fz+q)\\&+\zeta(gx+fy+cz+r)\\&+\ px+qy+rz+d=0,\end{aligned}\qquad\text{or}\qquad\begin{aligned}&x(a\xi+h\eta+g\zeta+p)\\&+y(h\xi+b\eta+f\zeta+q)\\&+z(g\xi+f\eta+c\zeta+r)\\&+\ p\xi+q\eta+r\zeta+d=0.\end{aligned}$$

5683 That is, the forms

$$\xi U_x + \eta U_y + \zeta U_z + U = 0 \quad \text{and} \quad xU_\xi + yU_\eta + zU_\zeta + U = 0$$

are convertible, U standing for $f(x, y, z)$ in the first, and for $f(\xi, \eta, \zeta)$ in the second.

5685 The intersection of the polar planes of two points is called the *polar line* of the points.

5686 The polar plane of the vertex is the plane of contact of the tangent cone.

PROOF.—If $\xi\eta\zeta$ be the vertex and xyz the point of contact, equation (5683) is satisfied. If x, y, z be the variables and $\xi\eta\zeta$ constant, the second form of that equation shows that the points of contact all lie on the polar plane of the point $\xi\eta\zeta$.

5687 Every line through the vertex is divided harmonically by the quadric and the polar plane.

PROOF.—In equation (5684) put $x = \xi + Rl$, $y = \eta + Rm$, $z = \zeta + Rn$ to determine R, the distance from the vertex to the polar plane. This gives $R = \dfrac{-2U}{lU_\xi + mU_\eta + nU_\zeta}$, employing (5680).

Now, if r, r' are the roots of the quadratic (5676), with ξ, η, ζ written for x, y, z, it appears that $\dfrac{2rr'}{r+r'} = R$, which proves the theorem.

5688 Every line (lmn) drawn through a point xyz parallel to the polar plane of that point is bisected at the point, and the condition of bisection is

$$lU_x + mU_y + nU_z = 0.$$

PROOF.—The equation is the condition for equal roots of opposite signs in the quadratic (5676). Since l, m, n are the dir. cos. of the line and U_x, U_y, U_z those of the normal of the polar plane (5683), the equation shows that the line and the normal are at right angles (5532).

5689 The last, when x, y, z are the variables, is also the equation of the diametral plane conjugate to the direction lmn. Expanded it becomes

$$(al+hm+gn)\,x + (hl+bm+fn)\,y + (gl+fm+cn)\,z$$
$$+pl+qm+rn = 0.$$

For the point xyz moves, when x, y, z are variable, so that every diameter drawn through it parallel to lmn is bisected by it, and the locus is, by the form of the equation, a plane.

If the origin be at the centre of the quadric, p, q, and r of course vanish.

5690 The coordinates of the centre of the general quadric $U = 0$ (5673) are

$$x = \frac{\Delta'_p}{2\Delta}, \quad y = \frac{\Delta'_q}{2\Delta}, \quad z = \frac{\Delta'_r}{2\Delta}.$$

PROOF.—Every line through xyz, the centre, is bisected by it. The condition for this, in (5688), is $U_x = 0$, $U_y = 0$, and $U_z = 0$, in order to be independent of lmn. The three equations in full are

$$\left.\begin{array}{l} ax+hy+gz+p=0 \\ hx+by+fz+q=0 \\ gx+fy+cz+r=0 \end{array}\right\}; \text{ and } \Delta' = \begin{vmatrix} a & h & g & p \\ h & b & f & q \\ g & f & c & r \\ p & q & r & d \end{vmatrix}; \quad \Delta = \begin{vmatrix} a & h & g \\ h & b & f \\ g & f & c \end{vmatrix}.$$

Solve by (582).

5691 The quadric transformed to the centre becomes

$$a\xi^2 + b\eta^2 + c\zeta^2 + 2f\eta\zeta + 2g\zeta\xi + 2h\xi\eta + \frac{\Delta'}{\Delta} = 0.$$

PROOF.—By the last theorem, the terms involving ξ, η, ζ in (5675) vanish. The value of U or $f(x, y, z)$, when xyz is the centre, appears as follows:—

$$U = \tfrac{1}{2}U_t \text{ (5680)} = px+qy+rz+d = \frac{p\Delta'_p + q\Delta'_q + r\Delta'_r}{2\Delta} + d \text{ (5690)} = \frac{\Delta'}{\Delta} \text{ (1647)}.$$

The last equation, being again transformed by turning the axes so as to remove the terms involving products of coordinates, becomes

5692 $$\alpha x^2 + \beta y^2 + \gamma z^2 + \frac{\Delta'}{\Delta} = 0,$$

5693 where α, β, γ are the roots of the discriminating cubic

$$R^3 - R^2(a+b+c) + R(bc+ca+ab-f^2-g^2-h^2) - \Delta = 0,$$

or $$(R-a)(R-b)(R-c) - (R-a)f^2 - (R-b)g^2 - (R-c)h - 2fgh = 0.$$

PROOF.—It has been shown, in (1847–9), that the roots of the discriminating cubic $\left(\text{multiplied in this case by } -\frac{\Delta}{\Delta'}\right)$ are the reciprocals of the maximum and minimum values of $x^2+y^2+z^2$. But such values are evidently the squares of the axes of the quadric surface. Let the central equation of the surface be $\frac{x^2}{a^2}+\frac{y^2}{b^2}+\frac{z^2}{c^2}=1$. Therefore $\frac{1}{a^2}=-\frac{\Delta a}{\Delta'}$, &c., producing the equation above.

5694 The equations of the new axis of x referred to the old axes of ξ, η, ζ are

$$(F+\alpha f)\,x=(G+\alpha g)\,y=(H+\alpha h)\,z;$$

and similar equations with β and γ for the y and z axes.

PROOF.—When lmn, in (5689), is a principal diameter of the quadric, the diametral plane becomes perpendicular to it, and therefore the coefficients of x, y, z must be proportional to l, m, n. Putting them equal to Rl, Rm, Rn respectively, we have the equations

$$\left.\begin{aligned}(a-R)\,l+hm+gn&=0\;\ldots\ldots\;(1)\\ hl+(b-R)\,m+fn&=0\;\ldots\ldots\;(2)\\ gl+fm+(c-R)\,n&=0\;\ldots\ldots\;(3)\end{aligned}\right\}.$$

The eliminant of these equations is the discriminating cubic in R already obtained in (5693).

From (1) and (2), $l:m=hf-g\,(b-R):gh-f\,(a-R)$,

and from (2) and (3), $m:n=fg-h\,(c-R):hf-g\,(b-R)$;

therefore $(gh-af+Rf)\,l=(hf-bg+Rg)\,m=(fg-ch+Rh)\,n$,

which establish the equations, since $x:y:z=l:m:n$ and $F=gh-af$, &c., as in (4665).

5695 The direction cosines of the axes of the quadric.

If the discriminating cubic be denoted by $\phi\,(R)=0$, and its roots by α, β, γ; the direction cosines of the first axis are

$$\sqrt{-\frac{\phi_a(\alpha)}{\phi_\alpha(\alpha)}},\qquad \sqrt{-\frac{\phi_b(\alpha)}{\phi_\alpha(\alpha)}},\qquad \sqrt{-\frac{\phi_c(\alpha)}{\phi_\alpha(\alpha)}}.$$

For the second and third axes write β and γ in the place of α.

PROOF.—Let $F+\alpha f=L$, $G+\alpha g=M$, $H+\alpha h=N$ .... (i.),
$(\alpha-b)(\alpha-c)-f^2=\lambda$, $(\alpha-c)(\alpha-a)-g^2=\mu$, $(\alpha-a)(\alpha-b)-h^2=\nu$...(ii.).
Then the equation $\phi\,(\alpha)=0$ may be put in either of the forms

$$L^2=\mu\nu,\quad M^2=\nu\lambda,\quad N^2=\lambda\mu\;\ldots\ldots\ldots\ldots\ldots\;\text{(iii.)}.$$

Now the dir. cos. of the first axis are, by (5694), proportional to

$$\frac{1}{L}:\frac{1}{M}:\frac{1}{N}=\surd\lambda:\surd\mu:\surd\nu,\ \text{by (iii.)}.$$

Their values are, therefore,

$$\frac{\surd\lambda}{\surd(\lambda+\mu+\nu)},\quad \frac{\surd\mu}{\surd(\lambda+\mu+\nu)},\quad \frac{\surd\nu}{\surd(\lambda+\mu+\nu)}.$$

But $\lambda=-\frac{d\phi\,(a)}{da}$ and $\lambda+\mu+\nu=\frac{d\phi\,(\alpha)}{d\alpha}$, by actual differentiation of the cubic in (5693).

5696 Cauchy's proof that the roots of the discriminating cubic (5693) are all real will be found at (1850).

5697 The equation of the enveloping cone, vertex xyz, of the general quadric surface $U=0$ (5673) is

$$4\,(abcfgh\between lmn)^2 U = (lU_x+mU_y+nU_z)^2,$$

with $\xi-x$, $\eta-y$, $\zeta-z$ substituted for l, m, n.

PROOF.—The generating line through xyz moves so as to touch the quadric. Hence the quadratic in r (5676) must have equal roots. The equation admits of some reduction.

5698 When U takes the form $ax^2+by^2+cz^2=1$, equation (5697) becomes

$$(al^2+bm^2+cn^2)(ax^2+by^2+cz^2-1) = (alx+bmy+cnz)^2.$$

5699 The condition that the general quadric equation may represent a cone is $\Delta'=0$; that is, the discriminant of the quaternary quadric, (5674) or (1644), must vanish.

PROOF.—By (5692). Otherwise $\Delta'=0$ is the eliminant of the four equations $U_x=0$, $U_y=0$, $U_z=0$, $U=0$, the condition that equation (5675) may represent a cone.

5700 The condition that the plane $lx+my+nz=0$ may touch the cone $(abcfgh\between xyz)^2=0$ is the determinant equation on the right.

$$\begin{vmatrix} a & h & g & l \\ h & b & f & m \\ g & f & c & n \\ l & m & n & \end{vmatrix} = 0.$$

PROOF.—Equate the coefficients l, m, n to those of the tangent plane (5681), p, q, r being zero, and xyz the point of contact. A fourth equation is $lx+my+nz=0$, which holds at the point of contact. The eliminant of the four equations is the determinant above.

5701 The condition that the plane $lx+my+nz+t=0$ may touch the quadric

$$(abcdfghpqr\between xyz1)^2=0$$

(5673) is the determinant equation on the right.

$$\begin{vmatrix} a & h & g & p & l \\ h & b & f & q & m \\ g & f & c & r & n \\ p & q & r & d & t \\ l & m & n & t & \end{vmatrix} = 0.$$

PROOF.—As in (5700).

5702 If the origin is at the centre, $p = q = r = 0$. In that case, transposing the last two rows and last two columns, the determinant becomes

$$\begin{vmatrix} a & h & g & l & 0 \\ h & b & f & m & 0 \\ g & f & c & n & 0 \\ l & m & n & 0 & t \\ 0 & 0 & 0 & t & d \end{vmatrix} = 0, \quad \text{or,} \quad d \begin{vmatrix} a & h & g & l \\ h & b & f & m \\ g & f & c & n \\ l & m & n & \end{vmatrix} = t^2 \begin{vmatrix} a & h & g \\ h & b & f \\ g & f & c \end{vmatrix}.$$

5703 The condition that the line of intersection of the planes

$$lx+my+nz+t = 0 \ldots \text{(i.)} \quad \text{and} \quad l'x+m'y+n'z+t' = 0 \ \ldots \ \text{(ii.)}$$

may touch the general quadric $(abcdfghpqr \between xyz1)^2 = 0$, is the determinant equation deduced below.

Multiply equation (i.) by ξ and (ii.) by η to obtain the plane

$$(l\xi+l'\eta)\,x+(m\xi+m'\eta)\,y+(n\xi+n'\eta)\,z+t\xi+t'\eta = 0 \ \ldots\ldots\ldots \text{(iii.)},$$

passing through the intersection of (i.) and (ii.). The line of intersection will touch the quadric if (iii.) coincides with the tangent plane at a point xyz, and if xyz be also on (i.) and (ii.). Therefore, equating coefficients of (iii.) and the tangent plane at xyz (5681), we get the six following equations, the eliminant of which furnishes the required condition,

$$\left.\begin{aligned} ax+\ hy+\ gz+pw &= \ l\xi+\ l'\eta \\ hx+\ by+\ fz+qw &= m\xi+m'\eta \\ gx+\ fy+\ cz+rw &= \ n\xi+\ n'\eta \\ px+\ qy+\ rz+dw &= \ t\xi+\ t'\eta \\ lx+my+nz+tw &= 0 \\ l'x+m'y+n'z+t'w &= 0 \end{aligned}\right\}, \quad \therefore \begin{vmatrix} a & h & g & p & l & l' \\ h & b & f & q & m & m' \\ g & f & c & r & n & n' \\ p & q & r & d & t & t' \\ l & m & n & t & & \\ l' & m' & n' & t' & & \end{vmatrix} = 0.$$

RECIPROCAL POLARS.

5704 The method of reciprocal polars explained at page 665 is equally applicable to geometry of three dimensions.

Taking poles and polar planes with respect to a sphere of reciprocation, we have the following rules analogous to those on page 666.

RULES FOR RECIPROCATING.

5705 *A plane becomes a point.*

5706 *A plane at infinity becomes the origin.*

5707 *Several points on a straight line become as many planes passing through another straight line. These lines are called reciprocal lines.*

5708 *Points lying on a plane become planes passing through a point, the pole of the plane.*

5709 *Points lying on a surface become planes enveloping the reciprocal surface.*

5710 *Therefore, by rules* (5708) *and* (5709), *the points in the intersection of the plane and a surface become planes passing through the pole of the plane and enveloped both by the reciprocal surface and by its tangent cone.*

5711 *When the intersecting plane is at infinity, the vertex of the tangent cone is the origin.*

5712 *Therefore the asymptotic cone of any surface is orthogonal to the tangent cone drawn from the origin to the reciprocal surface. The cones are therefore reciprocal.*

5713 *The reciprocal surface of the quadric is a hyperboloid, an ellipsoid, or a paraboloid, according as the origin is without, within, or upon the quadric surface.*

5714 *The angle subtended at the origin by two points is equal to the angle between their corresponding planes.*

5715 *The reciprocal of a sphere is a surface of revolution of the second order.*

5716 *The shortest distance between two reciprocal lines passes through the origin.*

5717 The reciprocal surface of the general quadric $(abcdfghpqr \between xyz1)^2 = 0$ (5674), the auxiliary sphere being $x^2+y^2+z^2 = k^2$, is

$$\begin{vmatrix} a & h & g & p & \xi \\ h & b & f & q & \eta \\ g & f & c & r & \zeta \\ p & q & r & d & -k^2 \\ \xi & \eta & \zeta & -k^2 & \end{vmatrix} = 0, \qquad \text{or, if } p=q=r=0, \quad d\begin{vmatrix} a & h & g & \xi \\ h & b & f & \eta \\ g & f & c & \zeta \\ \xi & \eta & \zeta & 0 \end{vmatrix} - k^4 \begin{vmatrix} a & h & g \\ h & b & f \\ g & f & c \end{vmatrix}.$$

PROOF.—The polar plane of the point $\xi\eta\zeta$ with respect to the sphere is $\xi x+\eta y+\zeta z-k^2=0$. This must touch the given surface, and the condition is given in (5701).

5718 The reciprocal surface of the central quadric $\frac{x^2}{a^2}+\frac{y^2}{b^2}+\frac{z^2}{c^2}=1$, when the origin of reciprocation is the point $x'y'z'$, is

$$(\xi x'+\eta y'+\zeta z'-k^2)^2 = a^2\xi^2+b^2\eta^2+c^2\zeta^2,$$

or, with the origin at the centre,

5719 $$a^2\xi^2+b^2\eta^2+c^2\zeta^2 = k^4.$$

PROOF.—Let p be the perpendicular from $x'y'z'$ upon a tangent plane of the quadric, and $\xi\eta\zeta$ the point where p produced, intersects the reciprocal surface at a distance ρ from $x'y'z'$. Then

$$k^2\rho^{-1} = p = lx'+my'+nz'-\sqrt{(a^2l^2+b^2m^2+c^2n^2)}. \qquad (5630)$$

Multiplying by ρ produces the desired equation.

THEORY OF TORTUOUS CURVES.

5721 DEFINITIONS.—The *osculating plane* at any point of a curve of double curvature, or *tortuous curve*,* is the plane containing either two consecutive tangents or three consecutive points.

5722 The *principal normal* is the normal in the osculating plane. The radius of *circular curvature* coincides with this normal in direction.

5723 The *binormal* is the normal perpendicular both to the tangent and principal normal at the point.

5724 The *osculating circle* is the circle of curvature in the osculating plane, and its centre, which is the centre of circular curvature, is the point in which the osculating plane intersects two consecutive normal planes of the curve.

5725 The *angle of contingence*, $d\psi$, is the angle between two consecutive tangents or principal normals. The *angle of torsion*, $d\tau$, is the angle between two consecutive osculating planes.

* Otherwise named "space curve."

5726 The *rectifying plane* at any point on the curve is perpendicular to the principal normal; and the intersection of two consecutive rectifying planes is the *rectifying line* and axis of the osculating cone.

5727 The *osculating cone* is a circular cone touching three consecutive osculating planes and having its vertex at their point of intersection.

The *rectifying developable* is the envelope of the rectifying planes, and is so named because the curve, being a geodesic on this surface, would become a straight line if the surface were developed into a plane.

5728 The *polar developable* is the envelope of the normal planes, being the locus of the line of intersection of two consecutive normal planes. Three consecutive normal planes intersect in a point which is the centre of *spherical curvature:* for a sphere having that centre may be described passing through four consecutive points of the curve.

5729 The *edge of regression* is the locus of the centre of spherical curvature.

5730 The *rectifying surface* is the surface of centres (5773) of the polar developable.

5731 An *evolute* of a curve is a geodesic line on the polar developable. It is the line in which a free string would lie if stretched between two points, one on the curve and one anywhere on the smooth surface of the polar developable.

5732 In Figure (180) A, A', A'', A''' are consecutive points on a curve. The normal planes drawn through A and A' intersect in CE; those through A' and A'' in $C'E'$, and those through A'' and A''' in $C''E''$. CE meets $C'E'$ in E, and $C'E'$ meets $C''E''$ in E'. The principal normals in the normal planes are AC, $A'C'$, $A''C''$, and these are also the radii of curvature at A, A', A'', while C, C', C'' are the centres of curvature. $\angle ACA' = d\psi$ and $CA'C' = d\tau$.

The surface $ECC'C''E'$ is the polar developable, $CC'C''$ being the locus of the centres of curvature, and $EE'E''$ is the edge of regression.

EA is the radius and E the centre of spherical curvature for the point A. hH, HH', $H'H''$ are elemental chords of an evolute of the curve, AhH being a normal at A, and $A'HH'$ a normal at A', and so on. The first normal drawn is arbitrary, but it determines the position of all the rest.

PROPERTIES OF A TORTUOUS CURVE.

5733 The equation of the osculating plane at a point xyz on the curve is

$$(\xi - x)\lambda + (\eta - y)\mu + (\zeta - z)\nu = 0.$$

5734 λ, μ, ν are the direction cosines of the binormal, and their complete values are

$$\rho(y_s z_{2s} - y_{2s} z_s), \quad \rho(z_s x_{2s} - z_{2s} x_s), \quad \rho(x_s y_{2s} - x_{2s} y_s).$$

5735 The angle of contingence

$$d\psi = \sqrt{\{(y_s z_{2s} - y_{2s} z_s)^2 + (z_s x_{2s} - z_{2s} x_s)^2 + (x_s y_{2s} - x_{2s} y_s)^2\}}\, ds.$$

PROOF.—Let the direction of a tangent be lmn, and that of a consecutive tangent $l+dl$, $m+dm$, $n+dn$. Since the normal of the plane must be perpendicular to both these lines, we shall have, by (5532),

$$l\lambda + m\mu + n\nu = 0 \quad \text{and} \quad (l+dl)\lambda + (m+dm)\mu + (n+dn)\nu = 0,$$

therefore $\lambda : \mu : \nu = mdn - ndm : ndl - ldn : ldm - mdl,$

and the denominator in the complete values of λ, μ, ν is

$$\sqrt{\{(mdn - ndm)^2 + \&c.\}} = \sin d\psi,$$

by (5521); that is, $= d\psi$. Also $l, m, n = x_s, y_s, z_s$ and $dl = x_{2s} ds$, &c. Therefore $\lambda = (y_s z_{2s} - y_{2s} z_s)\dfrac{ds}{d\psi}$. Similarly, μ and ν; and $s_\psi = \rho$, by (5146).

5736 The radius of curvature ρ at a point xyz.

$$\frac{1}{\rho^2} = x_{2s}^2 + y_{2s}^2 + z_{2s}^2 = \frac{x_{2t}^2 + y_{2t}^2 + z_{2t}^2 - s_{2t}^2}{s_t^4}.$$

PROOF: $\quad d\psi = \sqrt{\{(y_s z_{2s} - y_{2s} z_s)^2 + \&c.\}}\, ds$, in (5735),

therefore $\psi_s = \sqrt{\{(x_s^2 + y_s^2 + z_s^2)(x_{2s}^2 + y_{2s}^2 + z_{2s}^2) - (x_s x_{2s} + y_s y_{2s} + z_s z_{2s})^2\}}$

$= \sqrt{(x_{2s}^2 + y_{2s}^2 + z_{2s}^2)}$; since $x_s^2 + y_s^2 + z_s^2 = 1$;

and differentiating this equation makes $x_s x_{2s} + \&c. = 0$.

Otherwise, geometrically, precisely as in the proof of (5141), we find the direction cosines of the principal normal to be

5737 $\quad \cos\alpha = \rho x_{2s}, \quad \cos\beta = \rho y_{2s}, \quad \cos\gamma = \rho z_{2s}.$

Therefore $\quad \rho^2(x_{2s}^2 + y_{2s}^2 + z_{2s}^2) = \cos^2\alpha + \cos^2\beta + \cos^2\gamma = 1.$

The change to the independent variable t is made by (1762).

5738 The angle of torsion, in terms of λ, μ, ν of (5734), is

$$d\tau = \sqrt{(\lambda_s^2 + \mu_s^2 + \nu_s^2)}\, ds = (\lambda x_{3s} + \mu y_{3s} + \nu z_{3s})\, \rho ds$$

$$= \sqrt{\{(\mu\nu_s - \mu_s\nu)^2 + (\nu\lambda_s - \nu_s\lambda)^2 + (\lambda\mu_s - \lambda_s\mu)^2\}}.$$

PROOF.—By (5745), we have $(d\tau)^2 = (d\lambda)^2+(d\mu)^2+(d\nu)^2$ .... (i.), which gives the first form. The third reduces to this by the method in (5736). For the second form put $u = y_s z_{2s} - y_{2s} z_s$, &c., then

$$\frac{\lambda}{u} = \frac{\mu}{v} = \frac{\nu}{w} = \frac{1}{K} = \frac{ds}{d\psi} \text{ (5734)}, \quad d\lambda = \frac{du}{K} - \frac{udK}{K^2}, \text{ \&c.}$$

Substitute in (i.), reducing by $K^2 = u^2+v^2+w^2$ and $KdK = udu+vdv+wdw$.

CURVATURE AND TORTUOSITY.

5739 Radius of curv., $\rho = \dfrac{ds}{d\psi}$; Curvature $= \dfrac{1}{\rho} = \dfrac{d\psi}{ds}$.

Radius of torsion, $\sigma = \dfrac{ds}{d\tau}$; Tortuosity $= \dfrac{1}{\sigma} = \dfrac{d\tau}{ds}$.

If τ_s changes sign while passing through the values zero or infinity, there is a point of *inflected torsion* or a *cuspidal point*, respectively. If τ_s, without changing sign, passes through zero or infinity, there is a point of *suspended torsion* or *infinite torsion* respectively.

If τ_s is zero, identically, the curve is plane.

5740 The radius of spherical curvature,

$$R = \sqrt{(\rho^2+\rho_\tau)}.$$

PROOF.—In Fig. (180) $R^2 = \rho^2+EC^2$ and $EC = \rho_\tau$ by analogy with $q = p_\psi$ n a plane curve (see proof of 5147).

5741 The element of arc of the locus of centres of circular curvature is

$$ds' = Rd\tau, \text{ and therefore } R = s'_\tau.$$

PROOF.—In Fig. (180) $ds' = CC' = \rho d\tau \sec\phi = Rd\tau$.

5742 The radius of curvature of the edge of regression

$$= S''_\tau = RR_\rho = \rho+\rho_{2\tau},$$

S'' being the arc of the edge of regression.

PROOF.—An inspection of Figure (180) shows that R and ρ stand in the same relation to the edge of regression that r and p occupy with regard to a curve in the standard formula. In fact we may substitute R for r, ρ for p,

ϕ for θ, τ for ψ, and ϕ remains ϕ. The chosen line of reference AB being always parallel to the tangent EC, then $AEC = BAE = \phi$. Also the angle of contingence $CEC' = CAC' = d\tau$, by the right angles at C and C'. Accordingly, we have the formula $\rho = s_\psi = rr_p = p + p_{2\psi}$ from (5146–8), and the values above corresponding to them.

5743 *A method of estimating the variation in direction of a right line whose position is given as depending upon the form of a tortuous curve at every point.*

Let x, y, z be the direction cosines of the line referred to a fixed principal normal, tangent, and binormal of the curve [x, y, z may either be constants with respect to the varying principal normal, tangent, and binormal, or they may be functions of the angle between the binormal and the spherical radius].

5744 The complete changes in x, y, z, with respect to the fixed origin and axes, will be

$$\delta x = dx + y\,d\psi - z\,d\tau,$$
$$\delta y = dy - x\,d\psi,$$
$$\delta z = dz + x\,d\tau.$$

PROOF.—In Figure (180) AC, AB are the fixed axes of x and z. Let a line AL of unit length be drawn always parallel to the line in question; then, if x, y, z be the coordinates of L, x, y, z will also be the direction cosines of AL, and therefore of the given line.

Now, suppose A to move to A', and consequently AL to take the position $A'L'$. Then the changes in x, y, z will be the changes dx, dy, dz relatively to the moving axes, plus the changes due to the rotations $d\psi$ round the binormal and $d\tau$ round the tangent. With the usual notation, we shall have

$$\delta x = dx + \omega_2 z - \omega_3 y, \quad \delta y = dy + \omega_3 x - \omega_1 z, \quad \delta z = dz + \omega_1 y - \omega_2 x,$$

with $\omega_1 = 0$, $\omega_2 = -d\tau$, $\omega_3 = -d\psi$.

5745 If $d\chi$ be the angular change in the direction of the right line,

$$d\chi = \sqrt{\{(\delta x)^2 + (\delta y)^2 + (\delta z)^2\}}.$$

For $d\chi = LL'$ since AL is a unit length.

EXAMPLES.

5746 The angle between two consecutive radii of circular curvature being $d\epsilon$,

$$(d\epsilon)^2 = (d\psi)^2 + (d\tau)^2.$$

PROOF.—Here, in (5744), $x=1$, $y=0$, $z=0$, therefore $\delta x=0$, $\delta y=-d\psi$, $\delta z=d\tau$. Substitute these values in (5745).

5747 The angle, $d\eta$, between two consecutive radii of spherical curvature, ϕ being the inclination to the binormal,

$$(d\eta)^2=(d\psi.\sin\phi)^2+(d\phi-d\tau)^2.$$

PROOF.—In (5744) the direction cosines of R (Fig. 180) are $x=\sin\phi$, $y=0$, $z=\cos\phi$,
therefore $\delta x=\cos\phi\,(d\phi-d\tau)$, $\delta y=-d\psi\sin\phi$, $\delta z=-\sin\phi\,(d\phi-d\tau)$.
Substitute in (5745).

5748 The angle of contingence of the locus of the centres of circular curvature,

$$(d\chi)^2=(d\psi.\cos\phi)^2+(d\phi+d\tau)^2.$$

PROOF.—The dir. cos. of the tangent at C to the locus (Fig. 177) are

$$x=\cos\phi,\quad y=0,\quad z=\sin\phi\,;$$

therefore $\delta x=-\sin\phi\,(d\phi+d\tau)$, $\delta y=-d\psi\cos\phi$, $\delta z=\cos\phi\,(d\phi+d\tau)$.
Substitute in (5745).

5749 The osculating plane of the same curve has its direction cosines in the ratios

$$\frac{d\psi}{d\chi}\sin\phi\cos\phi:-\left(\frac{d\phi}{d\chi}+\frac{d\tau}{d\chi}\right):-\frac{d\psi}{d\chi}\cos^2\phi.$$

PROOF.—As in the Proof of (5735), the dir. cos. of the normal to this plane are proportional to $y\delta z-z\delta y$, $z\delta x-x\delta z$, $x\delta y-y\delta x$. Substitute the values in last proof.

5750 The angle of torsion of the same curve is found from (5745) and (5744) as above, x, y, z being in this case the dir. cos. of the normal of the osculating plane as given in (5749).

5751 The direction cosines of the rectifying line are

$$0,\quad\frac{d\tau}{d\epsilon},\quad\frac{d\psi}{d\epsilon}.$$

PROOF.—The rectifying plane at A' (Fig. 180) is perpendicular to the normal $A'C'$. Therefore its equation is $x-yd\psi+zd\tau=0$. The ultimate intersection of this plane with the rectifying plane at A (that is, the plane of yz) is the rectifying line. Hence the equation of the latter is $yd\psi=zd\tau$; and the dir. cosines reduce to the above by (5746).

5752 Cor.—The vertical angle of the osculating cone

$$= 2\tan^{-1}\frac{d\psi}{d\tau}.$$

5753 The angle of torsion of the involute of the curve is

$$= \sqrt{(\psi_{2\epsilon}^2+\tau_{2\epsilon}^2)}\, d\epsilon.$$

Proof.—This angle is also the angle between two consecutive rectifying lines. Therefore, taking the dir. cosines from (5751), we must put in (5744)

$$x = 0,\quad y = \frac{d\tau}{d\epsilon},\quad z = \frac{d\psi}{d\epsilon};$$

therefore $\delta x = \frac{d\tau}{d\epsilon}\,d\psi - \frac{d\psi}{d\epsilon}\,d\tau = 0;\quad \delta y = \tau_{2\epsilon}\,d\epsilon;\quad \delta z = \psi_{2\epsilon}\,d\epsilon.$

5754 The angle of torsion of an evolute of the curve

$$= d\psi \sin(\alpha-\tau).$$

Proof.—(Fig. 180.) Let $HH'H''$ be an evolute of the curve, AH the tangent to it in the normal plane of the original curve at A, and let $\alpha = CAH$, the inclination of AH to the principal normal. At any other point H'' of the evolute, where its tangent is $A''H'H''$, let the corresponding angle be $\theta = C''A''H''$. Then $\theta = \alpha-\tau$, τ being the sum of the angles of torsion between A and A'', or the total amount of twist of the osculating plane. Now the normal of the osculating plane of the evolute at H'' is perpendicular to HH' and $H'H''$, two consecutive tangents. Therefore its dir. cosines in (5744) must be

$$x = -\sin(\alpha-\tau),\quad y = 0,\quad z = \cos(\alpha-\tau);$$

therefore $\delta x = \cos(\alpha-\tau)\,d\tau + 0 - \cos(\alpha-\tau)\,d\tau = 0,$

$$\delta y = \sin(\alpha-\tau)\,d\psi;\quad \delta z = \sin(\alpha-\tau)\,d\tau - \sin(\alpha-\tau)\,d\tau = 0.$$

Hence the angle required $= \delta y = d\psi\sin(\alpha-\tau)$.

5755 Approximate values of the coordinates of a point on a tortuous curve near to the origin in terms of the arc, the axes of x, y, z being the principal normal, tangent, and binormal, and the arc s being measured from the origin:

$$x = \frac{s^2}{2\rho} - \frac{s^3\rho_s}{6\rho^2} - \frac{s^4}{24\rho^3}(1-2\rho_s^2+\rho\rho_{2s}) + \&c.,$$

$$y = s - \frac{s^3}{6\rho^2} + \frac{s^4\rho_s}{8\rho^3} + \&c.,\qquad z = \frac{s^3}{6\rho\sigma} - \frac{s^4}{24}\left(\frac{\sigma_s}{\rho\sigma^2} + \frac{2\rho_s}{\rho^2\sigma}\right) + \&c.,$$

ρ and σ being respectively the radii of circular curvature and torsion.

PROOF.—By Taylor's theorem (1500), since x, y, z, s are the same as dx, dy, dz, ds initially, we have $x = x_s s + \frac{1}{2}x_{2s}s^2 + \frac{1}{6}x_{3s}s^3 + \&c.$, and similar expansions for y and z. The dir. cosines of the principal normal at the point xyz will be, from (5737),

$$\cos(-\psi) = \rho x_{2s}, \quad \cos\left(\frac{\pi}{2}+\psi\right) = \rho y_{2s}, \quad \cos\left(\frac{\pi}{2}-\tau\right) = \rho z_{2s};$$

$\psi \equiv d\psi$ and $\tau \equiv d\tau$ being estimated positive as drawn in Figure (180) for positive values of x, y, z.

Differentiate these equations for s, and in the results put the initial values

$$x_s = z_s = 0, \quad y_s = 1, \quad \psi = \tau = 0, \quad \psi_s = \frac{1}{\rho}, \quad \tau_s = \frac{1}{\sigma}, \quad \&c.,$$

to determine the derivatives in the above expansions.

THE HELIX.

5756 The *helix* is a curve traced on a cylinder of radius a, so that its tangent preserves a constant inclination, $= \frac{1}{2}\pi - \alpha$, to the axis. Taking the axis of the cylinder for the z axis of coordinates, the equations of the helix are

$$x = a\cos\theta, \qquad y = a\sin\theta, \qquad z = a\theta\tan\alpha.$$

5757 The radius of curvature $\rho = a\sec^2\alpha$.

5758 The radius of torsion $\sigma = 2a\operatorname{cosec} 2\alpha$.

PROOF.—ρ from (5806); since $\rho_1 = a$, $\rho_2 = \infty$, and $\theta = \alpha$ at every point. By (5739), $\sigma = s_\tau$. But $dz = ds\sin\alpha$ and $ad\tau = dz\cos\alpha$.

5759 The helix of closest contact with a given curve may be found as follows.

Determine the constants a and α from equations (5757-8), with the known values of ρ and σ for the given curve; then place the helix to have a common tangent with the curve at the point, and make the osculating planes coincide.

GENERAL THEORY OF SURFACES.

5770 DEFINITIONS.—A *tangent plane* passes through three consecutive points on a surface which are not in the same right line.

5771 The *normal* at any point of a surface is perpendicular to the tangent plane.

5772 A *normal plane* is any plane through the normal.

5773 A *line of curvature* on a surface is a line along which consecutive normals to the surface intersect. At every point of a surface there are usually two lines of curvature at right angles to each other; and to these correspond two principal radii of curvature. The two lines of curvature coincide with the principal axes of the indicatrix at the point. See (5778).

5774 The *surface of centres* is the locus of the centres of principal curvature. There are two such surfaces, for there are two centres on each normal, and the normal is a tangent to both surfaces. Either surface may be regarded as generated by the evolutes of the lines of principal curvature.

5775 A *geodesic* is a line traced on a surface along which the osculating plane at every point contains the normal to the surface. See (5779).

5776 The *radius of geodesic curvature* * of a curve traced on a surface is measured by the ratio of the element of arc of the curve to the angle between consecutive normal sections of the surface drawn through consecutive tangents of the curve. *Geodesic curvature*, being the reciprocal of this, is therefore the rate of angular deviation of the normal section per unit length of the curve.

5777 An *umbilicus* is a point on a surface where a section parallel to and close to the tangent plane is a circle; in other words, the indicatrix is a circle.

For a definition of *Indicatrix*, see (5795).

5778 In Figure (182) OCD is the normal at O to a curved surface; AOA', BOB' are the lines of curvature, therefore the normals to the surface at A and O intersect in the centre of curvature radius ρ_1 (5773), and the normals at B and O, in the centre, radius ρ_2. The normals to the line of curvature BOB' at B and O, *drawn in the osculating plane* BOB', intersect in K, and those at B' and O intersect in H. HOD is the angle between the osculating plane of the line of curvature and the plane of normal section. Similarly for the line of curvature AOA'.

5779 If POP' be a geodesic, its osculating plane POP' contains OD the normal to the surface at O, and therefore $\rho = OD$, the radius of curvature of this section at O; but PE, the normal to the surface at P, does not intersect OD, the consecutive normal at O, unless the geodesic coincides with one of the lines of curvature, OA or OB. The angle DPE is the angle of torsion which vanishes in the latter case.

---

* Not to be confounded with the radius of curvature of a geodesic.

GENERAL EQUATION OF A SURFACE.

5780 Let the general equation of a surface be represented by $\phi(x, y, z) = 0$.

5781 The equations of any tangent at a point xyz are

$$\frac{\xi-x}{l} = \frac{\eta-y}{m} = \frac{\zeta-z}{n}, \quad \text{with} \quad l\phi_x+m\phi_y+n\phi_z = 0.$$

PROOF.—At an adjacent point $x+rl$, $y+rm$, $z+rn$, we have

$$\phi(x+rl,\ y+rm,\ z+rn) = 0,$$

therefore, by (1514), $\phi(x, y, z)+r(l\phi_x+m\phi_y+n\phi_z) = 0$, the rest vanishing in the limit. But $\phi(x, y, z) = 0$, therefore

$$l\phi_x+m\phi_y+n\phi_z = 0.$$

But l, m, n are the direction cosines of the line joining the two points, which becomes a tangent in the limit; and if $\xi\eta\zeta$ be any point on this line distant ρ from xyz, $\xi-x = \rho l$, $\eta-y = \rho m$, $\zeta-z = \rho n$, &c.

5782 The equation of the tangent plane at xyz is

$$(\xi-x)\,\phi_x+(\eta-y)\,\phi_y+(\zeta-z)\,\phi_z = 0.$$

PROOF.—Eliminate l, m, n from $l\phi_x+m\phi_y+n\phi_z = 0$ by $\xi-x = \rho l$, &c., as above.

TANGENT LINE AND CONE AT A SINGULAR POINT.

5783 If, in the expansion in (5781) by Taylor's theorem, all the derivatives of $\phi(x, y, z)$ of an order up to n inclusive vanish, we have

$$\phi(x+rl, y+rm, z+rn) = \phi(x, y, z) + \frac{r^{n+1}}{\underline{|n+1}}(ld_x+md_y+nd_z)^{n+1}\phi(x, y, z) = 0.$$

There are in this case $n+2$ coincident points at xyz in the direction lmn, and since the equation $(ld_x+md_y+nd_z)^{n+1}\phi(x, y, z) = 0$ is of the $n+1^{\text{th}}$ degree in l, m, n; $n+1$ tangents to the surface at xyz can, in general, be drawn in any given plane through that point. This equation now takes the place of the conditional equation in (5781).

5784 Equation (5782) is now replaced by

$$\{(\xi-x)\,d_x+(\eta-y)\,d_y+(\zeta-z)\,d_z\}^n\phi(x, y, z) = 0,$$

the equation of the locus of all tangents at the point xyz, and representing a conical surface generated by the motion of those tangents.

5785 The equation of the normal at xyz is

$$\frac{\xi-x}{\phi_x} = \frac{\eta-y}{\phi_y} = \frac{\zeta-z}{\phi_z}. \qquad (5782)$$

5786 The equation of the tangent at a point $x'y'z'$ on the curve of intersection of the tangent plane at xyz with the surface is

$$\frac{\xi-x'}{\lambda}=\frac{\eta-y'}{\mu}=\frac{\zeta-z'}{\nu},$$

with the two conditions

$$\lambda\phi_x+\mu\phi_y+\nu\phi_z=0, \qquad \lambda\phi_{x'}+\mu\phi_{y'}+\nu\phi_{z'}=0.$$

For these are the conditions of perpendicularity to the normals of the tangent planes at xyz and $x'y'z'$ respectively.

There are three exceptional cases in which the ratios $\lambda:\mu:\nu$ have more than one set of values; namely—

5787 I.—When ϕ_x, ϕ_y, ϕ_z vanish simultaneously, there is a tangent cone at xyz.

5788 II.—When $\phi_{x'}$, $\phi_{y'}$, $\phi_{z'}$ vanish simultaneously, $x'y'z'$ is a singular point on the surface.

5789 III.—When $\frac{\phi_x}{\phi_{x'}}=\frac{\phi_y}{\phi_{y'}}=\frac{\phi_z}{\phi_{z'}}$. In this case the point $x'y'z'$ coincides with xyz, and the tangent there meets the curve in more than two coincident points, the condition for which is

$$(\lambda d_x+\mu d_y+\nu d_z)^2\,\phi\,(x,y,z)=0 \quad\text{............ (i.)},$$

with

$$\lambda\phi_x+\mu\phi_y+\nu\phi_z=0 \quad\text{.................... (ii.)}.$$

These equations furnish two sets of values of the ratios $\lambda:\mu:\nu$, giving thereby the directions of two *inflexional tangents* (tangents to the curve of intersection) at xyz, each meeting the surface in three coincident points. If all the derivatives of an order less than n vanish at xyz, equation (i.) will be replaced by $(\lambda d_x+\mu d_y+\nu d_z)^n\phi\,(x,y,z)=0$, which, together with (ii.), will determine n inflexional tangents at the point.

5790 The polar equation of the tangent plane at the point $r\theta\phi$, r', θ', ϕ' being the variables, is, writing u for r^{-1},

$$u'=(u\cos\theta-u_\theta\sin\theta)\cos\theta'+(u\sin\theta+u_\theta\cos\theta)\cos(\phi'-\phi)\sin\theta'$$
$$+u_\phi\operatorname{cosec}\theta\sin(\phi'-\phi)\sin\theta'.$$

PROOF.—Write the polar equation of the plane through $p\alpha\beta$, the foot of the perpendicular on the plane from the origin; thus

$$pu = \cos\theta \cos\alpha + \sin\theta \sin\alpha \cos(\phi - \beta).$$

Differentiate for θ and ϕ to find pu_θ and pu_ϕ, and eliminate p, α, and β. This elimination is troublesome.

5791 The length of the perpendicular from the origin upon the tangent plane at xyz,

$$p = \frac{x\phi_x + y\phi_y + z\phi_z}{\sqrt{\{\phi_x^2 + \phi_y^2 + \phi_z^2\}}} \quad \text{or} \quad \frac{nc}{\sqrt{\{\phi_x^2 + \phi_y^2 + \phi_z^2\}}}, \quad (5782, 5549)$$

the second form being the value of p when the equation of the surface is $\phi(x, y, z) = c$, a constant, and when ϕ is a homogeneous function of the nth degree (1624).

5793 In polar coordinates,

$$\frac{1}{p^2} = u^2 + u_\theta^2 + u_\phi^2 \operatorname{cosec}^2\theta = \frac{r^2 + r_\theta^2 + r_\phi^2 \operatorname{cosec}^2\theta}{r^4}.$$

PROOF.—Add together the values of the squares of pu, pu_θ, and pu_ϕ found in (5790).

For a geometrical proof, see *Frost and Wolstenholme*, Art. (314).

THE INDICATRIX CONIC.

5795 DEF.—The indicatrix at any point of a surface is the curve in which the surface is intersected by a plane drawn parallel to the tangent plane at that point and infinitely near to it.

5796 The following abbreviations will be employed—
The derivatives of $\phi(x, y, z)$, $\phi_{2x}, \phi_{2y}, \phi_{2z}, \phi_{yz}, \phi_{zx}, \phi_{xy}, \phi_x, \phi_y, \phi_z$,
will be denoted by $a, b, c, f, g, h, l, m, n.$

5797 PROP.—The *indicatrix* at a point xyz of a surface $\phi(x, y, z) = 0$ is the conic in which the elementary quadric surface

5798 I.

$$a\xi^2 + b\eta^2 + c\zeta^2 + 2f\eta\zeta + 2g\zeta\xi + 2h\xi\eta = -\frac{R^2}{\rho}\sqrt{l^2 + m^2 + n^2} \equiv N$$

is intersected by the tangent plane at xyz, whose equation is

5799 II. $$l\xi + m\eta + n\zeta + \tfrac{1}{2}N = 0.$$

The origin of coordinates is the point xyz in both equations. R is an indefinitely small radius from the centre of the quadric (I.) to a point $\xi\eta\zeta$ on the indicatrix, and ρ is the radius of curvature of the section of the surface ϕ by a normal plane drawn through R; the ratio $R^2 : \rho$ being constant for all such planes.

PROOF.—Let O, in Fig. (181), be the point xyz on the surface ϕ; $x+\xi$, $y+\eta$, $z+\zeta$ an adjacent point P. Then

$$\phi(x+\xi,\ y+\eta,\ z+\zeta) = \phi(x,\ y,\ z)+l\xi+m\eta+n\zeta+\tfrac{1}{2}(a\xi^2+\ldots+2h\xi\eta)+\&c.$$

With xyz for origin, draw the quadric surface

$$a\xi^2+b\eta^2+c\zeta^2+2f\eta\zeta+2g\zeta\xi+2h\xi\eta = N \quad\text{.....................(i.)}$$

and the plane

$$l\xi+m\eta+n\zeta+\tfrac{1}{2}N = 0 \quad\text{..........................(ii.).}$$

Since ξ, η, ζ are very small, N is likewise. Also the unwritten terms in the above expansion may be neglected in the limit. Hence, any point $\xi\eta\zeta$ lying on the intersection of the quadric (i.) and the plane (ii.) will also lie on the original surface $\phi(x+\xi,\ y+\eta,\ z+\zeta) = 0$.

To determine N, we have the perpendicular from xyz upon the plane (ii.), $p = \dfrac{-N}{2\sqrt{(l^2+m^2+n^2)}}$ (5549). The radius of curvature of the section of the surface ϕ made by a normal plane at O drawn through P being ρ, we have $\rho = \dfrac{R^2}{2p}$, and therefore $N = -\dfrac{R^2}{\rho}\sqrt{(l^2+m^2+n^2)}$.

In the Figure, $R = OP$, $p = OL$, and the intersection of (i.) and (ii.) is the conic PSQ. Since p is indefinitely small, we may put $N = 0$ in equation (ii.). This amounts to taking the parallel section of the quadric by the tangent plane at O instead of the section PSQ. But these two will be equal in all respects, since the section of the quadric is a *central* one.

5800 If $m = 0$, equation (II.) becomes $l\xi+n\zeta = 0$, and if the inclination of the indicatrix plane to the plane of xy be α, $\tan\alpha = -\dfrac{l}{n}$. To obtain, in this case, the equation of the indicatrix in its own plane, put $\xi = \xi'\cos\alpha$, $\zeta = \zeta'\sin\alpha$, and $\eta = \eta'$, in equation (I.).

5801 When none of the three constants l, m, n are zero, the quadric (I.) simplifies as follows—

From (II.) we have $l\xi+m\eta = -n\zeta$ and two similar equations. Square these, and by the results eliminate the terms in $\eta\zeta$, $\zeta\xi$, $\xi\eta$ from (I.), which then becomes

5802 III. $$H\xi^2+K\eta^2+L\zeta^2 = N,$$

where $H = a+\dfrac{l}{mn}(lf-mg-nh)$, $\quad K = b+\dfrac{m}{nl}(mg-nh-lf)$,

$$L = c+\frac{n}{lm}(nh-lf-mg).$$

This is the equation of another quadric intersecting the plane (II.) in the indicatrix.

5803 The equation of a surface for points near an origin O (Fig. 182), the normal at O being taken for z axis, is

$$\frac{x^2}{\rho_1}+\frac{y^2}{\rho_2}=2z,$$

where ρ_1, ρ_2 are the radii of curvature of the normal sections through the x and y axes, and those sections will be proved to be the lines of curvature at O.

PROOF.—Let $AC=a$ and $BC=b$ be the semi-axes of the indicatrix conic at a small distance z from O (5795). The equation of the conic will therefore be $\frac{x^2}{a^2}+\frac{y^2}{b^2}=1$; but $\frac{a^2}{z}=2\rho_1$ and $\frac{b^2}{z}=2\rho_2$, giving the equation required.

Secondly, on a line of curvature, the normal to the surface at a point xyz will intersect the z axis (5773). The condition for this, by (5533) [with xyz for abc, the origin for $a'b'c'$,

$$L,\ M,\ N=\phi_x,\ \phi_y,\ \phi_z\ (5785)\ =\frac{2x}{\rho_1},\ \frac{2y}{\rho_2},\ -2,\ \text{and}\ L',\ M',\ N'=0,0,1\Big],$$

gives $xy\left(\frac{1}{\rho_2}-\frac{1}{\rho_1}\right)=0$, therefore $x=0$ or $y=0$ on a line of curvature.

Q. E. D.

5804 If the equation of the surface with the same axes be

$$z=ax^2+2hxy+by^2+2fyz+2gzx+cz^2+\text{higher powers},$$

then
$$\rho_1=\frac{1}{2a},\quad \rho_2=\frac{1}{2b}.$$

PROOF.—Put $y=0$ and divide by z, therefore $1=a\frac{x^2}{z}+2gx+cz+\&c.$ When x and z vanish, we have $1=2a\rho_1$.

5805 For a normal section making an angle θ with AC,

$$\frac{1}{\rho}=2\left(a\cos^2\theta+2h\sin\theta\cos\theta+b\sin^2\theta\right).$$

PROOF.—Turning the axes in (5804) through the angle θ by (4049), the coefficient of x'^2 becomes $a\cos^2\theta+$ as above.

5806 *Euler's Theorem.*—If ρ be the radius of curvature of any other normal section at O, making an angle $ACP=\theta$ with AC (Fig. 182),

$$\frac{1}{\rho}=\frac{\cos^2\theta}{\rho_1}+\frac{\sin^2\theta}{\rho_2}.$$

PROOF.—Let $r = CP$; then $x = r\cos\theta$, $y = r\sin\theta$, and $r^2 = 2\rho z$, which substitute in (5793).

5807 COR. — The sum of the curvatures of two normal sections at right angles to each other is constant; or, if ρ, ρ' be the radii of curvature for those sections, and ρ_a, ρ_b the radii for the principal sections,

$$\frac{1}{\rho}+\frac{1}{\rho'}=\frac{1}{\rho_a}+\frac{1}{\rho_b}.$$

5808 The radius of curvature of a normal section varies as the square of the radius of the indicatrix in that section.

PROOF.—From $r^2 = 2\rho z$, in Figure (182).

5809 *Meunier's Theorem.*—The radius of curvature of an oblique section of a surface is equal to the radius of curvature of the normal section through the same tangent multiplied by the cosine of the inclination of the planes.

PROOF.—(Fig. 183.) $\rho'=\frac{PN^2}{NC}$, $\rho=\frac{PN^2}{NO}$, therefore $\frac{\rho'}{\rho}=\frac{NO}{NC}=\cos\phi$, when NO and NC vanish.

5810 Quadratic for y_x at a point on the surface $z=\phi(x, y)$ giving the direction of the principal normal sections, and, therefore, of the lines of curvature (notation 1815).

$$\{pqt-(1+q^2)s\}\,y_x^2+\{(1+p^2)t-(1+q^2)r\}\,y_x$$
$$+\{(1+p^2)s-pqr\}=0.$$

PROOF.—(i.) The equations of the normals at the consecutive points xyz and $x+dx$, $y+dy$, $z+dz$ of the surface $\phi(x, y, z)=0$ are

$$\frac{\xi-x}{\phi_x}=\frac{\eta-y}{\phi_y}=\frac{\zeta-z}{\phi_z}\quad\text{and}\quad\frac{\xi-(x+dx)}{\phi_x+d\phi_x}=\frac{\eta-(y+dy)}{\phi_y+d\phi_y}=\frac{\zeta-(z+dz)}{\phi_z+d\phi_z}.$$

5811 The condition of intersection is, by (5533),

$$\begin{vmatrix} dx & dy & dz \\ \phi_x & \phi_y & \phi_z \\ d\phi_x & d\phi_y & d\phi_z \end{vmatrix}=0,\quad\text{or}\quad\begin{vmatrix} 1 & y_x & p+qy_x \\ p & q & -1 \\ r+sy_x & s+ty_x & 0 \end{vmatrix}=0,$$

by dividing the first row by dx, and putting $z_x = \phi_x + \phi_y y_x$, $d\phi_x = \phi_{2x} + \phi_{xy} y_x$, &c. The form of $\phi(x, y, z)$ being, in this case, $\phi(x, y) - z$, ϕ_z becomes -1, and $d\phi_z$ becomes zero. The determinant equation produces the quadratic.

(ii.) *Otherwise.*—Consider $\xi\eta\zeta$ the point of intersection of consecutive normals. The equations of a normal being

$$\frac{\xi - x}{p} = \frac{\eta - y}{q} = \frac{\zeta - z}{-1} \quad \text{or} \quad \xi - x = p(z - \zeta) \quad \text{and} \quad \eta - y = q(z - \zeta).$$

Differentiate both equations for x, considering ξ, η, ζ constant and p, q functions of x and y; the results are

$$1 + (r + sy_x)(z - \zeta) + p(p + qy_x) = 0 \quad \text{and} \quad y_x + (s + ty_x)(z - \zeta) + q(p + qy_x) = 0.$$

Eliminate $z - \zeta$ to obtain the quadratic in y_x.

5812 If the equation of the surface be in the form $\phi(x, y, z) = 0$, the quadratic for y_x may be obtained in the same way. The requisite substitutions in the first determinant are found from $\phi_x + \phi_y y_x + \phi_z z_x = 0$, giving z_x; $d\phi_x = \phi_{2x} + \phi_{xy} y_x + \phi_{xz} z_x$, &c., and with the notation of (5796) the determinant equation and quadratic for y_x becomes

$$\begin{vmatrix} n & ny_x & -(l + my_x) \\ l & m & n \\ an - gl + (hn - gm)\, y_x & hn - fl + (bn - fm)\, y_x & gn - cl + (fn - cm)\, y_x \end{vmatrix} = 0.$$

5813 The above determinant, or the corresponding one in (5810), is the differential equation of the lines of curvature.

5814 The radii of curvature of the principal normal sections of the surface $\phi(x, y, z) = 0$ at a point xyz are given by the following quadratic, in which Δ' is the bordered determinant in (5700), and the notation is that of (5796) and (1620).

$$\boldsymbol{\Delta'\rho^2 + \{(a+b+c)(l^2+m^2+n^2) - (abcfgh \between lmn)^2\}\, k\rho - k^4 = 0,}$$

where $k^2 = l^2 + m^2 + n^2$.

Proof.—The quadratic in (5653) applied to a section of the quadric (I.) (5798) by the plane (II.), becomes

$$\Delta' R^4 + \{(b+c)\, l^2 + (c+a)\, m^2 + (a+b)\, n^2 - 2fmn - 2gnl - 2hlm\}\, NR^2 \\ - (l^2 + m^2 + n^2)\, N^2 = 0,$$

whose roots, being the two values of R^2, are the squares of the principal semi-axes of the indicatrix. Put $R^2 = \frac{N\rho}{k}$, as in the Proof of (5797).

5815 Otherwise, the quadratic in (5651) might be applied to a section of the quadric (III.) (5802) by the plane (I.).

5816 If the equation of the surface be given in the form

$z = \phi(x, y)$, the quadratic becomes [writing, as in (1815), p, q, r, s, t for $z_x, z_y, z_{2x}, z_{xy}, z_{2y}$],

$$(rt-s^2)\,\rho^2 - \{(1+p^2)\,t - 2pqs + (1+q^2)\,r\}\,k\rho + k^4 = 0,$$

where $k^2 = p^2+q^2+1$.

Otherwise, this equation may be found from the two equations obtained in the second proof of (5810), by eliminating y_x instead of $z-\zeta$.

5817 The radius of curvature at a point xyz on the surface $\phi(x, y, z) = 0$ of the normal section whose tangent has the direction cosines λ, μ, ν is, with the notation of (5796) and (1620),

$$\rho = \frac{\sqrt{(l^2+m^2+n^2)}}{(abcfgh \between \lambda\mu\nu)^2}.$$

PROOF.—From equation (I.) (5798), since ξ, η, ζ are respectively equal to $R\lambda$, $R\mu$, and $R\nu$.

5818 The curvature at any point of a surface $\phi(x, y, z) = 0$ is termed *elliptic* or *synclastic*, *hyperbolic* or *anticlastic*, and *parabolic* or *cylindrical*, according as the indicatrix is an ellipse, hyperbola, or two parallel right lines, or according as the principal curvatures have the same signs, opposite signs, or one of them vanishes; and this will be according as the determinant Δ', in (5814), or s^2-rt, in (5816), is *negative*, *positive*, or *zero*.

PROOF.—The rule follows at once from the consideration that the two values of ρ in the quadratic of (5814) must have the same sign in the first case, different signs in the second, and that one value must be infinite in the third case.

5819 The condition for an umbilicus is that the indicatrix must be a circle; therefore, either (III.) (5802) must be a sphere, or, if it be a quadric surface, the plane (II.) must make a circular section of it, and therefore either l, m, or n must vanish.

5820 Otherwise, the quadratic in (5814) or (5816) must have equal roots.

5821 Otherwise, the conditions for an umbilicus on the surface $\phi(x, y, z) = 0$ are the two equations

$$\frac{bn^2+cm^2-2fmn}{m^2+n^2} = \frac{cl^2+an^2-2gnl}{n^2+l^2} = \frac{am^2+bl^2-2hlm}{l^2+m^2}.$$

PROOF.—The radius of the indicatrix, and therefore also ρ in (5817), is constant for all values of λ, μ, ν. Now, by (5817),

$$a\lambda^2 + \&c. = \frac{k}{\rho};$$

$$\therefore\quad \left(a-\frac{k}{\rho}\right)\lambda^2+\left(b-\frac{k}{\rho}\right)\mu^2+\left(c-\frac{k}{\rho}\right)\nu^2+2f\mu\nu+2g\nu\lambda+2h\lambda\mu = 0,$$

and $l\lambda+m\mu+n\nu = 0$, since $\lambda\mu\nu$ is always tangential, and lmn is normal to the surface. As these equations are true for all values of λ, μ, ν, the second expression must be a factor of the first. The quotient, by division, is therefore

$$\left(a-\frac{k}{\rho}\right)\frac{\lambda}{l}+\left(b-\frac{k}{\rho}\right)\frac{\mu}{m}+\left(c-\frac{k}{\rho}\right)\frac{\nu}{n}.$$

Equating to zero each of the three coefficients of the remainder, and eliminating ρ, we obtain the above conditions.

5822 If a common factor of the three fractions in (5821) exists, that factor equated to zero is the differential equation of a *line of spherical curvature* at every point of which there is an umbilicus. If the fractions are identically equal, the surface has an umbilicus at every point, and must therefore be a sphere.

5823 The number of umbilici on a surface of the nth degree cannot exceed $n(10n^2-25n+16)$. *Salmon*, p. 208.

5824 The condition that the indicatrix may be a rectangular hyperbola is

$$(a+b+c)(l^2+m^2+n^2) = (abcfgh \between lmn)^2.$$

PROOF.—The quadratic in (5814) must have equal roots of opposite signs.

Similarly, when $z = \phi(x, y)$ is the equation of the quadric, the condition becomes

$$(1+p^2)\,t-2pqs+(1+q^2)\,r = 0. \qquad (5816)$$

5825 The condition that the indicatrix may become two coinciding lines.

Here equation I. (5798) must represent a cone, and the plane (II.) must touch it. Hence $N = 0$, and, if ζ be eliminated, the quadratic for the ratio $\xi : \eta$ obtained is

$$(an^2+cl^2-2gnl)\,\xi^2+2\,(clm-fnl-gmn+hn^2)\,\xi\eta+(bn^2+cm^2-2fmn)\,\eta^2 = 0,$$

and this must have equal roots.

CURVATURE OF A SURFACE.

5826 DEFS.—*Integral curvature* of a closed surface is equal to the area of that part of the surface of a sphere of unit radius which is intercepted by radii drawn parallel to the normals at all points of the given surface. This area also measures the solid angle of the cone generated by the radii. The curve on the sphere is called the *horograph* of the curve on the original surface. In other words, *integral curvature* of a closed surface is the area of the horograph of its boundary.

5827 *Average curvature* is the integral curvature divided by the area of the surface.

Specific curvature is the average curvature of a small element at the point; *i.e.*, $\frac{(ds)^2}{\rho_1\rho_2} \div (ds)^2 = \frac{1}{\rho_1\rho_2}$.

5828 The last is the usual measure of curvature at a point, and its value in coordinates of the point is given by

$$\frac{1}{\rho_1\rho_2} = -\frac{\Delta'}{(l^2+m^2+n^2)^2} \quad \text{or} \quad \frac{rt-s^2}{(1+p^2+q^2)^2}, \qquad (5796)$$

according as $\phi(x, y, z) = 0$ or $z = \phi(x, y)$ is the form of the equation to the surface.

PROOF.—From the product of roots of the quadratics in (5814) and (5816).

5829 In a plane curve *integral curvature* is the plane angle contained by the terminal normals, and *average curvature* is the integral curvature divided by the length of the curve.

5830 Another measure of curvature at a given point of a surface is the ratio of the area of the indicatrix to the area of the indicatrix cut off by the same plane on a sphere of unit radius which touches the surface internally at the point. This measure is $= \sqrt{\rho_1\rho_2}$.

PROOF.—Putting $AC = R_1$, $BC = R_2$, in Fig. (182), and $OC = z$, the area of the indicatrix of the surface is $\pi R_1 R_2$ at an ellipsoidal point. But $R_1^2 = 2\rho_1 z$ and $R_2^2 = 2\rho_2 z$, therefore $\pi R_1 R_2 = 2\pi z\sqrt{(\rho_1\rho_2)}$. Also the indicatrix of the sphere $= 2\pi z$ since $\rho_1 = \rho_2 = 1$ for the sphere.

5831 The radius of curvature of any normal section at a point P of an ellipsoid (Fig. 184) is equal to the square of the semi-diameter parallel to the tangent of that section, divided by the perpendicular from P upon the diametral plane conjugate to OP.

PROOF.—Let AOB be the plane parallel to the tangent plane at P; $OA = d$, the semi-diameter in it parallel to the given tangent PT. Draw PR perpendicular to OA and $PN \equiv p$ perpendicular to the plane AOB. The radius of curvature at P of the elliptic section $PA = \frac{d^2}{PR}$ (4536). Therefore, by (5809), the radius of curvature of the normal section through the same tangent PT, will be $\rho = \frac{d^2}{PR} \times \frac{PR}{PN} = \frac{d^2}{p}$.

5832 The principal radii of curvature at P, viz. ρ_1, ρ_2, are found from their sum and product thus: putting γ for OP, and a, b, c for the semi-axes of the ellipsoid,

$$\rho_1+\rho_2 = \frac{a^2+b^2+c^2-\gamma^2}{p}, \quad \rho_1\rho_2 = \frac{a^2b^2c^2}{p^4}.$$

PROOF.—Let α, β be the semi-axes of the section AOB (Fig. 184), then $\alpha^2+\beta^2+\gamma^2 = a^2+b^2+c^2$ (5642) and $p\alpha\beta = abc$ (5648). By these values eliminate α, β from $\rho_1+\rho_2 = \frac{\alpha^2+\beta^2}{p}$ and $\rho_1\rho_2 = \frac{\alpha^2\beta^2}{p^2}$ (5831).

5833 The lines of curvature on a quadric surface are its intersections with the confocal quadrics.

PROOF.—Let the quadric and confocal be the ellipsoid and one-fold hyperboloid in (Fig. 178) intersecting in the line DPE, and let their equations be, as in (5656),

$$\frac{x^2}{a^2}+\frac{y^2}{b^2}+\frac{z^2}{c^2} = 1 \text{ ...... (i.) and } \frac{x^2}{a^2+\lambda}+\frac{y^2}{b^2+\lambda}+\frac{z^2}{c^2+\lambda} = 1 \text{ ...... (ii.).}$$

At any point P on the line of intersection x, y, z satisfy the three following equations:—

First, the differential of (ii.), $\frac{x\,dx}{a^2+\lambda}+\frac{y\,dy}{b^2+\lambda}+\frac{z\,dz}{c^2+\lambda} = 0$.

Second, the difference of (i.) and (ii.),

$$\frac{x^2}{a^2(a^2+\lambda)}+\frac{y^2}{b^2(b^2+\lambda)}+\frac{z^2}{c^2(c^2+\lambda)} = 0.$$

Third, the difference of their differentials,

$$\frac{x\,dx}{a^2(a^2+\lambda)}+\frac{y\,dy}{b^2(b^2+\lambda)}+\frac{z\,dz}{c^2(c^2+\lambda)} = 0.$$

The eliminant of these equations in x, y, z produces the determinant equation here annexed, which, by (5811), is the condition for the intersection of consecutive normals. Hence this condition holds for every point of the line of intersection of (i.) and (ii.).

$$\begin{vmatrix} dx & dy & dz \\ \frac{x}{a^2} & \frac{y}{b^2} & \frac{z}{c^2} \\ \frac{dx}{a^2} & \frac{dy}{b^2} & \frac{dz}{c^2} \end{vmatrix} = 0.$$

The general method of determining the lines of curvature of a surface from the differential equation in (5811) is here exemplified in the case of an ellipsoid.

5834 The determinant just written gives for the differential equation of the lines of curvature

$$(b^2-c^2)\,x\,dy\,dz+(c^2-a^2)\,y\,dz\,dx+(a^2-b^2)\,z\,dx\,dy=0 \quad\text{........ (i.).}$$

To solve this, multiply by $\frac{z}{c^2}$ and substitute for z and dz from the equation of the quadric. The result is of the form

$$Axy\,y_x^2+(x^2-Ay^2-B)\,y_x-xy=0,$$

in which $A=\frac{a^2(b^2-c^2)}{b^2(a^2-c^2)}$, $B=\frac{a^2(a^2-b^2)}{a^2-c^2}$; or, multiplying by $\frac{y}{x}$,

$$A\frac{yy_x}{x}(xyy_x-y^2)-B\frac{yy_x}{x}+(xyy_x-y^2)=0,$$

which is of the form in (3236). Solving by that method, we find that the two equations $\frac{yy_x}{x}=\alpha$ and $xyy_x-y^2=\beta$ have the common primitive $\alpha x^2-y^2=\beta$, which, with the relation $A\alpha\beta-B\alpha+\beta=0$, constitutes the solution. The result is that the projections of the lines of curvature upon the xy plane are a series of conics coaxal with the principal section of the ellipsoid, and having their axes $\mathfrak{a}$, $\mathfrak{b}$ varying according to the equation

$$\frac{\mathfrak{a}^2(a^2-c^2)}{a^2(a^2-b^2)}+\frac{\mathfrak{b}^2(b^2-c^2)}{b^2(b^2-a^2)}=1.$$

At an umbilicus $y=0$, therefore, equation (i.) becomes $[(b^2-c^2)\,xdz+(a^2-b^2)\,zdx]\,dy=0$. Here $dy=0$, being a solution, gives $y=C=0$, showing that the plane of zx contains a line of curvature. The other factor, equated to zero, taken with the differential equation of the curve $c^2x\,dx+a^2z\,dz=0$, gives the coordinates of the umbilicus, as in (5603).

OSCULATING PLANE OF A LINE OF CURVATURE.

5835 Let ϕ be the angle between the osculating plane and the normal section through the same line of curvature, ds an element of the other line of curvature, and ρ, ρ' their radii of curvature respectively: then

$$\tan\phi=\frac{d\rho}{ds}\cdot\frac{\rho'}{\rho'-\rho}.$$

PROOF.—Fig. (185). Let OA, OB be the lines of curvature; OP, AP consecutive normals along OA; and OS, BS the same along OB. Also, let BQ, CQ be consecutive normals along the line of curvature BC. Then, ultimately, $OP=\rho$, $OS=\rho'$, $BQ=\rho+d\rho$. Also, let QP produced meet the osculating plane of AO in R. Join RO and RA, and draw QN at right angles to PS. Since the tangent to AO at O is perpendicular to the plane $OBQP$ and that at A to $ACQP$, it follows that both tangents are perpendicular to QP, which must therefore be perpendicular to the osculating plane ARO. Hence ϕ or $ROP=PQN$.

Now $\frac{NQ}{ds}=\frac{SQ}{SB}=\frac{\rho'-\rho-d\rho}{\rho'}$, $\therefore\ \tan\phi=\frac{NP}{NQ}=\frac{d\rho}{ds}\cdot\frac{\rho'}{\rho'-\rho}$ ultimately.

5836 At every point on a line of curvature of a central conicoid pd is constant, where d is the semi-diameter parallel to the tangent at the point and p is the perpendicular from the centre upon the tangent plane.

Proof.—Let the first and third confocals in (5661) be fixed, and therefore a_1 and a_3 constant. Draw the second confocal through the point of contact P of the tangent plane (Fig. 178). Then, by (5663), p_1d_3 and p_3d_1 are constant along the line of intersection of the first and third surface, because, by (5661), $d_3^2=a_1^2-a_2^2$ and $d_1^2=a_2^2-a_3^2$.

GEODESIC LINES.

5837 The equations of a geodesic on the surface $\phi(x,y,z)=0$

are $$\frac{x_{2s}}{\phi_x}=\frac{y_{2s}}{\phi_y}=\frac{z_{2s}}{\phi_z}.$$

Proof.—The osculating plane of the curve contains the normal to the surface (5775); therefore, by (5737) and (5785).

5838 A geodesic is a line of maximum or minimum distance along the surface between two points.

Proof.—The curve drawn in the osculating plane from one point to a contiguous point is shorter than any other by Meunier's theorem (5809), for any oblique section has a shorter radius of curvature and therefore a longer arc. A succession of minimum arcs, however, may constitute a maximum curve distance between the extreme points; for example, two points on a sphere can be joined by either of two arcs of a great circle, the one being a minimum and the other a maximum geodesic.

5839 A surface of revolution such as the terrestrial globe affords a good illustration. A meridian and a parallel of latitude drawn through a point near the pole are the two lines of curvature at the point. The meridian is also a geodesic, but the parallel is evidently not, for its plane does not contain the normal to the surface.

5840 A geodesic is the line in which a string would lie if stretched over the convex side of a smooth surface between two fixed points.

Proof.—Any small arc of the string POP' (Fig. 182) is acted upon by tensions along the tangents at P and P', and by the normal reaction of the surface at O. But these three forces act in the osculating plane (5775); therefore the string will rest in equilibrium on the surface in that plane.

COR.—Two equal geodesics drawn from a point and indefinitely near to each other are at right angles to the line which joins their extremities.

5841 If a geodesic has a constant inclination to a fixed line, the normals along it will be at right angles to that line.

PROOF.—Let lmn be the fixed line and α the constant angle; then

$$lx_s + my_s + nz_s = \cos\alpha, \quad \text{and therefore} \quad lx_{2s} + my_{2s} + nz_{2s} = 0.$$

Therefore, by (5837), the principal normal is at right angles to lmn.

EXAMPLE.—The helix, the axis being the fixed line.

5842 On any central conicoid pd is constant along a geodesic, where p is the perpendicular from the centre upon the tangent plane and d is the semi-diameter parallel to the tangent of the geodesic.

PROOF.—(Fig. 186.) Let AT, BT be the tangents at the two extremities of a small geodesic arc AB, and let the tangent planes at A and B be ADC and BCD. AT and BT make equal angles with CD, by the property of shortest distance, for if the plane BCD be turned about CD until it coincides with the plane ADC, ATB will become a straight line, and therefore $\angle ATD = BTC = i$, say.

Let ω be the angle between the tangent planes; let the perpendiculars upon those planes from A, B be $AM = q$, $BN = q'$, and from the centre of the quadric p, p'; and let xyz and $x'y'z'$ be the points A, B. Then

$$q = AT \sin i \sin\omega, \quad q' = BT \sin i \sin\omega, \quad \therefore\ q : q' = AT : BT \ \ldots\ldots \text{(i.)},$$

$$q = p'\left(\frac{xx'}{a^2} + \frac{yy'}{b^2} + \frac{zz'}{c^2} - 1\right) \ (5628); \quad q' = p\left(\frac{x'x}{a^2} + \frac{y'y}{b^2} + \frac{z'z}{c^2} - 1\right),$$

therefore $$q : q' = p' : p \ \ldots\ldots \text{(ii.)}.$$

Again, let d, d' be the semi-diameters parallel to AT and BT. Then, by (5677), $AT : BT = d : d'$; therefore $p' : p = d : d'$ or $pd = p'd'$; that is, pd is constant.

5843 If a line of curvature be plane, that plane makes a constant angle with the tangent plane to the surface.

PROOF.—Let PQ, QR, RS be equal consecutive elements of the line of curvature. The consecutive normals to the surface bisect PQ and QR and meet in a point. Therefore they are equally inclined to the plane PQR. Similarly the second and third normals are equally inclined to the plane QRS, and so on. Hence, if the curve be plane, all the normals are equally inclined to its plane. Hence also the following theorem.

5844 *Lancret's Theorem.*—The variation in the angle between the tangent plane and the osculating plane of a line of curvature is equal to the angle between consecutive osculating planes.

5845 Cor.—If a geodesic be either a line of curvature or a plane curve, it is both, but a plane line of curvature, as in (5839), is not necessarily a geodesic.

GEODESIC CURVATURE.

Theorem.—The square of the curvature at any point of a curve traced on a surface is equal to the sum of the squares of the normal and geodesic curvatures (5776), or

5846 $$\frac{1}{\rho^2} = \frac{1}{\rho'^2} + \frac{1}{\rho''^2},$$

where ρ' is the radius of curvature of the normal section and ρ'' the radius of geodesic curvature. Also, if ϕ be the angle between the plane of normal section and the osculating plane,

5847 $$\rho = \rho'' \sin\phi = \rho' \cos\phi.$$

Proof.—Let $PQ = QR$ (Fig. 187) be consecutive elements of any curve traced on a surface. Produce PQ to S, making $QS = PQ$. Let $QT = PQ$ be the consecutive elements of the section of the surface drawn through PQS and the normal at Q. Join RS, ST, TR. $PQSR$ is the osculating plane of the curve PQR. $PQST$ is the plane of normal section, and therefore PQT is a geodesic. QRT is the tangent plane, and STR is a right angle.

Then, putting $SQR = d\psi$, $SQT = d\psi'$, $RQT = d\psi''$, $RST = \phi$, we have

$$\rho = \frac{ds}{d\psi}, \quad \rho' = \frac{ds}{d\psi'}, \quad \rho'' = \frac{ds}{d\psi''}. \tag{5776}$$

Therefore $$\frac{\rho}{\rho''} = \frac{ds\,.\,d\psi''}{ds\,.\,d\psi} = \frac{RT}{RS} = \sin\phi.$$

Also $$\frac{\rho}{\rho'} = \frac{ds\,.\,d\psi'}{ds\,.\,d\psi} = \frac{ST}{SR} = \cos\phi, \text{ as in (5809).}$$

Thus both theorems are proved. Note that ρ' is the radius of curvature of the geodesic PQT, while ρ'' is the radius of geodesic curvature of PQR.

RADIUS OF TORSION OF A GEODESIC.

5848 If θ be the angle between the geodesic and one of the lines of curvature; ρ_1, ρ_2 the principal radii of normal curvature, and σ the radius of torsion,

$$\frac{1}{\sigma} = \left(\frac{1}{\rho_1} - \frac{1}{\rho_2}\right) \sin\theta \cos\theta.$$

PROOF.—(Fig. 182.) Let $OP = ds$ be the geodesic, OA, OB the lines of curvature, and $\theta = ACP$. The angle of torsion $d\tau$ measures the rotation of the normal to the surface round $OP = ds$. But this angle is equal to the sum of the rotations of the normal round OA and OB resolved along ds. For, in travelling along each of the lines CN and NP, which are in the directions of the lines of curvature, the normal rotates only about the other. Therefore, if ω_1, ω_2 be the rotations round OA, OB, $d\tau = \omega_1 \cos\theta + \omega_2 \sin\theta$.

But $\omega_1 = -\dfrac{ds \sin\theta}{\rho_2}$, $\omega_2 = \dfrac{ds\cos\theta}{\rho_1}$; $\therefore\ \dfrac{1}{\sigma} = \dfrac{d\tau}{ds} = \left(\dfrac{1}{\rho_1} - \dfrac{1}{\rho_2}\right)\sin\theta\cos\theta$.

5849 The product pd has the same value for all geodesics which touch the same line of curvature.

PROOF.—By theorems (5836) and (5842), since the product where they touch it must be the same as that for the line of curvature.

5850 The product pd has the same value for all geodesics drawn through any umbilicus on a conicoid.

PROOF.—The semi-diameter d, in this case, is the radius of a circular section, and therefore equal to the mean semi-axis b for all the geodesics; and p is the same for all.

5851 The geodesics drawn through any point on a conicoid to two umbilici make equal angles with either line of curvature through the point.

PROOF.—pd is the same for each geodesic, by the last, and p is the same for each; therefore d is the same, that is, the diameters parallel to the two geodesics at the point are equal; therefore they are equally inclined to each axis of their section; but these axes are parallel to the lines of curvature (5803); therefore, &c.

5852 Hence the geodesics joining any point to two opposite umbilici lying on the same diameter are continuations of each other.

5853 The sum of the distances of any point on a line of curvature from two interior umbilici is constant; and the difference of the distances from one interior and one exterior umbilicus is constant.

PROOF.—Geometrically, as in the analogous theorem for the focal distances in a conic, if r, r' are the distances and $r+dr$, $r'+dr'$ the distances for a consecutive point on the line of curvature, it follows from (5851) that $dr = -dr'$ for interior umbilici and $dr = dr'$ for exterior ones.

5854 A system of lines of curvature and the umbilici on a quadric surface has therefore analogous properties with a system of confocal conics and their foci in a plane, the geodesics corresponding to straight lines.

5855 In the same way, every surface has a geodesic geometry proper to itself; spherical trigonometry, for instance, being the geodesic geometry of the sphere.

INVARIANTS.

INVARIANTS OF A SINGLE FUNCTION.

5856 The constancy of the ratio $R^2 : \rho$ in equation (5798) gives rise to the following invariant forms. Since the quadric surface I and the tangent plane II are the same for all positions of the coordinate axes, they have been called respectively the *invariable quadric* and the *invariable plane*. As a consequence,

5857 $$\phi_x^2+\phi_y^2+\phi_z^2$$

is an invariant of $\phi(x, y, z)$.

PROOF.—By (5791), since the perpendicular from the origin upon the invariable plane is constant. Also, the coefficients of the discriminating cubic (5693) of the invariable quadric will not be altered by transformation of axes. Therefore the following are also invariant forms :—

5858 $\phi_{2x}+\phi_{2y}+\phi_{2z}$,

5859 $\phi_{2y}\phi_{2z}+\phi_{2z}\phi_{2x}+\phi_{2x}\phi_{2y}-\phi_{yz}^2-\phi_{zx}^2-\phi_{xy}^2$,

5860 $\phi_{2x}\phi_{2y}\phi_{2z}+2\phi_{yz}\phi_{zx}\phi_{xy}-\phi_{2x}\phi_{yz}^2-\phi_{2y}\phi_{zx}^2-\phi_{2z}\phi_{xy}^2$.

5861 A similar theorem applied to a function $\phi(x, y)$ of two variables gives the *invariable conic* and *invariable line;* namely,

$$\xi^2\phi_{2x}+2\xi\eta\phi_{xy}+\eta^2\phi_{2y}=1 \quad \text{and} \quad \xi\phi_x+\eta\phi_y=1;$$

and from these the invariants,

5863 $$\phi_x^2+\phi_y^2, \quad \phi_{2x}+\phi_{2y}, \quad \phi_{2x}\phi_{2y}-\phi_{xy}^2,$$

5866 $$x\phi_y-y\phi_x, \quad x\phi_y+y\phi_x.$$

PROOF.—The last two invariants are obtained from the cosine of the angle between the invariable line (5862) and the fixed line $y\xi - x\eta = 0$, joining the point xy with the origin, or the fixed line $x\xi + y\eta = 0$.

INVARIANTS OF TWO FUNCTIONS.

5868 An invariant of the two functions $\phi(x, y)$, $\psi(x, y)$ is

$$\phi_x \psi_x + \phi_y \psi_y.$$

PROOF.—Form the cosine of the angle between the invariable lines $\xi\phi_x + \eta\phi_y = 1$ and $\xi\psi_x + \eta\psi_y = 1$, observing (5863).

Also the two following expressions are invariants,

5869 $$\phi_{2y}\psi_{2x} + \phi_{2x}\psi_{2y} - 2\phi_{xy}\psi_{xy},$$

5870 $$\phi_{2x}\psi_{2x} + \phi_{2y}\psi_{2y} + 2\phi_{xy}\psi_{xy}.$$

PROOF.—From the invariable conics (5861) of ϕ and ψ, we get

$$(\phi_{2x} + \lambda\psi_{2x})\,\xi^2 + 2\,(\phi_{xy} + \lambda\psi_{xy})\,\xi\eta + (\phi_{2y} + \lambda\psi_{2y})\,\eta^2$$

invariable for any value of λ. Hence the coefficients of the several powers of λ in the invariant

$$(\phi_{2x} + \lambda\psi_{2x})(\phi_{2y} + \lambda\psi_{2y}) - (\phi_{xy} + \lambda\psi_{xy})^2$$

are also invariants. This gives (5869). Subtracting the latter from the invariant $(\phi_{2x} + \phi_{2y})(\psi_{2x} + \psi_{2y})$ produces (5870).

INTEGRALS FOR VOLUMES AND SURFACES.

5871 If V be the volume included between the surface $z = \phi(x, y)$, three rectangular coordinate planes, the cylindrical surface $y = \psi(x)$, and the plane $x = a$, Fig. of (1906)

5872 $$V = \iiint dx\,dy\,dz = \iint z\,dx\,dy.$$

For the limits and demonstration, see (1906).

5874 The area of the surface $\phi(x, y, z) = 0$ or $z = f(x, y)$ will be

$$S = \iint \frac{\sqrt{(\phi_x^2 + \phi_y^2 + \phi_z^2)}}{\phi_z}\,dx\,dy \quad \text{or} \quad S = \iint \sqrt{(1 + z_x^2 + z_y^2)}\,dx\,dy.$$

PROOF.—The area of the element whose projection is $dxdy$ will be $dxdy\sec\gamma$, where γ is its inclination to the plane of xy, and therefore the angle between the normal and the z axis. Therefore

$$\sec\gamma = \sqrt{(\phi_x^2+\phi_y^2+\phi_z^2)} \div \phi_z = \sqrt{(1+z_x^2+z_y^2)}, \quad \text{by (1708).}$$

5875 Let the equation of a surface APB (Fig. 188) in polar coordinates be $r=f(\theta,\phi)$, and let V be the volume of the sector contained by the planes AOB, AOP, including an angle $\phi = PHC$, the given surface APB, and the portion OPB of the surface of a right cone whose vertex is O, axis OA, and semi-vertical angle $\theta = AOB$ or AOP; then

$$V = \tfrac{1}{3}\int_0^\theta\int_0^\phi r^3 \sin\theta \, d\theta \, d\phi.$$

PROOF.—Through P, any point on the surface, describe a spherical surface PCD, with centre O and radius $r=OP$. The volume of the elemental pyramid, vertex O, base Pe, $=\frac{1}{3}r.Pf.Pg = \frac{1}{3}r.rd\theta.r\sin\theta d\phi$. Here the error of the small portions, like PE, ultimately disappears in the summation, since the volume of PE, being equal to $\frac{1}{2}dr.rd\theta.r\sin\theta d\phi$, is of the third order of small quantities; and so in similar instances.

5876 The area of the same surface APB (Fig. 188) is

$$S = \int_0^\theta\int_0^\phi r\sqrt{\{(r^2+r_\theta^2)\sin^2\theta + r_\phi^2\}}\, d\theta\, d\phi.$$

PROOF.—Let the perpendicular from O upon the tangent plane at P to the given surface be $ON=p$. The element of

$$\text{area } PE = \text{area } Pe.\frac{OP}{ON} = rd\theta.r\sin\theta d\phi.\frac{r}{p} = \frac{r^3\sin\theta}{p}d\theta d\phi.$$

Substitute the value of p in (5793).

SURFACE OF REVOLUTION.

If $y=f(x)$ (Fig. 90) be the generating curve, and the x axis the axis of revolution, V the volume, and S the surface included between the planes $x=a$, $x=b$;

5877 $$V = \int_a^b \pi y^2 dx, \qquad S = \int_a^b 2\pi y\sqrt{(1+y_x^2)}\,dx.$$

PROOF.—The volume of the elemental cylinder of radius y and height dx is $\pi y^2 dx$. The element of the surface of revolution is

$$2\pi y\,ds = 2\pi y\,s_x dx = 2\pi y\sqrt{(1+y_x^2)}\,dx. \qquad (5113)$$

Guldin's Rules.—When the generating curve of a surface of revolution is a closed curve, and does not cut the axis of revolution, a solid annulus, or ring, is formed.

5879 RULE I.—*The volume of the solid ring is equal to the area of the generating curve multiplied by the circumference of the circle described by the centroid* of the area.*

5880 RULE II. — *The surface of the ring is equal to the perimeter of the generating curve multiplied by the circumference described by the centroid of the perimeter.*

PROOF.—Let A be the area of the closed curve, and dA any element of A at a distance y from the axis of revolution. The volume generated

$$=\int 2\pi y\,dA = 2\pi\int y\,dA = 2\pi\bar{y}A,$$

by the definition of the centroid (5885), $\bar{y}$ being its distance from the axis. Similarly, if P be the perimeter, writing P instead of A.

Quadrature of surfaces bounded by lines of constant gradient.

5881 Defining the curve (γ) as the locus of a point on the given surface at which the normal has the constant inclination γ to the z axis; let $F(\gamma)$ be the projection of the area bounded by the curve (γ) upon the xy plane; then the area itself will be found from the formula,

$$S=\int_0^\gamma \sec\gamma F'(\gamma)\,d\gamma.$$

PROOF.—The element of area between two consecutive curves (γ) and $(\gamma+d\gamma)$ projected on the xy plane will be $dF(\gamma)=F'(\gamma)\,d\gamma$; and, since the slope is the same throughout the curve (γ), this projected element must be equal to the corresponding element of the surface multiplied by $\cos\gamma$.

5882 RULE. — *Equate coefficients of the equation of the tangent plane with those of* $l\xi+m\eta+n\zeta=p$, *and eliminate* l *and* m *from* $l^2+m^2+n^2=1$. *The result will be an equation in* x, y *and* $n=\cos\gamma$, *representing the projection of the curve* (γ) *upon the* xy *plane. From this* $F(\gamma)$ *must be found.*

5883 Ex.—Taking the elliptic paraboloid $\frac{x^2}{a}+\frac{y^2}{b}=2z$; the tangent plane at xyz is $\frac{\xi x}{a^2}+\frac{\eta y}{b^2}-\zeta=z$. Equating coefficients of the last with $l\xi+m\eta+n\zeta=p$, and substituting for l and m in $l^2+m^2+n^2=1$, we obtain for the projection on the xy plane, $\frac{x^2}{a^2}+\frac{y^2}{b^2}=\tan^2\gamma$. The area of this ellipse

* *Centre of mass*, or *gravity*.

is $F(\gamma) = \pi ab \tan^2\gamma$, and therefore $F'(\gamma) = 2\pi ab \tan\gamma \sec^2\gamma$. Consequently, by (5881), $S = 2\pi ab \int_0 \tan\gamma \sec^3\gamma\, d\gamma = \frac{2}{3}\pi ab\,(\sec^3\gamma - 1)$.

CENTRE OF MASS.

5884 DEFINITIONS.—The *moment* of a body with respect to a plane is the sum of the products of each element of mass of the body and the distance of the element from the plane.

5885 The distance (denoted by $\bar{x}$) of the *centre of mass* * from the same plane is equal to the *moment* of the body divided by its *mass*.

5886 NOTE.—If the body be of uniform density, as is supposed to be the case in all the following examples, assume unity for the density, and read *volume* instead of *mass* in the above definitions.

The definition gives the following formulæ for the position of the centre of mass of a uniform body:

5887 For a *plane curve*,

$$\bar{x} = \frac{\int x\,ds}{\int ds} = \frac{\int x\sqrt{(1+y_x^2)}\,dx}{\int\sqrt{(1+y_x^2)}\,dx} = \frac{\int r\cos\theta\sqrt{(r^2+r_\theta^2)}\,d\theta}{\int\sqrt{(r^2+r_\theta^2)}\,d\theta}. \quad (5116)$$

For $\bar{y}$, change x into y and $\cos\theta$ into $\sin\theta$; but observe that in all cases, if the body be symmetrical about the axis of x, $\bar{y}$ vanishes. The formula gives the centre of volume of the portion of the curve included between the limits of integration.

For a *plane area*,

5890
$$\bar{x} = \frac{\iint x\,dx\,dy}{\iint dx\,dy} = \frac{\int xy\,dx}{\int y\,dx}.$$

The area is bounded by the curve, the x axis, and the ordinates $x = a$, $x = b$, if such be the limits of integration.

For a *plane sectorial area* bounded by two radii $SP = r$, $SP' = r'$ (Fig. 28) and the curve $r = F(\theta)$;

* Also called *centre of gravity* or *inertia*, and more recently *centroid*.

5892 $$\bar{x} = \frac{\iint r^2 \cos\theta \, d\theta \, dr}{\iint r \, d\theta \, dr} = \frac{\frac{2}{3}\int r^3 \cos\theta \, d\theta}{\int r^2 d\theta}.$$

5894 $$\bar{y} = \frac{\iint r^2 \sin\theta \, d\theta \, dr}{\iint r \, d\theta \, dr} = \frac{\frac{2}{3}\int r^3 \sin\theta \, d\theta}{\int r^2 d\theta}.$$

The second forms for $\bar{x}$ and $\bar{y}$ give the centroid of an area like SPP' (Fig. 28). The double integrals applied to that figure require the limits of the integration for r to be from 0 to $F(\theta)$, and afterwards for θ, from $\theta_1 = ASP$ to $\theta_2 = ASP'$. But, if applied to the area in (Fig. 109), the order of integration must be reversed, as explained in (5209).

For a *surface of revolution* round the x axis,

5896 $$\bar{x} = \frac{\int xy\sqrt{(1+y_x^2)}\,dx}{\int y\sqrt{(1+y_x^2)}\,dx} = \frac{\int r^2 \sin\theta\cos\theta\sqrt{(r^2+r_\theta^2)}\,d\theta}{\int r\sin\theta\sqrt{(r^2+r_\theta^2)}\,d\theta}.$$

Proof.—By (5885), for the *moment* $= \int x \,.\, 2\pi y ds$ and the *area* $= \int 2\pi y ds$; the second form by (5116). If $x = a$, $x = b$ are the limits of integration, the surface is bounded by the parallel planes $x = a$, $x = b$; and in the second form, the corresponding values of θ are the limits defining the same parallel planes.

For any surface,

5898 $$\bar{x} = \frac{\iint x\sqrt{(1+z_x^2+z_y^2)}\,dx\,dy}{\iint \sqrt{(1+z_x^2+z_y^2)}\,dx\,dy}. \qquad (5874)$$

For a *solid of revolution* round the x axis,

5899 $$\bar{x} = \frac{\int xy^2 dx}{\int y^2 dx} = \frac{\iint r^3 \sin\theta\cos\theta\, d\theta\, dr}{\iint r^2 \sin\theta \, d\theta \, dr}.$$

Proof.—By (5885), for the *moment* $= \int x \,.\, \pi y^2 dx$ and the *volume* $= \int \pi y^2 dx$. The limits as in (5896).

5901 For *any solid figure* bounded as described in (5871), the coordinates of the centroid are given by

$$V\bar{x} = \iiint x\,dx\,dy\,dz = \iint xz\,dx\,dy,$$

$$V\bar{y} = \iiint y\,dx\,dy\,dz = \iint yz\,dx\,dy,$$

$$V\bar{z} = \iiint z\,dx\,dy\,dz = \tfrac{1}{2}\iint z^2\,dx\,dy,$$

where $$V = \iiint dx\,dy\,dz = \iint z\,dx\,dy,$$

as in (5872–3), and the limits are as defined in (1906).

5902 For the *wedge shaped solid* (*OAPB*, Fig. 188) defined by the polar coordinates r, θ, ϕ, as in (5875),

$$V\bar{x} = \tfrac{1}{4}\iint r^4 \sin^2\theta \cos\phi \, d\theta\, d\phi,$$

$$V\bar{y} = \tfrac{1}{4}\iint r^4 \sin^2\theta \sin\phi \, d\theta\, d\phi,$$

$$V\bar{z} = \tfrac{1}{4}\iint r^4 \sin\theta \cos\theta \, d\theta\, d\phi,$$

where $$V = \tfrac{1}{3}\iint r^3 \sin\theta \, d\theta\, d\phi.$$

PROOF.—By (5875); multiplying the elementary pyramid $\frac{1}{3}r^3 \sin\theta\, d\theta\, d\phi$ separately by the distances of its centroid from the coordinate planes; viz., $\frac{3}{4}r \sin\theta \cos\phi$, $\frac{3}{4}r \sin\theta \sin\phi$, and $\frac{3}{4}r \cos\theta$.

MOMENTS AND PRODUCTS OF INERTIA.

5903 DEFINITIONS.—The *moment of inertia* of a body about a given right line or axis is the sum of the products of each element of mass and the square of its distance from the line.

5904 The square of the *radius of gyration* of the body about the given line is equal to the *moment of inertia* of the body divided by its mass.

5905 The moment of inertia of a body with respect to a plane or point is the sum of the products of each element of mass and the square of its distance from the plane or point.

5906 The *product of inertia* of a body with respect to two rectangular coordinate planes is the sum of the products of each element of mass and its distances from the two planes.

5907 Let A, B, C be the moments of inertia of a body about three rectangular axes; A', B', C' the moments of inertia with respect to the three planes of yz, zx, and xy; and

F, G, H the products of inertia with respect to the second and third planes, the third and first, and the first and second respectively. F, G, H are more frequently called the products of inertia about the axes of yz, zx, and xy respectively.

By the definitions we have the values

5908 $$A = \Sigma m\,(y^2+z^2), \qquad F = \Sigma myz,$$
$$B = \Sigma m\,(z^2+x^2), \qquad G = \Sigma mzx,$$
$$C = \Sigma m\,(x^2+y^2), \qquad H = \Sigma mxy.$$

5914 $$\left.\begin{aligned} A' &= \Sigma mx^2 = S-A \\ B' &= \Sigma my^2 = S-B \\ C' &= \Sigma mz^2 = S-C \end{aligned}\right\} \quad \text{where } S = \frac{A+B+C}{2}, \; = \Sigma m\,(x^2+y^2+z^2), \; = \Sigma mr^2.$$

5920 *Theorem I.*—The M. I. of a lamina about an axis perpendicular to its plane is equal to the sum of the two M. I. about any two axes in its plane drawn through the foot of the perpendicular axis and at right angles to each other.

Proof.—By the definition (5903), and Euc. I. 47.

5921 *Theorem II.*—The M. I. of a body about a given axis, plane, or point is equal to the M. I. about a parallel axis or plane through the centroid, or about the centroid itself respectively, plus the M. I. of the whole mass, supposed collected at the centroid, about the given axis, plane, or point.

Proof.—In the figure, p. 168, let the given axis be perpendicular to the paper at B; let A be the centroid, and m an element of mass at C; then, for every thin section of the solid parallel to the paper,

$$\text{M. I.} = \Sigma m.BC^2 = \Sigma m\,(AC^2+AB^2-2AB.AD)$$
$$= \Sigma m.AC^2+\Sigma m.AB^2-2AB.\Sigma m.AD.$$

But $\Sigma m.AD = 0$, by (5885), since A is the centroid of the body, which proves the proposition. Similarly for the plane or point.

Cor. I.—Hence, if the M. I. about any axis is known, that about any parallel axis can be found without integration. For, let I_1 be the M. I. about a given axis, whose distance from the centroid is a, and let I_2 be the required M. I. about an axis whose distance from the centroid is b; then, by Theorem II.,

$$I_2 = I_1 - m\,(a^2-b^2).$$

COR. II.—The M. I. has the same value for all parallel axes at the same distance from the centroid.

5922 *Theorem III.*—The *product of inertia* for two assigned axes is equal to the product for two parallel axes through the centroid of the body plus the product taken for the whole mass collected at the centroid with respect to the assigned axes.

PROOF.—Let $x = \bar{x}+x'$, $y = \bar{y}+y'$ be the coordinates of an element of the body with respect to the assigned axes; $\bar{x}$, $\bar{y}$ being the coordinates of the centroid, and x', y' the coordinates of the same element with respect to parallel axes through the centroid, all axes being parallel to z. Then

$$\Sigma mxy = \Sigma m(\bar{x}+x')(\bar{y}+y') = \bar{x}\bar{y}\,\Sigma m+\Sigma mx'y'+\bar{x}\,\Sigma my'+\bar{y}\Sigma mx'$$
$$= \bar{x}\bar{y}\,\Sigma m+\Sigma mx'y'.$$

Since $\Sigma mx'$ and $\Sigma my'$ vanish by the definition of the centroid.

5923 The M. I. of a body with respect to a point is equal to the M. I. for any plane through the point plus the M. I. about the normal to the plane through the point.

PROOF.—For the origin and yz plane,

$$\Sigma mx^2+\Sigma m(y^2+z^2) = \Sigma mr^2. \qquad (5908, '14, '19)$$

5924 Given the moments and products of inertia, A, B, C, F, G, H, as above, about three rectangular axes, the moment of inertia of the body about a line through the origin, whose direction cosines are l, m, n, will be

$$I = Al^2+Bm^2+Cn^2-2Fmn+2Gnl+2Hlm.$$

PROOF.—(Fig. 11.) Let xyz be a point P of the body, OM the line lmn, and PM the perpendicular upon it. Then the M. I. about OM

$= \Sigma m(OP^2-OM^2) = \Sigma m\{(x^2+y^2+z^2)(l^2+m^2+n^2)-(lx+my+nz)^2\}$ (5530)

producing the above result, by (5908–13).

ELLIPSOIDS OF INERTIA.

5925 The equation of the Momental Ellipsoid is

$$Ax^2+By^2+Cz^2-2Fyz-2Gzx-2Hxy = M\epsilon^4,$$

obtained by putting $Ir^2 = M\epsilon^4$. M being the mass of the body, and ϵ^4 a constant to make the equation homogeneous. Hence *the square of the radius of the momental ellipsoid for any point varies inversely as the moment of inertia of the body about that radius.*

5926 If the products of inertia vanish, the axes are called the *principal axes* of the body.

5927 At every point of a body there are always three principal rectangular axes.

PROOF.—These are evidently the principal axes of the momental ellipsoid of the point; for if the coordinate axes be made to coincide with the former, F, G, H will vanish.

5928 The equation of the momental ellipsoid referred to its principal axes will be

$$Ax^2+By^2+Cz^2 = M\epsilon^4.$$

5929 The moment of inertia about a line lmn will now be

$$I = Al^2+Bm^2+Cn^2.$$

THE ELLIPSOID OF GYRATION.

5930 The equation of the Ellipsoid of Gyration referred to principal axes is

$$\frac{x^2}{A}+\frac{y^2}{B}+\frac{z^2}{C} = \frac{1}{M}.$$

It is the reciprocal surface of the momental ellipsoid (5719), and its property is—

5931 *The moment of inertia about the perpendicular from the origin upon the tangent plane varies as the square of the perpendicular.*

5932 For any other rectangular axes through the point, the equation of the ellipsoid of gyration is, by (5717),

$$\begin{vmatrix} A & -H & -G & x \\ -H & B & -F & y \\ -G & -F & C & z \\ x & y & z & \frac{1}{M} \end{vmatrix} = 0,$$

being the reciprocal surface of the momental ellipsoid,

$$(A, B, C, -F, -G, -H \between xyz)^2 = M,$$

with the radius of the sphere of reciprocation $= 1$. The equation when expanded becomes

5933 $$(BC-F^2)\,x^2+\ldots+2\,(FG+CH)\,xy = \begin{vmatrix} A & H & G \\ H & B & F \\ G & F & C \end{vmatrix}\frac{1}{M}.$$

LEGENDRE'S EQUI-MOMENTAL ELLIPSOID.

5934 The equation is

$$\frac{x^2}{A'}+\frac{y^2}{B'}+\frac{z^2}{C'}=\frac{5}{M},$$

with the values in (5914).

5935 The mass of this ellipsoid is taken equal to that of the body, and it has the same principal moments of inertia.

THE MOMENTAL ELLIPSOID FOR A PLANE.

5936 If A', B', C' be the moments of inertia for the three coordinate planes, as in (5914), the M. I. for a plane through the origin whose dir. cos. are l, m, n, will be

$$I'=A'l^2+B'm^2+C'n^2+2Fmn+2Gnl+2Hlm.$$

PROOF: $I'=\Sigma m\,(lx+my+nz)^2=\Sigma mx^2.\,l^2+\&c.=A'l^2+\&c.$

5937 The momental ellipsoid for this plane will be

$$A'x^2+B'y^2+C'z^2+2Fyz+2Gzx+2Hxy=M\epsilon^4,$$

and its property is—

5938 *The M. I. for any plane passing through the centre of the ellipsoid is equal to the inverse square of the radius perpendicular to the plane.*

5939 If r be a radius of this ellipsoid, and a, b, c its semi-axes, the M. I. about r

$$=\frac{1}{a^2}+\frac{1}{b^2}+\frac{1}{c^2}-\frac{1}{r^2}.$$

PROOF.—(Fig. 11.) M. I. about r, plus M. I. for the plane OM perpendicular to r

$$=\Sigma mOP^2=\Sigma mx^2+\Sigma my^2+\Sigma mz^2=\frac{1}{a^2}+\frac{1}{b^2}+\frac{1}{c^2}, \text{ by (5938)}.$$

EQUI-MOMENTAL CONE.

5940 The equation of the equi-momental cone at any point of a body, referred to principal axes of the body at the point, is $$(A-I)\,x^2+(B-I)\,y^2+(C-I)\,z^2=0,$$

its property being that

5941 *The generating line passes through the given point, and moves so that the M. I. about it is a constant* $= \mathrm{I}$.

PROOF.—Let lmn be the generating line in one position, then
$$Al^2+Bm^2+Cn^2 = I\,(l^2+m^2+n^2). \quad \text{Therefore, \&c.}$$

5942 *Theorem.*—If two systems have the same mass, the same centroid, principal axes and principal moments of inertia at the centroid, they have equal moments of inertia about any right line whatever, and are termed *equi-momental*. By (5906) and (5929).

5943 If two bodies are equi-momental, their projections are equi-momental.

PROOF.—If the projection be from the xy plane in the ratio $1:n$, the coordinates x, y, z of a particle become x, y, nz, and the mass m becomes nm. The conditions in (5942) will then be fulfilled.

MOMENT OF INERTIA OF A TRIANGLE.

5944 The M. I. of a triangle ABD (Fig. 190) about a side BD, distant p from the opposite vertex A, is

$$I=\frac{mp^2}{6}.$$

PROOF.—Let $BD\equiv a$ and $EF\equiv y$; $I=\int_0^p \frac{a\,(p-y)}{p}\,y^2dy=\frac{ap^3}{12}=m\,\frac{p^2}{6}$

5945 The M. I. of a triangle ABC (Fig. 190) about a straight line BD passing through a vertex B, and distant p and q from the vertices A and C, is

$$I=m\,\frac{p^2+pq+q^2}{6}.$$

PROOF.—By (5944), taking difference of M. I. of the triangles ABD, CBD.

5946 The M. I. of a triangle ABC about an axis through its centroid parallel to BD, is

$$I=m\,\frac{p^2-pq+q^2}{18}. \qquad \text{By (5921)}$$

5947 Cor.—If the triangle be isosceles, so that $p = q$, the last two moments of inertia become

$$\frac{mp^2}{2} \text{ and } \frac{mp^2}{18}.$$

5949 The M. I. of the triangle about axes perpendicular to ABC through B and through the centroid, respectively, are

$$m\frac{3(c^2+a^2)-b^2}{12} \text{ and } m\frac{(a^2+b^2+c^2)}{36}. \qquad (5920)$$

5951 The M. I. about GF of the trapezoid $ACGF$ (Fig. 190), is

$$m\frac{p^2+q^2}{6}.$$

5952 The moments and products of inertia of a triangle about any axes are the same for three equal particles, each one-third of the mass of the triangle, placed at the mid-points of its sides.

Proof.—(Fig. 190.) The M. I. of the three particles at the mid-points of AB, BC, CA about BD, any line through a vertex, will be

$$\frac{M}{3}\left\{\frac{(p+q)^2}{4}+\frac{p^2}{4}+\frac{q^2}{4}\right\},$$

which is equal to that of the triangle, by (5945).

MOMENTAL ELLIPSE.

5953 If α, β be the radii of gyration of a plane area to principal axes Ox, Oy, where O is the centroid, the equation of the momental ellipse for the point O will be

$$\alpha^2x^2+\beta^2y^2 = 2\alpha^2\beta^2.$$

5954 Also the area is equi-momental with three equal particles, each one one-third of its mass placed anywhere on the ellipse so that O may be their centroid.

Proof.—Let xy, $x'y'$, $x''y''$ be the coordinates of three equi-momental particles: then

$$\frac{m}{3}(x^2+x'^2+x''^2) = m\beta^2; \quad \frac{m}{3}(y^2+y'^2+y''^2) = m\alpha^2; \quad xy+x'y'+x''y'' = 0;$$

and the two systems have the same centroid; therefore

$$x+x'+x''=0 \quad \text{and} \quad y'+y''+y'''=0.$$

Eliminating x', y', x'', y'' between the five equations, we find the equation of (5953) for the locus of xy.

5955 The momental ellipse for the centroid of a triangle is the inscribed ellipse touching the sides at their mid-points.

Proof.—(Fig. 189.) The inscribed ellipse, which touches two sides at their mid-points, also touches the third side at its mid-point, by Carnot's theorem (4779). Now DF is parallel to AC, the tangent at E; therefore BE, which bisects DF, passes through the centre O of the ellipse. Similarly, AD passes through it; therefore O is the centroid of the triangle.

Let $OE \equiv a'$, and let b' be the semi-diameter parallel to AC; then $\frac{ON^2}{a'^2}+\frac{FN^2}{b'^2}=1$. But $ON=\frac{a'}{2}$, therefore $FN^2=\frac{3}{4}b'^2$. The M. I. about OE, by (5954), $=\frac{2}{3}m\frac{3}{4}b'^2\sin^2\omega = m\frac{a^2b^2}{2a'^2}$, where a, b are the semi-axes. Hence the M. I. about OD, OE, OF varies inversely as the squares of those lines, and therefore the ellipse in the diagram is a momental ellipse, since it has six points which fulfil the requirements.

5956 The projections of a plane area and its momental ellipse form another plane area and its momental ellipse. (5943)

5957 The M. I. of a tetrahedron $ABCD$ about any plane through A is

$$\frac{m}{10}(\alpha^2+\beta^2+\gamma^2+\beta\gamma+\gamma\alpha+\alpha\beta),$$

where α, β, γ are the perpendiculars on the plane from B, C, D.

5958 The tetrahedron is also equi-momental with four particles, each one-twentieth of the mass, placed at the vertices, and a particle equal to the remaining mass placed at the centroid (5942).

5959 The equi-momental ellipsoid of a tetrahedron has the same centroid, and touches each edge at its middle point.

Proof.—By projecting a regular tetrahedron and its equi-momental sphere (for the centroid) of radius $=\sqrt{3}\times$ radius of inscribed sphere.

5960 To find the point, if it exists, in a given right line at which the line is a principal axis, and to find the other principal axes at the point.

Let O be a datum point in the line. Take this for origin, the given line for axis of z, and OX, OY for the other axes. Then, if h be the distance from O to the required point O', and θ the angle between OX and the principal axis $O'X'$,

5961 $$h = \frac{\Sigma myz}{\Sigma my} = \frac{\Sigma mzx}{\Sigma mx} \quad \text{and} \quad \tan 2\theta = \frac{2H}{A-B},$$

where A, B, H are the moments and product of inertia about OX, OY.

PROOF.—At the point $0, 0, h$, $\Sigma m(z-h)x = \Sigma m(z-h)y = 0$, from which h is found; and the equation for θ is that for determining the principal axes of the elliptic section of the momental ellipsoid, whose equation is $Ax^2+2Hxy+By^2 = M\epsilon^4$, as in (4408).

5964 The equality of the two ratios in (5961) is the condition that the z axis should be a principal axis at *some* point of its length.

5965 If an axis be a principal axis at more than one point of its length, it passes through the centroid of the system; and, conversely, if it be a principal axis at the centroid, it is so at every point of its length.

PROOF.—For h must be indeterminate in (5961). Therefore $\Sigma myz = 0$, $\Sigma my = 0$, $\Sigma mzx = 0$, $\Sigma mx = 0$.

5966 The principal axes $O'X'$, $O'Y'$ are parallel to the principal axes of the projection of the body in the original plane of xy. By (5962–3).

5967 Given the principal axes of a body at its centroid, to find the principal axes and moments of inertia at any point in the principal plane of xy.

Let C in the Figure of (1171) be the centroid, CX, CY principal axes, A, B the M. I. about them, and P the given point. Find two points S, S', called *foci of inertia*, such that the X and Y moments of inertia there are equal, and therefore

$$B+m.CS^2 = A; \quad \text{giving} \quad CS = CS' = \sqrt{\frac{A-B}{m}} \ldots \text{(i.)}.$$

The internal and external bisectors of the angle SPS' will be two of the principal axes at P, and the third will be the normal to the plane.

PROOF.—The X and Y principal moments being equal at S, the moment about every line through S in this plane is the same. [For $I = Al^2+Bm^2+Cn^2$ and $n=0$ and $A=B$, therefore $I=A$.] Therefore the moments about SP and $S'P$ are equal. Therefore the bisectors PT, PG of the angles at P will be principal axes.

5968 Let SY, $S'Y'$ be the perpendiculars on PT, and SZ, $S'Z'$ those upon PG; then the M. I. about PT and PG will be respectively,

$$A+mSY.S'Y' = B+m\left(\frac{SP+S'P}{2}\right)^2.$$

$$A-mSZ.S'Z' = B+m\left(\frac{SP-S'P}{2}\right)^2.$$

PROOF.—Draw CR perpendicular to SY. The M. I. about CR $(\theta \equiv SCR)$

$$= A\cos^2\theta + B\sin^2\theta \ (5929) = A-(A-B)\sin^2\theta$$
$$= A-m\,CS^2\sin^2\theta \text{ (by i.)} = A-mSR^2.$$

Therefore M. I. about $PT = A-mSR^2+mRY^2$ (5921)

$$= A+m(RY+SR)(RY-SR) = A+mSY.S'Y'$$
$$= A+mBC^2 \ (1178) = B+mAC^2 \text{ (by i.)} = B+m\left(\frac{SP+S'P}{2}\right)^2.$$

Similarly for the M. I. about PG.

5969 Hence, if an ellipse or hyperbola be described with S, S' for foci, the tangent and normal at any point of the curve are principal axes, and the M. I. about either is constant for that curve.

5970 Similarly, for a point P in *any* plane through the centroid O, it may be shown that the same construction will give the axes PT, PG about which the product of inertia vanishes, OX, OY being the axes at O in the given plane about which the product of inertia vanishes.

5971 The condition for the existence of a point in a body at which the M. I. about every axis through it shall be the same, is—

There must be two principal axes of equal moment at the centroid, and the M. I. about each must be less than the third principal moment.

Two such points will then exist situated on the axis of unequal moment, and equi-distant from the centroid.

5972 Given the principal axes at the centroid of a body and the moments of inertia about them, to find the principal axes and moments at any other point.

[See (5975) for the result.]

Let A, B, C be the given principal moments, and let the mass of the body be unity. Then the ellipsoid of gyration at the centroid O, and a quadric confocal with it, will be

$$\frac{x^2}{A}+\frac{y^2}{B}+\frac{z^2}{C}=1 \quad \text{and} \quad \frac{x^2}{A+\lambda}+\frac{y^2}{B+\lambda}+\frac{z^2}{C+\lambda}=1.$$

5973 Prop. I.—The M. I. is constant for all tangent planes of the confocal, and is equal to the

$$\textit{M. I. for the origin } O+\lambda = S+\lambda. \qquad (5919)$$

Proof.—Let l, m, n be the dir. cos. of the tangent plane of the confocal, p the perpendicular on the plane from O. The M. I. for this plane

$= \text{M. I. for a parallel plane through } O+p^2$ (5921)

$= A'l^2+B'm^2+C'n^2+p^2$ (5936)

$= (S-A)\,l^2+(S-B)\,m^2+(S-C)\,n^2+(A+\lambda)\,l^2+(B+\lambda)\,m^2+(C+\lambda)\,n^2$ (5914, 5631) $= S+\lambda$, which is independent of l, m, n.

5974 Prop. II.—All these planes are principal planes at their points of contact, and if the three confocals be drawn through any point P, the tangent planes at P to the confocal ellipsoid, two-fold hyperboloid, and one-fold hyperboloid, are respectively the principal planes of greatest, least, and mean moments of inertia. The normal to the confocal ellipsoid is the axis of least moment, and the normal to the two-fold hyperboloid is the axis of greatest moment.

Proof.—Draw any other plane through P. The perpendicular on it from O is less than the perpendicular on the parallel tangent plane to the confocal ellipse, and greater in the case of the two-fold hyperbola. Then, by (5921).

The solution of the problem at (5972) is now given by Proposition III.

5975 Prop. III.—The principal moments of inertia at P are $OP^2-\lambda_1$, $OP^2-\lambda_2$, $OP^2-\lambda_3$, and the normals to the three confocals at P are the principal axes.

Proof.—The M. I. about the x axis at P

$= \text{M. I. for the origin } P - \text{M. I. for the } yz \text{ plane}$

$= S+OP^2-(S+\lambda_1) = OP^2-\lambda_1$ (5921–73).

5976 The principal moments of inertia above, expressed in terms of λ_1 of the confocal ellipsoid and d_2, d_3, its principal semi-diameters conjugate to OP, will, by (5661), become

$$OP^2-\lambda_1, \qquad OP^2-\lambda_1+d_3^2, \qquad OP^2-\lambda_1+d_2^2.$$

5977 The condition that the line abc, lmn, referred to principal axes at the centroid, may itself be a principal axis at some point of its length, is

$$\frac{\frac{a}{l}-\frac{b}{m}}{A-B} = \frac{\frac{b}{m}-\frac{c}{n}}{B-C} = \frac{\frac{c}{n}-\frac{a}{l}}{C-A} = \frac{1}{p}.$$

Here abc is *any* point on the line, and if a confocal quadric of the ellipsoid of gyration at the centroid be drawn through the stated principal point of the given line, p is the perpendicular from the origin upon the tangent plane of the confocal at that point.

PROOF.—The given line $\frac{x-a}{l} = \frac{y-b}{m} = \frac{z-c}{n}$ .... (i.)

must be a normal to the confocal $\frac{x^2}{A+\lambda}+\frac{y^2}{B+\lambda}+\frac{z^2}{C+\lambda} = 1$ .... (ii.).

Therefore, by (5629), $l = \frac{px}{A+\lambda}$, $m = \frac{py}{B+\lambda}$, $n = \frac{pz}{C+\lambda}$ .... (iii.).

Eliminate x, y, z from (i.) by means of (iii.), and from the resulting equations eliminate p, and the condition above is obtained.

Also, by (5631),

$$p^2 = (A+\lambda)\,l^2+(B+\lambda)\,m^2+(C+\lambda)\,n^2 = Al^2+Bm^2+Cn^2+\lambda \;\ldots\;\text{(iv.)}.$$

The principal point xyz is now found by eliminating λ and p from equations (iii.), by means of (iv.) and (5977).

INTEGRALS FOR MOMENTS OF INERTIA.

By the definition (5903), the following indefinite integrals for moments of inertia are obtained :—

5978 For a plane curve, $y=f(x)$, the M. I. about the x and y coordinate axes are

$$\int y^2 ds \text{ and } \int x^2 ds; \quad \text{and therefore } \int (x^2+y^2)\,ds = \int r^2 ds$$

is the M. I. about an axis perpendicular to both the former through the origin (5920).

5980 Observe that ds may be changed into dx, dy, or $d\theta$ by the substitution formulæ (5113, '16).

5981 For a plane area bounded by the coordinate axes, the ordinate y and the curve $y = f(x)$, the M. I. about the x and y axes are

$$\iint y^2 dx\,dy = \tfrac{1}{3}\int y^3 dx \quad \text{and} \quad \iint x^2 dx\,dy = \int x^2 y\,dx.$$

5983 And the M. I. about an axis perpendicular to both the former drawn through the origin,

$$= \iint (x^2+y^2)\,dx\,dy = \iint r^3 dr\,d\theta = \tfrac{1}{4}\int r^4 d\theta,$$

but in the last two integrals the area has the boundaries described in (5894).

5986 The M. I. of a solid bounded by three rectangular coordinate planes and the surface $z = f(x, y)$ about the z axis, will be

$$\iint (x^2+y^2)\,z\,dx\,dy = \iiint r^4 \sin^3\theta\,dr\,d\theta\,d\phi,$$

but in the last integral the solid is bounded as described in (5875).

5988 The volume, which represents the mass in all these cases, has already been expressed (5205, 5871); and by dividing by the volume, the square of the radius of gyration of the solid is found (5904).

Proofs.—Formulæ (5981–3) are directly obtainable by geometry from figures 90 and 91, and formulæ (5986–7) from figures 168 and 188. The transition to polar coordinates may also be effected by the formula of (2774).

5989 In expressing moments of inertia, the factor m will stand for the mass of the body, and the remaining factor will therefore be the value of the square of the radius of gyration.

PERIMETERS, AREAS, VOLUMES, CENTRES OF MASS, AND MOMENTS OF INERTIA OF VARIOUS FIGURES.

RECTANGULAR LAMINA AND RIGHT SOLID.*

For a *rectangle* whose sides are a, b, the *moments of inertia* about the sides, and an axis perpendicular to both where they meet, are respectively

6015 $$m\frac{b^2}{3},\quad m\frac{a^2}{3},\quad m\frac{a^2+b^2}{3}.$$

PROOF: $\int_0^b ax^2\,dx = \frac{ab^3}{3} = m\,\frac{b^2}{3}$. The third by (5920).

6018 Hence, for a *right solid*, whose dimensions are $2a$, $2b$, $2c$,

$$M.\,I.\text{ about the axis of figure }2c = m.\frac{a^2+b^2}{3}.$$

ARC OF A CIRCLE.

6019 Let AB (Fig. 191) be the arc of a circle whose centre is O and radius r. Let the angle $AOB \equiv \theta$; then

$$\textit{Length of arc } AB = r\theta. \qquad (601)$$

6020 Huygens' Approximation.—RULE.—*From* 8 *times the chord of half the arc take the chord of the whole arc, and divide the remainder by* 3.

PROOF.—The rule gives $\frac{r}{3}\left(16\sin\frac{\theta}{4} - 2\sin\frac{\theta}{2}\right)$.

Expand the sines by (764) as far as θ^3, and the result is $r\theta$.

6021 Taking an axis OX through the mid-point of the arc with origin O, the *centroid of the arc* is given by (5889)

$$\bar{x} = \frac{2r\sin\frac{1}{2}\theta}{\theta}. \quad \text{Hence for a semi-circle } \bar{x} = \frac{2r}{\pi}.$$

* For $M.\,I.$ of a triangle, see (5944-52).

6023 Also, for *the centroid of BX*, $\bar{y} = \dfrac{2r\sin^2\frac{1}{2}\alpha}{\alpha}$, where $\alpha = \angle XOB$.

6025 The *M. I. of the arc AB* about OX and OY are

$$\frac{mr^2}{2}\left(1-\frac{\sin\theta}{\theta}\right) \quad \text{and} \quad \frac{mr^2}{2}\left(1+\frac{\sin\theta}{\theta}\right). \qquad (5978)$$

6027 *M. I.* about axes perpendicular to XOY, through O and X the mid-point of the arc respectively, are

$$mr^2 \quad \text{and} \quad m.2r^2\left(1-\frac{\sin\frac{1}{2}\theta}{\frac{1}{2}\theta}\right). \qquad (5979)$$

6029 Cor.—The *M. I.* of a *circular ring* about a diameter $\Big\} = \dfrac{mr^2}{2}$.

SECTOR OF CIRCLE. *AOB* (Fig. 191)

6030 *Area* $= \dfrac{r^2\theta}{2}$, $\quad \bar{x} = \dfrac{4r\sin\frac{1}{2}\theta}{3\theta}$. For XOB, $\bar{y} = \dfrac{4r\sin^2\frac{1}{2}\alpha}{3\alpha}$.

Proof.—$\bar{x}$, $\bar{y}$ are respectively $\frac{2}{3}$ of $\bar{x}$, $\bar{y}$ in (6021, '3); since the centroid of each elemental sector is distant $\frac{2}{3}r$ from O. Otherwise, by (5893, '5).

6033 The *M. I.* about OX and OY are

$$\frac{mr^2}{4}\left(1-\frac{\sin\theta}{\theta}\right) \quad \text{and} \quad \frac{mr^2}{4}\left(1+\frac{\sin\theta}{\theta}\right).$$

Proof.—By (5981–2); or integrate (6025–6) for r from 0 to r.

SEGMENT OF CIRCLE. *ABX* (Fig. 191)

6035 *Area* $= \dfrac{r^2}{2}(\theta-\sin\theta)$, $\qquad \bar{x} = \dfrac{4r\sin^3\frac{1}{2}\theta}{3(\theta-\sin\theta)}$.

6037 For CBX, $\quad \bar{y} = \dfrac{r(2-3\cos\alpha+\cos^3\alpha)}{3(\alpha-\sin\alpha\cos\alpha)}$.

Proofs.—From the sector and triangle; otherwise, the centroid, by (5893, '5).

6038 *M. I.* about *OX* and *OY*, (5981-2)

$$\frac{r^4}{24}(3\theta-4\sin\theta+\sin\theta\cos\theta) \quad\text{and}\quad \frac{r^4}{8}(\theta-\sin\theta\cos\theta).$$

6040 Cor.—Hence, for a *semi-circle*, $\bar{x}=\frac{4r}{3\pi}$.

6041 Also, the *M. I. of a circle* about a diameter, and about a central axis perpendicular to its plane, are respectively

$$\frac{mr^2}{4} \quad\text{and}\quad \frac{mr^2}{2}. \qquad (5920)$$

THE RIGHT CONE.

If h be the height, r the radius of the base, and l the slant,

6043 *Curved surface* $= \pi rl$. *Volume* $= \frac{1}{3}\pi r^2 h$.

6045 Distance of *centroid* from vertex $= \frac{3}{4}h$.

6046 *M. I.* about axis of figure $= m\frac{3}{10}r^2$.

6047 *M. I.* about cross axes through the vertex and centroid respectively.

$$m\tfrac{3}{20}(r^2+4h^2) \quad\text{and}\quad m\tfrac{3}{80}(4r^2+h^2).$$

FRUSTUM OF CYLINDER.

Let θ be the inclination of the cutting plane to the base, and c the length of the axis intercepted.

6048 The distance of the *centroid* from the axis is

$$\bar{x}=\frac{a^2\tan\theta}{4c}.$$

6049 The *M. I.* about the axis $= m\frac{a^2}{2}$, being the same as that of a cylinder of height c. Hence, by (5921) and the value of $\bar{x}$ above, the M. I. about any line parallel to the axis can be found.

SEGMENT OF SPHERICAL SURFACE. (Fig. 191)

Let O be the origin of coordinates; $OA \equiv r$ the radius; and $OC \equiv x$ the abscissa of AB the plane of section.

6050 The *curved area of* $AB = 2\pi r\,(r-x)$ = the area of its projection on the enveloping cylinder of the sphere.

PROOF: $$\text{Area} = \int_x^r 2\pi y \frac{r}{y}\,dx = 2\pi r\,(r-x). \quad (5878)$$

6051 For centroid of surface, $\bar{x} = \frac{1}{2}(r+x)$.

6052 The M. I. about the axes OX, OY are

$$\frac{m}{3}(2r^2-rx-x^2) \quad \text{and} \quad \frac{m}{6}(4r^2+rx+x^2).$$

HEMISPHERICAL SURFACE.

6054 *Area* $= 2\pi r^2$, $\quad \bar{x} = \dfrac{r}{2}$. (6050–1)

6056 *M. I.* about OX or OY $= m\,\frac{2}{3}r^2$. (6052–3)

SEGMENT OF SPHERE.

6057 *Volume* $= \dfrac{\pi}{3}(2r+x)(r-x)^2$, $\quad \bar{x} = \dfrac{3\,(r+x)^2}{4\,(2r+x)}$.

6059 *M. I.* about $OX = \dfrac{\pi}{30}(r-x)^3\,(8r^2+9rx+3x^2)$.

6060 *M. I.* about OY

$$= \frac{\pi}{60}(r-x)^2(16r^3+17r^2x+18rx^2+9x^3).$$

PROOF.—As in (6146–7); or put $a=b=c$ in the results.

HEMISPHERE.

6061 *Volume* $= \frac{2}{3}\pi r^3$, $\quad \bar{x} = \frac{3}{8}r$. (6064)

PROOF.—Vol. = surface (6054) $\times \dfrac{r}{3}$, by elemental pyramids having their common vertex at the centre of the sphere. Otherwise, make $x=0$ in (6057).

6063 M. I. about OX or $OY = m\,\frac{2}{5}r^2$. (6059–60)

SECTOR OF SPHERE.

6064 $\textit{Volume} = \frac{2}{3}\pi r^2(r-h), \qquad \bar{x} = \frac{3}{8}(r+h).$

PROOF.—Vol. = surface (6050) $\times \frac{r}{3}$. $\bar{x} = \frac{3}{4}$ of $\bar{x}$ in (6051), since the centroid of each elemental pyramid is distant $\frac{3}{4}$ths of r from the centre.

6066 For the *M. I.* add together the *M. I.* of the cone and segment (6046, '59).

THE PARABOLA, $y^2 = 4ax$. (Fig. of 1220)

6067 *Rad. of curv.* $\rho = \frac{2SP^2}{SY} = 2a\left(1+\frac{x}{a}\right)^{\frac{3}{2}}$. (4542)

6069 Coordinates of *centre of curvature*

$$3x+2a, \qquad -\frac{y^3}{4a^2}. \tag{4545}$$

6071 $\textit{Arc } AP \equiv s = \sqrt{(ax+x^2)} + a\log\frac{\sqrt{x}+\sqrt{(a+x)}}{\sqrt{a}}.$

6072 $= a\left[\cot\theta\operatorname{cosec}\theta + \log\cot\left(\tfrac{1}{2}\theta\right)\right].$

PROOF: $s = \int\sqrt{\left(1+\frac{a}{x}\right)}\,dx.$ (5197, 4206)

Substitute $\sqrt{x}$, and integrate by (1931).

6073 $\textit{Arc } AL = a\sqrt{2} + a\log(1+\sqrt{2}).$

Centroid of arc AP with above value of s.

6074 $s\bar{x} = \frac{2x+a}{4}\sqrt{(x^2+ax)} + \frac{a^2}{8}\log\frac{2x+a+2\sqrt{(x^2+ax)}}{a}.$

6075 $s\bar{y} = \frac{4}{3}\left\{\sqrt{a(x+a)^3} - a^2\right\}.$

6076 For *centroid of arc AL*, putting $x = a$,

$$\bar{x} = \frac{6\sqrt{2}+\log(3+2\sqrt{2})}{8\{\sqrt{2}+\log(1+\sqrt{2})\}}\,a, \qquad \bar{y} = \frac{4}{3}\cdot\frac{(2\sqrt{2}-1)\,a}{\sqrt{2}+\log(1+\sqrt{2})}.$$

Half-segment of parabola *ANP*.

6078 $Area = \frac{2}{3}xy, \quad \bar{x} = \frac{3}{5}x, \quad \bar{y} = \frac{3}{8}y.$

6081 The *M. I.* about the x and y axes are

$$m\tfrac{4}{5}ax \quad \text{and} \quad m\tfrac{3}{7}x^2.$$

THE ELLIPSE.

6083 The equation being $b^2x^2+a^2y^2 = a^2b^2$, the *length of the arc AP* (Fig. of 1205), putting ϕ for the eccentric angle of P, is

$$s = a\int_0^\phi \sqrt{(1-e^2\cos^2\phi)}\,d\phi.$$

PROOF. — In $(ds)^2 = (dx)^2+(dy)^2$ (5113), substitute $dx = -a\sin\phi\,d\phi$, $dy = b\cos\phi\,d\phi$, by (4276), and use (4260).

6084 The length of the *elliptic quadrant AB* is

$$\frac{\pi a}{2}\left\{1-\frac{e^2}{4}-\frac{3e^4}{2!\,2!\,2^4}-\frac{3^2.5e^6}{3!\,3!\,2^6}-\frac{3^2.5^2.7e^8}{4!\,4!\,2^8}-\text{\&c.}\right\}.$$

PROOF.—Expand the binomial surd above, and employ (2454) and (2472). Similarly, from (5887) and (5978) the three following values are found.

6085 For the *centroid* of the same quadrant,

$$\bar{x} = \frac{2a}{\pi}\cdot\frac{1-\frac{1}{3}e^2-\frac{1}{15}e^4}{1-\frac{1}{4}e^2-\frac{3}{64}e^4} \quad \text{approximately}.$$

6086 The *M. I.* about the x and y axes are approximately,

$$\frac{mb^2}{2}\cdot\frac{1-\frac{1}{8}e^2-\frac{1}{64}e^4}{1-\frac{1}{4}e^2-\frac{3}{64}e^4} \quad \text{and} \quad \frac{ma^2}{2}\cdot\frac{1-\frac{3}{8}e^2-\frac{5}{64}e^4}{1-\frac{1}{4}e^2-\frac{3}{64}e^4}.$$

6088 *Fagnani's Theorem.*—(Fig. 192.) Let P be any point on the ellipse, CY the perpendicular on the tangent at P; $\angle ACY = \theta$; Q the point whose eccentric angle $= \frac{1}{2}\pi-\theta$. Then

6089 $$PY+AP = a\int\sqrt{(1-e^2\sin^2\theta)}\,d\theta = BQ;$$

and in the hyperbola (Fig. 193)

6090 $$PY-AP = a\int\sqrt{(1-e^2\sin^2\theta)}\,d\theta.$$

6091 Cor.—The difference between the lengths of the infinite curve and asymptote $= a\int_0^\alpha \sqrt{(1-e^2\sin^2\theta)}\,d\theta$, where $\tan\alpha = \frac{a}{b}$.

Proofs.—By (5203),

$$AP+PY \text{ or } s+q = \int_0^\theta p\,d\theta = a\int \sqrt{(1-e^2\sin^2\theta)}\,d\theta = BQ, \text{ by (6083).}$$

In the hyperbola we have $q-s = \int p\,d\theta$.

6092 Draw the tangent at Q and the perpendicular CU upon it. Let x, x' be the abscissæ of P, Q. The following relations subsist,

$$PY = \frac{e^2xx'}{a} = QU, \quad CY.CU = ab, \quad CP^2+CU^2 = a^2+b^2.$$

Proof.—Let ϕ = the eccentric angle of P, and let $ACU = \theta'$. Then

$$\tan\phi = \frac{ay}{bx} = \frac{b}{a}\tan\theta. \qquad (4276\text{–}80)$$

Similarly for Q, $\tan\left(\frac{\pi}{2}-\theta\right) = \cot\theta = \frac{b}{a}\tan\theta'$,

therefore $\tan\phi = \cot\theta'$ or $\phi = \frac{\pi}{2}-\theta'$ .... (i.).

The relation therefore between P and Q is reciprocal. Now $PY = e^2x\sin\theta$ (4295) and $x' = a\sin\theta$, therefore $PY = \frac{e^2xx'}{a} = QU$, by the reciprocity.

Again, $CU^2 = a^2\cos^2\theta'+b^2\sin^2\theta'$ (4372) $= a^2\sin^2\phi+b^2\cos^2\phi$ ...... (ii.). Put ϕ in terms of θ by the above, and we find

$$CU^2 = \frac{a^2b^2}{a^2\cos^2\theta+b^2\sin^2\theta} = \frac{a^2b^2}{CY^2}.$$

Lastly, $CP^2+CU^2 = x^2+y^2+a^2\sin^2\phi+b^2\cos^2\phi$, by (ii), $= a^2+b^2$ (4276–7).

6095 When P coincides with Q, the point is called "Fagnani's point," $CY = \sqrt{(ab)}$, $PY = a-b$, and $x = a^{\frac{3}{2}}(a+b)^{-\frac{1}{2}}$.

6096 *Griffiths' Theorem.**—If an ellipse of eccentricity e, and a hyperbola of eccentricity e^{-1}, be placed as in the figure of 1205 (the circle representing the ellipse), P, p being considered corresponding points; then, calling PQ, in (6088), a Fagnanian arc, we have the following theorem:—

* J. Griffiths, M.A., *Proc. Lond. Math. Soc.*, Vol. v., p. 95.

The ratio of the difference of two Fagnanian arcs on the ellipse to the difference of the two corresponding arcs on the hyperbola is equal to the product of e^2 and the four abscissæ of the points on the ellipse.

SECTOR AND SEGMENT OF ELLIPSE.

6097 The formulæ for the sector and segment of a circle may be adapted to the ellipse *by writing* a *for* r *and multiplying linear dimensions parallel to the minor axis by* $b : a$. But α will then represent the eccentric angle of the semi-arc, and θ twice that angle. Thus, in the figure of (1205), if ACP be the half sector, $\alpha = ACp$, $\theta = 2ACp$.

Sector of ellipse ($2ACP$ in fig. of 1205):

6098 $$Area = \frac{ab\theta}{2},\quad \bar{x} = \frac{4a\sin\frac{1}{2}\theta}{3\theta},\quad \bar{y} = \frac{4b\sin^2\frac{1}{2}\alpha}{3\alpha},\qquad (6030\text{–}2)$$

the last being for the half sector ACP. The M. I. about the x and y axes are

6101 $$\frac{mb^2}{4}\left(1-\frac{\sin\theta}{\theta}\right) \text{ and } \frac{ma^2}{4}\left(1+\frac{\sin\theta}{\theta}\right).\qquad (6033)$$

Segment of ellipse ($2ANP$ in same figure):

6103 $$Area = \frac{ab}{2}(\theta-\sin\theta),\quad \bar{x} = \frac{4a\sin^3\frac{1}{2}\theta}{3(\theta-\sin\theta)}.\qquad (6035\text{–}6)$$

6105 For $\bar{y}$ of the *half segment* ANP, and for the M. I. about the x and y axes, replace r by b in (6037–8) and by a in (6039).

6108 For the *whole ellipse*, the *area* $= \pi ab$. (6103)

6109 For the *half ellipse*, $\bar{x} = \dfrac{4a}{3\pi}$. (6104)

6110 The *M. I.* about the x and y axes, and a third central axis perpendicular to both,

$$\frac{mb^2}{4},\quad \frac{ma^2}{4},\quad \text{and}\quad \frac{m(a^2+b^2)}{4}.\qquad (6041\text{–}2)$$

6113 The area of the ellipse whose equation is

$$(a\,b\,c\,f\,g\,h \between x\,y\,1)^2 = 0, \quad \text{is} \quad = \frac{\pi\Delta}{\sqrt{(ab-h^2)^3}} \quad \text{or} \quad \frac{\pi\Delta}{\sqrt{C^3}}.$$

PROOF.—If α, β be the semi-axes of the conic, the area $\pi\alpha\beta$ takes this value, by (4414) and (4407).

6114 *Lambert's Theorem.*—The area of a focal sector of an ellipse, as PSP' (Fig. 28), in terms of ϕ, ϕ', the eccentric angles of P, P', is

$$\frac{ab}{2}\{\phi-\phi'-e(\sin\phi-\sin\phi')\} = \frac{ab}{2}\{\chi-\chi'-(\sin\chi-\sin\chi')\}.$$

In the second value, $\sin\frac{\chi}{2}$ and $\sin\frac{\chi'}{2}$ are $=\frac{1}{2}\sqrt{\frac{r+r'\pm c}{a}}$ respectively, where $r = SP$, $r' = SP'$, and $c = PP'$,* a result of use in Astronomy.

THE HYPERBOLA.

6115 The *length of an arc* of the hyperbola $b^2x^2-a^2y^2 = a^2b^2$ and the abscissa of its *centroid* may be approximated to, as in (6084) for the arc of an ellipse, by the substitutions from (4278),

$$\int ds = a\int \sec\phi\sqrt{(e^2\sec^2\phi-1)}\,d\phi$$

and
$$\int x\,ds = a^2\int \sec^2\phi\sqrt{(e^2\sec^2\phi-1)}\,d\phi.$$

6117 *Landen's Theorem.*—This theorem gives any arc of an hyperbola in terms of the arcs of two ellipses, as follows:

$$\int\sqrt{(a^2+b^2+2ab\cos C)}\,dC =$$

$$\int\sqrt{(a^2-b^2\sin^2 A)}\,dA+\int\sqrt{(b^2-a^2\sin^2 B)}\,dB+2a\sin B+\text{const.},$$

that is—*Arc of ellipse whose semi-axes are* $a+b$ *and* $a-b$ $=$ *Arc of ellipse whose major axis is* $2a$ *and eccentricity* $b:a$ $+$ *difference between a right line and the arc of an hyperbola whose major axis is* b *and eccentricity* $a:b$.†

* *Williamson's Integ. Calc.*, Art. 137. † *Ibid.*, Art. 157.

6118 *Area ANP* (Fig. of 1183) bounded by x, y, and the curve

$$= \frac{b}{2a}\left\{x\sqrt{(x^2-a^2)}-a^2\log\frac{x+\sqrt{(x^2-a^2)}}{a}\right\}. \qquad (1931)$$

6119 $$= \tfrac{1}{2}\left\{xy-ab\log\left(\frac{x}{a}+\frac{y}{b}\right)\right\}. \qquad (4271)$$

6120 *Area of sector* between CA, CP and the curve

$$= \frac{ab}{2}\log\left(\frac{x}{a}+\frac{y}{b}\right).$$

6121 *Area between two ordinates* y_1, y_2, when the asymptotes are the coordinate axes

$$= \frac{ab}{2}\log\frac{x_2}{x_1}.$$

PROOF: $$\sin 2ACO\int y\,dx = \frac{2ab}{a^2+b^2}\int\frac{a^2+b^2}{4}\frac{dx}{x}. \qquad (4387)$$

6122 The *centroid of ANP*, A being the area (6118), is given by

$$A\bar{x} = \frac{b}{3a}(x^2-a^2)^{\frac{3}{2}}; \qquad A\bar{y} = \frac{b^2}{2a^2}\left(\frac{x^3}{3}-a^2x+\frac{2a^3}{3}\right).$$

6124 The *M. I. of ANP* about the x and y axes are

$$\frac{b^3}{24a^3}(2x^3-5a^2x)\sqrt{(x^2-a^2)}+\frac{ab^3}{8}\log\frac{x+\sqrt{(x^2-a^2)}}{a}.$$

6125 $$\frac{b}{8a}(2x^3-a^2x)\sqrt{(x^2-a^2)}-\frac{a^3b}{8}\log\frac{x+\sqrt{(x^2-a^2)}}{a}.$$

THE ELLIPTIC PARABOLOID.

6126 Equation, $$\frac{x^2}{a}+\frac{y^2}{b} = 2z.$$

6127 *Vol. of segment* $= \pi\sqrt{(ab)}\,z^2$, $\quad \bar{z} = \frac{2}{3}z$.

6129 *M. I.* about the axes of x, y, and z respectively,

$$m\left(\frac{az}{3}+\frac{z^2}{2}\right), \quad m\left(\frac{bz}{3}+\frac{z^2}{2}\right), \quad m\,\frac{a+b}{3}\,z.$$

6132 The *surface* S of the same segment may be found from

$$S = 4\int_0^{\sqrt{(2az)}}\int_0^{\sqrt{\left(2bz-\frac{b}{a}x^2\right)}}\sqrt{\left(1+\frac{x^2}{a^2}+\frac{y^2}{b^2}\right)}\,dx\,dy. \quad (5874)$$

6133 If the surface of the paraboloid be bounded by a curve of constant gradient γ (5881), the area becomes

$$S = \tfrac{2}{3}\pi ab\,(\sec^3\gamma - 1). \quad (5883)$$

THE PARABOLOID OF REVOLUTION.

6134 Equation, $x^2+y^2 = 2az$ or $r^2 = 2az$.

6136 *Surface of segment*, $S = \frac{2}{3}\pi\sqrt{a}\left\{(2z+a)^{\frac{3}{2}} - a^{\frac{3}{2}}\right\}$. (5880)

6137 *Volume* $= \pi az^2 = \frac{1}{2}\pi r^2 z$, $\bar{z} = \frac{2}{3}z$. (5887, '99)

6140 *M.I.* about axis of figure $= \dfrac{mr^2}{3}$ (6131)

6141 For *M.I.* about OX and OY put $a = b$ in (6129–30).

THE ELLIPSOID.

6142 Equation, $\dfrac{x^2}{a^2}+\dfrac{y^2}{b^2}+\dfrac{z^2}{c^2} = 1$, semi-axes a, b, c (5600).

6143 The *surface of the segment* cut off by the plane whose abscissa is x, will be found from

$$S = 4\int_x^a\int_0^{\frac{b}{a}\sqrt{a^2-x^2}}\left(\frac{a^4b^4+b^4(c^2-a^2)x^2+a^4(c^2-b^2)y^2}{a^4b^4-a^2b^4x^2-a^4b^2y^2}\right)^{\frac{1}{2}}dx\,dy.$$

Proof.—By (5874) and (5629, '7), eliminating z by means of the equation of the surface.

6144 The *volume* of the solid segment and the *centroid* are given by

$$V = \frac{\pi bc}{3a^2}(2a+x)(a-x)^2,\quad \bar{x} = \frac{3(a+x)^2}{4(2a+x)}.$$

Proofs.—Let (Fig. 177) represent one octant of the ellipsoid; OA, OB, OC being the principal semi-axes. The elemental section

$$4PNQ = \pi NP.NQ\,dx = \pi\frac{b}{a}\sqrt{a^2-x^2}\,\frac{c}{a}\sqrt{a^2-x^2}\,dx.$$

Therefore $\text{Vol.} = \dfrac{\pi bc}{a^2}\displaystyle\int_x^a (a^2-x^2)\,dx = \dfrac{\pi bc}{3a^2}(2a^3-3a^2x+x^3) =$ &c.

The moment with respect to the plane of yz

$$= \frac{\pi bc}{a^2}\int_x^a (a^2x - x^3)\,dx = \frac{\pi bc}{4a^2}(a^2 - x\ )^2,$$

and division by the volume gives $\bar{x}$ as above.

6146 The $M.\,I.$ of the solid segment about the axis a

$$= \frac{\pi bc\,(b^2+c^2)}{60a^4}\,(a-x)^3\,(8a^2+9ax+3x^2).$$

PROOF.—(Fig. 177.)

$$M.\,I. = \int_x^a \pi\,NP.NQ.\frac{NP^2+NQ^2}{4}\,dx \ (6112) = \frac{\pi bc\,(b^2+c^2)}{4a^4}\int_x^a (a^2-x^2)^2\,dx = \&c.$$

6147 The $M.\,I.$ about the axis b

$$= \frac{\pi bc}{15a^2}\left\{\frac{c^2}{a^2}\,(a-x)^3\,(8a^2+9ax+3x^2)+2a^5-5a^2x^3+3x^5\right\}.$$

PROOF: $\quad M.\,I. = \int_x^a \pi\,NP.NQ\left(\frac{NQ^2}{4}+ON^2\right)dx \qquad (5921)$

$$= \frac{\pi bc}{4a^4}\int_x^a (a^2-x^2)^2\,dx + \frac{\pi bc}{a^2}\int_x^a (a^2-x^2)\,x^2\,dx = \&c.$$

6148 The *volume* of the whole ellipsoid $= \frac{4}{3}\pi abc.$

PROOF.—By making $x = 0$ in (6144).

Otherwise: Let $\xi\eta\zeta$ be the point on the auxiliary sphere of radius r corresponding to xyz on the ellipsoid. By (5638–9), $rx = a\xi$, $ry = b\eta$, $rz = c\zeta$.

Therefore $\quad \int dx\,dy\,dz = \frac{abc}{r^3}\int d\xi\,d\eta\,d\zeta = \frac{abc}{r^3}\,\frac{4}{3}\pi r^3. \qquad (6061)$

6149 For the *centroid* of the semi-ellipsoid $\bar{x} = \frac{3a}{8}$. (6145)

6150 The $M.\,I.$ about the axis $a = m\,\frac{(b^2+c^2)}{5}$. (6146)

6151 The *volume of a segment* cut off by *any* plane PNQ (Fig. 177), where $OA = d$ is the semi-conjugate diameter, and $AN = h$, is

$$V = \pi abc\,\frac{h^2\,(3d-h)}{3d^3}.$$

PROOF.—Taking the area of the section from (5655), the volume of the segment will be

$$\frac{\pi abc\sin\theta}{p}\int_x^d \left(1-\frac{x^2}{d^2}\right)dx, \quad \text{where} \quad \sin\theta = \frac{p}{d},$$

θ being the inclination of d to the cutting plane. Integrate, and put $x = d-h$.

PROLATE SPHEROID.

Put $c = b$ in equation (6142) of the ellipsoid; then a will be the semi-axis of revolution.

6152 The *surface of the zone* between the plane of yz and a parallel plane at a distance x is

$$S = \pi b \left\{ \frac{a}{e} \sin^{-1} \frac{ex}{a} + \frac{x}{a} \sqrt{(a^2 - e^2x^2)} \right\}.$$

PROOF.—By (5878). $S = \frac{2\pi be}{a} \int_0^x \sqrt{\left(\frac{a^2}{e^2} - x^2\right)}\, dx.$

Then by (1933). Otherwise, make $b = c$ in (6143), and reduce.

6153 COR.—The *whole surface* $= 2\pi b \left(b + \frac{a}{e} \sin^{-1} e\right).$

6154 The *centroid* of the surface of the zone in (6152) is given by

$$\bar{x} = \frac{2\pi be}{3aS} \left\{ \frac{a^3}{e^3} - \left(\frac{a^2}{e^2} - x^2\right)^{\frac{3}{2}} \right\}.$$

PROOF.—From $S\bar{x} = \frac{2\pi be}{a} \int_0^x x \sqrt{\left(\frac{a^2}{e^2} - x^2\right)}\, dx.$

6155 The *M. I.* of the same zone is

$$= \pi ab^3 \left(\frac{1}{e} - \frac{1}{4e^3}\right) \sin^{-1} \frac{ex}{a} + \frac{\pi b^3}{a} \left(1 + \frac{1}{4e^2} - \frac{x^2}{2a^2}\right) x\sqrt{(a^2 - e^2x^2)}.$$

6156 And for the whole surface, by making $x = a$ and doubling,

$$M.\,I. = \pi ab^3 \left(\frac{2}{e} - \frac{1}{2e^3}\right) \sin^{-1} e + \pi b^4 \left(1 + \frac{1}{2e^2}\right).$$

PROOF: $M.\,I. = 2\pi \int y^3 \sqrt{\left(1 + \frac{b^4x^2}{a^4y^2}\right)}\, dx = \frac{2\pi b^3 e}{a^3} \int (a^2 - x^2) \sqrt{\left(\frac{a^2}{e^2} - x^2\right)}\, dx$

$$= \frac{2\pi b^3 e}{a} \int \sqrt{\left(\frac{a^2}{e^2} - x^2\right)}\, dx - \frac{2\pi b^3 e}{a^3} \int x^2 \sqrt{\left(\frac{a^2}{e^2} - x^2\right)}\, dx.$$

The first integral by (1933). For the second, by Rule VI. 2048, we obtain the formula

6157 $$\int x^2 \sqrt{(a^2 - x^2)}\, dx = \frac{a^4}{8} \sin^{-1} \frac{x}{a} + \frac{2x^2 - a^2}{8} x \sqrt{(a^2 - x^2)},$$

in which $\frac{a}{e}$ must now be written for a.

6158 For the *volume*, *moment of inertia*, and *abscissa of centroid* of the solid prolate spheroid, make $c = b$ in (6144–51), a being the axis of revolution.

OBLATE SPHEROID.

6159 Put $b=a$ in the equation (6142) of the ellipsoid; then c will be the semi-axis of revolution.

The surface of the zone between the plane of xy and a parallel plane at a distance z, is

$$S = \frac{\pi a}{c^2} z \sqrt{(c^4+a^2e^2z^2)} + \frac{\pi c^2}{e} \log \frac{aez+\sqrt{(c^4+a^2e^2z^2)}}{c^2}.$$

PROOF.—By (5878). $S = \frac{2\pi a^2 e}{c^2} \int_0^z \sqrt{\left(\frac{c^4}{a^2e^2}+z^2\right)}\, dz.$ Then by (1931).

6160 COR.—The whole surface $= 2\pi a^2 + \frac{\pi c^2}{e} \log \frac{1+e}{1-e}.$

6161 The centroid of the surface of the zone in (6159) is given by

$$\bar{z} = \frac{2\pi a^2 e}{3c^2 S} \left\{ \left(\frac{c^4}{a^2e^2}+z^2\right)^{\frac{3}{2}} - \frac{c^6}{a^3e^3} \right\}.$$

PROOF.—As in (6154). $\bar{z}$ for the surface of half the spheroid is obtained in this case by making $z=c$, but in (6154) put $x=a$.

6162 The M. I. of the same zone is

$$M.\,I. = \frac{\pi a^3}{c^2}\left(1-\frac{c^2}{4a^2e^2}-\frac{z^2}{2c^2}\right) z \sqrt{(c^4+a^2e^2z^2)}$$
$$+\frac{\pi c^2\,(4a^2-3c^2)}{4e^3} \log \frac{aez+\sqrt{(c^2+a^2e^2z^2)}}{c^2}.$$

6163 And for the whole surface, by making $z=c$ and doubling,

$$M.\,I. = \pi a^4 \left(1-\frac{c^2}{2a^2e^2}\right) + \frac{\pi c^2\,(4a^2-3c^2)}{2e^3} \log \frac{a\,(1+e)}{c}.$$

PROOF: $M.\,I. = 2\pi \int x^3 \sqrt{\left(1+\frac{a^4z^2}{c^4x^2}\right)}\, dz = \frac{2\pi a^4 e}{c^4} \int (c^2-z^2) \sqrt{\left(\frac{c^4}{a^2e^2}+z^2\right)}\, dz.$

The first integral involved is given at (1931), and the second is obtained in the same way as in the Proof of (6155), giving

6164 $\int x^2 \sqrt{x^2+a^2}\, dx = \frac{2x^3+a^2x}{8} \sqrt{(x^2+a^2)} - \frac{a^4}{8} \log \{x+\sqrt{(x^2+a^2)}\}.$

6165 For the *volume*, *moment of inertia*, and *abscissa of centroid* of the solid oblate spheroid, make $b=a$ in (6144–51), c being the axis of revolution.

JOINT INDEX

TO THE

SYNOPSIS

AND TO THE

PAPERS ON PURE MATHEMATICS,

CONTAINED IN THE UNDERMENTIONED

BRITISH AND FOREIGN JOURNALS

AND

TRANSACTIONS OF SOCIETIES.

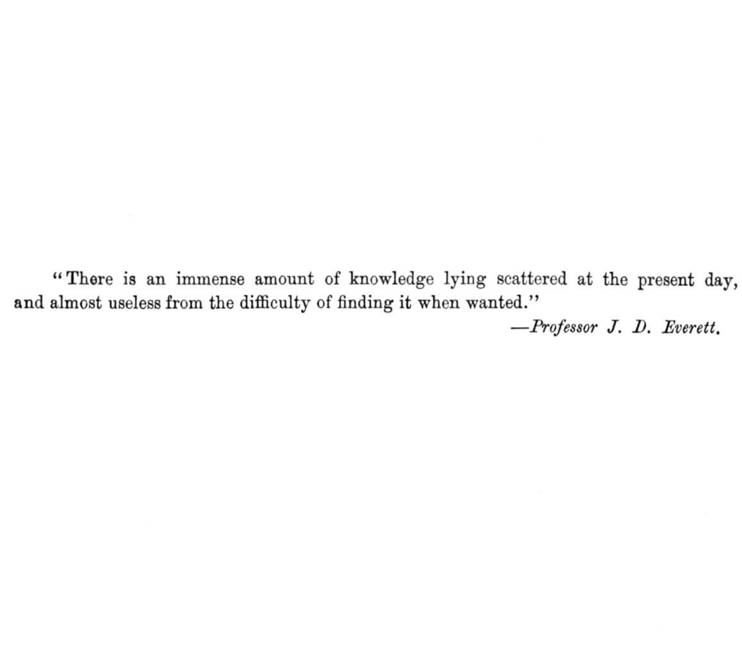

"There is an immense amount of knowledge lying scattered at the present day, and almost useless from the difficulty of finding it when wanted."

—*Professor J. D. Everett.*

KEY TO THE INDEX.

Prefixed to each title will be found the symbol by which the work is referred to in the Index. The words "*with Vol.*" or "*with Year,*" signify that any number following the symbol in the Index denotes, respectively, the Volume or Year of the journal. The year is given in all cases in which the work consists of more than one series of volumes. In order to connect the volumes with the years of publication, a Chronological Table is prefixed to the Index; in which table successive series of numbers in any column indicate successive series of volumes of the publication.

A. *with Vol.*—Archiv der Mathematik; 1843 to 1884; 70 vols. [B. M. C.: *P.P.* 1580.] *

Ac. *with Vol.*—Acta Mathematica, Zeitschrift Journal, herausgegeben von G. Mittag-Leffler; Stockholm, 1882 to 1885; 7 vols.

AJ. *with Vol.*—American Journal of Mathematics; Baltimore. Editor: J. J. Sylvester, F.R.S.; 1878 to 1885; 7 vols. [B. M. C.: *P.P.* 1575.*b.*]

An. *with Year.*—Annali di Scienze Matematiche e Fisiche, compilati da Prof. Barnaba Tortolini; Rome, 1850–57; *afterwards*—Annali di Matematiche pura et applica; Rome, 1858–65. Series II., Annali di Matematiche pura et applica, compilati da F. Brioschi e L. Cremona; Milan, 1868–85; 23 vols. in all. [B. M. C.: *P.P.* 1573 and 952.]

At. *with Year.*—Atti della Reale Accademia delle Scienze di Napoli; 1819 to 1878; 15 vols. [B. M. C.: for 1819–55, 8 vols., *Acad.* 2813; for 1863 to 1878, 7 vols., *Acad.* 96.]

C. *with Vol.*—Comptes rendus hebdomadaires des séances de l'Académie des Sciences; Paris, 1835 to 1885; 100 vols. [B. M. C.: *Acad.* 424 and *R.R.* 2099.*c.*] †

CD. *with Vol.*—Cambridge and Dublin Mathematical Journal. Editor, W. Thomson, B.A.; 1846 to 1854; 9 vols. [B. M. C.: *P.P.* 1565.]

CM. *with Vol.*— Cambridge Mathematical Journal; 1839 to 1845; 4 vols. [B. M. C.: *P.P.* 1565.]

CP. *with Vol.*—Cambridge Philosophical Transactions; 1822 to 1881; 13 vols. [B. M. C.: *Acad.* 3008.]

---

* *i.e.*, British Museum Catalogue, Press mark *P.P.* 1580.

† *R.R.* signifies *Reading-Room volumes within reach.*

E. *with Vol.*—Educational Times Reprint of Mathematical Questions and Solutions, with additional papers; London, half-yearly, 1863 to 1885; 44 vols. Editor: W. J. C. Miller, B.A. [B. M. C.: 2242.*c*.]

G. *with Vol.*—Giornale di Matematiche ad uso degli studenti delle università Italiane, pubblicato per cura del professore G. Battaglini; Naples, 1863–85; 23 vols. [B. M. C.: *P.P.* 1572.]

I. *with Vol.*—Journal of the Institute of Actuaries, or, The Assurance Magazine; London, 1850–84; 24 vols. [B. M. C.: *P.P.* 1423.*g.g.* and 126.]

J. *with Vol.*—Journal für die reine und angewandte Mathematik, herausgegeben von A. L. Crelle; 1826–1856; and Journal als Fortsetzung des von A. L. Crelle gregrundeten Journals von C. W. Borchardt; Berlin, 1856–1884; 97 vols. [B. M. C.: *P.P.* 1585 and *R.R.* 2022.*g*.]

JP. *with Vol.* — Journal de l'Ecole Polytechnique; Paris, 1796 to 1884; 34 vols. [B. M. C.: *T.C.* 1.*b*.]

L. *with Year.*—Journal de Mathématiques pures et appliquées, ou Recueil mensuel de mémoires sur les diverses parties des Mathématiques, publié par Joseph Liouville; Paris, 1836 to 1884; 49 vols. [B. M. C.: *P.P.* 1575 and *R.R.* 2022.*g*.]

LM. *with Vol.*—London Mathematical Society's Proceedings; 1866 to 1885; 16 vols. [B. M. C: *Acad.* 4265, 2.]

M. *with Vol.*—Mathematische Annalen, in Verbindung mit C. Neumann begründet durch R. F. A. Clebsch unter Mitwirkung der Herren Prof. P. Gordan, Prof. C. Neumann, Vols. 1–9; and Prof. K. V. Mühl, gegenwärtig herausgegeben von Prof. F. Klein und Prof. A. Mayer, Vols. 10, &c. Leipsig, 1869–1885; 25 vols. [B. M. C.: *P.P.* 1581.*b*.]

Man. *with Year.*—Manchester Memoirs, or, Memoirs of the Literary and Philosophical Society of Manchester; 1805 to 1884; 23 vols. [B. M. C.: 255.*d*., 9–12, and *Acad.* 1360.]

Me. *with Year.*—The Oxford, Cambridge, and Dublin Messenger of Mathematics; 1862 to 1871; 5 vols. *Continued as*—The Messenger of Mathematics. Editors: W. A. Whitworth, M.A., C. Taylor, D.D., R. Pendlebury, M.A., J. W. L. Glaisher, F.R.S.; Cambridge, 1872 to 1885; 14 vols. [B. M. C.: *P.P.* 1565.*b*. and 463.]

Mél. *with Vol.*—Mélanges mathématiques de l'Académie des Sciences de Saint Petersburg; 1849 to 1883; 6 vols. [B. M. C.: *Acad.* 1125/9.]

Mém. *with Year.*—Mémoires de l'Académie Impériale des Sciences de

Saint Petersburg; 1809 to 1883; 51 vols. [B. M. C.: 1809–30, *T.C.* 9.*a.*, 14–20; 1831–59, *Acad.* 1125/2; 1859–83, *Acad.* 1125/3.]

Mo. *with Year.*—Monatsbericht der Königlich-Preussichen Akademie der Wissenschaften zu Berlin; 1856 to 1881; *continued as*— Sitzungsberichte der Königlich, &c.; 1881 to 1885; 30 vols. [B. M. C.: *Acad.* 855.]

N. *with Year.*—Nouvelles Annales de Mathématiques. Journal des Candidats aux Ecoles Polytechnique et Normal. Rédigé par MM. Terquem et Gerono; Paris, 1842 to 1882; 41 vols. [B. M. C.: *P.P.* 1544.]

No. *with Year.*—Nova Acta Regiæ Societatis Scientiarum Upsaliensis; 1775 to 1884; 26 vols. [B. M. C.: for 1775–1850, 14 vols., *T.C.* 6.*b.*, 13–19; for 1851–1884, 12 vols., *Acad.* 1076.]

P. *with Year.*—The Philosophical Transactions of the Royal Society of London; 1781 to 1884; 104 vols: *i.e.*, vols. 71 to 174. [B. M. C.: *R.R.* 2021.*g.*]

Pr. *with Vol.*—Proceedings of the Royal Society of London; 1800 to 1885; 39 vols. [B. M. C.: for vols. 1 to 23, *Acad.* 3025, 21; for vols. 24, &c., *R.R.* 2101.*d.*]

Q. *with Vol.*—Quarterly Journal of Mathematics; 20 vols. Cambridge, 1857–78; Editors: J. J. Sylvester, F.R.S., N. M. Ferrers, F.R.S., G. G. Stokes, F.R.S., A. Cayley, F.R.S., M. Hermite, F.R.S.; 1878–85, N. M. Ferrers, F.R.S., A. Cayley, F.R.S., J. W. L. Glaisher, F.R.S. [B. M. C.: *P.P.* 1566 and 251.]

TA. *with Vol.*—Transactions of the American Philosophical Society; Philadelphia, 1818–71; 14 vols. [B. M. C.: *Acad.* 1830/3 and *T.C.* 18.*b.* 11.]

TE. *with Vol.*—Transactions of the Royal Society of Edinburgh; 1788 to 1880; 29 vols. [B. M. C.: *T.C.* 15.*b.* 1 and *R.R.* 2099.*g.*]

TI. *with Vol.*—Transactions of the Royal Irish Academy; 1786 to 1879; 26 vols. [B. M. C.: *Acad.* 1540 and *R.R.* 2099.*g.*]

TN. *with Vol.*—Transactions of the Royal Society of New South Wales; Sydney, 1867–83; 17 vols. [B. M. C.: *Acad.* 1971.]

Z. *with Vol.*—Zeitschrift für Mathematik und Physik, herausgegeben von Dr. O. Schlömilch und Dr. B. Witzschel; Dr. O. Schlömilch, Dr. E. Kahl, und Dr. M. Cantor; Leipzig, 1856–1883. 28 vols. [B. M. C.: *P.P.* 1581.]

Year.	A	Ac	AJ	An	At	C	CD	CM	CP	E	G	I	J	JP	Year.
1801	...	...	...	...	...	...	...	...	...	...	...	...	...	3	1801
2	...	...	...	...	...	...	...	...	...	...	...	...	...	...	2
3	...	...	...	...	...	...	...	...	...	...	...	...	...	...	3
4	...	...	...	...	...	...	...	...	...	...	...	...	...	4	4
5	...	...	...	...	...	...	...	...	...	...	...	...	...	...	5
6	...	...	...	...	...	...	...	...	...	...	...	...	...	5, 6	6
7	...	...	...	...	...	...	...	...	...	...	...	...	...	7	7
8	...	...	...	...	...	...	...	...	...	...	...	...	...	...	8
9	...	...	...	...	...	...	...	...	...	...	...	...	...	8	9
10	...	...	...	...	...	...	...	...	...	...	...	...	...	...	10
11	...	...	...	...	...	...	...	...	...	...	...	...	...	...	11
12	...	...	...	...	...	...	...	...	...	...	...	...	...	...	12
13	...	...	...	...	...	...	...	...	...	...	...	...	...	9	13
14	...	...	...	...	...	...	...	...	...	...	...	...	...	...	14
15	...	...	...	...	...	...	...	...	...	...	...	...	...	10	15
16	...	...	...	...	...	...	...	...	...	...	...	...	...	...	16
17	...	...	...	...	...	...	...	...	...	...	...	...	...	...	17
18	...	...	...	...	...	...	...	...	...	...	...	...	...	...	18
19	...	...	...	...	I. 1	...	...	...	...	...	...	...	...	...	19
20	...	...	...	...	...	...	...	...	...	...	...	...	...	11	20
21	...	...	...	...	...	...	...	...	...	...	...	...	...	...	21
22	...	...	...	...	...	...	...	...	1	...	...	...	...	...	22
23	...	...	...	...	...	...	...	...	...	...	...	...	...	12	23
24	...	...	...	...	...	...	...	...	...	...	...	...	...	...	24
25	...	...	...	...	2	...	...	...	...	...	...	...	...	...	25
26	...	...	...	...	...	...	...	...	...	...	...	...	1	...	26
27	...	...	...	...	...	...	...	...	2	...	...	...	2	...	27
28	...	...	...	...	...	...	...	...	...	...	...	...	3	...	28
29	...	...	...	...	...	...	...	...	...	...	...	...	4	...	29
30	...	...	...	...	...	...	...	...	3	...	...	...	5, 6	...	30
31	...	...	...	...	...	...	...	...	...	...	...	...	7	13	31
32	...	...	...	...	3	...	...	...	...	...	...	...	8, 9	...	32
33	...	...	...	...	...	...	...	...	4	...	...	...	10	14	33
34	...	...	...	...	...	...	...	...	...	...	...	...	11, 12	...	34
35	...	...	...	...	...	1	...	...	5	...	...	...	13, 14	15	35
36	...	...	...	...	...	2, 3	...	...	...	...	...	...	15	...	36
37	...	...	...	...	...	4, 5	...	...	...	...	...	...	16, 17	...	37
38	...	...	...	...	...	6, 7	...	...	6	...	...	...	18	16	38
39	...	...	...	...	4	8, 9	...	1	...	...	...	...	19	...	39
40	...	...	...	...	...	10	...	...	...	...	...	...	20, 21	...	40
41	...	...	...	...	...	11–13	...	2	...	...	...	...	22	17	41
42	...	...	...	...	...	14, 15	...	...	7	...	...	...	23, 24	...	42
43	1	...	...	...	...	16, 17	...	3	...	...	...	...	25, 26	...	43
44	2, 3	...	...	...	5	18, 19	...	...	...	...	...	...	27, 28	...	44
45	4, 5	...	...	...	...	20, 21	...	4	...	...	...	...	29, 30	...	45
46	6, 7	...	...	...	...	22, 23	1	...	...	...	...	...	31, 33	...	46
47	8, 9	...	...	...	...	24, 25	2	...	...	...	...	...	34, 35	18	47
48	10, 11	...	...	...	...	26, 27	3	...	...	...	...	...	36, 37	19	48
49	12, 13	...	...	...	...	28, 29	4	...	9	...	...	...	38	...	49
50	14, 15	...	...	I. 1	...	30, 31	5	...	...	...	...	1	39, 40	...	50
51	16, 17	...	...	2	6	32, 33	6	...	...	...	...	...	41, 42	...	51
52	18, 19	...	...	3	II. 1	34, 35	7	...	...	...	...	2	43, 45	...	52
53	20, 21	...	...	4	...	36, 37	8	...	...	...	...	3	46, 47	20	53
54	22, 23	...	...	5	...	38, 39	9	...	...	...	...	4	48	...	54
55	24, 25	...	...	6	2	40, 41	...	...	...	...	...	5	49, 50	...	55
56	26, 27	...	...	7	...	42, 43	...	...	9	...	...	...	51, 52	21	56
57	28, 29	...	...	8	...	44, 45	...	...	...	...	...	6	53, 54	...	57
58	30, 31	...	...	II. 1	...	46, 47	...	...	...	...	...	7	55	...	58
59	32, 33	...	...	2	...	48, 49	...	...	...	...	...	...	56, 57	...	59
60	34, 35	...	...	3	...	50, 51	...	...	...	...	...	...	58	...	60
61	36, 37	...	...	4	...	52, 53	...	...	...	...	...	8	59, 60	22	61
62	38, 39	...	...	...	...	54, 55	...	...	...	...	...	9	61	...	62
63	40	...	...	5	III. 1	56, 57	...	...	...	...	1	10	62	23	63
64	41, 42	...	...	6	...	58, 59	...	...	10	1, 2	2	11	63	...	64
65	43, 44	...	...	7	2	60, 61	...	...	...	3, 4	3	...	64	24	65
66	45, 46	...	...	...	...	62, 63	...	...	...	5, 6	4	12	65, 66	...	66
67	47	...	...	...	...	64, 65	...	...	...	7, 8	5	13	67	25	67
68	48	...	...	III. 1	3	66, 67	...	...	...	9, 10	6	...	68, 69	...	68
69	49, 50	...	...	2	4	68, 69	...	...	...	11, 12	7	14	70	...	69
70	51	...	...	3	...	70, 71	...	...	...	13, 14	8	15	71, 72	26	70
71	52	...	...	4	...	72, 73	...	...	11	15, 16	9	...	73	...	71
72	53	...	...	...	...	74, 75	...	...	...	17, 18	10	16	74, 75	...	72
73	54, 55	...	...	5	5	76, 77	...	...	...	19, 20	11	17	76	...	73
74	56, 57	...	...	...	...	78, 79	...	...	...	21, 22	12	...	77, 78	27	74
75	58	...	...	6	6	80, 81	...	...	...	23, 24	13	18	79, 80	...	75
76	59, 60	...	...	7	...	82, 83	...	...	...	25, 26	14	19	81	...	76
77	61	...	...	8	...	84, 85	...	...	...	27, 28	15	...	82–84	...	77
78	62	...	1	...	7	86, 87	...	...	...	29, 30	16	20	85, 86	...	78
79	63, 64	...	2	...	...	88, 89	...	...	12	31, 32	17	21	87, 88	28	79
80	65, 66	...	3	...	...	90, 91	...	...	...	33, 34	18	...	89, 90	29	80
81	67	...	4	...	...	92, 93	...	...	13	35, 36	19	22	91, 92	30, 31	81
82	68	1, 2	5	...	...	94, 95	...	...	...	37, 38	20	23	93, 94	32	82
83	69	3, 4	...	...	...	96, 97	...	...	...	39, 40	21	...	95	33	83
84	70	5–7	6	...	...	98, 99	...	...	...	41, 42	22	24	96, 97	34	84
1885	...	...	7	...	...	100	...	...	...	43, 44	23	...	...	...	1885
	A	Ac	AJ	An	At	C	CD	CM	CP	E	G	I	J	JP	

Year.	L	LM	M	Me	Man.	Mém.	Mél.	Mo	N	No	P	Pr	Q	TA	TE	TI	TN	Z	Year.
1801	...	...	...	...	...	...	...	...	...	...	91	...	...	...	...	...	...	...	1801
2	...	...	...	...	...	...	...	...	...	...	92	...	...	...	...	8	...	...	2
3	...	...	...	...	...	...	...	...	...	...	93	...	...	...	...	9	...	...	3
4	...	...	...	...	...	...	...	...	...	...	94	...	...	...	...	...	...	...	4
5	...	...	...	...	I. 1	...	...	...	...	...	95	...	...	...	5	...	...	...	5
6	...	...	...	...	...	...	...	...	...	...	96	...	...	...	...	10	...	...	6
7	...	...	...	...	...	...	...	...	...	...	97	...	...	...	...	...	...	...	7
8	...	...	...	...	...	...	...	...	...	...	98	...	...	...	...	...	...	...	8
9	...	...	...	...	...	I. 1	...	...	...	...	99	...	...	...	...	...	...	...	9
10	...	...	...	...	...	2	...	...	...	...	100	...	...	...	...	11	...	...	10
11	...	...	...	...	...	3	...	...	...	...	101	...	...	...	...	...	...	...	11
12	...	...	...	...	...	...	...	...	...	...	102	...	...	...	6	...	...	...	12
13	...	...	...	...	2	4	...	...	...	...	103	...	...	...	...	...	...	...	13
14	...	...	...	...	...	...	...	...	...	...	104	1	...	...	...	...	...	...	14
15	...	...	...	...	...	5	...	...	...	I. 7	105	...	...	...	7	12	...	...	15
16	...	...	...	...	...	...	...	...	...	...	106	...	...	...	...	...	...	...	16
17	...	...	...	...	...	...	...	...	...	...	107	...	...	...	...	...	...	...	17
18	...	...	...	...	...	6	...	...	...	...	108	...	...	1	8	13	...	...	18
19	...	...	...	...	3	...	...	...	...	...	109	...	...	...	...	...	...	...	19
20	...	...	...	...	...	7	...	...	...	...	110	...	...	...	...	...	...	...	20
21	...	...	...	...	...	...	...	...	...	8	111	...	...	...	...	...	...	...	21
22	...	...	...	...	...	8	...	...	...	...	112	...	...	...	...	...	...	...	22
23	...	...	...	...	...	...	...	...	...	...	113	...	...	...	9	...	...	...	23
24	...	...	...	...	4	9	...	...	...	...	114	...	...	...	...	...	...	...	24
25	...	...	...	...	...	...	...	...	...	...	115	...	...	2	...	14	...	...	25
26	...	...	...	...	...	10	...	...	...	...	116	...	...	...	10	...	...	...	26
27	...	...	...	...	...	...	...	...	...	9	117	...	...	...	...	...	...	...	27
28	...	...	...	...	...	...	...	...	...	...	118	...	...	...	...	15	...	...	28
29	...	...	...	...	...	...	...	...	...	...	119	...	...	...	...	...	...	...	29
30	...	...	...	...	...	11	...	...	...	...	120	2	...	3	...	16	...	...	30
31	...	...	...	...	5	II. 1	...	...	...	...	121	...	...	...	...	...	...	...	31
32	...	...	...	...	...	...	...	...	...	10	122	...	...	...	...	...	...	...	32
33	...	...	...	...	...	2	...	...	...	...	123	...	...	...	...	...	...	...	33
34	...	...	...	...	...	...	...	...	...	...	124	...	...	4	...	...	...	...	34
35	...	...	...	...	...	...	...	...	...	...	125	...	...	...	...	...	...	...	35
36	I. 1	...	...	...	...	...	...	...	...	...	126	...	...	...	...	...	...	...	36
37	2	...	...	...	...	...	...	...	...	...	127	3	...	5	...	17	...	...	37
38	3	...	...	...	...	3	...	...	...	...	128	...	...	...	...	...	...	...	38
39	4	...	...	...	...	...	...	...	...	11	129	...	...	6	...	18	...	...	39
40	5	...	...	...	...	...	...	...	...	...	130	...	...	...	...	...	...	...	40
41	6	...	...	...	...	4	...	...	...	...	131	...	...	7	...	...	...	...	41
42	7	...	...	...	6	...	...	...	I. 1	...	132	...	...	...	...	...	...	...	42
43	8	...	...	...	...	...	...	...	2	...	133	4	...	8	...	19	...	...	43
44	9	...	...	...	...	5	...	...	3	12	134	...	...	...	...	...	...	...	44
45	10	...	...	...	...	...	...	...	4	...	135	...	...	...	...	20	...	...	45
46	11	...	...	...	7	...	...	...	5	...	136	...	...	9	...	...	...	...	46
47	12	...	...	...	...	...	...	...	6	13	137	...	...	...	...	...	...	...	47
48	13	...	...	...	8	...	...	...	7	...	138	...	...	...	...	21	...	...	48
49	14	...	...	...	...	...	...	...	8	...	139	...	...	...	...	...	...	...	49
50	15	...	...	...	...	6	...	...	9	14	140	5	...	...	...	...	...	...	50
51	16	...	...	...	9	...	...	...	10	...	141	...	...	...	...	...	...	...	51
52	17	...	...	...	10	...	...	...	11	...	142	...	...	...	...	...	...	...	52
53	18	...	...	...	...	7	1	...	12	...	143	...	...	10	...	22	...	...	53
54	19	...	...	...	11	...	...	...	13	...	144	6	...	...	23	...	...	...	54
55	20	...	...	...	12	...	...	...	14	I. 1	145	7	...	...	...	...	...	...	55
56	II. 1	...	...	...	13	...	...	1	15	...	146	...	...	...	...	...	...	1	56
57	2	...	...	...	14	8	...	2	16	...	147	8	1	...	...	...	...	2	57
58	3	...	...	...	15	9	...	3	17	2	148	...	2	...	...	...	...	3	58
59	4	...	...	...	...	III. 1	2	4	18	...	149	9	...	...	...	23	...	4	59
60	5	...	...	...	...	2	...	5	19	...	150	10	3	11	...	...	...	5	60
61	6	...	...	...	...	3	...	6	20	3	151	...	4	...	...	...	...	6	61
62	7	...	...	I. 1	II. 1	4	...	7	II. 1	...	152	11	5	...	...	...	...	7	62
63	8	...	...	...	...	5, 6	...	8	2	4	153	12	...	12	...	...	...	8	63
64	9	...	...	2	...	7	...	9	3	...	154	13	6	...	...	...	...	9	64
65	10	...	...	...	2	8	...	10	4	5	155	14	...	...	...	...	...	10	65
66	11	1	...	3	...	9	3	11	5	...	156	...	...	...	...	...	...	11	66
67	12	...	...	...	...	10	...	12	6	...	157	15	8	...	24	...	1	12	67
68	13	...	...	4	3	11	...	13	7	6	158	16	9	...	...	...	2	13	68
69	14	2	1	...	...	12, 13	...	14	8	...	159	17	...	13	25	...	3	14	69
70	15	...	2	...	...	14, 15	...	15	9	7	160	18	...	...	...	...	4	15	70
71	16	3	3, 4	5	4	16	...	16	10	...	161	19	10	14	...	...	5	16	71
72	17	...	5	II. 1	...	17, 18	4	17	11	...	162	20	...	...	26	...	6	17	72
73	18	4	6	2	...	19, 20	...	18	12	8	163	21	...	...	...	...	7	18	73
74	19	5	7	3	...	21	...	19	13	...	164	22	...	...	...	24	8	19	74
75	III. 1	6	8	4	...	...	...	20	14	9	165	23	13	...	...	25	9	20	75
76	2	7	9, 10	5	5	22	...	21	15	...	166	24	14	...	27	...	10	21	76
77	3	8	11, 12	6	...	23, 24	...	22	16	...	167	25, 26	...	...	...	...	11	22	77
78	4	9	13	7	...	25	...	23	17	...	168	27	15	...	...	...	12	23	78
79	5	10	14, 15	8	6	26	...	24	18	10	169	28, 29	16	...	28	26	13	24	79
80	6	11	16, 17	9	...	27	...	25	19	...	170	30	...	...	29	...	14	25	80
81	7	12	18	10	...	28, 29	5	26	20	11	171	31, 32	17	...	...	...	15	26	81
82	8	13	19	11	7	30	...	27	III. 1	...	172	33	18	...	...	...	16	27	82
83	9	14	20, 21	12	...	31	6	28	...	...	173	34, 35	19	...	...	...	17	28	83
84	10	15	22, 23	13	8	...	...	29	...	12	174	36, 37	...	...	...	...	...	...	84
1885	...	16	24, 25	14	...	...	...	30	...	...	...	38, 39	20	...	...	...	...	...	1885
	L	LM	M	Me	Man.	Mém.	Mél.	Mo	N	No	P	Pr	Q	TA	TE	TI	TN	Z	

EXPLANATION OF ABBREVIATIONS, &c.

a. c.	= *areal coordinates.*	i. eq.	= *indeterminate equation.*
alg.	= *algebraic.*	imag.	= *imaginary.*
ap.	= *application.*	l. c. m.	= *lowest common multiple.*
anal.	= *analytical.*	num.	= *numerical.*
ar. p.	= *arith. progression.*	o. c.	= *oblique coordinates.*
c. c.	= *Cartesian coordinates.*	p. c.	= *polar coordinates.*
cn.	= *construction.*	p. d. e.	= *partial difference equations.*
cond.	= *condition.*	p. e.	= *polar equation.*
curv.	= *curvature.*	perp.	= *perpendicular.*
d. c.	= *differential calculus.*	pl.	= *plane.*
d. e.	= *differential equations.*	pr.	= *problem.*
d. i.	= *definite integral.*	q. c.	= *quadriplanar coordinates.*
eq.	= *equation.*	rad.	= *radius.*
ex.	= *example* or *exercise.*	sd.	= *solid or* 3-*dimensional.*
ext.	= *extension.*	sol.	= *solution.*
f.	= *formula.*	sym.	= *symmetrically.*
f. d. c.	= *finite difference calculus.*	ta.	= *table.*
f. d. e.	= *finite difference equation.*	t. c.	= *trilinear coordinates.*
geo.	= *geometrical.*	tg. c.	= *tangential coordinates.*
g. p.	= *geometrical progression.*	tg. e.	= *tangential equation.*
gn.	= *general.*	th.	= *theorem.*
gz.	= *generalization.*	tr.	= *treatise* (*i.e.*, more than 50 pages).
h. c. f.	= *highest common factor.*	transf.	= *transformation.*
i. c.	= *integral calculus.*		

15_3.—*The suffix means that three articles under the same heading will be found in the volume.*

$\overset{\cdot\ \cdot\ \cdot\ \cdot}{18—21}$.—*Means that one article on the subject will be found in each of the four consecutive volumes.*

References to the Synopsis stand first, and are the numbers of Articles, not of Pages. An asterisk (*) is prefixed where such numbers will be found.

The unclassified references following a principal title commonly refer to papers on the general theory of the subject; but some papers are occasionally included amongst these of which the titles are too long for insertion, and do not admit of abbreviation.

Subjects which might well have been included under the same heading appear sometimes under different ones, for the following reasons:—Exigencies of space have decided the insertion of the number of the volume only of the particular work in question, and a subsequent examination of the Index of that volume is required in order to find the page. It, therefore, became desirable not to change the original title of the paper when there was danger, by so doing, of making it unrecognizable. When, however, the same subject appears in two parts of this Index under different names, cross references from one to the other are given. Some changes, however, have been made when the synonym was perfectly obvious; for instance, when a reference to a journal, published fifty years ago, is found under the heading of "Binary Quantics," the actual title of the article will, in all probability, be "Homogeneous Functions of Two Variables," and so in a few other instances.

INDEX.

THEORIE DES OPERATIONS LINEAIRES

By S. BANACH

—1933-63. xii + 250 pp. 5⅜x8. 8284-0110-1. **$4.95**

DIFFERENTIAL EQUATIONS

By H. BATEMAN

CHAPTER HEADINGS: I. Differential Equations and their Solutions. II. Integrating Factors. III. Transformations. IV. Geometrical Applications. V. Diff. Eqs. with Particular Solutions of a Specified Type. VI. Partial Diff. Eqs. VII. Total Diff. Eqs. VIII. Partial Diff. Eqs. of the Second Order. IX. Integration in Series. X. The Solution of Linear Diff. Eqs. by Means of Definite Integrals. XI. The Mechanical Integration of Diff. Eqs.

—1917-67. xi + 306 pp. 5⅜x8. 8284-0190-X. **$4.95**

ERGODENTHEORIE, u.a.

By H. BEHNKE, P. THULLEN, and E. HOPF

TWO VOLUMES IN ONE.

THEORIE DER FUNKTIONEN MEHRERER KOMPLEXER VERAENDERLICHEN, by *H. Behnke* and *P. Thullen.*

ERGODENTHEORIE, by *E. Hopf.*

—1934/1937-68. 210 pp. 5½x8½. 8284-0068-7. Two vols. in one. **Prob. $4.95**

L'APPROXIMATION

By S. BERNSTEIN and CH. de LA VALLÉE POUSSIN

TWO VOLUMES IN ONE:

Leçons sur les Propriétés Extrémales et la Meilleure Approximation des Fonctions Analytiques d'une Variable Réelle, *by Bernstein.*

Leçons sur l'approximation des Fonctions d'une Variable Réelle, *by Vallée Poussin.*

—1925/1919-69. 363 pp. 6x9. 8284-0198-5. Two vols in one. **In prep.**

CONFORMAL MAPPING

By L. BIEBERBACH

Translated from the fourth German edition by F. STEINHARDT.

Partial Contents: I. Foundations; Linear Functions (Analytic functions and conformal mapping, Integral linear functions, $w = 1/z$, Linear functions, Groups of linear functions). II. Rational Functions. III. General Considerations (Relation between c. m. of boundary and of interior of region, Schwarz's principle). IV. Further Study of Mappings ($w = z + 1/z$, Exponential function, Trigonometric functions, Elliptic integral of first kind). V. Mappings of Given Regions (Illustrations, Vitali's theorem, Proof of Riemann's theorem, Actual constructions, Potential-theoretic considerations, Distortion theorems, Uniformization, Mapping of multiply-connected plane regions onto canonical regions).

"Presented in very attractive and readable form."—*Math. Gazette.*

—1952. vi + 234 pp. 4½x6½. 8284-0090-3. Cloth **$3.25**
8284-0176-4. Paper **$1.50**

BASIC GEOMETRY

By G. D. BIRKHOFF and R. BEATLEY

"is in accord with the present approach to plane geometry. It offers a sound mathematical development . . . and at the same time enables the student to move rapidly into the heart of geometry."—*The Mathematics Teacher.*

"should be required reading for every teacher of Geometry."—*Mathematical Gazette.*

—3rd ed. 1959. 294 pp. 5⅜x8. 8284-0120-9. **$6.00**

VORLESUNGEN UEBER DIFFERENTIALGEOMETRIE, Vols. I, II

By W. BLASCHKE

TWO VOLUMES IN ONE.

Partial Contents: VOL. I. 1. Theory of Curves. 2. Extremal Curves. 3. Strips. 4. Theory of Surfaces. 5. Invariant Derivatives on a Surface. 6. Geometry on a Surface. 7. On the Theory of Surfaces in the Large. 8. Extremal Surfaces. 9. Line Geometry.

VOL. II. *Affine Differential Geometry.* 1. Plane Curves in the Small. 2. Plane Curves in the Large. 3. Space Curves. 5. Theory of Surfaces. 6. Extremal Surfaces. 7. Special Surfaces.

—3rd ed. (Vol. I); 2nd ed. (Vol. II). 1930/23-67. 589 pp. 6x9. 8284-0202-7. Two vols. in one. **$12.00**

KREIS UND KUGEL

By W. BLASCHKE

Isoperimetric properties of the circle and sphere, the (Brunn-Minkowski) theory of convex bodies, and differential-geometric properties (in the large) of convex bodies. A standard work.

—x + 169 pp. 5½x8½. 8284-0059-8. Cloth **$3.50**
8284-0115-2. Paper **$1.50**

INTEGRALGEOMETRIE

By W. BLASCHKE and E. KÄHLER

THREE VOLUMES IN ONE.

VORLESUNGEN UEBER INTEGRALGEOMETRIE, Vols. I and II, by *W. Blaschke.*

EINFUEHRUNG IN DIE THEORIE DER SYSTEME VON DIFFERENTIALGLEICHUNGEN, by *E. Kähler.*

—1936/37/34-49. 222 pp. 5½x8½. 8284-0064-4. Three vols. in one. **$6.00**

FUNDAMENTAL EXISTENCE THEOREMS, *by G. A. BLISS.* See EVANS

VORLESUNGEN UEBER FOURIERSCHE INTEGRALE

By S. BOCHNER

—1932-48. 237 pp. 5½x8½. 8284-0042-3. **$3.50**

MODERN PURE SOLID GEOMETRY

By N. A. COURT

In this second edition of this well-known book on synthetic solid geometry, the author has supplemented several of the chapters with an account of recent results.

—2nd ed. 1964. xiv + 353 pp. 5½x8¼. 8284-0147-0. **$7.50**

MODERN PURE GEOMETRY FOR HIGH-SCHOOL MATHEMATICS TEACHERS

By N. A. COURT

—1969. Approx. 100 pp. 6x9. **In prep.**

SPINNING TOPS AND GYROSCOPIC MOTION

By H. CRABTREE

Partial Contents: Introductory Chapter. CHAP. I. Rotation about a Fixed Axis. II. Representation of Angular Velocity. Precession. III. Discussion of the Phenomena Described in the Introductory Chapter. IV. Oscillations. V. Practical Applications. VI-VII. Motion of Top. VIII. Moving Axes. IX. Stability of Rotation. Periods of Oscillation. APPENDICES: I. Precession. II. Swerving of "sliced" golf ball. III. Drifting of Projectiles. IV. The Rising of a Top. V. The Gyro-compass. ANSWERS TO EXAMPLES.

—2nd ed. 1914-67. 203 pp. 6x9. 8284-0204-3. **$4.95**

GESAMMELTE MATHEMATISCHE WERKE

By R. DEDEKIND

—1930/31/32-68. 1,359 pp. 6x9. 8284-0220-5.
Three vol. set. **In prep.**

THE DOCTRINE OF CHANCES

By A. DE MOIVRE

In the year 1716 Abraham de Moivre published his *Doctrine of Chances*, in which the subject of Mathematical Probability took several long strides forward. A few years later came his *Treatise of Annuities*. When the third (and final) edition of the *Doctrine* was published in 1756 it appeared in one volume together with a revised edition of the work on Annuities. It is this latter two-volumes-in-one that is here presented in an exact photographic reprint.

—3rd ed. 1756-1967. xi + 348 pp. 6x9. 8284-0200-0. **$7.95**

THEORIE GENERALE DES SURFACES

By G. DARBOUX

One of the great works of the mathematical literature.

An unabridged reprint of the latest edition of *Leçons sur la Théorie générale des surfaces et les applications géométriques du Calcul infinitésimal.*

—Vol. I (2nd ed.) xii+630 pp. Vol. II (2nd ed.) xvii+584 pp. Vol. III (1st ed.) xvi+518 pp. Vol. IV. (1st ed.) xvi+537 pp.
 In prep.

COLLECTED PAPERS

By L. E. DICKSON

—1969. 4 vols. Approx. 3,400 pp. 6½x9¼. **In prep.**

HISTORY OF THE THEORY OF NUMBERS

By L. E. DICKSON

"**A monumental work** . . . Dickson always has in mind the needs of the investigator . . . The author has [often] expressed in a nut-shell the main results of a long and involved paper *in a much clearer way than the writer of the article did himself.* The ability to reduce complicated mathematical arguments to simple and elementary terms is highly developed in Dickson."—*Bulletin of A. M. S.*

—Vol. I (Divisibility and Primality) xii + 486 pp. Vol. II (Diophantine Analysis) xxv + 803 pp. Vol. III (Quadratic and Higher Forms) v + 313 pp. 5⅜x8. 8284-0086-5.
Three vol. set. **$22.50**

STUDIES IN THE THEORY OF NUMBERS

By L. E. DICKSON

—1930-62. viii + 230 pp. 5⅜x8. 8284-0151-9. **$4.50**

ALGEBRAIC NUMBERS

By L. E. DICKSON, et al.

TWO VOLUMES IN ONE.

Both volumes of the *Report of the Committee on Algebraic Numbers* are here included, the authors being L. E. Dickson, R. Fueter, H. H. Mitchell, H. S. Vandiver, and G. E. Wahlen.

Partial Contents: CHAP. I. Algebraic Numbers. II. Cyclotomy. III. Hensel's p-adic Numbers. IV. Fields of Functions. I′. The Class Number in the Algebraic Number Field. II′. Irregular Cyclotomic Fields and Fermat's Last Theorem.

—1923/28-67. ii + 211 pp. 5⅜x9. 8284-0211-6.
Two vols. in one. **$4.95**

INTRODUCTION TO THE THEORY OF ALGEBRAIC EQUATIONS, *by L. E. DICKSON. See* SIERPINSKI

VORLESUNGEN UEBER ZAHLENTHEORIE

By P. G. L. DIRICHLET and R. DEDEKIND

The fourth (last) edition of this great work contains, in its final form, the epoch-making "Eleventh Supplement," in which Dedekind outlines his theory of algebraic numbers.

"Gauss' *Disquisitiones Arithmeticae* has been called a 'book of seven seals.' It is hard reading, even for experts, but the treasures it contains (and partly conceals) in its concise, synthetic demonstrations are now available to all who wish to share them, largely the result of the labors of Gauss' friend and disciple, Peter Gustav Lejeune Dirichlet (1805-1859), who first broke the seven seals . . . [He] summarized his personal studies and his recasting of the *Disquisitiones* in his *Zahlentheorie.* The successive editions (1863, 1871, 1879, 1893) of this text . . . made the classical arithmetic of Gauss accessible to all without undue labor."—*E. T. Bell,* in *Men of Mathematics* and *Development of Mathematics.*

—Repr. of 4th ed. 1893-1968. xv + 657 pp. 5⅜x8.
8284-0213-2. **$13.50**

AUTOMORPHIC FUNCTIONS

By L. R. FORD

"Comprehensive . . . remarkably clear and explicit."—*Bulletin of the A. M. S.*

—2nd ed. (Corr. repr.) 1929-51. x + 333 pp. 5⅜x8. 8284-0085-7. **$6.50**

THE CALCULUS OF EXTENSION

By H. G. FORDER

—1941-60. xvi + 490 pp. 5⅜x8. 8284-0135-7. **$6.50**

GRUNDGESETZE DER ARITHMETIK

By G. FREGE

TWO VOLUMES IN ONE.

—1893/1903-69. xxvi + 253 + xvi + 266 pp. 5⅜x8. Two vols. in one. **In prep.**

CURVE TRACING

By P. FROST

This much-quoted and charming treatise gives a very readable treatment of a topic that can only be touched upon briefly in courses on Analytic Geometry. Teachers will find it invaluable as supplementary reading for their more interested students and for reference. The Calculus is not used.

Partial Contents: Introductory Theorems. II. Forms of Certain Curves Near the Origin. Cusps. Tangents to Curves. Curvature. III. Curves at Great Distance from the Origin. IV. Simple Tangents. Direction and amount of Curvature. Multiple Points. Curvature at Multiple Points. VI. Asymptotes. VIII. Curvilinear Asymptotes. IX. The Analytical Triangle. X. Singular Points. XI. Systematic Tracing of Curves. XII. The Inverse Process. CLASSIFIED LIST OF THE CURVES DISCUSSED. FOLD-OUT PLATES.

Hundreds of examples are discussed in the text and illustrated in the fold-out plates.

—5th (unaltered) ed. 1960. 210 pp. + 17 fold-out plates. 5⅜x8. 8284-0140-3. **$4.50**

LECTURES ON ANALYTICAL MECHANICS

By F. R. GANTMACHER

Translated from the Russian by PROF. B. D. SECKLER, with additions and revisions by Prof. Gantmacher.

Partial Contents: CHAP. I. Differential Equations of Motion of a System of Particles. II. Equations of Motion in a Potential Field. III. Variational Principles and Integral-Invariants. IV. Canonical Transformations and the Hamilton-Jacobi Equation. V. Stable Equilibrium and Stability of Motion of a System (Lagrange's Theorem on stable equilibrium, Tests for unstable E., Theorems of Lyapunov and Chetayev, Asymptotically stable E., Stability of linear systems, Stability on basis of linear approximation, . . .). VI. Small Oscillations. VII. Systems with Cyclic Coordinates. BIBLIOGRAPHY.

—Approx. 300 pp. 6x9. 8284-0175-6. **In prep.**

COMMUTATIVE NORMED RINGS

By I. M. GELFAND, D. A. RAIKOV, and G. E. SHILOV

Partial Contents: CHAPS. I AND II. General Theory of Commutative Normed Rings. III. Ring of Absolutely Integrable Functions and their Discrete Analogues. IV. Harmonic Analysis on Commutative Locally Compact Groups. V. Ring of Functions of Bounded Variation on a Line. VI. Regular Rings. VII. Rings with Uniform Convergence. VIII. Normed Rings with an Involution and their Representations. IX. Decomposition of Normed Ring into Direct Sum of Ideals. HISTORICO-BIBLIOGRAPHICAL NOTES. BIBLIOGRAPHY.

—1964. 306 pp. 6x9. 8284-0170-5. **$6.50**

THE ALGEBRA OF INVARIANTS

By J. H. GRACE and A. YOUNG

An introductory account.

Partial Contents: I. Introduction. II. The Fundamental Theorem. III. Transvectants. V. Elementary Complete Systems. VI. Gordan's Theorem. VII. The Quintic. VIII. Simultaneous Systems. IX. Hilbert's Theorem. XI. Apolarity. XII. Ternary Forms. XV. Types of Covariants. XVI. General Theorems on Quantics. APPENDICES.

—1903-65. vii + 384 pp. 5⅜x8. 8284-0180-2. **$4.95**

DIE AUSDEHNUNGSLEHRE

By H. G. GRASSMANN

—2nd ed. 1862-1969. 388 pp. 6x9.

GYROSCOPIC THEORY

By G. GREENHILL

This work is intended to serve as a collection in one place of the various methods of the theoretical explanation of the motion of a spinning body, and as a reference for mathematical formulas required in practical work.

—1914-67. vi + 277 pp. + Fold-out Plates. 6½x10¾. 8284-0205-1. **$9.50**

LES INTEGRALES DE STIELTJES et leurs Applications aux Problèmes de la Physique Mathématique

By N. GUNTHER

The present work is a reprint of Vol. I of the publications of the V. A. Steklov Institute of Mathematics, in Moscow. The text is in French.

—1932-49. 498 pp. 5⅜x8. 8284-0063-6. **$6.95**

ONDES: Leçons sur la Propagation des Ondes et les Equations de l'Hydrodynamique

By J. HADAMARD

"[Hadamard's] unusual analytic proficiency enables him to connect in a wonderful manner the physical problem of propagation of waves and the mathematical problem of Cauchy concerning the characteristics of partial differential equations of the second order."—*Bulletin of the A. M. S.*

—Repr. 1949. viii + 375 pp. 5½x8½. 8284-0058-X. **$6.00**

REELLE FUNKTIONEN. Punktfunktionen

By H. HAHN

—1932-48. xi + 415 pp. 5½x8½. 8284-0052-0. **$6.00**

ALGEBRAIC LOGIC

By P. R. HALMOS

"Algebraic Logic is a modern approach to some of the problems of mathematical logic, and the theory of polyadic Boolean algebras, with which this volume is mostly concerned, is intended to be an efficient way of treating algebraic logic in a unified manner.

"[The material] is accessible to a general mathematical audience; no vast knowledge of algebra or logic is required . . . Except for a slight Boolean foundation, the volume is essentially self-contained."—*From the Preface.*

—1962. 271 pp. 6x9. 8284-0154-3. **$4.50**

LECTURES ON ERGODIC THEORY

By P. R. HALMOS

CONTENTS: Introduction. Recurrence. Mean Convergence. Pointwise Convergence. Ergodicity. Mixing. Measure Algebras. Discrete Spectrum. Automorphisms of Compact Groups. Generalized Proper Values. Weak Topology. Weak Approximation. Uniform Topology. Uniform Approximation. Category. Invariant Measures. Generalized Ergodic Theorems. Unsolved Problems.

"Written in the pleasant, relaxed, and clear style usually associated with the author. The material is organized very well and painlessly presented."

—*Bulletin of the A.M.S.*

—1956-60. viii + 101 pp. 5⅜x8. 8284-0142-X. **$3.25**

INTRODUCTION TO HILBERT SPACE AND THE THEORY OF SPECTRAL MULTIPLICITY

By P. R. HALMOS

A clear, readable introductory treatment of Hilbert Space.

—1957. 2nd ed. (C.r. of 1st ed.) 120 pp. 6x9. 8284-0082-2. **$3.50**

ELEMENTS OF QUATERNIONS

By W. R. HAMILTON

Sir William Rowan Hamilton's last major work, and the second of his two treatises on quaternions.

—2nd ed. 1899/1901-68. 1,185 pp. 6x9. 8284-0219-1. Two vol. set. **Prob. $29.50**

RAMANUJAN: Twelve Lectures on His Life and Works

By G. H. HARDY

The book is somewhat more than an account of the mathematical work and personality of Ramanujan; it is one of the very few full-length books of "shop talk" by an important mathematician.

—1940-59. viii + 236 pp. 6x9. 8284-0136-5. **$4.95**

GRUNDZUEGE DER MENGENLEHRE

By F. HAUSDORFF

The original, 1914 edition of this famous work contains many topics that had to be omitted from later editions, notably, the theories of content, measure, and discussion of the Lebesgue integral. Also, general topological spaces, Euclidean spaces, special methods applicable in the Euclidean plane, the algebra of sets, partially ordered sets, etc.

—1914-49. 484 pp. 5⅜x8. 8284-0061-X. **$7.50**

SET THEORY

By F. HAUSDORFF

Hausdorff's classic text-book is an inspiration and a delight. The translation is from the Third (latest) German edition.

"We wish to state without qualification that this is an indispensable book for all those interested in the theory of sets and the allied branches of real variable theory."—*Bulletin of A. M. S.*

—2nd ed. 1962. 352 pp. 6x9. 8284-0119-5. **$7.50**

VORLESUNGEN UEBER DIE THEORIE DER ALGEBRAISCHEN ZAHLEN

By E. HECKE

"An elegant and comprehensive account of the modern theory of algebraic numbers."

—*Bulletin of the A. M. S.*

—1923-69. 264 pp. 5½x8½. 8284-0046-6.

INTEGRALGLEICHUNGEN UND GLEICHUNGEN MIT UNENDLICHVIELEN UNBEKANNTEN

By E. HELLINGER and O. TOEPLITZ

"Indispensable to anybody who desires to penetrate deeply into this subject."—*Bulletin of A.M.S.*

—1928-53. 286 pp. 5⅜x8. 8284-0089-X. **$4.95**

THEORIE DER ALGEBRAISCHE FUNKTIONEN EINER VARIABELN

By K. HENSEL and G. LANDSBERG

Partial Contents: PART ONE (Chaps. 1-8): Algebraic Functions on a Riemann Surface. PART TWO (Chaps. 9-13): The Field of Algebraic Functions. PART THREE (Chaps. 14-22): Algebraic Divisors and the Riemann-Roch Theorem. PART FOUR (Chaps. 23-27): Algebraic Curves. PART FIVE (Chaps. 28-31): The Classes of Algebraic Curves. PART SIX (Chaps. 32-37): Algebraic Relations among Abelian Integrals. APPENDIX: Historical Development. Geometrical Methods. Arithmetical Methods.

—1902-65. xvi + 707 pp. 6x9. 8284-0179-9. **$9.50**

Grundzüge Einer Allgemeinen Theorie der LINEAREN INTEGRALGLEICHUNGEN

By D. HILBERT

—1912-53. 306 pp. 5½x8¼. 8284-0091-1. **$4.95**

THE DEVELOPMENT OF MATHEMATICS IN CHINA AND JAPAN

By Y. MIKAMI

"Filled with valuable information. Mikami's [account of the mathematicians he knew personally] is an attractive features."

—*Scientific American.*

—1913-62. x + 347 pp. 5⅜x8. 8284-0149-7. **$4.95**

GESAMMELTE ABHANDLUNGEN

By H. MINKOWSKI

TWO VOLUMES IN ONE.

Minkowski's Collected Works are issued under the editorship of David Hilbert, with the assistance of Andreas Speiser and Hermann Weyl.

—1911-67. 871 pp. 6x9. 8284-0208-6. Two vols. in one. **$17.50**

GEOMETRIE DER ZAHLEN

By H. MINKOWSKI

—1896-53. viii + 256 pp. 5⅜x8. 8284-0093-8. **$4.95**

DIOPHANTISCHE APPROXIMATIONEN

By H. MINKOWSKI

—1907-57. viii + 235 pp. 5⅜x8. 8284-0118-7. **$4.95**

FERMAT'S LAST THEOREM, *by L. J. MORDELL.* See KLEIN

INVERSIVE GEOMETRY

By F. MORLEY and F. V. MORLEY

—1937-54. xi + 273 pp. 5⅜x8. 8284-0101-2. **$3.95**

INTRODUCTION TO NUMBER THEORY

By T. NAGELL

A special feature of Nagell's well-known text is the rather extensive treatment of Diophantine equations of second and higher degree. A large number of non-routine problems are given.

—2nd ed. 1964. 309 pp. 5⅜x8. 8284-0163-2. **$5.50**

LEHRBUCH DER KOMBINATORIK

By E. NETTO

A standard work on the fascinating subject of Combinatory Analysis.

—2nd ed. 1927-58. viii + 348 pp. 5⅜x8. 8284-0123-3. **$6.00**

THE THEORY OF SUBSTITUTIONS

By E. NETTO

—2nd ed. (C.r. of 1st ed.) 1964. 304 pp. 5⅜x8. 8284-0165-9. **$4.95**

DIE GAMMAFUNKTION

By N. NIELSEN

TWO VOLUMES IN ONE.

HANDBUCH DER THEORIE DER GAMMAFUNKTION, by *N. Nielsen.* A standard, and very clearly written treatise on the gamma function and allied topics.

Part I (8 chapters) deals with the analytic theory of the gamma function; Part II (8 chapters) deals with definite integrals expressible in terms of the gamma function; and Part III deals with faculty series.

THEORIE DES INTEGRALLOGARITHMUS UND VERWANDTER TRANSZENDENTEN, by *N. Nielsen.* A treatise on certain transcendental functions. There are numerous references to the *Handbuch.*

—1906-65. 448 pp. 6x9. 8284-0188-8. Two vols. in one. **$7.50**

Vorlesungen über DIFFERENZENRECHNUNG

By N. H. NÖRLUND

—1924-54. ix + 551 pp. 5⅜x8. 8284-0100-4. **$6.50**

FUNCTIONS OF REAL VARIABLES
FUNCTIONS OF A COMPLEX VARIABLE

By W. F. OSGOOD

TWO VOLUMES IN ONE.

"*Well-organized courses, systematic, lucid, fundamental,* with many brief sets of appropriate exercises, and occasional suggestions for more extensive reading. The technical terms have been kept to a minimum, and have been clearly explained. The aim has been to develop the student's power and to furnish him with a substantial body of classic theorems, whose proofs illustrate the methods and whose results provide equipment for further progress."—*Bulletin of A. M. S*

—1936-58. 676 pp. 5x8. 8284-0124-1. Two vols. in one. **$6.50**

LEHRBUCH DER FUNKTIONENTHEORIE

By W. F. OSGOOD

THREE VOLUMES IN TWO.

Partial Contents: CHAP. I. The Calculus. II. Functions of Real Variables. III. Uniform Convergence. IV. Line Integrals and Multiply-Connected Regions. V. Set Theory. VI. Complex Numbers. Analytic Functions. Linear Transformations. VII. Rational Functions. VIII. Multiple-Valued Functions; Riemann Surfaces. IX. Analytic Continuation. X. Periodic Functions. XI. Infinite Series, Infinite Product Development . . . XIII. Logarithmic Potential. XIV. Conformal Mapping and Uniformization.

Vol. II: CHAPS. I-III. The Theory of Functions of Several Complex Variables. IV. Algebraic Functions and Abelian Integrals . . . VI. Abel's Theorem; Riemann-Roch Theorem; etc. VII. Periodic Functions of Several Complex Variables. VIII. Applications.

—Vol. I. 5th ed. 1928-65. 818 pp. 5⅜x8. 8284-0193-4. **$8.50**
—Vol. II. 2nd ed. 1932-65. 686 pp. 5⅜x8. 8284-0182-9. **$8.50**

EIGHT-PLACE TABLES OF TRIGONOMETRIC FUNCTIONS

By J. PETERS

Eight-place tables of natural trigonometric functions for every second of arc, with an appendix on the computation to twenty decimal places.

MAIN TABLE: The values of sine, cosine, tangent, and cotangent are given to 8 decimal places for every second of arc, from 0°0′0″ to 90°0′0″. For example, it can be read off directly from the table (without interpolation) that the value of Cos 89°19′33″ is 0.11622293.

TABLE II (*Supplementary Table*): 21-Place Tables of Sine and Cosine for every 10′ from 0°0′ to 90°0′.

TABLE III (*Supplementary Table*): 21-Place Tables of Sine and Cosine and their Differences, for every second from 0′0″ to 10′0″.

Several other supplementary tables are given.

"Peters' table is considered to be the standard (i.e., the best) 8-place trigonometric table."

—*Mathematics of Computation*

—1968. xii + 956 pp. 8x11. 8284-0174-8. Regular edition **$22.00**
8284-0185-3. Thumb-indexed **$25.00**

METHODS FOR GEOMETRICAL CONSTRUCTION, *by J. PETERSEN*. See BALL

IRRATIONALZAHLEN

By O. PERRON

—2nd ed. 1939-51. 207 pp. 5⅜x8. 8284-0047-4. Cloth **$3.75**
8284-0113-6. Paper **$1.50**

DIE LEHRE VON DEN KETTENBRUECHEN

By O. PERRON

Both the Arithmetic Theory and the Analytic Theory are treated fully.

"An indispensable work . . . Perron remains the best guide for the novice. The style is simple and precise and presents no difficulties."

—*Mathematical Gazette.*

—2nd ed. 1929-50. xii + 524 pp. 5⅜x8. 8284-0073-3. **$6.95**

THEORIE DES FONCTIONS ALGEBRIQUES DE DEUX VARIABLES INDEPENDANTES

By E. PICARD and G. SIMART

TWO VOLUMES IN ONE.

—1897/1906-69. 796 pp. Two vols. in one. **In prep.**

SUBHARMONIC FUNCTIONS

By T. RADO

TWO VOLUMES IN ONE: *Subharmonic Functions* and *The Problem of Plateau*, both by Tibor Rado.

—1937/33-70. 172 pp. 5½x8½. Two vols. in one. **In prep.**

COLLECTED PAPERS

By S. RAMANUJAN

Ramanujan's papers on Number Theory are edited by G. H. Hardy, P. V. Seshu Aiyar, and B. M. Wilson.

—1927-63. xxxvi + 355 pp. 6x9. 8284-0159-4. **$6.95**

RAMANUJAN, *by G. H. HARDY. See* HARDY

EINFUEHRUNG IN DIE KOMBINATORISCHE TOPOLOGIE

By K. REIDEMEISTER

—1932-50. 221 pp. 5½x8¼. 8284-0076-8. **$3.50**

KNOTENTHEORIE

By K. REIDEMEISTER

—1932-48. 78 pp. 5½x8½. 8284-0040-7. **$2.25**

ON THE HYPOTHESES . . . , *by B. RIEMANN.*
See CLIFFORD. *See also* WEYL

FOURIER SERIES

By W. ROGOSINSKI

Translated by H. Cohn. Designed for beginners with no more background than a year of calculus, this text covers, nevertheless, an amazing amount of ground. It is suitable for self-study courses as well as classroom use.

"The field covered is extensive and the treatment is thoroughly modern in outlook . . . An admirable guide to the theory."—*Mathematical Gazette.*

—2nd ed. 1959. vi + 176 pp. 4½x6½. 8284-0067-9. Cloth **$3.00**
8284-0178-0. Paper **$1.39**

ANALYTIC GEOMETRY OF THREE DIMENSIONS

By G. SALMON

A rich and detailed treatment by the author of *Conic Sections, Higher Plane Curves,* etc.

Partial Contents: Chap. I. Coordinates. III. Plane and Line. IV-VI. Quadrics. VIII. Foci and Focal Surfaces. IX. Invariants and Covariants of Systems of Quadrics. XI. General Theory of Surfaces. XII. Curves and Developables (Projective properties, non-projective properties, . . .).

Vol. II. Chap. XIII. Partial Differential Equations of Families of Surfaces. XIII (a). Complexes, Congruences, Ruled Surfaces. XIII (b). Triply Orthogonal Systems of Surfaces, Normal Congruences of Curves. XIV. The Wave Surface, The Centro-surface, etc. XV. Surfaces of Third Degree. XVI. Surfaces of Fourth Degree. XVII. General Theory of Surfaces. XVIII. Reciprocal Surfaces.

—Vol. I. 7th ed. 1927-58. xxiv + 470 pp. 5x8. 8284-0122-5. **$4.95**

—Vol. II. 5th ed. 1928-65. xvi + 334 pp. 5x8. 8284-0196-9. **$4.95**

LEHRBUCH DER TOPOLOGIE

By H. SEIFERT and W. THRELFALL

This famous book is a modern work on *combinatorial topology* addressed to the student as well as to the specialist. It is almost indispensable to the mathematician who wishes to gain a knowledge of this important field.

"The exposition proceeds by easy stages **with examples and illustrations at every turn.**"

—*Bulletin of the A. M. S.*

—1934-68. vii + 353 pp. 5⅜x8. 8284-0031-8. **$6.00**

TEXTBOOK OF TOPOLOGY

By H. SEIFERT and W. THRELFALL

A translation of the above.

—Approx. 380 pp. 6x9. **In prep.**

FROM DETERMINANT TO TENSOR, *by W. F. SHEPPARD. See* KLEIN

HYPOTHESE DU CONTINU

By W. SIERPINSKI

An appendix consisting of sixteen research papers now brings this important work up to date. This represents an increase of more than forty percent in the number of pages.

"One sees how deeply this postulate cuts through all phases of the foundations of mathematics, how intimately many fundamental questions of analysis and geometry are connected with it . . . a most excellent addition to our mathematical literature."

—*Bulletin of A. M. S.*

—2nd ed. 1957. xvii + 274 pp. 5⅜x8. 8284-0117-9. **$4.95**

CONGRUENCE OF SETS, and other monographs

By SIERPINSKI, KLEIN, RUNGE, and DICKSON

FOUR VOLUMES IN ONE.

ON THE CONGRUENCE OF SETS AND THEIR EQUIVALENCE BY FINITE DECOMPOSITION, by *W. Sierpinski.* 1. Congr. of Sets. 2. Translation of Sets. 3. Equiv. of Sets by Finite Decomposition. 4. D. into Two Parts. . . . 7. Paradoxical D's. . . . 10. The Hausdorff Paradox. 11. Paradox of Banach and Tarski. 12. Banach Measure. The General Problem of Measure. 13. Absolute Measure. 14. Paradox of J. von Neumann.

THE MATHEMATICAL THEORY OF THE TOP, by *F. Klein.* Well-known lectures on the analytical formulas relating to the motion of the top.

GRAPHICAL METHODS, by *C. Runge.*

INTRODUCTION TO THE THEORY OF ALGEBRAIC EQUATIONS, by *L. E. Dickson.* Dickson's earliest (1903) account of the subject, substantially less abstract than his later exposition. *From Dickson's Preface:* "The subject is here presented in the historical order of its development. The First Part (Chaps. I-IV) is devoted to the Lagrange-Cauchy-Abel theory of general algebraic equations. The Second Part (Chaps. V-XI) is devoted to Galois' theory of algebraic equations . . . The aim has been to make the presentation strictly elementary, with practically no dependence upon any branch of mathematics beyond elementary algebra. There occur numerous illustrative examples, as well as sets of elementary exercises."

—1954/1897/1912/1903-1967. 461 pp. 5¼x8. 8284-0209-4. Four vols. in one. **$6.50**

NON-DIFFERENTIABLE FUNCTIONS, *by A. N. SINGH.* See HOBSON

DIOPHANTISCHE GLEICHUNGEN

By T. SKOLEM

—1938-50. ix + 130 pp. 5½x8½. 8284-0075-X. **$3.50**

COLLECTED MATHEMATICAL PAPERS

By H. J. S. SMITH

Something of Smith's stature as a mathematician may be gathered from one of many references to him in Bell's *Development of Mathematics:* "such men as Lagrange, Legendre, Gauss, Eisenstein, Dirichlet, Hermite, H. J. S. Smith, Minkowski, and Siegal."

Smith's famous *Report on the Theory of Numbers* (also published separately; see below) is included in the present work.

—1894-1965. 1,422 pp. 6½x9¼. 8284-0187-X. Two vol. set. **$29.50**

REPORT ON THE THEORY OF NUMBERS

By H. J. S. SMITH

A little more than half a century after the publication of Gauss's *Disquisitiones*, H. J. S. Smith undertook to present an account of the status of number theory. The result was Smith's now classical *Report.*

Although it is a standard work and referred to frequently, the report has not been conveniently available to the practising mathematician. It was first published, in six parts, in the "Report of the British Association." It was then reprinted in the 1,400-page *Collected Mathematical Papers.* It is now, for the first time, available separately in book form.

—1894-1965. iv + 360 pp. 6½x9¼. 8284-0186-1. **$6.00**

INTERPOLATION

By J. F. STEFFENSEN

"A landmark in the history of the subject.

"Starting from scratch, the author deals with formulae of interpolation, construction of tables, inverse interpolation, summation of formulae, the symbolic calculus, interpolation with several variables, in a clear, elegant and rigorous manner . . . The student . . . will be rewarded by a comprehensive view of the whole field. . . . A classic account which no serious student can afford to neglect."—*Mathematical Gazette.*

—2nd ed. 1950-65. 256 pp. 5⅜x8. 8284-0071-7. **$4.95**

ALGEBRAISCHE THEORIE DER KOERPER

By E. STEINITZ

"Will always be considered as one of the classics of this branch of mathematics . . . I should like to recommend the book to students of algebra."—*O. Ore, Bulletin of the A.M.S.*

—1930-50. 176 pp. 5⅜x8. 8284-0077-6. **$3.75**

LAMESCHE, MATHIEUSCHE UND VERWANDTE FUNKTIONEN IN PHYSIK UND TECHNIK

By M. J. O. STRUTT

—1932-68. viii + 116 pp. 6x9. 8284-0203-5. **$3.95**

A HISTORY OF THE MATHEMATICAL THEORY OF PROBABILITY

By I. TODHUNTER

Introduces the reader to *almost every process and every species of problem which the literature of the subject can furnish.* Hundreds of problems are solved in detail.

—1865-1965. xvi + 624 pp. 5⅜x8. 8284-0057-1. **$7.50**

A HISTORY OF THE CALCULUS OF VARIATIONS IN THE 19th CENTURY

By I. TODHUNTER

A critical account of the various works on the Calculus of Variations published during the early part of the nineteenth century. Of the seventeen chapters, fourteen are devoted to the Calculus of Variations proper, two to various memoirs that touch upon the subject, and the seventeenth is a history of the conditions of integrability. Chapter Nine contains a translation in full of Jacobi's memoir.

—1862-1961. xii + 532 pp. 5⅜x8. 8284-0164-0. **$7.50**

SET TOPOLOGY

By R. VAIDYANATHASWAMY

In this text on Topology, the first edition of which was published in India, the concept of partial order has been made the unifying theme.

Over 500 exercises for the reader enrich the text.

CHAPTER HEADINGS: I. Algebra of Subsets of a Set. II. Rings and Fields of Sets. III. Algebra of Partial Order. IV. The Closure Function. V. Neighborhood Topology. VI. Open and Closed Sets. VII. Topological Maps. VIII. The Derived Set in T_1 Space. IX. The Topological Product. X. Convergence in Metrical Space. XI. Convergence Topology.

—2nd ed. 1960. vi + 305 pp. 6x9. 8284-0139-X. **$6.95**

[...]TURES ON THE GENERAL THEORY OF INTEGRAL FUNCTIONS

By G. VALIRON

—1923-49. xii + 208 pp. 5¼x8. 8284-0056-3. **$3.50**

L'APPROXIMATION, by *VALLÉE POUSSIN.*

See BERNSTEIN

GRUPPEN VON LINEAREN TRANSFORMATIONEN

By B. L. VAN DER WAERDEN

—(Ergeb. der Math.) 1935-48. 94 pp. 5½x8½. 8284-0045-8. **$2.50**

THE LOGIC OF CHANCE

By J. VENN

One of the classics of the theory of probability. Venn's book remains unsurpassed for clarity, readability, and sheer charm of exposition. No mathematics is required.

CONTENTS: PART ONE: Physical Foundations of the Science of Probability. CHAP. I. The Series of Probability. II. Formation of the Series, III. Origin, or Causation, of the Series. IV. How to Discover and Prove the Series. V. The Concept of Randomness. PART TWO: Logical Superstructure on the Above Physical Foundations. VI. Gradations of Belief. VII. The Rules of Inference in Probability. VIII. The Rule of Succession. IX. Induction. X. Causation and Design. XI. Material and Formal Logic . . . XIV. Fallacies. PART THREE: Applications. XV. Insurance and Gambling. XVI. Application to Testimony. XVII. Credibility of Extraordinary Stories. XVIII. The Nature and Use of an Average as a Means of Approximation to the Truth.

—4th ed. (Repr. of 3rd ed.) xxix + 508 pp. 5⅜x8.
8284-0173-X. Cloth **$6.00**
8284-0169-1. Paper **$2.25**

ANALYTICAL THEORY OF CONTINUED FRACTIONS

By H. S. WALL

Partial Contents: CHAP. I. The c. f. as a Product of Linear Fractional Transformations. II. Convergence Theorems. IV. Positive-Definite c. f. VI. Stieltjes Type c. f. VIII. Value-Region Problem. IX. J-Fraction Expansions. X. Theory of Equations. XII. Matrix Theory of c. f. XIII. C. f. and Definite Integrals. XIV. Moment Problem. XVI. Hausdorff Summability. XIX. Stieltjes Summability. XX. The Padé Table.

—1948-67. xiv + 433 pp. 5⅜x8. 8284-0207-8. **$7.50**